Asa Gray, Institution Smithsonian

Synoptical flora of North America

Vol. 1, Part 2: Caprifoliaceae - Compositae

Asa Gray, Institution Smithsonian

Synoptical flora of North America
Vol. 1, Part 2: Caprifoliaceae - Compositae

ISBN/EAN: 9783337271664

Printed in Europe, USA, Canada, Australia, Japan

Cover: Foto ©berggeist007 / pixelio.de

More available books at **www.hansebooks.com**

SYNOPTICAL

FLORA OF NORTH AMERICA.

SYNOPTICAL
FLORA OF NORTH AMERICA.

By ASA GRAY, LL.D.,

F.M. R.S. & L.S. Lond., R.I.A. Dubl., Phil. Soc. Cambr., Roy. Soc. Upsala, Stockholm, Gœttingen;
Roy. Acad. Sci. Munich, &c.; Corresp. Imp. Acad. Sci. St. Petersburg,
Roy. Acad. Berlin, and Acad. Sci. Instit. France.

FISHER PROFESSOR OF NATURAL HISTORY (BOTANY) IN HARVARD UNIVERSITY.

VOL. I. — PART II.

CAPRIFOLIACEÆ — COMPOSITÆ.

Published by the Smithsonian Institution, Washington.

NEW YORK:
IVISON, BLAKEMAN, TAYLOR, AND COMPANY.
LONDON: WM. WESLEY, 28 ESSEX ST., STRAND,
AND TRÜBNER & CO.
LEIPSIC: OSWALD WEIGEL.
JULY, 1884.

University Press:
JOHN WILSON AND SON, CAMBRIDGE.

SYNOPTICAL
FLORA OF NORTH AMERICA.

DIVISION II. GAMOPETALOUS DICOTYLEDONOUS PLANTS.

PERIANTH consisting of both calyx and corolla, the latter more or less gamopetalous. (Exceptions: A part of *Ericaceæ*, *Plumbaginaceæ*, *Styracaceæ*, and *Oleaceæ* have unconnected petals; some *Oleaceæ*, &c., are apetalous.)

GENERAL KEY TO THE ORDERS.

* Ovary inferior or mainly so: stamens borne by the corolla, alternate with its lobes, and
  +— Unconnected: leaves opposite or whorled.

69. CAPRIFOLIACE Æ. Stamens as many as corolla-lobes (one fewer in *Linnæa*, doubled by division in *Adoxa*). Seeds albuminous. Leaves opposite; stipules none, or rare as appendages to base of petiole.

70. RUBIACEÆ. Stamens as many as corolla-lobes, mostly four or five. Ovary with two or more cells or placentæ. Seeds albuminous. Leaves all simple and entire, with stipules between or within the petioles or bases, or whorled without stipules, the additional leaves probably representing them.

71. VALERIANACEÆ. Stamens fewer than corolla-lobes, one to four. Ovary with one cell containing a suspended ovule which becomes an exalbuminous seed, and commonly two empty cells or vestiges of them. No stipules.

72. DIPSACACEÆ. Stamens as many as or fewer than corolla-lobes, two or four. Ovary simple and one-celled, with a single suspended ovule, becoming an albuminous seed. Flowers capitate. Corolla-lobes imbricated in the bud.

  +— +— Stamens with anthers connate into a tube.

73. COMPOSITÆ. Syngenesious stamens as many as their corolla-lobes, five, sometimes four. Ovary one-celled, with a solitary erect ovule, becoming an exalbuminous seed in an akene. Lobes of the corolla valvate in the bud. Flowers in involucrate heads. No stipules.

1

* * Ovary either inferior or superior, two–several-celled: stamens free from the corolla or nearly so, inserted with it, as many or twice as many as its lobes or petals, when of same number alternate with them: no stipules. (Orders from these onward are in Vol. II. Part I.)

+— Juice milky except in the first order: corolla-lobes valvate or induplicate in the bud.

74. GOODENIACEÆ. Corolla irregular, epigynous. Stamens or at least filaments distinct. Stigma indusiate. Juice not milky.
75. LOBELIACEÆ. Corolla irregular, epigynous or perigynous. Stamens five, monadelphous or syngenesious, or both. Stigma not indusiate. Cells of ovary or placentæ two. Seeds numerous. Juice usually more or less milky and acrid. Inflorescence centripetal.
76. CAMPANULACEÆ. Corolla regular, epigynous. Stamens five, mostly distinct. Stigmas two to five, introrse, at the summit of the style, which below bears pollen-collecting hairs. Cells of ovary and capsule two to five, many-seeded. Juice milky and bland. (Exception: *Sphenoclea*.)

+— +— Juice not milky nor acrid: corolla-lobes or petals imbricate or sometimes convolute in the bud.

77. ERICACEÆ. Flowers mostly regular, symmetrical, and tetra–pentamerous throughout: corolla sometimes moderately irregular, epigynous or hypogynous. Stamens distinct, as many and oftener twice as many as petals or corolla-lobes. Cells of the ovary (with few exceptions) as many or even twice as many as the divisions of the calyx or corolla. Style and mostly stigma undivided.

* * * Ovary superior, many-celled: stamens five to eight, as many as the lobes of the hypogynous corolla, and borne in the throat of its long tube.

78. LENNOACEÆ. Root-parasites.

* * * * Ovary superior: stamens (or antheriferous stamens) of the same number as the proper corolla-lobes or petals and *opposite* them: flowers regular.

+— Ovary one-celled, with solitary ovule or free placenta rising from its base: seeds small.

80. PLUMBAGINACEÆ. Stamens and styles or lobes of the style five, except in *Plumbago*, the former hypogynous or borne on the very base of the almost or completely distinct unguiculate petals. Ovary uniovulate, in fruit becoming an akene or utricle. Herbs or somewhat shrubby.
81. PRIMULACEÆ. Stamens four or five, rarely six to eight, borne on the corolla (or in *Glaux*, which is apetalous, on the calyx alternate with its petaloid lobes): staminodia only in *Samolus*. Ovules several or numerous, sessile on the central placenta. Fruit capsular. Herbs.
82. MYRSINACEÆ. Shrubs or trees, with dry or drupaceous fruit and solitary or very few seeds, usually immersed in the placenta: otherwise as *Primulaceæ*.

+ + Ovary few-several-celled, with solitary ovules in the cells, usually only one maturing into a large bony-coated seed in a fleshy pericarp.

S3. SAPOTACEÆ. Shrubs or trees, mostly with milky juice and alternate simple leaves. Flowers small, hermaphrodite, tetra-heptamerous. Calyx and corolla much imbricated in the bud; the latter often bearing accessory lobes or appendages within, sometimes petaloid staminodia also.

* - * * * Ovary inferior or superior, few-several-celled: cells of the fruit one-seeded: stamens at least twice as many as the petals or lobes of the corolla, sometimes indefinitely numerous and borne on or united with their base or tube: flowers regular: shrubs or trees, with simple alternate leaves, sometimes a resinous but no milky juice.

S4. EBENACEÆ. Flowers dioecious or polygamous; the male ones polyandrous. Ovary superior and corolla hypogynous. Styles as many or half as many as the cells of the ovary, distinct or partly united. Fruit fleshy, containing solitary or few large seeds with bony testa and cartilaginous albumen.

S5. STYRACACEÆ. Flowers hermaphrodite, nearly pentapetalous and a numerous cluster of stamens adnate to base of each petal, or more gamopetalous and the fewer stamens monadelphous in a single series. Style and stigma entire. Corolla epigynous, in *Styrax* perigynous. Fruit dry or nearly so, one–four-seeded, when dehiscent the seed bony: albumen fleshy.

* * * * * * Ovary or gynoecium superior, dicarpellary, or in some monocarpellary, very rarely tri-pentacarpellary, sometimes appearing to be tetracarpellary by the division of the two ovaries: stamens borne on the corolla (in apetalous *Oleaceæ*. &c., on the receptacle), alternate with its divisions or lobes, of the same number or fewer.

+ Corolla not scarious and veinless.

++ Regular with stamens fewer than its lobes or petals, or no corolla: style one: seeds solitary or very few.

S6. OLEACEÆ. Trees or shrubs, with opposite (rarely alternate) leaves: no stipules, no milky juice. Stamens usually two, alternate with the carpels; these two-ovuled, or sometimes four-ovuled: seed mostly solitary, albuminous. *Forestiera* and part of *Fraxinus* apetalous and even achlamydeous.

++ ++ Corolla regular and stamens as many as its divisions, five or four.

= Ovaries two (follicular in fruit); their stigmas and sometimes styles permanently united into one: plants with milky juice: flowers hermaphrodite: leaves simple, entire.

S7. APOCYNACEÆ. Stamens distinct, or the anthers merely connivent or lightly cohering: pollen ordinary. Style single.

S8. ASCLEPIADACEÆ. Stamens monadelphous and anthers permanently attached to a large stigmatic body: pollen combined into waxy pollinia or sometimes granulose masses. Carpels united only by the common stigmatic mass.

4   GENERAL KEY TO THE GAMOPETALOUS ORDERS.

— — Ovaries two, with styles slightly united below or distinct. Vide 94.

═ ─ ─ Ovary one, compound, with two or three (very rarely four or five) cells or placentæ: stamens distinct (or anthers at most lightly connate).

a. Leaves opposite, simple, and mostly entire, with stipules or stipular line connecting their bases: no milky juice.

89. LOGANIACEÆ. Ovary dicarpellary, two-celled : style single, but stigmas occasionally four, usually only one. Seeds numerous : embryo rather small, in copious albumen.

b. Leaves with no trace of stipules: milky juice only in *Convolvulaceæ*.

90. GENTIANACEÆ. Leaves opposite, sessile, simple and entire, except in *Menyantheæ*. Ovary dicarpellary, one-celled, many-ovuled : placentæ or ovules parietal. Stigmas mostly two, introrse. Fruit capsular, septicidal, i. e. dehiscent through the placentæ or alternate with the stigmas. Seeds with minute embryo in fleshy albumen. Herbage smooth.

79. DIAPENSIACEÆ. Leaves alternate and simple, smooth. Ovary tricarpellary, three-celled, as also the loculicidal many-seeded capsule, which has a persistent columella. Stamens five, either borne in sinuses of the corolla or monadelphous : in some a series of petaloid staminodia alternate with the true stamens. Anthers inflexed on apex of the filament, or transversely dehiscent. Calyx and corolla imbricated in the bud. Style one : stigma three-lobed. Embryo small in fleshy albumen. Depressed or scapose and acaulescent perennials.

91. POLEMONIACEÆ. Leaves opposite or alternate, from entire to compound. Ovary tri-(very rarely di-)carpellary, with as many cells, becoming a loculicidal capsule, with solitary to numerous seeds borne on a thick placental axis. Stamens five, distinct, borne on the tube or throat of the corolla ; the latter convolute in the bud, the calyx imbricated. Style three-cleft or three-lobed at the summit : stigmas introrse. Seeds with comparatively large straight embryo in rather sparing albumen.

92. HYDROPHYLLACEÆ. Leaves mostly alternate, disposed to be lobed or divided. Inflorescence disposed to be scorpioid in the manner of the next order. Corolla five-lobed, imbricated or sometimes convolute in the bud. Stamens five, distinct. Ovary undivided, dicarpellary, and style (with one exception) two-parted or two-lobed : stigmas terminal. Capsule one-celled with two parietal or introflexed placentæ, each bearing two or more pendulous (or when very numerous horizontal) seeds, or sometimes two-celled by the junction of the placentæ in the axis. Seeds with reticulated or pitted or roughened testa : a small or slender straight embryo in solid albumen.

93. BORRAGINACEÆ. Leaves alternate, mostly entire, and with whole herbage apt to be rough, hirsute, or hispid. Inflorescence cymose, commonly in the scorpioid mode, the mostly uniparous or biparous cymes evolute into unilateral and often ebracteate false spikes or racemes. Corolla five-lobed, sometimes four-lobed, imbricate or convolute or sometimes plicate in the bud. Ovary dicarpellary, but usually seeming tetramerous, being of four (i. e. two biparted) lobes around the base of the style, maturing into as many separate or separable nutlets ; or ovary not lobed, two-four-celled, in fruit drupaceous or dry, containing or splitting into as many nutlets. Solitary seed with a mostly straight embryo and little or no albumen ; radicle superior or centripetal.

GENERAL KEY TO THE GAMOPETALOUS ORDERS. 5

94. CONVOLVULACEÆ. Leaves alternate and petioled. Stems usually twining or trailing, but some erect, many with milky juice. Flowers borne by axillary peduncles or cymose-glomerate. Calyx of imbricated sepals. Corolla with four-five-lobed or commonly entire margin, plicate and the plaits convolute in the bud, sometimes induplicate-valvate or imbricated. Ovary two-celled or sometimes three-celled, with a pair of erect anatropous ovules in each cell, becoming comparatively large seeds (these sometimes separated by spurious septa of the capsular fruit), with smooth or hairy testa. Embryo incurved, with ample foliaceous plaited and crumpled cotyledons (in *Cuscuta* embryo long and spiral without cotyledons) surrounded by little or no albumen : radicle inferior. *Dichondra* has two distinct ovaries.

95. SOLANACEÆ. Leaves alternate, sometimes unequally geminate. Inflorescence various, but no truly axillary flowers. Corolla in some a little irregular, its lobes or border induplicate-plicate or rarely imbricate in the bud. Ovary normally two-celled (occasionally three–five-celled) and undivided, with many-ovuled placentæ in the axis : style undivided : stigma entire or bilamellar. Seeds numerous, with incurved or coiled or rarely almost straight embryo in copious fleshy albumen : cotyledons seldom much broader than the radicle.

++ ++ ++ Corolla irregular, more or less bilabiately so ($\frac{2}{3}$); its lobes variously imbricate or convolute, or sometimes almost regular: stamens fewer than corolla-lobes, four and didynamous, or only two : style undivided : stigma entire or two-lobed or bilamellar; the lobes anterior and posterior : ovary in all dicarpellary ; the cells or carpels anterior and posterior.

= Pluriovulate or multiovulate.

96. SCROPHULARIACEÆ. Ovary and capsule completely two-celled : placentæ occupying the middle of the partition. Seeds comparatively small or minute, mostly indefinitely numerous, sometimes few. Embryo small, straight or slightly curved, in copious fleshy albumen : cotyledons hardly broader than the radicle.

97. OROBANCHACE.E. Ovary one-celled with two or four (doubled) parietal many-ovuled placentæ. Seeds very many in fleshy albumen, with minute embryo, having no obvious distinction of parts. Root-parasites, destitute of green herbage.

98. LENTIBULARIACE.E. Ovary one-celled, with a free central multiovulate placenta : globular capsule mostly bursting irregularly. Seeds destitute of albumen, filled by a solid oblong embryo. Bilabiate corolla personate and calcarate. Stamens two : anthers confluently one-celled. Aquatic or paludose plants, with scapes or scapiform peduncles, sometimes almost leafless.

99. BIGNONIACEÆ. Ovary and capsule two-celled by the extension of a partition beyond the two parietal placentæ, or in some genera simply one-celled. Seeds numerous, large, commonly winged, transverse, filled by the horizontal embryo : cotyledons broad and foliaceous, plane, emarginate at base and summit, the basal notch including the short radicle : no albumen. Trees or shrubs, many climbing, large-flowered : leaves commonly opposite.

100. PEDALIACEÆ. Ovary one-celled, with two parietal intruded placentæ, which are broadly bilamellar or united in centre, or two–four-celled by spurious septa from the walls. Fruit capsular or drupaceous, few–many-seeded. Seeds wingless, with thick and close testa, filled by the large straight embryo : cotyledons thickish. Herbs, with mainly opposite simple leaves : juice mucilaginous.

101. ACANTHACEÆ. Ovary two-celled, with placentæ in the axis, bearing a definite number of ovules (two to eight or ten in each cell), becoming a loculicidal capsule. Seeds wingless, destitute of albumen (or a thin layer in *Elytraria*), either globular on a papilliform funicle, or flat on a retinaculum. Embryo with broad and flat cotyledons.

= = Cells of the ovary uniovulate or biovulate.

102. SELAGINACEÆ. Ovary two-celled : ovule suspended. Embryo in fleshy albumen : radicle inferior. Leaves alternate.

103. VERBENACEÆ. Ovary two–four-celled, in fruit di–tetrapyrenous, not lobed, in *Phryma* one-celled and becoming an akene. Ovule erect from the base of each cell or half-cell. Seed with little or no albumen : radicle inferior.

104. LABIATÆ. Ovary deeply four-lobed around the style, the lobes becoming dry seed-like nutlets in the bottom of a gamosepalous calyx. Ovule erect. Seed with little or no albumen : radicle inferior. Commonly aromatic herbs or undershrubs.

+ + Corolla scarious and nerveless : flowers tetramerous, regular.

105. PLANTAGINACEÆ. Calyx imbricated. Corolla-lobes imbricated in the bud. Stamens four or fewer. Style entire. Ovary and capsule one–two-celled : cells sometimes again divided by a false septum. Seeds mostly amphitropous and peltate, with straight embryo in firm fleshy albumen. Chiefly acaulescent herbs, with one–many-flowered commonly spike-bearing scapes, arising from axils of the leaves.

Order LXIX. CAPRIFOLIACEÆ.

Shrubby, or a few perennial herbaceous plants, with opposite leaves normally destitute of stipules, and regular or (in the corolla) irregular hermaphrodite flowers; calyx-tube adnate to the 2–5-celled or by suppression 1-celled ovary; stamens as many as lobes of the corolla (in *Linnæa* one fewer, in *Adoxa* doubled) and alternate with them, inserted on its tube or base; embryo small in the axis of fleshy albumen. Corolla-lobes generally imbricated in the bud. Ovules anatropous, when solitary suspended and resupinate; the rhaphe dorsal. Seed-coat adherent to the albumen. Flowers commonly 5-merous.

TRIBE I. SAMBUCEÆ. Corolla regular, short, rotate or open-campanulate, 5-lobed. Style short or hardly any; stigmas 3 to 5. Ovules solitary in the (1 to 5) cells. Fruit baccate-drupaceous; the seed-like nutlets 1 to 5. Inflorescence terminal and cymose.

* Herb, with stamens doubled and flowers in a capitate cluster. Anomalous in the order.

1. **ADOXA.** Calyx with hemispherical tube adnate to above the middle of the ovary; limb about 3-toothed. Corolla rotate, 4–6-cleft. Stamens a pair below each sinus of the corolla, each with a peltate one-celled anther, and the short subulate filaments approximate or united at base (one stamen divided into two). Ovary 3–5-celled: style short, 3–5-parted. Ovule suspended from the summit of each cell. Fruit greenish, maturing 2 to 5 cartilaginous nutlets. Cauline leaves a single pair.

* * Frutescent to arborescent: inflorescence compound-cymose: flowers articulated with their pedicels: stamens as many as corolla-lobes: anthers 2-celled: calyx 5-toothed.

2. **SAMBUCUS.** Leaves pinnately compound. Corolla rotate or nearly so. Ovary 3–5-celled, forming small baccate drupes with as many cartilaginous nutlets. Embryo nearly the length of the albumen.

3. **VIBURNUM.** Leaves simple, sometimes lobed. Corolla rotate or open-campanulate. Ovary 1-celled and 1-ovuled, becoming a drupe with a single more or less flattened nutlet or stone. Embryo minute. Cymes in some species radiate.

TRIBE II. LONICEREÆ. Corolla elongated or at least campanulate, commonly more or less irregular. Style elongated: stigma mostly capitate. Fruit various. Stipules or stipular appendages seldom seen.

* Herbs, with axillary sessile flowers and drupaceous fruit.

4. **TRIOSTEUM.** Calyx-lobes 5. Corolla tubular-campanulate, somewhat unequally 5-lobed; tube gibbous at base. Stamens 5. Ovary 3- (sometimes 4–5-) celled, with a single suspended ovule in each cell: style slender: stigma 3-lobed. Fruit a fleshy drupe, crowned with the persistent calyx-lobes: putamen bony, costate, at length separable into 3 (rarely 4 or 5, or by abortion 2) thick one-seeded nutlets.

* * Fruticulose creeping herb, with long-pedunculate geminate flowers and dry one-seeded fruit, but a 3-celled ovary.

5. **LINNÆA.** Calyx with limb 5-parted into subulate-lanceolate lobes, constricted above the globular tube, deciduous from the fruit. Corolla campanulate-funnelform, not gibbous, almost equally 5-lobed. Stamens 4, two long and two shorter, included. Ovary 3-celled; two of the cells containing several abortive ovules; one with a solitary suspended ovule, forming the single seed in the dry and indehiscent coriaceous 3-celled small fruit. Style exserted: stigma capitate.

* * * Shrubs, with scaly winter-buds, erect or climbing: fruit 2–many-seeded: style slender: stigma capitate, often 2-lobed.

6. **SYMPHORICARPOS.** Calyx with a globular tube and 4–5-toothed persistent limb. Corolla regular, not gibbous, from short-campanulate to salverform, 4–5-lobed. Stamens as

many as the lobes of the corolla, inserted on its throat. Ovary 4-celled; two cells containing a few sterile ovules: alternate cells containing a single suspended ovule. Fruit a globose berry-like drupe, containing 2 small and seed-like bony smooth nutlets, each filled by a seed; sterile cells soon obliterated.

7. **LONICERA.** Calyx with ovoid or globular tube and a short 5-toothed or truncate limb. Corolla from campanulate to tubular, more or less gibbous at base; the limb irregular and commonly bilabiate (⅔), sometimes almost regular. Stamens 5, inserted on the tube of the corolla. Ovary 2-3-celled, with several pendulous ovules in each cell, becoming a few-several-seeded berry.

8. **DIERVILLA.** Calyx with slender elongated tube, and 5 narrow persistent or tardily deciduous lobes. Corolla funnelform (or in large-flowered Japanese species more campanulate), inconspicuously gibbous at base; a globular epigynous gland within occupying the gibbosity; limb somewhat unequally or regularly 5-lobed. Stamens 5, inserted on the tube or throat of the corolla: anthers linear. Ovary 2-celled. Fruit a narrow capsule, with attenuate or rostrate summit, septicidally 2-valved, many-seeded.

1. **ADÓXA,** L. (From ἄδοξος, obscure or insignificant.) — Single species, an insignificant small herb, of obscure affinity, now referred to the present order.

A. **Moschatéllina,** L. (MOSCHATEL.) Glabrous and smooth: stem and once to thrice ternately compound radical leaves a span high from a small fleshy-scaly rootstock: cauline pair of leaves 3-parted or of 3 obovate and 3-cleft or parted leaflets: flowers small, greenish-white or yellowish, 4 or 5 in a slender-pedunculate glomerule: corolla of the terminal one 4-5-cleft, of the others 5-6-cleft: drupe merely succulent: odor of plant musky. — Lam. Ill. t. 320; Gærtn. Fruct. t. 112; Schk. Handb. t. 109; Torr. & Gray, Fl. i. 648. — Subalpine, under rocks, Arctic America to N. Iowa, Wisconsin, and the Rocky Mountains to Colorado. (Eu., N. Asia, &c.)

2. **SAMBÚCUS,** Tourn. ELDER. (Classical Latin name, said by some to come from σαμβύκη, a stringed musical instrument.) — Suffrutescent to arborescent (in both Old and New World); with large pith to the vigorous shoots, imparipinnate leaves, serrate leaflets, small flowers (usually white and odorous) in broad cymes, and red or black berry-like fruits. Stems with warty bark. Stipule-like appendages hardly any in our species; but stipels not rare. Flowers occasionally polygamous, produced in summer.

* Compound cymes thyrsoid-paniculate; the axis continued and sending off 3 or 4 pairs of lateral primary branches, these mostly trifid and again bifid or trifid: pith of year-old shoots deep yellow-brown: no obvious stipule-like nor stipel-like appendages to the leaves: early flowering and fruiting.

S. **racemósa,** L. Stems 2 to 12 feet high, sometimes forming arborescent trunks: branches spreading; leaves from pubescent to nearly glabrous: leaflets 5 to 7, ovate-oblong to ovate-lanceolate, acuminate, thickly and sharply serrate: thyrsiform cyme ovate or oblong: flowers dull white, drying brownish: fruit scarlet (has been seen white), oily: nutlets minutely punctate-rugulose. — Spec. i. 270; Jacq. Ic. Rar. i. t. 59; Hook. Fl. i. 279; Gray, Bot. Calif. i. 278. *S. pubens,* Michx. Fl. i. 181; DC. Prodr. iv. 323; Torr. & Gray, Fl. ii. 13; Mechan, Nat. Flowers, ser. 2, ii. t. 21, flowers wrongly colored. *S. pubescens,* Pers. Syn. i. 328; Pursh, Fl. i. 204. — Rocky banks and open woods, Nova Scotia to the mountains of Georgia, in cool districts, west to Brit. Columbia and Alaska, and the Sierra Nevada, California. (Eu., N. Asia.)

Var. **arboréscens,** TORR. & GRAY, l. c. A form with leaflets closely serrate with strong lanceolate teeth. — Washington Terr. to Sitka.

Var. **laciniáta,** KOCH, with leaflets divided into 3 to 5 linear-lanceolate 2-3-cleft or laciniate segments, occurs on south shore of L. Superior, *Austin.*

S. **melanocárpa,** GRAY. Glabrous, or young leaves slightly pubescent: leaflets 5 to 7, rarely 9: cyme convex, as broad as high: flowers white: fruit black, without bloom: otherwise much like preceding. — Proc. Am. Acad. xix. 76. — Ravines of the Rocky Mountains of Montana (*Watson*) to those of E. Oregon (*Cusick*), south to the Wahsatch (*Watson*).

New Mexico (*Fendler*), and the Sierra Nevada, California (*Brewer, Bolander*): a plant with foliage not unlike that of *S. Canadensis*.

* * Compound cymes depressed, 5-rayed; four external rays once to thrice 5-rayed, but the rays unequal, the two outer ones stronger, or in ultimate divisions reduced to these; central rays smaller and at length reduced to 3-flowered cymelets or to single flowers: pith of year-old shoots bright white: "berries" sweet, never red: nutlets punctate-rugulose.

— **S. Canadénsis,** L. Suffrutescent or woody stems rarely persisting to third or fourth year, 5 to 10 feet high, glabrous, except some fine pubescence on midrib and veins of leaves beneath: leaflets (5 to 11) mostly 7, ovate-oval to oblong-lanceolate, acuminate, the lower not rarely bifid or with a lateral lobe: stipels not uncommon, narrowly linear, and tipped with a callous gland: fruit dark-purple, becoming black, with very little bloom. — Spec. i. 269; Michx. Fl. i. 281; Torr. & Gray, Fl. ii. 13. *S. nigra*, Marsh. Arbust. 141. *S. humilis*, Raf. Amer. Nat. 13. *S. glauca*, Gray, Pl. Wright. ii. 66 (not *Nutt.*), narrow-leaved form; Bot. Mex. Bound. 71. — Moist grounds, New Brunswick to the Saskatchewan, south to Florida, Texas, west to the mountains of Colorado, Utah, and Arizona; fl. near midsummer. Nearly related to *S. nigra* of Eu.

Var. laciniáta. Leaflets or most of them once or twice ternately parted into lanceolate divisions.— Indian River, Florida, *Palmer*. A still more dissected form, in waste places, Egg Harbor, *Mrs. Treat*, may be *S. nigra*, var. *laciniata*, of the Old World.

— **S. glaúca,** NUTT. Arborescent, 6 to 18 feet high; the larger forming trunks of 6 to 12 inches in diameter, glabrous throughout: leaflets 5 to 9, thickish, ovate to narrowly oblong; lower ones rarely 3-parted: stipels rare and small, subulate or oblong: fruit blackish, but strongly whitened with a glaucous mealy bloom, larger than in *S. Canadensis*. — Nutt. in Torr. & Gray, Fl. ii. 13; Wats. Bot. King Exp. 134; Gray, Bot. Calif. i. 278, in part.— Oregon and throughout California, common near the coast, eastward to Idaho and Nevada.

— **S. Mexicána,** PRESL. Arborescent, with trunks sometimes 6 inches in diameter: leaves and young shoots pubescent (sometimes slightly so, sometimes cinereous or tomentulose-canescent): leaflets, &c., nearly as preceding: fruit (as far as seen) destitute of bloom. — Presl. in DC. Prodr. iv. 323; Gray, Pl. Wright. ii. 66, & Bot. Mex. Bound. 71. *S. glauca*, Benth. Pl. Hartw. 313; Gray, Bot. Calif. l. c. in part. *S. velutina*, Durand in Pacif. R. Rep. v. 8. — California, from Plumas Co. southward to mountains of Arizona, and New Mexico on the Mexican border. Glabrate forms too near *S. Canadensis*. (Mex.)

3. **VIBÚRNUM,** L. (Classical Latin name of the WAYFARING-TREE, *V. Lantana*, of Europe.) — Shrubs or small trees (of various parts of the world); with tough and flexible branches, simple and not rarely stipulate or pseudo-stipulate leaves, and terminal depressed cymes of mostly white flowers, produced in spring or early summer. — *Viburnum* and *Opulus*, Tourn.

V. TINUS, L. (*Tinus*, Tourn., Œrst.), the LAURESTINUS, cultivated from Europe, with putamen not flattened and ruminated albumen, is left out of view in our character of the genus, as also the outlying forms with campanulate or more tubular corolla, upon which Œrsted (in Vidensk. Meddel. 1860) has founded genera, with more or less reason. The albumen in the N. American species is even, or obscurely ruminated in the first species.

§ 1. Cyme radiant; marginal flowers neutral, with greatly enlarged flat corollas as in *Hydrangea*: drupes coral-red turning dark crimson or purple, not acid: putamen sulcate: leaves pinnately straight-veined, scurfy: winter-buds naked.

— **V. lantanoídes,** MICHX. (HOBBLEBUSH.) Low and straggling, with thickish branches, sometimes 10 feet high, scurfy-pubescent on the shoots and inflorescence: leaves ample (when full grown 6 inches long), conspicuously petioled, rounded-ovate, abruptly acuminate, finely doubly serrate, membranaceous, minutely stellular-pubescent and glabrate above, rusty-scurfy beneath on the 10 or 12 pairs of prominent veins, and when young also on the very numerous transverse connecting veinlets: stipules small and subulate, or obsolete: fruit ovoid, flattish; the stone moderately flattened, 3-sulcate on one face, broadly and deeply sulcate on the other, and the groove divided by a strong median ridge, the edges also slightly sulcate: seed reniform in cross section and somewhat lobed; the albumen not ruminated. — Fl. i. 179; Torr. & Gray, Fl. i. 18; Audubon, Birds Amer. i. t. 148. *V. alnifolium*, Marsh. Arbust. 162. *V. Lantana*, var. *grandiflorum*, Ait. Kew. i. 372. *V. grandiflorum*, Smith in Rees Cycl. — Moist woods, New Brunswick and Canada to N. Carolina in the higher mountains; fl. spring. (Japan?)

§ 2. Cyme radiant, or not so: drupes light red, acid, edible, globose: putamen very flat, orbicular, even (not sulcate nor intruded or costate): leaves palmately veined: winter-buds scaly. — *Opulus*, Tourn.

V. Opulus, L. (HIGH CRANBERRY, CRANBERRY-TREE.) Nearly glabrous, occasionally pubescent, 4 to 10 feet high: leaves dilated, three-lobed, roundish or broadly cuneate at 3-ribbed or pedately 5-ribbed base; the lobes acuminate, incisely dentate or in upper leaves entire: slender petioles bearing 2 or more glands at or near summit, and usually setaceous stipules near base: cymes rather ample, terminating several-leaved branches, radiant. — Spec. i. 268; Ait. Kew. i. 373 (var. *Americanum*); Michx. Fl. i. 180 (vars.); Torr. & Gray, l. c. *V. trilobum*, Marsh. Arbust. 162. *V. opuloides*, Muhl. Cat. *V. Oxycoccus* & *V. edule*, Pursh, Fl. i. 203. — Swamps and along streams, New Brunswick to Saskatchewan, Brit. Columbia and Oregon, and in Atlantic States south to Pennsylvania. Variable in foliage; no constant difference from the European, which is cultivated, in a form with most flowers neutral, as SNOWBALL and GUELDER ROSE. (Eu., N. Asia.)

V. pauciflórum, PYLAIE. Glabrous or with pubescence, 2 to 5 feet high, straggling: leaves of roundish or broadly oval outline, unequally dentate, many of them either obsoletely or distinctly 3-lobed (the lobes not longer than broad), about 5-nerved at base, loosely veiny: cymes small, terminating short and merely 2-leaved lateral branches, involucrate with slender subulate caducous bracts, destitute of neutral radiant flowers: stamens very short: fruit nearly of preceding. — Pylaie, Herb.; Torr. & Gray, Fl. ii. 17; Herder, Pl. Radd. iii. t. 1, f. 3. *V. acerifolium*, Bong. Veg. Sitka, 144. — Cold moist woods, Newfoundland and Labrador, mountains of New England to Saskatchewan, west to Alaska and Washington Terr., southward in the Rocky Mountains to Colorado.

§ 3. Cyme never radiant: drupes blue, or dark-purple or black at maturity.

* Leaves palmately 3–5-ribbed or nerved from the base, slender-petiolate: stipules subulate-setaceous: pubescence simple, no scurf: primary rays of pedunculate cyme 5 to 7: filaments equalling the corolla.

+ Pacific species: drupe oblong-oval, nearly half-inch long, bluish-black.

V. ellípticum, HOOK. Stems 2 to 5 feet high: winter-buds scaly: leaves from orbicular-oval to elliptical-oblong, rounded at both ends, dentate above the middle, not lobed, at length rather coriaceous, 3–5-nerved from the base, the nerves ascending or parallel: corollas 4 or 5 lines in diameter: stone of fruit deeply and broadly sulcate on both faces; the furrow of one face divided by a median ridge. — Hook. Fl. i. 280; Gray, Bot. Calif. i. 278. — Woods of W. Washington Terr. and Oregon (first coll. by *Douglas*), to Mendocino and to Placer Co., California, *Kellogg, Mrs. Ames.*

+ + Atlantic species: drupe globular, quarter-inch long, bluish-purple or black when ripe: cyme mostly with a caducous involucre of 5 or 6 small and subulate or linear thin bracts.

V. acerifólium, L. (ARROW-WOOD, DOCKMACKIE.) Soft-pubescent, or glabrate with age, 3 to 6 feet high, with slender branches: winter-buds imperfectly scaly: leaves membranaceous, rounded-ovate, 3-ribbed from the rounded or subcordate base, and with 3 short and acute or acuminate divergent lobes (or some uppermost undivided), usually dentate to near the base (larger 4 or 5 inches long): cymes rather small and open: corolla 2 or 3 lines in diameter: stone of drupe lenticular, hardly sulcate on either side. — Spec. i. 268; Vent. Hort. Cels. t. 272; Michx. Fl. i. 180; Wats. Dendr. Brit. ii. t. 118 (poor); Hook. Fl. l. c. (partly); Torr. & Gray, l. c. 17; Emerson, Trees of Mass. ii. t. 19. — Rocky and cool woods, New Brunswick to Michigan, Indiana, and N. Carolina.

V. densiflórum, CHAPM. Lower, 2 to 4 feet high: leaves smaller (inch or two long), with mostly shorter lobes or sometimes none: cyme denser: involucrate bracts more conspicuous and less caducous: stone of the drupe undulately somewhat 2-sulcate on one face and 3-sulcate on the other. — Fl. ed. 2, Suppl. 624. — Wooded hills, W. Florida, *Chapman.* Also, Taylor Co., Georgia, *Neisler*, a glabrate form. Too near *V. acerifolium.*

* * Leaves pinnately and conspicuously veiny with straight veins (impressed-plicate above, prominent beneath and the lowest pair basal), thinnish, coarsely dentate: stipules subulate-setaceous: cymes pedunculate, about 7-rayed: stone of the drupe more or less sulcate. ARROW-WOOD.

+ Stone and seed flat, slightly plano-convex: leaves all short-petioled or subsessile.

— **V. pubéscens,** PURSH. Slender, 2 to 5 feet high; leaves oblong- or more broadly ovate, acute or acuminate, acutely dentate-serrate (1½ to 3 inches long, on petioles 2 to 4 lines long, or upper hardly any), soft-tomentulose with simple downy hairs beneath, but varying to slightly pubescent (and in one form almost glabrous with upper face lucidulous): peduncle generally shorter than the cyme: drupe oval, 4 lines long, blackish-purple, flattened when young; stone lightly 2-sulcate on the faces, margins narrowly incurved, no intrusion on ventral face. — Fl. i. 202 (excl. habitat, and syn. Michx.); Torr. Fl. i. 320; DC. Prodr. iv. 326; Hook. Fl. i. 280; Torr. & Gray, Fl. ii. 16; Gray, Man. ed. 5, 206; (Erst. l. c. t. 7, fig. 21), 22. *V. dentatum*, var. *pubescens*, Ait. Kew. i. 372 ? *V. dentatum*, var. *semitomentosum*, Michx. Fl. i. 179, in small part (spec. from L. Champlain). *V. villosum*, Raf. in Med. Rep. 1808, & Desv. Jonr. Bot. i. 228, uot Swartz. *V. Rafinesquianum*, Rœm. & Schult. Syst. v. 630. — Rocky ground, Lower Canada to Saskatchewan, west to Illinois, south to Stone Mountain, Georgia. (Not, as Pursh would have it, in the lower parts of Carolina.)

+ + Stone deeply sulcate-intruded ventrally: transverse section of seed about three-fourths annular, with flattish back: leaves rather slender-petioled.

— **V. dentátum,** L. Shrub 5 to 15 feet high, with ascending branches, glabrous or nearly so, no stellular pubescence: leaves from orbicular- to oblong-ovate, with rounded or sub-cordate base, acutely many-dentate (2 or 3 inches long); primary veins 8 to 10 pairs (some of them once or twice forked), often a tuft of hairs in their axil: peduncle generally longer than the cyme: drupe ovoid, three lines long, terete, bright blue, darker at maturity. — Spec. i. 268; Jacq. Hort. Vind. i. t. 36; Torr. l. c.; Wats. Dendr. Brit. t. 25; Torr. & Gray, l. c., excl. var.; Gray, Man. l. c. *V. dentatum*, var. *lucidum*, Ait. Kew. l. c. — Wet ground, chiefly in swamps, New Brunswick to Michigan, and south to the mountains of Georgia. Seems to pass into following, but the extremes widely different.

— **V. mólle,** MICHX. Young shoots, petioles, cymes, &c. beset with stellular pubescence: leaves orbicular or broadly oval to ovate, more crenately dentate, soft-pubescent at least beneath (larger 4 inches long); veins of the preceding or fewer; petioles shorter: drupe 4 lines long, more pointed by the style: calyx-teeth more conspicuous. — Fl. i. 180, but foliage only seen; Gray, Man. ed. 3 & ed. 5, 206. *V. dentatum*, var. *semitomentosum*, Michx. l. c. in large part; Ell. Sk. i. 365. *V. dentatum*, var. ? *scabrellum*, Torr. & Gray, Fl. ii. 16. *V. scabrellum*, Chapm. Fl. i. 72. — Coast of New England (Martha's Vineyard, *Bessey*) to Texas; flowers at the north in summer, later than *V. dentatum*.

* * * Leaves lightly or loosely pinnately veined, of firmer or somewhat coriaceous texture, petioled, mostly glabrous: stipules or stipule-like appendages none: mature drupes black or with a blue bloom, mealy and saccharine; the stone and seed flat or lenticular, plane: winter-buds of few and firm scales: petioles and rays of the cyme mostly lepidote with some minute rusty scales or scurf.

+ Cymes peduncled, about 5-rayed: drupes globose-ovoid, 3 lines long: stone orbicular, flattened-lenticular: shrubs 5 to 8 or 12 feet high, in swamps.

— **V. cassinoídes,** L. (WITHE-ROD.) Shoots scurfy-punctate: leaves thickish and opaque or dull, ovate to oblong, mostly with obtuse acumination, obscurely veiny (1 to 3 inches long), with margins irregularly crenulate-denticulate or sometimes entire: peduncle shorter than the cyme. — Spec. ed. 2, ii. 384 (pl. Kalm), excl. syn., at least of Mill. & Pluk.; Torr. Fl. i. 318; DC. l. c. *V. squamatum*, Willd. Enum. i. 327; Wats. Dendr. Brit. t 24. *V. pyrifolium*, Pursh, Fl. i. 201, not Poir. *V. nudum*, Hook. Fl. i. 279; Emerson, Trees of Mass. ed. 2, 411, t. 18. *V. nudum*, var. *cassinoides*, Torr. & Gray, Fl. ii. 14; Gray, Man. l. c. — Swamps, Newfoundland to Saskatchewan, New England to New Jersey and Pennsylvania: flowers earlier than the next.

— **V. núdum,** L. Obscurely scurfy-punctate: leaves more veiny, oblong or oval, sometimes narrower, entire or obsoletely denticulate, lucid above (commonly 2 to 4 inches long): peduncle usually equalling the cyme. — Spec. i. 268 (pl. Clayt.); Mill. Ic. t. 274; Willd. Spec. i. 1487; Michx. Fl. i. 178; Sims, Bot. Mag. t. 2281; Torr. & Gray, l. c., var. *Claytoni*. — Swamps, New Jersey or S. New York to Florida and Louisiana: fl. summer, or southward in spring.

Var. angustifólium, TORR. & GRAY, l. c. Leaves linear-oblong or oblong-lanceolate. — *V. nitidum*, Ait. Kew. i. 371, ex. char. — N. Carolina to Louisiana.
Var. grandifólium. Larger leaves 8 inches long, 4 wide. — E. Florida, *Mrs. Treat*.
Var. serótinum, RAVENEL, in Chapm. Fl. Suppl. 624. A strict or more simple-stemmed form, with foliage of the type, and smaller blossoms, produced in November! — On the Altamaha River, near Darien, Georgia, *Ravenel*.

+ + Compound cymes sessile, of 3 to 5 cymiferous rays, subtended by the upper leaves,

++ Many-flowered: trees or arborescent, 10 to 30 feet high: winter-buds minutely rusty-scurfy or downy, ovoid and acuminate: leaves ovate or oval, lucid, closely and acutely serrate, abruptly rather long-petioled : drupes comparatively large, oval, 5 to 7 lines long, when ripe sweetish and black or bluish from the bloom, with very flat stone. — BLACK HAW, SHEEP-BERRY, SWEET VIBURNUM.

— V. Lentágo, L. Often arboreous: leaves ovate, acuminate (larger 3 or 4 inches long), thickly beset with very sharp serratures : petioles mostly undulate-margined : larger winter-buds long-pointed, grayish. — Spec. i. 268; Michx. l. c.; Wats. Dendr. Brit. t. 21; Hook. l. c.; Torr. & Gray, l. c. 15. — Woods and banks of streams, Canada to Saskatchewan, Missouri, and mountains of Georgia; fl. spring.

— V. prunifólium, L. Seldom arboreous: leaves from roundish to ovate or oval with little or no acumination and finer serratures (larger ones 2 or 3 inches long) : petioles naked, or on strong shoots narrowly margined, these and the less pointed winter-buds often rufous-pubescent. — Spec. i. 268 (*Mespilus prunifolia*, &c., Pluk. Aln. t. 4, f. 2); Michx. l. c.; Duham. Arb. ii. t. 38 (Wats. Dendr. Brit. t. 23?); Torr. & Gray, l. c. *V. pyrifolium*, Poir. Dict. v. 658; Wats. Dendr. Brit. t. 22. — Dry or moist ground, New York (and Upper Canada?) to Michigan, Illinois, and south to Florida, Texas, and Kansas: flowering early.

++ ++ Cymes (3-4-rayed) and the lucid coriaceous commonly entire leaves small.

— V. obovátum, WALT. Shrub 2 to 8 feet high: leaves from obovate to cuneate-spatulate or oblanceolate, obtuse or retuse, with some obsolete teeth or none (half-inch to thrice that length), narrowed at base into very short petiole : flowering cymes little surpassing the leaves: drupes oval, 5 lines long, black; stone thickish-lenticular, the faces obscurely sulcate. — Walt. Car. 116; Pursh, Fl. i. 201; Ell. Sk. i. 366; Lodd. Bot. Cab. t. 1476; DC. Prodr. iv. 326. *V. cassinoides* (Mill. Ic. t. 83?); Willd. Spec. i. 1491; Michx. Fl. i. 179, not L. *V. laevigatum*, Ait. Kew. i. 371; Pursh, l. c.; DC. l. c. — Wooded banks of streams and swamps, Virginia to Florida in the low country.

4. TRIÓSTEUM, L. FEVERWORT, HORSE-GENTIAN. (Name shortened by Linnæus from *Triosteospermum*, Dill., meaning three bony seeds or stones to the fruit.) — Coarse perennial herbs (of Atlantic N. America, one Japanese and one Himalayan); with simple stems, ample entire or sinuate leaves more or less connate at base, and pinnately veiny; the dull-colored sessile flowers in their axils, either single or 2 to 4 in a cluster, produced in early summer, followed by orange-colored and reddish drupes. In our species the foliaceous linear calyx-lobes are as long as the corolla (about half-inch), and longer than the fruit. — Lam. Ill. t. 150; Gærtn. Fruct. t. 26. *Triosteospermum*, Dill. Elth. 394, t. 293.

— T. perfoliátum, L. Minutely soft-pubescent, or stem sometimes hirsute, stout, 2 to 4 feet high : leaves ovate to oblong, acuminate, narrowed below either to merely connate or more broadened and connate-perfoliate base : corolla dull brownish-purple : nutlets of the drupe 3-ribbed on the back. — Spec. i. 176; Schk. Handb. t. 41; Bigel Med. Bot. i. 90, t. 19; Bart. Veg. Mat. Med. t. 4; Sweet, Brit. Fl. Gard. ser. 2, t. 45; Torr. & Gray, Fl. ii. 12. *T. majus*, Michx. Fl. i. 107. — Alluvial or rich soil, Canada and New England to Illinois and Alabama. — Also called TINKER'S-WEED, WILD COFFEE, &c.

— T. angustifólium, L. l. c. Smaller : stem hirsute or hispid : leaves oblong-lanceolate or narrower, tapering above the more or less connate bases : corolla yellowish. — Torr. & Gray, l. c. *T. minus*, Michx. l. c. *Periclymenum herbaceum*, &c., Pluk. Aln. t. 104, f. 2. — Shady grounds, Virginia to Alabama, Missouri, and Illinois.

5. LINNÆA, Gronov. TWIN-FLOWER. (Dedicated to *Linnæus.*) — Gronov. in L. Gen. ed. i. 188. — Single species; fl. early summer.

— **L. boreális,** GRONOV. Trailing and creeping evergreen, with filiform branches, somewhat pubescent: leaves obovate and rotund, half-inch to inch long, crenately few-toothed, somewhat rugose-veiny, tapering into a short petiole: peduncles filiform, terminating ascending short leafy branches, bearing at summit a pair of small bracts, and from axil of each a filiform one-flowered pedicel, occasionally the axis prolonged and bearing another pair of flowers; pedicels similarly 2-bracteolate at summit, and a pair of larger ovate glandular-hairy inner bractlets subtending the ovary, soon connivent over it or enclosing and even adnate to the akene-like fruit: flowers nodding: corolla purplish rose-color, rarely almost white, sweet-scented, half-inch or less long. — L. Fl. Lapp. t. 12, f. 4, & Spec. ii. 631; Wahl. Fl. Lapp. 171, t. 9, f. 3; Fl. Dan. t. 3; Schk. Handb. t. 176; Lam. Ill. t. 536; Torr. & Gray, Fl. ii. 3. — Cool woods and bogs, New England to New Jersey and mountains of Maryland, north to Newfoundland and the Arctic Circle, westward in the Rocky Mountains to Colorado and Utah, the Sierra Nevada in Plumas Co., California, and northwest to Alaskan Islands; in Oregon, &c. Var. LONGIFLORA, Torr. in Wilkes S. Pacif. E. Ex. xvii. 287, with longer and more funnelform corolla. (N. Eu., N. Asia, &c.)

6. SYMPHORICÁRPOS, Dill. SNOWBERRY, INDIAN CURRANT. (Συμφορέω, to bear together, καρπός, fruit, the berry-like fruits mostly clustered or crowded.) — Low and branching shrubs (N. American and Mexican), erect or diffuse, not climbing; with small and entire (occasionally undulate or lobed, very rarely serrate) and short-petioled leaves, scaly leaf-buds, and 2-bracteolate small flowers, usually crowded in axillary or terminal spikes or clusters, rarely solitary, produced in summer; the corolla white or pinkish. — Dill., Elth. 371. t. 278; Juss. Gen. 211; DC. Prodr. iv. 338; Benth. & Hook. Gen. ii. 4; Gray, Jour. Linn. Soc. xiv. 9. *Symphoria,* Pers. Syn. i. 214.

§ 1. Short-flowered: corolla urceolate- or open-campanulate, only 2 or 3 lines long.

* Style bearded: fruit red: flowers all in dense and short axillary clusters: corolla 2 lines long, glandular within at base.

— **S. vulgáris,** MICHX. (CORAL-BERRY, INDIAN CURRANT.) Soft-pubescent or glabrate: branches slender, often virgate, flowering from most of the axils: leaves oval, seldom over inch long, exceeding the (1 to 4) glomerate or at length spiciform dense flower-clusters in their axils: corolla sparingly bearded inside: fruits very small, dark red. — Fl. i. 106: DC. Prodr. iv. 339; Torr. & Gray, Fl. ii. 4; Gray in Jour. Linn. Soc. l. c. 10. *Symphoricarpos,* Dill. l. c. *S. parriflora,* Desf. Cat., &c. *Lonicera Symphoricarpos,* L. Spec. i. 175. *Symphoria conglomerata,* Pers. l. c. *S. glomerata,* Pursh, Fl. i. 162. — Banks of streams and among rocks. W. New York and Penn. to Illinois, Nebraska, and Texas.

Var. **spicátus** (*S. spicatus,* Engelm. in Pl. Lindh. ii. 215) is a form with fructiferous spikes more elongated, sometimes equalling the leaves. — Texas, *Lindheimer.*

* * Style glabrous: fruit white, in terminal and upper axillary clusters, or solitary in some axils.

— **S. occidentális,** HOOK. (WOLF-BERRY.) Robust, glabrous, or slightly pubescent: leaves oval or oblong, thickish (larger 2 inches long): axillary flower-clusters not rarely pedunculate, sometimes becoming spicate and inch long: corolla 3 lines high, 5-cleft to beyond the middle, within densely villous-hirsute with long beard-like hairs: stamens and style more or less exserted. — Fl. i. 285; Torr. & Gray in Fl. ii. 4: Gray in Jour. Linn. Soc. l. c. *Symphoria occidentalis,* R. Br. in Richards. App. Frankl. Jour. — Rocky ground, Michigan to the mountains of Colorado, Montana (and Oregon?), north to lat. 64°.

—**S. racemósus,** MICHX. (SNOW-BERRY.) More slender and glabrous: leaves round-oval to oblong (smaller than in the preceding): axillary clusters mostly few-flowered, or lowest one-flowered: corolla 2 lines high, 5-lobed above the middle, moderately villous-bearded within, narrowed at base: stamens and style not exserted. — Fl. i. 107; Hook. l. c.; Torr. & Gray, l. c.; Gray, l. c. *Symphoria racemosa,* Pers. l. c.; Pursh, Fl. i. 169; R. Br. Bot.

Mag. t. 2211; Lodd. Bot. Cab. t. 230; Bart. Fl. Am. Sept. i. t. 19. *S. elongata* and *S. heterophylla*, Presl, ex DC.— Rocky banks, Canada and N. New England to Penn., Saskatchewan, and west to Brit. Columbia and W. California, even to San Diego Co.

Var. pauciflórus, ROBBINS. Low, more spreading: leaves commonly only inch long: flowers solitary in the axils of upper ones, few and loosely spicate in the terminal cluster.— Gray, Man. & in Jour. Linn. Soc. l. c. — Mountains of Vermont and Penn., Niagara Falls to Wisconsin and northward, in Rocky Mountains south to Colorado, west to Oregon.

S. móllis, NUTT. Low, diffuse or decumbent, soft-pubescent, even velvety-tomentose, sometimes glabrate: leaves orbicular or broadly oval (half to full inch long): flowers solitary or in short clusters: corolla open-campanulate and with broad base (little over line high), 5-lobed above the middle, barely pubescent within: stamens and style included.— Torr. & Gray, Fl. l. c.; Gray, l. c. & Bot. Calif. i. 279. *S. ciliatus*, Nutt. in Torr. & Gray, l. c., a glabrate form, from the char.— Wooded hills, California, both in the Coast Ranges and the Sierra Nevada, first coll by *Coulter* and *Nuttall*.

Var. acútus. Not improbably a distinct species, but materials incomplete: leaves very soft-tomentulose, oblong-lanceolate to oblong, acute at both ends or acuminate, sometimes irregularly and acutely dentate.— *S. mollis?* Torr. in Wilkes Pacif. E. Ex. xvii. 328.— Washington Terr. east of the Cascade Mountains, *Pickering & Brackenridge*, with the narrower and entire leaves. Lassen's Peak, N. E. California, *Mrs. Austin*, with broader leaves, commonly having 3 or 4 unequal serratures on each margin.

§ 2. Longer-flowered: corolla from oblong-campanulate to salverform, 5-lobed only at summit: fruit (in the Mexican *S. microphyllus* flesh color, ex Bot. Mag. t. 4975) in ours white: flowers mostly axillary: leaves small.

* Style glabrous: corolla with broad and short lobes slightly or merely spreading.

S. rotundifólius, GRAY. Tomentulose to glabrate: leaves from orbicular to oblong-elliptical, thickish (half to three-fourths inch long): corolla elongated-campanulate, 3 or 4 lines long; its tube pubescent within below the stamens, twice or thrice the length of the lobes: nutlets of the drupe oval, equally broad and obtuse at both ends.— Pl. Wright. ii. 66, Jour. Linn. Soc. l. c., & Bot. Calif. i. 279. *S. montanus*, Wats. Bot. King Exp. 132, partly.— Mountains of New Mexico and adjacent Texas to those of Utah, N. W. Nevada, adjacent California, and north to Mt. Paddo, Washington Terr., *Suksdorf*: first coll. by *Wright* and *Bigelow*.

S. oreóphilus, GRAY. Glabrous or sometimes with soft pubescence: leaves oblong to broadly oval, thinner: corolla more tubular or funnelform, 5 or 6 (rarely only 4) lines long; its tube almost glabrous within, 4 or 5 times the length of the lobes: nutlets of the drupe oblong, flattened, attenuate and pointed at base.— Jour. Linn. Soc. l. c. 12, & Bot. Calif. l. c. *S. montanus*, Gray in Am. Jour. Sci. xxxiv. 249, not HBK.— Mountains of Colorado, Utah, and Arizona, to the Sierra Nevada, California, and E. Oregon; first coll. by *Parry*.

* * Style bearded: corolla with oblong widely spreading lobes.

S. longiflórus, GRAY, l. c. Glabrous or rarely minutely pubescent, glaucescent: leaves spatulate-oblong varying to oval, thickish, small (quarter to half inch long): corolla white, salverform, slender; the tube 4 to 6 and lobes one and a half lines long, very glabrous within: anthers linear, subsessile, half included in the throat: nutlets of the fruit oblong.— Mountains of S. Nevada and Utah, *Miss Searls, Parry, Ward, Palmer*, &c. Apparently also S. W. Texas, *Havard*.

7. **LONÍCERA**, L. HONEYSUCKLE, WOODBINE. (*Adam Lonitzer*, Latinized *Lonicerus*, a German herbalist.) — Shrubs of the northern hemisphere, some erect, others twining; with normally entire leaves, occasionally on some shoots sinuate-pinnatifid; the flowers variously disposed, produced in spring or early summer.

§ 1. XYLÓSTEON, DC. Flowers in pairs (rarely threes) from the axils of the leaves, the common peduncle bibracteate at summit, the ovaries of the two either distinct or connate: ours (the genuine species of the section)- all erect and branching shrubs, with rather short corollas; the calyx-limb minute or obsolete. — *Xylosteon*, Tourn., Juss. *Xylosteum*, Adans., Michx., &c.

* Bracts at the summit of the peduncle small or narrow, often minute, sometimes obsolete or caducous: bractlets to the two flowers minute or none.

+ Leaves glaucescent or pale both sides, oblong-elliptical, very short-petioled, reticulate-venulose beneath: corolla ochroleucous, sometimes purplish-tinged, 4 to 6 lines long.

— **L. cærúlea,** L. A foot or two high, from villous-pubescent to glabrous or nearly so: leaves little over inch long, very obtuse: peduncles shorter than the flowers, usually very short: corolla moderately gibbous at base, not strongly bilabiate (sometimes glabrous, sometimes hairy): bracts subulate or linear, commonly larger than the ovaries; these completely united, forming a globular 2-eyed (black and with the bloom blue) sweet-tasted berry. — Spec. i. 174; Pall. Fl. Ross. t. 37; Sims, Bot. Mag. t. 1965; Jacq. Fl. Austr. v. Suppl. t. 17; Hook. Fl. j. 283; Torr. & Gray, Fl. ii. 9; Herder, Pl. Radd. iii. 15, t. 3. *L. villosa* (Muhl. Cat.) & *L. velutina*, DC. Prodr. iv. 337, excl. syn. in part. *Xylosteum villosum*, Michx. Fl. i. 106 (the very villous or hirsute form, *L. cærulea*, var. *villosa*, Torr. & Gray, l. c.); Bigel. Fl. Bost. ed. 2, 88; Richards. App. Frankl. Jour. *X. Solonis*, Eaton, Man. Bot. 518. — Moist ground, Newfoundland and Labrador, south to the cooler parts of New England, Wisconsin, &c., north to the Arctic Circle, west to Alaska, and south in the higher mountains to the Sierra Nevada, California. The American and E. Asian forms somewhat different from the European. (Eu., N. Asia.)

— **L. oblongifólia,** Hook. A yard or more high, minutely puberulent to glabrous, glaucescent: leaves 1 to 3 inches long: peduncles filiform, commonly inch long: corolla with conspicuous gibbosity at base, deeply bilabiate, the narrow lower lip separate far below the middle: bracts minute or caducous: ovaries either distinct, or united at base, or completely connate (even on the same plant): berries red or changing to crimson, mawkish. — Fl. i. 284. t. 100; Torr. & Gray, l. c. *L. villosa*, DC. l. c. in part. *Xylosteum oblongifolium*, (Goldie in Edinb. Phil. Jour. vi. 323. — Bogs, Canada and N. New England and New York to Michigan.

+ + Leaves bright green, thinnish, ovate or oblong: peduncles slender: berries red: shrubs with slender spreading or straggling branches.

++ Corolla dark dull purple, strongly bilabiate: calyx-teeth subulate: bracts subulate, caducous.

— **L. conjugiális,** Kellogg. Leaves pubescent when young, ovate or oval, often acuminate, short-petioled (1 to 2½ inches long): peduncles at least thrice the length of the flowers: corolla 4 or 5 lines long, gibbous-campanulate, with upper lip crenately 4-lobed; throat with lower part of filaments and style very hirsute: ovaries two-thirds or wholly connate. — Proc. Calif. Acad. ii. 67, fig. 15; Wats. Bot. King Exp. 133. *L. Breweri*, Gray, Proc. Am. Acad. vi. 537, vii. 349. — Woods of the Sierra Nevada, California and adjacent Nevada, at 6,000–10,000 feet, first coll. by *Veatch*. Also mountains of Washington Terr., *Howell, Suksdorf*.

++ ++ Corolla honey-yellow or ochroleucous, rarely a slight tinge of purple, oblong-funnelform, two-thirds to three-fourths inch long, with 5 short almost equal lobes: the tube with a small but prominent saccate gibbosity at base, merely pilose-pubescent within: calyx-limb barely crenate-lobed or truncate: divergent ovaries and mostly the berries quite distinct, subtended by very small subulate bracts, and each with minute rounded bractlets.

— **L. Utahénsis,** Wats. Leaves oval or elliptical-oblong, rounded at both ends, very short-petioled, glabrous or nearly so from the first, or soon glabrate, not ciliate, reticulate-venulose at maturity (inch or two long): peduncle seldom over half-inch long. — Bot. King Exp. 133. — Mountains of Utah, *Watson, Parry, Siler*. Montana, and Cascades from Oregon to Brit. Columbia.

— **L. ciliáta,** Muhl. (Fly-Honeysuckle.) Leaves ovate to oval-oblong. acutish or somewhat acuminate, loosely pilose-pubescent when young, especially the margins, 2 inches long at maturity, more distinctly petioled: full-grown peduncles two-thirds to nearly inch long: berries distinct, light red, watery. — Cat. 22; DC. Prodr. iv. 235; Hook. Fl. l. c.; Torr. & Gray, l. c. *L. Canadensis*, Rœm. & Schult. Syst. v. 260. *Xylosteum Tartaricum*, Michx.

Fl. i. 106. *X. ciliatum*, Pursh, Fl. i. 161, excl. var., which is *Symphoricarpos racemosus* according to Nutt. *Vaccinium album*, L. Spec. i. 350, specimen of Kalm. — Rocky moist woods, New Brunswick to the Saskatchewan, and New England to Penn. and Michigan. Flowering in spring, when the leaves are developing.

L. TARTÁRICA, L., of the Old World, with rose-colored flowers, is commonly planted as an ornamental shrub.

* * Bracts at the summit of the peduncle oblong to ovate or cordate and foliaceous: bractlets conspicuous and accrescent.

— **L. involucráta**, BANKS. Pubescent, sometimes glabrate, 2 to 10 feet high : leaves from ovate to oblong-lanceolate, from acutish to acuminate, 2 to 5 inches long, petioled : peduncles an inch or two long, sometimes 3-flowered : corolla yellowish, viscid-pubescent, half-inch or more long, tubular-funnelform, with 5 short hardly unequal lobes: bractlets 4 or united into 2, viscid-pubescent, at first short, obovate or obcordate, in fruit enlarging and enclosing or surrounding the two globose dark-purple or black berries. — Spreng. Syst. i. 759 ; DC. Prodr. l. c. 336; Lindl. Bot. Reg. t. 1119; Torr. & Gray, l. c.; Gray, Bot. Calif. i. 280. *L. Ledebourii*, Esch. Mem. Acad. Petrop. (1826) x. 284 ; DC. l. c. *L. Mociniana*, DC. l. c., probably from California, not Mexico. *L. intermedia*, Kellogg, Proc. Calif. Acad. ii. 154, fig. 47. *Xylosteum involucratum*, Richards. App. Frankl. Journ. 6. — Wooded grounds, from Gaspé Co., Lower Canada (*Allen*), and S. shore of Lake Superior northward, west to Alaska, southward in the Rocky Mountains to Colorado and Utah, and nearly throughout California.

§ 2. CAPRIFÓLIUM, DC. Flowers sessile in variously disposed terminal or axillary clusters, commonly quasi-verticillate-capitate : corolla more or less elongated : berries orange or red at maturity : stems climbing (twining) : upper leaves usually combined into a connate-perfoliate disk. — *Caprifolium*, Juss.

* Limb of corolla almost regular or slightly bilabiate, very much shorter than the elongated tube: stamens and style little exserted: flowers nearly scentless. — *Periclymenum*, Tourn. TRUMPET-HONEYSUCKLES.

— **L. sempérvirens**, L. Evergreen only southward, glabrous : leaves oblong, glaucous or glaucescent beneath, uppermost one or two pairs broadly connate : flowers in 2 to 5 more or less separated whorls of 6: the spike pedunculate : corolla scarlet-red varying to crimson and yellow inside, or sometimes wholly yellow; the narrow tube inch or more long; lobes sometimes almost equal, sometimes short-bilabiate, merely spreading, seldom over 2 lines long. — Spec. ii. 173 (Herm. Hort. Lugd. 484, t. 483); Ait. Kew, i. 230; Walt. Car. 131; Sims, Bot. Mag. t. 1781, & 1753; Bot. Reg. t. 556; Torr. & Gray, Fl. ii. 5; Meehan, Nat. Flowers, ser. 2, i. t. 45. *L. Virginiana* & *L. Caroliniana*, Marsh. Arbust. 80. *Caprifolium sempervirens*, Michx. Fl. 105; Pursh, Fl. i. 160; Ell. Sk. i. 271. — Low grounds, Connecticut and Indiana to Florida and Texas. Commonly cultivated. (There are indications of a nearly related species in Lower California.)

— **L. ciliósa**, POIR. Leaves ovate or oval, glaucous beneath, usually ciliate, otherwise glabrous; uppermost one or two pairs connate into an oval or orbicular disk : whorls of flowers single and terminal, or rarely 2 or 3, and occasionally from the axils of the penultimate pair of leaves, either sessile or short-peduncled : corolla glabrous or sparingly pilose-pubescent, yellow to crimson-scarlet, with thicker tube than the preceding, more ventricose-gibbous below; limb slightly bilabiate; lower lobe 3 or 4 lines long. — Dict. v. 612; DC. Prodr. iv. 333; Torr. & Gray, l. c. *Caprifolium ciliosum*, Pursh, Fl. i. 160. *C. occidentale*, Lindl. Bot. Reg. t. 1457. *Lonicera occidentalis*, Hook. Fl. i. 282. — Rocky Mountains in Montana to the coast of Brit. Columbia, the mountains of California and of Arizona. From mountains near Chico, California, comes a form which, by nearly naked margin of leaves and three-whorled pedunculate spike, makes transition to *L. sempervirens*.

* * Limb of corolla ringent; the spreading or recurved lips comparatively large, and stamens and style conspicuously exserted. — *Caprifolium*, Tourn. TRUE HONEYSUCKLES.

+— Tube of corolla elongated (fully inch long), wholly glabrous inside, as are stamens and style: flowers very fragrant: Atlantic species resembling the cultivated Italian or Sweet Honeysuckle of Middle and S. Europe, *L. Caprifolium*, L.

— **L. gráta**, AIT. Glabrous: leaves obovate or oblong and the upper one or two pairs connate, paler or somewhat glaucous beneath: flowers in terminal capitate cluster and from the axils of the connate-perfoliate leaves: corolla reddish or purple outside; the limb white within, fading to tawny yellow; lips over half-inch long; tube not gibbous: berries orange-red. — Kew. i. 231; Willd. Spec. i. 984; DC. Prodr. iv. 332: Darlingt. Fl. Cest. ed. 2. 159; Torr. & Gray, Fl. ii. 5. *Caprifolium gratum*, Pursh, Fl. i. 161. — Moist and rocky woodlands, N. New Jersey to Pennsylvania and mountains of Carolina according to Pursh, to "W. Louisiana, *Hale*," in Torr. & Gray, Fl. But it may be doubted if really different from *L. Caprifolium* of Europe, and if truly indigenous to this country.

+ + Tube of corolla less than inch long, but larger than the limb: the throat or tube below hairy within: Atlantic species.

++ Corolla bright orange-yellow: tube not gibbous, fully half-inch or more long: filaments and style glabrous: "flowers fragrant," produced early.

— **L. fláva**, SIMS. Somewhat glaucous, wholly glabrous: leaves broadly oval, 2 or 3 upper pairs connate into a disk: flowers in a terminal capitate cluster: corolla glabrous; the slender tube at upper part within or prolonged adnate base of filaments hirsute-pubescent. — Bot. Mag. t. 1318; Lodd. Bot. Cat. t. 338: DC. Prodr. iv. 332. *Caprifolium Fraseri*, Pursh, Fl. i. 160, excl. N. Y. habitat. *C. flavum*, Ell. Sk. i. 271. — "Exposed rocky summit of Paris Mountain in S. Carolina," in Laurens Co. *Fraser*. This very ornamental plant was first noticed in Drayton's View of South Carolina, published in 1802, p. 64, as growing on Paris Mountain, Greenville; afterwards it was collected by *Fraser*. Ell. l. c. Upper Georgia, *Boykin*, &c. It has not been found elsewhere; but it is still sparingly in cultivation.

++ ++ Corolla shorter, more or less hirsute within the throat: tube usually somewhat gibbous.

= Rather freely twining and high-climbing, little or not at all glaucous, pubescent: leaves deep green above.

— **L. hirsúta**, EATON. Leaves oval, conspicuously veiny and venulose both sides (3 or 4 inches long), soft-pubescent (as also usually the branchlets) and pale beneath; upper one or two pairs connate, lower short-petioled. corolla orange-yellow fading to dull purplish or brownish, more or less viscid-pubescent outside; tube half-inch long, little exceeding the limb; throat and lower part of filaments hirsute. — Eaton, Man. Bot. ed. 2. 307 (1818); Torr. Fl. i. 342; Hook. Bot. Mag. t. 3103, & Fl. i. 282; Torr. & Gray, Fl. ii. 6. *L. ciliosa* Muhl. Cat. 22, not DC. *L. Douglasii*, Hook. l. c., being *Caprifolium Douglasii*, Lindl. Trans. Hort. Soc. vii. 244; DC. l. c.; Loudon, Encl. Trees & Shrubs, 330, fig. 972. *L. parviflora*, var. ? Torr. & Gray, Fl. ii. 7. mainly. *L. pubescens*, Sweet, Hort. Brit. 194; DC. Prodr. iv. 332; Loudon, Encl. Trees & Shrubs, 529 (under *L. flava*). *L. Goldii*, Spreng. Syst. i. 758. *Caprifolium pubescens*, Goldie in Edinb. Phil. Jour. vi. 323; Hook. Exot. Fl. t. 27. — Rocky banks, &c., Northern New England and Canada to Penn., Michigan, and north shore of Lake Superior to the Saskatchewan.

= = Feebly twining or merely sarmentose or bushy, 2 to 6 feet high. conspicuously glaucous.

— **L. Sullivántii**, GRAY. At length much whitened with the glaucous bloom, 3 to 6 feet high, glabrous: leaves oval and obovate-oblong, thickish, 2 to 4 inches long, all those of flowering stems sessile, and most of them connate, the uppermost into an orbicular disk: corolla pale yellow, glabrous outside; tube half-inch or less long, little longer than the limb: filaments nearly glabrous. — Proc. Am. Acad. xix. 76. — *L.* n. sp.? Sulliv. Cat. Pl. Columb. 37. *L. flava*, var. Torr. & Gray, Fl. ii. 6; Gray, Man., mainly. — Central Ohio to Illinois, Wisconsin, and Lake Winnipeg, also Tennessee and apparently in mountains of N. Carolina.

— **L. glaúca**, HILL. Glabrous, or sometimes lower face of leaves tomentulose-puberulent, 3 to 5 feet high, generally bushy: leaves oblong, often undulate (glaucous, but less whitened than in the preceding, 2 or at most 3 inches long), 2 to 4 upper pairs connate: corolla quite glabrous outside, greenish yellow or tinged or varying to purple, short; the tube only 3 or 4 lines long, rather broad, nearly equalled by the limb, within and also style and base of filaments hirsute. — Hort. Kew. (1769) 446. t. 18: Gray, Proc. Am. Acad. xix. 77. *L. dioica*, L. Syst. Veg. 215; Ait. Kew. i. 230; Bot. Reg. t. 138. but not diœcious. *L. media*, Murr. in Comm. Gœtt. 1776. 28. t. 3. *L. parviflora*, Lam. Dict. i. 728 (1783); Torr. Fl. i. 245; DC.

l. c.; Hook. l. c.; Torr. & Gray, l. c. excl. var.; Gray, Man., and a part of var. *Douglasii.* *Caprifolium glaucum,* Mœnch, Meth. 502. *C. bracteosum,* Michx. Fl. i. 105. *C. parviflorum,* Pursh, Fl. i. 161. *C. dioicum,* Rœm. & Schult. Syst. v. 260. — Rocky grounds, Hudson's Bay ! and to Saskatchewan, Canada, New England, Penn., and mountains of Carolina ?

L. albiflóra, TORR. & GRAY. Wholly glabrous, or with minute soft pubescence, bushy, also disposed to twine, 4 to 8 feet high : leaves oval, inch long, or little longer, glaucescent both sides, usually only uppermost pair connate into a disk and subtending the simple sessile glomerule : corolla white or yellowish-white, glabrous; the tube 3 to 5 lines long, hardly at all gibbous : style and filaments nearly naked. — Fl. ii. 6; Gray, Pl. Lindh. ii. 213. *L. dumosa,* Gray, Pl. Wright. ii. 66, Bot. Mex. Bound. 71, the minutely pubescent form. — Rocky prairies and banks, W. Arkansas and Texas to New Mexico and Arizona, first coll. by *Berlandier, Leavenworth, Lindheimer,* &c. (Adj. Mex., *Palmer.*)

  + + + Tube of corolla only quarter-inch long, equalled by the limb, gibbous, more or less hairy within : Pacific species.

L. hispídula, DOUGL. Bushy and sarmentose, often feebly twining : leaves small (inch or so in length, or the largest 2½ inches), oval, or from orbicular to oblong, rounded at both ends, or lower and short-petioled ones sometimes subcordate, uppermost connate or occasionally distinct : spikes slender, commonly paniculate, of few or several whorls of flowers : corolla from pink to yellowish, barely half-inch long : filaments and especially style more or less pubescent at base. — Dougl. in Lindl. Bot. Reg. t. 1761 (the latter figured and published the species as *Caprifolium hispidulum*); Gray, Proc. Am. Acad. viii. 627, & Bot. Calif. i. 280. *L. microphylla,* Hook. Fl. i. 283. — Polymorphous species, of which the typical form (var. *Douglasii,* Gray, l. c.) is hirsute or pubescent with spreading hairs, disposed to climb : lower leaves mostly short-petioled and inclined to subcordate, not rarely a foliaceous stipule-like appendage between the petioles on each side : inflorescence and pink corollas glabrous. — Wooded region of Brit. Columbia to Oregon, first coll. by *Douglas.*

Var. **vacíllans,** GRAY, l. c. Stem and leaves either glabrous or pubescent, with or without hirsute hairs : inflorescence and corollas pubescent or glandular, varying to glabrous : otherwise like the Oregon type. — *L. Califórnica,* Torr. & Gray, Fl. ii. 7; Benth. Pl. Hartw. *L. ciliosa,* Hook. & Arn. Bot. Beech. 143, 349, not Poir. *L. pilosa,* Kellogg, Proc. Calif. Acad. i. 62. — From Oregon to Monterey, California.

Var. **subspicáta,** GRAY, l. c. Bushy, more or less pubescent or glandular-pubescent above, at least the pale pink or yellowish flowers : leaves small (half-inch to inch long), even uppermost commonly distinct : stipule-like appendages rare. — *L. subspicata,* Hook. & Arn. Bot. Beech. 349 ; Torr. & Gray, l. c.; Torr. Bot. Mex. Bound. 71, t. 29. — Common in California, from Monterey to San Diego.

Var. **interrúpta,** GRAY, l. c. Like the preceding, or sometimes larger-leaved and more sarmentose, but glabrous or minutely puberulent, more glaucous : spikes commonly elongated, of numerous capitellate whorls : corolla perfectly glabrous, pinkish or yellowish, less hairy inside. — *L. interrupta,* Benth. Pl. Hartw. 313. — Common in California : also Santa Catalina Mountains, Arizona, *Pringle, Lemmon.*

8. **DIERVÍLLA,** Tourn. BUSH HONEYSUCKLE. (*Dr. Dierville* took the original species from Canada to Tournefort in the year 1708.) — Low shrubs (of Atlantic N. America, Japan, and China) : with scaly buds, simply serrate membranaceous leaves, and flowers in terminal or upper axillary naked cymes, produced in early summer. — The E. Asian species, *Weigela,* Thunb. (of which *D. Japonica* is common in cultivation), have ampliate and mostly rose-colored corollas, herbaceous calyx-lobes deciduous from the beak of the fruit, and reticulate-winged seeds. Ours have small and narrow-funnelform corollas, of honey-yellow color, thin-walled capsule, and close coat to the seed, the surface minutely reticulated ; herbage nearly glabrous. — Torr. & Gray, Fl. ii. 10.

D. trífida, MŒNCH. Branchlets nearly terete ; leaves ovate-oblong, acuminate, distinctly petioled : axillary peduncles more commonly 3-flowered : limb of the corolla nearly equalling the tube, sometimes irregular, three of the lobes more united, the middle one deeper yellow and villous on the face: capsule oblong, with a slender neck or beak, crowned with slender-subulate calyx-lobes. — Meth. 492; Torr. & Gray, l. c. excl. var. *D. Acadiensis fruticosa*, &c., Tourn. Act. Acad. Par. 1706, t. 7, f. 1; L. Hort. Cliff. 63, t. 7; Duham. Arb. ed. 1. *D. Tournefortii*, Michx. Fl. i. 107. *D. humilis*, Pers. Syn. i. 214. *D. Canadensis*, Willd. Enum. 222; DC. Prodr. iv. 330; Hook. Fl. i. 281. *D. lutea*, Pursh, Fl. i. 162. *Lonicera Diervilla*, L. Mat. Med. 62, & Spec. i. 175. — Rocky and shady ground, Newfoundland and Hudson's Bay to Saskatchewan, south to Kentucky and Maryland, and in the mountains to N. Carolina.

D. sessilifólia, BUCKLEY. Branchlets quadrangular; leaves ovate-lanceolate, gradually acuminate, closely sessile, of firmer texture, more acutely serrulate: cymes several-flowered; corolla-lobes nearly equal, shorter than the tube, one of them obscurely pilose: capsule short-oblong, short-necked, and crowned with short lanceolate-subulate calyx-lobes. — Am. Jour. Sci. xlv. 174; Chapm. Fl. 170; Fl. Serres, viii. 292. — Rocky woods and banks, mountains of Carolina and Tennessee, first coll. by Curtis.

Order LXX. RUBIACEÆ.

Herbaceous or woody plants; with opposite entire and stipulate leaves, varying to verticillate, or in the *Stellatæ* the leaves in whorls without stipules (unless accessory leaves be counted as such); mostly hermaphrodite regular flowers, either 5-merous or 4-merous; calyx-tube adnate to the ovary: and stamens as many as and alternate with the lobes of the corolla, inserted on its tube or throat. Style single, sometimes with 2 or more lobes or stigmas. Fruit various: seeds in our genera albuminous.

Of this vast and largely tropical order 26 of the 140 recognized genera come within our limits, but more than half of them only in subtropical Florida. They rank under 14 of the 25 recognized tribes, — too large a scaffolding for a fragmentary structure. So they are here disposed under three series; of which the third is only a special modification in foliage of the second.

Series I. CINCHONACEÆ. Ovules numerous in each cell.

* Fruit capsular: seeds numerous, flat, winged all round.

1. **EXOSTEMA.** Calyx with clavate tube, 5-toothed. Corolla salverform, with long and narrow tube and 5-parted limb; lobes long-linear, imbricated in the bud. Stamens inserted near the base of the corolla-tube: filaments and style filiform, exserted: anthers slender-linear, fixed by the base. Capsule 2-celled, septicidal. Seeds downwardly imbricated on the placentæ.
2. **PINCKNEYA.** Calyx with clavate tube; limb of 5 subulate-lanceolate lobes, or in the outer flowers of the cyme one (or rarely two) of them an ample petaloid and petiolate leaf, all deciduous. Corolla salverform with somewhat enlarging throat, and 5 oblong recurved-spreading lobes, valvate or nearly so in the bud. Stamens inserted low down on the corolla: filaments filiform: anthers oblong, fixed by the middle, slightly exserted. Style exserted: stigma barely 2-lobed. Capsule didymous-globular, 2-celled, loculicidal, and valves at length 2-parted. Seeds horizontal, with small nucleus, broad and thin lunate-orbicular wing, and comparatively large embryo: cotyledons broad.
3. **BOUVARDIA.** Flowers heterogone-dimorphous. Calyx with turbinate or campanulate tube, and 4 subulate persistent lobes. Corolla tubular or salverform, the 4 short lobes valvate in the bud. Stamens inserted on the throat or on the tube below it: anthers sub-sessile, oblong or linear. Style filiform and more or less exserted in long-styled flowers, much shorter in the other sort: stigmas 2, obtuse. Ovary 2-celled. Capsule didymous-globose, coriaceous, loculicidal. Seeds peltate, somewhat meniscoidal, imbricated on the globular placentæ.

* * Fruit capsular or at least dry, 2-celled: seeds several or numerous in each cell, wingless: calyx-tube short; lobes persistent: corolla valvate in the bud: almost all herbs, with leaves no more than opposite: stipules not setose, or in one species setulose.

+— Summit or sometimes even three fourths of the capsule free from the calyx at maturity: flowers in most and probably in all heterogone-dimorphous: seeds peltate: albumen corneous.

4. **HOUSTONIA.** Flowers 4-merous. Calyx-lobes mostly distant. Corolla salverform to funnelform, with 4-parted limb. Stamens (according to the form) inserted either in the throat or lower down on the tube; anthers oblong or linear, fixed by near the middle. Style reciprocally long or shorter: stigmas 2, linear or oblong. Capsule usually somewhat didymous-globular, or emarginate at the free summit, there loculicidal, occasionally afterwards partially septicidal. Seeds few or moderately numerous in each cell, on usually ascending placentæ, acetabuliform, meniscoidal, or sometimes barely concave on the hilar face, not angulate; testa scrobiculate or reticulate.

+— +— Summit of capsule not extended beyond the adnate calyx-tube: flowers not heterogone-dimorphous, small: seeds numerous, angulate or globular, smooth or nearly so: albumen fleshy.

5. **OLDENLANDIA.** Flowers 4-merous. Corolla from rotate to short-salverform, 4-lobed. Stamens short: anthers oval. Capsule hemispherical, oval, or turbinate, loculicidal across the summit.

6. **PENTODON.** Flowers 5-merous. Calyx-tube turbinate or obpyramidal: limb of 5 deltoid-subulate teeth, in fruit distant. Corolla short-funnelform, 5-lobed. Stamens 5, short: anthers short-oblong. Capsule obconical, obscurely didymous, loculicidal across the truncate summit. Seeds very numerous, minute, reticulated. Stipules or some of them 2-4-subulate.

* * * Fruit baccate or at least fleshy and indehiscent, many-seeded (rarely few-seeded),

+— Five-celled: shrubby.

7. **HAMELIA.** Calyx 5-toothed, persistent. Corolla tubular, 5-lobed, imbricated in the bud. Stamens inserted low on the tube: filaments short: anthers linear. Style filiform: stigma fusiform, sulcate. Berry ovoid. Seeds very numerous in the cells, minute, angulate or flattened. Inflorescence scorpioid-cymose.

+— +— Ovary and fruit 2-celled, sometimes imperfectly so by the placentæ not meeting in the axis: shrubs.

8. **CATESBÆA.** Flowers 4-merous. Calyx-lobes subulate, persistent. Corolla funnelform; lobes short, ovate or deltoid, valvate in the bud. Stamens inserted low down on the tube: anthers linear. Ovary 2-celled: style filiform: stigma undivided. Berry coriaceous, globular. Seeds flattened.

9. **RANDIA.** Flowers 5-merous, rarely 4–7-merous. Corolla salverform or somewhat funnelform; the lobes convolute in the bud. Stamens inserted on the throat of the corolla: filaments short or none: anthers linear, acute or acuminate. Ovary completely 2-celled: style stout: stigma clavate or fusiform, entire or 2-lobed. Berry globose or ovoid. Seeds mostly imbedded in the pulpy placentæ, sometimes very few: testa thin, adherent to the corneous albumen.

10. **GENIPA.** Flowers 5-merous. Calyx-tube more or less produced beyond the summit of the ovary, the border truncate or sometimes bearing small teeth. Corolla salverform; the lobes convolute in the bud. Anthers linear, nearly sessile. Ovary one-celled, with two projecting parietal placentæ which almost meet in the centre. Berry large, becoming 2-celled by the junction or coalescence of the ample pulpy many-seeded placentæ in the centre. Seeds large, flat: albumen cartilaginous.

Series II. COFFEACEÆ. Ovules solitary in the cells of the ovary: leaves with obvious stipules, opposite or only casually in threes or fours.

* Shrubs: flowers compacted in pedunculate heads with a globose receptacle.

11. **CEPHALANTHUS.** Flowers 4-merous, crowded in a long-pedunculate head, but distinct, dry in fruit. Calyx oblong, soon obpyramidal: limb obtusely 4-lobed. Corolla

tubular-funnelform, with 4 short lobes imbricated in the bud, one lobe outside. Stamens included: filaments short, inserted in the throat: anthers 2-mucronate at base. Style long-exserted: stigma clavate-capitate. Ovary 2-celled, a solitary anatropous ovule pendulous from near the summit of each cell. Fruits akene-like, obpyramidal by mutual pressure, 1-2-seeded.

12. **MORINDA.** Flowers usually 5-merous, compacted and the ovaries or fruits confluent in a short-peduncled fleshy head. Calyx urceolate or hemispherical, with truncate or obscurely dentate limb. Corolla salverform or somewhat funnelform, mostly short; lobes valvate in the bud. Stamens short, inserted in the throat. Style bearing 2 slender stigmas. Ovary 4-celled, or rather 2-celled and the cells 2-locellate; an ascending ovule in each cell. Fruits drupaceous, maturing 2 to 4 bony seed-like nutlets, all confluent into a succulent syncarp.

    * * Shrubs: flowers distinct, in cymes or panicles: fruit drupaceous,

        + With 4 to 10 cells, at least in the ovary.

13. **GUETTARDA.** Flowers 4-9-merous (sometimes polygamo-diœcious). Calyx with ovoid or globular tube, continued above the ovary into a cupulate or campanulate limb; the border truncate, commonly irregularly denticulate or dentate. Corolla salverform, with elongated tube, and rounded or oblong lobes imbricated in the bud. Stamens inserted on the tube or throat of the corolla, included: filaments short or none: anthers linear. Style filiform: stigma subcapitate or minutely 2-lobed. Ovary 4-9-celled: an anatropous ovule suspended from the summit of each cell on a thickened funiculus. Drupe globular, with thin flesh, and a bony or ligneous 4-9-celled and lobed putamen; the cells and contained seed narrow. Embryo cylindrical: albumen little or none.

14. **ERITHALIS.** Flowers 5-merous, varying to 6-10-merous. Calyx with obovate or globular tube and a truncate or denticulate short limb or border. Corolla rotate, parted into 5 or more oblong-linear divisions, valvate, or at tips slightly imbricated in the bud. Stamens inserted on the base of the corolla: filaments hairy at base: anthers linear-oblong. Style thickish: stigma of 5 or more minute lobes. Ovary 5-10-celled, with solitary pendulous ovules. Drupe small, globose, 5-10-sulcate, containing as many bony seed-like nutlets. Embryo small in copious albumen.

       + + With 2 (rarely by variation 3) cells to the ovary: ovules anatropous.

15. **CHIOCOCCA.** Flowers 5-merous, in axillary panicles or racemes. Calyx with ovoid or turbinate tube and 5-toothed limb. Corolla funnelform, 5-cleft; the lobes valvate or at apex obscurely imbricated in the bud. Stamens inserted on the very base of the corolla: filaments monadelphous at base, somewhat hairy: anthers linear. Style filiform: stigma clavate. Ovules suspended. Drupe globular, small, containing two coriaceous seed-like nutlets.

16. **PSYCHOTRIA.** Flowers (small) 5-merous, sometimes 4-merous, in terminal naked cymes. Calyx short. Corolla from campanulate to short-tubular or funnelform, not gibbous; lobes valvate in the bud. Stamens short, inserted in the throat of the corolla, distinct. Stigma 2-cleft. Ovule erect from the base of each cell. Drupe globular, small, containing 2 flattened and commonly costate or cristate nutlets. Leaves mostly dilated and membranaceous. Flowers in some heterogone-dimorphous.

17. **STRUMPFIA.** Flowers (very small) 5-merous, in axillary thyrsiform cymes. Calyx short, 5-toothed. Corolla short, 5-parted; lobes oblong-lanceolate, lightly imbricated in the bud. Stamens inserted on the very base of the corolla: filaments very short, monadelphous: anthers oblong, with adnate introrse cells, connate by their broad coriaceous connectives into an ovoid tube. Style hirsute: stigmas 2, obtuse. Ovule erect from the base of each cell. Drupe small, with a 2-celled 2-seeded (or by abortion single-seeded) putamen. Leaves linear, rigid, Rosemary-like.

    * * * Suffruticose and procumbent plants: flowers axillary and sessile: fruit drupaceous, 2-celled: seeds peltate.

18. **ERNODEA.** Flowers 4-6-merous. Calyx-tube ovoid; lobes elongated, subulate-lanceolate, persistent. Corolla salverform; lobes valvate in the bud, linear, at length revolute. Stamens inserted on the throat of the corolla, much exserted: filaments filiform: anthers linear-oblong. Ovary 2-celled, with a peltate amphitropous ovule borne at the middle of the

cells. Style filiform, exserted: stigmas 2, obtuse. Drupe obovate, thin-fleshy, containing 2 cartilaginous plano-convex nutlets. Seed plano-convex. Embryo straight in fleshy albumen: cotyledons cordate, foliaceous: radicle inferior. Leaves fleshy-coriaceous, sessile.

* * * * Low herbs, with entire and naked interpetiolar stipules: ovules erect, anatropous: style filiform: stigmas filiform or linear.

19. **MITCHELLA.** Flowers (3–6-) generally 4-merous, heterogone-dimorphous, geminate at the summit of a peduncle and the ovaries of the two connate. Calyx-teeth persistent. Corolla between salverform and funnelform; lobes valvate in the bud, upper face densely villous-bearded within. Stamens inserted in the throat of corolla, with oblong anthers, on short filaments when the filiform style is exserted, on long exserted filaments when the style and stigmas are included. Style-branches 4, hirsute-stigmatose down the inner side. Fruit a globular baccate syncarp, containing 8 compressed roundish cartilaginous nutlets (4 to each flower). Albumen cartilaginous: embryo minute. Prostrate and creeping evergreen.

20. **KELLOGGIA.** Flowers (3–5-) generally 4-merous, singly slender-pedunculate. Calyx with obovate tube and minute persistent teeth. Corolla between funnelform and salverform; lobes naked, valvate in the bud. Stamens inserted in the throat of the corolla, more or less exserted: filaments flattened: anthers oblong-linear, fixed above the base. Style filiform, exserted: stigmas 2, linear-clavate, papillose-pubescent. Ovary 2-celled: ovules erect from the base, anatropous. Fruit small, dry and coriaceous, beset with uncinate bristles, separating at maturity into 2 closed carpels, which are conformed and adherent to the seed, somewhat reniform in cross section. Embryo comparatively large, in fleshy albumen: cotyledons elliptical, as long as the radicle.

* * * * Low herbs, with short-vaginate stipules setiferous or sometimes only 4–6-cuspidate: ovary 2–4-celled: solitary ovules borne on the septum and amphitropous: fruit dry: seed sulcate or excavated on the ventral face: embryo in corneous or firm-fleshy albumen; the radicle inferior: flowers small, sessile in terminal and axillary glomerules: corolla funnelform or salverform; lobes valvate in the bud.

+– Fruit circumscissile, upper part with persistent calyx-limb falling off, exposing the seeds.

21. **MITRACARPUS.** Flowers commonly 4-merous, capitate-glomerate. Calyx-lobes persistent, unequal, the alternate pair mostly shorter or minute and stipule-like. Stamens inserted on the throat of the corolla. Short style-branches or stigmas 2. Fruit didymous, membranaceous, 2-celled, a pyxidium, the upper half separating from the lower by transverse circular dehiscence. Seed cruciately 4-lobed on the ventral side.

+– +– Fruit septicidal into its 2 to 4 component carpels: calyx-limb gamophyllous at base and circumscissile-deciduous as a whole at or before dehiscence: stamens borne on the throat of the corolla.

22. **RICHARDIA.** Flowers (4–8-) commonly 5–6-merous and 2–4-carpellary. Calyx-lobes ovate-lanceolate or narrower. Corolla funnelform. Stigmas 2 to 4, linear or spatulate. Carpels separating from apex to base, coriaceous, roughish, closed or nearly so; no persistent axis.

23. **CRUSEA.** Flowers (3–5-) usually 4-merous and 2- (sometimes 3–4-) carpellary. Calyx-lobes subulate to triangular-lanceolate, sometimes very unequal or intermediate ones reduced to small teeth. Corolla salverform to narrow funnelform. Stigmas 2 to 4, linear to spatulate-oval. Fruit 2–4-lobed, separating from a persistent axis into obovoid or globular chartaceous carpels, which either open at the commissure or sometimes remain closed.

+– +– +– Fruit septicidal at summit or throughout, its 2 or rarely 3 carpels or valves bearing persistent and quite or nearly distinct calyx-teeth.

24. **SPERMACOCE.** Calyx-teeth, lobes of the short corolla, and stamens 4, or two of the former sometimes abortive. Fruit small, from membranaceous to thin-crustaceous, one or both the carpels opening ventrally to discharge the seed: no persistent carpophore, or sometimes a thin dissepiment remaining.

25. **DIODIA.** Calyx-lobes (1 to 6) usually 2 or 4, distinct, distant. Corolla funnelform or nearly salverform, with mostly 4-lobed limb, and stamens as many, inserted in its throat. Style filiform, entire or 2-cleft: stigmas 2. Fruit somewhat fleshy-drupaceous or crustaceo-coriaceous, tardily separating through the dissepiment into 2 closed carpels: no carpophore.

Series III. STELLATÆ. Ovules (peltate and) solitary in the cells of the ovary: embryo incurved, in corneous albumen: leaves verticillate without stipules, unless the supernumerary leaves be foliaceous stipules, which may in some cases be nearly demonstrated.

26. GALIUM. Flowers 4-merous (rarely 3-merous), 2-carpellary, sometimes diœcious. Calyx-tube globular; limb obsolete, a mere ring or obscure border. Corolla rotate; lobes valvate, and commonly acuminate or mucronate apex inflexed in the bud. Stamens with short filaments and anthers. Style 2-cleft or styles 2: stigmas capitellate. Ovary 2-celled, 2-lobed; a single amphitropous ovule borne on the middle of the dissepiment in each cell. Fruit didymous, dry, fleshy-coriaceous, or occasionally baccate, articulated on the pedicel, tardily separating into two closed carpels, or only one maturing. Seed deeply hollowed on the face; seed-coat adnate to the albumen within, and often also to the pericarp.

1. EXOSTÉMA, Rich. (Not *Erostemma*, to which later authors have changed the name, which is from ἔξω, on the outside, and στήμα, stamen, i. e. stamens exserted.) — Tropical American shrubs or trees, one reaching Florida. — Rich. in Humb. & Bonpl. Pl. Æquin. i. 131, t. 38. *Exostemma*, DC. Prodr. iv. 358; A. Rich. Rub. 200; Benth. & Hook. Gen. ii. 42. *Cinchona* § *Exostema*, Pers. Syn. i. 195 (1805), where the name first appears.

E. Caribǽum, ROEM. & SCHULT. Shrub 6 to 12 feet high, glabrous; leaves oblong-ovate to lanceolate, coriaceous; stipules subulate, small; flowers on short and simple axillary peduncles, fragrant; calyx-teeth very short; corolla white or tinged with rose; tube inch long and lobes hardly shorter; seeds narrowly winged.— Syst. v. 18; Torr. & Gray, Fl. ii. 36. *Cinchona Caribœa*, Jacq. Amer. t. 179; Lamb. Cinch. t. 4. *C. Jamaicensis*, Wright, in Phil. Trans. lxvii. t. 10; Andr. Bot. Rep. t. 481.— Keys of Florida. (W. Ind., Mex.)

2. PINCKNÉYA, Michx. GEORGIA BARK. (*Charles Cotesworth Pinckney*.) — Single species.

P. púbens, MICHX. Tall shrub or small tree, pubescent; leaves ample, oblong-oval to ovate, acute at both ends, petioled; stipules subulate, caducous; cymes terminal and from upper axils, pedunculate; petaloid calyx-lobe resembling the leaves in form, pink-colored, 2 inches or more long; corolla inch long, cinereous-pubescent, purplish; capsule half-inch in diameter.— Fl. i. 103, t 13; Michx. f. Sylv. t. 49; Bart. Fl. Am. Sept. t. 7; Audubon, Birds, t. 165; Torr. & Gray, Fl. ii. 37. *P. pubescens*, Gærtn. Fruct. iii. 80, t. 194. *Pinknea pubescens*, Pers. Syn. i. 197. *Cinchona Caroliniana*, Poir. Dict. vi. 40.— Marshy banks of streams in pine barrens of the low country, S. Carolina to Florida; fl. early summer.

3. BOUVÁRDIA, Salisb. (*Dr. Charles Bouvard*.) — Low shrubs or perennial herbs (from Texas to Central America, some cultivated for ornament); with mostly sessile and not rarely verticillate leaves, subulate interposed stipules, and handsome tubular flowers in terminal cymes.— Parad. Lond. t. 88; HBK. Nov. Gen. & Spec. iii. t. 288; Benth. & Hook. Gen. ii. 36.— Leaves in our species mostly verticillate and corolla not glabrous, its short lobes ascending or barely spreading. Flowers heterogone-dimorphous in the manner of *Houstonia*.

B. ováta, GRAY. Herbaceous, glabrous, obscurely scabrous; leaves mostly in fours, short-petioled, ovate, one or two inches long, costately 5-veined on each side of the midrib; corolla probably purple or reddish, inch long, minutely puberulent.— Pl. Wright. ii. 67.— S. Arizona, between San Pedro and Santa Cruz, *Wright*.

B. triphýlla, SALISB. Suffruticose or more shrubby, scabro-puberulent, 2 to 5 feet high; leaves in threes or fours (or on branchlets in pairs), from oblong-ovate to broadly lanceolate, usually hispidulous-scabrous, at least the margins, 3–4-veined each side of the midrib; corolla scarlet, about inch long, outside furfuraceous-pubescent.— Parad. Lond. l. c. (broad-leaved var., but not with villous-closed throat in any form); Ker, Bot. Reg. t. 107; Sims, Bot.

Mag. t. 1854; Lindl. Bot. Reg. xxvi. t. 37. *B. Jacquini*, HBK. l. c. 385; DC. Prodr. iv. 365; Gray, Pl. Wright. ii. 67. *B. quaternifolia*, DC. l. c. ? *B. coccinea*, Link, Enum. i. 139. *B. ternifolia*, Schlecht. in Linn. xxvi. 98. *B. splendens*, Graham in Bot. Mag. t. 3781. *Ixora ternifolia*, Cav. Ic. iv. t. 305. *I. Americana*, Jacq. Hort. Schœnb. iii. t. 257. *Houstonia coccinea*, Andr. Bot. Rep. t. 106. — Rocky ground, S. Arizona, &c., *Wright, Thurber, Rothrock, Pringle, Lemmon*. (Mex.)

Var. angustifólia. Cinereous-puberulent or hirtellous: leaves smaller (8 to 18 lines long), subsessile, less veiny, from oblong-lanceolate to almost linear. — *B. hirtella* & *B. angustifolia*, HBK. l. c. 384. *B. hirtella*, Gray, Pl. Wright. i. 80, ii. 67. — S. W. Texas to Arizona, *Wright*, &c. (Mex.)

4. HOUSTÓNIA, Gronov. (Named by Gronovius, as says Linnæus, in memory of *Dr. Wm. Houston*, who died in Jamaica in 1733.) — Low herbs, or one or two suffruticulose (Atlantic-American and Mexican), with heterogonedimorphous flowers; the corolla blue or purple to white, upper face of lobes sometimes puberulous. — L. Hort. Cliff. 35. & Gen. ed. 1 (1737); Juss. Gen. 197; Gray, Proc. Am. Acad. iv. 313, & Man. ed. 5. 212; Benth. & Hook. Gen. ii. 60. *Hedyotis* in part (Wight & Arn.), Torr. & Gray, Fl. ii. 36. (*Macrohoustonia*, Gray, Proc. Am. Acad. iv. 314, is a peculiar group of Mexican species, between this genus and *Bouvardia*.)

§ 1. EUHOUSTÓNIA. Low herbs, comparatively small-flowered: leaves not rigid: capsule more or less didymous or emarginate, sometimes septicidal as well as loculicidal across the broad summit.

* Delicate species, inch to span high: corolla salverform: anthers or stigmas included or only partially emerging from the throat: peduncles single, elongated and erect in fruit: seeds rather few acetabuliform with a deep hilar cavity: stipules a transverse membrane uniting the petioles, mostly entire or truncate and naked.

+ Perennial by delicate filiform creeping rootstocks or creeping stems: peduncles filiform, inch or two long: seeds subglobose with orifice of the deep hilar cavity circular.

H. cærúlea, L. (BLUETS of the Canadians, INNOCENCE.) Perennial by slender rootstocks, forming small tufts, erect, a span or more high, glabrous, and with lower leaves hispidulous: these spatulate to obovate and short-petioled; upper small and nearly sessile: corolla violet-blue to lilac, varying to white, with yellowish eye; tube (2 or 3 lines long) much exceeding calyx-lobes, longer than or equalled by those of corolla: capsule obcordate-depressed, half free. — Spec. i. 105 (Moris. Hist. sect. 15, t. 4, f. 1; Plnk. Alm. & Mant. t. 97, f. 9); Sims, Bot. Mag. t. 370; Barton, Fl. Am. Sept. t. 34, f. 1. *H. pusilla*, Gmel. Syst. i. 236 ! *H. Linnæi*, var. *elatior*, Michx. Fl. i. 85. *H. serpyllifolia*, Graham, Bot. Mag. t. 2822, from habitat and figure, but corolla-tube too short. *Hedyotis cærulea*, Hook. Fl. i. 286; Torr. & Gray, Fl. ii. 38. *H. gentianoides*, Endl. Iconogr. t. 89. *Oldenlandia cærulea*, Gray, Man. ed. 2, 174. — Low and grassy grounds, Canada to Michigan and the upper country of Georgia and Alabama; fl. early spring.

H. serpyllifólia, MICHX. Perennial by prostrate extensively creeping and rooting filiform stems, and some subterranean ones, glabrous or slightly and minutely hispidulous below: leaves orbicular to ovate or ovate-spatulate (2 to 4 lines long) and abruptly petioled, or upper ones on flowering stems oblong and nearly sessile: corolla deep violet-blue, rather larger than in *H. cærulea*. — Fl. i. 85; Pursh, Fl. i. 106. *H. tenella*, Pursh, l. c. *Hedyotis serpyllifolia*, Torr. & Gray, Fl. ii. 39. *Oldenlandia serpyllifolia*, Gray, Man. ed. 2; Chapm. Fl. 180; — Along streamlets and on mountain-tops in the Alleghanies, from Virginia to Tenn. and S. Carolina; flowering through early summer.

+ + Winter-annuals, branching from the simple root, glabrous or obscurely scabrous: peduncles a quarter-inch to at length sometimes an inch long: capsule somewhat didymous, less than half free: mature seeds generally as of the preceding.

H. pátens, ELL. An inch to at length a span high, with ascending branches and erect peduncles: leaves spatulate to ovate: corolla much smaller than that of *H. cærulea*: the tube twice the length of the calyx-lobes and more or less longer than its lobes, violet-blue or purplish without yellowish eye. — Sk. i. 191 ; Gray, Proc. Am. Acad. iv. 314. *H. Linnæi*, var. *minor*, Michx. Fl. i. 85. *Hedyotis minima*, Torr. & Gray, Fl. l. c. in part, & *H. cœrulea*, var. *minor*. — Dry or sandy soil, S. Virginia to Texas in the low country, also Illinois ? and Tennessee ; fl. early spring.

Var. pusílla. An inch or so high, more diffuse in age ; leaves narrowly spatulate (half a line or a line wide) ; upper ones nearly linear : seeds smoother, with more open and oval hilar cavity, and sometimes an elevated line within, as described in Proc. Am. Acad. l. c., a character not found in the larger and broader leaved form. Perhaps from the char. this is the true *H. patens*, Ell. But we have it only from Louisiana (*Hale*, *Drummond*) and Texas, *Drummond* and others : there passing into the other form.

H. mínima, Beck. More diffuse, commonly scabrous : leaves spatulate to ovate : flowers usually larger : calyx-lobes more foliaceous, oblong-lanceolate, sometimes 2 lines long, very much longer than the ovary, equalling the tube of the purple or violet corolla ; lobes of the latter 2 or 3 lines long : primary peduncles sometimes declined in fruit ! — Amer. Jour. Sci. x. 262 ; Gray, l. c. *Hedyotis minima*, Torr. & Gray, l. c., in part only. — Dry hills, Missouri and Arkansas to Texas, first coll. by *L. C. Beck* about St. Louis ; fl. early spring.

* * Slender leafy-stemmed annual, with lateral horizontal peduncles, and very small flowers: corolla short-salverform: seeds crateriform, with a medial hilar ridge.

H. subviscósa, Gray. A span or two high, minutely viscidulous-pubescent, with rather simple spreading branches : leaves narrowly linear, half-inch long : peduncle in first fork and from all following nodes, rather shorter than leaves, horizontally refracted in fruit : calyx and capsule a line high : corolla about same length, white : capsule didymous, only the summit free : seeds 10 in each cell. — Proc. Am. Acad. iv. 314. *Oldenlandia subviscosa*, Wright in Gray, Pl. Wright. ii. 68. — S. Texas, *Berlandier*, *Wright*.

* * * Depressed or low-tufted species : corolla salverform or in one species funnelform : filaments as well as anthers or summit of style reciprocally exserted quite out of the throat : *fructiferous peduncles all short and recurved*.

+— Annual, with small funnelform corolla ; seeds open-crateriform ; scarious stipules setulose-ciliate !

H. humifúsa, Gray. Much branched from the root, repeatedly dichotomous, forming a depressed turf, puberulent and viscid : leaves linear-lanceolate, thickish (half-inch or more long), mucronate : flowers in all the forks, crowded with the leaves at the ends of branchlets : calyx 4-parted into long setaceous-subulate spreading lobes : corolla pale purple or nearly white, open-funneliform, 3 lines long, hardly twice the length of the calyx ; the oblong lobes puberulous inside : capsule a line in diameter, globose-didymous, three-fourths free, only the base girt by the short accrete calyx-tube. — Proc. Am. Acad. iv. 314 (not of Heusl. Biol. Bot. which is *H. Wrightii*). *Hedyotis* (*Houstonia*) *humifusa*, Gray, Pl. Lindh. ii. 216. — Sandy or gravelly plains and hills, Texas, *Wright*, *Lindheimer*, *Reverchon*, &c. ; fl. spring.

+— +— Perennials, prostrate, with naked stipules and elongated salverform corolla, flowering conspicuously in early spring : later growth producing through the summer inconspicuous cleistogamous flowers, with short (yet mostly well-formed but unopening) corollas.

H. rotundifólia, Michx. Perennial by slender rootstocks or shoots, more or less creeping, glabrous or with some hispidulous pubescence : leaves somewhat orbicular, slightly petioled, not longer than the internodes : peduncles 2 to 4 lines long or in cleistogamous flowers very short : developed corollas bright white, with filiform tube (3 or 4 lines long) longer than the oblong lobes : capsule more than half free, somewhat didymous : seeds comparatively large (half-line in diameter), rough-scrobiculate, acetabuliform. — Fl. i. 85 ; Pursh, l. c. ; Ell. l. c. *Hedyotis rotundifolia*, Torr. & Gray, Fl. ii. 38. *Oldenlandia rotundifolia*, Chapm. Fl. 180, the later "apetalous fruiting" flowers noted. — Low sandy ground, S. Car. to Florida and Louisiana.

H. rúbra, Cav. Suffrutescent and multicipital from a deep root, forming a depressed tuft of 2 to 4 inches high, glabrous or minutely puberulent, densely leafy : leaves narrowly linear, an inch or more long, or earlier ones rather lanceolate and shorter : corolla "red " or rather purple, sometimes lilac or varying to white ; tube half-inch to nearly inch long, slender ; oblong acute lobes 2 or 3 lines long : capsule 2 lines wide, less high, didymous, fully three-fourths free : seeds open-crateriform. — Ic. vi. t. 374 ; Benth. Pl. Hartw. 15. *Hedyotis* (*Houstonia*) *rubra*, Gray, Pl. Fendl. 61. *Oldenlandia* (*Houstonia*) *rubra*, Gray, Pl. Wright. ii. 68. — Stony or gravelly hills, New Mexico and Arizona. (Mex.)

╋ ╋ ╋ Lignescent-rooted perennial, with small and short corolla and naked stipules.

— **H. Wrightii**, Gray. Many-stemmed from a deep root, a span or less high, erect or spreading, glabrous or very obscurely pruinose: branches quadrangular: leaves thickish, linear or lowest rather lanceolate (half-inch to inch long): flowers in terminal glomerate leafy cymes: corolla purplish or nearly white, between salverform and funnelform, 2 to hardly 4 lines long, with narrow oblong lobes: capsules on very short recurved peduncles, globose-didymous, about three-fourths free: cells 5–8-seeded: seeds crateriform, with a small hilar ridge. — Proc. Am. Acad. xvii. 202. *H. humifusa*, Gray, Pl. Wright. i. 82, & *Oldenlandia humifusa*, Pl. Wright. ii. 68, chiefly, not Pl. Lindh. — Hills, S. W. Texas and New Mexico to S. W. Arizona, first coll. by *Wright*. (Adj. Mex., *Parry & Palmer*.)

* * * * Erect perennials: corolla funnelform or in one species almost salverform, small: stamens and summit of style reciprocally exserted quite out of the throat: fructiferous peduncles erect: capsule from a third to nearly half free: seeds oval or roundish, barely concave on ventral face and with more or less of a medial hilar ridge: stipules entire, scarious, between and connecting the bases of the sessile cauline leaves: fl. mostly in summer.

— **H. purpúrea**, L. Forming small tufts or offsets by filiform rootstocks, a span to a foot high, hirsutulous-pubescent to glabrous; radical leaves ovate or oblong, short-petioled: flowers corymbosely cymose: corolla funnelform, light purple or lilac, varying to nearly white: capsule globular and obscurely didymous, upper half free. — Spec. i. 105; Pursh, Fl. i. 107; Gray, Man. ed. 5, 212. *H. varians*, Michx. Fl. i. 86. *H. pubescens*, Raf. Med. Rep. & Desv. Jour. Bot. i. 230, if of the genus. *Oldenlandia purpurea*, Gray, Man. ed. 2, 173. *Hedyotis lanceolata*, Poir. Suppl. iii. 14. *H. umbellata*, Walt. Car. 85 ? *Anotis lanceolata*, DC. Prodr. iv. 433. — Canada to Texas. — Truly polymorphous, of which the typical form "leaves ovate-lanceolate," L., or *latifolia*, is comparatively large, often a foot high and pubescent: leaves ovate to ovate-lanceolate, inch or two long, the larger with rounded closely sessile base: calyx-lobes subulate, sometimes slightly sometimes conspicuously surpassing the emarginate summit of the capsule. — *H. purpurea*, Torr. & Gray, Fl. ii. 40. This form from Maryland to Arkansas, and southward to Alabama, especially in and near the mountains.

Var. ciliolàta, Gray, Man. l. c. A span high: leaves only half-inch long, thickish: cauline oblong-spatulate; radical oval or oblong, in rosulate tufts, hirsute-ciliate: calyx-lobes a little longer than the capsule. — *H. ciliolata*, Torr. in Spreng. Syst. Cur. Post. 40, & Fl. i. 173. *H. longifolia*, Hook. Bot. Mag. t. 3099, not Gærtn. *Hedyotis ciliolata*, Torr. & Gray, Fl. ii. 40 (excl. syn. *H. serpyllifolia*, Graham). — Chiefly northward, on rocky banks along the Great Lakes and their tributaries, Canada to Michigan and south to Kentucky, passing into the next.

Var. longifólia, Gray, l. c. A span or two high, mostly glabrous, thinner-leaved: leaves oblong-lanceolate to linear (6 to 20 lines long); radical oval or oblong, less rosulate, not ciliate: calyx-lobes little surpassing the capsule. — *H. longifolia*, Gærtn. Fruct. i. 226, t. 49, f. 8; Willd. Spec. i. 583. *Hedyotis longifolia*, Hook. Fl. i. 286 ? Torr. & Gray, l. c. *H. angustifolia*, Pursh, Fl. i. 106, partly. — Rocky or gravelly ground, Canada to Saskatchewan, Missouri, and Georgia.

Var. tenuifólia. Slender, lax, diffuse, 6 to 12 inches high, with loose inflorescence, almost filiform branches and peduncles: cauline leaves all linear, hardly over a line wide: otherwise as preceding. — *H. tenuifolia*, Nutt. Gen. i. 95. *Hedyotis longifolia*, var. *tenuifolia*, Torr. & Gray, l. c. — S. E. Ohio, and through the mountains, Virginia to N. Carolina and Tennessee.

Var. calycósa. Near a foot high: leaves broadly lanceolate, thickish: calyx-lobes elongated (2 to 4 lines long), much surpassing the capsule. — *Hedyotis calycosa*, Shuttlew. in distrib. Pl. Rugel. — Mountains of Alabama (*Rugel*) to Arkansas (*Nuttall*), and Illinois (*E. Hall*); also coll. by *Drummond*.

— **H. angustifólia**, Michx. Rather rigid, becoming many-stemmed from a perpendicular root, glabrous: leaves narrowly linear or lowest somewhat spatulate, on the stems commonly fascicled in the axils: flowers corymbosely or paniculately cymose, short-pedicelled or subsessile: corolla nearly salverform, 2 or 3 lines long, mostly white, upper face of the lobes commonly villous-pubescent: capsule with turbinate or acutish base, only the summit free, and barely equalled by the short calyx-teeth, first opening across the tip, at length septicidal: seeds obscurely concave on the hilar face. (Transition to *Oldenlandia*.) — Fl. i. 85; Gray, l. c. *H. fruticosa* & *H. rupestris*, Raf. *Hedyotis stenophylla*, Torr. & Gray, l. c. *Olden-*

landia angustifólia, Gray, Pl. Wright. ii. 56, & Man. ed. 2. — Barrens, Illinois to Kansas, and Tennessee to Florida and Texas.

— Var. **filifólia**. Diffuse, disposed to be liguescent at base: cauline leaves mostly filiform: flowers and capsules smaller, more pedunculate. — *Oldenlandia angustifolia*, Chapm. Fl. 181. — Rocky pine barrens near the coast, Florida. In Texas passing into the ordinary form.

Var. **rigidiúscula**. A span to a foot high, stouter: leaves mostly rigid, from linear to lanceolate: flowers disposed to be glomerate and sessile, but some pedunculate. — S. and W. Texas, *Palmer, Havard*, &c. Coast of E. Florida, *Rugel*. (Mex.)

§ 2. ERICÓTIS. Fruticose or fruticulose: leaves setaceous or acerose-linear, rigid, fascicled: flowers (purplish) and seeds nearly as in the last preceding subdivision. — Gray, Proc. Am. Acad. xvii. 203.

H. fasciculáta, Gray, l. c. A span to a foot or more high, decidedly shrubby, with rigid and tortuous spreading branches, glabrous or birtello-puberulent: stipules very short: leaves subulate-linear, thickish, 2 to 4 lines long, much fascicled: flowers cymulose, shortpedicelled: corolla 2 or 3 lines long, between salverform and funnelform, the tube sometimes hardly or sometimes twice longer than the lobes: capsule barely a line long, about one-third free: seeds 4 or 5 in each cell, elongated-oblong, barely concave on the ventral face. — Includes some of *Hedyotis stenophylla* or *Oldenlandia angustifolia*, var. *parviflora* of Gray, Pl. Wright. i. & ii. — S. W. borders of Texas and adjacent New Mexico, *Bigelow, Wright, G. R. Vasey*. (Adj. Mex., *Palmer*.)

H. acerósa, Gray, l. c. A span or two high, fruticulose, tufted, with slender ascending branches, minutely hispidulous-pubescent or glabrate, very leafy throughout: stipules short, commonly with a median cusp: leaves acicular-setaceous, 3 to 5 lines long: calyx-lobes similarly setaceous: flowers sessile: corolla purplish, salverform with slightly dilated throat: its slender tube 3 or 4 lines long, much exceeding the ovate lobes: capsule globular, over a line long, about a quarter part free, much overtopped by the acicular calyx-lobes; cells 12-20-seeded: seeds roundish, with small ventral excavation. — *Hedyotis (Ereicotis) acerosa*, Gray, Pl. Wright. i. 81. *Oldenlandia acerosa*, Gray, Pl. Wright. ii. 67. *Moniostoma acerosa*, Hemsl. Biol. Centr.-Am. Bot. ii. 31. — High plains and hills, S. W. Texas, and adjacent New Mexico, *Wright*, &c. (Adj. Mex., first coll. by *Gregg*.)

5. **OLDENLÁNDIA**, Plum. (*Dr. H. B. Oldenland*.) — Mostly subtropical and humble herbs, with inconspicuous white or whitish flowers. — Nov. Gen. 42. t. 36, & Pl. Am. ed. Burm. t. 212. f. 1; L. Gen. ed. 1, 562; Benth. & Hook. Gen. ii. 58.

* Corolla salverform, surpassing the calyx: flowers cymose: calyx-lobes distant in fruit.

O. Greénei, Gray. Erect annual, paniculately branched, a span or more high, glabrous: leaves spatulate-linear or broadly linear with narrowed base (the larger ones inch long): flowers sessile in the forks and along the lax branches of the pedunculate cyme: calyx-teeth triangular-subulate, about the length of the turbinate tube: corolla less than 2 lines long with tube longer than its own lobes and those of the calyx: capsule quadrangular-hemispherical, or at first somewhat turbinate: seeds moderately angled. — Proc. Am. Acad. xix. 77. — Pinos Altos Mountains, New Mexico, *Greene*. Huachuca Mountains, S. Arizona, *Lemmon*.

* * Corolla rotate, shorter than the calyx-lobes, inconspicuous: capsule rounded at base: stipules mostly bimucronate or bicuspidate: calyx-teeth approximate at base: diffuse low herbs; fl. summer.

O. Bóscii, Chapm. A span or so high from a perennial root, diffusely spreading, slender, glabrous: leaves linear with attenuate base, inch or less long, obscurely one-nerved: flowers few or solitary and nearly sessile at the axils: calyx-teeth broadly subulate, rather shorter than the capsule. — Fl. 181. *Hedyotis Boscii* DC. l. c. 420; Torr. & Gray. Fl. ii. 41. — Low or wet ground, S. Carolina to Arkansas and Texas.

O. glomeráta, Michx. A span to a foot high from an annual root, erect or soon diffuse, freely branching, somewhat hirsutulous-pubescent: leaves from ovate to oblong, thinnish, half-inch long, contracted at base as if petioled: flowers in terminal or lateral sessile glomerules, rarely solitary: calyx-lobes ovate or oblong, foliaceous, longer than the subglobose or hemispherical hirsute capsule. — Fl. i. 83; Pursh, Fl. i. 102. *O. uniflora*, L. Spec. i. 119 name passed by as incorrect. *Hedyotis auricularia*, Walt. Car. 85, not L. *H. glomerata*, Ell. Sk. i. 187; DC. l. c.; Torr. & Gray, l. c. *H. glomerata* & *H. Virginica*, Spreng. Syst. i. 412. — Low grounds near the coast, Long Island, New York, to Florida and Texas. (Cuba.)

6. **PÉNTODON**, Hochst. (Πέντε, five, ὀδούς, tooth, differing from the preceding genus in 5-merous flowers, therefore five calyx-teeth.) — Tender and weak somewhat succulent annuals, glabrous; with 4-angular branching and diffusely spreading stems, ovate or oblong short-petiolate leaves. 2–3-flowered terminal peduncles, occupying the forks of the stem or becoming lateral, or by suppression of leaves bearing several quasi-racemose flowers: corolla white. — Flora, 1844, 522; Benth. & Hook. Gen. ii. 59. *Hedyotis* § *Pentotis*, Torr. & Gray, Fl. ii. 42. — Consists of an African species (*P. decumbens*, Hochst. l. c., *Oldenlandia pentandra*, DC.) and the following, which differs from the character of that plant in the points mentioned below.

P. **Hálei.** Leaves rather obtuse: peduncles shorter than the leaves, or hardly any: pedicels only twice the length of the flowering or fruiting calyx, soon clavate-thickened; corolla only a line long, not hirsute within. — *Hedyotis Halei*, Torr. & Gray, l. c. *Oldenlandia Halei*, Chapm. Fl. 181. — Low swampy grounds, W. Louisiana, *Hale*. Florida, *Rugel*, *Garber*, *Curtiss*. (Cuba.)

7. **HAMÉLIA**, Jacq. (*H. L. DuHamel du Monceau.*) Tropical American shrubs: with petiolate sometimes verticillate leaves, interpetiolar lanceolate-subulate stipules, and red or yellow flowers in naked and scorpioid terminal cymes. — Stirp. Amer. 71. t. 50. *Duhamelia*, Pers. Syn. i. 203.

— **H. pátens**, Jacq. l. c. Shrub 8 or 10 feet high, cinereous-pubescent on all young parts: leaves more commonly in threes, oval-oblong, acuminate: cyme 3–5-rayed, with flowers almost sessile along its branches: corolla crimson, puberulent, almost cylindrical, over half-inch long: fruits black, small. — Desc. Fl. Ant. t. 107. *H. coccinea*, Swartz, Prodr. 46. — Keys and shores of E. Florida. (W. Ind. to Brazil.)

8. **CATESBÆA**, Gronov. (*Mark Catesby*, author of Nat. Hist. of Carolina, Florida, etc., and of Hortus Brit.-Amer., etc.) — W. Indian spinose shrubs; one has reached the shores of Florida. — L. Gen. ed. 1, 356.

C. **parviflóra**, Swartz. Shrub 4 to 6 feet high, with rigid very leafy branches, glabrous, spinose from the axils: leaves mostly fascicled at the nodes, coriaceous, shorter than the spines (quarter to half inch long), roundish, lucid: flowers very small for the genus, solitary and sessile: corolla only half-inch long, white: berry small, white. — Prodr. 30. & Fl. i. 236; Vahl, Ecl. i. 12, t. 10; Griseb. Fl. W. Ind. 317; Chapm. Fl. ed. 2, Suppl. 625. — Bahia Honda Key, S. Florida, *Curtiss*. (W. Ind.)

9. **RÁNDIA**, Houst. ex L. (Dedicated by Houston, in a letter to Linnaeus, to *John Rand*, an English apothecary.) — As now received, an ample genus of tropical shrubs or trees, largely Asiatic and African, but the original species American, often spinose, and with sessile flowers in the axils or terminating short branchlets. — L. Hort. Cliff. 485, & Gen. ed. 1, 376; Benth. & Hook. Gen. ii. 88.

— **R. aculeáta**, L. Shrub 4 to 8 feet high, glabrous, with rigid spreading branches: axillary spines simple, sometimes few, not rarely wanting: leaves obovate to elliptical, at length coriaceous, from 2 inches down to half-inch long, many fascicled in the axils or on short spurs: calyx-teeth short and small: corolla white, 3 or 4 lines long: berries less than half inch long, subglobose, blue or black, not many-seeded. — P. Browne, Jam. t. 8. f. 1; Griseb. Fl. W. Ind. 318; Chapm. Fl. 179. *R. aculeata* & *R. mitis*, L. Spec. ii. 1192, the latter nearly a spineless form. *R. latifolia*, Lam. Dict. iii. 24, & Ill. t. 156. *Gardenia Randia*, Swartz. Fl. Ind. Occ. i. 526; Sims, Bot. Mag. t. 1841. — Coast and Keys of S. Florida. (W. Ind., &c.)

R. XALAPENSIS, Mart. & Gal. occurs not very far beyond the Mexican border.

10. GÉNIPA, Plum. (Altered from an aboriginal name.) — Shrubs or small trees of Tropical America; with ample coriaceous and mostly lucid leaves. deciduous interpetiolar stipules, no spines, but rather large white or whitish flowers which are more or less pedunculate in a terminal cyme, and a large firm-rinded berry. — Plum. Cat. 20. & Pl. Amer. ed. Burm. 127. t. 136; Tourn. Inst. 658. t. 436, 437; Griseb. Fl. W. Ind. 316.

G. clusiæfólia, GRISEB. l. c. Glabrous; plant blackening in drying: leaves obovate, very obtuse or retuse, mucronulate, slightly petioled, 2 to 5 inches long, nearly straight-veined, fleshy-coriaceous, lucid; truncate calyx-limb bearing 5 distant and slender subulate teeth; corolla inch long, fleshy, glabrous within and without; tube longer than the oblong-lanceolate lobes; acute tips of anthers exserted; stigmas 2, subulate; fruit 2 or 3 inches long, ovoid. — *Gardenia clusiæfolia*, Jacq. Coll. App. 37, t. 4; DC. Prodr. iv. 381. *Randia? clusiæfolia*, Chapm. Fl. 179. *Seven-years Apple*, Catesb. Cnr. i. 59. t. 59. — Keys and shores of S. Florida, first coll. by *Blodgett*. (Bahamas, Cuba.)

— GARDENIA FLORIDA, L., cult. as CAPE JESSAMINE, belonging to the genus most allied to *Genipa*, is planted out freely in the Southern Atlantic States.

11. CEPHALÁNTHUS, L. BUTTON-BUSH. (Κεφαλή, head, and ἄνθος, flower, the blossoms densely aggregated in a round head.) — Two or three American and as many Asiatic or African species.

C. occidentális, L. Shrub 3 to 15 feet high, glabrous or pubescent; stipules one on each side between the petioles, triangular, sphacelate, at length deciduous. Leaves ovate to lanceolate; flowers white; setiform bractlets between the flowers glandular-capitate; calyx not glandular, a little hairy around the base. — Spec. i. 95; Lam. Ill. t. 59; Michx. Fl. i. 87; Schk. Handb. t. 21; Bart. Fl. Am. Sept. iii. t. 91; Torr. & Gray, Fl. ii. 31; Gray, Bot. Calif. i. 282. — Swamps and along streams, Canada to Florida and Texas, Arizona, and California; fl. summer. Var. *brachypodus*, DC., of Texas, and var. *Californicus*. Benth. Pl. Hartw., are mere forms, with leaves short-petioled and often in threes. Var. *salicifolius* (*C. salicifolius*, Humb. & Bonpl. Pl. Æquin. t. 98) is an unusually narrow-leaved Mexican form. (Mex., Cuba.)

— **12. MORÍNDA**, Vaill. (Name contracted from *Morus Indicus*, the syncarp resembling a mulberry.) — Tropical shrubs or small trees, mostly glabrous; with oval to lanceolate leaves, their bases or petioles united by small scarious stipules, terminal or axillary peduncles, and white flowers. — *Roioc*, Plum. Nov. Gen. 11. t. 26.

M. Roioc, L. Low shrub, or sometimes climbing by twining; leaves oblong-lanceolate; stipules subulate-pointed; peduncles solitary, bearing single or sometimes geminate small heads — Spec. i. 176; Jacq. Hort. Vind. t. 16; Desc. Fl. Ant. t. 129. — Coast and Keys of S. Florida. (W Ind.)

13. GUETTÁRDA, L. (*Dr. J. E. Guettard.*) — Tropical and subtropical shrubs, chiefly American, and one widely diffused littoral species; leaves ovate to oblong, petioled, with prominent primary veins beneath; flowers in axillary pedunculate cymes; the corollas sericeous-canescent outside. — L. Gen. ed. 5. 428; Vent. Choix. t. 1; DC. Prodr. iv. 455, excl. § 4. *Mathiola*, Plum. Gen. 16; L. Gen. ed. 1, 49.

G. scábra, Lam. Arborescent: leaves obovate to oblong (4 or 5 inches long), mucronate, coriaceous, at length rugose, hispidulous-papillose and scabrous above, soft-pubescent beneath; primary veins (9 to 11 pairs) very prominent beneath and veinlets between well reticulated: peduncles elongated: corolla often inch long; tube retrorsely silky-villous; lobes 5, rarely 6 or 7: drupe quarter-inch in diameter, 4-6-celled. — Ill. t. 154, f. 3 ; Vent. Choix, t. 1 ; DC. l. c. 456 ; Griseb. Fl. W. Ind. 322. *G. ambigua*, Chapm. Fl. 178, not DC. *Mathiola scabra*, L. Spec. ii. 1192. — S. Florida, *Chapman, Garber*. (W. Ind.)

G. ellíptica, Swartz. Arborescent: leaves from broadly oval to elliptical-oblong (inch or two long), thinnish, pilose-pubescent, often glabrate, at least above; primary veins 4 to 6 pairs; transverse veinlets not prominent: peduncles and small cymes shorter than the leaves. flowers usually 4-merous. corolla quarter-inch long, externally canescent: drupe size of a pea, 4-8-celled, 4-2-seeded. — Prodr. 59, & Fl. Ind. Occ. i. 635 ; DC. l. c. 457 ; Torr. & Gray, Fl. ii. 35 ; Griseb. l. c. *G. Blodgettii*, Shuttlew. distrib. coll. Rugel ; Chapm. Fl. 178. — S. Florida, first coll. by *Blodgett*. (W. Ind., Mex.)

14. ERÍTHALIS, P. Browne.
(Ancient Greek name of some plant, from ἔρι, very much, and θαλλός, green shoot. Pliny applied it to some green *Sedum*, and P. Browne to this lucid green shrub.) — West Indian littoral shrubs or low trees, very smooth and resiniferous: the following is the principal species.

— **E. fruticósa,** L. Leaves mostly obovate, about 2 inches long, coriaceous : cymes pedunculate, many-flowered : border of the calyx repand-truncate : corolla white, quarter-inch long; lobes widely spreading : drupes not over 2 lines in diameter, purple. — Spec. ed. 2, ii. 251 ; DC. Prodr. iv. 465 ; Desc. Fl. Ant. t. 242 ; Torr. & Gray, Fl. ii. 35 ; Griseb. Fl. W. Ind. 336. *E. fruticulosa*, &c., P. Browne, Jam. 165, t. 17, f. 3. *E. odorifera*, Jacq. Stirp. Amer. 72, t. 173, f. 23. — Shores and Keys of S. Florida. (All W. Ind.)

15. CHIOCÓCCA, P. Browne. SNOWBERRY. (Χιών, snow, κόκκος, berry.)
— Tropical American shrubs, commonly sarmentose or twining, glabrous ; with coriaceous shining leaves on short petioles, and small yellowish-white flowers in axillary racemes or panicles; the small berry-like drupes at maturity white. — P. Browne, Jam. 164; Jacq. Stirp. Amer. 68; L. Gen. ed. 6, 92. — Some species are obviously heterogone-dimorphous !

— **C. racemósa,** L. Usually twining and climbing: leaves from ovate or oval to lanceolate-oblong, shining, about equalled by the racemiform panicles: corolla short-funnelform, at most 4 lines long : anthers included : mature drupe quarter-inch in diameter and globose ; only the immature flattened and when dried didymous. — Spec. ed. 2, i. 246 ; Andr. Bot. Rep. t. 284 ; Hook. Exot. Fl. t. 93 ; DC. Prodr. iv. 482 ; Torr. & Gray, Fl. ii. 32. *Lonicera alba*, L. Spec. ed. 1, 175. — Var. *parvifolia* (*C. parvifolia*, Griseb. Fl. W. Ind. 337) is a smaller-leaved and low form, mostly with simple and shorter racemes. — Coast and Keys of Florida. (W. Ind. to S. Am.)

16. PSYCHÓTRIA, L.
(Name changed by Linnæus from the original *Psychotrophum* of P. Browne, which was formed of ψυχή, soul, and τροφή, nourishment: seeds used as a substitute for coffee.) — A large genus of shrubs, of most tropical regions, commonly with membranaceous leaves, and small flowers in naked terminal cymes ; in some heterogone-dimorphous. — *Psychotrophum* & *Myrstiphyllum*, P. Browne.

— **P. undáta,** Jacq. Shrub 8 to 18 feet high, with woody spreading branches, glabrous or with some ferruginous pubescence : stipules rather large, broad, blunt, united and sheathing, sphacelate-scarious, caducous (the sheath usually splitting down one side) : leaves from oval to elliptical lanceolate, acuminate at both ends ; primary veins transverse or little ascending : cyme sessile, of about 3 primary rays and secondary divisions : corolla white or whitish, villous in the throat, with lobes shorter than tube : drupes red, ellipsoidal when dry (subrotund, Jacquin), the nutlets striate-costate on the back. — Hort. Schœnb. iii. 5, t. 260 ; DC. Prodr.

iv. 513; Griseb. Fl. W. Ind. 342. *P. nervosa,* Swartz, Fl. Ind. Occ. i. 403. *P. lanceolata,* Nutt. in Am. Jour. Sci. v. 290, ferruginous-pubescent form. in fruit, and glabrous form also mentioned: DC. l. c. 513. *P. chimarrhoides,* & *P. oligotricha,* DC. l. c. 514, glabrous or nearly glabrous forms. *P. rufescens,* HBK. ? Griseb. l. c., the ferruginous-pubescent form. — Woods of E. and S Florida along the coast. first coll. by *Michaux* and *Ware.* (W. Ind., Mex.?)

P. tenuifólia, SWARTZ. Shrub 1 to 4 feet high, with more simple and erect partly herbaceous flowering branches, glabrous or commonly with a very minute pruinose puberulence, no ferruginous hairiness: stipules distinct, ovate, often acute, sometimes setaceously-acuminate, caducous, leaves oblong-lanceolate or broader (3 to 6 inches long), acuminate at both ends: cyme either short-peduncled or sessile, compactly many-flowered: flowers nearly of the preceding: drupes not seen in the Florida plant, according to Swartz "oblong," (ellipsoidal, Grisebach,) in Cuban specimens globose. — Fl. Ind. Occ. i. 402 (ex char.); Griseb. l. c. 341. *P. lanceolata,* in distrib. coll. Rugel. in part. & coll. Curtiss: also Chapm. Fl. l. c. in part; Griseb. Cat. Cub. 135, not Nutt. (Near *P. pubescens,* Swartz, but has different stipules.) — Rich woods, S. Florida; Tampa and Manatee River, *Leavenworth, Rugel,* Indian River, *Curtiss.* (W. Ind.)

17. STRÚMPFIA, Jacq. *(C. C. Strumpf,* who edited the fourth edition of Linn. Genera Plantarum.) — Stirp. Amer. 218; Lam. Ill. t. 731; A. Rich. Mem. Rub. t. 9; Benth. & Hook. Gen. ii. 117. — Single species.

S. marítima, JACQ. Low shrub, much branched, erect, exceedingly leafy: branches where the leaves have fallen annulate-roughened by the square se remains of the stipules, which closely approximate: leaves more commonly in threes, firm-coriaceous and rigid, linear, with strongly revolute margins, glabrous or puberulent, at length shining, inch or less long, mostly exceeding the flower-clusters: corolla white: fruit white. — Desc. Fl. Ant. t. 205; DC. Prodr. iv. 470; Chapm. Fl. 178; Griseb. l. c. 336. *Tournefortia,* &c., Plum. Amer. ed. Burm. t. 251, f. 1. — Rocks on the sea-shore, Keys of Florida. (W. Ind.)

18. ERNÓDEA, Swartz. (Ἐρνώδης, sprouting or branching.) — Prodr. 29. & Fl. Ind. Occ. i. 223, t. 4. *Knoxia,* P. Browne. Jam. 140. *Thymelea,* Sloane, Hist. Jam. t. 169. — Single species.

E. littorális, SWARTZ, l. c. Procumbent, suffruticose, glabrous: leaves fleshy-coriaceous, lanceolate, acute, inch or less long, crowded on the branchlets, obscurely nerv se-veined: stipules short-vaginate, produced between the leaf-bases into cuspidate points: corolla yellowish, half-inch or less long: drupe yellow, pisiform, crowned by the conspicuous calyx-lobes. — A. Rich. Mem. Rub. t. 5, f. 2; Torr. & Gray, Fl. ii. 30; Griseb. Fl. W. Ind. 347. — Shores of S. Florida. (W. Ind.)

19. MITCHÉLLA. L. PARTRIDGE-BERRY. — *(Dr. John Mitchell* of Virginia, earliest N. American botanical author. founder of several new genera in 1741.) — Gen. ed. 5, 49; Lam. Ill. t. 63. *Chamædaphne,* Mitch. — Of a single species, for that of Japan seems not different.

M. répens, L Small creeping evergreen, glabrous or nearly so: leaves deep green, ovate or subcordate, half-inch to near an inch in length, slender-petioled: stipules triangular-subulate, minute: peduncle short, terminal: corollas white or tinged with rose outside; tube half-inch long, surpassing the oblong lobes: two-eyed "berry" rather dry and tasteless, bright red, sometimes white. — Spec. i. 111 *(Lonicera, &c., Gronov.: Sivinga baccifera,* &c., Pluk. Amalth. t. 444. Catesb. Car. t. 20. Lodd. Bot. Cab. t. 979; Bart. Fl. Am. Sept. t. 95, f. 1; Torr. & Gray, Fl ii. 34; Gray, Struct Bot. ed. 6, fig. 467-469. *M. undulata,* Sieb. & Zucc.; Miquel, Prolus. Jap. 275. — Woods, especially under Coniferæ, Nova Scotia and Canada to Florida and Texas. (Mex., Japan.)

20. KELLÓGGIA. Torr. *(Dr. Albert Kellogg,* of California.) — Wilkes. S. Pacif. Ex. Exped. xvii. 332 (1874), t. 6 (1862); Benth. & Hook. Gen. ii. 137;

Gray, Proc. Am. Acad. vi. 589, & Bot. Calif. i. 282. — Single species: most allied to *Galopina* of S. Africa.

— **K. galioides**, Torr. l. c. Slender and glabrous or puberulent perennial, a span to a foot high, with foliage of a *Houstonia* (leaves only opposite, lanceolate, sessile, with small and entire or 2-dentate interposed stipules), fruit and paniculate inflorescence of a *Galium*, and corolla (of *Asperula*) white or pinkish, 2 or 3 lines long, the lobes equalling or shorter than the tube. — Mountain woods, mostly under coniferous trees, Sierra Nevada, California (first coll. by *Brewer* and *Torrey*), south to mountains of Arizona, east to Utah, and north to Washington Terr. and N. W. Wyoming.

21. MITRACÁRPUS, Zuccarini. (Μίτρα, a girdle or head-band, evidently taken in the sense of mitre, and καρπός, fruit.) — Low annuals or perennials (American and one or two African); with the habit of *Spermacoce*, and with small white flowers. — Zucc. in Rœm. & Schult. Syst. Mant. iii. 210, name given only in the accusative case, "Mitracarpum," in index rightly under the nominative "*Mitracarpus*." Mistaken for a nominative, we have the ungrammatical *Mitracarpum*, by Cham. & Schlecht., followed by A. Rich., DC., Endl., Benth. & Hook., and wrongly corrected by Benth. Bot. Sulph. and Gray, Pl. Wright., into *Mitracarpium*. (Vide Proc. Am. Acad. xix. 77.) *Staurospermum*, Thonning in Schum. Pl. Guin. 73, is of same date (1827).

M. breviflórus, Gray. Annual, a span or two high, nearly glabrous and smooth, bearing 2 or 3 axillary verticillastrate-capitate clusters and a terminal one: leaves lanceolate, about inch long: stipules with few setiform appendages: two larger calyx-lobes lanceolate-subulate, longer than tube, equalling or surpassing the small (barely line long) glabrous white corolla; intermediate ones small and dentiform, hyaline. — Pl. Wright. ii. 68; Rothr. in Wheeler Rep. vi. 137. — Ravines and hillsides, S. Arizona, *Wright*, *Thurber*, *Rothrock*, &c. (Adj. Mex., *Berlandier*, &c.)

M. lineáris, Benth. Bot. Sulph., of Lower California, also coll. by *Xantus*, has narrow leaves, and tube of corolla at least twice the length of the calyx.

22. RICHÁRDIA, Houst., L. (*Dr. H. Richardson* of London, father of *Richard Richardson*, the correspondent of Gronovius, &c. See Smith's Corr. Linnæus and other Naturalists, ii. 173.) — Hispid or hirsute perennials or annuals, natives of Tropical America; with broadish subsessile leaves, setiferous stipules, and whitish flowers; these mostly in a terminal capitate cluster, involucrate by the one or two uppermost pairs of leaves. — Gen. Pl. ed. 1, 100; Gærtn. Fruct. t. 25; Ruiz & Pav. Fl. Per. & Chil. t. 279; Hiern in Fl. Trop. Afr. iii. 242. *Richardsonia*, Kunth in Mem. Mus. Par. iv. 430, & HBK. Nov. Gen. & Spec. iii. 350, t. 279: but it appears that this, which correctly indicates the naturalist to whom the genus was dedicated, cannot be allowed to supersede the original name, faulty as it is in this respect.

— **R. scábra**, L. Loosely branching and spreading: leaves ovate to lanceolate-oblong (inch or two in length), roughish: stipules with rather few setiform appendages: glomerules of flowers and fruit depressed: corolla 2 or 3 lines long. — Spec. i. 330. *R. pilosa*, Ruiz & Pav. l. c.; HBK. l. c. *Richardsonia scabra*, St. Hil. Pl. Us. Bras. 8, t. 8; DC. Prodr. iv. 567; Chapm. Fl. ed. 2, Suppl. 624. — Low or sandy grounds, abundantly naturalized in the low country, S. Carolina to Texas, called *Mexican Clover* in Alabama, and relished by cattle; the root in S. America used as an emetic and as a substitute for Ipecac. Sparingly occurs as a ballast-weed at Northern ports. (Nat. from Mex. & S. Am.)

23. CRÚSEA, Cham. (*Prof. Wm. Cruse*, of Kœnigsberg, who wrote on *Rubiaceæ*.) — Perennials or annuals (of Mexico and adjacent districts), with habit of *Diodia*, the rose-colored or white corollas elongated in the typical species : stamens and style usually exserted. — Linnæa, v. 165 ; DC. Prodr. iv. 566 ; Hook. & Arn. Bot. Beech. t. 99 ; Benth. & Hook. Gen. ii. 144 (calyx wrongly said to persist on the fruit) ; Gray, Proc. Am. Acad. xix. 77. — where the genus is extended.

* Corolla rose-purple, with slender almost filiform tube: erect annual.

C. Wrightii, GRAY. Sparsely hirsute, about a foot high, with long internodes : leaves oblong-lanceolate, nervose-veiny, upper attenuate-acute ; uppermost four or more involucrate around the solitary capitate glomerule : calyx-lobes 4, attenuate-subulate and almost equal, nearly equalling the corolla-tube, or two of them sometimes very short, hispid-ciliate toward the base : corolla salverform, 2 lines long ; stigmas 2, short-linear : ovary and immature fruit didymous. — Pl. Wright. ii. 68. — Plains and mountains of S. Arizona, *Wright, Lemmon*. Habit of *C. rubra*, but far smaller-flowered.

* * Corolla white or whitish, small (about 2 lines long): stamens and style little exserted: stigmas short: low and diffuse annuals or perennials.

C. subuláta, GRAY. Glabrous and smooth throughout : stems ascending from an annual root, a span or two high, somewhat paniculately branched : branches flowering from most of the axils : leaves narrowly linear becoming subulate (inch or less long) : clusters rather few-flowered : corolla almost salverform : calyx-lobes 2 or 3 lanceolate and foliaceous, one or two much smaller and partly scarious or reduced to stipule-like teeth : gynœcium 2-merous: fruit cuneate-obovate, slightly didymous, obscurely puberulent : carpels coriaceous, at maturity separating from a narrow linear and bifid persistent carpophore (not unlike that of some Umbelliferæ) and opening on the ventral face. — Proc. Am. Acad. xix. 78. not that of Hemsl. Biol. Centr.-Am., which is a slip of pen or type for *C. subulata*, Hook. & Arn. *Spermacoce subulata*, Pav. ex DC. (*Borreria subulata*, DC. Prodr. iv. 543) : Hemsl. l. c. 60. — S. Arizona, *Wright* (from seeds which were raised in Botanic Garden, Cambridge, in 1852), *Lemmon*. (Mex.)

C. allocócca, GRAY, l. c. Hirsute or hispidulous to almost glabrous, diffusely branched from a perennial root, low and much spreading or depressed, flowering from summit and uppermost axils : leaves from linear to oblong-lanceolate (half-inch to barely inch long) : corolla funnelform, 3-4-lobed : calyx-lobes 3 to 5, commonly 4 and equal, lanceolate, longer than the ovary and fruit : gynœcium 3-4-merous : stigmas short and broad : fruit obovate-globose, sometimes glabrous and smooth, sometimes partially or wholly hispidulous, 3-4-coccous, more commonly 3-coccous ; the carpels flattened on the ventral face, separating from a weak scarious carpophore, either closed or torn open ventrally. — *Diodia tricocca*, Torr. & Gray, Fl. ii. 30. *D. tetracocca*, Hemsl. Biol. Centr.-Am. Bot. ii. 56. t. 40. f. 10-15. *Spermacoce ? tetracocca*, Martens & Gal. Bull. Acad. Brux. xi. 132. fide Hemsl. — Prairies of Texas, first coll. by *Berlandier, Drummond*, &c. (Mex.)

24. **SPERMACÓCE**, Dill. (Σπέρμο. seed, ἀκωκή, point ; the carpels pointed or crowned with one or more calyx-teeth.) — Low herbs, with small and white sometimes bluish or purplish flowers, and small fruits in sessile glomerules at the nodes ; chiefly tropical, the greater number American. — Dill. Elth. ii. 370. t. 227 ; L. Gen. ed. 1. 25 ; Benth. & Hook. Gen. ii. 143. *Spermacoce & Borreria*, Meyer. Prim. Fl. Esseq. 79 ; Cham. & Schlecht. in Linn. iii. 310, 355 ; DC. Prodr. iv. 540, 552. Fl. summer : corolla in our species short and white.

S. INVOLUCRÁTA, Pursh, Fl. i. 105, appears to have been founded on *Crusea rubra*, Cham. & Schlecht. (notwithstanding the "flowers white"), and without much doubt was wrongly attributed to this country.

* Leaves from oval to oblong-lanceolate, contracted into a narrow base or short and margined petiole, obliquely more or less pinnate-veined, in ours smooth and glabrous or a little scabrous: fruit splitting into the two carpels, one broadly open on the ventral face and discharging its seed, the other closed (at least at first) by the membranaceous or coriaceous dissepiment. — *Spermacoce*, G. F. W. Meyer, l. c. ; DC.

+— Corolla very villous in the throat, very short: root apparently perennial.

S. glábra, MICHX. Spreading or decumbent, smooth and glabrous: stems a foot or so long: leaves oblong-lanceolate and oblong (inch or two long), not prominently veined: corolla more campanulate than funnelform, little surpassing the large calyx-teeth (only a line and a half long): subsessile anthers and style included: fruit somewhat turbinate, smooth (nearly 2 lines long), crowned by the 4 conspicuous at length triangular-lanceolate spreading calyx-teeth, their bases slightly united. — Fl. i. 82; Pursh, Fl. i. 105 (excl. the remark that corolla is longer than in the next); Torr. & Gray, Fl. ii. 27. *Diodia glabra*, Pers. Syn. i. 124. Probably *Spermacoce verticillis tenuioribus*, Dill. Elth. l. c., therefore *S. tenuior*, L. Spec. except as to syn. Pluk. — River-banks, S. Ohio to Florida, Arkansas, and Texas.

+— +— Corolla glabrous or merely pubescent in the throat: root annual.

S. tenúior, L. partly, LAM. Ascending or spreading: leaves oval-oblong to oblong-lanceolate, more or less scabrous, with 4 or 5 pairs of more prominent veins: corolla funnelform, twice or thrice the length of the calyx, and with more or less exserted stamens and style, yet in some plants nearly as short as in the preceding species, and with stamens and style included (probably dimorphous): fruit didymous-obovate, commonly pubescent or puberulent (only a line or so long), coriaceo-crustaceous, crowned with the four short deltoid or triangular-lanceolate distinct calyx-teeth. — L. Spec. i. 102, as to Pluk. Alm. t. 136, f. 4, perhaps also of Dill. Elth. l. c.; Pursh, Fl. l. c.; Lam. Ill. i. 273, t. 62, f. 1; Schk. Handb. t. 32; A. Rich. Mem. Rub. t. 4, no. 2, exel. fig. c?; Griseb. Fl. W. Ind. 349. *S. Chapmanii*, Torr. & Gray, Fl. ii. 27, form with the more conspicuous corolla, &c. — River-banks, Florida and Louisiana; rare. (W. Ind. to Brazil.)

S. Portoricénsis, BALBIS. Annual or perhaps perennial, diffusely spreading, wholly smooth and glabrous: leaves smaller than in the preceding (half-inch to inch long), drying blackish, with inconspicuous venation: glomerules mostly small and few-flowered: corolla only half-line long, short-campanulate, glabrous inside: subsessile anthers and style included: fruit globular (a line or less long), very smooth or rarely obscurely puberulent, thinnish, crowned with small narrowly subulate calyx-teeth, their bases distant: seed strongly scrobiculate-punctate. — DC. Prodr. iv. 552; Polak in Linn. xli. 373. *S. tenuior*, Torr. & Gray, Fl. ii. 27, & of distrib. Rugel, Curtiss, &c. *S. tenuior*, var. *Portoricensis*, Griseb. — Keys and near shores of Southern Florida, *Blodgett, Rugel, Garber, Curtiss*. (W. Ind., &c.)

* * Leaves in our species narrow and sessile: fruit septicidal through the thin dissepiment and both carpels ventrally dehiscent. — Species of *Borreria*, Meyer, l. c.; DC. l. c. *Bigelovia*, Spreng. Syst.

S. parviflóra. Annual, glabrous or a little hirtellous-pubescent: stems slender, spreading or erect, a span to a foot high: leaves from narrowly lanceolate to spatulate-oblong (inch or less long), obscurely veined: glomerules in many of the axils, globose: corolla salverform, about a line long: stamens and style included: fruit half a line long, didymous-globular, thinnish, surmounted by the four attenuate-subulate and longer nearly equal calyx-teeth, two on each carpel: seed scrobiculate. — *Borreria parviflora*, Meyer, Fl. Esseq. 83, t. 1, f. 1–3; DC. l. c. *B. micrantha*, Torr. & Gray, Fl. ii. 28. *B. Domingensis*, Griseb. Cat. Cub. 141. Hardly *Spermacoce hirta*, Swartz, referred here by Griseb. — Waste grounds, S. Florida, *Leavenworth, Garber, Curtiss*. (W. Ind., Mex., S. Am.)

S. podocéphala. Suffrutescent perennial, a span to a foot high in tufts, glabrous and smooth or sometimes obscurely hirtello-puberulent: stipular bristles few: leaves numerous, about the length of the internodes and usually axillary-fascicled, narrowly linear (inch or less long), seldom over a line wide, veinless, not rarely with revolute margins, 2 to 6 uppermost raised on a longer peduncle-like internode and involucrating the solitary terminal globose glomerule: corolla short-funnelform, a line or little more in length: fruit obovate and didymous, each carpel surmounted by a rather shorter subulate or obtuse calyx-tooth; intermediate teeth rudimentary or wanting: seed minutely scrobiculate. — *Borreria podocephala*, DC. Prodr. iv. 542; Chapm. Fl. 175 (var. *pumila*); Griseb. Fl. W. Ind. 350, the W. Indian specimens attenuate and with less fascicled leaves. *Spermacoce pygmaea*, Wright in Sauv. Fl. Cubana, 72. — S. Florida, *Blodgett*, &c. S. Texas, *Berlandier, Palmer*. (Mex., W. Ind.)

25. **DIÓDIA**, Gronov. (Δίοδος, a thoroughfare, wayside plants.) — Low herbs (nearly all American), usually decumbent; with white or bluish flowers

either solitary or few and sessile in the axils of the leaves, produced all summer; stipules long-setiferous.— L. Hort. Cliff. App. 493, & Gen. ed. 2, 291 (bad character, genus there attributed to Gronovius); Jacq. Ic. Rar. t. 29; Lam. Ill. i. 276. t. 63; Torr. & Gray, Fl. ii. 28, excl. spec. 3. *Diodia* § *Endiodia*, DC.

§ 1. Style 2-cleft and stigmas filiform, and anthers nearly linear; both exserted: fruit somewhat drupaceous-fleshy before maturity.

D. Virginiána, L. Diffusely spreading or procumbent from a perennial root, from nearly glabrous to hirsute: leaves from oblong to lanceolate, inch or two long, bright green, with 4 or 5 pairs of oblique veins: stipular bristles strong and flat, not very many, commonly sparingly hirsute: corolla about half-inch long, with slender tube: fruit 3 or 4 lines long, from glabrous to hirsute, crowned with 2 (or sometimes 3 or 4) lanceolate conspicuous calyx-teeth: carpels suberose-crustaceous, with a thin epicarp, 3-costate on the back. — Spec. i. 104, & Mant. ii. 330; Torr. & Gray, Fl. ii. 29, with vars. *D. Virginica*, Jacq. Ic. Rar. l. c.; Lam. l. c.; Michx. Fl. i. 181; Pursh, Fl. i. 105; DC. Prodr. iv. 562. *D. tetragona*, Walt. Car. 87. *D. hirsuta*, Pursh, Fl. i. 106. *Spermacoce Virginiana*, A. Rich. Mem. Rub. t. 4. no. 3. fruit only. — Low grounds, along streams, S. New Jersey to Florida, Texas, and Arkansas.

§ 2. Style entire: stigma capitate-2-lobed, and with the short anthers shorter than the purplish corolla-lobes: fruit wholly dry and thin-crustaceous.

D. téres, Walt. Diffusely spreading or ascending from an annual but sometimes lignescent root, rigid, from puberulent to hirsute: branches terete, rather quadrangular above: leaves from linear to lanceolate, commonly inch long, rather rigid, scabrous: bristles of the truncate stipules numerous, long and slender, usually equalling the flowers and surpassing the fruit: corolla only 3 lines long: fruit obovate-turbinate, commonly hispidulous, only 2 lines high, crowned with the mostly 4 shorter and equal or unequal deltoid-lanceolate or at length ovate calyx-lobes, often 3 on one carpel and one on the other. — Car. 87; DC. Prodr. iv. 562; Torr. & Gray, l. c. *Diodia*, Gronov. in Clayt. Fl. Virg. ed. 1, 71, ed. 2, 17, at least in part, fide herb. *Spermacoce diodina*, Michx. Fl. i. 82; Pursh, Fl. i. 85. — Sandy soil, New Jersey and Penn., toward the coast, to Florida, Texas, and in Mississippi Valley to W. Illinois. (Adj Mex., W. Ind., for it probably includes *D. prostrata*, Swartz.)

Var. angustáta. Slender: stem often simple, all the upper part quadrangular: leaves narrowly linear-lanceolate or linear: fruit hispidulous to puberulent, varying to quite smooth, and to smooth and glabrous herbage. — *D. teres*, var.? Gray, Pl. Wright. ii. 69. — S. Arizona, *Thurber*, *Wright*, *Lemmon*, the latter specimens a remarkably smooth form.

26. **GÁLIUM, L.** BEDSTRAW. CLEAVERS. (*Galium*, as written by the old herbalists, and even by Tournefort: supposed to come from γάλα, milk, which some species were used to curdle, in place of rennet.) — Very large genus, indigenous to all temperate regions: leaves sessile: flowers small, in summer. — Gen. ed. 5, 46. *Galium* & *Aparine*, L. Gen. ed. 1-3. *Gallium*, *Aparine*, & *Cruciata*, Tourn. *Galium* & *Relbunium*, Benth. & Hook. Gen. ii. 149.

RÉNIA (*R. tinctoria*, L., of the Old World, is the cultivated MADDER) has 5-merous flowers and baccate fruit, the latter a character of several species of *Galium*.

ASPÉRULA ODORÁTA, L., the WOODRUFF of Europe, sweet-scented in drying, has been occasionally found around German settlements.

§ 1. Species locally naturalized in the Atlantic States: fruit dry at maturity.

G. vérum, L. Perennial: stems smooth, erect: leaves 8 or sometimes 6 in the whorls, linear, roughish, soon deflexed: flowers very numerous, paniculate, yellow, rarely cream-color: fruit usually smooth. — Dry fields. E. Massachusetts. (Nat. from Eu.)

G. Mollúgo, L. Perennial, smooth throughout: stems erect or diffuse, 2 or 3 feet long: leaves 8 or on branchlets 6 in the whorls, oblanceolate to nearly linear: flowers very numerous in ample almost leafless panicles: fruit smooth. — Roadsides and fields, New York and Penn. (Nat. from Eu.)

G. Ánglicum, Huds. Annual, slender, diffuse, seldom a foot high, glabrous: leaves 5 to 7 in the whorls, oblanceolate to nearly linear (quarter-inch long), minutely spinulose-scabrous on margins and angles of stem: flowers rather few, cymulose on leafy branches, greenish-white, very small: fruit glabrous, but more or less tuberculate-granulate. — *G. Parisiense*, L. var. *Anglicum*, Hook. & Arn. Brit. Fl. &c. — Roadsides in dry soil, Bedford Co., Virginia, *A. H. Curtiss.* (Nat. from Eu.)

G. tricórne, With. Annual, resembling *G. Aparine*, rather stout, with simple branches, spreading or procumbent: leaves 6 or 8 in the whorls, oblanceolate, cuspidate-mucronate (inch or less long), retrorsely prickly-hispid on margins, as also on angles of stem: flowers usually only 3 in the umbelliform cymules, dull white: fruits comparatively large, tuberculate-granulate, not hairy, hanging on recurved stout pedicels (likened to the three balls of a pawnbroker's shop). — Rare in waste or cult. fields eastward. (Nat. from Eu.)

§ 2. Indigenous species: fruit dry.

* Annuals: fruit more or less uncinately hispidulous or hirsute, in one species sometimes naked: flowers hermaphrodite: corolla white or whitish.

    + Coarse, reclining: leaves 6 to 8 in the whorls.

G. Aparíne, L. (Cleavers, Goose-Grass.) Stems 1 to 4 feet long, retrorsely aculeolate-hispid on the angles, as also on the margins and midrib of the oblanceolate or almost linear cuspidate-acuminate leaves: peduncles rather long, 1 to 3 in upper axils or terminal, bearing either solitary or 2 or 3 pedicellate flowers: fruit not pendulous, rather large, granulate-tuberculate and the tubercles tipped with bristles. — Reichenb. Ic. Fl. Germ. t. 1597. — Shaded grounds, Canada to Texas, and Aleutian Islands to California; eastward mainly as an introduced plant, or appearing so. (Eu., Asia.)

    Var. **Vailléntii**, Koch. Smaller, more slender: leaves seldom inch long: flowers usually more numerous: fruit smaller (carpels when dry only a line or so in diameter), hirsute or hispidulous. — Fl. Germ. ed. 1, 330. *G. Aparine*, var. *minor*, Hook. Fl. i. 290. *G. Vaillantii*, DC. Fl. Fr. iv. 203. — Texas to California, Montana, and Brit. Columbia; certainly indigenous: perhaps so in Canada, &c. (Eu.)

    + + Small and low, more erect: leaves mostly 4 in the whorls.
    ++ Flowers on solitary naked peduncles.

G. bifólium, Watson. Smooth and glabrous, a span or two high, sparingly branched slender: leaves oblanceolate to nearly linear, four in whorls (larger half-inch long), the alternate ones smaller, or uppermost nearly reduced to a single pair: fructiferous peduncles about the length of the leaves, horizontal, and the minutely hispidulous fruit decurved on the naked tip. — Bot. King Exp. 134, t. 14. — Mountains of Utah, Nevada, and S. Montana, *Watson.* W. Colorado, *Brandegee*, and Sierra Nevada, California.

G. Texénse, Gray. Hispidulous-hirsute or upper part of stem glabrous, weak and slender, a foot or less high: leaves broadly oval, equal, in fours, thin, one-nerved (only 3 or 4 lines long), the sides and margins equally beset with straight bristly hairs: peduncles terminal and 1-flowered; the primordial ones naked and filiform, 4 to 10 lines long; single axils proliferous into a similar shoot which bears an unequally 4-leaved small whorl and a short peduncle or pedicel: bristles of the fruit much shorter than the carpels, barely uncinulate. — Proc. Am. Acad. xix. 80. *G. Californicum*, var. *Texanum*, Torr. & Gray, Fl. ii. 20. *G. uncinulatum*, Gray, Pl. Lindh. ii. 215 ? probably not DC., nor *G. obstipum*, Schlecht., which are perhaps perennial and have a different inflorescence, but are nearly allied. — Hills and riverbanks, Texas, *Drummond* (immature), *Lindheimer*, *Wright*, *Hall*, *Reverchon*.

    ++ ++ Flowers and fruit solitary and sessile between a pair of bracteal leaves which resemble the cauline ones: stem and leaves hispidulous, or sometimes nearly glabrous.

G. virgátum, Nutt. A span or two high, simple or with simple and strict branches from the base: leaves oblong-linear or oblong, thickish, 2 or 3 lines long; most of the axils floriferous: peduncles exceedingly short, reflexed in fruit, not proliferous: carpels copiously uncinate-hispid, shorter than the arrect bracteal leaves, which often appear as if belonging to the whorl itself. — Torr. & Gray, Fl. ii. 20 ; Gray, Pl. Lindh. ii. 215. *G. Texanum*, Scheele in Linn. xxi. 597, badly described. — Naked prairies of Arkansas, W. Louisiana, and Texas, first coll. by *Nuttall*.

    Var. **leiocárpum**, Torr. & Gray, l. c. Fruit quite smooth and glabrous ; herbage commonly almost so. — With the ordinary form.

G. prolíferum, Gray. More branching, less hispidulous or glabrate, weaker: leaves thinner, oval or oblong, alternate ones rather smaller: flowers solitary terminating a pedunculiform axillary branch of twice or thrice the length of the whorled leaves, and the fruit hardly surpassed by its pair of bracts, or one or even two more by prolification from the bracts: fruit of the preceding — Pl. Wright. ii. 67. *G. virgatum*, var. *diffusum*, Gray, Pl. Wright. i. 80. — Stony hills, along the Rio Grande between Texas and New Mexico, *Wright*, &c. Hills near Tucson, Arizona, *Pringle*. Perhaps S. Utah, *M. E. Jones*, specimen insufficient. (Adj. Mex., *Palmer*.)

　　* * Perennials, wholly herbaceous: slender roots of several species containing red coloring-matter (madder): flowers hermaphrodite (at least not diœcious): bristles on the fruit short and uncinate or none.

　　+— Leaves in fours throughout or rarely even fewer, comparatively large, either broad or inch or more long, none cuspidate-pointed,

　　++ Broad, one-nerved, with usually an obscure pair of lateral veins at base: flowers yellowish white to brown-purplish: fruit hispid.

G. pilósum, Ait. Commonly hirsutulous-pubescent: stems ascending, two feet long, paniculately branched above: leaves oval, callous-mucronulate, puncticulate (the largest hardly inch long): cymules few-flowered: flowers all short-pedicelled. — Ait. Kew. i. 145; Pursh, Fl. i. 104; Torr. & Gray, Fl. ii. 24. *G. Bermudense*, L. Spec. i. 105, as to syn. Pluk., from which also the specific name, but with the incongruous char. "foliis linearibus"; and the plant seems unknown from Bermuda. *G. purpureum*, Walt. Car. 87. not L. *G. puncticulosum*, var. *pilosum*, DC. Prodr. iv. 601. — Open woods in dry soil, S. New England to Indiana, Arkansas, Texas, and Florida. (*G. obovatum*, HBK., of S. America, is near to this.)

　　— Var. **puncticulósum**, Torr. & Gray, l. c. Almost glabrous: leaves varying to elliptical-oblong, hispidulous-ciliate. — *G. Bermudense*, L. l. c. as to syn. Gronov. *G. puncticulosum*, Michx. Fl. i. 80; DC. l. c. *G. Bermudianum*, Pursh, Fl. i. 104. *G. punctatum*, Pers. Syn. i. 128. — Virginia to Texas.

　　++ ++ Leaves broad, distinctly 3-nerved, pointless or merely callous-mucronate: flowers never bright white.

　　= Fruit hispid: cymes rather few-flowered, with divisions or peduncles in fruit divaricate or diverging: corolla from dull cream-color or greenish to brown-purplish: stems comparatively simple and low.

G. Kamtschátioum, Steller. A span to a foot high: stems weak, mainly glabrous: leaves orbicular to oblong-ovate, thin (half-inch to inch or so long), slightly pilose or hirsutulous, at least the nerves and margins: flowers few or several in the pedunculate cymules, all distinctly and rather slenderly pedicellate: corolla glabrous, yellowish white, not turning dark, its lobes merely acute. — Steller in Rœm. & Schult. Syst. iii. Mant. 186; Gray, Proc. Am. Acad. xix. 80. *G. obovatum*, Ledeb. Fl. Ross. ii. 412; Schmidt, Fl. Sachal. 203; Maxim. Mel. Biol. ix. &c., not of HBK., which is S. Amer. and has pinnately veiny leaves. *G. Littellii*, Oakes in Hovey Mag. vii. 177 (1841); Gray, Proc. Am. Acad. viii. 388. *G. circæzans*, var. *montanum*, Torr. & Gray, Fl. ii. 24. — Mountains of Gaspé, Lower Canada (*Dr. Allen*), higher mountains of New England (*Littell, Tuckerman, Oakes*, &c.); also those of Oregon and Washington Terr. (*Hall, Howell, Henderson, Suksdorf*, chiefly forms with oblong-ovate and acutish leaves), to Unalaska, *Eschscholtz*. (Adj. E. Asia, the Sachalin plant exactly that of N. New England and Canada.)

G. circǽzans, Michx. About a foot high, hirsutulous-pubescent or glabrate: leaves oval or oblong-ovate, obtuse (largest inch and a half long): flowers short-pedicelled or subsessile in the fork and along the simple branches of the cyme: fruit at length deflexed: corolla greenish, hirsutulous outside, the lobes acute or acuminate. — Fl. i. 80; DC. l. c.; Torr. & Gray, l. c. excl. vars. *G. brachiatum*, Muhl. Cat., not Pursh. *G. boreale!* Walt. Car. 237. *G. circæoides*, Rœm. & Schult. Syst. iii. 256. — Dry woods, Canada to Florida, N. W. Arkansas, and Texas. Leaves sweet-tasted, wherefore called *Wild Liquorice*.

G. lanceolátum, Torr. A foot or two high, simple-stemmed, nearly glabrous: leaves (except lowest) broadly lanceolate, verging to ovate-lanceolate, acute or acutish (2 inches long): corolla glabrous, larger and the lobes more acuminate than in preceding, yellowish turning dull purple: inflorescence similar: fruit less hispid. — Fl. N. & Mild. States. 168; Hook. Fl. i. 280; Gray, Man. *G. Torreyi*, Bigel. Fl. Bost. ed. 2, 56. *G. circæzans*, var.

lanceolatum, Torr. Cat. Pl. N. Y. 23; Torr. & Gray, Fl. ii. 24. — Dry woods, New England to Upper Michigan and Canada.

= = Fruit very smooth and glabrous, rather fleshy: corolla dark brown-purple; lobes acuminate.

G. latifólium, MICHX. A foot or more high, somewhat glabrous: leaves oblong- to ovate-lanceolate (mostly 2 inches long), hispidulous-ciliate, lineate-puncticulate, almost petiolate: cymes effusely paniculate, many-flowered; flowers on filiform pedicels, which are erect even in fruit. — Fl. i. 79; DC. Prodr. iv. 599; Torr. & Gray, Fl. ii. 25, excl. var. — Open woods in the Alleghany Mountains, Penn. (*Porter*) to Carolina and Tennessee, first coll. by *Michaux*.

++ ++ ++ Leaves narrow, with lateral nerves obscure or none: otherwise like *G. latifolium*.

G. Arkansánum, GRAY. Less than foot high: stem and branches glabrous, slender: leaves from lanceolate to linear (at most inch long, 1 to 3 lines wide), hispidulous-ciliate on the margins and midrib beneath: effuse cymes, flowers, &c. of the last preceding: fruiting pedicels minutely upwardly scabrous. — Proc. Am. Acad. xix. 80. *G. latifolium.* var., Torr. & Gray, l. c. — Arkansas, near the Hot Springs, *Engelmann, Dr. Foreman*.

++ ++ ++ ++ Leaves narrow, distinctly 3-nerved, blunt: flowers bright white, copious.

G. boreále, L. Erect, a foot or two high, mostly smooth and glabrous, very leafy: leaves from linear to broadly lanceolate, often with fascicles of smaller ones in the axils: flowers in numerous close cymules collected in a terminal and ample thyrsiform panicle; the uppermost leaves being reduced to pairs of small oblong or oval bracts: fruit small, hispidulous, or at first canescent and soon glabrous and smooth. — Spec. i. 108; Fl. Dan. t. 1024; Pursh, Fl i. 104; Hook. Fl. i. 289; Torr. & Gray, l. c. *G. septentrionale*, Rœm. & Schult. Syst. iii. 253; DC. Prodr. iv. 601. *G. strictum*, Torr. Cat. Pl. N. Y. 23. *G. rubioides* of Am. authors, form with smooth fruit and broadish leaves. (True *G. rubioides*, L., N. Asia to Kamtschatka, has evident reticulate venation between the ribs of the broader leaves, and enlarged vesicular as well as smooth fruit.) — Rocky banks of streams, Canada to Penn., New Mexico, California, and north to Arctic regions, in various forms. (Eu., N. Asia.)

+— + Leaves in fours, fives, or sixes, small, one-nerved, pointless: plants low, slender and weak, and slender-rooted: flowers very small, white: fruit smooth and glabrous.

G. Brandegéi, GRAY. Loosely cespitose-depressed, with the aspect of *Callitriche* or *Elatine*, smooth and nearly glabrous: branches or stems a span or less long: leaves in fours, obovate to spatulate-oblong, slightly succulent, 1 to 3 lines long, one or two of the whorl usually smaller than the others; midrib indistinct: peduncles solitary in upper axils or geminate and terminal, one-flowered, little longer than the leaves. — Proc. Am. Acad. xii. 58. — New Mexico, in valley of the upper part of the Rio Grande at Los Pinos, 9,000 feet, spreading on moist ground, *Brandegee*.

G. trífidum, L. Weakly erect, branching, 5 to 20 inches high, smooth and glabrous, except the retrorsely scabrous angles of the stem and usually more hispidulous and sparse roughness of the midrib beneath and margins of the leaves: these in sixes, fives, or not rarely fours, linear or oblanceolate, or lanceolate-oblong, obtuse, 4 to 7 lines long; the midrib evident: peduncles slender, scattered, 1-several-flowered; flowers often 3-merous (whence the specific name), as commonly 4-merous. — Spec. i. 105; Fl. Dan. t. 48; DC. Prodr. iv. 597; Torr. & Gray, Fl. ii. 22. *G. tinctorium*, L. l. c. 106; DC. l. c., larger form with leaves in sixes and flowers 4-merous. *G. Claytoni*, Michx. Fl. i. 78; Hook. Fl. i. 288. — Sphagnous bogs and wet ground, Newfoundland and Labrador to Aleutian Islands, and south to Texas, Arizona, and California. (Eu., N. Asia, Japan.)

Var. **pusíllum,** GRAY, Man., among the many forms of the species, is the smallest, a span or two high: leaves only in fours, 3 or 4 lines long, narrow, in age often reflexed: peduncles 1-flowered. — In cold bogs, a Northern form, and in the Rocky Mountains and Sierra Nevada to Colorado and California.

Var. **latifólium,** TORR. The larger and broadest-leaved form: leaves 6 or 7 lines long, often 2 lines wide: cymules few-several-flowered. — Fl. N. & Midd. States, 165; Gray, Man. *G. obtusum*, Bigel. Fl. Bost. ed. 2, 65. — Canada to Texas and mountains of California.

G. concínnum, TORR. & GRAY. Diffuse and erect, freely branching, about a foot high, smooth and glabrous, except the roughened angles of the stem and margins of the leaves: these all in sixes, oblanceolate-linear, mucronulate, veinless, rather lucid and firm (drying bright green), the midrib prominent beneath: flowers numerous in loose and open cymes on filiform peduncles or branchlets, and on filiform but rather short pedicels: corollas bright white. — Fl. ii. 23; Gray, Man. l. c. Perhaps *G. parciflorum*, Raf. in Med. Rep. v. 360, & Desv. Jour. Bot. i. 227 ? — Dry hills, Pennsylvania and Virginia to Michigan, Illinois, Kentucky, and Arkansas, first coll. by *Short*.

+ + + Leaves in sixes, sometimes fives or on the branchlets fours, cuspidately mucronate or acuminate.

++ Fruit smooth and glabrous: plant rough and adhesive by retrorse prickles: flowers bright white.

G. asprèllum, Michx. Glabrous, paniculately branched, erect and 2 feet high, or when supported by bushes 3 to 5 feet high, very floriferous: leaves lanceolate, about half-inch long, in sixes or on the branchlets fives or fours; their margins, midrib beneath, and prominent angles of the stem armed with strong retrorse prickles rather than bristles: cymes many-flowered: fruits small, like those of *G. trifidum*. — Fl. i. 78; DC. Prodr. iv. 598; Torr. & Gray, Fl. ii. 23. *G. Pennsylvanicum*, Muhl. Cat.; Willd. ex Rœm. & Schult. Syst. Mant. iii. 183. *G. spinulosum*, Raf. Prec. Decouv. 1814, 40. *G. micranthum*, Pursh, Fl. i. 103 ? by the char., except as to fruit. — Alluvial ground, especially low and shaded banks of streams. Canada, New England to Michigan and mountains of Carolina. (E. Asia ?)

++ ++ Fruit from scabrous or papillose to uncinately hispid: angles of the stem and midrib beneath minutely retrorse-hi-pidulous or scabrous or nearly naked in the same species; margins of leaves either antrorsely or retrorsely hispidulous-ciliolate, or naked in the same species, or even on different parts of same leaf.

G. aspérrimum, Gray. Stems erect or diffusely ascending, but weak, a foot or two high, probably from a perennial root: leaves lanceolate (about half-inch to inch long): cymes twice or thrice dichotomous, with filiform peduncles and pedicels: corolla white or turning purplish: ovary merely puberulent or scabrous: fruit granulate-scabrous, and sometimes minutely hispidulous. — Pl. Fendl. 60, & Bot. Calif. i. 284; Watson, Bot. King Exp. 134; Rothrock in Wheeler Rep. vi. 138. — Shady places in mountains, New Mexico (first coll. by *Fendler*) and Arizona to Nevada, California, and E. Oregon: mostly var. *asperulum*, Gray, Bot. Calif. l. c.; but the hispid or hispidulous roughness very variable.

G. triflórum, Michx. Diffusely procumbent, smoothish: herbage sweet-scented (as of *Asperula odorata*) in drying: stems a foot to a yard long: leaves elliptical-lanceolate to narrowly oblong (inch or two long): cymes once or twice 3-rayed: pedicels soon divaricate: corolla yellowish white to greenish, its lobes hardly surpassing the bristles of the ovary: fruit uncinate-hispid. — Fl. i. 80; Willd. Hort. Berol. i. 66; Pursh, Fl. i. 104; Hook. l. c.; Torr. & Gray, l. c. *G. cuspidatum*, Muhl. Cat., Ell. Sk. i. 197; DC. l. c. *G. brachiatum*, Pursh, l. c. 103. *G. suaveolens*, Wahl. Fl. Lapp. 48. *G. Pennsylvanicum*, Barton, Comp. Fl. Philad. 83. — Open and dry or moist woods, Canada to Alabama, Colorado, Rocky Mountains, W. California, and north to Alaskan Islands. (N. Eu., Japan.)

* * * Perennials with suffrutescent or suffruticose base: leaves 4 in the whorls: their margins, midrib, and angles of stem destitute of retrorse hispidness or roughness: fruit hirsute with long and straight (not at all uncinate-tipped) bristles: Western species of arid districts. — § *Trichogalium*, Gray.

+ Flowers hermaphrodite or monœcious-polygamous, paniculate and short-pedicelled, small: corolla only a line in diameter, brown-purple: stems numerous in tufts from the woody base, a foot or less high, slender, much branched: leaves narrow, 2 to 4 lines long, one-nerved, pointless.

G. Rothróckii, Gray. Glabrous, erect: leaves narrowly linear, rigid: bristles not very copious, not longer than the body of the fruit. — Proc. Am. Acad. xvii. 203. — S. Arizona, *Wright* (mixed with the following species), *Rothrock, Lemmon*. (Lower Calif., *Orcutt*.)

G. Wríghtii, Gray. Hirsute-pubescent throughout, diffuse: leaves linear to narrowly oblong, hardly at all rigid: bristles of fruit as long as its diameter. — Fl. Wright. i. 80, ii. 67. — Crevices of rocks in ravines, W. Texas to S. Arizona, *Wright, Lemmon*.

+ + Flowers diœcious: corolla greenish white or yellowish.

++ Leaves narrowly linear, with midrib little prominent and no lateral nerves or veins: stems elongated.

G. angustifólium, Nutt. Becoming shrubby at base, 1 to 4 feet high, with rigid virgate branches, smooth and glabrous or minutely pruinose-puberulent: leaves barely mucronulate (half-inch to inch long or on branches shorter, half-line to line wide) : cymes small, in narrow panicles, the fertile more or less condensed : corolla a line or two in diameter, dull white ; bristles of the fruit about the length of the body. — Gray, Bot. Calif. i. 285. *G. trichocarpum,* Nutt. (not DC.) & *G. angustifolium,* Nutt. in Torr. & Gray, Fl. ii. 82. — California, common from Santa Barbara to San Diego, Tejon, and apparently to the Mohave.

++ ++ Leaves narrowly-lanceolate to ovate, with midrib prominent beneath and continuous with stem-angles, sometimes a pair of lateral nerves : stems low or diffuse.

G. Matthéwsii, Gray. Glabrous and smooth, paniculately much branched, woody at base : leaves rigid, oblong- to ovate-lanceolate, veinless, with stout midrib, 2 or 3 lines long, some of the upper cuspidate-acute : flowers (of fertile plant) naked-paniculate : corolla barely a line in diameter : bristles of immature fruit rigid, not longer than the body. — Proc. Am. Acad. xix. 80. Arid district, Inyo Co , E. California, *Dr. Matthews.* Probably same from borders of S. W. Colorado and New Mexico, with rather longer and narrower leaves, *Brandegee.*

G. stellátum, Kellogg. Diffuse and bushy from woody base, a foot or two high, much branched, hispidulous-puberulent, sometimes nearly glabrous : leaves rigid, ovate-lanceolate (and 4 or 5 lines long) to narrow-lanceolate and small on flowering branches, acuminate-cuspidate, destitute of lateral nerves and veins ; margins either naked or hispidulous-ciliate : flowers paniculate and crowded : corolla white, little over a line in diameter : bristles of the fruit soft and flaccid at maturity, longer than the body. — Proc. Calif. Acad. ii. 97, fig. 26. *G. acutissimum,* Gray, Proc. Am. Acad. vii. 350, male plant. — Rocky cañons and dry hills, S. Utah and Arizona, first coll. by *Newberry.* (Islands off Lower California.)

G. multiflórum, Kellogg. A span to a foot high from a barely suffrutescent base, in tufts, glabrous, pruinose-puberulent or sometimes pubescent : leaves from broadly ovate to ovate-lanceolate, mucronate-apiculate, or minutely and abruptly acuminate, thickish, 4 to 7 lines long, a pair and sometimes two pairs of indistinct or obvious lateral nerves from the base ; uppermost leaves on flowering shoots usually only opposite : flowers short-pedicelled, thyrsoid-crowded in upper axils, or the fertile often solitary and sparse : corolla yellowish, a line or two in diameter : fruit when well formed densely clothed with hirsute bristles considerably longer than the body. — Proc. Calif. Acad. l. c., fig. 27. (Very poor name, the flowers not abundant for the genus and scattered.) *G. Bloomeri* & *G. hypotrichium,* Gray, Proc. Am. Acad. vi. 538, the latter founded on imperfect specimens with polygamous flowers and undeveloped fruit. *G. Bloomeri* & *G. multiflorum,* Watson, Bot. King Exp. 135 ; Gray, Bot. Calif. i. 285. — E. California to Utah, on the mountains of the drier districts, first coll. by *Bloomer, Veatch,* &c. Specimens east of the Sierra Nevada and vicinity mostly of the subjoined var.

Var. Watsóni. Mostly glabrous and smooth : leaves thinner, oblong-lanceolate (commonly about half-inch long and 2 lines wide), with lateral nerves either distinct or obsolete. — *G. multiflorum,* Watson, l. c. in great part.— Cañons and gulches, N. Arizona to E. Oregon and adjacent Idaho.

Var. hirsútum, *G. Bloomeri,* var. *hirsutum,* Gray, Bot. Calif. l. c., is an ambiguous form, with broad but thinnish leaves and whole herbage hirsute-pubescent. — Sierra Valley, California, *Lemmon.*

§ 3. Indigenous species, perennials : fruit baccate (leaves 4 in the whorls, one-nerved). — *Relbunium,* Endl.

* Pacific species, with ovate to oblong-linear (not rigid acerose) leaves : flowers of most and perhaps of all subdiœcious or polygamous, yellowish, purplish, or white ; sterile flowers in small loose cymes ; fertile somewhat solitary and scattered.

+ Berry so far as known purple or black, small.

G. púbens, Gray. Wholly herbaceous, somewhat cinereous with a fine and partly soft partly scabrous pubescence : stems much branched, diffuse, a foot or two long : leaves from roundish-oval to oblong, thickish, mostly pointless (largest half-inch long) ; margins at most hispidulous-scabrous : forming fruit glabrous and smooth ; mature fruit not seen, probably fleshy. — Proc. Am. Acad. vii. 350 ; Bot. Calif. i. 284, with var. *scabridum,* growing in more exposed situations. — California, in and near Yosemite Valley, first coll. by *Torrey* and *Bolander.*

G. Califórnicum, Hook. & Arn. Wholly herbaceous, from slender creeping rootstocks, often in low tufts, a span or two high, or diffuse, with slender stems a foot long, hispid or hirsute, rarely glabrate in age: leaves thinnish, ovate or oval, apiculate-acuminate (quarter-inch to half-inch long), margins and midrib hispid-ciliate; fruit glabrous, on recurved pedicels. — Hook. & Arn. Bot. Beech. 349; Torr. & Gray, l. c. 20 (excl. var. *T. ranum*); Gray, Bot. Calif. i. 283. — Shady ground, common in the western part of California, especially in the coast ranges.

G. Nuttállii, Gray. Tall and much branched from suffrutescent base, often supported by and as if climbing over bushes, or procumbent, mostly glabrous, except minutely aculeolate-hispidulous angles of stem and margins of leaves, these sometimes naked: leaves small, oval to linear-oblong, mucronate, macronulate, or obtuse: fruit smooth and glabrous. — Pl. Wright. i. 80, & Bot. Calif. i. 283. *G. suffruticosum,* Nutt. in Torr. & Gray, Fl. ii. 21, not Hook. & Arn. — California toward the coast, from San Diego to Humboldt Co.

+ + Berry white (blackening in drying), very smooth, juicy.

G. Bolánderi, Gray. Herbaceous from a woody root, diffuse, a foot or two high, glabrous, sometimes pubescent: angles of the stem not at all or hardly scabrous: leaves oblong-linear or lanceolate, rather acute, about half-inch long, thickish, with margins and midrib either smooth and naked or sparsely hispidulous; those of branchlets not rarely opposite: corolla dull purplish. — Proc. Am. Acad. vii. 350, xix. 80, & Bot. Calif. i. 284. male plant. *G. margaricoccum,* Gray, Proc. Am. Acad. xiii. 371, in fruit. — Dry ground, western side of Sierra Nevada, California, from the Yosemite northward, and apparently to Humboldt Co.; first coll. by *Bolander,* and the fruit by *Gray* and *Hooker.*

* * New Mexican, with linear leaves, diœcious: fruit unknown.

G. Féndleri, Gray. A span or two high from a tufted frutescent base, cinereous-puberulent and barely scabrous, slender: leaves hardly if at all rigid except the very small and squamaceous ones which are imbricated on the bases of the annual shoots; those above linear, about 4 lines long, less than line wide, rather acute, with midrib somewhat conspicuous beneath: flowers somewhat paniculate, short-pedicelled: corolla yellowish. — Pl. Fendl. 60. — Exposed mountain-sides, near Santa Fé, New Mexico, *Fendler,* male plant; and a female which is glabrous (also the ovary), or below barely pruinose-puberulent, perhaps not of the species. Santa Rita Mountains, Arizona, *Pringle,* male only.

* * * Texano-Californian, herbaceous, with very narrow and rigid small leaves, and very small white corollas.

G. Andréwsii, Gray. Depressed-cespitose and with slender creeping rootstocks, glabrous or nearly so: the matted tufts a span or less high: leaves very crowded, acerose-subulate, usually shining, either naked or sparsely spinulose-ciliate, 2 to 4 lines long: flowers diœcious; male slender-pedicelled in few-flowered terminal cymes; female solitary, subtended by a whorl of leaves which are longer than the fructiferous at length deflexed pedicel: berry dark-colored, smooth. — Proc. Am. Acad. vi. 538, & Bot. Calif. l. c. 286. — Dry hills, on the coast of California from Lake Co. to San Diego, and in the interior to Tejon, first coll. by *Dr. Andrews.*

G. microphýllum, Gray. Diffusely spreading or ascending, smooth and glabrous, but not shining; branches a span to a foot long: leaves shorter than the internodes and narrowly linear (or small, broader, and crowded at the base of stems), usually mucronate, with narrow midrib prominent beneath and callous naked margins, mostly 2 to 4 (rarely 5 or 6) lines long: flowers apparently all hermaphrodite, solitary on a very short or on a longer and peduncle-like axillary branchlet and sessile in its whorl of involucriform leaves, or this proliferous and bearing a second whorl and flower: ovary and young fruit scabro-puberulous or at length granulose, at maturity fleshy-baccate. — Pl. Wright. i. 80, ii. 66. *Relbunium microphyllum,* Hemsl. Biol. Centr.-Am. Bot. ii. 63. — Rocky ravines, &c., S. W. Texas to S. Arizona, first coll. by *Wright.* (Adj. Mex., where there is a pubescent variety, *Relbunium polyplocum,* Hemsl. l. c.)

* * * * Atlantic North American, herbaceous, with oval to linear leaves, and usually solitary hermaphrodite flowers: corolla white: berry purple, in our species naked-pedicellate beyond the ultimate involucriform whorl, mostly pendulous at maturity. — *Relbunium,* Benth. & Hook.

G. uniflórum, Michx. Smooth and glabrous: stems assurgent from filiform rootstocks, slender, rather simple: leaves linear (about inch long and a line wide), with somewhat scabrous margins : flowers solitary or in pairs from the pedunculiform axillary branchlet; the pedicels in fruit longer than or equalling the involucrate whorl, when in pairs one of the two commonly involucellate or nuibracteate; ovary and berry glabrous. — Fl. i. 79; Ell. Sk. i. 95; Torr. & Gray, Fl. ii. 21. — Woods in rich soil, S. Carolina to Florida and Texas.

— **G. hispidulum**, Michx. l. c. Hirsute-pubescent, hispidulous, scabrous, or sometimes almost smooth and glabrous, a foot or two high, diffusely branched and spreading : leaves oblong or oval, mucronate, a quarter to half an inch long: branchlets only floriferous : pedicels solitary or commonly 2 or 3 from the small involucral whorl, all naked, or one of them minutely bracteolate : ovary scabrous-puberulent : berry glabrate. — Ell. l. c.; Torr. & Gray, l. c. *G. hispidum*, Pursh, Fl. i. 104. *Rubia peregrina*, Walt., not L. *R. Brownei*, Michx. l. c., excl. syn. Browne. *R. Walteri*, DC. Prodr. iv. 590. — Dry or sandy soil, S. New Jersey to Florida, along the coast.

Order LXXI. VALERIANACEÆ.

Herbs (rarely suffruticose); with opposite leaves, no stipules, hermaphrodite or sometimes polygamo-diœcious flowers in cymose inflorescence, a 5-merous somewhat irregular epigynous corolla, bearing fewer (1 to 3, rarely 4) stamens on its tube, an ovary invested by the calyx-tube, and of one to three cells, but only one ovuliferous, a solitary suspended seed with a straight embryo and no albumen. Limb of calyx none, or of lobes or teeth, or evolved on the fruit into a kind of pappus. Corolla either obscurely or manifestly irregular (bilabiately, ⅖); lobes imbricated in the bud. Filaments and style filiform: stigma undivided and truncate, or minutely 3-cleft. Ovule anatropous. Fruit dry and indehiscent, a kind of akene.

1. **VALERIANA.** Calyx-limb of 5 to 15 setiform lobes, which are inrolled and inconspicuous until fruiting, when they are evolute and form a kind of plumose pappus. Corolla from campanulate-funnelform to salverform, the tube or body often gibbous or slightly saccate anteriorly. Stamens 3. Ovary 1-celled, and with mere vestiges of two lateral cells, ripening into a flattened akene, which is mostly 1-nerved on one face, 3-nerved on the other, and with a more or less evident nerve at each margin, which marks the position of a suppressed empty cell. — Perennials (with hardly an exception), the roots with a peculiar scent.

2. **VALERIANELLA.** Calyx-limb not pappose, in all ours more or less obsolete. Corolla from short-funnelform to salverform, with or without gibbosity, or sometimes a sac or spur at base; limb 5-parted, from nearly regular to obscurely or plainly bilabiate, or 4-parted with the posterior lobe notched or 2-cleft. Stamens 3, very rarely 2. Fruit various, the two abortive cells sometimes obsolete and nerviform at the lateral angles, commonly enlarged, sometimes converted into wings. Annuals, with entire or sparingly dentate or incised leaves; cauline sessile.

1. **VALERIÁNA**, Tourn. (Old herbalist's name, from *valeo*, to be strong, from use in medicine.) — Herbs (chiefly of northern temperate zone); with roots of peculiar scent, various leaves, and white or rose-colored flowers, in terminal cymes, produced in early summer. — L. Gen. 8, in part; DC. Prodr. iv. 632; Hœck in Engler, Bot. Jahrb. iii. 2.

PHYLLÁCTIS OBOVÁTA, Nutt. Gen. i. 21, is omitted, having been described from a plant of the Upper Missouri, not yet in flower, perhaps an undeveloped *V. edulis*.

* Erect from a large fusiform perpendicular stock branching below into deep and thickened roots: leaves thickish, nervosely veined, not serrate.

— **V. édulis**, Nutt. Glabrous or glabrate: the nascent herbage often tomentulose-puberulent, sometimes remaining so on the leaf-margins, a foot or at length 3 feet or more high: radical leaves oblanceolate to spatulate, tapering into a margined petiole, entire or some sparingly laciniate-pinnatifid ; cauline rarely none, commonly 1 to 3 pairs, sessile, and pinnately parted into 3 to 7 linear or lanceolate divisions, or terminal one spatulate: flowers polygamo-dioecious, yellowish white, sessile in the cymules, which form an elongated thyrsiform naked panicle: fruit ovate, puberulent or glabrous. — Nutt. in Torr. & Gray, Fl. ii. 48; Gray, Pl. Fendl. 61, & Man. Bot.   *V. ciliata*, Torr. & Gray, l. c.   *Patrinia ceratophylla*, Hook. Fl. i. 290.   *P. longifolia*, Macnab in Edinb. Phil. Jour. xix. — Wet plains and prairies, Ohio and W. Canada to Brit. Columbia, and south in the mountains of Colorado and Nevada to New Mexico and Arizona. Root a staple food of the Root-diggers and other Indians.

  * * Erect from creeping or ascending (but not vertical) rootstocks, which emit slender roots, glabrous or with a little sparse pubescence: leaves thinnish, loosely veiny, often with some simple and some divided and margins either entire or dentate on same plant; the radical ones on slender naked petioles; bracts of the cyme slenderly linear-subulate, mostly longer than the (usually quite glabrous) fruit; flowers hermaphrodite, but in the first species more or less dimorphous; corolla white to light rose-color.

  +— Tube of corolla from shorter than the throat and limb to less than twice their length; no sarmentose radical branches.

— **V. sylvática**, Banks. Stems from 8 to 30 inches high : radical leaves mostly simple and ovate to oblong, occasionally some 3–5-foliolate ; cauline more or less petioled, 3–11-foliolate or parted, the divisions entire or rarely few-toothed : fruiting cymes open, at length thyrsoid-paniculate: corolla 3 lines or in more fertile form only 2 lines long : the tube short: stigma nearly entire. — Richards. App. Frankl. Journ. ed. 2, 2; Hook. Fl. i. 291; Beck, Bot. 164; Torr. & Gray, Fl. ii. 47 (with var. *uliginosa*, a somewhat pubescent form); Gray, Man. & Bot. Calif. i. 287.   *V. dioica*, Pursh, Fl. ii. 727.   *V. dioica*, var. *sylvatica*, Gray in Proc. Acad. Philad. 1863, 63 ; Watson, Bot. King Exp. 136. — Wet ground, Newfoundland and Hudson's Bay country, south to S. New York, west to Brit. Columbia, and southward in the mountains to New Mexico and Arizona. In S. Utah it occurs with puberulent fruit, as collected by *Palmer*.

— **V. Sitchénsis**, Bong. More robust, from thicker and branching ascending rootstocks: leaves larger; cauline short-petioled, only 3–5-foliolate : the divisions orbicular to oblong-ovate, or in the upper leaves ovate-lanceolate, not rarely dentate or repand (larger 2 or even 3 inches long): cymes contracted : corolla funnelform, 4 lines long (but also a shorter form) : stigma entire. — Veg. Sitch. 145; Ledeb. Fl. Ross. ii. 438.   *V. pauciflora*, Hook. Fl. i. 292, t. 101, not Michx.   *V. capitata*, var. *Hookeri*, Torr. & Gray, Fl. ii. 48. — Moist woods, Sitcha, British Columbia, and through Washington Territory to S. Idaho and the northern Rocky Mountains.

V. capitáta, Pall. Stem rather slender from a creeping rootstock, 6 to 20 inches high, with long internodes; cauline all sessile (or lowest very short-petioled), only 2 or 3 pairs, all undivided and entire or few-toothed or some of them 3-parted, mainly ovate or oblong, an inch or two long : cyme capituliform or in fruit open-glomerate : corolla, &c. as of the preceding, 3 or 4 lines long : stigma 3-lobed. — "Link Jahrb. i. 3, 66," ex Ruem. & Schult. Syst. Mant. i. 257; DC. Prodr. iv. 367; Hook. l. c.; Torr. & Gray, l. c. (excl. var.); Ledeb. Ic. Pl. Ross. t. 346, & Fl. Ross. ii. 435; Trautv. Imag. t. 39. — Alaskan coast and Islands, north to Arctic region, first coll. by *Pallas*. (Adj. Asia to N. Eu.)

  +— +— Tube of corolla slender, much longer than the throat and limb.

— **V. Arizónica**, Gray.   A span or two high from tufted creeping rootstocks, glabrous, no sarmentose branches: leaves somewhat succulent; radical ovate (inch long), mostly entire and simple, some with one or two pairs of minute lobes on upper part of the rather long and margined petiole; cauline 2 pairs, subsessile, 3–5-parted, lobes oblong to lanceolate : cyme glomerate: corolla half-inch long, tubular, with gradually expanding throat: stigma minutely 3-cleft. — Proc. Am. Acad. xix. 81 — Arizona, in the mountains near Prescott, *Palmer*. Santa Catalina Mountains, *Lemmon*. Fruit not seen.

— **V. pauciflóra**, Michx. Stem 1 to 3 feet high from a slender creeping rootstock, erect, and with basal sarmentose branches or runners: leaves thin; radical and lowest cauline cordate and long-petioled, crenate or entire, not rarely with one and sometimes two pairs of small roundish lateral leaflets: upper cauline pinnate, with 3 larger leaflets ovate, one or two lower pairs smaller and more remote, lowest near base of petiole: cyme corymbiform and somewhat glomerate, commonly many-flowered (notwithstanding specific name): tube of corolla almost filiform, half-inch and more long, several times longer than the throat and limb. — Fl. i. 18; Nutt. Gen. i. 20; Torr. & Gray, l. c. — Alluvial river-banks, Pennsylvania to Illinois, Missouri, and Tennessee; first coll. by *Michaux*.

* * * Sarmentose-climbing or diffuse, with fibrous roots, glabrous: flowers very numerous in diffuse and compound paniculate cymes: bracts very small: corolla minute, seldom over a line long.

V. sorbifólia, HBK. A diffuse form of the Mexican species: stem weak, 2 or 3 feet long, springing from an annually produced small oblong tuber: leaves pinnate (except sometimes the radical), 5-13-foliolate; leaflets from rounded-ovate to oblong-lanceolate, coarsely serrate, or even laciniate: cymes loosely flowered in an elongated and naked (often foot long) terminal panicle. — Nov. Gen. & Spec. iii. 332. — Cañon in the Huachuca Mountains, S. Arizona, *Lemmon*, in a form with only 5 to 7 unusually large and broad leaflets, some almost 2 inches long, from rounded-ovate to oblong. (Mex.)

V. scándens, L. Root unknown: stem sarmentose and feebly twining, branching: leaves long-petioled; cauline 3-foliolate, with leaflets from deltoid- to oblong-ovate, acuminate, entire or repand, rarely with a few teeth, or lowest leaves simple and cordate: panicles effuse, axillary and terminal, elongated, the ultimate branches with the sessile flowers spicately disposed. — Spec. ed. 2, i. 27; Willd. Spec. i. 180; DC. l. c.; Torr. & Gray, Fl. i. 47. — Thickets in S. Florida, climbing several feet high. (W. Ind. to Brazil.)

2. VALERIANÉLLA, Tourn. CORN SALAD.

(Diminutive of *Valeriana*.) — Annuals, commonly winter-annuals (of the northern temperate zone), mostly low or slender and erect, ours glabrous or nearly so, except the fruit: leaves similar in all the species, from obovate to oblong and spatulate, entire or upper ones occasionally incised or toothed, radical rosulate, cauline sessile or even somewhat connate at base: flowers variously glomerate-cymose, the corolla from white to rose-color or rarely bluish. As in some species of *Valeriana*, so in some of these, the hermaphrodite flowers in different individuals are dimorphous as to size of corolla and exsertion of stamens and style, yet not as in heterogone dimorphism. — Vaill., Haller, &c.: Gray, Proc. Am. Acad. xix. 81. *Valerianella* & *Fedia*, Mœnch, Meth. 486, 493. *Fedia*, Gærtn. Fruct. ii. 36, t. 86; Woods in Trans. Linn. Soc. xviii. 23, t. 21; Torr. & Gray, Fl. ii. 50. *Valerianella, Dufresnia, Betckea,* & *Fedia*, DC. Prodr. *Valerianella, Plectritis,* & *Fedia*. Benth. & Hook. Gen. ii. 155, 156.

§ 1. VALERIANÉLLA proper. Corolla with nearly regular 5-parted limb, funnelform or more open throat with or without a small saccate gibbosity at its base anteriorly, and a short proper tube: stamens 3: fruit with the two empty cells manifest, or often enlarged and closed, sometimes at length confluent into one and rarely bursting: calyx-limb in American species none, or a mere tooth or oblique border: stem dichotomous above; the branches or pedunculiform branchlets terminated by corymbosely disposed glomerate cymes or cymules of small flowers. — *Valerianella*, Mœnch; Dufresne, Hist. Valer. 56; Krok, Monogr. Valer. in Svenska Vetensk. Acad. Handl. v. no. 1, 1864; Benth. & Hook. Gen. ii. 156, excl. § *Siphonella*.

* Introduced species: corolla bluish: a gibbous corky mass at the back of the fertile cell of the fruit.

V. olitória, Poll. Fruit flattish and obliquely roundish-rhomboidal: empty cells as large as fertile one and its corky back, contiguous, the thin partition between them at length breaking up. — Hist. Pl. Palat. i. 30; Mœnch, l. c.; Dufresne, Valer. 56, t. 3, f. 8; Krok, l. c. 88, t. 4, f. 40. *V. cœrulea* (& *rhombicarpa*), Aikin in Ent. Man. Bot. *Valeriana locusta, olitoria,* L. Spec. i. 33. *Fedia olitoria*, Vahl, Enum. i. 19; Woods, l. c. 430, t. 24, f. 1; Torr. & Gray, Fl. ii. 51; Porter in Am. Nat. vi. 386, fig. 102. — Old fields near dwellings, New York to Penn. and Louisiana; not common. (Nat. from Eu.)

* * Indigenous species: corolla white; no corky mass behind fertile cell of the fruit.

+ Fertile cell decidedly larger and broader than the two empty ones, and cross section of the fruit more or less triangular, the empty cells occupying the obtuser angle: tube of the corolla slender, commonly as long as the throat and limb.

V. chenopodifólia, DC. Stem a foot or two high, with long internodes and few forks: leaves comparatively large (1 to 3 inches long): glomerate small cymes few and slender-peduncled: bracts broadly lanceolate, narrowly scarious margined when dry: fruit glabrous or minutely pubescent, 2 lines long, ovate-triangular, the cross-section triquetrous or more or less rounded at the sterile angle, two empty cells about as deep but not as broad as the fertile, sometimes confluent into one when old. — Prodr. iv. 629; Gray, Proc. Am. Acad. xix. 82, founded on *Fedia chenopodifolia,* Pursh, Fl. ii. 727. from specimen in herb. Sherard. *V. triquetra,* Hochst. & Steud. ex Shuttlew. in Flora, 1837, 211, t. 3; Krok, l. c. 54, t. 2, f. 13. *Fedia radiata,* Torr. Fl. i. 35, not Michx. *F. Fagopyrum,* Torr. & Gray, Fl. ii. 51; Gray, Struct. Bot. ed. 5, fig. 881-884; Torr. Fl. N. Y. i. t. 46; Porter, l. c. fig. 103: name from likeness of fruit to buckwheat. *Valerianella Fagopyrum,* Walp. Repert. ii. 527. — Moist grounds, W. New York to Wisconsin, Kentucky, and Virginia.

V. amarélla, Krok. A span or two high, amply corymbosely branched above, bearing numerous and more open cymes: bracts lanceolate-linear, small: fruits very small (about half a line long), trigonous-ovate, densely white-hirsute, with rather obtuse lateral angles and that of the empty cells rounded, these decidedly shorter as well as much smaller than the fertile, almost filiform or sometimes almost obliterated. — Monogr. l. c. 55, t. 2, f. 14; Gray, l. c. *Fedia amarella,* Lindheimer, ex Eugelm. in Pl. Lindh. ii. 217. Spec. name from a peculiar bitterness of the herbage. — Low grounds, Texas, *Lindheimer, Wright, Hall, Reverchon,* &c. Gibbosity at base of corolla-throat sometimes very prominent and saccate, almost spur-like.

+ + Fertile cell fully as broad as the two introrse and parallel contiguous and more or less inflated empty ones, occupying the whole back of the fruit, apex projecting in a short obtuse tooth of a line long, the cross section quadrate, a conspicuous groove down the anterior face: stem a foot or two high, twice or thrice forked and spreading, the pedunculiform branchlets bearing one to three glomerate cymes: fruit in same species either pubescent or glabrous.

V. radiáta, Dufr. Fruit ovate-tetragonal, downy-pubescent or sometimes glabrous on one or all sides: fertile cell oblong-ovate, flattish; sterile cells as thick as or thicker than the fertile, a broad shallow groove between them. — Hist. Valer. 57; Krok, l. c. 64, t. 2, f. 22. not DC., who seems to have had *V. olitoria. Valeriana locusta, radiata,* L. l. c. (the plant of Clayton): Walt. Car. 166. *V. radiata,* Willd. Spec. i. 184. *Fedia radiata,* Michx. Fl. i. 118: Ell. Sk. i. 42; Torr. & Gray, Fl. ii. 52; Porter, l. c. fig. 164. — Low grounds, Penn. to Michigan, Florida, and Texas. Var. *leiocarpa* is a smooth-fruited form of it.

V. stenocárpa, Krok. Fruit tetragonal-oblong, commonly glabrous, sometimes pubescent; fertile cell oblong, obscurely narrowed upward, flattish and straight, thicker than the linear-oblong approximate sterile cells, and the groove between the latter narrow. — Monogr. Valer. l. c. t. 2, f. 1. *Fedia stenocarpa,* Engelm. Pl. Lindh. ii. 216. — Texas, *Berlandier* (part of no. 334), *Lindheimer, Hall.*

+ + + Fertile cell much narrower or smaller than the ampliate empty ones, one-nerved on the back, the fruit of orbicular or round-ovate circumscription, glabrous or with slight sparse pubescence.

V. Woodsiána, Walp. Habit of *V. radiata* and *V. chenopodifolia.* Fruit a line or more long; fertile cell oblong-ovate or ovate-lanceolate, tipped by a small soft and blunt tooth; empty cells introrse, either contiguous or somewhat diverging, inflated, with an oblong depression in the middle, sometimes an open concavity. — Walp. Repert. ii. 527; Krok, l. c. 66, t. 3, f. 23; Gray, l. c. 82. *Fedia Woodsiana,* Torr. & Gray, l. c. 52. *F. radiata,* var., Porter, l. c. fig. 105. — Moist grounds, New York and Penn. to Texas.

Var. umbilicáta, Gray, l. c. Empty cells ampliate and in age confluent, vesicular by incurvation of circular margin, forming a deep and rounded or obscurely cruciform umbilication. — *V. umbilicata,* Krok, l. c. 67, t. 3, f. 25. *Fedia umbilicata,* Sulliv. in Am. Jour. Sci. xlii., & Gray, Man. ed. 1, 183. *F. radiata,* var. *umbilicata,* Porter, l. c. 387, fig. 108. — New York & Penn. to Ohio and southward, first coll. by *Sullivant.*

Var. patellária, Gray, l. c. Empty cells divergent and obcompressed-dilated, so that the sterile face becomes open-concave, emarginate at top and bottom, and the whole fruit meniscoidal or saucer-shaped, the expanded and flattened sterile cells forming a kind of wing, or at length this incurved at base, or also at summit, and so nearly passing into var. *umbilicata*. — *V. patellaria*, Krok, l. c. 67, t. 3, f. 24. *Fedia patellaria*, Sulliv. l. c. *V. radiata*, Shuttlew. in Flora, 1837, 209, t. 3. *Fedia radiata*, var. *patellaria*, Porter, l. c. 387, fig. 106. — Ohio (*Sullivant*), Pennsylvania, &c.

§ 2. SIPHONÉLLA, Krok. Corolla salverform; the slender tube double or quadruple the length of the obscurely bilabiate-irregular limb, commonly bearing a minute boss or incipient spur near middle or base, sometimes with none; lobes oblong, the two posterior slightly more united and averse from the three anterior: stamens 3: fruit with divergent empty cells much larger than the fertile: habit and inflorescence of the preceding: bracts ciliate with gland-tipped denticulations. — Gray, l. c. 82. *Fedia* § *Siphonella*, Torr. & Gray, Fl. ii. 50. (Species referred to *Plectritis* by Nutt. in herb.)

V. longiflóra, WALP. Leaves ligulate-oblong and lower ones spatulate: cymes glomerate, many-flowered, corymbosely disposed: tube of corolla nearly filiform, 4 or 5 lines long, not rarely with a small boss at base, purplish or pink, 3 or 4 times the length of the lobes: fruit nearly orbicular in outline, somewhat meniscoidal, the semioval empty cells coriaceous with membranous face, parallel-contiguous and separated by a narrow partition, but widely diverging, each larger than the oblong obtusely short-tipped fertile one. — Repert. ii. 527; Krok, Monogr. l. c. 97, t. 4, f. 46. *Fedia longiflora*, Torr. & Gray, l. c. — Low rocky grounds, N. W. Arkansas, *Nuttall, Engelmann, Harvey*.

V. Nuttállii, WALP. l. c. Tube of the white or cream-colored corolla only about twice the length of the limb, bearing a little boss near the middle: fertile cell with a narrow soft projecting tip. — Krok, l. c. *Fedia Nuttallii*, Torr. & Gray, l. c. — Plains of Arkansas, *Nuttall, Engelmann*. Handsome in cultivation, a span or two high, very floriferous.

§ 3. PLECTRÍTIS, Lindl. Corolla with either manifestly or very obscurely bilabiate limb; proper tube shorter than the broadly or narrowly funnelform throat, which bears a descending spur at its base, this in one species obsolete or wanting: fruit one-celled, its body triangular or nearly so, one angle dorsal, the lateral angles bearing wings (these in place of the two empty cells, and apparently formed by their early separation and evolution from the middle of the ventral face), or in one species wingless: calyx-limb none: plants all nearly alike in herbage and thyrsoid-glomerate inflorescence; the cymules condensed into a capituliform or interrupted spiciform glomerule terminating stem or branches, and commonly one to three others verticillastrate at the nodes or axils below: flowers rose-color or white: all Pacific-American. — Bot. Reg. t. 1094; Gray, Proc. Am. Acad. xix. 83. *Plectritis & Betckea*, DC. l. c. *Plectritis*, Benth. & Hook.

* Fruit somewhat meniscoidal, only obtusely angled dorsally: cotyledons incumbent, i. e. parallel to the ventral face and expanded wings.

V. macrócera, GRAY. Flowers small, commonly in 2 to 4 somewhat distant and spicately disposed verticillastrate clusters: corolla narrow, white or pinkish, only a line or two long, with spur sometimes as long as the throat or body, sometimes only half its length; limb somewhat equally spreading and hardly at all bilabiate, or equally 4-lobed and posterior lobe emarginate-bifid: fruit commonly glabrous or puberulent, obtuse or even lightly lineate-sulcate on the dorsal angle, the broad wing of orbicular circumscription, sometimes spreading or very open, so that the ventral face is saucer-shaped, sometimes incurved so that it is acetabuliform. — Proc. Am. Acad. xix. 83. *Plectritis congesta*, var. *minor*, Hook. Fl. i. 291. *P. macrocera*, Torr. & Gray, Fl. ii. 50; Gray, Bot. Calif. i. 287, excl. syn. Fisch. & Meyer. — Dry ground, Washington Terr. to S. California, Nevada, and Arizona. Varies much in length and thickness of the spur, also in that of the tubular part of the corolla below the spur, which is sometimes slender and stipe-like, sometimes short.

* * Fruit strongly carinate-angled dorsally: cotyledons accumbent (transverse) to the ventral face.
+ Wings conspicuous, more or less introrse, in the last species small.

V. congésta, LINDL. Commonly rather stout: flowers in a capituliform or oblong simple or interrupted thyrsus, or sparingly verticillastrate below: corolla rose or flesh-colored, 3 or 4 lines long or in some individuals smaller, with obviously hilabiate limb, and spur half or less the length of the very gibbous throat; fruit broadly winged, and with prominent but rather obtuse keel, from glabrous to puberulent or sometimes thickly short-villous either on fertile cell or on wings also. — Bot. Reg. t. 1094; Gray, l. c. *Plectritis congesta*, DC. Prodr. iv. 631; Hook. l. c.; Torr. & Gray, l. c.; Gray, Bot. Calif. i. 287. *P. brachystemon*, Fisch. & Meyer, Ind. Sem. Petrop. 1835, Suppl. 47 (22), a form with smaller flowers (the state with included stamens and style) and villous-pubescent fruit, according to specimen from St. Petersb. garden; but the char. of flowers, four times smaller than in *P. congesta* and white, would be that of *V. macrocera*. — Low and moist ground, Brit. Columbia to W. California.

V. anómala, GRAY. Either slender or rather stout, freely branching: corolla only a line long, white or flesh-colored, wholly *destitute of spur*, at most a small mammæform gibbosity near the base of the short and broadly funnelform throat; limb small, obscurely bilabiate (usually 4-lobed and posterior lobe emarginate or 2-cleft): fruit comparatively large (mostly a line and a half long), acutely angled with sharp edge on the back, with broad wings usually inflexed at base and expanding above, but some fruits wingless. — Proc. Am. Acad. xix. 83. — Wet grounds on and near the Columbia River; Multnomah Co., Oregon, *Howell*, and Klickitat Co., Washington Terr., *Suksdorf*.

V. aphanóptera, GRAY, l. c. Slender, with aspect and inflorescence of the next: corolla only a line long, white, with obviously bilabiate limb and short basal spur: fruit puberulent or glabrate, trigonous; dorsal angle salient but rather obtuse; lateral angles with distinct but narrow incurved wings. — Springy ground on hillsides, along the Columbia River, Washington Terr., *Suksdorf*. Columbia Plains, *Nuttall*, under unpublished name of *Plectritis capitata*, appears to be the same; specimen insufficient.

+ + Wings wholly wanting to the triquetrous fruit, the lateral angles of which resemble the dorsal. — *Betckea*, DC.

V. samolifólia, GRAY, l. c. A span to a foot high: verticillastrate clusters 2 to 4, small: bracts slender-subulate (not pinnately parted as Hœck states, but uppermost sometimes palmately 3-parted): corolla a line or so in length, obscurely bilabiate, with short conical-saccate spur: akene-like fruit of the shape of buckwheat, glabrous or a little pubescent, in Chilian plants hardly, in ours rather over, a line long. — *Betckea samolifolia*, DC. l. c. 642. *B. major*, Fisch. & Meyer, l. c. (5) 30. *Plectritis samolifolia* & *P. major*, Hœck in Engler, Jahrb. iii. 37. — Low grounds on the Columbia River, Washington Terr., Oregon (*Suksdorf*), and coast of California, coll. by the Russian botanists. (Chili, smaller form.)

ORDER LXXII. DIPSACEÆ.

Herbs (all of the Old World); with opposite or verticillate leaves, no stipules, capitate and involucrate inflorescence; the flowers subtended by bracts, and each with a more or less obvious involucel, hermaphrodite; calyx-tube adnate to the one-celled simple ovary; corolla epigynous; stamens inserted on its tube alternate with its lobes, of equal number or fewer, wholly unconnected: style filiform and stigma simple; ovule solitary and suspended, anatropous; seed with a straight embryo in fleshy albumen. Corolla irregular or nearly regular; the lobes imbricated in the bud. Fruit an akene, more or less adnate to the involucel which embraces it.

SCABIÓSA ATROPURPÚREA, L., SWEET SCABIOUS of the gardens, is familiar; and one or two of the following genus have become spontaneous.

1. **DÍPSACUS**, Tourn. TEASEL. (Greek and Latin name of Teasel, said to come from δυψάς, thirsty.) — Flowers in a terminal head or short spike, in which,

however, the anthesis is not simply centripetal; the subtending bracts and those of the conspicuous polyphyllous and unequal involucre spinescent. Involucel calyx-like, prismatic, truncate and crenulate at the border, enclosing the whole ovary and fruit, at length adnate to its thin walls. Calyx-limb cup-like, 4-toothed or lobed. Corolla tubular-funnelform, 4-lobed, nearly regular. Stamens 4. Stigma lateral. Coarse and rough or prickly biennials, with cauline leaves mostly connate and cup-like at base.

— D. SYLVÉSTRIS, MILL. Prickly, 3 to 5 feet high: leaves lanceolate-oblong: involucre longer than the head: flowers flesh-colored: bracts of the receptacle tipped with a long and straight flexible awn. — Roadsides of the Eastern States: fl. summer. (Nat. from Eu.)

D. FULLÓNUM, L. (FULLER'S TEASEL.) Probably an ancient derivative of the preceding: involucre usually shorter than the cylindraceous head, at length reflexed: bracts of the receptacle rigid-spinescent, the tips recurved or hooked; whence useful for raising a nap on woollen cloth. — Escaped from cult. in some places, apparently established in S. California. (Nat. from Eu.)

ORDER LXXIII. COMPOSITÆ.

Flowers in an involucrate head on a simple receptacle, 5-merous or sometimes 4-merous; with lobes of the epigynous corolla valvate in the bud; stamens as many as corolla-lobes and alternate with them, inserted on the tube; anthers connate into a tube (syngenesious); style in all fertile flowers 2-cleft or lobed at summit and bearing introrse-marginal stigmas; ovary one-celled, a single anatropous ovule erect from the base, becoming an exalbuminous seed with a straight embryo, the inferior radicle shorter and narrower than the cotyledons; the fruit an akene. Tube of calyx wholly adnate to the ovary; its limb none, or obsolete, or developed into a cup or teeth, scales, awns, or capillary bristles. Corolla with nerves running to the sinuses, then forking and bordering the lobes, rarely as many intermediate nerves. Anthers commonly with sterile tip or appendage; the cells introrse, discharging the pollen within the tube; this forced out by the lengthening of the style, which in hermaphrodite and male flowers is commonly hairy-tipped or appendaged. Pollen-grains globose, echinulate, sometimes smooth, in *Cichoriaceæ* 12-sided. Leaves various: no true stipules. Development of the flowers in the head centripetal; of the heads, when clustered or associated, more or less centrifugal, i. e. heads disposed to be cymose. Juice watery, in some resinous, in the last tribe milky.

Heads *homogamous* when all its flowers are alike in sex; *heterogamous* when unlike (generally marginal flowers female or neutral, and central hermaphrodite or by abortion male); *androgynous* when of male and female flowers; *monœcious* or *diœcious* when the flowers of separate sexes are in different heads either on same or different plants; *radiate* when there are enlarged ligulate flowers in the margin; wholly *ligulate* when all the flowers have ligulate corollas; *discoid* when there are no enlarged marginal corollas. When these exist they are sometimes collectively called the *ray*: the other flowers collectively occupy the *disk*. The head (compound flower of the early botanists), in Latin *capitulum*, is also named *anthodium*. Its involucre (*periclinium* of authors) is formed of separate or sometimes connate reduced leaves, i. e. *bracts* (*squamæ* or scales of De Candolle, &c.): the innermost of these bracts subtend the outermost or lowest flowers. The axis within or above these is the *receptacle* (*clinanthium* of some), which varies from plane to conical, or oblong. or even cylindrical or subulate. When the receptacle bears flowers only, it is *naked*, although the surface may be *alveolate, foveolate*, or merely *areolate*, according as the insertion of the ovaries or akenes is surrounded or circumscribed by honeycomb-like or lesser elevations; or, when these project into bristles, slender teeth, or shreds, it is *fimbrillate*: it is *paleaceous* when the disk-flowers are subtended by bracts; these usually chaff-like, therefore called *paleæ, chaff*, or simply bracts of the receptacle. In place of calyx-limb there is more commonly a circle of epigynous bristles, hairs, or awns, the *pappus*, a name extended to the calyx-limb of whatever form or texture; its parts are bristles, awns, paleæ, teeth, &c., according to shape and texture. Corollas either all *tubular* (usually enlarging above the insertion of the stamens into the *throat*, and 4–5-lobed at summit, mostly regular); or the marginal ones strap-shaped, i. e. *ligulate*, the elongated limb (*ligule*) being explanate, and 3–5-toothed at the apex. Such are always female or neutral, or, when all the flowers of the head have ligulate corollas, then hermaphrodite. Anthers with basal auricles either rounded or acute, or sometimes produced into tails (*caudate*). Branches of the style in female flowers and in some hermaphrodite ones margined with stigma, i. e. stigmatic lines, quite to the tip; in most hermaphrodite flowers these lines shorter, occupying the lower portion, or ending at the appendage or hairy tip. — An immense order, comprising a tenth part of known phænogamous plants, an eighth of those of North America.

Key to the Tribes.

Ser. I. TUBULIFLORÆ. Corollas tubular and regular in all the hermaphrodite flowers.

Heads homogamous and discoid: flowers all hermaphrodite and never yellow: anthers not caudate at base.
Style-branches elongated filiform-subulate, hispidulous throughout; stigmatic lines only near the base: leaves alternate. . . . . I. VERNONIACEÆ.
Style-branches elongated, more or less clavate-thickened upward and obtuse, minutely papillose-puberulent, stigmatic only below the middle. II. EUPATORIACEÆ.
Heads homogamous or heterogamous, discoid or radiate: flowers not rarely yellow: style-branches of hermaphrodite flowers with stigmatic lines mostly prominulous and extending either to the naked summit or to a more or less distinct pubescent or hispidulous tip or appendage.
Anthers not caudate at base: style-branches in hermaphrodite flowers flattened and with a distinct (but sometimes very short) terminal appendage: disk-corollas generally yellow; rays of same or different color. . . . . . III. ASTEROIDEÆ.
Anthers caudate: style-branches of hermaphrodite flowers slender, destitute of any terminal appendage, the stigmatic lines extending quite to (or vanishing near) the naked obtuse or truncate summit: leaves alternate: heads in our genera discoid except in *Inula*. . . . . IV. INULOIDEÆ.
Anthers not caudate: style-branches with truncate or variously appendiculate pubescent or hispid tips: involucre not scarious: receptacle paleaceous, i. e. with chaffy bracts subtending at least the outer disk-flowers: pappus various or none, never of fine capillary bristles. . . . . V. HELIANTHOIDEÆ.
Anthers not caudate: receptacle naked: pappus from paleaceous to setiform or none: herbage often punctate with resinous or pellucid dots or glands: otherwise nearly as preceding. . . . . VI. HELENIOIDEÆ.

Anthers not caudate: receptacle naked or sometimes paleaceous: involucre of dry and scarious bracts: style-branches mostly truncate: pappus coroniform, or of short paleæ or squamellæ, or none. . . . . VII. ANTHEMIDEÆ.
Anthers not caudate: receptacle naked: involucre little or not at all imbricated, not scarious. Pappus of numerous soft-capillary bristles. VIII. SENECIONIDEÆ.
Anthers conspicuously caudate, and with elongated mostly connate cartilaginous appendages at tip: style-branches short or united, destitute of appendage, stigmatic quite to the obtuse summit, smooth and naked, but sometimes a pubescent or hispidulous ring or node below · involucre much imbricated: receptacle densely setose or fimbrillate, or favose: akenes thick and hard: pappus usually plurisetose. Heads never truly radiate. . . . . IX. CYNAROIDEÆ.
Also GOCHNATIEÆ of X. MUTISIACEÆ.

Ser. II. LABIATIFLORÆ. Corollas of all or only of the hermaphrodite flowers bilabiate.
Receptacle naked: anthers conspicuously caudate: style-branches short, smooth, not appendaged. . . . . X. MUTISIACEÆ.

Ser. III. LIGULIFLORÆ. Corollas all ligulate and flowers hermaphrodite.
Receptacle naked or paleaceous: anthers not caudate: style-branches filiform, naked, stigmatic only toward the base. Herbage with milky juice. XI. CICHORIACEÆ.

TRIBE I. VERNONIACEÆ. Heads homogamous, discoid, with flowers all hermaphrodite and corollas tubular, regular or nearly so, except *Stokesia*. Involucre imbricated. Anthers without tails at base. Style-branches slender, filiform or attenuate-subulate, acute, hispidulous or hispid; stigmatic lines only near the base. Leaves usually alternate. Flowers never yellow.

  * Anomalous genus, with enlarged and palmately quasi-ligulate outer corollas.

1. **STOKESIA.** Heads many-flowered. Involucre broad; its bracts in several series; outermost wholly foliaceous and spreading; inner with foliaceous pectinately spinulose-ciliate spreading appendage to an appressed coriaceous base. Receptacle fleshy, flat, naked. Central corollas tubular and deeply 5-lobed, slightly more cleft posteriorly, otherwise regular; outer successively more and more palmately ligulate and radiant, the marginal ones larger and wholly so, the narrowly cuneate-oblong ligule longer than the tube and (regularly or irregularly) 5-cleft. Akenes short, thick, 3-4-angled, slightly contracted at the callous base and apex. Pappus of 4 or 5 aristiform smooth and white paleæ, caducous. Flowers blue.

  * * Normal genera, with tubular 5-lobed corollas.

2. **ELEPHANTOPUS.** Heads 2-5-flowered, condensed into glomerules. Involucre narrow, compressed; the imbricated bracts dry and somewhat chaffy, alternately plane and conduplicate; the four outermost shorter. Receptacle small, naked. Corolla commonly a little irregular, being slightly deeper cleft on the inner side; the deeply 5-lobed limb therefore somewhat palmate. Akenes 10-striato, the apex truncate. Pappus of rigid bristles or awns, mostly with paleaceous base, persistent.

3. **VERNONIA.** Heads not glomerate, several-many-flowered, rarely one-flowered. Involucre of dry or partly herbaceous much imbricated bracts. Receptacle plane, naked. Corolla regularly 5-cleft into narrow lobes. Akenes mostly 10-costate, with truncate apex and a cartilaginous callous base. Pappus double, at least in all our species: the inner of rigid capillary hirtello-scabrous bristles, outer a series of small squamellæ or short and stout bristles, both more or less persistent.

TRIBE II. EUPATORIACEÆ. Heads homogamous, discoid, with flowers hermaphrodite and corolla tubular and regular. Receptacle in a few genera paleaceous, in most naked. Anthers without tails at base. Style-branches elongated, more or less clavate

COMPOSITÆ. 51

or thickened upward, minutely papillose or puberulous, or glabrous ; the stigmatic lines only near the base and inconspicuous. Leaves either opposite or alternate. Flowers never yellow, at most ochroleucous.

* Akenes 5-angled, destitute of intervening ribs: corolla-lobes or teeth usually short.
+ Pappus various, but never wholly capillary, sometimes obsolete or wanting.
++ Involucre 5-flowered (sometimes 3-4-flowered), cylindrical, of 5 or 6 mostly equal rather rigid bracts: receptacle small, naked: corolla narrow.

4. **STEVIA.** Akenes linear, slender, sometimes compressed. Pappus coroniform-paleaceous or aristiform, or composed both of awns (one or more) and short scales.

++ ++ Involucre many-flowered, lax, of 12 to 18 herbaceous or submembranaceous equal and nerveless bracts: receptacle naked: corolla abruptly much dilated above the narrow tube, rose-purple or flesh-color.

5. **SCLEROLEPIS.** Receptacle conical. Style-branches elongated, filiform-clavate. Akenes with an inane stipitiform base. Pappus conspicuous, of 5 broad and very obtuse or truncate cartilaginous nerveless scales. Leaves verticillate.

6. **TRICHOCORONIS.** Receptacle convex. Style-branches comparatively short, linear, flat or flattish-filiform, not enlarged upward. Akenes not contracted above the basal callus. Pappus minute or small, multisetulose-coroniform. Leaves opposite or alternate.

++ ++ ++ Involucre many-flowered, campanulate or hemispherical, of 2 or more series of striate-nerved bracts, more or less imbricated: receptacle flat or convex: corolla rather narrow: akenes not stipitate-attenuate, but with a strong basilar callus.

7. **AGERATUM.** Involucre of mostly narrow bracts in 2 or 3 series, not conspicuously unequal. Receptacle either naked or paleaceous. Pappus simple, paleaceous, and the paleæ either muticous or aristate, or coroniform; the crown sometimes almost obsolete.

8. **HOFMEISTERIA.** Involucre more imbricated; outer bracts successively shorter. Receptacle naked. Pappus of 2 to 12 slender or capillary bristles equalling the narrow corolla, and as many or fewer alternating or exterior short and thin paleæ.

+ + Pappus wholly of capillary and mostly uniserial bristles;
++ These merely scabrous, indefinitely numerous: receptacle naked.

9. **MIKANIA.** Involucre 4-flowered, composed of 4 or sometimes 8 similar and equal thin bracts, with or without a loose and somewhat herbaceous subtending one. Receptacle small. Stems (at least in ours) twining.

10. **EUPATORIUM.** Involucre few-many-flowered, of more than 4 bracts. Neither twining nor climbing.

++ ++ These long-plumose, rather few: receptacle naked.

11. **CARMINATIA.** Involucre several-flowered, cylindraceous, of several lanceolate-linear 3-5-striate thin imbricated bracts, the exterior shorter. Receptacle small, flat, naked. Corollas slender, 5-toothed. Style-branches filiform, acutish. Akenes slender, narrowish at the apex. Pappus of 10 to 18 bristles, which are slightly coherent at base in a single series, plumose with long arachnoid hairs, deciduous together.

* * Akenes 10-costate or striate.

+ Bracts of the involucre not herbaceous, striate-nerved, conspicuously so when dry, regularly imbricated; the outer ones successively shorter: receptacle naked: corollas slender, 5-toothed at summit; the teeth mostly glandular externally · pappus a single series of plumose or scabrous capillary bristles: heads few-many-flowered.

12. **KUHNIA.** Pappus conspicuously plumose. Bracts of the involucre narrow, in few series. Leaves nearly all alternate.

13. **BRICKELLIA.** Pappus from barbellate or subplumose to merely scabrous. Leaves opposite or alternate.

+ + Bracts of the involucre somewhat herbaceous or partly colored, inconspicuously when at all striate or nerved, even when dry,

++ Well imbricated, the outer successively shorter: leaves punctate, entire: flowers rose-color, or abnormally and rarely white.

= Pappus paleaceous-aristiform : leaves opposite.

14. **CARPHOCHÆTE.** Heads 4–6-flowered. Involucre cylindrical; the bracts acuminate, rather few. Receptacle small, naked. Corolla narrow and long, hypocrateriform; limb 5-parted into slender linear-lanceolate lobes. Akenes slender, barely puberulent. Pappus of long linear-subulate erose-denticulate scarious paleæ, the thickened costa continued into a barbellulate scabrous awn; and with 1 to 5 small nearly nerveless muticous paleæ.

= = Pappus of numerous capillary or stouter bristles, from plumose to barbellulate-scabrous: anther-tips emarginate or retuse: leaves alternate.

15. **LIATRIS.** Heads 4–many-flowered. Involucre spirally imbricate. Receptacle naked. Corolla narrow, with gradually dilated throat and elongated-lanceolate or linear spreading lobes. Akenes slender or tapering from apex to base, pubescent. Pappus about a single series of firm and mostly equal bristles, from plumose to barbellate. Herbs, with heads in a terminal reversed spike or raceme, sometimes becoming paniculate.

16. **GARBERIA.** Heads about 5-flowered. Involucre imbricate in 5 nearly vertical ranks (3 or 4 in each rank) of somewhat herbaceous acute bracts. Receptacle small, naked. Corollas with slender tube, abruptly cyathiform-ampliate throat, and lanceolate spreading lobes. Akenes, &c. of *Liatris*. Pappus copious, in two or more series of slender barbellate-scabrous bristles, the outer smaller and shorter. Broad-leaved shrub, with heads corymbosely cymose.

17. **CARPHEPHORUS.** Heads many-flowered. Involucre campanulate; the imbricated bracts all appressed. Receptacle chaffy; the chaff subtending the outer flowers, and mostly shorter than they, thin, deciduous with the fruit. Corolla-lobes ovate or short-lanceolate. Akenes of *Liatris*. Pappus of one or more series of barbellate or plumose bristles. Herbs, with heads corymbosely cymose.

++ ++ Little-imbricated involucre of bracts nearly all equal in length: receptacle plane, naked: corolla narrow, with short-ovate or oblong lobes: leaves broad, obscurely or not at all punctate: perennial herbs, fibrous-rooted from a small caudex.

18. **TRILISIA.** Heads 5–10-flowered. Pappus of rather rigid minutely barbellate bristles, nearly in a single series. Leaves entire; cauline sessile. Cymules paniculate or somewhat cymose.

TRIBE III. **ASTEROIDEÆ.** Heads either heterogamous and radiate, the ligulate ray-flowers feminine or rarely neutral, or homogamous with the flowers all hermaphrodite and tubular, or rarely the female flowers with filiform corolla and no ligule, or in *Baccharis* diœcious and the female corollas all filiform. Receptacle seldom paleaceous. Corolla of the hermaphrodite flowers regularly 5-lobed, rarely 4-lobed (obscurely palmate in *Lessingia*). Anthers obtuse and entire or barely emarginate at base. Style-branches of hermaphrodite flowers flattened, conspicuously margined by the stigmatic lines, and extended into a hispid or papillose (sometimes very short) appendage. Pappus various, or sometimes none. Leaves mostly alternate. Disk-flowers usually yellow. — Tribe of nearly 100 genera, the largest being *Aster* and *Solidago*. The characters of the subtribes fail in a few instances, either through absence of the rays, or as to their color.

Subtribe I. HOMOCHROMEÆ. Disk wholly of hermaphrodite flowers, of the same color as the ray when that is present, mostly yellow : these corollas tubular with more or less ampliate throat and 4–5-lobed limb. Receptacle not chaffy, flat or merely convex. Involucre closely imbricated, mostly in several series. (Flowers white in most species of *Lessingia*: rays often white in *Pentachæta* and in one *Solidago*.)

* Pappus none, or coroniform or paleaceous, or squamellate, or somewhat setose only in infertile disk-flowers: heads radiate : involucral bracts coriaceous or chartaceous, sometimes with herbaceous or greenish tips, the outer successively shorter. (See also *Pentachæta*. The four following genera are very close.)

19. **GYMNOSPERMA.** Heads several-flowered. Involucre ovoid or oblong; its bracts obtuse, concave. Receptacle small. Ligules very small, not surpassing the disk-corollas. Akenes oblong, slightly compressed, 4–5-costate, glabrous, destitute of pappus. Heads very small and numerous, in glomerate terminal cymes.

20. **XANTHOCEPHALUM.** Heads many-flowered, pluriradiate. Involucre broadly campanulate or hemispherical. Receptacle flat or convex. Ligules usually elongated, numerous. Akenes truncate and naked at summit, or edged with a ring bearing a minute coroniform or squamellate pappus, or none. Heads larger, solitary at the end of the branchlets or in open cymes.

21. **GUTIERREZIA.** Heads few-many-flowered. Involucre oblong-clavate or turbinate to campanulate. Receptacle from plane to conical, commonly alveolate or fimbrillate. Style-appendages mostly slender. Ligules 1 to 8. Akenes short, obovate or oblong, terete or 5-angled. Pappus pluripaleaceous, but the paleæ often reduced to a crown or minute, in the ray commonly smaller and sometimes wanting. Heads small, paniculate or cymose-clustered. Disk-flowers, or some of the central ones, occasionally infertile.

22. **AMPHIACHYRIS.** Heads few-many-flowered; the ray-flowers only fertile; those of the disk hermaphrodite or subhermaphrodite but infertile. Involucre, &c. of *Gutierrezia*. Style-appendages lanceolate; the stigmatic lines below them indistinct or obsolete. Fertile akenes pubescent. Pappus of the ray-flowers obsolete or coroniform, or of a few setiform squamellæ; of the disk setiform-paleaceous, viz. of 5 to 20 aristiform paleæ or weak bristles more or less dilated and united at base, nearly equalling the corolla.

33. **BRACHYCHÆTA.** Heads, &c. as in *Solidago*. Pappus both in the ray and disk of bristles abbreviated to squamellæ.

* * Pappus pauciaristate, i. e. of few (less than 10) elongated awns or rigid bristles, sometimes caducous or wanting: heads radiate or in same genus rayless, solitary at the end of the branches.

23. **GRINDELIA.** Involucre many-flowered, hemispherical or at first globose; its bracts numerous and narrow, imbricated in many series, firm and rigid, with more or less herbaceous tips. Style-appendages lanceolate or linear. Akenes short and thick, compressed or turgid, or the outer triangular, truncate, glabrous. Pappus of 2 to 8 caducous nearly smooth awns or corneous bristles.

24. **PENTACHÆTA.** Involucre many-several-flowered, hemispherical or campanulate, of thin and scarious-margined appressed bracts, destitute of herbaceous tips, commonly tipped with a delicate mucro. Style-appendages filiform-subulate, hispid. Akenes oblong to fusiform-obovate, somewhat compressed and villous. Pappus of mostly 5 (rarely 3 to 6 or even 8) slender and persistent scabrous bristles with somewhat enlarged base, sometimes reduced to this base, or even this obsolete. Receptacle short-alveolate. Rays either yellow, or white or pink!

* * * Pappus multisetose in either the disk or ray, but not in both, and akenes of disk and ray unlike.

25. **BRADBURIA.** Involucre campanulate, of rather broad and thin scariously margined and mucronate-acuminate appressed bracts. Ray-flowers about 12, fertile; the style very short. Disk-flowers about the same number, infertile; their style-branches destitute of stigmatic lines, filiform, barbellate-hispid. Akenes of the ray sparsely villous, trigonal-turbinate with a strong rib at each angle; the pappus of numerous unequal rigid capillary bristles, little longer than the akene: those of the disk abortive, with pappus of very few (usually 2) bristles which are somewhat chaffy-dilated at base.

26. **HETEROTHECA.** Involucre hemispherical or broadly campanulate, of narrow rather rigid bracts. Receptacle alveolate. Ray- and disk-flowers numerous, both fertile. Style-branches of the hermaphrodite flowers tipped with a lanceolate or sometimes (in the same species?) ovate-triangular appendage. Akenes of the ray thickish, often triangular, with no pappus or an obsolete crown, rarely a bristle or two; of the disk compressed, sericeous-hirsute, and with a double pappus; the inner of copious and long capillary scabrous bristles, outer of numerous short and stout bristles or setiform squamellæ.

* * * Pappus multisetose and double, both in disk and (when present) in the ray; the inner capillary; outer very short and setulose or squamellate.

27. **CHRYSOPSIS.** Heads many-flowered, with rays numerous, or rarely wanting. Involucre campanulate or hemispherical, of narrow regularly imbricated bracts. Style-appendages from linear-filiform to slender-subulate. Akenes compressed, but often turgid, from obovate to linear-fusiform. Principal pappus of numerous capillary scabrous bristles;

this surrounded at base by a series of minute short bristles or squamellæ; these sometimes inconspicuous or obsolete. (One or two species of *Erigeron* with ochroleucous and even yellow rays may be sought here.)

* * * * * Pappus multipaleolate or aristate rather than capillary-setose: involucre subglobose or hemispherical, of very broad bracts, all or inner ones scarious-margined; receptacle alveolate-fimbrillate; the fimbrillæ little shorter than the akenes.

28. **ACAMPTOPAPPUS.** Heads 12-36-flowered, discoid or radiate: flowers all fertile. Bracts of the involucre chartaceo-coriaceous, round-oval or broadly oblong and concave, bordered by an erose-fimbriate thin-scarious margin, closely imbricated in about 3 series. Disk-corollas funnelform. Style-branches tipped with a thickish subulate appendage. Akenes globular-turbinate, very densely long-villous and at length tomentose, 5-nerved under the wool. Pappus hardly longer than the akene, equalling the corolla, of 15 to 18 flattened and rigid paleaceous awns, the tips of which are mostly a little dilated, and of as many shorter unequal setiform awns or bristles, persistent.

29. **XANTHISMA.** Heads many-flowered, radiate; the flowers all fertile. Bracts of the involucre closely imbricated and appressed, mainly ovate or obovate, with short coriaceous base and herbaceous upper portion, more or less scarious-margined; some smaller inner ones. Disk-corollas narrowly funnelform. Style-branches tipped with a filiform-subulate appendage. Akenes turbinate, 4-5-costate-angled, and with intermediate less conspicuous ribs, sericeous-pubescent. Pappus paleaceous-aristate, persistent, composed of 10 or 12 rigid bristles which are minutely scabrous above, gradually paleaceous-dilated towards the base, and longer than the disk-corolla, as many more one-half shorter, and usually 5 still smaller and shorter external ones.

* * * * * Pappus capillary-multisetose and mostly alike in all the flowers, simple, consisting of numerous capillary scabrous bristles in one or more series: receptacle more or less alveolate and the alveoli often dentate.

+— Disk-corollas 5-toothed or 5-lobed, the lobes from ovate to oblong or narrower: flowers yellow, with rare exceptions: style-appendages from ovate-lanceolate to filiform.

27. **CHRYSOPSIS.** Species with outer pappus obscure or wanting would be sought here.
30. **APLOPAPPUS.** Heads usually many-flowered, radiate, rarely discoid, or with infertile rays. Disk-corollas narrow, 5-toothed. Involucre usually (but not always) broad; the bracts with or without herbaceous tips. Akenes from turbinate to linear.
31. **BIGELOVIA.** Heads 3-30-flowered, destitute of rays, small. Involucre narrow; the bracts chartaceous or coriaceous, mostly destitute of foliaceous or herbaceous tips. Akenes narrow, terete or angled, hardly compressed, mostly at least 5-nerved. Pappus of somewhat equal bristles. Inflorescence not racemiform.
32. **SOLIDAGO.** Heads few- or several-, rarely many-flowered, mostly radiate, small, commonly in racemiform or spiciform clusters, sometimes fastigiate-cymose or in a thyrsus. Involucre narrow; its bracts mostly not herbaceous-tipped. Akenes terete or angulate, 5-12-nerved or costate. Pappus of equal elongated bristles. Leaves not cordate.
33. **BRACHYCHÆTA.** Heads (very small) and flowers of *Solidago*. Pappus of mere rudimentary bristles, shorter than the akene. Lower and radical leaves cordate.

+— +— Disk-corollas with limb 5-parted into long and narrow lobes, in the same genus either yellow, white, or violet-purple: no true rays: style-tips comose-bearded.

34. **LESSINGIA.** Heads homogamous, 5-25-flowered: flowers all perfect. Corollas with slender tube divided into long and narrow lobes; the marginal ones sometimes larger and with a deeper cleft on the inner side, somewhat imitating a palmate ligule. Involucre campanulate or turbinate: its bracts well imbricated and appressed, mostly with herbaceous tips. Receptacle flat, alveolate. Anthers included, tipped with a slender-subulate appendage. Style-branches tipped with a very short and obtuse or truncate densely hispid appendage, which usually bears either a setiform cusp among the tufted bristles, or a more conspicuous subulate prolongation. Akenes turbinate or cuneiform, silky-villous, 2-5-nerved. Pappus of numerous and unequal rigid and scabrous capillary bristles.

Subtribe II. HETEROCHROMEÆ. Disk of hermaphrodite and mostly fertile flowers; their corolla yellow, or rarely cream-color, sometimes changing to purple; the ray not yellow, wanting only in certain species (much reduced and inconspicuous in a section of *Erigeron* and one of *Aster*). Receptacle naked (not paleaceous), with an occasional exception.

* Pappus of both disk and ray none or coroniform.
+ Involucre broad, many-flowered: rays numerous, fertile, conspicuous.

35. **BELLIS.** Bracts of the involucre nearly equal in length, herbaceous or somewhat membranaceous. Receptacle conical or hemispherical. Style-branches tipped with a short triangular appendage. Akenes obovate and compressed, nerveless except at the margins. Pappus none.

36. **APHANOSTEPHUS.** Bracts of the involucre imbricated in few series, broadly lanceolate, somewhat herbaceous with scarious apex and margins, the outer shorter. Receptacle conical or hemispherical. Style-branches with a very short and obtuse appendage. Akenes prismatic or terete, truncate: the broad apex bordered with a short coroniform and either dentate or entire and minutely setulose-ciliate pappus. Base of corolla-tube often skirrous-thickened in age.

37. **GREENELLA.** Bracts of the involucre imbricated in a few series, the outer shorter, all oblong, coriaceo-chartaceous, with scarious margins and an herbaceous dorsal tip. Receptacle flat, barely convex. Style-branches with a linear rather obtuse appendage, which is four times the length of the quadrate stigmatiferous portion and much exserted. Akenes short, somewhat turbinate, obscurely 8-nerved, hispidulous. Pappus a hyaline-scarious crown cleft into numerous setulæ or denticulations.

+ + Involucre narrower and flowers less numerous.

38. **KEERLIA.** Involucre narrowly campanulate or turbinate; its bracts imbricated in few series, of unequal length, oblong, smooth, thin-membranaceous with scarious margins, mostly setaceously mucronate. Receptacle small and flat. Rays 5 to 15, with oblong ligule on a slender tube. Style-appendages either short and obtuse or long and slender. Akenes obovate and compressed or subclavate, 2-3-nerved, and with very small epigynous disk; those of the disk-flowers mostly sterile. Pappus minute and coroniform or evanescent from the mature akenes.

* * Pappus of solitary or few setiform awns or bristles and of a few paleæ or a crown: rays conspicuous, fertile: akenes without wings or callous margins: receptacle flat or nearly so.

39. **CHÆTOPAPPA.** Involucre several-many-flowered, campanulate or narrower; its bracts oblong or lanceolate, thin-herbaceous, with scarious margins and tip, imbricated in two or more ranks, the outer shorter. Rays 5 to 20; the ligule oblong, raised on a slender tube. Disk-flowers often sterile: their style-appendages short. Akenes either fusiform or compressed, 2-5-nerved. Pappus of five or fewer thin and nerveless short paleæ, alternating with as many or fewer mostly long and setiform scabrous awns, or the latter sometimes wanting.

40. **MONOPTILON.** Involucre many-flowered, broad; its numerous bracts equal and almost in a single series, narrowly linear, somewhat herbaceous. Rays numerous: the ligule obovate or oblong and with a rather long tube: this and the tube of the disk-corolla sparsely villous. Style-appendages triangular and obtuse. Akenes oblong-obovate, compressed, one-nerved at each margin, or the outermost also on one face. Pappus both in disk and ray a short and cupuliform barely denticulate crown, and a single setiform awn which is barbellate or plumose toward the apex.

* * * Pappus short-setulose or squamellate and mostly biaristate, i. e. a ring of very short bristles or setiform squamellæ and a pair of naked upward-tapering awns, one over each edge of the broad and flat winged or callous-margined akene: involucre many-flowered, hemispherical or broader; the bracts somewhat herbaceous and thin-margined: receptacle strongly convex or low conical: rays conspicuous, fertile; their akenes occasionally 3-winged.

41. **DICHÆTOPHORA.** Involucre somewhat uniserial; the lanceolate bracts of equal length. Style-appendages of the disk-flowers triangular-lanceolate. Akenes surrounded by an almost orbicular firm wing, its edge and the body of the akene glochidiate-hispid. Pappus of two divergent awns about half the length of the akene, and of several minute

squamellæ, which are shorter than and concealed by the bristly hairs of the akene. Low annual.

42. BOLTONIA. Involucre imbricated, appressed; the outer bracts shorter. Style-appendages short-lanceolate. Akenes obovate, very flat, with callous or winged margin, glabrous or minutely and sparsely hispidulous. Pappus of several short-setulose squamellæ, and usually of 2 (rarely 3 or 4) elongated rigid awns. Leafy-stemmed perennials.

 * * * * Pappus a single series of long awns (or only 2 or 3) or of coarse and rigid bristles, or in the (fertile and conspicuous) ray reduced to squamellæ or palæa: receptacle flat.

43. TOWNSENDIA. Involucre broad, many-flowered, imbricated; the bracts lanceolate, with scarious margins and tips, outer usually shorter and inner more membranaceous. Receptacle broad, merely areolate. Style-appendages lanceolate. Akenes obovate or oblong, much compressed, and with thickish or mostly callous margins, those of the ray sometimes triangular. Awns or bristles of the pappus from hispidulo-scabrous to barbellulate.

 * * * * * Pappus of numerous capillary bristles, at least in the disk, with or without a short setulose or squamellate outer series: receptacle flat or barely convex: akenes mostly compressed.

 +— Style-appendages of disk-flowers comose-bearded: anthers tipped with slender-subulate appendages, as in *Lessingia*: rays neutral.

44. CORETHROGYNE. Involucre broad, imbricated; its bracts with herbaceous or green tips. Receptacle foveolate, rarely with a few chaffy bracts toward the margin. Style-appendages short-lanceolate, dorsally beset with long hispid hairs forming a bearded tuft. Akenes of the ray abortive and with reduced scanty pappus or none; those of the disk narrow, silky-villous or pubescent, few-nerved; their pappus of rather rigid and unequal capillary bristles.

 +— +— Style-appendages merely hispidulous or puberulous, not comose.
 ++ Pappus none or a mere vestige in the ray-flowers: these often sterile but styliferous.

45. PSILACTIS. Involucre hemispherical; its bracts imbricated in 2 or 3 series, and with herbaceous tips, or the outer herbaceous. Rays in a single series, sometimes short. Akenes pubescent, narrow; those of the ray sometimes with an obscure ring in place of pappus; those of the disk bearing a single series of soft capillary bristles. Leafy-stemmed annuals.

 ++ ++ Pappus present and mostly similar in ray and disk: flowers (with rare exceptions) all fertile. Genera of difficult limitation.

46. EREMIASTRUM. Involucre broad, many-flowered, of numerous lax and linear bracts, all as long as the disk and nearly in a single series, herbaceous, with hyaline-scarious and erose-fimbriate margins, their back setose-hispid. Receptacle broad and flat. Rays numerous in a single series; ligules broad, their base and tube villous-barbate. Style-appendages lanceolate. Akenes obovate-oblong, compressed, 2-nerved. Pappus of rather few very unequal and somewhat rigid bristles; the stronger ones considerably shorter than the disk-corolla and only 8 to 12; the smallest and outermost setulose and scarcely longer than the hirsute pubescence of the akene, sometimes coalescent irregularly at base; intermediate ones of various length and more numerous. Depressed winter-annual.

47. SERICOCARPUS. Involucre several-flowered, narrow, of closely imbricated and appressed whitish and coriaceous or cartilaginous bracts, with green-herbaceous abrupt and looser or spreading tips. Receptacle small, foveolate. Corollas both of ray and disk white or cream-color; the rays seldom over 5 or 6, rather broad; disk-corollas (8 to 20) with 5-cleft limb. Style-appendages lanceolate-subulate. Akenes narrow, little compressed, 2-nerved, sericeous-pubescent. Pappus of numerous unequal scabrous bristles. Leafy-stemmed perennials.

48. ASTER. Involucre from hemispherical to campanulate, sometimes oblong or turbinate, imbricated in several or few series of unequal bracts, mostly in part herbaceous. Rays numerous, not very narrow. Style-appendages from slender-subulate to ovate-acute, commonly lanceolate. Akenes mostly compressed, 10-4-2-nerved, and the pappus mostly simple and copious, rarely distinctly double. Leafy-stemmed herbs, the greater part perennials.

49. ERIGERON. Differs from *Aster* in the more naked-pedunculate heads, simpler involucre of narrow and erect equal bracts, which are never coriaceous, nor foliaceous or with distinct herbaceous tips, narrower and usually very numerous rays often occupying more than one series, very short and roundish or obtuse style-appendages, small akenes for the greater part 2-nerved, and more scanty or fragile pappus, in many with a conspicuous short outer series.

Subtribe III. CONYZEÆ. Characters of the preceding subtribe; but corolla of the numerous female flowers reduced to a filiform or short and narrow tube, wholly destitute of ligule.

50. **CONYZA.** Heads small, many-flowered. Bracts of the campanulate involucre narrow, inappendiculate, in 1 to 3 series. Female flowers much more numerous than the hermaphrodite; their filiform or slender tubular corolla shorter than the disk and style, truncate or 2–4-toothed at the apex. Akenes small, compressed. Pappus a single series of soft capillary bristles, sometimes an added outer series of short bristles or squamellæ.

Subtribe IV. BACCHARIDEÆ. Heads discoid and unisexual. Corolla of the fertile flowers filiform. Pappus of capillary bristles.

51. **BACCHARIS.** Heads completely diœcious, many-flowered. Involucre regularly imbricated, of squamaceous bracts. Receptacle mostly flat and naked, rarely chaffy. Flowers of the male heads with tubular-funnelform 5-cleft corolla, and style-branches as of *Aster* or *Solidago*, but the stigmatic portion obsolete and ovary abortive; the female with corolla reduced to a slender truncate or minutely toothed tube, shorter than the filiform style. Akenes 5–10-costate. Pappus of the male flowers of a series of scabrous and often tortuous and more or less clavellate bristles; of the fertile flowers of usually more numerous and fine bristles, and often elongated in fruit. Shrubby or some herbaceous.

TRIBE IV. INULOIDEÆ. Heads heterogamous and either radiate or discoid; the female flowers being either ligulate or filiform (rarely open-tubular), or sometimes homogamous and tubuliflorous. Anthers sagittate, and the base of the lobes produced into more or less of a tail (*caudate*) or other appendage. Style-branches of the hermaphrodite flowers filiform or flattish, not appendaged; the stigmatic lines running to or vanishing near the roundish or truncate tip, which is at most papillose or somewhat penicillate: style of staminate-sterile flowers commonly entire. Pappus usually capillary or none. Leaves mostly alternate and heads homochromous; the involucre commonly dry or scarious, rarely foliaceous. See also *Senecionideæ*, subtribe *Tussilagineæ*. (No North American species has conspicuous rays, except a naturalized *Inula*.)

Subtribe I. PLUCHEINEÆ. Heads discoid, heterogamous and mostly androgynous. Involucre more or less dry, but hardly scarious. Receptacle not paleaceous. Female flowers with filiform corolla. Adjacent anther-tails or acuminate bases connate, at least in our genera.

52. **PLUCHEA.** Heads many-flowered, largely of female flowers, a few hermaphrodite but usually sterile ones in the centre. Involucre imbricated, of coriaceous to submembranaceous bracts; the outer broad, all but the innermost persistent. Receptacle flat, naked and glabrous. Corolla of the female flowers reduced to a slender truncate or 2–3-toothed tube, shorter than the style; of the hermaphrodite-sterile ones regularly 5-cleft, the style either entire or 2-cleft at apex. Akenes small, 4–5-angled or sulcate. Pappus a series of capillary and soft or rigid bristles. Heads cymosely clustered or scattered.

53. **PTEROCAULON.** Heads and flowers as in *Pluchea*, but involucre of fewer and linear or subulate bracts: these deciduous with the matured flowers, leaving a few short basal ones which are more persistent, mainly by their implexed wool. Receptacle small, naked, sometimes pilose. Heads glomerate and the glomerules spicate. Perennial herbs.

Subtribe II. FILAGINEÆ. Heads heterogamous, mostly androgynous, discoid. Involucre of few scarious or firmer bracts. Receptacle chaffy; a chaff (palea) or involucral bract enclosing or subtending each female flower or akene. Corolla of the female flowers a filiform tube, shorter than the style; of the few hermaphrodite commonly sterile flowers regularly 4–5-toothed; their anthers sometimes only acutely sagittate or auriculate at base, and the short style-branches or undivided style not truncate. Akenes (with one or two exceptions) smooth and even, small and seed-like, the very thin pericarp destitute of nerves or other markings, conformed to the seed and sometimes connate with the simple seed-coat, or evanescent. Low and floccose-woolly annuals. (Characterized to the exclusion of three outlying Indo-African genera.)

    * Akenes gibbous, so that the corolla- and style-bearing true apex is introrsely lateral, enclosed in the compressed and cucullate mostly indurating bracts or chaff.

  54. **MICROPUS.** Heads several- but hardly many-flowered; female flowers in one or two series on a small and rarely somewhat elevated receptacle, each enclosed (all but the style) in a conduplicate cucullate herbaceous lanate bract or chaff, only the tip of which is scarious-appendiculate; the few hermaphrodite-sterile ones in the centre mostly naked. Involucre outside of the flower-bearing bracts scanty and scarious. Ovary and akene obovate, laterally compressed, destitute of pappus, remaining enclosed in the usually indurating laterally compressed fructiferous bracts, which sooner or later fall away. Pappus of the sterile flowers none, or rarely a few caducous bristles.

    * * Akenes (straight or slightly curved) with the minute epigynous disk at the summit.

      +— Female flowers all bracteate and all destitute of pappus;

    ++ The fructiferous bracts enclosing the akenes and deciduous with them, tipped with a hyaline appendage.

  55. **STYLOCLINE.** Heads ovoid; the boat-shaped fructiferous bracts borne on a slender column or receptacle, their erect hyaline tips usually conspicuous, the involute or saccate-conduplicate body or base embracing the obovate or oblong more or less compressed akene: those subtending the 4 or 5 central hermaphrodite-sterile flowers barely concave or plane. Pappus of the latter flowers a very few caducous bristles or none.

  56. **PSILOCARPHUS.** Heads globose; the numerous fructiferous bracts heaped on the globular or oval receptacle, cucullate-saccate and utricular, half-obovate or half-obcordate in outline, very rounded at top, herbaceo-membranaceous, the apex introrse, and the ovate or oblong hyaline appendage inflexed or porrect, or sometimes erect. Akene loose in the comparatively ample utricular bract, oblong or narrower, straight, slightly compressed. No bracts and no pappus to the few hermaphrodite-sterile flowers. Leaves mainly opposite!

    ++ ++ Fructiferous bracts open and merely subtending the (usually numerous) akenes.

  57. **EVAX.** Akenes from obcompressed to terete, sometimes minutely papillose or puberulent. Bracts of the female flowers from scarious to firm-chartaceous, not hyaline-appendaged. Hermaphrodite flowers sometimes fertile, destitute of pappus. Receptacle from barely convex to subulate.

    +— +— Female flowers more or less of two kinds: the upper ones surrounding the hermaphrodites and like them with a capillary pappus; the others destitute of pappus.

  58. **FILAGO.** Receptacle from subulate to obconical or hemispherical; its naked summit or centre bearing several or somewhat numerous commonly fertile flowers, which are all provided with a pappus of rather copious capillary bristles; the few central ones hermaphrodite, the others female; the cluster subtended by open scarious bracts; the sides or base of the receptacle bearing several or numerous female flowers, each subtended or its calvous akene loosely enclosed by a concave or boat-shaped bract. Akenes terete or obscurely compressed, sometimes roughish-papillose.

Subtribe III. GNAPHALIEÆ (Veræ). Heads discoid, heterogamous and androgynous or diœciously homogamous; the hermaphrodite or staminate flowers when in the same head much fewer than the female; the latter with filiform-tubular corolla shorter than the style; the former with style or style-branches mostly truncate; all usually with capillary pappus. No bracts or chaff on the receptacle bracts of the involucre numerous, more or less scarious, or with scarious and often colored or petaloid summits. Anther-tails slender. Ours floccose-woolly herbs, with alternate leaves.

  59. **ANTENNARIA.** Heads diœcious, many-flowered. Involucre pluriserially imbricated. Receptacle convex or flattish. Male flowers with mostly undivided style and a rather scanty pappus of clavellate or apically barbellulate or crisped bristles. Female flowers with oblong

or narrower and terete or flattish akenes, and a copious fine-capillary pappus, the soft and naked bristles of which are commonly united at base, so as to fall in a ring. Low perennials.

60. **ANAPHALIS.** Heads diœcious, but usually with a few hermaphrodite-sterile flowers in the centre of the female heads. Pappus of male flowers of bristles little if at all thicker at the apex; of the female flowers not united at base but falling separately. Otherwise as in the preceding: the female plant differing from the following only in the sterility of the few central flowers.

61. **GNAPHALIUM.** Heads heterogamous, fertile throughout, of few or many series of female surrounding a smaller number of hermaphrodite flowers. Involucre pluriserially imbricated; the scarious and commonly partly woolly bracts with or without colored papery tips or appendages. Style of hermaphrodite flowers 2-cleft. Pappus of numerous merely scabrous capillary bristles, in a single series.

Subtribe IV. EUINULEÆ. Heads heterogamous, with the female ligulate and radiate, and the disk-flowers all hermaphrodite and fertile (or the former sometimes wanting). Receptacle naked. Style-branches of the latter linear, rounded at the apex. Akenes mostly coriaceous. Old-world genera, one naturalized.

62. **INULA.** Heads many-flowered; the rays numerous. Involucre imbricated, hemispherical or campanulate; outer bracts herbaceous. Receptacle flat or nearly so. Akenes more or less 4-5-costate. Pappus of capillary scabrous bristles. Flowers yellow.

Subtribe V. ADENOCAULEÆ. Heads heterogamous, discoid: both female and hermaphrodite (sterile or fertile) flowers with tubular more or less ampliate 4-5-toothed or lobed corollas. Involucre not scarious. Receptacle naked. Akenes elongated, striate or nerved: pappus none. Leaves alternate, partly floccose-woolly. (Here also *Carpesium*, L.)

63. **ADENOCAULON.** Heads several-many-flowered; the marginal flowers female; the more numerous central ones hermaphrodite-sterile. Involucre of few thin-herbaceous bracts. Receptacle flat, naked. Corollas all somewhat alike; of the sterile flowers broadly funnelform and deeply 4-5-cleft; of the fertile ones less ampliate, either regularly 4-lobed or (in the Chilian species) bilabiate, the outer lip 3-lobed; the style of the former undivided, somewhat clavate; of the latter with short and broad stigmatic branches. Anthers sagittate, and the auricles minutely but evidently caudate, connate. Akenes obovoid-oblong or clavate, very obtuse, lightly 4-5-nerved, much exceeding the involucre, the upper part beset with stout stipitate (nail-shaped) glands: ovary of the sterile flowers inane. Leaves floccose-woolly underneath.

TRIBE V. HELIANTHOIDEÆ. Heads heterogamous and the female flowers ligulate and radiate, or rarely with corolla wanting, and in the latter case some are monœcious: or sometimes homogamous by the absence of the ligulate ray-flowers; those of the disk all with regularly 4-5-toothed tubular corolla. Receptacle paleaceous, i. e. with bracts (*paleæ*, chaff) subtending either all the disk-flowers or the marginal ones only, except in the first subtribe and the peculiar uniflorous or 2-4-locular heads of the third. Anthers at most sagittate, not caudate at base. Style-branches of hermaphrodite or sterile flowers (or the undivided style in some of the latter) truncate or continued into a hairy (from conical to subulate) appendage. Pappus various or none, never of truly capillary bristles. Leaves more commonly opposite, at least the lower ones. — A varied tribe of more than 100 genera.

Subtribe I. MILLERIEÆ. Ray-flowers solitary or few and fertile: disk-flowers hermaphrodite-sterile. Receptacle naked. Pappus none. (Artificial group.)

64. **PLUMMERA.** Heads of 2 to 5 ray and 6 or 7 disk-flowers. Involucre obpyramidal, double: the outer composed of mostly 4 ovate-oblong obtuse carinate bracts, which are united to or above the middle into a coriaceous or cartilaginous cupule; the inner of as many barely equal cuneate-obovate plane alternating bracts, their rounded summit firmscarious. Receptacle flat, naked. Female flowers with a dilated-cuneate 3-lobed ligule on a

short tube. Hermaphrodite-sterile, with usually a stipitiform abortive ovary, tubular-funnelform 5-toothed corolla, its short proper tube thickened, and a truncate-bicapitellate style. Akene of the fertile flowers obovate, turgid, free, thinnish, nearly nerveless, the upper part villous with long soft hairs, wholly destitute of pappus; the epigynous areola small and slightly depressed. Flowers yellow.

156. **BLENNOSPERMA.** Heads many-flowered. Involucre simple, broad. Akenes pyriform. Perhaps belongs here rather than to the *Helenioideæ*, where it has no near relatives: perhaps it should stand next to *Crocidium*, no. 186.

Subtribe II. **MELAMPODIEÆ.** Ray-flowers ligulate (sometimes obscurely so) and fertile, the ligule mostly deciduous: disk-flowers hermaphrodite-sterile. Akenes usually with coriaceous or thicker pericarp, the style mostly entire. Receptacle paleaceous throughout, *Parthenice* excepted. (Artificial group.)

  * Involucre of the several-flowered heads cylindrical or fusiform, of membranaceous or thinnish and striate-nerved oblong bracts, not enclosing or embracing the few narrow and somewhat obcompressed akenes; those of the small and flat receptacle scarious. (Allied to the *Coreopsideæ*.)

65. **DICRANOCARPUS.** Involucre of 3 or 4 narrow bracts (and rarely 1 or 2 short and subtending herbaceous ones), at length loose, deciduous in age. Ray-flowers 3 or 4; the ligule very small, shorter than the style, 2-3-lobed, and with hardly any tube. Disk-flowers 3 or 4: corolla with rotately-spreading limb parted into 5 ovate lobes, rather longer than the narrow tube: style tipped with an entire conical appendage: ovary inane. Akenes dimorphous, one or two elongated to fully twice or thrice the length of the involucre, from subulate to oblong-linear, nearly smooth, puberulent, long persistent on the receptacle, tipped with two diverging or divaricate and stout persistent naked rigid awns or horns; the others shorter, comparatively thicker, often tuberculate-rugose, the truncate apex bearing a pair of very short divaricate horns or hardly any.

66. **GUARDIOLA.** Involucre of 3 concave and herbaceo-membranaceous bracts completely enclosing 3 or 4 narrower white-scarious ones; and there are narrower similar paleæ subtending the sterile flowers. Tube of the corolla both of ray and disk long and filiform; in the 1 to 5 fertile flowers equalling the oblong ligule; in the 5 to 20 sterile flowers commonly several times longer than the abruptly dilated deeply 5-cleft limb (its lobes linear). Filaments very villous! Style of the disk-flowers mostly 2-cleft at apex, its branches linear or subulate, hispid to the base: ovary inane. Akenes oblong, slightly obcompressed, smooth, the rounded apex wholly destitute of pappus; the base suberose-fleshy, sometimes as a boss upon one side.

  * * Involucre of the many-flowered heads broad; the inner bracts concave, embracing and half enclosing the thick and turgid obovoid akenes: pappus none.

67. **POLYMNIA.** Heads hemispherical or broader. Involucre of about 5 loose exterior bracts, and as many or more numerous smaller and thinner interior ones, subtending as many fertile flowers: sterile flowers subtended by mostly scarious chaffy bracts of the flat receptacle. Ray-flowers with a short hairy tube bearing either a short or a long ligule, or none at all. Corolla of the disk-flowers thin, ampliate above, and with ovate lobes. Sterile style commonly 2-cleft or 2-pointed at the hairy apex. Akenes very thick, short, smooth, marginless.

  * * * Involucre double, strongly dimorphous; exterior of 4 or 5 herbaceous or foliaceous plane bracts; interior of a single series of small bracts, which completely and permanently enclose the obovate or oblong more or less compressed smooth and glabrous akenes with a pericarp-like accessory covering, at length deciduous together: pappus none.

68. **MELAMPODIUM.** Fructiferous and transformed involucral bracts commonly indurated, naked or unarmed. Receptacle convex or conical. Akenes more or less obovate and incurved.

69. **ACANTHOSPERMUM.** Fructiferous involucral bracts armed with hooked prickles or spines, forming a kind of bur. Otherwise as *Melampodium*. Ligules minute, concave or cucullate.

  * * * * Involucre broad, of plane or barely concave bracts; innermost subtending obcompressed (mostly much flattened) akenes, but not enclosing nor embracing them.

COMPOSITÆ.

+— Ray-flowers, or rather their ovaries and akenes, in more than one series, and with elongated exserted deciduous ligules; the akenes falling free, or with only the subtending bract. (*Schizomeria* of Mexico is of this section, but with fewer ray-flowers.)

70. **SILPHIUM.** Heads large, many-flowered. Involucre of thickish more or less foliaceous imbricated bracts; the innermost (next to the akenes) small and chaffy. Receptacle comparatively small, the central part bearing the sterile flowers somewhat turbinate in age; its chaffy bracts linear, flat, or involute around the pedicilliform abortive ovaries. Corollas of the ray with a long and spreading ligule on a very short tube; of the disk cylindrical-tubular, the teeth very short and thickish. Sterile style entire, much elongated in anthesis, hispidulous. Akenes very flat and broad, imbricated in 2 or 3 series, completely free from the subtending bract and from those of adjacent male flowers, surrounded by a winged margin which is produced more or less beyond the summit on each side into a callous tooth or auricle; pappus none or sometimes a pair of short rigid awns or teeth, with which the wing is confluently united.

+— +— Ray-flowers and akenes in a single series; the latter with 2 or 3 bracts of sterile flowers (paleæ of the receptacle) attached to their base on the inner side, which they take with them, and commonly also the subtending involucral bract, when they fall.

++ Heads conspicuously radiate: the ligules plane and exserted, subsessile, yellow, oblong or oval: receptacle flat; its bracts or chaff, or at least the outer, of rather firm texture and more or less involving their sterile flowers or inane ovaries.

71. **BERLANDIERA.** Heads broad, with 5 to 12 radiate fertile and many hermaphrodite-sterile flowers. Involucre hemispherical or broader; its bracts in about three series: outermost smaller and more foliaceous: the succeeding larger and usually dilated-obovate; innermost thinner, becoming membranaceous or chartaceous and reticulated in age. Disk-corollas, style, &c. as in *Silphium*. Akene very flat or slightly meniscoidal, obovate, wingless, not toothed or notched at summit, carinately unicostate down the inner face, with evanescent or obsolete pappus (sometimes two minute and caducous bristly teeth or awns), at the back more or less coherent with the base of the subtending plane involucral bract, at length falling away with it, the ventral face partly covered by the spatulate bracts of the 2 or 3 attached sterile flowers. Alternate-leaved perennials.

72. **CHRYSOGONUM.** Heads of mostly 5 radiate fertile and rather numerous sterile flowers. Involucre campanulate, double; outer of 5 loose and obovate or spatulate foliaceous bracts which surpass the disk: inner of as many small oval firm-membranaceous erect bracts, each subtending a fertile flower. Chaffy bracts of the receptacle thin, linear, the outer broader and partly enclosing as many sterile flowers. Ligules oblong: disk-corollas cylindraceous, with 5 small spreading teeth. Akene obovate or oval, flat or slightly meniscoidal, with acute but wingless margins, one-nerved on the convex back, carinately 1-2-costate on inner face, crowned with a truncate or crenately emarginate semi-cupulate pappus, i. e. the crown wanting on the inner side; the involucral bract with which it falls little larger than its back, and closely applied to but not cohering with it; adjacent sterile bracts less persistent with the akene than in the following genus. Low and opposite-leaved perennial.

73. **LINDHEIMERA.** Heads of 4 or 5 radiate fertile and rather numerous sterile flowers. Involucre double; outer of 4 or 5 loose and foliaceous narrow lanceolate bracts; inner of as many larger ovate-oblong herbaceous bracts, becoming chartaceo-membranaceous, subtending the fertile flowers. Receptacle small; its inner bracts chaffy, narrowly linear and nearly plane; the outer more herbaceous, spatulate, and enclosing the filiform abortive ovary, which forms a long pedicel to the subtended sterile flower. Ligules oval or oblong, almost entire: disk-corollas funnelform. Akene obovate, flat, slightly meniscoidal, unicostate on the middle of each face (dorsally not adnate to the much larger subtending bract), surrounded by a cartilaginous entire wing, which is confluent at apex with two triangular-subulate rigid teeth or horns: a similar but smaller and naked tooth projecting from the summit of the ventral costa. Erect annual, with sessile leaves, the upper opposite.

74. **ENGELMANNIA.** Heads of 8 to 10 conspicuously radiate and of many sterile flowers. Involucre hemispherical, somewhat double; the outer of about 10 rather linear mainly foliaceous bracts in two series, i. e. the outermost smaller and wholly loose and foliaceous, the succeeding similar but with ovate-dilated and appressed coriaceous base; inner of firm-

coriaceous oval or obovate concave bracts with short and abrupt foliaceous tips, subtending the fertile flowers. Bracts of the receptacle all firm-coriaceous and persistent, linear or lanceolate, partly enclosing the sterile flowers or their pedicel-like inane ovary. Ligules oblong, almost entire; disk-corollas narrowly funnelform, 5-lobed. Sterile style somewhat clavate, very hispid. Akenes filling nearly the whole deeply concave face of the subtending bract (not adnate to it), and covered by the 2 or 3 rigid internally attached chaffy bracts of sterile flowers, obovate, wingless, both faces carinately unicostate; indurated bases of the bract, the chaff, and the akene firmly and inseparably united, tardily falling away from the receptacle. Pappus a conspicuous and persistent firm-scarious and hispid crown, more or less cleft into 3 or 4 irregular lobes or into a pair of lanceolate scales; that of the abortive ovaries somewhat similar but rudimentary, nearly as in *Silphium*. Alternate-leaved perennial.

++ ++ Heads small, mostly hemispherical, nearly or quite discoid; the corolla of the 5 to 8 female flowers with the ligule very short, often broader than long, or obsolete: receptacle conical or convex: flowers whitish: corollas of the sterile ones tubular-funnelform, and the style glabrous except the more or less dilated and truncate pubescent summit, their ovary obsolete: akenes more turgid dorsally: involucre of rather few ovate or orbicular appressed bracts, in about two series: leaves alternate.

75. **PARTHENIUM.** Fertile flowers 5, with obcordate or 2-lobed almost sessile concave ligule barely surpassing the disk, or a truncate emarginate cup. Bracts of the involucre chartaceous or partly herbaceous, and the inner more scarious: those of the usually conical receptacle cuneate, tomentose at summit, partly enclosing the sterile flowers. Akenes oval or obovate, commonly pubescent, surrounded by a filiform callous margin, which is firmly coherent at base with the bases of the bracts of the contiguous pair of sterile flowers and of the subtending bract, at length tearing away from the akene; the summit bearing the marcescent corolla, and a pappus of two chaffy awns or scales, or sometimes hardly any.

76. **PARTHENICE.** Fertile flowers 6 to 8, with corolla hardly equalling the disk, not longer than the style, an obliquely cleft tube, with ligule obsolete or reduced to 2 or 3 small teeth; sterile flowers 40 or 50, with funnelform corolla. Involucre of 5 somewhat herbaceous oval exterior bracts, and of 6 or 8 somewhat larger orbicular-obovate and more scarious interior ones, these subtending the fertile flowers. Receptacle convex, with linear-oblong or spatulate chaffy bracts subtending the outer series of sterile flowers, but mostly minute or wanting to the inner flowers. Akenes oblong-obovate, glabrous, wingless, but acute-margined, with an incurved apiculation terminated by a small sphacelate obtuse tip, inserted by a very small base, falling away at maturity with the involucral and two receptacular bracts, but these readily separating. Pappus none; and corolla deciduous.

Subtribe III. AMBROSIEÆ. Fertile flowers apetalous, or with corolla reduced to a tube or ring around the base of the 2-parted style: disk-flowers staminate, with funnelform or obconical 4-5-lobed corolla, anthers slightly united and their short terminal appendage inflexed, the abortive style hairy only at the somewhat enlarged and depressed summit, the ovary a mere rudiment. Pappus none (or a vestige in *Oxytenia* and *Dicoria*). Heads small; the flowers whitish or greenish.

* Head androgynous (rarely all male in *Dicoria*), having few (or rarely solitary) female flowers at the margin; the more numerous male flowers all or most of them subtended by slender and commonly spatulate chaffy bracts: anther-tips short and obtuse, rarely pointed: involucre open: akenes usually large for the size of the head, free. — *Ivea*, DC.

+- Akenes turgid, mostly obovate or pyriform, marginless; dilated summit of the sterile style hispidulous or radiately penicillate.

77. **IVA.** Female flowers 1 to 5, with or without the tube or cup representing a corolla. Akenes more or less obcompressed, glabrous, puberulent, or glandular; the terminal areola small.

78. **OXYTENIA.** Female flowers about 5, wholly destitute of corolla; their style with 2 oval or oblong and very obtuse stigmatic lobes. Involucre of about 5 coriaceo-herbaceous dilated-ovate and rather rigidly acuminate bracts. Receptacle convex, small; the 10 to 20 sterile flowers subtended by slender chaffy bracts with cuneate-dilated tips, or these wanting to the central ones. Akenes (immature) obovate and turgid, very villous, nearly pyriform

(sometimes with a single diaphanous and minute squamella to represent pappus!), with large terminal areola bearing around the base of the style a fleshy annular disk. Lower part of the disk-flowers and their chaff beset with some villous hairs, like the very long and soft ones which thickly clothe the akenes.

+ + Akenes flattened, obcompressed, wing-margined.

79. **DICORIA.** Female flowers one or two, wholly destitute of corolla: male flowers 6 to 12, with mere rudiments of ovary and style. Involucre of 5 oval or oblong herbaceous bracts; and within one or two larger and broad thin-scarious bracts, subtending the fertile flowers; or these wanting in male heads. Receptacle small, flat, with a few narrow and hyaline chaffy bracts among the flowers. Filaments almost free from the obconical corolla, monadelphous up to the lightly connected anthers! the tube dilated and 5-toothed at summit. Akenes much surpassing the outer involucre, oblong, anteriorly flat, convex or somewhat angled dorsally, abruptly bordered by a thin-scarious pectinate-dentate wing or edge. Pappus rudimentary, of several small and setiform squamellæ.

* * Heads unisexual, monœcious: the fertile with solitary or 2 to 4 completely or nearly apetalous female flowers in a closed nutlet-like or bur-like involucre, only the style-branches ever exserted; the sterile of numerous male flowers in an open involucre, the heads in a raceme or spike of centripetal evolution: akenes turgid-obovoid or ovoid, wholly destitute of pappus: flowers greenish or yellowish: male corollas obconical. — *Ambrosieæ*, DC.

+ Involucre of the sterile heads gamophyllous; the receptacle low, and abortive style with dilated apex radiately penicillate or fimbriate.

80. **HYMENOCLEA.** Involucre of the male flowers saucer-shaped and 4-6-lobed, rarely more cleft: bracts of the receptacle subtending the outer flowers obovate or spatulate; inner filiform or none: filaments distinct: anther-tips blunt. Involucre to the solitary fertile flower ovoid or fusiform, beaked at apex, the lower part furnished with 9 to 12 dilated and silvery-scarious persistent transverse wings.

81. **AMBROSIA.** Involucre of the male flowers from depressed-hemispherical to turbinate, 5-12-lobed or truncate, herbaceous. Receptacle flat or flattish, usually with some filiform chaff among the outer flowers. Anther-tips (at first inflexed, at length erect) setiferous-acuminate. Involucre to the solitary fertile flower nucumentaceous, apiculate or beaked at the apex, and usually armed with 4 to 8 tubercles or short spines in a single series below the beak. Sterile heads spicate or racemose above the fewer fertile ones.

82. **FRANSERIA.** Heads of male flowers as *Ambrosia*, or sometimes intermixed with the female. Fertile involucre 1-4-flowered, 1-4-celled, a single pistil to each cell, 1-4-rostrate, more or less bur-like, being armed over the surface with several or numerous prickles or spines (the spiny free tips of component bracts) in more than one series. Leaves mostly alternate.

+ + Involucre of the sterile heads polyphyllous, and the receptacle cylindraceous.

83. **XANTHIUM.** Involucre of the globular sterile heads one or two series of small narrow bracts: receptacle distinctly paleaceous, a cuneate or linear-spatulate chaffy bract partly enclosing each male flower: filaments monadelphous: anthers distinct but connivent; the inflexed apical appendage mucronate: sterile style unappendaged. Fertile heads a closed and ovoid bur-like 2-celled and 2-flowered involucre, 1-2-beaked at the apex, the surface clothed with uncinate-tipped prickles: each flower a single pistil, maturing a thick ovoid akene, the two permanently enclosed in the indurated prickly involucre. Leaves alternate.

Subtribe IV. ZINNIEÆ. Ray-flowers ligulate and fertile; the ligule with very short tube or none, persistent on the akene and becoming papery in texture! (but at length falling or decaying away in *Heliopsis lævis*): disk-flowers hermaphrodite and in our genera fertile, numerous, subtended or embraced by chaffy bracts; the corolla cylindraceous. Leaves opposite and heads singly terminating the stem or branches.

* Leaves all or mostly entire: akenes of the disk compressed, all or some of them (either of disk or ray) toothed or awned from the summit of the angles or edges.

84. **ZINNIA.** Involucre campanulate or cylindraceous; its closely appressed-imbricated bracts dry and firm, broad, with rounded summit often margined. Receptacle becoming

conical or cylindraceous ; the chaffy bracts conduplicate around the disk-flowers. Lobes of the disk-corolla mostly velvety-villous. Style-branches of disk-flowers with either truncate or subulate tips. Akenes wingless or nearly so ; of the ray obcompressed-triquetrous, of the disk much compressed. Pappus when present of erect awns or chaffy teeth. Leaves mostly sessile and rays showy.

85. **SANVITALIA**. Involucre short and broad, of dry or partly herbaceous bracts. Receptacle from flat to subulate-conical, at least in fruit; its chaffy bracts concave or partly conduplicate. Ligules (entire or 2-toothed at apex) often short and small. Disk-corollas with glabrous lobes. Style-branches of disk-flowers truncate or capitellate at tip. Akenes either all or the exterior thick-walled; of the ray commonly 3-sided, and the angles produced into as many thick and rigid divergent awns or horns; those of the disk often heterogeneous, from compressed-quadrangular to flat, some usually wing-margined, the pappus of one or two slender awns or teeth or none. Leaves commonly petioled.

* * Leaves commonly serrate, slender-petioled : akenes not compressed.

86. **HELIOPSIS**. Involucre short, of nearly equal oblong or lanceolate bracts, the outer herbaceous. Receptacle from high-convex to conical; the pointless chaffy bracts partly embracing the disk-flowers. Ligules large, oblong or narrower: disk-corollas glabrous. Style-branches tipped with a very short conical hirsute appendage. Akenes short and thick, obtusely 4-angular, or in the ray somewhat triangular, with broad truncate summit, wholly destitute of pappus, or sometimes with the annular border 1–4-dentate.

99. **BALSAMORRHIZA** § KALLIACTIS also has persistent ligules !

Subtribe V. **VERBESINEÆ**. Ray-flowers ligulate and either fertile or neutral, or not rarely wanting, the ligule not becoming papery and persistent on the fruit (with one exception), but sometimes marcescent : disk-flowers hermaphrodite and fertile (or some of the inner often failing to produce fruit), subtended and sometimes enwrapped by the bracts of the receptacle. Anthers often blackish. Akenes various but those of the disk never obcompressed : pappus cupulate or coroniform, or of teeth or awns from the 2 to 4 principal angles, or of some squamellæ, or of a few stout (but not capillary) bristles, or none. Leaves either opposite or alternate.

* Involucre 4- rarely 5-lobed and foliaceous, valvate and saliently 4-5-angled in the bud : akenes short and thick : pappus when present pluripaleaceous in the manner of *Helenioideæ*.

87. **TETRAGONOTHECA**. Heads many-flowered ; ray-flowers few or several, fertile. Principal involucre membranaceo-foliaceous, spreading in anthesis ; the 4 or 5 broadly ovate bracts connate at base ; within are 6 to 15 small chaffy bracts subtending ray-flowers, and similar to the thin-membranaceous and nervose lanceolate chaffy bracts of the at length conical receptacle. Ligules with short tube or almost nearly sessile, 5–8-nerved : disk-corollas with elongated cylindraceous throat, 5-lobed. Style-branches of the disk-flowers hispid above, and tipped with a rather long acute or acuminate appendage. Akenes more or less 4-sided, with a broad flat summit, destitute of pappus, or with a crown of numerous chaffy squamellæ. Leaves opposite.

* * Involucre of several or numerous distinct bracts.

+— Bracts of the receptacle permanently investing the akenes as an indurated accessory covering.

88. **SCLEROCARPUS**. Heads many-flowered; the ray-flowers several, neutral. Involucre of rather few more or less herbaceous bracts, the outer loose and spreading. Receptacle convex or conical ; its at length coriaceous or cartilaginous bracts closely investing the akenes and falling away with them by an articulation. Disk-corollas 4–5-cleft : style-branches mostly with subulate appendages. Akenes smooth, oblong or obovoid : pappus a short crown or ring, or none. Branching herbs.

+— +— Bracts of the receptacle mostly reduced to awn-shaped chaff or bristles subtending the naked akenes.

89. **ECLIPTA**. Heads many-flowered : ray-flowers numerous, small and short, fertile. Involucre broad, of one or two series of herbaceous bracts. Receptacle nearly flat. Disk-corollas 4-toothed, rarely 5-toothed ; their style-branches with short obtuse or triangular tips.

Akenes thick, in the ray mostly 3-sided and in the disk compressed, more or less margined, without pappus, or sometimes with 2 to 4 teeth or short awns. Leaves opposite and heads small.

+ + + Bracts of the many-flowered receptacle concave or complicate, loosely embracing or subtending the disk-akenes, mostly persistent.

++ Rays uniformly none, the flowers all hermaphrodite and fertile: involucre dry or partly so: akenes not flat nor margined: pappus of slender awns or none.

90. **MELANTHERA.** Involucre hemispherical; the disk in fruit globular, and squarrose with the mostly pointed rather rigid striate concave bracts of the convex or low-conical receptacle: bracts of the involucre ovate to lanceolate, thickish, nerveless, in 2 or 3 series, somewhat equal in length. Corolla 5-lobed, with campanulate-oblong ampliate throat. Style-branches tipped with a subulate hispid appendage. Akenes thick and short, compressed-quadrangular, somewhat obpyramidal, with broad truncate summit: pappus of 2 or more slender caducous awns. Leaves opposite, petioled.

91. **VARILLA.** Involucre short, of rather few and small linear-lanceolate appressed-imbricate and mostly few-striate bracts, similar to those of the at length high-conical or oblong receptacle. Corolla with narrow cylindraceous throat, 5-toothed. Style-branches with short and obtuse or minutely apiculate conical tips. Akenes narrow, linear-oblong, terete, rather thin-walled, smooth, evenly 8–15-nerved: pappus setulose or none. Shrubby or suffruticose.

92. **ISOCARPHA.** Involucre, receptacle, and dry bracts nearly of the preceding genus. Corolla similar but small. Style-branches with subulate tips. Akenes 4–5-angled, small, little compressed, destitute of pappus. Herbaceous.

93. **SPILANTHES.** Some (exotic) species have no ray-flowers, and akenes not flat, with pappus also wanting: these resemble *Isocarpha*.

++ ++ Rays present, but in several genera occasionally wanting: involucre commonly herbaceous or foliaceous, or partly so.

= Receptacle high, from conical to columnar or subulate, at least in fruit. (Here *Gymnolomia*, as to two species, would be sought.)

α. Rays fertile, or not rarely wanting: style branches of the disk-flowers truncate and sometimes penicillate at tip: akenes small: leaves opposite.

93. **SPILANTHES.** Involucre of a few somewhat herbaceous loosely appressed bracts. Bracts of the receptacle soft and chaffy, shorter than the flowers, more or less conduplicate and embracing the akenes, at length falling with them. Disk-corollas 4–5-toothed. Akenes of the ray triquetrous or obcompressed; those of the disk either moderately or much compressed and with acute or nerve-like margins, sometimes ciliate-fimbriate. Pappus a setiform awn from one or more of the angles, or none.

b. Rays sterile (imperfectly styliferous in *Echinacea*, otherwise completely neutral), soon drooping, sometimes marcescent, the ligule with very short tube or none: style-branches tipped with an acute or obtuse hispid appendage: leaves mostly alternate.

94. **ECHINACEA.** Involucre imbricated in 2 or 3 or more series and squarrose; its bracts lanceolate. Disk at first only convex, becoming ovoid and the receptacle acutely conical. Chaffy bracts of the latter firm and completely persistent, linear-lanceolate, carinate-concave, acuminate into a rigid and spinescent cusp, surpassing the disk-flowers. Ligules elongated and pendent in age, rose-colored or rose-purple, marcescent, usually imperfectly styliferous. Disk-corollas cylindraceous, with 5 erect teeth and almost no proper tube (a ring upon which the stamens are inserted). Akenes suberose-cartilaginous, acutely quadrangular, somewhat obpyramidal, with a thick coroniform pappus more or less extended into triangular teeth at the angles: the basal areola central.

95. **RUDBECKIA.** Involucre looser, spreading, more foliaceous. Disk from hemispherical or globose to columnar, and receptacle from acutely conical to cylindrical and subulate; its chaffy bracts not spinescent, but sometimes soft-pointed. Ligules yellow or partly (rarely wholly) brown-purple. Disk-corollas with a short but usually a manifest proper tube. Akenes 4-angled, prismatic, in some species quadrangular-compressed, or in one nearly terete. Pappus a coriaceous or firm-scarious and often 4-toothed crown, sometimes deep and cupuliform, sometimes obsolete, or none.

5

96. **LEPACHYS.** Akenes short and broad, compressed, acutely margined or sometimes winged at one or both edges, somewhat laterally or obliquely inserted on the slender-subulate receptacle: pappus a chaffy or aristiform tooth over one or both edges, or none, the crown minutely squamellate and evanescent or none. Chaffy bracts of the receptacle conduplicate or deeply navicular, with thickened and truncate or somewhat hooded summit, embracing and hardly surpassing the akenes, at length deciduous with them. Corollas of the disk with hardly any proper tube. Ligules, involucre, &c., of *Rudbeckia*.

== == Receptacle from flat to convex, or in certain species conical: akenes not winged nor very flat, when flattened not margined or sharp-edged.

a. Rays fertile: style-branches of the disk-flowers hispid for all or much of their length: receptacle flat or merely convex: ray akenes commonly triquetrous or obcompressed: pappus persistent or none.

97. **WEDELIA.** Akenes thick and turgid, cuneate-oblong or pyriform, with roundish summit; those of the disk obtusely if at all quadrangular, or flattened only at the inane base: pappus a paleaceous commonly lobed and at length indurated cup. Involucre rather simple and foliaceous. Leaves opposite: stem herbaceous.

98. **BORRICHIA.** Akenes equably and acutely quadrangular, or in the ray triangular: pappus a somewhat toothed cup or crown. Involucre imbricated; outer bracts sometimes foliaceous. Bracts of the receptacle concave, rigid. Leaves opposite: stem woody.

99. **BALSAMORRHIZA.** Akenes destitute of pappus, oblong; of the disk quadrangular and often with intermediate nerves (these and the angles usually salient). Ligules with a distinct tube. Involucre broad; the outer bracts foliaceous, sometimes enlarged. Bracts of the receptacle linear-lanceolate. Style-appendages filiform or slender-subulate. Tuberous-rooted low herbs.

100. **WYETHIA.** Akenes prismatic, large, 4-angled, or in the ray 3-angled and in the disk often flattened, also with intermediate salient nerves: pappus a lacerate chaffy or coriaceous crown, or cut into nearly distinct squamellæ, commonly produced at one or more of the angles into chaffy rigid awns or teeth. Involucre campanulate or broader, more or less imbricated; outer bracts often foliaceous. Bracts of the receptacle lanceolate or linear, partly embracing the akenes. Style-appendages slender-subulate or filiform, very hispid. Thick-rooted and large-headed herbs, with alternate leaves.

b. Rays sterile, rarely wanting: akenes quadrangular-compressed or more turgid, or flatter, but none margined or winged; those of the ray inane or sterile: chaffy bracts of the convex or conical receptacle either strongly concave or conduplicate and embracing the akenes: leaves either opposite or alternate.

101. **GYMNOLOMIA.** Pappus none or a minute denticulate ring; the truncate apex of the short akenes commonly at length covered by the base of the corolla, the tube of which is usually pubescent.

102. **VIGUIERA.** Pappus of two chaffy awns or paleæ, one to each principal angle of the akene, or occasionally one or two more, and of two or more intermediate shorter commonly truncate paleæ or squamellæ on each side, either persistent or deciduous. Akenes commonly pubescent. Peduncles slender.

103. **TITHONIA.** Pappus of *Viguiera* or more persistent: habit of the annual species of *Helianthus:* involucre somewhat peculiar, of about two series of bracts, with appressed and rigid usually striate base and loose foliaceous tip. Peduncles clavate and fistular under the head.

104. **HELIANTHUS.** Pappus promptly deciduous, of two scarious and pointed or somewhat awned paleæ, mostly no intermediate squamellæ or paleæ, except sometimes as detached or partly united portions of the principal paleæ. Akenes usually glabrous or glabrate. Proper tube of disk-corollas short, and the throat cylindrical and elongated.

== == == Receptacle flat, convex, or sometimes becoming conical: akenes (of the ray or margin often triquetrous) of the disk either flat-compressed and margined or thin-edged, or if turgid some of them winged: pappus not caducous.

a. Truly shrubby, rayless, alternate-leaved: akenes wingless.

105. **FLOURENSIA.** Rays none in the Mexican (several and neutral in the Chilian) species. Involucre of 2 or 3 series of oblong or lanceolate bracts, at least the outer herba-

ceous or foliaceous. Receptacle flat; its chaffy bracts scarious-membranaceous, conduplicate around the akenes and tardily deciduous with them. Proper tube of the corolla fully half the length of the oblong-campanulate throat. Appendages of the style-branches from oblong to dilated-spatulate, obtuse. Akenes compressed, narrowly oblong-cuneate, callous-margined, very villous, bearing a nearly persistent pappus of a subulate somewhat chaffy awn from each angle of the truncate summit, and commonly some intermediate smaller ones or squamellæ.

 b. Herbaceous, or sometimes shrubby: leaves never decurrent on the stem: rays neutral, rarely wanting: mature akenes all wingless or nearly so, emarginate or truncate at summit, the margins either villous-ciliate or naked.

106. ENCELIA. Pappus none, or an awn or its rudiment answering to each margin of the wingless akene: no intermediate squamellæ.

107. HELIANTHELLA. Pappus of delicate squamellæ between the two chaffy teeth or awns which surmount the two acute margins of the akene (and sometimes the lateral angles when there are any), or these obsolete in age, but not caducous. Ovary often wing-margined, but mature akene not so.

 c. Herbaceous, or rarely suffruticose: rays fertile or sometimes neutral in *Verbesina*, or occasionally wanting: akenes or some of them developing winged margins, or sometimes all wingless, none villous-ciliate: style-appendages acute.

108. ZEXMENIA. Involucre campanulate or hemispherical, imbricated; the bracts commonly broad, erect, and dry, or the outermost sometimes loose and foliaceous or with spreading herbaceous tips. Rays fertile. Receptacle flat or convex. Akenes of the ray or outermost of the disk triquetrous; of the disk more or less compressed, sometimes flat, truncate at summit, variably and narrowly winged or acutely margined, awned from one or more of the margins or angles, the awns either connected by dilated bases or with intermediate and separate or confluent persistent squamellæ. Leaves opposite, rarely alternate.

109. VERBESINA. Involucre campanulate or hemispherical and more or less imbricated, rarely more spreading, from somewhat herbaceous to foliaceous. Rays fertile, or styliferous but infertile, or sometimes neutral, sometimes none. Receptacle from convex to conical: disk from convex to ovoid, not squarrose in fruit. Akenes usually winged and flat or much compressed, 2-awned, or in the ray triquetrous and 1-3-awned, with no intermediate squamellæ, and even the awns sometimes obsolete or wanting. Leaves opposite or alternate, apt to be decurrent as wings on the stem.

110. ACTINOMERIS. Involucre simple, of few and small herbaceous and loose bracts, deflexed under the globular fruiting disk, which is globose even in anthesis, and echinate-squarrose in fruit by the spreading of the akenes in all directions on the small and soon globular receptacle. Rays neutral, few and irregular or none. Akenes flat, obovate, winged or wingless in the same head. Pappus of 2 slender-subulate naked awns, at length divergent, sometimes with 2 or 3 intermediate awns or awn-like squamellæ.

Subtribe VI. COREOPSIDEÆ. Akenes obcompressed or sometimes terete, and the subtending chaffy bracts flat or hardly concave: otherwise as in *Verbesineæ*. Heads many-flowered. Leaves mostly opposite. Style-tips of the disk-flowers produced into a cusp or cone, or sometimes capitellate-truncate.

 * Involucre single: habit of the preceding group.

111. SYNEDRELLA. Heads with few or several fertile ray-flowers and more numerous disk-flowers; the latter with slender tube to the corolla. Involucre ovoid or oblong, of rather few bracts; the outer larger than the inner, erect, mostly foliaceous. Bracts of the receptacle scarious-membranaceous. Style-appendages of the disk-flowers slender. Akenes or some of them wing-margined, and the wings commonly lacerate or undulate, in the ray often triquetrous, the angles or wings surmounted each by a rigid naked awn. Annuals.

 * * Involucre double, rarely indistinctly so: receptacle flat or merely convex; the thin chaffy bracts of the receptacle mostly deciduous with the akenes. Base of style not rarely bulbous-dilated.

 +— Rays always neutral (rarely wanting): akenes never rostrate-attenuate nor with retrorsely barbed awns: no ring at the junction of tube and throat of disk-corolla.

112. COREOPSIS. Involucre of two distinct series of bracts, all commonly united at the very base; outer foliaceous, narrower, and usually spreading; inner erect or incurved after anthesis, more membranaceous, each series commonly 8 in number. Rays about 8, wanting in one or two species. Disk-corollas with slender tube and funnelform or campanulate 5-lobed or 5-toothed limb. Akenes flat, or becoming meniscoidal, orbicular to linear-oblong, winged or wingless, truncate or emarginate at summit, bearing 2, rarely 3 or 4 naked (or upwardly hispid) awns, or naked scales, or teeth, or sometimes wholly destitute of pappus.

++ ++ Rays fertile or neutral, or wanting: awns of the pappus when present retrorsely barbed or hispid.

++ Bracts of the involucre distinct, or united only at the common base.

113. BIDENS. Akenes neither winged nor beaked, 2–5-awned; the awns retrorsely hispid or aculeolate, mostly persistent. Rays neutral (in one Mexican species styliferous), yellow or white, sometimes wanting: no ring to the disk-corollas.

114. COSMOS. Akenes slender and beaked: rays purple or rose color, in one species orange-yellow: otherwise as *Bidens;* the awns apt to be deciduous.

115. HETEROSPERMUM. Akenes dimorphous; the outer with winged or callous margin, mostly cymbiform; inner narrower, attenuate upward, marginless; these and sometimes the outer with 2 retrorsely barbed awns. Rays fertile: no ring to the disk-corollas. Heads rather few-flowered.

116. LEPTOSYNE. Akenes oval or oblong, truncate or emarginate, some of them usually wing-margined or bordered. Rays pistillate and often fertile, occasionally neutral. Disk-corollas with slender tube girt at summit or near it by a bearded or naked ring, a dilated throat, and 5-lobed limb.

++ ++ Bracts of the inner involucre united into a cup.

117. THELESPERMA. Involucre of *Coreopsis;* but the bracts of the inner connate to or above the middle, fleshy below, their free summits more membranaceous and scarious-margined; outer of shorter and narrow somewhat foliaceous spreading bracts, connate at base with the inner. Chaffy bracts of the flat receptacle wholly white-scarious, with a 2-nerved midrib, otherwise nerveless, deciduous with the akenes. Rays about 8, neutral, cuneate-obovate, or in some species wanting. Disk-corollas with long and slender tube, abrupt campanulate or cylindrical throat, and linear to ovate spreading lobes. Anthers wholly exserted. Style-appendages tipped with a cusp or cone. Akenes slightly obcompressed or terete, narrowly oblong to linear, marginless, beakless, attached by a broad callus, at least the outer ones tuberculate, papillose, or rugose; the abrupt summit crowned with a pair of persistent and stout awns or rather scales, the margins of which are retrorsely hispid-ciliate, or sometimes pappus obsolete or wanting. Leaves opposite.

Subtribe VII. GALINSOGEÆ. Pappus pluripaleaceous, and akenes commonly turbinate and 5-angled: otherwise nearly as *Verbesineæ.* Receptacle chaffy throughout: otherwise as *Helenioideæ.* Ours all herbs, and leaves except in *Galinsoga* alternate and entire.

* Bracts (chaff) of the receptacle concreted, coriaceous or cartilaginous, persistent, forming deep alveoli, resembling honeycomb, in which the akenes are enclosed: rays neutral.

118. BALDWINIA. Heads many-flowered, conspicuously radiate. Involucre imbricated, shorter than the convex disk; its bracts small, coriaceous and partly herbaceous. Disk-corollas with a short soon indurated tube, above cylindraceous, 5-toothed; the teeth glandular-puberulent. Style-appendages truncate and penicillate, with a subulate tip. Akenes turbinate, silky-villous: pappus of 7 to 12 nerveless thin-scarious paleæ.

* * Bracts of receptacle distinct, linear or filiform, rigid: rays none: paleæ of the pappus thin-scarious, nerveless.

119. MARSHALLIA. Heads many-flowered, homogamous. Involucre one or two series of narrow and equal herbaceous bracts. Receptacle at length conical. Corollas with a filiform tube and the limb 5-parted into linear lobes. Style-branches truncate at apex. Akenes turbinate, 5-costate: paleæ of the pappus 5 or 6, ovate or lanceolate-deltoid, acute or acuminate, nearly entire and naked.

* * * Bracts of the receptacle distinct, chaffy-membranaceous or scarious, mostly deciduous with the fruit: rays fertile. 2–3-lobed: paleæ of the pappus firmer, with a thickish axis and fimbriate or barbellate margins, or sometimes wanting.

120. **GALINSOGA.** Heads small, with 4 or 5 short rays and rather numerous disk-flowers. Involucre broadly campanulate or hemispherical, of ovate and thin nearly equal bracts in two series. Receptacle conical. Disk-corollas short, 5-toothed: style-tips acute. Akenes turbinate, 4–5-angled. Pappus of several thickish oblong or obovate paleæ, with fimbriate-barbellate or almost plumose margins or summit, or wanting. Leaves opposite, serrate.

121. **BLEPHARIPAPPUS.** Heads with 3 to 6 exserted fertile rays, and 7 to 12 disk-flowers; the central of these commonly infertile. Bracts of the involucre linear-lanceolate, erect, nearly equal, in one or two series. Receptacle convex; the chaff thin or scarious and narrow. Rays 3-cleft: disk-corollas 5-cleft. Style of fertile disk-flowers filiform, 2-cleft at apex only, and the short branches merely truncate; of the central and infertile ones entire. Akenes turbinate, silky-villous. Pappus of rather numerous narrow linear or aristiform paleæ, with thickish axis, and hyaline margins which are mostly lacerate-fimbriate so as to appear pectinate-plumose, sometimes abortive or wanting.

Subtribe VIII. MADIEÆ. Ray-flowers ligulate and fertile (rarely wanting), each subtended by a bract of the mostly uniserial involucre which partly or completely encloses its akene: disk-flowers hermaphrodite, but some or all of them sterile (sometimes all fertile); their style-branches subulate and hispid. Bracts of the receptacle always present between ray- and disk-flowers, generally none to the central ones. Pappus none (or a mere rudiment or crown) to the ray-akenes, paleaceous or aristiform or else none to the disk-flowers. Pacific-American herbs, commonly glandular-viscid and heavy-scented: such in California called *Tarweeds*.

* Akenes laterally compressed, those of the ray particularly so, and enclosed in conduplicate-infolded laterally-compressed involucral bracts.

122. **MADIA.** Heads many–several-flowered. Involucre ovoid or oblong, few–many-angled by the salient narrow or carinate backs of the involucral bracts. Receptacle flat or convex, bearing a single series of bracts enclosing the disk-flowers as a kind of inner involucre, either separate or connate into a cup. Ray-flowers 1 to 20, with cuneate or oblong 3-lobed ligules; their akenes more or less oblique and with flat sides: disk-flowers with or without a pappus, either sterile or fertile.

* * Akenes of the ray from obovate or triangular with broad rounded back to clavate-oblong, more commonly obcompressed, never laterally compressed with narrow back,

+— Arcuate-incurved and obcompressed, completely invested by the whole of the conformed at length coriaceous involucral bracts.

123. **HEMIZONELLA.** Heads few-flowered; the ray-flowers only 4 or 5: disk-flowers solitary or rarely 2 to 4; both fertile and destitute of pappus. Involucre as in *Madia* § *Harpæcarpus*, but the 4 or 5 arcuate infolded bracts broad on the back and rather obcompressed; those of the receptacle 3 to 5 and connate into a cup. Ligules minute. Akenes glabrous or sparsely pilose, obovate or somewhat fusiform; of the disk straight but oblique. Leaves mostly opposite.

+— +— Ray akenes thick and short, turgid, partly enclosed by the lower part of the involucral bract.

124. **HEMIZONIA.** Heads many- or sometimes few-flowered: bracts of the involucre rounded on the back. Ray-akenes more or less oblique: those of the disk abortive or infertile, or in the later sections some or even most of them fertile, with or without pappus. Leaves mainly alternate.

+— +— +— Ray akenes mostly obcompressed, never laterally compressed, wholly enclosed in an obcompressed basal portion of the subtending involucral bract, the dilated margins of which are abruptly infolded.

125. **ACHYRACHÆNA.** Heads many-flowered: ray-flowers 6 to 10, with 3-cleft ligule much shorter than its filiform tube, little surpassing the disk: disk-corollas slender, 5-

toothed. Involucre oblong-campanulate, of lanceolate thin-herbaceous bracts, deciduous at maturity: bracts of the nearly flat receptacle similar but thinner, only between the disk and ray, distinct. Akenes all clavate, with attenuate base, symmetrical, 10-costate, the ribs or the alternate ones tuberculate-scabrous at maturity; those of the ray slightly obcompressed, rounded at apex and with slightly protuberant areola, not rarely an abortive pappus in the form of a minute denticulate crown; those of the disk chiefly fertile, the truncate apex bearing a large pappus of 10 elongated-oblong obtuse silvery-scarious paleæ, the 5 inner as long as the corolla and akene, the alternate outer ones shorter.

126. **LAGOPHYLLA.** Heads several-flowered: ray-flowers about 5, with 3-parted or deeply 3-cleft ligules: disk-flowers sterile, with 5-lobed corollas. Bracts of the involucre thin-herbaceous, deciduous with the enclosed akene: bracts of the small receptacle 5 to 12 between the ray and disk. Akenes of the disk slender, abortive, destitute of pappus or with some caducous bristles; of the ray obcompressed, oblong-obovate, smooth and glabrous, nearly straight, the areola not protuberant, rarely a saucer-shaped cup in place of pappus.

127. **LAYIA.** Heads many-flowered, broad: ray-flowers 8 to 20, with 3-lobed or toothed ligules: disk-flowers fertile, or the central sometimes infertile; their corollas cylindraceous-funnelform and 5-lobed. Bracts of the involucre flattened on the back below, with abruptly dilated thin margins infolded so as to enclose the ray-akene. Receptacle broad and flat, bearing a series of thin chaffy bracts between the ray- and disk-flowers, sometimes additional more scarious ones among the flowers. Akenes of the ray obcompressed, obovate-oblong or narrower, almost always smooth and glabrous, destitute of pappus (or rarely a crown or vestige), the terminal areola somewhat protuberant and disciform; those of the disk similar or more linear-cuneate, mostly pubescent, bearing a pappus of 5 to 20 bristles, awns, or paleæ, or rarely none.

TRIBE VI. **HELENIOIDEÆ.** Heads heterogamous and the ligulate ray-flowers mostly fertile, or homogamous; the disk-flowers hermaphrodite and fertile, rarely some infertile, with regular 4–5-toothed tubular corolla. Receptacle naked, i. e. destitute of bracts (*paleæ*), but rarely fimbrillate. Bracts of the involucre herbaceous or membranaceous, not scarious. Style-branches of hermaphrodite flowers with either truncate or appendiculate tips. Pappus paleaceous or aristiform, or sometimes plurisetose, but the bristles when capillary always more or less rigid. — A peculiarly American tribe, differing from the preceding in the total absence of receptacular bracts; some genera with setose pappus making transition to the *Senecionideæ;* others, with short pappus or none, to the *Anthemideæ.*

Subtribe I. JAUMIEÆ. Involucre of broad bracts imbricated in two or more series. Ligules not persistent. Akenes 5-angled or terete and several-nerved. Many-flowered heads in ours radiate, and the ray-flowers fertile. No oil-glands.

* Receptacle setose-fimbrillate, convex: pappus plurisetose.

128. **CLAPPIA.** Involucre hemispherical, of rather few oval and very obtuse somewhat striate coriaceous bracts, imbricated in 2 or 3 series. Rays 12 to 15, linear, 3-denticulate at apex. Disk-corollas with slender tube and campanulate 5-cleft limb. Style-branches conical-tipped. Akenes equalled by the very slender fimbrillæ of the receptacle, oblong-turbinate, terete, 8–10 nerved, hirtellous on the nerves. Pappus of 20 to 25 rigid and somewhat paleolate hispidulous-scabrous distinct bristles, broader toward the base, longer than the akene. Fruticulose, with alternate fleshy leaves.

* * Receptacle naked: pappus in ours none.

129. **JAUMEA.** Involucre campanulate, its bracts fleshy or membranaceous, the outer shorter. Corollas glabrous. Receptacle in ours conical. Style-branches papillose or hairy, truncate or short-conical at tip. Akenes 10-nerved: pappus in exotic species of narrow and pointed or awned strongly 1-nerved paleæ, in ours none.

130. **VENEGASIA.** Involucre very broad, of 2 or 3 series of roundish membranaceous erect bracts, some innermost narrower and scarious, and a series of outer and loose narrower herbaceous ones. Receptacle flat. Rays numerous, elongated, entire or 3-toothed at the narrow apex: tube of corollas glandular-bearded, especially at base. Style-branches very obtuse. Akenes many-nerved, destitute of pappus.

Subtribe II. RIDDELLIEÆ. Involucre of narrow equal erect bracts. Ligules persistent and becoming papery on the usually striate-nerved akenes. Herbage more or less white-woolly: no oil-glands.

      * Pappus paleaceous: rays very broad, few.

131. **RIDDELLIA.** Heads with 3 or 4 ray- and 5 to 12 disk-flowers, all fertile. Involucre cylindraceous-campanulate, of 4 to 10 linear-oblong coriaceous woolly bracts, and a few smaller scarious ones within, sometimes an additional narrow outer one. Receptacle small, flat. Ligules as broad as long, abruptly contracted at base into a short tube, truncate and 2-3-lobed, 5-7-nerved, the nerves uniting in pairs within the lobes. Disk-corollas elongated-cylindraceous, with very short proper tube, and short externally glandular-bearded teeth. Style-branches truncate-capitate. Akenes narrow, terete, obscurely striate or angled. Pappus of 4 to 6 hyaline nerveless and pointless paleæ.

      * * Pappus none: rays several or numerous: disk-flowers numerous.

132. **BAILEYA.** Involucre hemispherical, of numerous thin-herbaceous linear bracts in 2 or 3 series, very woolly on the back. Receptacle flat or barely convex. Ray-flowers 5 to 50; the ligules from round-oval to oblong-cuneate, 3-toothed at apex, 7-nerved, tapering into a narrow but not tubular base, becoming scarious-papery but thin, persistent on the truncate summit of the akene. Disk-flowers fertile; their corollas tubular-funnelform above the short proper tube, 5-toothed; the teeth glandular-bearded. Style-branches short, with truncate-capitate tips. Akenes oblong-linear or clavate, somewhat angled, pluricostate or striate; the truncate apex obscurely toothed by extension of the ribs, or in the ray callous-thickened.

133. **WHITNEYA.** Involucre campanulate, of 9 or 12 oblong or broadly lanceolate equal thin-herbaceous bracts, nearly in a single series, in fruit somewhat cymbiform-carinate near the base, not villous. Receptacle narrow-conical, villous. Ray-flowers 7 to 9; ligule elongated-oblong, minutely 3-toothed at apex, 10-16-nerved (the nerves also prominent on the short tube), becoming thin-papery, persistent. Disk-flowers numerous, infertile, the tubular-funnelform obtusely 5-toothed corollas persistent on the sterile akenes: style-branches linear, pubescent externally, with rather obtuse tips. Ray-akenes only maturing, oblong, slightly obcompressed, obtuse at both ends, lightly nerved.

Subtribe III. PERITYLEÆ. Involucre of equal and narrow erect bracts, in only one or two series. Ray-flowers female or none; the ligule deciduous: disk-corollas narrow, 4-toothed. Akenes flat, with only marginal callous nerves, usually much ciliate. Style-branches and their appendages slender. Receptacle flat or convex. Plants not floccose-tomentose, and with no oil-glands. (*Hulsea*, 154, might be sought here. *Eatonella*, 137, and *Crockeria*, 137ª, also have flat and ciliate akenes with strong marginal nerves.)

134. **LAPHAMIA.** Head several- to many-flowered. Bracts of the hemispherical involucre distinct, more or less overlapping. Style-tips setaceous subulate, hirsute. Margin of akenes naked or not much ciliate. Pappus none, or of one or two, or sometimes about 20 bristles. Suffruticulose perennials, or herbaceous from a thick woody base, mostly yellow-flowered.

135. **PERITYLE.** Head many-flowered. Involucre of preceding, or the bracts more carinate-concave and partly embracing outer akenes. Style-branches with either short (acute or obtuse) or slender hirsute tips. Akenes at maturity cartilaginous-margined, usually strongly ciliate. Pappus a squamellate or cupulate crown, and commonly a slender awn from one or both angles. Mostly annuals, white- or yellow-flowered.

136. **PERICOME.** Head many-flowered, homogamous. Involucre a strictly single series of numerous narrow bracts, which are lightly connate by their edges into a campanulate cup. Disk-corollas slender, with viscous-glandular tube nearly the length of the cylindrical throat, from which the anthers are much exserted. Style-tips filiform, rather obtuse. Akenes strongly villous-ciliate. Pappus a squamellate lacerate-ciliate crown, and sometimes a pair of short awns, one from each angle of the akene. Perennial, yellow-flowered, with long-acuminate leaves.

Subtribe IV. HELENIEÆ. (*Baericæ* & *Euheleniæ*, Benth. & Hook., excl. gen.) Involucre hardly at all imbricated; its bracts when broad nearly equal or in a single series. Ligules not persistent. Disk-flowers numerous except in *Schkuhria*, with 5 or rarely 4 teeth or lobes. Akenes few-nerved or angled, or more numerously striate-angled only when turbinate or pyriform. No oil-glands. (*Raillardella*, 190, might be sought here.)

* Anomalous: akenes (as in *Perityle*) flat-compressed, with no lateral nerves, the callous or nerved margins densely ciliate-fringed; rays fertile or none: disk-corolla with dilated limb: style-tips truncate-capitate, with or without a slight cusp.

137. **EATONELLA.** Involucre of 5 to 8 oval or oblong obtuse and distinct bracts. Receptacle hardly convex. Disk-corollas short. Akenes callous-margined, ciliate with dense very long villosity, outermost obcompressed. Aspect of *Eriophyllum*.

137a. **CROCKERIA.** Involucre and other characters of *Lasthenia* § *Hologymne*. Akenes obovate-oval, very densely fringed with clavate glandular hairs.— See p. 445.

* * Baeria type: receptacle conical, mostly high-conical and acute, beset after the akenes have fallen by projecting points (as if pedicels, on which they were inserted): bracts of the involucre herbaceous, in one or rarely two series and commonly broad, sometimes cupulate-connate: female flowers ligulate, or sometimes wanting: akenes narrow and from oblong (or in one *Monolopia* somewhat obovate) to linear, usually tapering to the base, few-nerved and angled or nerveless, not callous-margined: herbage not impressed-punctate nor resinous-atomiferous.

+— Involucre (almost always) gamophyllous and simple, hemispherical or campanulate: disk-corollas with rather slender tube and dilated throat or limb: anther-tips ovate or oblong: style-tips capitate-truncate or obtuse.

143. **ERIOPHYLLUM.** Akenes slender, and usually with a paleaceous pappus.

138. **MONOLOPIA.** Head conspicuously radiate, with broad ligules: inner disk-flowers often infertile. Receptacle high-conical. Involucre broad, of one or rarely two series of bracts, which are normally connate by their edges into a several-toothed or lobed hemispherical cup, but sometimes distinct even to the base. Lobes of disk-corollas somewhat bearded. Akenes obovate or obovate-oblong, quadrangular-compressed or the outer obcompressed-triangular, sometimes acute-margined, with small terminal areola, and no pappus. Floccose-tomentose and alternate-leaved annuals.

139. **LASTHENIA.** Head radiate, or discoid by diminution of the ligules: disk-flowers all fertile. Involucre a single series of bracts connate by their edges into a 5-15-toothed glabrous green cup. Disk-corollas 4–5-lobed. Akenes linear or narrowly oblong, compressed, slightly 2–3-nerved or nerveless, nearly marginless, scabro-puberulent or glabrous. Pappus of 5 to 10 firm and subulate-tipped paleæ, or none. Glabrous and smooth annuals, with opposite entire sessile leaves

+— +— Involucre of few or several distinct and thinnish herbaceous bracts in a single series, loose, open at maturity of fruit, not rarely deciduous: disk-corollas with slender tube which equals or exceeds in length the campanulate or cyathiform 5-lobed (rarely 4-lobed) limb: leaves all opposite, sometimes connate at their sessile bases.

140. **BURRIELIA.** Head few-flowered, discoid, the 1 to 3 female flowers with ligule wanting or shorter than the style. Involucre cylindraceous, of 3 or 4 narrowly oblong plane bracts. Receptacle slender-subulate. Style-tips short-ovate, rather obtuse. Akenes slender, fusiform-linear, flattish. Pappus of 2 to 4 long attenuate-subulate paleæ.

141. **BAERIA.** Head mostly many-flowered, radiate: rays 5 to 15, conspicuous. Bracts of the campanulate or hemispherical involucre as many, ovate or oblong, plane or becoming somewhat carinate at middle, at least below. Receptacle subulate to high-conical. Style-tips from truncate-capitate, with or without a central apiculation, to ovate and sometimes with a cuspidate appendage. Akenes clavate-linear to linear-cuneate. Pappus a few paleæ or paleaceous awns, or both, often wanting.

* * * Bahia type: receptacle flat or convex (rarely obtusely conical): akenes from linear to obpyramidal, rarely 5-angled, occasionally with intermediate nerves: flowers (with few exceptions) all fertile.

COMPOSITÆ. 73

+ Involucre many-flowered, from hemispherical to cylindraceous; the bracts strictly erect, not membranaceous, persistent, from oblong to oval, more or less carinate-concave in fruit and partly receiving the subtended akene: herbage mostly floccose-woolly, not impressed-punctate nor resinous-atomiferous: leaves alternate or opposite.

142. **SYNTRICHOPAPPUS.** Involucre narrow, of about 5 equal and oblong carinate-concave thinnish-herbaceous bracts, which are partly wrapped around the ray-akenes. Receptacle flat. Ray-flowers about 5, with oval ligules 3-lobed or toothed at summit. Disk-corollas with very short proper tube, and elongated funnelform or cylindraceous throat, the stamens therefore inserted near the base; lobes 5, ovate-oblong. Anther-tips slender, long-lanceolate or linear. Style-tips elongated-lanceolate, acute, flattened, of the Asteroid type. Akenes linear-turbinate, 5-costate or angled, hairy. Pappus of numerous barbellulate white bristles in a single series, rather shorter than the disk-corollas, paleaceously and somewhat unequally united into a ring at base, deciduous, or in the second species wanting. Low and branching annuals, short-peduncled.

143. **ERIOPHYLLUM.** Involucre from hemispherical-campanulate to oblong, commonly equalling the disk, of one or sometimes two series of oblong or narrower firm-herbaceous or coriaceous permanently erect bracts, either distinct or sometimes partially united into a cup, at least in fruit concave or concave-carinate at centre, into which concavity the subtended akenes are partially received. Receptacle from convex or rarely conical to plane. Ray-flowers usually with broad ligules, very rarely none. Disk-corollas with distinct and sometimes slender proper tube. Style-tips truncate, obtuse, or obscurely capitellate-conical. Akenes narrow, from clavate-linear to cuneate-oblong, mostly 4-angled. Pappus of nerveless and mostly pointless (rarely awned or setiform) paleæ. Floccose-tomentose or rarely glabrate herbs, rarely suffruticose.

+ + Involucre many- (at least 12-20-) flowered; the bracts wholly herbaceous, not colored nor scarious-tipped, broad or broadish, plane or merely concave, equal and in a single or hardly double series, not embracing akenes: receptacle small; corolla-lobes or teeth short: herbage destitute of impressed punctures and resinous atoms, not floccose-lanate.

144. **BAHIA.** Involucre hemispherical or obovate and lax or open in fruit; the plane bracts distinct to and commonly narrower at the base. Receptacle small, mostly flat. Female flowers with exserted ligules, or rarely none. Style-tips truncate or obtuse. Akenes narrow, quadrangular. Pappus (rarely wanting) of several scarious paleæ, with callous-thickened opaque base, which is sometimes extended into a strong midnerve (costa).

145. **AMBLYOPAPPUS.** Characters of *Bahia*; but involucre of only 5 to 6 broadly obovate bracts, their centre in age more or less carinate-concave; small receptacle conical; head discoid; corollas all short-tubular, and in the few female flowers minutely 2-3-toothed, shorter than the style, in the hermaphrodite flowers 5-toothed, the teeth soon connivent. Akenes elongated-obpyramidal, pubescent. Pappus of 8 to 12 oblong obtuse rather firm paleæ, with merely thickened base and no costa, nearly equalling the corollas.

+ + + Involucre 3-9-flowered; its bracts few, equal, broad and with roundish more or less scarious-petaloid summit, concave-carinate: corollas only 5-toothed: herbage minutely impressed-punctate and resinous-atomiferous.

146. **SCHKUHRIA.** Heads effusely paniculate. Involucre clavate-turbinate or obpyramidal, of 4 or 5 erect bracts and sometimes an accessory bractlet at base. Receptacle very small. Female flowers only one or two, with a short or sometimes obsolete ligule not exceeding the hermaphrodite flowers, or altogether wanting. Akenes obpyramidal-tetragonal, the faces not rarely 2-3-striate. Pappus of 8 scarious paleæ, the larger often equalling the short corolla, either nerveless with callous-thickened base, or with a prominent costa. Leaves or their divisions filiform.

+ + + + Involucre many- (rarely 12-15-) flowered; its bracts mostly appressed, with scarious-membranaceous and usually colored tips and sometimes margins: disk-corollas deeply 5-cleft: anthers partly or wholly exserted: leaves alternate, not impressed-punctate except in *Hymenopappus*: receptacle small and flat: heads except in two species homogamous: flowers seldom yellow, but sometimes so.

147. **HYMENOTHRIX.** Involucre turbinate-campanulate, or in age more open, about 30-flowered, shorter than the disk; its principal bracts 7 to 10, obovate or lanceolate-oblong.

thin, half or more scarious-petaloid, plane; commonly one or more accessory outer bracts. Ray-flowers 6 to 10 and with oblong exserted ligule 3-cleft at the apex, or none. Disk-corollas with narrow tube and lobes, one or two of the sinuses a little deeper than the others. Style-branches flattish, with subcapitate tips, with or without a central cusp. Akenes 4-5-angled, tapering from broad summit to attenuate base. Pappus about the length of the akene, of 12 to 20 narrow lanceolate hyaline paleæ, traversed by a strong costa which is excurrent into a scabrous awn.

148. **HYMENOPAPPUS.** Involucre broadly campanulate; its bracts 6 to 12, equal, obovate to broadly oblong, thin, the rounded summit and usually the margins scarious-colored or petaloid. Ray-flowers none. Disk-flowers numerous, all alike. Corolla with narrow tube, abruptly dilated throat, and ovate reflexed or widely spreading lobes. Style-branches with short and thick conical appendages. Akenes obpyramidal, 4-5-angled, with attenuate base, the faces 1-3-nerved; the nerves at maturity sometimes as prominent as the angles, except in one species. Pappus of 10 to 20 thin-scarious and mostly hyaline obtuse paleæ, with or without a costa or central opacity, sometimes very short and small or quite obsolete.

149. **FLORESTINA.** Involucre turbinate-campanulate, 15-25-flowered; its bracts 6 to 8 in a single series, equal, obovate-spatulate, thin-herbaceous, with scarious-colored (whitish or purplish) rounded tips. Ray-flowers none. Disk-flowers with corolla widely dilated above the short narrow tube, deeply 5-cleft into oblong spreading lobes. Style-branches terminated by a rather long attenuate-subulate hispid appendage. Akenes narrowly obpyramidal, 4-5-angled, pubescent. Pappus of 6 to 8 obovate pointless paleæ, hyaline-scarious from a callous thickened narrow base or axis.

150. **POLYPTERIS.** Involucre from broadly campanulate to turbinate; its bracts from spatulate to linear-lanceolate, commonly in two series and equal, rarely with some accessory shorter ones, the tips or (in the original species) a larger portion membranaceous and colored or petaloid. Rays in one species distinctly evolute into a palmate ligule and fertile; in the others wanting. Corolla of the disk-flowers with filiform tube abruptly dilated into a 5-parted limb, the long lobes lorate-linear. Stamens wholly exserted. Style-branches filiform, wholly hispidulous, acutish or barely obtuse. Akenes from linear and downwardly attenuate to clavate-obpyramidal, 4-sided, only minutely pubescent. Pappus of 6 to 12 equal paleæ, with a strong percurrent costa, otherwise hyaline-scarious, rarely abortive or wanting; in the outermost flowers usually shorter.

+ + + + + Involucre many- (or 12-30-) flowered; its bracts linear (rarely broader), erect, equal and similar in a single or hardly in two series, herbaceous to the tip, inclined to embrace subtended akenes: receptacle flat, mostly small: akenes slender, linear-tetragonal or more compressed, merely pubescent: head discoid (rarely an inconspicuous ligule): corollas with short lobes or teeth and long throat: leaves alternate.

++ Leaves simple, entire: flowers never yellow.

151. **PALAFOXIA.** Heads homogamous and flowers all alike, except in the pappus. Involucre oblong or campanulate. Corolla with tube and narrow lobes shorter than the cylindraceous throat. Style-branches elongated, filiform and obtuse or obscurely thickened toward the summit, puberulent for the whole length (altogether of the Eupatoriaceous type, but the stigmatic lines traceable nearly to the apex). Pappus of 4 to 8 usually unequal paleæ, with strong costa.

152. **RIGIOPAPPUS.** Heads heterogamous, inconspicuously radiate. Involucre turbinate-campanulate, of numerous narrowly linear rather rigid herbaceous bracts, which are somewhat involute at maturity. Ray-flowers 5 to 15; the corolla with slender tube, and oblong entire or 2-toothed ligule, not surpassing the disk. Disk-flowers more numerous; corolla small, with short proper tube, elongated narrow throat, and 3 to 5 short erect teeth. Anthers included. Style-branches with short and linear glabrous stigmatic portion, and a larger slender-subulate hispidulous appendage. Pappus nearly similar in disk and ray, of 3 to 5 rigid and wholly opaque paleaceous naked awns (smooth, flat, gradually tapering from base to apex), rarely obsolete.

++ ++ Leaves mostly cleft or compound: flowers in some species yellow.

153. **CHÆNACTIS.** Heads homogamous and tubuliflorous; but the marginal flowers commonly with ampliate limb to the corolla. Involucre campanulate or hemispherical.

Receptacle flat, naked, in one species bearing a few setiform bracts or fimbrillæ among the flowers. Corollas with short tube, long and narrow throat, and short teeth, or in the marginal flowers of some species with larger lobes or even imperfect palmate ligules, forming a kind of ray. Anthers usually partly exserted. Style-branches pubescent nearly throughout, slender, filiform or with attenuate-subulate tips. Pappus of hyaline nerveless paleæ (or rarely with the vestige of a costa), in one species wanting.

+ + + + + + Involucre many-flowered, hemispherical; its bracts in 2 or 3 series, thin-herbaceous, rather loose, sometimes unequal, from linear to oblong, plane: receptacle flat, corneous-scrobiculate: disk-corollas with long and narrow throat and 5 short lobes or teeth: style-branches with short and thickened obtuse tips: akenes linear-clavate or cuneate-oblong, villous: pappus of 4 or 5 wholly hyaline paleæ; these erose or lacerate at summit, or dissected into capillary bristles: leaves mostly alternate, woolly or glabrate.

154. **HULSEA.** Bracts of the involucre linear or lanceolate. Ray-flowers numerous (10 to 60) and ligulate, but sometimes short and inconspicuous. Disk-corollas with proper tube slender or narrow, but shorter than the cylindraceous throat. Akenes linear-cuneate, compressed or somewhat tetragonal, soft-villous, especially the margins. Pappus of mostly 4 truncate paleæ, from erose or lacerate at summit to nearly entire.

155. **TRICHOPTILIUM.** Bracts of the involucre about 20, equal; those of the outer series ovate-lanceolate; those of the inner narrowly spatulate or lanceolate and membranaceous. Ray-flowers none. Disk-flowers 30 to 40: the corollas with very short tube, cylindraceous-funnelform throat, and 5 short ovate lobes, those of the marginal flowers slightly enlarged after the manner of *Chænactis*, but regular, the nerves deeply intramarginal. Anther-tips oblong-lanceolate. Style-branches linear, glabrous, and with stigmatic lines continued up to the obtuse tip. Akenes oblong-turbinate, 5-nerved or angled, hirsute-villous. Pappus of 5 ovate or oblong hyaline nerveless paleæ, which are resolved above into numerous slender bristles, the middle ones rather shorter than the corolla.

* * * * Receptacle flattish or convex, many-flowered: ray-flowers female and fertile; those of the disk sterile: involucral bracts few in a single series, broad and plane, membranaceous: akenes pyriform.

156. **BLENNOSPERMA.** Involucre hemispherical or depressed: its bracts 5 to 12, equal, oblong, plane, herbaceous or partly membranaceous, the tips sometimes colored, the bases somewhat united. Ray-flowers 5 to 12: some of them in our species not rarely apetalous, the others with ligule oblong or elliptical, entire, sessile on the ovary, being destitute of tube: style-branches flat, linear or oblong. Disk-flowers numerous (20 or more): corollas with narrow tube, abruptly expanded into a broadly campanulate 4–5-lobed limb: anthers oval: style undivided, with capitate or disk-shaped apex: ovary abortive, a mere rudiment. Akenes (of the ray) obscurely 8–10-ribbed, with small areola, wholly destitute of pappus; the surface powdered with papillæ which develop mucilage when wet.

* * * * * Receptacle from convex to oblong: involucre many-flowered, various, of more than one series of bracts, or irregular: akenes short, obpyramidal or turbinate, sometimes more oblong, 5–10-costate or angled, mostly silky-villous or hirsute: disk-flowers all fertile; the corolla 4–5-toothed: leaves alternate, in many minutely impressed-punctate or resinous-atomiferous.

+ Receptacle naked, i. e. destitute of awn-like fimbrillæ among the flowers: style-branches of the disk-flowers dilated-truncate and somewhat penicillate at tip.

++ Involucre erect, at least not spreading or reflexed.

148. **HYMENOPAPPUS**, with turbinate or obpyramidal costate akenes, might be sought here.

64. **PLUMMERA** is like *Actinella* § *Hymenoxys*, without pappus, and disk-flowers sterile.

157. **ACTINELLA.** Heads radiate (except in S. American species). Involucre campanulate or hemispherical, or sometimes broader; its bracts in two or more series, somewhat herbaceous or coriaceous, often rigid; outer sometimes united. Receptacle from conical to convex. Rays fertile. Pappus of 5 to 12 thin and mostly hyaline paleæ, with more or less manifest costa or none; these sometimes truncate, more commonly acuminate or aristate at tip. Mostly low herbs, and bitter-aromatic.

++ ++ Involucre spreading or soon reflexed, herbaceous, usually with some inconspicuous short scarious interior bracts: akenes turbinate, 8–10-costate: heads mostly radiate: receptacle more or less elevated.

158. **HELENIUM.** Bracts of the involucre subulate or linear. Rays fertile or sterile, rarely none. Disk-corollas commonly with short or almost obsolete proper tube (the stamens inserted close to the base), and 4–5-toothed limb; the teeth obtuse, glandular-pubescent. Pappus of usually 5 or 6 thin scarious paleæ. Leaves commonly impressed-punctate, mostly decurrent.

159. **AMBLYOLEPIS.** Principal bracts of the involucre foliaceous, lanceolate; an inner hyaline-scarious series resembling the conspicuous blunt nerveless paleæ of the pappus. Rays fertile, ample. Disk-corollas glabrous throughout, and with a distinct tube as long as the ampliate throat, 5-cleft; the lobes lanceolate, attenuate-acute. Akenes broadly turbinate and with 10 thick ribs. Leaves neither punctate nor decurrent.

+– +– Receptacle (from convex to globular) beset with setiform or subulate or rarely small dentiform fimbrillæ among the flowers.

160. **GAILLARDIA.** Involucre broad; the bracts in 2 or 3 series, all but the short inner series largely foliaceous or herbaceous and lax. Ray-flowers neutral, rarely styliferous and fertile, sometimes none: ligules 3-toothed or 3-cleft. Disk-corollas with short narrow tube, enlarged cylindraceous throat, and 5 ovate-triangular to subulate teeth or short lobes, which are beset with jointed hairs. Style-branches with penicillate tuft at summit of the stigmatic portion, thence produced into a filiform or shorter appendage. Akenes turbinate, 5-costate, covered with long villous hairs which sometimes rise only from the base of the akene. Pappus conspicuous, longer than the akene, of 5 to 10 hyaline-scarious paleæ, with a costa mostly excurrent into an awn, which about equals disk-corollas.

Subtribe V. **FLAVERIEÆ.** Involucre of the small heads composed of a few equal connivent bracts in a single series, sometimes one or two small additional ones at base. Ligules small (little or not at all surpassing disk-flowers), not persistent. Akenes terete, oblong or linear, 8–10-striate-costate. Style-branches truncate. Leaves opposite. No oil-glands, nor resinous atoms.

161. **SARTWELLIA.** Heads with about 5 ligulate female and rather numerous hermaphrodite tubular flowers. Bracts of the involucre 5, oval or oblong, somewhat fleshy, in fruit somewhat carinate-concave and subtending ray-akenes. Receptacle convex. Ligules mostly entire, obovate or roundish. Disk-corollas narrow, 4–5-toothed. Pappus a deep paleaceous cupule with minutely fimbriolate edge (doubtless composed of 4 or 5 truncate paleæ which are completely connate), or of 4 or 5 narrowly oblong fimbriolate-truncate nerveless paleæ alternating with as many setiform awns, all united only at the base.

162. **FLAVERIA.** Heads one-several-flowered; the flowers all fertile, homogamous and tubular, or one female and short-ligulate. Disk-corollas 5-toothed. Involucre of 2 to 5 mostly carinate-concave bracts. Pappus none.

Subtribe VI. **TAGETINEÆ.** Involucre a series of equal bracts, either distinct or united into a cup or tube, dotted or striped with oil-glands, not rarely subtended or calyculate by some loose accessory bracts, several-many-flowered. Rays when present fertile; ligules not persistent. Akenes mostly narrow and striate. Pappus various. — Mostly glabrous and smooth herbs or undershrubs, strong-scented, the herbage like the involucre commonly dotted with some oil-glands.

* TRUE TAGETINEÆ. Style-branches of hermaphrodite flowers more or less elongated, appendiculate or truncate.

+– Pappus simple, of copious capillary scabrous bristles: akenes linear: receptacle small, naked and smooth: bracts of the involucre distinct.

163. **POROPHYLLUM.** Ray-flowers none. Disk-flowers numerous or several. Involucre of 5 to 10 bracts. Style-branches tipped with long filiform-subulate hispid appendages. Akenes slender.

164. CHRYSACTINIA. Ray-flowers conspicuous, with linear ligules. Disk-flowers numerous; their corolla narrow and 5-toothed, and style-branches tipped with short obtuse or conical appendages. Involucre of 10 or more short bracts. Akenes short-linear, not attenuate upward. Flowers all yellow.

— + Pappus of distinct bristles and distinct paleæ; bracts of the many-flowered involucre distinct.

165. NICOLLETIA. Involucre oblong or cylindraceous, of 8 to 12 thinnish bracts, nearly naked at base. Receptacle quite naked. Disk-corollas narrow-tubular, 5-toothed. Style-branches tipped with long filiform-subulate appendages. Akenes filiform-linear, with tapering base. Pappus double; outer of indefinitely numerous capillary bristles like those of *Porophyllum*; inner of 5 lanceolate long hyaline paleæ, with costa excurrent into a scabrous awn.

— + + Pappus either wholly paleaceous, or some or all of the paleæ bearing or largely resolved into awns or capillary bristles; bracts of the involucre gamophyllous or sometimes distinct; receptacle variously fimbrillate, alveolate-dentate, or more strictly naked.

166. DYSODIA. Pappus multi-setose-polyadelphous, i. e. all or most of the 10 or more paleæ resolved, except a basal portion, into several (9 or more) or indefinitely numerous capillary but rather stiff bristles. Involucre hemispherical or campanulate, usually calyculate with a series of loose accessory bracts, the proper bracts generally gamophyllous at base, rarely quite separate, rarely united to near the summit. Style-appendages sometimes slender, sometimes an abrupt apiculation or short obtuse cone.

167. HYMENATHERUM. Pappus of several or numerous paleæ, either 1–5 aristate or pointed, or partly resolved into as many bristles, or some or all of them entire and even truncate (rarely even concreted). Involucre campanulate, copulately gamophyllous high up, with or without some loose accessory bracts. Style-branches truncate or very obtuse, sometimes tipped with a minute apiculation. Akenes mostly terete, and striate.

168. TAGETES. Paleæ of the pappus 3 to 6, firm, commonly unequal, entire, not setiferous, but one or more of them frequently subulate-pointed or aristiform. Involucre naked at base, gamophyllous nearly throughout into an oblong or more elongated cup or tube. Akenes compressed or angulate, hardly striate. Herbs.

* * PECTIDEÆ. Style of hermaphrodite flowers slender, hispidulous, terminated by two very short obtuse and inappendiculate stigmatic branches.

169. PECTIS. Heads radiate, several-many-flowered. Involucre naked at base, or nearly so, cylindrical or campanulate, of few or several equal carinate bracts in a single series. Receptacle small, naked. Disk-corollas 5-lobed, one or two sinuses often deeper, thus becoming bilabiate. Akenes linear, terete or angled. Pappus of few or numerous bristles or awns, sometimes paleaceous-dilated at base, or of paleæ, or reduced to paleaceous-coroniform, rarely obsolete. Opposite-leaved herbs.

TRIBE VII. **ANTHEMIDEÆ.** Heads homogamous with flowers all tubular and hermaphrodite, or more commonly heterogamous, with the female flowers ligulate and radiate, or sometimes with corolla reduced to a tube or obsolete. Receptacle either naked or with some chaffy bracts. Bracts of the involucre imbricated, wholly or partly dry and scarious or scale-like, not foliaceous, seldom herbaceous. Anthers without tails at base. Style-branches of the hermaphrodite flowers truncate, and sometimes with obscure conical tips. Akenes usually small and short, with no pappus or a paleaceous crown, or a circle of squamellæ. — Strong-scented or bitter-aromatic herbs or undershrubs, the greater part of the Old World; with alternate leaves; distinguished from the preceding tribe by the scarious imbricated involucre; from the *Asteroideæ*, by the truncate style-tips, &c. The first genus would go with *Helenioideæ*, except for the paleæ of the receptacle.

* Receptacle paleaceous, i. e. with chaffy bracts subtending some or all the disk-flowers; heads radiate, or the rays wanting in certain species.

— Anomalous, with involucre (of comparatively few and broad thin bracts and aspect of *Hymenopappus*.

170. LEUCAMPYX. Involucre broadly hemispherical; its bracts broadly oval, equal, in 2 or 3 series of 4 or 5 each, membranaceous, their margins white-scarious. Receptacle somewhat convex, with oblong-lanceolate wholly scarious bracts subtending disk-flowers and partly folded round the akenes. Ray-flowers 8 or 10, fertile; ligule cuneate-obovate, ample, on a slender glandular tube, somewhat persistent on the akene. Disk-flowers numerous: corolla with narrow tube, ampliate-campanulate throat, and 5 spreading lobes: style-branches linear, with an obscure obtuse tip slightly produced beyond the stigmatic portion. Akenes large for the tribe, obovate-trigonous, with narrowed base and rounded summit, lightly 5-nerved, glabrous, slightly incurved. Pappus an obscure squamellate crown, soon obsolete.

+ + Involucre of comparatively small imbricated bracts, the outer successively shorter: receptacle convex to oblong: style-branches truncate-penicillate.

171. ANTHEMIS. Involucre hemispherical, many-flowered. Chaffy bracts of receptacle sometimes hyaline, sometimes aristiform. Akenes terete or 4-10-angled or ribbed, not flattened, glabrous; the truncate summit naked, or with a very short coroniform or auriculate pappus. Heads comparatively large.

172. ACHILLEA. Involucre campanulate or obovate. Chaffy bracts of the receptacle membranaceous, like the innermost bracts of the involucre. Rays few or several, short and broad. Akenes oblong or obovate, obcompressed, callous-margined, glabrous, destitute of pappus.

* * Receptacle naked, i. e. destitute of bracts or chaff among the flowers.

+ Heads comparatively large, radiate, or rarely discoid and homogamous by the absence of ligulate female flowers, pedunculate, solitary at the summit of the branches, or sometimes corymbosely cymose, never racemosely paniculate: akenes glabrous: tube of disk-corolla either terete or ancipital.

173. MATRICARIA. Receptacle conical or ovoid, or rarely lower when young. Akenes 3-5-ribbed or nerved on the face or sides, rounded on the back.

174. CHRYSANTHEMUM. Receptacle from flat to hemispherical. Akenes (at least of the disk) 5-10-ribbed or nerved all round; of the ray in certain species triquetrous.

+ + Heads sessile, discoid, heterogamous: female flowers most numerous, apetalous; their akenes pointed or armed with indurated persistent style.

175. SOLIVA. Heads many-flowered, largely of female flowers: a few hermaphrodite but mostly sterile ones in the centre; these with a short and thick 2-6-toothed corolla and usually undivided style. Involucre of 5 to 12 nearly equal bracts in not more than 2 series. Receptacle flat. Akenes obcompressed, with rigid wings or callous margins, which are commonly spinulose-pointed at summit, and the apex armed by the spiniform persistent style. Pappus none.

+ + + Heads slender-peduncled, discoid, heterogamous: female flowers apetalous: style deciduous.

176. COTULA. Heads many-flowered: female flowers in one or two rows: disk-flowers with 4-toothed corolla, fertile or infertile. Bracts of the involucre greenish, in about 2 ranks. Akenes raised on pedicels at maturity (these remaining on the flat or convex receptacle), obcompressed, commonly thick-margined or narrowly winged, in our species destitute of pappus or nearly so.

+ + + + Heads discoid, heterogamous, and the few or uniserial female flowers with a tubular 2-3-toothed or lobed corolla (in one species imperfectly radiate), or sometimes homogamous, the female flowers wanting and the hermaphrodite rather few: style deciduous: akenes truncate or obtuse: receptacle quite naked or sometimes hirsute: involucre imbricated in few or several ranks.

177. TANACETUM. Heads corymbosely cymose or glomerate, rarely solitary, many-flowered: female flowers with tubular 3-5-toothed corolla, either equal or oblique or imperfectly ligulate. Akenes 5-ribbed or 3-5-angular, with broad truncate summit, bearing a coroniform pappus or none. Anther-tips broad and mostly obtuse.

178. ARTEMISIA. Heads paniculately disposed, few-many-flowered, small, wholly discoid, heterogamous, the female flowers with small and slender tubular corolla, and the hermaphrodite either sterile or fertile; or homogamous, with the flowers all hermaphrodite and fertile. Anther-tips slender and pointed. Akenes obovate or oblong, mostly with small epigynous disk or summit, and no pappus.

TRIBE VIII. SENECIONIDEÆ. Heads heterogamous or homogamous. Involucre mostly one or two series of equal (and not scarious) bracts, sometimes unequal or even imbricated, with or without loose and short accessory ones at base. Receptacle naked. Anthers without tails at base, but not rarely sagittate. Style-branches of hermaphrodite flowers most commonly truncate or obtuse, tipped with short appendages or none. Pappus of numerous capillary bristles, sometimes caducous. Leaves usually alternate. (Copious capillary pappus, comparatively simple involucre, short or conical if any style-tips, tailless anthers, and naked receptacle, are the marks of this tribe, no account being here taken of the tropical American subtribe *Liabeæ*.)

* Style-branches of hermaphrodite fertile flowers roundish-obtuse, or at least not truncate, and wholly without appendage or hispidity at summit, simulating *Inuloideæ* or *Eupatoriaceæ*: pappus-bristles merely denticulate: receptacle naked, flat. — Subtribe *Tussilagineæ*, Benth. & Hook.

+– Heads submonœcious or subdiœcious; the hermaphrodite flowers (with rather deeply 5-cleft corolla) essentially sterile: akenes narrow, 5-10-costate, with elongating soft and white pappus: involucre a series of soft herbaceous bracts, with few or no loose accessory ones at base. — True *Tussilagineæ*.

179. TUSSILAGO. Head solitary, yellow-flowered, monœcious: female flowers in several series in the ray, slenderly ligulate: numerous subhermaphrodite flowers in the centre, with undivided style and sterile ovary.

180. PETASITES. Heads racemosely or corymbosely disposed, white- or purplish-flowered, subdiœcious: heads in the truly fertile plant wholly or chiefly of female flowers, with slender-tubular and irregularly 2-5-toothed or distinctly ligulate corolla; in the substerile with few of these in the margin, and numerous hermaphrodite-infertile flowers, like those of *Tussilago*, but their style commonly with 2-cleft or 2-toothed apex.

+– +– Heads homogamous, discoid, of wholly hermaphrodite and fertile flowers: style-branches very minutely granular-puberulent.

++ Corollas yellow, rather deeply 5-cleft, the lobes lanceolate: anthers much exserted and with lanceolate tips: akenes linear, glabrous: involucre hardly herbaceous, simple, of carinately one-nerved narrow bracts, and with few and small or no accessory bracts.

181. CACALIOPSIS. Heads very many-flowered. Involucre broadly campanulate, of 14 to 30 lanceolate-linear mostly acuminate bracts. Corolla with the cylindraceous throat rather longer than the slender tube. Anthers entire at base. Style puberulent for some distance below the slightly flattish branches. Akenes 10-striate. Pappus very copious, soft and white, equalling the corolla. Leaves palmately lobed, petioled.

182. LUINA. Heads about 10-flowered. Involucre oblong-campanulate, of 8 to 10 linear bracts. Corolla of the preceding, or the throat more ampliate. Anthers sagittate at base. Style glabrous, its flattened and linear branches obscurely papillose on the back, truncately obtuse. Akenes (immature) obscurely 10-striate. Pappus of the preceding, but less copious. Leaves entire, veiny, sessile.

++ ++ Corollas yellowish, obtusely 5-toothed: anthers little exserted, with oval obtuse tips: involucre mostly foliaceous!

183. PEUCEPHYLLUM. Heads 12-25-flowered. Involucre campanulate, of numerous subulate-linear or almost filiform nerveless bracts which resemble the leaves, in about 2 series, some of the outer looser and similar to the uppermost leaves. Corolla with very short proper tube and long cylindrical throat; the 5 teeth short, ovate, obtuse, erect, obscurely puberulent. Anthers minutely sagittate at base. Style-branches linear, flattish or semiterete, obscurely papillose-puberulent, the very obtuse tip wholly destitute of appendage. Akenes turbinate-oblong, obscurely 10-striate, very hirsute. Pappus shorter than the corolla, of very numerous and unequal rather sordid and roughish capillary bristles. Leaves short-filiform, crowded.

* * Style-branches of hermaphrodite flowers either truncate or capitellate at tip, which is either naked or penicillate or hirsute, and not rarely bearing a short conical or flattened appendage. — Subtribe *Eusenecioneæ*, Benth. & Hook.

+ Involucre lax (not erect-connivent), commonly of much overlapping or unequal bracts, 10–many-flowered.

++ Herbs, with alternate well-developed leaves and many-flowered heads.

184. PSATHYROTES. Heads homogamous; the flowers all hermaphrodite and fertile. Involucre of somewhat numerous bracts in two series, at least the outer more or less herbaceous. Receptacle flat. Corollas with extremely short proper tube (the filaments therefore inserted near the base), elongated cylindrical throat, and 5 very short obtuse teeth. Style-branches flattish, very obtuse or truncate, and with obscure appendage if any. Akenes terete, more or less turbinate, obscurely striate, villous or hirsute. Pappus copious, shorter than the corolla, of very unequal rather rigid obscurely denticulate bristles, at least in age fuscous or ferrugineous.

185. BARTLETTIA. Heads heterogamous, radiate: flowers all fertile. Involucre broadly campanulate, of 12 to 14 oblong-lanceolate bracts in 2 or 3 series, rather lax; the inner and larger membranaceous, 2 or 3 outermost short and more herbaceous. Receptacle convex, tuberculate. Corollas with long and slender pubescent tube; of the ray with narrowly oblong exserted ligule; of the disk with dilated-funnelform throat longer than the 5 ovate lobes. Anthers with ovate obtuse tips. Style-branches rather short, linear, flat, truncate, minutely hairy at the broad summit, and usually with a central setula. Akenes (at maturity) compressed, cuneate-oblong, with a strong salient nerve to each margin and usually on the middle of one face, these densely long-hirsute, the faces glabrate. Pappus equalling the disk-corolla; its numerous somewhat unequal bristles in a single series, rather rigid, barbellulate, fuscous.

186. CROCIDIUM. Heads heterogamous, radiate: flowers all fertile. Involucre hemispherical or more open, of 9 to 12 nearly equal and similar oblong-ovate or oblong-lanceolate thin-herbaceous bracts; no external calyculate ones. Receptacle conical. Ray-flowers about 12, with oval or oblong rather ample ligules: disk-corollas with slender tube rather longer than the campanulate throat; lobes 5, spreading. Anthers with deltoid-ovate acute tips. Style-branches short and broad, terminated by large deltoid appendages. Akenes fusiform-oblong, obscurely 3–5-costate, beset with hyaline oblong papillæ, which, detaching when wetted, throw out a pair of spiral threads, in the manner of *Senecio*, &c. Pappus a single series of equal white barbellate bristles, which are very deciduous, in the ray commonly wanting.

++ ++ Herb, with opposite leaves and many-flowered heads.

187. HAPLOESTHES. Heads heterogamous, many- (at least 20-) flowered, radiate: flowers all fertile. Involucre short-campanulate, of 4 or 5 nearly equal and similar rather fleshy orbicular or broadly oval bracts, the outer strongly overlapping the inner. Receptacle flat. Corollas with somewhat slender tube: ligules of the rather few and short ray-flowers oval: disk-corollas narrowish, deeply 5-toothed. Anther-tips lanceolate. Style-branches of *Senecio*. Akenes linear, terete, striate-costate, glabrous. Pappus a single series of rather rigid and scabrous whitish bristles, about equalling the disk-corolla.

++ ++ ++ Shrub, with alternate leaves reduced to scales, and 10–18-flowered heads with imbricated involucre.

188. LEPIDOSPARTUM. Heads homogamous. Involucre oblong-campanulate; its bracts scarious-chartaceous, regularly imbricated in 3 or 4 series, oblong, obtuse; the outer successively shorter; outermost ovate, passing into similar scaly bracts on the pedicel. Receptacle naked. Corolla with elongated tube, and lanceolate-linear spreading lobes, which much exceed the open campanulate throat. Anthers wholly exserted, slenderly and almost caudately sagittate at base, the tips lanceolate. Style-branches flattish, ending in short acutish pubescent tips. Akenes oblong, terete, obscurely 8–10-nerved, with large epigynous disk. Pappus very copious, of soft and whitish minutely scabrous capillary bristles.

+ + Involucre of 4 to 6 firm and concave close and strongly overlapping bracts, 4–9-flowered: shrubs, with alternate leaves.

189. TETRADYMIA. Heads homogamous. Involucre cylindrical to oblong, naked, i. e. no accessory bracts. Receptacle flat, small. Corollas with elongated tube, and lanceolate or linear spreading lobes longer than the short-campanulate throat. Anthers wholly exserted, acutely and even caudately sagittate at base: the tips triangular-lanceolate. Style-branches flattish, the truncate and minutely penicillate tips terminated by a very short and low obtuse cone. Akenes terete, short, obscurely 5-nerved, from extremely long-villous to glabrate or even glabrous. Pappus of fine and soft minutely scabrous capillary long bristles, white or whitish.

+ + + Involucre of numerous or several connivent-erect herbaceous equal bracts (with or without short accessory ones at base), many-flowered, or in some species of *Cacalia* of few bracts and few-flowered: ours herbs, the flowers all fertile; heads either homogamous or heterogamous with ligulate rays.

++ Pappus of comparatively few and unusually stout plumose bristles. (Transition to *Helenioideæ*.)

190. **RAILLARDELLA.** Heads 15-many-flowered (fewer-flowered only in depauperate plants), homogamous or heterogamous. Involucre cylindraceous or campanulate, a single series of linear equal bracts, their edges lightly connate below the middle, or not manifestly overlapping. Receptacle flat. Ray-flowers (when present) with irregular and cuneate deeply 3–4-cleft fertile ligules. Disk-corollas with rather short proper tube, elongated and narrow-funnelform throat, and 5 ovate obtuse naked teeth. Style-appendages flattish, hispidulous, tapering into lanceolate or cuspidate tips. Akenes linear, somewhat terete, obscurely several-nerved, pubescent. Pappus of 12 to 25 equal aristiform but soft and plumose bristles, nearly equalling the disk corollas.

++ ++ Pappus a single series of numerous rather rigid capillary bristles, from scabrous to barbellate: leaves chiefly opposite.

191. **ARNICA.** Heads many-flowered, conspicuously radiate, or the rays rarely wanting. Involucre campanulate, not calyculate-bracteolate at base, of several thin-herbaceous oblong-lanceolate to linear equal bracts in a single or somewhat double series. Receptacle flat, sometimes fimbrillate or villous. Corollas of the disk-flowers with a commonly elongated hirsute tube, a funnelform or cylindraceous throat, 5-lobed at summit. Style-branches flattish, at least above, there hirsute, with obtuse or acute tips. Akenes linear, more or less 5–10-costate or angled.

++ ++ ++ Pappus of soft-capillary and merely scabrous very numerous bristles: style-branches narrow, truncate or capitellate and often bearing a bearded ring at tip, which sometimes is produced into a short central cusp or obscure cone: leaves in our genera all alternate.

192. **SENECIO.** Heads heterogamous and radiate, or by the absence of ray homogamous and discoid, usually many-flowered. Corollas yellow, those of the disk 5-toothed, occasionally 5-lobed.

193. **CACALIA.** Heads homogamous, the flowers all hermaphrodite, few or numerous. Corollas white, rarely flesh-colored, with 5-cleft or 5-parted limb, the lobes usually with a midnerve.

194. **ERECHTITES.** Heads heterogamous and discoid, many-flowered; numerous outer flowers female; central ones hermaphrodite. Corollas all slender-tubular; those of the female flowers filiform and with usually slightly dilated and 2–4-toothed summit; of the hermaphrodite flowers with long filiform tube and short cyathiform 4–5-lobed limb. Receptacle flat, naked. Bristles of the pappus very soft and fine, elongated. Flowers whitish or yellowish.

TRIBE IX. **CYNAROIDEÆ.** Heads homogamous and tubiflorous, the flowers all hermaphrodite and with equally or sometimes rather unequally 5-cleft corollas, the lobes long and narrow; or sometimes radiatiform (falsely radiate) and heterogamous by enlargement of limb of corollas of marginal flowers, which are commonly neutral. Involucre much imbricated. Receptacle mostly flat or convex, often fimbrillate or densely setose. Anthers with tails at base, and commonly with elongated and connate cartilaginous apical appendages, their tips distinct. Style-branches destitute of appendage, short, sometimes distinct or partly so, more commonly united up to the simply obtuse tips, not hirsute or hispid, but sometimes an hispidulous or pubescent ring or node below. Akenes thickish and hard. Pappus setose or rarely paleaceous. Leaves alternate, the teeth or margins often prickly. (Nearly all the indigenous American representatives are Thistles.)

CRYPTOSTEMMA CALENDULACEA, of S. Africa, of the tribe *Arctotideæ* (lying between this tribe and *Anthemideæ*, and to which belongs *Gazania* of the gardens), is a ballast-weed at some ports in California, which it has reached via Australia.

Subtribe I. CARDUINEÆ. Akenes attached by their very base, mostly very glabrous: flowers all perfect (one Thistle diœcious), in ours numerous or in the first genus rather few in the head.

* Filaments distinct.

+— Leaves never prickly: style-branches partly distinct, slender; akenes oblong: filaments glabrous.

195. SAUSSUREA. Involucre obovoid to oblong; bracts appressed, muticous. Receptacle with setiform chaff among the flowers, or rarely naked. Pappus of numerous plumose bristles, more or less connate in an indurated ring at base, so falling from the akene in connection; with commonly some outer and smaller bristles, either less plumose or naked, which are separately deciduous.

196. ARCTIUM. Involucre globular; bracts slender-subulate or aristiform and spreading above the broader appressed base, hooked at tip. Receptacle densely setose. Pappus of numerous short and rigid or chaffy bristles, separately deciduous.

+— +— Leaves more or less prickly: style-branches concreted to or near the tip into a filiform or rarely short-cylindrical body; a pubescent ring below this either manifest or quite obsolete akenes obovate or oblong, compressed or somewhat turgid: pappus simple; its numerous bristles connate into a ring at base and falling from the akene in connection: filaments bearded or papillose-pubescent, rarely glabrous: involucre of numerous much imbricated and often prickly-tipped bracts, many-flowered.

197. CARDUUS. Bristles of the pappus naked, or at most barbellulate, not plumose: otherwise like *Cnicus*.

198. CNICUS. Bristles of the pappus long- and soft-plumose, or only their tips naked, or those of some marginal flowers occasionally almost naked to the base. Receptacle densely villons-setose.

199. ONOPORDON. Receptacle fleshy, alveolate, not setose: pappus not plumose: otherwise like *Cnicus*.

CYNARA, *Artichoke*, *Cardoon*, is sparingly cultivated, but not naturalized.

* * Filaments monadelphous below, glabrous: otherwise as preceding subdivision.

200. SILYBUM. Involucre depressed-globose, of rather large and rigid bracts in a few series; their upper portion herbaceous, spinose along the margins, and tapering into a rigid prickle, widely spreading. Receptacle and flowers nearly as in common Thistles. Bristles of the pappus numerous in more than one series, flattish, barbellulate-ciliolate or scabrous.

Subtribe II. CENTAURINEÆ. Akenes obliquely attached by one side of the base or more laterally.

201. CENTAUREA. Involucre ovoid or globose, many-flowered, mostly firm or rigid; bracts appressed and variously appendaged. Receptacle densely setose. Flowers sometimes all hermaphrodite and with corollas equally or obliquely 5-cleft into narrow lobes; more commonly the marginal ones neutral or sterile, and their corollas sometimes enlarged and widely spreading, forming a kind of false ray. Style-branches either concreted or partly separate. Akenes obovoid or oblong, turgid or compressed, usually smooth and glabrous, with a large epigynous disk, commonly surrounded by an elevated entire or denticulate margin. Pappus various, setose or partly paleaceous, occasionally obsolete or wanting.

TRIBE X. MUTISIACEÆ. (Ser. LABIATIFLORÆ, DC.) Heads in one subtribe homogamous, the hermaphrodite flowers all with regularly 5-cleft corollas; otherwise either homogamous or heterogamous and corollas bilabiate in the hermaphrodite flowers, sometimes simply ligulate in female ray-flowers. Anthers with long tails at base. Receptacle naked. Style-branches of hermaphrodite flowers not appendaged, usually short or very short, and like those of *Cynaroideæ* (but no node below) or of *Inuloideæ*. Leaves alternate. (Mostly South American, a few in other parts of the world: our five genera belong to three subtribes.)

Subtribe I. GOCHNATIEÆ. Heads homogamous; the corollas almost or quite regularly and deeply 5-cleft into linear lobes: style-branches usually rounded at tip. Ours shrubs. (Transition to *Cynaroideæ* and *Inuloideæ*.)

202. HECASTOCLEIS. Heads one-flowered, in a fascicle, surrounded by an involucriform cluster of leaves. Involucre cylindraceous, of several narrowly lanceolate rather rigid and cuspidate-acuminate bracts, appressed-imbricated. Flower hermaphrodite. Corolla rather chartaceous, narrow, equally cleft to the middle; the linear lobes widely spreading, not revolute. Anthers wholly exserted, subcoriaceous, bearing naked tails: the linear terminal appendages lightly connate, as long as the polliniferous portion. Style glabrous and even, not cleft, but terminated by an emarginate-2-lobed stigma. Akene (immature) cylindraceous, glabrous. Pappus coroniform, laciniate-dentate, corneous.

203. GOCHNATIA. Heads few–many-flowered, fasciculately paniculate or cymose. Involucre campanulate or oblong, of dry or coriaceous regularly imbricated bracts. Receptacle flat, naked. Corolla-lobes mostly revolute. Style-branches sometimes very short, sometimes fully twice longer than broad, flat, roundish-obtuse or nearly truncate at summit. Akenes oblong, silky-villous. Pappus of copious rather rigid capillary scabrous or barbellulate bristles, nearly equalling the corolla.

Subtribe II. GERBEREÆ, & III. NASSAUVIEÆ. Heads heterogamous or homogamous: corollas either all bilabiate ($\frac{3}{2}$), or marginal ones simply ligulate.

* Heads heterogamous and radiate: ray-flowers female and simply ligulate.

204. CHAPTALIA. Heads many-flowered: female flowers in two or more series and fertile; hermaphrodite flowers in the disk, all or some of them sterile. Involucre campanulate or turbinate, of narrow appressed-imbricated bracts, outer successively shorter. Corolla of the marginal flowers simply ligulate and 3-toothed at the end, or entire; those of an inner series more filiform, the ligule reduced to less than the length of the style; those of the hermaphrodite flowers more or less bilabiate, outer lip 3-toothed, inner 2-lobed or parted. Style in hermaphrodite flowers obtusely 2-lobed at apex, or when sterile entire. Akenes oblong or fusiform, 5-nerved, attenuate or rostrate at apex, bearing a copious pappus of very soft and fine capillary bristles. Scapigerous and monocephalous herbs.

* * Heads homogamous, of hermaphrodite and fertile flowers, all of them with bilabiate ($\frac{3}{2}$) corollas, the lower lip larger in marginal flowers, not rarely more elongated and radiatiform: style-branches comparatively long, mostly dilated or flattened above and truncate, rarely somewhat penicillate.

205. PEREZIA. Involucre few–many-flowered, imbricated in few to several series: bracts dry, chartaceous or coriaceous. Receptacle flat, naked, rarely pilose or fimbrillate. Akenes commonly papillose-puberulent, elongated-oblong, terete or obscurely angled, sometimes narrowed at apex, not rostrate. Pappus of copious capillary scabrous bristles, either rather rigid or soft. Flowers never yellow.

206. TRIXIS. Involucre several–many-flowered; proper bracts 8 to 12, equal in a single series, or in two unequal series, little if at all imbricated, usually subtended by a few foliaceous loose accessory ones or by bracteiform leaves. Receptacle in genuine species pilose. Akenes more slender, with a tapering or rostrate summit. Pappus soft. Flowers yellow.

TRIBE XI. CICHORIACEÆ. (Ser. LIGULIFLORÆ, DC.) Heads homogamous and ligulate; the flowers all hermaphrodite and with ligulate corolla; ligule 5-toothed at the truncate apex. Anthers sagittate-auriculate at base, not caudate: pollen-grains dodecahedral. Style-branches filiform, minutely papillose, not appendaged, but stigmatic lines evident only toward base. Receptacle almost always plane. Herbs (except a few insular genera), mostly with milky and bitter juice: leaves alternate. (Natural and well-definable subtribes being still a desideratum, artificial sections based primarily on the pappus are here employed.)

Series I. Pappus none : receptacle naked.

* Akenes truncate at base and apex, short, smooth: leaves all radical: involucre of nearly nerveless bracts, nearly unchanged in fruit, rather many-flowered.

207. PHALACROSERIS. Involucre of 12 to 16 equal and nearly herbaceous lanceolate bracts, naked or loosely unibracteate at base. Akenes short-oblong, slightly incurved, obscurely quadrangular: pericarp thin-coriaceous. Scape naked, monocephalous: flowers yellow.

208. ATRICHOSERIS. Involucre of 12 or more equal lanceolate bracts, and calyculate with a few minute ones. Akenes oblong, with corky pericarp, more or less 8-10-costate, the alternate ribs thicker. Scape bracteate and polycephalous: flowers white and purplish.

211. KRIGIA, & 219. MICROSERIS, very rarely want the pappus, or nearly so.

* * Akenes with rounded or somewhat contracted apex and small areola, narrow at base: involucre of several one-nerved equal bracts, unchanged or concavo-convex in fruit, 8-20-flowered : corollas yellow.

209. LAMPSANA. Involucre narrow, minutely calyculate-bracteate at base; the true bracts carinate, at least in fruit, then erect. Akenes narrowly obovate-oblong and somewhat obcompressed, minutely nervoso-striate, smooth. Leafy-stemmed and branching Old World annuals.

210. APOGON. Involucre not calyculate, of usually 8 oblong-lanceolate herbaceous bracts, in fruit becoming rather ovate by broadening of the base, concave and the tips conniving. Akenes terete, obovoid, merely rounded at summit, 10-costate, obscurely scabrous-lineolate transversely, rarely an obsolete vestige of pappus. Low annuals, becoming caulescent.

Series II. Pappus paleaceous or partly so, or aristiform, or plumose.

* Involucre simple and naked, i. e. of equal bracts and no short calyculate ones at base: akenes truncate: pappus of paleæ and (usually) of bristles: receptacle naked.

211. KRIGIA. Heads several-many-flowered. Bracts of the involucre thin-herbaceous. Akenes short-columnar or turbinate, pluricostate, terete or somewhat angular, with broad truncate summit. Pappus double; outer of pointless thin paleæ; inner of delicate naked bristles, these rarely wanting in one species. Flowers yellow.

* * Involucre either calyculate or imbricated, i. e. principal bracts equal and some short ones at base, or of less unequal bracts in two or more series, simple only in *Tragopogon*.

+— Akenes usually short, with truncate summit (sometimes a little narrowed beneath it, not rostrate): receptacle not chaffy: flowers never yellow: caulescent, with small or reduced leaves on the rigid stems or branches: flowers matutinal.

212. CICHORIUM. Heads several-many-flowered. Involucre double; its bracts herbaceous with coriaceous and indurating base, those of the inner series partly enclosing the subtended akenes, the 4 or 5 outer more spreading and herbaceous. Akenes somewhat angled; the broad summit bordered with a crown-like pappus of numerous short and blunt paleæ, in 2 or more series. Flowers normally blue.

213. STEPHANOMERIA. Heads 5-12-flowered, rarely 3-20-flowered. Involucre cylindraceous or oblong, of several appressed and equal plane membranaceous bracts and some short calyculate ones, not rarely with 2 or 3 of intermediate length, thus becoming imbricate. Akenes 5-angled or ribbed, sometimes with intermediate ribs. Pappus a series of plumose bristles, or rarely chaffy awns, not rarely naked toward the bases, which sometimes are lightly connate in phalanges. Flowers pink or rose color.

214. CHÆTADELPHA. Heads about 5-flowered. Involucre of *Stephanomeria*, cylindraceous, the accessory calyculate bracts very small, the membranaceous proper ones 5. Akenes short-linear, 5-angled, very smooth. Pappus of 5 rigid upwardly tapering awns, which bear on each side toward the base 3 to 5 rather shorter and slender rigid bristles. Flowers rose-color.

+— +— Akenes long-rostrate, base more or less excavated at insertion: receptacle naked: heads rather many-flowered: pappus a series of long-plumose bristles or awns.

215. **RAFINESQUIA.** Involucre conical or cylindraceous, of 7 to 15 linear acuminate equal bracts, somewhat fleshy-thickened at base, and some loose calyculate ones. Akenes terete, somewhat fusiform, obscurely few-ribbed, attenuate into a slender beak, not callous-thickened at the insertion. Pappus (white) of 10 to 15 slender bristles, softly long-plumose from base to near the tip. Leafy-stemmed and branching annuals; flowers white or tinged with rose-color.

216. **TRAGOPOGON.** Involucre campanulate or oblong, of several lanceolate and upwardly attenuate equal herbaceous bracts; no calyculate ones. Akenes somewhat fusiform, 5-10-costate, more or less excavated at insertion, tapering into a long beak, except perhaps the outermost. Pappus a series of numerous stout bristles, somewhat connate at base into a ring, long-plumose to near the apex, the plumes arachnoid and more or less interlacing. Simple-stemmed or branching biennials or perennials, with gramineous leaves, and large solitary heads of yellow or purple flowers.

+ + + Akenes either truncate or inner ones rostrate; receptacle paleaceous: soft slender chaff among the flowers: head rather many-flowered: involucre sparingly imbricated: flowers yellow.

217. **ANISOCOMA.** Involucre cylindraceous, of thin and very obtuse appressed bracts, somewhat herbaceous in centre and with broad white-scarious margins; innermost linear-oblong, 2 or 3 intermediate ones oblong; outer ones short-oval and orbicular. Chaffy bracts of receptacle long, linear-filiform or setiform. Akenes terete, linear-turbinate, 10-nerved, pubescent, short-attenuate at base, the truncate summit crowned with a narrow entire cuplike border or ring, within which is inserted the bright white pappus, of 10 or 12 rather rigid long bristles, in two series; the 5 longer ones (equalling the involucre) long-plumose above the middle; the others much shorter, less plumose, sometimes naked. Scapes monocephalous.

218. **HYPOCHŒRIS.** Involucre campanulate, of somewhat herbaceous marginless bracts. Chaffy bracts of the receptacle narrow and scarious. Akenes glabrous or scabrous, 10-ribbed, oblong or fusiform, tapering upward, at least the inner ones, into a beak. Pappus a series of fine plumose bristles, with or without some naked and shorter outer ones. Leaves chiefly radical and scapes bracteolate, often branching.

+ + + + Akenes either truncate at summit or upwardly attenuate, yet with no distinct or prolonged beak: receptacle not chaffy: pappus of awned or pointed scarious paleæ or of awns or bristles with paleaceous base, or plumose: flowers yellow, open in morning and dull weather.

219. **MICROSERIS.** Heads several-many-flowered, on naked simple scapes or peduncles. Corollas mostly with a hairy tube. Akenes 8-10-costate, with a basal callosity which is hollowed at the insertion. Pappus simple; its bristles or awns naked, in one or two species plumose (and then white) or barbellate.

220. **LEONTODON.** Heads many-flowered, on simple or branching scaly-bracteolate scapes. Involucral bracts narrow. Akenes minutely striate or rugulose, fusiform and tapering to the narrow summit, sometimes by more or less of a beak. Pappus one or two series of plumose (sordid) bristles, which are more or less lanceolate-widened at base, persistent.

220ª. **PICRIS.** Heads many-flowered, terminating leafy stems. Outer bracts of involucre loose or spreading. Akenes terete, 5-10-costate; the ribs rugose. Pappus one or two series of slender plumose bristles, not paleaceous at base.

Series III. Pappus of capillary bristles, scabrous, rarely barbellulate, never plumose nor paleaceous-dilated.

* Receptacle paleaceous, i. e. bearing narrow chaffy bracts among the flowers: corollas rose-color or rose-tinged.

221. **PINAROPAPPUS.** Involucre many-flowered, campanulate; its bracts imbricated and outer successively shorter, thinnish, the tips sphacelate. Chaff of the receptacle attenuate-linear, deciduous with the akenes. Akenes glabrous, slender, terete, 10-15-costate, tapering from the callous base into a short slender beak. Pappus sordid, of copious soft-capillary bristles, one or two outer series shorter, rather persistent.

* * Receptacle bearing some capillary bristles among the flowers: pappus all or the greater part deciduous in connection: akenes not flattened.

222. CALYCOSERIS. Involucre many-flowered, oblong-campanulate, of numerous erect linear-lanceolate scarious-margined bracts in a single series, and of a short and loose calyculate outer series. Delicate capillary bristles of the receptacle, one to each flower, as long as the akenes and deciduous with them. Akenes fusiform or oblong, 5-costate, attenuate into a short beak, which terminates in a shallow and denticulate scarious pappus-like crown, surrounding the base of a copious and white soft-capillary pappus; its bristles equal, deciduous all together.

223. MALACOTHRIX. The species with bristle-bearing receptacle belong here. Akenes short-columnar, truncate at both ends.

230. TROXIMON. One species sometimes bears chaffy bracts among the flowers: akenes short-rostrate.

* * * Receptacle naked.

+– Akenes not flattened: pappus promptly deciduous, mainly altogether, soft and white.

223. MALACOTHRIX. Involucre many-flowered, either imbricated or only calyculate. Receptacle sometimes with or sometimes without delicate capillary bristles interposed among the flowers. Akenes short, oblong or columnar, glabrous, terete and striately 5-15-costate, or 4-5-angled by the prominence of stronger ribs, slightly or not at all narrowed either way, with broad truncate apex having an entire or denticulate border or sharp edge. Pappus a series of soft and scabrous or near the base barbellulate bristles, which are deciduous more or less in connection, and commonly 1 to 8 outer and stronger ones which are more persistent and smoother.

228. CREPIS. One or two species incline to have most of the pappus-bristles fall in connection, also a few less deciduous.

224. GLYPTOPLEURA. Involucre 8-18-flowered, cylindraceous, of 7 to 12 nearly membranaceous linear-lanceolate equal hardly scarious-margined bracts, which are partly connate below, and some loose foliaceous ones or subtending leaves at base. Akenes narrowly oblong, often somewhat incurved, slightly tapering downward, with 5 thick obtuse ribs or angles, and the intervals conspicuously cancellate-sculptured, so as to form single rows of pits, at summit a short thick and 5-ribbed hollow beak exserted from a cupulate shoulder, and slightly dilated to bear the pappus: this bright white, of very numerous and fine hardly scabrous capillary bristles, in more than one series, caducous, outermost falling separately, inner mostly in connection at base.

+– +– Akenes not flattened: pappus persistent, or bristles tardily falling quite separately, never in connection (except, perhaps, by the breaking of the summit of an attenuate beak).

++ Beak to the akenes none or a mere attenuation.

= Heads solitary, terminating simple bractless scapes: flowers yellow.

225. APARGIDIUM. Involucre rather many-flowered, cylindraceous-campanulate; bracts somewhat herbaceous, lanceolate, acuminate, one-nerved, rather few in 2 or 3 series, or outer and broader ones more calyculate. Akenes linear-oblong, columnar, glabrous and smooth, truncate, not tapering at either end. Pappus sordid or brownish, of rather copious minutely barbellulate and rather fragile capillary bristles, with some outer and smaller ones merely scabrous. Perennial.

230. TROXIMON. Involucre many-flowered. Akenes tapering, 10-costate, beakless in original species.

= = Heads seldom solitary, borne by leafy stems or more or less bracteate scapes.

a. Flowers yellow (in an adventive species red-orange), or in one species white.

226. HIERACIUM. Involucre several-many-flowered, of narrow equal bracts and some short calyculate ones, or sometimes imbricate, having those of intermediate length, not thickened at base nor with thickened midribs. Akenes oblong or columnar, smooth and glabrous, mostly 10-ribbed or striate, either terete or 4-5 angular, slightly contracted at very base, commonly of same thickness to the truncate top, but in several species tapering to a narrower summit. Pappus of rather rigid scabrous fragile bristles, sordescent or fuscous, rarely white and soft, then passing into *Crepis*. Perennials, commonly with hispid or hirsute, or often glandular pubescence.

227. **CREPIS.** Involucre few-many-flowered, somewhat imbricated, or more commonly a series of equal bracts and some short calyculate ones, sometimes thickened at base after anthesis. Akenes from columnar to fusiform, 10-20-costate. Pappus of copious white and usually soft capillary bristles. Annuals or perennials.

 b. Flowers from whitish or cream-color to violet or rose-red; involucre narrow, unchanged in age, a series of equal erect bracts, and a few short calyculate ones at base; styles usually long and slender; akenes columnar or linear, or even fusiform, mostly truncate at summit.

228. **PRENANTHES.** Heads 5-30-flowered, mostly nodding before or during anthesis. Akenes terete or 4-5-angled, commonly striate, sometimes striately pluricostate, with truncate summit. Pappus of copious rather rigid capillary bristles, in the section *Nabalus* from whitish to ferruginous. Leafy-stemmed perennials, with paniculate or racemiform-thyrsoidly disposed heads. leaves dilated.

229. **LYGODESMIA.** Heads 3-12-flowered, erect. Akenes terete, obscurely few-striate or angled, commonly linear or slender-fusiform, in the larger species concave at insertion. Pappus of copious and usually unequal capillary bristles, either soft or rigidulous, from sordid-whitish to white. Stems mostly rush-like and striate, in one species spinescent, and leaves narrow-linear or reduced to scales. Flowers rose-colored.

 ++ ++ Beak to the akenes distinct and slender, except in one or two species of *Troximon*: heads erect before and during anthesis; involucre unchanged in age: akenes oblong or obovate to linear.

230. **TROXIMON.** Heads many-flowered, solitary, terminating simple naked scapes. Involucre campanulate or oblong, more or less imbricated. Akenes 10-costate or 10-nerved, smooth, not muricate nor sculptured, with or without a small callus at insertion; the beak various, or in two species wanting. Pappus white or whitish. Flowers yellow, orange, or rarely purple.

231. **TARAXACUM.** Heads many-flowered, solitary, terminating simple and fistulous naked scapes. Involucre campanulate or oblong, a single series of nearly equal narrow bracts, a little connate at base, and several or numerous calyculate bracts at the base. Style-branches slender and nearly filiform, as in most genera. Akenes oblong-obovate to fusiform, 4-5-costate or angled, and usually with some intervening nerves, muricate or spinulose, at least near the summit, which is abruptly contracted into a filiform beak. Pappus soft and capillary, dull white, no woolly ring at its base. Flowers yellow.

232. **PYRRHOPAPPUS.** Heads and involucre nearly of *Taraxacum*, terminating scapose or leafy stems or branches. Style-branches short, oblong, very obtuse. Akenes oblong or linear-fusiform, about 5-costate or sulcate, muriculate-rugulose or hirsutulous—scabrous, tapering abruptly into a long filiform beak. Pappus copious, soft and capillary, fulvous or rufous, its base usually surrounded by a soft-villous ring. Flowers yellow.

233. **CHONDRILLA.** Heads several-flowered, sessile or short-peduncled on slender branches. Involucre cylindrical, of several linear equal bracts, and some short calyculate ones. Akenes 4-5-angled and with intervening nerves or ribs, muricate toward the summit, which is abruptly produced into a filiform beak. Pappus fine and soft, bright white. Flowers yellow.

 + + + Akenes flattened: pappus of copious fine and soft capillary bristles: leafy-stemmed plants, with more or less paniculate heads.

234. **LACTUCA.** Involucre cylindraceous, or in fruit somewhat conoidal, several-many-flowered, either calyculately or more regularly imbricated. Akenes obcompressed, and with a beak or narrowed summit, which is more or less expanded at apex into a pappiferous disk. Pappus of bright white or rarely sordid bristles, falling separately.

235. **SONCHUS.** Involucre campanulate or broader, in age usually broadened and deshy-thickened at base, and becoming conical. Akenes obcompressed, destitute of beak or neck or dilated pappiferous disk. Pappus of very soft and fine flaccid bristles, which fall more or less in connection, and commonly one or more stronger ones, which fall separately.

Tribe I. VERNONIACEÆ, p. 50.

1. STOKÉSIA, L'Her. (*Jonathan Stokes*, a British botanist, coadjutor of Withering: some say *Dr. Wm. Stokes* of Dublin.) — A most peculiar genus, of a single species, of local habitat; a perennial, flowering in early summer; the large and showy head of flowers having considerable resemblance to that of a China Aster. — Benth. & Hook. Gen. ii. 234.

S. cyánea, L'Her. A foot high: stem stout, at first floccose-lanate; the few branches terminated by solitary heads: leaves glabrous, bright green, puncticulate, thickish; radical and lower cauline entire, oblong-lanceolate, tapering into a margined petiole; upper becoming ovate lanceolate, partly clasping, and bearing toward their base some spinulose-aristiform teeth; some subtending the head and passing into the bracts of the involucre: head, with the radiant marginal corollas (of an inch long), 3 inches in diameter: flowers bright purplish-blue. — L'Her. Sert. Angl. 27; Ait. Kew. ed. 2, iv. 491; Torr. & Gray, Fl. ii. 60; Hook. Bot. Mag. t. 4966; Meehan, Nat. Flowers, ii. t. 13. *Carthamus lævis*, Hill, Hort. Kew. 57, t. 5. *Cartesia centauroides*, Cass. Bull. Philom. 1816. *Centaurea Americana*, Hook. Comp. Bot. Mag. i. 48, by mistake. — Moist ground, in the low country, from southwestern part of S. Carolina to E. Louisiana: rare.

2. ELEPHÁNTOPUS, Vaill., L. (Greek for Elephant's foot, which is a translation of a Malabarian name of the original species.) — Perennial herbs, of warm regions, extending northward almost through the Atlantic U. S.; with undivided pinnately-veined leaves and usually bluish-purple flowers. — Benth. & Hook. Gen. ii. 237. *Elephantopus*, *Elephantosis*, & *Distreptus* (Cass.), Less., DC. — Our species all belong to the typical section of the genus; with stem dichotomously branching; heads capitately glomerate at the summit of pedunculiform branches, the compound glomerule involucrate by two or three cordate and closely sessile bracteiform leaves; and simple pappus of about 5 awns or rigid bristles, with chaffy-dilated base: fl. late summer. Of the nearly related species (with glabrous corolla) *E. scaber* belongs to the extra-American and *E. mollis* to the American tropics. Schultz Bip., in Linnæa, xx. 514, too hastily combined all the American species.

    * Stem leafy: upper cauline leaves very similar to the basal.

E. Caroliniánus, Willd. Rather softly hirsute or pubescent, sometimes 3 feet high: leaves thin, oval-obovate or ovate, crenate or repand-dentate, not rugose, nor prominently veined (the larger 4 to 8 inches long and 2 to 4 wide); uppermost oblong: chaffy base of awns of the pappus decidedly longer than the diameter of the akene, lanceolate-subulate and very gradually attenuate into the awn. — Spec. iii. 2390 (excl. syn.); Nutt. Gen. ii. 187; Ell. Sk. ii. 480; Torr. & Gray, Fl. ii. 60. *E. scaber*, Walt. Car. 217, &c., not L. — Dry soil in open woods, Pennsylvania to Illinois, Kansas, Texas, and Florida.

    * * Stem usually naked and scapiform: its few leaves small and bract-like; principal leaves radical and flat on the ground.

E. tomentósus, L. Somewhat canescently hirsute and villous; leaves silky-villous beneath (rather than tomentose), varying from obovate or rarely oval to narrowly-spatulate; veins of the lower surface prominent: scapiform stem a foot or two high: involucre of the large glomerules rigid: pappus-scales about the length of the breadth of the akene, triangular-subulate, attenuate into the bristle. — Spec. ii. 814, & ed. 2, excl. syn. Browne; Torr. & Gray, Fl. l. c. *E. Carolinianus*, var. *simplex*, Nutt. Gen. ii. 187. *E. nudicaulis*, Ell. Sk. ii. 481. *E. elatus*, Bertol. Misc. xi. 21, t. 5. — Virginia and Kentucky to Florida and Louisiana.

E. nudátus, Gray. Minutely strigose-pubescent: leaves membranaceous, green, at most somewhat hirsute beneath, from spatulate-obovate to oblanceolate, not prominently veined: glomerules smaller: pappus-scales very short, broadly deltoid, abruptly terminated by the bristle.— Proc. Am. Acad. xv. 47. (*Echinophoræ affinis Mariana*, etc., Pluk. Mant. 66, t. 388, fig. 6 ?) *E. scaber*, Michx. Fl. in part; Torr. & Gray, Fl. l. c., not L. *E. nudicaulis*. Ell. in herb. Hook., not of Sk. l. c. — Low and sandy woodlands, Delaware (*Canby*) to Georgia, W. Louisiana, and Arkansas (*Harvey*).

3. **VERNÓNIA**, Schreb. IRON-WEED. (*Wm. Vernon*, an early collector in Virginia, &c.) — Perennial herbs (or some in the tropics shrubs); with alternate and pinnately-veined leaves, and usually purple or rose-colored flowers, occasionally varying to white. — Gen. 541; DC. Prodr. v. 15; Torr. & Gray. Fl. ii. 57; Benth. & Hook., Gen. ii. 227. — A huge genus, of nearly 400 species, the greater part S. American, some S. African and S. Asian: the N. American species all of the section *Lepidaploa*, Benth. & Hook. l. c. (*Lepidaploa*, &c., Cass.), having somewhat spherical heads in terminal cymes or terminating corymbiform branches. Ours all many-flowered; the (fuscous or even ferruginous) pappus persistent or nearly so, and double; akenes commonly sprinkled or beset with resinous atoms between the salient ribs: foliage often puncticulate. Fl. late summer and autumn. The species are extremely difficult: there are spontaneous hybrids between such very different species as *V. Arkansana* and *V. Baldwinii*, *V. fasciculata* and *V. Baldwinii*, and even between *V. Baldwinii* and *V. Lindheimeri!*

* Stems leafy throughout: short outer pappus conspicuous, and squamellate rather than setose.
+ Heads large, sometimes an inch high, 50-70-flowered.

V. Arkansána, DC. Tall (8 or 10 feet), rather glabrous: leaves all linear-lanceolate (4 to 12 inches long and lines wide), attenuate-acuminate, runcinately denticulate: heads all on simple and somewhat clavate peduncles, nearly hemispherical: involucre green, very squarrose; its bracts all equalling the disk, and with long filiform tips (those of the upper reddish), the outer and loose ones filiform nearly or quite to the base: akenes minutely hispid on the ribs. — Prodr. vii. 264; Nutt. in Trans. Am. Phil. Soc. n. ser. vii. 283; Torr. & Gray, Fl. ii. 59; Torr. in Sitgreaves Exped. t. 2. — Plains and alluvial banks of streams, Missouri and Kansas to E. Texas.

+ + Heads smaller, half-inch high or less, 15-40-flowered, rarely only 10-flowered.

++ Leaves slightly or not at all scabrous, and without revolute margins, most of them acutely denticulate or serrate with rigid or somewhat spinulose teeth, varying from linear-lanceolate to oblong-ovate, acuminate or very acute, pinnately veined: stems leafy up to the inflorescence; cymes mostly compound. (Species not clearly limited.)

= Akenes under a lens more or less hispidulous on the ribs.

V. Noveboracénsis, WILLD. Somewhat glabrous or pubescent, 3 to 6 feet high: leaves from elongated- to oblong-lanceolate (3 to 9 inches long): heads in an open cyme, 20-40-flowered: involucre commonly brownish or dark purplish; the ovate and ovate-lanceolate bracts (or at least the upper ones) abruptly acuminate into a slender cusp or slender tortuous awn, usually some of the lower wholly aristiform and loose. — Spec. iii. 1632; DC. Prodr. v. 63; Torr. & Gray, Fl. ii. 57. *Serratula Noveboracensis* (founded on Herm. Parad. Bot., & Dill. Elth. 355. t. 263) and *S. præalta* (in herb. and of Dill. Elth. t. 264, bracts more aristate than the figure shows), L. Spec. ii. 818. *V. præalta*, Less in Linn. iv. 264; Hook. Fl. i. 304. *V. tomentosa*, Ell. Sk. ii. 288 (*Chrysocoma tomentosa*, Walt. Car. 196), a form with tomentulose pubescence. Varies with pale or sometimes white instead of pink-purple corollas, the involucre then greenish. — Low grounds, coast of New England to Georgia, west to Wisconsin and Missouri, but mostly an eastern species.

Var. latifólia. Lower, 2 to 5 feet high: leaves oblong-ovate or broadly lanceolate, pale or glaucescent beneath, the larger more coarsely serrate: heads fewer: involucre varying from hemispherical (of fewer bracts) to somewhat turbinate, and its bracts merely acute, acuminate, mucronate, or some with a short filiform cusp. — *Serratula glauca*, L. l. c., founded on Dill. Elth. 354. t. 262; the specimen has many aristate-tipped bracts. *Vernonia glauca* (and nearly *V. præalta*), Willd. Spec. iii. 1633. *V. ovalifólia*, Torr. & Gray, l. c.; Chapm.

Fl. 187, extreme form, mostly with muticous involucral bracts. — In shady places, Penn. and Ohio to Florida.

V. Baldwínii, Torr. Tomentulose, 2 to 5 feet high: leaves oblong- or ovate-lanceolate: involucre (a quarter-inch high) when young globose, hoary-tomentose, greenish, squarrose by the spreading or recurved acute or acuminate tips of its bracts. — Ann. Lyc. N. Y. ii. 211; Torr. & Gray. l. c. *V. sphæroidea*, Nutt. in Trans. Am. Phil. Soc. l. c. — Prairies and barren hills, E. Missouri to Texas; flowering early, in July and August. Passes into the next.

V. altíssima, Nutt. Nearly glabrous, or sometimes cinereous-pubescent, 5 to 10 feet high: leaves thinnish, veiny, obscurely if at all puncticulate, lanceolate or lanceolate-oblong: cyme usually loose or open: involucre of wholly appressed obtuse or merely mucronate-acute bracts; ribs of the akenes minutely or sparsely hispidulous. — Gen. ii. 134; Ell. Sk. ii. 289; Less. in Linn. vi. 639, partly. *V. præalta*, Michx. l. c., partly; DC. l. c., partly. *V. fasciculata*, var., Torr. & Gray, Fl. ii. 59; Chapm. Fl. 188. *Chrysocoma gigantea*, Walt. l. c. Varies much, especially in the size of the heads: the form *parviflora*, with involucre only 2 or 3 lines high and rather pauciseriate, being Nuttall's original. — Low or wet grounds, W. Penn. to Illinois, Louisiana and Florida.

Var. **grandifióra.** Less tall: heads larger: involucre mostly 4 lines high; the bracts 35 to 40 and in more numerous ranks. — Nutt. in Herb. Acad. Philad. — Low prairies and along streams, Illinois and Kentucky to Texas.

== == Akenes smooth and glabrous on the ribs, or nearly so: bracts of the involucre all closely appressed and inappendiculate, coriaceo-chartaceous.

V. fasciculáta, Michx. Glabrous, or the cyme puberulent, 2 to 5 feet high: leaves thickish, when dry puncticulate, from linear (and with obscure veins or veinlets) to oblong-lanceolate (and more evidently veined), conspicuously spinulose-denticulate: heads numerous and crowded on the branches of the compound cyme: involucre (3 or 4 lines high) 20-30-flowered; its bracts all obtuse, or some of the uppermost abruptly mucronate-acute. — Fl. ii. 94: Torr. & Gray, l. c., excl. vars. *V. corymbosa*, Schweinitz, in Keating, Narr. Long Exped. Mississ., the form with broad and short leaves. *V. altissima*, DC. l. c. partly, & excl. syn. Dill., &c. — Low grounds, prairies and river-bottoms, Ohio and Kentucky to Dakota and south to Texas.

++ ++ Leaves perfectly glabrous and smooth, veinless, commonly entire, narrowly linear, plane: heads narrow, few-flowered.

V. Lettermáni, Engelm. Habit of the preceding, 2 to 4 feet high, fastigiately and cymosely much branched at summit: leaves 3 or 4 inches long, only a line wide, the margins not revolute: heads numerous, pedunculate, clavate-cylindraceous, 10-14-flowered, half-inch long: bracts of the involucre all appressed and inappendiculate, but acute or acuminate; outermost ovate-subulate, innermost narrowly lanceolate and purple: ribs of the glandular akenes obscurely scabrous. — Proc. Am. Acad. xvi. 78. — Arkansas, on Cooper's Creek, *Bigelow*. Gravelly banks and sand-bars of the Washita, *Letterman*.

V. Jamésii, Torr. & Gray. Glabrous or nearly so, a foot or two high: leaves linear-lanceolate or linear, like those of narrowest forms of *V. fasciculata*, but smaller and less or obsoletely denticulate; veins and veinlets obscure: heads few or numerous in a loose and open corymbiform cyme, all pedunculate: involucre (4 or 5 lines high) 15-25-flowered, from hemispherical-campanulate to turbinate-oblong; its bracts all or mostly obtuse, or (in the larger form of involucre) acute or acuminate. — Fl. l. c.; Gray, Pl. Wright. i. 82. *V. altissima*, var. *marginata*, Torr. Ann. Lyc. N. Y. ii. 210. — Plains of Nebraska and Arkansas to W. Texas and E. New Mexico, first coll. by *Dr. James*.

++ ++ ++ Leaves with upper face scabrous and margins often revolute, then entire, not canescent.

V. angustifólia, Michx. Stem a foot to a yard high, slender, from roughish-hirsute to nearly glabrous: leaves from narrowly linear or approaching filiform to lanceolate, the broader ones sparsely denticulate and also veiny: cyme loose, simple or compound, sometimes paniculate, sometimes umbelliform, mostly naked: heads 15-25-flowered: involucre about 3 lines high, commonly somewhat turbinate; its bracts or most of them mucronate, sometimes cuspidate-acuminate: akenes minutely hirsute, at least on the ribs. — Fl. ii. 94; Ell. Sk. ii. 87; Torr. & Gray, l. c. *V. fasciculata*, DC. l. c., not Michx. *Chrysocoma graminifolia*, Walt. Car. 196. *Liatris umbellata*, Bertol. Misc. v. t. 4. — Dry pine barrens, N. Carolina to Florida, Arkansas, and Texas.

Var. scabérrima. Leaves mostly short and sparsely denticulate or toothed, from linear to oblong-lanceolate, scabrous to rough-hispidulous above: bracts of the involucre or some of them produced into long and loose or spreading subulate or filiform tips. — Torr. & Gray, l. c. *V. scaberrima*, Nutt. Gen. ii. 134; Ell. l. c. — South Carolina to Florida.

Var. Texána. Stem virgate, rather tall: lower leaves large, lanceolate (3 to 6 inches long); upper ones linear or subulate: cyme naked: bracts of the involucre all pointless or merely mucronate. — Torr. & Gray, l. c., character, without name. — Pine woods, Arkansas, Louisiana, and Texas.

Var. púmila, CHAPM. Glabrous and hardly at all scabrous, even the leaves; these small, mostly linear and entire: stem slender, a span to 18 inches high: cyme of few heads: bracts of the involucre pointless. — Bot. Gazette, iii. 5. — Wet pine barrens, S. E. Florida, *Blodgett, Garber*.

++ ++ ++ ++ Leaves with revolute entire margins, not scabrous, veinless, lanose beneath.

— **V. Lindheiméri,** GRAY & ENGELM. About a foot high, excessively leafy up to the corymbiform cyme, lanose-canescent, even to the obtuse and pointless bracts of the involucre; leaves narrowly linear (1½ to 3 inches long, a line or two wide), glabrate and green above; heads all pedunculate: akenes glabrous: pappus purple. — Proc. Am. Acad. i. 46, & Pl. Lindh. ii. 217. — Rocky hills and plains, W. Texas, *Lindheimer, Wright*, &c. *Berlandier* collected an apparent hybrid between this most distinct species and *V. Baldwinii*.

* * Outer pappus inconspicuous and rather setose than squamellate; cauline leaves few and small.

V. oligophýlla, MICHX. Minutely scabrous-pubescent: stem about 2 feet high, slender, bearing a few heads in a very loose naked cyme: radical leaves ample (4 to 8 inches long) in a rosulate tuft, oblong; cauline lanceolate, few and small, the uppermost reduced to subulate bracts; all veiny and denticulate: heads 15–30-flowered: bracts of the involucre subulate (mostly from a broad base), loose: hristles of the pappus slender: akenes hirtellous on the ribs. — Fl. ii. 94; DC. Prodr. v. 62; Torr. & Gray, Fl. ii. 57. *Serratula Carolinensis*, Dill. Elth. t. 261. *Chrysocoma acaulis*, Walt. Car. 196. — Low pine barrens, N. Carolina to Florida, near the coast. — Varies with foliage soft cinereous-pubescent: S. Carolina, *J. Donnell Smith*.

TRIBE II. EUPATORIACEÆ, p. 50.

4. STÉVIA, Cav. (*Dr. Pedro Esteve.*) — Herbs, rarely suffrutescent plants; with mostly opposite and triplinerved leaves, small and narrow heads usually corymbosely crowded in terminal naked cymes or fascicles, and flowers white or rose-purple: pappus variable; the awns when present barbellate-scabrous. — A large Mexican genus (a few species reaching our borders), also well developed on the eastern side of South America in corresponding latitudes. — Cav. Ic. iv. 32, t. 354–356; Schultz Bip. in Linn. xxv. 268.

* Branches and heads paniculate, loose: root annual.

S. micrántha, LAG. Puberulent and somewhat viscid: stem slender, a foot or two high, bearing short flowering branches almost from the base: leaves thin, ovate with subcuneate or rarely subcordate base, serrate (inch long), petioled: heads pedicellate in the loose clusters, 3 and 4 lines long: pappus of 3 awns with short paleaceous-dilated base, or in one or two flowers occasionally awnless. — Elench. Hort. Madrid, 1815, & Nov. Gen. & Spec. 27. *S. macella*, Gray, Pl. Wright. ii. 70. — Shady cliffs, New Mexico, *Wright*. Southern Arizona, *Lemmon*, by which is generally meant *Mr. J. G.* and *Mrs. Sara Plummer Lemmon*, associates in exploration. (Mex.)

* * Heads loosely cymose-paniculate and pedunculate: root perennial.

S. amábilis, LEMMON. Stem slender and virgate, or with long virgate branches, about 2 feet high: leaves all alternate, linear with narrowed base, or the lowest oblanceolate, entire, thinnish: involucre slender, glandular-viscid: flowers purple: pappus of 5 long awns and with extremely short (broader than long) intermediate paleæ. — Gray, Proc. Am. Acad. xix. 1. — Plains near Cave Cañon, S. Arizona, *Lemmon*.

* * * Heads subsessile and fasciculate; the fascicles corymbosely cymose; root perennial.

+— Herbaceous, leafy up to the dense fastigiate clusters of heads: leaves subsessile, serrate.

S. serráta, Cav. Pubescent or somewhat hirsute: leaves often alternate, crowded, from spatulate-linear to oblong-spatulate, irregularly and sometimes coarsely serrate or some entire, loosely veiny, strongly punctate: flowers white or pale rose; pappus 1–5-aristate or in some flowers reduced to a crown of short obtuse paleæ. — lc. iv. t. 355 ; DC. Prodr. v. 118. *S. ivæfolia*, Willd. Mag. Naturf. Berl. 1807, 137, & Enum. 855. *S. canescens*, HBK. Nov. Gen. & Spec. iv. 143; Benth. Pl. Hartw. 19; Gray, Pl. Wright. ii. 71. *S. virgata*, HBK. l. c. *S. punctata*, Schultz Bip. in Linn. xxv. 286. *Ageratum punctatum*, Jacq. Hort. Schœnbr. iii. t. 300. (Variable species.) — New Mexico and Arizona, *Wright* and later collectors. (Mex., Venezuela.)

S. Plúmmeræ, Gray. Puberulent and almost glabrous: leaves nearly all opposite, less crowded, oblong-lanceolate or broader, acute, incisely serrate, bright green, very conspicuously nervose-veiny and reticulated, hardly punctate (2 inches long): flowers rose-color: pappus of 4 broad and truncate fimbriate-denticulate paleæ. — Proc. Am. Acad. xvii. 204. — S. Arizona, Rucker Valley of the Chiricahua Mountains, *Mrs. Lemmon*, born *Plummer*.

Var. **álba**. Flowers white: leaves less serrate and not so strongly veiny. — S. Arizona, in Ramsey's Cañon, *Lemmon*.

+— +— Shrubby: leaves subsessile, mostly entire and opposite.

S. Lemmóni, Gray. Fruticose, puberulent throughout, leafy up to the dense clusters of very numerous heads: leaves linear-oblong, obtuse, thinnish, obscurely triplinerved: involucre somewhat viscid-pubescent: flowers apparently white: pappus a cupulate and nearly entire or merely lacerate crown. — Proc. Am. Acad. l. c. — S. Arizona, cañons in the Santa Catalina Mountains, *Lemmon, Pringle*.

S. salicifólia, Cav. Frutescent, low, nearly glabrous: leaves coriaceous, linear or linear-lanceolate, occasionally serrate, commonly glutinous-lucid: heads in small and more open fascicles: flowers white: pappus 1–3-aristate, or sometimes of obtuse paleæ. — Ic. l. c. t. 354; Schultz Bip. l. c. 290; Gray, Bot. Mex. Bound. 73. *S. angustifolia*, HBK. l. c. (awnless pappus). — S. border of Texas, *Parry*, a low and very narrow-leaved form. (Mex.)

5. **SCLERÓLEPIS**, Cass. (Σκληρός, hard, and λεπίς, scale, from the cartilaginous paleæ of the pappus.) — Genus of a single species, peculiar to the Atlantic coast. Fl. summer.

S. verticilláta, Cass. Subaquatic perennial, nearly glabrous, stoloniferous from the base: stems slender, usually simple, above the water bearing many whorls of narrowly linear one-nerved entire sessile leaves (half-inch to an inch long), and terminated by a solitary pedunculate small head (rarely branching at top and 3–4-cephalous): flowers rose-purple. — Dict. xxv. 365; DC. Prodr. v. 114; Torr. & Gray, Fl. ii. 65. *Ethulia uniflora*, Walt. Car. 195. *Sparganophorus verticillatus*, Michx. Fl. ii. 42. — Low pine-barren ponds and streams, in shallow water, New Jersey to Florida. Leaves 4 to 6 in the whorls.

6. **TRICHOCORÓNIS**, Gray. (Θρίξ, τριχός, hair, and κορωνίς, top or apex.) — Texano-Mexican herbs, fibrous-rooted, aquatic or paludose; with stems creeping at base or spreading, branching, leafy, pubescent with somewhat viscid and weak multicellular hairs: leaves of soft texture, opposite or the upper alternate, sessile and partly clasping, glabrate: heads slender-peduncled, terminating the branches: flowers flesh-color or rose-purple. — Pl. Fendl. 65; Benth. & Hook. Gen. ii. 240.

T. Wrightii, Gray, l. c. Stems assurgent from an annual root, paniculately-branched above: leaves undivided, sparingly serrate, half-inch or more long; the lower opposite and oblong; upper alternate and cordate-lanceolate: heads diffusely panicled, only two lines high and wide: involucral bracts about 18, oblong-lanceolate: receptacle convex: tube of the corolla shorter than the expanded throat and limb: style-branches narrow: pappus a minute but evident crown of more or less concreted setuliform squamellæ, or some of them aristellate. — *Ageratum?* (*Microgeratum*) *Wrightii*, Torr. & Gray, Proc. Am. Acad. i. 46.

Margacola parvula, Buckl. in Proc. Acad. Philad. 1861, 1862. — Wet ground in prairies, Texas. *Wright, Buckley*, &c. (Mex., *Palmer*.)

T. rivuláris, Gray, l. c. Stems floating, in shallow water rooting, and flowering branches emersed and ascending: leaves succulent, mostly opposite, an inch or two in length, cuneate-obovate, sparingly incised or palmately 3-lobed, contracted into a narrow connate-clasping auriculate base: heads fewer or solitary on simple peduncles, 3 or 4 lines in diameter: involucral bracts about 12, oval, obtuse: receptacle highly convex: tube of corolla slender, equalling the hemispherical throat and limb: style-branches flat and linear, acutish: pappus a minute and evanescent or obscure setulose crown. — In springs and streamlets, S. W. Texas, *Wright*, &c. (Adjacent Mex., *Gregg*, &c.)

7. AGÉRATUM, L. (Ancient Greek and Latin name of some aromatic plant of this order, probably an *Achillea*, from α privative and γῆρας, γήρατος, not waxing old, transferred by Linnæus to an American genus.) — Chiefly tropical, herbaceous, and with opposite petiolate leaves; heads small in terminal corymbiform cymes or rarely paniculate; flowers blue, purple, or white, in summer. — Benth. & Hook. Gen. ii. 241, excl. syn. *Oxylobus*. *Ageratum* & *Cælestina*, Cass., DC.; to which should be added *Alomia*, HBK., differing only in the want of pappus.

§ 1. EUAGERATUM. Pappus of distinct aristate or sometimes muticous paleæ: receptacle naked.

A. conyzoïdes, L. Annual, pubescent: leaves ovate or deltoid-subcordate, crenately serrate: pappus of 5 to 7 lanceolate rigid scales, mostly tapering into a scabrous awn which nearly equals the blue or white corolla. — Schk. Handb. t. 238; Hook. Exot. Fl. t. 15. *A. Mexicanum*, Sims, Bot. Mag. t. 2524, &c., a more pubescent form, common in ornamental cultivation. — Sparingly naturalized near towns in the S. Atlantic States. (Nat. from Trop. Amer., &c.)

§ 2. CŒLESTINA. Pappus coroniform or cupulate (by the union of the paleæ into an entire or toothed cup or border), sometimes obsolete. — *Cælestina*, Cass., DC., &c. (In our species the receptacle is naked, duration of root uncertain, and flowers usually blue or violet.)

A. corymbósum, Zuccagni. Scabrous-puberulent, erect: leaves short-petioled, ovate to oblong-lanceolate, irregularly few-several-toothed: floriferous branches naked above: corolla-tube glanduliferous: pappus prominently cupulate, more or less dentate. — Zuccagni ex Balb. in Hort. Taur. 1806; Pers. Syn. ii. 402. *A. cælestinum*, Sims, Bot. Mag. t 1730; Lodd. Bot. Cab. t. 623. *Cælestina ageratoides*, HBK. Nov. Gen. & Spec. iv. 151; Gray, Pl. Wright. ii. 70. *C. cærulea*, Cass. Dict. vi. suppl. 8. t. 93. *C. corymbosa*, DC. Prodr. v. 108. — New Mexico, *Wright*, &c. (Mex.)

A. littoràle, Gray. Glabrous, decumbent or assurgent: leaves rather succulent, long-petioled, ovate with cuneate base, serrate: corolla glabrous: pappus an extremely short crown, with or without several minute narrow teeth, or reduced to a mere ring. — Proc. Am. Acad. xvi. 78. — *Cælestina maritima*, Torr. & Gray, Fl. ij. 64; not *Ageratum maritimum*, HBK., which is a true *Ageratum* with diminutive pappus. — Key West, S. Florida, *Bennett, Blodgett, Palmer, Garber*.

8. HOFMEISTÉRIA, Walp. (*W. Hofmeister*, a vegetable histologist.) — Low suffrutescent plants; with heads terminating slender peduncles, small incised leaves either opposite or alternate on long petioles, and whitish flowers; the style-branches clavate. — Two species, the original one (*H. fasciculata*, Walp. Rep. vi. 106; *Helogyne*, Benth. Bot. Sulph. 20, t. 14), of Lower California, with 2-3-awned pappus.

H. pluriséta, Gray. Slightly puberulent and viscidulous, much branched: leaves with small (2 to 5 lines long) deltoid to oblong blade very much shorter than the petiole: heads about

20-flowered, 4 or 5 lines long: bracts of involucre with pointed somewhat spreading tips: akenes rather short: pappus of 10 or 12 bristles and about as many small and narrow acute squamellæ. — *Pacif. R. Rep.* vi. 96, t. 9, & *Bot. Calif.* i. 299. — Cañons, San Bernardino desert, Southeast California to Arizona and S. Utah, *Bigelow, Parry, Newberry*, &c.

9. MIKÁNIA, Willd. (*Prof. J. G. Mikan*, of Prague, or his son and successor, *J. C. Mikan*, who collected in Brazil.) — Twining perennials, or many erect and shrubby in tropical America, where most of the numerous species occur; with opposite leaves and small variously clustered heads. Our species, confined to the Atlantic States, have slender-petioled angulate-cordate leaves, corymbosely cymose heads of pale flesh-colored and more or less fragrant flowers, produced in summer and autumn; the throat of the corolla abruptly dilated from the narrow tube, and broadly campanulate. — Willd. Spec. iii. 1472; Benth. & Hook. Gen. ii. 246.

M. scándens, WILLD. Glabrous or puberulent: herbaceous stems high-twining: leaves somewhat hastately or deltoidly cordate, acuminate, irregularly and obtusely angulate-dentate or repand, rarely almost entire: heads crowded, about 3 lines long: involucral bracts lanceolate, acuminate or slender-apiculate: corolla-lobes ovate, much shorter than the very wide throat: akenes a line long, resinous-atomiferous. — Torr. & Gray, Fl. ii. 91: Baker in Fl. Bras. vi. 248, in part. *Eupatorium scandens*, L.; Jacq. Ic. Rar. t. 169; Michx. Fl. ii. 97. — Moist ground along streams, New England and W. Canada to Florida and Texas. (Mex. and W. Ind. to S. Brazil, mostly in peculiar forms, if not species.)

——**Var. pubéscens,** TORR & GRAY, l. c. From slightly to densely puberulent. — *M. pubescens*, Muhl. Cat. 71; Nutt. Gen. ii. 136. *M. menispermea*, DC. Prodr. v. 200. — Southern Atlantic States to Texas.

M. cordifólia, WILLD. Puberulent or pubescent, frutescent at base: branchlets often striate-angulate: leaves broadly cordate and augulate: inflorescence more compound: heads 4 or 5 lines long: involucral bracts oblong-linear, obtuse or muticous: corolla-lobes oblong-lanceolate, fully as long as the campanulate throat: akenes 1½ to 2 lines long, glabrous. — *Cacalia cordifolia*, L. f. Suppl. 351, & herb. Mutis, fide Baker, l. c. 253. *M. cordifolia* (and according to Baker also *M. rubiginosa*), Smith. *M. suaveolens*, HBK. Nov. Gen. & Spec. iv. 135. *M. gonoclada*, DC. l. c. 199. *M. convolvulacea*, DC. l. c. — W. Louisiana, *Hale*. (Mex., W. Ind., Brazil.)

10. EUPATÓRIUM, Tourn. THOROUGHWORT, &c. (*Mithridates Eupator*, king of Pontus.) — Perennial herbs, a few annuals, and some shrubby in the warmer regions; with commonly opposite leaves, mostly resinous-atomiferous and bitter; the small heads corymbosely cymose, or sometimes paniculate, rarely solitary. Fl. late summer and autumn. A vast genus as received in DC. Prodr. v. 141, and more extended by Benth. & Hook. Gen. ii. 245; chiefly American. The sections are too confluent for good subgenera.

§ 1. ÓSMIA, Benth. Involucre cylindrical or cylindraceous: the bracts squamaceous, coriaceous or firm-chartaceous, striate, pluriseriate, closely imbricated, the exterior successively shorter, obtuse: receptacle of the flowers flat or rarely convex: heads mostly clustered in corymbiform cymes: branching shrubs, or rarely herbs with suffrutescent base, tropical or subtropical: leaves all opposite. — *Osmia*, Schultz Bip. § *Cylindrocephala*, DC.

* Involucral bracts abruptly appendiculate with short foliaceous or partly colored squarrose tips: heads pedunculate. — § *Phyllocerocephala*, Gray, Pl. Wright. i. 88.

E. sagittátum, GRAY, l. c. Probably suffruticose, puberulent: leaves (inch long) slender-petioled, sagittate or hastate, otherwise entire, acute or acuminate: heads nearly half-inch long, in threes terminating divergent branchlets: involucre 30-40-flowered, its bracts firm-coriaceous, hardly striate, prominently appendaged by deltoid spreading foliaceous tips: flowers probably purplish. — "California," *Coulter*, no. 294. But the same as 253 of Upper Sonora in the Mexican collection, doubtless the real habitat. Yet may reach into Arizona. (Adj. Mex., *Coulter, Gregg, Palmer*.)

E. ivæfólium, L. Herbaceous or merely suffrutescent, somewhat hirsute or pubescent, strictly erect, 2 to 5 feet high: leaves lanceolate or the upper ones linear, hardly petioled, 3-nerved, sparsely and often coarsely serrate at the middle, mostly obtuse, roughish, an inch or two long: heads small (3 or 4 lines long), 10-20-flowered, in small and loose cymes: bracts of the cylindraceous involucre oblong, striate, with the very short somewhat truncate tips purple or greenish and slightly squarrose-spreading: flowers light purplish-blue or reddish. — Amœn. Acad. v. 405, & Spec. ed. 2, 1174; Torr. & Gray, Fl. ii. 81; Griseb. Fl. W. Ind. 359; Baker in Fl. Bras. l. c. 290. (*E. obscurum*, DC., & *E. concinnum*, Hook. & Arn., ex Baker.) *E. calocephalum*, Nutt. in Trans. Am. Phil. Soc. n. ser. vii. 286. *Liatris oppositifolia*, Nutt. in Am. Jour. Sci. v. 299. — Old fields, &c., Lower Mississippi, Louisiana, and Texas; the var. *Ludovicianum*, Torr. & Gray, l. c., a form with less serrate leaves and less squarrose involucre, the tips of the upper scales mostly petaloid and purple. (W. Ind. & Mex. to S. Brazil.)

* * Involucral bracts wholly inappendiculate and appressed.

E. heteroclínium, Griseb. Herbaceous, with somewhat ligneous base, 2 or 3 feet high, rather strong-scented, pubescent: branches ascending: leaves rather short-petioled, ovate-lanceolate with cuneate or truncate base to deltoid, obtusely serrate, 3-nerved, about an inch long: heads scattered, 5 or 6 lines long, 20-25-flowered, short-peduncled: involucre cylindraceous, glabrous, smooth and somewhat shining, pale; the bracts very obtuse, about 7-striate, more than usually deciduous: receptacle of the purple or bluish flowers convex. — Fl. Brit. W. Ind. 358. *Conoclinium rigidum*, Chapm. in Bot. Gazette, iii. 6, not DC. — Keys of S. Florida, *Blodgett, Chapman, Curtiss*. (Jamaica.)

E. conyzoídes, Vahl. Shrubby, with herbaceous divergent flowering branches. 4 to 10 feet high, from villous-pubescent to glabrate: leaves slender-petioled, ovate-lanceolate, varying to ovate, acuminate, mostly cuneate at base, sparsely and acutely serrate or sometimes entire, 3-nerved or triplinerved (larger 3 to 5 and smaller 1 or 2 inches long): heads numerous in the corymbiform open cymes, a third to half-inch long, 12-30-flowered: involucre cylindraceous or cylindrical, glabrous; the bracts 3-5-striate, rounded and somewhat greenish at the tip: receptacle of the pale blue or white flowers flat. — Symb. iii. 96; Schrank, Hort. Monac. t. 85; Baker, l. c. *E. odoratum*, L. in part. — Along the Rio Grande on the Mexican border of Texas, *Berlandier, Schott, Bigelow*, &c. Mouths of the Mississippi, *Trecul. E. Sabeanum*, Buckley in Proc. Acad. Philad. 1861, 456. The form with stouter heads and firmer greenish-tipped involucral bracts, common in Mexico, &c. (*E. floribundum*, HBK., *E. divergens*, Less., *E. Maximiliani*, Schrader, *E. conyzoides folio molli et incano*, etc., Pluk. t. 177, fig. 3), not the W. Indian form with more slender and pallid fewer-flowered involucre, and innermost bracts often acute, which approaches *E. odoratum*. (Trop. Amer.)

§ 2. EUPATORIUM proper. Involucre various; the bracts from thin-membranaceous or scarious to herbaceous, nerveless or few-nerved, mostly lax, either imbricated or equal and nearly uniseriate: receptacle flat, not hairy.

* Involucre cylindrical and imbricate in the manner of § 1, but thin-membranaceous and somewhat scarious when dry, faintly 3-striate: heads very numerous, corymbiform-cymose, mostly 5-10-flowered: leaves verticillate: stem herbaceous: herbage nearly destitute of resinous globules. — § *Verticillata*, DC.

E. purpúreum, L. (JOE-PYE WEED, TRUMPET WEED.) From pubescent to nearly glabrous: stems simple, 3 to 9 feet high, usually lineolate-punctate, often fistular: leaves commonly 3-6-nate, from oval-ovate to oblong-lanceolate, acuminate, coarsely serrate, reticulate-veiny, the base narrowed into a short petiole: cymes polycephalous, compound-corymbose and numerous: involucre (3 or 4 lines long) whitish and flesh-colored: flowers dull flesh-color or purple, rarely almost white. — Spec. ii. 838 (Corn. Canad. t. 72; Herm. Parad. t. 159; Moris. Hist. vii. t. 18); Torr. & Gray, Fl. ii. 81. *E. trifoliatum*, L. l. c., pl. Gronov. Virg. *E. maculatum*, L. Amœn. iv. 288, & Spec. ed. 2, 1174; Bart. Fl. Am. Sept. t. 102. *E. verti-*

cillatum, Muhl. in Willd. Spec. iii. 760. *E. ternifolium*, Ell. Sk. ii. 306; DC. Prodr. v. 151. — Low or wet ground, New Brunswick to Saskatchewan, Florida, and westward in wooded districts to New Mexico, Utah, and Brit. Columbia. Varies greatly, yet manifestly one species. The typical form very tall, growing in shady places, with smooth stem (usually purple above the nodes), large and thin leaves and loose inflorescence, its branches slender-peduncled. A narrow-leaved and attenuated form (var. *angustifolium*, Torr. & Gray, l. c.) is *E. falcatum*, Michx. Fl. ii. 99, and *E. lævigatum*, Torr. Cat. Pl. N. Y. The best marked of the variations are the following.

Var. **maculátum**, Darl. Common in open ground, 3 or 4 feet high, often roughish-pubescent: stem commonly purple, striate or sulcate; leaves somewhat rugose, 3-5-nate: inflorescence more compact and depressed. — Fl. Cest. 453; Torr. & Gray, l. c. *E. maculatum*, L. l. c. *E. fusco-rubrum*, Walt. Car. 199 ? *E. punctatum*, Willd. Enum. ii. 853. *E. dubium*, Poir. Suppl. ii. 606. — The most widely distributed form.

Var. **amœnum**. Leaves opposite or at most 3-4-nate, ovate or oblong, smoothish: stem slender, 2 feet high: heads fewer and only 3-5-flowered. — *E. amœnum*, Pursh, Fl. ii. 514. — An attenuate or depauperate form, growing in rather dry woods, mountains of Virginia to New York.

＊ ＊ Involucre imbricated, rather lax; the bracts of at least three or seldom only two lengths, the outer successively shorter. — § *Subimbricata*, DC.

╋ Heads as many as 20-flowered, large (about half-inch long): bracts of the involucre of 4 or 5 lengths, striate-nervose in the way of *Brickellia*: perennial herbs, of a Mexican type.

╋╋ Leaves entire, tomentose beneath.

E. Bigelóvii, Gray. Cinereous-pubescent, paniculately branched: leaves all opposite, ovate-lanceolate with a rounded or obscurely cordate base, acute, entire, short-petioled, puberulent above, soft-tomentose beneath, 3-5-ribbed at base: inflorescence somewhat paniculate: peduncles 3-5-cephalous: involucre turbinate, tomentulose, regularly imbricated; outer bracts ovate-lanceolate, acute, coriaceous, the innermost linear: flowers purplish: akenes nearly glabrous. — Bot. Mex. Bound. 75. — Arizona, on the Gila, *Bigelow*.

╋╋ ╋╋ Leaves acutely serrate, narrowed at the pinnately veined base, very short-petioled.

E. Brúneri. Minutely puberulent, apparently only a foot or two high: leaves opposite, ovate-oblong, acute, loosely veiny (2 or 3 inches long): paniculate rather slender peduncles bearing 3 or more sessile or short-peduncled heads: involucre campanulate, of comparatively few obscurely striate obtuse bracts; the outer oval, puberulent; inner ones scarious and glabrous, flesh-color (as probably are the flowers): akenes glabrous. — Damp ground, in the Rocky Mountains at Fort Collins, N. Colorado, *Dr. Bruner.*

╋╋ ╋╋ ╋╋ Leaves coarsely and often obtusely dentate, 3-5-ribbed at the cordate or sometimes truncate dilated base, slender-petioled, thin, bright green, acute or acuminate: flowers white or whitish: bracts of the campanulate involucre conspicuously striate-nerved: akenes minutely pubescent, not rarely 6-nerved, or with one or two of the nerves double !

E. Féndleri, Gray. A foot or two high, leafy, obscurely puberulent: leaves opposite or the upper alternate, deltoid-subcordate, tapering gradually to an acute or acuminate point: heads comparatively small and numerous, paniculate, all peduncled: bracts of the involucre all obtuse, the outer oblong. — Proc. Am. Acad. xvii. 205. *Brickellia Fendleri*, Gray, Pl. Fendl. 63, & Pl. Wright. ii. 73. (Some secondary or double ribs on many of the akenes connect this with *Brickellia*.) — Mountains of New Mexico and Arizona, *Fendler, Wright, Greene, Lemmon, Rusby.*

E. Párryi, Gray. Hirsutely pubescent (the spreading hairs of the stem somewhat glandular and viscid), loosely branched: leaves (so far as known) alternate, broadly ovate and rather deeply cordate, crenately dentate: heads rather few and large in an open naked panicle, slender-pedunculate: bracts of the involucre thin, oblong-lanceolate, acuminate, the innermost produced into a setiform tip. — Bot. Mex. Bound. 75. — Sierra de Carmel, S. border of Texas, on the Mexican side of the Rio Grande, *Parry.* (Mex.)

╋ ╋ Heads 3-9-flowered, small (only 2 or 3 lines long), paniculate: leaves (at least the lower) pinnately dissected, many of them alternate: involucral bracts 6 to 10, narrow, acute or abruptly pointed, narrowly scarious-margined, nerveless: flowers white or whitish: herbs very leafy, much branched, with habit of *Conyza* and *Artemisia.*

++ Very numerous heads in corymbosely paniculate cymules, 5-9-flowered.

E. pinnatifidum, ELL. Pubescent, 3 or 4 feet high. cauline leaves mainly opposite, sometimes 4-nate; lower 2-3-pinnately parted and incised into oblong or lanceolate divisions and lobes; upper once or twice parted into linear lobes: involucral bracts obtuse with a mucronate cusp. — Sk. ii. 295; DC. Prodr. v. 176 (not of 149, which is the earlier *E. brunii-folium*, Hook. & Arn., & *E. pinnatifissum*, Buck.); Torr. & Gray, l. c. 83. — Low grounds, near the coast, N. Carolina to Florida.

++ ++ Very numerous heads racemosely and thyrsoidly paniculate, 3-6-flowered: autumnal.

— **E. coronopifólium**, WILLD. Puberulent and sometimes pubescent, somewhat glutinous and balsamic-aromatic, 3 or 4 feet high: lower leaves more commonly opposite, twice 3-7-parted into linear entire or sparingly incised lobes; upper less compound, uppermost often entire, from broadly to narrowly linear: heads from over 2 to 3 lines long, in close spiciform panicles which are usually collected in an oblong thyrsus. — Spec. iii. 1750; DC. l. c. 176: Torr. & Gray, l. c. 83. *E. compositifolium*, Walt. Car. 199. *E. racemosum*, Bertol. Misc. v. 26, t. 1, from specimen with upper cauline and rameal leaves all entire. — *Chrysocoma coronopifolia*, Michx. Fl. ii. 102. — Sandy or dry soil, N. Carolina to Florida and Texas. Narrow-leaved forms too nearly approach the next.

— **E. fœniculáceum**, WILLD. l. c. (DOG-FENNEL.) Herbage fennel-scented when bruised, and slightly acrid: stem villous below with many-jointed slightly viscid hairs, 4 to 10 feet high, extremely leafy: leaves mostly glabrous, nearly all alternate, more compound than of the preceding and the lobes very narrowly linear or filiform: heads 2 lines long, loosely racemose-paniculate at the ends of the upper branches. — *E. faniculoides*, Walt. l. c. *E. leptophyllum*, DC. l. c. *Artemisia procerior*, etc., Dill. Elth. i. 38, t. 37. *A. capillifolia*, Lam. Dict. i. 267. *Mikania artemisioides*, Cass. Dict. Sci. Nat. liv. 130. *Traganthes*, Wallr. Sched. Crit. i. 456, ex Cass. l. c. — Moist pine barrens and low fields, common from N. Carolina to Florida. The varieties, *glabrum* and *lateriflorum*, Torr. & Gray, Fl., have no permanence. *E. leptophyllum*, DC., is only the more slender form. (W. Ind.)

++ ++ ++ Heads 3-15-flowered, 3 to 5 lines long: leaves undivided: flowers white (rarely purplish): involucre of rather few (8 to 12 or rarely 15) bracts.

++ Thyrsoid-paniculate, suffruticose: involucral bracts 3-nerved.

E. solidaginifólium, GRAY. A foot or two high, with simple branches, glabrate or minutely pubescent: leaves opposite, very short-petioled, oblong- or narrowly ovate-lanceolate from a rounded base, acute, entire or obscurely dentate, 3-nerved at or near the base, 10 to 18 lines long: thyrsus small (2 or 3 inches long), leafy at base, oblong or interrupted: heads few and crowded in each short-pedunculate cymule, 3-5-flowered: involucral bracts about 8, almost in two ranks, linear-lanceolate, acute: akenes pubescent. — Pl. Wright. i. 87, & ii. 74. — Dry hills between the Limpio and the Rio Grande in W. Texas, and near Santa Cruz, Arizona, *Wright, Pringle*, &c.

++ ++ Corymbosely cymose or fastigiate inflorescence: herbaceous perennials, mostly copiously resinous-atomiferous, some species becoming balsamic-glutinous: involucral bracts nerveless or nearly so.

= Leaves conspicuously petioled from a mostly truncate or abrupt base, strongly serrate: cymes broad: involucre cinereous-pubescent.

E. mikanioídes, CHAPM. Tomentose-pubescent when young, soon glabrate: stems simple, a foot or two high from a creeping base: leaves opposite, deltoid-ovate or the uppermost oblong, obtuse, thickish and rather fleshy, glandular-punctate, obtusely dentate (an inch or two long): heads 5-flowered: involucral bracts linear, rather obtuse. — Fl. 195. *E. crassifolium*, Shuttleworth in distrib. coll. Rugel. — Low and sandy ground, coast of Florida, *Chapman, Rugel*, &c.

E. serótinum, MICHX. Puberulent: stems 5 to 7 feet high, corymbosely branched above: leaves oblong- or ovate-lanceolate, acute or acuminate, thinnish, acutely serrate (3 to 6 inches long), many of the upper alternate, some of these cuneate at base: heads 7-15-flowered, very numerous: involucral bracts rather narrow (10 or 12) linear-oblong, very obtuse. — Fl. ii. 100; Torr. & Gray, Fl. ii. 89. *E. ambiguum*, Hook. Comp. Bot. Mag. i. 96, as to 'Covington' plant, is either this species or a (hybrid?) form between it and *E. semiserratum*, DC.. the *E. parviflorum*, Ell. — Low grounds, Maryland to Iowa, Florida, and Texas; Sept. to Nov. (Adj. Mex.)

= = Leaves from linear to oblong, sessile or some short-petioled from a narrowed base, chiefly opposite: heads mostly 5-flowered, occasionally 6-7-flowered.

a. Involucral bracts with conspicuous white-scarious acute tips; the inner equalling the flowers.

— **E. álbum,** L. Pubescent with jointed spreading hairs; stem 2 feet high; leaves oblong-lanceolate or narrowly oblong, commonly obtuse, coarsely serrate, veiny, sessile (2 to 4 inches long); cymes fastigiate; involucre (4 or 5 lines long) mostly bright white and glabrous throughout, well imbricated; its bracts slender-mucronate, the outer sometimes pubescent and dark-dotted with resinous globules. — Mant. 111; Walt. Car. 199. *E. glandulosum,* Michx. Fl. ii. 98. *E. stigmatosum,* Bertol. Misc. v. 15, t. 5. — Sandy fields and pine barrens, Long Island, N. Y., and Penn. to Florida and Louisiana.

Var. subvenósum. More minutely roughish-pubescent: leaves smaller, only an inch or two long, mostly acute, with smaller and more appressed serratures, less veiny and more manifestly 3-nerved at base, where the upper cauline are not narrower; involucral bracts not so white. — Long Island (*E. S. Miller*) and New Jersey. Burke Co., N. Carolina ?

— **E. leucólepis,** Torr. & Gray. Puberulent: stem slender, about 2 feet high: leaves lanceolate or linear, minutely and sparingly appressed-serrate, thickish, obscurely 3-nerved at base, closely sessile (1 to 3 inches long); involucre (3 lines long) canescently pubescent; the narrowed tips of the bracts white-scarious. — Fl. ii. 84. *E. linearifolium,* Michx., Pursh, &c., partly. *E. hyssopifolium,* Ell. Sk. ii. 296; Hook. Comp. Bot. Mag. i. 96. *E. glaucescens,* var. *leucolepis,* DC. l. c. 177. — Moist pine barrens, New Jersey to Florida and Louisiana, in the low country.

b. Involucral bracts obscurely if at all scarious, mostly obtuse, at length shorter than the flowers.

— **E. hyssopifólium,** L. Merely puberulent: stems about 2 feet high, very leafy, commonly with fascicles in the axils, simple, corymbosely branched at summit: leaves occasionally verticillate, linear, obtuse, entire or sparingly dentate, narrowed at base, ¾ to 2 inches long, the broader forms with lateral nerves: cymes crowded: involucre (3 lines long) canescently pubescent and glandular; bracts rather few, the inner with somewhat scarious margins and tips, obtuse, sometimes apiculate. — Spec. ii. 836 (Dill. fig. & Pluk.); Torr. & Gray, Fl. ii. 84. *E. linearifolium,* Walt. Car. 199; Michx. l. c. (partly); Willd. l. c. *E. linearifolium & hyssopifolium* (chiefly), DC. l. c. — Dry and sterile soil, Mass. to Florida and Texas, along and toward the coast. Varies greatly in the foliage, the extreme forms being, on one hand, that with very narrowly linear and much fascicled leaves; on the other, the

Var. laciniátum. Leaves lanceolate and linear-lanceolate, irregularly and coarsely dentate, even laciniate. — Penn. and Kentucky to Carolina and Louisiana.

Var. tortifólium. Leaves oblanceolate or spatulate-linear, mostly short, all entire, inclined to be vertical by a twist at base, many of them alternate. — *E. tortifolium,* Chapm. in Bot. Gazette, iii. 5. *E. cuneifolium,* A. H. Curtiss, distrib. 1194. — Sandy pine barrens, S. Carolina, Georgia, and Florida. The lower leaves resemble the uppermost of *E. cuneifolium,* but are all entire, often reflexed as well as vertical.

E. cuneifólium, Willd. Habit, involucre, and pubescence of the preceding: leaves short (half to a full inch long), oblanceolate to cuneate-spatulate, obtuse, glaucescent, few-toothed toward the extremity, or the upper entire, uppermost very small and oblong-linear. — Spec. iii. 1759, excl. syn. (not DC.); Torr. & Gray, l. c. 85; Chapm. l. c. *E. linearifolium,* Michx. l. c., in part. *E. glaucescens,* Ell. l. c. 297; DC. l. c., excl. var. *E. hyssopifolium,* DC. l. c., in part. *E. cassinifolium,* Bertol. Misc. v. 17, t. 6. — Dry ground, South Carolina, Georgia, Alabama, and Florida.

— **E. semiserrátum,** DC. Tomentulose-pubescent: stems 2 or 3 feet high, much branched above: leaves oblong-lanceolate, mostly acute or acuminate (commonly 2 or even 3 inches long), serrate with numerous unequal teeth from above or below the middle to the apex, triplinerved, rather veiny, narrowed at base, the lower into a short mostly distinct petiole: cymes numerous; heads small: involucre (2 lines long) canescently pubescent, of few bracts; the longer linear-oblong, very obtuse, the others much shorter. — *E. semiserratum & E. cuneifolium,* DC. Prodr. v. 177. *E. parviflorum,* Ell. Sk. ii. 299; Torr. & Gray, l. c., not Swartz. *E. ambiguum,* Hook. Comp. Bot. Mag. i. 96 (1835), in part only, the Jacksonville plant, but heads not "8-10-flowered." — Virginia to Florida, Arkansas, and Texas. In dry and open ground, plants with smaller and firmer leaves pass into

Var. lancifólium. Glabrate: leaves lanceolate and verging to linear, 5 to 2 lines wide, rather rigid, 3-nerved from near the base.—*E. parviflorum*, var. *lancifolium*, Torr & Gray, l. c.— W. Louisiana and Texas, *Drummond*, *Leavenworth*, *Hale*.

— **E. altíssimum**, L. Pubescent: stems 4 to 7 feet high, branched at summit, very leafy: leaves lanceolate, tapering gradually to both ends, acuminate, acutely serrate above the middle, 2 to 4 inches long, with 3 conspicuous parallel nerves (giving the aspect of a triplinerved *Solidago*); uppermost entire: cymes numerous and irregular: heads fully 3 lines long: involucre canescently pubescent; its bracts oblong and very obtuse.— Jacq. Hort. Vind. t. 164; Michx. Fl. ii. 97; Torr. & Gray, l. c. *Kuhnia glutinosa*, DC. Prodr. v. 127, not Ell.— Dry ground, Penn. to Iowa, N. Carolina, and Texas.

= = = Leaves sessile or very short-petioled with a broad base, normally opposite, occasionally 3-nate: involucre pubescent.

a. Heads mostly 5-flowered, in one species 6-8-flowered: herbage roughish-pubescent: inner bracts of involucre acutish or acute, or sometimes acuminate at the thin tip.

— **E. teucrifólium**, WILLD. Stem 2 or 3 and even 8 feet high, not very leafy: leaves oblong, coarsely and irregularly serrate, rarely somewhat incised, slightly petioled (2 to 4 inches long); the upper small and few-toothed, sometimes hastately 1-2-toothed near the broad sessile base, or lanceolate and entire, usually alternate, as are the branches of the corymbiform general inflorescence: cymes rather small and dense. — Spec. iii. 1753, & Hort. Berol. t. 32; Torr. & Gray, l. c. *E. pilosum*, Walt. Car. 199 ? *E. verbenæfolium*, Michx. Fl. ii. 98. *E. lanceolatum*, Muhl. in Willd. l. c. *E. pubescens*, Bigel. Fl. Bost. ed. 2, 296, not Muhl. — Moist and shady ground, Mass. to Florida and Louisiana.

— **E. rotundifólium**, L. Stem a foot to a yard high, strict, corymbosely branched at summit: leaves in the typical form round-ovate, obtuse or abruptly acute, sessile or nearly so from a truncate or obscurely cordate base, regularly and closely crenate-dentate, veiny (larger 2 inches long): cymes corymbosely fastigiate, dense. — Spec. ii. 837 (Pluk. Alm. 141, t. 88, fig. 4); Torr. & Gray, l. c. *E. Marrubium*, Walt. Car. 199 ? — Dry and sterile soil, especially in pine barrens, Canada ! and New Jersey to Florida and Texas.

Var. scábridum. A form with smaller (an inch or two long) and more scabrous or cinereous leaves, the upper and sometimes all with cuneate base; affecting drier and more sterile soil. — *E. scabridum*, Ell. Sk. ii. 298; Chapm. Fl. 196. — Lower part of S. Carolina to Florida and Texas.

Var. ovátum, TORR. Commonly taller and larger: leaves ovate (often 2 or 3 inches long), acute, hardly truncate at base, more strongly serrate, sometimes laciniately so, either roughish-pubescent or smoother and glabrate: heads sometimes but not generally 7-8-flowered. — Torr. in DC. Prodr. v. 178. *E. pubescens*, Muhl. in Willd. Spec. iii. 1155; Willd. Enum. ii. 852; Torr. & Gray, l. c. *E. obovatum*, Raf. in Med. Rep. hex. 2, v. 359 ? *E. ovatum*, Bigel. Fl. Bost. ed. 2, 296. — Massachusetts to S. Carolina, near the coast.

b. Heads 5-flowered: herbage glabrous: narrow involucre more imbricated; its bracts obtuse.

— **E. sessilifólium**, L. Corymbosely branched above, 2 to 6 feet high: leaves oblong- or ovate-lanceolate, tapering from near the rounded or truncate closely sessile base into a narrow acumination, finely serrate, pinnately veiny (3 to 6 inches long): cymules small and crowded, few-headed, numerous in effusely compound cymes. — Spec. ii. 837; Torr. & Gray, l. c. *E. truncatum*, Ell. Sk. ii. 298, not Willd. — Dry and wooded ground, Mass. to Illinois, Virginia, and along the mountains to Alabama.

c. Heads 10-15-flowered (or by confluence sometimes many-flowered), much crowded: leaves perfoliate or connate-clasping, divaricate, narrow and elongated, one-ribbed: stems 2 to 4 feet high.

— **E. perfoliátum**, L. (THOROUGHWORT, BONESET.) Stem villous-pubescent, fastigiately branched above, stout: leaves lanceolate, connate-perfoliate, tapering gradually to an acuminate apex, finely and closely crenate-serrate, rugose, soft-pubescent, or almost tomentose beneath, 4 to 8 inches long: heads small (3 lines long) but very numerous, in dense compound-corymbose cymes, mostly 10-flowered: bracts of the involucre linear-lanceolate, with slightly scarious acutish tips. — Spec. ii. 838 (Pluk. Alm. 140, t. 87, fig. 6); Bart. Veg. Med. Mat. t. 37; Bigel. Med. Bot. i. 38, t. 2; Raf. Med. Bot. t. 36; Torr. & Gray, l. c. 88. — Wet ground, New Brunswick to Dakota, south to Florida and Louisiana. Varies with purple flowers (Penn. *Porter*), and with leaves in threes (Virginia, *Curtiss*, &c.); also into

Var. truncátum, with the upper or even all of the leaves disjoined and truncate at base; some of them alternate.— *E. truncatum*, Muhl. in Willd. Spec. iii. 1751. *E. salviæfolium*, Sims, Bot. Mag. t. 2110. — With the normal form.

—— Var. **cuneátum**, ENGELM. (*E. cuneatum*, Engelm. in Torr. & Gray, l. c.), with smaller leaves narrowed as well as disjoined at base, and fewer-flowered heads, has the appearance of being a hybrid between *E. semiserratum* and *E. perfoliatum*. — Eastern Arkansas and Missouri, *Engelmann*. Also Louisiana, *Hale*, a form between this state and the preceding.

— **E. resinósum**, TORR. Puberulent, glutinous with resinous atoms: stem slender, 2 or 3 feet high, fastigiate-corymbose at summit: leaves linear-lanceolate (4 to 6 inches long, 4 to 6 lines wide), half-clasping or slightly connate, finely serrate, glabrate above, canescent beneath;. cymules numerous in compound cymes: bracts of the 10–15-flowered involucre narrowly oblong, obtuse. — DC. Prodr. v. 176 ; Torr. & Gray, Fl. ii. 88.— Wet pine barrens, New Jersey, where it was first collected by *Bartram*.

+ + + + Heads 24–30-flowered, hardly over 2 lines long : bracts of the involucre of three lengths, obtuse, thin, conspicuously few-nerved : habit of the following section.

E. pycnocéphalum, LESS. Pubescent or nearly glabrous : stems slender, erect or spreading from a perennial root, a foot or two high : leaves membranaceous, deltoid-ovate or subcordate, acute or acuminate, coarsely serrate or dentate, slender-petioled : cymes small and compact, solitary or corymbosely clustered at the end of naked branches : heads very short-pedicelled : involucre campanulate; the bracts mostly glabrous, oblong and oblong-linear, very obtuse; innermost equalling the white flowers.— Less. in Linn. vi. 404. *E. Schiedeanum*, Schrad. Ind. Sem. Hort. Gœtt. 1832, 3; DC. Prodr. v. 159. *E. multinerve*, Benth. Pl. Hartw. 76. *E. Sonoræ*, Gray, Pl. Wright. ii. 76.— Rocky ravines, S. Arizona and along the Mexican borders of Texas; a form with small and deeply dentate leaves, and comparatively few and small heads. *E. Schiedeanum*, var. *grosse-dentatum*, Gray, Bot. Mex. Bound. 76. (Mex., &c.)

* * * Involucre (campanulate or oblong) of bracts all of the same length or nearly so, in one or two series, or with only a few accessory and shorter ones at base: leaves mainly opposite, petioled. — § *Eximbricata*, DC.

+— Shrubby, freely branched: flowers white, sometimes purplish-tinged.

— **E. Wríghtii**, GRAY. A foot or two high, puberulent : branches very leafy : leaves small (half-inch long), ovate, obtuse, entire or obscurely few-toothed, thickish, scabrous, abruptly contracted into a short margined petiole : heads (3 or 4 lines long), about 12-flowered, rather few in a somewhat leafy terminal cyme : involucre half the length of the flowers, of about 10 oblong-lanceolate acute or obtusish greenish obscurely 3-nerved and equal bracts in a double series, sometimes one or two small accessory ones. — Pl. Wright. i. 87, ii. 73.— Guadalupe Mountains, western borders of Texas, *Wright*.

E. villósum, SWARTZ. Shrub 4 to 6 feet high, rusty-pubescent: leaves ovate or somewhat deltoid, rather obtuse, sparingly serrate or some entire, tomentulose beneath (1 to 3 inches long), on short slender petioles : heads small (2 or 3 lines long), 8–15-flowered, numerous and crowded in corymbiform cymes : involucre half the length of the fully developed flowers, of 8 to 10 oblong-lanceolate obtuse and nerveless equal bracts.— DC. Prodr. v. 172 ; Chapm. Fl. 196. *E. Cubense*, DC. l. c. ?— S. Florida, *Blodgett, Garber, Curtiss*, &c. (W. Ind.)

E. ageratifólium, DC. Shrub 3 to 7 feet high, with slender and spreading mostly herbaceous branches, green and nearly glabrous : leaves deltoid-ovate, obtusish or obtusely acuminate, coarsely and rather obtusely dentate (2 or 3 inches long), slender-petioled : heads (5 lines long), pedicelled, numerous in corymbiform cymes, 10–30-flowered : involucral bracts 8 to 12, narrowly lanceolate or linear, acutish, greenish, nerveless above, somewhat 2-ribbed at base. — Prodr. v. 173; Torr. & Gray, Fl. ii. 90 (var. *Texense*, which does not differ); Gray, Pl. Lindh. ii. 219 ; Griseb. Fl. W. Ind. 360. *E. Berlandieri*, DC. l. c. 167. *E. Lindheimerianum*, Scheele in Linn. xxi. 599. *Bulbostylis deltoides*, Buckley in Proc. Acad. Philad. 1861, 456.— Rocky shaded hills and ravines, Texas, *Lindheimer, Wright*, &c.; fl. Nov. to May. (W. Ind., Mex.)

+ +— Herbaceous perennials, or the first species barely lignescent at base.

++ Corolla wholly glabrous even in the bud.

— **E. occidentále**, HOOK. Minutely puberulent, glabrate : stems 8 to 20 inches high, strict, simple or with few ascending branches : leaves ovate with truncate base, rarely subcordate or roundish, obtuse or acute, sparingly dentate, sometimes merely repand or entire, an inch or two long, rather short-petioled: cymes small and rather compact, somewhat paniculate; heads (4 or 5 lines long) 15-25-flowered: involucre hardly longer than the mature akenes; its bracts about 15 in two series, nearly equal, lanceolate, rather firm, nearly nerveless; corolla white or flesh-color. — Fl. i. 305; Torr. & Gray, Fl. ii. 91. *E. Oreganum*, Nutt. Trans. Am. Phil. Soc. vii. 286. — Crevices of rocks, Washington Territory and Oregon east of the Cascade Mountains, N. Nevada, and through the Sierra Nevada of California.

— Var. **Arizónicum.** Larger (2 feet high), more branching and floribund: leaves sometimes 2½ inches long. — *E. ageratifolium*, var. *? herbaceum*, Gray, Pl. Wright. ii. 74. *E. Berlandieri*, Gray, Bot. Mex. Bound. 76, not DC. — Mountains of Arizona and New Mexico: also California, *Bridges*. The opposite extreme from the plant of Oregon, which has small and thinnish leaves, but not unlike plants from the Sierra Nevada.

━ **E. incarnátum,** WALT. More or less pubescent: stems 2 or 3 feet long, slender and weak, loosely or diffusely branched: leaves thin, deltoid, or ovate-lanceolate with broad truncate or cordate base, tapering to a mostly acuminate apex, coarsely crenate or serrate (an inch or two long), veiny, slender-petioled: cymes small and lax: heads (2 or 3 lines long) about 20-flowered: involucre nearly equalling the pale purple or sometimes white corolla; its bracts unequal, narrow, thin and 2-nerved when dry, the inner linear, a few external ones much shorter. — Car. 200; Ell. Sk. ii. 306; DC. Prodr. v. 175; Torr. & Gray, l. c. — N. Carolina to Florida, Louisiana, and Texas. (Adj. Mex.)

++ ++ Lobes of the pure white corolla more or less bearded outside in the bud, sometimes very sparsely and minutely so, or the beard fugacious: heads 15-30- or sometimes 8-14-flowered, cymose.

== Involucre 2 or 3 lines long, rather narrow; the linear bracts nearly equal, green externally and nerveless when fresh, but more or less 2-nerved when dried: cymes corymbiform and naked, usually ample.

━ **E. ageratoídes,** L. f. Nearly glabrous, sometimes pubescent: stems 1 to 3 feet high, branching above: leaves bright green, membranaceous, long-petioled, ovate, with truncate or subcordate or broadly cuneate base, acuminate, coarsely and rather sharply dentate-serrate, conspicuously veiny, 3 to 5 inches long: cymes ample, corymbose-cymose. — Suppl. 355; DC. Prodr. v. 175; Torr. & Gray, Fl. ii. 89. *E. urticæfolium*, Reich. Syst. iii. 719; Michx. Fl. ii. 100, not L. f. *E. altissimum*, L. Syst. Veg. 614. *E. odoratum?* Walt. Car. 200? *E. Fraseri*, Poir. Suppl. ii. 600 (Lam. Ill. t. 672, fig. 4). *Ageratum altissimum*, L. Spec. ii. 839 (Corn. Canad. t. 21; Moris. Syst. sect. 7, t. 18, f. 11). — Moist woodlands and rich soil, Canada to Minnesota, Arkansas, Georgia, and Louisiana. A state with viscid-villous stem and petioles, Bedford Co., Virginia, *Curtiss*.

Var. **angustátum.** Smaller, slender: leaves from ovate-lanceolate to broadly lanceolate, much acuminate, coarsely serrate with only 3 to 6 teeth on each margin, commonly cuneate at base: cymes looser: heads only 8-12-flowered. — W. Louisiana, *Hale.* Texas, *Wright, Lindheimer.*

━**E. aromáticum,** L. Herbage not aromatic, minutely puberulent: stems more simple, a foot or two high: leaves dull green, thicker, mostly short-petioled, ovate, often subcordate, acutish or obtuse, crenate-serrate, 1½ to 3 inches long: cymes simpler. — Spec. ii. 839, fide herb. & syn. Pluk. & Gronov.; DC. l. c.; Torr. & Gray, l. c. *E. cordatum*, Walt. Car. 199 ? *E. ceanothifolium*, Muhl. in Willd. Spec. iii. 1755; Ell. Sk. ii. 303; DC. l. c. — Dry woods and pine barrens, mostly in sterile soil, coast of Massachusetts to Florida. Passes on the one hand almost into the preceding; on the other, into

Var. **melissoídes.** Slender, roughish-puberulent, strict, somewhat paniculately cymose at summit: heads 5-12-flowered: leaves subcordate-ovate or oblong, ¾ to 2 inches long, obtuse, crenulate-dentate or with few coarser teeth, very short-petioled or even subsessile, somewhat scabrous, most of them much shorter than the internodes. — *E. melissoides*, Willd. l. c. *E. cordiforme*, Poir. Suppl. ii. 600. *E. cordatum*, DC. l. c., & var. *Fraseri.* — Sterile soil, Penn. ? to Florida and Louisiana.

Var. **incísum.** An insufficiently known plant, with the straggling habit and glabrous involucre of *E. incarnatum*; probably a form either of this or the preceding species: leaves slender-petioled, thickish, coarsely or laciniately dentate, broadly cuneate at base, considerably like those of *E. cælestinum*, var. *salinum*, Griseb.: "flowers very fragrant." — *E. suaveolens*, Chapm. Bot. Gazette, iii. 5, not of HBK. — Manatee, &c., S. Florida, *Chapman.*

═ ═ Involucre less than 2 lines long; the bracts broader, green externally, 2-3-nerved when dry: inflorescence somewhat paniculate and leafy.

E. paupérculum, GRAY. A foot or two high, nearly glabrous: leaves ovate-lanceolate (mostly inch long), roundish at base, obtusely serrate, on rather short slender petioles: heads 25-flowered, small (2 lines high), few in the numerous small cymes, which are paniculately disposed, terminating short leafy branches: bracts of the involucre 10 or 12, oblong-lanceolate, puberulous, little over half the length of the white flowers: corolla-lobes slightly hirsute outside or becoming naked: pappus soft and white. — Proc. Am. Acad. xvii. 205. — Santa Rita Mountains, S. Arizona, along brooks and on dripping rocks, *Pringle*.

═ ══ ═ Heads comparatively large and few in the cymes, 25-35-flowered: involucre 3 or 4 lines high, rather broad.

E. Rothróckii. Glabrous (or peduncles somewhat pubescent): stems a foot or two high, simple or brachiately branched above: leaves bright green, ovate or deltoid-ovate, usually acuminate, coarsely and sharply serrate, sometimes irregularly or doubly serrate, and the teeth tipped with a callous gland (the larger 2 inches long, with petiole half-inch or less, smaller in depauperate plants nearly sessile): bracts of the involucre 15 to 17, equal and similar, linear-lanceolate, mostly acute, glabrous, when dry pale and somewhat scarious and conspicuously 2-3-nerved, nearly equalling the white and soft barbellulate-scabrous pappus: corolla-lobes rather strongly bearded externally. — Mountains of S. Arizona: on Mount Graham, *Rothrock* (740, 741); Chiricahua Mountains, *Lemmon*. Heads larger and fewer than in the Mexican *E. grandidentatum*, DC.; the involucre not imbricated as in *E. Fendleri*.

§ 3. CONOCLÍNIUM, Benth. Receptacle of the flowers conical or hemispherical: otherwise as in the *Eximbricata* subsection of the preceding: habit of *Ageratum* § *Cœlestina*: flowers blue or violet (sometimes white), sweet-scented: bristles of the pappus rather scanty in a single series: leaves opposite: perennial herbs. — *Conoclinium*, DC. Prodr. v. 135.

E. cœlestínum, L. (MIST-FLOWER.) Somewhat pubescent: stems erect, branched at summit: leaves deltoid-ovate or subcordate, obtuse or acutish, obtusely serrate, rarely with some coarser salient teeth, slender-petioled: cymes rather compact: receptacle obtusely conical. — Spec. ii. 838 (Dill. Elth. t. 114; Pluk. Mant. t. 394); Michx. Fl. ii. 100. *Calestina cærulea*, Spreng. Syst. iii. 446, not Cass. *Conoclinium cœlestinum*, DC. l. c.; Torr. & Gray, Fl. ii. 92. — Moist shady ground, New Jersey to Florida and Texas, and west to Arkansas and Illinois. *Conoclinium dichotomum*, Chapm. in Bot. Gazette, iii. 5, appears to be a lax and more branched form, of Florida and Texas, found only on the coast, approaching the var. *salinum*, Griseb. Cat. Cub. 146. (Cuba.)

E. betónicum, HEMSL. From tomentose-villous to glabrate: stems lax, loosely branching: branches naked and pedunculiform at summit, bearing some small corymbose or paniculate cymes: leaves oblong, mostly obtuse, with cordate base, crenate, petioled: receptacle low-conical. — Biol. Centr.-Am. Bot. ii. 93. *Conoclinium betonicum*, DC. Prodr. v. 135; Gray, Bot. Mex. Bound. 76. — Southern border of Texas on the Rio Grande, *Schott*, a glabrate form. (Adjacent Mex.)

Var. **subíntegrum**. Leaves sometimes truncate, commonly obtuse or cuneate at base, obscurely crenate, denticulate, repand or entire, from villous or cinereous-tomentulose to nearly glabrous. — *Conoclinium betonicum*, var. *integrifolium*, Gray, Pl. Wright. i. 88. *Eupatorium Hartwegi*, Benth. Pl. Hartw. 19? — Southern border of Texas, *Wright*, *Bigelow*, &c. (Mex.)

E. Gréggii. Minutely puberulent: stems erect, a foot or two high, bearing one or few small and dense cymes at the naked pedunculiform summit: leaves nearly sessile, palmately 3-5-cleft or parted; the divisions laciniate-pinnatifid into narrow lobes: receptacle low-conical. — *Conoclinium dissectum*, Gray, Pl. Wright. i. 88; Bot. Mex. Bound. 76. *Eupatorium dissectum*, Gray, Proc. Am. Acad. xviii. 100 (name only), not Benth. Bot. Sulph. 113, with which Hemsley has confounded it. — Low ground, S. Texas to Arizona near the Mexican border, *Wright*, &c. (Adjacent Mex., first coll. by *Gregg*.)

E. LÚTEUM, Raf. in Med. Rep. N. Y., is doubtless a false species. E. CRASSIFÓLIUM and E. VIOLÁCEUM, Raf. Fl. Ludov., are fictitious, as are all the species of that work.

11. **CARMINÁTIA**, Moçino. (*Prof. B. Carminati*, of Pavia, wrote on the materia medica.) — Single species, an annual; with opposite or partly alternate broad and long-petioled thin leaves, and racemiform-paniculate heads of whitish flowers. — DC. Prodr. vii. 267; Deless. Ic. Sel. iv. 98.

C. tenuiflóra, DC. l. c. Sparsely pubescent or hirsute: stems a foot to a yard high, terminating in a leafless virgate panicle: leaves broadly deltoid-ovate, as wide as long, repand-dentate, veiny, often shorter than the petiole: heads half-inch long: soft pappus bright white. — Gray, Pl. Wright. ii. 71. — New Mexico and Arizona, *Wright*, &c. (Mex., first coll. by *Moçino*.)

12. **KÜHNIA**, L. (*Dr. Adam Kuhn*, of Philadelphia, took the original species to Linnæus.) — Perennials of Atlantic U. S. and Mexico; with chiefly alternate leaves (more or less sprinkled with resinous atoms, as in allied genera), usually with scattered or corymbosely cymose heads, these of 10 to 30 whitish or at length purple flowers, produced in late summer or autumn: pappus mostly tawny. — Spec. ed. 2, ii. 1662 (excl. syn. Pluk.), & Gen. ed. 6, 95 (the anthers wrongly described, from imperfect or monstrous blossoms). *Critonia*, Gærtn., not Browne. *Kuhnia* § *Strigia*, DC. Prodr. v. 126.

K. SCHÄFFNERI, Gray, Proc. Am. Acad. xvii. 207, of Mexico, is a peculiar species, with scapiform monocephalous peduncles and tuberous roots. The rest of the genus is the following.

K. eupatorioídes, L. Stems wholly herbaceous, 2 or 3 feet high: leaves from oblong- (or even ovate-) lanceolate to linear, irregularly few-toothed or upper ones entire, the lower narrowed at base and sometimes short-petioled: pubescence minute or soft and cinereous, or hardly any: heads more or less cymose-clustered. — L. f. Dec. ii. 21, t. 11; Torr. & Gray, Fl. ii. 78. *K. eupatorioides* & *K. Critonia*, Willd. Spec. iii. 1773. *K. dasypia, glutinosa, elliptica, tuberosa, fulva (media, glabra),* & *pubescens*, Raf. *Critonia Kuhnia*, Gærtn. Fruct. ii. 411, t. 174, f. 7; Michx. Fl. ii. 101. — Dry ground, New Jersey and Penn. to Montana, and south to Texas. Very variable; the extreme forms are

Var. **corymbulósa**, Torr. & Gray, l. c. A foot or two high, stouter, somewhat cinereous-pubescent or tomentulose: leaves rather rigid and sessile, from oblong to lanceolate, coarsely veiny: heads rather crowded. — *K. glutinosa*, Ell. Sk. ii. 292, not DC. *K. suaveolens*, Fresen. Ind. Sem. Fraucf. 1838. *K. Maximiliani*, Sinning in Neuwied. Trav. *K. macrantha*, Buckley in Proc. Acad. Philad. 1861, 456. — Prairies and plains, Illinois to Dakota and Nebraska, and south to Alabama and Texas.

Var. **grácilis**, Torr. & Gray, l. c. Leaves from lanceolate to very narrowly linear, few-toothed or all but lower entire, minutely puberulent or nearly glabrous: general inflorescence more open and paniculate. — *K. paniculata*, Cass. Dict. xxiv. 516; DC. l. c. *K. Critonia*, Ell. l. c. — Carolina to Florida, Alabama, &c. Seems to pass into the following.

K. rosmarinifólia, Vent. Perhaps more lignescent at base, a foot or two high: leaves all entire, linear or linear-lanceolate, mostly with revolute margins, and the upper almost filiform, from a quarter of a line to 2 lines wide, somewhat scabrous: heads more scattered or paniculate: plume of the bristles of the pappus perhaps a little shorter. — Hort. Cels. t. 91 (poor figure of a broadish-leaved form, with too much imbricated involucre): DC. l. c. (excl. syn. Ort. ?), but surely from Mexico, not "Cuba." *K. frutescens*, Hornem. Hort. Hafn. ii. 791. *K. leptophylla*, Scheele in Linn. xxi. 598. *K. eupatorioides*, var. *gracillima*, Gray, Pl. Lindh. ii. 218, a very slender-leaved form, which connects with the slenderest of the preceding. — Rocky open ground, Texas to Arizona. (Mex.)

13. **BRICKÉLLIA**, Ell. (*Dr. John Brickell* of Georgia, correspondent of Muhlenberg and Elliott.) — Herbs or undershrubs; with opposite or alternate veiny leaves, and variously disposed heads of white, ochroleucous, or rarely flesh-colored flowers, in late summer. A genus of about 40 species, of the warmer parts of the U. S. and Mexico. · A single annual species (*B. diffusa*, which may

reach Florida) is widely tropical American, and there is an anomalous species in Brazil. — Sk. ii. 290; Benth. Bot. Sulph. 22; Gray, Pl. Wright. i. 84, & Proc. Am. Acad. xvii. 206; Benth. & Hook. Gen. ii. 247. *Coleosanthus*, Cass. Dict. x. 36. *Rosalesia*, Llave & Lex.? *Clavigera* & *Bulbostylis*, DC. Prodr. v. 127, 138.

B. HASTÁTA, Benth., is a well-marked species of Lower California, described in Bot. Sulph. 21. In that work the genus was first extended to its proper limits, but made to rest on the bulbous base of the style (which is of little account) instead of the 10-costate akene.

* Heads 35–50 flowered, large or middle-sized: pappus-bristles merely scabrous or densely serrulate.
  + Herbage white-tomentose: leaves rounded, pointless.

B. incána, GRAY. A foot or two high, loosely branched from a suffrutescent base; dense and fine tomentum somewhat deciduous: leaves alternate, sessile, subcordate-rotund or ovate, entire (less than inch long): heads solitary terminating the branchlets, inch high, pedunculate: involucre broadly campanulate, pluriserial; its bracts firm-chartaceous; short outer ones ovate, inmost lanceolate-linear: akenes (5 lines long) cinereous-pubescent. — Proc. Am. Acad. vii. 350, & Bot. Calif. i. 300. — S. E. California along the Mohave River, *Cooper, Parry, Parish.*

  + + Puberulent to almost glabrous: leaves sessile or subsessile, all alternate, not cordate,
  ++ Rigid-coriaceous, spinulose-pointed and toothed: fruticulose.

B. atractyloídes, GRAY. A foot or less high, woody except the new shoots, much branched: leaves ovate, acuminate, bright green, minutely scabrous-atomiferous, 3-nerved and reticulate-veined (an inch or less long): branchlets terminating in a solitary (half-inch long) and slender-pedunculate head: involucre campanulate; its bracts firm-chartaceous; outer ovate, acuminate, little shorter than the linear-lanceolate innermost. — Proc. Am. Acad. viii. 290. — Rocky ravines, St. George, S. Utah, *Palmer, Parry.*

  ++ ++ Leaves not coriaceous, pointless, seldom an inch long, sometimes viscidulous: stems herbaceous from a lignescent base or stock, a span to a foot or so high, leafy to the top: heads mostly singly terminating corymbose leafy branches.
  = Heads three fourths of an inch long, involucrately surrounded or subtended by small uppermost leaves.

B. Greénei, GRAY. Very viscid: leaves ovate, obtuse, minutely more or less serrate, and the lower short-petioled; upper oblong and often entire, uppermost forming accessory loose bracts to the involucre: proper involucral bracts lanceolate and linear, acuminate, glabrous: akenes not glandular, glabrous, or at the upper part hirtellous with a few scattered short bristles on the ribs. —Proc. Am. Acad. xii. 58. — N. E. California, on Scott River, *Greene.*

  = = Heads two-thirds or over half inch long, naked at base, commonly somewhat peduncled: leaves entire, rarely with a tooth or two, obscurely 3-nerved, puberulent and minutely somewhat granadular-granulose or atomiferous, graveolent, becoming slightly viscidulous.

B. oblongifólia, NUTT. Leaves oblong or some upper ones lanceolate, obtuse or mucronate: involucral bracts all acute or mucronate-pointed; outer and short ones oblong-lanceolate; inner narrowly linear: akenes sprinkled with minute sessile and stipitate glands, toward summit often a few bristles. — Trans. Am. Phil. Soc. l. c. 288; Torr. in Wilkes Pacif. Exp. xvi. t. 9; Eaton, Bot. King Exp. 137. — In gravelly or dry soil, E. Oregon to Brit. Columbia, first coll. by *Nuttall.*

Var. abbreviáta, GRAY. Dwarf: leaves seldom half-inch long: involucral bracts less acute: akenes minutely and sparsely glandular on the ribs, otherwise glabrous. — Eaton, l. c, t. 15, f. 7–10. — W. Humboldt Mountains, Nevada, *Watson.*

B. linifólia, EATON. Rather more pubescent: leaves oblong-lanceolate to almost linear: involucre of the preceding, or bracts more attenuate-acute: akenes minutely hispid on the ribs, not glandular. — Bot. King Exp. 137, t. 15, f. 1–6. — Sandy banks of streams, &c., Sierra Nevada, California, to Utah and borders of Arizona; first coll. by *Watson.*

B. Mohavénsis, Low, more cinereous-pubescent, brachiately branched: leaves narrowly oblong; bracts of the involucre obtuse, rather broadly linear, outermost oblong: akenes cinereous-hispidulous: pappus-bristles approaching barbellulate. — Rocky washes in the Mohave Desert, S. E. California, *Parish.*

⊢ ⊢ ⊢ Barely pubescent or glabrate perennial herbs, not viscid: leaves slender-petioled, at least all the lower ones opposite, deltoid-ovate or cordate, serrate, mostly acuminate or attenuate-acute, thinnish; heads half to two-thirds inch long; involucre subtended by some loose linear-subulate accessory bracts. Typical species.

B. cordifólia, ELL. l. c. Minutely soft-pubescent; stem branching. 3 feet high; leaves deltoid-cordate or the upper deltoidly ovate-lanceolate, crenate-serrate; heads rather few, loosely corymbosely cymose, 40-50-flowered; involucral bracts somewhat coriaceous, linear, mostly obtuse; pappus rufous or tawny. — Torr. & Gray, Fl. ii. 80. *Eupatorium Brickellia*, DC. Prodr. v. 182. — Wooded hills, W. Georgia and adjacent parts of Alabama and Florida: rare, first coll. by *Dr. Brickell*.

B. grandiflóra, NUTT. Puberulent or almost glabrous; stem 2 or 3 feet high, paniculately branched; the numerous heads paniculate-cymose and drooping; leaves broadly or narrowly deltoid-cordate, or the upper deltoid-lanceolate, coarsely dentate-serrate and with an entire gradually acuminate apex (the larger 4 inches long); involucre about 40-flowered; its bracts papery and scarious-margined when dried; the short outer ones ovate; inner oblong-linear, obtuse or acutish, or some exterior ones with loose subulate acumination; pappus white, inclined to deciduous. — Trans. Am. Phil. Soc. n. ser. vii. 287; Torr. & Gray, l. c. *Eupatorium? grandiflorum*, Hook. Fl. ii. 26. — Hills along streams of the Rocky Mountains and the Sierra Nevada, from Montana to the borders of Oregon, and south to New Mexico and Arizona. Name of the species not appropriate.

Var. **petioláris**, GRAY. Heads and leaves commonly smaller; the latter inclined to hastate-deltoid, and equalled or even surpassed by the slender petiole! — Proc. Am. Acad. xvii. 207. — Mountains of Arizona, *Lemmon*, and the borders of New Mexico, *Rusby*. Passes into the following and into the typical form.

Var. **mínor**, GRAY (Proc. Acad. Philad. 1863, 67), is a smaller form, with leaves only an inch or two long, heads proportionally small, involucre 30–35-flowered. — Clear Creek, Colorado, to California in the Sierra Nevada above Lake Tahoe, and mountains of Arizona.

B. símplex, GRAY. Resembles the preceding; stem a foot or two high, slender, simple, bearing a single terminal or 3 or 4 racemose slender-pedunculate comparatively large heads, or producing numerous simple floriferous branches; involucre about 30-flowered, of less imbricated and acute bracts, most of them linear, the outer series very short, as are the few loose subtending ones; leaves 10 to 20 lines long, from deltoid-cordate to deltoid-oblong, mostly obtuse. — Pl. Wright. ii. 73. — Shaded hills, Arizona, *Wright, Thurber, Lemmon*.

* * Heads 9–25-flowered (or in the penultimate species 3–5-flowered), not over half an inch long; pappus-bristles scabrous or not manifestly barbellate, except in the penultimate subdivision.

⊢ Leaves slender-petioled, all opposite, deltoid-cordate or triangular-hastate, small; heads pedunculate, in naked cymes terminating the stem or branches; bracts of the involucre thin, smooth and glabrous; shrubby.

B. Coúlteri, GRAY. A foot to a yard high, with numerous spreading slender branches, only the flowering ones herbaceous, minutely puberulent to glabrous; leaves from sparingly laciniate-dentate to nearly entire, acute or acuminate (larger ones an inch long, smaller less than half-inch); heads rather few in the naked and very open cymes, slender-pedunculed, half-inch high; involucre about 12-flowered; its bracts linear-lanceolate, subulately acuminate or acute; akenes pubescent. — Pl. Wright. i. 86. — Common in Arizona, in cañons, first coll. by *Coulter*. (Adj. Mex., *Gregg, Palmer,* &c.)

⊢ ⊢ Leaves distinctly petioled, all or mostly alternate; stems shrubby at base; inflorescence thyrsiform,

⊢⊢ Naked when well developed; the heads distinctly peduncled or in pedunculate small corymbiform cymes, forming an ample nearly leafless open paniculate thyrsus.

B. floribúnda, GRAY. Glabrate or barely puberulent below, but the branches with the inflorescence and outer involucral bracts glandular-pubescent and viscid; stem 4 feet high, woody only at base, much branched; leaves slender-petioled, deltoid-ovate or the lower subcordate, irregularly dentate (2 and 3 inches long); veins loosely reticulated; heads (5 lines long) 15-22-flowered; bracts of the involucre broadly linear and obtuse, with some oblong-ovate acutish short ones, and often 2 or 3 loose and herbaceous ones subtending the head. — Pl. Wright. ii. 73. *B. Wrightii*, Rothrock in Wheeler Rep. vi. 140, not Gray, l. c. — Ravines and river banks, S. Arizona, *Wright, Palmer, Rothrock, Lemmon, Pringle*.

++   +-   Foliose, i. e. the heads sessile or short pedunclcd, terminating short leafy branchlets or in axillary clusters, forming a spiciform, paniculate, or interrupted leafy thyrsus.

=   Involucre naked at base, all the bracts dry and chartaceous, glabrous and smooth, the outermost very short and appressed, wholly destitute of green tips.

    a. Leaves mainly with truncate or subcordate base, crenate or dentate, but not laciniate: involucral bracts all obtuse, or innermost linear ones abruptly acute; short outermost oval and ovate: heads 10–20-flowered, 4 or 5 lines high.

B. Rúsbyi. Tall, copiously branched, largely herbaceous, amply floriferous, with the habit of *B. floribunda*, except that the inflorescence is thyrsoid-paniculate, minutely puberulent: leaves (2 to 4 inches long) from deltoid-ovate to ovate-lanceolate, with truncate or some with more or less cuneate base, gradually tapering to an acute or acuminate apex, unequally dentate to or above the middle. — Mountains of New Mexico, *Greene*, *Rusby*, *G. R. Vasey*, and of S. Arizona, *Lemmon*.

—— B. Wríghtii, GRAY. Usually much branched from a woody base, 2 to 4 feet high, puberulent, sometimes a little scabrous. leaves broadly deltoid-ovate, or rounded-cordate and obtuse, or at most acute (but not prolonged upward), more or less crenate-dentate (larger cauline an inch and a half long, smaller only half-inch): heads glomerate-paniculate, the clusters shorter than or little surpassing the subtending leaves: involucre often purple. — Pl. Wright. ii. 72. *B. Californica*, var., Gray, Pl. Fendl. 64. — W. borders of Texas to Colorado and Arizona, where it is not clearly distinguishable from *B. Californica*.

    Var. ténera. A form with thin dilated-ovate leaves, fewer heads, and pale involucre, evidently growing in shade. — *B. tenera*, Gray, Pl. Wright. ii. 72. — Mountain ravines, S. Arizona, *Wright*, *Lemmon*.

    Var. renifórmis. Leaves also thin, broader than long, some of them quite reniform, coarsely crenate, mostly surpassing the glomerules of heads. — *B. reniformis*, Gray, Pl. Wright. i. 86; an older name than *B. Wrightii*, but inappropriate for the species, of which this is an extreme form. — Mountain valley near the western border of Texas, *Wright*.

—— B. Califórnica, GRAY. Moderately and virgately branched, 2 or 3 feet high, minutely puberulent: leaves ovate, obtuse, rarely subcordate, somewhat crenate-dentate, commonly an inch or less long, mostly surpassed by the small clusters of heads, these rather spicately glomerate, forming an interrupted strict thyrsus. — Pl. Fendl. 64, Pl. Wright. i. 85, & Bot. Calif. i. 300. *Bulbostylis Cavanillesii*, DC. Prodr. v. 38, as to Calif. plant. *B. Californica*, Torr. & Gray, Fl. ii. 79. — California, from Mendocino Co. southward to adjacent parts of Nevada and Arizona, and Utah?

    b. Leaves cuneate at base, tapering into the petiole, very numerous, incised or deeply toothed, seldom an inch long, the upper about equalling the glomerate heads in their axils: involucre narrow, 4 or 5 lines long; bracts mostly obtuse, the outer oblong, innermost linear: much branched and shrubby, 2 to 5 feet high.

B. baccharídea, GRAY. Leaves coriaceous, resinous-atomiferous and very glutinous, rhombic-ovate or oblong, and with 2 to 5 strong teeth to each margin, much reticulated: heads 15–18-flowered. — Pl. Wright. i. 87. — Mountains of S. W. Texas, east of El Paso, *Wright*. San Francisco Mountains, N. E. Arizona, *Greene*.

B. laciniáta, GRAY. Leaves thin, puberulent and somewhat scabrous, ovate-cuneate and oblong, laciniate-toothed or lobed, obscurely veiny: heads 9–12-flowered. — Pl. Wright. i. 87. *B. dentata*, Schultz Bip. Bot. Herald, 301, excl. syn. DC. — S. W. Texas, east of El Paso, *Wright*. S. Arizona, *Thurber*. (Mex., first coll. by *Berlandier*.)

=   =   Involucre of firmer bracts, the outer with greenish and somewhat spreading tips, outermost loose and herbaceous and passing into the small leaves of the branchlets.

B. microphýlla, GRAY. Glandular-puberulent or pubescent and viscid, a foot or two high from a partly woody base, paniculately much branched; the short leafy branchlets terminated by 1 to 3 heads: leaves subcordate or ovate to oblong, when old somewhat scabrous, obtuse or apiculate, sparingly denticulate or nearly entire, the larger half-inch long, those of flowering branchlets a line or two long; heads nearly half-inch long, about 15-flowered. — Pl. Wright. i 85; Bot. Calif. i. 300. *Bulbostylis microphylla*, Nutt. Trans. Am. Phil. Soc. n. ser. vii. 287; Torr. & Gray, Fl. i. 79. — Dry interior of Oregon and California in the eastern part of the Sierra Nevada to Idaho, the mountains of Utah, and S. W. Colorado; first coll. by *Nuttall*.

+ + + Leaves sessile, subsessile, or the lower short-petioled: heads not pendulous.
++ Leaves mainly opposite, veiny: heads mostly pedunculate: bristles of the pappus merely scabrous or barbellulate-serrulate under a lens: last two species with much imbricated involucre.

= Stems ligne-cent at base, slender.

B. oligánthes, GRAY. Cinereous-puberulent, a foot or two high: leaves coriaceous, from oblong to linear, obtuse, obtusely and often obscurely serrate, an inch or two long, canescent and the veins very prominently reticulated beneath: peduncles mostly elongated, axillary and terminal, 1–3-cephalous, racemosely or somewhat corymbosely disposed: heads half-inch long, 10–12-flowered: bracts of the involucre mostly acute or acuminate; the short outermost ovate, innermost linear. — Pl. Wright. i. 84. & ii. 71. *Eupatorium oliganthes*, Less. in Linn. iv. 137. *Bullostylis oliganthes*, DC. Prodr. v. 139. — S. Arizona, *Wright, Thurber, Lemmon*, mostly a narrow-leaved form. (Mex.)

B. párvula, GRAY. Minutely scabro-puberulent, low: leaves deltoid-ovate, coarsely few-toothed, green both sides, barely half-inch long, the upper oblong, sparse and much smaller: peduncles few and slender, monocephalous, corymbosely disposed at the summit of the stems: head 5 lines long, about 12-flowered: bracts of the involucre few-ranked; innermost linear, rather obtuse; outer broader and mucronate-acute. — Pl. Wright. i. 87. — Mountains of S. W. Texas near the pass of the Rio Limpio, *Wright*.

== Stems herbaceous to the base: leaves reticulate-veiny.

B. Wislizéni, GRAY. Glandular-hirsute. 2 or 3 feet high: cauline leaves lanceolate-oblong with a subcordate closely sessile base, acute, acutely and numerously serrate, thin, loosely veiny, 1½ to 3 inches long; those of the branches mainly obtuse at base: peduncles axillary and simple and as long as the leaves, or exceeding them on axillary branches, filiform: heads 5 or 6 lines long, 12–20-flowered: bracts of the involucre all lanceolate and gradually acuminate, or the innermost linear. — Pl. Fendl. 64; Pl. Wright. i. 84. & ii. 71. — S. Arizona, on mountain-sides, *Wright*. (Heads rather smaller and fewer-flowered than in the original of adjacent Mex.)

Var. **lanceoláta**. Loosely paniculate-branched and floribund, the numerous heads smaller: leaves broadly lanceolate, the cauline half-inch wide, those of the branches small, or the upper minute. — San Francisco Mountains, N. E. Arizona, *Greene*.

B. betonicæfólia, GRAY. More minutely glandular-hirsute: stems 1 to 3 feet high, virgate: leaves subcordate-oblong, obtuse, crenate or obtusely dentate, rugosely veiny; the lower mostly with short but distinct petioles: inflorescence virgate-racemiform: peduncles mostly shorter and the 12-flowered heads rather smaller: otherwise nearly as the preceding. — Pl. Wright. ii. 72. — Hills, New Mexico and Arizona, *Wright, Thurber, Greene*, &c.

B. Lemmóni, GRAY. Cinereous-puberulent, not glandular, slender, a foot or two high: leaves lanceolate, all acute at base and as if with short margined petiole, remotely or obscurely serrate, lightly triplinerved (inch or two long): heads (5 or 6 lines long) numerous in a rather loose narrow leafy thyrsus, on slender short peduncles, 10–12-flowered: bracts of the involucre nearly all acute; the rather few and short outer ones ovate- to oblong-lanceolate, inner linear: akenes canescent. — Proc. Am. Acad. xvii. 206. — Chiricahua Mountains, S. Arizona, *Lemmon*.

B. Prínglei, GRAY, l. c. Cinereous-puberulent and the foliage scabrous: stem strict, rather stout, 2 feet high: leaves oblong-lanceolate, acute, mostly obtuse at base, on very short but distinct petioles, somewhat serrate, nearly coriaceous, 3-nerved from just above the base, conspicuously and beneath saliently reticulated: heads (half-inch long) in a loose and narrow leafy thyrsus, 20–24-flowered: involucre glabrous, rather longer than the scaly-bracteate peduncles; their rounded bracts passing above into the ovate and obtuse or barely mucronulate outer bracts of the involucre, the innermost of which are lanceolate and acute: akenes canescent. — Rocky cañons, S. Arizona, *Pringle, Lemmon*.

B. cylindrácea, GRAY & ENGELM. Cinereous-pubescent, somewhat scabrous, stem commonly stout and strict, 2 to 4 feet high: leaves oblong-ovate to ovate-lanceolate, mostly obtuse at both ends, obtusely serrate, thickish, 3-nerved or triplinerved from near the subsessile base (about 2 inches long): heads (6 to 8 lines long) numerous in a virgate racemiform thyrsus, short-peduncled, sometimes almost sessile, 10-flowered: involucre cylindrical, closely imbricated; the broadly ovate outer bracts in several ranks, mucronate, multistriate, mostly villous when young; inner broadly linear, obtuse or mucronulate: akenes pubescent. —

Proc. Am. Acad. i. 46, Pl. Lindh. ii. 218, Pl. Wright. l. c. — Hillsides and thickets, Texas, *Berlandier, Wright, Lindheimer,* &c. Varies into

Var. láxa, GRAY. Paniculately branched, and the branches bearing numerous smaller (5 or 6 lines long) loosely disposed and sometimes slender-peduncled heads, having fewer bracts to the involucre: leaves of the branches either subsessile or abruptly petioled. — Proc. Am. Acad. xvii. 207. — S. W. Texas, *Palmer.*

++ ++ Leaves alternate, veiny; stems herbaceous from a perennial root: pappus barbellate.

B. Riddéllii, GRAY. Minutely cinereous or puberulent, glabrate: stem strict and stout, 2 to 4 feet high, simple or fastigiately branched above, exceedingly leafy to the summit: leaves oblong-lanceolate, rather acute, sparingly denticulate, occasionally more dentate, often entire, thickish, obscurely veiny, 8 to 18 lines long: heads subsessile, numerous, crowded in a leafy spiciform thyrsus, 15-20-flowered, 4 or 5 lines long: involucre campanulate, somewhat pubescent; the bracts few-striate, obtuse or mucronate; the outer ovate, inner oblong-lanceolate: pappus barbellulate under a lens. — Pl. Wright. i. 83. *Clavigera dentata,* DC. Prodr. v. 128, but the character does not well agree, and the specific name is inappropriate. *C. Riddellii,* Torr. & Gray, Fl. ii. 77. — River banks, middle and southern parts of Texas, *Berlandier, Riddell, Wright, Lindheimer,* &c.

B. brachyphýlla, GRAY. Minutely puberulent: stems a foot or two high from a lignescent caudex, slender, simple, and bearing a few racemosely paniculate slender-pedunculate heads, or paniculately branched and polycephalous: leaves oblong-lanceolate, entire or sparingly serrate, half-inch or the larger an inch long: heads 5 lines long, 9-12-flowered: involucral bracts few, acute, short outermost ovate or oblong, inner linear: pappus-bristles almost plumose under a lens. — Pl. Wright. i. 84. *Clavigera brachyphylla,* Gray, Pl. Fendl. 63. — Rocks and ravines, western border of Texas, New Mexico, and Arizona, *Fendler, Bigelow, Greene, Lemmon,* &c.

++ ++ ++ Leaves mostly nervose and narrow, entire, the lower opposite: stems paniculately much branched: heads very numerous, thyrsoid-paniculate: akenes usually glabrous: pappus merely scabrous: plants nearly glabrous. (*B. spinulosa,* Gray, of Mexico, is of this group.)

B. squamulósa, GRAY. Suffrutescent at base, 2 or 3 feet high: stems or shoots of the first year bearing narrowly linear (2 or 3 inches long, less than 2 lines wide) obscurely 3-nerved deciduous leaves; flowering shoots the next year bearing only minute squamiform obtuse leaves of a line or less in length, and closely imbricated on short branchlets and thence passing into the bracts of the involucre: heads 10-12-flowered, turbinate, about 5 lines long: involucral bracts pluriseriate, thickish, obscurely nerved, green with whitish margins, externally somewhat canescent; the short outer ones ovate or oblong and obtuse, inner narrow and acutish. — Proc. Am. Acad. xv. 30. — New Mexico near Santa Rita del Cobre, *Greene.* S. Arizona, near Fort Huachuca, *Lemmon.* (San Luis Potosi, Mex.)

B. longifólia, WATSON. Suffruticose: flowering branches leafy; leaves lanceolate-linear (1 to 3 inches long, 2 or 3 lines wide), 3-nerved; upper gradually diminished in the open-paniculate leafy thyrsus: heads subsessile in small clusters, 3-5-flowered, only 3 lines long: bracts of the involucre about 10, of 2 or 3 lengths, conspicuously striate, obtuse. — Am. Nat. vii. 301; Rothr. in Wheeler Rep. vi. 139, t. 5. — S. Utah and S. Nevada, *Wheeler, Mrs. Thompson, Palmer.*

B. multiflóra, KELLOGG. Suffruticose: cauline leaves ovate-lanceolate and with divergent lateral nerves, an inch or two long; those in the crowded panicle from lanceolate to linear, small, and with obscure lateral nerves: heads 3-5-flowered: akenes sparsely hairy: otherwise much resembling *B. longifolia.* — Kellogg in Proc. Calif. Acad. vii. 49; Greene, Bull. Calif. Acad. i. 8. — On rocks, in a cañon of King's River, southern part of the Sierra Nevada, California, *Kellogg.*

++ ++ ++ ++ Leaves all alternate, spatulate, veinless: stems shrubby: heads sparse or solitary.

B. frutéscens, GRAY. Rigid undershrub with divaricate branches, cinereous-pubescent: leaves spatulate or obovate, entire, 3 to 5 lines long, including the attenuate petiole-like base: heads (half-inch long) terminating the branchlets, about 20-flowered: involucral bracts rather obtuse; outer ones somewhat greenish-tipped: akenes hispidulous-scabrous: bristles of the pappus very minutely but densely serrulate. — Proc. Am. Acad. xvii. 207. — Southern borders of California, *Sutton Hayes, Palmer, G. R. Vasey, Parish.*

14. **CARPHOCHÆTE**, Gray. (Κάρφος, scale or chaff, and χαίτη, bristle, from the pappus.) — Perennial herbs or suffrutescent plants (of New and Northern Mexico), glabrous or nearly so; with opposite and entire sessile thickish 1-3-nerved but nearly veinless leaves, and solitary or somewhat clustered heads, terminating leafy or pedunculiform branches: the flowers (about an inch long) much exceeding the involucre: this and the corolla rose-colored: nearly of *Liatris* habit, and pappus somewhat of *Stevia*. — Pl. Fendl. 65; Pl. Wright. i. 89, ii. 71.

C. **Bigelóvii**, Gray, l. c. A span to a foot high, woody at base, fasciculately branched: lower leaves spatulate-oblong, inch long, and fascicles of smaller ones in the axils, upper oblong or linear: heads sessile or very short-peduncled, mostly terminating very leafy somewhat paniculate short branchlets: aristiform paleæ of the pappus 11 to 14, and a few very small exterior squamellæ. — N. New Mexico, *Bigelow, Wright, Greene.* Arizona, *Pringle.* S. W. Texas, *Girard.* The one or two other species are more herbaceous, slender, and with loose pedunculate heads.

15. **LIÁTRIS**, Schreb. BLAZING STAR, BUTTON SNAKEROOT. (Name of unknown derivation.) — Perennial Atlantic N. American herbs; with simple virgate very leafy stems from a tuberous or mostly globose and corm-like stock, bearing reversely racemose or spicate heads of handsome rose-purple flowers (rarely also white), in late summer and autumn; the leaves all alternate, narrow, entire, rigid or with cartilaginous margins, mostly glabrous or glabrate. — Gen. 542 (where Gærtner's name is mentioned; but Gærtner takes up the genus, like Schreber, from the *Anonymos*, Walt., under the name *Suprago*, confusing it with *Vernonia*, and in a volume two years later than Schreber's); Torr. & Gray, Fl. ii. 67 (excl. § 2 & 3); Benth. & Hook. Gen. ii. 248.

* Pappus very plumose: heads 4-5-flowered; inner involucral bracts with prolonged petaloid tips. — *Calostelma,* Don.

- L. **élegans**, WILLD. Partly pubescent, 2 to 3 feet high: linear upper leaves commonly soon reflexed: spike or raceme virgate, dense, 3 to 20 inches long: heads either sessile or on bracteolate pedicels, about half-inch long: bracts of the involucre few-ranked, the inner dilated at tip into an oblong or lanceolate mucronate-acuminate rose-red spreading appendage, which surpasses the flowers and pappus. — Spec. iii. 1065; Michx. Fl. ii. 91; Ker, Bot. Reg. t 267; DC. Prodr. v. 129; Torr. & Gray, l. c. *Stæhelina elegans*, Walt. Car. 202. *Serratula speciosa*, Ait. Kew. iii. 138. *Eupatorium speciosum*, Vent. Cels. t. 79. *Liatris radians*, Bertol. Misc. v. 9, t. 1. — Dry pine barrens, Virginia? to Florida and Texas.

* * Pappus very plumose: heads 16-60-flowered, cylindraceous with turbinate base: bracts of involucre much imbricated, with herbaceous tips if any: lobes of the corolla pilose inside: leaves all linear and rigid, hardly punctate: the lower elongated and graminiform.

L. **squarrósa**, WILLD. Pubescent or partly glabrous: stem stout, 6 to 20 inches high: heads few (even solitary), or sometimes numerous in a leafy spike or raceme, rarely somewhat paniculate, the larger an inch or more long: bracts of the involucre all herbaceous and acuminate, or with foliaceous or herbaceous (or innermost slightly colored) lanceolate rigid and somewhat pungent tips: these usually squarrose-spreading and prolonged. — Torr. & Gray, l. c., incl. vars. *floribunda & compacta. Cirsium tuberosum,* etc., Dill. Elth. t. 71. fig. 82. *Serratula squarrosa,* L. Spec. ii. 818. *Pteronia Caroliniana*, Walt. Car. 292. — Dry gravelly or sandy soil, Upper Canada to Florida, Nebraska, and Texas. Passes into

Var. **intermédia**, DC. Heads narrow: bracts of the involucre erect or little spreading, less prolonged. — Prodr. v. 129; Torr. & Gray, l. c., with var. *compacta.* L. *intermedia*, Lindl. Bot. Reg. t. 948. — Upper Canada to Nebraska, Louisiana, and Texas.

L. **cylindrácea**, MICHX. Mostly glabrous, a foot high: heads few or several, 16-20-flowered, an inch or less long: bracts of the involucre all appressed, barely herbaceous, rounded and abruptly mucronate at tip, the outermost very short. — Fl. ii. 93; Ell. Sk.

ii. 275; Torr. & Gray, l. c., not Pursh. *L. graminifolia*, Willd. Spec. iii. 1636, excl. syn. Walt. & hab.; Muhl. Cat. 73. *L. stricta*, Macnab in Edinb. Phil. Jour. xix. 60. *L. flexuosa*, D. Thomas, in Am. Jour. Sci. xxvii. 338? — Dry prairies and open woodlands, Upper Canada and Michigan to Minnesota and Missouri.

* * * Pappus distinctly plumose to the naked eye: heads 3–6-flowered: bracts of the involucre acuminate or mucronate, coriaceo-herbaceous, not appendaged: corolla-lobes naked: leaves all narrowly linear or the upper acerose.

L. punctáta, Hook. Stems a span to 30 inches high from a thick and branching or sometimes globular stock, stout: leaves as well as bracts commonly punctate, rigid: head 4–6-flowered, oblong or cylindraceous, thickish, from half to three-fourths inch long, mostly numerous and crowded in a dense (below leafy) spike: bracts of the involucre oblong, abruptly or sometimes more gradually cuspidate-acuminate, often lanuginous-ciliate: pappus almost as plumose as in the preceding. — Fl. i. 306, t. 55; Torr. & Gray, l. c. *L. cylindrica*, Torr. Ann. Lyc. N. Y. ii. 210. *L. resinosa*, DC. Prodr. v. 129 (pl. Arkans.), not Nutt. — Dry prairies and plains, Saskatchewan and Minnesota west to Montana and Colorado, south to Texas and New Mexico. (Mex.)

L. acidóta, Engelm. & Gray. Stem a foot or two high from a globose or at length elongated tuber: leaves very slender: heads 3–5-flowered, three-fourths to half an inch long, numerous in a slender and strict naked spike: bracts of the involucre rather few, thinnish, mostly glabrous, ovate- and oblong-lanceolate, gradually or abruptly acuminate or cuspidate-mucronate: pappus short-plumose. — Pl. Lindh. i. 10; Gray, Pl. Wright. i. 83. *L. mucronata*, Torr. & Gray, Fl. ii. 70, not DC. — Prairies of Texas, *Drummond, Wright, Lindheimer*, &c. Var. *vernalis*, Engelm. & Gray, l. c., is a depauperate vernal form.

Var. **mucronáta**. Heads and flowers smaller; involucral bracts abruptly mucronate-pointed. — *L. mucronata*, Engelm. & Gray, Pl. Lindh. i. 10. — Texas, *Lindheimer*.

L. Boýkini, Torr. & Gray. Glabrous: stem very slender, a foot or two high: leaves punctate; lower narrowly linear, upper acerose: heads rather numerous in a strict naked spike, 3–4-flowered, hardly half-inch long: bracts of the involucre only about 8, thin, lanceolate, acuminate, the inner somewhat scarious at margins and tip: pappus short-plumose. — Fl. ii. 70. — Near Columbus, Georgia, *Boykin*. Not since found.

* * * * Pappus from barbellulate to minutely short-plumose under a lens, not to the naked eye.

+ Heads subglobose or hemispherical, 15–40-flowered: involucral bracts mostly spatulate, many-ranked, somewhat spreading: corolla-lobes comparatively short.

L. scariósa, Willd. Pubescent or glabrate: stem stout, 1 to 5 feet high: leaves spatulate- or oblong-lanceolate and tapering into a petiole (4 to 6 inches long, half-inch to inch and a half wide); upper narrowly lanceolate; uppermost small, linear, sessile: heads racemose or spicate, few or numerous (3 to 50), mostly 25–40-flowered and about an inch high and wide: involucral bracts broadest and rounded at summit, there either herbaceous or scarious edged and tinged with purple (rarely white-scarious): pappus-bristles minutely barbellate. — Willd. Spec. iii. 1635; Sims, Bot. Mag. t. 1709; Ker, Bot. Reg. t. 590; Lindl. Bot. Reg. t. 1654; Meehan, Nat. Flowers, ser. 2, ii. t. 29. *L. aspera* & *sphæroidea*, Michx. Fl. ii. 92. *L. scariosa* & *L. sphæroidea*, DC. l. c. *L. sphæroidea*, Brit. Fl. Gard. t. 87. *L. borealis*, Nutt. in Paxt. Mag. v. t. 27. *L. squarrosa*, Sweet, Brit. Fl. Gard. t. 44? *Serratula scariosa*, L. Spec. ii. 818. — Dry and usually sandy ground, Upper Canada and New England to the Saskatchewan, west to the Rocky Mountains, south to Florida and Texas. Varies greatly; in the involucre, which is either herbaceous or with the tips largely scarious and colored; in the size of the heads, &c., passing into the extreme microcephalous form (which except for the transitions would be regarded as a distinct species), viz.: —

Var. **squarrulósa**. Comparatively small and slender: heads merely half or two-thirds inch long, 14–20-flowered: involucral scales narrower, innermost sometimes linear or lanceolate and acutish. — *L. squarrulosa*, Michx. l. c. *L. heterophylla*, R. Br. in Ait. Kew. ed. 2, iv. 509; Pursh, Fl. ii. 508; Nutt. Gen. ii. 131. — Open woods, N. Carolina to Texas. The heads of ordinary *L. scariosa*, when abnormally numerous and paniculate, are sometimes reduced to the smallest size.

+ + Heads oblong, 5-flowered: involucre squarrose by the spreading colored tips of the bracts.

L pycnostáchya, Michx. Hirsute, or below glabrous: stem stout, 3 to 5 feet high: leaves crowded throughout; the lower lanceolate and the upper very narrowly linear: spike dense, cylindrical (5 to 18 inches long); heads (4 to 6 lines long) all sessile: bracts of the involucre 14 to 16, oblong or the inner narrower; the more or less scarious squarrose tips purple or purplish, usually acute: pappus copious, minutely barbellate. — Fl. ii. 91; Pursh, Fl. ii. 507 (excl. syn. Dill. & Walt. ?); Torr. & Gray, l. c. *L. brachystachya*, Nutt. in Jour. Acad. Philad. vii. 72, a glabrous form. — Prairies, Illinois and Iowa to Arkansas and Texas. Apparently this hybridizes with *L. spicata*: at least specimens occur which are intermediate between the two species.

+ + + Heads from short-oblong to cylindraceous: bracts of the involucre all appressed,
++ Obtuse and mostly rounded at the pointless apex.

= Leaves narrowly linear, or the lowermost larger and broader; upper ones gradually reduced to linear-subulate bracts.

━ **L. spicáta**, WILLD. Glabrous, or with some sparse hirsute pubescence: stem stout or tall, usually 2 to 5 feet high, very leafy: heads 8–13- (sometimes 5–7-) flowered, half-inch long, almost erect, closely sessile and numerous in a dense spike of a span to a foot or more in length: involucre obtuse or rounded at base; its bracts obscurely if at all glandular-punctate, but not rarely glutinous; the tips of the inner usually with narrow colored scarious margin. — Sims. Bot. Mag. t. 1411; Torr. & Gray, l.c. 73; Torr. Fl. N. Y. 35, t. 47. *Cirsium tuberosum*, &c., Dill. Elth. t. 72. fig. 83. *Serratula spicata*, L. Spec. ii. 819 (excl. syn. Gronov.); Andr. Bot. Rep. t. 401. *S. compta*, Dryander in herb. Banks, cited by Pursh under the next. *Liatris macrostachya*, Michx. Fl. ii. 91; Pursh, l. c. *L. resinosa*, Nutt. Gen. ii. 131, a small form with 5-flowered heads. *L. sessiliflora*, Bertol. Misc. v. 10, t. 2 (but our specimen from coll. Alabama, *Gates*, has hirsute foliage), a form with slender and looser spike. — Moist or rich soil, Mass. and New York to Wisconsin and Arkansas, and south through the upper country to Florida and Louisiana.

Var. **montána**. Low and stout, 10 to 20 inches high: leaves broader, lower ones half to two-thirds inch wide, obtuse: spike proportionally short and heads large. — *L. macrostachya*, Michx. l. c., in part. *L. pumila*, Loddiges. *L. spicata*, Sweet. Brit. Fl. Gard. t. 49. *L. pilosa*, in part, Torr. & Gray, Fl. ii. 74. — Rocky mountain-tops in Virginia and N. Carolina, where it abounds.

━ **L. graminifólia**, PURSH. More slender than the preceding, mostly only 2 or 3 feet high: leaves usually ciliate toward the base with scattered hispid hairs, rigid, often sparse: heads more sparsely spicate or scattered, not rarely becoming racemose or paniculate, mostly half-inch long: involucre acutish at base; its bracts firmer, oval and oblong, glandular-punctate on the herbaceous back, the rounded (or sometimes slightly herbaceous-apiculate) tip hardly at all scarious-edged. — (Willd. Spec. iii. 1636, only as to name & syn. of *Anonymos graminifolia*, Walt., which is also uncertain.) Pursh, Fl. ii. 308 (excl. portions of char. taken from Willd.); Nutt. Gen. ii. 131; Ell. Sk. ii. 274; DC. Prodr. v. 130, chiefly; Torr. & Gray, Fl. ii. 72. *L. pilosa*, var. *gracilis*, Nutt. Gen. ii. 131. *L. virgata*, Nutt. Jour. Acad. Philad. vii. 72, & Trans. Am. Phil. Soc. n. ser. vii. 284, a form running into compound-paniculate inflorescence, with small heads. *Serratula foliis linearibus*, Gronov. Virg. ed. 1, 92; cited by Linnæus under *S. spicata*. — Dry or moist ground, Virginia to Florida.

━ Var. **dúbia**, GRAY. Spike strict and virgate, with many approximate rather large heads, or occasionally racemiform, or abnormally paniculate: bracts of the involucre narrower and thinner, sometimes obscurely scarious-margined. — Man. 224 (Torr. & Gray, l. c.). *L. pilosa*, Willd. (*Serratula pilosa*, Ait. Kew. ed. 1, iii. 138?, apparently a state with unusually narrow involucral scales, and like Lodd. Cab. t. 356, the only character being "*S. foliis linearibus pilosis, floribus axillaribus longe pedunculatis*"); Pursh, l. c.; Ell. l. c.; Lindl. Bot. Reg. t. 595. *L. pilosa*, var. *lævicaulis*, & *L. spicata*, var. *racemosa*, DC. l. c. *L. dubia*, Bart. Mat. Med. ii. 222. t. 49. *L. propinqua*, Hook. Bot. Mag. t. 3829. — Sandy pine barrens, New Jersey to Florida and Alabama, near the coast, in wet or dry soil.

━ **L. grácilis**, PURSH. Cinereous-pubescent or glabrate: stem slender, 1 to 3 feet high: leaves rather short, mostly spreading; lower usually oblong-linear or oblanceolate, upper small and narrow: heads small (4 or 5 lines long), 3–5- or rarely 6–7-flowered, numerous in a virgate raceme, on spreading or horizontal slender pedicels, or rarely spicate, often loosely compound-paniculate: bracts of the involucre lax, rather few (7–10), thinnish, commonly glandular-puberulent, not scarious at tip. — Fl. ii. 508; Torr. & Gray, l. c. *L. pauciflosculosa*, Nutt. in Jour. Acad. Philad. l. c. 71. *L. lanceolata*, Bertol. Misc. v. 11, t. 3. — Dry pine barrens, Georgia, Alabama and Florida.

= = Leaves all very slender: heads 4 or 5 lines long.

L. tenuifólia, NUTT. Glabrous or with a few bristles below : stem strict and slender, 2 to 4 feet high : leaves rigid, attenuate-linear and when dry with revolute margins ; radical and lower cauline very numerous and crowded, a foot or less long, a line or two wide ; upper cauline short, becoming acerose or filiform and reduced to setaceous bracts : heads about 5-flowered and 4 lines long, very numerous in a strict virgate raceme (of a foot or two in length), which occasionally develops into a panicle : involucre of about 10 oblong bracts, not punctate, the inner more or less scarious and purplish : pappus strongly barbellate. — Gen. ii. 131 ; Ell. Sk. ii. 275 ; Torr. & Gray, l. c. *L. lævigata*, Nutt. Trans. Am. Phil. Soc. l. c. 285, a large form with coarser radical leaves. — Dry pine barrens, N. Carolina to Florida.

++ ++ Involucral bracts or most of them acuminate or mucronate-tipped,
= = Hirsute with short many-jointed hairs.

L. Gárberi, GRAY. A foot or two high, hirsute with many-jointed spreading hairs, or the linear and rigid strongly punctate leaves glabrate : upper leaves very short, linear-subulate, erect : heads 6-7-flowered, 5 or 6 lines long, crowded in a dense spike : involucre campanulate ; its bracts (about 10) greenish and very glandular-punctate, villous-hirsute, in age glabrate ; outer ones ovate, inner oblong, all obtuse and conspicuously mucronate-pointed ; pappus minutely barbellate. — Proc. Am. Acad. xv. 48. — Tampa, Florida, *Garber*.

= = Involucre glabrous or nearly so, narrow, indistinctly glandular-punctate, 3-5-flowered (bracts variable): pappus more distinctly barbellate toward the base.

L. Chapmánii, TORR. & GRAY. Tomentulose-puberulent, glabrate : stem a foot or two high, strict and rigid : leaves short, linear, or the lower oblong-linear and obtuse (1 to 3 inches long) and the upper small and narrow : heads numerous, mostly 3-flowered, erect in a strict and dense virgate spike : involucre cylindrical ; its bracts thinnish, lanceolate or the short outer ones oblong, mostly acute and mucronate or short-acuminate, sometimes pointless : flowers large for the size of the head, two thirds of an inch long · pappus half-inch long. — Fl. ii. 502 ; Chapm. Fl. 191. — Dry sandy ridges, Middle Florida, first coll. by *Chapman*.

L. pauciflóra, PURSH. Glabrous or minutely puberulent : stem slender, often weak and declining : leaves rigid, linear, mostly narrow : heads numerous in a virgate often secund spiciform raceme (of 6 to 24 inches in length), when secund on short spreading or recurving pedicels : involucre cylindraceous ; its bracts thinnish, oblong, or the short outermost oval and the inner lanceolate, mostly mucronate-acute or acuminate : flowers 5 or 6 and pappus 4 or 5 lines long. — Fl. ii. 510 ; Chapm. l. c. *L. secunda*, Ell. Sk. ii. 278 ; Torr. & Gray, Fl. ii. 71. — Sandy pine woods, S. Carolina to Florida.

16. GARBÉRIA, Gray. (The late *Dr. A. P. Garber*, the re-discoverer.) — Proc. Acad. Philad. Nov. 1879, 379, & Proc. Am. Acad. xvi. 79. *Liatris* § *Leptoclinium*, Nutt. in Trans. Am. Phil. Soc. n. ser. vii. 285 ; Torr. & Gray, Fl. ii. 76. *Leptoclinium*, Gray, Proc. Am. Acad. xv. 48, not Benth. & Hook.

G. fruticósa, GRAY, l. c. Shrub 4 to 6 feet high, branching, leafy : branchlets and involucre puberulent : leaves with base of a short petiole articulated with the stem, vertical by a twist, glabrous, pale and of the same hue both sides, nearly veinless, obovate, retuse (barely inch long) : heads (half-inch long) numerous in fastigiate naked terminal cymes : involucre much shorter than the pappus. — *Liatris fruticosa*, Nutt. in Am. Jour. Sci. v. 299. *Leptoclinium fruticosum*, Gray, l. c. — S. Florida, *Warr. Garber*. Found by the latter on dry sand-ridges of the western coast, at Tampa Bay. Lower leaves opposite according to Nuttall.

17. CARPHÉPHORUS, CASS. (Κάρφος, chaff, and φόρος, bearing.) — Perennials, with no bulbiform stock or tuber ; the rose-purple or white flowers in cymosely disposed heads ; all N. American, late-flowering. — Bull. Philom. 1816, & Dict. vii. 148 ; DC. Prodr. v. 132 (one species) ; Torr. & Gray, Fl. ii. 65.

§ 1. Pappus of copious and unequal minutely barbellate bristles, occupying more than one series : flowers purple : stem simple, leafy : even the lowest leaves alternate, cauline ones sessile : Atlantic-States species, herbs.

* Leaves all acerose, erect or appressed.

C. Pseudo-Liátris, Cass. l. c. Cinereous-pubescent, glabrate below, glaucescent; stems a foot or two high, very strict; leaves with base half-clasping the stem, rigid, somewhat carinate; lowest 8 or 10 inches long, a line or less broad; cauline gradually reduced to subulate appressed bracts; heads few or numerous in a small compact terminal cyme; involucral bracts ovate-lanceolate, acuminate, densely pubescent. — Torr. & Gray, l. c.; Bertol. Misc. v. 14. *Liatris squamosa*, Nutt. Jour. Acad. Philad. vii. 73; Hook. Comp. Bot. Mag. i. 95. — Grassy pine barrens, Alabama, Middle Florida, and Mississippi to Louisiana.

* * Leaves plane, thickish; radical ones spatulate, tapering into a margined petiole; cauline oblong, short, closely sessile; bracts of the involucre pluriserial.

C. tomentósus, Torr. & Gray, l. c. About 2 feet high. tomentulose, or below hirsute and glabrate; heads numerous in the cyme (over half-inch long); bracts of the involucre canescently hirsute and viscid, mostly acute. — *Liatris tomentosa*, Michx. Fl. ii. 73. *L. Walteri*, Ell. Sk. ii. 285, excl. syn. Walt. — Low pine barrens, N. Carolina to Florida.

C. corymbósus, Torr. & Gray, l. c. Stouter and taller, minutely hirsute or pubescent; cauline leaves broadly oblong; heads numerous in the compound cyme; involucre glabrous; the bracts all very obtuse or truncate, inner ones scarious-margined and erose at apex. — *Liatris corymbosa*, Nutt. Gen. ii. 132, excl. syn. *L. tomentosa*, Ell. l. c., not Michx. — Margin of swamps in pine barrens, N. Carolina to Florida.

C. bellidifólius, Torr. & Gray, l. c. About a foot high, rather slender, often branched below the middle, almost glabrous, cauline leaves narrowly oblong or oblanceolate; heads fewer and scattered, more pedunculate; involucre of looser bracts; the lower rather spreading, innermost thin and linear, all very obtuse. — *Liatris bellidifolia*, Michx. l. c.; Nutt. l. c. *Anonymos uniflorus*, Walt. Car. 198 ? — Sandy woods and pine barrens, from Wilmington, N. Carolina, to Georgia.

§ 2. KUHNIOIDES. Gray. Pappus a single series of about 15 plumose bristles; flowers white or ochroleucous; bracts of the involucre fewer, in about 3 ranks; stems much branched, shrubby at base, few-leaved; lower leaves opposite: Pacific species.

C. júnceus, Benth. Minutely hispid or glabrate, or above somewhat canescent, 2 or 3 feet high; branches slender and rigid, junciform; the branchlets often leafless, terminated by solitary or 2 or 3 hemispherical heads (of half-inch length); leaves linear, sometimes sparingly lobed, upper ones filiform or reduced to subulate bracts, or early deciduous; bracts of the involucre obtuse or acutish; outer ones canescently hirsute and ovate or oblong; inner ones thin and narrower. — Bot. Sulph. 21; Gray, Proc. Am. Acad. viii. 632, & Bot. Calif. 301. — Sandy banks of streams, southern borders of California to Arizona, where the involucral bracts are narrower. (S. Calif., first coll. by *Hinds*.)

C. ATRIPLICIFÓLIUS, Gray, Proc. Am. Acad. v. 159, from Cape San Lucas. S. California, *Xantus*, is possibly a form of the last, with oblong laciniate-toothed or somewhat hastate leaves, on distinct petioles, and rather oblong heads : specimens insufficient.

18. TRILÍSIA. Cass.

(*Tres*, three, *licia*, threads or girdles; application obscure.) — Atlantic U. S. perennials; with simple and erect rather tall leafy stems, terminating in a thyrsus or panicle of cymules of small heads: leaves entire, oval to lanceolate; cauline partly clasping, radical much larger and tapering at base into a margined petiole. Flowers rose-purple, autumnal. Involucre of few oval or oblong somewhat herbaceous equal bracts, usually with 2 or 3 small and loose exterior ones. — Bull. Philom., 1818, & Dict. iv. 310; Benth. & Hook. Gen. ii. 248. *Liatris* § *Trilisia*. DC., excl. spec.; Torr. & Gray, Fl. ii. 76.

T. odoratíssima, Cass. l. c. (VANILLA-PLANT. HOUNDSTONGUE.) Glabrous; stem 2 or 3 feet high; leaves thickish, pale, often glaucous, obscurely-veined, vanilla-scented in drying; radical and lower cauline 4 to 10 inches long, oval or oblong, upper ones becoming very small; heads (3 or 4 lines long) rather numerous in open cymules, and these cymosely paniculate; akenes glandular. — *Anonymos odoratissimus*, Walt. Car. 198. *Liatris odoratissima*,

8

Willd. Spec. iii. 1637; Michx. Fl. ii. 93; Andr. Bot. Rep. t. 633; Don in Brit. Fl. Gard. ser. 2, t. 184; Torr. & Gray, Fl. ii. 76. *Eupatorium glastifolium*, Bertol. Misc. v. 16, t. 4. — Low pine barrens, near the coast, Virginia ! to Florida and Louisiana.

- T. paniculáta, CASS. l. c. Viscid-pubescent or the foliage glabrate, a foot or two high: leaves smaller, green; radical lanceolate-spatulate; small cauline ones oblong-lanceolate: cymules short-peduncled, crowded in a narrow, panicle or thyrsus: akenes minutely pubescent. — *Anomynos paniculatus*, Walt. l. c. *Liatris paniculata*, Michx. Fl. ii. 93; Willd. Spec. iii. 1637; Torr. & Gray, Fl. ii. 76. — Damp pine barrens, Virginia to Florida, near the coast.

TRIBE III. ASTEROIDEÆ, p. 52.

19. GYMNOSPÉRMA, Less. (Γυμνός, naked, σπέρμα, seed, having no pappus.) — Perennial herbaceous or suffrutescent plants, erect, glabrous, mostly glutinous; with alternate entire narrow leaves, and small heads of yellow flowers in fastigiately corymbose glomerate cymes. Involucre about 2 lines long: ligules very small and short. — Syn. 194; DC. Prodr. v. 312, excl. § 2. — Founded on *Selloa glutinosa*, Spreng., said to come from S. Brazil, with infertile disk-flowers, to which DeCandolle added three Mexican species; but these are all reducible to one, viz. : —

G. corymbósum, DC. Woody at base, 2 or 3 feet high: leaves from oblong-lanceolate to linear; lower ones distinctly 3-nerved: flowers of the ray 5 to 9, of the disk mostly fewer, all fertile. — Torr. & Gray, Fl. ii. 192; Gray, Pl. Wright. ii. 94. *G. corymbosum, multiflorum*, & *scoparium*, DC. l. c. — Rocky soil, S. Texas; fl. autumn. (Mex.)

20. XANTHOCÉPHALUM, Willd. (Ξανθός, yellow, and κεφαλή, head.) — Herbaceous or suffruticose plants (chiefly Mexican); with alternate entire or lobed leaves, and yellow flowers in scattered or loosely cymose heads; the smaller-flowered species approaching the following genus. — Willd. in Gesel. Nat. Fr. Berl. 1807, 140; Benth. & Hook. Gen. ii. 249. *Xanthocoma*, HBK. Nov. Gen. & Spec. iv. 310, t. 412; DC. Prodr. v. 311.

X. SERICOCÁRPUM, Gray, Proc. Am. Acad. xv. 31, from San Luis Potosi, Mexico, has canescent akenes: in all other species they are glabrous or only sparsely pubescent. Our species are annuals.

X. Wrightii, GRAY. Very glabrous, not glutinous: stems slender, a foot or two high, corymbosely paniculate at summit: leaves linear, entire: heads rather numerous, terminating pedunculiform branchlets: involucre barely 3 lines high and wide; the bracts broad, obtuse, or apiculate with a short green tip: rays 12, oblong: style-appendages linear-lanceolate, acute: akenes all surmounted by an entire or obscurely denticulate coroniform border, without proper pappus. — Proc. Am. Acad. viii. 632. *Gutierrezia Wrightii*, Gray, Pl. Wright, ii. 78. — S. Arizona and New Mexico, *Wright*, *Thurber*, *Bigelow*, *Greene*.

X. gymnospermoídes, BENTH. & HOOK. l. c. Glutinous when young, occasionally with some deciduous tomentum: stem stout, 2 to 4 feet high: leaves oblong-lanceolate with a tapering base, sometimes sparingly denticulate; the lowest often broader, petioled, occasionally incised and even pinnatifid: heads corymbosely cymose, crowded: involucre hemispherical, 4 lines high, very many-flowered; the bracts narrow and with acute green tips, not very unequal: flowers deep golden-yellow: rays 30 to 50, only 2 lines long: style-appendages ovate: pappus in the ray none; in outer disk-flowers setulose-coroniform; in central and less fertile flowers of several unequal awns and mostly coroniform-concreted at base. — Rothrock in Wheeler Rep. vi. 140; Hemsl. Biol. Centr.-Am. Bot. ii. 111. *Gutierrezia ? gymnospermoides*, Gray, Pl. Wright. l. c.; Hook. Bot. Mag. t. 5155. — Banks of streams, Arizona, first coll. by *Wright*. (Mex., *Parry* & *Palmer*, which has been wrongly referred to the larger-flowered very serrate-leaved *X. Benthamianum*, Hemsl.)

21. GUTIERRÉZIA, Lag. (Named for some member of the noble Spanish family, *Gutierrez*.) — Herbs or suffrutescent plants (N. & S. American), glabrous, often glutinous; with narrowly linear and entire alternate leaves, and small heads of yellow flowers, either solitary terminating the branchlets, or in dense cymes in the manner of *Gymnosperma*, from which it is distinguished mainly by the pappus. — Nov. Gen. & Spec. 30; Hook. & Arn. Comp. Bot. Mag. ii. 51; Torr. & Gray, Fl. ii. 193. *Brachyris*, Nutt. Gen. ii. 103. *Brachyachyris*, Spreng. Syst. iii. 574. *Brachyris* (excl. spec. and § 2), *Hemiachyris*, & *Odontocarpha*, DC. Prodr. v. 312, 71.

G. LINEARIFÓLIA, Lag., the original species (of which no specimen named by Lagasca is extant), on account of the oblong involucre with bracts loose at apex, enclosing only about 8 or 10 flowers, may with the highest probability be referred to a Chilian species, the *Brachyris paniculata*, DC. Prodr. v. 313; and this, although not traceable at Madrid, was collected by Née, and has been communicated to herb. DC. and herb. Boissier, to the latter by Pavon.

§ 1. Pappus of ray and disk similar, or in the former shorter; ligules mostly short: involucral bracts in N. American species all appressed. — *Brachyris*, Nutt.

* Suffruticose, and the woody base much branched; heads fastigiately or paniculately cymose; receptacle plane or small; paleae of the pappus conspicuous, from narrowly oblong to linear-subulate.

— **G. Euthámiæ**, TORR. & GRAY. Bushy, from glabrous to hirtellous-puberulent, 6 to 18 inches high, with mostly strict and fastigiately polycephalous branches: leaves narrowly linear, verging to filiform: heads mostly clavate-oblong, few-several-flowered, not over 2 lines long, some short-pedunculate, others 3 to 5 in a glomerule (in the manner of *Solidago* § *Euthamia*): flowers of disk and ray not numerous (commonly 3 or 4 each, or the latter 5 or 6, sometimes only one or two each); akenes sericeous-pubescent. — *Solidago Sarothra*, Pursh, Fl. ii. 540. *Brachyris Euthamiæ*, Nutt. Gen. ii. 163; Hook. Fl. ii. 23. *B. Euthamiæ & B. divaricata*, Nutt. Trans. Am. Phil. Soc. vii. 313, the latter an open form. *Brachyachyris Euthamiæ*, Spreng. Syst. iii. 574. *Gutierrezia Euthamia & G. divaricata*, Torr. & Gray, Fl. ii. 193. — Arid plains and rocky hills, Saskatchewan to Montana, south to New Mexico, Arizona, and borders of California. (Adj. Mex., where there is also a form with rather broadly linear leaves, coll. *Berlandier*, *Thurber*.)

— **Var. microcéphala.** Heads smaller, narrower, few-flowered, commonly oblong-cylindraceous and the involucre of fewer and narrower bracts; flowers of disk and ray mostly reduced to one or two each; leaves either narrowly linear or nearly filiform: pappus, as in the species, varying from short-oblong and obtuse (as in Berlandier's Saltillo specimens) to linear-lanceolate, and even attenuate-acute (as in Parry & Palmer's); certainly passes into *G. Euthamiæ*. — *G. microcephala*, Gray, Pl. Fendl. 74. Pl. Wright., &c. *G. microphylla*, Durand & Hilgard. Pl. Heerm. 40. *Brachyris microcephala*, DC. Prodr. v. 313. — S. Texas and New Mexico to S. California. (Adj. Mex., first coll. by *Berlandier*.)

G. Califórnica, TORR. & GRAY. l. c. More loosely branched; heads seldom glomerate-fascicled, obovate-turbinate, 2 or 3 lines long: involucral bracts (except small outermost) broad, oblong to obovate: rays 8 to 10, short: disk-flowers 6 to 12: akenes more villous.— *Brachyris Californica*, DC. l. c. *Gutierrezia linearifolia* (with some of *G. Euthamiæ*), Gray, Bot. Calif. i. 302, not Lag. — Hills, California near the coast, from San Francisco Bay southward (first coll. by *Douglas*); also San Bernardino Mountains (*Parish*), and Mesas of Arizona, *Palmer*, *Lemmon*, *Pringle*.

* * Annual herbs, loosely much branched: heads singly terminating the branchlets and paniculate: involucre hemispherical or obscurely obovate, about 2 lines in diameter, many-flowered: rays 9 to 13: disk-flowers 20 to 30: receptacle more or less elevated and hirsute-fimbrillate: akenes very short, obovate or turbinate, 10-costate; the ribs very silky-villous.

— **G. sphærocéphala**, GRAY. Low: receptacle of the flowers obtusely conical or hemispherical: pappus of 5 or 6 ovate short coroniform-concreted paleæ, barely half the length of the akene. — Pl. Fendl. 73. Pl. Wright. ii. 79. — S. W. Arkansas, E. New Mexico, and S. W. Texas, *Fendler*, *Wright*, &c.

G. eriocárpa, Gray. Low or taller (a foot or two high): receptacle obtusely high-conical: pappus of 12 or more linear-lanceolate or subulate and mostly distinct paleæ, about half the length of the akene. — Pl. Wright. i. 94. — Plains and prairies, S. and W. Texas, *Wright, Havard.* (Mex.)

— G. BERLANDIÉRI, Gray, Proc. Am. Acad. xv. 31, is an allied species of the northern part of Mexico, with a pappus of numerous minute paleæ, which do not surpass the silky hairs of the akene.

§ 2. Pappus wanting in the ray-flowers: ligules comparatively long: habit of the preceding subsection. — *Hemiachyris*, DC.

— **G. Texána**, Torr. & Gray. Annual, effusely much branched, 2 or 3 feet high: branches slender, bearing the very numerous pedunculate heads in open compound panicles: involucre turbinate-campanulate, a line or two long: rays 8 to 10 (3 or 4 lines long); disk-flowers as many: akenes minutely pubescent; those of the disk with a minute pappus of ovate or subulate paleæ, of length less than the breadth of the akene. — Fl. ii. 194. *Hemiachyris Texana*, DC. Prodr. v. 314. *Brachyris microcephala*, Hook. Ic. t. 147. not DC. — Sterile plains, W. Arkansas to Texas. (Adjacent Mex.)

22. AMPHIÁCHYRIS, Nutt. (*Brachyris § Amphiachyris*, DC.) —

('Αμφί, about, or on both sides, and ἄχυρον, chaff.) — As here constituted, the genus consists of two rather low and fastigiately or diffusely much-branched and erect glabrous plants, with entire leaves; the first with the habit of *Gutierrezia*, the second sufficiently different to form a subgenus (AMPHIPAPPUS, Torr. & Gray): fl. yellow in late summer and autumn.

— **A. dracunculoídes**, Nutt. Annual, rather low, effusely corymbiform, the slender branches and branchlets terminating in single pedunculate heads: leaves narrowly linear or the uppermost filiform: involucre hemispherical or short-campanulate; the bracts 10 or 12, firm-coriaceous and whitish with abrupt green tips, mostly ovate or oval: rays 5 to 10, oval or oblong, nearly as long as the involucre; disk-flowers 10 to 20, wholly sterile, the ovary quite abortive; their pappus of 5 to 8 scarious almost aristiform smooth paleæ, cupulately united at base and slightly dilated upward: akenes (of the ray) with a minute or obscure coroniform pappus. — Traus. Am. Phil. Soc. vii. 313; Torr. & Gray, Fl. ii. 192. *Brachyris dracunculoides*, DC. Pl. Rar. Genev. vii. 1, t. 1, & Prodr. v. 313. *Brachyris ramosissima*, Hook. Ic. t. 142; DC. Prodr. vii. 278. *Gutierrezia Lindheimeriana*, Scheele in Linn. xxii. 351. — Plains, Kansas to Texas.

A. Fremóntii, Gray. Shrubby, a foot or two high, with rigid tortuous branches: leaves short (half or quarter-inch long), obovate or spatulate, commonly narrowed at base into a margined petiole: heads mostly sessile and glomerate in small corymbosely disposed cymes: involucre campanulate or oblong, 2 lines long: the bracts 7 to 9, thin, mostly destitute of green tips: rays 1 or 2, short: disk-flowers 3 to 6, with infertile glabrous ovary, and a pappus of about 20 flattish denticulate-hispid tortuous bristles, some of them branching or irregularly paleaceous-concreted at base: ray-akenes with a pappus of fewer and short bristles or squamellæ, more united at base. — Proc. Am. Acad. viii. 633, & Bot. Calif. i. 302. *Amphipappus Fremontii*, Torr. & Gray in Jour. Bost. Soc. Nat. Hist. v. 4; Torr. Pl. Frem. 17, t. 9. — Arid deserts on the Mohave, S. E. California, *Fremont*, to S. W. Utah, *Palmer*.

23. GRINDÉLIA, Willd. (*Prof. Hieronymus Grindel*, of Riga and Dorpat.) — Herbs, or some species shrubby, of coarse habit (American, mostly of the U. S. west of the Mississippi); with sessile or partly clasping and usually serrate rigid leaves, and rather large heads of yellow flowers terminating the branches; the narrow rays usually numerous, occasionally wanting; central disk-flowers not rarely infertile. Herbage often balsamic-viscid, the heads especially so before and during anthesis (whence called GUM-PLANT in California): fl. all

summer. — Gesel. Nat. Fr. Berl. Mag. 1807, 259; Dunal, Mem. Mus. Par. v. 48; DC. Prodr. v. 314; Benth. & Hook. Gen. ii. 250. *Demetria*, Lag. *Donia*, R. Br. *Aurelia*, Cass.

G. coronopifólia, Lehm., of Mexico, is *Xanthocephalum centauroides*, Willd., the original of that genus.

G. angustifólia, DC. in Dunal, founded on a drawing only, is not identified; probably of some other genus.

G. costáta, Gray, Proc. Am. Acad. xvii. 208. is a northern Mexican species, allied to *G. squarrosa* and *G. subdecurrens*, with lunate-gibbous 10-ribbed akenes. It may reach the U. S. borders.

* Stem or branches (at least above) and sometimes the leaves pubescent: rays very numerous: awns of the pappus 2 or 3, sometimes solitary: plants a foot to a yard high.

+— Atlantic and Mexican species: root in U. S. annual or biennial, perhaps more enduring in Mexico: akenes with no terminal border or teeth.

G. inuloides, Willd. l. c. Pubescence minute or short: leaves from oblong to lanceolate or almost ovate, serrate down to the partly clasping or broad base with close-set and often gland-tipped salient teeth: involucre glabrous (half-inch or more in diameter), at length squarrose: akenes short and turgid (the length barely double the breadth), with rounded-truncate summit and small areola, smooth or becoming corky-rugose transversely. — Dunal, l. c. 50, t. 5; Bot. Reg. t. 248; DC. Prodr. v. 315; Hook. Bot. Mag. t. 3737; Torr. & Gray, l. c., excl. var. β. *G. pubescens*, Nutt. Jour. Acad. Philad. vii. 74. *Inula serrata*, Pers. Syn. ii. 451. *Demetria spathulata*, Lag. Elench. Madr. 1814, 20. — Plains of Arkansas and Texas; common. (Mex.)

Var. microcéphala, Gray. Smaller, more branching: heads only half as large: akenes more commonly rugose-thickened but sometimes smooth: involucral bracts usually shorter and closer: the extreme forms seeming very distinct from the type, but connected by intermediate states. — Bot. Mex. Bound. 81. *G. microcephala*, DC. Prodr. v. 315. — S. Texas, first coll. by *Berlandier*. (Mex.)

+— +— Pacific species: root perennial but sometimes flowering the first year: akenes truncate and with a prominulous irregularly undulate or obscurely 3-5-toothed border around the terminal areola: pappus-awns stouter and more corneous, flattish: involucre in the same species either naked or surrounded by spreading foliaceous bracts passing into leaves.

G. hirsútula, Hook. & Arn. A foot or two high, simple or sparingly branched, villous-hirsute, or glabrate, sometimes even tomentose when young: leaves rather rigid and commonly serrate with rigid salient teeth, in the typical plant oblong, or lower ones spatulate and obtuse (cauline inch or two long and about half-inch wide), upper with partly clasping but not widened base, varying however to lanceolate and acute: heads solitary or few: involucre half-inch in diameter; its proper bracts with or without subulate-attenuate squarrose tips, and with or without the surrounding loose foliaceous bracts, which may surpass the disk. — Bot. Beech. 147, 351; DC. Prodr. vii. 278; Torr. & Gray, l. c.; Gray, Bot. Calif. i. 103. *G. rubricaulis*, DC. Prodr. v. 316. — Hills and open grounds, California from Monterey northward, where it seems to pass into or is not well discriminated from the following; first coll. by *Douglas*.

G. integrifólia, DC. A foot to a yard high, the taller plants corymbosely branching at summit and bearing several or numerous heads: pubescence soft-villous, sometimes sparse or vanishing: leaves of soft texture, commonly entire, occasionally serrate; cauline lanceolate, 3 or 4 inches long, mostly tapering from a broad base to an acute or acuminate apex; radical spatulate and obtuse: bracts of the involucre with mostly elongated setaceous-subulate points to the bracts. — Prodr. v. 315; Torr. & Gray, l. c. *G. stricta*, DC. Prodr. vii. 278. *G. virgata*, Nutt. Trans. Am. Phil. Soc. vii. 314, slender form. *Donia inuloides*, var., Hook. Fl. ii. 25. — Moist or shady ground, Oregon to British Columbia, chiefly toward the coast. Varies greatly in open ground having leaves of firmer texture, the lower sometimes coarsely serrate, even the upper barely acute: on the shores of British Columbia occurs a low form, glabrate and thickish-leaved, which perhaps too nearly approaches *G. cuneifolia*.

* * Whole herbage glabrous: stems equably leafy, a foot or two high: root mostly short-lived perennial, but sometimes annual in the same species: leaves firm or rigid.

← Akenes squarely truncate and even at the summit, not bordered nor toothed: pappus-awns only 2 or 3.

G. Arizónica, GRAY. Rather low and slender: cauline leaves oblong-linear or narrowly oblong, obtuse, mostly spinulose-denticulate or dentate: heads small (half-inch high): bracts of the involucre short and rather broad, the acute or subulate-acuminate tips not prolonged nor spreading. — Proc. Am. Acad. xvii. 208. *G. microcephala*, Rothr. in Wheeler Rep. vi. 141, not DC. — Mesas of Arizona and New Mexico, *Wright, Rothrock, Brandegee.* (Adj. Mex.)

G. squarrósa, DUNAL. Commonly only a foot or two high and branched from the base: leaves rigid; cauline from spatulate- to linear-oblong and with either broadish or narrowed half-clasping base, acutely and often spinulosely serrate or denticulate; sometimes radical and even cauline laciniate-pinnatifid: involucre strongly squarrose with the spreading and recurving short-filiform tips of the bracts: outer akenes commonly (but not always) corky-thickened and with broad truncate summit, those toward the centre narrower and thinner-walled and with smaller areola. — DC. Prodr. v. 315; Torr. & Gray, l. c. *Donia squarrosa*, Pursh, Fl. ii. 559; Sims, Bot. Mag. t. 1706; Nutt. Gen. ii. 163. *Aurelia amplexicaulis*, Cass. Dict. xxxvii. 468. *Grindelia subdecurrens*, DC. l. c. *G. arguta*, Gray, Pl. Wright. ii. 81, not Schrader. — Plains and prairies, Minnesota and Saskatchewan to Montana and south to Missouri and Texas, west to Nevada, Arizona, and borders of California. (Mex.) — Heads small or middle-sized: involucre half to two-thirds inch in diameter, very glutinous. Varies much: the following are the most marked forms.

Var. **núda**, GRAY. Rays wanting. — *G. squarrosa*, Gray, Pl. Fendl. 77. *G. nuda*, Wood in Bot. Gazette, iii. 50. — With the usual radiate form in New Mexico, Colorado, and recently about St. Louis, Missouri.

Var. **grandiflóra**, GRAY. Heads larger and with very numerous rays (of an inch in length): stem 2 to 4 feet high, strict and simple below: upper leaves from ovate to oblong, broader or not narrowed at base, more numerously and equally serrate either with obtuse or spinulose teeth. — Pl. Wright. i. 98. *G. grandiflora*, Hook. Bot Mag. t. 4628. *G. Texana*, Scheele in Linn. xxi. 60. — Texas, in two forms; one by *Berlandier, Wright*, &c., with heads no larger than is common in *G. squarrosa*, and the leaves elliptical or oval and obtuse, closely beset with obtuse callous teeth; the other collected by *Lindheimer, Reverchon*, &c., with spinulose or almost aristate teeth.

G. Oregána. Stem rather stout and tall, branched above: leaves thickish, not rigid, sparsely denticulate or entire, mostly obtuse, oblong-spatulate or lingulate, or the upper lanceolate (the larger cauline 4 inches long and an inch wide): heads large (rays an inch long): bracts of the involucre with erect or spreading slender linear-subulate tips: akenes minutely striate. — *G. virgata*, in part, Nutt. Trans. Am. Phil. Soc. vii. 314. *G. integrifolia*, in part, Nutt. l. c.; Torr. & Gray, l. c., not DC. *Donia glutinosa*, Hook. Fl. ii. 25, not R. Br. — Oregon to Idaho, in dry soil.

← ← Akenes all or some outer ones 1-2-dentate or auriculate-bordered at the summit, except perhaps in *G. glutinosa*.

++ Atlantic species: pappus-awns mostly 2.

G. lanceoláta, NUTT. Stem 2 feet high, slender: leaves lanceolate or linear, acute, spinulose-dentate or denticulate (lower sometimes laciniate): heads as in *G. squarrosa* but the subulate-attenuate elongated tips of the involucral bracts straight and erect or the lower spreading: summit of the akene produced from each or the outer margin into a short tooth. — Jour. Acad. Philad. vii. 73; Torr. & Gray, Fl. ii. 248. — Prairies and barrens, Texas, Louisiana, Arkansas, and Tennessee. (Barrens near Nashville, *Gattinger*, where it is probably indigenous.)

++ ++ Pacific species.

G. cuneifólia, NUTT. Suffrutescent, stout, 3 or 4 foot high, mostly maritime, much branched: leaves thick, from cuneate-spatulate to linear-oblong, almost all with narrowed base, denticulate-serrate or entire: involucre half-inch or more high, little glutinous, the tips of the bracts either scarcely or decidedly squarrose: pappus-awns 5 to 8. — Trans. Am. Phil. Soc. l. c. 315; Torr. & Gray, l. c.; Greene in Bot. Gazette, viii. 256. *G. robusta*, var. *angustifolia*, Gray, Bot. Calif. i. 304, chiefly. — Salt marshes and shores, California, from Santa Barbara Bay northward; flowering in October. Woody base of stem becoming an inch or two thick.

G. glutinósa, Dunal. Herbaceous nearly or quite to the base ("fruticose," Cav.), a foot or two high: leaves rather large, obovate or spatulate, mostly rounded at summit and with partly clasping (broad or narrowish) base, more or less serrate: heads large: involucre half to three-fourths inch high, its bracts close, acute or acuminate, with no prolonged squarrose tips: akenes obscurely if at all bordered at summit: pappus-awns 5 to 8, stout and flattened, sparingly ciliolate-scabrous or nearly smooth. — Mem. Mus. l. c. 49; DC. Prodr. v. 314; Gray, l. c. 303. *Aster glutinosus,* Cav. Ic. ii. 53, t. 168. *Doronicum glutinosum,* Willd. Spec. iii. 2115. *Inula glutinosa,* Pers. Syn. ii. 452. *Donia glutinosa,* R. Br. in Ait. Kew. ed. 2, v. 82. *Demetria glutinosa,* Lag. Nov. Gen. & Spec. 30. *Aurelia decurrens,* Cass. Dict. xxxvii. 468. (The pappus-awns in old-time cultivated specimens sparsely hirtello-ciliolate indeed, but not as figured by Cavanilles; in Californian specimens varying from obscurely so to smooth.) *Grindelia latifolia,* Kellogg, Proc. Calif. Acad. v. 36. — Shore of California, from Humboldt Co. (*Bolander*) and San Francisco Bay to Santa Barbara Islands, whence a very large-leaved and robust form was described by *Kellogg*. Fl. summer. ("Mexico," Cavanilles. "Peru," Bentham in Gen. Original habitat seemingly quite unknown, but doubtless it came from the Pacific shores.)

—— **G. robústa,** Nutt. Herbaceous to the base, rigid, branching, usually glutinous in the manner of *G. squarrosa,* which it resembles in the attenuate-acuminate and squarrose spreading or recurved tips to the involucral bracts: leaves more rigid and larger, oblong, varying to lanceolate, rigidly spinulose-serrate or denticulate, or uppermost entire: heads usually half-inch high: akenes (at least outer ones) obliquely auriculate or broadly nuidentate at summit: pappus-awns 2 or 3, rarely more. — Trans. Am. Phil. Soc. l. c. 314; Torr. & Gray, l. c.; Gray, Bot. Calif. l. c., excl. vars. *latifolia* & *angustifolia* in part, incl. var. *rigida*. *G. squarrosa,* Hook. & Arn. Bot. Beech. 147, not Dunal. — The common GUM-PLANT of California, common throughout the western part of the State, on dry hills, &c.. fl. summer.

—— **G. nána,** Nutt. Rather low and slender, 6 to 30 inches high, the larger plants corymbosely and freely branched above: leaves thinnish, lanceolate and linear, or the lower spatulate, entire or spinulose serrate: heads small (a quarter to a third of an inch high): bracts of the involucre with slender and squarrose soon revolute tips, in the manner of *G. squarrosa* (which this species represents northwestward): rays 16 to 30: akenes narrow, excisely truncate or bidentate at summit: pappus-awns mostly 2. — Trans. Am. Phil. Soc. l. c. 314. *G. humilis,* Torr. & Gray, Fl. ii. 248, not Hook. & Arn. *G. Pacifica,* Marcus E. Jones in Bull. Torrey Club, ix. 31, in a habitat much out of range; namely, at Santa Cruz, California. — Washington Terr. and east to N. W. Wyoming, south to Shasta, California. Some Oregon specimens have heads as large as those of *G. squarrosa,* but the akenes are different.

Var. discoídea, a rayless state of the species. — *G. discoidea,* Nutt. l. c. 315, not Hook. & Arn. — Oregon and Washington Terr., *Nuttall,* &c.

* * * Anomalous and obscure species, wholly glabrous: cauline leaves all very small and narrow, almost filiform.

G. húmilis, Hook. & Arn. Not glutinous, apparently perennial: stem simple, slender, 7 inches high, 2-cephalous at summit: radical leaves linear, 2 inches long, 2 lines wide at the obtuse obscurely denticulate apex, thence gradually tapering to base; cauline nearly all small and bract-like, all but lowest half-inch long, not over one third of a line wide, attenuate-acute. involucre half-inch high; bracts lanceolate, acute, largely green, erect, the outer successively shorter: rays rather long: bristles of the pappus apparently 3 or 4, slender. — Bot. Beech. 147. — Single specimen known, "California, *Beechey,*" therefore probably from Monterey. Very unlike any other.

24. PENTACHÆTA,

Nutt. (Πέντε, five, χαίτη, bristle; from the pappus of the original species.) — Californian annuals, low and slender. often depauperate, glabrous and smooth or with some pubescence; with filiform-linear and entire alternate leaves, heads terminating the pedunculiform summit of the stem and loose branches, with either homochromous or heterochromous flowers. produced in spring. — Trans. Am. Phil. Soc. n. ser. vii. 336; Gray, Proc. Am. Acad. viii. 633, & Bot. Calif. i. 305. *Pentachæta* & *Aphantochæta,* Torr. & Gray; Benth. & Hook. Gen. ii. 251. (See p. 445.)

§ 1. Flowers of both ray and disk golden yellow: involucre of comparatively numerous and regularly imbricated bracts.

P. aúrea, Nutt. l. c. At length diffusely branched, 3 to 12 inches high: heads mostly large for the size of the plant and many-flowered, but greatly varying: rays 7 to 40 (2 to 5 lines long): bracts of the involucre broadly lanceolate, mostly setaceous-acuminate, with green centre and broad scarious margins: akenes villous-pubescent: pappus-bristles 5, sometimes 6 to 8, as long as disk-corollas. — Gray, Bot. Mex. Bound. 81, Bot. Calif. l. c. — Open and dry ground, in the southernmost counties of California; first coll. by *Nuttall*.

§ 2. Flowers of the ray white or purple-tinged, sometimes wanting or else few and wanting the ligule: disk-corollas yellow or yellowish, or changing to purple in age: bracts of involucre somewhat equal and fewer, mostly obtuse and narrowly scarious-margined.

P. exílis, Gray, l. c. A span or so high, with simple or from the base simply branched monocephalous erect stems: heads in the larger form (here taken as the type) many-flowered, with hemispherical or broadly campanulate involucre (3 lines high), and 8 to 14 oblong rays, these 2 lines long: akenes oblong-turbinate, villous: pappus-bristles 5, shorter than disk-corollas, in some plants abortive or obsolete. — Bot. Calif. l. c.; Greene in Bot. Gazette, viii. 256. — Dry hills, middle part of California, from Santa Clara Co. northward.

Var. aphantochǽta, Gray, l. c. More or less depauperate, 2 to 4 inches high: heads narrower, from rather few- to 25-flowered, discoid, mostly having 3 to 5 female flowers with corolla destitute of ligule, sometimes these wanting: pappus reduced to 3 or 5 short cusps or obsolete. — *P. aphantochæta*, Greene in Bot. Gazette, l. c. *Aphantochæta exilis*, Gray, Pacif. R. Rep. iv. 99, t. 11, a delicate and few-flowered form. — Dry ground, from the Salinas Valley to El Dorado Co., first coll. by *Bigelow*. Var. *discoidea*, Gray, l. c., is partly a small form of this without female flowers, and partly the following, into which it may pass.

P. alsinoídes, Greene. A span high, at length diffusely and several times branched from the base, with pedunculated discoid heads in the forks: involucre only 2 lines long, of only 5 to 7 bracts, "3–5" or 6–7-flowered: flowers apparently all hermaphrodite: pappus-bristles 3 or 4, fully equalling the corolla and as long as the obovate-clavate pubescent akenes, rarely obsolete in some flowers. — Bull. Torrey Club, ix. 109, & Bot. Gazette, l. c. — Hills or dry ground around San Francisco Bay to El Dorado Co., first coll. by *Kellogg* and *Bolander*.

P. gracilis, Benth. in Hook. Ic. t. 1101, from Mexico, is *Oxypappus*, Benth.

25. **BRADBÚRIA,** Torr. & Gray. (In memory of *John Bradbury*, who collected plants on the Missouri which were published in Pursh's Flora.) — Fl. ii. 250; Benth. & Hook. Gen. ii. 251. — Single species.

B. hirtélla, Torr. & Gray, l. c. Annual, branched from the base, a foot or so high, hispid: slender branches terminated by single rather small heads of yellow flowers: radical and lower cauline leaves narrowly spatulate; those of the flowering branches small, spatulate-linear to nearly filiform, mucronate-pointed: rays 3 or 4 lines long. — Dry ground, Texas, *Drummond, Wright, Lindheimer*, &c.

26. **HETEROTHÉCA,** Cass. ("Ετερος, different, θήκη, case, from the unlike akenes of ray and disk.) — N. American and Mexican herbs (probably only three species, two of them very variable), with the aspect of *Chrysopsis*, hirsute or scabrous: flowers yellow: pappus reddish or ferruginous: lower leaves at base of petiole commonly with a foliaceous stipuliform dilatation, upper partly clasping. Peduncles and involucre more or less glandular. A bristle or two of pappus rarely found on ray-akenes. — Bull. Philom. 1817, & Dict. xxi. 130; DC. Prodr. v. 316; Torr. & Gray, Fl. ii. 251.

H. Lamárckii, Cass. l. c. Biennial or sometimes annual, 1 to 3 feet high, somewhat heavy-scented, branching, usually bearing numerous corymbiform-paniculate rather small heads: radical leaves oval or oblong, slender-petioled; cauline oblong, the upper mostly with subcordate-clasping base: involucre 3 to 5 lines high; rays about 20; their akenes mostly glabrous and obscurely crowned: outer pappus of the disk-flowers conspicuous. — *H. Lamarckii* & *H. scabra* (also apparently *H. Chrysopsidis* & *H. leptoglossa*), DC. l. c. 317. *H. scabra* (var. *Calycium* & var. *nuda*, which are confluent), Torr. & Gray, Fl. ii. 251. *H. latifolia*, Buckley in Proc. Acad. Philad. 1861, 459. *Inula subaxillaris*, Lam. Dict. iii. 259, fide Cass. *I. scabra*, Pursh, Fl. ii. 531. *Chrysopsis scabra*, Nutt. Gen. ii. 151; Ell. Sk. ii. 339; Bertol. Misc. vii. t. 4. — Sandy or barren dry soil, coast of Carolina to Texas, Arkansas, S. Arizona, and perhaps within the borders of California. (Mex. In original specimens of *H. Chrysopsidis*, DC., and others from Saltillo, &c., a setose pappus to the ray-flowers only abnormally occurs. *H. leptoglossa*, DC., has the crown of the ray-akenes with a sharp and sometimes undulate edge. In Parry & Palmer's no. 373 the crown is more salient and setulose-denticulate!)

— **H. grandiflóra**, Nutt. Villous-hispid or hirsute; stem stout, from a foot to 6 feet high, bearing rather large (sometimes rather small) heads: cauline leaves not clasping, or hardly so, and clasping base of petioles of the lowest occasionally wanting; involucre 4 or 5 lines high: rays about 30; their akenes minutely pubescent or glabrate; outer pappus of the disk-flowers less conspicuous; style-appendages shorter, mostly obtuse. — Trans. Am. Phil. Soc. vii. 315; Torr. & Gray, l. c. *Diplopappus scaber*, Hook. Fl. ii. 22. *Heterotheca floribunda*, Benth. Bot. Sulph. 24. *H. floribunda* (excl. pl. Coulter, which belongs to the preceding and is probably from Arizona) & *H. grandiflora*, Gray, Bot. Calif. i. 308. — California from Santa Barbara southward and east to the borders of Nevada. — Heads always smaller than those of *H. inuloides*, sometimes no larger than of the preceding species.

27. **CHRYSÓPSIS**, Nutt. (Χρυσός, ὄψις, of golden aspect, from the color of the blossom.) — Herbs (N. American, extending into Mexico), mostly perennials; with silky, lanate, hirsute or hispid pubescence, or rarely glabrous, entire or sometimes few-toothed leaves, the cauline sessile, and middle-sized heads of yellow flowers terminating the stem and branches; in late summer and autumn: pappus commonly fuscous or ferruginous. — Gen. ii. 150, under *Inula*: DC. Prodr. v. 326; Torr. & Gray, Fl. ii. 252.

§ 1. EUCHRYSOPSIS. Heads radiate: outer short pappus mostly manifest.

* Leaves narrow, elongated and nervose, gramineous or rather Luzula-like: whole herbage sericeous-lanate, or in age glabrate: root perennial: akenes compressed-fusiform: outer pappus squamellate-setulose. — *Pityopsis*, Nutt. Trans. Am. Phil. Soc. n. ser. vii. 317.

— **C. graminifólia**, Nutt. Stem a foot or two high, slender, generally leafy, stoloniferous underground: leaves 3–5-nerved or striate, silvery-sericeous, at least when young, lanceolate to linear; radical a span to a foot long: cauline successively shorter and becoming linear-subulate, erect: heads few or several and paniculate: involucre (half-inch or less high) somewhat turbinate; its regularly imbricated bracts many-ranked, glabrate, sometimes granulose-glandular on the back: peduncles when glabrate often hirtellous-glandular. — Torr. & Gray, Fl. ii. 252; Bertol. Misc. Bot. vii. t. 3. *C. graminifolia* & *C. argentea*, Nutt. Gen. ii. 151; Ell. Sk. iii. 234; DC. Prodr. v. 326. *C. oligantha*, Chapm. Fl. 216, an early-flowering form with few leaves and heads. *Inula graminifolia*, Michx. Fl. ii. 122. *I. argentea*, Pers. Syn. ii. 452. *Erigeron nervosum*, Willd. Spec. iii. 1953. *E. glandulosum*, Poir. Dict. viii. 487. *Diplopappus graminifolius*, Less. in Linn. v. 310. *D. sericeus*, Hook. Comp. Bot. Mag. i. 97. *Pityopsis* (*Sericophyllum*) *graminifolia* & *argentea*, Nutt. Trans. Am. Phil. Soc. l. c. — Dry pine barrens or sandy ground, Maryland to Florida and Texas; fl. autumn. A characteristic but variable species: leaves from 1 to 9 lines wide, and heads when numerous smaller than when few. (Mex. Probably *Hectorea rillosissima*, DC.)

Var. **áspera** (*C. aspera*, Shuttlew. in distrib. coll. Rugel), a glabrate rigid and polycephalous state, near St. Marks, Florida (probably on the very coast), the stem and leaves sparsely glandular-hispidulous!

C. pinifólia, Ell. A foot high, slender (the flowering branches almost filiform); very early glabrate, appearing glabrous, smooth throughout; lowest leaves narrowly linear and 2–3-nerved (at most a line and a half wide, 2 to 6 inches long); cauline filiform: heads solitary terminating the branches, or corymbose pedunculate, nearly as large as the average in

C. graminifòlia: akenes the same. — Sk. ii. 335. *Pityopsis pinifòlia*, Nutt. l. c. — Georgia, on sand-hills between the Flint and Chattahoochee, *Jackson* (*Ell.*), *Baldwin*.

C. falcáta, ELL. Low, seldom a foot high, branched from the base, very leafy to the top, loosely lanate, at length glabrate, not glandular: leaves from narrowly to oblong-linear, obscurely few-nerved, rigid (1 to 3 inches long); the cauline spreading and sometimes falcate-recurving: heads mostly numerous and cymose, small: involucre campanulate (3 or 4 lines long). — Sk. ii. 336 (note); DC. l. c.; Torr. Fl. N. Y. i. t. 56. *Inula falcata*, Pursh, Fl. ii. 532. *Pityopsis falcata*, Nutt. Trans. Am. Phil. Soc. l. c. — Barren land along the coast, Cape Cod to New Jersey.

* * Leaves not nervose or gramineous: involucre hemispherical: akenes turbinate-obovate and turgid-flattish (or in the last species more oblong), 3–5-nerved: outer pappus squamellate or setulose.

+ Pubescence arachnoid-lanate or cottony-villous and flocculent, deciduous, leaving a glabrous or minutely scabrous and glandular surface, sometimes glabrate from the first except on rosulate tufts of radical leaves: Atlantic species.

++ Heads comparatively small, seldom half-inch high, commonly cymose: arachnoid hairs sparse or wanting: stems very leafy: root no more than biennial.

C. scabrélla, TORR. & GRAY. Glandular-scabrous even to the rather obtuse bracts of the involucre, destitute of cobwebby hairs: stem rather stout: leaves oblong-lanceolate or spatulate: outer pappus setiform. — Fl. ii. 255. — Pine woods, Tampa, Florida, *Leavenworth*, *Garber*. Too near the broad-leaved form of the next.

C. trichophýlla, NUTT. Villous when young with very long and soft usually scattered hairs which mostly have a stouter base: stem slender, 1 to 3 feet high: leaves oblong-spatulate or oblanceolate and obtuse, or upper linear: bracts of the involucre smooth, acute: outer pappus squamellate-setulose. — Gen. ii. 150; Torr. & Gray, l. c. *Diplopappus trichophyllus*, Hook. Comp. Bot. Mag. i. 97. — Dry ground, N. Carolina to Florida and Louisiana, in the low country, chiefly on and near the coast. Broad-leaved form approaches *C. Mariana*: narrower comes too near the next.

C. hyssopifólia, NUTT. Glabrate and smooth, but the rosulate linear-spatulate or sometimes broadly spatulate (barely inch long) radical leaves floccose-woolly when young: stem slender, virgate, 2 or 3 feet high, very leafy with spatulate-linear to almost filiform (inch or so long) glabrous leaves: heads often numerous and cymose: otherwise as the preceding. — Jour. Acad. Philad. vii. 67. *C. trichophylla*, var. *hyssopifolia*, Torr. & Gray, Fl. ii. 254, excl. syn. Hook. — Sand-hills and dry pine barrens of Florida, on the coast.

++ ++ Heads larger: wool floccose: akenes often with 2 to 4 salient and glandular-thickened nerves or ribs: outer pappus more squamellate: leaves occasionally with a few serratures or denticulations, oblong, or the lower spatulate or obovate and uppermost lanceolate.

C. Mariána, NUTT. A foot or two high from a perennial root, loosely silky-villous with arachnoid hairs, glabrate in age: leaves thinnish, green: heads several in a corymbiform cluster: involucre glabrous but granulate-glandular. — Gen. l. c. (under *Inula*); Torr. & Gray, l. c.; Bertol. Misc. vii. t. 2. *Inula Mariana*, L. Spec. ed. 2, ii. 1240. *Aster Carolinianus pilosus*, etc., Mill. lc. t. 57. *Diplopappus Marianus*, Hook. l. c. — Pine barrens and sandy soil, coast of New York to Florida and Louisiana.

C. gossýpina, NUTT. l. c. A foot or two high from a biennial root, densely lanate, the wool becoming tomentose-floccose: leaves all obtuse, mostly short and spatulate or oblong: heads terminating pedunculiform branches or loosely corymbose: involucre very woolly, or becoming glabrate or even glandular. — Torr. & Gray, l. c.; Bertol. Misc. vii. t. 1. *Inula gossypina*, Michx. Fl. ii. 122. *I. glandulosa*, Lam. Dict. iii. 259? *Erigeron pilosum*, Walt. Car. 206. *Chrysopsis dentata*, Ell. Sk. ii. 337, a form with lower leaves few-toothed. *C. decumbens*, Chapm. Fl. 217, a coast form with glandular peduncles and involucre. — Sandy pine barrens, N. Carolina to Florida and Alabama, in the low country.

+ + Pubescence from hispid to silky-villous, persistent: root perennial. Includes a multitude of forms, seemingly not distinguishable into species.

C. villósa, NUTT. l. c. A foot or two high: leaves from oblong to lanceolate, rarely few-toothed, usually cinereous or canescently strigose or hirsute and sparsely hispid along the margins and midrib, an inch or two long: heads mostly terminating leafy branches sometimes rather clustered, naked at base or foliose-bracteate: involucre campanulate, 4 or 5 lines high; its bracts commonly strigulose-canescent, sometimes almost smooth, acute: akenes oblong-obovate, villous: outer pappus setulose-squamellate. — *Amellus villosus*, Pursh, Fl. ii. 564. *Diplopappus villosus* & *D. hispidus*, Hook. Fl. ii. 22. *Chrysopsis villosa, hispida, foliosa, mollis,* & *sessiliflora,* Nutt. Trans. Am. Phil. Soc. l. c.; Torr. & Gray, l. c.; also *C. canescens,* Torr. & Gray, *C. echioides,* Benth. Bot. Sulph. 25 & Pl. Hartw. 316. — Prairies, plains, and other open grounds, from Illinois and W. Alabama north to Saskatchewan, south to Arizona, and west to British Columbia and the coast of California; in various forms. The typical eastern and northern plant is rather large, with cinereous and roughish but not canescent pubescence. Westward, extending to the southern part of California, it usually becomes more canescent and villous as well as hirsute and hispid; the size and fulness of the heads greatly varying. The more marked but quite unlimited forms are the following: —

━━━ **Var. híspida,** Gray. Small and low, with hirsute and hispid pubescence, not canescent: heads particularly small: involucre not canescent, sometimes glabrous. — Proc. Acad. Philad. 1863, 65. *Diplopappus hispidus,* Hook. Fl. ii. 22. *Chrysopsis hispida,* DC. Prodr. vii. 279; Nutt. l. c. — Saskatchewan to Idaho, south to W. Texas, Nevada, and Arizona. And forms between this and the next in California.

━━━ **Var. víscida.** Low: leaves small, oblong to spatulate, green, sparingly if at all hispid, not rough, but viscid-hirtellous or with viscid points, and the involucre commonly viscidulous. — Utah and Arizona, in the mountains, *Jones, Greene, Pringle, Lemmon.*

Var. discoídea. Heads destitute of rays: involucre somewhat canescent: otherwise nearly as var. *hispida.* — Cañons, W. Montana, *Watson.*

Var. stenophýlla, Gray. Low and rough-hispid, rigid: leaves spatulate-linear, only a line or two wide: heads small. — Pl. Lindh. ii. 223. — Crevices of rocks, W. Texas, *Lindheimer,* and S. W. Arkansas, *Bigelow.*

━━━ **Var. canéscens.** Wholly canescent with short and appressed sericeous pubescence, and with some spreading hispid bristles along the stem and margins of the narrow mostly oblanceolate leaves: heads small: involucre also canescent: outer pappus less distinct. — *Aplopappus?* (*Leucopsis*) *canescens,* DC. Prodr. v. 349. *Chrysopsis canescens,* Torr. & Gray, Fl. ii. 256. — Texas, *Berlandier, Drummond, Wright, Lindheimer,* &c. Stems a foot, sometimes "2 to 5 feet," high; very leafy and branching.

Var. foliósa, Eaton. Canescent with appressed sericeous pubescence, mostly soft and destitute of hispid bristles; but stem often hirsute or villous: leaves short, oblong or elliptical: heads small, rather numerous and clustered. — Bot. King Exp. 164. *C. foliosa* & *C. mollis,* Nutt. l. c. *C. foliosa,* Gray, Pl. Wright. i. 99, & ii. 81, a small-leaved and somewhat hispid form, between this and var. *hispida.* — Rocky Mountains of Wyoming to Utah and Arizona.

Var. Rútteri, Rothrock. Most like the preceding, equally sericeous-canescent with usually longer soft hairs: heads of double the size, fully half-inch high and wide, solitary or few in a cluster, foliose-bracteate: rays 30 to 40, half-inch long. — Wheeler Rep. vi. 142. *C. foliosa,* var. *sericeo-villosissima,* &c., Gray, Pl. Wright. ii. 81. — S. Arizona, *Wright, Rothrock, Lemmon.* — Seemingly the most distinct form of all; but connected with the eastern type by one with slightly canescent leaves, Colorado, *Greene.*

━━━ **Var. sessiliflóra.** From hirsute and hispid or greenish to villous-canescent: leaves oblong or spatulate: heads mostly large, solitary and foliose-bracteate at base: outer pappus more conspicuous and squamellate. — *C.* (*Phyllotheca*) *sessiliflora,* Nutt. Trans. Am. Phil. Soc. l. c. 317; Gray, Bot. Calif. i. 309, partly, especially var. *Bolanderi. C. Bolanderi,* Gray, Proc. Am. Acad. vi. 543, which is a well-developed form. — California, near the coast, from Mendocino Co. to San Diego and Arizona. Disk-corollas in the bud tipped with some scattered very slender hairs.

━━━ **Var. echioides.** A branching form, with rather numerous and naked heads of small size, and usually small leaves, commonly canescently hispid, sometimes greener: passes into var. *foliosa,* var. *hispida,* &c. — *C. echioides,* Benth. Bot. Sulph. 25 (from Bodegas, a form nearer the foregoing) & Pl. Hartw. 316, form with small and scattered heads. *C. sessiliflora,* var. *echioides,* Gray, Bot. Calif. i. 309. — California, common from the Sacramento southward to Arizona.

* * * Leaves not nervose, somewhat veiny: involucre hemispherical: akenes turgid-obovate and flattish, indistinctly 10-nerved, minutely pubescent: outer pappus paleolate and conspicuous; inner not very copious: root annual.

C. pilósa, Nutt. A foot or two high, branching: branchlets terminated by solitary middle-sized heads: pubescence soft-hirsute or villous, also a minute glandulosity: leaves oblong-lanceolate, occasionally denticulate or toothed; the lower sometimes incised: bracts of the involucre acuminate, glandular-viscid: rays almost half-inch long. — Jour. Acad. Philad. vii. 66, & Trans. Am. Phil. Soc. l. c. (§ *Phyllopappus*); Torr. & Gray, l. c., § *Achyrœa*. — Open pine and oak woods, N. W. Arkansas, W. Louisiana, and Texas, first coll. by *Nuttall*.

§ 2. AMMÓDIA, Gray. Rays none: outer pappus slender-setulose, inconspicuous or obscure: somewhat viscid and pubescent perennials, with bracts of the involucre thinner and more scarious. — Proc. Am. Acad. vi. 543. *Ammodia*, Nutt. Trans. Am. Phil. Soc. n. ser. vii. 321.

C. Oregána, Gray. A foot or two high, paniculately branched: leaves oblong or lanceolate, sometimes hirsute or almost hispid; midrib conspicuous: involucre nearly equalling the flowers; its bracts pluriseriate: corollas slender: akenes oblong. — Proc. Am. Acad. l. c. & Bot. Calif. i. 309. *Ammodia Oregana*, Nutt. l. c.; Torr. Bot. Wilkes Exped. t. 9, A. *Brickellia Cumingii*, Klatt, in Abh. Nat. Gesells. Halle, xv. 5. — Sandy or gravelly banks of streams, Oregon and W. California.

Var. scabérrima. Leaves (of branches) small, these and the branches very hispid scabrous. — Dry creek, Tulare Co., California, *Congdon*.

C. Bréweri, Gray, l. c. A foot or more high, more slender, less pubescent or almost glabrous: leaves shorter, ovate-lanceolate or lanceolate, 3-nerved at base (an inch or two long): heads naked-pedunculate: involucre shorter; its bracts fewer-ranked and somewhat broader: corollas funnelform: akenes obovate. — California, in the Sierra Nevada, from 4,500 to 11,000 feet, in open woods, first coll. by *Brewer*.

28. ACAMPTOPÁPPUS, Gray. ("Ακαμπτος, unbending or stiff, and πάππος, pappus.) — Low shrubs, of the Arizona-Nevadan desert region, a foot to a yard high, glabrous or obscurely puberulous, not glandular nor resinous: leaves small, entire, sessile, nearly veinless except midrib, lower spatulate, upper linear-oblong to linear: heads terminating pedunculiform branchlets, yellow-flowered. — Proc. Am. Acad. viii. 634, xvii. 208.

A. sphærocéphalus, Gray, l. c. Branches striate, corymbosely polycephalous: heads discoid, homogamous, depressed-globular, 4 or 5 lines high: bracts of involucre whitish, outer ones commonly with a pale greenish spot. — *Aplopappus* (*Acamptopappus*) *sphærocephalus*, Gray, Pl. Fendl. 76; Torr. in Pacif. R. Rep. vii. t. 6. — Arizona and S. Utah to the Mohave desert in California, first coll. by *Coulter*.

A. Shockléyi, Gray. Branchlets simpler, monocephalous: head hemispherical, radiate: rays 10 to 12, elongated, linear-oblong, bright yellow: outer bracts of involucre more conspicuously green on the back. — Proc. Am. Acad. xvii. 208. — Mountains of S. W. Nevada, at Candelaria, Esmeralda Co., *W. S. Shockley*.

29. XANTHÍSMA, DC. (Ξάνθισμα, dyed yellow, alluding to the bright yellow flowers of the showy head.) — Prodr. v. 94; Benth. & Hook. Gen. ii. 253. *Centauridium*, Torr. & Gray, Fl. ii. 246. — Single species, near to *Aplopappus*, showy in cultivation.

X. Texánum, DC. l. c. Nearly glabrous, biennial or annual, 1 to 4 feet high, with virgate branches terminated mostly by solitary large heads: leaves from narrowly oblong to lanceolate; radical and lower cauline not rarely laciniate-pinnatifid and even bipinnately parted; cauline sessile, sparsely serrate or denticulate, or the upper entire: outer bracts of the involucre commonly narrowed below the green body or appendage; this whitish-margined, and sometimes with rounded barely mucronate summit, oftener either gradually or abruptly acuminate and cuspidate: rays about 20, an inch or less long, the ligule borne on a very short tube, and the style short. — Gray, Pl. Wright. i. 98 (var. *Berlandieri*, with rounded obtuse involucral bracts, and var. *Drummondii*, with pointed ones); Torr. Bot. Marcy Rep.

t. 10. *Centauridium Drummondii*, Torr. & Gray, Fl. ii. 246; Gray, Pl. Lindh. ii. 223. *Machaeranthera grandiflora*, Buckley in Proc. Acad. Philad. 1861. 456. — Open woods, Texas, *Berlandier, Drummond, Lindheimer*, &c.; fl. all summer.

30. **APLOPÁPPUS**, Cass. ('Απλόος, πάππος, simple pappus.) — A large American genus (chiefly W. North American and Chilian) the analogue of *Aster* in the heterochromous division and equally polymorphous; mostly herbaceous perennials, some suffruticose or even shrubby, a few annual: the flowers all yellow, produced in summer and autumn. — Dict. lvi. 168. *Haplopappus* & *Ericameria*, Benth. & Hook. Gen. ii. 253, 255. — Note that one or two species occasionally and certain species uniformly want the ray-flowers, obliterating the distinction between this genus and the following!

§ 1. PRIONÓPSIS. Gray. Heads very large and broad: involucre depressed-hemispherical, of lanceolate acuminate bracts, the outer mostly foliaceous and spreading: rays very numerous: disk-corollas narrow, merely 5-toothed: style-appendages short and rather obtuse: akenes very glabrous; those of the ray short, turgid-elliptical; of the disk oblong or narrower, and the central ones inane: pappus of very rigid and unequal bristles and comparatively little numerous; the innermost and larger ones somewhat flattened toward the base and their margins scabrous-ciliolate; the outermost very small and short: root annual or biennial. — Pl. Wright. i. 98. *Prionopsis*, Nutt. Trans. Am. Phil. Soc. n. ser. vii. 329. — (Connects with *Xanthisma* and has the foliage of *Grindelia*.)

A. **ciliátus**, DC. Very glabrous: stem 2 to 5 feet high, bearing few or several somewhat cymose-clustered heads (with the disk an inch in diameter), equably leafy to the top: leaves oval or the lower obovate (1 to 3 inches long), very obtuse, veiny, evenly and somewhat pectinately dentate with bristle-pointed teeth: pappus of the fertile akenes disposed to be deciduous in a ring. — Prodr. v. 346; Gray, Pl. Wright. i. 98. *Donia ciliata*, Nutt. Jour. Acad. Philad. ii. 118; Hook. Exot. Fl. i. t 45. *Prionopsis ciliata*, Nutt. l. c.; Torr. & Gray, Fl. ii. 245. — Hillsides and river-banks, Missouri and Kansas to Texas.

§ 2. APLOPAPPUS proper. Heads large or middle-sized, or sometimes small, commonly broad and with involucre of firm well-imbricated or rigid bracts: rays numerous, several, or rarely wanting: disk-corollas narrow, merely 5-toothed: style-appendages from ovate to linear-subulate: pappus commonly fuscous or rufous, and more or less rigid. (Habit and special characters various, but the groups too confluent and indefinite for first-class sections.)

* Heads rayless: bracts of the involucre rigid, appressed-imbricate with the outer successively shorter, all with abrupt and more or less spreading herbaceous tips: style-appendages ovate- or oblong-lanceolate: pappus rather rigid: leaves coriaceous, mostly oblong and spinulose-dentate. — *Aplopappus* § *Aplodiscus*, Torr. & Gray, Fl. ii. 242, excl. the first species, which is § *Aplodiscus*, DC. § *Haplodiscus* & *Eriocarpæa*, Benth. & Hook. l. c. — (One of the transitions to *Bigelovia* § *Aplodiscus*.)

A. **squarrósus**, Hook. & Arn. Suffruticose, 2 or 3 feet high, somewhat pubescent, glandular and glutinous: leaves thickly dentate (about inch long): heads numerous and spicately thyrsoid at the end of the branches, half-inch long: involucre elongated-turbinate; its bracts imbricated in many ranks, the lower usually imbricated on the peduncle, their tips mostly squarrose and glandular: akenes fusiform, glabrous, or sparsely pubescent. — Bot. Beech. 146; Torr. & Gray, Fl. ii. 242. *Pyrrocoma grindelioides*, DC. Prodr. v. 350. *Homiopappus squarrosus*, Nutt. Trans. Am. Phil. Soc. l. c. 332. — Dry hills on the coast of California, from Monterey to San Diego: first coll. by *Douglas*. Also on the foot-hills of the San Bernardino Mountains, *Parish*, &c.

A. **Nuttállii**, Torr. & Gray, l. c. Herbaceous from a ligneous stock, a span to a foot high: leaves from spatulate-oblong to almost lanceolate, rather sparsely pectinately dentate;

heads few terminating the branches, one-third inch high: involucre hemispherical; the bracts fewer-ranked and with slightly spreading greenish tips: akenes short, sericeous-canescent. — *Eriocarpum grindelioides*, Nutt. Trans. Am. Phil. Soc. l. c. 321. — Rocky Mountains and adjacent plains, north to Idaho and Saskatchewan, south to New Mexico and Arizona; first coll. by *Nuttall*.

 * * Heads radiate, with rays not rarely neutral or sterile, or in one species commonly discoidal by the diminution of the ligules: involucre well imbricated, of firm texture, the bracts either coriaceous with herbaceous tips or coriaceo-foliaceous: akenes (with two exceptions) glabrous and narrow: pappus capillary but rigid: style-appendages long and slender, acute or acutish: perennials, rigid-leaved. — § *Pyrrocoma*, Gray, Pl. Wright. i. 98. *Pyrrocoma* & *Homopappus*, in part, Nutt. Trans. Am. Phil. Soc. vii. 330, 333.

 +- Shrubby: rays conspicuous but sterile: appendage of the slender style-branches of the length and breadth of the stigmatic portion: akenes very glabrous, narrow, compressed, 4-nerved.

A. Berbéridis. Suffruticose, a foot or two high: flowering branches somewhat virgate, when young tomentose-pubescent, equably leafy, bearing numerous and racemose or sometimes solitary heads: leaves oval, very obtuse, spinulosely and evenly multidentate, half-clasping by an abrupt somewhat adnate base (half to full inch long), coriaceous, with conspicuous midrib but obscure veins: involucre broadly turbinate; its bracts numerous, in successively shorter ranks, broadly linear or outermost oblong, smooth, all with very obtuse and short rather appressed green tips: rays numerous, a quarter to nearly half an inch long, seldom styliferous: pappus merely sordid. — All Saints Bay, Lower California, so near that it may be expected within the U. S. border, *Parry*, *Miss Fish*.

 +- +- Herbaceous: style-appendages from subulate-filiform to narrowly subulate, much longer than the stigmatic portion.

 ++ Heads large and discoid, the sterile rays being hardly apparent or very small for the size of the head (when styliferous the style-branches sometimes tipped with a hispid appendage!): akenes completely glabrous and smooth, slender but flattish, 4-costate or nerved, often finely striate: rigid leaves commonly spatulate or lanceolate, on the same plant either entire or sparsely spinulose-toothed. — *Pyrrocoma*, Hook.

A. carthamoídes, Gray. Commonly a foot high, rather stout and leafy, scabro-puberulent when young, becoming smooth, bearing a solitary terminal large head and sometimes one or two in axils: leaves from spatulate to oblong or lanceolate: involucre hemispherical, half to three-fourths inch high, often leafy-subtended at base; its proper bracts coriaceous-rigid, from oblong to broadly lanceolate or innermost linear, more or less scarious-margined, most of them tipped with an abrupt mucro or cusp, the outer commonly loose and becoming leaf-like, either entire or spinulose-denticulate: rays almost always present and rather numerous; but their ligules inconspicuous, being short, involute, and concealed in the at length rufous or fulvous pappus. — Proc. Acad. Philad. 1863, 65. *Pyrrocoma carthamoides*, Hook. Fl. i. 306, t. 107; Torr. & Gray, Fl. ii. 243. — Dry plains and hills, Oregon, Washington Terr., and Idaho; first coll. by *Douglas*. Polymorphous species: the extremes are

Var. **máximus.** Robust, leafy, sometimes 2 feet high: radical leaves obovate or oval, 3 to 7 inches long; cauline oblong, with partly clasping base: heads ample, in fruit an inch high and broad: involucre of very numerous and broad or broadish bracts: rays sometimes more evolute, but small. — *Pyrrocoma radiata*, Nutt. Trans. Am. Phil. Soc. vii. 333; Torr. & Gray, l. c. — Of the same district, first coll. by *Nuttall*.

Var. **Cusickii.** Smaller: stems only a span or two high, ascending, few-leaved: leaves mostly spatulate-lanceolate: head three-fourths to nearly inch high in fruit, but narrow and much fewer-flowered: bracts of the involucre correspondingly fewer, lanceolate, mostly acute or acuminate. — Union Co., Oregon, flowering earlier (in June), *Cusick*. Perhaps a distinct species, but appears to pass into the smaller forms of the type.

 ++ ++ Heads middle-sized to small, evidently radiate; the exserted rays often infertile but styliferous; plants comparatively slender and more capituliferous.

 = Pubescence either cottony-tomentose and deciduous or none: leaves firm-coriaceous or rigid; cauline and mostly the radical lanceolate, the former disposed to be sparse or small at the upper part of stem: akenes or ovaries not rarely with some villous pubescence. — *Homopappus*, Nutt., excl. *H. uniflorus*.

A. racemósus, Torr. Stems usually virgate and simple, rigid, a foot or two high, leafy: leaves lanceolate or radical, sometimes oblong-spatulate (4 to 6 inches long, tapering into a petiole), entire or denticulate or on same plant spinulose-serrate; heads several or rather numerous, racemosely or spicately disposed along naked upper part of the stem or (either singly or 2 or 3 together) in axils of upper leaves: involucre (half-inch or less high) from hemispherical to turbinate-campanulate; its bracts rigid, well-imbricated, and with short abrupt mostly mucronate-pointed or apiculate green tips, these either erect or somewhat squarrose: rays (8 to 20) 2 or 3 lines long. — Torr. in Sitgreaves Rep. 162, as to syn., &c., probably not as to the specimen. *Homopappus racemosus*, Nutt. Trans. Phil. Soc. l. c. 332. *Pyrrocoma racemosa*, Torr. & Gray, Fl. ii. 244. The type is a form with virgate stem, bearing 3 to 9 racemosely or spicately disposed and approximate or remote heads, of the larger size, with involucre half or two-thirds inch broad as well as high, and akenes (or at least ovaries) more or less beset with villous hairs. *A. lanceolatus*, var. *strictus*, Gray, Proc. Am. Acad. viii. 389, is a form with more villous akenes. — Plains of Oregon, *Nuttall*, *Hall*, &c.; also Northern California, *Greene*, where it varies with many and correspondingly smaller heads, these glomerate in numerous axils, and the campanulate involucre disposed to be squarrose. Also it evidently passes into

    Var. **glomeréllus**. Heads narrower and smaller, disposed to be fascicled in twos or threes in the axils of small upper leaves, or at summit of stem or short peduncles: involucre often turbinate: akenes glabrate or sometimes glabrous: herbage somewhat more disposed to be balsamic-viscid. — *Homopappus glomeratus*, *paniculatus*, & *argutus*, Nutt. l. c. 331. *Pyrrocoma glomerata*, *paniculata*, & *arguta* (the latter a stouter and more leafy state), Torr. & Gray, l. c. *Aplopappus paniculatus*, Gray, Bot. Calif. i. 311. — Plains of the Columbia, E. Oregon, *Nuttall*, *Nevius*, *Cusick*, &c. N. W. Nevada, *Anderson*, *Lemmon*.

    Var. **virgátus**. Slender and smaller, with strict virgate stems and narrow leaves: heads as in the type, but only half the size, few, or in depauperate plants solitary. — *A. paniculatus*, var. *virgatus*, Gray, Bot. Calif. i. 312. — Eastern part of the Sierra Nevada, California, *Bolander*, *Lemmon*, &c. Passes into

    Var. **stenocéphalus**. This is to var. *glomerellus* what var. *virgatus* is to the type: it has scattered heads; these narrow, comparatively few-flowered; the bracts of the oblong-turbinate involucre rigid and more pointed. — *A. paniculatus*, var. *stenocephalus*, Gray, Bot. Calif. l. c. — With preceding var., *Lemmon*.

— **A. apargioídes**, Gray. Low, with numerous ascending or diffuse few-leaved or sometimes scapiform stems from a thick caudex, a span to a foot high, bearing solitary or few pedunculate heads: leaves lanceolate or the radical broader, from denticulate to laciniate-dentate or even pinnatifid: involucre hemispherical (a third to half an inch high); its bracts lanceolate to narrowly oblong, mostly obtuse, imbricated in few rather loose ranks, outer sometimes equalling the inner: rays 20 or more, oblong, comparatively large, commonly fertile: pappus softer. — Proc. Am. Acad. vii. 354, & Bot. Calif. i. 311. — Eastern parts of the Sierra Nevada, California and adjacent Nevada, from Sierra Co. to San Bernardino Mountains; first coll. by *Bolander*.

    = = Pubescence not tomentose nor floccose, but rather villous and persistent: leaves thinnish, oblong, more regularly and closely spinulose-serrate, numerous and approximate on the stem or branches up to the heads or nearly: rays fertile.

A. hírtus. A foot or less high, hirsutely pubescent and villous, even to the involucre, or at base lanuginous: stems rather simple, ascending, bearing few or scattered pedunculate heads: leaves membranaceous, pectinately serrate with long and salient slender-subulate teeth; cauline an inch or two long, radical sometimes 4 inches long and with margined petioles: involucre hemispherical, half-inch or more high; its bracts rather loose, linear, acuminate or acute, all about equalling the disk, the outer mainly foliaceous: rays 20 or more, conspicuous: akenes rather short, sericeous-pubescent; pappus soft, whitish. — Baker Co., Oregon, *Cusick*. Washington Terr., *Brandegee*. Might be arranged in a following subdivision, with *A. uniflorus*, but has the habit of the next.

A. Whitnéyi, Gray. About a foot high, somewhat minutely villous-pubescent, or foliage glabrous, branching, bearing rather numerous fasciculate-panicled and mostly sessile heads: leaves inch or less long, spinulose-dentate, those subtending the lower heads hardly smaller than the main cauline ones: involucre narrow, oblong-turbinate (about half-inch long), glabrous; its bracts lanceolate, acute, appressed, subcoriaceous, with short and sometimes indistinct green tips, well imbricated, outer successively shorter: rays 5 to 8, with oblong and small ligules, little surpassing the 10 to 20 disk-flowers: akenes oblong-linear, glabrous, striate: pappus rigidulous, rufous. — Proc. Am. Acad. vii. 354, & Bot. Calif. i. 312. — Sierra Nevada, California, at 8,500 to 9,000 feet, in open woods, Sonora Pass, *Bolander*, and on bleak summits in Siskiyou Co., *Greene*, *Pringle*. Involucre rather of the *Ericameria* section.

   * * * Heads conspicuously radiate, large and showy: rays fertile, very numerous, half-inch to inch long: involucre well imbricated, of numerous oblong to lanceolate firm bracts: akenes (and ovaries) wholly glabrous, flat and rather broad: pappus pale: style-appendages broadish, oblong to lanceolate, shorter or not longer than the stigmatic portion: wholly herbaceous perennials, smooth and glabrous, except some soft-villous pubescence or tomentum when young: leaves coriaceous, entire.

   + Stems equably and very leafy up to the sessile or subsessile heads.

A. **Fremónti**, Gray. A foot or less high, from slender lignescent rootstocks, simple or fastigiately branched above: leaves lanceolate (2 to 4 inches long, 3 to 8 lines wide), obscurely 3-5-nerved; lower narrowed and upper partly clasping at base: involucre (inch or less high) broadly campanulate; its bracts broadly lanceolate, conspicuously and often cuspidately acuminate: rays half-inch long: style-appendages ovate-oblong, obtuse: akenes obovate, striate-nerved, almost as long as the rigid pappus. — Proc. Acad. Philad. 1863, 65; Porter & Coulter, Fl. Colorad. 67. *Pyrrocoma foliosa*, Gray in Jour. Bost. Soc. Nat. Hist. v. 109. — Plains and rocky hills, Colorado, common on the Arkansas from Pueblo upward; first coll. by *Fremont*.

   Var. **Wárdi**. Dwarf: fascicled stems only a span high: leaves proportionally small, linear-lanceolate, destitute of lateral nerves: heads one-half smaller, 2 or 3 in a terminal glomerule: akenes double the length of the scanty pappus. — Wyoming (probably in southwestern part), *L. F. Ward*.

   + + Stems simple, solitary or several from a thick caudex, above with decreasing or sparse leaves and solitary or few naked and usually pedunculate heads, at base a tuft of ample lanceolate- or spatulate-oblong radical leaves (in the manner of the preceding and succeeding subdivisions): involucre hemispherical or broader: rays 30 to 50.

A. **cróceus**, Gray. Stem stout and erect, commonly a foot or two high, and with radical leaves a foot or less long (including the petiole), cauline leaves ovate-oblong to lanceolate, partly clasping (upper an inch or two long): head mostly solitary: involucre a full inch in diameter; its bracts ovate to spatulate-oblong, very obtuse, lax, inner with scarious erose-denticulate margins: rays saffron-yellow, sometimes inch long: akenes narrowly oblong, nearly the length of the pappus. — Proc. Acad. Philad. l. c. — Rocky Mountains of Colorado, especially in Middle Park, first coll. by *Parry*. A dwarf form in N. Arizona, *Rusby*.

A. **integrifólius**, T. C. Porter. Stems several from the caudex, ascending, a foot or less high: radical leaves 3 to 8 inches (including short petiole or tapering base), cauline lanceolate, or small uppermost linear: heads solitary or 2 or 3 in axils, smaller than in foregoing: involucral bracts narrowly oblong to linear-lanceolate, some loose outer ones usually equalling the disk and more foliaceous: rays bright yellow, half-inch long: immature akenes oblong. — Gray, Proc. Am. Acad. xvi. 79. — Mountain meadows, Wyoming and Montana, *Burke* (in herb. Hook.), *J. M. Coulter*, *Watson*, *Canby*. Verges to the larger-flowered form of the next species.

   * * * * Heads conspicuously radiate, smaller: rays fertile, half to barely quarter inch long: akenes turbinate or oblong, silky-pubescent or villous: style-appendages from ovate to subulate, shorter or rarely longer than the stigmatic portion. (Here *A. Whitneyi* might be sought.)

   + Perennial herbs, with mostly simple stems and a tuft of radical leaves from a thickened somewhat fusiform caudex: leaves coriaceous and when dry rigid, entire or spinulose-serrate, the cauline diminished upward: heads solitary or rather few, pedunculate: involucre hemispherical or broader, of firm and herbaceous-tipped or foliaceous bracts: rays 20 to 50: pappus pale or merely sordid, rather soft and fine: herbage more or less flocculent-tomentose when young, glabrate in age and smooth. — § *Arnicella*, Torr. & Gray, partly of Benth. & Hook.

A. **uniflórus**, Torr. & Gray. Stems a span to barely a foot high, ascending or erect, sometimes 5-6-leaved, sometimes rather scapiform or upper leaves reduced and bract-like, bearing a solitary head, rarely one or two from lower axils: leaves lanceolate or sometimes broader; radical 2 or 3 inches long and usually petioled: involucre commonly half-inch high and the linear or oblong-linear bracts all of same length, rather loose, outer all foliaceous: rays in larger heads 40 or 50. — *A. uniflorus* & *A. inuloides*, Torr. & Gray, Fl. ii. 241. *Donia uniflora*, Hook. Fl. ii. 25, t. 124. *Homopappus inuloides*, Nutt. Trans. Am. Phil. Soc. l. c. 333, a woolly form. — Plains of the Saskatchewan to Montana, and along the mountains to Utah and Colorado; first coll. by *Drummond*. Varies much in size, especially of the head; in the larger forms much broader than high, and very many-flowered.

— **A. lanceolátus**, TORR. & GRAY, l. c. Habit of the preceding: stems generally more leafy and bearing 3 to 15 heads; these when few subcorymbose, when more numerous racemosely or paniculately disposed: involucre in the type fully half-inch high; its bracts rather closely imbricated in 3 or 4 unequal series, lanceolate, acutish, with short green tips and whitish coriaceous base; outer successively shorter, occasionally some of them longer and more herbaceous. Such forms, when heads are very few or solitary, effect a transition to the foregoing species. — Eaton, Bot. King Exp. 160. *Donia lanceolata*, Hook. l. c. *Homopappus* (*Aetanaphoria*) *multiflorus*, Nutt. l. c. — Plains of Saskatchewan to the borders of Brit. Columbia, Idaho, and N. Nevada; first coll. by *Drummond*. The more robust form, with few and large heads, usually corymbosely disposed, and rays 30 or 40 in number and half-inch long, passes freely into

Var. Vaséyi, PARRY in Eaton, l. c., with heads a third or quite half smaller, disposed to be racemose, and involucre closer. — Saskatchewan to Wyoming, Utah, and Colorado.

Var. tenuicaúlis (*A. tenuicaulis*, Eaton, l. c.), is an extreme very slender and marked variety sometimes a foot high and bearing several racemose heads, sometimes more depauperate and only a span high: heads only 3 or 4 lines high: rays correspondingly reduced: involucre close, with short green tips. — Alkaline meadows, Nevada and Utah, first coll. by *Watson*. Apparently a form with laciniate leaves, in alkaline soil, E. Oregon, *Cusick*.

+ + Perennial herbs from a lignescent multicipital caudex or suffruticose base, with slender and branching stems, leafy up to the small heads: leaves all narrow and quite entire; involucre turbinate or obovate (4 or 5 lines high); its bracts well imbricated, appressed, chartaceo-coriaceous, with short and abrupt acute green tips, or these wanting in some: rays 7 to 10, with oblong ligules: disk-flowers not numerous: style-appendages ovate to narrow-lanceolate (thus distinguished from the *Ericameria* section, to which there is an approach).

A. multicaúlis, GRAY. Very dwarf, tufted, tomentulose, but early glabrate and smooth: stems 1 to 3 inches high from a ligneous caudex, simple or forked, bearing 3 or 4 leaves and few heads: leaves narrowly linear, or the lowest obscurely spatulate (about inch long): bracts of the involucre large and rather few (9 to 14), from ovate to oblong-lanceolate, cuspidate-acuminate, marked with a green spot below the slender cusp, or the outermost with a larger foliaceous tip: rays few: style-appendages ovate-triangular, half the length of the stigmatic portion: pappus scanty, somewhat fulvous. — Am. Nat. viii. 213. *Stenotus multicaulis*, Nutt. Trans. Am. Phil. Soc. l. c. 335; Torr. & Gray, Fl. ii. 238 — On rocks, Rocky Mountains of N. W. Wyoming, *Nuttall, Geyer, Parry*.

A. Hállii, GRAY. A foot or two high, paniculately branched from a suffrutescent or even more woody base, glabrous, very leafy: leaves lanceolate or linear, short (larger over inch long, 3 lines wide and spatulate-lanceolate), rather rigid, mostly scabrous (at least the margins); midrib prominent beneath and commonly some lateral veins: heads paniculate, terminating short branchlets or sometimes rather congested: involucral bracts broadish-linear, imbricated in several ranks, the outer successively shorter, the short tips merely mucronate-acute: rays about 10: style-appendages lanceolate, rather obtuse, about the length of stigmatic portion: pappus barely sordid. — Proc. Am. Acad. viii. 389, first described from mere branchlets, and these not well developed. — Base of the Cascade Mountains, Oregon and Washington Terr., *Hall, Howell, Suksdorf, Pringle*.

+ + + Annual or perennial herbs, branching, leafy: leaves not rigid, spinulosely dentate or pinnatifid, the teeth and tips commonly bristle-tipped: heads middle-sized or small: involucre hemispherical, of well-imbricated narrow bracts, the outer successively shorter: rays conspicuous, mostly numerous: pappus rather rigid, its bristles very unequal in size and strength. (Analogue of *Machæranthera* in *Aster*.) — § *Blepharodon*, DC., excl. spec.

++ Akenes short-turbinate, not compressed, obscurely 5-10-nerved under the canescent villosity: style-appendages short and broad, ovate or deltoid: rays 18 to 25, deep golden yellow: leaves not deeply cleft.

A. aúreus, GRAY. Perennial? and branched from the base, at first lightly lanuginous, minutely scabrous-glandular, a span or two high: leaves all narrowly linear, sparingly pinnatifid-dentate, at least toward the base (an inch or less long): heads 4 lines high: bracts of the involucre linear-oblong, mostly obtuse and muticous; the outer ones with short deltoid-ovate green tips, the longer innermost nearly scarious: stronger bristles of the pappus only 10 or 12. — Pl. Fendl. 76. — Low prairies, near Houston, Texas, *Wright*. Not since collected.

A. rubiginósus, TORR. & GRAY. Annual, 1 to 3 feet high, viscid-glandular and pubescent or puberulent: leaves lanceolate or narrowly oblong, incisely pinnatifid or dentate with salient narrow teeth: heads somewhat cymosely paniculate, 5 or 6 lines high, usually nakedpedunculate: bracts of the involucre linear-subulate and with slender spreading green tips: stronger bristles of the fulvous or at length rufous pappus more numerous. — Fl. ii. 240. — Low grounds from S. Texas to plains of Colorado up to the base of the Rocky Mountains; first coll. by *Drummond*.

Var. **phyllocéphalus**. A lower form, spreading, leafy up to the heads, which singly terminate the branches, and are accordingly larger or broader, leafy-involucrate and therefore sessile, or at least some of outermost bracts loose and foliaceous, inner less imbricated. — *A. phyllocephalus*, DC. Prodr. v. 347; Gray, Bot. Mex. Bound. 80. Without much doubt a state of *A. rubiginosus* (in which case a misleading name for the species); but may hold distinct. — Sea-beaches, S. Texas, also S. Florida. (Adj. Mex. *Berlandier*.)

++ ++ Akenes compressed, obscurely striate at maturity: style-appendages lanceolate, rather long: rays 15 to 30: involucre of numerous small and narrow short-tipped and wholly appressed bracts: leaves 1-2-pinnatifid.

A. grácilis, GRAY. Annual or becoming lignescent at base and more enduring, canescently pubescent, occasionally glabrate and glandular-scabrous: stems a span to a foot high, much branched: leaves linear or the lowest spatulate, pinnatifid, or the upper few-toothed or entire, tipped or also sparsely fringed with long and slender bristles: heads 4 or 5 lines high: bracts of the involucre mostly setaceous-tipped: pappus rigid; its larger bristles manifestly dilated below. — Pl. Fendl. 76, & Bot. Calif. i. 613. *Dieteria (Sideranthus) gracilis*, Nutt. Pl. Gamb. 177. — Plains, W. Texas to S. Utah, Arizona, and the southern border of California; first coll. by *Gambel*.

A. spinulósus, DC. Perennial, canescently puberulent or tomentulose, or glabrate: stems a span to a foot high, commonly spreading, cymosely branched at summit: leaves broader in outline than the preceding, pinnately and the lower often bipinnately parted into rather numerous lobes; lobes and teeth mucronate-setigerous: heads and involucre of the preceding: pappus more capillary and soft. — Prodr. v. 347; Torr. & Gray, Fl. ii. 240. *Amellus? spinulosus*, Pursh, Fl. ii. 564. *Starkea? pinnata*, Nutt. Gen. ii. 169. *Diplopappus pinnatifidus*, Hook. Fl. ii. 22. *Dieteria spinulosa*, Nutt. Trans. Am. Phil. Soc. l. c. 301. — Plains from Saskatchewan to Texas, and west to Dakota, Colorado, and Arizona. Varies in Texas to nearly glabrous throughout, also sometimes with divisions of the leaves nearly filiform. (Mex.)

A. ARENÁRIUS, Benth. Bot. Sulph., from Cape Lucas, Lower California, may have heterochromous heads, and be an *Aster*.

§ 3. ISOPÁPPUS, Benth. Heads small and narrow, loosely paniculate: involucre of subulate-lanceolate bracts, destitute of distinct green tips, appressed and imbricated in few ranks, the outer shorter: rays 5 to 15: disk-flowers 10 to 25; their corolla slightly ampliate upward, 5-toothed: style-appendages linear-subulate, much longer than the stigmatic portion: akenes narrow, sericeous-canescent: pappus ferruginous, of rather scanty fine and soft bristles: annuals, or sometimes more enduring, narrow-leaved. — *Isopappus*, Torr. & Gray.

A. divaricátus, GRAY. A foot or two high, with somewhat the aspect of *Chrysopsis graminifolia*, more slender and effusely paniculate, scabrous-pubescent or glandular, sometimes glabrate: leaves rigid, linear-lanceolate or lower spatulate-lanceolate, mucronate-acute or cuspidate, entire or beset with a few spinulose teeth, more or less setose-ciliate toward the base; the upper small and subulate and in the diffuse naked usually polycephalous panicle minute: heads 3 or 4 lines high: peduncles sometimes filiform, sometimes short: involucral bracts subulate-attenuate. — Proc. Am. Acad. xviii. 102. *Isopappus divaricatus*, Torr. & Gray, Fl. ii. 239. *Chrysopsis (Inula) divaricata*, Nutt. Gen. ii. 152. *C. Lamarckii*, Nutt. in Trans. Am. Phil. Soc. l. c. 315. *Heterotheca Lamarckii*, DC. Prodr. v. 317, as to char. & syn. Nutt. & Ell., excl. syn. Cass., Lam., & Pluk. Alm. — Dry and sandy ground, Georgia and Florida to Arkansas and Texas; flowering late. A rigid and rough-hispidulous form with less open inflorescence (*Lindheimer*, 254, *Drummond*, 157) is *Aplopappus Hookerianus*, Gray, Pl. Lindh. i. 40.

A. Hookeriánus. Low, loosely branched from the base, barely hirsute, not glandular: leaves not rigid, entire; upper linear or attenuate-lanceolate, sparingly hispidly ciliate; lower spatulate, short, naked: involucral bracts subulate-lanceolate, with less attenuate points. — *Isopappus Hookerianus*, Torr. & Gray, Fl. ii. 239. — Gonzales, Texas, *Drummond* (184 of coll. 3); not since found; perhaps an unusual state of *A. divaricatus*.

§ 4. STENÓTUS, Gray. Heads middle-sized, mostly broad: bracts of the involucre from ovate to lanceolate or even linear, not rigid, all of equal or moderately unequal length: rays several or numerous: disk-corollas somewhat ampliate upward and deeply 5-toothed: style-appendages various: pappus soft and white or whitish: perennials (herbaceous or fruticulose), of the Rocky Mountains and westward, with leaves all entire. — Proc. Am. Acad. vii. 353. *Stenotus*, Nutt. Trans. Am. Phil. Soc. l. c. 334.

* Solidaginiform herb: heads corymbiform-cymose or glomerate at the summit of a leafy stem: involucre campanulate: rays 12 to 20, small and narrow: akenes short and glabrous or nearly so.

A. Párryi, GRAY. Green and almost glabrous, puberulent and somewhat viscid above: stems 6 to 18 inches high from slender rootstocks: leaves oblong-obovate and spatulate, or the upper oblong-lanceolate, thinnish, loosely veiny (2 to 4 inches long): heads nearly half-inch high, rather numerous (in a dwarf form reduced to a glomerule of 2 or 3): involucral bracts oblong, obtuse, pale and chartaceous or the outer partly herbaceous, in about three moderately unequal ranks: flowers pale-yellow: style-appendages lanceolate, rather longer than the stigmatic portion. — Am. Jour. Sci. ser. 2, xxxiii. 10; Eaton, Bot. King Exp. 162. — Rocky Mountains, from those of Colorado to the Wahsatch, in open woods, 6,000 to 10,000 feet; first coll. by *Parry*. Has somewhat the aspect and character of a large corymbose *Solidago*. — Var. *minor* is a reduced subalpine form (Wahsatch Mountains, Utah, at 12,000 feet, *M. E. Jones*), with leaves only an inch or two long, and 2 or 3 narrower heads.

* * Typical species, herbaceous or suffruticulose and dwarf: heads solitary, terminating simple stems or branches: rays conspicuous.

+- Wholly herbaceous, chiefly alpine, disposed to be cespitose or multicipital, a span or less in height; leaves soft, not persistent: involucre hemispherical: rays 15 to 20: style-appendages oblong to subulate, shorter or not longer than the stigmatic portion.

++ Green, not woolly, mostly equably leafy up to the (half-inch) head.

A. pygmǽus, GRAY. Less than a span high, soft-pubescent or glabrate, not viscid nor glandular: leaves from linear-spatulate to spatulate-oblong: involucral bracts oblong, outer ones foliaceous and loose, very obtuse, equalling the thinner innermost: akenes pubescent. — Am. Jour. Sci. ser. 2, xxxiii. 239. *Stenotus pygmæus*, Torr. & Gray, Fl. ii. 237. — Rocky Mountains, Colorado, strictly alpine; first coll. by *James*.

A. Lyálli, GRAY. Rather taller, larger-leaved, viscid-puberulent: leaves obovate-spatulate to oblanceolate: involucre glandular; its bracts lanceolate, acute, sometimes 2 or 3 outermost oblong and more foliaceous: akenes and ovaries glabrous or nearly so. — Proc. Acad. Philad. 1863, 64. — Alpine region of Colorado Rocky Mountains, first coll. by *James*. Also in northern Rocky and Cascade Mountains, Montana to Oregon and Brit. Columbia; first coll. by *Lyall*.

++ ++ Woolly or tomentose, at least the involucre, above less leafy, or head pedunculate.

A. lanuginósus, GRAY. Fully a span high from creeping rootstocks, floccose-tomentose; leaves soft, narrowly spatulate or upper linear (inch or two long); the sparse uppermost almost filiform: involucre half-inch high; its bracts lanceolate, acute or acuminate, thin, nearly equal, in two series, outer barely greenish: style-appendages elongated-subulate: akenes sericeous-canescent. — Wilkes Ex. Exped. xvii. 347. — Mountains of Washington Terr.; first coll. by *Pickering* and *Brackenridge*, recently by *Nevius*, *Howell*, *Brandegee*; and Montana, *Watson*.

A. Brandegéi. A span high from a tufted caudex, einereous-pubescent or puberulent, and the involucre lanuginous-tomentose: radical leaves obovate or spatulate or roundish (half-inch long), contracted into a slender petiole; cauline few and sparse, small (quarter-inch long), oblong or lanceolate : head one-third inch high and broad : bracts of involucre loose, lanceolate, nearly equal: young akenes hirsute-pubescent: pappus rather scanty: style-appendages triangular-subulate. — Mountains of Washington Terr., in the Yakima district, *Brandegee.* — Aspect of an alpine *Erigeron*; but rays deep yellow and style-appendages acute.

+ + Depressed-cespitose from a multicipital lignescent caudex, glabrous or puberulent-scabrous: leaves rigid and persistent, crowded on the crowns of the caudex or on short shoots, and a few on the lower part of the scapiform flowering stems: rays 6 to 15, rather broad: style-appendages subulate: akenes canescently villous. — *Stenotus,* Nutt.

A. acaúlis, GRAY. Leaves from spatulate (and inch or less long) to oblanceolate or linear (and 2 or 3 inches long), mucronate, more or less 3-nerved and the broader ones veiny, commonly scabrous: scapiform flowering stems an inch to a span high, mostly monocephalous: bracts of the involucre from ovate to ovate-lanceolate, mucronately acute or acuminate, destitute of greenish tips; the outer a little shorter than the inner. — Proc. Am. Acad. vii. 353; Eaton, Bot. King Exp. 161. *Chrysopsis acaulis,* Nutt. in Jour. Acad. Philad. vii. 33, t. 3. *Stenotus acaulis,* Nutt. in Trans. Am. Phil. Soc. l. c. 334 ; Torr. & Gray, l. c. — Dry rocks on the mountains (at 6,000 to 8,000 feet, and extending to the alpine region), from Saskatchewan and N. Wyoming to E. Oregon, and south to Utah and the Sierra Nevada, California. Passes into

Var. glabrátus, EATON, l. c. Glabrous and smooth or almost so: flowering stems disposed to be leafy above and to branch, so bearing 2 or 3 heads. — *Chrysopsis cespitosa,* Nutt. in Jour. Acad. Philad. l. c. *Stenotus cespitosus,* Nutt. in Trans. Am. Phil. Soc. l. c.; Torr. & Gray, l. c. — Wyoming to Nevada and N. Arizona.

A. armerioides, GRAY. Smooth and glabrous: flowering stems naked above (for 1 to 3 inches), sometimes nearly scapiform: bracts of the campanulate involucre broadly oval, rounded-obtuse or retuse, muticous, of about three lengths; the outermost much shorter, most of them greenish at apex. — *Stenotus armerioides,* Nutt. l. c.; Torr. & Gray, l. c. — Rocks on mountains, from Wyoming to New Mexico and S. Utah; first coll. by *Nuttall.* Too near the preceding.

A. stenophýllus, GRAY. More suffruticulose, hirtellons-scabrous: leaves very narrow, linear-spatulate to filiform-linear (commonly inch or less long and half a line wide), one-nerved: scapiform peduncles inch or two long: involucral bracts linear, glandular-puberulent, equal, in one or two series. — Wilkes Ex. Exped. xvii. 347. — Mountains and stony hills, W. Idaho and Washington Terr. to northeastern borders of California, *Pickering* and *Brackenridge, Burke, Nevius, Howell, Lemmon.*

* * * Anomalous species, shrubby, a transition to the following section, of which it has the foliage and habit, but with broad rather large heads and little-imbricated involucre.

A. linearifólius, DC. Undershrub, a foot to a yard or more high, fastigiately much branched, with herbage often resinous-dotted and balsamic-viscid : branches thickly leafy: leaves all narrowly linear (an inch or less long, a line or less wide), sometimes almost filiform, many in axillary fascicles: heads solitary terminating the corymbiform branchlets, on peduncles bearing one or two setaceous-subulate bracts: involucre fully half-inch high ; its bracts thin, lanceolate, acute or acuminate, somewhat scarious-margined (at least when dry), in about 2 series of nearly equal length: rays about 12, oblong or broadly lanceolate, in largest heads nearly three-fourths inch long, in smaller only half that length; style-appendages from ovate- to lanceolate-subulate: akenes densely silvery-villous: pappus white, rather deciduous. — Prodr. v. 347; Gray, Bot. Calif. i. 311. *Stenotus linearifolius,* Torr. & Gray, Fl. ii. 238. — Dry hills, coast ranges of California from San Francisco Bay southward; and mountains of San Bernardino Co. to S. Utah and adjacent Arizona. Southward it bears more numerous and smaller heads than at the north.

§ 5. ERICAMÉRIA, Gray. Heads small or barely middle-sized, paniculately or corymbosely disposed: involucre oblong or campanulate, of well-imbricated bracts; these all chartaceous or thinner, appressed, and wholly destitute of herbaceous tips, or some outer looser ones foliaceous or foliaceous-tipped: rays few, sometimes only one (which alone definitely separates the group from *Bigelovia*, and even this fails in one or two species!): disk-corollas commonly somewhat ampliate upward and rather deeply 5-toothed: style-appendages (with some exceptions) filiform or slender-subulate: akenes slender: pappus fine and soft: all W. North American shrubby or fruticulose plants, very leafy, mostly with Heath-like foliage, glabrous or almost so, except in one species, disposed to be resinous-dotted and balsamic-viscid. — Pl. Wright. ii. 80. *Ericameria*, Nutt. Trans. Am. Phil. Soc. l. c.; Benth. & Hook. Gen. ii. 255.

* Anomalous for its broad although small leaves, also in the frequent absence of the scanty rays: involucral bracts (as of the next following group) all close and unappendaged, the outer successively shorter.

A. cuneátus, GRAY. Shrub a foot or so high, intricately branched and spreading, balsamic-glutinous: leaves thick, cuneate or rarely obovate, retuse, sometimes apiculate, entire but inclined to be undulate, usually resinous-punctate, 2 to 4 lines long, larger ones petioled: heads corymbosely fasciculate, 5 or 6 lines long: involucre turbinate; bracts lanceolate or nearly linear, rather obtuse: rays 2 or 3, or solitary and small, or as commonly wanting: style-appendages slender-subulate, not longer than the stigmatic portion: akenes pubescent. — Proc. Am. Acad. viii. 638, & Bot. Calif. i. 212. *Bigelovia spathulata*, Gray, Proc. Am. Acad. xi. 74, & Bot. Calif. i. 613, also *B. rupestris*, Greene in Bot. Gazette. vi. 184, the rayless state! — Cañons and cliffs in the Sierra Nevada, California, from Placer Co. and the Yosemite to the Mexican border below San Diego, and in Arizona; first coll. by *Bolander* and southward by *Palmer*, &c.

* * Typical species: leaves from filiform to very narrowly linear, thick: proper bracts of the involucre obtuse or barely acute and close: shrubs a foot to a yard or more high.

+ Heads only 3 or 4 lines high, in close cymose clusters terminating fastigiate branchlets: bracts of the involucre in only 2 or 3 series, no loose outer ones: leaves half-inch or less long: akenes villous: style-appendages shorter than the linear stigmatic portion, not attenuate.

A. laricifólius, GRAY. About a foot high: leaves linear-acerose, rigid, mucronate, conspicuously resinous-punctate and becoming viscid, crowded but seldom axillary-fascicled: larger ones narrowed downward and flatter: involucral bracts subulate-linear, acute: rays 3 to 6, with rather conspicuous oblong ligules: disk-flowers 10 or 12: style-appendages linear, rather obtuse. — Pl. Wright. ii. 80, & Pacif. R. Rep. iv. 99; Rothrock in Wheeler Rep. vi. 144. — Western borders of Texas to mountains of Arizona, first coll. by *Wright, Bigelow*, &c.

A. monáctis, GRAY. A foot to a yard or more high, hardly becoming viscid: leaves not punctate, mostly obtuse or pointless, more disposed to have axillary fascicles, otherwise not unlike those of the foregoing: involucral bracts only 8 or 10, oblong or linear-oblong, obtuse, thin-chartaceous: ray-flower solitary with an elongated-oblong ligule, wanting to some heads: disk-flowers 5 or 6: style-appendages oblong-ovate, acute. — Proc. Am. Acad. xix. 2. — S. E. California, on the San Bernardino Mountains and Mohave Desert, *Palmer, Parish, Pringle*.

+ + Heads 4 or 5 lines high, paniculate: involucral bracts imbricated in several ranks: style-appendages filiform-subulate: leaves all filiform or nearly terete, excessively numerous and axillary-fascicled.

++ Involucre narrow, 7-20-flowered; its bracts all erect, more or less obtuse, somewhat tomentulose-ciliolate when young; outer successively shorter, becoming greenish and passing into the very short leaves of the ultimate branchlets; cauline leaves short: shrubs 2 to 5 feet high, bearing very numerous heads: young parts disposed to be cinereous-pruinose or puberulent.

A. ericoídes, HOOK. & ARN. Fastigiately much branched: cauline leaves only half-inch and those of the dense fascicles 2 or 3 lines long: rays 3 to 5, short: akenes glabrous. — Bot. Beech. 146; DC. Prodr. v. 346; Gray, Bot. Calif. i. 313. *Diplopappus ericoides*, Less. in Linn. vi. 117. *Ericameria microphylla*, Nutt. Trans. Am. Phil. Soc. l. c.; Torr. & Gray, l. c. — California along the coast, especially on sand-hills near the sea: first coll. by *Chamisso*.

A. Pálmeri, GRAY. Paniculately much branched: cauline leaves often inch long: lower bracts of involucre more greenish-tipped; rays 3 or 4 and disk-flowers 5 to 15: akenes pubescent. — Proc. Am. Acad. xi. 74, & Bot. Calif. i. 613. — S. California, on hills, Los Angeles to the Mexican border (*Palmer, Nevin, Lyon*, and mountains of San Bernardino Co. to the desert on the Colorado River, *Parry, Lemmon, Parish, Pringle*. Heads of the plant in the interior districts very numerous in ample and rather naked panicles, at Los Angeles sparse and racemosely disposed along the elongated and intricate branches.

++ ++ Involucre larger, campanulate, 15–30-flowered, subtended by several loose outer bracts having elongated-subulate herbaceous tips: leaves longer.

A. pinifólius, GRAY. Shrub 2 to 5 feet high, rather stout, with rigid erect branches: cauline leaves from very narrowly linear to filiform, an inch or more long, mucronate; those of the fascicles and branchlets much shorter: heads not very numerous in a contracted panicle, or scattered: proper bracts of the involucre broadly lanceolate and with a greenish keel or midrib; loose outer ones normally subulate, shorter than the innermost, and passing into the small leaves of the flower-bearing branchlet, or in a vernal state (with solitary larger heads) developed into an involucriform rosette of acerose-filiform leaves: rays commonly 6 to 10, short; akenes almost glabrous. — Proc. Am. Acad. viii. 636, & Bot. Calif. i. 312, there described from the abnormal vernal state, in which the large and many-flowered head, terminating a very leafy branch, seems to consist of two or three confluent ones. In autumn the normal paniculate and naked heads are developed. — S. California, from Los Angeles Co. to the foot-hills of the San Bernardino, *Bolander, Parry, Nevin, Parish*, &c.

* * * Leaves from narrowly linear to lanceolate-spatulate, not rigid nor punctate, mostly plane, seldom with axillary fascicles: low and suffruticose, not at all or very slightly balsamic or viscidulous: at least the outer involucral bracts acute or acutely herbaceous-tipped: akenes pubescent to glabrate.

+— Glabrous throughout: leaves narrow.

A. Bloómeri, GRAY. A foot or two high, with erect and rigid usually virgate branches, sometimes lower, very leafy: leaves from narrowly spatulate-linear to filiform-linear, an inch or two long: heads showy, half to three-fourths inch high, in dwarf plants solitary terminating fastigiate branches, commonly several and racemosely clustered, or more numerous and thyrsoid-paniculate: involucre oblong; its inner bracts oblong-lanceolate or linear, chartaceous with thin-scarious and erose-ciliate margins, some obtuse, some acute or tipped with a soft cusp, most of the outer bearing a filiform foliaceous tip: rays 2 to 4, rarely solitary, oblong, deep yellow, half-inch or less long: disk-flowers 8 to 20: their style-appendages long and much exserted, setaceous-subulate: akenes 3 or 4 lines long, sparsely pubescent. — Proc. Am. Acad. vi. 541, vii. 354, viii. 356, & Bot. Calif. i. 313, with var. *angustatus*, the narrower-leaved form, passing freely into the broader, and to this belongs *A. resinosus*, Gray in Wilkes Ex. Exp. xvii. 346, t. 10. *Ericameria erecta*, Klatt in Abh. Naturf. Gesel. Halle, xv. 6, from the char. & habitat. — California and adjacent Nevada, along the Sierra Nevada from Kern Co. northward to Washington Terr.; first coll. by *Pickering* and *Brackenridge*, next by *Bloomer* and *Anderson*.

A. nánus, EATON. A span to a foot high, in depressed tufts, fastigiately branched, disposed to be balsamic-glutinous: leaves from narrowly linear to narrowly spatulate (the largest less than inch long): heads solitary or fastigiate-clustered at summit of branchlets, 3 or 4 lines high, narrow: bracts of the involucre lanceolate, acute or acuminate, pale, wholly destitute of green tip or midrib, except one or two looser and subulate outermost: flowers all pale or ochroleucous, or even "white": rays small, 3 to 6 or in some heads wanting; disk-flowers 8 to 12, with deeply 5-*cleft corolla*: style setaceous-subulate and hispid: akenes either pubescent or glabrous. — Bot. King Exp. 150. *A. resinosus*, Gray, Bot. Calif. 313. *Ericameria nana* & *E. resinosa*, Nutt. Trans. Am. Phil. Soc. l. c. 318; Torr. & Gray, Fl. ii. 236. — Rocky hills and cliffs, eastern borders of Washington Terr., and Oregon, adjacent Idaho, and W. Nevada; first coll. by *Nuttall*. Flowers said by *Cusick* and *Brandegee* to be white; by Nuttall, in his *E. resinosa*, ochroleucous.

Var. **cervínus**. Leaves broader; lower ones from oblanceolate to obovate-spatulate: heads more scattered. — *A. cervinus*, Watson, Am. Nat. vii. 30; Rothrock in Wheeler Rep. vi. 142, t. 6. — Cañons, S. W. Utah and adjacent Arizona, *Wheeler, Palmer*.

+— +— Minutely viscidulous-pubescent.

A. Watsóni, GRAY. A span or two high, like the broader-leaved variety of the foregoing, but coarser and manifestly pubescent: leaves from lanceolate with narrowed base to obovate-spatulate, thinnish: heads half-inch or less high, loosely fastigiate-clustered: involucral bracts linear-lanceolate, attenuate-acute, usually one or two outer ones loose and foliaceous, these sometimes equalling the head and resembling uppermost leaves of the branchlets; rays 4 to 8, about 3 lines long: disk-flowers hardly more numerous: young akenes pubescent. — Proc. Am. Acad. xvi. 79. Part of *A. suffruticosus*, Eaton, l. c., which, indeed, it approaches, but is nearer the preceding. — Mountains of Nevada, *Watson, Palmer*, and of E. Utah, *M. E. Jones*.

§ 6. MACRONĒMA, Gray. Heads middle-sized or rather large, solitary or few, terminating leafy branches: involucre campanulate, of lanceolate or linear bracts in few ranks and of somewhat equal length; innermost thin-chartaceous or partly scarious; outer with conspicuous foliaceous tips, or loose and foliaceous, passing into leaves: rays few and conspicuous, or in the typical species wanting: style-appendages long and attenuate-filiform, much exserted: akenes slender, compressed, few-nerved, soft-pubescent: pappus soft and slender: low and many-stemmed from a suffrutescent base, not resinous-punctate: stems or branches leafy to the summit, but no axillary fascicles: leaves soft, spatulate-oblong to broadly linear, sessile, entire, but margins sometimes undulate. — Proc. Am. Acad. vi. 542. xvi. 79. *Macronema*, Nutt. Trans. Am. Phil. Soc. l. c. 322.

   * Connecting with preceding group; the involucre being somewhat imbricated.

A. Greénei, GRAY. About a foot high, branching from a decidedly shrubby base, not viscidulous, or above very obscurely viscid-puberulent: the typical form otherwise quite glabrous: leaves spatulate-oblong or somewhat lanceolate (half-inch to barely inch long, 2 or 3 lines wide), obtuse or mucronate: heads solitary or few and crowded, half-inch high: bracts of the involucre in about 3 series, lanceolate to linear, all but the innermost with conspicuous and spreading mostly elongated-subulate foliaceous tips: rays 2 to 7, 3 or 4 lines long: disk-flowers 10 to 16. — Proc. Am. Acad. xvi. 80. — Mountains of N. California, about the heads of the Sacramento, *Greene, Pringle*. Also mountains of Oregon and Washington Terr., *Cusick*. Passes freely into

   Var. **móllis**, GRAY, l. c. From cinereous-puberulent to canescent-tomentose, even to the more foliaceous involucre. — *A. mollis*, Gray, Proc. Am. Acad. xvi. 80. — N. California (the intermediate form), *Pringle*. Mountains of Oregon and Washington Territory, *Cusick, Brandegee*, &c.

   * * Low, a span or two high, viscidly glandular-puberulent: heads commonly solitary, terminating the leafy simple stems or branches: involucre simpler and loose outer bracts more foliaceous, often enlarged: species probably confluent.

A. suffruticósus, GRAY. Destitute of tomentum: stems glandular-pubescent or puberulent: heads two-thirds to three-fourths inch high: rays 2 to 5 and somewhat exserted, or none: disk-flowers 10 to 30. — Proc. Am. Acad. vi. 542, & Bot. Calif. i. 313. *Macronema suffruticosa*, Nutt. l. c. — Alpine or subalpine region of the Sierra Nevada, California, from Mariposa Co. and Humboldt Mountains, Nevada, northward to Oregon and N. Wyoming; first coll. by *Nuttall*.

A. Macronéma, GRAY, l. c. Stems stouter, whitened by a dense and close tomentum: head commonly larger (inch long): rays always wanting. — *Macronema discoidea*, Nutt. l. c. — Rocky Mountains in Colorado and Wyoming, and higher mountains in Nevada and eastern border of California; first coll. by *Nuttall*.

31. **BIGELÓVIA**, DC. (*Dr. Jacob Bigelow*, author of Florula Bostoniensis, Medical Botany of U. S., &c.) — The original a perennial herb, most related to *Solidago*; as now extended a large genus (N. American, mainly western, with an anomalous Andean representative), mostly of suffrutescent or more shrubby plants, the genuine species with few-flowered heads of marked habit and character, while others are only artificially and not definitely distinguished from *Aplopappus*, especially from § *Ericameria*, by the total want of ray-flowers. Yet some genuine *Aplopappi* are rayless. — DC. Mem. Comp. t. 5, & Prodr. v. 329

(excl. § 3); Gray, Proc. Am. Acad. viii. 638; Benth. & Hook. Gen. ii. 1232. *Chrysothamnus*, Nutt. Trans. Am. Phil. Soc. ser. 2, vii. 323; Benth. & Hook. Gen. ii. 255. *Linosyris*, Torr. & Gray, Fl. ii. 232, not Cass., which Old-World genus differs in that its heads when perchance heterogamous are heterochromous. The various types in the genus are connected by gradations, so that the sections are not very distinct.

§ 1. CHRYSOTHAMNÓPSIS, Gray, l. c. Heads comparatively large but narrow, at least half-inch long, 5-20-flowered: bracts of the involucre comparatively large, chartaceous and acuminate, and some outer ones prolonged into a slender herbaceous tip or appendage; when numerous the vertical ranks become more or less apparent: corollas 5-toothed or barely 5-lobed at summit: low and suffrutescent, with linear entire leaves, not punctate nor viscid or resinous, except that the first species is slightly so.

* Genuine: style-appendages setaceous-subulate or filiform, conspicuously exserted out of the corolla: akenes slender, sericeous-pubescent: anther-tips oblong-lanceolate: involucre cylindraceous, shorter than the developed (5 to 15) flowers and pappus: stems or branches whitened (at least when young) by a close pannose tomentum: heads thyrsoidly paniculate or glomerate. Connects on one hand with *Aplopappus* § *Macronema*, on the other with § *Chrysothamnus*. *B. ceruminosa*, of the latter, approaches this group.

+ Bracts of the involucre comparatively few, not showing obvious vertical ranks.

B. Bolánderi, GRAY, l. c. Leaves oblanceolate-linear or narrower, green and glabrate, somewhat viscidulous (about inch long), very obscurely 3-nerved: heads few and clustered, sometimes three-fourths inch long, 5-11-flowered: bracts of the involucre only about 10, thin-chartaceous, lanceolate, with a soft acuminate apex, or one or two outermost herbaceous-tipped: alveoli of the receptacle paleaceous-dentate. — *B. Bolanderi* & *B. Howardii* in part (sp. Bolaud.), Gray, Bot. Calif. i. 315, 316. *Linosyris Bolanderi*, Gray, Proc. Am. Acad. vii. 354. — Mono Pass in the Sierra Nevada, California, at 10,000 feet, *Bolander*.

B. Párryi, GRAY, l. c. Not viscidulous: stems rather strict, leafy to the summit: leaves linear (2 or 3 inches long, 2 lines or less wide), obscurely 3-nerved, glabrous; upper ones hardly diminished in size and overtopping all the heads of the strict and narrow thyrsiform-virgate panicle: heads little over half-inch long, 10-15-flowered: bracts of the involucre about 12, lanceolate and gradually acuminate, rather prominently 1-nerved, thin-chartaceous, a few exterior tapering into a prolonged subulate-linear herbaceous appendage: alveoli of receptacle short and nearly entire. — *Linosyris Parryi*, Gray in Proc. Acad. Philad. 1863, 66. — Parks of the Colorado Rocky Mountains, *Parry, Hall & Harbour*, &c.

+ + Bracts of the involucre more numerous and disposed to fall into 4 or 5 vertical ranks, in the manner of § *Chrysothamnus*.

B. Nevadénsis. Rigid, more branching, cinereous-puberulent or tomentulose when young: leaves coriaceous, oblanceolate to linear, mucronate (the mucro generally recurving), obscurely or not at all 3-nerved, at most inch long: heads few and glomerate at the naked summit of the branches, often three-fourths inch long, 5-flowered: bracts of cylindrical involucre more imbricated and numerous (15 to 18), rigid-chartaceous, pubescent or somewhat hirsute-ciliate, all with prolonged slender acumination, outer broadly lanceolate or oblong and with prolonged slender-subulate tip or appendage recurving and rigid. — *B. Howardi*, var. *Nevadensis*, Gray, Proc. Am. Acad. viii. 641, & Bot. Calif. i. 316. *Linosyris Howardi*, var. *Nevadensis*, Gray, l. c. vi. 541. — Eastern and arid portion of the Sierra Nevada, &c., on the borders of California and Nevada, *Bloomer, Anderson, Brewer, Watson*, &c. Is the analogue of *Aplopappus Bloomeri*.

B. Howárdi, GRAY. Lower, more tufted, canescently tomentulose when young: leaves narrowly linear, rigid (an inch or two long, barely a line wide), obscurely 1-nerved; upper mostly overtopping the glomerate (about half-inch long) narrow heads: involucre 5-flowered, glabrous; its bracts thinnish, lanceolate, apiculate-acuminate, or some loose outer ones with prolonged subulate-filiform appendage. — Proc. Am. Acad. viii. 641, excl. var. *Linosyris Howardi*, Parry, in Gray, Proc. Am. Acad. vi. 541, excl. var. — Parks of the Rocky Mountains in Colorado to the borders of New Mexico and Utah; first coll. by *Parry*. Forms approach *B. graveolens*.

* * Style-appendages short-subulate, these and the deltoid-ovate obtuse anther-tips hardly exserted: akenes linear-oblong, glabrous: involucre campanulate-cylindraceous, equalling the 15 to 20 flowers: herbage glabrous throughout.

B. Engelmánni, GRAY. A span or two high, in tufts from a suffrutescent subterranean branching caudex or rootstock: stems simple, very leafy up to the cymose glomerate heads: leaves all narrowly linear (inch or two long, only a line wide), rigid: heads (few or rather numerous in the cluster) barely half-inch long: bracts of the involucre firm-chartaceous, oblong or innermost lanceolate, regularly imbricated and appressed, outer similar but short, all abruptly mucronate or short-cuspidate, slightly greenish below the tip.—Proc. Am. Acad. xi. 75.—Plains of Colorado at Hugo Station, *Engelmann, Parry, Patterson*.

§ 2. CHRYSOTHÁMNUS, Gray, l. c. Heads narrow or small, 5-flowered (in *B. Douglasii* sometimes 6–7-flowered), mostly numerous and crowded: involucre (anomalous first species excepted) of dry and chartaceous more or less carinate bracts imbricated so as to form 5 conspicuous vertical ranks (less manifestly so when the bracts are less numerous): corollas narrow: style-appendages with exserted subulate- or setaceous-filiform appendages: akenes slender: fruticose or suffruticose and branching, with entire narrow leaves.—*Bigelovia*, § 2, DC. l. c. *Chrysothamnus*, Nutt. Trans. Am. Phil. Soc. l. c.

* Transition to preceding section: involucral bracts comparatively large, not carinate nor obviously 5-stichous, some outer ones foliaceous-acuminate or appendaged: anther-tips very short and obtuse: corollas said to be even "white."

B. álbida, M. E. JONES. Shrubby, a foot or two high, more or less resinous-viscid, fastigiately branched, very leafy: leaves all filiform, mucronate, not obviously punctate: heads fastigiate-glomerate at the summit of the branchlets, 5 or 6 lines long: involucre oblong-turbinate or cylindraceous: its bracts rather few and coriaceo-chartaceous, lanceolate; outer with rather rigid subulate-acuminate and recurved or spreading foliaceous tip or appendage; inner wholly chartaceous and pointless: corollas probably ochroleucous; lobes of the deeply cleft limb linear-lanceolate: akenes pubescent.— Gray, Proc. Am. Acad. xvii. 209.—Arid districts, east of the Sierra Nevada: Owens Valley, California, coll. 1875, *Kellogg*. Wells Station, W. Nevada, *Marcus E. Jones*, who states that the flowers are white.

* * Genuine species, with thinner more chartaceous and carinate involucral bracts, none foliaceous-tipped: anther-tips lanceolate or narrowly oblong.

+– Akenes and ovaries glabrous, 4–6-angled and with broad epigynous disk: pappus rigidulous: corollas 5-toothed or short-lobed: bracts of the involucre acute or acuminate, numerous and strictly 5-stichous, 5 or 6 in each vertical rank: herbage not punctate, slightly or not at all balsamic-resinous: heads half to three-fourths inch long, somewhat fastigiately glomerate.

—— **B. depréssa**, GRAY, l. c. Obscurely scabro-puberulent and pale, a span or two high from a decumbent woody base: branches leafy up to the glomerule or fasciculate cyme of few heads: leaves short (about half-inch or less long), lanceolate or lowest rather spatulate, rigid, mucronate-acute, with carinate midrib and no veins: heads half-inch long: involucral bracts lanceolate, gradually acuminate into an almost setaceous tip.— *Chrysothamnus depressus*, Nutt. Pl. Gamb. 171. *Linosyris depressa*, Torr. in Sitgreaves Rep. 161.— Plains of S. Colorado to adjacent New Mexico and S. Utah; first coll. by *Gumbel*.

B. pulchélla, GRAY, l. c. Glabrous and green, shrubby, 2 or 3 feet high, fastigiately much branched, very leafy up to fastigiate-cymose heads: leaves narrowly linear, plane (inch or less long), rather obtuse, with ciliolate-scabrous margins and midrib not prominent: heads two-thirds to three-fourths inch long: involucral bracts rigid-chartaceous and lower ones obscurely herbaceous on the back, much carinate, acute and cuspidate-mucronate.— *Linosyris pulchella*, Gray, Pl. Wright. i. 96; Torr. in Sitgreaves, l. c. t. 4.— W. borders of Texas to adjacent New Mexico and Colorado; first coll. by *Wright*.

B. Bigelóvii, GRAY, l. c. Canescent with fine close tomentum when young, glabrate, shrubby, a foot to a yard high, fastigiately much branched, rigid: branches less leafy, bearing a few fastigiate-clustered heads (these half to two-thirds inch high): leaves nearly filiform: involucral bracts lanceolate, acute, thinnish, all pale: receptacle sometimes bearing a prominent chaffy cusp.— *Linosyris Bigelovii*, Gray, Pacif. Ex. Exp. iv. 98, t. 12. — N. New Mexico and adjacent Colorado; first coll. by *Bigelow*.

+ + Akenes (smaller) canescently pubescent or villous (*B. leiosperma* excepted!): herbage commonly graveolent, and in most species becoming more or less resinous-pruinose or balsamic-viscid.

++ Leafless or sparsely leaved, shrubby, with rush-like or broom-like branches, 2 feet or more high: leaves when present filiform, not punctate: heads fasciculate-clustered: involucre somewhat clavate, 4 or 5 lines long, very glabrous; the bracts wholly thin-chartaceous and pale, very strictly pentastichous and about 5 in each vertical rank, all muticous; the inner ones linear, outer successively and regularly shorter, outermost minute: akenes slender, appressed-villous.

B. júncea, Greene. Strict, fastigiately very much branched: branches slender and junciform, mostly leafless, greenish and minutely canescent, apparently not becoming viscid: bracts of the involucre acutish, at least the innermost: corolla-lobes short-lanceolate, in the bud externally beset with delicate long hairs. — Bot. Gazette, vi. 184. — E. Arizona, on calcareous bluffs of the Gila, near the New Mexican boundary, *Greene*.

B. Mohavénsis, Greene. Stouter, with fewer and looser sometimes flexuous rigid branches, canescent with a fine pannose tomentum, or in age glabrate and becoming viscidulous: sparse leaves often present, an inch or less long: bracts of the involucre obtuse: corolla-lobes narrowly lanceolate, wholly glabrous. — Bull. Torr. Club, incd. *B. juncea*, Gray in distrib. Pringle, not Greene. — On the Mohave Desert, *Greene, Parry, Pringle*. Host-plant of *Pholisma*, according to Pringle.

++ ++ Leaves numerous, filiform or nearly so, not obviously punctate: heads shorter: involucral bracts 3 or 4 in each vertical rank, some or most of them with small setaceous or subulate spreading or recurving tips: lobes of 5-cleft limb of corolla linear or linear-lanceolate: stems fastigiately branched.

B. ceruminósa, Gray. Shrubby, a foot or two high, glabrate, balsamic-viscidulous or pruinose-resinous: leaves rather scattered on the slender branches, spreading or recurving: heads cymose-fascicled, about 5 lines long, narrow: bracts of the viscidly lucid involucre narrowly lanceolate, abruptly produced into a spreading setiform tip or short awn, or the much shorter outermost muticous. — Proc. Am. Acad. viii. 643, & Bot. Calif. i. 316. *Linosyris ceruminosa*, Durand & Hilgard, Pl. Heerm., & Pacif. R. Rep. v. 9, t. 6. — S. California in Tejon Pass, *Dr. Heermann*. Not since seen.

B. Greénei, Gray. Suffruticose, about a foot high, green and glabrous, more or less balsamic-viscid: leaves very numerous on the branches, filiform-acerose, but flat and margins minutely ciliolate-scabrous: heads numerous and fastigiate-cymose, 3 or 4 lines high: bracts of the subclavate involucre fewer and firmer-chartaceous, oblong, abruptly subulate-tipped or short outermost mucronate, only about 3 in each vertical rank, these ranks comparatively indistinct: anthers and stigmas less exserted. — Proc. Am. Acad. xi. 75. — Colorado; on the Huerfano Plains, *Greene*. Near Twin Lakes in the Colorado Mountains, and Cottonwood Cañon, Utah, *M. E. Jones*.

++ ++ ++ Leaves numerous, all involute-filiform, resinous-punctate and glabrous, as are the branchlets, but at length balsamic-viscid or pruinose-waxy: no tomentum: heads open-paniculate, 4 or 5 lines high: bracts of the cylindraceous involucre less numerous, only 3 or 4 in each vertical rank, from oblong to linear, obtuse and pointless, little carinate: corolla with short oblong lobes or teeth: pappus soft: low-shrubby, fastigiately or paniculately much branched, very leafy: leaves an inch or less long.

B. teretifólia, Gray. Branches rigid, fastigiate: involucral bracts narrowly oblong to broadly linear, rather firm-chartaceous, in about 4 vertical ranks, all but innermost tipped with a greenish and glandular subapical spot. — Proc. Am. Acad. viii. 644, & Bot. Calif. i. 316. *Linosyris teretifolia*, Durand & Hilgard, Pl. Heerm., & Pacif. R. Rep. v. 9, t. 7. — Arid hills, S. E. California, bordering the Mohave Desert; first coll. by *Dr. Heermann*. Perhaps also in Arizona.

B. paniculáta, Gray, l. c. Less woody, more paniculate: involucral bracts broader, thinner, about 3 in each vertical rank, pale and wholly naked. — *Linosyris viscidiflora*, var. *paniculata*, Gray, Bot. Mex. Bound. 80. — Desert wastes, San Bernardino Co. to S. Utah! First coll. by *Schott*, later by *Parry, Parish, Palmer*.

++ ++ ++ ++ Leaves numerous, from filiform-linear or involute-filiform (but mostly plane or only canaliculate) to broadly linear or lanceolate, not resinous-punctate but sometimes viscidulous: heads fastigiate-cymose or somewhat thyrsoid: bracts of the involucre obtuse or somewhat acute and muticous (in one ambiguous form even pointed!): slender style-appendages well exserted, especially in the first species.

═ At least the branches when young, and commonly in age, whitened by a close pannose tomentum: subulate-filiform style-appendages longer than the stigmatic portion: pappus soft.

B. gravéolens, GRAY. A foot to a yard or more high, bearing numerous crowded heads: these half or two-thirds inch high: leaves mostly flocculent-tomentose when young, often glabrate in age, not rigid: the larger spatulate-linear, or linear-lanceolate (2 inches long and fully 2 lines wide, obscurely if at all 3-nerved); the narrowest almost filiform, at least when dry, and margins involute: involucre thin-chartaceous when dry: corolla-lobes or teeth short, from lanceolate to nearly ovate: akenes linear: pappus soft. — Proc. Am. Acad. viii. 644. — The typical form of this polymorphous species has the bracts of sometimes viscidulous involucre narrowly oblong to linear-lanceolate, rather obtuse to acutish or even quite acute: short corolla-lobes commonly oblong-lanceolate, varying to nearly ovate and shorter, the tube naked or nearly so. — *Chrysocoma dracunculoides*, Pursh, Fl. ii. 517, not Lam. *C. graveolens*, Nutt. Gen. ii. 136. *Bigelovia dracunculoides*, DC. Prodr. v. 329. *Chrysothamnus dracunculoides* & *C. speciosus*, Nutt. Trans. Am. Phil. Soc. l. c. *Linosyris graveolens*, Torr. & Gray, Fl. i. 234. — Sterile and especially alkaline soil, Dakota to British Columbia, and south to S. California and New Mexico. Heads sometimes cymose, sometimes thyrsoid-glomerate. Forms of the latter occur with firmer involucral bracts, some of them even acuminate, as if connected with *B. Howardi*.

Var. glabráta, GRAY, l. c. Includes forms of the above with the usually narrow leaves early glabrate or perhaps glabrous from the first, sometimes balsamic, sometimes not. — Includes *Linosyris viscidiflora*, Hook. Lond. Jour. Bot. vi. 243, in part, no. 102, *Geyer*, from the northern Rocky Mountains, and *Bigelovia Douglasii*, var. *stenophylla*, Gray, Bot. Calif. i. 614, from the southern borders of California, *Palmer*. Not rare in Colorado, where even the branches sometimes early lose their light tomentum.

Var. albicaúlis, GRAY, l. c. Branches for the most part permanently and very densely white-tomentose and leaves floccose-tomentose: involucre either tomentulose or glabrate; its bracts commonly acutish: corolla-lobes more or less lanceolate and the tube villous- or arachnoid-pubescent. — *Chrysocoma nauseosa*, Pursh, l. c., Nutt. Gen. l. c., therefore *Bigelovia Missouriensis*, DC. l. c., but chiefly found west of the Rocky Mountains. *Chrysothamnus speciosus*, var. *albicaulis*, Nutt. Trans. Am. Phil. Soc. l. c. *Linosyris albicaulis*, Torr. & Gray, Fl. ii. 234. — Rocky Mountains of Wyoming to Brit. Columbia, and the eastern side of the Sierra Nevada to San Bernardino Co., California.

Var. latisquámea, GRAY, l. c. Rather stout, white-tomentose or partly glabrate: heads numerous in the corymbiform cymes: bracts of the glabrous involucre mostly elliptical-oblong, very obtuse: lobes or teeth of the corolla short, somewhat lanceolate, the tube glabrous. — S. E. Colorado to adjacent New Mexico, and S. Utah, *Fendler* (no. 341), *Bigelow, Dr. Henry Ward*.

Var. hololeúca, GRAY, l. c. Slender, white-tomentose even to the heads; these rather small, numerous in corymbiform cymes terminating sparsely-leaved branches: leaves very narrowly linear, inch long, and uppermost short and bract-like: involucral bracts small, linear-oblong, very obtuse: corolla merely 5-toothed, its tube bearing cobwebby hairs: akenes (as in the species) villous-pubescent. — Owens Valley in the southeastern part of the Sierra Nevada, California, *Dr. Horn*.

B. leiospérma. A foot or two high, with rigid slender branches, bearing small glomerate cymes, white-tomentose, or in age somewhat glabrate: leaves sparse, and uppermost very small, involute-filiform: involucre glabrous; its bracts small, oblong, or innermost linear-oblong, very obtuse: corolla glabrous and with 5 short ovate teeth: ovary and akenes completely glabrous! — St. George, Southern Utah, *Palmer*, coll. 1875. Candelaria, S. W. Nevada, *W. H. Shockley*.

══ Green, no tomentum, either smooth and glabrous or scabro-puberulous: style-branches less exserted, thicker, shorter than the stigmatic portion: pappus rigidulous: akenes shorter.

B. Douglásii, GRAY, l. c. From 6 inches to 6 feet high, fastigiately branched, sometimes resinous-viscid, often slightly or not at all so leaves from very narrowly linear or almost filiform (but plane or canaliculate) to lanceolate-oblong, mostly 3-nerved: heads few or numerous and fastigiate-cymose, 3 or 4 lines high: bracts of the involucre comparatively few, only 2 to 4 in each vertical rank (these ranks therefore less conspicuous), from broadly to linear-oblong or lanceolate, obtuse (rarely acute), firm-chartaceous, not rarely some of the outer with firmer and indistinctly greenish apex: corollas rather deeply cleft into oblong-lanceolate lobes. — (*Crinitaria viscidiflora*, Hook. Fl. ii. 24, apparently, in part: this founded on two specimens, both with heads undeveloped, one puberulent, one glabrous, to be referred either to this species or to *Aplopappus, Ericameria, nanus*.) *Chrysothamnus viscidiflorus* & *C. pumilus*, Nutt. Trans. Am. Phil. Soc. l. c. *Bigelovia viscidiflora*, DC. Prodr. vii. 279. *Linosyris viscidiflora*, Torr. & Gray, l. c. *Brickellia linearifolia*, Klatt, Abh. Nat. Gesells. Halle, xv. 5. — Plains and mountains, in sterile soil, Dakota to Washington Terr. and border of Brit. Columbia, dry eastern border of California, and south to Arizona and New Mexico; in various forms. Taking the forms with linear and lanceolate smooth leaves as the type, the marked variations are

    Var. púmila (*Chrysothamnus pumilus*, Nutt. l. c., with his var. *euthamioides*), a dwarf northern and mountain state, a span or two high, glabrous or minutely puberulent and disposed to be viscidulous; the simple branches bearing very few heads in a close cluster: outer involucral bracts either somewhat greenish-tipped or passing into bract-like leaves. — N. Montana to Washington Terr. and mountains of Utah.

    Var. serruláta, Gray, l. c. Taller: leaves linear or narrowly lanceolate, serrulate-ciliolate, sometimes scabrous and rigid. — *Linosyris serrulata*, Torr. in Stansbury Rep. 389; Eaton, Bot. King Exp. 157. — Common through the whole dry interior region.

    Var. tortifólia, Gray, l. c. Leaves twisted: otherwise like the preceding. — Plains of Colorado to the Sierra Nevada, California. Here *Linosyris lanceolata*, Hook. Lond. Jour. Bot. vi. 243.

    Var. stenophýlla, Gray, l. c. Leaves slender, at most a line wide by an inch or two long, or narrower and varying to filiform, smooth: flowers sometimes only 4. — N. W. Nevada to S. E. California, Utah, and New Mexico.

    Var. latifólia, Gray, l. c. Stouter and taller, smooth and glabrous, or puberulent: leaves lanceolate to narrowly oblong (the broadest even half-inch wide by thrice that length), often obtuse, 3-5-nerved: flowers sometimes 6 or 7 in the head. — *Linosyris viscidiflora*, var. *latifolia*, Eaton, Bot. King Exp. l. c. — S. Idaho, Nevada, and Utah.

    Var. lanceoláta. Low, but bearing compact cymes of numerous (5-7-flowered) heads: leaves short, lanceolate or broadly linear, scabro-puberulent. — *Chrysothamnus lanceolatus*, Nutt. Trans. Am. Phil. Soc. l. c. *Linosyris lanceolata*, Torr. & Gray, Fl. ii. 233. *Bigelovia lanceolata*, & *B. Douglasii*, var. *puberula*, in part, Gray, l. c. 639, 644. (*Linosyris viscidiflora*, var. *puberula*, Eaton, l. c., is mainly a scarcely puberulent narrow-leaved form of the type.) — Head-waters of the Platte, Wyoming and Montana, &c. Passes into var. *serrulata* and var. *tortifolia*.

+ + + Akenes glabrous, as also the ovaries, nearly terete: bracts of involucre rounded-obtuse, not prominently pentastichous: anthers and style-tips little exserted: suffrutescent, green and glabrous, not punctate.

B. Vaséyi, Gray. A span or two high, somewhat balsamic-viscid but wholly glabrous, leafy up to the fastigiate-cymose cluster of heads: leaves linear or spatulate-linear, obtuse, plane (at most inch long), with obscure midrib: involucre cylindraceous, 3 or 4 lines long; its bracts narrowly oblong, firm-chartaceous, and all but innermost with a thickened greenish spot at the very obtuse apex: lobes of the corolla short-linear: style-appendages narrowly subulate, rather obtuse, half the length of the stigmatic portion: pappus fine and soft, rather short. — Proc. Am. Acad. xii. 58. — Colorado Rocky Mountains, in Middle Park and Gunnison Valley, *Vasey, Parry*. Utah, *Ward*. — Transition to *Solidago* and to § *Aplodiscus*.

§ 3. EUBIGELÓVIA, Gray, l. c. Heads as of preceding section, very narrow, 3-4-flowered: alveoli of the receptacle prolonged into subulate teeth or at the centre into a chaff-like cusp: limb of corolla enlarging and 5-cleft: style-appendages ovate-subulate, shorter than the stigmatic portion: akenes short, somewhat turbinate: pappus rigidulous: wholly herbaceous perennial, with entire narrow leaves: habit of *Solidago* § *Euthamia*.

—— **B. nudáta**, DC. Glabrous: stems slender, a foot or two high from a small caudex, strict and simple up to the compound-fastigiate and corymbose cyme of numerous heads: leaves not punctate nor obviously viscid, spatulate to nearly filiform, uppermost small and bract-like: heads barely 3 lines high, subclavate: bracts of the involucre about 3 in each rather indistinct vertical rank, oblong-linear, obtuse and firm-chartaceous, or at least outermost with short greenish tips. Leaves in the original of the species spatulate-linear, or uppermost narrower, lowest and radical commonly broader (sometimes half-inch wide) and rounded-obtuse. — Prodr. v. 329. & Mem. Comp. t. 5   *B. nudata*, var. *spathulæfolia*, Torr. & Gray. Fl. ii. 232. *Chrysocoma nudata*, Michx. Fl. ii. 101; Nutt. Gen. ii. 137. — Low pine barrens, New Jersey to Florida and Louisiana: fl. autumn.

Var. **virgáta**, Torr. & Gray, l. c.   Cauline leaves linear-filiform, or lowest and the radical linear-spatulate. — *B. virgata*, DC. l. c.   *Chrysocoma virgata*, Nutt. l. c. — New Jersey to Texas.   Passes into the broader-leaved form.

§ 4. EUTHAMIÓPSIS.   Heads (small) 7–25-flowered: bracts of the involucre wholly chartaceous, or in some obscurely greenish at tip, hardly carinate, obtuse or nearly so and muticous, appressed-imbricated in 3 or 4 series, but vertical ranks inconspicuous: style-appendages hardly exserted out of the 5-lobed limb of the corolla, subulate-oblong to short-filiform, shorter or not longer than the stigmatic portion: akenes mostly short and turbinate, sericeous-pubescent: shrubby, becoming more or less balsamic-viscid, and with entire *punctate* leaves: corollas of outermost flowers sometimes deformed. — § *Aplodiscus, Euthamioideæ*, mainly, Gray, Proc. Am. Acad. viii. 639.

* Stems simple below and fastigiately branched above, 3 to 12 feet high, bearing numerous heads in close and ample corymbiform cymes: leaves plane: involucral bracts small, lanceolate, wholly chartaceous and pale, or midnerve obscurely greenish.

—**B. Paríshii**, Greene.   Leaves thickish, lanceolate or oblong-linear (inch or two long, quarter to nearly half an inch wide), mucronate, strongly punctate: heads 10–12-flowered, fully 3 lines long). — Bull. Torr. Club, ix. 62 — Mountains near San Bernardino, S. E. California. *Parish*, &c.   Stems sometimes 2 or 3 inches in diameter.

B. arboréscens, Gray.   Leaves narrowly linear, very numerous (1 to 3 inches long, a line wide), moderately punctate: heads 20–25-flowered, barely 3 lines long: outer flowers often deformed. — Proc. Am. Acad. viii. 640.   *Linosyris arborescens*, Gray, Bot. Mex. Bound. 79. — Dry ground, common in the Coast Ranges of California, sparingly in the Sierra Nevada; first coll. by *Fitch* and *Kellogg*.

* * Branched from the base: heads paniculate or more scattered: leaves filiform, thickish: bracts of involucre larger and rather few, oblong, obtuse.

B. Coóperi, Gray.   Apparently low, with leaves half-inch or less long, balsamic-viscid: heads few in a cluster at the end of the branchlets, 6–8-flowered: bracts of involucre narrowly oblong, chartaceous, pale to the apex: style-appendages ovate-subulate. — Proc. Am. Acad. viii. 640, & Bot. Calif. i. 315. — S. E. California, on eastern slope of Providence Mountains, *Cooper*. Not again found: only branchlets known.

B. brachýlepis, Gray.   Shrub 4 to 6 feet high: leaves inch or half-inch long, balsamic-viscid, conspicuously resinous-punctate: heads loosely paniculate or solitary terminating paniculate branchlets, 8–12-flowered, 4 or 5 lines high: bracts of the campanulate involucre oblong, more or less carinate by a glandular thickened midnerve, innermost not surpassing the linear-oblong akenes, outermost passing into small commonly imbricated scales on the peduncle: style-appendages subulate-filiform. — Bot. Calif. i. 614. — S. California, along the southern borders of San Diego Co., near the Mexican frontier, *Palmer, Cleveland, Nevin*. (Adj. Lower Calif.)

B. DIFFÚSA, Gray, Proc. Am. Acad. viii. 640 (*Linosyris Sonoriensis*, Gray, l. c. 231, *Ericameria diffusa*, Benth. Bot. Sulph. 29. *Solidago diffusa*, Gray, Proc. Am. Acad. v. 159), of Lower California (*Hinds, Xantus*) and Sonora *Palmer*, is a species of this group, with filiform leaves obscurely punctate, and involucral bracts of firmer texture, the tips greenish, verging therefore to the next section.

§ 5. APLODÍSCUS, Gray, l. c. Heads several-many-flowered: bracts of the involucre either coriaceous or firm-chartaceous, and usually somewhat herbaceous or thickened at the obtuse or barely acute apex, all strictly appressed and well imbricated, but the vertical ranks inconspicuous: style-appendages subulate-lanceolate or broader, shorter than the stigmatic portion: akenes short, sericeous-pubescent: herbaceous or suffruticose, commonly more or less balsamic-viscid: leaves not punctate, sometimes dentate or pinnatifid. — *Aplopappus* § *Aplodiscus*, DC. Prodr. v. 350, excl. *A. ramulosus*, which is a *Baccharis*.

* Herbaceous down to suffrutescent base: leaves linear: bracts of the involucre thin-coriaceous or almost chartaceous, and with obscure if any greenish tips.

B. pluriflóra, GRAY, l. c. Leaves narrowly linear, entire: heads 15-18-flowered, 4 lines high: involucre somewhat turbinate, very smooth; its thinnish bracts lanceolate, acute: otherwise like the next, of which it is perhaps a mere form, but is insufficiently known. — *Chrysocoma graveolens*, Torr. Ann. Lyc. N. Y. ii. 211, not Nutt. *Linosyris pluriflora*, Torr. & Gray, Fl. ii. 233. — Colorado? probably on the Arkansas or South Fork of the Platte, *James* in Long's expedition.

B. Wrightii, GRAY, l. c. Commonly glabrous or nearly so: stems rather strict and slender, a foot or two high from the lignescent base: leaves thickish, narrowly linear, entire, sometimes lower ones sparingly laciniate-dentate, margins either smooth or sparingly hirtello-scabrous: heads (4 or 5 lines high) 7-15-flowered, usually numerous and crowded in a corymbiform cyme: bracts of the involucre oval-oblong to broadly lanceolate, obtuse; the back at or near the apex usually greenish, but no definite tip. — *Linosyris Wrightii* & *L. heterophylla*, Gray, Pl. Wright. i. 95, ii. 80. — Banks of streams and in saline, soil, W. Texas to S. Colorado and Arizona; first coll. by *Wright*.

Var. **hirtélla**. Leaves cinereous-hirtellous or hirsute-pubescent and roughish, but often glabrate in age or only ciliolate: stems sometimes pubescent. — *Linosyris hirtella*, Gray, Pl. Wright. i. 95. — Same range; first coll. by *Wright*.

* * Suffrutescent: leaves linear-filiform and pinnately parted: involucre nearly of the preceding.

B. coronopifólia, GRAY, l. c. Glabrous: stems freely branching, slender, a foot or two high, leafy: divisions of the leaves 3 to 9, often half-inch long, not thicker than the filiform rhachis, setulose-mucronate: heads somewhat thyrsoid-glomerate (4 or 5 lines long), 10-12-flowered. — (Excl. pl. Arizona, Palmer.) *Linosyris coronopifolia*, Gray, Pl. Wright. i. 96. — S. Texas along the Rio Grande, *Wright, Bigelow, Havard, Palmer.*

* * * Suffruticose: bracts of involucre more coriaceous and more definitely greenish-tipped.

+– Leaves all entire (or rarely a tooth or two), linear or spatulate-linear: branches partly herbaceous: glabrous.

B. Drummóndii, GRAY, l. c. About a foot high, with many slender erect or ascending branches or stems from a woody base: leaves all narrowly linear, with tapering base (inch or two long, seldom over a line wide): heads 5 or 6 lines high, rather numerous in a corymbiform cyme, 18-30-flowered: involucre campanulate; its bracts linear-oblong, with obtuse or obtusish and short green or greenish tips; pappus rather soft. — *Linosyris Drummondii*, Torr. & Gray, Fl. ii. 233. — Coast of Texas, and Lower Rio Grande; first coll. by *Berlandier, Drummond, Trécul.* E. Arizona, *Rusby.* (Adj. Mex., *Palmer.*)

B. acradénia, GREENE. A foot or so high, very many slender stems or branches forming broad tufts from a woody base: leaves spatulate-linear (half-inch to inch long, entire or rarely some small teeth: heads glomerate-cymose, 4 lines high, 10-20-flowered: involucre campanulate, of more rigid oblong bracts, the back at the obtuse apex bearing a protuberant rounded resiniferous gland: pappus rigid, of very unequal bristles. — Bull. Torr. Club, x, 126. — Mohave Desert, S. E. California, *Greene, Parry, Jared,* &c. S. Utah, *Palmer.* Transitions apparently occur between this and the next.

+– +– Leaves serrate, dentate, or pinnatifid, occasionally entire: shrubby, 2 to 4 feet high.

B. venéta, GRAY. Glabrous, or the herbage when young loosely pubescent, or almost tomentose: leaves short (half-inch or lower twice or thrice this length), spatulate or oblanceolate, or sometimes cuneate-oblong, sparsely or irregularly spinulose-dentate or serrate, or denticulate with spinulose teeth, sometimes incised, some upper or fascicled ones varying toward linear and entire : heads more or less glomerate at the end of the branchlets, 15-35-flowered, 4 or 5 lines high; bracts of the turbinate or campanulate involucre with obtuse or sometimes acutish or mucronate-acute green tips (these occasionally bearing an indistinct resinous gland); pappus of rather rigid and very unequal bristles. — *B. veneta* & *B. Menziesii*, Gray, Proc. Am. Acad. viii. 638. & Bot. Calif. i. 315.   *Baccharis veneta*, HBK. Nov. Gen. & Spec. iv. 68.   *Linosyris Mexicana*, Schlecht. Hort. Halens. 7, t. 4.   *Aplopappus discoideus*, DC. Prodr. v. 350.   *A. Menziesii*, Torr. & Gray, Fl. ii. 242.   *Pyrrocoma Menziesii*, Hook. & Arn. Bot. Beech. 351.   *Isocoma vernonioides*, Nutt. Trans. Am. Phil. Soc. vii. 320.   (*B. tridentata*, Greene, Bull. Torr. Club, 1 c., *Linosyris dentata*, Kellogg, Proc. Calif. Acad. ii. 16, is apparently a form of this, from Cedros Island off Lower California.) — Southern part of California (first coll. by *Menzies*) to borders of Arizona. (Mex.)

— **B. Hartwégi**, Gray.   Cinereous-puberulent or glabrate, a foot or two high : leaves from linear to narrowly oblong, pinnatifid ; the lobes 5 to 11, oblong-linear, short (only a line or two long) : heads smaller than in the preceding, into which it may pass. — Hemsl. Biol. Centr.-Am. Bot. ii. 115. — S. Arizona, *Palmer* (taken for *B. coronipifolia* in Proc. Am. Acad. viii. 639), *Lemmon*. (Mex.)

32. SOLIDÁGO, L. Golden-rod. (*Solidus* and *ago*, to make solid or draw together, in allusion to reputed vulnerary properties.) — Perennial herbs (one species somewhat shrubby) ; with mostly strict stems, entire or serrate alternate leaves, the cauline sessile or nearly so, the radical tapering into margined petioles (never cordate) ; the small heads thyrsoid-glomerate, or sometimes corymbosely cymose, or more commonly in racemiform secund clusters ; the flowers yellow, or in one species whitish in the disk and white in the ray ; rarely the rays wanting. — Gen. ed. 1, 258 (name from Vaill.) ; Torr. & Gray, Fl. ii. 195. — A large genus, of nearly 100 species, mostly Atlantic N. American, but with several Pacific species, a few Mexican or S. American, one or two European and N. Asiatic : fl. late summer and autumn. — For notes on the species in the older herbaria, and a synopsis, see Proc. Am. Acad. xvii. 177-199.

S. LATERIFLÓRA, L. Spec. ii. 879, is *Aster diffusus*, Ait.

S. NOVEBORACÉNSIS, L. l. c., is probably *Aster Tartaricus*, and not North American.

Species founded on garden plants and not identified with, or obviously referable to, North American originals, are the following : —

S. AMBÍGUA, Ait. Kew. iii. 217, cult. 1759 by Miller, of unknown source, appears to have been some European form of *S. Virgaurea*, although later plants cultivated under this name may be derivatives of *S. latifolia*, L.

S. ELLÍPTICA, Ait.   Glabrous and smooth up to the flowering branches, 2 or 3 feet high, equably leafy : leaves of rather firm texture, oval or oblong, acuminate at both ends, the larger 3 or 6 inches long, 1½ or 2 wide, more or less serrate with fine acute teeth, somewhat veiny : thyrsus somewhat leafy ; the heads (3 lines long) racemose-paniculate on erect branches, little or not at all secund : bracts of the involucre oblong-lanceolate, acutish or obtuse : rays 7 to 9 : akenes villous-pubescent. — Kew. iii. 214 ; DC. Prodr. v. 334 ; Gray, Proc. Am. Acad. xvii. 181. *S. plantaginea*, Desf. Cat. ed. 3, 402. — Cultivated from early times in European gardens, not identified as indigenous. The typical form is here taken to be that of the Banksian herbarium, cult. hort. t. 1778.   A second original specimen, to which the syn. Mill. Dict. belongs, is

Var. ILLIFLÓRA, Gray, l. c.   Leaves of somewhat firmer texture, from oval to broadly lanceolate : heads rather larger, in short or somewhat elongated and racemiform erect or spreading clusters, which are mostly axillary and shorter than the leaves. — *S. latifolia*, L., as to Pluk. Alm. 389, t. 235, f. 3.   *S. latissimifolia*, Mill. Dict. ed. 7.   *S. lateriflora*, Willd. Spec. iii. 2057, &c., not L. nor Ait.   *S. fragrans*, Willd. Enum. Suppl 331, a narrower-leaved form.   *S. verrucosa*, Schrad. Hort. Goett. 12, t. 6 ?   *S. Clelia*, DC. Prodr. v. 331, & perhaps *S. dubia*, Scop. Del. Insub. ii. 19, t. 10. — Cultivated from ante-Linnæan times in European collections, not identified in N. America, but doubtless of American origin.

S. RECURVÁTA, Willd., and S. LÍVIDA, Willd. (Enum. 889 & 491), described from cultivated plants, are referred to under *S. cæsia*, L., p. 145.

S. LITHOSPERMIFÓLIA, Willd. Enum. 891, referred to under *S. sempervirens*, L., was probably derived from that, under cultivation becoming pubescent and duller green. But without the transitions as seen in *S. integrifolia*, Desf., this would seem improbable.

S. CORYMBÓSA, Poir. Suppl. v. 461 (not Ell.), is only the European *S. Virgaurea*, L.

S. GRANDIFLÓRA, Desf. Cat. ed. 3, 403, of unknown source, is evidently a tall cultivated state of the Italian *S. littoralis*, Savi.

S. FUSCÁTA, Desf. l. c. 402. Glabrous and very smooth (the inflorescence barely puberulent): stem 3 or 4 feet high, with numerous ascending purplish branches, very leafy: leaves oblong-lanceolate, acute at both ends or acuminate, entire, or the lower (3 or 4 inches long) with a few minute and obscure teeth, of somewhat firm texture, and minutely reticulated inconspicuous venation, a pair or two of primary veins more evident: heads hardly 3 lines long, numerous in a narrow or virgate thyrsus, not secund: bracts of the involucre rather broad (outer oblong) and obtuse: rays 6 to 8, short: young akenes puberulent. — Of unknown source; cult. in Paris Garden from 1828. Habit somewhat of *S. puberula*.

Of species founded on indigenous specimens there remains wholly obscure only the following: —

S. PAUCIFLÓRA, Raf. in Med. Rep. (hex. 2), v. 359. "Stem simple, smooth: leaves oblong-lanceolate, acute, entire: flowers 1 to 5, terminal. — Gloucester Co., New Jersey, and Kent Co., Delaware," *Rafinesque*. Not to be identified.

§ 1. VIRGAÚREA, DC. (*Virga-aurea*, Tourn.) Receptacle of the head alveolate: rays commonly fewer or not more numerous than disk-flowers: herbs.

* Involucre *squarrose*, the bracts having herbaceous recurving or spreading tips (yet occasionally erect in all the species): general inflorescence thyrsiform or reversed racemiform-paniculate, not unilateral: leaves pinnately veined, from ovate to lanceolate; the lower ones commonly petioled, and acutely more or less serrate; the upper often entire. (*Chrysastrum*, Torr. & Gray.) — SQUARROSÆ.

+ Rays none: ovaries hirsute: bristles of the pappus unequal, all with clavellate tips.

S. discoídea, TORR. & GRAY. Pubescent or hirsute, somewhat cinereous: stem branching above: lower leaves ovate, coarsely serrate, on slender and margined petioles, 3 inches long: upper small, often entire, oval or oblong: heads (3 or 4 lines long) rather scattered in the racemiform thyrsus, 10–20-flowered: disk-corollas deeply 5-cleft: pappus often tinged with purple. — Fl. ii. 195. *Aster? discoideus*, Ell. Sk. ii. 358. — Dry soil, Georgia to Florida and Louisiana; first coll. by *Elliott*.

+ + Rays present and conspicuous, rather numerous: bristles of the pappus not evidently clavellate-thickened: akenes glabrous or nearly so.

S. squarrósa, MUHL. Green, pubescent or glabrate: stem stout and simple, 2 to 5 feet high: lower leaves ovate or oblong, 6 to 10 inches long: heads (5 or 6 lines long) numerous and crowded at least on the lower branches of the (foot or two long) leafy thyrsus: green squarrose tips of the involucral bracts short and broad, obtuse or abruptly acute. — Cat. 79; Nutt. Gen. ii. 161; Torr. & Gray, l. c. *S. confertiflora*, Nutt. in Jour. Acad. Philad. vii. 102; Hook. Fl. ii. 4, not DC. — Rocky soil, New Brunswick and Canada to Ohio and upper part of Virginia.

S. petiolaris, AIT. Puberulent or pubescent with fine short hairs, somewhat pale or cinereous: stem slender, a foot to a yard high: leaves comparatively small, elliptical-oblong to broadly lanceolate, scabrous-ciliate; the lower 2 inches or so in length, serrate with a few coarse teeth toward the apex, narrowed at base, obscurely or sometimes distinctly and abruptly short-petioled, mostly glabrous or glabrate above, minutely hairy at least along the veins beneath; upper smaller, sessile, entire: heads (3 to 5 lines long) loosely or sometimes more compactly disposed in a narrow or irregular thyrsus: involucral bracts narrow and acute; the outer green or with green tips, and more or less squarrose; inner ones appressed. — Ait. Kew. iii. 216; Smith in Rees Cycl.; Torr. & Gray, Fl. ii. 203, not Muhl., Ell., or Less. *S. erecta*, Nutt. Gen. ii. 161. *S. elata?* Ell. Sk. ii. 389. *S. squarrosa*, Nutt. in Jour. Acad. Philad. vii. 102, not Nutt. Gen., nor Muhl. *S. petiolaris*, var. *squarrulosa*, Torr. & Gray, l. c. — Dry soil, especially in pine barrens. N. Carolina to Florida, Kansas, and Texas. The specific name quite inappropriate; and the squarrose tips of the bracts are sometimes obsolete, thus invalidating the rather marked character of this group.

— Var. angústa, GRAY. Leaves greener, glabrate, narrower, nearly all entire; the lower sometimes 3 or 4 inches long and half-inch or less wide, tapering into a margined petiole. — Proc. Am. Acad. xvii. 189. *S. angusta*, Torr. & Gray, Fl. ii. 204. — W. Louisiana. *Hale*, and Fredericksburg, Texas, *Thurber*.

* * Involucre of inappendiculate and wholly appressed bracts in this and all the following divisions: heads small (at most 3 lines long), disposed more or less in axillary glomerate or short-racemiform clusters along the leafy stem, or not rarely with some or most of the clusters in an almost naked thyrsus: leaves unicostate, pinnately veiny. — GLOMERULIFLORÆ, Torr. & Gray.

+ Akenes canescently hirsute-pubescent: leaves normally thin and membranaceous, very sharply serrate, acuminate, bright green, usually surpassing the short clusters in their axils, except where these become confluent into a thyrsus at the summit.

++ Stem and branches terete, often glaucous.

— **S. cǽsia**, L. Slender, commonly branching and glabrous or nearly so up to the peduncles, smooth, a foot or two high: leaves lanceolate or the lower from ovate- to oblong-lanceolate, sessile, serrate with erect or ascending teeth, the venation not prominent: heads small, few-flowered: bracts of the involucre all obtuse. — Spec. ii. 879 (founded on Dill. Elth. 414, t. 307, & *Virga-aurea Marilandica*, etc., Ray); Ait. Kew. iii. 217; Torr. & Gray, Fl. ii. 199. *S. flexicaulis*, L. l. c., as to herb., excl. char. & syn. — Shaded banks, or in wooded grounds, Canada to N. W. Arkansas, Georgia, and Texas.

——Var. axilláris, GRAY. Proc. Am. Acad. l. c. (*S. axillaris*, Pursh, Fl. ii. 542), is a common form in shade, with elongated-lanceolate thin leaves, all much exceeding the short clusters of rather few heads. — Chiefly northern, in woods.

Var. paniculáta, GRAY, l. c. Paniculately branched above, smaller-leaved, floribund; the clusters of heads becoming racemose-paniculate toward the end of the branches: stems often purple and branchlets occasionally pubescent. — *S. gracilis*, Poir. Dict. viii. 476; DC. Prodr. v. 336. *S. Schraderi*, DC. l. c.! (abnormal form), & of the Gardens. *S. arguta*, Spreng. Syst., not Ait. *S. arguteu*, Hornem. ex Spreng. — A form of drier and open grounds, commoner in S. States, and of European cultivation, where it is much altered, and appears to pass into

S. RECURVÁTA, Willd. Enum. 889 (not Mill. Dict.). Tall, more paniculate, and the heads in more loosely crowded clusters on spreading (but hardly recurved) or ascending flowering branches, few if any in the axils of cauline leaves: usually some pubescence. — European gardens. May be a hybrid between *S. cæsia* and *S. ulmifolia* or *S. rugosa*.

S. LÍVIDA, Willd. l. c. 491. Stouter, purple-stemmed, with thyrsiform-paniculate inflorescence of more crowded heads; apparently a cultivated modification of *S. cæsia*. var. *paniculata*, with a large-flowered indigenous form of which (from Monticello, Georgia, *Porter*) it is congruous. It is *S. flabellata*, Schrader ex Spreng. (*S. arguta*, Spreng.), and *S. flabelliformis*, Wendl. in DC. Prodr. v. 336.

++ ++ Stem and branches angled, manifestly so in dried specimens, green, not glaucous.

— **S. latifólia**, L. Stem much angled, often flexuous, glabrous, 1 to 3 feet high: leaves ample and normally thin, broadly ovate or the upper ovate-lanceolate, conspicuously acuminate, abruptly and acuminately contracted at base into as it were a winged petiole of usually about the length of the axillary clusters, mostly pilose-pubescent beneath, thickly and coarsely serrate with salient subulate teeth: rays 3 or 4: disk-flowers 6 or 7: akenes very hirsute. — Spec. ii. 879 (ex herb. & habitat, excl. syn. Pluk.); Torr. & Gray, Fl. ii. 198. *S. flexicaulis*, L. l. c. ex syn. & char. (not of herb.); Ait. Kew. iii. 217; DC. Prodr. v. 335. *S. flexicaulis*, var. *latifolia*, Willd. Spec. iii. 2064. *S. macrophylla*, Bigel. Fl. Bost. ed. 2, 305, not Pursh. — Moist woods and shaded banks, Nova Scotia to Minnesota, south to Missouri and along the mountains to Tennessee and Georgia. In grounds exposed to the sun, the clusters of heads are often extended and spiciform, or the whole inflorescence becomes a terminal thyrsus.

— **S. lancifólia**, TORR. & GRAY. Nearly glabrous: stem strict and stout, 3 or 4 feet high, sulcate-angled: leaves elongated-lanceolate or the lower broader, sessile by a gradually narrowed entire base, above sharply serrate with the teeth ascending, 4 to 8 inches long: heads

10

larger than in the preceding (3 lines long), usually more spicately clustered and with more numerous flowers (rays about 8): involucre of more imbricated and broader very obtuse narrowly oblong bracts, externally granular-puberulent when young: akenes canescently hirsute. — Chapm. Fl. 209. *S. ambigua*, var. ? *lancifolia*, Torr. & Gray, Fl. ii. 200. — Damp woods of the higher Alleghanies in N. Carolina and Tennessee; first coll. by *Curtis*.

S. Curtísii, Torr. & Gray. Glabrous or somewhat pubescent: stem commonly branching, slender, moderately angled, 2 feet high: leaves from oblong to elongated-lanceolate, with gradually attenuate entire base, subsessile, serrate with ascending subulate teeth, 3 to 5 inches long: heads in looser clusters, smaller and fewer-flowered (rays 4 to 7): bracts of the involucre much fewer, linear, obtuse. — Fl. ii. 200 (excl. var.); Chapm. l. c. *S. flexicaulis*, in part, in herb. Michx. — Open woods, mountains of Virginia to Georgia, at low or moderate elevations; first coll. by *Michaux*, next by *Curtis*.

Var. **púbens**, Gray, l. c. From sparsely to somewhat densely pubescent: leaves from ovate with tapering base to lanceolate. — *S. pubens*, M. A. Curtis in Torr. & Gray, Fl. ii. 198; Chapm. l. c. — Common in the mountains of Carolina, Tennessee, and Georgia; first coll. by *Curtis*.

+ + Akenes glabrous: inflorescence less axillary-clustered, more virgately thyrsoid.

S. montícola, Torr. & Gray. Nearly glabrous: stem slender, a foot or two high: leaves from oblong-ovate to oblong-lanceolate, thinnish, acuminate or acute at both ends, 1 to 4 inches long; the lower rather sparingly serrate with acute teeth: heads small: involucral bracts linear, acutish: rays 5 or 6, yellow. — Chapm. Fl. 209. *S. Curtisii*, var. ? *monticola*, Torr. & Gray, Fl. ii. 200. — Alleghany Mountains, from Maryland to Georgia and Alabama; first coll. by *Curtis*.

S. bícolor, L. Puberulent, commonly cinereous: stem often hirsute below, strict, a foot to a yard (rarely a span) high: leaves oblong or the lower obovate and ovate, short, mostly obtuse; lower slightly or obtusely serrate: clusters crowded in a simple or compound often elongated thyrsus; involucral bracts linear-oblong, very obtuse: rays from 5 to 14, small, *white, and the disk-corollas also white or yellowish*. — Mant. 114; Ait. Kew. iii. 216; Torr. & Gray, Fl. ii. 197. *S. alba*, Mill. Dict. *Virga-aurea flore albo*, etc., Pluk. Alm. t. 114, fig. 8. *S. viminea*, Bosc in herb. Poiret, therefore *S. erecta*, DC. Prodr. v. 340. *Aster bicolor*, Nees, Ast. 283. — Dry ground, Nova Scotia to Virginia and the upper part of Georgia.

Var. **cóncolor**, Torr. & Gray, l. c. Flowers both of ray and disk yellow (or some rays yellow, others white): foliage sometimes greener, sometimes lanate-hirsute. — *S. hispida*, Muhl. in Willd. Spec. iii. 2063. *S. hirsuta*, Nutt. in Jour. Acad. Philad. vii. 103, & Trans. Am. Phil. Soc. n. ser. vii. 327. — New Brunswick and Maine to Lake Superior, Missouri, and Tennessee.

Var. **lanáta**, Gray, l. c. Low, villous-lanate: heads simply spicately crowded at the summit of the stem or branches. — *S. lanata*, Hook. Fl. ii. 4. — Plains of the Saskatchewan toward the Rocky Mountains, *Drummond*.

* * * Heads mostly large for the genus (in some 6 and seldom less than 4 lines long, smaller in forms of *S. humilis*, &c.), many-flowered, collected in thyrsoidal inflorescence which is not at all secund nor strictly racemiform (but in two species approaches corymbiform): rays 6 to 14: leaves veiny from a simple midrib, in most species bright green: stems commonly low or not tall. (From the inflorescence a few other species, such as *S. speciosa*, might be sought here.) — Thyrsiflor.æ.

+ Southwestern species, fully 2 feet high: leaves very numerous up to or into the inflorescence, uniform in size and shape, short (inch or two long), closely sessile, of rather firm texture, entire, rough-margined, somewhat scabrous: pubescence minute and somewhat cinereous: heads 4 lines long: bracts of the involucre narrow, obtusish, or in some acute.

S. Bigelóvii, Gray. Cinereous-puberulent: leaves oval and oblong, mostly obtuse at both ends and hispidulous on the margin; thyrsus simple or compound, rather dense or at length open: involucre broadly campanulate, puberulent: akenes minutely pubescent or glabrate. — Proc. Am. Acad. xvi. 80, xvii. 190. *S. petiolaris*, Gray in Bot. Mex. Bound. 79, not Ait. — Mountains of New Mexico and Arizona, *Bigelow, Wright, Parry, Greene, Lemmon*. (Adj. Mex.)

Var. **Wríghtii**, Gray, l. c. Leaves sometimes narrower: thyrsus simple and short, of comparatively few heads, or corymbiform almost in the manner of the *Corymbosæ*. —

S. petiolaris, var., Gray, Pl. Wright. i. 94. *S. Californica*, var., Rothrock in Wheeler Rep. vi 145. — S. W. Texas and New Mexico to Arizona; same collectors.

S. Lindheimeriána, SCHEELE. Obscurely puberulent and glabrate, strict, more rigid, especially the broadly lanceolate or oblong more acute and greener leaves: heads densely glomerate in an oblong spiciform thyrsus: involucre oblong-campanulate, its bracts more unequal; akenes glabrous. — Linn. xxi. 599. *S. speciosa*, var. *rigidiuscula*, Gray, Pl. Lindh. ii. 222, not Torr. & Gray. — Texas, on rocky bluffs and in exsiccated beds of streams, *Lindheimer*, *Reverchon*.

++ ++ Southern Alleghanian species: leaves thinner, mostly ample, bright green, tapering to both ends, some of them acutely serrate; pubescence loose and somewhat hirsute.

S. Buckléyi, TORR. & GRAY. Stem 2 or 3 feet high, glabrous below: leaves ovate-oblong to oblong-lanceolate (the larger 3 to 6 inches long): thyrsus loose and elongated, nearly naked: heads 4 or 5 lines long, mostly pedunculate: bracts of the involucre narrowly oblong with rounded-obtuse green tips: akenes glabrous. — Fl. ii. 198. — Lincoln Co., N. Carolina, *Curtis*. Middle Alabama, *Buckley*. Jasper Co., Georgia, *Porter*.

S. glomeráta, MICHX. Mostly glabrous up to the inflorescence: stem stout, 1 to 3 feet high, leafy to the top: leaves ample, from oblong-ovate to lanceolate-oblong, acuminate (the lower 5 to 12 inches long): heads 5 or 6 lines long, in a leafy interrupted thyrsus, or often in remote axillary clusters, all or most of them much shorter than the subtending leaves: involucral bracts oblong, obtuse: akenes glabrate. — Fl. ii. 117; Torr. & Gray, Fl. ii. 209. — Moist wooded sides of the high mountains of Carolina and Tenn., especially Grandfather and Roan. The well-developed inflorescence hardly ever glomerate, therefore the name of this most marked species is misleading.

S. spithamǽa, M. A. CURTIS. Stems a span to a foot high, roughish-pubescent, leafy to the top: leaves glabrate; lower obovate-spatulate; upper oblong (an inch or two in length), acute: heads (barely 4 lines long) somewhat corymbosely glomerate at the summit, also (in cult.) in low axillary clusters: involucral bracts acute or acutish: rays short, hardly surpassing the disk: akenes pubescent. — Gray in Am. Jour. Sci. xlii. 42; Torr. & Gray, Fl. ii. 208. — Rocky summits of the higher mountains in N. Carolina, especially on Grandfather and Roan; first coll. by *Curtis*.

++ ++ ++ Boreal and montane, of difficult and uncertain limitation: rays usually numerous.

++ Bracts of the involucre acute.

S. macrophýlla, PURSH. Glabrous or a little pubescent: stem stout, 8 inches to 3 or even 4 feet high, leafy to or near the summit: leaves thin, ovate or the upper ovate-lanceolate, acuminate, acutely serrate; the lower (3 or 4 inches long) rounded at base or abruptly contracted into a long winged petiole: heads (5 or 6 lines long) mostly pedunculate, few or loose in the clusters, which in smaller specimens form a simple oblong or racemiform thyrsus, and in the larger occupy the axils of many of the cauline leaves: bracts of the involucre narrowly lanceolate-linear, thin and when dry somewhat scarious: rays rather long and narrow: akenes glabrous or rarely a little pubescent at summit. — Fl. ii. 542; Gray in Proc. Am. Acad. xvii. 187, 191. *S. thyrsoidea*, E. Meyer, Pl. Labrad. (1830), 63; Torr. & Gray, Fl. ii. 207. *S. leiocarpa*, DC. Prodr. v. 339. *S. Virgaurea*, Bigel. Fl. Bost ed. 2, 306, excl. var. *S. multiradiata*, Nutt. Trans. Am. Phil. Soc. n. ser. vii. 328, not Ait. — Mountain woods of N. New England, extending upward to the limit of trees, north to L Superior, Hudson's Bay, and Labrador. (Approaches *S. Virgaurea*, var. *leiocarpa*, of E. Asia.)

S. multiradiáta, AIT. Villous-pubescent above or glabrate, a span to a foot or so high: leaves of rather firm texture and fine venation, minutely and sparingly serrate above, sometimes entire; cauline spatulate to lanceolate, all tapering gradually to sessile base, or the radical into a slender margined petiole: heads (mostly 4 lines long) generally few and glomerate in a single terminal roundish or oblong compact often corymbiform cluster, occasionally with one or two looser axillary clusters or branches: bracts of the involucre narrowly lanceolate, thinnish or thin-edged: rays numerous and narrow: akenes pubescent. — Ait. Kew. iii. 218; Pursh, Fl. ii. 542; Hook. Fl. ii. 5. *S. compacta*, Turcz. in Bull. Mosq. 1840, 73, ex char. *S. Virgaurea*, var. *arctica*, DC. Prodr. v. 239. *S. Virgaurea*, var. *multiradiata*, Torr. & Gray, Fl. ii. 207. — Labrador and Hudson's Bay to Behring Strait and Unalaska. The original high northern form very near to forms of *S Virgaurea*. Bracts of the involucre attenuate. On the northern Rocky Mountains passes into

Var. scopulórum, Gray, l. c. More glabrous, 3 to 18 inches high, commonly strict: heads when numerous in a more open or compound cluster, mostly smaller: bracts of the involucre closer, shorter, and merely acute. — *S. corymbosa*, Nutt. l. c. (*S. heterophylla* in herb.). — Along the higher Rocky Mountains to New Mexico, Utah, &c., the Cascade Mountains, and rare (in a dwarf state) along the Sierra Nevada.

Var. Neo-Mexicana. Two feet high, with numerous heads more loosely disposed in approximate axillary as well as terminal clusters, composing a narrow elongated thyrsus, somewhat like that of *S. macrophylla*. — High summits of the Mogollon Mountains, N. Mexico, *Rusby*. A doubtful plant.

S. Virgaúrea, L. Of this Old World and polymorphous or confused species, the var. *alpestris* (of which *S. macrophylla* is the American representative) reaches the Asiatic side of Behring Strait, and seems to pass into *S. multiradiata*. — The var. *Cambrica* is represented by

Var. alpína, Bigel. Dwarf, 2 to 8 inches high, obscurely pubescent or glabrous: leaves few, thickish, spatulate or obovate, mostly obtuse; cauline sessile, the uppermost lanceolate, lowest and radical narrowed into a margined petiole: heads (4 lines long) 3 to 7 in a terminal cluster, or also subsolitary in uppermost axils: involucre broad; its bracts rather broadly lanceolate, barely acute: akenes pubescent. — Fl. Bost. ed. 2, 307; Torr. & Gray, l. c. — Alpine summits of the mountains of N. New York, New England, and Lower Canada, on Anticosti, and Hudson's Bay? Seems nearly to pass into *S. humilis*, and like that to be somewhat viscid.

++ ++ Bracts of the involucre obtuse.

S. húmilis, Pursh. Glabrous, disposed to be glutinous, bright green: stems strict, a span to a foot high, leafy: leaves of firm texture and fine venation, smooth; cauline all sessile; upper lanceolate to nearly linear, entire; lower and radical becoming spatulate with long attenuate base, sparingly appressed-serrate above the middle: heads (3½ or 4 lines long), rather crowded in a narrow racemiform paniculate simple or sparingly branched thyrsus (which is leafy below and naked above): bracts of the involucre oblong-linear: akenes pubescent. — Fl. ii. 543 (the Newfoundland plant, in herb. Banks, where Solander indicated the species); Hook. Fl. ii. 5; Torr. & Gray, Fl. ii. 206, not of Desf. & DC. *S. stricta*, Hook. l. c., partly. *S. Virgaurea*, var. *humilis*, Gray, Man. 241. — Rocky ground, Newfoundland to Saskatchewan and Rocky Mountains, Northern New England, and at two remarkable southern stations in the Atlantic States (viz. on the Susquehanna, York Co., Penn., *Porter*, and Great Falls of the Potomac, *Robbins*, *Vasey*): in the Rocky Mountains south to New Mexico and Utah, perhaps also Sierra Nevada in California, there too like *S. multiradiata*, var. *scopulorum*. The typical plant is narrow-leaved, with slender but rigid stems and virgate inflorescence: it often becomes larger, broad-leaved, and with ample compound thyrsus; and on mountains occurs a dwarfer broad-leaved form, passing to

Var. nána. A western alpine form, analogous to *S. Virgaurea*, var. *alpina*, 2 to 5 inches high, with spatulate to obovate leaves, and few heads in a close glomerule, or more numerous in a spiciform thyrsus. — *S. Virgaurea*, var. *humilis*, Gray, Proc. Am. Acad. viii. 389. *S. Virgaurea*, var. *alpina*, Rothrock in Wheeler Rep. vi. 145. — High Rocky Mountains, Colorado (first coll. by *Parry*), and the Cascades of Oregon and Washington Terr., *Hall*, *Howell*, *Suksdorf*.

Var. Gillmáni, Gray, Proc. Am. Acad. xvii. 191. Large, 2 feet high, rigid, in cultivation with compound ample panicle, and laciniate-dentate leaves. — Sand-hills of the Lake shores, N. Michigan, *Gillman*, *W. Boott*.

S. confertiflóra, DC. A foot or two high, strict, rigid, sometimes strikingly glutinous or resiniferous: leaves nearly of the preceding: heads smaller and numerous, fewer-flowered, crowded in a virgate or pyramidal compound thyrsus. — Prodr. v. 339; Fisch. & Meyer, Ind. Sem. Petrop. (1840), vii. 57. *S. glutinosa*, Nutt. Trans. Am. Phil. Soc. l. c. 328. — Coast of Brit. Columbia to Oregon, first collected by *Henke*, with inflorescence incompletely evolute. Shoalwater Bay, *Cooper*. Sauvie's Island, *Howell*. Near Portland, *Pringle*, in a form too near *S. humilis*.

+ + + + Californian coast species: rays inconspicuous, shorter than the disk.

S. spathuláta, DC. Glabrous, glutinous: stem a foot high, few-leaved, terminated by a single spiciform thyrsus, the upper clusters of which are monocephalous, the lower 2-5-cephalous, and about equalled by the small subtending leaves: lower and radical leaves spatulate, rounded at apex, these sharply serrate, below long-attenuate into a margined petiole: heads 4 lines long: involucral bracts oblong and broadly linear: akenes silky-pubescent. — Prodr. v. 339; Gray, Proc. Am. Acad. xvii. 191. *S. petiolaris*, Less. in Linn. vi. 502; Hook. & Arn. Bot. Beech. 145, in part. *S. spiciformis*, Torr. & Gray, Fl. ii. 202. *Homopappus? spathulatus*, Nutt. Trans. Am. Phil. Soc. vii. 332 — Monterey, California, first coll. by *Hanke, Chamisso, Coulter*. Not "Mexico," where, however, is the somewhat related *S. simplex*, HBK., recently rediscovered by Schaffner.

* * * Heads small or middle-sized (2, 3, or rarely 4 lines long), not in a terminal corymbiform cyme, but in paniculate or racemiform clusters, which when well developed are collected in a terminal and more or less naked compound panicle or set of panicles (a few species tend to have axillary clusters, or the panicle leafy below); when the clusters are racemiform and spreading they are apt to be secund: stems generally simple or branching only at summit. — (*Erectæ* in part and *Unilaterales*, DC.) PANICULATÆ.

+ Confined to the sea-coast or the vicinity of brackish water, very smooth and glabrous, and with firm and thickish or even somewhat fleshy bright-green foliage; but occasionally varying with some minute pubescence in and toward the inflorescence, &c. (*S. lithospermifolia* is manifestly pubescent, but that species is not known as an indigenous plant): leaves obscurely puncticulate, entire, or some lower ones a little serrate, with a prominent midrib, but inconspicuous veins and veinlets in a fine reticulation; the lower leaves sometimes with one or two pairs of low-lateral or basal ribs or veins: inflorescence thyrsoidal, but the clusters sometimes racemiform and even secund. — *Maritimæ*.

++ Flowering rather early, commonly stout and middle-sized or tall: general inflorescence paniculate or hardly strict, leafy at the base: upper leaves not notably unlike the lower ones, and not appressed.

S. confínis, GRAY. Apparently pale green: leaves lanceolate and rather short (cauline 2 to 3 inches long), or the radical obovate: heads small (2 lines long), crowded in a dense oblong panicle, not secund, on glabrous pedicels: rays small, not surpassing the disk-flowers: akenes canescently pubescent. — Proc. Am. Acad. xvii. 191. *S. sempervirens*, Gray, Bot. Calif. i. 319, as to pl. coll. Palmer. — S. California; in San Diego Co., *Palmer, Cleveland, Vasey.* San Bernardino Co., at hot springs on the lower mountains, *Parish.*

S. sempérvirens, L. Bright green, leafy to the top, 2 to 8 feet high: leaves lanceolate or varying to linear and mostly acute or the lower obtuse, lowest often oblong and spatulate, of firm or rather fleshy texture: heads commonly large (4 or 5 lines long, or in slender forms only 3 lines long) and showy, numerous in short racemiform or corymbulose and somewhat secund clusters collected in an open thyrsus, or when fewer loosely paniculate: flowers golden yellow; rays 7 to 10, mostly large — Spec. ii. 878; Ait. Kew. iii. 214; DC. Prodr. v. 335; Torr. & Gray, Fl. ii. 211. *S. Mexicana*, L. l. c. 879, & authors. *S. carnosa* & *Noveboracensis*, Mill. Dict. *S. lævigata*, Ait. Kew. l. c. 215; Nutt. Gen. ii. 159. *S. limonifolia*, Pers. Syn. ii. 249. *S. Azorica*, Hochst. in Seubert, Fl. Azor. 31. — Along and near the sea-coast and tidal streams, New Brunswick and Canada to Florida, in wet or dry soil. Also San Francisco Bay, &c., on the Pacific. Inflorescence occasionally pubescent, and when away from salt water not rarely the upper part of the stem also, and leaves duller, so approaching the following cultivated variety. (Mex., Bermuda, Azores.)

Var. **vimínea**, GRAY, Proc. l. c. 192. Cultivated form, with duller-green leaves, which have lost the somewhat fleshy-coriaceous texture: upper part of stem and the inflorescence appressed-puberulent: racemiform clusters hardly developed, but the heads more scattered in a leafy panicle. — *S. integerrima*, Mill. Dict. *S. viminea*, Ait. Kew. l. c. 215; Willd. Spec. iii. 2064. *S. integrifolia*, Desf. Cat. 1804, 103, & ed. 3, 402; DC. Prodr. l. c., excl. syn. Nutt. *S. carinata*, Schrad. in DC. l. c. 337. — Common in European Botanic Gardens; passes into *S.* LITHOSPERMIFÓLIA, Willd. Enum. 891, and *S.* ELÁTA, Pursh (Solander, mss.), Fl. ii. 543). Taller, robust, larger-leaved, even the leaves somewhat puberulent. Unknown as indigenous, obviously *S. sempervirens*, var. *viminea*, more altered; but so unlike the species that it demands separate mention.

++ ++ Late-flowering, wholly glabrous, virgate; the upper portion of the stem beset with small appressed leaves: heads (commonly 3 lines long) in a strict and narrow naked panicle.

S. strícta, AIT. Stem simple, slender, very strict, 3 to 8 feet high: leaves all entire or the lowest cauline and radical rarely a little serrate; these oblong or spatulate and very obtuse; cauline very numerous, approximate, small and becoming bract-like, appressed, from oblong or spatulate to linear-lanceolate, obtuse but mucronate-apiculate: heads commonly in a simple and very narrow virgate panicle of a span or two in length, but not rarely fastigiate compound: rays 5 to 7. — Ait. Kew. iii. 216 (as to the true original, cult., with inflorescence branched); Pursh, Fl. ii. 540; Gray, Proc. Am. Acad. xvii. 182, 192, not of subsequent authors. *S. virgata*, Michx. Fl. ii. 117; Ell. Sk. ii. 384; Torr. & Gray, Fl. ii. 201. *S. linoides*, Solander, in herb. Banks, not Boott & A. Gray. *S. genistoides*, Bertol. Misc. Bot. vii. 37. — Low and sandy pine barrens. New Jersey to Florida and Louisiana. (Cuba.)

Var. angustifólia, GRAY, l. c. Leaves narrower and the lower longer, all entire; radical mostly lanceolate or narrowly spatulate (4 to 7 inches long, 4 to 9 lines wide); cauline lanceolate gradually reduced to subulate-linear: clusters of the strict panicle often more racemiform and secund. — *S. angustifolia*, Ell. Sk. ii. 388; Torr. & Gray, ii. 212. — Forms in brackish soil not clearly distinguished from the most slender and narrow-leaved *S. sempervirens*. — Carolina to Florida and Texas, along the coast.

S. flavóvirens, CHAPM. Stem 2 to 6 feet high: radical and lower cauline leaves oblong-ovate or oblong, obtusely serrate, ample (4 to 6 inches long besides the winged petiole); upper oblong (gradually reduced to half or quarter inch), all obtuse and yellowish green: inflorescence and heads of the preceding, but the short racemiform clusters at length more spreading and secund: rays few, mostly 3. — Fl. 211. — Florida, in brackish marshes at Apalachicola, *Chapman*. Robust and largest-leaved specimens of *S. stricta* seem to pass into this.

+ + Not maritime, nor alpine, nor canescently pubescent, and leaves not triple-ribbed. Yet in some a pair of stronger primary veins in the larger lower leaves gives nearly the character of the *Triplinerviæ*, and the Pacific species, *S. lepida* and *S. elongata*, referred to the latter, would rather be sought here. — *Unicostatæ*.

++ Slender, wholly glabrous and smooth, with narrow obscurely veined leaves, rayless!

S. gracíllima, TORR. & GRAY. Stem simple or with long and very slender branches, 2 or 3 feet high: leaves thickish; radical and lower cauline spatulate-lanceolate with long tapering base, 3 or 4 inches long, obscurely serrulate; upper mostly linear and becoming small, entire: heads comparatively large, 3 lines long, more or less secund in a long and slender and virgate racemiform or sometimes paniculate inflorescence (its apex often recurving): involucre broad; its bracts oblong, very obtuse, thickish, mostly greenish at the tip: flowers 9 to 12, one sometimes imperfectly ligulate: akenes pubescent. — Fl. ii. 215; Chapm. Fl. 212. — Dry pine barrens, Middle Florida, *Chapman*, &c.

++ ++ Rather small-leaved, *minutely puberulent*, but with no other pubescence: leaves not at all triple-ribbed, the small upper ones only obscurely venulose: heads (small) in a narrow thyrsoid panicle, never secund.

S. pubérula, NUTT. Smooth, the soft puberulence nearly imperceptible to the naked eye: stem rather slender, 2 feet or more high, very leafy, strict: leaves obovate and oblong, or the lower (1 to 3 inches long) spatulate, these more or less serrate; upper entire, from oblong to lanceolate: heads crowded on the short branches of the thyrsus; involucral bracts subulate-lanceolate: rays small, about 10: akenes glabrous. — Gen. 162; Torr. & Gray, Fl. ii. 202. *S. pubescens*, Ell. Sk. ii. 381. — Sandy ground, New Brunswick (glabrate and ambiguous form) and New England (chiefly along the coast, occasionally on the mountains) to Florida and Mississippi. Southward the characteristic minute puberulence is more manifest in

Var. pulverulénta, CHAPM. Almost cinereous-puberulent; the upper cauline leaves shorter and broader, gradually diminished to half or quarter inch in length. — Fl. 210. *S. pulverulenta*, Nutt. l. c. 161; Ell. l. c.; Torr. & Gray, Fl. l. c. *S. obovata*, Bertol. Misc. Bot. vii. 36. — Lower Georgia, Alabama, Florida.

++ ++ ++ Obscurely-veined and mainly entire-leaved species; the cauline leaves closely sessile or partly clasping by a broad base, with midrib prominent beneath, but veins and veinlets usually very inconspicuous: heads (about 2 lines long) crowded in slender spreading or recurving racemiform and secund clusters, which are all collected in a mostly short and broad naked terminal panicle: involucre of narrow and rather obtuse few-ranked bracts: rays 3 to 5 or rarely more: disk-flowers hardly more numerous.

= Leaves all entire and glabrous, smooth, except the margins, usually more or less pellucid-punctate.

S. odóra, AIT. Stem simple, 2 or 3 feet high, rather slender, often reclining, glabrous, or above minutely pubescent in lines: leaves commonly anise-scented when bruised, narrowly or linear-lanceolate, acute, spreading (1½ to 4 inches long, half-inch or much less in width): rays rather small. (A form, var. *inodora*, Gray, Man., growing with the ordinary plant, is scentless.) — Ait. Kew. iii. 214 (Pluk. Alm. t. 116, f. 6, & 236. f. 1); Pursh, Fl. ii. 539; Bigel. Med. Bot. i. 188, t. 20; Torr. & Gray, Fl. ii. 219. *S. retrorsa*, Michx. Fl. ii. 117. *S. puncticulata*, DC. Prodr. v. 332. — Dry or sandy soil, Canada to Florida and Texas, chiefly near the coast, but as far interior as Kentucky. (Mex.)

S. Chapmáni, Gray. Rather stouter and more rigid: stem roughish-puberulent above: leaves oblong or elliptical, obtuse or even rounded at the apex, about an inch long; those next the broad expanding thyrsus very small, often roundish. — Proc. Am. Acad. xvi. 80, xvii. 193. *S. odora*, Chapm. Fl. 213, in part. — Pine barrens of Florida, *Chapman, Garber, Curtiss*, distributed as *S. tortifolia*. Between *S. odora* and *S. pilosa*.

== Lower leaves more or less serrulate and all scabrous or pubescent, not punctate, more evidently veiny than the preceding, spreading: stem very leafy: rays small, hardly surpassing the disk-flowers.

S. tortifólia, Ell. Stem slender, 2 or 3 feet high, scabrous-puberulent: leaves all linear (an inch or two long, 1½ to 3 lines wide), acutish, commonly twisted, especially in age, hirtello-puberulent or glabrate, the lower with a few sharp denticulations: heads small, few-flowered. — Sk. ii. 377; Hook. Comp. Bot. Mag. i. 97; Torr. & Gray, Fl. ii. 220. *S. odora*, Michx. Fl. ii. 118. not Ait. *S. retrorsa*, Pursh, Fl. ii. 539; Nutt. Gen. ii. 159, not Michx. — Dry sandy soil, coast of Virginia to Florida and Texas.

S. pilósa, Walt. Stem stouter, 3 to 7 feet high, hirsute with short spreading hairs: leaves lanceolate-oblong (2 or 3 inches long), or the upper elliptical or oblong (8 to 18 lines long), these mostly obtuse, the midrib beneath and margins scabrous- or hirsute-ciliate; the lower with some acute small teeth: rays several or few and trifid, very small. — Car. 207 (not Mill. Dict.); Torr. & Gray, Fl. ii. 219. *S. fistulosa*, Mill. Dict. *S. altissima*, Michx. l. c., ex herb. *S. pyramidata*, Pursh, Fl. ii. 537; Nutt. Gen. ii. 118. *S. villosa*, Ell. Sk. ii. 372; DC. Prodr. v. 333. — Moist ground, New Jersey (pine barrens) to Florida and Louisiana, in the low country: flowering late.

++ ++ ++ ++ Leaves not small for the size of the plant, not prominently veiny, of firm texture, entire or little serrate, glabrous and smooth, but sometimes with ciliolate-scabrous margins: heads (middle-sized) crowded in thyrsoid inflorescence, not secund.

= Pacific species: rays rather numerous (8 to 15) and small: akenes pubescent.

S. Tolmieána. Low, a foot or less high, leafy up to the short and rather broad inflorescence of spiciform somewhat corymbosely disposed clusters: leaves thickish and veins very inconspicuous, linear or lanceolate (2 or 3 inches long), entire, rarely with some minute serratures, the margins usually scabrous-ciliolate: heads about 3 lines high: involucral bracts lanceolate, acutish, thin. (Has been taken for a form of *S. Missouriensis*, var. *montana*.) — Idaho, Washington Territory and Oregon; first coll. by *Tolmie*, then by *Spalding* and later collectors.

S. Guiradónis, Gray. Slender, 2 feet high, bearing rather few heads in a simple virgate thyrsus: leaves all quite entire. thickish, bright green, attenuately cuspidate-acuminate; radical and lower cauline linear-lanceolate (2 to 4 lines wide, about 4 inches long); upper more attenuate: bracts of the involucre small, lanceolate or linear, acutish. — Proc. Am. Acad. vi. 543, & Bot. Calif., in part. — California, along brooks, base of San Carlos Peak, Fresno Co., *Guirado*, an assistant of *Prof. Brewer*.

S. spectábilis, Gray. A foot or two high: heads numerous and crowded in a narrow or compound and broader thyrsus: leaves paler, sometimes thinner; cauline lanceolate, or the small uppermost becoming linear, acute: lower and radical spatulate-lanceolate or oblong, acutish or obtuse, often an inch wide and obscurely triple-ribbed; radical rarely with a few serratures: involucral bracts lanceolate or broader, mostly obtuse. — Proc. Am. Acad. xvii. 193. — *S. Guiradonis*, var. *spectabilis*, Eaton, Bot. King Exp. 154. *S. Guiradonis*, in part. Gray, Bot. Calif. i. 319; Rothrock in Wheeler Rep., &c. — From the western slopes of the Sierra Nevada, California, to the interior mountains of Nevada, *Bloomer, Watson*, &c.

= = Atlantic species: akenes glabrous or sometimes slightly and sparsely pubescent: rays conspicuous, 5 or 6.

S. uliginósa, Nutt. Stem 2 or 3 feet high, strict: leaves lanceolate and oblong-lanceolate, mostly acute or acuminate, acutely and sparsely serrulate or else entire; radical and lower cauline 4 to 8 inches long, tapering gradually into a margined petiole; some ascending primary veins obvious: thyrsus narrowly oblong or virgate, dense, the short clusters appressed: heads 3 lines long: bracts of the involucre narrowly oblong or nearly linear. — Jour. Acad. Philad. vii. 101, mainly; Gray, Proc. Am. Acad. xvii. 193. *S. stricta*, Hook. Fl. ii. 4, in part; Torr. & Gray, Fl. ii. 204, not Ait. — Bogs and wet ground, Newfoundland and Canada to L. Superior, south to New England and the mountains of Pennsylvania.

— **S. speciósa**, Nutt. Commonly 3 to 6 feet high and robust: leaves thickish and generally ample, oval, ovate, or oblong, entire or little serrate, rather abruptly narrowed into a sessile base, or the larger into a winged petiole (these often 4 to 6 inches long and 2 or 3 wide); uppermost small and lanceolate or oblong; primary veins spreading and obscure, seldom more obvious than the finely reticulated veinlets: thyrsus narrow, composed of numerous short or rarely elongated spiciform clusters, rigid, rather showy: heads 3 or 4 lines long: bracts of the well-imbricated involucre of firm texture, narrowly oblong, very obtuse, and with a greenish midnerve. — Gen. ii. 160 (excl. syn. Pers.); Torr. & Gray, Fl. ii. 205. *S. sempervirens*, Michx. Fl. ii. 119, in part. *S. petiolaris*, Muhl. Cat. 79, not Ait. — Margin of woodlands, in moist or rather fertile soil, Canada and E. New England to N. Carolina and west to Arkansas.

Var. **angustáta**, Torr. & Gray, l. c. Stem 2 or 3 feet high: leaves smaller; the lower spatulate-oblong or oblanceolate, 2 to 4 inches long, seldom an inch wide, sometimes entire; upper an inch or two long: thyrsus commonly more simple and virgate, sometimes racemosely compound. — *S. erecta?* Ell. Sk. ii. 385; DC. Prodr. v. 340? — Sandy open ground or prairies, New Jersey to Minnesota and south to Florida and Texas.

Var. **rigidiúscula**, Torr. & Gray, l. c. A form of the var. *angustata*, growing in dry open places, with more rigid and rougher-edged small leaves. — Minnesota to Nebraska and Texas.

++ ++ ++ ++ ++ Leaves veiny, at least the lower serrate (except sometimes in *S. juncea*): heads racemosely paniculate; the racemiform clusters when well developed secund and commonly scorpioid-recurving, sometimes not so in the earlier species. Atlantic species.

= Leaves (the lower ample, those of the branches small) *shagreen-scabrous* on the upper face: involucral bracts broadish: heads many-flowered, rather large.

— **S. pátula**, Muhl. Stem strongly angular and striate, rather stout, 2 to 4 feet high, with rigid elongated branches: leaves pale green, loosely venose and venulose, sharply and rather finely serrate, smooth and glabrous (as is the stem), except the upper face which is strikingly scabrous when rubbed from point to base (being thickly set with minute sharp callosities antrorsely directed); cauline oval or oblong, 4 to 8 inches long besides the abruptly narrowed base or winged petiole of the lowest; the uppermost and those of the flowering branches sometimes equalling the at length spreading clusters of the narrow or virgate thyrsus: heads 3 or even 4 lines long: bracts of involucre linear-oblong, very obtuse: rays 6 or 7, small, light yellow: disk-flowers 8 to 12: akenes minutely pubescent. — Willd. Spec. iii. 2059; Torr. & Gray, Fl. ii. 213. *S. asperata*, Pursh, Fl. ii. 538, as to herb. Lamb. "*S. angulata*, Muhl.," ex Spreng. in herb. Willd.; Schrad. in DC. Prodr. v. 331. *S. Frankii*, Hochst. & Steud. in distrib. Frank. — Wet soil, Canada to Wisconsin, south to Georgia, Missouri, and Texas. Flowering rather early.

Var. **strictula**, Torr. & Gray, l. c. Narrower leaved, and with simpler and strict inflorescence. — *S. salicina*, Ell. Sk. ii. 389, ex char. *S. scabra*, Hook. Comp. Bot. Mag. i. 97. — Commoner southward to Louisiana and Florida.

= = Leaves not scabrous, both faces minutely cinereous-puberulent: heads small, many-flowered, in an erect panicle, hardly secund, vernal!

— **S. vérna**, M. A. Curtis. Stem 2 feet high, cinereous-pubescent, bearing a loose naked panicle: leaves of rather firm texture; radical oval or ovate, minutely serrate (3 or 4 inches long), abruptly contracted into a long margined petiole, the primary veins (2 or 3 pairs) conspicuous; upper cauline small and sparse, oblong, entire: heads barely 3 lines long: bracts of the involucre rather narrow and thin: rays 10 or 12: akenes pubescent. — Torr. & Gray. Fl. ii. 205. — Open and sandy pine woods, near Wilmington, N. Carolina, *Curtis*. Flowering in May! Obscure species, not since met with.

= = = Leaves commonly thin and membranaceous, loosely veiny (if firmer the veinlets of the lower face conspicuously reticulated), small or not large: heads small (about 2 lines long): bracts of the involucre rather few and narrow, obtuse: akenes pubescent.

a. Rays from uone to 3: leaves clasping!

S. amplexicaúlis, TORR. & GRAY. Minutely soft-pubescent or glabrate: stem slender, 1 to 3 feet high, with spreading branches: leaves ovate, acute or acuminate, acutely serrate, rather scabrous above and soft-pubescent beneath; the upper slightly narrowed above the dilated auriculate-clasping base; lower cauline with longer contracted portion; lowest and radical wing-petioled below the truncate or subcordate base of the lamina (this about 2 inches long): racemiform clusters of the thyrsus slender, secund, often simple; pappus shorter than the disk-corolla. — Fl. ii. 218 (not Martens, which is *S. Riddellii*); Chapm. Fl. 213. — Open dry woods, Florida to Louisiana, *Leavenworth*, *Chapman*, *Rugel*, distributed by Shuttleworth as *S. auriculata*. Makes the nearest approach to *Brachychæta*.

b. Rays 4 to 6 or rarely more, small, and disk-flowers little more numerous: leaves sessile by a narrow base, pinnately veiny: pubescence of spreading hairs, or hardly any.

S. rugósa, MILL. Stem hirsute or pubescent with spreading hairs, low or moderately tall (1 to 6 feet high), mostly slender, very leafy to the top: leaves thin and soft, or in dry open ground becoming thicker and firmer, from oval-ovate to oblong-lanceolate (1 to 4 inches long), mostly acute or acuminate, sometimes obtuse, usually hirsute on the veins and veinlets beneath; these conspicuous and often rugose-reticulated, sometimes scabrous above: inflorescence when well developed recurved-spreading, but sometimes erect; bracts of the involucre linear. — Dict. ed. 8; Willd. Spec. iii. 2058; Ait. Kew. ed. 2, v. 66; Gray, Proc. Am. Acad. xvii. 194. *Virga-aurea* sp., Dill. Elth. 406–411, t. 304, 305, 308, appended in L. Spec. 878 to *S. altissima*, but not referred to it. *S. altissima* & *S. aspera*, Ait. Kew. iii. 212; Willd. l. c. *S. scabra*, Muhl. in Willd. l. c. *S. villosa*, Pursh, Fl. ii. 537. *S. humilis*, Desf. Cat. ed. 3, 402; DC. l. c., a low form, commonly with the racemiform clusters erect, or hardly spreading and secund. *S. asperula*, Desf. Cat. l. c. 403 ? *S. hirta*, Willd. Enum. 891. *S. rigidula*, Bosc, in hort. Paris ? *S. asperata*, Soland. mss., and so of Pursh as to herb. Banks. *S. altissima*, Torr. & Gray, Fl. ii. 216 (incl. *altissima*, *pilosa*, *recurvata*, *Virginiana*, Mill. Dict.), not L. — Moist or dry ground, Newfoundland and Labrador to Texas; very common eastward in the Atlantic States. Polymorphous, not readily sorted into definable varieties; in shade thin-leaved; in open and dry soil has small and broader, firmer, more scabrous, and rugose-reticulated leaves. *S. rugosa*, Mill., is the best of the old names to take up.

S. ulmifólia, MUHL. Resembles the thinner-leaved and least pubescent forms of the preceding (into which it appears to pass), but with stem smooth and glabrous, except perhaps the summit: leaves bright green, nearly smooth and glabrous, or pubescent, membranaceous, acute or acuminate at both ends, usually coarsely serrate, the larger veins conspicuous but veinlets inconspicuous: thyrsus more naked: bracts of the involucre of firmer texture and more obtuse. — Willd. Spec. iii. 2060; Darlingt. Fl. Cest. 457; Torr. & Gray, Fl. ii. 217. *S. lateriflora*, Ait. Kew. iii. 211, not L. — Moist woodlands and copses, Maine to Iowa, Arkansas, and Texas. *S. multiflora*, Desf. (in Poir. Suppl. v. 462) Cat. l. c. 402, DC. Prodr. v. 336, appears to be a state of this, altered by cultivation.

Var. microphýlla. A reduced and rather rigid form: with lower leaves 2 inches long; upper reduced to half an inch, obtuse, obscurely serrate. — Texas, *Lindheimer*, *Wright*.

= = = = Leaves usually of firm texture and inconspicuous reticulation, occasionally thin and membranaceous or more veiny, not scabrous above, commonly glabrous as also the stems: bracts of the involucre from broadly linear to narrowly oblong, obtuse.

a. Stem equably and very leafy up to or into the pyramidal compound thyrsus: leaves comparatively short and broad, even the lower not much narrowed downward, the secondary veins often manifest.

S. Ellióttii, TORR. & GRAY. Smooth and glabrous throughout, or the thyrsus somewhat pubescent: stem tall, rigid: leaves from ovate-oblong to oblong-lanceolate, apiculate-acuminate or acute, minutely and sparsely serrate with appressed teeth, scabrous on the margin, mostly closely sessile by a broadish base (1 to 4 inches long): heads (3 lines long) crowded on the secund and spreading or sometimes ascending and straight racemiform or spiciform branches of the pyramidal panicle: bracts of the involucre rather broadly linear: rays 8 to 12, short: akenes pubescent. — Fl. ii. 218. and *S. elliptica* of the same, as to the plant of New York. *S. elliptica* ? Ell. Sk. ii. 376. *S. elongata*, Hort. Par. 1832. — Moist ground near the coast, Massachusetts to New York and through the low country south to Georgia.

b. Less leafy, or leaves toward the naked panicle small compared with the lower, which are contracted or tapering into a conspicuous narrowed base or winged petiole: veins inconspicuous; panicle commonly narrow, or its branches short; plants wholly smooth and glabrous, except the somewhat ciliolate-scabrous margins to the leaves, in drier ground sometimes obscurely scabrous.

S. neglécta, TORR. & GRAY. Stem strict and simple, 2 to 4 feet high: leaves bright green, lanceolate or the larger oblong-lanceolate, acute, mostly serrate or serrulate; radical ones ample (often a foot or more long, including the elongated petiole): panicle generally thyrsoid and narrow, of short and crowded more or less secund clusters, or in larger plants more compound with spreading racemiform branches: heads at most 3 lines long: involucral bracts oblong-linear: rays 3 to 7 and disk-flowers 5 to 7: akenes from sparsely puberulent to glabrous. — Fl. ii. 213; Gray, Man. ed. 2, 204. — In swamps, especially in sphagnous bogs, or on their borders, Lower Canada to Maryland, west to Illinois and Wisconsin. Forms with almost entire leaves and strict panicle too nearly approach *S. uliginosa,* Nutt., while some with large and serrate leaves are more like *S. arguta.* The most slender is

Var. linoídes. Stem simple, commonly 2 feet high, slender: radical leaves 4 to 8 inches long, a third to half inch wide; upper cauline very small and erect: panicle of rather few and approximate racemiform secund clusters: heads rather smaller: rays only 2 or 3. — *S. uliginosa,* Nutt. in Jour. Acad. Philad. vii. 101, in part, but not of his own herb. nor descr. *S. linoides,* Torr. & Gray, Fl. ii. 216, not of Soland. in herb. Banks, which is *S. stricta,* Ait. *Bigelovia? uniligulata,* DC. Prodr. v. 329. *Chrysoma uniligulata,* Nutt. Trans. Am. Phil. Soc. vii. 325. — Sphagnous swamps, Massachusetts to New Jersey.

S. Terræ-Nóvæ, TORR. & GRAY. Still obscure species, probably a form of *S. neglecta,* somewhat dwarfed and with a corymbosely paniculate thyrsus: involucral bracts rather thinner and narrower. — Fl. ii. 206. — Sphagnous bogs, Newfoundland, *Pylaie, Miss Brenton.*

c. Stems not strict, disposed to branch below the inflorescence: racemiform clusters of the inflorescence often leafy-bracteate, rather rigid, sparse and ascending, or forming a loose elongated thyrsus: leaves more veiny and serrate; cauline commonly abruptly contracted into a petiole-like or narrow base: rays not numerous, sometimes wanting: bracts of the involucre rather firm, obtuse, mostly greenish toward the tip.

S. Boóttii, HOOK. Sometimes minutely scabrous-pubescent, or below hirsute with jointed hairs, often quite glabrous: stem slender, 2 to 5 feet high: leaves rather finely serrate with ascending teeth; radical and lower cauline from ovate to oblong-lanceolate, acuminate (the larger 3 to 5 inches long, besides the petiole-like base) upper small, oblong to narrowly lanceolate, often entire: heads (2 and 3 lines long) rather loosely racemose: bracts of the campanulate involucre oblong-linear: rays 2 to 4 or rarely 5, sometimes solitary or none: akenes pubescent. — Comp. Bot. Mag. i. 97; Torr. & Gray, Fl. ii. 215. *S. juncea,* DC. Prodr. v. 334, not Ait. — Dry wooded ground, Virginia to Florida, Louisiana, and Texas. The larger forms northward nearly approach the next species. Southward the smaller ones pass into

Var. brachyphýlla, GRAY. More slender; the flowering branches even filiform: larger leaves an inch or two long, all from ovate to oblong, seldom acuminate, commonly obtuse, upper reduced to half or quarter inch, sessile by a broad base: heads sparse, 4-7-flowered; rays none or an imperfect one. — Proc. Am. Acad. xvii. 195. *S. brachyphylla,* Chapm. in Torr. & Gray, Fl. ii. 215, & Fl. 213. — Dry woodlands, Georgia and Florida, *Chapman,* &c.

Var. Ludoviciána, GRAY, l. c. Perhaps a distinct species, stouter, tall, rather large-leaved: lower leaves and lower part of the stem sometimes roughish-hirsute or hispidulous with many-jointed hairs, or glabrous: heads larger, even 4 lines long! — *S. Boottii,* var. *e,* partly, Torr. & Gray, l. c. — W. Louisiana, *Hale.*

S. argúta, AIT. Glabrous, sometimes slightly pilose-pubescent: stem 2 to 4 feet high: leaves thinnish (in shade membranaceous), usually ample; the lower and larger 5 to 9 inches long, ovate or oval, acuminate, very strongly and sharply (or even doubly) serrate with salient teeth; upper reduced to oblong-lanceolate, only the small ones of the branches entire: heads somewhat crowded on the branches of the irregular panicle, fully 3 lines long: involucral bracts oblong-linear: rays 5 to 7, rather large: disk-flowers 10 to 12: akenes glabrous or sometimes slightly pubescent. — Ait. Kew. iii. 213; Pursh, Fl. ii. 538; Muhl. Cat.; Darlingt. Fl. Cest. 458; DC. Prodr. v. 333; Gray, Proc. Am. Acad. xvii. 180, 195; not Torr. & Gray, who followed a wrong determination. *S. verrucosa*, Schrad. Hort. Goett. 12, t. 6 ? *S. Muhlenbergii*, Torr. & Gray, Fl. ii. 214. — Moist woodlands, New England and Canada to Ohio, through Pennsylvania to the mountains of Virginia.

Var. Caroliniána. Leaves of firmer texture, simply serrate as in *S. Boottii*, but larger: heads thicker, with 4 or 5 short rays and 10 to 14 disk-flowers; involucral bracts firmer, oblong: akenes pubescent. — Mountains of N. Carolina and of adjacent S. Carolina and Georgia, *G. R. Vasey, J. Donnell Smith*. Perhaps distinct both from this and the preceding species.

d. Stems not strict, simple or corymbosely branched at summit: inflorescence an open spreading panicle, usually as broad as high, composed of recurving naked and minutely subulate-bracteate secund-racemiform clusters of crowded small heads, the rhachis and pedicels slender: rays numerous and small.

S. júncea, Ait. Mostly smooth and nearly glabrous: stem 1 to 3 feet high, rigid, commonly simple up to the mostly crowded branches of the wide panicle: leaves of rather firm texture; radical oval to oblong-spatulate, tapering into a winged petiole, usually large and sharply serrate; cauline from narrowly oblong to lanceolate (larger 3 or 4 inches long), not rarely almost entire or sparsely serrulate, the small upper not much narrowed at base: paniculate racemes slender: heads seldom over 2 lines long: bracts of the involucre small and pale : rays 7 to 12, hardly surpassing and little fewer than the disk-flowers : akenes glabrous or slightly pubescent. — Kew. iii. 213; Pursh, l. c.; Hook. Fl. ii. 3; Gray, Proc. l. c. *S. ciliaris*, Muhl. in Willd. Spec. iii. 2056 : Darlingt. l. c.; DC. l. c. 331 (excl. syn. *S. glabra*). *S. arguta*, Torr. & Gray, Fl. ii. 214, not Ait., &c., as was wrongly supposed. — Common in dry or rocky ground, Hudson's Bay and Saskatchewan to Wisconsin, and through the Northern States to the upper country of Carolina and Tennessee. — The original type by Solander is a small form from Hudson's Bay. The specific name alludes to the inflorescence, remotely resembling that of some species of *Juncus*. *S. ciliaris* is a common broad leaved form, the larger leaves a little ciliate. — Var. SCABRELLA (*S. arguta*, var. *scabrella*, Torr. & Gray, l. c.) is a form with rigid and roughish leaves, growing in arid soil. Wisconsin and Illinois to Kentucky; in which district the leaves become more or less triple-ribbed and rigid, and seemingly pass into *S. Missouriensis*.

+ + + Not maritime: leaves more or less triple-ribbed, or with a pair of lateral veins continued by inosculation parallel to the midrib, yet these sometimes obscure or evanescent. — *Triplinerviæ*.

++ Smooth and glabrous, at least as to the stem and bright green leaves (the latter sometimes a little pilose-pubescent in *S. serotina*), not cinereous or canescent: inflorescence when well developed of naked and secund commonly recurving racemiform clusters, collected in a terminal compound panicle: akenes more or less pubescent.

= Leaves of firm texture, rather rigid, lanceolate, acute or acuminate, the slender lateral ribs not rarely evanescent in the upper leaves: bracts of the involucre rather firm; the short outermost ovate or oval and the inner oblong-linear, all obtuse. A form of the first species connects with the last preceding.

a. Rays rather small: stems leafy to the summit: leaves commonly with scabrous margins, the larger mostly with some scattered teeth or denticulations.

S. Missouriénsis, Nutt. Low or middle-sized, smooth : leaves thickish, mostly tapering to both ends, and the serratures when present sharp and rigid, somewhat nervose; lower spatulate-lanceolate (larger 4 to 6 inches long); upper mostly linear and entire, acute; sometimes all entire: racemiform clusters approximated in a short and broad panicle (like those of *S. juncea*, but usually shorter), recurving in age : rays 6 to 13, small. — Jour. Acad. Philad. vii. 32, & Trans. Am. Phil. Soc. vii. 327 (excl. hab. N. Carol.); Torr. & Gray, Fl. ii. 322. *S. serotina*, Hook. Comp. Bot. Mag. i. 97, not Ait. *S. glaberrima*, Marteus in Bull. Acad. Brux. viii. (1841), 68. — Dry prairies, Indiana and Tennessee to Texas, and westward to the Rocky Mountains; in the more eastward stations passing into or else hybridizing with *S. juncea*.

Var. montána, Gray. Dwarf, 6 to 15 inches high : leaves entire or with few small serratures; cauline obscurely triplinerved, an inch or two long: panicle small and compact (at most 2 or 3 inches long); its clusters short, crowded, seldom recurved or much secund. — Proc. Am. Acad. xvii. 195. *S. Missouriensis*, Nutt. Jour. Acad. Philad. l. c., as to the original from "upper branches of the Missouri, *Wyeth*." — Dakota to the Saskatchewan and west to Idaho.

Var. extrária, GRAY, l. c. A foot or two high, robust: leaves broader (the largest sometimes an inch wide), sparingly serrate or entire: heads rather larger: rays more conspicuous. — Dry ground, in the mountains, Colorado to S. Arizona, *Parry, Hall & Harbour, Greene, Pringle, Lemmon,* &c.

S. Gattíngeri, CHAPM. ined. Slender, mostly strict and barely 2 feet high: branches and inflorescence perfectly smooth and glabrous: leaves ciliolate; lowest cauline and radical lanceolate-spatulate, appressed-serrulate, obviously triplinerved; upper canline mainly entire and without lateral ribs, oblong-lanceolate and an inch or so long, and the upper reduced to half or quarter inch, but near the inflorescence very small and bract-like: racemiform clusters of small heads open and spreading, not recurving, disposed to form a corymbiform very naked panicle: involucral bracts oblong, very obtuse, yellowish in the dried plant: flowers 15 to 20 in the head: akenes appressed-puberulent or the lower part glabrous. — *S. Missouriensis,* var. *pumila,* Chapm. Fl. Suppl. 627. — Rocky barrens and cedar glades, Rutherford Co., Tennessee, *Gattinger.* Between the preceding and the following.

S. Shórtii, TORR. & GRAY. Slender, 2 to 4 feet high: upper part of stem and flowering branches scabrous with minute appressed puberulence: leaves bright green, oblong-lanceolate, rather short (longer only 2 or 3 inches long, toward the inflorescence moderately reduced), acute, mostly with a few small serratures: panicle oblong or pyramidal; its racemiform clusters commonly slender and soon recurving: heads narrow, 10-14-flowered: involucral bracts narrowly oblong: akenes pubescent. — Fl. ii. 222. — Rocks, at the Falls of the Ohio, near Louisville, *Rafinesque, Short.* N. W. Arkansas, *F. L. Harvey.*

b. Leaves with entire and smooth margins: rays larger.

S. Marshálli, ROTHR. Tall (only the upper part of stem known), slender: leaves linear-lanceolate, acute; the lateral ribs mostly obscure: panicle naked, of loose recurving racemes; the rhachis and slender pedicels setaceously bracteate: heads 3 lines long, rather broad: bracts of the involucre broadish, of firm texture, mostly greenish on the back: rays about 8, and disk-flowers more numerous: akenes pubescent. — Rothrock in Wheeler Rep. vi. 146. — Mountains of S. Arizona, near the Chiricahua Agency, *Lieut. Marshall.*

= = Leaves thinner, sometimes membranaceous: bracts of the involucre chiefly linear, obtuse: branches and upper part of the stem not rarely scabrous-puberulent or minutely hairy.

S. Leavenwórthii, TORR. & GRAY. Stem strict, slender, rigid, 2 to 4 feet high, scabro-puberulent even to below the middle: leaves mostly linear (3 or 4 inches long and as many lines wide), very sharply and finely serrate, both ribs and veins inconspicuous: heads 3 lines long, in an ample open panicle: involucral bracts thin, linear, obtuse: rays 10 or 12, small. — Fl. ii. 223; Chapm. Fl. 214. — Damp soil, Florida to S. Carolina, near the coast, *Leavenworth, Chapman.*

S. rupéstris, RAF. Stem lax, 2 or 3 feet high, smooth nearly to the small panicle: leaves membranaceous, linear-lanceolate, sparsely and sharply serrulate or denticulate, or the upper entire (1 to 3 inches long): heads very small (barely 2 lines long): rays 4 to 6, small. — Ann. Nat. 14; Torr. & Gray, Fl. ii. 225. — Rocky banks of streams, along the Ohio River, Kentucky, Indiana, and Western Virginia. Probably only an extreme glabrous form of *S. Canadensis.*

S. serótina, AIT. Stem stout, 2 to 7 feet high, very smooth and glabrous up to or near the ample panicle, sometimes glaucous: leaves commonly ample, lanceolate or broader (3 to 6 inches long), sharply and saliently serrate, in the typical plant glabrous both sides: heads crowded, rather large and full (3 lines long): rays 7 to 14, moderately large and conspicuous: bracts of the involucre broadly linear or linear-oblong. — Kew. iii. 211; Gray, Proc. Am. Acad. xvii. 179, 196. *S. gigantea,* Willd. Spec. iii. 2056, and subsequent authors. *S. glabra,* Desf. Cat. ed. 3, 402; DC. Prodr. v. 331. *S. fragrans,* Hort. Par., not Willd. *S. Pitcheri,* Nutt. Jour. Acad. Philad. vii. 101, & Trans. Am. Phil. Soc. l. c. 326, forms with broad and comparatively short leaves and rather smaller heads. *S. elongata,* var., Torr. & Gray, l. c., in part. — Moist or rich soil, Newfoundland to Brit. Columbia, Oregon, and south to Texas. Passes insensibly into

Var. gigantéa, GRAY, l. c. Commonly tall, 5 to 8 feet high: leaves with the lateral ribs more prominent beneath, and these more or less pilose-pubescent or hispidulous, sometimes the veins or even the whole under surface pubescent. — *S. gigantea,* Ait. l. c. *S. serotina,* Willd.; Torr. & Gray, etc. — Chiefly in the Atlantic States, from Canada to

Texas. From Willdenow to the latest authors this has passed as the true *S. serotina*, and that for this.

++ ++ Minutely pubescent or glabrate, not cinereous nor scabrous, thinnish-leaved, and the lateral ribs commonly obscure: panicle mostly erect and thyrsiform, often compact, and the heads little if at all secund: involucre of small and thin narrow bracts: rays 12 to 18, small. (Related to the preceding and following, also to *S. rugosa*.)

S. lépida, DC. A foot or two high: leaves from oblong to broadly lanceolate, acute, 3 or 4 inches long, very sharply and mostly coarsely serrate, sometimes for most of their length, sometimes only above the middle, in some the teeth almost none: thyrsus very short and compact, an inch or two long, little surpassing the upper leaves, not at all secund: heads fully 3 lines long: bracts of the involucre subulate-linear, attenuate-acute. — Prodr. v. 339. *S. gigantea*, Hook. Fl. ii. 2, in part. — Alaska, coast and islands, *Hawks, Kellogg*, &c., and Brit. Columbia.

S. elongáta, NUTT. Like the preceding, or taller, sometimes a yard high: leaves commonly narrower: thyrsus more developed and compound, 3 to 8 inches long, its branches occasionally spreading: bracts of the involucre linear, acutish or obtuse. — Trans. Am. Phil. Soc. l. c.; Torr. & Gray, Fl. 223, mainly. *S. stricta*, Less. in Linn. vi. 502. *S. elata*, Hook. Fl. ii. 5, not Solander. — Along streams, Brit. Columbia to California, and east to Montana, Slave Lake, &c. Seemingly passes on the northwest coast into *S. lepida*, and eastward into *S. Canadensis*.

++ ++ ++ Pubescent (at least the stem), either hirsutely or canescently, or hispidulous-scabrous: branches of the panicle when well developed secund.

= Leaves tapering gradually to an acute or acuminate point, generally thin or thinnish: panicle open, of naked and secund mostly recurving racemiform clusters: bracts of the involucre narrow and thin: rays small and short.

S. Canadénsis, L. Stem 2 to 6 feet high, from scabrous- or cinereous-puberulent to hirsute: leaves mostly lanceolate, puberulent, pubescent, or nearly glabrous, sharply serrate or the upper entire, veiny, and with lateral ribs prolonged parallel to the midrib: heads small, ordinarily only 2 lines long: bracts of the involucre small and pale, narrowly linear, acutish or obtuse: rays 9 to 16, more numerous than the disk-flowers. — Spec. ii. 878 (excl. syn. Pluk.); Ait. Kew. iii. 210; Torr. & Gray, Fl. ii. 221. *S. altissima*, L. l. c., that is *Virgaaurea altissima*, etc., Martyn, "Cent." (Hist. Pl.) 14, t. 14; not of most subsequent authors, who have followed the conjectural references to Dill. Elth. *S. reflexa*, Ait. l. c. 211; Willd. Spec. iii. 2056. *S. nutans*, Desf. Cat. ed 3, 402. *S. longifolia*, Schrader, in DC. Prodr. v. 330. — Moist or dry and shady ground, New Brunswick to Brit. Columbia (and north to Slave Lake), south to Florida and mountains of Arizona: flowering rather early. — The more marked forms varying from the ordinary are the following.

Var. prócera, TORR. & GRAY, l. c. Leaves less serrate or the upper entire, at least the lower face and upper portion of the stem cinereous-pubescent or tomentulose with very short and fine pubescence: inflorescence less open or the branches ascending in less developed or cultivated plants: heads sometimes larger. — *S. procera*, Ait. l. c.; Willd. l. c. *S. eminens*, Bischoff, hort. Heidelb. — Open ground, Canada and Saskatchewan to Idaho and Texas, the northwestern forms commonly dwarf.

Var. scábra, TORR. & GRAY, l. c. Like the foregoing, but the short pubescence rough or hispidulous: leaves shorter, oblong-lanceolate to oblong-ovate, more entire, more veiny (approaching rough-leaved forms of *S. rugosa*): heads sometimes 3 lines long. — *S. scabra*, Muhl. Fl. Lancast. ined., not Willd., which is *S. rugosa*. — Drier and sunnier places, Penn. to Florida and Texas. (*S. scabrida*, DC. Prodr. v. 331, of Mexico, appears to be a form of this.)

Var. canéscens, GRAY. Stem and both faces of the narrow and commonly entire leaves canescent with soft and fine pubescence: bracts of the involucre broader and more obtuse. — Proc. Am. Acad. xvii. 197. — S. W. Texas, *Berlandier, Lindheimer, Bigelow*, and *S. New Mexico, Thurber*.

Var. Arizónica, GRAY, l. c. Minutely cinereous-pubescent or puberulent, hardly scabrous: stems low: heads mostly 3 lines long: thin bracts of the involucre commonly acutish. — *S. mollis*, Rothr. in Wheeler Rep. vi. 146. — Mountains of S. Utah, *Ward*, and of New Mexico & Arizona, *Bigelow, Rothrock*. (Heads, &c., nearly of *S. velutina*, DC., a Mexican species, which approaches this and the preceding ambiguous forms of *S. Canadensis*.)

== Leaves obtuse or abruptly apiculate, or acutish, of firm or coriaceous texture, upper ones entire: pubescence all close, cinereous or canescent, or scabro-hispidulous; lateral ribs commonly incomplete, often obscure or wanting: panicle mostly compact, naked: bracts of the involucre broadish and obtuse, of firm texture: rays fewer and larger, golden yellow. The species are confluent.

a. Cinereous to canescent with fine and soft or at length minutely scabrous pubescence: leaves firm but seldom very rigid.

S. Califórnica, NUTT. Stem rather stout, either low or tall, canescently puberulent or pubescent: leaves oblong or the upper oblong-lanceolate and the lower obovate, obtuse or apiculate, entire or the lower with some small teeth, canescently puberulent or beneath more pubescent: thyrsus virgate, 4 to 12 inches long, dense; the racemiform clusters erect or barely spreading in age, when elongated mostly secund, and even with the apex at length recurved; heads 3 or 4 lines long: bracts of the involucre lanceolate-oblong or oblong-linear, mostly obtuse, externally somewhat puberulent: rays 7 to 12, fewer than the disk-flowers: akenes minutely pubescent. — Trans. Am. Phil. Soc. l. c.; Torr. & Gray, Fl. ii. 203; Gray, Bot. Calif. i. 319. *S. puberula*, Cham. & Schlecht. in Linn. vi. 502, not Nutt. *S. petiolaris*, Hook. & Arn. Bot. Beech. 145, partly. *S. velutina*, DC., var. "panicula contracta," DC. Prodr. v. 332, Hænke, whose "Real del Monte" is Monterey, California. — Dry ground, California to the borders of Nevada and Mexico.

Var. **Nevadénsis**, GRAY. Thyrsus and its clusters more secund: heads rather smaller: involucre mostly glabrous. — Bot. Calif. l. c. — Sierra Nevada, California, and Nevada from Plumas Co. to Owens Valley, &c. Transition to *S. nemoralis*.

S. nemorális, AIT. Mostly low, with the fine and uniform close pubescence either soft or (in age and in dried specimens) minutely scabrous: leaves from spatulate-obovate to oblanceolate or somewhat linear; upper entire and small (half-inch or more long); radical and lower cauline sparingly serrate: thyrsus and its compact racemiform clusters secund, commonly recurved-spreading: heads 2 or 3 lines long: bracts of the involucre oblong-linear or narrower, obtuse, smooth and glabrous: flowers (appearing rather early) deep yellow: rays 5 to 9, usually more numerous than the disk-flowers: akenes closely pubescent. — Kew. iii. 213; Pursh, Fl. ii. 537; DC. Prodr. v. 333; Torr. & Gray, Fl. ii. 220. *S. hispida*, Muhl. in Willd. Spec. iii. 2063; Pursh, Fl. ii. 541. *S. conferta*, Poir. Dict. viii. 459. *S. cinerascens*, Schwein. in Ell. Sk. ii. 375. *S. decemflora*, DC. Prodr. v. 322. *S. puberula*, DC. l. c. 333, not Nutt. — Dry hills or sterile soil, throughout Canada and Saskatchewan to Florida and Texas, and west to Arizona, Utah, and Nevada; in the eastern region soft-cinereous; beyond the Mississippi often greener and more scabrous; or in Utah and New Mexico greenish and hardly scabrous. In the Rocky Mountains and northward mostly occur low and more canescent forms. (Adj. Mex.)

Var. **incána**, GRAY, Proc. l. c. Dwarf, a span to a foot high: leaves oval or oblong, rigid, more or less canescent, sometimes rather strongly serrate, sometimes mostly entire: racemiform clusters erect or the lower somewhat spreading, collected in a dense oblong or conical thyrsus. — *S. mollis*, Bartl. Ind. Sem. Hort. Gœtt. 1836, 5; DC. Prodr. v. 279; in cult. specimens the involucral bracts are narrowish and somewhat acute, as also in one form of *S. incana*, Torr. & Gray, Fl. ii. 221 (excl. var.), while in a similar one, collected with it by Nicollet, they are linear-oblong and obtuse. — Plains of Minnesota and Dakota (*Nicollet*, &c.) to the Rocky Mountains of Montana and Colorado. (Adj. Mex.)

S. nána, NUTT. A span to a foot high, canescent with minute dense puberulence, not scabrous in age: leaves mostly obovate or spatulate and entire, small: heads (3 lines long) broad, few or rather numerous in an oblong or corymbiform panicle, not at all secund: bracts of the involucre oval or oblong, very obtuse: otherwise nearly as *S. nemoralis*. — Nutt. Trans. Am. Phil. Soc. l. c. 327 (in herb. "*S. pumila*"); Torr. & Gray, l. c. — Rocky Mountains and high plains, Wyoming to N. Arizona and N. E. Nevada; first coll. by *Nuttall*.

b. Hispidulous-scabrous, rigid, green!

S. rádula, NUTT. Stem a foot or two high, scabro-puberulent: leaves rigidly coriaceous, short, loosely reticulate-veined, occasionally with well-developed lateral ribs, obtuse, sparsely serrate or entire, from oval or obovate to oblong-spatulate (lowest 2 or 3 inches long, uppermost an inch or less, or rounded ones on the branches reduced to half or quarter inch), very hispidulous-scabrous at least on the veins, the midrib and margins often hispid: branches of the thyrsus secund and when well developed recurved-spreading: heads 2 and at most 3 lines long: bracts of the involucre rather rigid, glabrous, oval to linear-oblong: rays 3 to 6. rather fewer than disk-flowers: akenes minutely pubescent. — Jour. Acad. Philad. vii. 327; Torr. & Gray. Fl. ii. 220. *S. rotundifolia*, DC. Prodr. v. 332, & *S. scaberrima*, Torr. & Gray, l. c., broad-leaved form. *S. decemflora*, Gray. Pl. Lindh. ii. 223, not DC. — Dry hills and prairies, S. W. Illinois to Arkansas, W. Louisiana, and Texas; first coll. by *Berlandier* and *Nuttall*.

c. Scabro-puberulent, somewhat cinereous, small-leaved; the lateral ribs obsolete.

S. sparsiflóra, GRAY. Founded on incomplete specimens (branches), of doubtful affinity, scabrous rather than puberulent, leafy into the narrow and strict branches of the panicle: leaves all small (the larger hardly an inch long), lanceolate-linear, rather acute at both ends, rigid, entire, with lateral ribs and veins almost obsolete: heads somewhat scattered or few in the short imperfectly racemiform and somewhat secund clusters. 3 lines long: bracts of the involucre rather small, oblong-linear, barely obtuse: rays 6 to 10, little surpassing the disk. — Proc. Am. Acad. xii. 58; Rothr. in Wheeler Rep. vi. 146. — S. Arizona, near Camp Lowell, *Rothrock*. Llano Estacado, N. W. Texas on the borders of New Mexico, *Bigelow*. — To which must be added

Var. **subcinérea**, GRAY. Quite cinereously puberulent, the leaves scabro-puberulent: heads more crowded and secund in the virgate panicles: rays more conspicuous. — Proc. Am. Acad. xvii. 197. — Rucker Valley, S. Arizona, *Lemmon*. Base of stem and lower leaves unknown: the affinity decidedly with *S. nemoralis*. Also a form between this and *S. Canadensis*, var. *canescens*, with larger heads, &c., coll New Mexico in the Mogollon Mountains, 1881, *Rusby*.

= = = Leaves thinnish, puberulent but green, broad, acute, divergently triplinerved and veiny: branches of the loose panicle racemiform, secund, leafy: bracts of the involucre narrowly oblong, obtuse, outer with greenish tips: rays few.

S. Drummóndii, TORR. & GRAY. Soft-puberulent: stem 3 feet high, freely branched: leaves ovate or broadly oval, nearly or quite glabrous above; cauline copiously serrate, commonly acute at both ends, almost petioled (lower 3 or 4 inches long and 2 or more broad); those of the flowering branches numerous even through the inflorescence, from 2 inches down to a quarter-inch long, obtuse, sparingly denticulate or entire: rays 4 or 5, often 3-lobed, rather large. — Fl. ii. 217. *S. ulmifolia*, Hook. Comp. Bot. Mag. i. 97. — S. W. Illinois and Missouri to Louisiana, flowering late; first coll. by *Drummond*. Allied in some respects to *S. rugosa* and *S. amplexicaulis*.

* * * * * Heads in a compact and corymbiform thyrsus or cyme: radical leaves mostly long-petioled and with prominent midrib: akenes except in the first species wholly glabrous. — CORYMBOSÆ.

+- Leaves, even the radical, not triplinerved, flat; cauline sessile, very numerous: involucre of oblong-linear to oval faintly striate bracts: akenes very glabrous.

S. rígida, L. Somewhat cinereous with a short and dense, either soft or (in age) rather scabrous pubescence: stem stout, 2 to 5 feet high (rarely more dwarf): leaves rigid, obscurely serrate or entire; radical and lowest cauline oval or oblong, rounded at both ends or acute at base, 3 to 7 inches long; upper cauline ovate-oblong, gradually smaller upward, with slightly clasping or decurrent base: clusters dense: heads about 5 lines long, campanulate, many- (over 30-) flowered: involucral bracts broad: rays 7 to 10, rather large: akenes turgid, 12-15-nerved. — Spec. ii. 880; Ait. Kew. iii. 216; Michx. Fl. ii. 118; Ell. Sk. ii. 390; Torr. & Gray, Fl. ii. 208. *S. grandiflora*, Raf. in Med. Rep. hex. 2, v. 359, & Desv. Jour. Bot. i. 226. — Dry and gravelly or sandy soil, Canada to the Saskatchewan, south to the upper part of Georgia, southwest to Texas and W. Colorado. Varies with smaller heads, looser inflorescence, and greener more scabrous leaves, in Texas, &c.

S. corymbósa, ELL. Stem and leaves (except their margins) quite smooth and glabrous, green: heads (3 to 5 lines long) in looser inflorescence: akenes short, turgid, 10-nerved: otherwise as in the preceding, of which it may be a glabrous variety. — Sk. ii. 378; Torr. & Gray, l. c.; not of Poir. Suppl. v. 461, which is a form of *S. Virgaurea*. — Upper and middle Georgia and Alabama; first coll. by *Mr. Jackson*: apparently also in Texas.

S. Ohioénsis, RIDDELL. Glabrous and smooth throughout: stem slender, 2 or 3 feet high: radical and lower cauline leaves lanceolate or elongated-oblong, 5 to 9 inches long, half-inch to an inch or more wide, attenuate at base, almost entire; upper lanceolate, sessile by a narrowed base: cyme fastigiate: heads pedicellate, small (3 lines long), narrow, 16-24-flowered: bracts of the involucre narrower: rays 6 to 9, small: akenes slightly 5-nerved.— Synop. 57; Torr. & Gray, l. c.— Low prairies or meadows, W. New York to Ohio and Indiana; first coll. by *Riddell.*

+ + Leaves somewhat conduplicate; lower slightly triplinerved.

S. Riddéllii, FRANK. Glabrous and smooth, or the inflorescence puberulent: stem a foot or two high, very leafy: leaves elongated-lanceolate, entire; radical 8 to 12 inches long, attenuate at both ends; cauline rather long, erect at the base which nearly sheathes the stem, partly conduplicate above, and the upper part falcately arcuate: heads densely cymose, 3 or 4 lines long, 20-30-flowered: rays 7 to 9, small and narrow: akenes faintly 5-nerved.— Riddell, Synops. l. c.; Torr. & Gray, Fl. ii. 210. *S. amplexicaulis*, Martens in Bull. Acad. Brux. viii. (1841) 68.— Wet prairies, Ohio (first coll. by *Riddell*) to Iowa and Missouri. (Also Fort Monroe, Virginia, *Vasey* and *Chickering*, these adventive?)

S. Houghtóni, TORR. & GRAY. Stem slender, 10 to 20 inches high: leaves indistinctly nerved, rather rigid, scattered (3 or 4 inches long, 2 to 4 lines wide): heads rather few in a corymbiform cyme, 20-30-flowered: rays 7 to 10, rather large: bracts of the involucre oblong-linear: akenes 4-5-nerved.— Gray, Man. ed. 1, 211, ed. 5, 242.— Swamps, north shore of L. Michigan, *Houghton.* Genessee Co., New York, *Paine.* Flowering early.

+ + + Leaves flat, smooth, and glabrous, linear or linear-lanceolate, entire, more or less triplinerved or 3-nerved, or nervose: heads only 3 or 4 lines long.

S. nítida, TORR. & GRAY. Stem 2 or 3 feet high, very smooth except the summit and inflorescence, which are minutely hirsute: leaves coriaceous and rigid, evidently nervose, punctate (the larger 4 to 6 inches long, 3 to 5 lines wide): heads numerous in the corymbiform cyme, about 14-flowered: rays 2 or 3, large: bracts of the involucre narrowly oblong: akenes 10-nerved.— Fl. ii. 210.— Dry pine woods and barrens, W. Louisiana and Texas; first coll. by *Drummond* and *Leavenworth.*

S. púmila, TORR. & GRAY. Dwarf, a span or more high, many-stemmed from a woody branching and cespitose caudex, glabrous throughout, punctate, somewhat resinous: leaves rigid, 3-nerved, acute; radical 2 or 3 inches long: cyme glomerate-fastigiate: heads narrowly oblong, 5-8-flowered: rays 1 to 3, short: involucral bracts rigid, somewhat carinate, and with small green (sometimes mucronulate) tips: mature akenes flattish and unusually broad, rather longer than the rigid pappus: akenes 5-nerved.— Fl. ii. 210. *Chrysoma pumila*, Nutt. in Trans. Am. Phil. Soc. vii. 325.— Rocky dry places, N. W. Texas to S. W. Utah, Nevada, and Idaho, mostly in the mountains; first coll. by *Nuttall.*

§ 2. EUTHÁMIA, Nutt. Receptacle of the flowers fimbrillate or the alveoli pilose: rays very small, almost always more numerous than the disk-flowers and never surpassing them in height: heads glomerately and fasciculately cymose, small: leaves very numerous, all linear, entire, 1-5-nerved, somewhat punctate, sessile: akenes villous-pubescent, short and turbinate: filiform rootstocks extensively creeping.— *Euthamia*, Cass. Dict. xxxvii. 471; Nutt. Trans. Am. Phil. Soc., l. c.

* Taller and paniculately branched Pacific species.

S. occidentális, NUTT. Stems 2 to 6 feet high: the branches terminated by small clusters of mostly pedicellate heads: leaves usually 3-nerved, glabrous and smooth even on the midrib, and margins obscurely scabrous: bracts of the involucre rather narrow: rays 16 to 20: disk-flowers 8 to 14.— Torr. & Gray, Fl. ii. 226; Eaton in Bot. King Exp. 156. *S. lanceolata*, Cham. & Schlecht. in Linn. vi. 502; Hook. Fl. ii. 6, partly. *Euthamia occidentalis*, Nutt. in Trans. Am. Phil. Soc. n. ser. vii. 326. *Aplopappus bacchuroides*, Benth. Bot. Sulph. 24.— Moist ground, British Columbia to S. California, extending eastward to New Mexico, Colorado, and Montana.— Long rootstocks tuberous-thickened at the extremity.

* * Comparatively low, a foot or at most a yard high, cymosely much branched above and flat-topped: heads mostly glomerate-sessile: Atlantic species.

S. lanceoláta, L. Leaves lanceolate-linear, distinctly 3-nerved and the larger with an additional outer pair of more delicate nerves, minutely scabrous-pubescent on the nerves beneath: outer bracts of the involucre ovate or oblong, and the inner linear: rays 15 to 20: disk-flowers 8 to 12. — Mant. 114; Ait. Kew. iii. 214; Torr. & Gray, Fl. ii. 226. *S. graminifolia*, Ell. Sk. ii. 391. *Chrysocoma graminifolia*, L. Spec. ii. 841. *Euthamia graminifolia*, Nutt. Gen. ii. 162 (subgen.), & Trans. Am. Phil. Soc. l. c. — Low ground, Canada to Georgia, and northwest to Montana.

— **S. tenuifólia,** PURSH. Lower (a foot or two high), slender, more resinous-atomiferous and glutinous, but glabrous: leaves all narrowly linear, one-nerved or with a pair of indistinct lateral nerves: heads smaller: rays 6 to 12: disk-flowers 5 or 6. — Fl. ii. 540; Ell. Sk. ii. 392; Torr. & Gray, l. c. *S. lanceolata*, var. *minor*, Michx. Fl. ii. 116. *Erigeron Carolinianum*, L. Spec., being *Virgaurea Carol.*, &c., Dill. Elth. 412, t. 306, f. 394. *Euthamia tenuifolia*, Nutt. l. c. — Sandy or gravelly and moist or dry ground, coast of New England to Florida and Texas.

— **S. leptocéphala,** TORR. & GRAY. A foot or two high, with more simple branches, wholly smooth and glabrous except the margin of the leaves; these with prominent midrib, very obscure lateral nerves, and no apparent veins: bracts of the involucre and the head narrower: rays 8 or 10: disk-flowers 3 or 4. — Fl. ii. 226. — Low ground, W. Louisiana and Texas; first coll. by *Leavenworth* and *Drummond*. Also, in a narrow-leaved form, N. W. Arkansas, *F. L. Harvey*.

§ 3. CHRYSÓMA, Torr. & Gray. Suffruticose: leaves fleshy-coriaceous, peculiarly areolate-venulose in the dried state: otherwise as § *Virgaurea.* — *Chrysoma,* Nutt., in part.

S. pauciflosculósa, MICHX. A foot or two high, much branched from the shrubby base, glabrous, somewhat viscid: leaves from spatulate-oblanceolate to linear, very obtuse, entire, an inch or two long and with a contracted petiole-like base, one-nerved or obscurely 3-nerved, not venose, but minutely and uniformly venulose, the impressed veinlets forming microscopic quadrate or roundish meshes over both surfaces: thyrsus somewhat corymbosely paniculate; the clusters only obscurely secund: heads 3 or 4 lines long: rays 1 to 3, rather long: disk-flowers 3 to 5, deep yellow: akenes pubescent; pappus brownish. — Fl. ii. 116; Torr. & Gray. Fl. ii. 224. *Chrysoma solidaginoides*, Nutt. in Jour. Acad. Philad. vii. 67, & Trans. Am. Phil. Soc. vii. 325. — Dry hills and sand-banks on the sea-shore, S. Carolina to Florida and Alabama; flowering late. (Bahamas.)

33. **BRACHYCHÆTA,** Torr. & Gray. (Βραχύς, short, χαίτη, bristle, from the very abbreviated setose pappus, which, with the cordate leaves, somewhat artificially distinguishes the genus from *Solidago*.) — Single species, flowering in late summer and autumn. — Fl. ii. 194.

— **B. cordáta,** TORR. & GRAY, l. c. Soft-pubescent: stems 2 or 3 feet high from a perennial root: leaves membranaceous, veiny, mostly acutely serrate; radical rather large, round-cordate, on long and nearly wingless petioles; cauline ovate, the lower on winged petioles: heads 2 or 3 lines long, narrow, solitary or fascicled in the racemiform and secund clusters or narrow thyrsus: bracts of the involucre with greenish tips, inner ones linear-oblong: flowers golden yellow, those of the disk and short ray each 4 or 5: pappus shorter than the akene and shorter than the proper tube of the corolla. — *Solidago sphacelata*, Raf. Am. Nat. (1820), 14. *S. cordata*, Short, Cat. Pl. Kentucky, Suppl. *Brachyris ovatifolia*, DC. Prodr. v. 313. — Open woods, &c., W. North Carolina and E. Kentucky to the upper part of Georgia; apparently first coll. by *Rafinesque*.

34. **LESSÍNGIA,** Cham. (Dedicated to the eminent German author, *G. E. Lessing*, and to his grand-nephews, *Karl Lessing* the painter, and *Christian Fr. Lessing*, author of Syn. Gen. Compositarum.) — Californian annuals or biennials, flocculent-woolly when young; with alternate leaves and rather small heads of flowers, either of the xanthic or cyanic series; the pappus becoming fuscous or rufous. Nerves of the corolla-lobes deeply intramarginal, the æstivation indu-

11

plicate up to the nerve. — Linnæa, iv. 203; Gray in Benth. Pl. Hartw. 315, Proc. Am. Acad. vii. 351, viii. 364, & Bot. Calif. i. 306. — Flowering spring and summer.

* Flowers yellow, sometimes purplish in age; some of the marginal ones with conspicuously larger and more or less irregular and radiatiform corolla: bracts of the involucre with herbaceous tips: akenes narrow, compressed, 2–3-nerved: style-branches truncate-obtuse, bearing a brush-like tuft of bristles, in which a minute or obscure setiform tip is partly or wholly hidden: heads about 3 lines high, terminating spreading slender branchlets.

L. Germanórum, CHAM. l. c. Low and diffusely spreading from the base, or procumbent, arachnoid-lanate with appressed white tomentum, glabrate with age; filiform flowering branches sparsely leafy or naked: lower leaves spatulate and usually pinnatifid or incised, with long tapering entire base; those of the branches becoming linear and entire, all narrowed at base: involucre hemispherical; its bracts with loose and foliaceous tips or the outer foliaceous, all glandless. — Torr. in Wilkes Exped. xvii. 326, t. 7 (style bad); Gray in Pl. Hartw. l. c., & Bot. Calif. 307, only in part. — Open dry ground, near San Francisco and in adjacent parts of California; first coll. by *Chamisso*. Corollas said by Chamisso to be "croceous."

L. glandulífera, GRAY. Diffusely much branched from an erect stem, more rigid, above glabrous or early glabrate: leaves more commonly entire, sometimes spinulose-dentate; those of the branches small and very numerous (3 to 1 lines long), or minute and almost covering flowering branchlets, ovate-lanceolate or oblong, thick and rigid, commonly beset along the margins with yellowish tack-shaped glands: involucre campanulate to turbinate; its bracts more appressed, the outer successively shorter, and some or all of them glanduliferous. — Proc. Am. Acad. xvii. 207. *L. Germanorum* in part, & *L. ramulosa*, var. *tenuis*, Gray, Bot. Calif. l. c., in part. — Arid grounds, from Monterey to San Diego, San Bernardino, &c.; common. The glands are like those of *Calycadenia* on a smaller scale, sometimes copious and strongly marked, sometimes few and inconspicuous.

* * Flowers purple or white; the corollas all alike and regular or nearly so: bracts of the involucre with appressed or erect tips: akenes less or hardly at all compressed, 4–5-nerved.

+— Stems slender and loosely branching, erect, a span to a foot or two high: white wool deciduous in age: leaves oblong to lanceolate or the lower spatulate, entire or sparingly dentate, the small upper with partly clasping or adnate base: involucral bracts mostly herbaceous-tipped.

L. ramulósa, GRAY, l. c. Somewhat granulose- or hirtellous-glandular on the glabrate branches and upper leaves, occasionally with some minute tack-shaped glands: stem usually stout at base: heads (3 or 4 lines long) terminating diffuse slender branchlets: involucre campanulate or somewhat turbinate, 10–20-flowered: corollas short (purple): style-appendages with minute setiform tip. — On dry hills, not rare through the northwestern part of California to Bay of San Francisco; first coll. by *Pickering* and *Brackenridge*.

Var. **ténuis**, GRAY. A slender and ambiguous form, not thickened at base of stem, low and diffuse, analogous to the depauperate states of the next species. — Bot. Calif. i. 307, as to pl. of Rothrock in Wheeler Rep. vi. 364. — Southeastern California, at head of Peru Creek, *Rothrock*.

L. leptóclada, GRAY. Glabrous after denudation of the floccose wool: stem slender (the taller forms 2 feet or more high, the most depauperate only 3 or 4 inches), and with long virgate or filiform branches bearing solitary or few heads: upper leaves commonly with sagittiform-adnate base: involucre turbinate, from 20-flowered down (in depauperate plants) to 5-flowered; its bracts in numerous ranks: corolla conspicuously exserted: style-appendages with a conspicuous subulate tip. — Proc. Am. Acad. vii. 351, & Bot. Calif. l. c. — Dry ground, common through the western and central parts of California, in very diverse forms; sometimes with numerous heads spicately crowded along the summit of the branches, and too nearly approaching the next.

L. virgáta, GRAY. More densely woolly: stem and virgate branches more rigid: upper leaves appressed, concave, carinately one-nerved: heads spicately sessile, each in the axil of a leaf of nearly the same length: involucre cylindrical, woolly, 5–7-flowered: style-branches with a conspicuous subulate tip. — Pl. Hartw. l. c.; Bot. Calif. l. c. — On the Sacramento, probably in the northern part of the State, *Pickering* and *Brackenridge*, *Newberry*.

++ Depressed or dwarf, flowering from the ground; inner bracts of involucre cartilaginous-aristate!

L. nána, GRAY, l. c. Usually stemless, a very woolly and pellet-like tuft from a slender root, an inch or two high, a cluster of sessile (half-inch long) heads, each surrounded by a rosulate cluster of spatulate or lanceolate leaves: involucre 10-12-flowered; its outer bracts linear-lanceolate, mucronate-acute or cuspidate, little herbaceous; inner ones pearly white, scarious-chartaceous, tapering into a rigid subulate acumination or awn which equals the flowers and very rufous pappus: akenes short and turgid: tip to the tufted style-appendages wanting. — Torr. in Wilkes Exped. xvii. 338. t. 7, poor. — Dry ground, foot-hills of the Sierra Nevada, from Siskiyou Co. to Kern Co., *Pickering, Fitch, Muir, Canby, Rothrock.*

Var. cauléscens. Leaves larger; radical ones much surpassing the sessile heads in their axils; also several developed stems, of an inch to 4 inches high, sparsely leaved, and bearing either solitary or 3 or 4 spicately disposed heads. — S. California, at Tehachipi Pass, *Parry.*

— 35. **BÉLLIS,** Tourn. DAISY. (Latin name, from *bellus*, pretty.) — Low herbs, of the northern hemisphere: the typical species perennial and stemless: radical leaves obovate; rays white, rose-colored, or purple. The akenes in the two perennial Mexican species, viz. *B. xanthocomoides (Brachycome,* Less.) and

— *B. Mexicana,* Gray. Pl. Wright. i. 93 (coll. Wright and Bourgeau), as also in our annual species, are less flat, and marginal nerves slender or less thickened, than in the Old World species — Benth. & Hook. Gen. ii. 265.

— B. PERÉNNIS, L., the common European DAISY, is escaping from cultivation and beginning to be spontaneous in a few places.

— **B. integrifólia,** MICHX. Annual, sparsely pilose-pubescent, diffusely branched and leafy, a span to a foot high: leaves spatulate-obovate and the upper narrower, entire: peduncles terminating the branches: bracts of the involucre ovate-lanceolate, acuminate, scarious-margined: rays half-inch or less in length, usually pale violet. — Fl. ii. 131; Hook. Bot. Mag. t. 3455; Torr. & Gray, Fl. ii. 189. *Eclipta integrifolia,* Spreng. Syst. iii. 602. *A. xanthium integrifolium,* Nutt. Trans. Am. Phil. Soc. ser. 2, vii. 312. — Low grounds, Kentucky to Arkansas and Texas; fl. spring and summer.

36. **APHANÓSTEPHUS,** DC. ('Aφανής, vanishing or inconspicuous, and στέφος, crown; from the pappus.) — Texano-Mexican annuals or biennials, sometimes perhaps of longer duration, pubescent, leafy-stemmed and branching; with rather showy heads, resembling those of DAISY, on solitary peduncles terminating the branches, and nodding before anthesis: leaves from entire to pinnately lobed: rays from white to violet-purple: akenes almost or quite glabrous. Fl. summer. — Gray, Pl. Wright. i. 93; Benth. & Hook. Gen. ii. 262; Gray, Proc. Am. Acad. xvi. 80. *Aphanostephus, Keerlia* (excl. one species, which is a *Xanthocephalum*), & *Leucopsidium,* DC. Prodr. v. 309, 310, vi. 43.

* Pappus a very short crown with a ciliate-fringed edge, which commonly is obsolete in age: base of the corolla-tube seldom thickened.

A. Arizónicus, GRAY. Erect, a foot high, minutely soft-pubescent, not cinereous: upper leaves linear and entire; lower linear-spatulate, 3-5-lobed or laciniate: heads small, on at length clavate-thickened peduncles: akenes narrow, terete, evenly striate with about 10 narrow ribs. — Proc. Am. Acad. xvi. 81. *A. ramosissimus,* Rothrock in Wheeler Rep. vi. 147. — Arizona, on the Gila River, *Rothrock.*

— **A. ramosíssimus,** DC. Erect or at length diffuse, slender, a foot or less high, hispidulous-pubescent: upper leaves linear or lanceolate, entire or few-toothed; lower laciniate-pinnatifid or incised: heads on slender peduncles: rays 3 to 5 lines long: akenes almost terete and even, the ribs or nerves few and mostly obscure, except on some outermost. — Prodr. v. 310; Gray, Pl. Wright. l. c.; Torr. in Marcy Rep. t. 9. *A. Riddellii,* Torr. & Gray, Fl. ii. 189. *A. pilosus,* Buckley in Proc. Acad. Philad., a remarkably hispid form. *Egletes* *ramosissima*, Gray, Pl. Fendl. 71, & Pl. Lindh. ii. 220. — Rocky and sandy prairies, Texas. (Adjacent Mex.)

A. húmilis, GRAY, l. c. Low and diffuse, soft-pubescent and cinereous: leaves rarely entire, often pinnatifid: heads on slender peduncles: rays 3 or 4 lines long: akenes shorter and more distinctly costate-angulate. — *Leucopsidium humile*, Benth. Pl. Hartw. 18. *Egletes humilis*, Gray, Pl. Fendl. 71. — Southern and western borders of Texas, *Wright*, *Palmer* (but his plant, no. 494, doubtful), *Reverchon*. (Mex.)

A. RAMÓSUS, Gray, Proc. Am. Acad. xvi. 90 (*Keerlia ramosa*, DC.), Mexico, *Keerl*, is imperfectly known.

* * * Pappus more conspicuous and dentate or laciniate: base of the corolla-tube in age prominently thickened and indurated, long persistent on the strongly angulate-costate akene.

A. Arkansánus, GRAY, l. c. A foot high, cinereous-pubescent: leaves from oblong-spatulate to broadly lanceolate; lower often toothed or sinuate-lobed: heads larger: rays commonly half-inch long: outer akenes usually suberose-angled or ribbed: pappus mostly obtusely 4-5-lobed or pluridentate. — *Leucopsidium Arkansanum*, DC. Prodr. vi. 43. *Keerlia skirrobasis*, DC. Prodr. v. 310; Deless. Ic. iv. t. 18; Hook. Ic. t. 240. *Egletes Arkansana*, Nutt. Trans. Am. Phil. Soc. vii. 394; Torr. & Gray, Fl. ii. 411. — Plains of Arkansas, Kansas, and Texas; first coll. by *Berlandier*.

Var. Hállii, GRAY, l. c. Somewhat smaller: leaves varying from entire to pinnately parted: crown of the pappus more conspicuous, deeply cleft into 4 or 5 unequal subulate-acuminate lobes! — Texas, *E. Hall* (no. 303, 304), *Palmer*.

37. GREENÉLLA, Gray. (*Rev. Edward Lee Greene*, the discoverer.)
— Slender and low winter annuals; the typical species (analogous to *Gutierrezia*) diffuse and conspicuously radiate; an ambiguous species rayless, and perhaps not truly congeneric. — Proc. Am. Acad. xvi. 81.

G. Arizónica, GRAY, l. c. Smooth and glabrous, diffusely branched from the base: leaves small (inch or less long), entire, veinless, sessile, alternate; radical ones lanceolate or obscurely spatulate, hispidulous-ciliolate; cauline narrowly linear and gradually reduced to subulate: heads solitary at summit of divergent filiform branchlets: involucre 2 or 3 lines high and wide; bracts with a conspicuous subapical green spot: rays 10 to 16, oblong or obovate, white: mature akenes densely white-villous, the hairs tipped with a capitellate gland: border of the pappus-crown multisetulose-dissected. — Mesas of Arizona, *Greene* (1877), *Lemmon*, *Pringle*. The root obviously not perennial.

G. discoidea, GRAY. Stems or branches numerous from a probably monocarpic but lignescent root, strict, very leafy: leaves all narrowly linear, acute; the lower (over an inch long) with obscurely ciliolate-scabrous margins: heads somewhat corymbose: involucre barely 2 lines high; the bracts more scarious and with indistinct green spot: rays none: ovaries glabrous; pappus pluridenticulate. — Proc. Am. Acad. xix. 2. — S. Arizona, in Tanner's Cañon, *Lemmon*.

38. KEÉRLIA, Gray. (*F. W. Keerl*, a German traveller in Mexico.) —
Diffusely and slenderly branched Texan herbs, leafy-stemmed; with small paniculate heads on almost capillary peduncles, white or purple rays, and oblong entire sessile leaves; the style-appendages in one species much elongated (in the manner of the preceding genus), and this has only sterile ovaries in the disk. — Pl. Lindh. ii. 220, & Pl. Wright. i. 92, not DC., whose genus of this name was founded on two species of *Aphanostephus* and a *Xanthocephalum*, to which was added a synonyme belonging to a *Bellis*.

K. bellidifólia, GRAY & ENGELM. Annual, pubescent, effusely branched from near the base, a span or two high; when young with the habit of *Bellis integrifolia*: lower leaves obovate or spatulate; uppermost somewhat linear: involucre only 2 lines long: rays 4 to 15, blue: style-appendages in the disk-flowers short and very obtuse: akenes obovate-clavate and moderately compressed. — Proc. Am. Acad. i. 47; Pl. Lindh. l. c.; Pl. Wright, l. c. — Fertile soil, Texas, *Lindheimer*, *Wright*.

K. effúsa, Gray. Perennial, often 2 feet high, with simple stem branching above into an effuse ample panicle: leaves (an inch or less long) bispid as well as the stem, rigid and scabrous, oblong, mostly with a broad sessile base: heads very numerous: involucre more turbinate: rays 4 to 7, white: disk-flowers somewhat more numerous, apparently always sterile, and with elongated linear-lanceolate style-appendages: fertile akenes obovate, flat, callous-nerved at the margins (or with one margin 2-nerved). — Pl. Lindh. ii. 221; Pl. Wright. i. 93. — Hillsides, central parts of Texas, *Berlandier*, *Lindheimer*.

39. CHÆTOPÁPPA, DC. (Χαίτη, bristle, and πάππος, pappus.) — Low
and small Texano-Mexican winter annuals, diffusely branched; the branches terminated by small heads: rays white or purple: leaves entire, the lower spatulate, upper gradually becoming linear or reduced to subulate bracts. Fl. spring and early summer. — *Chætanthera*, Nutt. Jour. Acad. Philad. vii. 111. *Chætophora*, Nutt. in DC. *Chætopappa* & *Distasis*, DC. Prodr. v. 301, 279; Benth. & Hook. Gen. ii. 268. *Diplostelma*, Gray. Pl. Fendl. 72.

C. asteroídes, DC. l. c. Slender, 2 to 10 inches high, pubescent: involucre (2 lines long) rather narrow, of 12 to 14 bracts: rays 5 to 12: disk-flowers 8 to 12: style-appendages very obtuse: akenes slender, little compressed, obscurely few-nerved, pubescent, all the central ones sterile and often awnless: paleæ of the pappus very thin and hyaline, narrowly oblong, not rarely lacerate or cleft. — Torr. & Gray, Fl. ii. 187. *Chætanthera asteroides*, Nutt. l. c. — Dry ground, Texas to Arkansas and the borders of Missouri. (Adjacent Mex.)

Var. imbérbis, Gray. Awns of the pappus wanting in all the flowers: the paleæ rather broader and sometimes coroniform-concreted. — Proc. Am. Acad. xvi. 82. — E. Texas, *Wright*.

C. Párryi, Gray. More rigid, 9 inches or more high: leaves subcoriaceous, hispidulous and glabrate: involucre (3 lines long) turbinate: rays 6 or 7: style-appendages short and very obtuse: akenes quite glabrous; the fertile ones fusiform and somewhat compressed, 4-nerved, with a pappus of 4 or 5 firmer and cuneiform-quadrate paleæ which are laciniately fimbriate at the truncate apex, and of few or sometimes solitary more delicate awns, these occasionally little longer than the paleæ, sometimes wanting; disk-akenes mostly inane and awnless. — Proc. Am. Acad. xvi. 82. *Distasis modesta*, var., Gray, Bot. Mex. Bound. 78. — Mt. Carmel, on the Rio Grande, between Texas and Mexico, *Parry*.

C. modésta, Gray, l. c. Less slender and pubescence more hirsute than in *C. asteroides*: involucre broadly campanulate: its bracts obtuser and more numerous: rays 9 to 20: disk-flowers 40 to 60, all but the central fertile; their style-appendages narrower and acutish: akenes oblong or linear, much compressed, pubescent when young, with merely marginal nerves or occasionally a facial one, only the central ones sterile: pappus of 5 oblong erose-truncate at length subcoriaceous paleæ, alternating with as many rather rigid awns. — *Distasis modesta*, DC. Prodr. v. 279. *Diplostelma bellioides*, Gray, Pl. Fendl. 73 — Dry ground, Texas, *Berlandier*, *Wright*, &c. (Adjacent Mex.)

Dístasis ? heterophýlla, Hemsl. Biol. Centr.-Am. Bot. ii. 119, of Mexico, is hardly of this genus, probably not of the tribe.

40. MONÓPTILON, Torr. & Gray. (Μόνος, single, πτίλον, feather, al-
luding to the solitary plumose bristle of the pappus.) — Jour. Bost. Nat. Hist. Soc. v. 106, t. 13; Benth. & Hook. Gen. ii. 307; Gray, Bot. Calif. i. 306. — Single species.

M. bellidifórme, Torr. & Gray, l. c. A small but pretty annual, much branched from the very base, depressed, villous-hirsute: heads terminating the numerous leafy branchlets, half-inch in diameter, inclusive of the white or violet-purple rays: leaves small, spatulate or linear-spatulate, the uppermost involucrate around the head. — Arid or desert plains, S. E. California to S. W. Utah, *Fremont*, *Parry*, *Palmer*, *Parish*.

41. DICHÆTÓPHORA, Gray. (Δίς, χαίτη, φορά, bearing two bristles,
i. e. pappus-awns.) — Pl. Fendl. 73. — Single species; in Benth. & Hook. Gen.

ii. 209, referred (along with a species of *Perityle* and an *Achætogeron*) to a section of *Boltonia*.

D. campéstris, GRAY. A small and Daisy-like winter annual, at first acaulescent with a scapiform peduncle (1 to 3 inches high), at length with leafy branches terminated by a slender monocephalous peduncle: leaves spatulate, entire, somewhat hirsute: head 2 or 3 lines high, the ovate disk soon surpassing the involucre; rays 16 to 20, apparently white or rose-color. — Pl. Fendl. 73, perhaps excl. syn. *Brachycome? xanthocomoides*, Torr. & Gray, Fl. ii 190, the specimen of which is too young for determination. — Southern borders of Texas, *Berlandier* (no. 1465, specimen too young), *Havard*, in fruit. (Adj. Mex., *Gregg, Palmer*.)

42. **BOLTÓNIA**, L'Her. (*James Bolton*, an English botanical author.) — Perennial and leafy-stemmed herbs (wholly of the United States), Aster-like, glabrous, glaucescent, mostly tall; with striate-angled stems, entire sessile leaves commonly becoming vertical by a twist at base, rarely decurrent; and with rather showy heads; the numerous rays white, purplish, or violet; fl. autumn. — Sert. Angl. 27 (with figures cited which were never published); DC. Prodr. v. 301; Benth. & Hook. Gen. ii. 209, excl. § *Asteromœa*, Blume, which passes into *Calimeris*, and also § 3, which is a mixture. Wings of the akene broadish and thin, narrow and thickish, or obsolete in the same species, or even in the same head.

* Stems (2 to 7 feet high) paniculately much branched and slender: heads small; the disk only about 2 lines high and wide.

B. diffúsa, ELL. Lower leaves lanceolate; upper linear, those of the loose and almost filiform flowering branches or branchlets becoming linear-subulate and minute: rays mostly white, barely 2 lines long involucre as in the next, but the bracts more numerous and unequal. — Sk ii. 400; Hook. Comp. Bot. Mag. i. 97; DC. l. c. & Torr & Gray, l. c., excl. syn. Bot Mag. — Low grounds, South Carolina to Texas and along the Mississippi region north to Illinois.

* * Stems (2 to 8 feet high) simple and more cymose-paniculate at summit, leaves broadly lanceolate or the uppermost linear-lanceolate; heads short-peduncled, larger; the disk in fruit a third to half an inch in diameter; rays 4 to 6 lines long.

B. asteroídes, L'HER. Bracts of the involucre lanceolate, acute, mostly greenish · rays from white to purplish or pale violet-color· setulose squamellæ of the pappus mostly numerous and conspicuous, the two awns sometimes wanting or obsolete, more commonly present and little shorter than the akene. — *Matricaria asteroides*, L Mant. 116. *M. glastifolia*, Hill, Hort. Kew. 19, t. 3. *Chrysanthemum Carolinianum*, Walt. Car. 204. *Boltonia glastifolia & B. asteroides*, L'Her. l. c.; Michx Fl. ii. 132; Willd. Spec. iii. 2162, Sims, Bot. Mag t. 2381 & 2554; DC. l. c. — Moist or wet ground along streams, Pennsylvania to Illinois and Florida. The awnless form (*B. asteroides*) is not constant to this character, but is commonly smaller, and with fewer and smaller heads.

Var. **decúrrens**, ENGELM in herb. A large form (in cultivation 7 or 8 feet high), with leaves alate-decurrent on the stem and even the branches; the wings sometimes ending below in a free and subulate point: pappus-awns slender. — Missouri, *Eggert*.

B. latisquáma, GRAY. Heads rather larger and more showy rays blue-violet. bracts of the involucre oblong to ovate, obtuse or mucronate-apiculate, awns of the pappus uniformly present and conspicuous, the setulose squamellæ small — Am Jour. Sci. ser 2, xxxiii 238. — Kansas and W. Missouri, near the mouth of the Kansas River, *Parry* Now not rare in cultivation, the handsomest species.

Var. **occidentális**. Heads rather smaller: rays white. — River-bottoms of Union Co., Eastern Oregon, *Cusick*.

43. **TOWNSÉNDIA**, Hook. (*David Townsend*, botanical associate of Dr. Darlington of Penn.) — Depressed or low many-stemmed herbs (of the Rocky Mountains); with from linear to spatulate entire leaves, and comparatively large heads, resembling those of *Aster*; the numerous rays from violet or rose-

purple to white; fl. from early spring to summer. Akene commonly beset with bristly "duplex" hairs, having a forked or glochidiate-capitellate apex. Involucral bracts mostly ciliate. — Fl. ii. 16. t. 119; Torr. & Gray, Fl. ii. 185; Gray, Proc. Am. Acad. xvi. 82. For structure of the achenial hairs, see Macloskie in Proc. Am. Nat. xvii. 31, xviii. 1102.

* Bracts of the involucre conspicuously attenuate-acuminate: head large; the involucre half-inch or more high, and rays half-inch long: fl. summer.
+ Caulescent biennials or annuals, somewhat hirsute-pubescent, but the foliage at length glabrate: involucre naked; its bracts from lanceolate to ovate-lanceolate: rays showy, bright blue or violet. (Pappus of the first species anomalous!)

T. exímia, GRAY. Stems erect, simple or sparingly branching, 6 to 14 inches high: leaves spatulate or the upper lanceolate: head sparingly leafy-bracted or naked at base: involucral bracts ovate-lanceolate and somewhat rigidly cuspidate-acuminate, whitish-scarious with green centre: akenes broadly obovate, almost cartilaginous, glabrate (sprinkled with a few short and obscure glochidiate-tipped hairs): pappus wholly persistent, of 2 subulate at length corneous stout awns which are rather shorter than the akene (sometimes wanting in the ray), and a circle of rigid squamellæ which are mostly coroniform-concreted at base and rigid in age. — Pl. Fendl. 70; Pacif. R. R. Exp. iv. 98; Proc. Am. Acad. l. c. 83. — Mountain sides, New Mexico and adjacent part of Colorado, *Fendler, Bigelow,* &c.

T. grandiflóra, NUTT. Stems spreading from the base, sometimes divergently branched above, a span or two high: upper leaves often linear, 2 or more uppermost subtending the head, involucre nearly of the preceding: akenes narrowly obovate, sprinkled with glochidiate-capitellate hairs: pappus in the ray reduced to a crown of short squamellæ, in the manner of the genus, and of the disk plurisetose and longer than the akene. — Trans. Am. Phil. Soc. n. ser. vii. 306; Torr. & Gray, l. c. — Plains and hills, Wyoming and W. Nebraska to the borders of New Mexico, first coll. by *James* and *Nuttall*.

T. Párryi, EATON. Stems erect, simple, stout, naked and pedunculiform above, 2 to 6 inches high (the taller forms sometimes branching): leaves mostly spatulate: bracts of the very broad involucre lanceolate, thinner, with softer and less attenuate tips, or the outer barely acuminate: akenes narrowly obovate, canescently pubescent, the hairs acute and simple or many of them 1-2-dentate at tip: pappus of the ray plurisetose like that of the disk, or somewhat more scanty, rays "blue" or violet. — Am. Naturalist, viii. 212; Gray, Proc. Am. Acad. xvi. l. c. — Wyoming, Montana, and E. Idaho, *Hayden, Parry,* &c.

Var. **alpína,** GRAY, l c. A dwarf and alpine form, more pubescent and cinereous: leaves very small, at most half-inch long: flowering stem about the same length or hardly any: involucral bracts less pointed: "rays pink." — Wyoming on the high divide between Stinking Water and the Yellowstone (confounded with *T spathulata*), *Parry*.

+ + Depressed-stemless and monocephalous perennial.

T. condensáta, PARRY. Very lanuginous with long and soft arachnoid hairs, the spatulate obovate leaves (with blade 2 or 3 lines long and tapering into a very much longer petiole) rosulate-crowded around the large and broad sessile head, the whole forming a globular or hemispherical woolly tuft, an inch and a half high and surmounting a slender stoloniform caudex: bracts of the involucre linear and soft, with a weak attenuate apex, all nearly equal in length: rays 100 or more, narrow: disk-flowers also very numerous: pappus of ray and disk similarly and slenderly plurisetose and long. — Am. Nat. viii. 213 (description by Eaton). — Wyoming, on a high alpine peak of the Owl Creek range, July, *J. D. Putnam*.

* * Bracts of the involucre not prominently if all acuminate: heads mostly smaller or narrower: pappus of the disk and often of the ray plurisetose.
+ Hairs on the akene mostly copious and slender, some simple, others bifid or bi- (rarely tri-) dentate at the apex, the teeth or lobes ascending or merely spreading and usually acute: heads middle-sized, more or less naked-pedunculate, the pink or rarely white rays and the involucre each from a third to barely half an inch long, bracts of the latter few-ranked: annuals or biennials. (The most western species in range.)
++ Pappus of the ray like that of the disk, but somewhat shorter.

T. flórifer, GRAY. A span or more high, cinereous-hirsute: stems rather slender from an annual root, leafy, leaves linear or the lowest lanceolate-spatulate, acute, mostly apiculate-

acuminate : involucral bracts linear-lanceolate, little unequal. — Proc. Am. Acad. xvi. 84.
T. strigosa, Gray, in Wilkes Exped. xvii. 344, not Nutt. *Erigeron florifer*, Hook. Fl. ii. 20.
Aplopappus florifer, Hook. & Arn. Bot. Beech. 351. *Stenotus florifer*, Torr. & Gray, Fl.
ii. 238. — Sandy banks of the Columbia River and its tributaries, east of the Cascades, Montana to Washington Terr. and Oregon ; first coll. by *Douglas*.

T. scapígera, Eaton. Low (2 to 4 inches high), hirsutely pubescent : heads on scapiform 1-2-leaved stems : radical leaves spatulate (often broad and short, with a long narrowed base or petiole) : involucral bracts rather broadly lanceolate. — Bot. King Exp. 145, t. 17. *Aplopappus florifer*, var., Hook. & Arn. Bot. Beech. 351 ? — Rocky ridges in the mountains, Nevada and Modoc Co., N. E. California, *Watson, Lemmon, Mrs. Austin*. Flowering early : a winter annual or biennial.

 Var. cauléscens, Eaton, l. c. A summer form, more slender and sparingly leafy-stemmed, with rather smaller heads. — Nevada, in Monitor Valley, *Watson*.

 Var. ambígua, Gray, l c. More leafy-stemmed from a slender root, fully a span high : rays white : pappus of the ray sometimes little more than half the length of that of the disk. — Rabbit Valley, Utah, at 7,000 feet, *Ward*.

++ ++ Pappus of the ray setose-squamellate, shorter than the breadth of the akene.

T. Watsóni, Gray, l. c. Somewhat cinereous with a close short pubescence : stems 4 to 7 inches high from a slender root, spreading, nearly all branching above and sparsely leafy, therefore bearing numerous short-pedunculate heads : leaves narrowly spatulate and oblanceolate : involucral bracts oblong-lanceolate : hairs of the akene rather shorter and obtuse or at length 2-3-dentate at tip. — *T. strigosa*, Eaton, Bot. King Exp. 145, not Nutt. — Great Salt Lake, Utah, on the shore of Stanbury Island, June, *Watson*.

+ + Hairs on the akene, or most of them, glochidiate-capitellate, i. e. obtusely bidentate at apex, and the apparently somewhat glandular lobes recurved or revolute, thus appearing to be minutely depressed-capitate under a lens.

++ Head large, three-fourths to a full inch long (without the rays) : plants green and glabrate, depressed-acaulescent from a perennial root, with habit of *T. sericea :* leaves large, much surpassing the head, minutely sericeous-pubescent when young, in age with only some ciliate or other hairs toward the attenuate petiole-like base, plane and coriaceous : involucre well imbricated.

T. Wilcoxiána, Woon. Leaves linear-spatulate, elongated (1 to 3 inches long including the petiole-like base) : head mostly solitary, sometimes an inch long, short-pedunculed or subsessile : bracts of the involucre lanceolate or the inner linear, barely acutish : ray and disk-pappus of similar slender and elongated bristles. — Bull. Torr. Club, vi. 163, & Bot. Gazette, iii. 50. — Colorado, in the San Luis Mountains, *E. K. Smith*. Indian Territory, *Dr. Wilcox*. Patagonia Mountains, Arizona, *Lemmon*.

T. Rothróckii, Gray. Leaves more broadly spatulate and shorter (inch or less long), rosulate around the solitary head which is closely sessile at the surface of the ground, or at length with one or two additional heads and tufts from the same crown : involucre shorter and broader ; its bracts oblong, mostly obtuse : ray-pappus of squamellate bristles not longer than the breadth of the akene, or with one or two more elongated. — Rothrock, in Wheeler Rep. vi. 148, t. 7 ; Proc. Am. Acad. xvi. 85. — Mountains of South Park, Colorado, in the alpine district, at 13,500 feet, *Rothrock, J. D. Allen*.

++ ++ Heads from three fourths down to one third of an inch long, sessile, or rarely on a very short naked peduncle : plants sericeous- or strigulose-pubescent, depressed-acaulescent or low-caulescent : involucre well imbricated : ray-pappus mostly plurisetose.

— **T. serícea,** Hook. Depressed-acaulescent perennial, with closely sessile solitary or few heads on the crown next the ground, surrounded and surpassed or equalled by the linear or linear-spatulate leaves, at length multicipital and pulvinate-tufted, an inch or two high : head an inch or less long : involucral bracts narrowly lanceolate, mostly acute : rays white or purplish-tinged : pappus of the ray plurisetose like that of the disk (forma *papposa*, Gray, Pl. Fendl.), or of fewer but similar bristles, or (in the northern part of its range) with most of the bristles short and aristiform, and even reduced to squamellæ little longer than the width of the akene.—Fl. ii. 16, t. 119 ; Torr. & Gray, Fl. ii. 185 ; Gray, Pl. Fendl. 69 ; Meehan, Nat. Flowers, ser. 2, i. t. 47 ; Gray, Proc. l. c. 85. *Aster? exscapus*, Richards. Appx. Frankl. Journ. 32. — On dry hills, plains, or mountains, Saskatchewan to Rocky Mountains in lat. 54°, thence south to New Mexico and Arizona ; fl. April and May. — Varies from large-headed, comparatively broad-leaved, and glabrate forms (which may almost pass into the two preceding species), to a narrowly leaved and more sericeous form with head barely two-thirds inch long, as in the original northern specimens (both grow together in Colorado, "the latter flowering two weeks later"), or sometimes even yet more reduced, so that the heads are barely half-inch long.

Var. **leptótes**, Gray, is an ambiguous form from Middle Park, Colorado (*Parry*), with heads less than half-inch long, and all but the primary ones somewhat distinctly pedunculate: leaves narrowly linear with attenuate base. Perhaps a distinct species.

T. Arizónica, Gray. Depressed subcaulescent and multicipital, or branching from a perennial root, forming a lax pulvinate tuft of 2 or 3 inches high, minutely sericeous-canescent: leaves spatulate, short (about half-inch long), seldom surpassing the barely sessile and mostly foliose-fulcrate hemispherical heads (these merely half-inch high): bracts of the involucre lanceolate, mostly obtuse: pappus of ray and disk alike and of equal length, rather rigid, about the length of the akene (2 or 3 lines long). — Proc. Am. Acad. xvi. 85. — Arizona, adjacent Utah (*Palmer*), and N. W. New Mexico, *Coues & Palmer*, *M. E. Jones*, *Matthews*.

T. incána, Nutt. Depressed-caulescent or subcaulescent from a winter annual or perennial root, an inch to a span high, branching, strigulose-cinereous or canescent: leaves from narrowly spatulate to almost linear; uppermost fulcrate around the sessile (about half-inch) heads and seldom surpassing them: involucral bracts more sericeous and ciliate and less obtuse than in the foregoing: pappus of the ray from a third to half the length of that of the disk. — Trans. Am. Phil. Soc. l. c.   *T. Fremontii*, Torr. & Gray, in Jour. Bost. Nat. Hist. Soc. v. 108, where the heads are wrongly said to be larger than those of *T. sericea*. — Mountains of Wyoming to S. Utah and the borders of S. Nevada; first coll. by *Nuttall*.

++ ++ ++ Heads about one-third inch long, sessile among the rosulate leaves: herbage soft-lanate: pappus deciduous in a ring! — § *Crophorus*, Nutt.

T. spathuláta, Nutt. Depressed and multicipital from a slender perennial root, forming a tuft an inch or so high: leaves crowded, spatulate, densely villous-lanate; the upper about equalling the heads: bracts of the involucre oblong-lanceolate, acute: rays rather short, pinkish: pappus of ray and disk similar and of the same length, of slender bristles. — Trans. Am. Phil. Soc. l. c. — Rocky Mountains in Wyoming; on the Black Hills of the Platte, *Nuttall*, and Wind-River Mountains, *Parry*.

++ ++ ++ Heads small, only a quarter-inch high (exclusive of the rays), mostly short-pedunculate, hemispherical: involucre of few-ranked broadly lanceolate and barely acute bracts: caulescent and branching (at least in age) and summer-flowering: pappus of the (sometimes infertile but feminine) ray shorter, commonly setose-squamellate.

= Green and glabrate, perennial.

T. glabélla, Gray. An inch or two high from a slender rootstock, nearly simple, sparsely pilose-pubescent when young: leaves thickish, soon glabrous, spatulate (an inch or less long, including the usually slender petiole); the uppermost usually surpassed by the slender and naked (sometimes inch-long) peduncle: involucre glabrous: pappus of the ray in one specimen plurisetose and nearly half the length of that of the disk, in another reduced to short squamellæ. — Proc. Am. Acad. xvi. 86. — Pagosa, S. W. Colorado, *Newberry*.

= = Cinereous with fine and close somewhat strigulose pubescence, flowering from near the ground at first, but becoming taller (4 to 10 inches high) and loosely branching: pappus of ray-akenes always reduced to a crown of short squamellæ, with rarely one or two short bristles. (Species hardly distinct.) — § *Nanodia*, Nutt.

T. Féndleri, Gray. Root slender, but apparently perennial: leaves linear: bracts of the involucre unequal, in about 3 ranks, acute. — Pl. Fendl 70. & Proc. Am. Acad. l. c. — Gravelly hills, New Mexico and S. Colorado, fl. May to Sept.; first coll. by *Fendler*.

T. strigósa, Nutt. Winter annual, with slender root, flowering when only half-inch high, often attaining a span in height: early leaves spatulate; later ones linear: heads rather smaller: bracts of the involucre broader, acutish, in about 2 ranks, the outer shorter. — Trans. Am. Phil. Soc. l. c.; Gray, Pl. Fendl. l. c. — Gravelly hills and plains, Wyoming to New Mexico and Arizona; first coll. by *Nuttall*.

T. Mexicána, Gray, Pl. Fendl. 70 (from about Saltillo, &c., Mexico, *Gregg*, *Parry*, *Palmer*, and from farther south, *Galeotti*), differs slightly from the last in having the two ranks of involucral bracts of equal length and all very obtuse.

44. CORETHRÓGYNE, DC. (Κόρηθρον, γυνή, besom-style, from the brush-like tuft of bristles on the style-appendages.) — Rather low and Aster-like Californian perennials, whitened, at least when young, with cottony tomentum; the stem or branches terminated by solitary somewhat large and showy heads: rays violet-blue or purple: disk yellow, often changing to purplish: pappus tawny or ferruginous: peduncles, with the bracts, &c., usually glandular under the wool: leaves sessile, entire, or merely serrate. Fl. summer. — Prodr. v. 215; Nutt. in Trans. Am. Phil. Soc. vii. 290; Torr. & Gray, Fl. ii. 97; Gray, Bot. Mex. Bound. 76, Proc. Am. Acad. vii. 351, & Bot. Calif. i. 320.

* Heads pretty large and broad, disposed to be solitary, terminating a simple stem or simple branches: involucre hemispherical, half-inch or more in diameter; its bracts little unequal and outer ones largely herbaceous: style-appendages strongly comose.

C. obováta, BENTH. Stems decumbent from spreading rootstocks, a foot or two long: leaves obovate or spatulate, obtuse, sparsely serrate or dentate above; those of the ascending branches small, from oblong to linear-lanceolate: rays violet, varying to white suffused with pink. — Bot. Sulph. 23. *C. spathulata*, Gray, Proc. Am. Acad. vii. 317, & Bot. Calif. l. c. — Coast of California, from Bodegas (where it was first coll. by *Hinds*) to Humboldt Co., *Bolander, Kellogg*, &c.

C. Califórnica, DC. Stems erect or ascending: leaves linear and entire, or the lowest lanceolate-spatulate and few-toothed: sometimes a few bracts on the receptacle, like the innermost of the involucre, subtending outer flowers: rays violet or purple. — Prodr. v. 215; Gray, Bot. Calif. i. 321. *C. Californica* & *C. incana* (the common state, with no bracts on the receptacle), Nutt. l. c.; Torr. & Gray, l. c. — California, along and near the coast, Monterey to San Diego; first coll. by *Douglas*.

* * Heads smaller, solitary and terminating the branches, or often more numerous and loosely paniculate: involucre camp nulate or broadly turbinate, much imbricated; the bracts mainly appressed, outer ones successively shorter; all with short green tips: style-appendages scantily comose.

C. filaginifólia, NUTT. Stems slender, erect or ascending, a foot or two high, commonly bearing few or (when depauperate) even solitary heads: leaves oblanceolate-spatulate and few-toothed or entire; upper often linear or reduced and bract-like on the branchlets; the white tomentum usually persistent, or when deciduous the branchlets and involucre little if at all glandular: rays violet. — Trans. Am. Phil. Soc. vii. 290; Gray, Bot. Mex. Bound. 76, & Bot. Calif. l. c. *Aster? filaginifolius* & *A. tomentellus*, Hook. & Arn. Bot. Beech. 146. *Diplopappus leucophyllus*, Lindl. in DC. v. 278. *Aplopappus (Pyrochaeta) Hankei*, DC. l. c. 349. *Corethrogyne filaginifolia* & *C. tomentella*, Torr. & Gray, Fl. ii. 99. — Common through the western and southern parts of California from Monterey southward; flowering at almost all seasons, varying greatly. The following are the more extreme forms.

—— **Var. virgáta**, GRAY. Slender, becoming glabrate and greener in age, often bearing numerous heads in a very open panicle: involucre and naked branchlets disposed to be glandular-viscidulous. — Bot. Calif. l. c. *C. virgata*, Benth. Bot. Sulph. 23. — Common from Monterey southward; first coll. by *Hinds* and by *Fremont*.

—— **Var. rígida**. A stouter and rigid form, either very white-tomentose or in age glabrate, then viscidulous-glandular: leaves from spatulate-lanceolate to oval or obovate: heads commonly numerous and paniculate. — *C. incana*, var. *rigida*, &c., Benth. Pl. Hartw. 316. *C. tomentella*, Durand, Pacif. R. Rep. v. App. 8. *C. filaginifolia*, var. *tomentella*, Gray, Bot. Calif. l. c., in part. — Dry and open ground, Monterey to San Bernardino Co.

C. DETÓNSA, Greene in Bull. Torr. Club, x. 41, of uncertain origin, probably Lower California, may be a form of the last variety.

45. PSILÁCTIS, Gray. (Ψιλός, naked, ἀκτίς, ray; no pappus to ray-flowers.) — Texano-Mexican annuals, minutely pubescent, or glandular, or glabrate; with slender and loosely paniculate-branching stems, pinnatifid or incised lower leaves tapering into a petiole, and narrow often entire upper ones; the heads small, terminating the branches, with violet or purplish or white rays; these usually infertile: fl. summer. — Pl. Fendl. 71; Benth. & Hook. Gen. ii. 269.

P. Coúlteri, GRAY, l. c. Branched from near the base, glabrous or obscurely hispidulous-puberulent, or the rigid spreading flowering branchlets granulose-glandular: leaves all short, rigid, mostly incisely dentate, those of the branchlets minute: involucral bracts oblong or broadly lanceolate: rays conspicuous (quarter-inch long) and rather broad: pappus copious. — S. Arizona, *Palmer, Lemmon, Pringle*, &c. Mohave Desert, California, *Parish, Coulter*, &c. (Adj. Mex.)

P. asteroídes, GRAY, l. c. Scabro-puberulent, a foot to a yard high from a plainly annual root. lower leaves spatulate or oblong, sometimes laciniate-pinnatifid, sometimes barely dentate; upper mostly linear and entire: involucral bracts lanceolate or linear: rays smaller and narrower: pappus less copious. — S. W. Texas to Arizona, *Wright*, &c. (Adj. Mex.)

P. BREVILINGULATA, Schultz Bip., of Mexico, the remaining species, resembles *P. asteroides*, but is more slender, with narrower leaves, smaller heads, and small rays which hardly surpass the pappus.

46. EREMIÁSTRUM, Gray. ('Ερῆμια, desert, ἄστρον, star. i. e. Aster of the desert.) — Pl. Thurb. in Mem. Acad. v. 320; Benth. & Hook. Gen. ii. 270. — Single species.

E. bellioídes, GRAY, l. c. Small winter annual, diffusely branched from the very base and depressed, hirsutely hispid throughout: leaves linear-spatulate, entire (half-inch long); those at the summit of the flowering branches loosely rosulate-involucrate around the solitary heads, and passing into involucral bracts: rays oblong-linear, white, acutely 2-3-dentate at the apex, 4 lines long: disk yellow. — On the desert near the Rio Colorado, borders of California and Arizona; fl. January and February, *Thurber, Newberry, Schott, Palmer, W. G. Wright*, &c., and borders of S. Utah, *Parry*. — Seldom, if ever, are the bristles of the pappus combined in clusters so as to form laciniate paleæ.

47. SERICOCÁRPUS, Nees. (Σηρικός, silky, καρπός, fruit, the akenes sericeous-pilose.) — Perennial herbs, of low or moderate stature (all N. American); with alternate commonly entire and sessile leaves, and small heads usually fascicled in a terminal compact cyme; both disk and ray white or whitish, or the latter changing in age to purplish: fl. midsummer. — Nees, Ast. 148; DC. Prodr. v. 261; Torr. & Gray, Fl. ii. 101; Benth. & Hook. Gen. ii. 270.

* Pappus ferruginous: leaves sparingly serrate, comparatively thin and veiny, and the radical ones large.

S. conyzoídes, Nees, l. c. A foot or two high, and slightly pubescent or glabrate: radical and lower cauline leaves spatulate (2 to 5 inches long, tapering into a margined petiole), obtuse; upper ones oblong-lanceolate: involucre turbinate, 18-20-flowered: rays rather short and broad. — *Conyza asteroides*, L. Spec. ii. 861. *Aster Marylandicus* (Pluk. Mant.), Michx. Fl. ii. 108. *A. conyzoides*, Willd. Spec. iii. 2043. — Dry woodlands, common from Maine to Ohio and south to Florida.

* * Pappus white: leaves entire, firmer, smaller, obscurely veined, disposed to be vertical, mostly obtuse: green tips of involucral bracts short, seldom squarrose: stems more leafy.
+ Atlantic species: akenes short, canescent-sericeous.

S. solidagíneus, NEES, l. c. Green, almost glabrous: stems strict and slender, 2 feet high, acutely striate-angled: leaves from linear to spatulate-lanceolate, an inch or two long: heads mostly glomerate and sessile, narrow, rather few-flowered: involucral bracts oblong, very smooth and rigid: rays at length elongated. — *Conyza linifolia*, L., l. c. *Aster Americanus albus*, &c., Pluk. Alm. t. 79, fig. 2. *A. solidagineus*, Michx l. c. *A. solidaginoides*, Willd. l. c. *Galatella obtusifolia*, Lehm. Ind. Sem. Hamb. 1837. — Moist woodlands, Canada to Tennessee, Alabama, and Louisiana.

S. tortifólius, Nees, l. c. Cinereous-puberulent and somewhat scabrous: stems not angled, a foot or two high: leaves conspicuously vertical by a twist, obovate or spatulate, short: heads more corymbosely cymose and most of them pedicellate, somewhat turbinate: involucral bracts more numerous and narrower, the innermost linear, puberulent: rays small. — *Conyza bifoliata*, Walt. Car. 204. *Aster tortifolius*, Michx. l. c.; Ell. Sk. ii. 341. *A. scabrosus*, Bertol. Misc. Bot. vii. t. 5. *A.* (*Leucocoma*) *Collinsii*, Nutt. Jour. Acad. Philad. vii. 82, a form with leaves sparingly crenate-serrate (*S. tortifolius*, var. *Collinsii*, Torr. & Gray, Fl. ii. 102). — Dry pine woods and barrens, Virginia near the coast to Florida and Louisiana, flowering later than the others.

+ + Pacific species: akenes more slender, less canescent.

S. rígidus, Lindl. Green, from scabrous-pubescent to nearly glabrous: stems a foot or two high, somewhat striate-angled: leaves oblong-lanceolate (an inch or two long), acute or obtuse, hardly rigid: heads crowded, either short-pedicelled or glomerate, somewhat turbinate, about 15-flowered: involucral bracts rather narrow and one-nerved: rays small and short, but sometimes surpassing the disk. — Hook. Fl. ii. 14, & DC. l. c.; Gray, Bot. Calif. i. 320. *S. Oregonensis*, Nutt. Trans. Am. Phil. Soc. n. ser. vii. 302, a form with larger rays and rather large leaves. — Woods, Washington Terr. to the Sierra Nevada, California.

48. **ÁSTER**, Tourn. Starwort, Aster. ('Αστήρ, a star, late Latin name for these and similar radiate-flowered and Daisy-like plants.) — Herbs, mainly of the northern hemisphere, and especially of North America, the larger part perennial: fl. late summer and autumn. The yellow disk-flowers in several species change to purple or rose-color. — L. Gen. ed. 1, 254; DC. Prodr. v. 261; Torr. & Gray, Fl. ii. 103; Benth. & Hook. Gen. ii. 271, at least as to American species; Gray, Proc. Am. Acad. xvi. 97, & xvii. 164.

The line between *Aster* and *Erigeron* is arbitrary. Aster is far the most difficult of our genera, both for the settlement of the names of the species and for their limitation, in respect to which little satisfaction has been attained as the result of prolonged and repeated studies. For full reasons why the following Linnæan species subside, see Proc. Am. Acad. xvii. supra cit.

A. divaricátus, L. Founded, as to the specimen in the herbarium, upon the upper part of a plant of *A. corymbosus*, Ait., wanting the cordate serrate leaves; as to the plants of Plukenet and Gronovius, upon *A. infirmus*, Michx.

A. rígidus, L. Founded wholly upon Clayton's plant as described in Fl. Virg.; is a mere synonym of *A. linariifolius*, L.

A. linifólius, L., and A. hyssopifólius, L. Mant., belong to *Galatella*, and are not American. The plant of Gronov. Fl. Virg., referred to, is *A. tenuifolius*, L.

A. ánnuus, L., is well known as *Erigeron annuus*.

A. vérnus, L., is *Erigeron nudicaulis*, Michx.

A. mutábilis, L. Not to be found in the Linnæan herbarium. In the first ed. of the Spec. Pl. was compared with an "*A. serotinus*," which was neither published nor preserved in the herbarium, has as a synonym a plant figured by Plukenet which is not preserved in his herbarium: finally, in ed. 2, the character was reconstructed in a manner incompatible with the original one, and before the Plukenetian synonym "Herm Hort. Lugd. t. 67" was introduced, which, on the authority of the herbaria of Royen and of Sherard, is *A. lævis*, L., and the comparison was changed to one with *A. Tradescanti*, L. *A. mutabilis* of later authors belongs to older-named species.

A. míser, L. Founded wholly upon a much exaggerated figure in Dillenius, with fictitious character, drawn from the plate. See *A. vimineus*, var. *foliolosus*.

A. ciliátus, Walt. Car. 209, is not at all identified.

Series I. Perennials, with multicipital rootstocks or caudex, usually multiplying by creeping subterranean shoots.

§ 1. Amellástrum. Bracts of the broad involucre 2-3-serial, of slightly unequal length; outer foliaceous or herbaceous-tipped: akenes broad, obovate, compressed, mostly flat and with callous marginal nerves only: pappus simple, or with an indistinct short outer series. — Gray, Proc. Am. Acad. xvi. 97. Consists of *A. Amellus* and the following, of less marked character.

↠ **A. alpínus,** L. A span or more high, with simple erect stems from a thickish caudex, bearing a single large head, cinereous-pubescent: leaves entire, somewhat 3-nerved, spatulate; upper small and linear: involucral bracts loosely erect, oblong-linear, herbaceous, all nearly equalling the disk: rays violet: akenes hirsute: usually a manifest short outer pappus. — Spec. ii. 872; Jacq. Fl. Austr. t. 88; Bot. Mag. t. 199; Hook. Fl. ii. 6. — Arctic shores, *Pelly*. Alpine region of northern Rocky Mountains, down to about lat. 49°, *Drummond, Burke, Bourgeau, Macoun*. The American plant mostly with fewer and shorter rays. (En., N. Asia.)

§ 2. MEGALÁSTRUM. Head very large (an inch in diameter exclusive of the large and numerous rays), pedunculate and solitary, terminating rigid branches): bracts of the hemispherical involucre imbricated in 2 or 3 unequal series, somewhat herbaceous, gradually attenuate-acuminate: akenes oblong, compressed, 2–4-nerved: pappus-bristles unusually coarse and rigid, rather scanty. — Gray, Pl. Wright. ii. 75. (Related to subsection *Xylorrhiza* and to *Townsendia*.)

A. Wrightii, Gray, l. c. Viscous-pubescent, a foot or more high: leaves oblong-spatulate, setulose-mucronate, entire, with one or two setiferous teeth, an inch and a half long including the margined petiole, thickish, obscurely veined: bracts of the involucre ovate-lanceolate or the inner narrower, rather lax, viscid, slightly scarious-margined, the caudate-acuminate tips surpassing the disk: rays purple, narrowly oblong, 30 to 40 (8 or 9 lines long): akenes (young) loosely pubescent: pappus white, of unequal strongly denticulate bristles, the larger almost aristiform. — *Townsendia* (*Megalastrum*) *Wrightii*, Gray, Bot. Mex. Bound. 78. — Rocks and stony hills on the Rio Grande, S. W. Texas, from the mouth of the great cañon, *Wright, Bigelow*.

A. tortifólius, Gray. Tomentose-pubescent, at length glabrate, 2 feet high: leaves from oblong to narrowly lanceolate, rigid, spinulosely dentate and acuminate, sometimes incisely pinnatifid, veiny: involucral bracts narrower and more numerous, lanceolate-subulate, rigid, the longest barely equalling the disk: rays pale purple or violet, often an inch long: akenes sericeous-canescent: pappus at length ferruginous, deciduous in a ring. — Proc. Am. Acad. vii. 353. & Bot. Calif. i. 323, not Michx. *Aplopappus tortifolius*, Torr. & Gray, Jonr. Bost. Soc. Nat. Hist. v. 109. — Mountains bordering the Mohave Desert, S. E. California to S. Utah, *Fremont, Newberry, Cooper*, &c.

§ 3. HELEÁSTRUM. Heads mostly smaller: pappus (simple) unusually coarse and rigid, the stronger bristles somewhat clavellate at tip: bracts of the involucre partly or the outer wholly foliaceous and linear-lanceolate, rigid, imbricated in several series, but the outermost little shorter than the inner: receptacle alveolate or fimbrillate: style-appendages filiform-subulate: akenes narrow and mostly slender, little compressed, 8–10-nerved, nearly glabrous: leaves all linear and entire or with some spinulose teeth, rigid, one-nerved and with obscure lateral nerves: rays numerous and elongated: pappus ferruginous or tawny. — Gray, Proc. Am. Acad. xvi. 97. *Heleastrum*, DC. Prodr. v. 263.

* Leaves as if gramineous, and bearing some spinuliform serratures or denticulations, mostly cuspidate at tip, smooth and glabrous: stems simple from a tuberiform rootstock, strict and slender, rigid, 2 feet high, pilose below, or the base of the leaves ciliate with some long and jointed hairs: flowering in summer.

A. eryngiifólius, Torr. & Gray. Stems bearing solitary or 2 or 3 large heads: leaves mostly lanceolate-linear (2 or 3 lines wide, the larger 4 inches long, upper ones gradually reduced to erect bracts): involucre hemispherical, very many-flowered; its bracts linear-lanceolate, attenuate into an almost setiform cusp: rays an inch long, pale blue or white. —

Fl. ii. 502; Chapm. Fl. 199. *Prionopsis? Chapmanii*, Torr. & Gray, l. c. 245. — Low pine barrens on the coast, Florida, *Chapman, Rugel, Mohr.*

A. spinulósus, CHAPM. Stem bearing few or several spicately disposed smaller heads: leaves narrowly linear, attenuate (half to 2 lines wide); the lower and radical 6 to 12 inches long, upper gradually reduced to setaceous-subulate appressed bracts; the margins merely spinulose-denticulate or mostly entire: involucre campanulate, its bracts mostly subulate from a broad base: rays half-inch long, pale violet. — Fl. 199. — Damp pine barrens, W. Florida near the coast, *Chapman.*

* * Leaves all entire, thickish: pubescence if any short and scabrous: flowering in autumn.

◂── **A. paludósus**, AIT. Stems sometimes branching, a foot high, bearing few or several often racemosely or spicately disposed heads (of half-inch height): leaves from broadly to narrowly linear (1 to 4 lines wide, 2 to 4 inches long); involucre nearly hemispherical; its bracts more unequal, the outer lanceolate-subulate and lax, inner linear-spatulate with herbaceous merely acute tips. rays rather short, deep violet — Hort. Kew iii. 201; Ell. Sk. ii. 343; Torr. & Gray, Fl. ii. 109; Chapm. l. c. *A. grandiflorus*, Nutt. Gen. ii. 156, not L. *Tripolium paludosum*, Nees, Ast. 155. *Heleastrum paludosum*, DC. Prodr. v. 264. — Wet pine barrens in the low country, N. Carolina to Texas, Arkansas, and Missouri.

§ 4. HESPERÁSTRUM. Heads with *neutral rays:* bracts of the campanulate involucre well imbricated and unequal, the outer with short herbaceous spreading tips style-appendages slender-subulate: akenes narrow, hardly at all compressed, 5-nerved and with intermediate striæ: pappus simple and soft. — Gray, Bot. Calif. i. 323, & Proc. Am. Acad. xvi. 97. (Resembling on the one hand § *Machæranthera*, and *Corethrogyne* on the other.)

A. Shasténsis, GRAY, l. c. A span or two high, in small tufts from a perennial root, paniculately branched, slender, canescently puberulent: leaves entire, an inch or less long; lower spatulate: uppermost linear and reduced to subulate bracts: heads rather numerous, scattered: involucre (nearly half-inch high) somewhat viscid-glandular; its bracts lanceolate or linear, mostly with acute and spreading green tips rays 12 to 20, violet; 3 to 5 lines long, occasionally (var. *eradiatus*) wanting — *Machæranthera (Hesperastrum) Shastensis*, Gray. Proc. Am. Acad vi. 539 — California· on Mount Shasta, above and below the limit of trees, first coll. by *Brewer*, and on Lassen's Peak, *Mrs. Austin*. The rayless state on Scott Mountains at 9,000 feet, *Greene.*

§ 5. BIÓTIA. Heads (small or middle-sized) corymbosely cymose: bracts of the campanulate well-imbricated involucre subcoriaceous and wholly appressed, obtuse and merely greenish or thickish but not spreading at the tip (transition to § *Orthomeris*, but passing into the succeeding subsection): outer successively shorter: rays not numerous (6 to 18), white or purplish-tinged: style-appendages subulate-lanceolate: akenes 3-several-ribbed or nerved, hardly or moderately compressed, mostly linear: pappus slightly rigid, simple: radical and lower cauline *leaves cordate, on long naked petioles*, ample, conspicuously serrate and acuminate: fl. midsummer and early autumn. (Other Asters with cordate petiolate leaves are only the *Heterophylli*.) — Torr. & Gray, Fl. ii. 104. *Biotia*, DC. Prodr. v. 264.

── **A. corymbósus**, AIT. Stem slender. 2 feet high, sometimes flexuous, terete: leaves membranaceous, much longer than wide, gradually or very prominently acuminate and acuminately serrate: involucre only one-fourth inch high, little surpassing the rather broadly compressed fusiform akenes: rays 6 to 9, white — Kew iii. 207; Willd. Spec. iii. 2036; Torr. & Gray, l. c. *A. divaricatus*, L. Spec ii 873, as to herb., excl. syn. Gronov & Pluk. (which relate to *A. infirmus*), and cordate leaves not described; name to subside *A. cordifolius*, Michx. Fl. ii. 114, in part. *Eurybia corymbosa*, Cass Dict. xxxvii 487; Nees, Ast. 143; Lindl. Bot. Reg. t. 1532. *Biotia corymbosa*, DC. l. c. 265. — Woodlands, Canada to upper part of Georgia.

― A. macrophýllus, L. Stem stout, somewhat striate-angled, 2 or 3 feet high: leaves thickish, serrate with proportionally smaller and broader less salient teeth, abruptly acuminate; radical and lowest usually broadly ovate- or even reniform-cordate (4 to 10 inches long); upper ovate to oblong, often wing-petioled, and uppermost sometimes sessile by a broad base. involucre commonly 5 lines high, often viscid-puberulent, in fruit much surpassing the fusiform-linear obscurely compressed akenes: rays 10 to 15, white or tinged with bluish purple. — Spec. ed. 2, ii. 1232; Willd. l. c.; Torr. & Gray, l. c. *Euryhia macrophylla* & *E. Jussiei*, Cass. Dict. l. c. *E. macrophylla* (larger-leaved and more scabrous form), *glomerata*, & *Schreberi* (thinner-leaved form in shade and drier soil), Nees, Ast. 140. *Lisia Schreberi*, *latifolia* (*A. latifolius*, Desf Cat., form approaching *A. corymbosus*), *glomerata*, & *macrophylla*, DC. l. c. — Woodlands, commonly in damp or rich soil, from Canada and Manitoba to the mountains of Georgia. Variable species: of which forms with smaller heads and thinner leaves appear to pass into *A. corymbosus*. A robust form, with large heads, more glandular involucre and peduncles, upper leaves ovate and sessile, lower and petioled cauline leaves all rounded at base, and most of the radical ones little cordate (*A. macrophyllus* of Willd. herb. chiefly, the rays perhaps violet), comes near the next following.

§ 6. ASTER proper. Heads various: bracts of the involucre (or at least the outer ones) with green herbaceous tips or appendages, or wholly or partly foliaceous, imbricated or pluri-serial, their margins not scarious: akenes from obovate-oblong to linear, 3–several-nerved: pappus rather fine and soft, or in the first subsection more coarse and rigid, simple, i. e. with no exterior setulose series. — § *Aster* proper, with part of *Oritrophium* & *Calliastrum*, Torr. & Gray, Fl. Probable hybrids abound.

* 1. Involucre well imbricated; the bracts appressed and coriaceous, with short and abrupt mostly obtuse herbaceous or foliaceous spreading tips (the outermost sometimes loose and more foliaceous): akenes narrow, 5–10-nerved, from minutely pubescent to glabrous: pappus mostly more rigid than in any of the following: rays showy, blue or violet: leaves of firm texture, more or less scabrous (the last species excepted), none of them cordate or clasping (§ *Calliastrum*, Torr & Gray. Fl. ii. 106, excl. spec.). — SPECTABILES.

    ← Radical and lowest cauline leaves ovate or ovate-oblong, some with rounded base, or even subcordate heads half-inch high: involucre nearly hemispherical; the green tips of the involucral bracts very short and either indistinct and erect or abruptly spreading: stems a foot or two high. Transition to *Biotia*, possibly hybrids of the preceding species with true Asters, being local and rare, but if so the other parent not at all determinable.

A. mirábilis, TORR & GRAY. Scabrous-pubescent, bearing few or several somewhat paniculate heads: leaves all ovate or oval, finely and acutely serrate, hispidulous-scabrous above: upper small and roundish, lower abruptly contracted into margined petioles (true radical not seen): involucre nearly smooth and glabrous, neither glandular nor viscid; its bracts with roundish-obtuse abrupt and very short squarrose-spreading tips: rays about 20, half-inch long violet: pappus ferruginous. — Fl. ii. 165. — Near Columbia, S. Carolina, *Gibbes*, Sept., 1835, not since collected.

A. Hervéyi, GRAY. Slightly scabrous, the corymbose branches and short peduncles glandular-puberulent, leaves minutely or obscurely serrate: radical and lowest cauline ovate on slender naked petioles; upper lanceolate: heads loosely corymbiform cymose: involucral bracts all erect and with less distinct close tips, pulvernlent-glandular: the short outer oblong-linear rays 15 to 24, narrow, half-inch long, lilac or violet. — Man. ed. 5, 230. *Eurybia commixta*, Nees. Ast. 143. *Biotia commixta*, DC. Prodr v. 265 (excl. syn. ?), is a robust cultivated form of this, which has long been in the gardens, of unknown origin. — Borders of oak woods, E. Mass. and Rhode Island, near the coast, *Hervey, Sargent*, &c. Grows in company with *A. macrophyllus* and *A. spectabilis*, evidently most related to the former, both in foliage and involucre: fl. Aug, Sept.

    ← ← Radical leaves all tapering at base into winged or margined petioles.

    ←← Involucre squarrose by the spreading or recurving herbaceous tips of the bracts: akenes slender, slightly pubescent, leaves obscurely veined, slightly scabrous: rootstocks slender and creeping, stems low, bearing few or several (rarely solitary) short-peduncled and showy heads. Atlantic U. S. species.

— **A. spectábilis**, AIT. A foot or more high, bearing several somewhat paniculate or cymose heads · leaves oblong-lanceolate or the lower spatulate- or oval-oblong, obscurely serrate or the upper entire. involucre hemispherical, half-inch high; the bracts glandular-puberulent and somewhat viscid, upper half of most of them herbaceous and recurved-spreading: rays numerous, three-fourths inch long or more, bright violet. — Kew. iii. 209; Nutt. Gen. ii. 157; Nees, Ast. 42; Lindl. Bot. Reg. t. 1527; Torr. Fl. N. Y. i. 336, t. 51. *A. elegans*, Willd. Spec. iii. 2042 (ex. char. and mainly herb.); Wenderoth in Act. Soc. Nat. Marb. ii. 17. *A. speciosus*, Hornem. Hort. Hafn. ii. 816, fide DC. — Sandy soil, Massachusetts to Delaware, near the coast, and perhaps farther southward, where it is replaced by the next.

— **A. surculósus**, MICHX. A foot or less high from long filiform rootstocks, bearing solitary or few pedunculate heads, which are generally smaller than those of *A. spectabilis*, but not dissimilar: leaves entire or nearly so, rigid, lanceolate or the upper linear and the radical oblong-lanceolate: involucre sometimes puberulent, but hardly glandular. — Fl. i. 112; Nutt. Gen. ii 157; Ell. Sk. ii. 354, ex char.; Nees, Ast. 40; Torr. & Gray, Fl. ii. 109. — Moist rocky and gravelly ground, or sometimes in sand, coast of New Jersey to Georgia, and on the Blue Ridge in North and South Carolina, where it was first collected by *Michaux*, and where it abounds.

A. grácilis, NUTT. About a foot high, slender, from slender and occasionally tuberous-thickened rootstocks, smoothish, not glandular nor viscid, bearing few or several cymosely disposed small heads: leaves oblong-lanceolate, entire or nearly so, small (an inch or two long, 2 to 6 lines wide): involucre turbinate, a quarter or third of an inch long, glabrous, coriaceous and whitish, with very short deltoid or ovate green tips, only about 30-flowered: rays 9 to 12, a quarter to half an inch long: akenes rather short. — Gen. ii. 158; Torr. & Gray, l. c. — Pine barrens, New Jersey to N. Carolina, also E. Kentucky and Tennessee, according to Nuttall. The larger forms closely related to *A. spectabilis*, with which it is associated; the more slender ones nearly approach *Sericocarpus*.

++ ++ Involucre hardly if at all squarrose; the tips of the bracts less definite and less spreading: stems very leafy: leaves pinnately veiny and reticulated, acutely serrate, more or less scabrous: heads middle-sized, corymbosely cymose or rarely solitary: style-appendages rather short and thick: Northern and Western species.

— **A. rádula**, AIT. Nearly glabrous or with some scattered hairs: stem slender and strict, a foot or two high, bearing few or solitary mostly slender-pedunculate heads: leaves veiny, oblong-lanceolate or narrower, acuminate, somewhat hispidulous-scabrous, thinnish (inclined to be rugulose in drying, about 2 inches long, 3 to 9 lines wide), each margin with 3 to 7 serratures toward the middle; upper cauline sometimes oblong-ovate with subcordate sessile base · involucre nearly hemispherical, 3 or 4 lines high; its bracts in few series, obtuse, ciliolate · the outermost oblong, inner narrower, shorter than the disk: rays half-inch long, pale violet; akenes glabrous, striate-nerved. — Kew. iii. 210; DC. Prodr. l. c. 230; Torr. & Gray, Fl. ii. 106; Torr. Fl. N. Y. i. t. 50. *A. nudiflorus*, Nutt. Gen. ii. 157, a broader-leaved and most luxuriant southern form Was cultivated in 1839 in the Berlin Garden as *Biotia commixta*, var. *stricta*. — Swamps, Delaware to E. Massachusetts, west to the mountains of Pennsylvania (Pocono), thence north to Nova Scotia and Newfoundland. Passes into

Var. strictus. Reduced boreal form, a span to a foot high, with either oblong or narrowly lanceolate barely serrulate leaves, and solitary or rarely 2 or 3 heads. — *A. biflorus*, Michx Fl. ii. 111; Torr. & Gray, l. c. *A. strictus*, Pursh, Fl. ii. 556. — Higher mountains of New England to Labrador.

A. Sibíricus, L. A span to a foot high, often fastigiately branched from the base, somewhat cinereous-pubescent or puberulent, or the foliage scabrous: heads solitary, terminating the stem or corymbiform branches: leaves oblong-spatulate to broadly lanceolate, acutely more or less serrate (an inch or more, or in largest form even 3 inches long): involucre broadly campanulate, 3 lines high, shorter than the disk; its bracts narrowly lanceolate, with mostly acute and loose herbaceous tips. rays 3 or 4 lines long, violet: akenes piloso-pubescent. — Spec ii. 872 (Gmel. Fl Sibir. ii. t. 80, f. 2), larger than American form; Herder in Radde, Reis. iii. 11, *A. montanus*. Richards. App Frankl. Jour. 32; Torr. & Gray, Fl. ii. 107. *A. Richardsonii*, Spreng. Syst. iii. 528; Hook. Fl. ii. 7. *A. Espenbergensis*, Nees, Ast. 36. *A Prescottii*, Lindl in DC Prodr. v. 231 — Arctic coast and Alaskan Islands to Rocky Mountains in Wyoming and Montana. (N. E. Asia to Arctic Eu.)

Var. gigantéus, *A. Richardsonii.* var. *giganteus,* Hook. l. c., and *A. montanus,* var. *giganteus,* Torr. & Gray, l c., is a stout and large form, of the Arctic regions, nearly answering to the original *A. Sibiricus,* L., of Siberia.

A. radulínus, GRAY. Between the preceding and the following. 10 to 20 inches high: leaves from oval-obovate to broadly lanceolate (2 to 4 inches long), serrate with numerous sharp teeth, scabrous; heads numerous, corymbosely cymose: involucre broadly turbinate, 3 or 4 lines high; its bracts regularly imbricated and onter successively shorter, cinereous-pubescent or glabrate, not glandular nor viscid, from broadly lanceolate or oblong to linear, abruptly acutish or obtuse: rays 3 to 5 lines long, pale violet, sometimes whitish; akenes minutely pubescent. — Proc. Am. Acad. viii. 388, & Bot. Calif. i. 323. *A. radula,* Less. in Linn. vi. 125, not Ait. — Dry ground, California, from Monterey northward, and in the Sierra Nevada to Oregon and Washington Terr.

A. conspícuus, LINDL. Scabrous; stem 2 feet high, stout, rigid, bearing several or numerous corymbosely cymose heads; leaves rigid, ovate, oblong, or the lower obovate, acute, ample (commonly 4 to 6 inches long and 1½ to 4 inches broad), acutely serrate, rigid, reticulate-venulose as well as veiny: involucre broadly campanulate, about equalling the disk, 5 to 6 lines high; its bracts in several series, minutely glandular-puberulent or viscidulous, lanceolate, acute, the greenish tips little spreading: rays half-inch long, violet: akenes minutely pubescent. — Hook. Fl. ii. 7, & DC. Prodr. v. 230; Torr. & Gray, Fl. ii. 107. — Saskatchewan to British Columbia, and south to the Yellowstone in the Rocky Mountains; first coll. by *Drummond.*

++ ++ ++ Involucre very squarrose by the foliaceous widely spreading tips of the bracts, smooth and glabrous, as is also the foliage: heads large and paniculate: Alleghanian.

A. Curtísii, TORR. & GRAY. Almost wholly glabrous and smooth: stems 2 or 3 feet high, rather slender, the larger loosely paniculately branched; branches bearing scattered large heads: radical and lower leaves (3 or 4 inches long) ovate-lanceolate, acuminate, sparingly serrate, gradually or abruptly contracted into winged petioles; upper ones lanceolate and sessile, becoming entire: involucre hemispherical, equalling the disk, half-inch high; the much imbricated coriaceous bracts very conspicuously appendaged with foliaceous ovate or short-lanceolate tips, or the outer more than half foliaceous: rays rather broad, half-inch long or more, deep violet-blue: akenes compressed, broader upward and with narrowed apex, glabrous. — Fl. ii. 110. — Margin of woodlands, in dry soil, through the Alleghanies in N. Carolina and adjacent borders of Tennessee; very showy.

* 2. Involucre and usually branchlets viscidly or pruinose-glandular, therefore more or less graveolent, either well imbricated or loose: rays showy, violet to purple: akenes mostly several-nerved and narrow: pubescence not sericeous: leaves all entire or lower with few and rare teeth, except in the last species: cauline all sessile or partly clasping: true perennials, mostly multiplying by subterranean rootstocks or other shoots. (Glandular involucre also in species of § *Machæranthera,* some of which are short-lived perennials.) — GLANDULOSI.

+ Bracts of involucre rather well imbricated, commonly with more or less rigid appressed base and foliaceous or herbaceous tips: rays not extremely numerous, from 15 to 40.

++ Stem simple: leaves and heads proportionally large: Rocky Mountain alpine or subalpine species.

A. integrifólius, NUTT. Stem mostly a foot or more high, stout, sparsely leafy, villous-pubescent but glabrate, bearing few or several racemosely or thyrsoidly disposed heads: leaves of firm texture, oblong or spatulate (the larger 4 to 7 inches long) or the smaller upper ones lanceolate, sometimes obsoletely repand-serrulate, apiculate, traversed by a strong midrib, venulose-reticulated, glabrate, half-clasping: lowest tapering into a long stout wing-margined petiole with clasping base: heads fully half-inch high, hemispherical: involucre and branchlets viscid-glandular; its bracts few-ranked, linear, ascending, not squarrose; the outer sometimes short and rather close, commonly violet and more foliaceous, nearly equalling the inner; these equalling the disk: rays 15 to 25, bluish-purple, half-inch long: akenes compressed-fusiform, 5-nerved, and sometimes with intermediate nerves, sparsely pubescent: pappus decidedly rigid. — Trans. Am. Phil. Soc. n. ser. vii. 291; Torr. & Gray, Fl. ii. 111; Gray, Bot. Calif. i. 324. — Open and moist subalpine woods or meadows, Montana to the Cascade Mountains in Oregon, south to Colorado, and along the Sierra Nevada, California, in the Yosemite, &c.; first coll. by *Nuttall.*

12

A. Kíngii, EATON. A span or less high, cespitose : leaves mainly radical, spatulate, entire, or with few sharp teeth, mucronate, thinnish, glabrous or nearly so (1 to 3 inches long) : flowering stems pubescent and above glandular, bearing solitary or 3 to 5 middle-sized heads: involucre somewhat campanulate, 4 or 5 lines high, merely puberulent-glandular, hardly at all viscid ; the bracts linear-lanceolate with attenuate and squarrose-spreading green tips : rays less than 30, barely half-inch long, white : akenes narrow, pubescent. — Bot. King Exp. 141, t. 16. — Utah, in the Wahsatch Mountains at 7,000 to 11,000 feet, *Watson, Parry, M. E. Jones.*

++ ++ Stems branching: leaves comparatively small : species neither alpine nor subalpine.

= Involucre of the small and scattered or somewhat racemosely disposed heads not squarrose; the green tips of the bracts more or less erect : slender and low species, a span to a foot or less high, of the Rocky Mountain and interior western region.

— **A. campéstris,** NUTT. Pruinose-puberulent and viscidulous, somewhat heavy-scented : leaves linear (about an inch long, a line or two wide) or lower narrowly lingulate-spatulate (radical "serrulate," Nuttall), mostly glabrate, some obscurely 3-nerved : involucre 3 or 4 lines high, hemispherical, of rather few-ranked and little unequal linear acute bracts, pruinose-glandular : rays 3 or 4 lines long, light violet or purple. — Trans. Am. Phil. Soc. (n. ser.) vii. 293. — Low grounds and plains, interior of Washington Terr. and Idaho to Montana (first and sparingly coll. by *Nuttall* and *Spalding*, recently by *Watson, Suksdorf, Forwood,* &c.), E. Oregon (*Cusick*) to N. California (*Greene*).

Var. Bloómeri. More-rigid (in drier more exposed situations) : stem and leaves hirsutulous : involucral bracts sometimes more unequal. — *A. Bloomeri,* Gray, Proc. Am. Acad. vi. 539, & Bot. Calif. 323. — High slopes, &c., W. Nevada, *Bloomer, Lemmon,* in specimens of the latter, from Carson, passing into *A. campestris.*

— **A. Féndleri,** GRAY. Rigid, a span to a foot high, sparsely hispidulous : the linear one-nerved firm leaves hispid-ciliate, otherwise usually smooth and glabrous : involucre somewhat campanulate (3 lines high) ; outer bracts shorter, linear-oblong, obtuse, pruinose-glandular, inner acute or apiculate : rays violet, 4 lines long. — Pl. Fendl. 66. *A. Nuttallii,* var. *Fendleri,* Gray, Pacif. R. Rep. vi. 97. — Plains and sand-hills, from W. Kansas to S. Colorado and N. New Mexico ; first coll. by *Fendler.*

= = Involucre of the large heads very squarrose-foliaceous : leaves proportionally very small, rigid, recurved or reflexed.

A. grandiflórus, L. About 2 feet high, with long and slender spreading rigid branches, hispid with short spreading bristles, not viscid : leaves oblong-linear or obscurely spatulate, rough-hispidulous ; cauline rarely 2 inches long ; of the branches half to less than quarter inch long ; uppermost passing into bracts of the (half-inch high) many-ranked obscurely granulose-viscid involucre ; the green tips oblong-linear or shorter, or the inner linear : rays three-fourths inch long, deep violet, large and numerous, rather broad : akenes little compressed, canescent, 7–10-costate. — Spec. ii. 877 (Martyn, Hist. Pl. Rar. t. 191 ; Dill. Elth. t. 36, fig 41) ; Mill. Ic. t. 282 ; Bot. Reg. t. 273 ; Hoffm. Phyt. Blatt. 65, t. A, f. 1. *A. asperrimus,* Nutt. Trans. Phil. Soc. vii. 293. — Dry and gravelly soil, Virginia to Georgia in the middle country.

= = = Involucre of middle-sized (a third to half inch) heads well imbricated : the unequal bracts with loose squarrose-spreading tips : leaves not rigid, spreading.

— **A. Nóvæ-Ángliæ,** L. Stem stout and strict, 2 to 8 feet high, very leafy to the top, coarsely hirsute with many-jointed hairs, also with glandular pubescence : leaves lanceolate or broadly linear, pubescent (2 to 5 inches long), entire, slightly if at all narrowed below, half-clasping by a strongly auriculate-cordate base : heads crowded : rays 50 to 60 or more, fully half-inch long, purple. — Spec. ii. 875 (Hort. Cliff. 408 ; Herm. Par. Bot. t. 98) ; Bot. Reg. t. 183 ; Torr. & Gray, Fl. ii, 145. *A. amplexicaulis,* Lam. Dict. i. 304, excl. syn. Tourn. *A. spurius,* Willd. iii. 2032, a low and branching form with scattered heads. *A. concinnus,* Colla, Hort. Rip. App. iii. t. 12, not Willd. — Low grounds, Canada and Saskatchewan to S. Carolina and Colorado. A peculiar and handsome species.

—**Var. róseus,** DC. Rays rose-colored. — (Bot. Reg. l. c. fig. d.) *A. roseus,* Desf. Cat. ed. 3, 401, not Stev. — With the ordinary form occasionally, permanent in cultivation.

A. oblongifólius, NUTT. About 2 feet high : stem hirsute-pubescent, very leafy, corymbosely branched : leaves from narrowly oblong to broadly linear (larger cauline 2 inches long, 3 or 4 lines wide), somewhat hispidulous-puberulent; those of flowering branchlets not rarely glandular: involucre campanulate, aromatic-scented, the linear bracts granulose-glandular and viscidulous: rays 25-30, bright violet, 5 or 6 lines long: akenes cinereous-pubescent. — Gen. ii. 156, & Trans. Am. Phil. Soc. l. c. 294; Torr. & Gray, Fl. ii. 143. *A. biennis*, Torr. Ann. Lyc. N. Y. ii. 122, not Nutt. *A. multiceps*, Lindl. in DC. Prodr. v. 237, from St. Louis. not "Louisiana." — Rocky banks and bluffs, Penn. and Virginia, from the Alleghanies westward to Wisconsin, Kansas, and Texas.

Var. **rigídulus.** Low, more fastigiate, with more rigid and hispidulous scabrous leaves. — *A. Kumleini*, Fries, in distrib. Mus. Ups. no. 5. — In drier and more exposed places, Illinois and Wisconsin, to Texas and Colorado.

+ + Bracts of the involucre loose and more or less herbaceous (or somewhat colored) almost from the base, linear-attenuate, all equalling the disk: heads hemispherical, numerous and usually thyrsoidly or cymosely congested at the summit of the simple very leafy stem: rays numerous and narrow: style-appendages lanceolate: akenes hirsute.

A. **modéstus,** LINDL. Merely pubescent or glabrate: stem more slender, 2 feet high: leaves thinnish, lanceolate or broader (2 to 4 inches long), sparingly and acutely serrate or denticulate, very acute, mostly narrowed to a sessile or partly clasping but not auriculate base: heads fewer and smaller than in the preceding: bracts of the involucre and rays less numerous; these "pale blue." — Hook. Fl. ii. 8, & DC. l. c. 231; Torr. & Gray, l. c. *A. Seyanus*, Nutt. Trans. Am. Phil. Soc. l. c. 294. *A. Unalaschkensis?* var. *major*, Hook. Fl. ii. 7. *A. mutatus*, Torr. & Gray, Fl. ii. 142. — Moist woods, Oregon to Brit. Columbia on the Pacific, and east to Saskatchewan and Pembina (*Macoun*).

∗ 3. Involucre imbricated, hemispherical, not glandular nor viscid, squarrose with ovate or lanceolate foliaceous tips to the bracts: *pubescence wholly soft and cinereous: cauline leaves all with sagittate-auriculate clasping base*, both sides of the same hue, entire; base of stem said to be somewhat woody! — SAGITTIFERI.

A. **Caroliniánus,** WALT. Minutely and softly cinereous-pubescent, not glandular nor viscid: stem diffusely branched, often *reclining*, 4 to 10 feet long, with straggling slender branches: leaves oblong-lanceolate, an inch or two long, contracted above the *sagittately auriculate* insertion: heads terminating small-leaved branchlets: bracts of the involucre well imbricated; the outer shorter and somewhat spatulate, with ovate-lanceolate green tips or more foliaceous; inner linear: rays 5 or 6 lines long, narrow, pale purple or rose-color: akenes narrow, glabrous, 10-nerved. — Car. 208; Michx. Fl. ii. 211; Ell. Sk. ii. 353; Chapm. Fl. 205. *A. scandens*, Jacq. f. Ecl. t. 125. — Marshes and river-banks near the coast. S. Carolina to Florida.

∗ 4. Involucre imbricated: the bracts with coriaceous base and foliaceous or herbaceous loose tips: *pubescence soft-sericeous and sometimes silvery: leaves of the same hue both sides*, all entire, disposed to become vertical by a twist near the sessile base: heads middle-sized or smaller: rays violet: akenes narrow, 8-10-nerved. — SERICEO-CONCOLORES.

+ Heads terminating open branches, middle-sized (about half-inch high): involucre loose and foliaceous, of comparatively large bracts; the outermost passing into leaves of the branchlets: leaves mucronate: akenes glabrous.

A. **seríceus,** VENT. A foot or two high, paniculately branching: leaves silvery-white with soft silky pubescence, oblong (an inch or less long), or the lowest oblanceolate (3 inches long): involucre oblong; foliaceous tips of the bracts from ovate to lanceolate, sericeous-canescent: rays 18 to 25, fully half-inch long, rather broad, deep violet. — Hort. Cels. t. 33; Pursh, Fl. ii. 348; Torr. & Gray, l. c. 112. *A. argenteus*, Michx. Fl. ii. 111. — Prairies and dry banks, Minnesota and Illinois to Tennessee and Texas.

Var. **montánus.** Less silvery, merely canescent: leaves commonly narrower: upper leaves and bracts of the shorter involucre sometimes glabrate and villose-ciliate: approaching the next species. — *A. montanus*, Nutt. Gen. ii. 155. — In the mountainous district from Buncombe Co., N. Carolina, to N. W. Georgia.

A. **phyllólepis,** TORR. & GRAY. More slender and with long simple branches, merely canescent: leaves small; lower cauline inch or more long, oblong; the branches elliptical to oblong-lanceolate, half to quarter inch long; uppermost and the large ovate or ovate-lanceolate foliaceous portion of the involucral bracts cuspidate-acuminate, glabrate, conspicuously hirsute-ciliate: rays less than half-inch long. — Fl. ii. 113. *A. sericeus*, var. *microphyllus*,

DC. Prodr. v. 233. *A. ciliatus*, Nutt. Trans. Am. Phil. Soc. vii. 295, not Walt. — Prairies, W. Louisiana and Texas; first coll. by *Drummond*.

+ + Heads smaller, usually numerous and racemosely disposed on virgate simple stems; involucre closer and of small bracts; akenes silky-villous.

A. cóncolor, L. Stems slender, 2 feet high, sometimes from a tuberous-thickened rootstock, very leafy: leaves small, canescent with minute pubescence, rarely glabrate, from oblong to short-linear; the lower on fertile stems only inch long, above gradually reduced in the inflorescence to small bracts: heads rather narrow (4 lines high): bracts of the involucre lanceolate, erect, sericeous-canescent; the tips short and narrow, or sometimes more prolonged: rays 10 to 15, 3 or 4 lines long, violet-purple. — Spec. ed. 2, ii. 1228; Walt. Car. 209; Torr. & Gray, Fl. ii. 113; Bertol. Misc. Bot. vii. t. 6. — Sandy or gravelly soil, mostly in pine barrens, toward the coast, Rhode Island to Florida and Louisiana.

* 5. Involucre turbinate, pluriserial, not glandular; the appressed coriaceous whitish bracts with definite and short (mostly ovate) and slightly squarrose green tips, outer successively shorter: heads rather small, but large in proportion to the *minute (line or two long) crowded and uniform cauline leaves;* radical leaves rosulate, subsessile, abruptly larger and very unlike the cauline, sometimes an inch long: herbage scabrous: rays violet, 3 or 4 lines long: akenes short, pubescent: flowering late in autumn. — BRACHYPHYLLI.

A. squarrósus, WALT. Stems rigid, slender, paniculately much branched, a foot or two high, bearing scattered heads: branches throughout uniformly squarrose with the minute recurved-spreading rigid leaves; these mostly ovate-subulate and a line long; lowest on sterile shoots 2 or 3 lines long, lanceolate-subulate, mucronate-cuspidate: bracts of the obovate-turbinate involucre with very obtuse or roundish green tips. — Car. 209; Michx. Fl. ii. 112; Ell. Sk. ii. 530; Torr. & Gray, Fl. ii. 114. — Dry pine barrens, N. Carolina to Florida.

A. adnátus, NUTT. More hispidulous-scabrous and virgately branched: leaves almost imbricated on the stem and branches, lanceolate-oblong, with clasping base, appressed and by the midrib adnate to the stem for most of their length, only the lowest larger and free: heads rather smaller and involucral bracts acutish. — Jour. Acad. Philad. vii. 82; Hook. Comp. Bot. Mag. i. 97; Torr. & Gray, l. c. *A. microphyllus*, Torr. ex Lindl. in DC. Prodr. v. 244; Bertol. Misc. Bot. vi. t. 5. — Pine barrens of Alabama and W. Florida.

* 6. Involucre ovoid with turbinate base or campanulate, appressed-imbricated, pluriserial; the bracts narrow, coriaceous, minutely granulose or scabrous, but not glandular, acute, with indistinctly marked green tips, the outer successively shorter: *whole herbage scabrous-pubescent: cauline leaves all with sessile and completely cordate-clasping base,* the basal lobes generally meeting or overlapping around the stem; radical not cordate; all entire: heads showy: akenes many-striate, sericeous-pubescent, narrow. — PATENTES.

A. pátens, AIT. Stems 2 or 3 feet high, with long and slender rigid divergent branches, mostly bearing single heads: leaves from oblong to broadly lanceolate, rather rigid, scabrous, rarely with obscure serratures, roughly hispidulous-ciliolate; the cauline an inch or two long, sometimes narrowed above the broad auriculate clasping base; those of the branchlets gradually reduced to small subulate bracts: heads half-inch or less high: rays a third to half an inch long, about 24, deep violet. — Ait. Kew. iii. 201; Pursh, Fl. ii. 551; Torr. & Gray, Fl. ii. 114. *A. undulatus*, Ell. Sk. ii. 361, not L. *A. amplexicaulis*, Michx. Fl. ii. 114; Bigel. Fl. Bost. ed. 2, 312, not Lam. *A. patentissimus*, Lindl. in DC. Prodr. v. 232, a rigid and long-branched form. — Dry open grounds, Massachusetts to Florida, west to Michigan, Arkansas, and Texas.

Var. **grácilis,** HOOK. A foot or two high, more slender: heads and oblong to oval leaves smaller and more scabrous. — Comp. Bot. Mag. i. 97. — Alabama to Texas, &c., a common Southwestern form.

Var. **phlogifólius,** NEES. The other extreme: leaves larger (cauline 3 or 4 inches long, an inch or more wide, softer and membranaceous), hardly scabrous, sometimes contracted above an auriculate-dilated base: heads paniculate on shorter branches: involucral bracts in fewer ranks, almost glabrous. — Ast. 49; Torr. & Gray, l. c. *A. phlogifolius*, Muhl. in Willd. iii. 2034; Nutt. Gen. ii. 156. *A. alatus*, Aikin in Eaton & Wright, Man. 146 ? *A. auritus*, Lindl. in DC. l. c., cultivated form, with thinner and lax involucre. — In woods or shady moist ground, New York to North Carolina and Tennessee. A part of *A. undulatus*, L., may belong here.

* 7. Heads small, or in one species middle-sized, paniculate: lower cauline and radical *leaves cordate and petioled:* no glandular or viscid pubescence: akenes compressed, short, 3–5-nerved: rays violet, purplish, or sometimes almost white: bracts of the involucre with short and appressed green tips, except in the first. — HETEROPHYLLI.

+ Anomalous species, with middle-sized heads, many rays, and squarrose foliaceous involucre!

— **A. anómalus,** ENGELM. Pubescent and somewhat scabrous, a foot to a yard high, paniculately or virgately branched above, bearing numerous loosely disposed heads: leaves veiny, thinnish, entire, mostly oblong- to lanceolate-ovate with narrow and often deep cordate base, those of branchlets reduced and lanceolate to subulate: heads half-inch high: involucre pluriserially imbricated, hirsutulous-pubescent, of attenuate-linear bracts; their foliaceous upper half recurved or widely spreading: rays bright violet, about 40, quarter to half an inch long: akenes glabrous. — Torr. & Gray, Fl. ii. 503. — Limestone cliffs, Illinois, Missouri, and Arkansas; first coll. by *Engelmann.*

+ + True *Heterophylli,* with smaller heads, 10 to 20 rays, and a close involucre of appressed or erect bracts. Occasional specimens with only the radical leaves cordate.

++ Leaves all entire or nearly so (lower sometimes with a few teeth), of rather firm texture, all much longer than wide, none clasping: heads showy: rays violet, 5 or even 6 lines long, 15 to 20 in number: involucre 3 or 4 lines high; its bracts all appressed and with mostly definite short green tips, outer successively shorter.

— **A. Shórtii,** HOOK. Stem 2 to 4 feet high, rather slender, leafy to the summit, bearing racemose-paniculate heads: leaves minutely soft-pubescent, mostly glabrate and smooth above, thin-veiny, nearly all petioled; radical and principal cauline ovate-lanceolate with distinctly cordate base and on slender naked petioles, tapering-acute (3 to 5 inches long), only on ultimate short branchlets or peduncles reduced to subulate bracts. involucre sometimes puberulent; its bracts narrow, less rigid and less definitely green-tipped than in the next: rays light violet. — Fl. ii. 9; Torr. & Gray, Fl. ii. 118. — Border of thickets and shaded banks, Ohio to Georgia in the upper country, west to Kentucky and Illinois; first coll. by *Short.*

— **A. azúreus,** LINDL. Stem 2 to 4 feet high, paniculately or racemosely compound at summit: branches slender and rigid: leaves hirtello-scabrous both sides; radical and lowest cauline ovate-lanceolate with subcordate base, on slender petioles (3 to 6 inches long): cauline oblong or lanceolate with winged petiole or attenuate base, verging to linear, and on the branchlets reduced to numerous small and slender-subulate rigid bracts: involucre glabrous and smooth; green tips of the bracts ovate or deltoid: rays deep violet-blue. — Hook. Comp. Bot. Mag. i. 98, & DC. Prodr. v. 244 (incompletely described for want of lower leaves); Torr. & Gray, Fl. ii. 118. *A. Oolentangiensis,* Riddell, Synops. 55. — Prairies and border of woods, W. New York and Ohio to Minnesota, and southwest to Arkansas and Texas, where there are forms with hardly a cordate leaf!

++ ++ Leaves some entire; but lower almost always somewhat serrate, rather firm in texture, longer than broad; the base, or that of wing-margined petiole of lower cauline, *cordate-clasping:* greenish tips of the less rigid involucral bracts short and rather obtuse.

— **A. undulátus,** L. Pale or dull with a minute somewhat cinereous and sometimes scabrous pubescence: branches rather rigid, racemosely or paniculately bearing several or rather numerous racemosely disposed heads: leaves at most inconspicuously or obtusely serrate; upper mainly entire, lanceolate or oblong with partly clasping base, above diminished to subulate bracts; middle ones ovate or ovate-lanceolate, abruptly contracted below and with dilated cordate-clasping base, sometimes panduriform, below subcordate on margined petioles; lowest cauline and radical cordate on slender naked petioles: heads 4 lines high: rays bright violet or sometimes paler. — Spec. ii. 875 (Hort. Cliff. & Herm. Parad. t. 96, whence the name, & Moris. Hist. 120); Ait. Kew. iii. 206; Hoffm. Phyt. Blatt. 77, t. C, f. 1: Torr. & Gray, l. c.; Spragne, Wild Flowers, t. 4. *A. diversifolius,* DC. Prodr. v. 234. *A. paniculatus,* Nutt. Gen. ii. 56, not Ait., nor Lam. *A. sagittifolius,* Ell. Sk. ii. 262, not Willd. — Dry ground, margin of woods, &c., Canada to Florida, Kentucky, and Arkansas. Southward in the low and middle country the common form is

— **Var. diversifólius.** More rigid, scabrous or scabro-puberulent, and with longer virgate flowering branches, which are beset with minute subulate or lanceolate (or below oblong) leaves, only the lower cauline having a narrowed base or winged petiole. — *A. diversifolius,* Michx. Fl. ii. 113 *A. scaber,* Ell. Sk. ii. 262. *A. asperulus,* Torr. & Gray, Fl.

ii. 120, not Wall. *A. Baldwinii*, Torr. & Gray, Fl. ii. 127, from specimens with rameal leaves only. — Dry ground, S. Carolina to Florida and Louisiana. (Named as distinct new species in herb. Poiret and herb. Banks.)

+++ +++ +++ Leaves nearly all sharply serrate, thinner, none cordate-clasping either by base of upper sessile ones or by appendage at base of petiole.

━━ Involucral bracts obtuse or obtusish with conspicuously marked and definite very short green tips: heads small and numerous.

A. cordifólius, L. Green, slightly pubescent to nearly glabrous, paniculately much branched above into thyrsoidal inflorescence of very many small heads: leaves membranaceous, acutely serrate, cordate-ovate on nearly naked petioles, or uppermost lanceolate and sessile, acuminate: heads 2 or 3 lines high: rays only 10 to 12, pale violet or whitish. — Spec. ii. 875, mainly (Hort. Cliff., & syn. Cornuti & Morison); Ait. l. c. 207; Lindl. Bot. Reg. t. 1597 (unusual form); Pursh, Fl. ii. 552; Torr. & Gray, Fl. ii. 120; Gray, Proc. Am. Acad. xvii. 165. *A. paniculatus* & *A. heterophyllus*, Willd. Spec. iii. 2035. *A. paniculatus, heterophyllus,* & *cordifolius,* Nees, Ast., chiefly. — Common in woodlands, New Brunswick to the mountains of Georgia, west to Wisconsin and Missouri. A singular abnormal state, collected by Moser on the Pocono in Pennsylvania, has some of the lower cauline leaves lanceolate and laciniate; others oblong-ovate, simply serrate, barely subcordate-contracted into a winged petiole: perhaps a hybrid with *A. diffusus.*

══ Involucral bracts acute or acutish, with the green tips more indefinite.

a. Heads rather larger than in the last preceding (3 or 4 lines high), numerous, thyrsoid-paniculate on the rather rigid branches: stems rather stout, 2 to 6 feet high: leaves mostly gradually acuminate.

A. Drummóndii, LINDL. Pale with a fine and mostly soft cinereous pubescence: leaves from cordate to cordate-lanceolate and mostly on margined petioles, or the small uppermost lanceolate and sessile by a narrow base, obtusely or acutely serrate (the large 4 inches, smaller about an inch long), sometimes scabrous above: bracts of the involucre acute or acutish: rays violet-blue or paler, 3 to 5 lines long. — Hook. Comp. Bot. Mag. i. 97, & DC. l. c. 294; Torr. & Gray, l. c. — Open grounds and border of woods, Illinois and Minnesota to Texas. Forms pass into the next.

A. sagittifólius, WILLD. Green, from glabrous to sparsely pilose-pubescent: stem strict, 2 or 3 feet high: leaves oblong- and ovate-lanceolate, acutely more or less serrate; radical and lowest cauline narrowly cordate, on naked petioles; upper subcordate or truncate at base and contracted into a winged petiole (3 to 5 inches long); uppermost linear-lanceolate and sessile: heads densely thyrsoid-paniculate: bracts of the involucre subulate-linear and mostly attenuate, the tips rather loose: rays purplish, pale violet, or bluish, sometimes nearly white. — Spec. iii. 2035; Nees, Ast. 56; Hook. Fl. ii. 9; Torr. & Gray, Fl. ii 121. *A. cordifolius,* Willd. l. c., as to spec. from Muhl. *A. paniculatus,* Muhl. Cat.; Darlingt. Fl. Cest. 464. *A. hirtellus* & *A. urophyllus,* Lindl. in DC. Prodr. v. 233, from "St. Louis, *Drummond,*" not "Louisiana." — Open grounds, Canada and W. New York to Dakota, Missouri, and Florida.

b. Heads larger (4 or 5 lines high) and comparatively few in a loose thyrsus or panicle terminating the simple stem: only lowest leaves cordate or subcordate: involucral bracts looser and less imbricated.

A. Lindleyánus, TORR. & GRAY. Green, sparsely pilose or nearly glabrous: stem 10 to 20 inches high, rather stout, the smaller plants bearing few heads: radical and lowest cauline leaves ovate, moderately or some obscurely cordate, on winged or margined petioles; upper ovate- to oblong-lanceolate; uppermost sessile and acuminate at both ends: bracts of the involucre linear-attenuate; the outer little shorter: rays pale violet, 3 to 5 lines long. — Fl. ii. 122. *A. paniculatus,* Ait. Kew. iii. 207; Hook. Fl ii. 8 (chiefly), not Lam. *A. pracox,* Lindl. in Hook. Fl. ii. 9, not Willd. — Labrador to Upper Canada, Lake Superior, Saskatchewan, and the borders of British Columbia. The origin aloft the species was raised by Gordon from Labrador seeds, and has more extended inflorescence of smaller heads than is common in the wild plant.

Var. **ciliolátus** (*A. ciliolatus,* Lindl. in Hook. Fl. & DC. Prodr. l. c. 235) is a dwarf arctic form, a span to 8 inches high, bearing few (not cordate) leaves and only 2 to 7 heads; from Slave Lake, *Richardson.*

* 8. Heads and inflorescence various: no cordate petioled leaves; radical leaves all acute or attenuate at base: not glandular nor viscid, nor silky-canescent: akenes compressed, few-nerved. — HOMOPHYLLI. Nees.

+— Whole plant *very smooth and glabrous* (sometimes hispidulous roughness on leaf-margins or a little pubescence on branchlets or peduncles): involucre of middle-sized or rather large heads pluriserial, from turbinate to campanulate, of rather firm closely imbricated appressed bracts with short green tips, outer successively shorter: leaves of firm texture, entire, or sometimes with a few teeth: rays of the showy heads violet or blue, rarely pale. — *Læves.*

++ Typical species, usually pale and glaucescent or glaucous; with involucral bracts whitish-coriaceous below and abruptly green-tipped (most conspicuous in dried specimens): akenes 4-5-ribbed: leaves on flowering branchlets commonly much reduced to rigid subulate bracts.

— **A. turbinéllus,** LINDL. Slender, 3 feet high, diffusely paniculate above: leaves light green, not rigid, from oblong to narrowly lanceolate, and all with narrow base (2 or 3 inches long), scabrous-ciliolate: heads (half-inch or more high) terminating divergent and minutely bracteolate slender branchlets: involucre elongated-turbinate or subclavate; its many-ranked bracts with very short and obtuse green tips: rays a third to half inch long. bright blue-violet: akenes minutely pubescent. — Comp. Bot. Mag i. 98. & DC. Prodr. v. 244; Torr. & Gray, Fl. ii. 117. — Hillsides and plains, Illinois and Missouri to W. Arkansas and Louisiana. Handsome species, flowering late.

— **A. virgátus,** ELL. Slender, strict and simple, with few or several racemose heads, or with virgate branches terminated with single heads: these and the flowers nearly as of *A. lævis*: cauline leaves lanceolate or linear, of firm texture, little if at all dilated at base: lower ones usually long and narrow; those of the branchlets subulate-acute and rigid. — Sk. ii. 553; Torr. & Gray, Fl. ii. 116; Chapm. Fl. 201. *A. vimineus,* Willd. Spec. iii. 2046 ? (fide herb., but a peculiar and imperfect specimen), not Lam. nor Nees. *A. purpuratus,* Nees. Ast. 118, & *A. miser,* Lam. Dict. i. 308. *A. attenuatus,* Lindl. in Comp. Bot. Mag. i. 97. *A. gracilentus,* Torr. & Gray, l. c. 166. — Upper N. Carolina to Louisiana and Texas. Form with narrow and linear leaves (lower 3 or 4 inches long by 2 or 3 lines wide) seems very distinct: broader-leaved forms pass into the next.

— **A. lǽvis,** L. Stouter, 2 to 4 feet high, rigid: leaves from ovate or oblong to lanceolate (4 or 5 inches long, decreasing upward); radical and lowest cauline contracted below into a winged petiole; upper all with auriculate or subcordate partly clasping base: heads sparsely thyrsoid-paniculate, on short and rigid branchlets: involucre campanulate or obscurely turbinate; the whitish coriaceous bracts bearing abrupt rhomboid or deltoid short green tips: rays 20 or 30, broadish, sky-blue verging to violet: akenes glabrous or nearly so. — Spec. ii. 876; Ait. Kew. iii. 206; Nees, Ast. 128. partly; Torr. & Gray, Fl. ii. 216 (the var. β is the typical plant); Gray, Proc. Am. Acad. xvii. 166. *A. rubricaulis,* Lam. Dict. i. 305: Nees, Ast. 131. *A. amplexicaulis,* Muhl. in Willd. Spec. iii. 2046; Nees. l. c., not of others. *A. Pennsylranicus,* Poir. Suppl. i. 498. *A. cyaneus,* Hoffm. Phyt. Blatt. 71, t B, f. 1; Nees, l. c.; Lindl. Bot. Reg. t. 1495. *A. glaucescens* & *A. politus,* Nees, Synops. 23. *A. lævigatus,* Hook. Bot. Mag. t. 2595. not Lam. nor Willd. — Borders of woodland, in dry or barely moist ground, Canada to Louisiana and west to the Rocky Mountains from Saskatchewan to New Mexico. A form from Fort Edward, N. Y. (*Vanderberg*), bore white rays changing to rose.

Var. **Géyeri.** A foot or two high: involucre broader and less imbricated; its bracts of thinner texture, mostly attenuate-acute, the green tip less definite. — Valleys of the Northern Rocky Mountains to Idaho, south to Wyoming, &c.

++ ++ Ambiguous species, green, at least not glaucous: involucre greener and somewhat looser.

A. versícolor, WILLD. Leafy up to the more corymbosely disposed inflorescence: leaves thinner than in preceding, bright green, oblong-lanceolate, obscurely if at all auriculate and not broadened at insertion, lower with some sharp serratures: involucre short-campanulate: rays "changing from white to deep violet," or commonly pale or bright violet from the first. — Spec. iii. 2045 & Enum. ii. 885; Nees. Ast. 128. *A. lævigatus,* Willd. l. c. 2046 (in part); Nees. l. c. 129. not Lam. *A. lævis* of the same authors, Lindl. Bot. Reg. t. 1500. *A. mutabilis,* Willd. l. c. 2045; Nees. l. c. 125. *A. confertus,* Nees. Ast. 146, white-fl. state. — Common in European gardens, doubtless from Atlantic N. America; but decisive indigenous specimens hardly known.

A. concínnus, WILLD. Stem and paniculate branches slender, 1 to 3 feet high (above often showing traces of pubescence in lines): leaves pale green, lanceolate, mostly somewhat serrate or serrulate; upper ones an inch or two long; lowest and radical spatulate-lanceolate and tapering into a winged petiole: heads much smaller than in preceding, numerous: rays 4 or 5 lines long, violet. — Enum. ii. 884; Nees, Ast. 121 (excl. β?); Lindl. Bot. Reg. t. 1619. *A. elegans*, Hort. Par. 1814, not Willd. — North America, received by Willdenow from Muhlenberg. An indigenous specimen from Pennsylvania, *Minn*, in herb. Cosson. This and perhaps that of N. Carolina, *Schweinitz* in herb. Ell. (now lost), and Arkansas, *Harvey*, seem to be the only indigenous ones seen.

+ + Involucre of the small or barely middle-sized and paniculately or racemosely disposed heads (3 or 4 lines high) pluriserially imbricated; its bracts rather rigid, narrow, with subulate or acute green nearly erect tips; rays white, sometimes turning purplish or violaceous: leaves mostly narrow and entire, narrowed at base: akenes minutely pubescent. — *Ericoidei*.

++ Heads disposed to be corymbosely or open-paniculate on erect branches: involucre nearly hemispherical: rays numerous, bright white, disposed to turn rose-purplish, 4 lines long.

—**A. Pórteri**, Gray. A foot or less high, glabrous and smooth (except hirsute ciliation of lowest leaves), either simple or branching above, bearing several or numerous thyrsoidly or corymbosely disposed heads: leaves linear or lower spatulate-linear (2 to 4 inches long, 1 to 3 lines wide), radical spatulate: heads broad: involucral bracts linear-subulate; outer little shorter than inner. — Proc. Am. Acad. xvi. 99. *A. ericoides*, var.? Proc. Acad. Philad. 1863, 64. *A. ericoides*, var. *strictus*, Porter & Coult. Fl. Colorad. 56. — Common in the Colorado Rocky Mountains at middle elevations, *Fremont, Parry, Hall & Harbour*, &c.

A. polyphýllus, Willd. Mostly tall (4 or 5 feet high), with virgate branches, glabrous: cauline leaves narrowly lanceolate or linear (4 or 5 inches long, quarter to half inch wide); those of flowering branchlets small and subulate-linear: heads paniculate (4 lines high): involucral bracts lanceolate-subulate, outer successively shorter. — Enum. 888; Spreng. Syst. iii. 336. *A. Americanus Beloideefolius*, &c., Pluk. Alm. t. 78. f. 5? *A. tenuifolius*, Nees, Ast. 119, in part; Torr. & Gray, Fl. ii. 132, in part. — N. Vermont to Wisconsin, south to Penn. and N. Carolina. Showy in cultivation, flowering much earlier than *A. ericoides*.

++ ++ Heads disposed to be racemose along spreading branches or branchlets: rays 15-25, and smaller, bright white, rarely purplish-tinged.

—**A. ericoídes**, L. Glabrous or nearly so in the typical form (but with hirsute varieties), rather rigid, a foot to a yard high, with lateral branches spreading or ascending and commonly unilaterally capituliferous: radical leaves oblanceolate and spatulate, often sparingly serrate; cauline narrowly lanceolate or linear and narrowed at both ends, entire; those of the branches and branchlets gradually diminished to setaceous-subulate: heads usually 3 lines high: tips of the involucral bracts somewhat abruptly subulate-acute or acuminate from a rigid or coriaceous base: akenes little compressed, scarcely nerved. — Spec. ii. 875 (specimen in herb. an attenuate cultivated form, not of syn. Dill. & Gronov., which are of *A. multiflorus*); Ait. Kew. iii. 202; Spreng. Syst. iii. 531; Torr. & Gray, Fl. ii. 123; Gray, Proc. Am. Acad. xvi. 165, not of Lam. nor Michx. *A. tenuifolius*, Willd. Spec. iii. 2026 (excl. syn.); Nutt. Gen. ii. 155; Nees, Ast. 119, partly. *A. dumosus*, Hoffm. Phyt. Blatt. t. A, f. 2. *A. ericoides & A. glabellus*, Nees, Ast. 107. *A. pauciflorus*, Martens, Bull. Acad. Brux. viii. (1841), 67. — Dry and open ground, Canada to Florida and the Mississippi.

Var. Reevésii (*A. Reevesii* of the gardens) is the most rigid form, comparatively stout, glabrous except that the leaves are often hispidulous-ciliate toward the base; the heads and rays as large and the latter about as numerous as in *A. polyphyllus*. It is *A. virgatus*, A. H. Curtiss, distrib. no. 1279, from dry river-banks near Nashville, Tenn.

Var. villósus, Torr. & Gray, l. c. Stem (generally low) with branches and not rarely the leaves villous-hirsute or hispid-hirsute. — *A. villosus*, Michx. Fl. ii. 113. *A. pilosus*, Willd. Spec. iii. 2055; Nees, Ast. 109. — Ohio to Iowa and Missouri, south to W. North Carolina. The var. *platyphyllus*, Torr. & Gray. l. c., is a very hirsute state of this, with leaves broader, some even an inch wide and sparingly serrate.

Var. pusíllus. Slender, a span to a foot high, glabrous: cauline leaves mostly slender-subulate or filiform: heads small (2 lines high), narrow, few-flowered: involucre turbinate; its bracts less rigid: rays 2 lines long. — Serpentine barrens, Lancaster, Penn., *Porter*. A singular form, probably dwarfed by sterility of soil.

Var. Pringlei. A slender and strict glabrous form, seldom over a foot high, simple or with few erect branches, rather small heads, and shorter tips to the involucre, mentioned in Proc. Am. Acad. xvi. 99. — Rocky islands and shores, northern part of Lake Champlain, *Pringle, E. Brainard.*

+ + + Involucre of the numerous small and *racemosely disposed heads with squarrose or at least spreading herbaceous tips* to the well-imbricated unequal bracts, these tips obtuse or merely mucronate-apiculate: cauline leaves small, all linear and entire, not at all or scarcely narrowed at the abrupt closely sessile or partly clasping base: akenes canescent-hirsute: herbage with somewhat cinereous or hirtellous pubescence. — *Multiflori.*

++ Rays amethystine-violet or purple: leaves not rigid.

A. amethýstinus, NUTT. Cinereously puberulent or the stems hirsutulous, 2 to 5 feet high, paniculately much branched: heads 3 lines high: tips of involucral bracts merely spreading, acutish, not ciliate: rays rather numerous, 3 lines long. — Trans. Am. Phil. Soc. (n. ser.) vii. 294; Torr. & Gray. Fl. ii. 144; Gray, Man. ed. 5. 234. — Rather low grounds, E. Massachusetts to Illinois and Iowa. This has been cult. in European gardens under the names of *A. pilosus* and *Bostoniensis.* It has much the habit of *A. oblongifolius*, but is destitute of viscidity and aroma.

++ ++ Rays white, rarely bluish or purple-tinged.

A. multiflórus, AIT. Low (a foot or two high), bushy-branched, cinereous or green: leaves rigid, scabrous- or hispidulous-ciliate: uppermost of the branchlets passing into involucral bracts; these mostly with obtuse tips: heads in the ordinary forms little over 2 (at most 3) lines long, and with only 10 to 15 or 20 rays. — Kew. iii. 203; Willd. Spec. iii. 2027; Torr. & Gray. Fl. ii. 124, with var. *stricticaulis*, a slender strict form of the North. *A. ericoides, dumosus*, Dill. Elth. t. 36. *A. ericoides*, L. spec. as to syn. Dill.; Michx. Fl. ii. 113; Selak Handb. t. 245, & (var. *multiflorus*) Pers. Syn. ii. 443. *A. ciliatus*, Muhl. in Willd. Spec. iii. 2027. *A. scoparius*, DC. Prodr. v. 242, a rather strict slender-leaved Texan form. *A. hebecladus*, DC. l. c., a very small-leaved hirtellous Texano-Arizonian form. — Dry or sterile ground, Canada to Georgia and Texas, common throughout Atlantic States, southwest to Arizona, northwest to Saskatchewan and Brit. Columbia. (Mex.) The most wide-spread species.

A. commutátus. A foot or so high, with divergent branches: heads more scattered and twice or even thrice the size of those of *A. multiflorus* (3 or 4 lines high and broad): rays 20 to 30: otherwise nearly as the preceding. — *A. multiflorus*, var. *commutatus*, Torr. & Gray, l. c., excl. syn. *A. biennis*, Torr. Ann. Lyc. N. Y., at least mainly. *A. ramulosus*, var. *incanopilosus*, Lindl. in DC. Prodr. l. c. & Hook. Fl. ii. 12. — Plains and river-banks, Dakota and Saskatchewan, to Utah and E. Oregon. Seems to pass into the preceding on one hand, and into *A. adscendens* on the other.

A. falcátus, LINDL. Much like a strict and simple-stemmed *A. multiflorus*, perhaps a high northern form of it: leaves all narrowly linear, glabrate or sparingly and minutely (and the stem more obviously) pubescent with soft somewhat appressed hairs. involucre broader, glabrous; its bracts thinner and looser; outer herbaceous to near the base and as long as the attenuate innermost. — Torr. & Gray, Fl. ii. 126. *A. falcatus* & *A. ramulosus* (as to the type), Lindl. in DC. Prodr. v. 241, 243, & Hook. Fl. ii. 12 — Subarctic America, from Cumberland House to Fort Franklin, near the Arctic Circle and Arctic coast, *Richardson*.

+ + + + Involucre of the small (2 or 3 lines high) and numerous heads mostly of the *Heterophylli*, pluriserial; the bracts not coriaceous, regularly and closely imbricated (outer successively shorter), smooth and glabrous, mostly whitish below and with definite short green tips, these not spreading; stems usually slender and not very tall; the branches divergent or divaricate (except in *A. racemosus*), and racemosely branched or racemosely capitulierous: leaves from lanceolate to subulate, not cinereous nor more than minutely scabrous, commonly spreading; all Atlantic species. — *Divergentes.*

++ Heads more scattered and singly terminating the racemose or compound-paniculate minutely foliose slender branches.

A. dumósus, L. Mostly quite glabrous and smooth, 1 to 3 feet high: leaves all entire and obtuse, commonly reflexed or widely spreading; the cauline linear (1 to 3 inches long and as many lines wide), of rather firm texture; those of branches and branchlets gradually smaller and shorter; ultimate ones reduced to minute bracts: involucre campanulate or short-turbinate (2 or 3 lines long), well imbricated and with very definite and broadish oval or oblong green tips to the obtuse or sometimes barely acutish bracts: rays from violet to nearly white, 2 lines long. — Spec. ii. 873 (with syn. mainly); Ait. Kew. iii. 202; Torr. & Gray, Fl.

ii. 128. *A. sparsiflorus*, Michx. Fl. ii. 112; Willd. Enum. 880 (with unusually large lower leaves); Nutt. Gen. ii. 155. *A. fragilis*, Lindl. Bot. Reg. t. 1537, & herb. — Border of moist or dry woods, New England and W. Canada to Florida and Texas. Runs into various forms, such as

Var. **coridifólius**, Torr. & Gray, l. c. A more rigid and effuse Southern form, with rather coriaceous leaves, especially the very small ones of the elongated branches and branchlets; involucral bracts also more rigid. — *A. coridifolius*, Michx. Fl. ii. 112; Willd. Spec. iii. 2028; Nees, Ast. 104; Lindl. Bot. Reg. t. 1487. *A. foliolosus*, Ell. Sk. ii. 345, not Ait. *A. foliolosus*, var. *coridifolius*, Nutt. Gen. ii. 155. *A. multiflorus*, Bertol. Misc. v. t. 5, fig. 3. — Pine barrens, S. Carolina to Florida and Louisiana.

Var. **subulæfólius**, Torr. & Gray, l. c. Rather rigid form, with ascending flowering branches, on which the somewhat large heads are often subracemosely paniculate and bearing erect or little spreading subulate-linear or linear-oblong very small leaves. — Open woods and pine barrens, Carolina to Texas; also N. W. Arkansas. The var. *subracemosus*, Torr. & Gray, l. c., was made up of specimens, some fairly referable to the present form, others to *A. vimineus*, var. *foliolosus*, or of intermediate forms.

++ ++ Heads racemosely unilateral, usually numerous or crowded along the flowering branches; the branchlets or minutely leafy peduncles shorter or little longer than the involucre: disk-flowers apt to turn purple.

A. racemósus, Ell. Minutely scabrous-pubescent along the numerous slender erect or ascending branches, probably rather tall (base of stem unknown): leaves rigid, linear, small, acute, entire: heads small (little over 2 lines high), somewhat spicately or more loosely racemose: involucre hemispherical and of narrower and acuter bracts than in the following: rays only a line or two long, purplish. — Sk. ii. 348; Torr. & Gray, Fl. ii. 127. — S. Carolina to Florida and Louisiana, in the low country or along the coast. Perhaps also Texas, but specimens (of *Lindheimer*) insufficient. Species insufficiently understood.

A. vimíneus, Lam. Nearly glabrous: stem 2 to 5 feet high, slender, simple, with numerous usually horizontal foliolose flowering branches, bearing numerous usually crowded heads: leaves linear or narrowly lanceolate, entire, or the lower with few serratures (the longer cauline 3 to 4 or 5 inches long, 2 to 4 lines wide): heads 2 or 3 lines high: bracts of involucre linear, usually acutish: rays commonly pure white (not rarely changing to purplish, even on a part of the plant), about 2 lines long. — Dict. i. 306 (1783, form with somewhat lanceolate cauline leaves); Gray, Proc. Am. Acad. xvii. 169. *A. Tradescanti*, L., as to one specimen in herb. Cliff., of very doubtful authority, also of hort. Par. in early days, of Nees, DC., Torr. & Gray, Fl. ii. 129, with var. *fragilis*; not of herb. L. (hort. Ups.), nor *A. Tradescanti*, Morison. *A. secundiflorus*, Desf. Hort. Par. 1815, & *A. diffusus*, DC. Prodr. v. 242, partly. *A. multiflorus*, Nutt. Gen. ii. 155, excl. syn. *A. tenuifolius*, Ell. Sk. ii. 347, not L. *A. fragilis*, Nees, Ast. 101, in part, not Willd. — Moist ground, Canada to Florida and west to Arkansas, most common northward; flowering rather early.

Var. **foliolósus**. Leaves linear, entire: branches ascending, bearing sparse or more paniculate heads: consists of forms intermediate between *A. vimineus* and *A. dumosus*, but with smaller heads than is usual in the latter, and thinner as well as narrower involucral bracts. — *A. foliolosus*, Ait. Kew. iii. 202. *A. ericoides*, *Meliloti agrariæ umbone*, Dill. Elth. 39, t. 35, a coarsely exaggerated figure: from which figure solely the char. & descr. of *A. miser*, L. Spec. ii. 877, were made, neither these nor the figure answering at all well to the dried specimen in herb. Sherard; which is said to have been raised from New England seeds. *A. dumosus*, var. *subracemosus*, Torr. & Gray, Fl. ii. 128. — New England to Illinois.

A. diffúsus, Ait. Either pubescent or almost glabrous, a foot to 4 or 5 feet high; the larger plants widely and divergently branched: leaves thinnish, mostly broadly lanceolate or wider, with much narrowed base, acute or acuminate; lower cauline generally 3 to 5 inches long, sparingly and acutely serrate; those of the flowering branches becoming small and entire, some of them surpassing the crowded or sometimes more scattered heads, which are usually disposed along the length of the flowering branches, either singly or in clusters; radical leaves ovate and slender-petioled: involucre campanulate; its bracts linear, obtuse or sometimes acutish, and with a definite short green tip: rays small, white, or sometimes tinged with purplish or violet. — Ait. Kew. iii. 205; Nees, Ast. l. c., &c. *A. divergens*, Ait. l. c.; Nees, l. c. *A. pendulus*, Ait. l. c. 204 (a form with narrowish and less serrate leaves, verging to or connecting with the preceding species); Nees, Ast. 100. *A. Tradescanti*,

Michx. Fl. ii. 118, not of L., &c. *A. miser*, Nutt. Gen. ii. 158 (a cinereous-pubescent variety or state, of sun-burnt situations, short-leaved and glomerate-clustered, partly the var. *glomerellus*, Torr. & Gray, under this name): Torr. & Gray, Fl. ii. 129; not L. (pl. Dill.), nor Ait. *A. parviflorus*, Darlingt. Fl. Cest. 446, not Nees. *Solidago lateriflora*, L. Spec. ii. 879. — Dry or barely moist ground, Canada to Texas and west to Missouri. *A. diffusus* is, on the whole, the best of three names of same date.

— Var. horizontális. A robust, very bushy-branched and exceedingly floriferous cultivated form; the leaves thickish, those of the widely spreading flowering branches small and short, entire; white rays more conspicuous. — *A. horizontalis*, Desf. Cat. ed. 3, 402. *A. recurvatus*, Willd. Spec. iii. 2047. — A plant of the gardens, not exactly matched by indigenous specimens, but evidently of this species.

— Var. thyrsoídeus. From slightly to distinctly cinereous-pubescent: leaves from ovate-oblong to lanceolate: flowering branches ascending, rather rigid, either short or somewhat elongated: heads thyrsoid-paniculate or spicate-glomerate, less secund. — Part of *A. diffusus*, var. *glomerellus*, Torr. & Gray, l. c. — New York to Illinois and Upper Canada. Western forms connect with the next species.

— Var. hirsuticaúlis (*A. hirsuticaulis*, Lindl. in DC. Prodr. v. 242, and *A. miser*, var. *hirsuticaulis*, Torr. & Gray, Fl. l. c.), founded only on specimens from Albany, N. Y., Beck, in herb. Torr. & Lindl., is a singular form, probably growing in much shade, with long and narrow leaves, as of *A. vimineus*, the midrib of these beneath and the stem very hirsute. Other forms in Torr. & Gray, Fl., are ambiguous between this and *A. vimineus*.

Var. bifrons. A luxuriant form, growing in shady and moist grounds, with large and thin elongated-lanceolate leaves, and spreading branches with loosely disposed and mostly larger heads: a transition to the following section and to *A. paniculatus*, Lam. — *A. bifrons*, Lindl. in DC. Prodr. v. 243. — Shady banks of Kentucky River (*Short*) to Illinois.

+ + + + + Involucre various, in some imbricated and with short close tips, as in the last preceding section, in others more loose and herbaceous: heads when numerous either thyrsoid- or open-paniculate on erect or ascending branches. — *Vulgares*.

++ Cauline leaves sessile, but neither with cordate or auriculate base (except in forms of *A. Novi-Belgii* and *A. foliaceus*), nor with abrupt winged-petiole-like lower portion.

= Atlantic United States species, with branching stems or several or many heads when well developed, none alpine or subalpine: herbage disposed to be glabrous, but branches often pubescent in lines.

a. Involucre of the small or middle-sized heads close and erect; its bracts narrow, imbricated in successive lengths, the small green tips not dilated nor spreading. Species seemingly confluent in a series.

— A. Tradescánti, L., partly. Stem slender, 2 to 4 feet high, with numerous erect or ascending branches and branchlets: leaves lanceolate or linear, slightly serrate or entire, thinnish: small heads numerous, corymbosely or somewhat racemosely paniculate; only two or three lines high: bracts of the involucre linear, acutish, partly green at tip and down the back: rays white, or sometimes tinged with lilac, only about 2 lines long. — Spec. ii. 876 (as to Hort. Cliff. in part, if herb. Cliff. is of any authority, and as to syn. *A. Virginianus parvis floribus Tradescanti*, Moris. Hist. iii. 121, whence the name); Ait. Kew. iii. 294 (var. fl. albis); Gray, Proc. Am. Acad. xvii. 166; not of L. Hort. Ups. & herb., only in small part of Torr. & Gray, Fl. *A. miser*, Ait. Kew. iii. 205, not L. *A. fragilis*, Willd. Spec. iii. 2051; not *A. Tradescanti fragilis*, Torr. & Gray. *A. leucanthemus*, Desf. Cat. 102; Poir. Suppl. i. 500. *A. artemisiæflorus*, Poir. l. c., ex char. *A. parviflorus*, Nees, Ast. 99, a rather strict form. *A. tenuifolius*, var., in part, Torr. & Gray, l. c. 132, not of others. — Open grounds, Canada to Virginia, Illinois, and Saskatchewan. Cult. from earliest days in European gardens. Some forms, both cult. and wild, show affinity to *A. dumosus*, *vimineus*, and *diffusus*; others differ from the next following species only in the smaller heads and flowers.

— A. paniculátus, Lam. Stem 2 to 8 feet high, freely and paniculately branched: leaves from elongated oblong to narrowly lanceolate, mostly attenuate-acuminate, sharply serrate or denticulate, or upper entire, thin: heads about a third of an inch high, in loose and open mostly leafy panicles: bracts of the involucre narrowly linear, with tapering acute or acuminate green tips (or outermost wholly green on back): rays 3 or 4 lines long, white varying to purplish or pale violet (in drying often turning blue). — Lam. Dict. i. 306 (1783, the char. not good for the involucre, but it is the *A. serotinus procerior*, &c., Tourn., cited by Lam.);

not Ait. 1789, nor of Willd. *A. Tradescanti*, L. Spec. ii. 876, as to herb. ("H. U.") & Hort. Ups. 262, not Hort. Cliff. & syn. Morison (whence the name *Tradescanti*); Ait. Kew. l. c. 204, as to var. *floribus coruleis*. *A. junceus*, Ait. l. c., as to pl. H. Kew. 1777 only. *A. dracunculoides*, Willd. Spec. iii. 2050, a form nearest to the preceding species, not Lam. *A. recurvatus*, Willd. Herb. fol. 1, but hardly of Spec. iii. 2047. *A. lanceolatus*, Willd. l. c., & *A. bellidiflorus*, Willd. Enum., are cultivated forms. *A. Lamarckianus*, Nees, Ast. 100, at least as to syn. Lam. *A. tenuifolius* (Nees in part), and *A. simplex*, Torr. & Gray, l. c., mainly, and *A. carneus*, Nees, Syn. 27 & Ast. 96, by the char., belong to this rather than to the next species, as do some indigenous (but not original) specimens named by Nees. *A. salicifolius*, Scholler, Fl. Barb. Suppl. (1785), 328, — to which belong *A. salignus*, Willd. Spec. iii. 240, *A. simplex*, Willd. Enum. 887, and probably *A. strictus*, Poir. Suppl. 498, — represents a form of this same species, either very early naturalized in Hungary and Germany, or possibly indigenous. *A. laxus*, Willd. Enum. 886, seems to be a similar form. — Low or moist ground, New Brunswick to Saskatchewan, E. Montana, and Louisiana, abundant in the Northern States, and polymorphous. A small and slender form, in Northern sphagnous bogs, with linear leaves, resembles *A. longifolius* in habit and foliage.

A. salicifólius, (LAM. ?) AIT. Resembles the preceding, equally branching: leaves commonly less elongated, less serrate or entire, of firmer texture, apt to be scabrous, and the fine reticulation of the veinlets manifest: involucre more imbricated; its bracts firmer, linear, with shorter and more definite green tips, these acute or obtusish: heads (as large as in preceding or broader) disposed to be thyrsoid or racemose-glomerate on the ascending branches: rays purplish to violet, rarely white. — Lam. Dict. l. c. ? (no Lamarckian specimens seen); Ait. Kew. iii. 203; Muhl. Cat.; Darlingt. Fl. Cest. 467. *A. praealtus*, Poir. Suppl. i. 493, merely a change of Aiton's name, not Nees. *A. eminens*, Willd. Enum. 886, is either this or the preceding. *A. rigidulus*, Desf. Cat. (1815), 122. *A. obliquus*, Nees, Ast. 76, cult. form. *A. onustus?* and perhaps *A. carneus*, Nees, Ast. 122, 96, on cult. forms, but char. and some specimens of latter are of the preceding species. *A. stenophyllus*, Lindl. in DC. Prodr. v. 242, narrow-leaved form. *A. carneus*, in part, & *A. Greenei*, Torr. & Gray, Fl. ii. 134. — Low grounds, Canada and New England to Saskatchewan, E. Montana, and Texas: most abundant in the Mississippi valley. The original of Ait. Kew., in the Banksian herbarium, is of flowering branches only, with small leaves.

Var. subásper. A rigid and commonly scabrous form, with thyrsoid-contracted and foliose inflorescence: broad heads commonly foliose-bracteate: bracts of involucre broader and firmer, often obtuse: rays violet. — *A. subasper*, Lindl. in Comp. Bot. Mag. i. 97, & DC. Prodr. v. 257. *A. carneus*, var. *subasper*, Torr. & Gray, l. c. — Illinois to Texas.

Var. cæruléscens. A strict and rigid form, with the rather large heads in a more naked inflorescence, and leaves all entire: involucral bracts with narrower acute or acutish tips. — *A. cærulescens*, DC. Prodr. v. 235. — Rocky banks, E. to W. Texas, *Berlandier, Lindheimer*, &c.

b. Involucre of the small or barely middle-sized heads looser and less imbricated; but its bracts erect or hardly at all spreading, narrow and linear, with acute and not at all dilated green tips, or outermost wholly herbaceous, these little shorter or equalling the inner: leaves linear or lanceolate, not rigid, not dilated at base, sparingly denticulate or entire.

A. júnceus, AIT. Slender, a foot to a yard high, the smaller plants simple-stemmed and with few heads, smooth and nearly glabrous: leaves linear or nearly so (3 to 5 inches long, 2 to 4 lines wide), entire, or lower with rare denticulations: involucre 3 lines high; its bracts all small, narrowly linear and erect, thinnish, manifestly imbricated in 2 or 3 series, and the outer more or less shorter (thus connecting with *A. paniculatus* of the preceding subdivision): rays light violet-purple, 4 or 5 lines long. — Hort. Kew. iii. 204, the indigenous specimen Halifax, *Hallgren*. *A. salicifolius*, Richards. App. Frankl. Journ. ed. 1, 478, ed. 2, 20, not Ait. *A. laxifolius*, Lindl. in Hook. Fl ii. 10, mainly; hardly of Nees, Ast., who had a cult. plant of Leyden Garden, and in herb. Lindl. so named a plant of *A. paniculatus?* *A. laxifolius*, var. *borealis*, & var. *latiflorus*, Torr. & Gray, Fl. ii. 138. *A. æstivus*, Gray, Man. mainly. *A. borealis*, Provancher, Fl. Canad. i. 308. — Wet meadows and cold bogs, Nova Scotia to Ohio, Michigan, Wisconsin, and north to Hudson's Bay, Saskatchewan, and Rocky Mountains, &c. Appears to pass into the next.

A. longifólius, LAM. A foot to a yard high, glabrous or pubescent, leafy: leaves elongated-lanceolate to linear-lanceolate, entire or sparingly serrulate, 3 to 7 inches long, tapering to both ends: involucre 4 or 5 lines high, little or not at all imbricated; its bracts all of nearly equal length, some looser outermost not rarely quite herbaceous: rays 3 or 4 lines long, violet or purplish, rarely almost white. — Dict i. 306, chiefly (and partly *A. paniculatus*, Lam.), fide herb. Par. *A. æstivus*, Ait. Kew. iii. 203: Willd. Spec. iii. 2030; Nees, Ast. 74; a shorter-leaved cultivated form. *A. eminens*, Nees, Ast. 87, in part, perhaps also *A. laxifolius*, Nees, certainly Hook. Fl., in part. *A. salicifolius*, Willd. ! herb. (not Ait.), therefore seemingly *A. hiemalis*, Nees, Ast. 77, said to blossom late. *A. floribundus*, Willd. fide spec. cult. herb. Par. 1814, hardly of Spec. Pl. *A. virgineus*, Nees, Ast. 88. *A. squarrulosus*, Nees, Ast. 86 ? — Low grounds or along streams, Labrador to Montana, Slave Lake, south to Canada and N. New England. Like other boreal species, flowers early when cultivated in lower latitudes.

Var. villicaúlis. A small and low form, with simple stem (a foot or less high) and midrib of narrow leaves beneath densely white-villous: heads few or solitary: rays deep violet. — Northern Maine, at Fort Kent, *Miss Furbish*.

c. Involucre of the middle-sized heads of firmer and more herbaceous or foliaceous-tipped and linear to spatulate bracts, imbricated in few to several series, of more or less unequal length, their summits from slightly to squarrose-spreading: leaves of rather firm texture: rays violet: comparatively late-flowering.

— **A. Novi-Bélgii, L.** Rather low, rarely tall, glabrous and smooth, or pubescent in lines on the branches: leaves from oblong to linear-lanceolate, entire or sparsely or obscurely serrate; upper with sessile base partly clasping and not rarely somewhat auriculate: heads mostly 4 or 5 lines high and bright blue-violet rays of equal length. — The commonest later-flowered blue Aster of the Atlantic border, in low or wet grounds, truly polymorphous, both in wild forms and in those of long European cultivation, many of which are not identified with indigenous originals. — Spec. ii. 877 (truly founded on the *A. Novi-Belgii*, etc., Herm. Hort. Lugd. 67, t. 69, raised from seed collected about the year 1680 in the vicinity of New York, whence the name, and probably represented by the plant of Hort. Cliff. 408; not by indigenous specimen in herb. Linn. from Kalm, which is *A. puniceus*, L., nor by plant in herb. from Upsal garden); Nees, Ast. 79; Gray, Proc. Am. Acad. xvii. 167. *A. serotinus*, Mill. Dict., probably. *A. floribundus*, Willd. Spec. iii. 2048. Assume as most normal, if not the original Leyden type, the common form away from influence of salt water, and with leaves not thickish; these from narrowly to oblong-lanceolate, their upper surface not rarely scabrous, and linear involucral bracts with narrow and acute spreading or recurving upper portion. — Common in wet grounds, New Brunswick and Canada to Georgia, chiefly eastward, but extending to Ohio and Illinois. *A. eminens*, var. *virgineus*, Lindl Bot. Reg. t. 1656, appears to be a nearly white-rayed form. *A. laxus*, Torr. & Gray, Fl. ii. 134, a very narrow-leaved form, and *A. præaltus* (Poir. ?), Torr. & Gray, l. c., one with broader leaves. *A. longifolius*, Gray, Man. ed. 5. 233 ; Sprague, Wild Flowers, 49, t. 10.

Var. lævigátus. Smooth and glabrous throughout or nearly so: leaves mostly oblong-lanceolate, little if at all thickened ; upper cauline disposed to be half-clasping by an abrupt or obscurely auriculate base: involucral bracts in few ranks, rather short, all not far from same length, loosely erect, and with comparatively short acutish herbaceous tips; thus resembling *A. versicolor* except that the involucral bracts are much less imbricated and little unequal. — *A. lævigatus*, Lam. Dict. i. 306; Poir. Suppl. i. 495, not Willd. &c. *A. mutabilis*, Ait. Kew. iii. 205 (cult. hort. Collinson & Kew, 1777, & herb. Jacq.); not L. by char., syn. Pluk., nor syn. Herm. *A. serotinus* & *Novi-Belgii*, in part, Willd. Spec. iii. 2048; Nees, Syn. Ast. 24. *A. brumalis* (also *A. onustus*, partly, & *A. eminens*, var. *lævigatus*). Nees, Ast. 88, &c. *A. argutus*, Nees, Ast. 69, file spec. Schultz Bip., hort. Bonn.; but char. does not accord. — Newfoundland to New England: hardly any wild specimens exactly answering to the plant cultivated and even naturalized in Europe ; but many that connect with the following, viz. : —

Var. litóreus. Stems rigid, low, or sometimes 3 or 4 feet high and then paniculately much branched, very leafy: leaves thickish and firm, very smooth (rarely upper face somewhat scabrous), oblong to lanceolate, upper partly clasping and sometimes auriculate ; bracts of the involucre loosely imbricated in several ranks, outer commonly spatulate, all but innermost with broadish or obtuse herbaceous and mostly thickish tips. — *A. Novi-Belgii*, L., as to Hort. Cliff., at least herb. Cliff. *A. tardiflorus*, Willd. Spec. iii. 2049, and of most later authors, not L. *A. adulterinus*, Willd. Enum. 884; Lindl. Bot. Reg. t. 1571. *Symphyotrichium unctuosum*, Nees, Ast. 135. The synonyms all from cultivated plants, less showy than the wild; and a dwarf form, very floriferous, with small leaves and small heads, is *A. cœspitosus* of the gardens (as cited by Lindley under *A. adulterinus*), probably the parent also of *A. Novi-Belgii*, var. *minimus*, of the gardens, with rose-purple rays. — Saline marshes and shores, Canada and New England to Georgia: evidently passes into the thinner-leaved form taken as the type, wherever it recedes from the influence of brackish water. The old cultivated forms evidently much altered in the European gardens.

Var. elódes. Slender, a foot or two high and simple, sometimes taller and with ampler panicle: leaves thickish, long and narrowly linear (2 to 5 inches long, 2 or 3 lines wide), entire; those of flowering branches or open panicle small and bract-like: involucre of rather well-imbricated narrow bracts, with short and mostly spreading acutish tips. — *A. elodes*, Torr. & Gray, Fl. ii. 136, chiefly. *A. longifolius*, Gray, Man. 233, in part, not Lam. — Swamps near the coast, New Jersey to Virginia. Would seem to be a most distinct species; but passes by gradations into forms of the type of the species, with narrow-lanceolate denticulate leaves of thinner texture; and the broad-leaved var. of Torr. & Gray, l. c., into the preceding variety.

Var. thyrsiflórus. Very leafy, smooth: cauline leaves narrowly lanceolate or nearly linear (2 to 4 inches long, 2 to 4 lines wide below the middle), attenuate-acuminate, commonly serrulate, of rather firm texture: heads numerous in a narrow thyrsoid panicle, or somewhat racemosely paniculate on elongated branches, rather large: involucre of the narrow bracts with attenuate and spreading or squarrose-recurving tips, as in the typical form. — *A. thyrsiflorus*, Hoffm. Phyt. Blatt. i. 83, t. D, f. 1 (yet figure and description answer rather to a broader-leaved form, either of the type or of the var. *lævigatus*); Poir. Suppl. i. 502; Nees, Ast. 65; DC. Prodr. v. 235, with var. *squarrosus*, Lindl. in DC. (*A. eminens*, Lindl. Bot. Reg. t. 1614, with abnormally foliose involucre) *A. spectabilis*, Willd. Spec. iii. 2048, as to descr. (not char.) & herb., not Ait. — Said to come from Virginia: cultivated plants not matched by indigenous specimens.

= = Pacific or Rocky Mountain species.

∗. Involucre of the middle-sized or small heads conspicuously and regularly pluriserially imbricated; outer bracts successively shorter; all loosely erect or little spreading, and with obtuse or obtusish mostly short and broadish herbaceous tips (occasionally in early or less developed heads some outer bracts foliaceous): leaves entire, or lower sometimes slightly serrate.

1. Heads mostly half-inch high, hemispherical, loosely paniculate: leaves comparatively large, none broadened at the insertion.

A. Chamissónis, Gray. Rather tall (2 to 4 feet high), with loosely spreading branches and branchlets, pubescent with spreading hairs or glabrate: leaves bright green, broadly lanceolate (larger cauline 3 to 6 inches long and an inch or less broad, those of flowering branchlets small): bracts of the broad involucre all but inner with obtuse and oval or apparently spatulate obtuse green tips (coarser and looser than in the next): rays bright violet, 4 to 6 lines long. — Torr. in Wilkes Exped. xvii. 341; Bot. Calif. i. 324. *A. radula*, Less. in Linn. vi. 125, fide Nees. *A. Chilensis*, Nees, Ast. 133; Torr. & Gray, Fl. ii. 112, but not at all Chilian. *A. spectabilis*, Hook. & Arn. Bot. Beech. 146, not Ait. *A. Neesii*, Schultz Bip. in Flora, 1856, 354, name merely suggested. — Moist thickets and along streams, California toward the coast, for nearly the whole length of the State, apparently reaching Oregon; first coll. by *Hænke* and *Chamisso*.

2. Heads smaller, 3 to 5 lines high.

A. Menziésii, Lindl. A foot or two high, strict, from cinereous-pubescent throughout to almost glabrous, bearing mostly numerous or thyrsoidly racemose-paniculate and rather small heads on rigid erect branchlets or peduncles; leaves lanceolate or the lower spatulate-oblong (2 or 3 inches long), on the branches small and linear or reduced to linear-subulate, so that the well-developed panicle is comparatively naked: involucre seldom over 3 lines high, short-turbinate, of linear slightly spatulate bracts in several rather closely imbricated ranks, nearly all obtuse: rays violet or purple, 3 lines long. — Hook. Fl. ii. 12, & DC. Prodr. v 243; Torr. & Gray, Fl. ii. 113 (described from the original, starved and arid, cinereous-canescent specimens, collected by *Menzies* in California, not "Oregon"); Torr. in Wilkes Exped. xvii. t. 8 (a similar form, collected on the Sacramento); Gray, Bot. Calif. i. 324. *A. Durandii*, Nutt. ex Durand in Pacif. R. Rep. v. 8. Has been sometimes taken for *A. falcatus*. — Dry or moist ground, throughout California to W. Nevada. There are connecting forms between this and the preceding, and others verging to the following.

◂ **A. adscéndens,** LINDL. A span to a foot or two high, rather rigid, from nearly glabrous to strigulose-hirsutulous: stems ascending or erect from creeping rootstocks, commonly branching, bearing few or rather numerous loosely paniculate or subcorymbose heads (these 4 or 5 lines high): leaves of firm and thickish texture (veins obscure), linear to spatulate-lanceolate, entire, with margins commonly hispidulous-ciliate or scabrous: bracts of the hemispherical involucre oblong-linear or obscurely spatulate, moderately unequal and in comparatively few ranks: the green tips looser, either glabrous, puberulent, or ciliolate; inner often mucronulate: rays 3 or 4 lines long, violet or purple. — DC. Prodr. v. 231, & Hook. Fl. ii 8; Torr. & Gray, Fl. ii. 111. *A. denudatus,* var. *ciliatifolius,* Nutt. Trans. Am. Phil. Soc. vii. 293. *A. falcatus,* Eaton in Bot. King Exp. 140, mainly. *A. multiflorus,* var. *commutatus,* Gray, Pl. Wright. ii. 76, a large form. — Plains and moist banks, Saskatchewan and Montana, to Colorado, New Mexico, N. Arizona, and W. Nevada, ascending the mountains to 10,000 feet; first coll. in Brit. America by *Drummond.*

Var. **denudátus,** TORR. & GRAY, l. c. A low or slender form, smoother, less leafy, or raineal leaves much reduced in size, and smaller heads and rays. — *A. denudatus* (& *A. ramulosus,* in part, as to specimens), Nutt. l. c. 292. *A. Nuttallii,* Torr. & Gray, l. c. 126; Eaton in Bot. King Exp., l. c. — Plains of Utah to S. Idaho; first coll. by *Nuttall.*

Var. **Yosemitánus.** Greener, less rigid, with comparatively large heads and looser involucre. — Sierra Nevada, from Summit to the Yosemite Valley.

A. Hállii, GRAY. Stem strict, a foot or two high, leafy to the top, bearing numerous short racemosely disposed and ascending flowering branches: these minutely pubescent; leaves (1 or 2 inches long, barely 2 lines wide) entire, scabrous-ciliolate, otherwise smooth and glabrous, neither dilated nor contracted at base: heads small (3 lines high) and numerous, somewhat racemosely paniculate and crowded: involucre campanulate, glabrous; the bracts subspatulate-linear with oval or oblong green tips rather close and erect: rays 2 or 3 lines long, white or whitish. — Proc. Am. Acad. viii. 388, name only. — Dry ground, Oregon, *E. Hall* (distrib. no. 243), *Lobb* (289), *Henderson.* Perhaps this is *A. bracteolatus,* Nutt. Trans. Am. Phil. Soc. vii. 293, no specimens of which seem to have been preserved, and which is compared with *A. campestris,* but is said to have a smooth involucre.

§ 8. Involucre of the middle-sized heads more or less imbricated but looser; the bracts all narrow (linear or approaching subulate), thinnish, from moderately to hardly unequal, loosely erect, all acute or acutish, with not at all dilated tips, nor are the outermost normally enlarged-foliaceous: leaves mostly entire.

1. Low, or only a foot or two high, chiefly of the mountains and high northward, mostly glabrous or a little pubescent.

A. Andínus, NUTT. Dwarf, with decumbent stems 2 or 3 inches long from filiform creeping rootstocks, bearing a solitary comparatively large head: leaves only half-inch long; radical and lower cauline spatulate; cauline (2 or 3) linear-lanceolate: involucre hemispherical, 4 lines high: its linear acutish bracts of almost equal length, nearly glabrous: rays violet, 4 lines long (35 or 40): style-tips short-lanceolate, acute. — Trans. Am. Phil. Soc. vii. 290; Torr. & Gray, Fl. ii. 154. — Rocky Mountains at Thornberg's Ridge, Wyoming, lat. 42°, near perpetual snow, *Nuttall.* Not since found; perhaps a high alpine state of *A. Fremonti.*

A. spathulátus, LINDL. Low, a span or two high, with ascending stems sparingly branched above and bearing 3 to 5 corymbosely disposed pedunculate rather large (half-inch high) heads: leaves (1½ to 3 inches long) linear-spatulate or upper linear-lanceolate with half-clasping base, and radical broader: involucre hemispherical; its linear bracts acutish, nearly equal: rays rather short, 3 lines long. — DC. Prodr. v. 231, & Hook. Fl. ii. 8; Torr. & Gray, l. c. — Subarctic America, between Bear Lake and Fort Franklin, on the Mackenzie River, *Richardson.* Approaches the next; but not matched.

◂ **A. Fremónti.** A span to a foot (rarely 2 feet) high, glabrous or some minute soft pubescence along the upper part of the slender erect stem: leaves thinner and with margins either quite naked and smooth or obscurely ciliolate-scabrous; radical and lowest cauline oblong or oblanceolate, or somewhat obovate (inch or two long), and tapering into a slender margined petiole; cauline from oblong-lanceolate to linear, commonly half-clasping at base; heads solitary in the smaller specimens, several in the larger, one third to half an inch high (and the numerous violet rays 4 lines long), somewhat naked-peduncled: bracts of the involucre narrowly linear, obtuse or acutish, or the inner acute, some of the outer shorter, all loose and similar. — *A. adscendens*, var. *Fremonti*, Torr. & Gray, Fl. ii. 503. *A. adscendens?* partly, Gray, Bot. Calif. i. 324. *A. laxifolius*, in part, Hook. Lond. Jour. Bot. vi. 160. — Rocky Mountains, from Montana to Colorado and Utah, in wet ground below the alpine region, west to the Cascade Mountains, lat. 49°, and along the Sierra Nevada, California.

Var. **Paríshii.** A dubious form (connecting with the next species?), with more imbricated and acute involucral bracts, their margins ciliolate. — Bear Valley in the San Bernardino Mountains, S. E. California, *Parish*.

A. occidentális, NUTT. A span to a foot or more high, smooth and glabrous (except some minute pubescence below the head), slender; smaller plants simple, bearing solitary or few heads; larger with slender branches and several or more numerous corymbose or paniculate heads (these 4 or 5 lines high): leaves mainly linear and narrow; cauline 1 to 3 inches long and only a line or two wide, rarely lanceolate and larger, occasionally (in Nuttall's specimens) bearing one or two salient lateral teeth or lobes; radical sometimes lanceolate-spatulate with long tapering base: involucre of narrowly or subulate-linear acutish or acute thinnish loose bracts, obviously imbricated, of 2 or 3 lengths: rays light violet, about 4 lines long. — Torr. & Gray, Fl. ii. 164 (*Tripolium occidentale*, Nutt. Trans. Am. Phil. Soc. vii. 296), a small and weak alpine form, apparently of a species which at lower elevations becomes taller, rather freely branched, and in Oregon passes into a diffusely much branched and paniculate polycephalous form. — Moist grounds and along streams, Idaho to Washington Terr., and along the Sierra Nevada, California, to Kern Co. (*A. œstivus*, Rothrock in Wheeler Rep.); first coll. in Oregon by *Douglas*.

Var. **scabriúsculus.** More strict, rather rigid, probably in drier soil with more exposure to aridity, stem and leaves scabrous-puberulent. — *A. œstivus*, Eaton in Bot. King Exp. 141. — Mountains of N. E. Nevada and Utah, *Watson*, *Wood*.

Var. **intermédius.** Ambiguous between *A. occidentalis* and a glabrous variety of *A. Menziesii* or of *A. adscendens*, a foot or two high, rather rigid, somewhat sparingly leafy, with paniculate flowering branches; short outer bracts of the involucre often quite obtuse, but narrower than in the two last-mentioned species: radical and sometimes cauline leaves lanceolate. — Wet meadows, Falcon Valley, &c., Washington Terr., *Suksdorf*, *Howell*, *Braudeyee*, and N. California, *Pringle*.

2. Tall (3 to 8 feet high) and branching, leafy to the top, paniculately polycephalous: Southwestern.

A. hespérius. Resembles *A. paniculatus* and *A. salicifolius* of the East, equally variable, from nearly glabrous and smooth to scabrous-pubescent. leaves lanceolate, entire or the larger with a few denticulations (2 to 5 inches long, 3 to 8 lines wide): heads rather crowded, 4 or 5 lines high: involucre of narrowly linear or more attenuate acute or gradually acuminate erect bracts, either unequal and imbricated, or with some loose and slender herbaceous exterior ones which equal the inner: rays either white or violet, 3 or 4 lines long. — Damp soil and along streams, S. Colorado and New Mexico to Arizona and S. California. Has been variously taken for *A. longifolius*, *Novi-Belgii*, *œstivus*, &c., and coll. by *Wright*, *Greene*, *Rothrock*, *Cleveland*, *Parish*, *Lemmon*, &c.

c. Involucre loose and foliaceous-bracteate at least some of the outer bracts herbaceous or foliaceous to the base or nearly so, equalling the inner, and more or less enlarged, either ascending or squarrose-spreading: the involucre of primary or early heads is more foliaceous; but, when the heads are more numerous, the enlarged outer bracts are not rarely wanting.

1. Heads small.

A. Oregánus, NUTT. Nearly glabrous: stem rather slender, 2 feet high, paniculately branched at summit, or bearing several to many paniculate heads: these about 3 lines high: leaves linear-lanceolate, entire (2 to 4 lines wide): outer and herbaceous involucral bracts lanceolate, acute, not longer than the thin and narrow inner ones (in some heads few or none): rays about 2 lines long, white or purplish. — Torr. & Gray, Fl. ii. 163, viz. *Tripolium Oreganum*, Nutt. Trans. Am. Phil. Soc. vii. 296, on small and hardly developed specimens. *A. simpler* and perhaps *A. carneus*, Eaton in Bot. King Exp. l. c. *A. laxifolius*, in part, Hook. Lond. Jour. Bot. vi. 240, not Nees. — Wet banks of streams and boggy meadows, Idaho and N. Nevada to Oregon and Washington Terr.; probably also N. California.

2. Heads middle-sized or large: rays violet or purple. (Species confluent.)

A. Douglásii, LINDL. Smooth, glabrous or nearly so: stems 2 or 3 feet high, with erect or ascending branches, bearing several or numerous paniculate heads; these 5 or 6 lines high: cauline leaves (either thinnish or rather firm) lanceolate (2 to 6 inches long, 3 to 5 lines broad in the middle), tapering to both ends, inserted by a narrow base, commonly serrate along the middle by acute and appressed or erect teeth: bracts of the involucre linear and acute, loosely imbricated and the small green tips commonly spreading; outer foliaceous ones few and not dilated, often wanting: rays 5 or 6 lines long. — DC. Prodr. v. 239 (not of herb. DC.), & Hook. Fl. ii. 11; Torr. & Gray, Fl. ii. 138. *A. subspicatus*, Nees, Ast. 74, from Cape Mulgrave, Alaska, is doubtless a form of this or of the next, and the name might apply to some specimens of either with contracted inflorescence. — Moist ground, commonly in shade, Northern Brit. Columbia to Oregon and N. California.

A. foliáceus, LINDL. Smooth and glabrous, or upper part of stem tomentulose or pubescent: leaves from broadly lanceolate to oblong and the lower spatulate, entire or nearly so; upper cauline very commonly with partly clasping and sometimes even subcordate-auriculate base: heads half-inch high, when few or solitary fully as broad, when more numerous less ample: involucre mostly with conspicuous loose foliaceous lanceolate or broadly linear outer bracts, which equal the inner, or sometimes more imbricated and squarrose: rays violet or purple, in the larger heads nearly half-inch long. — DC. Prodr. v. 228. Here made to include very various forms. The originals, from Unalaska and Sitka, are rather low, simple, or simple-stemmed with short monocephalous branches, leafy about the heads; farther south it becomes more branching, 2 or 3 feet high; generally differing from the preceding species in the ampler and broader as well as entire leaves, disposed to be half-clasping at base, and the leafy-bracted or much greener involucre. *A. Douglasii*, Eaton, Bot. King Exp. 141, & Gray, Bot. Calif. i. 324, mainly. — Wet ground, Alaska, Brit. Columbia, and along the mountains to eastern part of California and Nevada. Eastward it passes into

Var. fróndeus. Stem simple or with sparing erect flowering branches, sparsely leaved: leaves comparatively ample, 4 or 5 inches long; lower tapering into winged petioles, upper often with clasping base: heads solitary or few, naked-pedunculate, broad: involucral bracts linear-lanceolate, loose and not imbricated, all equalling the disk, occasionally the outermost broader and leaf-like. — *A. adscendens*, var. *Parryi*, Eaton, Bot. King Exp. 139. — Subalpine on the Cascade and Rocky Mountains, from the borders of Brit. Columbia to those of Colorado and the Wahsatch in Utah.

Var. apricus. Like a dwarf state of the preceding variety, grown in exposed places, somewhat rigid, thicker-leaved; stems ascending from tufted rootstocks, a span or two high, bearing solitary or 2 to 3 broad heads: involucral bracts all alike, somewhat spatulate-linear, obtuse or acutish: rays "deep blue-violet and reddish-purple intermixed." — High mountains of Colorado, at Union Pass, *Rothrock*, and near Gray's Peak, at 11,000–12,000 feet, in open and very dry places, *Patterson*. On Mount Paddo, Washington Terr., *Suksdorf, Howell*, the latter in a taller form, and looking toward *A. spathulatus*.

Var. Párryi. Includes some ambiguous forms, seemingly between the preceding variety and *A. Fremonti*, with stems a span to a foot high, with smooth and thickish rather large leaves, mostly naked heads; the involucre sometimes foliaceous-bracteate in the manner of the present species, sometimes wholly of the narrow and closer bracts of *A. Fremonti*. With that species this has been referred to *A. adscendens*. — Rocky Mountains of Colorado, subalpine, *Parry* (417), *Hall & Harbour* (253), *Vasey* (251), &c., and S. Wyoming, *H. Engelmann*.

Var. Búrkei. A foot or two high, rather stout, simple or branched above, leafy to the top: leaves thickish, very smooth, ample; upper cauline mostly oblong, and with broadly half-clasping usually auriculate insertion: heads solitary or several, very broad: involucre of oblong or spatulate and obtuse loosely imbricated bracts, the outer commonly shorter, or ontermost sometimes more foliaceous and equalling the disk. — Rocky Mountains, *Burke* in herb. Hook. Simcoe Hills, Washington Terr., *Howell*. Wahsatch Mountains at Alta, Utah, *M. E. Jones*. Mogollon Mountains, New Mexico, and Arizona, *Rusby*.

Var. Cánbyi. Like the preceding form in foliage, apparently tall and stout (base of stem and lower leaves wanting), leafy throughout the thyrsoid panicle of numerous subsessile heads: these comparatively small: upper leaves (only ones seen) rather broadly oblong and with broad half-clasping base obscurely auriculate: bracts of the involucre imbricated, with small and erect lanceolate green tips, only in some heads a few of the outermost loose and foliaceous, but seldom equalling the disk. — On White River in Western Colorado, *Vasey*, 1868, distributed under the name of *A. Canbyi*, Vasey: perhaps a distinct species.

13

Var. Eatóni. Rather tall (2 or 3 feet high), branching, bearing numerous and smaller paniculate or glomerate heads, and comparatively narrow lanceolate leaves: involucre loosely imbricated; outer and sometimes inner bracts foliaceous, either erect or squarrose-spreading: transitional between *A. foliaceus* and *A. Oregunus*, and some specimens approaching *A. Chamissonis.* — *A. Douglasii* mainly, Eaton, Bot. King Exp. 141. — Open ground or woods and along streams, Brit. Columbia to California along the borders of Nevada, and northeastward to Montana.

A. ámplus, LINDL. Glabrate: stem over 2 feet high, strict, robust, remotely leafy: leaves thinnish, acutely and saliently serrate or serrulate, or some entire, oblong or oval-lanceolate; cauline 2¼ to 5 inches long, mostly with narrowed partly clasping base; radical larger (over inch and a half wide), tapering into very long wing-margined petioles: heads several on rather naked peduncles: bracts of the involucre lanceolate and linear, of about two series, loose, of equal length, all rather shorter than the developed disk. — Hook. Fl. ii. 10, & DC. l. c. 236; Torr. & Gray, Fl. ii. 137. — Northern Rocky Mountains, *Drummond.* Not since collected, seen only in herb. Hook., perhaps rather of the following group.

++ ++ Cauline leaves either conspicuously contracted at base, some as it were into a winged petiole, or with auriculate-clasping insertion, or with both: involucre lax.

= Narrowed base of leaves not cordate- or auriculate-clasping at insertion.

A. Ellióttii, TORR. & GRAY. Glabrous, or the stout (2 or 3 feet high) stem minutely pubescent, very leafy to the corymbosely paniculate inflorescence: leaves thickish, oblong-lanceolate, serrate with small and appressed rather obtuse teeth, tapering below into the narrowed and as if wing-petioled base; upper 4 to 6 inches and lowest a foot or less long, including the channelled winged petiole: heads numerous, nearly half-inch high: bracts of involucre all of nearly equal length, loose, very narrowly subulate-linear, their green tips mostly spreading: rays narrow, "bright purple," 5 lines long. — Fl. ii. 140; Chapm. Fl. 204. *A. puniceus,* Ell. Sk. ii. 355, by the detailed descr. and specimen, excl. char. from *Willd.* — Swamps in the low country near the coast, S. Carolina to Florida.

A. pátulus, LAM. Glabrous or somewhat pubescent, either low or 2 to 4 feet high, with loose flowering branches: leaves ovate- or oblong-lanceolate, sharply serrate in the middle, acuminate at both ends, the lower into wing-margined petiole or attenuate base, even the upper with obscure if any auriculate insertion: heads loosely paniculate, about 4 lines high: bracts of involucre linear, erect or nearly so, loosely imbricated, the outer more or less shorter: rays light violet or purple, varying to white. — Dict. i. 308; DC. Prodr. v. 234. *A. Tradescanti,* Hoffm. Phyt. Blatt. 86, t. D, f. 2, not L. *A. pallens* & probably *A. præcox,* Willd. Enum. Suppl. 58. *A. Cornuti* (Wendl. ex Nees, where published, and why *Cornuti?*) & *A. acuminatus,* Nees, Ast. 58 & 60. *A. abbreviatus,* Nees, Syn. Ast. 16. — Canada and New Brunswick to E. New England, chiefly known in cultivation: introduced into the Paris garden in the days of Tournefort and Vaillant. There is a low form in the gardens, early flowering, having weak and often decumbent stems, as Lamarck characterized his species. The taller plants flower later.

= = Base of most cauline leaves auriculate- or cordate-clasping at insertion: involucral bracts loose, disposed to be equal in length and the outer foliaceous.

a. Atlantic species, chiefly Northern.

A. tardiflórus, L. A foot or two high, glabrous or stem somewhat pubescent (not hispid), bearing corymbosely disposed heads: leaves lanceolate or oblong-lanceolate, gradually acuminate, mostly with contracted or tapering base and with auriculate or obliquely semi-auriculate insertion: heads about 5 lines high: rays pale violet. — Spec. ed. 2, ii. 1231 (founded on plant cult. in hort. Upsal., low, with weak stems, which grew for 18 years before it flowered, and then late, whence the name: represented in the herb. by two specimens of the non-flowering, with the semi-amplexicaul spatulate-lanceolate leaves well marked, and one flower-bearing), not of later authors and gardens. *A. vimineus,* Nees, Ast. 68, in part, not Lam. nor Willd. — Along streams, Lower Canada and New Brunswick to Labrador. Nearly related to *A. patulus* on the one hand, to the succeeding and to *A. puniceus,* var. *laricaulis,* on the other. Ordinarily not a late-flowered species.

A. prenanthoídes, MUHL. A foot or two high, nearly glabrous, or the slender stem pubescent in lines, bearing loosely corymbiform cymose heads: leaves thin and elongated

(4 to 8 inches long), oblong- or ovate-lanceolate, saliently serrate in the middle, attenuate-acuminate, and lower half or third narrowed as if into a broadly winged petiole, which is more or less (in most cases conspicuously) dilated into an auriculate-clasping base; upper surface minutely scabrous, lower smooth: heads (mostly 4 lines high) on short rather rigid and divergent peduncles: rays not very numerous, about 5 lines long, pale violet or in shade whitish: bracts of involucre narrow and outer more or less spreading — Willd. Spec. iii. 2046; Nees, Ast. 61; Darlingt. Fl. Cest. 465; Torr. & Gray, Fl. ii. 142. — Moist ground especially along streams, W. New England to Penn. and Wisconsin, and throughout Canada. The var. *scaber*, Torr. & Gray, l. c. (New York & Penn. in few specimens, with stem almost hispid in the upper part, or else tall and branching), is probably a hybrid with *A. puniceus*.

A. puníceus, L. Stem commonly 3 to 7 feet high, loosely branching above, rather stout, often red or purple (whence the name), hispid with spreading bristles which are taper-pointed from a thickened rigid base (but sometimes these are few and sparse): leaves not rigid (3 to 6 inches long), oblong-lanceolate, acuminate, from coarsely and irregularly serrate to sparingly denticulate or sometimes entire, not at all or slightly narrowed toward the subcordate-semiamplexicaul base, commonly scabrous above and often hispid along the midrib beneath: heads (4 to 6 lines high) subsessile, either sparsely paniculate or thyrsoid-crowded: involucre of loose and thin soft and narrowly linear merely herbaceous bracts, with or without some larger and more foliaceous accessory ones: rays half-inch long, violet, varying to purple or occasionally white. — Spec. ii. 875 (Hort. Cliff., Herm. Lugd. t. 651, &c.); Ait. Kew. iii. 208; Michx. Fl. ii. 115; Lindl. Bot. Reg. t. 1636 (var. *demissus*), Torr. & Gray, Fl. ii. 140. *A. hispidus* & *A. amœnus*, Lam. Dict. i. 306. *A. blandus*, Lodd. Bot. Cab. t. 959. — Swamps and low thickets, Nova Scotia and Canada, west to Dakota, and common in the Atlantic States as far south as N. Carolina and the upper part of Georgia. A common species in cool districts, generally well marked, but running into some peculiar varieties, which may mostly be grouped under the following.

Var. **lævicaúlis**. Usually lower, a foot to a yard high: stem mostly green, smooth and naked below, above with mere traces of the characteristic hispid or hirsute pubescence: leaves serrate. — *A. blandus*, Pursh, Fl. ii. 555 (Solander in herb. Banks), appears to be this, but may be *A. tardiflorus*. *A. firmus*, Nees, Ast. 66, a low form, certainly of *puniceus*, with few-flowered branches. *A puniceus*, var. *firmus*, Torr. & Gray, l. c. *A. confertus*, Hort. Par. 1835-1869, probably Nees, Ast. 126, a form with numerous thyrsoid-crowded heads. *A. rimineus*, Nees, Ast. 68 (form with longer and nearly glabrous branches), not of Willd., nor of Lam. — New England, Canada, &c.

Var. **lucídulus**. A foot to a yard high, very leafy: stems glabrous, or with vestiges or even conspicuous traces of hispidulous pubescence: leaves lanceolate, entire or sparingly denticulate, somewhat lucid, wholly glabrous, but upper surface more or less scabrous: heads commonly numerous and thyrsoid-paniculate: involucral bracts less loose and less attenuate. — *A. lucidus*, Wenderoth, Ind. Sem. Marb., ex DC. Prodr. v. 247. *A. puniceus*, var. *rimineus*, Torr. & Gray, l. c., chiefly. — Low ground, New England to Illinois, Wisconsin, and northward.

b. Rocky Mountain and Western species.

A. Cusickii, GRAY. Soft-pubescent throughout, or sometimes approaching to glabrous: stems a foot or so high, simple or corymbosely branched, leafy to summit: leaves thin, nearly entire, oblong-lanceolate or oblong; upper ones moderately contracted above the deeply cordate-clasping base; lower with more elongated narrow lower portion or winged petiole with dilated but smaller auriculate-clasping insertion: heads large (over half-inch high) and broad, terminating simple or leafy short branches: involucre very foliaceous or foliose-subtended and loose; the larger and broader-lanceolate outer bracts fully equalling the inner; rays numerous, narrowly linear, nearly half-inch long, pale violet: akenes glabrous. — Proc. Am. Acad. xvi. 99. — Along subalpine streams, in mountains of E. Oregon, *Cusick*. Mountain meadows of W. Idaho, *Watson*. The latter seemingly connects with

Var. **Lyálli**. Villous with soft pubescence: stem over 2 feet high, rather stout: radical leaves not seen; cauline mostly narrowed below and with more or less auriculate half-clasping base, but even lower and larger (5 inches long and inch broad) not petiolar-contracted: heads terminating simple leafy branches: rays long for the size of the head (8 or 9 lines). — Between the Kootenay and Pend Oreille, Washington Terr., Aug., 1861, *Lyall*. Perhaps a distinct species and more allied to *A. amplus*, seen only in herb. Kew.

§ 7. ERIGERÁSTRUM. Involucre of *Erigeron*, i. e. broad, of very many and narrow acute or attenuate bracts, all of the same length, herbaceous, with no distinction of body and tip: rays numerous and narrow: pappus simple: heads solitary, or rarely two, large, terminating the simple stem: this leafy to the top, in which and in the acute style-tips the section differs from *Erigeron*, to which it makes transition : arctic and subarctic species.

A. peregrínus, PURSH. Tomentose-pubescent and glabrate, a span to 20 inches high from a thickish creeping rootstock: leaves oblong-lanceolate or upper ovate-lanceolate, these closely sessile by partly clasping base (inch or two long), either entire or sharply denticulate-serrate: head half-inch high and broader: bracts of the involucre attenuate, tomentose-pubescent or villous, not at all viscid or glandular: rays half-inch long, violet-purple.— Fl. ii 556; Torr. & Gray, Fl. ii. 155; Herder in Pl. Radd. ii. 10, in part. *A. Unalaschensis*, Less. in Linn. vi. 122. *A. Tilesii*, Wikstr. in Act. Holm. 1822, 13 ? *A. salsuginosus*, Hook. Fl. ii. 7, in part. *A. consanguineus*, Ledeb. Fl. Ross. ii. 473 ? — Alaskan Islands to Arctic coast; first coll. by *Nelson*. (Arct. E. Asia.) Has been confused with *A. salsuginosus*, Richards., now removed to *Erigeron*, which is naked-stemmed above, its involucre viscidulous-glandular and not villous.

A. pygmǽus, LINDL. Villous-pubescent and below glabrate, a span or less high and loosely cespitose: stems assurgent from a slender rootstock or creeping base: leaves lingulate-lanceolate to linear, entire, obtuse, nearly veinless (mostly an inch long): head about 4 lines high, solitary: bracts of the involucre spreading, linear, acute or obtuse, flaccid, densely or sparsely villous: rays 30 or more, apparently violet. — Hook. Fl. ii. 6, & DC. Prodr. v. 228; Torr. & Gray, Fl. ii. 154. — Arctic sea-coast, *Richardson, Rae*, &c. Seemingly connects with *Erigeron grandiflorus;* but has subulate and very acute style-tips.

§ 8. DŒLLINGÉRIA. *Pappus manifestly double;* outer setulose. i. e. of numerous rigid and short bristles or squamellæ in a distinct series, inner of long capillary bristles, some of which are usually clavellate-thickened at the tip: involucre of § *Orthomeris*, i. e. bracts destitute of herbaceous tips and thin-coriaceous. shorter than the disk: rays not numerous (8 to 13), always white: disk-corollas barely yellowish: akenes mostly obovate, several-nerved: heads corymbosely cymose (rarely solitary) at summit of stem or sparing branches, not large: leaves mostly entire, not rigid, veiny: pappus becoming tawny. — Gray, Proc. Am. Acad. xvi. 98. *Diplostephium*, Cass. Dict. xxxvii. 486, not HBK. *Dœllingeria*, Nees, Ast. 176, excl. spec. *Diplostephium*, § 1, DC. Prodr. v. 272, excl. spec. *Diplopappus* § *Triplopappus*, Torr. & Gray, Fl. ii. 182. (The most distinct subgenus, even worthy of generic rank, except for some transitions. *A. oboratus*, Meyer, *Rhinactina*, Less., has similar pappus, but is otherwise as *Xylorrhiza*.)

* Leaves acute or acuminate, all entire, generally green and almost glabrous, with loose veins and beneath a minute reticulation of veinlets (visible only under a lens): bracts of the short involucre mostly obtuse: akenes turgid-obovate at maturity, glabrate or glabrous: pappus rather rigid, at least some of the longer bristles clavellate: disk-corollas deeply 5-lobed.

A. umbellátus, MILL. Stem 2 to 7 feet high, generally tall and corymbose at summit, very leafy, bearing numerous rather crowded cymosely disposed heads: leaves lanceolate to oblong-lanceolate (3 to 6 inches long), acuminate and with tapering base: involucre hardly longer than the akenes; its bracts lanceolate-linear, rather obtuse: style-appendages deltoid-ovate, acutish: stronger pappus-bristles delicately clavellate. — Dict. ed. 7, no. 2; Ait. Kew. iii. 199; Hoffm. Phyt. Blatt. 74, t. B, f. 2. *A. amygdalinus*, Lam. Dict. i. 305; Michx. Fl. 109; Lindl. Bot. Reg. t. 1517. *Chrysopsis amygdalina*, Nutt. Gen. ii. 153. *Diplostephium umbellatum* & *D. amygdalinum*, Cass. l. c.; DC. l. c. 272; *Diplopappus umbellatus*, and *D. amygdalinus*, partly, Torr. & Gray, l. c. 183. — Low grounds, Newfoundland, S. Labrador, and Saskatchewan to Arkansas and Georgia; the typical form commoner northward: low forms with broader and more scabrous leaves common southward.

—— **Var. púbens.** Lower face of the oblong-lanceolate leaves tomentulose-pubescent, also usually the flowering branchlets. — Saskatchewan to Upper Michigan.

Var. latifólius. Stems 2 to 5 feet high: leaves from ovate-lanceolate to ovate, comparatively short, less narrowed or sometimes even rounded at base. — *A. humilis*, Willd. Spec. iii 2038, as to char. and indigenous specimen in herb., from Muhl., not Hort Berol. t. 67. *A. amygdalinus*, Bertol. Misc. vi. t. 5, f. 1. *Dallingeria amygdalina*, Nees, Ast. 179, chiefly, excl. syn. *D. cornifolia*, Lindl. in Hook. Comp. Bot Mag. i. 98. *Diplopappus amygdalinus*, Torr. & Gray, l. c. — Low pine barrens, &c., Penn. and New Jersey to Florida and Texas. Extreme forms seem very different from *A. umbellatus*, having leaves even 2 inches wide by 3 in length. In specimen from Georgia, *J. Donnell Smith*, style-appendages (abnormally?) rounded-obtuse.

—— **A. infírmus,** Michx. Stem slender, often flexuous, a foot to a yard high, less leafy, simple or with diverging flowering branches, bearing several or few (or even solitary) pedunculate heads: leaves obovate to ovate or oblong (rarely lanceolate, lower small and scattered), with attenuate base and bispidulous-ciliolate margin and midrib, more copious primary and some loosely reticulated secondary veins: involucre more imbricated, of thicker and broader obtuse bracts: style-appendages linear-subulate: pappus more rigid; bristles of the longer pappus nearly all clavellate, rather scanty. — Fl. ii. 109. *A. divaricatus*, L. Spec., as to syn. Gronov. & Pluk. Alm. t. 79, not of herb., nor char. *A. cornifolius*, Muhl. in Willd. Spec. iii. 2039. *A. humilis*, Willd. Hort. Berol. t. 67 (not herb. nor Spec. l. c.); Pursh, Fl. ii. 548 ; Ell. Sk. ii. 366. *Chrysopsis humilis*, Nutt. Gen. ii. 153, at least partly. *Dallingeria cornifolia*, Nees, Ast. 181. *Diplostephium cornifolium*, DC. l. c. *Diplopappus cornifolius*, Torr. & Gray, Fl. ii. 182. — Open woodlands, Massachusetts and Penn. to Upper Georgia, Tennessee, and Louisiana?

* * Leaves obtuse, occasionally toothed, both veins and veinlets conspicuously reticulated beneath: akenes oblong, pubescent: pappus softer and finer, inner bristles not clavellate: disk-corollas with short lobes.

—— **A. reticulátus,** Pursh. Canescently puberulent: stems strict, 1 to 3 feet high, simple or fastigiately branched at summit, bearing few or numerous slender-pedunculate heads: leaves oval or oblong, or lowest obovate (larger 3 inches long and 2 wide): involucral bracts lanceolate: rays 10 to 13, rather long and narrow. — Fl. ii. 548. *Chrysopsis obovata*, Nutt. Gen. ii. 152. *Aster obovatus* & *A. dichotomus* (the latter a slender and paniculately branching state), Ell. Sk. ii. 368, 366. *Diplostephium boreale*, Spreng. Syst. iii. 544. *D. obovatum* & *D. dichotomum*, DC. l. c. *Dallingeria obovata*, Nees, Ast. 182. *Diplopappus obovatus*, Torr. & Gray, Fl. ii. 184. — Low pine barrens, S. Carolina to Florida.

§ 9. IÁNTHE. *Pappus less distinctly double;* outer setulose (in one species obscure), inner not clavellate: otherwise as in § *Orthomeris:* involucre about equalling the disk, of narrow and appressed well-imbricated bracts: rays 10 to 18, violet: akenes narrow, villous: low and tufted plants, with rigid stems, which are thickly beset with the small linear or lanceolate entire and rigid one-nerved and veinless leaves. — Gray, Proc. Am. Acad. xvi. 98. *Diplostephium* § *Amelloidea*, Nees, Ast. 199. *Diplopappus* § *Amelloidei*, DC. Prodr. v. 277, partly. *Diplopappus* § *Ianthe*, Torr. & Gray, Fl. ii. 181.

* Head rather large (half-inch high) and broad: style-appendages elongated, subulate-linear or narrower: akenes flat, with strong marginal nerves and sometimes a single lateral nerve.

—— **A. linariifólius,** L. Stems 6 to 20 inches high, puberulent, strict, very leafy up to the heads: leaves widely spreading (except the small ones on the branchlets), narrowly linear, mucronate, about an inch long, green, smooth except the hispidulous-ciliolate or scabrous acute margins ; uppermost more or less passing into the rigid acutish bracts of the pluriserial campanulate or somewhat turbinate involucre: rays deep violet. — *A. linariifolius* & *A. rigidus*, L. Spec. ii. 874 ; Bart. Fl. Am. Sept. iii. t. 104 ; Bertol. Misc. Bot. v. t. 6. *A. pulcherrimus*, Lodd. Bot. Cab. i. t. 6. *Chrysopsis linariifolia*, Nutt. Gen. ii. 152. *Diplostephium linariifolium*, Nees, Ast. 199. *Diplopappus linariifolius* (Hook. Fl., Torr. & Gray, Fl.) & *D. rigidus*, Lindl. in DC. Prodr. v. 277. — Dry sandy or gravelly soil, Newfoundland to Wisconsin and Texas. A variety with white rays is occasionally seen.

A. scopulórum, GRAY. Scabro-puberulent and somewhat cinereous: stems tufted, rigid, only a span high, terminated by a solitary pedunculate head: leaves short (3 to 6 lines long), rigid, from oblong to linear or the lowest spatulate, the broader obtuse with an abrupt mucro, callous-margined: involucre broadly campanulate; its bracts imbricated in about 3 series, scabro-puberulent, lanceolate, acuminate: rays half-inch long, light violet: outer pappus sometimes distinctly squamellate. — Proc. Am. Acad. xvi. 98. *Chrysopsis alpina*, Nutt. Jour. Acad. Philad. vii. 34, t. 3, fig. 2. *Diplopappus alpinus*, Nutt. Trans. Am. Phil. Soc. vii. 304; Torr. & Gray, Fl. l. c. — Rocky Mountains, Montana and Wyoming, to W. Nevada and the border of California; first coll. by *Wyeth*.

A. stenómeres, GRAY. More slender, 6 to 10 inches high, green, minutely scabrous: solitary naked pedunculate head larger: leaves all linear (half to full inch long, a line wide), acutely mucronate, hardly margined: involucre broad; its bracts barely in two moderately unequal series, linear, acute or acuminate, thinnish, often pubescent: rays pale violet, over half-inch long: outer pappus setulose. — Proc. Am. Acad. xvii. 209. — Rocky Mountains of Montana and Idaho, *Burke, Watson*.

 * * Head smaller (a third to a quarter inch high) and narrow: the disk-flowers sometimes hardly more numerous than those (12 to 15) of the ray: style-appendages ovate and obtuse: akenes less compressed, lightly few-nerved: outer pappus of few or indistinct unequal short bristles.

A. ericæfólius, ROTHROCK. About a span high, strigosely canescent or hispidulous and glandular-scabrous, much branched: branches erect or diffuse, terminated by somewhat pedunculate heads: leaves commonly hispid-ciliate, erect or little spreading, 3 to 6 lines long; lowest spatulate and tapering into a petiole; upper from linear to nearly filiform, piliferous-mucronate: bracts of the involucre in about 3 series, lanceolate, acute or apiculate, thinnish, scarious-margined: rays purple or violet, sometimes white. — Rothrock in Bot. Gazette, ii. 70, & Wheeler Rep. vi. 152. *Inula? ericoides*, Torr. in Ann. Lyc. N. Y. ii. 212. *Eucephalus ericoides*, Nutt. in Trans. Am. Phil Soc. l. c. 299. *Diplopappus ericoides*, Torr. & Gray, Fl. ii. 182; Gray, Pl. Fendl. 69, var. *hirtella*, a hispid form. — Dry hills, Kansas and Texas to Utah, Arizona, and border of California; first coll. by *James*. (Adj. Mex.)

Var. ténuis, GRAY. Much less or not at all hispid: branches filiform and diffuse: all the upper leaves minute. — New Mexico, *Wright*, &c. (Adjacent Mex. to San Luis.)

§ 10. ORTHÓMERIS. Pappus simple: bracts of the involucre imbricated and appressed, destitute of foliaceous or herbaceous tips, often scarious-edged or more or less dry: rays fertile. — Gray, Proc. Am. Acad. xvi. 98. § *Orthomeris* with part of § *Oxytripolium*, Torr. & Gray, Fl.; Benth. & Hook. Gen. l. c.

 * Involucre well imbricated, of small and narrow bracts, greener than in others of this section (much as in *Aster* proper): low and slender herbs (a foot or less high), leafy-stemmed, branching above; with mostly linear erect and entire leaves, and several small white-rayed heads: akenes somewhat 4–5-angled or nerved.

A. ptarmicoídes, TORR. & GRAY. Rather rigid, 6 to 20 inches high in a tuft from short and thickish rootstocks, from smooth or minutely scabrous to hirtellous-puberulent, bearing a corymbiform cyme of several or numerous heads: leaves of firm texture, linear or the lower spatulate-lanceolate, lucid both sides, the broader ones nervose: bracts of the campanulate or somewhat turbinate involucre oblong-lanceolate, obtuse, thickish, rather rigid: rays 2 to 4 lines long, bright white, broadish: style-appendages acutely lanceolate-subulate: pappus white, of rather rigid bristles, longer ones manifestly clavellate at tip: akenes very glabrous, hardly at all compressed. — Fl. ii. 160. *Chrysopsis alba*, Nutt. Gen. ii. 152. *Inllingeria ptarmicoides*, Nees, Ast. 183. *Diplopappus albus*, Hook. Fl. ii. 21. *Hebecsteum album*, DC. Prodr. v. 264, excl. syn. Willd. *Aster albus*, Eaton & J. Wright. Man. Bot. 146, not Willd. herb. & Spreng. Syst. (which is *A. Amellus*). *Eucephalus albus*, Nutt. Trans. Am. Phil. Soc. vii. 299. — Rocky banks and bluffs, W. New England (S. Hadley, Mass.), to Illinois, the Saskatchewan, and the mountains of Colorado; first coll. by *Nuttall*. Depauperate plants sometimes only 2 or 3 inches high, and monocephalous.

Var. Georgiánus, GRAY. Taller and slender, over 2 feet high: lowest leaves 5 or 6 inches long, sometimes with 2 or 3 coarse denticulations: heads and rays rather small — Proc. Am. Acad. xvi. 98; Chapm. Fl. Suppl. 627. — Upper Georgia, near Rome, *Chapman*. Nearly the same from open woods of N. W. Arkansas, *F. L. Harvey*.

Var. lutéscens. Rays pale yellow, small. — *A. lutescens*, Torr. & Gray, Fl. l. c. *Diplopappus albus*, var. *lutescens*, Hook. l. c. *D. lutescens*, Lindl. in DC. l. c. — Saskatchewan, on dry limestone rocks of Red River, *Douglas*, a broadish-leaved scabrous-puberulent form. Englewood, N. Illinois, *E. J. Hill*, a slender and smooth form, with numerous and unusually small heads.

A. Lemmóni. Slender, from filiform rootstocks, somewhat strict, smooth and glabrous, bearing a few rather scattered heads: leaves not rigid nor lucid, not nervose; cauline somewhat gramineous, narrowly linear and attenuate (larger 4 or 5 inches long, a line or two wide), on flowering branches gradually reduced to subulate-attenuate; radical shorter, lanceolate-oblong or spatulate: involucre (3 lines high) equalling the disk, of about 3 series of thin linear and acute or acuminate bracts: rays 2 lines long: pappus of soft and slender bristles: akenes minutely canescent. — Along mountain streams in S Arizona: Santa Rita Mountains, *Pringle*, and Huachuca Mountains, *Lemmon*.

  * * Involucre rather loosely imbricated, of thin narrowly linear-lanceolate attenuate-acute bracts in not more than 3 series: akenes glandular, several-nerved: stems leafy, a foot or two high from filiform creeping rootstocks, bearing several or sometimes solitary long-peduncled middle-sized heads: leaves mostly pinnately veined, thin or thinnish, from lanceolate to oblong-ovate. Northern Atlantic species.

A. acumminátus, Michx. Somewhat pubescent or puberulent: stem leafless below, leafy and somewhat corymbosely branched above, or often simple, sometimes flexuous: leaves membranaceous, 3 to 6 inches long, mostly oblong with cuneiform-attenuate base and slender acuminate apex, sharply and coarsely dentate, primary veins abundant and conspicuous: heads usually several and corymbiform-paniculate, barely half-inch high: rays linear, white, or tinged purplish: style-appendages lanceolate-subulate, slender: akenes narrow. — Fl. ii. 109; Hook. Bot. Mag. t. 2707, & Fl. ii. 9; Torr. & Gray, Fl. ii. 157, not Nees. *A. divaricatus*, Lam. Dict. i. 305 (herb. Juss.), not L. *A. diffusus*, var. *acuminatus*, Pers. Syn. ii. 447. *Diplostephium acuminatum*, DC. Prodr. v. 273. — Deep and cool woods, S. Labrador to Pennsylvania, and along the mountains to Georgia.

A. nemorális, Ait. Somewhat puberulent: stem slender, very leafy above, sometimes simple and bearing a single slender-pedunculate head, often corymbosely or somewhat umbellately branched above, the branches similarly monocephalous: leaves from oblong-lanceolate to broadly linear, an inch or two long, acutish or obtuse, tipped with a callous point, entire or slightly few-toothed, scabrous above; those of the flowering branches or peduncles linear-subulate and scattered: involucre of more numerous linear-subulate puberulent bracts: rays broadly linear, lilac-purple: style-appendages broadly lanceolate: akenes broader. — Kew. iii. 198; Nutt. Gen. ii. 154; Torr. & Gray, l. c. *A. uniflorus*, Michx. Fl. ii. 110, small and simple-stemmed form. *A. ledifolius*, Pursh, Fl. ii. 544. *Galatella nemoralis*, Nees, Ast. 173. — Bogs and swamps, Newfoundland and Hudson's Bay to New Jersey.

  * * * Involucre closely and regularly appressed-imbricated in several series of ovate or ovate-lanceolate dry and chartaceous (sometimes purplish-tinged) bracts: akenes oblong, compressed, more or less pubescent: stems leafy to the top, bearing several or rarely solitary pedunculate heads: leaves mostly pinnately veined, sessile, from lanceolate to oblong-ovate, commonly entire. Rocky Mountain and Northwestern species. — *Eucephalus*, Nutt.

  ← Style-appendages subulate, equalling or longer than the stigmatic portion: involucral bracts all thin and dry, acute or acutish, commonly tomentose-ciliate, at least when young: akenes rather broad and flat, hirsute, becoming glabrate at maturity: stems mostly simple and 2 or 3 feet high, striate-angled: heads showy: rays purple or violet.

A. Engelmánni, Gray. Commonly rather tall and robust, green, slightly puberulent to glabrous: leaves thin, ovate-oblong to broadly lanceolate (2 to 4 inches long), loosely veined, the larger sometimes with a few small acute teeth, upper commonly tapering at apex into a slender or cuspidate acumination: heads (fully half-inch high), hemispherical, either racemosely disposed on slender axillary peduncles or somewhat thyrsoid-cymose: involucral bracts mostly acute or acuminate; some outer ones loose, narrow and partly herbaceous, or with loose pointed tips; inner purplish: rays about half-inch long: style-appendages attenuate-subulate: akenes obovate-oblong with narrowish summit. — Am. Jour. Sci. ser. 2, xxxiii. 238, without char. *A. elegans*, var. *Engelmanni*, Eaton, Bot. King Exp. 144. — Rocky Mountains, Utah and Wyoming to the Brit. boundary and in the Cascades, Washington Terr.; first coll. by *H. Engelmann* and *Lyall*.

Var. ledophýllus, GRAY. Stem lower (not over 2 feet high), rather strict: leaves smaller (inch or two long), cottony-tomentulose beneath, at least when young; lower obtuse and merely mucronate, uppermost with slender cuspidate point usually developed. — Proc. Am. Acad. viii. 388. *A. ledophyllus,* Gray, Proc. Am. Acad. xvi. 98, without char. — Mount Hood at 4,000 feet and upward, *Hall, Howell,* Mount Paddo (Adams), *Suksdorf.* Seemingly distinct, but passes into the type and into the following variety.

Var. glaucéscens. Stem mostly slender, 1 to 3 feet high, in the larger plants more branched above and bearing rather numerous corymbosely disposed heads: leaves somewhat glaucous, wholly glabrous (except the minute ciliolation of the species), lanceolate, 2 or 3 inches long, 3 to 7 lines wide, uppermost usually attenuate-cuspidate: heads smaller or less broad: involucre of fewer and closer bracts. — Washington Terr.: on Mount Paddo, and Simcoe Mountains, *Suksdorf; Howell.* N. California, mountains of Siskiyou Co., *Greene, Pringle,* distributed as *A. elegans,* which it approaches.

A. élegans, TORR. & GRAY. Slender, 1 to 3 feet high, mostly scabro-puberulent: leaves thickish, pale, lanceolate (inch or two long), erect, the upper apiculate-mucronate, the veins inconspicuous: heads several at summit of simple stem or branches, comparatively small and few-flowered (4 or 5 lines high): involucral bracts all close and conspicuously woolly-ciliate, barely acute, outer ovate, none with pointed tips: rays rather few, about 4 lines long: style-appendages linear-subulate, hardly acute. — Fl. ii. 159; Eaton, l. c. (a somewhat ambiguous form). *Eucephalus elegans,* Nutt. Trans. Am. Phil. Soc. vii. 298. — Mountains of Wyoming and Montana to N. Nevada and E. Oregon; first coll. by *Nuttall.*

+ + Style-appendages obtuse and short-oblong, shorter than the stigmatic portion: involucral bracts firmer; all the outer obtuse, not ciliate nor scarious-margined: akenes narrower, merely pubescent.

A. glaúcus, TORR. & GRAY. Throughout smooth and glabrous, glaucescent or pale: stems a foot high from extensively creeping filiform rootstocks, branching, bearing several or numerous paniculate heads: leaves thickish, lanceolate (1 to 3 inches long, a quarter to half inch broad), rather obtuse, when dry reticulate-venulose both sides: involucre 3 lines high, imbricated in about 3 ranks: rays bright violet, 4 to 6 lines long. — Fl. ii. 150; Eaton, l. c. *Eucephalus (Lagatea) glaucus,* Nutt. l. c. — Rocky Mountains, Wyoming to Colorado and Utah.

* * * Involucre less imbricated, hemispherical; the bracts in few ranks and in the typical species somewhat equal, partly greenish, with or without scarious margins: pappus-bristles not clavellate-thickened at tip: low-stemmed or acaulescent, from a thick and sometimes ligneous caudex or rootstock, with solitary or few pedunculate heads, and rather large and numerous rays: leaves thickish, narrow, one-nerved or nervose, entire. — *Xylorrhiza,* Nutt. Trans. Am. Phil. Soc. vii. 298. § *Orthomeris, Xylorrhiza,* Torr. & Gray, l. c.; also Benth. & Hook. Gen. ii. 273, excl. syn. *Rhinactina* (which has a double pappus) & *Arctogeron* (which has the characters of *Erigeron* with somewhat too imbricated involucre). Western montane species.

+ Genuine species, with comparatively large (half-inch high or more) and showy heads, and thickish leaves: pappus-bristles rather rigid.

++ Heads terminating short leafy stems which arise from creeping and ligneous rootstocks: involucral bracts acuminate and mucronate-tipped: style-appendages triangular- or lanceolate-subulate, not attenuate, shorter than the stigmatic portion: akenes oblong, very villous. — *Xylorrhiza,* Nutt.

A. Párryi, GRAY. Tomentose-pubescent and cinereous, a span high: leaves mostly spatulate and obtuse with a mucronate point, an inch or more long: heads usually solitary on peduncle surpassing the leaves, very broad: bracts of the involucre oblong-lanceolate, densely cinereous-pubescent: rays white, over half-inch long. — Am. Nat. viii. 212. — Rocky Mountains in Wyoming, on marshy flats of Sandy Creek, Green River, &c., *Parry, A. J. McCosh.*

A. Xylorrhíza, TORR. & GRAY. Less pubescent and glabrate, 4 to 8 inches high: leaves from narrowly spatulate-lanceolate to linear (1 or 2 inches long, 1 to 3 lines wide): the upper commonly equalling the 1 to 3 peduncles: heads smaller: involucral bracts more attenuate: rays "pale red" or "pale rose-color," 4 lines long. — *Xylorrhiza villosa* & *X. glabriuscula,* Nutt. Trans. Am. Phil. Soc. n. ser. vii. 297, 298. *Aster Xylorrhiza* & *A. glabriuscula,* Torr. & Gray, l. c. 158; the latter a more leafy-stemmed form. — Clayey soil and on rocks, Rocky Mountains of Wyoming, toward the sources of the Platte; first coll. by *Nuttall.* Laramie Plains, *Parry.*

++ ++ Head (broad and large for the plant) solitary on the simple and scapiform few- and small-leaved stems, which with the cluster of narrow radical leaves rise from a thickened caudex: involucral bracts linear, acutish, rather loose, often tomentulose when young: the plants otherwise glabrous and smooth; rays numerous, purple or violet: style-appendages slender and acute, usually more than twice the length of the stigmatic portion: akenes narrow: pappus strongly denticulate.

A. Andersóni, GRAY. Scapiform stems a span to a foot high, erect: radical leaves ligulate-linear or slightly broader upward, gramineous, mostly acute (2 to 10 inches long. 2 or 3 lines wide), nervose when dry; upper cauline reduced to scattered subulate bracts: head broad (fully half-inch high and wide): style appendages filiform: akenes oblong-linear, soft-villous. — Proc. Am. Acad. vii. 352. & Bot. Calif. i. 325. *Erigeron Andersoni,* Gray, l. c., vi. 540. — Wet subalpine meadows, along the whole length of the Sierra Nevada, California and borders of Nevada; first coll. by *Anderson.*

A. pulchéllus, EATON. Scapiform stems spreading and assurgent, 2 to 4 inches long: radical leaves from lingulate-spatulate to narrowly linear an inch or two long, obtuse, nerveless, in the larger western form often 3 or 4 lines wide near apex, and heads as large as those of *A. Andersoni:* in the smaller more eastern form only a line wide and heads smaller: style-appendages linear-subulate: akenes linear, striate, glabrate, at least below. — Bot. King Exp. 143. t. 16, the small and slender form, published in 1871. *A. alpigenus,* Gray, Proc. Am. Acad. viii. 389 (1873), the larger form first collected by *Tolmie,* and published as *Aplopappus alpigenus,* Torr. & Gray, Fl. ii. 241, the rays supposed to be yellow, whereas they are violet. — On the higher mountains, viz. Ranier, Paddo, and Hood, of Washington Terr. and Oregon, rediscovered by *Hall, Howell, Suksdorf, Mrs. Barratt,* and the smaller form on Blue Mountains, E. Oregon, *Cusick,* those of N. Nevada, *Watson,* also Rocky Mountains of Wyoming and Montana, *Hayden, Parry, Scribner.*

++ ++ Ambiguous species, with small heads (2 or 3 lines high) few or solitary, terminating very slender leafy stems or branches; and leaves small and slender: style-appendages ovate-subulate, about the length of the stigmatic portion: akenes compressed, hispidulous-pubescent, 2-3-nerved: pappus rather scanty and fragile (therefore near to *Erigeron,* but with the style-tips of *Aster*): small and many-stemmed from a somewhat ligneous caudex, nearly glabrous.

A. Watsóni, GRAY. Cespitose, 2 to 4 inches high; the filiform stems mostly monocephalous: leaves filiform-linear, or the lower and larger (inch long) with spatulate-dilated apex; upper very small: bracts of the involucre lanceolate, acute, commonly purplish-tinged, 2-3-seriate: rays white or purplish: style-appendages ovate or triangular and acuminate-subulate. — *A. glacialis,* in part, Eaton, Bot. King Exp. 142 (no. 500), also mixed with specimens of *A. pulchellus.* — Mountains of N. Nevada, Wahsatch Mountains at the head of American Fork; first coll. by *Watson.*

A. arenarioídes, EATON. Stems tufted on a woody caudex, almost filiform, 6 to 9 inches high, sparingly branched above, or bearing 2 to 4 heads: leaves filiform-linear, even the lower (inch or two long) only obscurely dilated upward: uppermost reduced to minute subulate bracts: bracts of the involucre linear, rather rigid, unequal and 3-seriate: rays white or bluish: style-appendages ovate-subulate, merely acute. — Proc. Am. Acad. viii. 647. *Erigeron stenophyllum,* Eaton, Bot. King Exp. 152. t. 17, not Gray. — Wahsatch Mountains, above Cottonwood Cañon, 8,000–9,000 feet, *Watson.*

* * * * * Involucre (except in *A. pauciflorus*) well imbricated and with short outer bracts disposed to pass into scale-like bracts of the peduncle: herbs or shrubby plants, maritime or of alkaline soil; the leaves more or less fleshy or reduced to scales. — § *Oxytripolium* in part (the perennial species), Torr. & Gray.

+ Heads rather large (about half-inch high), with showy violet rays: involucre well imbricated in several ranks: leaves long and narrow, entire, moderately fleshy; very glabrous herbs of the Atlantic coast. (Here also *A. imbricatus.* Walp. Rep. ii. 574. *Tripolium imbricatum,* Nutt., and the true *T. conspicuum,* Lindl. in DC., of Chili; see Proc. Am. Acad. xvii. 210.)

A. Chapmáni, TORR. & GRAY. Stem simple and slender, 2 or 3 feet high, from a thickish caudex, bearing a few simple slender monocephalous branches at summit: leaves rigid when dry, linear, or radical spatulate-linear (these 5 to 9 inches long, including the long attenuate base), obscurely nerved when dry; cauline becoming subulate-filiform and erect, and reduced on the branches to minute bracts: involucre campanulate, equalling the disk; its rather firm bracts mostly oblong-lanceolate, acute or mucronate: style-appendages ovate-subulate: akenes oblong, 7-10-nerved: pappus rather rigid. — Fl. ii. 161; Chapm. Fl. 205. — Pine-barren swamps, W. Florida, *Chapman, Curtiss.*

A. tenuifólius, L. Stem simple or paniculately branched above, a foot or two high from a weak and slender rootstock, often flexuous, somewhat sparsely leafy: leaves rather fleshy, at least thickish, linear, tapering to both ends, acute; the lower (2 or 3 lines wide) with long tapering base; upper subulate-attenuate: involucre turbinate; its bracts lanceolate-subulate and attenuately very acute: style-appendages linear-subulate: akenes narrow, 5-ribbed, hispidulous-pubescent: pappus soft. — Spec. ii. 873 (excl. syn. Pluk.) & herb.; Gray, Proc. Am. Acad. viii. 647. *A. flexuosus*, Nutt. Gen. ii. 154; Torr. & Gray, l. c. *A. sparsiflorus*, Pursh, Fl. ii. 547; Ell. Sk. ii. 346, not Michx. .1. *Tripolium*, Walt. Car. 154. — Salt or brackish marshes, coast of Mass. to Florida. This is one of the plants of Clayton which by the character in Gronov. Fl. Virg. was referred by Linnæus to *A. linifolius.*

+ + Heads rather small (quarter-inch high), with conspicuous violet or purple rays: little imbricated involucre with peduncles and upper part of stem *viscid-glandular:* wholly herbaceous, western, might be sought among the *Glandulosi* of true *Aster.*

A. pauciflórus, Nutt. Stem 6 to 20 inches high from a slender creeping rootstock, simple and bearing few heads, or branching above and with several corymbosely disposed short-peduncled heads: leaves moderately fleshy, linear, or radical subspatulate or elongated-lanceolate, entire, uppermost reduced to short sparse bracts: bracts of short hemispherical involucre rather fleshy and green, moderately unequal and rather loose, in only 2 or 3 ranks: style-appendages lanceolate-subulate: akenes narrow, compressed, striate-nerved, appressed-pubescent. — Gen. ii. 154, & Trans. Am. Phil. Soc. vii. 292; Torr. & Gray, Fl. ii. 164. *A. caricifolius*, HBK. Nov. Gen. & Spec. iv. 92, t. 333. *Tripolium subulatum*, Nees, Ast. 167; Lindl. in Hook. Fl. ii. 15, & DC Prodr. v. 254. *T. caricifolium*, Schauer in Linn. xix. 721. — Wet saline soil, Saskatchewan and Dakota to New Mexico, Utah, and Arizona. (Mex.)

Var. **gracíllimus,** Gray, Pl. Wright. ii. 76, a very slender form, with leaves almost filiform; from New Mexico, *Wright.*

+ + + Heads small or rather small, with close imbricated involucre and who'e herbage smooth and glabrous: branching plants with lignescent base, or even shrubby, all of the Southwestern borders and Mexican, and in saline soil.

++ Low and spreading or tufted, with merely liguescent base, leafy: rays purple or violet, rather conspicuous, about 3 lines long.

A. blepharophýllus, Gray. Loosely surculose-tufted, with ascending flowering stems a span or two high: leaves fleshy, conspicuously hispid-ciliate with strong bristles; those of creeping sterile shoots and rosulate tufts linear-spatulate, half-inch long; of the branching flowering stems much smaller, short-linear, and upper ones reduced to minute and merely bristle-tipped scales: heads 3 lines high: involucre turbinate; its bracts dry and pale, ovate-oblong to lanceolate, rather obtuse, carinate-one-nerved: rays 10 to 14: style-appendages short-subulate: akenes obscurely striate-nerved, not compressed, sericeous. — Pl. Wright. ii. 77. — Las Playas Springs, New Mexico, *Wright.*

A. ripárius, HBK. A foot or two high from a somewhat liguescent base, diffusely branched: branches terminated by solitary heads (of 4 or 5 lines in height and equally broad): leaves linear and entire, or lowest spatulate and incisely few-toothed, an inch or less long, on the branches toward the heads gradually reduced to small subulate bracts: involucre shorter than the disk; its numerous well-imbricated bracts narrowly lanceolate and with subulate-acuminate greenish tips: style-appendages subulate, rather short: akenes pubescent, obscurely striate: pappus rufous. — Nov. Gen. & Spec. iv. 92, the rays said to be white, which is probably a mistake, and the involucre subsquarrose, but it is not so, though the outer may be a little loose. *A. Sonoræ*, Gray, Pl. Wright. ii. 76. — S. Arizona, west of the Chiricahui Mountains, *Wright.* (Mex., *Humboldt.*)

++ ++ Taller, much branched, rigid, woody at base, with small heads terminating the branchlets: rays small (a line or two long) and white or none: anomalous species.

A. carnósus, Gray. Glaucescent or pale, 2 or 3 feet high; the rigid slender stems diffusely and at length intricately much branched: lower leaves linear and very fleshy, an inch or less long; upper and those of the branchlets reduced to small or minute subulate scales: heads 3 or 4 lines high: involucre campanulate or turbinate, of lanceolate acute chartaceous bracts: *rays wanting:* style-appendages linear-subulate: akenes sericeous-pubescent. — *Lino-*

syris? carnosa, Gray, Pl. Wright. ii. 80. *Linosyris intricata*, Gray, Proc. Am. Acad. xvii. 208, a slender form, with smaller heads. — Saline arid region, S. Arizona, *Wright*, to California, in the Mohave Desert, *Parry, Greene, Pringle, Parish*, and near Visalia, *Congdon*.

A. spinósus, BENTH. Base of stem usually persistent and woody, sending up (3 to 5 feet long) slender and lithe striate green branches, resolved into paniculate branchlets, terminated by small heads: cauline leaves small, linear or spatulate-lanceolate, entire, mostly few and fugacious, some of them with soft subulate spines in or above their axils; these of the branchlets reduced to subulate scales or wanting: involucre hemispherical, 2 lines high, of small and thinnish subulate-lanceolate bracts, imbricated in about 3 series: rays white, 2 lines long: style-appendages subulate-triangular, much shorter than the stigmatic portion: akenes glabrous. — Pl. Hartw. 20; Torr. & Gray, Fl. ii. 165; Gray, Pl. Lindh. ii. 219. — Banks of streams, or in moist ground, S. W. Texas to Arizona and S. California, common; first coll. by *Berlandier*. (Mex.)

A. Pálmeri, GRAY. Decidedly *shrubby*, with the habit of a small-leaved *Baccharis*, 3 or 4 feet high, very much branched throughout: branchlets slender, striate-angled, terminated by the small heads: leaves apparently not fleshy, narrowly linear (of the branches an inch or less long), entire: involucre equalling the disk, barely 3 lines high, of closely imbricated narrowly oblong obtuse rather dry bracts: rays 6 to 10, a line long: disk-flowers about 20: akenes sericeous-pubescent. — Proc. Am. Acad. xvii. 209. Perhaps rather of the W. Indian genus *Gundlachia*, Gray, Proc. Am. Acad. xvi. 100. — S. Texas, at Corpus Christi Bay, *Palmer*.

Series II. Biennials and annuals.

§ 11. OXYTRIPÓLIUM. Involucre of § *Orthomeris*; the bracts thin and narrow, linear-lanceolate or linear-subulate, gradually very acute or acuminate, commonly greenish above or in the centre, but without herbaceous tips, imbricated in few series, the outer successively shorter, all erect-appressed: rays at least equalling the disk, numerous, often more numerous than the disk-flowers (revolutely coiled in drying): style-appendages lanceolate-subulate: akenes narrow, more or less pubescent, few-nerved: pappus fine and soft: glabrous and smooth annuals, chiefly of saline soil, paniculately branched, bearing numerous small heads, with bluish or purplish rays, and with entire narrowly lanceolate or linear leaves, on the branchlets reduced to subulate bracts. — Gray, Proc. Am. Acad. xvi. 98. *Tripolium* § *Oxytripolia*, DC. Prodr. v. 253, excl. spec. *Tripolium* § *Astropolium*, Nutt. in Trans. Am. Phil. Soc. n. ser. vii. 296. *Aster* § *Oxytripolium*, Torr. & Gray, Fl. ii. 161, in part. The two species are quite distinct in the Atlantic U. S., but seemingly confluent in Mexico and S. America.

A. exílis, ELL. Mostly slender and diffusely branched above: principal cauline leaves linear (3 or 4 inches long, 1 or 2 lines wide, lowest sometimes broader and lanceolate, rarely with a few serratures): heads 3 lines high: bracts of the involucre linear-subulate or more lanceolate and acuminate: rays 15 to 40, bluish or purple, rather conspicuous (about 2 lines long), usually much surpassing the pappus: disk-flowers generally more numerous. — Ell. Sk. ii. 344; Torr. & Gray, Fl. ii. 163: believed to be the species here described; but the original of herb. Ell. is now lost. *A. divaricatus*, Torr. & Gray, l. c., not L., &c. *A. subulatus*, Michx. Fl. ii. 112, in part. *Tripolium subulatum*, Nees. Ast. 157, in part; DC. Prodr. l. c. 254, excl. var. *boreale*. *Tripolium divaricatum*, Nutt. Trans. Am. Phil. Soc. l. c. 296. — Subsaline or even not at all brackish moist soil, S. Carolina to Texas, Arizona, and California: on the southern borders occurs with very short ligules. (Mex., W. Ind., &c.)

Var. austrális, the commoner Mexican and S. American form of this polymorphous and widely diffused species, is less diffuse, less slender, often broader-leaved, and with larger heads, the involucral bracts broader, less acute, and greener or purplish-tinged. — *A. subulatus*, Less. in Linn. vi. 120. *Erigeron multiflorum*, Hook. & Arn. Bot. Beech. 87. *Tripolium conspicuum* of authors, but not the original of Lindley. — Coast of Oregon and California (at Visalia, in the interior, *Congdon*, a form with unusually large heads), &c. (Mex. to Chili, Brazil, &c.)

A. subulátus, MICHX. Stouter, only a foot or two high, with short usually purplish stems and branches: leaves somewhat fleshy, linear-lanceolate (lower 4 to 6 inches long, 2 to 4 lines wide), or the upper linear passing into subulate: heads narrower, cylindraceous, 4 lines high: bracts of the involucre linear-subulate with much attenuate apex: rays 25 to 30, purplish, very small and inconspicuous, hardly surpassing the disk, with ligule very much shorter than the tube, often surpassed by the (not very copious) mature pappus, more numerous than the (10 to 15) disk-flowers. — Fl. ii. 111, partly (char. "ligulis minimis," & hab.); Nutt. Gen. ii. 154. *Tripolium subulatum*, Nees, DC., &c., in part. *Aster linifolius*, Torr. & Gray, Fl. ii. 162, not L., not even as to the syn. "Gron. Virg" cited (which belongs to *A. tenuifolius*, p. 202). — Salt marshes, from New Hampshire to Florida. Closely connects with the following section.

§ 12. CONYZÓPSIS. Involucre campanulate, of 2 or 3 series of linear or oblong bracts, nearly equal in length; the outer foliaceous or herbaceous and loose, resembling the rameal leaves; the inner more membranaceous or scarious: rays small and not longer than the mature pappus, or the ligule wanting; the female flowers mostly in more than one series and more numerous than the hermaphrodite; these with slender corolla, its limb 4-5-toothed: style-appendages lanceolate: akenes narrow, not compressed, 2-3-nerved, appressed-pubescent: pappus simple, very soft: low and branching leafy-stemmed annuals (of W. North America and N. E. Asia, and of moist subsaline soil), nearly glabrous, except that the linear (or the lowest spatulate) chiefly entire leaves are more or less hispidulous-ciliate; the numerous rather small heads in well-developed plants disposed to be racemose-paniculate. (Char. from the two genuine species, which are intermediate between the *Oxytripolium* section, *A. subulatus* connecting them, and *Conyza*.) — Gray, Proc. Am. Acad. xvi. 99. *Aster* § *Oxytripolium*, subsect. *Conyzopsis*, Torr. & Gray, Fl. ii. 162. *Brachyactis*, Ledeb. Fl. Ross. ii. 495; Benth. in Hook. Ic. Pl. xii. 6 (excl. spec.), & Gen. Pl.; Gray, Proc. Am. Acad. viii. 647, & Bot. Calif. i. 326.

A. frondósus, TORR. & GRAY. A span to a foot or more high, branching from the base, when low usually spreading, when taller the branches bearing numerous spicately paniculate heads (of 4 lines in height): outer bracts of the involucre linear-oblong, obtuse, wholly foliaceous and loose, numerous: rays in anthesis exserted, a line long, linear, pinkish-purple, always longer than the style, but equalled or surpassed by the mature copious pappus. — Fl. ii. 165. *Tripolium frondosum*, Nutt. Trans. Am. Phil. Soc. n. ser. vii. 296. *A. angustus*, Gray, Pl. Wright. ii. 76; Eaton, Bot. King Exp. 144, not Torr. & Gray. *Brachyactis ciliata*, var. *carnosula*, Benth. in Hook. Ic. Pl. xii. 6. *B. frondosa*, Gray, Proc. Am. Acad. l. c.; Bot. Calif. l. c. — Borders of springs, pools, &c., Rocky Mountains of Idaho to the Sierra Nevada, California, and the Rio Grande in New Mexico.

A. angústus, TORR. & GRAY. Leaves commonly narrower: bracts of the involucre all linear, acute: corolla of the ray-flowers reduced to the tube and much shorter than the elongated style, or rarely with a rudimentary ligule? — Fl. ii. 162. *Crinitaria humilis*, Hook. Fl. ii. 24. *Linosyris? humilis*, Torr. & Gray, l. c. 234. *Erigeron ciliatus*, Ledeb. Fl. Alt. iv. 92, & Ic. t. 100. *Conyza Altaica*, DC. Prodr. v. 380. *Tripolium angustum*, Lindl. in Hook. Fl. ii. 15, & DC. l. c. 254. *Brachyactis ciliata*, Ledeb. Fl. Ross. ii. 495; Benth. l. c. (excl. var.); Gray, Proc. Am. Acad. viii. 647. (The poor figure in Ledeb. Ic. l. c. represents a ligulate female flower, which accords with neither specimens nor character.) — Saline wet ground, Saskatchewan to Utah and Colorado, eastward to Minnesota, and now extending to Chicago, &c. (N. Asia.)

§ 13. MACHÆRANTHÉRA. Involucre pluriserially imbricated, hemispherical or campanulate; the bracts linear, coriaceous below, and with herbaceous or foliaceous spreading tips: rays numerous and conspicuous, violet or bluish purple: akenes narrowed downward, compressed, few-nerved, and the faces somewhat striate: receptacle alveolate. the alveoli toothed or lacerate: style-appendages from linear-lanceolate to filiform-subulate: pappus copious and simple, of rather rigid unequal bristles: leafy-stemmed and branching biennials (sometimes more enduring, but no rootstocks, stolons or buds below the crown), or occasionally annuals (W. N. American and Mexican): the showy heads terminating the branches: involucre either canescent or somewhat viscid or glandular: leaves from sparingly dentate to bipinnately parted, the teeth or lobes apt to be bristle-tipped. — Gray. Proc. Am. Acad. viii. 647, & Bot. Calif. i. 322. *Machæranthera*, Nees, Ast. 224; Gray, Pl. Wright. i. 90. *Dieteria*, Nutt. Trans. Am. Phil. Soc. vii. 300; Torr. & Gray, Fl. ii. 99.

 * Anomalous, seemingly perennial and multicipital, but otherwise of this section.

 A. Coloradoénsis, GRAY. A span or less high, forming a tuft of short few-leaved stems on a strong tap root, canescently pubescent, not at all glandular: leaves spatulate or oblanceolate (about an inch long), coarsely dentate, the teeth tipped with conspicuous bristles: heads solitary, broadly hemispherical, half-inch high: involucral bracts small and numerous, well imbricated, subulate-lanceolate, rather close: rays 35 to 40, violet-purple, barely half-inch long: akenes turbinate, short, densely canescent-villous, half the length of the comparatively rigid pappus. — Proc. Am. Acad. xi. 76; Rothrock in Wheeler Rep. vi. 149, t. 7. — Common in South Park, Colorado, *Porter, Canby, Greene*, &c. Also San Juan Pass, at 12,000 feet, *Brandegee*.

 * * Genuine species, with annual or biennial but never truly perennial root.

 +- Involucre densely hispidulous as well as viscid, very squarrose: akenes glabrous or glabrate: pappus slender: heads large and broad (the disk two-thirds to full inch in diameter): herbage green, not canescent, glabrate: leaves from incisely dentate to entire, their teeth or tips obscurely if at all mucronate-setigerous: rays bright violet, showy: root biennial or somewhat more enduring.

 A. Pattersóni, GRAY. A span or two high, branched from the summit of the tap root: stems or branches with soft or cottony-tomentulose pubescence, or glabrate: leaves thickish, spatulate or lingulate, entire or coarsely few-toothed, none widened at base: heads solitary or few: involucral bracts lanceolate: rays about 30, fully half-inch long. — Proc. Am. Acad. xiii. 272, excl. var. *Machæranthera canescens*, var. *alpina*, Porter & Coulter, Fl. Colorad. 59. — Moist ground along streams, Gray's Peak, Colorado; first coll. by *Parry*, then by *Patterson*, &c.

 A. Bigelóvii, GRAY. A foot or two high, robust: stem leafy, branching above, roughish-hirsute to glabrate; the flowering branches or peduncles glandular-hirsute, terminated by showy large heads: leaves oblong or lanceolate, irregularly and sometimes incisely dentate, sometimes entire; radical lanceolate-spatulate; cauline oblong to lanceolate, usually with broadish partly clasping base: involucral bracts very numerous, linear-attenuate or the prolonged and much recurved tips about filiform: rays very many, an inch or less long. — Pacif. R. Rep. iv. 97, t. 10. *A. Townshendii*, Hook. f. Bot. Mag. t. 6430 (wrong as to the broadly obovate style-appendages figured and described); Robinson, Garden, xvii. t. 228. — Southern Colorado and New Mexico, *Bigelow, Brandegee, Rusby*, &c. Very handsome in cultivation.

 +- +- Involucre from nearly glabrous to glandular-puberulent or canescent, not rarely viscid, but not hirsute or hispidulous: heads less ample: akenes densely pubescent.

 ++ Leaves at most incisely dentate, rather rigid: root disposed to be biennial or somewhat more enduring. — *Dieteria*, Nutt.

 A. gymnocéphalus, GRAY. Stem erect, simple or branched from a rather slender root, commonly hirsute or hispidulous, equably leafy to the top: branches bearing solitary usually naked-pedunculate middle-sized heads: leaves spatulate-oblong to lanceolate; cauline short (inch or less long), usually obtuse, copiously serrate or denticulate with spinulose-setigerous teeth: involucre depressed-hemispherical, half-inch or less high; its bracts linear-subulate with the tips squarrose: rays purple, 4 or 5 lines long: receptacle fimbrillate. — Proc. Am. Acad. xv. 32; Hook. f. Bot. Mag. t. 6549. *Aplopappus gymnocephalus*, DC. Prodr. v. 346, & *A. blephariphyllus*, Gray, Pl. Wright. i. 97; the ray-flowers having been thought to be yellow

Machæranthera setigera, Nees in Linn. xix. 722. — Dry ground S. W. borders of Texas, *Wright, Howard.* (Mex.)

A. canéscens, Pursh. Commonly a foot or two high and loosely much branched, even from the indurated root, bearing numerous paniculate heads, sometimes dwarf and with simple contracted inflorescence, pale and cinereous-puberulent or minutely canescent, or greener and glabrate: leaves lanceolate to linear or the lower spatulate, from entire to irregularly dentate, or occasionally laciniate, the rigid teeth mostly with mucronate-setulose tip: heads when numerous 4 or 5 lines and when fewer half inch high: involucre turbinate to hemispherical, of rigid usually well-imbricated bracts: rays violet, 4 or 5 lines long: akenes narrow, canescent. — Fl. ii. 547; Gray, Bot. Calif. i. 322. *A. biennis*, Nutt. Gen. ii. 155. *Dieteria canescens, pulverulenta, divaricata, viscosa,* & *sessiliflora,* Nutt. Trans. Am. Phil. Soc. vii. 300; Torr. & Gray, Fl. ii. 100. *Machæranthera canescens,* Gray, Pl. Wright. ii. 75; Eaton, Bot. King Exp. 146. — The forms taken as the type of this polymorphous species are cinereous, rigid, when well developed bearing numerous heads: bracts of the involucre regularly imbricated in numerous ranks, coriaceous and appressed, with the green tips short and spreading, seldom at all viscous or glandular. — Open and sterile ground and sandy banks of streams, Saskatchewan to the eastern parts of Brit. Columbia, on the plains south to W. Texas, also eastern side of the Sierra Nevada to Arizona. (Adj. Mex.)

Var. víridis. A green form, hardly rigid, of less arid situations, either sparsely scabro-puberulent or almost glabrous: involucral bracts looser, either with short and ascending or longer and squarrose-spreading tips, sometimes rather hispidulous-glandular. — *Machæranthera canescens,* var. *glabra*, Gray, Pl. Wright. i. 89, &c. *Aster Pattersoni*, var. *Hallii*, Gray, Proc. Am. Acad. xiii. 272, is rather a subalpine form of this. — W. borders of Texas to Utah.

Var. latifólius. Green, minutely soft-pubescent, 2 feet or more high: leaves thinnish, nearly membranaceous, comparatively large, sometimes spatulate-oblong and over half-inch wide: heads large and few: involucre hemispherical; tips of its bracts mostly attenuate-subulate and squarrose-spreading, canescent and obscurely glandular. — *Dieteria asteroides*, Torr. in Emory Rep. 142. *Machæranthera canescens,* var. *latifolia*, Gray, Pl. Wright. ii. 75. — New Mexico and Arizona, in moist ground; passes into var. *tephrodes*.

Var. viscósus. Canescent or cinereous: leaves narrow, rather rigid; the upper mostly entire and the lower coarsely dentate: involucre campanulate or turbinate, squarrose; the prominent (either short or elongated) foliaceous tips of the bracts viscid-glandular, either spreading or recurved. — *Dieteria viscosa* & *D. sessiliflora* (rays probably only abnormally if ever at all "ochroleucons"), Nutt. l. c.; Torr. & Gray, l. c. *D. incana*, Torr. & Gray, l. c. *Diplopappus incanus*, Lindl. Bot. Reg. t. 1693 (form with little viscidity to involucre); Hook. Bot. Mag. t. 3882 (involucre very foliaceous-squarrose). — Arid districts, Wyoming to California.

Var. tephródes. More or less canescent, especially the hemispherical involucre of the large heads; the bracts with elongated and subulate-attenuate foliaceous tips, not glandular; the hoary pubescence sometimes looser. — *A. incanus*, Gray, Bot. Calif. i. 322. — S. California, Arizona, and New Mexico.

++ ++ Leaves 1-3-pinnately cleft or parted, not rigid: involucre hemispherical, its bracts mostly looser; akenes more strongly striate: root commonly annual: stem diffusely branched. — *Machæranthera*, Nees, l. c. *Dieteria* § *Pappochroma*, Nutt., excl. spec.

A. tanacetifólius, HBK. Pubescent, often rather viscid, very leafy, commonly a foot or two high: lowest leaves 2-3-pinnately parted; uppermost simply pinnatifid or on the flowering branchlets entire; lobes short, setulose-mucronate: heads half-inch high: bracts of the involucre narrowly linear, with slender mostly linear-subulate spreading foliaceous tips, or the outermost almost wholly foliaceous: rays numerous (half-inch long or more), bright violet: akenes rather broad, villous. — Nov. Gen. & Spec. iv. 95. *A. chrysanthemoides*, Willd. in Spreng. Syst. iii. 538. *Machæranthera tanacetifolia*, Nees, Ast. 224; Hook. Bot. Mag. t. 4624; Gray, Pl. Wright. i. 90. *Chrysopsis* (*Pappochroma*) *coronopifolia*, Nutt. Jour. Acad. Philad. iv. 34. *Dieteria coronopifolia*, Nutt. Trans. Am. Phil. Soc. vii. 300; Torr. & Gray, l. c. — Moist ground, Nebraska to Texas, Arizona, and borders of California. (Mex.)

Var. pygmǽus, a low and small form, seemingly a precocious state, with less dissected leaves, rather smaller heads, and much shorter foliaceous tips to the involucral bracts, seems to connect this with the following. — *Machæranthera canescens,* var. *humilis* & var. *pygmæa*, Gray, Pl. Wright. ii. 74. — New Mexico, *Wright.*

A. parviflórus, Gray. Glabrous, somewhat viscid, low and slender: leaves narrow, simply pinnatifid, barely inch long; the lobes short-linear, obtuse, hardly mucronate: heads 3 or 4 lines high: involucre closer; the bracts with short oblong or ovate-lanceolate acute green tips: rays 3 lines long: akenes canescently sericeous. — Bot. Calif. i. 322, note. *Machæranthera parviflora*, Gray, Pl. Wright. i. 90. — New Mexico from the Rio Grande to W. Arizona, *Wright*, *Thurber*, &c.

49. ERÍGERON, L. FLEABANE. ('Ηρ and γέρων, old man in spring.)

— A rather large genus of herbs or barely suffrutescent plants, verging on the one hand to *Aster*, on the other to *Conyza*, and only arbitrarily to be separated on the lines of junction; the heads disposed to be solitary and long-pedunculate; rays (occasionally absent in certain species, uniformly wanting in two or three others) violet, purple, white, rarely ochroleucous (or in anomalous species even clear yellow!); disk-flowers yellow, not changing to purple: akenes commonly 2-nerved. — L. Gen. ed. 2. 400 (*Erigerum* in ed. 1, after Dodoens, who had Groundsel in view, and this form may explain how the name was taken for neuter by Linnæus); Torr. & Gray. Fl. ii. 166: Benth. & Hook. Gen. ii. 280, excl. § *Oritrophium* (which must belong either to *Celmisia* or *Aster*): Gray, Proc. Am. Acad. xvi. 86. *Erigeron, Trimorphæa, Phalacroloma, Stenactis*. &c., Cass. *Erigeron, Stenactis, Phalacroloma, Polyactidium* (*Polyactis*, Less.), *Heterochæta*, & *Woodvillea*, DC. Prodr. (Genera founded on the pappus and number of the rays, mostly unavailable even for good subgenera.) The series here commences with Asteroid and ends in Conyzoid forms.

§ 1. EUERÍGERON. Rays elongated and conspicuous, or in a few species uniformly wanting, in one or two (*E. compositus, E. concinnus*) occasionally abortive: no rayless female flowers between the proper ray and disk.

* Perennials, commonly dwarf from a multicipital caudex, alpine or rarely alpestrine, with comparatively large and mostly solitary heads: *involucre loose or spreading, and copiously lanate with long multiseptate hairs*; rays about 100, narrow: leaves entire.

+ Whole herbage gnaphalioid-lanate: pappus double; the short outer multisquamellate.

E. Muírii, Gray. A span high, densely clothed with long and soft white (apparently persistent) floccose wool; stems simple and monocephalous, rather leafy: leaves lanceolate-spatulate (an inch or two long), or uppermost narrowly lanceolate: involucre squarrose, as of the following species: rays white, a third of an inch long. — Proc. Am. Acad. xvii. 210. — Cape Thompson, Alaska, *John Muir*.

+ + Herbage green, with or without villous or hirsute pubescence: pappus nearly simple.

E. uniflórus, L. Stems an inch to a span or two high, strictly monocephalous, few-leaved, often naked and pedunculiform at summit: radical leaves spatulate or oblanceolate (inch or two long); cauline lanceolate to linear: involucre usually hirsute as well as lanate, occasionally becoming naked; the linear acute bracts rather close, or merely the short tips spreading: rays purple or sometimes white, 2 or 3 or rarely 4 lines long. — Fl. Lapp. t. 9, f. 3, & Spec. ii. 864; Hook. Fl. ii. 17; Torr. & Gray, Fl. ii. 169; Ledeb. Fl. Ross. ii. 490; Reichenb. Ic. Fl. Germ. xvi. 914. *E. alpinus* & *Hieracium pusillum*, Pursh, Fl. ii. 532, 502. *E. pulchellus*, var., & *E. alpinus*, in part, DC. Prodr. v. 287. *E. eriocephalus*, J. Vahl, Fl. Dan. t. 2298, is either this or possibly a form of the next. — Labrador to Arctic coast, and Unalaska, south to the Sierra Nevada, California, and mountains of Colorado, in the alpine region. Forms with a comparatively hirsute involucre occur in the Rocky Mountains; and some are not well distinguished from the next. (Greenland, Eu., N. Asia to Kamts.)

E. lanátus, Hook. Stems about a span high from a multicipital caudex, scapiform or few-leaved, monocephalous: radical leaves spatulate to obovate, about half-inch long, tapering into a narrowed base or into a slender margined petiole; some primary ones occasionally palmately 3-lobed; cauline one or two, small and linear, or hardly any; head not larger than that of *E. uniflorus*, and involucre similar, but densely soft-lanate: rays rather broader, 3 lines long, white. — Fl. ii. 17, t. 121; Torr. & Gray, Fl. ii. 168. *E. grandiflorus*, var. *lanatus*, Gray, Proc. Am. Acad. xvi. 92. *Aster glacialis*, Torr. & Gray, Fl. ii. 503, not Nutt. — Alpine summits of the Rocky Mountains in Montana and Brit. America, *Drummond*, *Fremont*, *Bourgeau*, and of the Cascades, *Lyall*. (Lyall's plant may have yellowish rays, and pass into *Aplopappus Brandegei*.)

E. grandiflórus, Hook. Stems a span or two high, rather stout, usually several-leaved and monocephalous: radical leaves obovate-spatulate, an inch or so long; cauline oblong to lanceolate, usually half-inch or less long: heads larger: involucre half-inch high, very woolly; its linear and attenuate-acuminate bracts squarrose-spreading or the tips recurved: rays violet or purple, a third to half inch long. — Fl. ii. 18, t. 123; Torr. & Gray, l. c.; Eaton, Bot. King Exp. 184 (a somewhat abnormal form, in the Uinta Mts.) — Rocky Mountains, in or near the alpine region, from British Columbia (*Drummond*) to Colorado, where it sometimes has fewer and linear cauline leaves, and approaches *E. uniflorus*.

Var. **elátior**, Gray. A foot or two high, leafy up to the 1 to 4 pedunculate heads, pubescent, but hardly hirsute: leaves oblong to ovate-lanceolate, 2 to 4 inches long; cauline closely sessile by a broad base: involucre fully half-inch high: rays half-inch long. — Am. Jour. Sci. ser. 2, xxxiii. 237, & Proc. Am. Acad. xvi. 92. — Subalpine and lower, in the Rocky Mountains of Colorado; first coll. by *Parry*.

* * *Submaritime* perennial: heads of the largest, the disk a full inch in diameter: involucre rather loose, villous with long multiseptate hairs: rays about 100, rather broad, Aster-like: pappus simple: leaves obovate or spatulate, ample, mostly entire, graveolent. — *Woodvillea*, DC., color of ray-flowers mistaken.

E. glaúcus, Ker. A span to a foot high, viscidulous and more or less pubescent, producing a tuft of radical leaves from a rather fleshy crown, and some ascending monocephalous or occasionally branching stems: leaves glaucescent or pale green, but hardly glaucous, somewhat succulent; larger radical 3 or 4 inches long and an inch wide, rarely 2-3-toothed; upper cauline few and small: rays half-inch long, bright violet: akenes 4-nerved. — Bot. Reg. t. 10; DC. Prodr. v. 284; Torr. & Gray, Fl. ii. 172; Gray, Bot. Calif. i. 330. *E. maritimum* & *E. hispidum*, Nutt. Trans. Am. Phil. Soc. vii. 310. *E. Bonariensis*, Spreng. Syst. iii. 528. *Aster Californicus*, Less. in Linn. vi. 121; Hook. & Arn. Bot. Beech. 146; DC. l. c. 228. *Stenactis glauca*, Nees, Ast. 275. *Woodvillea calendulacea*, DC. l. c. 318. — Banks or bluffs of the Pacific coast, within the influence of salt water, Oregon to Santa Barbara, California, flowering for most of the year; probably first coll. by *Menzies*.

* * * True perennials from rootstocks or a caudex, neither stoloniferous-surculose nor flagelliferous: involucre from hispid or villous to glabrous, but not lanate, in the first species loose and spreading: all Western or Northern species. — Part of § *Phœnactis*, Nutt. Trans. Am. Phil. Soc. vii. 310; Benth. & Hook. Gen. ii. 280.

+— Comparatively tall and large, a foot or more high except in alpine or depauperate forms, leafy-stemmed, glabrous to soft-hirsute: leaves rather ample, entire or occasionally few-toothed: heads pretty large, with usually very numerous rays: montane or alpestrine.

++ Aster-like; the rays comparatively broad: involucre rather loose: heads solitary, or on larger plants few and corymbosely disposed: pappus simple.

E. salsuginósus, Gray. Rootstocks short and thickish: stem commonly 12 to 20 inches high, the summit or peduncles lanate-pubescent or puberulent: no bristly or hirsute hairs: leaves very smooth and glabrous or glabrate, bright green, thickish; radical and lower cauline leaves spatulate to nearly obovate, with base attenuate into a margined petiole; upper cauline ovate-oblong to lanceolate, sessile, conspicuously mucronate or apiculate-acuminate; uppermost small and bract-like: bracts of the involucre loose or even spreading, linear-subulate or attenuate, *viscidulous*, at most puberulous (or at some northern stations sometimes pubescent): disk over half an inch in diameter: rays 50 to 70, purple or violet, half-inch or more long. — Proc. Am. Acad. xvi. 93. *Aster salsuginosus*, Richards. in Frankl. Journ. App. ed. 2, 32; Hook. Bot. Mag. t. 4942; DC. Prodr. v. 229; Torr. & Gray, Fl. ii. 150. *A. Unalaschensis*, Less. in Linn. vi. 124; DC. l. c. 228. — Wet ground, Kotzebue Sound and Unalaska, and along the higher mountains southward to California, Utah, and New Mexico; first coll. by *Richardson*.

Var. **angustifólius**, Gray. A span to a foot high: radical and lower cauline leaves from narrowly spatulate to lanceolate (only 3 or 4 lines broad), somewhat scabrous on margins; upper cauline linear-lanceolate, small; rays about 40. — Proc. Am. Acad. l. c. *Aster salsuginosus*, var. *angustifolius*, Gray, Bot. Calif. i. 325. — Mountains of Washington Terr. (*Brandegee*) to the Sierra Nevada, California, as far south as Kern Co., *Lemmon, Mrs. Austin, Matthews*, &c. Passes into

Var. glaciális. A span high, few-leaved, monocephalous: leaves as of the type (of which this is a reduced alpine form), but smaller. — *Aster glacialis*, Nutt. Trans. Am. Phil. Soc. vii. 291; Torr. & Gray, Fl. ii. 155. — Alpine region of the Rocky Mountains; first coll. by *Nuttall* in Wyoming.

E. Howéllii. Rootstock filiform: stem a foot high, equably leafy, monocephalous: leaves membranaceous, glabrous and smooth; radical obovate, slender-petioled; cauline mostly ovate and with broad half-clasping base (larger ones 2 inches long and an inch wide), sometimes one or two sharp denticulations, mucronate-acuminate: involucre, &c., nearly of the foregoing: rays only 30 to 35, two-thirds inch long, *a line or two wide*, white. — *E. salsuginosus*, var. *Howellii*, Gray, Proc. Am. Acad. xvi. 93. — Oregon, in the Cascade Mountains, *Howell*.

E. Coúlteri, T. C. PORTER. Rootstock slender; stem 6 to 20 inches high, equably leafy, bearing solitary or rarely 2 or 3 rather slender-pedunculate heads: leaves membranaceous, obovate to oblong, either entire or serrate with several sharp teeth, pilose-pubescent to glabrous, cauline inconspicuously mucronulate: disk of the head about half an inch wide: involucre less attenuate and spreading than that of *E. salsuginosus*, obscurely viscidulous but hirsute (as also the peduncle) with spreading hairs: rays 50 to 70, rather narrowly linear, half-inch or more long, white, varying to purplish. — Porter & Coulter, Fl. Colorad. 61; Rothrock in Wheeler Rep. vi. 154; Gray, Proc. Am. Acad. xvi. 93. — Rocky Mountains of Colorado, at about 10,000 feet, *Coulter*, &c., of Utah, *Ward, Jones*, &c., and Sierra Nevada, California, *Brewer, Bolander, Greene*.

++ ++ Less Aster-like: rays 100 or more and narrow: involucre closer: pappus more or less double, but the exterior minute, setulose or subulate-squamellate: stems chiefly erect, tufted, generally leafy to the summit, and bearing few or several heads: leaves entire. (Species hard to discriminate, montane, but never alpine.) — § *Phœnactis*, Nutt. Trans. Am. Phil. Soc. l. c., in part.

E. speciósus, DC. Sparingly and loosely hirsute or with a few scattering hairs: stems mostly 2 feet high, very leafy to the top: leaves lanceolate, acute (3 to 8 lines wide), sparsely ciliate; lowest more or less spatulate: involucre hirsute-pubescent, or sometimes almost glabrous: rays half-inch to almost an inch long, violet. — Prodr. v. 284, & vii. 274; Torr. & Gray, Fl. ii. 173. *E. glabellus*, var. *mucronatus*, Hook. Fl. ii. 19. *Stenactis speciosa*, Lindl. Bot. Reg. t. 1577; Hook. Bot. Mag. t. 3067. — British Columbia to Oregon and perhaps N. California, near the coast.

E. macránthus, NUTT. From hirsute-pubescent to nearly glabrous: stem 10 to 20 inches high: leaves from lanceolate to ovate; upper often reduced in size: involucre glabrous or nearly so, but commonly minutely glandular: rays half-inch long (heads not larger, as the name would imply, but rather smaller than those of the preceding): short outer pappus more conspicuous, sometimes nearly squamellate. — Trans. Am. Phil. Soc. l. c.; Torr. & Gray, l. c. *E. grandiflorum*, Nutt. Jour. Acad. Philad. vii. 31, not Hook. — Rocky Mountains, from Wyoming to New Mexico and S. W. Utah, at 8,000 to 10,000 feet in the southern portions of its range.

E. glabéllus, NUTT. From partly glabrous to copiously hirsute, disposed to be naked above: stems 6 to 20 inches high: leaves lanceolate or the lowest somewhat spatulate; upper linear-lanceolate and gradually reduced to subulate bracts: heads in the typical forms considerably smaller than those of the two preceding species: involucre strigosely hirsute or pubescent: rays violet, purple, and rarely white, a third to half an inch long: outer pappus setulose. — Gen. ii. 147, & Jour. Acad. Philad. l. c.; Hook. Bot. Mag. t. 2923, & Fl. ii. 19 (excl. var. γ); Torr. & Gray, l. c., with vars. *asper* & *pubescens*. *E. asper*, Nutt. l. c., a somewhat roughish-hirsute form. *E. pulchellus*, Hook. Fl. ii. 19, partly. — Minnesota and Saskatchewan to the Rocky Mountains, and southward to Colorado and Utah. Occurs in various forms; the small or slender northern forms of the plains naked-stemmed and simple; some of the larger more equably leafy and approaching the preceding, others by the copious pubescence leading to the ambiguous

Var. móllis, GRAY. Somewhat cinereous with a soft and short spreading pubescence, a foot or two high, leafy to the top: leaves oblong-lanceolate: cinereous pubescence of the involucre soft and spreading.—Proc. Acad. Philad. 1863, 64, & Proc. Am. Acad. l. c.—Colorado Rocky Mountains, at 8,000 to 9,000 feet; first coll. by *Parry*, *Hall*, &c. Perhaps a distinct species.

++ ++ Low, rarely a foot high, conspicuously hispid or hirsute with spreading bristly hairs: leaves entire, narrow: involucre close: rays numerous, occasionally wanting in one species: *pappus conspicuously double*, but least so in the first species.

++ Sparingly branched stems several or numerous from the crown of a tap root, more or less leafy: heads middle-sized: disk a third to half an inch in diameter: involucre hispid: rays 50 to 80, long and narrow, soon deflexed, occasionally wanting in the second species.

E. púmilus, NUTT. Radical and lower cauline leaves from spatulate-linear to lanceolate (a line or two wide); upper linear: rays white (4 lines long): outer pappus of short bristles little or not at all thicker than the inner ones and more or less intermixed with them.—Gen. ii. 147; Torr. & Gray, Fl. ii. 174. *E. hirsutus*, Pursh, Fl. ii. 742, not Lour.—Dry upper plains, Dakota to Colorado, and in the Rocky Mountains, west to Utah.

E. concínnus, TORR. & GRAY. Like the preceding, but usually with more dense and shaggy hirsuteness and less rigid leaves: stems not rarely somewhat copiously branched: rays violet or blue, rarely white: outer pappus conspicuous and squamellate or paleaceous (the paleæ varying from subulate to oblong!).—Fl. ii. 174; Eaton, Bot. King Exp. 151, with var. *condensatus*, a dwarf and condensed form with monocephalous stems, and commonly wide (but fewer) paleæ to the pappus. *E. strigosus*, var. *hispidissimus*, Hook. Fl. ii. 18, chiefly. *Distasis? concinna*, Hook. & Arn. Bot. Beech. 350.—Arid regions between the western slopes of the Rocky Mountains and the Sierra Nevada and Cascades, from Wyoming to New Mexico and Brit. Columbia to Arizona.

Var. aphanáctis, GRAY. Discoid, the rays being nearly destitute of ligule or wanting.—Proc. Am. Acad. vi. 540.—Colorado to Nevada and the borders of California.

++ ++ More branched and leafy, over a span high; with smaller heads, fewer rays, and somewhat naked involucre more imbricated: anomalous Texano-New-Mexican species.

E. Bigelóvii, GRAY. Cinereous-hispidulous, diffusely branched from the base, leafy up to the short-pedunculate scattered heads: leaves small, spatulate-lanceolate or upper linear (less than inch long), lowest more spatulate and petioled: bracts of the hemispherical involucre rather rigid, lanceolate, acuminate, obviously of 2 or 3 lengths, the outer sparingly hispidulous: rays 40 to 50, purple or violet (3 lines long): outer pappus of slender-subulate squamellæ, about a third the length of the inner bristles.—Bot. Mex. Bound. 78.—On the Rio Grande near Fronteras, at the borders of Texas, New Mexico, and Chihuahua, *Wright*, *Bigelow*.

E. Brandegéi. A very anomalous and imperfectly known plant, green, sparsely hispidulous-hirsute, less branched: radical leaves spatulate-linear; cauline linear and small, or upper minute bracts of involucre short-linear, almost naked: rays 30 or more, white: outer pappus of coriaceous squamellæ which are commonly confluent with the scanty bristles of the inner, perhaps abnormal: only one specimen seen.—Adobe plains, S. W. Colorado, on the borders of New Mexico, *Brandegee*.

++ ++ ++ Tufted stems very short and densely leafy, bearing simple and monocephalous scapiform or few-leaved flowering stems (about a span high): head proportionally large: rays 25 to 50, not very narrow, 3 or 4 lines long: leaves narrowly spatulate-linear.

E. poliospérmus. Leaves hispid throughout, an inch or more long, filiform-spatulate, the broader summit a quarter or half a line wide: head half-inch high: involucre of rather loose and slender hispidulous bracts: rays about 25, blue-violet or white: akenes densely white-villous outer pappus slender-squamellate, fully as long as the breadth of the akene, covered by the copious white silky hairs of the latter.—Umatilla, Oregon, *Howell*, and Washington Terr., in the Wallawalla region, *Brandegee*, *Tweedy*. Resembles the next.

E. Chrysópsidis. Hispid, also with some minuter pubescence: leaves spatulate-linear, an inch or two long, commonly a full line wide at summit: involucre rather hirsute: rays 40 to 50, "golden yellow": akenes barely pubescent or hirsutulous: outer pappus less conspicuous, merely setulose, otherwise very like the preceding — *Chrysopsis hirtella*, DC. Prodr. v. 327. *E. ochroleucus*, var. *hirtellus*, Gray, Proc. Am. Acad. xvi. 90.—Stony hills and in wet clay on mountain sides, E. Oregon and adjacent Washington Terr., *Douglas*, *Cusick*, *Nevius*, *Howell*. Must be retained in *Erigeron* (of which it has the involucre and style), notwithstanding the pure yellow rays, which also occur in *E. peucephyllus*. It can hardly pass into *E. ochroleucus*.

+ + + Dwarf, cespitose from a multicipital caudex, with monocephalous flowering stems, often scapose: radical leaves dissected: pappus simple.

E. compósitus, Pursh. From hirsute to glabrate, with slender margined petiole setose-ciliate: radical leaves much crowded on the crowns of the caudex, usually 1-3-ternately parted into linear or short and narrow spatulate lobes, the few on the erect flowering stems 3-lobed or entire and linear: involucre (3 or 4 lines high) sparsely hirsute: rays from 40 to 60, not very narrow, white, purple, or violet, mostly 3 or 4 lines long. — Fl. ii. 535: Fl. Dan. xii. 1999; Hook. in Trans. Linn. Soc. xiv. 374, t. 13, & Fl. ii. 17; DC. Prodr. v. 288; Torr. & Gray, Fl. ii. 167. *E. pedatus*, Nutt. Trans. Am. Phil. Soc. vii. 308. *Cineraria Lewisii*, Richards. in Frankl. Journ. App. ed. 2, 32. — Alpine and alpestrine districts of the Rocky Mountains, and of the Sierra Nevada, from S. Colorado and California to Brit. Columbia and arctic sea-coast. (Greenland and Spitzbergen.)

—— **Var. discoídeus**, Gray. Rays wanting or abortive: head commonly smaller. — Am. Jour. Sci. ser. 2, xxxiii. 237; Eaton, Bot. King Exp. 148. — Same range as the radiate form, often accompanying it; first coll. by *Parry*, &c.

—— **Var. trifídus**, Gray. Small blade of leaves simply 3-5-fid: the lobes from oblong to obovate. — Proc. Am. Acad. xvi. 90. *E. trifidus*, Hook. Fl. ii. 17, t. 120. — Rocky Mountains, N. Colorado to Brit. Columbia; first coll. by *Drummond*, later by *J. M. Coulter* and *Canby*.

—— **Var. pinnatiséctus**, Gray, l. c. Usually a large form: numerous violet-purple rays 5 lines long: leaves pinnately parted into 9 to 11 linear and entire or rarely 2-3-cleft divisions. — Mountains of Colorado, from South Park to the Sierra Blanca; first coll. by *Hall*.

E. Prínglei, Gray. Smooth and glabrous, densely cespitose from a lignescent multicipital caudex: radical leaves laciniate-pinnatifid into 3 to 5 short-lanceolate or broadly subulate pointed lobes; those of the ascending (2 or 3 inches long) flowering stems linear, entire, 5 or 6 in number: involucre hardly 3 lines high, glabrous: rays 20 or 30, purple or whitish, 3 lines long. — Proc. Am. Acad. xvii. 210. — Cliffs of Mount Wrightson, Santa Rita Mountains, Arizona, *Pringle*.

+ + + + Dwarf or low species, alpine or alpestrine, entire-leaved, cespitose from multicipital caudex, no fine or cinereous pubescence, monocephalous: leaves few on the simple stems, at least the radical broader than linear: rays rather numerous and not very narrow: pappus simple or nearly so.

++ Involucre glabrous but pruinose-glandular, brownish-purple: alpine and Aster-like, smooth and green.

E. leiómerus. A span high from the somewhat surculose branches of the caudex, smooth and very glabrous (or some minute hairiness at least on the petioles): leaves bright green, mainly radical and spatulate, very obtuse (larger about inch long, with tapering base or petiole of at least equal length), from 2 to 6 lines wide; cauline only 2 or 3 and smaller: involucre 3 lines high, not unlike that of *E. salsuginosus*, but close, the bracts lanceolate and not attenuate: rays about 40, linear, violet, 3 or 4 lines long. — *Aster glacialis*, Eaton, Bot. King Exp. 142, but hardly that of Nuttall (which is rather a high alpine form of *A. salsuginosus*, to which this is related). Comes close to the next species, to which it has been referred. — Rocky Mountains of Colorado, Utah, and Nevada, in the alpine region; first coll. by *Parry, Hall & Harbour, Watson*.

++ ++ Involucre hirsute or pubescent, greenish: herbage not strigulose nor cinereous.

E. ursínus, Eaton. A span or two high, loosely cespitose: leaves duller green, mostly smooth and glabrous, but their margins more or less hirsute-ciliate, spatulate to narrowly oblanceolate; cauline ones lanceolate or linear and acute: involucre (3 lines high) and naked summit of flowering stem hirsute-pubescent: rays 40 or 50, purple, narrowly linear, 3 lines long. — Bot. King Exp. 148; Gray, Bot. Calif. i. 327. — Alpine and subalpine region, Rocky Mountains, Wyoming to S. Colorado, Uinta Mountains, Utah, and on Mount Dana, California; first coll. by *Watson*.

E. radicátus, Hook. A span high or less, densely tufted: leaves all spatulate-linear or somewhat wider (broadest only a line or two wide), hirsute or hirsutely ciliate, or sometimes almost naked, then glabrous; no glandular roughness: involucre more or less villous-pubescent (barely 3 lines high); rays white or purplish, 2 or 3 lines long. — Fl. ii. 17. *E. nanus* & *E. radicatus*, Nutt. Trans. Am. Phil. Soc. vii. 308. — Alpine or subalpine in the Rocky Mountains, from British America (*Drummond, Macoun*) to Wyoming, S. Colorado, and Utah, *Nuttall, Parry*, &c.

E. glandulósus, T. C. Porter. Cespitose from a stout caudex, a span to almost a foot high, rigid, minutely granulose-glandular or glandular-scabrous (but sometimes obsoletely so), and with sparse hirsute or hispid hairs, especially on the margins of the leaves: these thickish, spatulate to linear-oblanceolate, 1 to 3 inches long; upper cauline small: head comparatively large, 4 or 5 lines high: involucre glandular or viscid as well as pubescent: rays 40 or 50, violet or purple, 4 to 6 lines long: an obscure outer setulose pappus. — Porter & Coulter, Fl. Colorad. 60; Gray, Proc. Am. Acad. xvi. 90. — Bleak mountain-tops, alpestrine and subalpine, and sometimes descending to lower levels, Colorado, *J. M. Coulter, Hall & Harbour, Greene*, &c. Some forms approach *E. pumilus*.

++ ++ ++ ++ ++ Various Rocky Mountain to Pacific species, with entire leaves, none truly alpine, none hispidly hirsute (except very rarely some spreading bristly hairs fringing base of leaves): involucre close, disposed to be somewhat imbricated and rigid: rays not very numerous, in several species uniformly wanting.

++ A span or two high from a simple or multicipital caudex: leaves only few and narrow on the weak and ascending simple or sparingly branched flowering stems; but radical ones with obovate or spatulate blade, only half-inch long, contracted into a petiole of at least equal length, cinereously puberulent or canescent: heads only 3 or 4 lines high: rays 18 to 30, pale violet or purple: akenes compressed, 2-3-nerved: pappus nearly simple.

E. asperugíneus, Gray. Cinereous with minute roughish pubescence: stems commonly simple from the slender caudex, monocephalous: involucre obscurely hirsute, a single series of equal bracts: rays 18 or 20. — Proc. Am. Acad. xvi. 91. *Aster asperugineus*, Eaton, Bot. King Exp. 142. — Utah, in the E. Humboldt Mountains, *Watson, M. E. Jones*.

E. téner, Gray, l. c. Canescent with very fine and close or almost imperceptible pubescence (either silvery-whitish or becoming greener): stems several from a stouter caudex, weak and ascending, bearing single or 2 or 3 heads: involucre minutely canescent; its narrow and close bracts unequal, somewhat in 2 or 3 ranks: rays 25 to 30. — *E. cæspitosum*, var. *tenerum*, Gray, Bot. Calif. i. 328 — High mountains of Utah, N. W. Nevada, and of the Sierra Nevada on the borders of California, *Watson, Brewer*, &c., to those near the sources of the Sacramento, *Pringle*, Red Rock Creek, and of Wind River, Montana, *Watson, Dr. Forwood*.

++ ++ A span to near a foot high, cespitose on a stout multicipit 1 caudex, silvery-canescent, with simple and monocephalous or rarely somewhat branching stems: leaves from narrowly spatulate to linear: rays 40 or 50, white or purple changing to white: *akenes slender and nearly terete, 5-10-nerved or striate:* pappus double; the outer subulate-setulose and conspicuous.

E. cánus, Gray. Silvery appressed pubescence obviously strigulose under a lens, that of the involucre loose and spreading: stems 4 to 9 inches high, leafy: linear cauline leaves gradually diminishing upward; radical spatulate lanceolate or narrower: head 4 lines high: rays narrow, 3 lines long: akenes glabrous, striately 8-10-nerved. — Pl. Fendl. 67, & Proc. Am. Acad. viii. 650. — Dry and gravelly hills, Northern New Mexico and Colorado; first coll. by *Fendler*. Also on the Platte in Wyoming, *Geyer*.

E. argentátus, Gray. Silvery white pubescence throughout very close and fine, the separate hairs undistinguishable: stems 6 to 12 inches high: radical leaves very densely clustered, linear-spatulate or broader, inch or two long; cauline scattered and much smaller: head broad, fully half-inch high: rays rather broad and large, half-inch long: immature akenes sericeous-pubescent or villous, 5-8-nerved. — Proc. Am. Acad. viii. 649. *E. cæspitosum*, Eaton, Bot. King Exp. 153, in small part (no. 549), not Nutt. — Arid interior region, Utah and Nevada, *Watson, Miss Searls, Ward, Palmer, M. E. Jones*.

++ ++ ++ A foot or less high from a thick multicipital caudex, more or less branching and leafy, minutely silvery-canescent (the pubescence fine and short): leaves all narrowly linear: rays 30 to 50, elongated (large for the involucre of about 3 lines high), purple or sometimes white: akenes narrow, 4-nerved, disposed to be tetragonal.

E. Paríshii. Rigid and rather stout, at length somewhat corymbosely branched: leaves spatulate-linear (largest 2 lines wide or nearly so), rather short: heads short-peduncled:

involucre cinereous-puberulent and glandular: rays nearly half-inch long, purple: disk-corollas beset with some sparse and short minute hairs; akenes sparsely hirsute: pappus conspicuously double; outer setose-squamellate. — Rocky cañons, borders of the Mohave Desert, S. E. California, *Parish.*

E. Utahénsis, GRAY. Slender, but rigid, with sparse branches from dense clumps: leaves narrowly linear or almost filiform (larger 2 inches long and barely a line wide): heads slender-peduncled: involucre cauescent: rays fully half-inch long: disk-corollas sparsely hirsute toward the base: immature akenes villous: pappus almost simple; the outer being scanty and setulose, hardly distinguishable from the villous hairs of the akene. — Proc. Am. Acad. xvi. 99. *E. stenophyllus*, var. ? *tetrapleurus*, Proc. Am. Acad. viii. 650. — Rocky hills in the arid region of S. Utah, *Mrs. Thompson, Parry, Palmer.* This and the preceding are showy species, nearly related to *E. argentatus*, all with a close and somewhat imbricated involucre.

++ ++ ++ ++ Either low or comparatively tall, leafy-stemmed or subscapose: akenes compressed, 2-nerved, rarely 3-nerved.

= Heads radiate: leaves all narrowly linear to filiform, the broadest not over a line wide: pubescence either cinereous or obscure. (Also one or two of the following subdivision are sometimes very narrow-leaved.)

a. Involucre of the ample head half-inch high, of linear and equal bracts; and rays half-inch long.

E. stenophýllus, GRAY. Green and glabrate, but obscurely strigulose-puberulent when young: stems simple and monocephalous, less than a foot high, naked and pedunculiform at summit: leaves mostly 2 inches long, hardly widening upward; upper ones sparse and smaller; bracts of involucre somewhat hirsute-pubescent and glandular: ovary villous; pappus simple or nearly so. — Pacif. R. Rep. iv. 42; Proc. Am. Acad. viii. 650. & xvi. 89. — Hills on the Pecos, N. W. Texas, *Bigelow.* A smaller plant from Fort Wingate, New Mexico, *Matthews*, may belong here, but has merely hirsutulous young akenes.

b. Involucre only 2 or 3 lines high, of shorter and unequal somewhat imbricated bracts: rays 2 to 4 lines long.

E. filifólius, NUTT. Canescent or cinereous throughout with very fine close pubescence, no loose hairs: stems slender, a span to two feet high from lignescent slender base or branched rootstock, leafy, usually paniculately branched and bearing several or rather numerous heads: leaves linear-filiform or quite filiform (some lower ones occasionally dilated upward to a line in width and flat): involucre canescent: rays 30 to 50, rarely even 80, purple, violet, or white, 3 or 4 lines long: akenes slightly pubescent or glabrate: pappus simple, of fragile and indistinctly scabrous bristles. — Trans. Am. Phil. Soc. vii. 328; Torr. & Gray, Fl. ii. 177; Gray, Proc. Am. Acad. xvi. 89. *Diplopappus filifolius*, Hook. Fl. ii. 21, is either this or the next. *Chrysopsis canescens*, DC. Prodr. v. 328. — Rocky or dry ground, from Brit. Columbia, mostly east of the Cascades, and Idaho, to the Sierra Nevada in California and Nevada; first coll. by *Douglas.*

E. peucephýllus, GRAY. Low, with flowering stems a span or two high from broad depressed tufts, simple and with naked summit or peduncle monocephalous or occasionally forking and 2-3-cephalous, cinereous-puberulent or glabrate: leaves filiform or lowest slightly dilated upward (to not over half a line in breadth): involucre hirsute-pubescent or glabrate: rays 20 or 30, usually short (2 or 3 lines long), pale blue to cream-color or pure yellow: pappus manifestly double, the outer squamellate. — Proc. Am. Acad. xvi. 89. — Dry hills, from Brit. Columbia (and east to Cypress Hills, *Macoun*) to the Sierra Nevada in California and adjacent Nevada, east to Idaho.

c. Involucre 3 or 4 lines high, of equal bracts: rays of equal length.

E. ochroleúcus, NUTT. Low, a span or two high, somewhat cespitose on the caudex, from cinereous-pubescent to glabrate, and attenuate lower part of the leaves not rarely sparsely hirsute-ciliate: stems usually simple, naked above and monocephalous, occasionally with one or two additional heads: leaves rather rigid, narrowly linear, the radical (2 or 3 inches long) often a line wide at the upper part: involucre tomentose or hirsute-pubescent: rays 40 to 60, "ochroleucous," white, or purplish (not known to be yellow): outer pappus setulose. — Trans. Am. Phil. Soc. vii. 309; Torr. & Gray, Fl. ii. 178; Gray, Proc. Am. Acad. xvi. 89, excl. var. *E. pumilus*, Hook. Lond. Jour. Bot. vi. 242, in part, not Nutt. *E. canescens*, Parry in Jones Exp. no. 139, canescent form. *Diplopappus linearis*, Hook. Fl.

ii. 21 ? — Gravelly hills and plains, N. Wyoming and Montana to Idaho, *Nuttall, Spalding, Geyer, Parry*, &c.

= = Heads rayless: leaves filiform to narrowly spatulate-linear, chiefly from the multicipital caudex: dwarf flowering stems more or less scapiform and monocephalous.

E. Bloómeri, GRAY. Densely cespitose, cinereous-puberulent or glabrate and pale: radical leaves 1 to 3 inches long, the larger dilated upward sometimes to a line or more in width; cauline few and nearly filiform: scapiform flowering stems 2 to 6 inches high: head almost half an inch high: involucral bracts equal, linear-lanceolate, soft-villous or canescent: akenes glabrate, oblong-linear, flat: pappus whitish, simple. — *Proc. Am. Acad.* vi. 40, & *Bot. Calif.* i. 328: *Eaton, Bot. King Exp.* 148. — Stony ground, mountains of Nevada to Idaho, and from the Sierra Nevada, California, to E. Oregon; first coll. by *Bloomer*. Habit of the last preceding species, to which it is most allied.

= = = Heads radiate: leaves from narrowly linear to oblong.

a. Stems naked above, more commonly simple and monocephalous, only a span or two high: pappus simple.

E. Nevadénsis, GRAY. Stems numerous from a multicipital caudex, erect, a span to nearly a foot high: leaves all lanceolate or linear; radical 1 to 4 inches long, 1 to 4 lines wide, strigulose-cinereous; uppermost small and subulate: head always solitary, half-inch high: involucre villous-pubescent, sometimes glabrate; its bracts equal: rays rather broadly linear, white or pale blue, 4 to 6 lines long: akenes comparatively large: pappus rather coarse. — *Proc. Am. Acad.* viii. 649; *Bot. Calif.* i. 328. *E. cæspitosus*, var. *grandiflorus*, Eaton, *Bot. King Exp.* 153, in part. — Sierra Nevada, California, and W. Humboldt Mountains, Nevada, at 5,000 to 8,000 feet; first coll. by *Bloomer, Watson*, &c. Appears to pass into the somewhat doubtful

Var. **pygmǽus**, GRAY, l. c. Dwarf, subcaulescent: leaves spatulate-linear, an inch or more long, a line or so wide, more minutely pubescent or cinereous, and glabrate: head considerably smaller: involucre slightly hirsute: rays purple. — Sierra Nevada, California, above Ebbett's Pass (at 9,500 feet) and Mono Pass (10,750 feet), *Brewer*. Also Mount Dyer, Plumas Co. (a connecting form), *Mrs. Austin*.

E. Eatóni, GRAY. Stems several from the crown of a strong tap root, slender and weak, diffuse, 3 to 9 inches long, simple or with 2 or 3 monocephalous branches: leaves all linear, thickish, minutely strigulose-pubescent; radical about 2 inches long and the broadest 2 lines wide: heads only 3 lines high: bracts of the sparsely hirsute involucre little unequal: rays seldom over 20, at most 3 lines long, white or purplish. — *Proc. Am. Acad.* xvi. 91. *E. ochroleucus*, Eaton, *Bot. King Exp.* 152, not Nutt. — Rocky Mountains of Colorado and Wyoming, and the Uinta and Wahsatch Mountains in Utah; first coll. by *Watson* and *Eaton*.

b. Stems more leafy and disposed to branch, but sometimes monocephalous: pubescence cinereous: outer pappus setulose, sometimes rather manifest, sometimes obscure or none.

E. cæspitósus, NUTT. Low, a span to rarely a foot high, many-stemmed and ascending or spreading from a stout multicipital caudex, from cinereous to canescent with dense and fine short pubescence (this generally spreading and soft, sometimes hispidulous, rarely fine and appressed, at least on young parts): stems of smaller plants monocephalous: radical leaves spatulate to lanceolate, and cauline lanceolate-oblong to linear (half-inch to 2 inches long): heads short-peduncled, 3 or 4 lines high: bracts of the involucre rather unequal: rays 40 or 50, linear, 3 or 4 lines long, white, sometimes tinged with rose-color. — *Trans. Am. Phil. Soc.* vii. 307 (a small and low form); Torr. & Gray, *Fl.* ii. 179. *Diplopappus canescens* (*Erigeron canescens*, Torr. & Gray, l. c.) & *D. grandiflorus* (*E. cæspitosus*, var. *grandiflorus*, Torr. & Gray, l. c.), Hook. *Fl.* ii. 21, the latter a large form. — Mountains and high plains, Saskatchewan and Montana to Utah and borders of New Mexico, and eastern part of the Sierra Nevada, California. A variable species. Western forms come near to the next.

E. corymbósus, NUTT. Taller, often a foot or two high, erect from creeping rootstocks, soft-cinereous or sometimes hispidulous with the mostly spreading short pubescence: radical leaves narrow-lanceolate or spatulate-lanceolate (largest 3 or 4 inches long and 3 or 4 lines wide), 3-nerved; cauline linear and narrow: heads sometimes solitary, usually several and corymbosely disposed on short slender peduncles: involucre 3 lines high, canescently pubescent: rays 30 to 50, mostly narrow and 3 to 5 lines long, blue or violet, apparently sometimes white. — *Trans. Am. Phil. Soc.* vii. 308; Torr. & Gray, l. c.; Gray, *Bot. Calif.* i. 329.

— Mountains of Montana to those of Washington Terr. and sparingly of California; first coll. by *Nuttall*. A soft-pubescent form, subalpine in Washington Terr. and E. Oregon, *Cusick*, *Brandegee*, has white rays; a similar one, coll. by *Lyall* near the British boundary, has blue rays. Nuttall's character of achenium, "nearly smooth and striate," does not accord with his specimens.

E. Bréweri, GRAY. A span to a foot high from slender rootstocks, slender, erect or ascending, leafy up to the solitary or several and corymbosely disposed heads, scabrous-cinereous with minute spreading pubescence: leaves small (the largest barely inch long), narrowly spatulate or uppermost nearly linear, obtuse: heads 3 or 4 lines high: involucre glabrous or minutely granuloso-glandular; its bracts unequal, obtuse: rays 12 to 20, violet, 3 lines long. — Proc. Am. Acad. vi. 541, & Bot. Calif. l. c. — Open woods of the Sierra Nevada, California, from Kern Co. to Shasta; first coll. by *Brewer* and *Torrey*.

  c. Stems (commonly from slender rootstocks) leafy, mostly branched above and bearing few or several heads: pubescence not cinereous nor spreading, either strigose or none: pappus essentially simple.

E. decúmbens, NUTT. Slender, commonly low or spreading, 6 to 18 inches high, strigulose-pubescent or puberulent, or glabrate: leaves linear or sometimes linear-spatulate (radical not rarely 4 to 6 inches long and only a line or two wide, sometimes 3 lines wide): involucre minutely hirsute or pubescent: rays 15 to 40, white, purplish, or violet-tinged. — Trans. Am. Phil. Soc. l. c. 309; Torr. & Gray, l. c. — Mountains, from Montana and Utah to Oregon and northern part of the Sierra Nevada, California; first coll. by *Douglas* and *Nuttall*.

E. foliósus, NUTT. A foot or two high, smooth and glabrous, or with some minute roughish hairs, usually branched above, and bearing scattered or loosely corymbose heads: leaves linear, obtuse, the larger an inch or two long and 2 or 3 lines wide, but often much narrower: heads hemispherical, 3 or 4 lines high: involucre of somewhat unequal bracts, either minutely puberulent-strigose or glabrous, rarely hirsute: rays 20 to 40, narrow, 3 to 5 lines long, violet or purple, rarely white. — Trans. Am. Phil. Soc. l. c., & Fl. Gamb. 117; Gray, Bot. Calif. i. 329 (excl. var. *inornatus*), & Proc. Am. Acad. xvi. 88. *E. Douglasii*, Torr. & Gray, Fl. ii. 177. *E. decumbens*, Benth. Pl. Hartw. 316, not Nutt. *Diplopappus occidentalis*, Hook. & Arn. Bot. Beech. 350. — Sparsely wooded ground, common nearly throughout California, especially in the western parts; first coll. by *Douglas*. Nuttall's name was given to the broader-leaved form. This passes freely into

  Var. **stenophýllus**, GRAY, l. c. A common form, with leaves from only a line wide to slender and filiform. — *E. stenophyllus*, Nutt. Pl. Gamb. 176, not Gray. — Same range, and equally common.

  Var. **tenuíssimus**. Slender, small-leaved: leaves nearly all filiform, erect or ascending; the longest only an inch long; upper gradually shorter, becoming setaceous-subulate: heads much smaller. — San Diego Co. on the Mexican border, and within Lower California, *Parry*, *Palmer*, *Orcutt*.

= = = = Heads wholly rayless: stems leafy to the summit: pappus simple.

E. inornátus, GRAY. Commonly glabrous throughout and smooth, or with some sparse hirsute pubescence: stems 10 to 20 inches high, erect: leaves from broadly to narrowly linear (an inch or two long, a line or two wide): heads usually several and cymosely disposed at the summit of the stem, short-peduncled, 3 lines high: involucre campanulate; its bracts unequal and somewhat imbricated, very glabrous. — Proc. Am. Acad. xvi. 88. *E. foliosus*, var. *inornatus*, Gray, Bot. Calif. i. 330. — Pine woods, Sierra Nevada and coast ranges of California to those of E. Oregon and Washington Territory; first coll. by *Newberry*. Comes near some forms of *E. foliosus*, but rayless.

  Var. **angustátus**. Leaves very narrowly linear or almost filiform: heads few or scattered and paniculate. — Red Mountain, Mendocino Co., California, *Kellogg* & *Harford*, and Napa Co., *Greene*.

  Var. **viscídulus**. Low and stouter: heads fewer and larger (4 lines high): leaves spatulate-linear, shorter (seldom an inch long): stems and peduncles occasionally hirsute-pubescent, and as well as the leaves commonly more or less viscid. — Mountains of northern part of California, *Kellogg* & *Harford*, *Pringle*.

E. súpplex, GRAY. Villous-hirsute: stems decumbent or ascending from a slender rootstock, mostly simple, a span to a foot long, terminated by single and very broad (5 to 6 lines high) short-peduncled heads: leaves spatulate-lanceolate or uppermost linear, mucronate-apiculate (an inch or two long): involucre villous; its bracts linear-lanceolate, equal. — Proc. Am. Acad. vii. 353, & Bot. Calif. i. 330. — N. E. California, in Humboldt and Mendocino Co., *Bolander, Pringle*; the latter a nearly erect form.

E. míser, GRAY. Cespitose from a thickish caudex or rootstock, canescently villous: stems ascending, 3 to 5 inches high, leafy up to the solitary or few and small (3 lines high) heads: leaves from oblong-spatulate to short-linear (4 to 8 lines long): involucre glabrate or minutely glandular, short; its bracts lanceolate or linear, acute: flowers comparatively few. — Proc. Am. Acad. xiii. 372, & Bot. Calif. ii. 445. — On Mount Stanford and vicinity, in the Sierra Nevada, California, *Lobb, Kellogg, Greene*, &c.: fl. late.

+ + + + + + + S. Arizonian, with diffusely branched and trailing stems, very leafy branches, bearing very small heads, soft-cinereous pubescence, and lower leaves commonly 3–5-lobed or toothed: pappus simple.

E. Lemmóni, GRAY. Stems a foot or two long, apparently from slender creeping rootstocks: leaves half-inch long or less, spatulate; upper all entire, lower tapering into more or less of a petiole, many of them 1–5-toothed or incisely lobed: heads terminating short branchlets, short-peduncled: involucre 2 lines high: rays of about same length, 40 or 50, light purple. — Proc. Am. Acad. xix. 2. — Tanner's Cañon, Huachuca Mountains, S. Arizona, *Lemmon*.

+ + + + + + + Northeastern species, smooth and slender, erect, from filiform rootstocks, leafy-stemmed, entire-leaved; with small and Aster-like heads of only 20 or 30 rays: pappus quite simple.

▬ **E. hyssopifólius**, MICHX. Nearly glabrous, a span to a foot high, sparingly branched: branches terminated by a solitary slender-peduncled head: leaves small and numerous, linear or lower somewhat spatulate, thinnish, entire, an inch or less long: rays 3 lines long, white or tinged with purple. — Fl. ii. 123; Gray, Proc. Am. Acad. xvi. 87. *Aster graminifolius*, Pursh, Fl. ii. 545; DC. Prodr. v. 227; Torr. & Gray, Fl. ii. 156. *Galatella graminifolia*, Hook. Fl. ii. 15. — Moist and rocky banks, Newfoundland and New Brunswick to Hudson's Bay, northern borders of New England to Lake Superior and Slave Lake; first coll. by *Michaux*.

* * * Perennial by rosulate offsets, producing a scapiform stem from a rosette of radical leaves: heads small and Aster-like, bearing only 20 or 30 rays: disk convex, only 3 lines broad: akenes mostly 4-nerved: pappus quite simple: S. Atlantic species. — § *Erigeridium*, Torr. & Gray.

▬ **E. nudicaúlis**, MICHX. Glabrous or glabrate: scapiform stems solitary or occasionally several from the rosette of obovate or spatulate thickish and sparingly denticulate leaves: cauline leaves few and small, or merely bracts: heads several, corymbosely cymose: rays white and pinkish, 2 or 3 lines long. — Pursh, l. c. *Erigeron* (*Erigidium*) *vernus*, Torr. & Gray, Fl. ii. 176. *E. integrifolius*, Bertol. Misc. Bot. vi. t. 3, not Bigel. *Aster vernus*, L. Spec. ii. 876. *Doronicum bevifolium*, Walt. Car. 205? *Stenactis verna*, Nees, Ast. 275; DC. l. c. 299. — Low pine barrens near the coast, Virginia to Florida and Louisiana: fl. spring.

* * * * Perennial by biennial rosulate offsets borne on apex of stoloniform creeping rootstocks, or some species probably biennial: leaves membranaceous, commonly serrate or dentate: heads middle-sized or small, with glabrate involucre: rays numerous: pappus quite simple: species not montane.

+ Rays not very narrow, not more than 60 or 70.

▬ **E. bellidifólius**, MUHL. Stoloniferous-cespitose, making rosulate offsets from slender subterranean shoots, villous-hirsute: flowering stems usually a foot or more high, simple, naked above and bearing 3 to 9 (or, when depauperate, only single) umbellately cymose middle-sized heads: radical leaves cuneiform-obovate or spatulate, mostly coarsely few-toothed, on very short-winged petioles; cauline few, oblong or lanceolate: bracts of the involucre appressed: rays violet or bluish-purple, a third to half inch long: akenes almost glabrous. — Willd. Spec. iii. 1958; Sims, Bot. Mag. t. 2402; Torr. & Gray, Fl. ii. 170. *E. pulchellus*, Michx. Fl. ii. 124, excl. syn. Gronov.; Darl. Fl. Cest. ed. 2, 492; Hook. Fl. ii. 19, excl. var. — Damp ground, borders of woodlands, Canada to Illinois and Louisiana: fl. spring.

E. Oregánus, GRAY. Perhaps only biennial, pubescent: rosulate tufts many-leaved, sending up weak or diffuse leafy stems of a span or two in length, bearing solitary or few rather small heads: leaves spatulate, or the radical cuneate-obovate; these an inch or two long, coarsely 3-5-toothed or incised; cauline more entire, inch long: rays pale purple, quarter-inch long. — Proc. Am. Acad. xix. 2. — Oregon, along the Columbia River under overhanging cliffs in Multnomah Co., *Howell*.

+ + Rays very narrow, almost filiform, and numerous (much over 100): disk only 3 or 4 lines broad: stems scattered, erect, either from a biennial root or from a biennial or winter-annual offset.

E. Philadélphicus, L. Soft-hirsute, a foot or two high: stem striate-angled: leaves oblong, or lowest spatulate or obovate; upper cauline half-clasping, obtuse, sparingly and coarsely serrate or entire: peduncles thickened under the head: rays pink, about 3 lines long. — Spec. ii. 863; Willd. Spec. iii. 1957; Torr. & Gray, Fl. ii. 171. not Michx., Ell., &c. *E. purpureum*, Ait. Kew. iii. 186, D!!. l. c. *E. pulchellus*, var., Hook. Fl. ii. 19 (N. W. Am.). — Moist fields and border of woodlands, Hudson's Bay to Florida, Texas, California, and Brit. Columbia: fl. summer.

E. quercifólius, LAM. Pubescent with short spreading hairs, sometimes cinereous, about a foot high: radical and lowest cauline leaves obovate or spatulate, from repand to sinuate-pinnatifid: heads smaller than in the preceding: rays barely 2 lines long, from bluish or purplish to white. — Ill. t. 681, f. 4; Poir. Dict. vii. 490; Reichenb. Ic. Exot. t 134 (?); Torr. & Gray, l. c. *E. Philadelphicus*, Michx. Fl. ii. 123; Ell. Sk. ii. 396; DC. l. c., not L. — Low grounds, S. Carolina to Florida and Texas; fl. spring.

* * * * * * Perennial by rooting from decumbent or creeping leafy stems or stolons: rays very numerous and narrow: heads solitary, slender-peduncled.

E. répens. Cinereous-pubescent: stems prostrate or ascending from the slender root; prostrate ones rooting at the nodes: leaves obovate or broadly spatulate with cuneate base tapering into a petiole, obtusely and deeply 5-9-toothed or almost lobed: peduncles scapiform, 4 to 8 inches long: involucre 4 lines high: rays 3 lines long, white: pappus simple. — *E. scaposus*, Torr. & Gray, Fl. ii. 170; Gray, Pl. Lindh. i. 11, but hardly the Mexican *E. scaposus* nor *E. longipes*, DC. *E. scaposus*, var.? *cuneifolius*, Gray, Proc. Am. Acad. xvi. 94. — Sandy sea-coast, Texas, *Berlandier, Drummond, Lindheimer, Wright*, &c. (Probably also on the Mexican side of the Rio Grande.)

E. flagelláris, GRAY. More or less cinereous with fine appressed pubescence: stems slender, diffusely decumbent and flagelliform but leafy, some prostrate, many at length rooting at the apex and proliferous: leaves small, entire; radical spatulate and petioled; those of the branches passing to linear (from an inch to 3 lines long): peduncles 2 to 3 inches long: head barely 3 lines high: rays white or purplish: pappus double, the outer subulate-setulose. — Pl. Fendl. 69; Rothrock in Wheeler Rep. vi. 153. *E. divergens*, Hook. Lond. Jour. Bot. vi. 242. *E. divergens*, var., Gray, Pl. Wright. — Banks of streams, W. Texas and New Mexico to Colorado and S. W. Utah; also north to the Upper Platte; first coll. by *Fendler*.

* * * * * * * Annuals or sometimes biennials, leafy-stemmed and branching: heads conspicuously radiate, except in one species.

+ Akenes narrow, little compressed, with a broad and whitish truncate apex and a simple capillary pappus: heads small (only 3 lines high): rays 40 to 70, not very narrow.

E. Bellidiástrum, NUTT. A diffusely or loosely branched annual, a span or two high, cinereous-pubescent: leaves entire, spatulate-linear or the lowest broader (an inch or less long): heads paniculate, short-peduncled: rays light purple. — Trans. Am. Phil. Soc. vii. 307; Torr. & Gray, Fl. ii. 170; Gray, Proc. Am. Acad. viii. 648. — Low grounds, plains of Nebraska to New Mexico; first coll. by *Nuttall*.

+ + Akenes compressed, 2-nerved: pappus more or less double; outer short and subulate-squamellate or sometimes coroniform; inner often fragile or deciduous.

++ Leaves entire, sometimes dentate or lower incisely lobed, not dissected. — *Phalacroloma*, Torr. & Gray, Fl. ii. 175.

= Rays of the middle-sized or rather large heads numerous, well exserted, and with pappus like the disk-flowers: leaves all entire: Southwestern species.

E. Rúsbyi. Hirsute-pubescent or hispidulous, but green: stems a foot high from probably annual or biennial root, sparingly branched, somewhat diffuse or spreading, equably leafy: cauline leaves oblong-lanceolate, acute, closely sessile by a broad base, about an inch long;

radical not larger, obovate or spatulate, slender-petioled: heads solitary, terminating the branches, on rather slender peduncles: involucre broad, 3 lines high, slightly pubescent: rays about 50, apparently white, 4 lines long, not very narrow: pappus indistinctly double, the outer short and setulose. — Mogollon Mountains, N. Arizona, *Rusby*.

E. Arizónicus, Gray. Cinereous-hirsute throughout: stem 2 feet high from an annual root, strict, with simple branches, leafy: leaves oblong-lanceolate and sessile, or lower obovate-oblong and petioled, an inch or two long: heads solitary and short-peduncled, terminating the branches, half-inch high and broad: involucre hirsute: rays 80 to 100, white, 4 or 5 lines long: outer pappus very conspicuous, setose-squamellate. — Near Tanner's Cañon in the Huachuca Mountains, S. Arizona, *Lemmon*.

= = Rays of the small heads rather numerous but small, shorter than or barely equalling the flowers of the convex disk. Verges to § *Cœnotus*.

E. incómptus. A foot or two high, branched from the base, slender and erect, hirsute with short spreading pubescence, leafy: leaves narrowly linear (half-inch or inch long, a line or less wide), or lower narrowly spatulate-lanceolate and attenuate into slender petiole: heads slender-peduncled: involucre 2 lines high, shorter than the hemispherical disk: rays either very numerous or fewer, slender, with ligule only a line long, bluish or purplish: outer pappus conspicuous, subulate-squamellate, longer than the breadth of the glabrate akene; inner scanty and rather deciduous. — Carysito, Lower California, near the U. S. border, within which it probably occurs, *C. R. Orcutt*.

= = = Rays of the small heads only 30 or 40, well exserted, white, not very narrow, barely 3 lines long, and with pappus as in the disk-flowers: leaves narrow, entire.

E. modéstus, Gray. A foot or less high and much branched from an indurated but annual root, slender, rigid, cinereous-hirsute or hispid: branches terminated by the small (2 lines high) slender-pedunculate heads: upper leaves linear and lower narrowly spatulate, about an inch long. — Pl. Fendl. 68 (excl. syn.) & Pl. Lindh. ii. 220. — Dry and sterile rocky plains, W. and N. W. Texas, *Lindheimer*, *Wright*, &c.

= = = = Rays of small or barely middle-sized heads very numerous (about 100), narrow, with pappus like the disk-flowers; the inner of rather scanty bristles; outer of short subulate squamellæ: leaves from entire to sparingly lobed.

E. divérgens, Torr. & Gray. Diffusely branched and spreading, a foot or less high, cinereous-pubescent or hirsute: leaves linear-spatulate or the upper linear and the lowest broader (these 2 to 4 lines wide, sometimes laciniately toothed or lobed): heads 2 or 3 lines high, and the white or purplish or sometimes violet rays equally long: involucre hirsute: receptacle in age commonly very convex. — Fl. ii. 175: Gray, Pl. Fendl., Pl. Wright, &c. *E. strigosus*, var., Hook. Fl. ii. 18, in part. *E. (Oligotrichium) divaricatus*, Nutt. Trans. Am. Phil. Soc. l. c. 311. — Low plains and river-banks, Nebraska to W. Texas, Washington Terr., and California. (Adj. Mex.)

Var. **cinéreus**, Gray, l. c. Dwarf and flowering almost from the root, with the earliest heads on slender almost scapiform peduncles; or leafy and later heads shorter-peduncled: pubescence soft and cinereous. — *E. cinereus*, Gray, Pl. Fendl. 68. *E. nudiflorus*, Buckley in Proc. Acad. Philad. 1861, 456. — W. Arkansas to Arizona. (Adj. Mex.)

E. ténuis, Torr. & Gray. Branched from the annual or biennial root, ascending or erect, a span or two high, somewhat hirsute or pubescent: leaves oblong-spatulate or lanceolate, and the lowest obovate (4 to 6 lines wide), occasionally few-toothed or sinuate-lobed: heads little over 2 lines high: involucre nearly glabrous: rays white and purplish. — Fl. ii. 175. *E. quercifolium*, Nutt.; DC. Prodr. v. 285, not Lam. *E. Brazoensis*, Buckley, l. c. — Low grounds, Arkansas, Louisiana, and Texas.

= = = = Rays of the small heads not excessively numerous, nor very narrow (2 or 3 lines long), white or barely purplish-tinged; the bristles of their pappus commonly wanting or very few: outer pappus a short crown of distinct or partly united slender squamellæ, persistent after the fragile inner pappus has fallen: tall and erect winter annuals or biennials, leafy, branched above, bearing corymbosely cymose or paniculate heads, commonly produced all summer: leaves green, sometimes serrate or the lower incised: weedy species, of wide distribution; the two generally distinct in the Atlantic States, hardly so on the Pacific side. — *Phalacroloma*, Cass. Dict. xxxix. 404.

E. ánnuus, Pers. Sparsely hirsute with spreading hairs, 2 to 5 feet high: leaves membranaceous, from ovate to broadly lanceolate, mostly serrate, lower often very coarsely so:

involucre commonly beset with some bristly hairs. — Syn. ii. 431; Hook. Fl. ii. 20; Torr. & Gray, Fl. ii. 175. *E. heterophyllus*, Muhl. in Willd. iii. 1956; Pers. l. c.; Pursh, Fl. ii. 148; Bart. Veg. Mat. Med. t. 21. *E. strigosus*, Bigel. Fl Bost ed. 2. 302, not Muhl. *Aster annuus*, L. Hort. Cliff. & Spec. ii. 875. *Pulicaria annua*, Gaertn. Fruct. ii. 462. *Diplopappus dubius*, Cass. Bull. Philom. 1817 & 1818. *Stenactis dubia*, Cass. Dict. xxxvii. 485. *S. annua & S. strigosa* (excl. syn.), DC. Prodr. v. 299. *Phalacroloma acutifolium*, Cass. Dict. xxxix. 405. — Fields and open grounds, common from Canada to Virginia: also in Oregon, &c., in a form quite intermediate between this and the following. (Nat. in Eu.)

E. strigósus, MUHL. Pubescence appressed, either sparse and strigose or close and minute: stem seldom over 2 feet high: leaves of firmer texture, lanceolate and the upper entire; lower from spatulate-lanceolate to oblong, often sparingly serrate: involucre with few or no bristly hairs. — Willd. Spec. l. c.; Ell. Sk. ii. 394; Hook. l. c.; Torr. & Gray, l. c. *E. nervosum*, Pursh, l. c., not Willd. *E. ambiguus*, Nutt. Gen. ii. 147. *E. Philadelphicus*, Bart. Veg. Mat. Med. t. 20. *E. integrifolius*, Bigel. l. c. *Doronicum ramosum*, Walt. Car. 205. *Phalacroloma obtusifolium*, Cass. Dict. xxxix. 405. *Stenactis ambigua*, DC. l. c. — Dry open grounds, Canada and Saskatchewan to Texas, Oregon, and California. Passes into or mixes with the preceding. Occurs rarely with abortive rays, var. *discoideus*, Robbins, in Gray, Man. ed. 5, 237.

Var. **Beyríchii**. A slender form, with minute and sometimes almost cinereous pubescence, smaller heads, and rays from white to pale rose-color. — Torr. & Gray, l. c. *E. Beyrichii*, Hort. Berol. *Stenactis Beyrichii*, Fisch. & Meyer, Ind. Sem. Petrop. v. 27. *Phalacroloma Beyrichii*, Fisch. & Meyer, l. c. vi. 63. — Nebraska to Arkansas and Texas, perhaps first coll. by *Beyrich*.

++   ++ Leaves pinnately parted into narrow divisions: rays very numerous (100 or more) and narrow: pappus alike in ray and disk; the bristles of the inner very deciduous; the short squamellæ of the outer more or less confluent into a multidentate crown. — Original of *Stenactis*, Cass. ex Benth. *Polyactis*, Less. Syn. Comp. 188. *Polyactidium*, DC. Prodr. v. 281.

E. Neo-Mexicánus, GRAY. A foot or two high from a biennial or winter-annual root, leafy, paniculately branched, hispidulous or hispid with spreading bristly hairs; divisions of the cauline leaves 3 to 9, linear or linear-spatulate, obtuse, of the radical shorter and broader: rays white or purplish-tinged, narrowly linear, 4 or 5 lines long. — Proc. Am. Acad. xix. 2. *E. delphinifolius*, Gray, Pl. Wright. ii. 77; Rothrock in Wheeler Rep. vi. 153 (where the root is said to be perennial, which needs confirmation), not Willd. — Hillsides, New Mexico and Arizona, *Wright, Thurber, Palmer, Rothrock, Lemmon*.

E. DELPHINIFOLIUS, Willd. (*Stenactis*, Cass., *Polyactidium*, DC.), from which Bentham first distinguished our very similar species, appears to be wholly Mexican, has appressed pubescence and more numerous as well as more slender rays.

§ 2. TRIMORPHÆA. Rays inconspicuous or slender, numerous, sometimes not exceeding the disk: within them a series of rayless filiform female flowers (commonly none in the last species): leaves entire or nearly so. — *Trimorphæa*, Cass. Dict. xxxvii. & liv.

* Stems low from a truly perennial rootstock, mostly simple and monocephalous: ray-corollas bearing a few long and articulated hairs on the upper part of the tube: short outer pappus manifest.

E. alpínus, L. A span or so high, 1–3-cephalous: herbage and involucre more or less hirsute: leaves entire; lowest spatulate, uppermost usually linear: rays purple, about twice the length of the pappus. — Spec. ii. 864; Engl. Bot. t. 464; Fl. Dan. t. 292; Hook. Fl. ii. 18. excl. vars.; Reichenb. Fl. Germ. xvi. t. 914. — High region of Northern Rocky Mountains, *Drummond*, only specimen seen is not certain. (Eu., N. Asia.)

* * Stems a span to a foot or more high from a biennial or sometimes more enduring root, the larger plants branching and bearing several or numerous somewhat paniculately disposed heads: pappus nearly or quite simple.

E. ácris, L. More or less hirsute-pubescent, varying towards glabrous (not glandular): cauline leaves mostly lanceolate, the lower and radical spatulate: involucre hirsute: rays slender, equalling or moderately surpassing the disk and pappus, purple: filiform female flowers numerous. — Spec. ii. 863; Engl. Bot. t. 1158; Reichenb. l. c. t. 917; Blytt, Norg.

Fl. 561. *E. alpinus* & *E. glabratus*, in part, Hook. Fl. l. c. *Trimorphœa vulgaris*, Cass. Dict. liv. 324. — Anticosti to Labrador, Saskatchewan, &c., to Brit. Columbia and Oregon, and in the Rocky Mountains south to Colorado and Utah. (Eu., N. Asia.)

Var. **Drœbachénsis**, Blytt, l. c. Somewhat glabrous, or even quite so: involucre also green, naked, at most hirsute only at the base, often minutely viscidulous: slender rays somewhat slightly exserted, sometimes minute and filiform and shorter than the pappus. — *E. Drœbachensis*, O. Mueller, Fl. Dan. t. 874; Fries, Summa Scand. 182; Reichenb. Ic. Fl. Germ. xvi. t. 916. *E. elongatus*, Ledeb. Fl. Alt. iv. 91, & Fl. Ross. ii. 487. *E. Kamtschaticus*, DC. Prodr. v. 290. *E. glabratus*, Hook. Fl. ii. 18, mainly, not Hoppe. — New Brunswick and the north shore of Lake Superior to the Arctic Circle and Kotzebue Sound, south along the Rocky Mountains to Colorado and Utah, at about 10,000 feet. Clearly passes into the other form. (Eu., N. Asia.)

Var. **débilis**. Sparsely pilose: stems a span to a foot high from an apparently perennial root, slender, 1-3-cephalous: leaves bright green; radical obovate or oblong; cauline spatulate to lanceolate, short: involucre sparsely hirsute or upper part glabrate, the attenuate tips of the bracts spreading: rays in flower rather conspicuously surpassing the disk. — Northern Rocky and Cascade Mountains, Montana, *Canby*, *Sargent*, at Woodruff's Falls, the tips of involucral bracts strongly recurved. Mount Paddo, *Suksdorf*, *Howell*. Also Hudson's Bay, *Burke*, and N. Labrador, named by *Steetz*, *E. Drœbachensis*, var. *hirsutus*. Passing into that species or form.

E. armeriæfólius, Turcz. Sparsely hispid-hirsute or the leaves glabrous and most of the (narrowly linear and elongated) cauline bristly-ciliate: inflorescence more racemose and strict: involucre sparsely hirsute: rays filiform, extremely numerous, slightly surpassing the disk, whitish, no filiform rayless flowers seen (even in Siberian specimens, though described by Turczaninow). — Cat. Baik. & DC. Prodr. v. 291; Ledeb. Fl. Ross. ii. 489; Gray, Proc. Am. Acad. viii. 648, & Bot. Calif. i. 326. *E. lonchophyllus*, Hook. Fl. ii. 18. *E. glabratus*, var. *minor*, Hook. l. c., partly. *F. racemosus*, Nutt. Trans. Am. Phil. Soc. l. c. 312. — Saskatchewan and along the Rocky Mountains to Colorado, mountains of S. Utah, Nevada, and the Sierra Nevada, California. (N. Asia.)

§ 3. Cænótus, Nutt. Rays of the small and narrow seemingly discoid (and mostly thyrsoid-paniculate) heads inconspicuous, little if at all surpassing the disk or pappus; the narrow ligule always shorter than its tube, often shorter than the style-branches, or even obsolete: disk-flowers sometimes few, with usually 4-toothed corolla: annuals or biennials, with the aspect of *Conyza*, and passing into that genus: the pappus in the genuine species simple: bracts of the involucre not rarely somewhat unequal and imbricated. — Gen. ii. 148; Benth. & Hook. Gen. ii. 281.

*. Floccose-lanuginous with white wool, destitute of either hirsute or viscid pubescence.

E. eriophýllus, Gray. A foot or two high, bearing few heads on almost leafless branches: lower leaves spatulate-oblong, obtuse, serrate near the apex (inch long); upper linear, entire: involucre glabrate (3 lines high); corollas purplish, not exceeding the pappus: akenes oblong-obovate, flat, callous-margined: pappus completely simple, somewhat deciduous in a ring. — Pl. Wright. ii. 77. — S. Arizona, on the Sanoita, *Wright*.

* * Lightly arachnoid, but green and at length naked, somewhat viscid-pubescent.

E. subdecúrrens, Schultz Bip. A foot or two high, strict, bearing numerous heads in a virgate racemiform leafy thyrsus: leaves oblong-linear or lanceolate (inch or less long), sparingly dentate, or the lower sometimes sinuate-laciniate, the base partly adnate-clasping: involucre (2 lines high) sparsely hirsute with viscid hairs: flowers whitish: ligules very short: disk-flowers 6 to 10: pappus scanty, somewhat deciduous in a ring. — *Conyza subdecurrens*, DC. Prodr. v. 379. *C. Coulteri*, Rothrock in Wheeler Rep. vi. 155, not Gray. — Arizona, on Mount Graham at 9,000 feet, *Rothrock*. (Mex., *Schaffner*, *Parry* & *Palmer*, &c.)

* * * Pubescence hirsute or hispid, neither lanate nor viscid, very leafy.

+ Introduced weed: heads fully 3 lines high.

E. linifólius, Willd. A foot or two high, rather strict, bearing loosely paniculate heads, hirsute, also somewhat scabrous with minute appressed pubescence: upper leaves narrowly linear, mostly entire, narrowed downward; lowest broader, incisely toothed or laciniate-involucre cinereous-pubescent: ligules very small, shorter than the style and the at length ferruginous pappus. — Spec. iii. 1955; Benth. Fl. Austr. iii. 495. *E. ambiguus*, Schultz Bip. in Phyt. Cauar. ii. 208. *E. Bonariensis*, DC. Prodr. v. 289, in part. *Conyza ambigua*, DC. Fl. Franc. & Prodr. l. c. *C. sinuata*, Ell. Sk. ii. 323. — Waste grounds, coast of S. Carolina to Florida. (Intr. from tropics.)

+ + Indigenous weeds; but the common species now cosmopolitan: heads only 2 lines high: involucre almost glabrous: leaves commonly more or less hispid-ciliate.

E. Canadénsis, L. From sparsely hispid to almost glabrous: stem strict, 1 to 4 feet high, with numerous narrowly paniculate heads, or in depauperate plants only a few inches high and with few scattered heads: leaves linear, entire, or the lowest spatulate and incised or few-toothed: rays white, usually a little exserted and surpassing the style-branches. — Spec. ii. 863; Fl. Dan. t. 292; Torr. & Gray, Fl. ii. 167. *E. paniculatus*, Lam. Fl. Franc. *E. pusillus*, Nutt. Gen. ii. 148, a depauperate form. *E. strictum*, DC. Prodr. v. 289, a strict and setose-hispid form. *Senecio ciliatus*, Walt. Car. 208. — Open or waste grounds, throughout temperate N. America, especially the warmer parts. (Nat. in Eu., &c.)

E. divaricátus, Michx. Low (a span to a foot high), diffusely much branched, somewhat fastigiate: leaves all narrowly linear or subulate, entire: rays purplish, rarely surpassing the style-branches or the pappus. — Fl. ii. 123; Nutt. l. c.; Torr. & Gray, l. c. — Open grounds and river banks, Indiana to Minnesota, Nebraska, and Texas.

50. CONYZA (Tourn., L. in part), Less. (Name used by Dioscorides and Pliny for some kind of *Fleabane*, supposed to come from κώνωψ, a flea.) — Herbaceous or some shrubby, of various habit; what were the original species belong to *Inula*, &c., those now referred to it are of warm regions, and approach the *Cænotus* section of *Erigeron*. — Benth. & Hook. Gen. ii. 283.

C. Coúlteri, Gray. Apparently annual, a foot or two high, commonly branched, bearing numerous small heads in a mostly crowded thyrsoid leafy panicle, viscidly pubescent or partly hirsute with many-jointed hairs; cauline leaves linear-oblong, the lower spatulate-oblong and with partly clasping base, from dentate to laciniate-pinnatifid (an inch or two long); involucre 1 or 2 lines high, hirsute with rather soft spreading hairs, considerably shorter than the soft pappus: flowers whitish; the numerous female with an entire corolla-tube barely half the length of the style; hermaphrodite flowers only 5 to 7. — Proc. Am. Acad. vii. 355, & Bot. Calif. i. 332. *C. subdecurrens*, Gray, Pl. Fendl. 78, & Pl. Wright. i. 102, not of DC. *Erigeron discoideus*, Kellogg, Proc. Calif. Acad. v. 55. *E. subdecurrens*, Gray, Bot. Mex. Bound. 78. — River-bottoms, &c., W. Texas and Colorado to Arizona and California. Much resembling *C. subdecurrens*, DC., which, from the more developed corolla of the ray, is referred to *Erigeron*, but has also a different pubescence. (Adj. Mex.)

Var. tenuisécta. Greener, extremely leafy: leaves pinnately or even somewhat bipinnately parted into linear lobes: heads smaller and very numerous in an ample panicle. — S. Arizona, near Fort Huachuca, *Lemmon*. Apparently growing with the ordinary form.

51. BÁCCHARIS, L. (Named after *Bacchus*, unmeaningly.) — Shrubs, under-shrubs, or some perennial herbs; with alternate simple leaves, sometimes reduced to scales, and the branches commonly striate or sulcate-angled, bearing small heads of white or whitish or yellowish flowers. A huge American genus, chiefly tropical and S. American. — Benth. & Hook. Gen. ii. 286; Gray, Proc. Am. Acad. xvii. 212.

§ 1. Pappus of the fertile flowers very copious and pluriserial, elongated in fruiting, soft: akenes 5-10-costate: stems herbaceous from a lignescent or more woody base: leaves linear, 1-nerved: receptacle flat and broad, naked. Here also *B. juncea*, of S. Brazil (*Archenachne*, Cass., *Stephananthus*, Lehm.), and *B. Seemanni*, of Mexico. — Gray, Proc. Am. Acad. xvii. 211.

B. Wríghtii, GRAY. Very smooth and glabrous, a foot or two high, diffusely branching, sparsely leaved: slender branches terminated by solitary heads: leaves small; uppermost linear-subulate: involucre campanulate, 4 or 5 lines high; its bracts lanceolate, gradually acuminate, conspicuously scarious-margined, with a green back: pappus fulvous or sometimes purplish, four times the length of the scabrous-glandular 8–10-nerved akene. — Pl. Wright. i. 101, & ii. 83. — W. Texas to S. Colorado and Arizona. (Adj. Mex.)

B. Texána, GRAY. Glabrous, a foot or more high, with many nearly simple rigid stems from a woody base, leafy to the top, where it bears a few somewhat corymbosely disposed heads: leaves an inch or two long, rather rigid: involucre 3 lines long, of firmer and narrower merely acute bracts: akenes smoother. — Pl. Fendl. 75, & Pl. Wright. l. c. *Linosyris Texana,* Torr. & Gray, Fl. ii. 232, male plant. *Aplopappus linearifolius,* Buckley in Proc. Acad. Philad. 1861, 457. — Texas, forming large patches in dry prairies, *Berlandier, Drummond, Wright,* &c.

§ 2. Pappus of the fertile flowers more or less copious, but uniserial or nearly so, conspicuously elongating in fruiting, soft and fine, mostly flaccid and bright white: akenes 10-nerved: branching shrubs, glabrous or nearly so, usually viscous with a resinous exudation: leaves sometimes lobed or angulate-dentate: heads glomerate or paniculate: receptacle naked and flat.

* Eastern species, of the coast or along streams in subsaline soil: shrubs 3 to 12 feet high.

B. halimifólia, L. Cauline leaves from dilated-obovate to oblong with cuneate base, attenuate into a petiole, laciniately or angulately 3–9-toothed, those of the flowering branchlets becoming lanceolate and mostly entire: heads in pedunculate and paniculate glomerules (3 to 5 together): involucre of the male heads only 2 lines long, of oblong-ovate obtuse bracts; of the female rather longer and narrower, the inner bracts linear-lanceolate and acute. — Spec. ii. 860; Michx. Fl. ii. 125; Duham. Arb. i. t. 60. — Sea-coast, New England to Florida and Texas. (W. Ind.)

B. glomerulifióra, PERS. Brighter green: leaves mostly cuneate-obovate or the uppermost spatulate, less petioled or sessile, merely angulate-toothed: heads larger, sessile or in very short-peduncled glomerules in the axils of the upper leaves: involucre of both sexes campanulate, pluriserially imbricate, of obtuse bracts. — Syn. ii. 423; Pursh, Fl. ii. 523. *B. sessiliflora,* Michx. Fl. ii. 125; Ell. Sk. ii. 320, not Vahl. — Swamps near the coast, N. Carolina to Florida. (Bermuda.)

B. salícina, TORR. & GRAY. Leaves mostly subsessile, from oblong to linear-lanceolate, sparingly toothed, rarely entire: heads or glomerules pedunculate: involucre of both sexes campanulate (nearly 3 lines long), of mainly ovate and acutish bracts. — Fl. ii. 258. *B. salicifolia,* Nutt. Trans. Am. Phil. Soc. vii. 337. — Colorado (banks of the Arkansas, &c.) to W. Texas, on the Rio Grande, near El Paso.

B. angustifólia, MICHX. Rather strict: leaves narrowly-linear (larger 2 or 3 inches long, a line or two wide), entire or with few denticulations; and some lower ones broadly lanceolate and more serrate: heads or glomerules short-pedunculate, amply paniculate: involucre 2 lines long, of oblong-ovate or lanceolate bracts, the outer obtuse, innermost acute. — Fl. ii. 125; Ell. l. c.; Torr. & Gray, l. c. *B. salicina,* Gray, Pl. Wright. i. 101, not of ii., nor Nutt. — Brackish marshes, &c., S. Carolina to Florida, and to Texas on the Rio Grande; also S. Arizona, *Lemmon.* (Adj. Mex.)

* * Western species (Pacific coast to Arizona): branches smooth or nearly so, striate-angled.

B. piluláris, DC. Either depressed, spreading on the ground, or more erect and sometimes 4 feet high, leafy up to the glomerate sessile heads: leaves short (seldom over inch long), obovate and cuneate or roundish, very obtuse, sessile, coarsely few-toothed or some entire: involucre nearly hemispherical, 2 lines long; its bracts oval and oblong, all but the innermost very obtuse: flowers bright white. fertile pappus not over 4 lines long. — *B. pilularis & B. consanguinea,* DC. Prodr. v. 407, 408; Torr. & Gray, Fl. ii 259; Benth. Bot. Sulph. 25. *B. glomeruliflora,* Less. in Linn. vi. 506; Hook. & Arn. Bot. Beech. 147. — Near the coast, Monterey, California, to Oregon.

B. Emóryi, GRAY. Erect, with slender branches, 2 to 15 feet high: cauline leaves mostly oblong or the lower broader, with attenuate or cuneate base and the larger somewhat petioled, more or less triplinerved, often with 2 to 4 short lobes or teeth; those of the branches from oblanceolate to linear, mostly entire, 1-nerved: heads somewhat nakedly paniculate on the branchlets, short-pedunculate or the glomerules more or less pedunculate: involucre campanulate or oblong, 3 or sometimes 4 lines long, mostly of firm coriaceous and obtuse bracts; the outermost oval, inner oblong, the innermost thin, linear and acutish: pappus of male flowers bearded towards the tip; of the female in fruit half-inch long. — Bot. Mex. Bound. 83, & Bot. Calif. i. 333, described from mere branches. *B. pilularis*, Nutt. Trans. Am. Phil. Soc. l. c., partly, not DC. *B. salicina*, Rothr. in Wheeler Rep. vi. 156, & Bot. Calif. ii. 456, partly. — Along watercourses, from Los Angeles southward, through Arizona and in S. Nevada and Utah.

B. sarothroides, Gray. Erect, fastigiately much branched, 10 to 15 feet high: leaves all nearly linear, entire, 1-nerved, rigid, small; the larger (less than inch long and 2 lines wide) narrowed at base; those of the slender and strongly striate-angled branchlets commonly sparse and minute: heads loosely paniculate, terminating ultimate naked branchlets, small: involucre of the male campanulate, hardly 2 lines long; of the female rather oblong, only about 10-flowered; short outer bracts ovate or oval, very obtuse, innermost thin and broadly linear: clavellate tips of male pappus naked; female pappus in fruit 3 lines long. — Proc. Am. Acad. xvii. 212. — S. California, from San Diego to the Mexican line, *Sutton Hayes, Palmer*. Has been confounded with *B. Emoryi* and *B. sergiloides*. (Adj. Mex.)

* * * Species of Mexican border, with branchlets terete, less striate, pruinose-scabridous.

B. pteronioides, DC. Diffusely branched: leaves small (rarely half-inch long), crowded and fascicled on the branchlets, from lanceolate-spatulate to linear, thickish, nearly veinless, the larger 2–6-dentate: heads singly terminating very short densely leafy branchlets, which are crowded in a virgate or racemose way along the branches: involucre 3 lines long, campanulate: the outer bracts ovate or oblong: pappus of the male flowers not at all clavellate; of the female in fruit 4 lines long, not much surpassing the corolla. — Prodr. v. 410. *B. ramulosa*, Gray, Pl. Thurb. 301, & Bot. Mex. Bound. 84. *Aplopappus ramulosus*, DC. l. c. 350. *Linosyris (Aplodiscus) ramulosa*, Gray, Pl. Wright. i. 97, & ii. 80. — New Mexico and Arizona. (Mex.)

§ 3. Pappus rather rigid and scanty, short, not elongated with age, of the fertile flowers even in fruit not surpassing the style: akenes 10-nerved (the 5 primary nerves sometimes the more prominent): fertile corollas regularly cleft at apex into 5 subulate lobes: some chaff among the flowers on the sometimes elevated receptacle similar to the innermost involucral bracts: branches broom-like.

B. sergiloides, Gray. Suffruticose, glabrous, 3 to 5 feet high, very much branched: the slender and partly herbaceous branches and branchlets strongly striate-angled and naked, bearing a few small leaves and paniculate mostly short-pedunculate heads: larger leaves spatulate, entire, rarely 2–4-toothed (the larger seldom over half-inch long): heads 2 or 3 lines long: bracts of the involucre small, oblong or lanceolate, rather obtuse, of firm texture: fertile pappus barely twice the length of the mature akenes. — Bot. Mex. Bound. 83, Pacif. R. Rep. iv. 101, & Bot. Calif. i. 333, partly, and not well characterized. — Arid districts of S. E. California and adjacent Nevada to S. W. Utah, *Bigelow, Wheeler, Palmer, Parish,* &c. Varies in the amount of imbrication of the involucre, and the number of chaffy scales; when these are numerous the receptacle becomes conical and the disk very convex.

§ 4. Pappus of the fertile flowers not flaccid, little if at all elongated in fruit, not very copious: akenes only 5-nerved, sometimes 4-nerved. Southwestern, chiefly Pacific species.

* Scabro-puberulent or pubescent throughout, not glutinous: fruiting pappus manifestly surpassing the style: heads loosely paniculate: bracts of the involucre scarious-margined from a green or greenish back or centre, acute or acuminate: stems herbaceous from a woody or merely lignescent base, 2 or 3 feet high: leaves not rigid.

B. brachyphýlla, Gray. Minutely scabro-puberulent, diffusely much branched, slender: leaves small, entire, mostly linear, 1-nerved, the larger cauline seldom over half-inch long, on the branchlets mostly becoming minute and scale-like: heads 3 lines long, 12–15-flowered:

involucre of oblong-lanceolate or broader bracts: pappus in fruit 3 lines long. — Pl. Wright. ii. 83. — Rocky ground, S. Arizona to San Bernardino Co., California, *Wright, Palmer, Parry.*

B. Plúmmeræ, GRAY. Loosely pubescent, moderately branched: leaves linear-oblong, obtuse, irregularly and acutely serrate, the larger an inch or two long, obscurely 3-nerved: heads 4 lines long: involucre of linear bracts: akenes somewhat compressed and puberulent, obscurely 5-nerved: pappus in fruit 4 lines long. — Proc. Am. Acad. xv. 48, & Bot. Calif. ii. 456. — S. W. California, in mountain ravines behind Santa Barbara and Santa Monica, *Miss Plummer, Parish,* &c.

* * Glabrous or nearly so and smooth, sometimes glutinous: pappus in fruit slightly if at all surpassing the style.

+— Bracts of the 15–30-flowered involucre from oblong to linear, rather firm and with green centre or costa: receptacle flat: leaves comparatively small and rather rigid, serrate with rigid or spinulose teeth.

B. thesioides, HBK. A foot or two high from a woody base: branches rigid and slender: leaves linear-lanceolate or sometimes broader and narrowed to base, nearly or quite sessile, rather closely and evenly ciliately spinulose-serrate (the larger an inch or rarely 2 inches long), prominently 1-nerved, sometimes with obscure lateral nerves: heads 2 lines long, numerous in a corymbiform or an oblong naked panicle: pappus of the male flowers obscurely if at all thickened upward. — Nov. Gen. & Spec. iv. 61; DC. Prodr. v. 419. *B. ptarmicæfolia,* DC. l. c. *B. ptarmicæfolia?* or *thesioides,* Gray, Pl. Wright. ii. 83. — S. Arizona, *Wright, Lemmon.* (Mex.)

B. Bigelóvii, GRAY. Stems more copiously and loosely branched: leaves less rigid, from linear to oblong and the broader ones sometimes petioled, irregularly serrate, commonly obtuse: heads larger, more cymose: bristles of the male pappus thickened and barbellate at the tip. — Bot. Mex. Bound. 84. *B. ptarmicæfolia?* Gray, Pl. Wright. ii. 83. — Woods and shaded hillsides, in Arizona and New Mexico, *Bigelow, Wright, Thurber, Lemmon, Rusby.* (Adj. Mex.)

B. Havárdi. Stems copiously branched, slender: leaves hardly at all rigid; lower linear-oblanceolate and tapering into a slender petiole, laciniate-pinnatifid into several irregular slender-subulate lobes; those of the branchlets narrowly linear, 2–3-toothed or entire: heads loosely paniculate, only the male known, these barely 2 lines high, about 15-flowered: involucral bracts oblong: bristles of the pappus rigid, clavellate. — Guadelupe Mountains, western borders of Texas, *Havard.*

+— +— Bracts of the involucre narrowly oblong or linear-lanceolate, thin and pale with greenish centre: heads short and broad, many-flowered: receptacle hemispherical or broadly conical!

B. Douglásii, DC. Herbaceous nearly or quite to the ground, 3 to 5 feet high, loosely branched: leaves glutinous, not rigid, either entire or serrulate with minute and very acute denticulations, triplinerved from near the base, ovate-lanceolate (the larger 4 to 6 inches long) or the upper lanceolate, with attenuate-acute apex, the base contracted into a short margined petiole: heads numerous and densely cymose at the summit of naked branchlets, 3 lines long: involucral bracts erose-ciliate: female pappus barely 2 lines long, soft; male somewhat clavellate and barbellate above. — Torr. & Gray, Fl. ii. 259, excl. syn. Nutt., &c.; Gray, Bot. Calif. i. 333. *B. Douglasii & B. Hænkei,* DC. l. c. 400 & 401, the latter from Monterey in California. *B. glutinosa,* Hook. & Arn. Bot. Beech. 147? — Moist or wet ground, California, from San Francisco southward, and southeastward to San Bernardino.

+— +— +— Bracts of the involucre broader, thin-chartaceous, rather dry, with narrow scarious margins, at least the inner ones yellowish or tawny, destitute of green centre or distinct costa: the outer bracts ovate, inner oblong: heads many-flowered: receptacle flat: stems very leafy up to the corymbosely paniculate or cymose inflorescence, more terete than in the preceding species: leaves lanceolate, willow-like, acute at both ends, either denticulate-serrate or entire, subsessile.

B. glutinósa, PERS. Stems herbaceous above but woody toward the base, 3 to 10 feet high: branches somewhat striate-angled: leaves elongated-lanceolate, serrate with few or several scattered teeth on each side, more or less distinctly 3-nerved from near the base (3 or 4 and the larger 5 or 6 inches long): heads mostly 3 lines long or the male smaller, numerous and corymbosely cymose at the summit of comparatively simple stems or branches: involucre stramineous. — Syn. ii. 425; Hook. & Arn. Bot. Beech. i. 31 (*Molina viscosa*, Ruiz & Pav. Syst. 207). *B. glutinosa*, *B. cærulescens*, & *B. Alamani*, DC. Prodr. l. c. 402. 403. *B. Pingræa*, Nutt. Trans. Am. Phil. Soc. l. c. 337, excl. syn. *B glutinosa* & *B. cærulescens*, Gray, Bot. Calif. i. 333. — Along streams and in moist ground, S. California, from Los Angeles southward, and through Arizona to S. Colorado and the borders of Texas: fl. late in autumn. (Mex. to Chili.)

B. vimínea, DC. Stems truly shrubby, 6 to 12 feet high, producing short lateral flowering branches, these terete and minutely striate: leaves lanceolate, entire or some sparingly denticulate, obscurely 3-nerved, 2 or 3 inches long, or much smaller on the flowering shoots: heads usually 4 lines long, hemispherical, in small cymose clusters terminating numerous lateral branchlets: involucre tawny. — Prodr. v. 400; Gray, Bot. Calif. i. 333. — Along streams, California, from Monterey southward and to San Bernardino Co.: flowering in early spring; the foliage persisting all winter.

Tribe IV. INULOIDEÆ, p. 57.

52. PLÚCHEA, Cass. (For the Abbé N. A. *Pluche*, an amateur naturalist of the latter part of the eighteenth century.) — Warm-temperate or tropical plants; with alternate pinnately veined leaves, and heads of flesh-colored or dull purple flowers, cymosely or paniculately disposed or rarely solitary at the summit of the stem or branches. — Cass. Bull. Philom. 1817, & Dict. Sci. Nat. xlii.; Benth. & Hook. Gen. ii. 290.

§ 1. **Berthelótia**. Pappus of the hermaphrodite-sterile (or rarely fertile) flowers of more rigid bristles with clavellate-dilated tips: involucre chartaceo-coriaceous; the innermost narrowly linear and deciduous with the flowers. Very leafy sericeous-canescent shrubs. — Gray, Proc. Am. Acad. xvii. 212. *Berthelotia*, DC. Prodr. v. 375.

P. boreális, Gray, l.c. (Cachimilla, Arrow-wood.) Shrub several feet high, much branched, willow-like, very leafy up to the cymulose-glomerate heads, silvery with the very close and fine appressed pubescence: leaves entire, linear-lanceolate, sessile, acute at both ends: involucre campanulate; its outer bracts ovate, obtuse, tomentose: bristles of the pappus of the central flowers little stouter than of the others, but with abruptly enlarged tips, not united at base: style of the same entire. — *Tessaria borealis*, Torr. & Gray, in Emory Rep. (Notes of Reconnoissance, 1848) 143; & Sitgreaves Rep. 162, t. 5; Gray, Pl. Fendl. 75, Pl. Wright. i. 102 (§ *Phalacrocline*), & Bot. Calif. i. 334. *Polypappus sericeus*, Nutt. Pl. Gamb. 178. — Sandy banks of streams, from the Rio Grande on the western borders of Texas to S. California: fl. summer.

§ 2. **Stylimnus**. Pappus of both kinds of flowers fine and similar, more or less soft, none of the bristles at all thickened at tip: bracts of the involucre thin or thinnish: corolla of the hermaphrodite flowers somewhat enlarged upward: heavy-scented herbs, or in the tropics shrubby, somewhat pubescent and glandular, with membranaceous or slightly succulent pinnately-veiny leaves, commonly with some callous-mucronate teeth: heads cymose-clustered: flowers dull purple, in late summer or autumn. — *Stylimnus* & *Gynema*, Raf. in Jour. Phys. 1819, & Ann. Nat. 1820, 15. *Leptogyne*, Ell. Sk. ii. 322, as subgenus. *Pluchea* (*Stylimnus*) § 3, DC. Prodr. v. 451.

The first of the following species may fairly retain the now established name, rather than have a new one made; but *Conyza bifrons* was founded by Linnæus on European *Inulæ*, viz. on Hermann's figure, which in ed. 2 he refers to that genus, and on one of Plukenet's (mistaken for Canadian), which is certainly *I. bifrons*, as his herbarium shows. Of the many names for our second species, one of De Candolle's, which continues the principal Linnæan specific name, is to be preferred. *Conyza Carolinensis*, Jacq., is *Pluchea odorata*, wrongly attributed to Carolina.

P. bífrons, DC. Stems nearly simple, 2 or 3 feet high from a perennial root: leaves veiny, acutely denticulate, from oblong to lanceolate, commonly obtuse at both ends (2 to 4 inches long), partly clasping or closely aduate-sessile: heads glomerate in leafy-bracted sessile clusters: involucral bracts lanceolate. — Prodr. l. c.; Torr. & Gray, Fl. ii. 260, excl. syn. L.! *Baccharis fœtida*, L. Spec. 861, as to pl. Gronov. *B. viscosa*, Walt. Car. 202. *Conyza bifrons*, Pursh, Ell. l. c., &c., not L. *C. amplexicaulis*, Michx. Fl. ii. 126. *C. uliginosa*, Pers. Syn. ii. 427. — Wet soil, Cape May, New Jersey, and through the low country to Keys of Florida (where is a very narrow-leaved variety, *Conyza angustifolia*, Nutt. Jour. Acad. Philad. vii. 109 ?), and Texas. (W. Ind.)

P. camphoráta, DC. Stems 2 to 5 feet high from an annual (not perennial) root: leaves from oblong-ovate to oblong-lanceolate, acute or acuminate at both ends, denticulate or dentate (3 to 8 inches long), the larger distinctly or indistinctly petioled; primary veins often evident, but veinlets obscure: heads numerous and crowded in naked convex or corymbiform cymes, commonly short-pedicelled: involucral bracts from ovate to lanceolate, often tinged with purple. — *Erigeron camphoratum*, L. Spec. ed. 2, 1212 (Gronov. Virg. ed. 1, 96, Clayt. no. 165). *Baccharis fœtida*, L. Spec. ed. 1, 861, as to syn. Dill., not as to Gronov. *Conyza Marilandica*, etc., Dill. Elth. t. 88, fig. 104, & *C. Americana frutescens*, etc., Dill. l. c. t. 89. *C. Marylandica*, Michx. l. c. *C. Marilandica* & *C. camphorata*, Pursh, Fl. ii. 523; Ell. l. c. *Gynema dentata, viscida*, &c., Raf. Aun. Nat. 159. *Pluchea Marilandica* & *P. petiolata*, Cass. Dict. l. c. *P. Marilandica, fœtida, camphorata*, also (W. Ind.) *P. purpurascens* & *P. glabrata*? DC. l. c. *P. fœtida, camphorata*, & *purpurascens*, Torr. & Gray, Fl. ii. 261. — Salt marshes and moist saline soil, Mass. to Florida, Texas, Arizona, and coast of California: in shady places or less saline soil, with leaves thinner and more petioled, and involucre almost glabrous, when it is *P. petiolata*, Cass. (Adj. Mex., W. Ind.)

53. PTEROCAÚLON, Ell. BLACK-ROOT. (Πτερόν, wing, and καυλός, stem.)

— Mostly perennial herbs, the typical species American; with one exception all tomentose-canescent except the upper face of the sessile pinnately veined leaves, these decurrent on the whole stem, forming wings; small sessile heads spicate at the summit of the stem and virgate branches; the flowers usually white or whitish, in summer. — Ell. Sk. ii. 323; DC. Prodr. v. 453. *Chænolobus*, Cass. Dict. Sci. Nat. xlix. 348.

P. pycnostáchyum, ELL. l. c. Roots fasciculate and tuber-like or fusiform, black: stem 2 feet high, mostly simple: leaves from oblong to lanceolate, minutely denticulate: heads crowded in a dense and continuous spiciform naked thyrsus (of 3 to 8 inches in length): involucre lanato-tomentose. — Torr. & Gray, Fl. ii. 262. *Conyza pycnostachya*, Michx. Fl. ii. 126. *Chænolobus pycnostachyus*, Cass. l. c. *Gnaphalium undulatum*, Walt. Car. 203. — Dry pine barrens, near the coast, N. Carolina to Florida.

P. virgátum, DC. Root fusiform and fibrose (perhaps biennial): stem slender, simple or with virgate branches: leaves linear and very acute, entire, or the lower cauline lanceolate and obscurely serrulate, the venation hardly apparent: heads narrow, in separated glomerules; these forming a virgate and elongated interrupted spike-like inflorescence: involucre appressed-tomentose, or the subulate inner bracts glabrate. — Prodr. v. 454. *Gnaphalium virgatum*, L. Amœn. Acad. v. 405. *Conyza virgata*, L. Spec. ed. 2, 1206, with syn. *Chænolobus virgatus*, Cass. l. c. — Open pine woods near Houston, Texas, *Lindheimer*. (W. Ind., Mex.)

54. MÍCROPUS, L. (Μικρός, small, πούς, foot, the soft-woolly small heads or clusters like Leontopodium, or Lion's-foot, on a small scale.)

— Low floccose-woolly annuals, with alternate entire leaves, belonging to the Old World, except our Pacific coast species. — Gærtn. Fr. t. 164; Schkuhr, Handb. t. 267;

Benth. & Hook. Gen. ii. 297 (excl. § 3, 4); Gray, Proc. Am. Acad. viii. 651, & Bot. Calif. i. 335. — Our species, of the section BOMBYCILÆNA (with woolly fructiferous bracts smooth and crestless), approach *Stylocline* and *Filago* in the points which distinguish them from the European species.

M. califórnicus, FISCH. & MEYER. Slender, erect, 6 to 12 inches high: leaves mostly linear: fructiferous bracts 3 or 6, at length firm-coriaceous, somewhat half-obcordate or half-obovate in outline, straight anteriorly, and with the soon erect beak-like tip largely scarious. — Ind. Sem. Petrop. 1835, 42; Torr. & Gray, Fl. ii. 264; Gray, Bot. Calif. l. c. *M. (Rhyncholepis) angustifolius,* Nutt. Trans. Am. Phil. Soc. n. ser. vii. 339. — Plains and open ground from S. California northward, toward the coast, to Oregon. Heads vary in the wool, from long and copious, as in *M. bombycinus,* to short, as in the subjoined

Var. subvestítus, GRAY, Bot. Calif. l. c. Small; the wool of the bracts all short and wholly appressed. — Arroyo Grande near Monte Diablo, California, *Brewer.*

M. amphíbolus, Gray. Resembles the more loose-woolly forms of the preceding: female flowers about 10, somewhat imbricated at base on an oblong receptacle; their fructiferous bracts membranaceous or merely chartaceous at maturity, the beak an ovate almost wholly hyaline appendage which in flower is almost as long as the body and inflexed, at maturity porrect: sterile flowers subtended by some linear thin chaff, and with a pappus of a few bristles. — Proc. Am. Acad. xvii. 214. — California, Walnut Creek near Martinez, *Brewer,* and coll. (probably in same district) *Kellogg & Harford,* distrib. 416.

55. STYLÓCLINE, Nutt.
(Στῦλος, a column, and κλίνη, a bed, or receptacle, from the form of this.) — Floccose-woolly annuals, a span or less in height, branched from the base, erect or spreading, with entire alternate leaves and more or less glomerate heads. — Nutt. Trans. Am. Phil. Soc. n. ser. vii. 338; Gray, Proc. Am. Acad. viii. 652, & Bot. Calif. i. 336. *Micropus* § 3 & 4, Benth. & Hook. Gen. ii. 296. — Includes, besides the following species, *S. (Diplocymbium) Griffithii* of Afghanistan, most related to our second and very distinct section.

§ 1. EUSTYLÓCLINE, Gray, l. c. Fertile flowers numerous; their chaffy bracts pluriserially and closely imbricated in an ovoid head, thin, with at least the broad tips hyaline (barely a green midrib or centre), ovate in outline, promptly falling from the receptacle after maturity along with the loosely enclosed akene; those subtending the sterile flowers all scarious-hyaline and deciduous. Pappus of a very few capillary bristles generally present with the sterile flowers. — *Stylocline,* Nutt. l. c.

S. gnaphalioídes, NUTT. l. c. Leaves broadly linear or the upper oblong, obtuse: fructiferous bracts broadly ovate, moderately woolly on the back, almost wholly hyaline-scarious, a firmer central portion at base saccate-conduplicate and enclosing the narrowly obovate oblique laterally compressed akene. — Gray in Pacif. R. Rep. iv. 101, t. 13, &c. — Open grounds, California from the Stanislaus southward; first coll. by *Nuttall.*

S. micropoídes, GRAY. Leaves somewhat narrower and rather acute: heads more woolly, appearing less scarious and imbricated; fructiferous bracts having a narrower oblong-ovate hyaline tip, the oblong body densely long-woolly, without hyaline expanded margins, but wholly enwrapping the nearly straight and slightly compressed akene. — Pl. Wright. ii. 84, & Bot. Calif. l. c. *Micropus Grayana,* Hemsl. Bot. Centr.-Amer., name only. — Arid plains, S. California through S. Nevada and Arizona to New Mexico; first coll. by *Wright.*

§ 2. ANCISTROCÁRPHUS, Gray. Fertile flowers 5 to 9, loosely disposed on the slender receptacle; their enclosing bracts cymbiform, of firm texture except the narrow hyaline tip, tardily if at all deciduous at maturity; the few sterile flowers involucrately subtended by about 5 larger open bracts; these herbaceo-coriaceous, ovate-lanceolate, tapering into a rigid and incurved-uncinate cusp, persistent and at length stellately spreading: akene obovate-fusiform and obscurely obcompressed (the pericarp distinct from the seed and obsoletely few-nerved!), loosely enclosed in the involutely closed bracts: no pappus to sterile flowers: no involucre outside the fructiferous bracts. — Proc. Am. Acad. viii. 652.

—— **S. filagínea**, Gray, l. c. Erect or diffuse, appressed-lanate: leaves from linear to spatulate: heads capitate-glomerate, the hooked empty bracts at maturity 2 lines long. — *Ancistrocarphus filagineus*, Gray, Proc. Am. Acad. vii. 356. — Open ground, California from Mendocino Co. (*Bolander*) to the Mohave Desert (*Parry, Lemmon*), and northward to Union Co., Oregon, *Cusick*. Between *Stylocline* and *Evax*.

56. PSILOCÁRPHUS, Nutt. (Ψιλός, bare, κάρφος, chaff, not an appropriate name.) — Small and diffuse or depressed and much branched annuals (Pacific American), floccose; with most of the leaves opposite, and globose heads comparatively large and apt to be solitary at the forks and ends of the branches. Fructiferous chaff at length deciduous with the enclosed akene, or opening ventrally so that this is shed. Uppermost leaves involucrate around the sessile head. — Nutt. Trans. Am. Phil. Soc. l. c. 340; Torr. & Gray, Fl. ii. 265; Gray, l. c. *Bezanilla*. C. Gay, Fl. Chil. iv. 109, t. 46.

—— **P. Oregánus**, Nutt. Loosely lanate, erect or spreading, becoming a span high, but beginning to flower close to the ground: leaves mostly linear: heads when well formed 3 lines in diameter, and fructiferous bracts a line and a half long. — Gray, Bot. Calif. ii. 336. *P. Oreganus, globiferus* (excl. syn.), & *brevissimus* (excl. syn.), Nutt. l. c., the last two depauperate early-flowered states. — W. California, from Los Angeles to Oregon, and even to Boise City, Idaho, *Wilcox*.

—— Var. **elátior**, Gray, l. c. A robust well-developed form, 5 or 6 inches high, with larger leaves three-fourths inch long, and heads 4 lines broad. — Near Portland, Oregon, *Hall, Kellogg*.

—— **P. tenéllus**, Nutt, l. c. Canescent with a finer and closely appressed wool, slender, diffusely much branched, usually depressed and matted: leaves commonly spatulate, sometimes all linear, 3 or 4 lines long: heads 2 lines in diameter, the more vesicular fructiferous bracts a line long. — Gray, Bot. Calif. ii. 336. — Common through W. California to Washington Territory.

57. ÉVAX, Gærtn. (Name, unexplained by Gærtner, used as a joyous exclamation in Plautus, said by Wittstein, on the authority of old editions of Pliny, to be the name of an Arabian chief who wrote to the Emperor Nero about simples.) — Mostly dwarf and depressed annuals, or some typical species of the Old World perennials, floccose-woolly, represented in N. America by the following aberrant groups.

§ 1. HESPERÉVAX, Gray. Bracts of the oblong involucre and those of the receptacle subtending the female flowers from oblong to obovate, chartaceous, becoming coriaceous, persistent, barely concave: receptacle at length slender-columnar from a broader base, sparsely villous: the female flowers and bracts crowded at its base; the summit bearing a circle of 3 to 5 or 7 more herbaceous or coriaceous obovate or rotund tomentulose open bracts, subtending a few male flowers; these with a 2-cleft style but no ovary: akenes pyriform-obovate, somewhat obcompressed, very smooth. — Pacif. R. Rep. iv. 101, t. 11, Proc. Am. Acad. viii. 651, & Bot. Calif. i. 337.

—— **E. cauléscens**, Gray, l. c. Either annual or biennial, canescent with appressed or somewhat flocculent wool: leaves spatulate. Occurs under various forms, of which the typical, var. PETIOLATA, is comparatively stout and large; the blade of the leaves 6 to 15 lines long, tapering into petioles of an inch or two in length: the heads in their axils glomerate at the root, therefore stemless (*E. acaulis*, Greene, in Bot. Gazette, vii. 256), or on the summit of a simple stem or simple branches from the base, an inch or two high. — *Psilocarphus caulescens*, Benth. Pl. Hartw. 319. — Gravelly or alluvial soil, California, on the Sacramento, &c., *Hartweg, Bigelow, Kellogg, Parry,* and others.

Var. **sparsiflóra.** More caulescent and branching; leaves similar but smaller, rarely inch long (including the slender petiole), scattered on branching stems of at length 2 inches high, none rosulate at the base: heads in their axils accordingly scattered, narrowly oblong. — Southeastern part of California, San Luis Obispo and San Diego, *Brewer, Parry, Cleveland,* &c.

Var. **brevifólia.** Either depressed and rosulate, or with stems an inch or two high: leaves small and short-petioled, seldom over a quarter to half inch long. — Northern part of the State, Humboldt and Mendocino Co., *Bolander, Kellogg,* &c.

Var. **mínima.** A very exiguous form of the preceding variety, in the early and depressed state, but tending to be subcaulescent; the largest leaves barely half-inch long and hardly a line wide. — *Stylocline acaule*, Kellogg in Proc. Calif. Acad. vii. 112, exceedingly starved specimens, just coming into flower, coll. *Dr. Eisen*, at Fresno. The whole structure exactly of *E. caulescens*, and sterile flowers not "single," but 6 or 7, surrounded by 5 to 7 firm but not yet enlarging bracts.

§ 2. DIAPÉRIA. Bracts of the involucre thin; of the female flowers scarious, from oval to oblong-linear, barely concave, at maturity deciduous from the merely convex receptacle; those of the 2 to 5 staminate flowers (which have an undivided style and no ovary) similar or with woolly tips, or partly herbaceous, and somewhat embracing the flowers; no central prolongation to the receptacle: akenes obcompressed, smooth or very minutely papillose: heads small, aggregated in terminal foliose-involucrate glomerules. — *Diaperia*, Nutt. Trans. Am. Phil. Soc. vii. 337; Benth. & Hook. l. c. 298, extended. *Diaperia & Filaginopsis*, Torr. & Gray, Fl. ii. 263, 264.

E. prolífera, NUTT. Rather stout: stem often a span high, simple and erect, or with ascending branches from the base, bearing numerous small spatulate leaves and a capituliform glomerule (commonly half-inch in diameter), whence proceed 1 to 3 nearly leafless branches similarly terminated, sometimes again proliferous: heads cylindraceous or oblong-fusiform: fructiferous bracts chartaceo-scarious, oval or oblong, mainly naked; those embracing staminate flowers more herbaceous and woolly-tipped, of firmer or more herbaceous texture: staminate flowers each on a filiform stipe representing an abortive ovary: habit of *Filago Germanica*. — DC. Prodr. v. 459. *Diaperia prolifera*, Nutt. Trans. Am. Phil. Soc. l. c. 337; Torr. & Gray, Fl. ii. 264. — Dry or exsiccated ground, Arkansas to Texas, Colorado, and north to Dakota; first coll. by *Nuttall*.

E. multicaúlis, DC. Diffusely branched from the base, rather slender: capituliform glomerules much smaller and less foliose-involucrate: leaves oblanceolate or spatulate (3 or 4 lines long): heads globular or ovoid (only a line or two in diameter): involucre and apex of the receptacular bracts densely implexed-lanate; those of the female flowers narrowly oblong, of the male spatulate; these sessile without vestige of ovary. — Prodr. v. 459. *Filaginopsis multicaulis*, Torr. & Gray, l. c., & Pacif. R. Rep. ii. t. 3. *Micropus minimus*, DC. l. c. 461, a depauperate form. *Diaperia multicaulis*, Benth. & Hook. Gen. ii. 298. — Low or exsiccated alluvial ground, common from Texas and the borders of New Mexico even (coll. *Lemmon*) to the Mohave Desert in S. E. California. (Adj. Mex., *Berlandier, Gregg*.)

Var. **Drummóndii.** A slender form, commonly with some long woolly hairs on the limb or on the tube of the staminate corollas. — *Filaginopsis Drummondii*, Torr. & Gray, l. c. *Diaperia Drummondii*, Benth. & Hook. l. c. — E. Texas and Louisiana, in moist ground, *Drummond, Hale,* &c.

§ 3. CALYMMÁNDRA. Bracts of the simple involucre and of the female flowers mostly scarious, narrowly spatulate-oblong, plane, externally villous-lanate; of the five central hermaphrodite flowers shorter and broad, very woolly, involute around the lower half of the flower; all at length deciduous: hermaphrodite flowers also fertile, with funnelform 4-toothed corolla and linear-oblong obtuse style-branches: receptacle hemispherical: akenes of both kinds of flowers very smooth, obovate-oblong, obscurely obcompressed, the terminal areola larger and more evident, at least in the perfect flowers: heads small, axillary. — *Calymmandra*, Torr. & Gray, Fl. ii. 262.

E. cándida. A span or two high, slender, and with commonly simple branches, silvery white throughout with appressed wool: heads usually few in a foliose-involucrate cluster and sessile or nearly so in the axils of the spatulate or lanceolate leaves. — *Calymmandra candida*, Torr. & Gray, l. c., & Pacif. R. Rep. ii. t. 2. *Diaperia candida*, Benth. & Hook. Gen. ii. 298. — Alluvial or sandy ground, E. Texas, *Drummond, Berlandier, Wright*, &c., and in the northwestern part of the State, *Pope*.

58. FILÁGO, Tourn.

(*Filum*, a thread, in allusion to the cottony wool.) — Low annuals, mainly of the Old World, mostly erect and with the habit of the preceding. Ours have no pappus to the outer flowers. — DC. Prodr. vi. 247; Benth. & Hook. Gen. 299.

§ 1. Receptacle subulate; its pluriserial and well-imbricated bracts merely concave, subtending the loose akenes. — *Gifola*, Cass.

— **F. Germánica, L.** (COTTON-ROSE, HERBA IMPIA of the old herbalists.) A span to a foot high, erect, thickly beset with lanceolate commonly erect leaves, terminated by a capituliform globose glomerule of many heads, whence proceed a similar branch or branches, as if proliferous: heads ovate-oblong; its bracts ovate-lanceolate, acuminate. — Fl. Dan. t. 2787. — Dry fields, New York to Virginia. (Nat. from Eu.)

§ 2. Receptacle somewhat obconical or convex; its deeply concave or boat-shaped fructiferous bracts rather few, more or less enclosing the somewhat oblique akenes, loose or stellately spreading at maturity: glomerules smaller and looser. — *Oglifa*, Cass. (The indigenous Pacific American species are peculiar, as connecting with *Stylocline* and *Micropus*.)

— **F. Califórnica**, NUTT. Erect, leafy throughout, a span or two high, with the habit of *F. arvensis*: heads ovate, somewhat angular: bracteate female flowers 8 to 10; their bracts (or all but the innermost) broadly ovate and deeply boat-shaped, somewhat arcuate-incurved, very woolly, not herbaceous on the back, with a broadish and obtuse hyaline tip; inner bracts oblong, merely concave, nearly glabrous: akenes narrowly oblong, almost terete, minutely and obscurely papillose-granular: pappus of the upper female and hermaphrodite flowers copious, of the embraced akenes none. — Nutt. Trans. Am. Phil. Soc. l. c. 405, with var. *tomentosa*. *F. Califórnica* & *F. parvula*, Torr. & Gray, Fl. ii. 432. *Gnaphalium filaginoides*, Hook. & Arn. Bot. Beech. 359. — Open ground throughout California and to S. Utah.

F. depréssa, GRAY. Diffusely branched from the root, depressed-spreading: internodes all short, even the lower little longer than the glomerules of the oblong-ovate heads: bracteate female flowers 5 or 6; their bracts narrower and straighter, somewhat herbaceous on the back: akenes obovate, smooth, sometimes the uppermost bracteate ones also with pappus. — Proc. Am. Acad. xix. 3. — San Bernardino Co., S. California, in the desert at Hot Springs, *Parry, Parish*.

F. Arizónica, GRAY. Diffuse or at first erect, with widely spreading branches; the proliferous ones of elongated filiform internodes, widely separating the glomerules: bracteate female flowers 10 to 15; their bracts of firmer texture, ovate, open on the face: akenes clavate-oblong and arcuate, very smooth. — Proc. Am. Acad. viii. 652. — Arizona to San Diego Co., California, and Lower California, *Smart, Parry*, &c.

— **F. Gállica, L.**, forming the section LOEFIA (*Logfia subulata*, Cass.; characterized by the low and nearly plane receptacle, pentagonal-conical heads, about two-ranked female flowers, the akenes of the outer ones completely and firmly enclosed in the at length indurated base of the subtending bract), has been found at Newcastle, California, by *Mrs. Curran*, probably a weed of grain-fields or a waif.

F. RÉPENS and F. TEXÀNA. Scheele in Linn. xxii. 164, of Rœmer's collection in Texas, are not identified; probably are not of this genus.

59. ANTENNÁRIA, Gærtn., R. Br. (Bristles of male pappus likened to the *antennæ* of certain insects.) — Perennial, mostly low, canescently and often floccosely woolly herbs (occasionally glabrate), of north temperate and arctic zones, with whitish or purplish flowers; the bracts of the involucre pearly white or rose-color, or brownish, never yellow. — Gærtn. Fruct. ii. 410, t. 167 (excl. spec.); R. Br. in Linn. Trans. xii. 122; Benth. & Hook. Gen. ii. 301.

§ 1. Bristles of the male pappus hardly at all thickened but minutely barbellulate near the apex: akenes (in the first species) oblong-linear, obscurely 2–3-nerved, puberulent; the short hairs with 2-lobed and at length biuncinate tip, after the manner of *Townsendia:* bracts of the campanulate or somewhat turbinate involucre brownish, not radiant.

A. dimórpha, TORR. & GRAY. Depressed, cespitose from a stout multicipital caudex, bearing rosulate clusters of spatulate leaves; heads solitary and subsessile at the crown, or raised on a sparsely-leaved stem of an inch or less in height; male head 4 lines high, with broad and obtuse involucral bracts; female becoming half to three-fourths inch long, the inner bracts narrow and long-attenuate into a hyaline acuminate tip: pappus of the fertile flowers of long and flue smooth (not denticulate) bristles. — Fl. ii. 431; Eaton, Bot. King Exp. 186 (var. *Nuttallii* & var. *macrocephala*); Gray, Bot. Calif. i. 339. *Gnaphalium* (*Omalotheca, Heterophania*) *dimorphum*, Nutt. Trans. Am. Phil. Soc. vii. 405. — Dry hills, from eastern base of the Rocky Mountains, in Wyoming, &c., to the Sierra Nevada in California, and north to Brit. Columbia.

A. flagelláris, GRAY. Simpler, from a small caudex or biennial root, bearing smaller and fewer-flowered heads than the preceding and in the same manner, also copious naked and filiform stolons of a span or less in length, either as declined scapes bearing at their apex a head rosulately involucrate by small leaves, or rooting and forming a rosulate offset: leaves small, all narrowly linear; bracts of the female involucre less attenuate. — Proc. Am. Acad. xvii. 212. *A. dimorpha,* var. *flagellaris,* Torr. & Gray in Wilkes Exped. xvii. 366. — Dry rocks, Washington Territory and throughout E. Oregon, *Pickering & Brackenridge, Cusick, Howell.*

A. stenophýlla, GRAY, l. c. Stems erect from a subterranean caudex, slender, 4 to 6 inches high, without stolons, leafy, terminated by a capituliform glomerule of 2 to 4 heads: leaves very narrowly linear or almost filiform, attenuate to both ends (the larger 3 inches long), silvery-woolly; heads barely 3 lines long: involucral bracts in both sexes broadish and obtuse, dark brown, or in the male the inner ones with white tips: akenes (two thirds of a line long), minutely hirtellous-scabrous: female pappus scanty, only a line long; male pappus nearly of the preceding. — *A. alpina?* var. *stenophylla,* Gray in Bot. Wilkes Exped. l. c. — Banks of the Spipen River, Washington Terr., *Pickering & Brackenridge.* High hills, Union Co., E. Oregon, *Cusick.*

§ 2. Bristles of the male pappus stouter, with thickish and clavate or scarious-dilated tips.

* Not surculose by stolons, a span or more high: female heads narrow, cylindraceous or clavate: akenes glandular.

A. Géyeri, GRAY. Branched from a lignescent base, commonly stout, thickly woolly; stems very leafy to the top, bearing few or several somewhat spicately or cymosely disposed rather large heads: leaves spatulate or oblanceolate, less than inch long: involucre very woolly at base; of the female heads commonly 4 lines long, of the male shorter; the inner in both with conspicuous rose-purple or sometimes ivory-white tips, which in the latter are obtuse, in the former narrower and rather acute: bristles of the male pappus moderately clavate. — Pl. Fendl. 107, & Bot. Calif. i. 340. *Gnaphalium alienum*, Hook. Lond. Jour. Bot. vi. 251. — Hills, Washington Terr. to N. California; first coll. by *Geyer*.

A. microcéphala, GRAY. Simple-stemmed, slender, silvery-woolly: lower leaves spatulate; uppermost small and linear: heads rather numerous, small, loosely paniculate: involucre nearly glabrous throughout, fuscous, of the narrow female heads 3 lines long, of the broader male heads 2 lines long, the somewhat colored (whitish or purplish) tips scarious and inconspicuous: bristles of the male pappus with much dilated tips. — Proc. Am. Acad. x. 74, & Bot. Calif. l. c. — Dry eastern slope of the Sierra Nevada, in California and Nevada, *Stretch*, *Lemmon*, &c.

* * Not surculose-stoloniferous: stems simple from the subterranean branching caudex, rather strict, leafy, naked at summit, and bearing a mostly compound-cymose cluster of broad heads: inner bracts of the male involucre all with conspicuous ivory-white papery obtuse tips; those of the female with hardly any tips and more scarious: herbage silvery-lanate: larger lower leaves 3-nerved.

A. luzuloídes, TORR. & GRAY. Closely silky-woolly: stems slender, a span to a foot high: leaves all narrowly linear, or some of the lowest narrowly lanceolate-spatulate, small uppermost linear-subulate: heads small (2 lines, or the female barely 3 lines long), several or numerous: involucre glabrous nearly or quite to the base; the inner bracts in the female heads obtuse: akenes glandular: the spatulate and as it were petaloid tips of the male pappus obtuse. — Fl. ii. 430; Gray, Bot. Calif. l. c., excl. var. — Oregon, Washington Terr., and borders of Brit. Columbia, east to Wyoming.

A. argéntea, BENTH. Larger, 8 to 16 inches high: lower leaves all spatulate (the larger 4 or 5 lines wide): heads numerous in a more compound cyme, broader (fully 3 lines long): involucre in both sexes whiter than in the preceding species; innermost bracts of the female acutish: tips of male pappus even more dilated. — Pl. Hartw. 319. *A. luzuloides*, var. *argentea*, Gray in Pacif. R. Rep. iv. 54, & Bot. Calif. l. c. — California, in the Sierra Nevada, from Siskiyou Co. to the Yosemite district.

A. Carpáthica, R. BR. Floccosely white-woolly, rather stout: lower leaves spatulate-lanceolate and the upper linear: heads broad, 3 or 4 lines long: involucre conspicuously woolly at base, more or less livid, except the white tips to the bracts of the male; the inner bracts of the female commouly acutish and thin-scarious: akenes smooth and glabrous. The typical plant 2 to 6 inches high, with a simple close cluster of 3 to 7 heads, or even a solitary head: bristles of the male pappus gradually and moderately enlarged upward. — Hook. Fl. i. 329; DC. Prodr. vi. 269; Torr. & Gray, Fl. ii. 430; Reichenb. Ic. Fl. Germ. xvi. t. 951. *Gnaphalium Carpathicum*, Wahl. Fl. Carp. 258, t. 3. — Labrador (a monocephalous form!) and Anticosti, and from the northern Rocky Mountains to mountains of Oregon and Washington Terr. (Eu., N. Asia.)

Var. **pulchérrima**, HOOK. l. c. Stems 6 to 18 inches high: leaves mostly larger, the radical often half-inch or even almost an inch wide: heads more numerous, often in a compound cyme: bristles of the male pappus with more strongly and abruptly or even scariously dilated tips! — Rocky Mountains at lower elevations, extending to New Mexico, Oregon, and Brit. Columbia; first coll. by *Drummond*. Passes into the typical form as to stature, and even as to pappus.

* * Surculose-proliferous by either subterranean or humifuse and leafy shoots or stolons, in the first species least so.

+— Heads in a cymose cluster, sometimes solitary: involucre woolly at base.

A. alpína, GÆRTN. Somewhat cespitose: radical shoots few and short: flowering stems 1 to 4 inches high, bearing 2 to 5 heads, sometimes (var. *monocephala*, Torr. & Gray) a single head: radical leaves spatulate, half-inch long: involucre 3 lines high, livid-brownish; the inner of the male heads with whitish oblong tips, of the female wholly livid and scarious and from acutish to acuminate: akenes glandular. — Less. in Linn. vi. 221; Hook. l. c.; DC. l. c.; Torr. & Gray, l. c.; Fl. Dan. t. 2786. *A. monocephala*, DC. l. c., depauperate form. *A. Labradorica*, Nutt. Trans. Am. Phil. Soc. l. c. 406. *Gnaphalium alpinum*, L.; Reichenb. Ic. Pl. Crit. viii. t. 750. — Labrador and northward to Behring Strait and Aleutian Islands, and southward on the high mountains to Colorado and to California beyond the Yosemite. (Greenland, Eu.)

A. dioíca, GƷNTN. l. c. Freely surculose and forming broad mats: flowering stems 2 to 8 or even 12 inches high, bearing few or numerous heads: radical leaves from obovate to spatulate (half-inch to nearly inch long), rarely glabrate above: bracts of the involucre in both sexes with colored (white or rose-colored) and obtuse papery tips: akenes smooth and glabrous or sometimes minutely glandular. (Polymorphous.) — Hook. l. c.; Torr. & Gray, l. c. *Gnaphalium dioicum*, L., &c. *A. hyperborea*, Don, Engl. Bot. t. 2640, a glabrate form. *A. parvifolia*, Nutt. l. c. (*A. dioica*, var. *parcifolia*, Torr. & Gray, l. c.); form with small and very silvery leaves, and involucral bracts rarely of yellowish tinge. — Moist or dry ground, Newfoundland and Labrador, and through the Rocky Mountain region (alpine, subalpine, and lower along the streams), thence southward to New Mexico and S. California, and northwestward to Alaska. (Eu., Asia.)

Var. **congésta**, DC. l. c. A form too little developed, with heads sessile in a rosulate tuft of leaves terminating depressed stems, like the sterile creeping ones, occurs on Sierra Blanca, S. Colorado, at 13,000 feet: and similar but more caulescent forms, from mountains of S. Utah, California, Wyoming, &c.

A. plantaginifólia, HOOK. l. c. Freely surculose by long and slender sparsely leafy stolons, the offsets biennial; flowering stems more scapiform, 6 to 18 inches high, bearing small linear or lanceolate leaves and a cluster of several heads: radical leaves from roundish ovate to obovate and spatulate, the larger an inch or two long (besides the petiole), soon glabrate and green above, silvery-canescent beneath with a completely pannose coating, 3-5-nerved (but the nerves not rarely obsolete): involucre very woolly at base: inner bracts of the male heads with oval or oblong obtuse ivory-white tips, of the larger (4 to 6 lines long) female heads with white or whitish narrow and acute tips: akenes minutely glandular. — Torr. & Gray, Fl. l. c. 431. *A. plantaginea*, DC. l. c. *Gnaphalium plantaginifolium*, L. *G. plantagineum*, Murr. Syst. 748: Pursh, Fl. ii. 525. *G. dioicum*, var. *plantaginifolium*, Michx. Fl. ii. 128. — Dry hills and shaded grounds, Hudson's Bay to Florida, Texas, and New Mexico, and northwestward to British Columbia and Washington Territory. — Var. *monocephala*, Torr. & Gray, is an occasional form, with a single head; from Louisiana. On the Blue Ridge in Virginia, A. H. Curtiss collected the male of a remarkably small-headed and small-leaved form.

++ ++ Heads loosely paniculate: involucre almost glabrous.

A. racemósa, HOOK. l. c. Stoloniferous in the manner of the preceding, lightly woolly, becoming glabrate: flowering stems 6 to 20 inches high, slender, sparsely leafy, bearing few or numerous racemosely or paniculately disposed heads, nearly all slender-pedunculate: leaves thin; the radical broadly oval, an inch or two long, obscurely 3-nerved at base, rather veiny; lower cauline oblong; upper small and lanceolate: involucre scarious, brownish; the male 2 or 3 lines long, of obtuse bracts, the inner obscurely white-tipped; female 3 or 4 lines long, of narrow and mostly acute bracts: akenes glabrous. — Torr. & Gray, l. c. — Moist woods, Rocky Mountains along the British border, south to Wyoming, and west to the Cascade Mountains, &c.; first coll. by *Drummond*.

60. **ANÁPHALIS**. DC. EVERLASTING. (Said by DC. to be an ancient Greek name of some Gnaphalioid plant, and that it may be taken as an anagram of the very similar genus *Gnaphalium*.) — Chiefly perennial herbs, all but our species Asiatic: fl. late summer. — Benth. & Hook. Gen. ii. 303. *Anaphalis & Antennaria* § *Margaripes*, DC. Prodr. vi. 270, 271.

A. margaritácea, BENTH. & HOOK. l. c. Commonly a foot or two high, in tufts, very leafy, the white floccose wool rarely becoming tawny: leaves (2 to 5 inches long) from rather broadly to linear-lanceolate, soon glabrate and green above, the broader ones indistinctly 3-nerved: heads numerous, corymbosely cymose: bracts of the involucre very numerous, almost wholly pearly white, radiating in age. — *Gnaphalium margaritaceum*, L.; Engl. Bot. t. 2018. *G. Americanum*, &c., Clusius, Hist. i. 327, fig. 3. *Antennaria margaritacea*, R. Br. in Linn. Trans., &c. — Dry fields and open woods, Newfoundland to the Aleutian Islands, and through the northern and cooler portions of the United States, extending south to the higher mountains in Colorado (a var. *subalpina*, dwarf, broad-leaved, and with few glomerate heads), and the mountains of California. (N. E. Asia. Nat. in Eu.)

61. GNAPHÁLIUM, L. Cudweed, Everlasting. (Γναφάλιον, the Greek and also Latin name of these or similar plants). — Floccose-woolly herbs (of most parts of the world); with sessile and sometimes decurrent entire leaves, and cymosely clustered or glomerate heads of whitish or yellowish flowers. Involucre not rarely colored, but seldom yellow. Receptacle usually flat. Akenes terete or flattish, mostly nerveless. Fl. summer and autumn. — Torr. & Gray, Fl. ii. 426; Benth. & Hook. Gen. ii. 305.

§ 1. Eugnaphálium. Bristles of the pappus not at all united at the base, falling separately.

* Involucre woolly only at base, mainly scarious, in ours from white to brownish straw-color or rarely tinged with rose, not yellow: heads paniculately or corymbosely cymose or glomerate at summit of the leafy stem and branches: more or less fragrant herbs, erect, a foot or two high from an annual or biennial or sometimes perennial root: akenes in our species smooth and glabrous.

+ Leaves not at all decurrent, narrowed at base: hermaphrodite flowers very few: akenes sometimes lightly 3–4-nerved: stems freely branching, rather slender, 1 to 3 feet high.

— **G. polycéphalum**, Michx. Erect from an annual root, somewhat aromatic: branches either glabrous when the white wool is detached, or minutely viscid-pubescent when it is caducous: leaves thinnish, lanceolate or sometimes linear, mucronately acute or acuminate, often with finely undulate margins, soon bare and green and commonly viscid-puberulent or glandular above: heads in numerous rather close paniculately or cymosely disposed glomerules; involucre dull white, soon with a rusty tinge; its thin bracts oblong, obtuse. — Fl. ii. 127; DC. Prodr. vi. 227; Torr. & Gray, l. c. *G. obtusifolium*, L. Spec. ii. 851, a false name taken from the char. and figure of the doubtful plant of Dill. Elth. (but the figure of Morison is good and its leaves acute), changed in Lam. Dict. ii. 755 to *G. conoideum*, founded on the same ambiguous figure. — Open woods and dry ground, Canada to Wisconsin and south to Texas. (Mex.)

— **G. Wrightii**, Gray. Diffusely much branched from an apparently perennial root, persistently white-woolly, not glandular: leaves from spatulate to lanceolate (an inch or two long): heads (2 lines long) very numerous in small cymosely paniculate glomerules on loose spreading or divergent branchlets: involucre turbinate, grayish-white, very woolly at base; its bracts thin, oblong, obtuse, but most of them (at least the inner) with an acute apiculation. — Proc. Am. Acad. xvii. 214. *G. microcephalum*, Gray, Pl. Wright. i. 124, & ii. 99, not Nutt. — Dry ground, W. Texas and Arkansas to New Mexico and Arizona; first coll. by *Wright*. (Adj. Mex.)

+ + Leaves more or less adnate-decurrent at base, persistently white-woolly, slightly if at all glandular or heavy-scented.

— **G. Arizónicum**, Gray. Grayish-woolly: stems slender, strict, a foot high from an annual root: cauline leaves narrowly linear (inch and a half long, a line wide), slenderly decurrent; lowest short and somewhat spatulate: heads (2 lines or more long) very numerous and glomerate, the clusters fastigiate-cymose: involucre narrowly oblong, brownish; its thin bracts mostly lanceolate and acute. — Proc. Am. Acad. xix. 3. — *S.* Arizona, in dried beds of streams near Fort Huachuca, *Lemmon*.

— **G. microcéphalum**, Nutt. Slender, more loosely branched from an apparently perennial root: leaves linear or lower spatulate-lanceolate, with slenderly decurrent base: heads (2 or 3 lines long) rather few or loose in the paniculately or cymosely disposed glomerules: involucre from turbinate to campanulate, bright white; its bracts ovate or oblong (except the innermost), obtuse, though described by Nuttall as "acute." — Trans. Am. Phil. Soc. n. ser. vii. 404. — Along water-courses, S. California to Oregon; first coll. by *Nuttall*.

— **G. Sprengélii**, Hook. & Arn. Stems usually stout, 6 to 30 inches high from an annual or biennial root: leaves lanceolate or linear, or the lowest narrowly spatulate, densely white-woolly, or sometimes more thinly floccose, the short decurrent bases or adnate auricles rather broad: heads (3 lines long and wide) in single or few (rarely numerous and cymose) close glomerules terminating the stem or few branches: involucre hemispherical, white or with barely greenish-yellowish tinge, becoming slightly rusty in age; its bracts thin, oval and oblong, obtuse. (Slender forms resemble *G. luteo-album* of the Old World, which has duller or sordid heads and scabrous-pubescent akenes. A slender form in New Mexico, &c., nearly approaches the Mexican *G. gracile*, HBK., which has yellowish involucre.) — Bot. Beech. 150; Torr. & Gray, Fl. ii. 427; Gray, Bot. Calif. i. 341. *C. Chilense*, Spreng. Syst. iii. 480. ex Less. in Linn. vi. 525, but not Chilian. *G. luteo-album*, Hook. Fl.; Nutt. Trans. Am. Phil. Soc. l. c. (var. *occidentale*), &c. — Moist or dry ground, from N. Oregon to S. California, and eastward to W. Texas. (Mex.)

++ ++ ++ Leaves obviously adnate-decurrent, the upper face at least becoming naked and green in age and with the stem glandular-pubescent or glandular-viscid: herbage strongly balsamic-scented.

++ Root apparently annual or biennial.

G. decúrrens, Ives. Stem stout, 2 or 3 feet high, corymbosely branched at summit, and bearing crowded cymosely disposed glomerules of broad heads: leaves very numerous, lanceolate or the upper linear, white-woolly beneath or rarely glabrate: involucre broadly campanulate, white, usually becoming rusty-tinged; the thin-scarious bracts ovate and oblong, acutish, only the innermost linear-lanceolate and acute. — Am. Jour. Sci. i. 380, t. 1; Torr. Compend. 288; Hook. Fl. i 328; DC. Prodr. vi. 236; Torr. & Gray, l. c.; Gray, Bot. Calif. i. 346. — Rather open and dry ground, New England to Pennsylvania, Upper Michigan, Colorado, also Texas, New Mexico, and to Brit. Columbia and Washington Terr.

Var. **Califórnicum,** Gray, l. c. Bracts of the involucre more pearly white: leaves usually shorter. — *G. Californicum*, DC. l. c.; Torr. & Gray, l. c., excl. var. — Throughout the western part of California, and to San Bernardino Co. Foliage sometimes wholly green.

G. ramosíssimum, Nutt. Greener than *G. decurrens*, soon glabrate, and more glandular-viscid: stem 2 to 6 feet high, paniculately and fastigiately much branched above: leaves smaller, linear: heads amply and rather loosely paniculate, small (commonly 2 lines long), comparatively few-flowered · involucre turbinate; its bracts fewer, narrower, white or tinged with rose. — Pl. Gamb. 172; Gray in Wilkes Exped. xvii. 363, & Bot. Calif. 342. *G. Sprengelii*, var. *erubescens*, Nutt. Trans. Am. Phil. Soc. l. c., a form with rosy bracts. *G. Californicum*, var., Torr. & Gray, l. c. — Thickets, &c., W. California, from the Sacramento to Los Angeles; first coll. by *Nuttall*.

++ ++ Root lignescent-perennial.

G. leucocéphalum, Gray. Very white with close wool, except the upper face of the leaves: stems a foot or two high, strict, mostly simple, very leafy: cauline leaves all narrowly linear, small (not over 2 inches long, a line or two wide), attenuate-acute, commonly erect, hardly broader at the short-decurrent base, viscid-glandular above: heads in a rather close cyme: involucre broadly campanulate, much imbricated, pure pearly white; the bracts thin-papery, ovate and oblong, obtuse. — Pl. Wright. ii. 99. — Dry water-courses, western borders of Texas to Arizona and S. California, *Wright, Thurber, Parish*, &c.

* * Involucre less imbricated, more involved in wool, the scarious tips of the nearly equal bracts comparatively inconspicuous and dull-colored: heads glomerate and leafy-bracteate, only a line or so in length: low and branching annuals, a few inches or rarely a foot high: akenes in the same species either smooth or scabrous. Species perhaps confluent.

G. palústre, Nutt. Loosely floccose with long wool, erect, at length diffuse or weak: leaves (3 to 5 lines wide) spatulate or the uppermost oblong or lanceolate: tips of the linear involucral bracts white, obtuse. — Gray, Bot. Calif. i. 342. *G. palustre & G. gossypinum*, Nutt. Trans. Am. Phil. Soc. l. c.; Torr. & Gray, Fl. ii. 427. 428. — Common in all moist grounds, from Washington Terr. to S. California, east to Wyoming and New Mexico.

G. uliginósum, L. (Cudweed.) Appressed-woolly, soon diffusely branched: leaves spatulate-linear or the lower spatulate-oblanceolate: involucral bracts brownish to the tip or soon becoming so, acutish or obtuse, the outermost oblong. — Fl. Dan. 859: Engl. Bot. 1194; DC. Prodr. vi. 230. — Low or wet ground, a common weed, from Newfoundland to Virginia and west to the Mississippi; seemingly introduced from Eu. Also in Oregon and Brit. Columbia, where the preceding appears to pass into this. (Eu., N. Asia.)

G. strictum, Gray. Appressed-woolly: stem strict and simple, a span to a foot high, sometimes branching or with ascending stems from the base: leaves all linear, seldom a line wide: heads in spicately disposed glomerules in the axils or on short lateral branches: involucral bracts with brownish or somewhat whitish tips, obtuse. — Pacif. R. Rep. iv. 110; Rothr. in Wheeler Rep. vi. 157. — Eastern Rocky Mountain region, from Wyoming to New Mexico, and to Mount Agassiz, Arizona; first collected by *Fremont*.

* * * Involucre of the few and naked heads nearly glabrous, brown: female flowers comparatively few, only twice or thrice the number of the hermaphrodite: akenes broader and flatter: small and low alpine perennial. — *Omalotheca*, Cass., DC.

G. supínum, VILL. Cespitose: leaves white-woolly, mainly in radical tufts, linear, less than inch long, 2 or 4 on the (inch to span high) simple slender flowering stems, which bear 2 to 7 spicately disposed heads. — Delph. iii. 192; Engl. Bot. t. 1193; Hook. Fl. i. 329; Torr. & Gray, l. c. *G. pusillum*, Hænke; Schkuhr, Handb. t. 267. *Omalotheca supina*, DC. Prodr. vi. 245. — Alpine region of the White Mountains, N. Hampshire, and Labrador. (Greenland, Eu., Asia.)

§ 2. GAMOCHÆTA. Bristles of the pappus united in a ring at base and deciduous together from the akene : heads spicately or sometimes capitately glomerate, the lower glomerules leafy bracteate : involucre brownish, purplish, or sordid. — *Gamochæta*, Wedd. Chlor. And. i. 151.

* Northern (also European) species, perennials: stems strict and simple: akenes fusiform, hispidulous-pubescent.

G. Norvégicum, GUNNER. A span or two high, silvery-woolly throughout: leaves spatulate-lanceolate, acute; the radical often 6 inches long and half-inch wide; cauline sparse: heads in the upper axils and in an oblong spike: involucre 3 lines long, dark brown or the bracts with a lighter centre. — Fl. Norveg. 105 (Fl. Dan. t. 254); Syme, Engl. Bot. t. 744. *G. sylvaticum*, Smith (not L.); Torr. & Gray, Fl. ii. 429. *G. sylvaticum*, var. *fuscatum*, Wahl.; DC. Prodr. vi. 232. *G. fuscum*, Lam. Dict., not Scop. — Labrador; Mount Albert, Lower Canada, *Allen*. (Greenland, Eu.)

G. sylváticum, L. A span to a foot or more high, more leafy: leaves linear or the lowest linear-oblanceolate, glabrous or glabrate above: heads numerous in an elongated and leafy virgate-spiciform inflorescence: involucre light-colored; the bracts usually only brownish-tipped or with a brown spot below the hyaline-scarious tip. — Spec. ii. 856; Fl. Dan. t. 1229; Syme, l. c., t. 743. *G. rectum*, Smith, Engl. Bot. — New Brunswick, roadsides and muddy shores of the Bay of Chaleurs, *Fowler*, *Macoun*. Perhaps introduced. (Eu., N. Asia.)

* * More southern, and wholly American, annuals or biennials, chiefly of the sea-coast or near it.

G. purpúreum, L. Canescent with a silvery dense and close coating of white wool, sometimes becoming flocculent, simple or branched from the base: stems erect or ascending, 6 to 20 inches high: leaves spatulate, obtuse, usually becoming glabrate and green above: heads in a cylindraceous or oblong or in a more elongated spiciform inflorescence: involucre (2 lines long) brownish, often tinged with purple: akenes sparsely scabrous. — Spec. ii. 854 (pl. Gronov. & Dill. Elth.); Michx. Fl. ii. 127; Torr. & Gray, Fl. ii. 428; Klatt in Linn. xlii. 140. *G. spicatum*, *spathulatum*, *stachydifolium*, & *falcatum* (narrow-leaved form), Lam. Dict. ii. 757, &c. *G. Americanum* (Mill. Dict.?), Willd. Spec. iii. 1887. *G. Pennsylvanicum*, Willd. Enum. 867. *G. hyemale*, Walt. Car. 203. *G. Chamissonis*, DC. Prodr. vi. 233. *G. ustulatum*, Nutt. Trans. Am. Phil. Soc. l. c., ex char. And many other extra-North-American synonyms: polymorphous. — Coast of Mass. to Texas, and interior of Arkansas; also Washington Terr. to S. California and Arizona, in saline soil. (Mex., S. Amer.)

62. ÍNULA, L. (Old Latin name of Elecampane.) — A large and varied Old World genus, chiefly of perennial yellow-flowered herbs, with alternate simple leaves, sometimes tomentose, but not floccose-woolly. Section CORVISARTIA (*Helenium*, Adans., not L., *Corvisartia*, Mérat & Cass.) consists mainly of

I. HELÉNIUM, L. (ELECAMPANE, i. e. *Enula campana* of the herbalists.) A coarse and tall herb, in tufts from large perennial (bitter-mucilaginous) roots: leaves large, especially the petioled ovate radical ones, denticulate, tomentose beneath; cauline sessile, partly clasping: heads very large, solitary or few terminating the stem or flowering branches: outer bracts of the involucre ovate and foliaceous; inner smaller, obovate and spatulate, obtuse: rays very many, long and slender: akenes 4-sided, glabrous. — Fl. Dan. t. 728; Lam. Ill. t. 680; Sibth. Fl. Græc. t. 873. — Roadsides and pastures, escaped from gardens, and well established in the older States. (Nat. from Eu.)

63. ADENOCAULON. Hook. ('Αδήν, a gland, and καυλός, stem.) —
Perennial herbs; with alternate and dilated leaves on long and margined petioles, slender stems naked and paniculately branched above, and bearing very small heads of whitish flowers: the peduncles, &c., beset with stalked glands (whence the name) like those of the akenes but less stout. Floccose wool caducous, except on the lower face of the leaves. — Hook. Bot. Misc. i. 119. t. 15. & Fl. i. 308; Maxim. Fl. Amur. 152; Gray. Proc. Am. Acad. viii. 653. & xvii. 214.

— **A. bícolor**, Hook. l. c. Stem 1 to 3 feet high, leafy below: leaves ample, deltoid-cordate, coarsely sinuate-dentate or repand or slightly lobed, early glabrate and green above, white with the thin cottony wool beneath: bracts of the involucre 4 or 5 in a single series, ovate, reflexed in fruit, several times shorter than the (4 to 6) club-shaped akenes. — Gray, Bot. Calif. i. 335. — Damp woods, California to Brit. Columbia and east to Lake Superior: fl. summer. Quite distinct from the Chilian, less so from the Amur-Himalayan species.

TRIBE V. HELIANTHOIDEÆ. p. 59.

64. PLÚMMERA, Gray. (*Sara Plummer*, now *Mrs. J. G. Lemmon*, the discoverer. She and her husband have shared together the toils, privations, and dangers of arduous explorations in the wilds of Arizona and California, as well as in the delights of very numerous discoveries: so that wherever the name of *Lemmon* is cited for Arizonian plants, it in fact refers to this pair of most enthusiastic botanists.) — Proc. Am. Acad. xvii. 215. — Single species.

P. floribúnda, Gray, l. c. Erect and rather stout herb, apparently from a biennial root, 2 or 3 feet high, nearly glabrous, with bitter-aromatic odor and savor, fastigiately and corymbosely much branched above: branches terminating in loose cymes of numerous pedunculate heads: leaves all alternate, 1-3-ternately parted into filiform lobes, impressed-punctate: involucre only 2 lines long: corollas golden-yellow; those of the ray nearly glabrous, of the disk densely puberulent-glandular. — S. Arizona, in Apache Pass, *Mr. & Mrs. Lemmon*. — Corollas, involucre, odor, &c., nearly of *Actinella*, sect. *Psiladenia*.

65. DICRANOCÁRPUS, Gray. (Δίκρανον, a pitchfork, καρπός, fruit.) —
Mem. Am. Acad. v. 322 (Pl. Thurb.). & Bot. Mex. Bound. 85. — Single species.

D. parviflórus, Gray, l. c. Branching annual, a foot or less high, nearly glabrous: leaves all opposite, 1-2-ternately divided into filiform lobes, or the uppermost nearly simple: heads more or less pedunculate and paniculate, terminating slender branches, in flower a line long, yellowish: longest akenes 4 lines and their horns often 3 lines in length. — *Heterospermum dicranocarpum*, Gray, Pl. Wright. i. 109. — W. Texas, near the Pecos, *Wright*. (Adj. Mex., *Perry, Palmer*.)

66. GUARDÍOLA, Humb. & Bonpl. (The name of a Spanish naturalist.) — Perennial herbs (of Mexico and its northern borders), glabrous, branching; with merely serrate and commonly petiolate veiny leaves; the branches terminated by the cymulose-clustered heads of white flowers. — Pl. Æquin. i. 144, t. 41; Gray. Pl. Wright. i. 110; Benth. & Hook. Gen. ii. 347. *Tulocarpus*, Hook. & Arn. Bot. Beech. 208. t. 63.

G. platyphýlla, Gray. Somewhat glaucous, 2 or 3 feet high, corymbosely branched: leaves roundish-ovate, very obtuse, rigidly denticulate or dentate, commonly subcordate (the larger 3 or 4 inches long), very short-petioled: involucre narrow, 5 lines long: corollas pure white and anthers bright green. — Pl. Wright. ii. 91. — S. Arizona, *Wright*, *Thurber*, *Schott*, *Lemmon*, &c.: fl. summer.

67. POLÝMNIA, L. (Name of the muse *Polyhymnia*, shortened.) — Perennial herbs (Atlantic-American), or some S. American species shrubby or arborescent, commonly viscid-pubescent and heavy-scented, of coarse habit; with mostly opposite ample and membranaceous lobed or angulate leaves, commonly with margined petioles, or auriculate-appendaged at the insertion, and loosely paniculate heads of yellow or yellowish flowers, or the rays sometimes white; in summer. — Gen. ed. 4, 396; Lam. Ill. t. 711; Gærtn. Fruct. t. 174; DC. Prodr. v. 514. *Alymnia*, Neck. *Polymniastrum*, Lam.

§ 1. EUPOLÝMNIA. Akenes somewhat obcompressed and trigonous-obovoid, tricostate (namely with marginal and ventral nerves or ribs), not striate: heads rather small.

P. Canadénsis, L. Viscid-pubescent, 2 to 5 feet high: slender branches bearing loosely paniculate somewhat nodding heads of honey-yellowish flowers: leaves thin-membranaceous; uppermost (sometimes alternate) deltoid-ovate or somewhat hastate; lower variously pinnately lobed or the larger ones parted, acuminate, sharply denticulate, occasionally sinuate-dentate: disk of the head about 4 lines in diameter: loose outer bracts ovate-lanceolate or narrower: flowers yellowish; those of the ray 5, their ligule commonly minute or abortive, so that the head is discoid: akenes smooth and glabrous or sparsely puberulent, and with a narrow apiculate-protuberant epigynous disk: disk-corollas with abruptly much dilated campanulate throat and ovate lobes. — Amœn. Acad. iii. 15, t. 1, fig. 5, & Spec. ii. 926; Lam. l. c.; Michx. Fl. ii. 147; DC. Prodr. v. 515. *P. Canadensis*, var. *discoidea*, Gray, Man. ed. 3, 248. — Shaded and damp hillsides along streams, Canada to Pennsylvania and Missouri and in the higher Alleghanies to Carolina. Southward commonly with more evident rays and passing to

Var. radiáta. Ligules developed, dilated-cuneate, a fourth to a third of an inch long, 3-lobed, but seldom surpassing the disk, nearly white. — *P. Canadensis*, Torr. & Gray, Fl. ii. 272, mainly, from the character, excl. syn. Poir. & Lam. (which belong to a S. American species). — Extends to Hot Springs, Arkansas, *F. L. Harvey*.

§ 2. UVEDÁLIA, DC. Akenes somewhat laterally compressed, very stout, rather oblique, and their whole surface closely and strongly striate-nerved. (Here *P. variabilis*, Poir., *Polymiastrum*, Lam., i. e. all of § *Alymnia*, DC., excepting the original *P. Canadensis*.)

P. Uvedália, L. Commonly pubescent, not viscid, stout, 4 to 10 feet high: leaves ample (the larger a foot or two long and nearly as broad), of deltoid-ovate outline and 3-ribbed above the cuneate-decurrent base, 3-5-lobed, or the smaller only angulate-sinuate: heads somewhat cymosely paniculate, short-peduncled: the disk half-inch or more in diameter: outer involucral bracts broadly ovate: rays 10 to 14, with ligules bright yellow, linear-oblong to oval, usually half-inch in length, but sometimes hardly developed: akenes 3 lines long, glabrous: disk-corollas with cylindraceous throat and short lobes. — Spec. ed. 2, ii. 1303 (Pluk. Alm. t. 83, f. 3; Moris. Hist. iii. 6, t. 7, f. 55); Torr. & Gray, l. c. *Osteospermum Uvedalia*, L. Spec. ed. 1, ii. 923. — Fertile or moist grounds, New York to Florida and west to Missouri and Texas.

68. MELAMPÓDIUM, L. (Μέλας, black or dark, and πούς, foot, i. e. black-footed, an ancient name of Black Hellebore, from the root: unmeaningly transferred to these plants.) — Branching herbs, of the warm parts of America, the greater number Mexican; with opposite mostly sessile leaves, and pedunculate heads terminating the branches or in the forks. Rays in some short, in

others conspicuous. — Gaertn. Fruct. t. 169; R. Br. in Linn. Trans. xii. 104;
DC. Prodr. v. 517; Benth. & Hook. Gen. ii. 349.

> * Perennial and the base slightly lignescent: akene with apex exposed at the hooded orifice of the fructiferous bract: rays plane, conspicuously exserted, comparatively ample, tardily deciduous from the akene, white!

M. cinéreum, DC. Branched from the base, a span to a foot high, rather slender, cinereous or even silvery-canescent with a fine and mostly close pubescence, or greener and becoming strigulose: leaves linear or the lower lanceolate or spatulate, entire or undulate, or even sinuate-pinnatifid: peduncles slender: ligules 5 to 9, cuneate-oblong, 2-3-lobed at apex, 3 to 6 lines long: bracts of the involucre ovate, appressed, slightly united at base: fructiferous bracts (2 lines long, including the hood) turbinate, nearly terete, somewhat incurved, muricate with sharp tubercles; its hood about the length of the body and very much wider, imperfectly cupuliform, nearly smooth, callous-thickened or becoming suberose, its truncate and usually even margin commonly incurved. — Prodr. v. 518 (excl. habitat); Gray, Pl. Fendl. 75. Pl. Wright., &c. *M. leucanthum,* Torr. & Gray, Fl. ii. 271. — Open ground, W. Arkansas and Texas to Arizona. (Adj. Mex.)

—— Var. **ramosissimum.** (*M. ramosissimum,* DC. l. c. by the char., but habitat and number of distribution of this and of *M. cinereum* were interchanged in the Prodromus!) More loosely pubescent and diffusely branching: heads mostly smaller: hood of the fructiferous bracts with the less thickened margin little or not at all involute, sometimes erose or denticulate and bearing a mucro or short (seldom "uncinate") cusp. — Southern borders of Texas, *Berlandier, Palmer.* (Adj. Mex.)

> * * Annuals, commonly low, erect, branching, with linear or oblong mostly entire leaves: akene with merely the apex exposed at the summit of the enclosing fructiferous bract: ray and disk-corollas yellow. Our species all quite alike in foliage and habit.

M. híspidum, HBK. Hispidulous-hirsute, sometimes a foot high: rays very small, barely a line long: outer involucral bracts oval, distinct to the base; fructiferous bracts truncate and not at all appendaged at the somewhat oblique summit, more or less tuberculate on the back and sides. — Nov. Gen. & Spec. iv. 273, t. 399; DC. l. c. 520; Gray, Pl. Wright. ii. 85. — S. Arizona, *Wright, Lemmon.* (Mex.)

M. cupulátum, Gray. Somewhat hispidulous-pubescent: rays small but exserted, 2 lines long: outer involucral bracts connate to above the middle into an obtusely 5-lobed hemispherical or saucer-shaped cup: fructiferous bracts nearly of the preceding. — Proc. Am. Acad. viii. 291. — Borders of S. Arizona, on the Mexican side, *Schott.* (Mex.)

M. longicórnu, Gray. Sparsely hispidulous: rays exserted, oblong, when well developed 3 lines in length and as long as the involucre, the outer bracts of which are distinct: fructiferous bracts more nervose, little tuberculate or smooth, the summit cupulately produced and gradually extended exteriorly into a circinnate or revolute horn or rigid awn, fully as long as the body, longer and more attenuate than in *M. sericeum,* and sericeous-pubescent along the outside. — Mem. Amer. Acad. (Pl. Thurb.) v. 321, & Bot. Mex. Bound. 85. — S. Arizona, *Thurber, Schott.* (Adj. Mex.)

69. **ACANTHOSPÉRMUM,** Schrank. (Ἄκανθα, a prickle or thorn, and σπέρμα, seed, i. e. prickly-fruited.) — Homely annual weeds, much branched from the base; with opposite dentate leaves, in their axils and in the forks small subsessile or short-peduncled heads of yellowish flowers: the (4 to 7) bur-like involucral bracts enlarging in age. Natives of the tropics, one or two species becoming naturalized. — Pl. Rar. Hort. Monac. t. 53; DC. Prodr. v. 521. *Centrospermum,* HBK. Nov. Gen. & Spec. iv. 270. t. 397.

—— **A. xanthioídes,** DC. l. c. Diffusely procumbent or creeping; stems pubescent: leaves small (about inch long), mostly obovate, narrowed at base into a short petiole: fructiferous involucral bracts narrowly oblong, longitudinally sulcate, truncate, thickly beset especially along the angles with uniform and small hooked prickles. — *A. Brasilum,* Schrank, l. c.? a hirsute form. *Melampodium australe,* Loefl., L. *Centrospermum xanthioides,* HBK. l. c. — Roadsides and waste grounds, S. Carolina to Florida, &c. (Nat. from S. Amer.)

A. húmile, DC. Larger, commonly erect, hirsute: leaves wing-petioled or sessile by a cuneate base: fructiferous bracts somewhat 3-angled, not grooved, armed (besides the prickles) with one or two long spines from the truncate summit. — *A. humile* & *A. hispidum,* DC. l. c. *Melampodium humile,* Swartz, Prodr. 114. *Centrospermum humile,* Less. Syn. 217. — Ballast-weed, about Philadelphia and New York; naturalized at Pensacola. (Nat. from W. Ind.)

70. SÍLPHIUM, L. ROSIN-WEED. (Σίλφιον, ancient name of an Umbelliferous plant in N. Africa which produced a gum-resin, transferred by Linnæus, in his accustomed way, to an American genus.) — Tall and coarse perennials (all of Atlantic U. S.); with resinous juice, thick roots, commonly large leaves, and ample pedunculate heads of yellow flowers (one species with white rays!), produced in summer and autumn. — Torr. & Gray, Fl. ii. 275; Benth. & Hook. Gen. ii. 350.

S. brachiatum, A. Gattinger

* Stem square, leafy to the top: bases of the leaves or of their winged petioles cupulate-connate.

— **S. perfoliátum,** L. (CUP-PLANT.) Stem 4 to 8 feet high, commonly very smooth and glabrous: leaves either smooth or scabrous, sometimes hirsute-pubescent beneath, ovate or the upper ovate-lanceolate (the larger a foot or more long), dentate or denticulate with mucronate teeth; upper ones united by their broad bases and lower by winged petioles into a perfoliate cup: heads terminating the loosely cymosely disposed flowering branches, on naked peduncles: involucre short-campanulate, half or two-thirds inch high: outer bracts ovate, from erect to somewhat squarrose-spreading: rays inch long: akenes either with deep or shallow notch, the narrow wings being produced either into very small obsolete or prominent triangular teeth. — Spec. ed. 2, ii. 1301; Gouan, Hort. Monsp. 462; Hook. Bot. Mag. t. 3354; Torr. & Gray, l. c. *S. connatum,* L. Mant. 574, a form with branches somewhat hispid. *S. tetragonum* & *S. scabrum,* Moench, Meth. 606. *S. conjunctum,* Willd. Enum. 633. *S. Hornemanni,* Schrad. Hort. Gœtt.; DC. Prodr. v. 514. *S. erythrocaulon,* Bernh. in Spreng. Syst. iii. 630. — Alluvial soil, Michigan and Wisconsin to Upper Georgia and Louisiana. Common in cultivation; variable but characteristic.

◦ * Stem from obtusely quadrangular to terete, leafy: leaves all or some of them opposite, entire or serrate, not connate-perfoliate.

+— All but the lower sessile, and either all opposite or the upper pairs occasionally disjoined: akenes with a broad wing and a deep narrow notch: stems 2 to 4 feet high, rigid, very leafy to the top.

— **S. integrifólium,** MICHX. Stem smooth or scabrous, sometimes rough-hispidulous: leaves entire or denticulate, lanceolate-ovate or ovate-lanceolate; all the upper ones closely sessile by a broad and roundish or subcordate partly clasping base, and tapering from below the middle to an acute apex, scabrous above, from nearly glabrous and smooth to cinereous-pubescent beneath, 3 to 5 inches long, commonly of firm texture: heads somewhat corymbose, nearly all short-peduncled: involucre over half-inch high; its bracts mostly ovate and spreading: akenes broadly obovate, the body 4 lines long, the scarious wing a line or so wide, at least toward the summit. — Fl. ii. 146; Torr. & Gray, Fl. ii. 279, hardly of Ell. *S. lævigatum,* Pursh, as to char. *S. speciosum,* Nutt. Trans. Am. Phil. Soc. vii. 341, a very smooth form, the var. *læve,* Torr. & Gray, l. c. — Prairies, &c., Wisconsin and Illinois to Arkansas and Texas, and possibly to W. Georgia.

— **S. aspérrimum,** HOOK. Commonly taller: stem rough-hispid: leaves of the preceding but more scabrous: heads generally larger: akenes with broader wings, the triangular apical portions 2 or 3 lines high. — Comp. Bot. Mag. i. 99. *S. radula,* Nutt. Trans. Am. Phil. Soc. l. c. *S. scaberrimum,* Torr. & Gray, Fl. ii. 279, var. γ, hardly Ell. — Plains of Arkansas, Louisiana, and Texas.

— **S. scabérrimum,** ELL. Stem and commonly both sides of the leaves hispid: leaves in remoter pairs, thinner, oblong or ovate, all but the uppermost rather coarsely serrate and with narrowed or even short petiole-like base (the larger 4 to 6 inches long): heads fewer, more pedunculate: rays inch long: outermost involucral bracts smaller: akenes including broad wing nearly orbicular in outline, half-inch in diameter. — Sk. ii. 462; Torr. & Gray, Fl. ii. 279, excl. var. γ. — W. Georgia to Louisiana and E. Texas.

+ + Leaves rather few on the slender stem, the lower slender-petioled, often alternate: akenes with the broad wings of the preceding.

S. grácile, GRAY. Hispidulous: stem 12 to 30 inches high, rather naked, terminated by solitary or few mostly long-pedunculate heads: leaves membranaceous, ovate-oblong or oblong-lanceolate, acute at both ends, denticulate; radical and lower cauline ample (5 to 9 inches long); upper cauline from 2 inches to half-inch long: involucre of nearly equal and rather few oblong bracts: akenes orbicular or very broadly oval, broadly winged, and with a comparatively shallow notch. — Proc. Am. Acad. viii. 653. — Prairies of Middle Texas, *Drummond, Lindheimer, Hall,* &c. Sometimes the leaves are all alternate and the petioles of the one or two principal cauline 2 or 3 inches long, equalling the blade.

+ + + Leaves numerous on the stem, varying from opposite to alternate or the middle ones verticillate, only upper and alternate ones (if any) strictly sessile by a broad base: akenes with narrow wings and a comparatively shallow open notch; awn-like pappus-teeth usually evident and not rarely partly separate from the wing.

S. Asteríscus, L. Stem 2 to 4 feet high, commonly hispid: leaves from ovate-oblong to oblong-lanceolate, coarsely and irregularly dentate or serrate, or some entire, scabro-hispidulous or hispid, all the upper not rarely alternate, seldom any verticillate; upper commonly sessile by a rounded or partly clasping base; lower short-petioled: heads solitary or few on leafy branches: involucre foliaceous and squarrose (half-inch high), hirsute or hispidulous: akenes obovate-oval. — Spec. ii. 920 (Dill. Elth. t. 37, f. 42); Michx. Fl. ii. 146; DC. Prodr. v. 512; Torr. & Gray, Fl. ii. 278. — Dry sandy soil, common from Virginia and Tennessee to Florida and Louisiana.

Var. lævicaúle, DC., l. c. Stem smooth and glabrous, either throughout or up to the branches. — Torr. & Gray, l. c. *S. scabrum,* Walt. Car. 217. *S. reticulatum,* Muench, Meth. 607, fide syn. L. *S. Asteriscus,* var. *scabrum,* Nutt. Gen. ii. 183. *S. dentatum,* Ell. Sk. ii. 468; Torr. & Gray, l. c. *S. lanceolatum,* Nutt. Trans. Am. Phil. Soc. vii. 341, a narrow-leaved form. *S. Asteriscus,* var. *dentatum,* Chapm. Fl. 221. — *S.* Carolina to Georgia and Alabama.

S. trifoliátum, L., l. c. Stem 4 to 7 feet high, very smooth and glabrous, terminated by naked corymbiform panicles of numerous usually slender-peduncled heads: leaves lanceolate, varying to oblong-ovate, and from entire to sparsely serrate, from almost glabrous and smooth to scabrous or hispidulous-pubescent, seldom alternate, a part of them usually 3–4-nately verticillate, commonly acute at base and the upper subsessile, lower tapering into margined petioles: involucre somewhat campanulate, narrower and usually smaller than in the foregoing; the bracts hardly foliaceous, smooth and glabrous, except the ciliate margins: akenes broadly obovate-oval. — Hook. Bot. Mag. t. 3355; Torr. & Gray, l. c. *S. trifoliatum, ternatum, & otropurpureum,* Retz in Willd. Spec. iii. 2333. *S. ternifolium,* Michx. Fl. ii. 146. — Dry woodlands, Penn. and Ohio, and through the upper country to Alabama.

Var. latifólium. Stem 2 to 4 feet high: leaves broader, seldom more than opposite: heads fewer and broader. — *S. lævigatum,* Ell. Sk. ii. 465 (perhaps Pursh. Fl. ii. 578, but his character points to *S. integrifolium*); Torr. & Gray, l. c. — W. Georgia, Alabama, and low country of S. Carolina; and broad-leaved northern forms are similar.

* * * Stem terete, almost leafless and scapiform, bearing a loose panicle of slender pedunculate heads: involucre nearly hemispherical, of rounded mostly erect and coriaceo-herbaceous bracts: radical leaves ample, long-petioled, cordate at base: cauline when present all alternate and slender-petioled: herbage almost glabrous and smooth, or the leaves hispidulous and papillose-scabrous. (True ROSIN-WEEDS.)

S. compósitum, MICHX. Stem slender, 2 to 6 feet high, commonly glaucous: radical leaves of roundish-cordate or reniform or more ovate circumscription, 6 to 12 inches long or broad: heads small and numerous: involucre a third or rarely half inch high: rays small and scattered, 4 lines long: akenes roundish-obovate and with wing broadened above, so as to form a deep notch, with which the two subulate pappus-teeth are confluent, sometimes narrowly winged so that the summit is barely emarginate and minute pappus-teeth nearly free. — Torr. & Gray, Fl. ii. 276. The first-described form (var. *Michauxii,* & var. *ovatifolium,* Torr. & Gray, l. c.) has the leaves varying from deeply sinuate-pinnatifid to pinnately or somewhat ternately divided into 3 to 7 divisions, which are again sinuate-lobed; for this the specific name is appropriate, being *S. compositum,* Michx. Fl. ii. 145; Ell. l. c.; DC. l. c., and *S. laciniatum,* Walt. Car. 217, not L. *S. nudicaule,* M. A. Curtis in Bost. Jour. Nat. Hist. i. 127,

16

a form passing into var. *reniforme*, Torr. & Gray, l. c., has rounder leaves, some only sinuate-dentate, others deeply palmately cleft. *S. elatum*, Pursh, Fl. ii. 579. *S. terebinthinaceum*, Ell. Sk. ii. 463, not Jacq. *S. reniforme*, Raf. Med. Fl. ii. 283; Nutt. Trans. Am. Phil. Soc. vii. 341. — Pine woods and barrens, N. Carolina to Florida.

S. terebinthináceum, JACQ. (PRAIRIE DOCK.) Stem 4 to 9 feet high, bearing several or numerous large heads: leaves of thick and firm texture, cordate-oblong or sometimes ovate-oblong, a foot or two long (besides the long petiole), dentate with very many small teeth, becoming rough in age: involucre nearly an inch high: rays an inch or more in length: akenes obovate, narrowly winged, merely emarginate and obscurely 2-toothed at summit. — Hort. Vindob. i. t. 43; L. f. Suppl. 383; Gœrtn. Fruct. ii. 445, t. 171; Schk. Handb. t. 262; Hook. Bot. Mag. t. 3525; Torr. & Gray, l. c. — Prairies and dry open woodlands, Ohio and Michigan to Iowa and south to W. Georgia and Louisiana.

Var. **pinnatifidum**, GRAY. Leaves laciniately or sinuately pinnatifid. — Man. ed. 1, 220. *S. pinnatifidum*, Ell. l. c. — Ohio and W. Georgia, not common.

* * * * Stem terete (striate when dried), bearing alternate deeply pinnatifid or bipinnatifid coriaceous leaves, and sessile or short-peduncled large heads racemosely disposed along the naked summit, and bracteate: involucre rigid; its bracts ovate, thickened and at length coriaceous at base, with equally long or longer and spreading foliaceous acumination: rays numerous: herbage scabrous-hispidulous or hispid, very rough when dried. — COMPASS-PLANTS.

S. laciniátum, L. Stem 3 to 6 and even 12 feet high: radical leaves (a foot or two long) long-petioled, once or twice pinnately parted or below divided, the divisions and lobes lanceolate to linear; cauline with petiole simply dilated at base, or with stipuliform and sometimes palmatifid appendages; upper sessile and reduced to bracts: involucre inch or more high and broad: rays numerous, inch or two long, bright yellow: akenes half-inch long, oval, glabrous or nearly so, with narrow wing widening upward and an open shallow notch; no awns. — Spec. ii. 919; L. f. Dec. 5, t. 3; Jacq. f. Ecl. 1, t. 90; Torr. & Gray, l. c.; Mechan, Nat. Flowers, ser. 2, ii. t. 46; Hook. f. Bot. Mag. t. 6534. *S. spicatum*, Poir. Suppl. v. 157. *S. gummiferum*, Ell. Sk. ii. 460. — Prairies, Wisconsin to Dakota and south to Alabama, Kansas, and Texas. Leaves vertical and, especially the radical ones, disposed to place the edges north and south, — in respect to which there is abundant literature. See Alvord in Am. Naturalist, xvi. 626.

S. albiflórum, GRAY. Low, a foot to barely a yard high, very scabrous: leaves rigid, as broad as long, more disposed to pedate division; dilated base of petiole entire: tips of involucral bracts seldom surpassing the disk: rays *white*, about inch long: akenes puberulent; the narrow wing produced and dilated at summit into somewhat triangular teeth which are adnate to a pair of subulate and more or less projecting awns, the notch narrow. — Proc. Am. Acad. xix. 4. — On cretaceous rocks, W. & N. Texas, *Reverchon*.

71. **BERLANDIÉRA**, DC. (*J. L. Berlandier*, a Genevese botanist and collector, explored parts of Texas and Mexico, died at Matamoras in 1851.) — Perennial herbs (of the southeastern borders of the U. S.); with canescent or cinereous herbage, thick roots, alternate leaves, and pedunculate heads: the rays yellow: involucre radiately expanding in fruit. Fl. spring and summer. — Prodr. v. 517; Benth. Pl. Hartw. 17; Torr. & Gray, Fl. ii. 280.

* Stems leafy up to the inflorescence of mostly rather numerous and short-peduncled heads: leaves crenate, some or all the cauline cordate; radical oblong.

B. Texána, DC. Hirsute-tomentose; the pubescence not pannose, that of the (2 or 3 feet high) very leafy stem commonly hirsute or villous, the coarser hairs many-jointed: cauline leaves from oblong-cordate to subcordate-lanceolate, greenish, merely cinereous beneath, somewhat scabrous above; upper closely sessile, lower short-petioled: heads usually fastigiate-cymose. — Prodr. l. c.; Deless. Ic. Sel. iv. t. 26; Torr. & Gray, l. c. *B. longifolia*, Nutt. Trans. Am. Phil. Soc. vii. 342. — Margin of woods and hillsides, Texas (first coll. by *Berlandier*), W. Louisiana and Arkansas to S. W. Missouri. Leaves of *Betonica*.

Var. **betonicifólia**, TORR. & GRAY, l. c. A form with most of the cauline leaves petioled, and the peduncles hirsute with purplish hairs. — *Silphium betonicifolium*, Hook. Comp. Bot. Mag. i. 99. — Louisiana, *Drummond*.

— **B. tomentósa**, Nutt. l. c. Canescent throughout with soft and close pannose tomentum, no hirsute or villous hairs, when glabrate hardly at all scabrous: stem a foot or two high, rarely only a span high: leaves all obtuse, green above, generally whitish beneath; radical and lower cauline elongated-oblong and petioled; upper cauline usually ovate-oblong or oval, sometimes subcordate-ovate, short-petioled or sessile heads fewer, in low specimens almost solitary and longer-peduncled. — Torr. & Gray, Fl. ii. 282. *B. pumila*, Nutt. l. c. *Silphium pumilum*, Michx. Fl. ii. 146. *S. tomentosum, pumilum,* & *reticulatum?* Pursh, Fl. ii. 578, 579. *S. Asteriscus*, var. *pumilum*, Wood, Bot. 442. *Polymnia Caroliniana*, Poir. Dict. v. 505. — Dry pine barrens, N. Carolina to Florida, Arkansas, and Missouri.

Var. **dealbáta**, Torr. & Gray, l. c. More robust and leafy, 2 or 3 feet high, branching at summit and bearing more numerous and shorter-peduncled heads: cauline leaves broader and more sessile, densely white-tomentose beneath; lower broadly cordate, upper often deltoid (with or without a subcordate base), either obtuse or acute. — Texas, *Drummond, Hall, Reverchon*, a very soft-canescent form. Varies into a less canescent state, approaching *B. Texana*, the leaves scabrous above (var. γ, Torr. & Gray, l. c.), Arkansas, Louisiana, and Texas.

* * Stems commonly low and with long monocephalous peduncles; the earliest often produced from near the root, and scapiform, the later from leafy stems or branches: leaves variable, all attenuate at base, disposed to be pinnatifid or lyrate.

B. subacaúlis, Nutt. l. c. Barely cinereous with minute often hispidulous pubescence (or the peduncles sometimes hirsute), soon green, becoming a foot or so high and leafy: leaves of oblong-linear or oblong-spatulate outline, irregularly sinuately or lyrately pinnatifid, with short obtuse lobes: akenes narrowly obovate-oval, merely carinately costate on the inner face. — Torr. & Gray, Fl. ii. 282. *Silphium subacaule*, Nutt. in Am. Jour. Sci. v. 301; DC. Prodr. v. 512. *S. Nuttallianum*, Torr. Ann. Lyc. N. Y. ii. 216, as to syn. — Florida, in dry pine barrens; first coll. by *Ware*.

— **B. lyráta**, Benth. Canescent with minute white or gray tomentum: leaves at length greenish above, variously lyrate-pinnatifid; the lateral lobes oblong or narrower, obtusely dentate, sometimes incised: akenes obovate, the costa of the inner face strongly carinate. — Pl. Hartw. 17; Gray, Pl. Fendl. 78, & Pl. Wright. i. 103. *B. incisa*, Torr. & Gray, Fl. ii. 282. *Silphium Nuttallianum*, Torr. Ann. Lyc. N. Y. ii. 216, excl. syn. — Plains and hills, W. Texas and Arkansas to Arizona. (Mex.)

Var. **macrophýlla**. Radical leaves often a foot long, lanceolate-oblong or spatulate, either merely crenate or pinnatifid at base: later flowering stems sometimes 2 or 3 feet high. — S. Arizona, *Lemmon*.

72. **CHRYSÓGONUM**, L. (Greek name of some plant in Dioscorides. Linnæus gives the derivation of his genus from χρυσός, golden, and γόνυ, knee; of no obvious application.) — Gærtn. Fruct. ii. 436. t. 174; Lam. Ill. t. 713; DC. Prodr. v. 510; Torr. & Gray, Fl. ii. 274; Benth. & Hook. Gen. ii. 350, excl. syn. *Moonia*, &c.; Gray, Proc. Am. Acad. xvii. 216. *Diotostephus*, Cass. Dict. xlviii. 543. — Single species: fl. spring and summer.

— **C. Virginiánum**, L. Perennial from creeping rootstocks and sometimes by runners, pubescent, often hirsute, flowering acaulescently from the ground, also with stems a span to a foot high, bearing 3 or 4 pairs of long-petioled leaves; these ovate, mostly obtuse and crenate; cauline rarely subcordate and equalling or shorter than their petioles, or the radical obovate with cuneate attenuate base: peduncles solitary in the forks and terminal, all but the radical ones elongated: involucre one-third and yellow rays half inch long. — Spec. ii. 920 (Pluk. Alm. t. 83, f. 4. & 242. f. 3); Walt. Car. 217; Michx. Fl. ii. 148; Torr. & Gray, l. c. *C. Virginianum* & *C. Diotostephus*, DC. l. c. *Diotostephus repens*, Cass. l. c. — Dry ground, S. Pennsylvania to Florida. Varies considerably according to age and season, usually low when blossoming begins.

Var. **dentátum**, Gray. Leaves deltoid-ovate, acute, coarsely dentate-serrate, the tip and teeth, also the tips of the bracts of the outer involucre, terminated by a more conspicuous callous mucro. — Bot. Gazette, viii. 31. — High Island at the falls of the Potomac above Washington, *J. Donnell Smith, Ward, Vasey*.

73. LINDHEÍMERA, Gray & Engelm. (*Ferdinand Lindheimer*, the discoverer of this neat plant, now prized in cultivation, and remarkable for its golden yellow rays simulating a 5-petalous flower.) — Proc. Am. Acad. i. 47, Jour. Bost. Nat. Hist. vi. 225, & Pl. Lindh. ii. 225. Single species.

L. Texána, Gray & Engelm. l. c. At length 2 feet high from an annual root, hirsute or hispid, branching above, bearing loosely cymose-paniculate usually slender-pedunculate heads: lower leaves spatulate to cuneate-ovate, alternate, coarsely sinuate-dentate; upper ovate to ovate-lanceolate, with a broad closely sessile base, acuminate, commonly entire, mainly opposite, their edges and also the peduncles usually beset with some small tack-shaped glands: ligules half-inch or more long. — Open woods and bottoms of the upper Guadalupe River, &c., Texas, *Lindheimer*, *Wright*.

74. ENGELMÁNNIA, Torr. & Gray. (*George Engelmann*, an eminent botanist, died while this volume was printing, Feb. 4, 1884, æt. 75.) — Torr. & Gray in Nutt. Trans. Am. Phil. Soc. vii. 343, & Fl. ii. 283. *Angelandra*. Endl. Gen. Suppl. iii. 69. — Single species, in structure nearer to *Parthenium* than to *Silphium*. Fl. summer.

— **E. pinnatífida**, Torr. & Gray, l. c. A foot or two high from a stout perennial root, roughish-hirsute or hispid, branching above, and bearing somewhat paniculately disposed heads of golden-yellow flowers on mostly slender naked peduncles: leaves all alternate, deeply pinnatifid; radical and lower cauline short-petioled and their linear or oblong lobes sometimes sparingly lobulate; upper cauline sessile and with broad base: head about 4 lines high: rays half-inch or more long: akene rough-hispidulous. — Torr. in Marcy Rep. t. 11; Meehan, Nat. Flowers, ser. 2, i. t. 2. *E. texana*, Scheele in Linn. xxii. 155. — Prairies and rocky hills, Arkansas and Louisiana to Texas and Arizona. (Adj. Mex.)

75. PARTHÉNIUM, L. (Ancient name of some plant, from παρθένος, virgin.) — Herbaceous or suffruticose (all E. American), bitter-aromatic; with small heads of whitish flowers; in summer. — Gærtn. Fruct. t. 168; DC. Prodr. v. 531; Torr. & Gray, Fl. ii. 284; Benth. & Hook. Gen. ii. 351.

§ 1. PARTHENIÁSTRUM (Nissole), DC. Ligule more or less evident: caulescent, usually branching, with alternate leaves either dentate or variously lobed or divided: heads corymbosely or paniculately cymose.

* Herbaceous, with membranaceous once or twice pinnatifid leaves, and habit of *Ambrosia*.

— **P. Hysteróphorus, L.** A foot or two high, from an annual root, diffuse, strigosely pubescent, sometimes also hirsute, generally green: heads in a loose and open naked panicle: cauline leaves of broadly ovate outline, pinnately parted into 5 to 9 mostly narrow again pinnatifid lobes; of the flowering branches linear or lanceolate and entire or few-lobed: pappus of 2 rather large and roundish scales. — Spec. ii. 988; Bot. Mag. t. 2275. *Argyrochæta bipinnatifida*, Cav. Ic. iv. 54, t. 378. *Villanova bipinnatifida*, Ort. Dec. iv. 48, t. 6. (*P. lobatum*, Buckley in Proc. Acad. Philad. 1861, 457, should be this, by its "annual root," rather than the following.) — Waste grounds, Florida to Texas, where it may be indigenous, but probably introduced from within the tropics: also an imported ballast-weed as far north as Philadelphia. (Mex., Trop. Am.)

P. lyrátum. A foot high from a truly perennial root, canescent or cinereous with fine and close sometimes also loose hirsute pubescence, erect: heads corymbosely crowded, more pubescent: leaves of obovate or oblong outline, lyrately pinnatifid, the lobes short and oblong. — *P. Hysterophorus*, var. *lyratum*, Gray, Proc. Am. Acad. xvii. 216. — Texas, in the southern and western parts, *Berlandier*, *Lindheimer*, *Wright*, *Reverchon*, &c. Equally allied to the preceding species and to the Mexican *P. confertum*, Gray. (Adj. Mex.)

* * Fruticose or suffrutescent, with firmer and more simply lobed leaves.

— **P. incánum, HBK.** Decidedly shrubby, 1 to 3 feet high, much branched, canescent with fine tomentum: leaves mostly obovate in outline, sinuately pinnatifid into 3 to 7 oblong or roundish and obtuse lobes: heads numerous, paniculate-cymose: ligules commonly longer than broad: pappus a pair of short-subulate erect or at length spreading awns. — Nov Gen. & Spec. iv. 260, t. 391; Gray, Pl. Wright. i. 103. *P. incanum* & *P. ramosissimum*, DC. Prodr. v. 532. — Dry hills, W. Texas to Arizona. (Mex.)

P. argentátum, GRAY. Suffrutescent, a foot high, silvery-canescent with close tomentum: branches erect, rather leafless above, bearing comparatively large and few heads (of 2 lines in diameter): leaves lanceolate to spatulate in outline, some entire or incisely 2–3-toothed; the larger incisely pinnatifid into 2 to 7 acute lateral lobes: pappus a pair of lanceolate chaffy awns. — Bot. Mex. Bound. 86. — S. W. borders of Texas, *Bigelow*. (Adj. Mex., *Parry, Palmer*. Produces a gum or resin in Mexico.)

* * * Perennial herb, with larger heads and leaves; the latter undivided, thickish.

P. integrifólium, L. Stout, 1 to 3 feet high, minutely pubescent, corymbosely branched above, the branches terminated by a dense cyme of many heads (these a quarter-inch high): leaves ovate-oblong or narrower, thickly crenate-dentate, rarely doubly dentate or somewhat incised, hispidulous-scabrous, prominently veiny from a strong midrib; radical a foot or more long and tapering into a petiole; upper cauline closely sessile and broad at base: pappus a pair of small chaffy teeth or scales. — Spec. ii. 988 (Dill. Elth. t. 225; Pluk. Alm. t. 53 & 219); Lam. Ill. t. 766; Willd. Hort. Berol. t. 4; Torr. & Gray, l. c. — Dry ground, Maryland to Illinois and Texas.

§ 2. BOLÓPHYTUM, Torr. & Gray. (*Bolophyta*, Nutt.) Ligule wanting, the corolla being reduced to a truncate tube, which is obscurely notched at back and front: acaulescent cespitose perennial.

P. alpínum, TORR. & GRAY, l. c. Densely tufted on a thick branching caudex, depressed, rising only an inch or two high: leaves crowded, silvery-canescent with a fine appressed pubescence, and villous in the axils, spatulate-linear, barely inch long, entire: heads solitary and nearly sessile among the leaves: pappus a pair of oblong-lanceolate membranaceous scales. — *Bolophyta alpina*, Nutt. Trans. Am. Phil. Soc. n. ser. vii. 347. — Rocky Mountains in Wyoming (at 7,000 feet), on rocks near the Three Buttes, *Nuttall*.

76. **PARTHÉNICE**, Gray. (Παρθενική, a poetical form of the word from which the name of the preceding nearly related genus is derived.) — Pl. Wright. ii. 85; Benth. & Hook. Gen. 352. — Genus of a single species, allied also to the succeeding genus.

P. móllis, GRAY, l. c. Annual, with odor and savor of *Artemisia*, 4 to 6 feet high, paniculately branched, minutely puberulent-cinereous throughout, wholly destitute of any coarser pubescence: leaves membranaceous, all alternate, ovate, some of the larger (as much as 10 or 12 inches long) subcordate, acuminate, irregularly or doubly dentate, long-petioled: heads small (2 lines broad), numerous in loose axillary and terminal somewhat leafy panicles, mostly pedicellate: flowers greenish-white. — Hillsides and along streams, S. Colorado to Arizona, *Wright, Thurber, Lemmon*, &c. Fl. autumn.

77. **IVA**, L. (An unexplained name.) — American herbs or shrubs; with entire or dentate or dissected leaves, at least the lower ones opposite, and small spicately or racemosely or paniculately disposed or scattered and commonly nodding heads: fl. summer. — Lam. Ill. 766; Gaertn. Fruct. t. 164; DC. Prodr. v. 529. *Iva* & *Cyclachæna* (Fresen.), Torr. & Gray, Fl. ii. 285; Benth. & Hook. Gen. ii. 352.

§ 1. CYCLACHÆNA. Heads naked-paniculate, inconspicuously bracteate: corolla of the 5 fertile flowers a very short rudiment or none: leaves membranaceous, from incisely serrate to dissected, mostly petioled: flowers somewhat inclined to polygamo-diœcious through abortion of the ovaries: annual herbs. — *Cyclachæna*, Fresen. Ind. Sem. Hort. Franc. 1836, & Linn. xii.; Torr. & Gray. l. c. 285.

* Heads nearly sessile, crowded in narrow spiciform clusters which are aggregated in a panicle.

I. **xanthiifólia**, Nutt. Tall and coarse (3 to 5 feet high), pubescent, at least when young: leaves mainly opposite, long-petioled, broadly ovate, ample, coarsely or incisely serrate, acuminate, 3-ribbed at base, puberulently scabrous above, and when young canescent beneath: panicles axillary and terminal: involucre depressed-hemispherical, biserial; outer of 5 broadly ovate herbaceous bracts; inner of as many membranaceous dilated-obovate or truncate ones, which are strongly concave at maturity and half embrace the obovate-pyriform and glabrate akenes (on the apex of which sometimes persists a minute crown answering to the obsolete corolla, or this wholly absent). — Gen. ii. 185. *I. (Pirrotus) xanthiifolia* & *paniculata*, Nutt. Trans. Am. Phil. Soc. vii. 347. *Cyclachaena xanthiifolia*, Fresenius, l. c.; Torr. & Gray, Fl. ii. 286. *Euphrosyne xanthiifolia*, Gray, Pl. Wright. ii. 85. — Alluvial ground or along streams, Saskatchewan and Nebraska to New Mexico, Utah, and Idaho; first coll. by *Nuttall*.

I. **dealbáta**, Gray. A foot or two high, canescent with floccose wool except the elongated and narrow terminal panicle: leaves in greater part alternate, soft-tomentose, reticulate-veiny (1½ to 3 inches long), from obscurely angulate-toothed to laciniately pinnatifid, cuneately or abruptly contracted at base into a short winged petiole: heads only a line long: involucre of only 5 obovate concave somewhat herbaceous bracts: corolla of fertile flowers a short cup or ring: akenes pyriform, roughish and glandular. — Pl. Wright. i. 104. — Valleys of S. W. Texas, *Wright, Bigelow*. (Adj. Mex., *Thurber*, &c.)

* * Heads pedicellate, in looser panicles, more or less leafy-bracteate: habit and foliage of *Euphrosyne*.

I. **ambrosiæfólia**. A foot or two high, hirsute or villous-hispid, paniculately branched: leaves almost all alternate, thin, twice or thrice pinnately parted into small oblong lobes: involucre of 6 broadly ovate herbaceous outer bracts, and as many smaller obovate thin-scarious inner ones: corolla of fertile flowers a mere vestige: akenes turgid-obovate. — *Euphrosyne ambrosiæfolia*, Gray, Pl. Wright. i. 102, ii. 85. — W. borders of Texas and adjacent New Mexico, *Wright*. (Mex.)

§ 2. IVA proper. Heads spicately or racemosely disposed in the axils of leaves or foliaceous bracts, and nodding: fertile flowers with evident corolla. — *Iva*, Torr. & Gray, Fl. ii. 286; Benth. & Hook. Gen. l. c.

* Heads in terminal and solitary or paniculate compact squarrosely-bracteate spikes: leaves not coriaceous: root annual.

I. **ciliáta**, Willd. Rather stout, 2 to 6 feet high, strigose-hispidulous and hispid: leaves nearly all opposite, ovate, acuminate, sparsely serrate, the base abruptly contracted into a hispid petiole: spikes strict, 3 to 8 inches long; their bracts lanceolate and ovate-lanceolate, foliaceous, surpassing the at length deflexed heads, hispid-ciliate, as are the 3 or 4 (rarely 5) herbaceous and unequal distinct or partly united bracts of the involucre: akenes about 3, obovate, moderately flattish. — Spec. iii. 2386; Pursh, Fl. ii. 580; Torr. & Gray, Fl. ii. 287. *I. annua*, Michx. Fl. ii. 184, not L., unless possibly the detailed illustration by Schmidel should represent a state of it much altered in cultivation. *Ambrosia Pitcheri*, Torr. in Hook. Comp. Bot. Mag. i. 99, with a var. having linear and much elongated bracts to the spike. — Moist alluvial ground, Illinois to Nebraska, and south to Louisiana, Texas, and New Mexico. (Adj. Mex., *Berlandier*.)

* * Heads more loosely disposed in the axils of ordinary leaves, or upper ones commonly in the axils of foliaceous bracts.

+ Rather many-flowered; the fertile flowers 5 or rarely fewer: perennials or shrubby, with thickish and firm somewhat fleshy or coriaceous leaves.

++ Bracts of the fleshy-herbaceous involucre 6 to 9, imbricated in two or more ranks; and those among the numerous sterile flowers linear-spatulate.

I. **imbricáta**, Walt. A foot or two high from a suffrutescent base, honey-scented, smooth and glabrous or nearly so: stems thickish, ascending: leaves mainly alternate, fleshy, from spatulate-oblong to narrowly lanceolate, sessile, some of the larger (1 or 2 inches long) sparingly serrate: heads large for the genus (3 or 4 lines long), commonly pedunculate, the lower surpassed by and the upper surpassing the subtending leaves: involucre hemispherical-campanulate, the outer bracts orbicular; sterile flowers many, the fertile 2 to 4: akenes obovate-oval, turgid. — Car. 232; Michx. Fl. ii. 184; Torr. & Gray, l. c. — Sands of the sea-shore, Virginia to Florida and Louisiana. (W. Ind.)

++ ++ Bracts of the simpler involucre 5 or 4; those among the several or rather numerous sterile flowers reduced to linear filiform chaff: herbage minutely or sparsely strigulose or nearly glabrous, rarely more pubescent: leaves opposite and alternate.

I. frutéscens, L. (MARSH ELDER, HIGH-WATER SHRUB.) Shrubby, or on the northern coast nearly herbaceous, erect, 3 to 8 feet high, much branched: cauline leaves oval or oblong, 3 to 5 inches long, serrate, 3-nerved at base, petioled; those of the branches lanceolate and tapering to each end, and in the upper part of the inflorescence reduced to linear bracts mostly surpassing the heads: bracts of the involucre distinct, orbicular-obovate. — Amœn. Acad. iii. 25, & Spec. ii. 989; Walt. Car. 232; Lam. Ill. t. 166, f. 2; Michx. Fl. ii. 184; Torr. & Gray, Fl. ii. 287. — Brackish muddy shores and beaches along the sea-coast, from Massachusetts to Texas.

I. Hayesiána, GRAY. Suffrutescent, 2 or 3 feet high, with ascending rather simple branches: leaves obovate-oblong or spatulate, or the small uppermost lanceolate, obtuse, entire, nearly sessile; the larger 2 inches long; upper little or not at all surpassing the heads: involucral bracts distinct, roundish. — Proc. Am. Acad. xi. 78, & Bot. Calif. i. 614. — Brackish soil, San Diego Co., California, *Sutton Hayes, Palmer, G. R. Vasey.*

I. axilláris, PURSH. Herbaceous from somewhat woody creeping rootstocks: the stems or branches nearly simple, ascending, a foot or two high: leaves from obovate or oblong to nearly linear, obtuse, entire, sessile, rarely over inch long, even the uppermost usually much surpassing the mostly solitary heads in their axils: bracts of the hemispherical involucre connate into a 4-5-lobed or sometimes parted and sometimes merely crenate cup. — Fl. ii. 743; Nutt. Gen. ii. 185; Hook. Fl. i. 309, t. 106; Torr. & Gray, l. c. — *I. axillaris* (bracts almost separate) & *I. foliolosa* (bracts much united), Nutt. Trans. Am. Phil. Soc. l. c. 346. — Sandy saline soil, Saskatchewan and Dakota to New Mexico, and west to Brit. Columbia and California.

Var. **pubéscens**, GRAY. Villous with lax spreading hairs: involucre turbinate and almost entire. — Bot. Wilkes Exped. xvi. 350, & Bot. Calif. i. 343. — California, along the Bay of San Francisco.

+ + Heads 3-6-flowered, small (about a line long), very numerous, subsessile, all surpassed by the narrow-linear or filiform mostly alternate subtending leaves: slender erect annuals, with elongated or virgate flowering branches: chaffy bracts filiform. — § *Monnchæna*, Torr. & Gray, l. c.

I. microcéphala, NUTT. Nearly glabrous, 2 or 3 feet high, even the lower leaves narrowly linear (an inch or two long, a line wide), those subtending the loosely disposed hemispherical heads spreading: involucre of 4 or 5 distinct bracts: fertile and sterile flowers each about 3. — Trans. Am. Phil. Soc. l. c.; Torr. & Gray, l. c. — Dry pine barrens, E. and Middle Florida, *Baldwin, Chapman, Palmer, Curtiss.*

I. angustifólia, NUTT. Strigulose-scabrous or somewhat hirsute, 2 to 4 feet high: lower leaves lanceolate, acute at both ends (larger inch and a half long, 3 or 4 lines wide), some of them sparingly serrate; those of the branches from linear to filiform, the bracteal ones ascending: heads more crowded and spicate, turbinate: involucral bracts united by scarious edges into a cup: fertile flowers usually solitary; the sterile 2 to 5: anther-tips cuspidate-apiculate. — DC. Prodr. v. 529, & Trans. Am. Phil. Soc. l. c.; Torr. & Gray, l. c. — Gravelly banks or beds of streams, Arkansas, Louisiana, and Texas. (Adj. Mex.)

§ 3. CHORISÍVA. Heads scattered, lateral and ebracteate on leafy branches: fertile flowers with evident corolla.

I. Nevadénsis, M. E. JONES. Low and diffusely branched annual, leafy to the top, cinereously hirsute-pubescent: leaves obovate in outline, pinnately 3-7-parted into oblong or obovate obtuse lobes: heads small, sessile along the branches or rarely in the axil of a leaf: involucre of 3 nearly distinct ovate-oblong and very obtuse foliaceous bracts, considerably surpassing the 8 to 10 male and 3 or 4 female flowers; the latter subtended and akene partly enwrapped by as many roundish and hyaline interior bracts; their truncate corolla beset and fringed by long hairs. — Am. Naturalist, xvii. 973, but akenes not "finely striate." — Near Hawthorne, Nevada, *M. E. Jones.* — Insignificant but singular species, with the aspect of *Franseria Hookeriana.*

78. **OXYTÉNIA**, Nutt. ('Οξυτενής, pointed, "in allusion to the rigid narrow foliage.") — Pl. Gamb. 172; Benth. & Hook. Gen. ii. 353; Gray, Bot. Calif. i. 343. — Single species, Artemisia-like in habit; fl. autumn.

O. acerósa, Nutt. l. c. Shrubby, but soft-woody, 3 to 5 feet high, canescent, with erect branches sometimes leafless and rush-like: leaves when present alternate, pinnately 3-5-parted into long filiform divisions, or uppermost entire: heads numerous (2 lines long), in dense panicles. — Dry plains, S. W. Colorado to S. E. California, *Gambel, Wheeler, Brandegee*, &c.

79. **DICÓRIA**, Torr. & Gray. (Δίς, twice, used for two, and κόρις, a bug, from the aspect of the two akenes of the original species.) — Emory Rep. 143, & Bot. Mex. 86, t. 30; Gray, Proc. Am. Acad. xi. 76, & Bot. Calif. i. 615.

—— D. canéscens, Torr. & Gray, l. c. Herb a foot to a yard high, with annual root, stem becoming ligneseent at base and widely branched, herbage canescent with appressed pubescence and the branches hispid, becoming green and scabrous in age: lower leaves opposite, lanceolate and oblong, coarsely toothed or laciniate; upper alternate, ovate or roundish, all petioled: heads sparsely and irregularly racemose-paniculate, along slender nearly leafless branchlets, nodding in fruit: fertile flowers 2: inner bracts of the involucre peltoid-scarious (yellowish white), orbicular and deeply concave, accrescent in fruit (becoming 3 or 4 lines long), then inflated-saccate and loosely or partly enclosing the laciniately wing-margined akene, falling with it. — Desert washes, S. E. California and adjacent Arizona to S. Utah.

D. **Brandegéi**, Gray, l. c. Strigulose-canescent, diffusely and alternately branched (base of stem unknown): leaves of the branches oblong-lanceolate or partly spatulate, obtuse, mostly entire, an inch or less long and with slender petiole: heads sparse, racemose-paniculate; some all male: corollas sparsely hirsute: fertile flower solitary; its dilated-cuneate hyaline subtending bract hardly accrescent or surpassing the outer involucre: akene naked and exserted, bordered with pectinate callous teeth connected by an indistinct scarious margin. — Sandy bottoms of the San Juan, near the boundary between Colorado and Utah, *Brandegee*. Little Colorado, N. Arizona, *Rusby*, in flower only.

80. **HYMENOCLÉA**, Torr. & Gray. ('Υμήν, membrane, used for wing, and κλείω, to enclose.) — Two known species, of low and much branched shrubby plants, minutely canescent, or else glabrous and smooth; with slender diffuse branches, bearing profuse scattered or glomerate paniculate small heads, the two sexes intermixed, or the female in lower axils: leaves all alternate and linear-filiform; the lower sparingly and irregularly pinnately parted: fl. summer and autumn. — Pl. Fendl. 79; Gray, Pl. Wright. i. 104, Bot. Calif. i. 343.

—— H. **Sálsola**, Torr. & Gray, l. c. Fructiferous involucre fusiform, strobilaceous; the ample orbicular silvery-scarious wings spirally alternate, imbricated over each other, radiately spreading when mature and dry. — Torr. Pl. Fremont (Smiths. Contrib.) 14, t. 8. — Saline soil in the desert region, S. California, adjacent Arizona, and Nevada; first found by *Fremont*.

H. **monogýra**, Torr. & Gray, l. c. Fructiferous involucre smaller (2 lines long), winged only at the middle by a whorl of obovate or rhombic-reniform radiating scales of smaller size. — S. California through Arizona to S. W. Texas; first coll. by *Coulter*. (Adj. Mex.)

81. **AMBRÓSIA**, Tourn. Ragweed. (Ancient Greek and also Latin name of several plants, as well as of the food of the gods.) — Weedy or coarse herbs; with mostly lobed or dissected opposite and alternate leaves, and dull inconspicuous flowers; in summer. Sterile heads racemose or spicate, and with no subtending bracts; the fertile below, commonly in small clusters in the axils of leaves or bracts: fl. summer and autumn. — Lam. Ill. t. 765; Gaertn. Fruct. t. 161; Schk. Handb. t. 292; Torr. & Gray, Fl. ii. 354. — Fructiferous nut-like involucre called for shortness "fruit."

§ 1. CERCÓMERIS, Torr. & Gray, l. c. Sterile heads densely spicate, closely sessile; the involucre turbinate and half-truncate, the inner margin bearing a large lanceolate-acuminate hispid lobe, which by the deflexion of the head is strongly recurved and partly covers the orifice of the involucre, the bractless spike thus appearing as if retrorsely bracteate; fertile heads commonly solitary in axils below; leaves closely sessile by partly clasping base.

A. bidentáta, MICHX. Roughish-hirsute annual, 1 to 3 feet high, fastigiately branched above, very leafy up to the stout (span long) spikes: leaves mostly alternate, lanceolate, commonly with au acute lobe or tooth on each side near the broad base, thence tapering gradually to a point, usually entire: fertile involucre in fruit oblong, somewhat prismatic, the 4 strong angles or ribs terminating in acute strong spines of half the length of the spine-like beak: sterile heads about 10-flowered. — Fl. ii. 182; Pursh, Fl. ii. 581; Torr. & Gray, Fl. ii. 292. — Prairies and alluvial ground, Illinois and Missouri to Texas. (Adj. Mex.)

§ 2. AMBROSIA proper. Sterile heads racemose or spicate: sterile involucre commonly saucer-shaped or open-campanulate, with a several-toothed or truncate border: fertile flowers usually glomerate in axils below.

* Involucre of sterile heads unilaterally 3-ribbed: no chaff on the receptacle: leaves palmately cleft, ample, petioled.

A. trífida, L. Tall and stout annual, 3 to 12 feet high, or even higher, roughish-hispidulous, or partly hispid or hirsute, sometimes almost glabrous: leaves all opposite, very deeply 3-lobed or the lower 5-lobed; the lobes ovate-lanceolate, acuminate, serrate (in the larger leaves a span or more in length); petioles of the upper commonly wing-margined: sterile racemes long and dense: fertile heads clustered and as if involucrate by short bracts: fruit (matured fertile involucre) very thick and indurated, 4 or 5 lines long, obovoid-turbinate or obpyramidal, with 5 or sometimes 6 or 7 strong ribs or angles terminating above in spinous tubercles around the base of the conical beak. — Spec. ii. 987 (Moris. Hist. iii. sect. 6, t. 1, f. 4); Michx. l. c.; DC. Prodr. v. 527; Torr. & Gray, l. c. — Moist alluvial banks of streams, Canada and Saskatchewan to Florida, Missouri, Nebraska, &c.

Var. **integrifólia**, TORR. & GRAY, l. c. A depauperate form, with oblong or oval-lanceolate undivided leaves, and mostly solitary sterile racemes: spinous tubercles of the fruit less developed. — A. integrifolia, Muhl. in Willd. Spec. iv. 375. — New York to Illinois and Virginia.

A. áptera, DC. Very like the preceding, equally tall: petioles not margined; larger leaves commonly 5-lobed, and the middle lobe often 3-cleft: sterile racemes more numerous and paniculate: fruit smaller, 2 or 3 lines long, more obovoid, 4-8-ribbed, and with 4 to 6 short or obsolete tubercles. — Prodr. v. 527; Gray, Pl. Lindh. ii. 226. A. trifida, var. Texana, Scheele in Linn. xxii. 156. — Low grounds, Texas to New Mexico and S. W. Arizona; first coll. by Berlandier.

* * Involucre of sterile heads not costate, indistinctly radiate-veined: receptacle with some filiform or sometimes more dilated chaff: leaves opposite and alternate (in the adjacent Mexican A cheiranthifolia, Gray, Bot. Mex. Bound. 87, entire and canescent), mostly 1-3-pinnatifid or dissected.

A. artemisiæfólia, L. (ROMAN WORMWOOD, RAGWEED, BITTER WEED.) Annual, variously pubescent or hirsute, paniculately branched, a foot or two high, sometimes taller: leaves thinnish, bipinnatifid or pinnately parted with the divisions irregularly pinnatifid or sometimes nearly entire, on the flowering branches often undivided: sterile heads more or less pedicelled: fruit not 2 lines long, short-beaked, armed with 4 to 6 short acute teeth or spines. (Varies much, occasionally the sterile inflorescence abnormally fertile.) — Torr. & Gray, Fl. ii. 291. A. artemisiæfolia & A. elatior, L. Spec. 987, 988. A. absynthifolia & A. paniculata, Michx. Fl. ii. 183. A. heterophylla, Muhl. in Willd. Spec. iv. 378. Ira monophylla, Walt. Car. 232. — Dry ground, a weed of cultivated and waste grounds, Nova Scotia to Saskatchewan, Texas, California, and Washington Terr. (W. Ind. & Mex. to Brazil.)

A. longístylis, NUTT. Trans. Am. Phil. Soc. l. c. 344, — known only from Nuttall's specimen from "Rocky Mountains," described as having pinnatifid leaves, and conglomerate fertile flowers with styles about an inch long, — needs verification.

A. híspida, PURSH. Perennial, spreading from a suffrutescent base, strigose-hispidulous or hispid and hirsute: leaves all petioled, twice and thrice pinnatifid or interruptedly pinnately divided into numerous short and small oblong ultimate lobes: sterile raceme commonly solitary and elongated: fruit with a stout short beak and commonly 4 short acute tubercles. — Fl. ii. 743, the original in herb. Sherard was probably from Bahamas. *A. crithmifolia*, DC. Prodr. v. 525; Torr. & Gray, l. c. — Sandy sea-shore, Florida. (W. Ind.)

A. psilostáchya, DC. Perennial from slender running rootstocks, stouter than *A. artemisiæfolia*, 2 to 6 feet high, with strigose and some loose hirsute pubescence: leaves thickish; upper simply and lower twice pinnatifid; the lobes mostly lanceolate and acute: sterile heads commonly short-pedicelled: fruit mostly solitary in the axils below, turgid-obovoid, less than 2 lines long, rugose-reticulated, obtusely short-pointed, either wholly unarmed or (sometimes on the same plant) with four short either blunt or acute tubercles. — Prodr. v. 526; Gray, Pl. Wright. ii. 86, & Bot. Calif. i. 344. *A. Peruviana*, DC. l. c., as to pl. Mex., hardly of Willd. *A. coronopifolia*, Torr. & Gray, Fl. ii. 291. *A. Lindheimeriana* & *A. glandulosa*, Scheele in Linn. xxii. 156, 158. — Moist prairies and beds of streams, Illinois and Saskatchewan to Texas, Arizona, and California. (Mex.)

A. púmila, GRAY. Perennial, a span or two high from slender running rootstocks, canescent throughout with a dense and close silky pubescence, very leafy: leaves nearly all alternate and long-petioled, 2-3-pinnately parted into linear-oblong crowded lobes: sterile heads in a short spike: fruit obovoid, pubescent, muticous, a line long (rarely two are connate at base). — Proc. Am. Acad. xvii. 217. *Franseria pumila*, Nutt. Trans. Am. Phil. Soc. l. c. 344; Gray, Bot. Calif. i. 344, ii. 615. *Hemiambrosia*, Delpino, Stud. Comp. Artemis. 57. — San Diego, California, *Nuttall*, &c., recently coll. by *Cleveland* in fruit.

82. **FRANSÉRIA,** Cav. (*Ant. Franser*, a physician and botanist in Madrid in the time of Cavanilles.) — Herbs or shrubby plants (all American); with chiefly alternate leaves, some species with habit of *Ambrosia* and near it in character, others with the fruiting involucre nearly that of *Xanthium*. — Cav. Ic. ii. 78, t. 200; Willd. Hort. Berol. i. t. 2; DC. Prodr. v. 224; Torr. & Gray, Fl. ii. 292. *Franseria, Hemixanthidium,* & *Xanthidium*, Delpino, Stud. Comp. Artemis. 58–67.

§ 1. Spines of the fruiting and 1-2-flowered involucre comparatively few, conical, subulate, or flattened with the inner face more or less concave, usually straight or merely incurved. — § *Acantholæna*, DC.

* Herbaceous perennial: fruiting involucre seldom over a line long, in the same plant bearing either one or two flowers.

F. tenuifólia, GRAY. Erect, 1 to 5 feet high, leafy to the top, hispid, variously pubescent, or glabrate: leaves mostly 2-3-pinnately parted or dissected into narrowly oblong or linear lobes, and the narrow primary rhachis often with some interposed small lobes, the terminal elongated: sterile racemes commonly elongated and paniculate: fertile heads in numerous glomerules below, in fruit minutely glandular, usually 2-flowered, obovate with narrow obpyramidal base, armed with 6 to 18 short and stout incurving spines, their tips almost always hooked, and* an excavated cartilaginously bordered areola above each. (Larger leaves often 5 inches long or more.) — Pl. Fendl. 80, Pl. Wright. i. 104 (var. *tripinnatifida*), Bot. Mex. Bound. 87, & Bot. Calif. i. 346. *Ambrosia longistylis*, Gray, Pl. Fendl. 79, as to no. 407, perhaps of Nutt. *Ambrosia tenuifolia*, Spreng. Syst. iii. 851 ? *A. confertiflora* & *A. fruticosa* (excl. var.), DC. Prodr. v. 525, 526. *Xanthidium tenuifolium*, Delpino, l. c. 62. — Moist grounds, from Texas to N. Colorado, S. California, and southward. (Mex., Hawaii, &c.)

* * Herbaceous, with fruiting involucre 3 or 4 lines long at maturity, and longer stout or broad spines: stems low.

F. Hookeriána, NUTT. Diffusely spreading from an annual (or perennial?) root, freely branched, hirsute-pubescent or hispid, sometimes canescent with strigose-sericeous pubescence when young: leaves of ovate or roundish circumscription (1 to 3 inches broad) and bipinnatifid, or the upper oblong and pinnatifid: sterile racemes solitary or paniculate: fruiting involucre armed with flat and thin lanceolate-subulate smooth and glabrous long and straight spines, seemingly always 1-flowered. — Torr. & Gray, Fl. ii. 294. *F. Hookeriana* & *montana*, Nutt. l. c. *Hemizanthidium*, Delpino, l. c. 60. *Ambrosia acanthicarpa*, Hook. Fl. i. 309. — Plains and along streams, Saskatchewan to Washington Terr., California, Arizona, and W. Texas.

— **F. bipinnatífida**, Nutt. Procumbent, with stems 2 or 3 feet long from a perennial root, somewhat hirsute: leaves of ovate circumscription (an inch or two long), 2-3-pinnately parted into oblong-linear divisions and small oblong lobes, canescent with soft tomentum or fine hirsute-sericeous pubescence: sterile spike or raceme dense, of rather large heads: fruiting involucre ovate-fusiform, armed with rather short and thick but flattish tubercule-like spines, their acute tips sometimes incurving. — Trans. Am. Phil. Soc. vii. 507; Torr. & Gray, l. c. *F. Lessingii*, Walp. Rel. Meyen. 268. — Sands of the sea-coast, Washington Terr. to S. California.

F. Chamissónis, Less. Leaves cuneate-obovate or oblong-ovate with a cuneate base, 3-5-nerved at base, obtusely serrate, the lower often laciniate-incised; otherwise as the preceding, or the 2-flowered fruiting involucre rather thicker, the spines broader and more canaliculate. — Torr. & Gray, l. c. (with var. *cuneifolia*); Gray, Bot. Cal. i. 345. *F. Chamissonis*, var. *maleæfolia*, Less. in Linn. vi. 507; DC. l. c. *F. cuneifolia*, Nutt. l. c. — Sandy sea-beaches, Brit. Columbia to California.

— **F. díscolor**, Nutt. l. c. A foot or less high, erect from perennial slender creeping rootstocks: leaves canescently tomentose beneath, green and glabrate above, interruptedly bipinnatifid, oblong in outline, comparatively large (the lowest often 6 inches long); the lobes usually short and broad: sterile racemes commonly solitary: fruiting involucre ovoid, 2-flowered, canescent, armed with rather short conical-subulate very acute and straight spines. — Torr. & Gray, l. c. From station and char. probably *Ambrosia tomentosa*, Nutt. Gen. ii. 186. *Xanthidium discolor*, Delpino, l. c. — Plains, &c., Nebraska to Wyoming, Colorado, and New Mexico.

— **F. tomentósa**, Gray. A foot high, rather stout, erect from an apparently perennial base or rootstock, canescent with a dense sericeous tomentum: leaves very white beneath, cinereous above, pinnately 3-5-cleft or parted; the terminal division large, oblong or broadly lanceolate, serrate; upper lateral similar but smaller: lowest commonly very small and entire: fruiting involucre 3 lines long, turgid-ovoid, 2-flowered, nearly glabrous; the short spines conical-subulate, very acute, and the very tip usually uncinate-incurved. — Pl. Fendl. 80, & Pacif. R. Rep. iv. 102. — Along streams or river-beds, Kansas and E. Colorado, *Fendler*, *Bigelow*, *Hall*.

* * * Shrubby, low (1 to 3 feet high), much branched, canescent with a fine and close white tomentum, which is sometimes partly deciduous with age: sterile heads and fertile glomerules not rarely intermixed in short racemes or clusters.

+ Fruiting involucre mostly 2-flowered, smooth and glabrous, or barely puberulent; its spines flattened and dilated at base, mostly straight-pointed.

— **F. dumósa**, Gray. Divergently much branched, very canescent: leaves small, 1-3-pinnately parted into oblong or roundish (1 or 2 lines long) obtuse lobes: fertile involucre globular; its spines long, tapering from a broadish flat base to a slender aristiform point. — Frem. 2d Rep. 316, Bot. Mex. Bound. 86, & Bot. Calif. i. 345. *F. albicaulis*, Torr. Fl. Frem. 16. — Arid region, from S. E. California to S. Utah and S. Arizona; first coll. by *Coulter*.

F. deltoídea, Torr. Somewhat less woody, and less densely canescent-tomentulose: branches erect or spreading: leaves all undivided, from rhomboid-ovate or oblong to deltoid or obscurely hastate, minutely and often doubly crenate-serrate, an inch or less long, rather slender-petioled: fruiting involucre of the preceding, but the spines shorter and broader, flatter, lanceolate-subulate. — Pl. Frem. 15; Bot. Mex. Bound. 87; Bot. Calif. l. c. — *Xanthidium rhombophyllum*, Delpino, l. c. ? — Arid regions of Arizona, and perhaps adjacent part of California, *Fremont*, *Parry*, *Schott*, *Palmer*, &c. (Can hardly be *F. chenopodiifolia*, Benth., of Lower California.)

+ + Fruiting involucre only one-flowered, villous-lanate!

F. eriocéntra, Gray. Rigidly much branched, canescent with very minute tomentum: leaves soon green and glabrate above, cuneate-oblong to lanceolate, from sinuately few-toothed or lobed to sparingly and irregularly laciniate-pinnatifid, nearly sessile by attenuate base:

fruiting involucre with single subulate beak as long as the body, the latter bearing about 10 rather long rigid subulate-acerose spines, these nearly equalled by the long whitish wool. — Proc. Am. Acad. vii. 355, & Bot. Calif. ii. 345. Here also belongs the flowerless specimen coll. by Newberry, mentioned under *F. artemisioides* in the Colorado Expedition of Ives; and this is probably the nearest relative of *F. chenopodiifolia*, Benth. — Arid region, S. E. California and adjacent Nevada, *Cooper, Newberry.* Arizona and S. Utah, *Parry, Palmer, Lemmon.*

§ 2. Spines of the larger and 2–4- (commonly 3-) flowered involucre very numerous, comparatively slender, and conspicuously uncinate-tipped in the manner of *Xanthium.* (But the S. American *F. artemisioides* has stout spines.) — § **Xanthiopsis**, DC. l. c.

F. **ambrosioides**, Cav. Shrubby, 4 or 5 feet high, cinereous-pubescent: leaves rather long-petioled, oblong-lanceolate, mostly truncate or subcordate at base, acuminate, irregularly dentate or serrate, 2 to 4 inches long; petiole naked: fruit ovoid, nearly half-inch and slender prickles 2 lines long. — Ic. ii. 79, t. 200 (excl. syn); Gray, Bot. Calif. i. 346. *Xanthidium ambrosioides*, Delpino, Stud. Comp. Artemis. 63. — Arizona, *Bigelow, Palmer, Pringle,* &c. (Lower Calif., Mex.)

F. **ilicifólia**, Gray. Shrubby, at least the branches hirsute, very leafy: leaves rigidly coriaceous, scabrous, reticulate-veiny, sessile, somewhat clasping, oblong-ovate, coarsely dentate, the teeth and apex spinose: heads ovoid, those seen only 2-celled. — Proc. Am. Acad. xi. 77. — Cañons beyond the southern border of San Diego Co., California, *Palmer.* Gila Desert, Arizona, *Lemmon,* foliage only.

83. **XÁNTHIUM**, Tourn. Cockle-bur, Clot-bur. (Old Greek name of some plant the fruit of which, in the time of Dioscorides, was used to dye the hair yellow.) — Coarse annuals, chiefly American, of the warmer region, but now widely dispersed weeds; with branching stems, alternate and usually lobed or toothed leaves, and mostly clustered heads of greenish or yellowish flowers, in terminal and larger axillary clusters of both sexes, the male uppermost; the lower of few or solitary female heads in axils of leaves: fl. summer and autumn. — Gærtn. Fruct. t. 164; Schkuhr, Handb. t. 291; Benth. & Hook. Gen. ii. 355.

§ 1. Leaves cordate or ovate, 3-ribbed from the base, with dentate margins and often incised or lobed, on long petioles: axils unarmed: fruiting involucre with two prominent indurated beaks. — § *Euxanthium*, DC. Prodr. v. 523. — Perhaps all derivatives of a single species.

X. **strumárium**, L. A foot or two high: fruiting involucre half to two-thirds inch long, glabrous or puberulent; the beaks straight and rarely at all hooked at maturity, and spines rather slender. — Spec. ed. 2, ii. 1400; Fl. Dan. t. 270; Schkuhr, Handb. t. 291. — A weed of barnyards and in cult. grounds. (Sparingly nat. from Eu.? or Ind.?)

X. **Canadense**, Mill. Stouter: stem often punctate with brown spots: fruiting involucre about an inch long, densely beset with rather long prickles, the stout beaks at maturity usually hooked at the tip or incurved, the surface and base of the prickles more or less hispid, sometimes glabrate. — Dict. ed. 8, first after L. Spec. *X. majus Canadense,* Herm. Lugd. 635. *X. elatus Americanum,* etc., Moris. Hist. iii. 604, sect. 15, t. 2, fig. 2. *X. Carolinense,* etc., Dill. Elth. ii. 432, t. 231. *X. orientale,* L. l. c., in part. *X. Americanum,* Walt. Car. 231. *X. macrocarpum,* var. *glabratum,* DC. Prodr. l. c. *X. strumarium,* var. *Canadense,* Torr. & Gray, Fl. ii. 294. — Alluvial shores and waste grounds, from Texas to Saskatchewan, Nevada, and California: perhaps extended northward by man's indirect agency. In brackish soil it becomes

Var. **echinátum**. A form, usually dwarf, with still denser and longer prickles, these conspicuously hirsute or hispid. — *X. echinatum,* Murr. Comm. Gœtt. vi. 32, t. 4; Torr. & Gray, Fl. ii. 294. *X. maculatum,* Raf. in Am. Jour. Sci. i. 151. *X. macrocarpum,* DC. Fl. Fr. Suppl. 356, & Prodr. l. c. Sandy sea-shores and on the Great Lakes. (S. Am.)

§ 2. Leaves attenuate to both ends and short-petioled; their axils triply spiniferous. — § *Acanthoxanthium*, DC.

X. spinósum, L. A foot or two high, much branched: leaves ovate-lanceolate with cuneate base, the larger 3-lobed or incisely pinnatifid, glabrate and green above, white-tomentose beneath: axils bearing long and slender 3-parted yellow spines: fertile involucres solitary or few in upper axils, cylindraceous, half-inch long, obtuse, armed with short weak prickles, inconspicuously 1-2-beaked or pointless. — Lam. Ill. t 655, f. 4; DC. l. c. — A weed of S. Atlantic States and Pacific coasts, occasionally about seaports northward to Massachusetts. (Nat. from Trop. Am.)

84. ZÍNNIA, L. (Dr. *J. G. Zinn*. of Göttingen, who figured the original species as a *Rudbeckia*.) — American, chiefly Mexican, herbs or suffruticulose plants; with opposite and mostly sessile entire leaves, single heads terminating the branches, and showy flowers, the bright-colored rays long enduring: fl. summer. — Gen. ed. 6, 437; Gærtn. Fruct. t. 172; Gray, Pl. Wright. i. 105. *Zinnia* & *Diplothrix*, DC. Prodr. v. 534, 611.

§ 1. EUZÍNNIA. Herbs, mostly annual (some species perennial): leaves from ovate to linear: ray-flowers several or numerous, usually without pappus. — Pl. Wright. l. c.

Z. pauciflóra, L. Erect annual: leaves from lanceolate to oblong-ovate, commonly with subcordate base, scabrous: peduncle sometimes enlarging and hollow: involucre narrow-campanulate: ligules from obovate to narrowly spatulate, red, purple, or yellow: akenes of the disk 1-awned, sometimes with a rudiment of a second awn or tooth. — Webb, Spic. Gorg. 141. *Z. pauciflora* & *Z. multiflora*, L. Spec. ed. 2, 1269 (L. f. Dec. t. 12). *Z. tenuiflora*, Jacq. Ic. Rar. t. 590, with narrow ligules. *Z. revoluta*, Cav. Ic. iii. 251. *Z. leptopoda* & probably *Z. bicuspis*, DC. Prodr. v. 535. *Z. intermedia*, Engelm. Bot. Wisliz. 23. — Louisiana to Texas, but probably introduced, Arizona, apparently indigenous. (Mex., S. Am., and now widely dispersed.)

§ 2. DIPLÓTHRIX. Suffruticulose and tufted perennials: leaves narrow and rigid, connate-sessile, usually crowded: ray-flowers commonly few, and their akenes 2-4-aristate: head conspicuously pedunculate only in *Z. juniperifolia*. — Pl. Wright. l. c. *Diplothrix*, DC.

* Ligules shorter than or little surpassing the disk, sometimes wanting: stems mainly herbaceous.
— § *Heterogyne*, Gray, Pl. Wright. l. c.

Z. anómala, GRAY. Scabrous-hispid: stems or branches very numerous from a ligneous base and root, 4 to 8 inches high: leaves linear (half-inch to inch long, less than 2 lines wide), one-nerved, obscurely 3-nerved at base: peduncle shorter than the uppermost leaves: involucre oblong or campanulate (half-inch long): ligules 4 to 6, oval or oblong, 1 to 3 lines long, yellow or orange, occasionally the whole corolla wanting: hispid style-branches of the disk-flowers acuminate-subulate. — Pl. Wright. i. 106, t. 10, & ii. 86. — S. W. Texas, *Wright*. (Mex. near Saltillo, *Palmer*, with broader involucre.)

* * Ligules (4 or 5) ample, dilated-obovate or roundish, at maturity much surpassing the disk, light yellow or sulphur-color, becoming white in age: involucre narrow: stems or branches a span or more high from the stout woody base or branching caudex.

Z. grandifióra, NUTT. Scabro-hispidulous: leaves linear, 3-nerved at base: involucre usually 4 lines long: ligules at maturity 5 to 8 lines long: style-branches of the disk-flowers attenuate-subulate. — Trans. Am. Phil. Soc. n. ser. vii. 348; Torr. & Gray, Fl. ii. 298; Torr. in Emory Rep. t. 4 (style incorrect); Gray, l. c. — Plains and bluffs, E. Colorado to S. W. Texas and Arizona.

Z. púmila, GRAY. Cinereous-puberulent: leaves very narrowly linear (hardly half-line wide, half-inch or less long), one-nerved: involucre 2 or 3 lines long, and ligules 2 to 4 lines: style-branches of disk-flowers with short triangular-subulate tips. — Pl. Fendl. 81, Pl. Wright. l. c. — High plains and table-lands, S. W. Texas to Arizona. (Adj. Mex.)

— **Z. acerósa,** Gray. Cinereous-pubescent or glabrate: leaves acerose-filiform, very obscurely one-nerved, half-inch or more long: ligules 3 to 6 lines long: style-branches with subulate-ovate tips. — Pl. Wright. l. c. *Diplothrix acerosa,* DC. Prodr. v. 611.— Hills, S. W. Texas, *Wright.* (Adj. Mex.)

85. **SANVITÁLIA,** Lam. (*Sanvitali,* name of a noble Italian family.) — Mostly low and branching herbs, of Mexico and its border; with opposite and more or less petioled leaves, almost always entire, and rather small heads terminating the branches, ours and most of the species annuals. — Jour. Hist. Nat. ii. (1792), 176, t. 33; Ill. t. 686; Cav. Ic. iv. 31, t. 351; DC. Prodr. v. 628. *Lorentea,* Ort. Dec. iv. 42, t. 5.

§ 1. Involucre of 2 or 3 series of bracts, their tips commonly herbaceous: fructiferous receptacle from flat to strongly conical; its chaffy bracts soft or shorter than the flowers: disk commonly dark purple or brownish: rays yellow or turning whitish in age: ray-akenes mostly triangular; the comparative smoothness, granulation, or murication of disk-akenes inconstant.

— **S. ocymoídes,** DC. A span or two high, diffusely spreading, hispidulous or hirsute: leaves oval, obtuse, abruptly contracted into the petiole: ligules shorter than the akene and shorter than the three slender-subulate diverging awns: disk-akenes all wingless, quadrangular-compressed, sometimes 1-2-awned. — *S. ocymoides* & *S. tragiæfolia,* DC. l. c. — Southernmost border of Texas on the Rio Grande, *Berlandier, Schott.* (Adj. Mex.)

S. procúmbens, Lam. l. c. (*S. villosa,* Cav. Ic. l. c.), a Mexican species not uncommon in cultivation, has conspicuous ligules much exceeding the awns at their base, and flattened disk-akenes, some of them winged and 1-2-aristellate, some not; and the receptacle, at first barely convex, may become even acutely conical in age. *S. acinifolia,* DC. l. c., appears to be only a form of it.

S. angustifólia, Engelm. in Gray, Pl. Wright. i. 112, of Northern Mexico, has similar akenes and similar receptacle, but rays nearly as short as those of *S. ocymoides;* the chaff of the receptacle disposed to be rigid-tipped as in the following.

§ 2. Involucre a single series of dry bracts: fructiferous receptacle strongly and acutely conical; its chaffy bracts conspicuous and with rigid cuspidate tips: rays white: disk pale: leaves rarely denticulate.

— **S. Abérti,** Gray. Erect, at length a foot high, with ascending branches, minutely pubescent or hispidulous, glabrate: leaves lanceolate or nearly linear, 3-nerved, narrowed into a margined petiole: rays 1 to 3 lines long: akenes all corky-thickened; those of the ray almost terete, narrowly 4-sulcate, bearing 3 very short and stout nearly conical awns or tubercles; of the disk compressed-quadrangular, wingless, awnless, or sometimes minutely uniaristellate. — Pl. Fendl. 87, & Pl. Wright. i. 111. — S. W. Texas, New Mexico, and S. Arizona.

86. **HELIÓPSIS,** Pers. ("Ηλιος, the sun, ὄψις, likeness, from resemblance to the Sunflower.) — American perennials (or a Mexican and South American species annual); with loosely branching stems, ovate or oblong and veiny mostly serrate 3-ribbed or triple-ribbed leaves, on naked petioles, and pedunculate showy heads; the rather numerous rays yellow, and the disk yellowish. Fl. summer and autumn. — Syn. ii. 473; Torr. & Gray, Fl. ii. 302; Benth. & Hook. Gen. ii. 358.

— **H. lǽvis,** Pers. Smooth and glabrous or nearly so throughout, 3 or 4 feet high: leaves bright green, thinnish, oblong-ovate or ovate-lanceolate from a truncate or slightly cuneate-decurrent base, acuminate, coarsely and sharply serrate with numerous teeth (3 to 5 inches long): heads somewhat corymbose: rays broadly linear, an inch long, at length marcescent and decaying away: akenes wholly glabrous and smooth, the summit wholly truncate or obscurely 2-4-toothed. — Pursh, Fl. ii. 563; Dunal, Mem. Mus. v. 55; Hook. Bot. Mag. t. 3372;

Torr. & Gray, l. c., excl. vars.; Meehan, Nat. Flowers, ser. 2, ii. 42. *Buphthalmum helianthoides*, L. Hort. Ups., & Spec. ii. 904; Michx. Fl. ii. 130; L'Her. Stirp. t. 45. *Silphium helianthoides*, L. Spec. ii. 920, pl. Gronov. *S. solidaginoides*, L. Spec. 1. c. *Rudbeckia oppositifolia*, L. Spec. l. c. 907, pl. Gronov. *Helianthus lævis*, L. Spec. ed. 2, 1278. excl. syn. Gronov., which is *Bidens chrysanthemoides*. *Helepta grandiflora*, &c., Raf. Neog. — Dry or moist ground, Canada to Florida.

H. scábra, DUNAL. Hispidulous-scabrous, especially the leaves, 2 to 4 feet high: leaves from broadly ovate and subcordate to ovate-lanceolate, the upper occasionally entire; rays oblong, nearly or quite an inch in length: akenes smooth, but the angles above pubescent when young, the summit usually bearing an obscure or evident and irregular coroniform chaffy pappus, or sometimes 2 or 3 conspicuous and rigid teeth! Otherwise as in the foregoing, into which it may pass. — Mem. Mus. l. c. 56, t. 4; Hook. Fl. i. 310; DC. Prodr. v. 550, excl. syn. *H. canescens*, Dou. *H. lævis*, var. *scabra*, Torr. & Gray, Fl. ii. 303. — W. New York to Saskatchewan, Missouri, and Arkansas.

H. grácilis, NUTT. Slender, a foot or two high: leaves ovate-lanceolate to lanceolate (2 or 3 inches long), and with somewhat cuneate base, hispidulous-scabrous or almost smooth: heads very much smaller, barely 3 or 4 lines high, and the fewer (5 to 10) rays 5 to 8 lines long: akenes of *H. scabra*. — Trans. Am. Phil. Soc. n. ser. vii. 353. *H. lævis*, var. *minor*, Hook. Comp. Bot. Mag. i. 98. *H. lævis*, var. *gracilis*, Torr. & Gray, Fl. ii. 303. — Dry and shaded ground, Georgia and W. Florida to Louisiana and Arkansas, *Gates & Jewett, Drummond, Curtiss*, &c.

H. parvifólia, GRAY. Slender, a foot or two high, from cinereous-puberulent and somewhat scabrous to nearly glabrous and green: leaves deltoid-lanceolate or rhomboid-lanceolate or approaching deltoid-ovate, irregularly dentate with few or several teeth, or some entire, an inch or two long: head barely half-inch high: rays oval or oblong, half or two-thirds inch long: akenes glabrous but dull and rugulose-scabrous, the summit evenly truncate. — Pl. Wright. ii. 86; Rothr. in Wheeler Rep. vi. 159. *H. buphthalmoides*, Gray in Bot. Mex. Bound. 88, not Dunal. — Cañons and beds of streams, Arizona, *Wright, Thurber*, &c. (Lower Cal. & adj. Mex.)

H. BUPHTHALMOIDES, Dunal, of Mexico and S. America, and H. ANNUA, Hemsley (which may be the *H. canescens*, HBK., that being said to have an annual root, as has Parry & Palmer's no. 431), have puberulent or papillose-pubescent akenes.

87. TETRAGONOTHÉCA, Dill. (Τετράγωνος, four-angled, θήκη, case, i. e. involucre.) — Erect perennial herbs, all N. American, with striate stems; the leaves all opposite, mostly sessile or connate-amplexicaul, thinnish, dentate or sinuate-pinnatifid; heads rather large, on slender peduncles terminating the stem or branches; both disk- and ray-corollas light yellow, nervose (the ligules 5 to 10-nerved, and the usual 5 nerves of the throat of the disk-corollas not rarely doubled), marcescent and persisting almost to the maturity of the akenes. Fl. summer. — Dill. Elth. ii. 378. t. 283; Linn.; Benth. & Hook. Gen. ii. 367. *Tetragonotheca* & *Halea*, Torr. & Gray, Fl. ii. 303, 304.

§ 1. Akenes very thick, obscurely 4-sided or almost terete, wholly destitute of pappus: tube of the corolla villous below: stem simple: involucre very saliently 4-angled in the bud.

T. heliánthoides, L. Villous with somewhat viscid hairs: stem a foot or two high: leaves ovate or rhomboid-oblong, closely sessile by a narrow base, dentate, 4 to 6 inches long: lobes of the involucre and 6 to 9 rays about an inch long. — Spec. ii. 903; DC. Prodr. v. 532; Torr. & Gray, l. c. *Polymnia Tetragonotheca*, L. Syst.; Abbot, Ins. & Pl. Georg. ii. t. 69; Schkuhr, Handb. t 263. *Silphium Tetragonotheca*, Gærtn. Fruct. t. 171. — Dry ground, Virginia to Florida.

§ 2. Akenes distinctly 4-sided and the sides striate (somewhat pubescent), moderately narrowed from the truncate summit to base: pappus plurisquamellate or sometimes wanting: stems more branching: involucre ovoid and less angled in the bud. — *Halea*, Torr. & Gray, l. c.

T. Texána, GRAY & ENGELM. Minutely pubescent or glabrate: stems slender, a foot or two high, sometimes freely branched: cauline leaves laciniately pinnatifid or incised, 2 or 3 inches long; the lower tapering into margined connate petioles; upper with winged petioles or bases dilated at insertion and usually connate around the stem into a toothed disk: peduncles elongated (4 to 9 inches long); lobes of the involucre and 7 to 9 rays half-inch long: tube of the corollas glandular: pappus none, or very minute, or sometimes of numerous subulate squamellæ of length nearly equalling the breadth of the akene. — Proc. Am. Acad. i. 48. *Halea Texana*, Gray, Pl. Fendl. 83, & Pl. Lindh. ii. 227. *Tetragonosperma lyratifolium*, Scheele in Linn. xxii. 167. — Rocky ground, Texas, *Lindheimer*, *Wright*, &c. (Adj. Mex., *Berlandier*, *Palmer*.)

T. Ludoviciána, GRAY. Glabrous or nearly so: stem rather stout, 2 to 4 feet high, usually leafy to the top: leaves ovate or oblong, ample (the larger 4 to 7 inches long), saliently and acutely dentate, the lowest on winged petioles, upper all connate by mostly broad bases into a large perfoliate disk: peduncles mostly longer than the leaves: corollas with tube somewhat pubescent: ligules 10 to 12, oval, less than half-inch long: akenes (over 2 lines long) crowned with a conspicuous pappus of rigid oval or oblong chaffy scales in length equalling the breadth of the truncate summit. — E. Hall, List Pl. Tex. 13, no. 328. *Halea Ludoviciana*, Torr. & Gray, l. c. — Sandy soil, Louisiana (*Hale*, *Leavenworth*) & Texas.

Var. **repánda**. Depauperate or dwarf form; flowering sometimes from near the ground; the leaves therefore petioled, and the upper with perfoliate disk of united bases of the petioles, nearly as in *T. Texana*: peduncles elongated as in that species; so that it is as it were intermediate between the two. — *Halea repanda*, Buckley in Proc. Acad. Philad. 1861, 458. — Texas, near Corpus Christi, *Buckley*. W. of San Antonio, *Palmer*; an autumnal state, flowering as seedlings.

88. **SCLEROCÁRPUS**, Jacq. (Σκληρός, hard, καρπός, fruit, referring to the indurating enclosing bracts.) — Strigose-pubescent herbs (the original species African, the others mostly Mexican); with branching stems, terminal pedunculate heads of yellow flowers, and alternate or opposite leaves: fl. summer. — Act. Helv. ix. 34, t. 2, & Ic. Rar. t. 176; Benth. & Hook. Gen. ii. 364. *Aldama*, Llav. & Lex. Nov. Gen. Descr. i. 14. *Gymnopsis*, DC. Prodr. v. 561, in part.

S. uniseriális, BENTH. & HOOK. l. c. Annual, a foot or two high, loosely branched: leaves all alternate, slender-petioled, deltoid- or rhombic-ovate, or uppermost lanceolate, coarsely dentate, the strigose pubescence of the lower face canescent: loose involucral bracts nearly in a single series; corollas orange; ligules 5 to 9, oval or oblong: fructiferous bracts cartilaginous or bony, terete, roughish, in age often tuberculate. — *Gymnopsis uniserialis*, Hook. Ic. t. 145; Torr. & Gray, Fl. ii. 316; Revue Hort. 1853, t. 14; Belg. Horticole, 1854, t. 20. *Aldama uniserialis*, Gray, Pl. Lindh. ii. 228. — Moist or shady ground, Texas, *Berlandier*, *Drummond*, &c. (Mex.)

89. **ECLÍPTA**, L. (Name from ἐκλείπω, to be deficient, i. e. in pappus.) — Insignificant herbs, of the warmer regions, chiefly of shores; with opposite leaves, and scattered small heads of whitish or yellowish flowers; in summer. — Mant. Alt. 157; DC. Prodr. v. 489.

E. álba, HASSKARL. Annual, 1 to 3 feet high, or often procumbent and smaller, minutely strigose-pubescent: leaves lanceolate or oblong, sparingly serrate, sessile or the lower somewhat petioled: peduncles from the upper axils, sometimes equalling the leaves, sometimes shorter than the heads: ligules not surpassing the disk, white: akenes of the disk at length corky-margined, truncate at summit or 4-denticulate when young. — Pl. Jav. Rar. 528. *E. erecta* & *E. prostrata*, L. Mant. Alt. 286. *E. procumbens* & *E. brachypoda*, Michx. Fl. ii. 129. *E.* species 1–8, DC. l. c. *Cotula alba*, L. Syst., & *Verbesina alba*, L. Spec. *Eupatoriophalacron*, Dill. Elth. t. 113. *Amellus Carolinianus*, Walt. Car. 313. — Shores and river-banks, New Jersey to Texas. (All subtropical countries.)

90. **MELANTHÉRA**, Rohr. (Μέλας, black, and ἀνθηρά, used for anther.) — Scabrous herbs (chiefly tropical American); with quadrangular branching stems, opposite and sometimes lobed petioled leaves, and pedunculate heads: corolla white and anthers blackish in the genuine (rayless) species. Fl. summer. — Rohr, Skriv. Nat. Selsk. Kiob. 1792, ii. 213; DC. Prodr. v. 544. *Melananthera*, Michx. Fl. ii. 106.

M. hastáta, Michx. l. c. Stem 3 to 6 feet high from a perennial root, spotted: leaves from ovate to ovate-lanceolate, or uppermost lanceolate, some of them commonly and variously hastately 3-lobed, unequally serrate: bracts of the involucre broadly lanceolate, of the receptacle spinescently acuminate: heads in fruit half-inch in diameter. — DC. Prodr. v. 545. *M. trilobata, panduræformis*, &c., Cass. Dict. xxix. 485. *Bidens nivea*, L. Spec. ii. 833 (Dill. Elth. t. 46, 47). *Athanasia hastata*, Walt. Car. 201. — Moist ground, near the coast, S. Carolina to Louisiana. (Mex., W. Ind., &c.)

M. deltoídea, Michx. l. c. Leaves ovate to deltoid or obscurely hastate: heads smaller: bracts of the involucre ovate, of the receptacle only mucronate. — DC. l. c. *M. urticæfolia*, Cass. l. c. *M. Linnæi*, HBK. *Bidens nivea*, L. l. c. as to Dill. Elth. t. 47, f. 3. *Calea aspera*, Jacq. lc. Rar. t. 583. — S. Florida. (W. Ind. to S. Am.)

M. lanceoláta, Benth. A foot or two high: leaves lanceolate (1 to 3 inches long, 2 to 5 lines wide), somewhat serrate: bracts of the involucre oblong-ovate, of the receptacle cuspidately mucronate, short: disk about 4 lines in diameter. — Vidensk. Medel. 1852-3, 88. *M. microphylla*, Steetz in Seem. Bot. Herald, 156 (same year?). *M. angustifolia*, A. Rich. ex Griseb. Cat. Cub. 154. — S. Florida, *Garber*, &c. (W. Ind., Centr. Am.)

91. **VARÍLLA**, Gray. (Native Mexican name of this and some similar plants.) — Shrubby or suffrutescent, glabrous; with linear and narrow entire and sessile thickish or fleshy leaves, and pedunculate rather small heads, either corymbosely cymose or solitary; the flowers yellow. — Pl. Fendl. 106, & Pl. Wright. i. 123. — Two known species.

V. Mexicána, Gray, l. c. Shrub about 5 feet high, much branched: branches very leafy, terminated by a cyme of numerous short-peduncled heads: leaves not succulent, linear (1 to 3 inches long, at most 2 lines wide), attenuate to both ends, opposite: involucre somewhat turbinate, 2 lines long, half the length of the rather narrow head: pappus of 5 to 10 or 15 slender short bristles (which commonly bear 3 or 4 salient setulose denticulations), somewhat irregular, in length fully equal to the diameter of the akene. — Coahuila, near Parras, *Gregg, Wislizenus, Palmer*, &c., not yet found within U. S. (Mex.)

V. Texána, Gray. Low, suffrutescent, much branched and very leafy at base: leaves very succulent, terete, mostly alternate, obtuse: head larger, solitary on a long terminal and minutely bracteate peduncle: involucre not turbinate, very much shorter than the broadly ovoid conical disk: pappus none. — Pl. Wright. i. 103. — Saline soil, from the Nueces to the Rio Grande, S. Texas, *Wright, Trécul, Bigelow, Palmer*. (Adj. Mex.)

92. **ISOCÁRPHA**, R. Br. (From ἴσος, equal, κάρφος, chaff, the chaffy bracts of the receptacle and of the involucre similar.) — Tropical American herbs; with small heads of white or whitish flowers, either solitary or glomerate at the summit of a naked peduncle. — Trans. Linn. Soc. xii. 110; Benth. & Hook. Gen. ii. 365. *Dunantia*, DC. Prodr. v. 626.

I. oppositifólia, R. Br. l. c. Pubescent: stems slender, 1 to 3 feet high from a perennial (?) root, paniculately branched: leaves opposite, lanceolate, narrowed to both ends, triplinerved, entire or sparingly denticulate: heads commonly in threes, in fruit 4 or 5 lines long, narrow, with turbinate involucre: bracts of the involucre and receptacle pointed, becoming rigid and the receptacle columnar. — *Calea oppositifolia*, L. *Dunantia Achyranthes*, DC. Prodr. v. 672; Deless. Ic. Sel. iv. t. 37. — S. borders of Texas on the Rio Grande, *Schott*. (Adj. Mex., W. Ind.)

17

93. **SPILÁNTHES**, Jacq. (Σπῖλος, a spot or stain, ἄνθος, flower; name ordinarily without application.) — Usually spreading or creeping herbs (mainly tropical); with opposite and merely serrate leaves, rather small heads on peduncles terminating the stem and branches, the rays when present yellow or white, the disk-flowers yellow: herbage of some species acrid to the taste. Fl. summer. — Jacq. Amer. t. 214, Hort. Vind. t. 135, & Ic. Rar. t. 584; Schreb. Gen. 1266; DC. Prodr. v. 620. *Spilanthus*, L. Mant. 475; Gærtn. Fruct. ii. t. 167. — Our species is of the section *Acmella*, DC. (*Acmella*, Pers. Syn. ii. 472), having evident ligules.

S. répens, Michx. Perennial by the creeping base, slender, spreading or ascending, from hirsute-pubescent to almost glabrous: stems slender, a foot or two long: leaves from lanceolate to oblong-ovate, an inch or two long, from sparsely denticulate to serrate, abruptly or sometimes gradually contracted at base into a petiole: peduncles 2 to 4 inches long: bracts of the involucre oblong-lanceolate, mostly obtuse: rays 8 to 12, yellow, rather shorter than the obtusely ovoid disk· receptacle at length subulate-conical: akenes oblong, less than a line long. not flat, most of them tuberculate-roughened in age and minutely hispidulous, the margins not more so than the sides: pappus none or occasionally one or two minute awns. — Fl. ii. 131; DC. Prodr. v. 623. *S. repens* & *S. Nuttallii*, Torr. & Gray, Fl. ii. 356. *Anthemis repens*, Walt. Car. 211; Pursh, Fl. ii. 562. *Acmella repens*, Pers. Syn. l. c. *A. repens* & *A. occidentalis*, Nutt. Gen. ii. 171. — Low or wet ground, S. Carolina to Florida, Arkansas, and Texas.

94. **ECHINÁCEA**, Mœnch. ('Εχῖνος, hedgehog, or sea-urchin, in allusion to the spinescent bracts of the receptacle.) — Atlantic N. American perennial herbs; with thick and black roots of pungent taste (used in popular medicine under the name of *Black Sampson*), rather stout erect stems, undivided somewhat nervose leaves, the lower long-petioled, and solitary large heads on long peduncles terminating the stem and few branches; in summer. Rays from flesh-color to rose-purple or crimson, much elongating with age: disk purplish. — Meth. 591; Cass. Dict. xxxv., xlvii., &c.; DC. Prodr. v. 554, excl. sp. Mex. *Brauneria*, Necker. *Heliochroa*, Raf. Neog. 1825, no. 35, &c.

E. purpúrea, Mœnch. Commonly smooth and glabrous, or the leaves hispidulous and rough, sometimes the stem also hispid, 2 feet or more high: leaves ovate-lanceolate or the lower ovate from a broad base, commonly denticulate or acutely serrate, most of them abruptly contracted into a margined petiole, some of the middle occasionally opposite; lower often 3-5-plinerved involucre well imbricated: ligules (rarely almost white), at first an inch long and broadish, in age often elongated to 2 inches or more. — Torr. & Gray, Fl. ii. 305, with varieties. *E. purpurea* & *E. serotina*, DC. Prodr. v. 554. *Rudbeckia purpurea*, L. Spec. ii. 907 (Catesb. Car. t. 59; Pluk. Alm. t. 21, &c.); Bot. Mag. t. 2; Schkuhr, Handb. t. 259; Bart. Fl. Am. Sept ii. t. 64. *R. serotina*, Sweet, Brit. Fl. Gard. t. 4, & Lodd. Cab. t. 1539 (*R. purpurea*, var. *serotina*, Nutt. Gen. ii. 178), the hirsute or hispid form, which is *R. hispida*, Hoffm., and *R. speciosa*, Link. Enum., ex DC. *Heliochroa Linnæana*, *elatior*, *amœna*, *furcata*, &c., Raf. Neog. l. c. — Rich or deep soil, Virginia and Ohio to Illinois and Louisiana.

E. angustifólia, DC. Hispid, either sparsely or densely, a foot or two high, mostly simple: leaves from broadly lanceolate to nearly linear, entire, 3-nerved, all attenuate at base, the lower into slender petioles; bracts of the involucre in only about 2 series: heads and flowers nearly of the preceding (the fruiting disk often an inch high), or sometimes very much smaller. — Prodr. l. c.; Torr. & Gray, Fl. ii. 306; Hook. Bot. Mag. t. 5281; Sprague, Wild Flowers of Amer. t. 25. *E. pallida* & *E. sanguinea*, Nutt. in Trans. Am. Phil. Soc. n. ser. vii. 354. *Rudbeckia pallida*, Nutt. in Jour. Acad. Philad. vii. 77. — Prairies and barrens, Saskatchewan and Nebraska to Texas, and east to Illinois, Tennessee, and Alabama; in several forms; some too near the preceding.

95. RUDBÉCKIA, L. CONEFLOWER. (The two Professors *Rudbeck*, father and son, predecessors of Linnaeus at Upsal.) — N. American herbs, chiefly perennial; with alternate leaves, either simple or compound, and commonly showy pedunculate heads terminating stem and branches; the rays yellow, rarely with brown-purple base, in one species wholly crimson, the disk from fuscous to purplish black. Fl. summer. — Gaertn. Fr. t. 172. *Rudbeckia* & *Dracopis*, Cass. l. c.; DC. l. c.; Torr. & Gray, Fl. ii. 307, 316.

§ 1. EURUDBÉCKIA. Akenes prismatic-quadrangular, when laterally compressed yet with a salient angle or rib on the lateral faces: bracts persisting on the receptacle. — *Rudbeckia*, Cass., &c.

* Disk from hemispherical to globose or oblong-ovoid, dark-purple (at least the corollas) or brown: akenes (not rarely becoming somewhat curved) inserted by a central or slightly oblique basal areola.

+ Leaves elongated-linear, as it were gramineous, but rigid, nervose, shining, entire: chaffy bracts of the receptacle firm or rigid, carinate-concave, commonly mucronate from the thickish obtuse summit, rather shorter than the subtended flowers: style-tips conical-capitate: disk dark brown, globular, becoming ovoid in fruit: stems rush-like and striate, 2 feet or more high from a perennial root, bearing solitary rather small heads on long naked peduncles: rays in one species dark crimson!

R. atrórubens, NUTT. Either glabrous or sparsely and minutely strigulose: stems rigid, nearly simple, few-leaved: leaves rather obtuse, often purplish; radical and lowest cauline often a foot long, a quarter to half an inch wide: involucre a few small subulate-linear bracts: rays 9 or more, oblong, half-inch long, dark crimson; fructiferous disk two thirds of an inch long, its receptacle fusiform-conical; its chaffy bracts thick and firm, oblong, tipped with a short rigid mucro: akenes equably quadrangular, straight and with centrally basal insertion, a line and a half long, inclusive of the short cupulate and obscurely 4-toothed pappus. — Jour. Acad. Philad. vii. 80. *Echinacea atrorubens*, Nutt. Trans. Am. Phil. Soc. l. c. 354; Torr. & Gray, Fl. ii. 306 (with var. *graminifolia*); Chapm. Fl. 226. — Borders of pine-barren ponds, Georgia and Florida, in the low country (also Arkansas, according to *Nuttall*), *Wray*, *Chapman*, *Mohr*, &c.

R. bupleuroídes, SHUTTL. Perfectly glabrous and smooth, divergently branching: leaves pale green, attenuate-acute; the larger 7 or 8 inches long, 2 or 3 lines wide: heads smaller; disk even when fructiferous hemispherical or globular: rays bright sulphur-yellow, over half-inch long: chaffy bracts of the receptacle less rigid, obtuse with obscure or blunt mucro: akenes somewhat curved and with rather oblique insertion, 2 lines long, inclusive of the deep cupulate and irregularly dentate pappus. — Coll. Rugel distrib. by Shuttleworth; Chapm. Fl. Suppl. 629. *R. Mohrii*, Gray, Proc. Am. Acad. xvii. 217. — W. Florida. Wet pine barrens near St. Marks, *Rugel*, 1843. Margin of the Dead Lakes, near Iola, *C. Mohr*. — Makes approach to *R. nitida*, var. *longifolia*.

+ + Leaves broad, various in form, thinnish, veiny: chaffy bracts of the receptacle merely concave, thinnish, not rigid, acuminate into a slender almost awn-like cusp, about equalling the flowers; the whole disk black-purple: style-tips conical-capitate: root biennial.

R. tríloba, L. Bright green, sparsely hirsute or hispidulous, or the freely branching stem glabrous and smooth, 2 to 5 feet high: radical leaves commonly cordate, slender-petioled; cauline ovate-lanceolate or broader, with cuneate subsessile base, coarsely serrate, acuminate, or the upper lanceolate and nearly entire, the lower divergently 3-lobed or 3-parted: heads short-peduncled: involucre foliaceous, soon reflexed; its bracts linear or mostly so, unequal, nearly in a single series: rays 8 to 10, half-inch to inch long, deep yellow, sometimes particolored, the basal portion orange or even brown-purple: disk depressed-globular, becoming ovoid at maturity (about half-inch in diameter), glabrous, the upper part of the chaffy bracts and the flowers dark purple: akenes equably quadrangular: pappus a minute crown or border. — Spec. ii. 907 (pl. Gronov., Pluk., &c.); Michx. Fl. ii. 144 (excl. var.); Bot. Reg. t. 525; Bart. Fl. Am. Sept. i. t. 24; Torr. & Gray, l. c. *R. triloba, subtomentosa* (as to herb. & pl. Virg.), & *aristato*, Pursh, Fl. 575. *Peramibus hirtus*, Raf. Ann. Nat. 14. *Centrocarpha triloba* (at least as to "paleis acuminato-aristatis," though the rest of the character refers to

R. subtomentosa) & *C. aristata*, Don in Sweet, Brit. Fl. Gard. ser. 2, under t. 87. — Dry or moist ground, Penn. and Michigan to Illinois, and south to Georgia and Louisiana, but mostly affecting the mountains.

Var. rupéstris. Large; cauline leaves often 4 or 5 inches long: rays 9 to 13, an inch to inch and a half long, pure orange-yellow to the base: in habit approaching *R. subtomentosa*. — *R. rupestris*, Chickering in Bot. Gazette, vi. 188. — Rocky slopes of the Roan and other mountains on the borders of N. Carolina and Tenn., *Chickering*, &c.

Var. pinnatíloba, TORR. & GRAY, l. c. A peculiar form, slender: leaves small; many of the radical and lowest cauline pinnately 5-7-parted; upper ones seldom inch long: heads small, with rays at most half-inch and disk a quarter-inch long. — W. Florida, *Chapman*.

+– +– +– Leaves from lanceolate to ovate or broader: chaffy bracts of the receptacle pointless (obtuse or rarely acute), linear, concave or carinate-canaliculate, somewhat shorter than the disk-flowers: akenes nearly equably quadrangular, or in a few species moderately compressed: involucre foliaceous and variable, soon reflexed: disk very obtuse.

++ Cauline leaves or some of them 3-cleft or parted: disk of the head dull brownish: rays yellow, sometimes with dark base: root perennial: receptacle anisate-scented.

R. subtomentósa, PURSH. Cinereous with short and mostly soft pubescence, 2 to 5 feet high, branching above, leafy; leaves nearly all petioled, acutely serrate, veiny, ovate, or the terminal lobe ovate and the lateral oblong or lanceolate: peduncles not much elongated: rays numerous, becoming inch and a half long: disk hemispherical, becoming higher, half-inch broad; its bracts cinereous-puberulent and somewhat glandular at the obtuse tips: pappus a short crenately toothed crown. — Fl. ii. 575; Torr. & Gray, l. c. *R. triloba*, var., Michx. Fl. ii. 144. *R. odorata*, Nutt. Jour. Acad. Philad. vii. 78. *R. tomentosa*, Ell. Sk. ii. 453, as to syn. & char. *Centrocarpha triloba*, Don in Sweet, l. c., as to syn. and part of the char. — Prairies and open moist grounds, Illinois to Arkansas and Texas.

++ ++ Leaves undivided (rarely laciniate-dentate): stems more simple.

= Style-tips slender-subulate: bracts of the receptacle hispid or hirsute at and near the acutish summit: akenes small, equably quadrangular, wholly destitute of pappus: annuals or biennials, hispid with spreading bristly hairs.

R. bícolor, NUTT. A foot or two high from an annual root, simple or branching, slender or not very stout: leaves from lanceolate to oblong or the lower obovate, mostly obtuse and nearly entire, an inch or two long, indistinctly triplinerved, nearly all sessile: peduncles rarely elongated: rays half-inch to barely inch long, either pure yellow, or with brown purple spots at base, or the lower half deep blackish-purple: disk black. — Jour. Acad. Philad. vii. 81; Torr. & Gray, l. c. — Pine woods or sandy soil, Arkansas, Texas, and sparingly E. to Georgia. Often confounded with small forms of the next, and with *R. fulgida*. (Adj. Mex.)

R. hírta, L. Stouter and larger, 1 to 3 feet high from a biennial or sometimes annual root, rough-hispid and hirsute: leaves from oblong to lanceolate, sparingly serrate or nearly entire, slightly triplinerved, 2 to 5 inches long, the lower narrowed into margined petioles: rays when well developed an inch or two long, golden yellow, sometimes deeper colored toward the base: disk at first nearly black, in age dull brown, becoming ovoid in fruit. — Spec. ii. 907 (Dill. Elth. t. 218); Michx. Fl. ii. 143, mainly; Sweet, Brit. Fl. Gard. t. 82; Torr. & Gray, l. c., chiefly. *R. gracilis* (Herb. Banks.?), Nutt. Gen. ii. 178? a depauperate form. *R. discolor*, Ell.? not Pursh. *R. serotina*, Nutt. Jour. Acad. Philad. vii. 80, at least the cult. plant described, fide Herb. Acad. Philad. *R. strigosa*, Nutt. Trans. Am. Phil Soc. vii. 354, a hairy and short-rayed form. — Dry and open ground, Saskatchewan and W. Canada to Florida, Texas, and Colorado: naturalized in grass-fields in Eastern States: flowering early as a biennial.

== Style-tips short and thickened, obtuse (in *R. mollis* narrower and sometimes acutish): pappus more or less manifest: perennials.

a. Chaffy bracts of the receptacle obtuse and glabrous or nearly so, with blackish-purple tips of the same hue as the corollas, so that the hemispherical at length globose-ovoid disk is deep black-purple: rays golden yellow, not rarely orange toward the base: akenes small, equably quadrangular: pappus a very short commonly 4-toothed crown.

R. fúlgida, AIT. Hispid or hirsute, a foot or two high: leaves from narrowly to oblong-lanceolate, mostly entire, lowest and radical spatulate-lanceolate and tapering into slender petioles: foliaceous bracts of the involucre often ample and equalling or sometimes half the length of the 12 to 14 fully inch-long rays: disk over half-inch in diameter. — Ait. Kew. iii. 251; Bot. Mag. t. 1996; Bart. Fl. Am. Sept. i. t. 54, & iii. t. 98 (both figures doubtful); Torr. & Gray, l. c., partly. *R. chrysomela*, Michx. Fl. ii. 143. *R. discolor*, Pursh, Fl. ii. 574, DC. l. c., hardly of Elliott. — Dry soil, Pennsylvania ? and Virginia to Louisiana and Texas, west to Missouri; flowering rather late.

R. spathuláta, Michx. Strigulose: stem slender, 8 inches to 3 feet high: leaves obovate or spatulate, or the uppermost lanceolate, denticulate or sparingly serrate, their pubescence wholly appressed and short; radical and lowest cauline leaves mostly roundish at summit, at base abruptly contracted into a winged petiole, or even subcordate: peduncle usually elongated: involucre commonly shorter and rays fewer and broader than in the preceding, and disk smaller. — Fl. ii. 144; Nutt. Gen. ii. 178. *R. Heliopsidis*, A. H. Curtiss, coll. no. 1427, not Torr. & Gray. *R. fulgida*, Torr. & Gray, l. c., var. γ, & β in part. — Pine woods, Virginia to Tennessee and Florida.

R. speciósa, Wenderoth. Sparsely strigulose or hispid, or glabrate: stem 1 to 3 feet high, usually with spreading branches terminating in long naked peduncles: leaves ovate-lanceolate or the upper elongated-lanceolate, bright green, irregularly serrate or some laciniately dentate, acute or acuminate; radical and lower cauline oblong or ovate, 3-5-nerved, abruptly contracted into long margined petioles: rays 12 to 20, elongated, at length inch and a half long: disk two-thirds to three-fourths inch high at maturity, the tips of the purple chaffy bracts sparingly or obscurely ciliate: akenes larger and longer than in the related species (line and a half long), more curved. — Ind. Sem. Hort. Marb. 1828, & in Flora, 1829, i. Suppl. 30; Schrad. in DC. l. c.; Torr. & Gray, l. c.; Gard. Chron. 1881, ii. 372, fig. 72. Probably *R. aspera*, Pers. Syn. ii. 477. *R. fulgida*, Meehan, Nat. Flowers, ser. 2, i. t. 14. — Moist ground, Penn. to Michigan, Arkansas, and upper part of Alabama. Long cultivated in gardens as *R. fulgida*, &c.

b. Chaffy bracts of the receptacle with the obtuse tips canescently puberulent or pubescent, and the flowers duller purple; the disk therefore browner.

1. Cauline leaves all closely sessile or partly clasping, not nervose: bristly style-tips little thickened: akenes small: pappus very short or obsolete.

R. móllis, Ell. Cinereous, the leaves with fine and close pubescence, the (2 or 3 feet high and usually branching) stem with hirsute or villous hairs, leafy: leaves spatulate-oblong, obtuse, obscurely serrate, somewhat triplinerved (1 to 3 inches long): rays 12 to 20, at length inch and a half long and disk fully half-inch high. — Sk. ii. 453; Torr. & Gray, l. c. *R. spathulata*, Pursh, Fl. ii. 574. — Dry soil, Georgia and Florida.

2. Cauline leaves mostly petioled: heads small: quadrangular akenes only a line long: pappus an obscure crown or hardly any.

R. Heliópsidis, Torr. & Gray. Almost glabrous, 2 feet high, rather slender, branched above: leaves oblong-ovate, somewhat serrate, triplinerved and with a pair of nearly basal nerves, abruptly contracted, the upper into short and wing-margined, the lower into long and naked petioles: peduncles rather short and corymbose: involucre much shorter than the at length globular disk (which is hardly half-inch high): rays light yellow, 10 or 12. an inch or less long. — Fl. ii. 310. — Pine woods, Columbus, Georgia, *Boykin*. Cherokee Co. and Lee Co., Alabama, *Buckley, J. Donnell Smith*.

3. Cauline leaves mostly petioled and like the radical 3-5-nerved; the veinlets reticulated: heads large and showy: the soon drooping light yellow rays 1 or 2 inches long, and the hemispherical at length somewhat conical receptacle becoming three fourths of an inch high: involucre rather small: akenes somewhat compressed: pappus a conspicuous cup-shaped irregularly dentate or crenate crown: stem 2 or 3 feet high, usually simple, and head long-peduncled.

R. alismæfólia, Torr. & Gray. Glabrous or minutely scabrous: leaves oval, obtuse or sometimes acute, obscurely repand-dentate or entire, 3 to 6 inches long, abruptly contracted into the petiole: rays 10 to 15. — Fl. ii. 310. — Plains and open pine woods, S. Arkansas, W. Louisiana, and adjacent Texas, *Leavenworth, Hale, Drummond.*

R. grandiflóra, C. C. Gmelin. Hispidulous and scabrous throughout: leaves more rigid, ovate to oval-lanceolate or uppermost lanceolate, commonly acute or acuminate at both ends, sparingly serrate or denticulate, 4 to 9 inches long: rays 20 or more. — Hort. Bad. Carlsr. 1811; DC. l. c. 556 (with some erroneous characters as to chaff and pappus, taken from a plant of *R. hirta*); Torr. & Gray, l. c. *Centrocarpha grandiflora*, Don in Sweet. Brit. Fl.

Gard. ser. 2, t. 87, but has not the character of his genus, which was founded on *R. triloba*. — Dry plains, Arkansas and W. Louisiana.

* * Disk from globular to cylindrical, greenish, fuscous, or yellowish; its chaffy bracts navicular or more conduplicate, truncate or obtuse, little surpassing the mature akenes, sometimes deciduous from the receptacle at full maturity: style-branches with short and truncate-capitate or obtuse tips: akenes comparatively large and somewhat compressed, inserted by a more or less oblique or lateral areola, the more lateral when the receptacle is elongated: root in all perennial.

+ Rays several or numerous, an inch or two long, drooping, pure yellow: bracts of receptacle pubescent at summit.

++ Leaves entire or barely dentate: disk when well developed at length columnar, an inch or two long, three-fourths inch thick; the receptacle bodkin-shaped: akenes about 3 lines long: pappus a conspicuous irregularly toothed or denticulate cup: herbage completely glabrous and smooth, or sometimes slightly scabrous in age: stems simple or nearly so, and the long-pedunculed heads solitary or few: involucre comparatively small. — § *Macrocline*, Torr. & Gray, Fl. ii. 312.

R. nítida, NUTT. Stem 2 to 4 feet high: leaves bright green, commonly lucid, thin-coriaceous, nervose-ribbed, mostly acute, denticulate or entire; radical and lower cauline ovate-spatulate to lanceolate-oblong, tapering into long margined petioles, upper cauline sessile, oblong to lanceolate, 3 to 6 inches long. — Jour. Acad. Philad. vii. 78; Torr. & Gray, Fl. ii. 313. *R. lævigata?* Nutt. Gen. 178, not Pursh. — Wet ground, lower part of Georgia to Florida and Texas; first coll. by *Nuttall*.

Var. longifólia. Leaves elongated-lanceolate or broader, attenuate to both ends, sparingly dentate or repand-denticulate, more nervose-veiny, in age sometimes minutely scabrous; radical and lowest cauline 8 or 10 inches long, an inch or more broad in the middle. — *R. glabra*, DC. Prodr. v. 556. — Near Savannah, Georgia, according to herb. *DC.* Tuskegee, Alabama, *Beaumont.* Manatee, Florida, *Garber.*

R. máxima, NUTT. Stem 4 to 9 feet high, and whole plant smooth and glaucous: leaves from broadly ovate to oblong, mostly obtuse, repand-denticulate or entire, with numerous pinnate veins, the larger a foot or less long; upper cauline subcordate-clasping. — Trans. Am. Phil. Soc. vii. 354; Torr. & Gray, l. c. — Moist pine woods and plains, Arkansas, Louisiana, Texas; first coll. by *Nuttall.*

++ ++ Leaves more or less dentate, sometimes 2-lobed at base: pappus a conspicuous crown deeply cleft into four irregular chaffy lobes: Pacific species!

R. Califórnica, GRAY. Pubescent, slightly scabrous: stem simple, 2 to 4 feet high, bearing a solitary long-pedunculed head: leaves from ovate to oblong-lanceolate, the upper sessile by a narrow base: rays from half-inch to 2½ inches long, surpassing the loose linear bracts of the involucre: disk from short-oblong to cylindraceous (becoming sometimes 2 inches long); its bracts canescent at summit: akenes flattish. — Proc. Am. Acad. vii. 357, & Bot. Calif. i. 347. — California; moist ground in the Sierra Nevada; first coll. by *Bridges.*

++ ++ ++ All or most of the cauline leaves 3-7-cleft or divided: pappus a short 4-toothed or nearly entire crown: disk from globular or even hemispherical to oblong-cylindraceous in age, dull yellowish; the tip of the chaffy bracts canescent.

R. laciniáta, L. Glabrous and smooth, sometimes minutely hispidulous-scabrous, at least on the margins and upper face of the leaves: stem 2 to 7 feet high, branching above; leaves veiny, broad, incisely and sparsely serrate; radical commonly pinnately 5-7-foliolate or nearly so, and divisions often laciniately 2-3-cleft; lower cauline 3-5-parted, upper 3-cleft, and those of the branches few-toothed or entire: involucre loose and irregular, foliaceous: rays soon drooping, few or several, oblanceolate. — Spec. ii. 906 (Cornuti, Canad. t. 179, &c.); Michx. Fl. ii. 144; Bart. Fl. Am. Sept. i. t. 16; Torr. & Gray, l. c. *R. laciniata, quinata, & digitata*, Mill. Dict. ed. 6. *R. laciniata & digitata*, Ait. Kew. iii. 251; DC. l. c. — Moist ground, commonly in thickets, Canada to Florida, and westwardly from Montana to New Mexico and Arizona. A variable species, of which an extreme form is

Var. húmilis. A foot or two high, simple or branching, commonly slender, glabrous: radical leaves diverse, some of them undivided or with roundish divisions: heads smaller; the rays seldom inch long and globular disk barely half-inch high. — Probably *R. lævigata*, Pursh, l. c. — Alleghany Mountains from Virginia to Georgia and Tennessee, common in open woods, &c., at 4,000 to 6,000 feet.

R. heterophýlla, Torr. & Gray. Cinereous-pubescent: stem 2 to 4 feet high, slender, bearing several somewhat corymbose short-peduncled small heads: leaves coarsely and rather obtusely serrate; some of the radical cordate-obicular and undivided, others with 3 ovate undivided leaflets, the terminal petiolulate, lower cauline 3–5-parted; upper all ovate, coarsely toothed, nearly sessile: rays an inch or less long: disk in fruit globose and barely half-inch high. — Fl. ii. 312; Chapm. Fl. 228. — Swamps, Middle Florida, *Chapman*.

+ + Rays wholly wanting: proper tube of disk-corollas very short: disk brownish, from ovoid to columnar; its chaffy bracts puberulent at tip: receptacle bodkin-shaped: akenes rather large: scarious cupulate-coroniform pappus very conspicuous: stem stout, nearly simple, 2 or 3 feet high: involucre foliaceous, variable. — § *Acosmia*, Nutt.

R. occidentális, Nutt. Nearly glabrous and smooth, or somewhat scabrous-puberulent: leaves undivided, ovate or ovate-lanceolate, acuminate, entire or irregularly and sparingly dentate (4 to 8 inches long); upper sessile by a rounded or subcordate base; lower abruptly contracted into a short winged petiole, rarely a pair of obscure lateral lobes: disk in age becoming inch and a half long, and akenes 2 lines long. — Trans. Am. Phil. Soc. vii. 355; Torr. & Gray, l. c. — Woods along streams, Rocky Mountains of Wyoming to Idaho and Oregon: first coll. by *Nuttall*. Sierra Nevada in Butte Co., California (*Bidwell*), &c.

R. montána, Gray. Smoother, somewhat glaucous, tall and very stout: leaves (8 to 12 inches long) pinnately parted into 3 to 9 oblong-lanceolate divisions, or the lanceolate uppermost cauline with 2 to 4 narrow lateral lobes: disk cylindraceous or cylindrical, at length often 3 inches long and an inch in diameter: akenes with the deep coroniform pappus 3 or 4 lines long. — Proc. Am. Acad. xvii. 217. — Rocky Mountains of Colorado, *E. Hall, Brandegee*, the latter in the Elk Mountains.

§ 2. **Dracópis.** Akenes nearly terete, not angled, minutely striate, destitute of pappus, inserted by an obliquely lateral areola, and subtended by navicular bracts, which are more or less deciduous in age. — *Dracopis*, Cass., DC., &c.

R. amplexicaúlis, Vahl. A foot or two high from an annual root, smooth and glabrous, somewhat glaucous, leafy; the branches terminated by solitary rather showy heads: leaves strictly one-ribbed, reticulate-veiny, from entire to sparingly serrate: lower oblong-spatulate and sessile by a tapering base; upper oblong and ovate with cordate-clasping base. involucre of a few small foliaceous bracts: rays oblong, half-inch or more long, yellow, often with a brown-purple base. disk brownish, cylindraceous in age: receptacle slender: akenes small, minutely rugulose-roughened transversely between the sulcate striæ. — Act. Hafn. ii. 29, t. 4 (1793); Schkuhr, Handb. t. 259; Pursh, Fl. ii. 573. *R. amplexifolia*, Jacq. lc. Rar iii. t. 592 (1793). *R. perfoliata*, Cav. Ic. t. 252. *R. spathulata*, Nutt. Gen. ii. 178 (excl. hab.), not Michx. *Dracopis amplexicaulis*, Cass. Dict. xxxv. 273; DC. Prodr. v. 558; Torr. & Gray, Fl. ii. 316. — Low grounds, Louisiana and Texas. (Adj. Mex.)

96. **LÉPACHYS,** Raf. (λεπίς, a scale, and παχύς, thick, the upper part of the bracts of the receptacle thickened.) — Herbs (Atlantic N. American); with pinnately divided or parted alternate leaves, and terminal long-peduncled showy heads. the drooping rays mostly broad, yellow or partly brown-purple; the disk at first grayish. the truncate inflexed tips of the chaff canescently pubescent; disk-corollas yellowish turning fuscous. Heads redolent with anisate odor when bruised. Chaffy bracts commonly marked with an intra-marginal purple line or spot, containing volatile oil or resin. Fl. summer. — Less. Syn. 225; Torr. & Gray, Fl. ii. 313. *Lepachys* & *Ratibida*, Raf. in Jour. Phys. 1819, 100. *Obeliscaria*, Cass. Dict. xlvi. 401 (1825); DC. Prodr. v. 558.

§ 1. Akenes with convex or obscurely angled faces: root perennial. — *Obeliscaria*, Cass.

* Style-tips lanceolate-subulate: rays large and long.

L. pinnáta, Torr. & Gray, l. c. Strigulose-pubescent and scabrous, 3 to 5 feet high, slender: leaves 3–7-foliolate, and the leaflets lanceolate or broader, usually sparsely serrate, sometimes lobed, the uppermost commonly confluent: rays pure yellow, oblong-lanceolate, often 2 inches long or more, very much exceeding the at length short-oblong disk: chaffy bracts of the receptacle becoming much corky-thickened at the enlarging summit: ovary not rarely wing-margined; akenes subcuneate-oblong, the inner margin acute and salient, and produced at summit into a short rounded tooth, which is occasionally aristellate-pointed. — *L. pinnatifida* & *L. angustifolia*, Raf. l. c. *Rudbeckia pinnata*, Vent. Cels. t. 71; Smith, Exot. Bot. i. t. 38; Bot. Mag. t. 2310. *R. digitata*, Willd. Spec. iii. 2247, excl. syn. *R. tomentosa*, Ell. Sk. ii. 453, as to herb., hardly of char. *Obeliscaria pinnata*, Cass. l. c.; DC. l. c. — Dry prairies, W. New York to Michigan and Iowa, south to W. Florida and Louisiana.

* * Style-tips short and obtuse: rays oval or oblong, mostly shorter than the fruiting disk, not rarely particolored with brown purple: akenes commonly with a scarious and more or less ciliate margin or sometimes narrow wing to the inner edge: divisions or lobes of the leaves mostly entire.

— **L. Tagétes,** GRAY. A foot high, branching, leafy, strigulose-cinereous: leaves thickish, mostly with 3 to 7 narrowly linear rather rigid lobes: heads rather short-peduncled: rays few, a quarter to half an inch long: disk globose to barely oblong, half-inch high: pappus of one or sometimes two subulate or awn-like deciduous teeth, and no intermediate squamellæ. — Pacif. R. Rep. iv. 103. *Lepachys columnaris*, var. *Tagetes*, Gray, Pl. Wright. i. 106. *Rudbeckia Tagetes*, James in Long Exped. ii. 68. *R. globosa*, Nutt. Jour. Acad. Philad. vii. 19, & Traus. Am. Phil. Soc. l. c. 355. *Obeliscaria Tagetes*, DC. l. c. — Alluvial plains, Arkansas to W. Texas and New Mexico; first coll. by *James*.

— **L. columnáris,** TORR. & GRAY, l. c. Strigose-scabrous, a foot or two high, branching from the base, terminated by long peduncles bearing a showy head: divisions of the cauline leaves 5 to 9, from oblong to narrowly linear, sometimes 2-3-cleft: rays commonly an inch long or more, normally all yellow: disk at length columnar and inch or more long: pappus of the preceding, but usually a series of minute and delicate squamellæ around the broad flat summit. — *Rudbeckia columnaris*, Pursh, Fl. ii. 575; Bot. Mag. t. 1601; Hook. Fl. i. 311; Sprague, Wild Flowers of Amer., 43, t. 8. *Ratibida sulcata*, Raf. l. c. *R. columnaris*, Don, Brit. Fl. Gard. n. ser. iv. 361. *Obeliscaria columnaris*, DC. l. c. — Plains and prairies, Saskatchewan to the Rocky Mountains, and south to Texas and Arizona.

Var. **pulchérrima,** TORR. & GRAY, l. c. Differs only in having a part or even the whole upper face of the ray brown-purple; varies southward into more slender and branching forms, some with rays reduced to a quarter-inch. — *Obeliscaria pulcherrima*, DC. l. c. *Ratibida columnaris*, var. *pulcherrima*, Don, l. c. t. 361. — Nebraska to Arizona and Texas. (Adj. Mex.)

§ 2. Akenes completely flat: style-tips slender-subulate, very hispid: root probably annual or biennial. — § *Lophochæna*, Torr. & Gray, l. c.

L. pedunculáris, TORR. & GRAY. Strigose-scabrous or pubescent and somewhat cinereous, 2 or 3 feet high, including the naked peduncle of a foot or more: leaves rather large, irregularly bipinnately parted or pinnately parted and some of the lobes incisely pinnatifid or toothed, these oblong-linear or broader: rays obovate, an inch or less long and pure yellow, or sometimes only quarter-inch long and particolored: disk cylindrical, the largest an inch and a half long: akenes broadly and somewhat obliquely obovate, with no nerve or elevation on the face, from narrowly to broadly winged and squamellate-fimbriate on at least the inner edge, deeply notched at summit by an extension into two chaffy teeth, the inner one large and triangular-subulate, the outer smaller, and the notch fringed with small irregular squamellæ. — Fl. ii. 315. — Low ground, Texas, *Drummond, Wright*, &c.

Var. **pícta,** GRAY. Pubescence more cinereous: leaves simply and lyrately pinnately parted into fewer (5 to 7) divisions; these incised, the larger terminal one ovate-oblong or obovate: rays barely half-inch long, brown-purple with yellow edge: disk becoming inch and a half long. — Pl. Wright. i. 107. *L. serrata*, Buckley in Proc. Acad. Philad. 1861, 457. — Texas, near the coast, and in sandy woods, *Wright, Buckley, Hall*.

97. WEDÉLIA, Jacq.

(*Prof. G. W. Wedel,* of Jena, in the latter part of the 17th century.) — Tropical herbs or undershrubs, mostly of sea-shores; with opposite leaves, and lateral or terminal pedunculate heads of yellow flowers. One species has reached our southernmost coast.

— **W. carnósa**, Pers. Perennial herb, slightly strigose-hispidulous, glabrate: stem extensively creeping, sending up erect branches: leaves fleshy, mostly sessile, cuneate-oblong to obovate, somewhat serrate, often with some coarse teeth or 3 to 5 short lobes: rays golden yellow, 3-toothed, little surpassing the oblong foliaceous involucral bracts: akenes (3 lines long including the cupulate pappus) much thickened and muricate-scabrous at maturity, the attenuate base compressed and sharp-edged. — Syn. ii. 490; DC. Prodr. v. 538; Griseb. Fl. W. Ind. 371. *Silphium trilobatum*. L. Spec. ed 2. ii. 1302 (Plum. ed. Burm. t. 107, f. 2; Sloane, Jam. t. 155, f. 1). *Buphthalmum repens*, Lam. — Biscayne Bay, S. E. Florida, *Curtiss*. (W. Ind., S. Am.)

98. BORRÍCHIA, Adans.

(*Ole Borrich*, a Danish botanist of the 17th century.) — Shrubs or suffruticose and more or less fleshy plants of the sea-coast, canescent, or becoming glabrate and green; with opposite entire or denticulate leaves tapering somewhat into a petiole, and rather large heads of yellow flowers on terminal peduncles: fl. summer. — Fam. ii. 130; DC. Prodr. v. 488.

— **B. arboréscens**, DC. Shrub 4 feet or less high, fleshy, much branched: leaves spatulate-lanceolate, rigidly mucronate, veinless: involucre appressed: bracts of the receptacle obtuse or barely mucronate. — Prodr. l. c. *Asteriscus*, &c., Dill. Elth. t. 38. f. 43. *Coronu-solis frutescens*, &c., Plum. ed. Burm. t. 16, f. 2. *Buphthalmum arborescens*, L. Spec. ed. 2, ii. 1273. — Sandy shores and Keys. S. Florida. (W. Ind. to Peru.)

— **B. frutéscens**, DC. Less woody, more permanently canescent; the simpler stems 1 to 3 feet high: leaves fleshy-coriaceous, from obovate to spatulate-lanceolate, sometimes dentate: bracts of the involucre smaller and looser, spreading in age; of the receptacle spinulose-cuspidate. — Prodr. l. c. *Asteriscus frutescens*, &c., Dill. Elth. t. 38, f. 44. *Chrysanthemum fruticosum*, &c., Catesb. Car. i. t. 93. *Buphthalmum frutescens*, L. Spec. ii. 903; Walt. Car. 212; Torr. & Gray, Fl. ii. 268. — Sandy sea-coast, Virginia to Texas. (Mex., &c.)

99. BALSAMORRHÍZA, Hook.

(βάλσαμον, balsam, ῥίζα, root.) — Low perennials (all of Central and Western N. America); with thick and deep roots, which exude a terebinthine balsam, and send up a tuft of radical leaves, mostly on long petioles, and short simple few-leaved flowering stems or naked scapes, bearing large and mostly solitary heads of yellow flowers; the rays ample and numerous. Cauline leaves when present alternate or occasionally opposite, petioled. The root, when peeled (to get rid of the terebinthine rind) and baked, is an article of food to the aborigines, and the akenes are also eaten. — Fl. i. 310 (under *Heliopsis*); Torr. & Gray, Fl. ii. 300; Gray, Pl. Fendl. 81.

§ 1. Ligules becoming thin-papery, and persistent on or very tardily deciduous from the canescently pubescent akenes. — *Kalliactis*, Gray, l. c.

B. Careyána, Gray, l. c. Cinereous-pubescent, slightly scabrous: flowering stems a foot high, bearing 3 or 4 small lanceolate leaves and 2 to 7 racemosely disposed heads: leaves subcoriaceous, entire, reticulated; the radical cordate-lanceolate, a span or more in length: involucre half-inch or more high: ligules oval, hardly inch long, abruptly contracted into a very short but distinct tube: style-branches of the disk-flowers subulate and very hispid throughout. — Sandy plains on the Clearwater, Idaho, fl. May, *Spalding*. Rediscovered on the Wallawalla, Washington Terr., 1883, by *Brandegee*, with the rays deciduous from the mature fruit.

§ 2. Ligules deciduous in the ordinary manner: akenes glabrous; stems or scapes terminated by solitary or sometimes 2 or 3 heads.

  * Leaves entire or merely serrate; the principal ones cordate or with cordate base and long-petioled. — § *Artorhiza*, Nutt. Trans. Am. Phil. Soc. vii. 350. *Espeletia*, Nutt. Jour. Acad. Philad. vii. 39, not Humb. & Bonpl.

— **B. sagittáta**, Nutt. Silvery-tomentulose or canescent, and the involucre white-woolly: radical leaves from cordate-oblong to hastate, entire or nearly so (4 to 9 inches long, the base 2 to 6 inches wide, on petioles of greater length); the few and inconspicuous cauline from linear to spatulate: scape at length a foot or more high: rays 1 to nearly 2 inches long. — Gray, Bot. Calif. i. 348. *B. sagittata* & *B. helianthoides*, Nutt. Trans. Am. Phil. Soc. l. c.; Torr. & Gray, l. c. *Espeletia helianthoides* & *E. sagittata*, Nutt Jour. Acad. Philad. vii. 38, t. 4. *Buphthalmum sagittatum*, Pursh, Fl. ii. 564. — Rocky Mountains of Colorado to Montana and Brit. Columbia, the border of California, and S. Utah. Young stalks, root, and seeds used for food by the Indians. Outer bracts of the involucre sometimes oblong-lanceolate, foliaceous, and surpassing the disk (as in Pursh's original); or all more imbricate and conformed, the outer shorter.

B. deltoídea, Nutt. Trans. l. c. Green, more or less pubescent or glabrate: leaves broadly cordate to cordately ovate-lanceolate, sometimes nearly deltoid, from irregularly serrate to entire, 4 to 10 inches long: scape with small lanceolate or rarely ovate leaves, not rarely 2-3-cephalous: rays an inch or more long. — Torr. & Gray, l. c.; Gray, Bot. Calif. l. c. *B. glabrescens*, Benth. Pl. Hartw. 317. — Idaho and Brit. Columbia to S. California.

B. Bolánderi, Gray. Green, glabrate: stems stout, a span or two high, and bearing 2 or 3 subcordate nearly entire leaves, similar to and as large as the radical ones: principal involucre of the short-peduncled head a single or double series of ovate-lanceolate foliaceous bracts, over an inch long: apparently disk-akenes flattened. — Proc. Am. Acad. vii. 356, & Bot. Calif. l. c. — California, at Auburn, and on the Sacramento, *Fremont, Rich, Bolander.*

* * Leaves not cordate and entire, varying from laciniately dentate to pinnately or bipinnately divided: heads solitary on a naked scape, or scapiform stem bearing a pair of small opposite leaves towards the base: thick caudex or root exceedingly balsamic-resiniferous. Perhaps all forms of one polymorphous species. — § *Eubalsamorrhiza*, Nutt.

B. macrophýlla, Nutt. Green, not at all canescent, glabrate, except the ciliate margins of the leaves, usually minutely glandular-viscidulous: leaves ample, ovate or oblong in outline, a span to a foot long, some with only one or two lobes or coarse teeth, most of them pinnately parted into broadly lanceolate and commonly entire lobes (of 2 or 3 inches in length): scapes a foot or two high: bracts of the involucre from narrowly lanceolate to spatulate and foliaceous, an inch or two long, nearly equal, either half or fully the length of the rays. — Trans. Am. Phil. Soc. vii. 350; Torr. & Gray, Fl. ii. 301; Eaton, Bot. King Exp. 168. — Rocky and Wahsatch Mountains, Wyoming to Utah, *Nuttall, Fremont, Watson.*

B. terebinthácea, Nutt. Slightly and minutely if at all canescent: leaves from green and glabrate to minutely hispidulous-scabrous, or barely hirsutulous at margins, at length rigid and reticulate-veiny, oblong-lanceolate and with cuneate or truncate base (4 to 8 inches long, 1 to 3 wide), spinulosely dentate or sometimes crenate-dentate, or some laciniate-incised, or even pinnatifid: scapes a span to a foot high: involucre lanate-tomentose, of numerous and narrow linear-lanceolate and attenuate loose and nearly equal bracts, an inch long. — Trans. Am. Phil. Soc. vii. 349 (name only); Gray, Pl. Fendl. 82. *B. Hookeri*, var., Torr. & Gray, l. c. *Heliopsis? terebinthacea*, Hook. Fl. ii. 310? — W. Idaho to E. Oregon, in hard or stony ground, *Douglas, Spalding, Nevius, Cusick.*

B. Hoókeri, Nutt. l. c. Canescent with fine sericeous or more tomentose pubescence, but not at all hirsute: scapes and leaves a span to a foot high; the latter lanceolate or elongated-oblong in outline, pinnately or bipinnately parted into lanceolate or linear divisions or lobes, or some of them only pinnatifid or incised: involucre from canescently puberulent to lanate; its bracts from linear- to oblong-lanceolate, either unequal and well imbricated or sometimes the outermost foliaceous and enlarged. — Torr. & Gray, l. c., excl. var.; Eaton in Bot. King Exp. l. c. *Heliopsis? balsamorrhiza*, Hook. l. c. — Hills and rocky plains, eastern parts of Washington Terr. to Nevada and W. California; first coll. by *Douglas.*

Var. **incána**. Densely white-tomentose: leaves often of broader outline. — *B. incana*, Nutt. l. c. 350; Torr. & Gray, l. c. — Wyoming and Montana to northern parts of California; first coll. by *Nuttall.*

B. hirsúta, Nutt. l. c. Green, roughish-hirsute or hispidulous, not tomentose nor canescent: leaves lanceolate in outline, pinnately parted or divided, the divisions (9 to 15 lines in length) incisely toothed or again pinnatifid, soon rigid: scapes a span to a foot high: involucre hirsute-pubescent or glabrate, of narrowly lanceolate or more attenuate bracts. — Torr. & Gray, l. c.; Eaton, l. c. — Utah to Brit. Columbia and N. E. California, in the dry region; first coll. by *Douglas* and *Nuttall.*

100. WYÉTHIA, Nutt. (*Nathaniel J. Wyeth,* who collected the species on which the genus was founded, and with whom Nuttall subsequently crossed the continent.) — Stout and mostly low perennials (W. North-American); with more or less balsamic or resiniferous juice, ample and undivided pinnately veined alternate leaves (commonly entire), and large heads of mostly yellow flowers. (Thick roots and seeds were food of the Indians.) — Jour. Acad. Philad. vii. 38, & Trans. Am. Phil. Soc. vii. 351; Gray, Proc. Am. Acad. viii. 654. *Alarçonia,* DC. Prodr. v. 537.

 * Rays from "pale yellow" or dull straw-color to white. — The original *Wyethia,* Nutt.

W. helianthoídes, Nutt. A span to a foot and a half high, simple and with a single large head, or rarely 3 or 4, hirsute; leaves from oval to broadly lanceolate, denticulate or entire, 4 to 8 inches long, mostly narrowed at base into a short margined petiole; heads an inch high; bracts of the involucre narrowly lanceolate, numerous; rays nearly 2 inches long; akenes 4 lines long, either prismatic-quadrangular or flattish, 12-nerved; pappus shorter than the width of the akene, sometimes minute, chaffy-coroniform and cleft into few or several teeth. — Jour. Acad. Philad. l. c. t. 5; Gray, l. c. — Northern Rocky Mountains, in moist valleys, S. W. Montana to E. Oregon, *Wyeth, Nevius, Cusick, Watson, Scribner.*

 * * Rays bright yellow. — *Alarçonia,* DC. (Dedicated to the memory of Hernando de Alarçon, a noble Spanish navigator, who, in 1540, first visited and carefully surveyed the coast of California.)

 + Involucre of the very large and broad heads foliaceous; the spreading outer bracts ovate or oblong, commonly 2 inches or more in length, much surpassing the disk (which is of about equal breadth) and often exceeding the rays: akenes very stout and thick, half-inch long, with comparatively obtuse angles, crowned with a large chaffy-coriaceous calyciform pappus, which is cleft into unequal teeth or lobes: cauline leaves short-petioled.

W. helenioídes, Nutt. l. c. Very stout, 2 or 3 feet high, floccosely tomentose, glabrate in age: leaves oblong and ovate, mostly entire, radical a foot or two and upper cauline 6 to 8 inches long: akenes pubescent toward the summit. — Gray, Pl. Fendl. 82, Proc. Am. Acad. l. c., & Bot. Calif. i. 349. *Alarçonia helenioides,* DC. l. c. — Hillsides around and near San Francisco Bay, California; first coll. by *Douglas.*

W. glábra, Gray. A foot or two high, glabrous or nearly so, balsamic-viscid: leaves of the preceding in size and shape, or narrower, sometimes serrate: akenes glabrous. — Proc. Am. Acad. vi. 543, viii. 654, & Bot. Calif. l. c. — W. California, from Marin Co. southward, *Andrews, Brewer,* &c.

 + + Involucre of the smaller heads (about inch or less high) narrower and fewer-flowered, usually campanulate; the outer bracts even when foliaceous seldom surpassing the disk: akenes less thick, 3 to 5 lines long.

 ++ Glabrous and smooth throughout, usually balsamic-viscid: leaves thickish, lanceolate to oblong, upper sessile, lower tapering into margined petioles: outer bracts of the narrowish involucre disposed to be foliaceous.

W. amplexicaúlis, Nutt. A foot or two high, robust: leaves mostly lanceolate-oblong, entire or denticulate; radical often a foot or more long; upper cauline (a span or so long) partly clasping by a rounded or somewhat narrowed base: heads solitary or several, short-peduncled: involucral bracts broadly lanceolate, acute or obtuse, one or two outer ones occasionally foliaceous and larger: rays inch and a half long: akenes with a conspicuous crown cleft into acute teeth, and sometimes a small awn. — Trans. Am. Phil. Soc. l. c.; Torr. & Gray, l. c. *Espeletia amplexicaulis,* Nutt. Jour. Acad. Philad. vii. 38. *Silphium? læve,* Hook. Lond. Jour. Bot. vi. 244. — Moist valleys and plains, Rocky Mountains from Colorado to Montana, west to Nevada and Brit. Columbia. *Pe-ik* of the Indians.

W. longicaúlis, Gray. Nearly resembles preceding, taller, rather slender; leaves lanceolate, even uppermost with tapering base and not clasping: heads solitary or paniculate, on long and slender peduncles: outer series of involucral bracts oblong or somewhat spatulate, foliaceous, mostly surpassing the inner and the disk: rays only inch long: akenes with a short erosely denticulate crown. — Proc. Am. Acad. xix. 4. — Prairies of E. Humboldt Co., California, *Rattan.*

++ ++ Glabrous, but scabridous and balsamic-viscid: leaves ovate, abruptly petioled, coriaceous.

W. reticulata, Greene. Habit of *W. ovata*, only puberulent-hispidulous without tomentum, leafy up to the corymbosely disposed heads; cauline leaves ovate or subcordate, short-petioled (4 down to 2 inches long), 3-5-plinerved, and with veins and veinlets much reticulated, shining; those of flowering branches small, oblong, 3-nerved: heads hemispherical, little over half-inch high: bracts of involucre oblong-linear, obtuse, short; outer foliaceous and loose, sometimes one or two enlarged: rays apparently few and rather small: akenes compressed-quadrangular, glabrous (barely 3 lines long): pappus an extremely short erose-denticulate crown; no awn. — Bull. Calif. Acad. i. 9. — Banks of Sweetwater Creek, El Dorado Co., California, *Mrs. Curran*.

++ ++ Tomentose or woolly, but sometimes glabrate in age: leaves all petioled and becoming coriaceous, ample, even the cauline 4 to 7 inches long.

= Involucre hemispherical, of numerous broadly lanceolate bracts, not surpassing the disk: rays numerous, 20 to 24.

W. ováta, Torr. & Gray. Canescent with a soft not floccose tomentum, 2 or 3 feet high from running rootstocks, commonly branching: leaves ovate, the cauline subcordate and with acute apex, somewhat triplinerved; veinlets not much reticulated: pappus a chaffy, several-toothed crown. — Emory Rep. 143 (1848, wholly overlooked); Gray, Proc. Am. Acad. vii. 357, & Bot. Calif. l. c. — California, on the western side of the Sierra Nevada.

= = Involucre narrower, campanulate; the outer bracts larger than the inner and more or less surpassing the disk: rays fewer: leaves at length firm-coriaceous and the veinlets conspicuously reticulated.

W. móllis, Gray. White with floccose wool when young, more or less glabrate in age, 1 to 3 feet high, bearing solitary or few heads: leaves oblong and ovate, with either rounded or truncate or cuneate base: rays 10 to 15, over an inch long: akenes minutely pubescent at summit; pappus a truncate chaffy crown, and 2 or in the ray 3 to 5 subulate awns. — Proc. Am. Acad. vi. 544, viii. 655, &c. — Sierra Nevada, especially on the eastern side, from Sierra Valley to Virginia City, Nevada, and westward to the Yosemite; first coll. by *Anderson*.

W. coriácea, Gray. Sericeous-tomentose, stout, 1 to 3 feet high: leaves rigid, broadly ovate or oval, obtuse or apiculate, somewhat triplinerved, even the upper cauline (5 to 7 inches long) seldom longer than their petiole: rays 5 to 9, hardly surpassing the involucre: pappus a short obtusely 4-6-cleft crown. — Proc. Am. Acad. xi. 77, & Bot. Calif. i. 616. — San Diego Co., California, on the Mesa Grande, &c., *Palmer, Parish*.

++ ++ ++ ++ Hirsutely more or less pubescent, often somewhat balsamic-glutinous: leaves elongated-lanceolate, tapering to both ends, or the upper and sessile cauline broader: bracts of the involucre mostly foliaceous or herbaceous, lanceolate or broader, equalling the disk.

W. angustifólia, Nutt. A span to 2 feet high, and the radical leaves about as long, these occasionally denticulate or serrate, often undulate: involucre fully inch high, loose or spreading: head solitary; rays mostly numerous, inch and a half long: pappus a short and chaffy fimbriolate-cleft crown, and one or two or in the ray 3 or 4 elongated subulate awns, one of them about the length of the akene. — Gray, Proc. Am. Acad. viii. 655, & Bot. Calif. l. c. *W. angustifolia* & *W. robusta*, Nutt. l. c.; Torr. & Gray, Fl. ii. 299. *Helianthus longifolius*, Hook. Fl. i. 312; Hook. & Arn. Bot. Beech. 353. *H. Hookerianus*, DC. Prodr. *Alarconia angustifolia*, DC. Prodr. v. 537. — Plains and hills, commonly in moist ground, Washington Terr. to Monterey Bay, California.

W. Arizónica, Gray. A foot high, bearing a single or few and smaller heads: leaves oblong-lanceolate: involucre of fewer and more erect bracts: rays 8 to 12: pappus a very narrow crown, extended into 3 or 4 stout subulate teeth, or into one or two short awns. — Proc. Am. Acad. viii. 655; Rothrock in Wheeler Rep. vi. 161, t. 9; Meehan, Nat. Flowers, ser. 2, ii. t. 37. — Near streams and springs, S. Colorado to S. Utah and Arizona, *Palmer, Bishop, Siler, Rothrock*, &c.

++ ++ ++ ++ ++ Hispidulous, very scabrous, narrow-leaved: involucre more imbricated, squarrose.

W. scábra, Hook. A foot or two high (root unknown), rigid: cauline leaves linear, thick, 4 to 6 inches long, half-inch wide, sessile, attenuate-acute, the few veins confluent into lateral undulate nerves: involucre nearly hemispherical; its bracts imbricated in 3 or 4 series, all the outer with a coriaceous ovate-oblong appressed base, which is acuminate into a longer subulate filiform spreading very hispid-scabrous appendage: rays several, half-inch long: akenes acutely angled and with few or obscure intermediate nerves, very smooth, the 3 or 4 angles extended into a pappus of as many short and blunt teeth, which are barely coroniform-confluent at base. — Lond. Jour. Bot. vi. 245; Gray in Pacif. R. Rep. iv. 102, & Proc. Am. Acad. viii. 655. — S. Colorado and New Mexico to Utah and Wyoming, *Geyer, Bigelow, Parry, Ward*, &c.

101. **GYMNOLÓMIA**, HBK. (Γυμνός, naked, λῶμα, border, the pappus obsolete or none.) — Herbs or frutescent plants (of Mexico and adjacent countries), resembling the smaller-flowered species of *Helianthus;* with erect branching stems, alternate or opposite leaves, and heads of yellow flowers (or the disk brownish); the peduncles terminating the branches: fl. summer. — Nov. Gen. & Spec. iv. 217, t. 373, 374; Benth. & Hook. Gen. ii. 363. *Gymnopsis*, DC. Prodr. v. 561, in part.

   * Annuals: receptacle of the head conical and the disk high: bracts of the rather simple involucre linear. — *Heliomeris*, Nutt.

G. Pórteri, Gray. A foot or two high, slender, paniculately branched, sparingly hispid, otherwise nearly glabrous: leaves nearly all alternate, narrowly lanceolate or linear, entire: rays 5 to 8, oval or obovate (half-inch or more long), deep orange yellow: disk in age oblong-conical; its chaffy bracts oblong-lanceolate or the outer ovate, cuspidate-acuminate, striate, merely concave at maturity: fructiferous receptacle almost columnar: akenes turgid-obovate, very obscurely quadrangular, dull, somewhat puberulent, with small terminal areola, one of the angles or nerves sometimes slightly margined or umbonate at the summit: style-tips subulate and hispid. — Proc. Am. Acad. xii. 59; Meehan, Nat. Flowers, ser. 2, ii. t. 35. — *Rudbeckia? Porteri*, Gray, Pl. Fendl. 83. — Northern Georgia, known only on the isolated granite rock called Stone Mountain, near Atlanta, where it abounds; first coll. by *Prof. Porter*.

G. multiflóra, Benth. & Hook. A foot to a yard high, strigulose-pubescent or scabrous, sometimes also hispid, often much branched: leaves from narrowly linear to lanceolate, rarely broader, either alternate or mainly opposite, entire or obscurely denticulate: rays 10 to 15, golden yellow: disk hemispherical, in age little more elevated and receptacle obtusely conical; its bracts obtuse or the inner acute with soft acumination: akenes smooth, compressed, with convex or obtusely angulate sides: style-tips short and obtuse. — Benth. & Hook. ex Rothrock in Wheeler Rep. vi. 160, & Hemsl. Biol. Centr.-Am. ii. 162. *Heliomeris multiflora*, Nutt. l. c.; Gray, Pl. Fendl., Pl. Wright. ii. 87, with var. *hispida*, &c. — Sandy banks of streams, &c., W. Texas to Wyoming, Nevada, and Arizona. Very polymorphous: the root not perennial as was supposed. An indigenous specimen coll. by *Lemmon* in Arizona has disk-corollas all converted into rays or radiatiform ampliate lobes. (Mex.)

G. tríloba, Gray. Much branched, over 2 feet high (root not seen), obscurely puberulent, no hispid bristles: leaves roundish in general outline, 3-lobed, with subcordate or truncate base, short-petioled, the lobes short and broad: rays 12 or more, oblong-linear, elongated: disk hemispherical: receptacle low-conical: akenes of the preceding but more oblong. — Proc. Am. Acad. xvii. 217. — Mountains of S. Arizona south of Rucker's Valley, *Lemmon*.

   * * Perennial or frutescent: disk and receptacle low.

G. tenuifólia, Benth. & Hook. Shrubby, much branched, 2 or 3 feet high, scabrous-puberulent, very leafy: branches terminated by solitary long-peduncled heads: leaves alternate and the lower opposite, canescent beneath, pinnately or pedately parted into 3 to 7 narrow linear lobes, or the uppermost very narrow and entire, the margins mostly revolute: bracts of the involucre subulate-linear: rays 10 to 16: disk convex: chaffy bracts of the receptacle truncate-obtuse: akenes smooth, quadrangular-compressed. — Ex Hemsl. Biol. Centr.-Am. l. c. *Heliomeris tenuifolia*, Gray, Pl. Fendl. 84, Pl. Wright. &c. — S. W. Texas, *Wright, Havard*. (Adj. Mex., *Berlandier, Gregg*, &c.)

   * * * Annual: receptacle and disk barely convex: habit of *Encelia* and *Helianthus*.

G. encelioídes, Gray. A foot or two high from an annual root, strigose-canescent and the branching stem hispid: leaves ovate-oblong or obscurely deltoid, rather obtuse, nearly entire, mostly long-petioled, the lower opposite: heads barely half-inch high: involucre biserial; outer bracts all equal and equalling the disk, oblong-lanceolate, acute, white with soft but at length hispid pubescence, longer and larger than the nearly linear interior ones: rays 10 or 12, oval, showy, golden yellow, less than an inch long: disk-corollas with dark purple tips: akenes obovate-oblong, below sparsely and toward the summit thickly villous with slender hairs: pappus none, or a few very delicate setiform squamellæ shorter than the hairs of the akene. — Proc. Am. Acad. xix. 4. — S. E. California, at Aqua Caliente, in the Mohave Desert, *Parish.*

102. VIGUIÉRA, HBK. (*Dr. A. Viguier,* botanist, of Montpellier.) — Herbaceous or sometimes suffruticose plants (of the warm parts of America); with only the lower or rarely all the leaves opposite, yellow-flowered heads of only medium size (in our species), on peduncles at the summit of the branches, the akenes usually pubescent. — Nov. Gen. & Spec. iv. 224, t. 379; Benth. & Hook. Gen. ii. 375. *Viguiera, Leighia* (Cass.), & *Harpalium* (Cass.), in part, DC. Prodr. v. 578-584.

  * Disk of the head at maturity elevated or strongly convex (but at firs: often low) and the receptacle conical: root probably annual or biennial.

V. helianthoídes, HBK. Minutely hispidulous-pubescent or scabrous, green, or sometimes cinereous: stem 2 to 7 feet high, slender, paniculately branched above: leaves alternate or occasionally either upper or lower opposite, slender-petioled, mostly thin, ovate, acuminate, sometimes very broadly ovate (the larger 4 to 6 inches long and 3 or 4 wide), sometimes ovate-lanceolate, from slightly to coarsely serrate, triplinerved from near the base: heads paniculate, usually slender-peduncled: involucre only 3 lines high, shorter than the disk, nearly simple, of subulate or linear bracts: rays 7 to 10, obovate or oblong, over half-inch long: chaffy bracts of the receptacle somewhat cuspidately mucronate or acuminate: akenes villous-pubescent: paleæ a pair on each side between the chaffy awns, erose or fimbriolate at the truncate summit. — HBK. l. c.; Benth. & Hook. l. c. *V. helianthoides, Sayræana, laxo, brevipes,* and probably *V. microcline, triquetro,* also with little doubt *V. dentata* (Spreng., the *Helianthus dentatus,* Cav. Ic. iii. 10, t. 220), DC. Prodr. v. 579. *V. Texana,* Torr. & Gray, Fl. ii. 318. *Helianthella latifolia,* Scheele in Linn. xxii. 160. — Shady or more open grounds, Texas to Arizona. (Mex., Cuba.)

V. canéscens, DC. Less tall, more rigid, commonly cinereous: leaves coriaceous, entire or nearly so, from broadly ovate to oblong-lanceolate; the lower opposite: chaff of the receptacle more cuspidate: rays saffron-yellow: akenes canescently sericeous. — Prodr. v. 579. — S. Arizona, *Pringle,* a greener form, and in adjacent Mexico, *Palmer.* (Mex.)

  * * Disk flattish or convex: receptacle at maturity flat or hardly conical.

  +— Herbaceous to the base from a probably perennial root, not canescent nor tomentose.

V. cordifólia, Gray. Hispid or hispidulous and scabrous: stem rather stout, 2 or 3 feet high, leafy to the top, commonly branched above: leaves mostly all opposite. occasionally some alternate, subcordate-ovate or deltoid, acute, serrate or denticulate, 3-ribbed from the base, either sessile or short-petioled, rough; veinlets reticulated: heads mostly corymbose and short-peduncled: involucre campanulate, fully half-inch long, equalling the barely convex disk, commonly lanceolate and acuminate, erect, in 2 or 3 series: chaffy bracts of the receptacle gradually acuminate: akenes narrowly cuneate-oblong, almost equalled by the chaffy awns; the intermediate paleæ equalling the breadth of the akene, narrowly oblong, rigid. — Pl. Wright. i. 107, ii. 88. — Near water-courses, W. Texas to Arizona, *Wright, Schott, Lemmon,* &c. (Mex., *Schaffner.*)

  +— +— Shrubby or lignescent at base, low, not tomentose: leaves hispidulous-scabrous, mostly alternate, rigid.

V. laciniáta, Gray. Branching: leaves lanceolate or obscurely hastate, from laciniate-pinnatifid to nearly entire, abruptly petioled, an inch or two long, beneath with very prominent pinnate veins: branches bearing several cymosely disposed and pedunculate heads: involucre nearly half-inch high; its bracts lanceolate or the outermost ovate, acute or acuminate: rays half-inch long: akenes sparingly pilose, glabrate: pappus-awns chaffy; the intermediate chaffy paleæ laciniate or erose. — Bot. Mex. Bound. 89, & Bot. Calif. i. 354. — San Diego Co, California, *Schott, Newberry, Cleveland,* &c. (Lower Calif.)

V. Paríshii, GREENE. Branched from the base and diffuse; the nearly simple slender flowering branches (sometimes rather canescently pubescent) bearing mostly solitary pedunculate heads: leaves ovate, small (half-inch to barely inch long), somewhat serrate, short-petioled: involucre broad: its bracts lanceolate: akenes more villous: awns as long as the akene and chaffy-dilated only near the base; the paleæ much laciniate. — Bull. Torr. Club, ix. 15. — Desert region of W. Arizona and S. E. California to the coast at San Luis Rey, *Newberry, Palmer, Parry* & *Lemmon, Parish, W. G. Wright, Greene.*

+ + + Herbaceous? perennials, white-tomentose or canescent (at least the foliage): involucre of rather short imbricated bracts.

V. reticuláta, WATSON. Stem glabrate, few-leaved: leaves rigid and coriaceous, cordate, entire, strongly veined and reticulated beneath, 2 inches long, petioled, canescent with short rather silky pubescence: heads small (3 or 4 lines high), several in the corymbiform clusters: rays 3 lines long: subulate chaffy awns only twice the length of the laciniate paleæ of the pappus. — Amer. Nat. vii. 301. — Telescope Mountain, Nevada, *Wheeler.*

V. tephródes, GRAY. Silvery-white with close-pressed sericeous-hirsute (not tomentose) pubescence, which is probably somewhat deciduous: leaves alternate, ovate-oblong or the upper rather deltoid-lanceolate, entire, thickish, 3-ribbed at base and obscurely veiny, less than inch long, slender-petioled: heads few or solitary, less than half-inch high: akenes (or rather ovaries) short, with villous-ciliate margins and rather glabrous sides, about the length of the lanceolate awned paleæ, the short intermediate paleæ dissected into almost setiform squamellæ. — Proc. Am. Acad. xvii. 218. *Helianthus (Harpalium) tephrodes,* Gray, Bot. Mex. Bound. 90. *Viguiera nivea,* Gray, Bot. Calif. i. 354, excl. syn. Benth. & syn. Kellogg. — S. E. California, at Mirasol del Monte, in the Colorado Desert, *Schott.*

V. LANÁTA, Gray, Proc. Am. Acad. xvii. 218 (*Bahiopsis lanata,* Kellogg, Proc. Calif. Acad. ii. 35), of Lower California, with very pannose dense tomentum, is of the genus, but is not *Encelia nivea,* Benth.

V. TOMENTÓSA and V. DELTOÍDEA, Gray, Proc. Am. Acad. v. 161, are other species of Lower California.

103. TITHÓNIA, Desf.

(Τιθωνός, consort of Aurora.) — Robust annuals (all Mexican); with alternate petioled and 3-ribbed often 3-lobed ample leaves, and large heads of yellow flowers on long and stout upwardly thickened peduncles. Ligules entire or nearly so. Bracts of the receptacle rather rigid, striate, cuspidate or aristate. Akenes oblong or narrower, compressed-quadrangular: the pappus either deciduous or persistent. — Desf. Ann. Mus. i. 49. t. 4; Benth. & Hook. Gen. ii. 374. — The following was collected so near the southwestern boundary of the U. S. that it is here introduced.

T. Thúrberi, GRAY. Comparatively small and slender, 2 feet high, slightly hispid: leaves ovate, serrate, undivided: head only half-inch high, with little exserted orange-colored rays: bracts of the involucre lanceolate or oblong, with short foliaceous tips: akenes narrow: squamellæ of the pappus linear-oblong, coriaceous, the awns nearly smooth. — Proc. Am. Acad. viii. 655. *T. tubæformis,* Gray, Bot. Mex. Bound. 90, not Cass., which is a far larger and showy species. — Magdalena, State of Sonora, Mex., near Arizona, *Thurber.*

104. HELIÁNTHUS, L. SUNFLOWER.

(From ἥλιος, the sun, and ἄνθος, flower.) — Annual and perennial American herbs, almost all N. American, usually tall or coarse: with the lowest and sometimes all the leaves opposite and simple; heads pedunculate and terminating the stem or branches, produced in summer or autumn, with yellow rays (wanting in one or two species), and either yellow or dark purple disk-flowers: chaffy bracts of the receptacle either entire or 3-toothed: throat of the disk-corollas not rarely 10-nerved. Pappus normally of a thin and acuminate or awn-pointed palea from each of the two principal angles of the akene, or rarely an additional one: intermediate squamellæ present only in a few species, and then inconstant, or else mere appendages or lateral portions of the 2-paleaceous pappus. Juice of the stem resinous. — Schkuhr, Handb. t. 528; Torr. & Gray, Fl. ii. 318; Benth. & Hook. Gen. ii. 376, excl. syn. of *Flourensia* in part.

H. PAUCIFLORUS, Nutt. Gen. ii. 177, of "Lower Louisiana," with narrow serrate leaves, and ovate closely imbricated bracts to the involucre, has not been identified.

§ 1. Annuals: involucre spreading; its bracts attenuate to a point: disk brownish or dark purple: receptacle flat or nearly so: leaves petioled, 3-ribbed from or near the base, all but the lower usually alternate.

* Stem erect, commonly robust: chaffy bracts of the receptacle mostly 3-cleft at apex, the longer middle lobe lanceolate or linear and somewhat hirsute or hispid. Species of difficult limitation, apparently confluent.

H. argophýllus, TORR. & GRAY. White with soft and silky wool, which is sometimes floccose, in age more or less deciduous: leaves slightly serrate: otherwise as in the larger indigenous forms of the following. — Fl. ii. 318; Rev. Hort. 1857, 431 with figure. — Texas; first coll. by *Drummond*. Disk often inch and a half broad, and rays as long. Degenerates in cultivation apparently into

H. ánnuus, L. (COMMON SUNFLOWER.) Robust, when well developed tall, hispid, hispidulous, or scabrous: stem often spotted or mottled: leaves ovate and the lower cordate, serrate, the larger 6 to 12 inches long, the blade of the cauline ones longer than their petiole: bracts of the involucre from broadly ovate to oblong, aristiform-acuminate, below hispidly ciliate: disk in the wild plant commonly an inch or more in diameter. — Spec. ii. 904 (excl. habitat, for it came not from Peru, nor even from Mexico); Lam. Ill. 706; Gray, Bot. Calif. i. 353. *H. lenticularis*, Dougl. Bot. Reg. t. 1225; DC. Prodr. v. 586; Torr. & Gray, l. c. *H. tubræformis*, Nutt. Gen. ii. 177; Ind. Sem. Gœtt. 139. *H. ovatus*, Lehm. Ind. Sem. Hamb. 1828, & Linn. v. 376. *H. erythrocarpus*, Bartl. *H. macrocarpus*, DC. Prodr. l. c., a race of the garden Sunflower with larger and light-colored akenes, long cult. in Russia, &c., for food and oil. *H. multiflorus*, Hook. Fl. i. 313, excl. syn. (For history, &c., see Decaisne in Fl. des Serres, xxiii., and Gray & Trumbull in Am. Jour. Sci. ser. 3, xxiii. 245.) — Plains and alluvial grounds, Saskatchewan to Texas, and west to Washington Terr. and California. (Adj. Mex.) Fruit from early times collected by the N. American Indians for food and hair-oil; the plant cultivated for these uses. Gigantesque forms everywhere commonly cultivated for ornament.

H. petioláris, NUTT. A foot to a yard high, more slender, loosely branching, strigose-hispidulous, rarely hirsute: leaves oblong-lanceolate or ovate-lanceolate, entire or sparingly denticulate, barely acute, 1 to 3 inches long, cuneately attenuate or the lower abruptly contracted into a long and slender petiole: bracts of the involucre lanceolate or oblong-lanceolate, with acute and mucronate or sometimes more attenuate tips, seldom at all ciliate: disk half-inch or more in diameter. — Jour. Acad. Philad. ii. 115; Sweet, Brit. Fl. Gard. ser. 2, t. 75; DC. l. c.; Torr. & Gray, l. c. *H. patens*, Lehm. Ind. Sem. Hamb. 1828, & Ind. Schol. 1828, 19. *H. integrifolius*, Nutt. Trans. Am. Phil. Soc. vii. 366. — Dry plains, Saskatchewan to Texas, west to Oregon and Arizona: seemingly passes into the preceding species.

Var. **canéscens**, GRAY, Pl. Wright. i. 108. Leaves whitened with a fine and close strigulose-sericeous pubescence, the lowest ovate, all or most of them with blade longer than the petiole. — S. W. Texas and New Mexico; first coll. by *Wright*. A very similar variety from Nebraska, *H. Engelmanni*.

* * Stem erect, not tall: chaffy bracts of the receptacle entire or with a pair of small lateral teeth, and the apex prolonged into a naked cusp or awn: bracts of the involucre hirsute or hispid with long spreading hairs, oblong or lanceolate, mostly attenuate-acuminate.

H. scabérrimus, BENTH. A foot or two high: stem rather stout, branching, scabrous-hispid: leaves from ovate to oblong-lanceolate, from rather coarsely serrate to entire, 2 to 5 inches long, the base cuneately or more abruptly contracted into the petiole, both faces either slightly or strongly scabrous: disk about two-thirds inch in diameter, and rays of about equal length: cusp of the chaff mostly subulate-aristiform and equalling the developed disk-flowers. — Bot. Sulph. 28, not Ell. *H. Bolanderi*, Gray, Proc. Am. Acad. vi. 544, & Bot.

Calif. i. 353. — California, from San Francisco Bay northward, *Hinds* (who got it at Bodegas), *Bridges, Bolander, Mrs. Ames.*

H. exílis, GRAY. A foot or so high, slender, commonly hirsute: leaves lanceolate and ovate-lanceolate, sparingly denticulate, tapering into a slender petiole: heads from half to nearly full size of those of the preceding: cusp of the chaff a slender awn, surpassing the disk-flowers. — Proc. Am. Acad. vi. 545, & Bot. Calif. l. c. — Plains throughout the northern part of California. (The specimen from Owen's Valley, *Van Horn*, is probably a depauperate *H. petiolaris.*)

　　　　* * * Stems branched from the base, diffuse or decumbent; slender.

H. débilis, NUTT. Scabrous to hispidulous or hispid: stems a foot to a yard long: leaves from ovate to deltoid or obscurely hastate, occasionally subcordate, thinnish, 1 to 3 inches long, repand-denticulate to sparingly lobulate-dentate, slender-petioled: bracts of the involucre lanceolate and gradually subulate-acuminate: rays half-inch or more in diameter; its chaffy bracts with truncate or 3-toothed summit, the middle tooth aristiform-subulate: rays half-inch or more long. — Trans. Am. Phil. Soc. vii. 367; Torr. & Gray, l. c. 320; the coast form. *H. praecox,* Engelm. & Gray, Pl. Lindh. i. 13, form with more hispid stem. — Sandy shores of Florida, W. Louisiana, and E. Texas.

　　Var. **cucumerifólius.** A larger form, usually with purple-mottled stems, leaves irregularly serrate with salient teeth, more commonly subcordate, the larger 4 or 5 inches long, and the ampler (15 to 20) rays an inch or more long. — *H. cucumerifolius,* Torr. & Gray, l. c. 319. *H. Lindheimerianus,* Scheele in Linn. xxii. 159 ? — Sandy soil, often in woods, Texas, common westward.

§ 2. Perennials: receptacle convex, or in some at length low-conical: lower leaves almost always opposite.

　　* Involucre loose (about half-inch high), more or less squarrose in age, of subulate-lanceolate or narrower mostly attenuate-acuminate and almost equal bracts: disk (upper part of corollas) commonly but not always dark purple or turning brownish: all but the lower leaves linear or filiform and strictly one-nerved: slender creeping rootstocks, no tubers.

H. orgyális, DC. Stem smooth and glabrous, often 10 feet high, very leafy to the top: leaves mostly alternate, from long-linear (8 to 16 inches long, commonly 2 to 4 lines wide), or the lowest lanceolate, to almost filiform, slightly papillose-scabrous, the lower narrowed into a petiole and sometimes serrulate: bracts of the involucre filiform-attenuate, those of the receptacle entire: akenes oblong-obovate with a rounded summit, 3 lines long. — Notul. Pl. Rar. Genev. 12, & Prodr. v. 586, excl. syn.; Torr. & Gray, Fl. ii. 320. *H. giganteus,* var. *crinitus,* Nutt. Gen. ii. 177? — Dry plains, Nebraska to Arkansas and Texas, west to S. E. Colorado.

H. angustifólius, L. Scabrous, sometimes hispidulous: stems 2 to 6 feet high, rather sparsely leafy, slender: leaves thickish, entire, when dry with revolute margins: cauline sessile (the upper hardly narrowed at base), 3 to 7 inches long, mostly 2 or 3 lines wide, paler and smooth or sometimes canescent beneath, many of them opposite; radical somewhat spatulate or lanceolate: bracts of the involucre lanceolate and acute or attenuate-acuminate: rays numerous, inch- long: disk generally dark-purple: receptacular bracts entire or 3-toothed: akenes (barely 2 lines long) with broad truncate apex. — Spec. ii. 906; Walt. Car. 216; Michx. Fl. ii. 141; Bot. Mag. t. 2051; Bart. Fl. Am. Sept. t. 105; Torr. & Gray, Fl. ii. 320. *Coreopsis angustifolia,* L. l. c. 908; Mill. Ic. t. 224, f. 2; and *Rudbeckia angustifolia,* L. Spec. ed. 2, 1281. *Leighia bicolor,* Cass. Dict. xxv. 436. — Wet ground, pine barrens of New Jersey and Kentucky to Florida and Texas.

　　* * Involucre closer, of more imbricated and unequal inappendiculate bracts, none of them foliaceous: disk mostly dark-colored or dusky: leaves from lanceolate to ovate, rarely linear: herbage not tomentose nor conspicuously cinereous: Atlantic United States species, one of them reaching the Rocky Mountains.

　　+— Stems glabrous and very smooth or merely scabrous, leafy: leaves narrowly to broadly lanceolate: chaff of receptacle entire, merely mucronate.

H. Floridánus, GRAY. Stem from 2 to 6 feet high: leaves thinnish, bright green above, sparsely hispidulous-scabrous, lanceolate, sparingly or obscurely denticulate, somewhat triplinerved near the base, 2 to 4 inches long, 5 to 9 lines wide toward the base, often short-petioled; uppermost commonly alternate: bracts of the involucre from oblong-ovate to lanceolate and either acute or acuminate, glabrous, shorter than or sometimes equalling the yellowish or at length brownish disk; its bracts nearly glabrous: rays about an inch long, oblong. — Chapm. Fl. Suppl. 629. *H. angustifolius*, var. with broader leaves and yellow disk, Chapm. Fl. 229; Curtiss, distrib. 1437. — N. and E. Florida, *Chapman, Palmer, Garber, Curtiss*; flowering late.

H. ciliáris, DC. Glaucous: stems a foot or two high, very leafy: leaves nearly all opposite and sessile, lanceolate, varying to ovate-lanceolate or to linear, thickish, with undulate or repand margins, either very smooth and naked, or hispidulous with some scattered bristles, at least along the margins: bracts of the involucre ovate or oblong, obtuse or abruptly mucronulate, hirsutely ciliate: chaff of the brownish disk pubescent at tip: rays few or several, not surpassing the disk, sometimes none. — Prodr. v. 587; Gray, Pl. Fendl. 84, Pl. Wright. i. 108. — *Linsecomia glauca*, Buckley in Proc. Acad. Philad. 1861, 458. — Low and brackish ground, S. W. Texas, on the Rio Grande, and Arizona, *Berlandier, Wright, Thurber*, &c. (Adj. Mex.)

+ + Stems somewhat hirsute, scapiform and monocephalous: leaves roundish: chaff of receptacle with entire cuspidate-acuminate tips: rays commonly wanting: akenes rather flat, emarginate-2-toothed at summit: an anomalous species. — *Echinomeria*, Nutt.

H. rádula, TORR. & GRAY. Leaves hirsute or hispid, denticulate, triplinerved, mostly sessile; radical orbicular, 2 or 3 inches long, in a rosulate cluster; cauline 2 or 3 pairs near the base of the (foot or two high) simple stem, obovate with narrowed base, or above reduced to some narrow and minute ones: involucral bracts broadly lanceolate or oblong, acute, brownish-purple, as is the disk: the few rays when present little exserted: akenes with the unusually acute margins produced above more or less into a sort of tooth: pappus small. — Fl. ii. 321. *Rudbeckia radula*, Pursh, Fl. ii. 575. *R. apetala*, Nutt. Jour. Acad. Philad. vii. 77. *Echinomeria apetala*, Nutt. Trans. Am. Phil. Soc. vii. 356. — Low pine barrens, Georgia, Alabama, and Florida.

+ + + Stem and (mostly opposite and triplinerved) leaves more or less hispid, hirsute, or scabrous (or forms of the last species smoother): chaff of receptacle entire or some 3-toothed at the apex, pointless: rays numerous and conspicuous.

++ Disk of the head dark purple or brownish.

H. heterophýllus, NUTT. Stem slender, 1 to 3 feet high, naked above, bearing a single showy head: leaves hispid, entire; radical oval to spatulate-oblong; cauline 3 or 4 pairs and some minute ones above, narrowly lanceolate or linear, sessile, the lower with long tapering base: bracts of the involucre lanceolate, acuminate, ciliate: rays about 20, inch and a half long. — Jour. Acad. Philad. vii. 74; Hook. Comp. Bot. Mag. i. 98, partly; Torr. & Gray, l. c. — Low pine barrens, Georgia to Florida and Louisiana, toward the coast.

H. atrórubens, L. Stems stouter, sometimes leafy, sometimes few-leaved, 2 to 4 feet high, bearing few or several rather small heads: leaves hispid and scabrous, veiny, commonly thinnish, from roundish-ovate or rarely cordate to oblong-lanceolate, often serrate, contracted below into winged petioles, lower a span to a foot long, uppermost small and bract-like: bracts of the involucre oval or obovate, obtuse, cilioate: rays 10 to 16, rarely inch long. — Spec. ii. 906, & ed. 2, 1279 (Dill. Elth. t. 94; Martyn, Hist. Pl. Rar. t. 20); Ait. Kew. iii. 250; Michx. Fl. ii. 140, in part; Torr. & Gray, l. c., not Lam., nor Bot. Mag., Bot. Reg., &c. *H. sparsifolius*, Ell. Sk. ii. 415. *H. silphioides*, Nutt. Trans. Am. Phil. Soc. vii. 366. — Open woods, Virginia to Florida, Arkansas, and Louisiana.

H. rígidus, DESF. A foot or two (rarely 6 to 8 feet) high, rigid, sparingly branched: leaves very firm-coriaceous and thick, both sides hispidulous-scabrous, shagreen-like, entire or serrate, lightly triplinerved but indistinctly and sparingly veined; lower oblong and ovate-lanceolate, attenuate at base into short winged petioles; upper mostly lanceolate: heads comparatively large, showy (disk three-fourths inch high): involucre pluriserially imbricated; its bracts mainly ovate, obtuse or acutish, rigid, appressed, densely and minutely ciliate: rays numerous, generally inch and a half long: akenes oblong-obovate, 3 lines long: pappus of two large ovate-lanceolate paleæ, and sometimes two or four rather stout intermediate paleæ! more commonly none. — Cat. Hort. Par. ed. 3, 184; Torr. & Gray, Fl. ii. 322. *H. atrorubens*, Michx. l. c., in part; Bot. Reg. t. 508; Bot. Mag. t. 2668; DC. Prodr. v. 586. *H. diffusus*, Sims, Bot. Mag. t. 2020. *H. Missuricus*, Spreng. Syst. iii. 618, name in place of *diffusus.* *H. scaberrimus,* Ell. Sk. ii. 423. *H. Missouriensis* (Schweinitz) & *H. crassifolius,* Nutt. Trans. Am. Phil. Soc. l. c. *Harpalium rigidum,* Cass. Dict. Sci. Nat. xx. 200; DC. Prodr. v. 583, founded on the form with intermediate paleæ to the pappus. Plains and prairies, Saskatchewan and Michigan to W. Georgia, Texas, and eastern part of Colorado. Sometimes the disk-corollas are at first yellow!

++ ++ Disk yellow. (Here the Californian *H. gracilentus* would be sought.)

— **H. lætiflórus,** Pers. Resembles tall forms of the preceding, similarly scabrous or hispid, leafy: leaves commonly thinner, mostly oval-lanceolate, acuminate at both ends, 4 to 10 inches long, more or less serrate: heads usually several and rather short-peduncled: disk half-inch high: bracts of the involucre imbricated in only 2 or 3 series, from ovate- to oblong-lanceolate, acuminate or attenuate-acute, hirsutely ciliate or ciliolate, occasionally a little hirsute on the back: rays numerous, the larger inch and a half long.— Syn. ii. 476; DC. Prodr. v. 586, excl. syn. Ell.; Torr. & Gray, l. c. *H. atrorubens,* Lam. Dict. iii. 86, not L. — Prairies and barrens, Indiana, Illinois, Wisconsin.

Var. **tricúspis,** Torr. & Gray, l. c. Leaves less serrate: chaff of receptacle more commonly 3-toothed. — *H. tricuspis,* Ell Sk. ii. 422. W. Georgia, ex *Elliott.* Needs confirmation.

— **H. púmilus,** Nutt. Hispid and scabrous throughout: stems simple, a foot or two high, bearing 5 to 7 pairs of leaves and a few rather short-peduncled heads: leaves mostly ovate-lanceolate, acute, entire or nearly so (1½ to 4 inches long), rigid, abruptly contracted at base into a short margined petiole: involucre less than half-inch high, white-hirsute or scabrohispidulous; its bracts imbricated in about 3 series, oblong-lanceolate, acutish: rays about inch long.— Trans. Am. Phil. Soc. vii. 366; Gray in Am. Jour. Sci. ser. 2, xxxiii. 239. — Eastern Rocky Mountains and adjacent plains of the Platte, &c., from Wyoming to Colorado, *Nuttall, Hayden, Geyer, Parry, Hall & Harbour,* &c.

H. occidentális, Riddell. Stem slender, 2 or 3 feet high, sometimes smooth and glabrous, usually leafy only at and near the base: radical and lowest cauline leaves ovate to lanceolate-oblong, entire or denticulate, contracted at base into long margined petioles, minutely hirsute or hispidulous, moderately scabrous; upper cauline a few remote pairs, subsessile, lanceolate, and bract-like, of an inch or half-inch in length: heads few or sometimes solitary, small: bracts of the involucre ovate to lanceolate, acute or acuminate, glabrous, or the margins sometimes ciliate, sometimes naked: rays half-inch to nearly inch long: akenes when young and at summit pubescent. — Suppl. Cat. Ohio Pl. (1836), 13; Torr. & Gray, Fl. ii. 323. *H. heterophyllus,* Short, Cat. Kentucky Pl. Suppl. 3; Hook. Comp. Bot. Mag. i. 98, partly, not Nutt. — Prairies and oak barrens, in dry ground, Michigan to Kentucky and Missouri.

Var. **plantagíneus,** Torr. & Gray, l. c. Minutely puberulent and slightly or not at all scabrous: leaves rather more rigid: involucre obscurely ciliolate or naked. — Texas, *Drummond, Lindheimer, Wright.* (Adj. Mex.)

Var. **Dowelliánus,** Torr. & Gray. Like the preceding, but leafy to the middle or higher, the leaves larger and mostly ovate, and stem sometimes branching. — Fl. ii. 504. *H. Dowellianus,* Curtis in Am. Jour. Sci. xliv. 82. — Mountain region in the southwestern part of North Carolina, *Curtis, Buckley,* &c.

* * * Involucre looser and the bracts disposed to be more taper-pointed, or elongated, or foliaceous (closer and shorter in some species): disk except for the dark anthers yellow or yellowish.

+— Canescent or cinereous, at least the foliage, with soft and fine appressed (but not tomentose) pubescence: leaves all opposite, sessile, merely serrulate: heads middle-sized: bracts of the involucre imbricated; their attenuate tips seldom or little surpassing the disk: Atlantic species.

H. cinéreus, Torr. & Gray. A foot or two high, barely cinereous throughout with minute and slightly scabrous appressed pubescence: stem simple, somewhat equably leafy, bearing one or two slender-pedunculate small heads: leaves coriaceous, lanceolate-oblong, acute; lower (3 inches long) contracted into a rather long narrowed base; uppermost (about inch long) ovate-lanceolate with a broad sessile base: involucre half-inch high; its bracts lanceolate-subulate, canescent: rays 10 or 12, two-thirds inch long. — Fl. ii. 324, excl. var. — Texas, *Drummond.* Heads little larger than those of *H. occidentalis,* of which it may be a hybridized offspring.

← **H. móllis**, Lam. Canescent throughout: stems 2 or 3 feet high, very leafy, when young villous, in age often hirsute or hispid, simple and with solitary or few rather large heads, or branched above and more floriferous: leaves ovate-lanceolate or ovate with a cordate closely sessile or a clasping base, attenuate-acute or acuminate, 3 to 5 inches long, whitened with a soft pubescence, or the upper face becoming greener and scabrous: involucre two-thirds inch high, villous or sericeous: rays 15 to 25, an inch or more long. — Dict. iii. 85 (1789); DC. Prodr. v. 587; Torr. & Gray, l. c., not Willd., &c. *H. canescens*, Michx. Fl. ii. 140. *H. pubescens*, Vahl, Symb. ii. 92 (1791); Willd. Spec. iii. 2240; Ell. Sk. ii. 418; Hook. Comp. Bot. Mag. i. 98. — Dry barrens, Ohio to Iowa and south to W. Georgia and Texas. Well-marked species, but passes into a greener or less pubescent and somewhat scabrous variety.

+ + Soft-villous rather than tomentose (varying to merely pubescent) as to the lower face of the mostly alternate ample leaves, but the tall stem villous hirsute or even hispid: heads rather large: involucre loose and long: disk grayish: the corolla-lobes as well as the tips of the chaff externally hirsute!

H. tomentósus, Michx. Stems stout, 4 to 9 feet high, branching: leaves thinnish, ample (the larger cauline a foot long), from ovate to oblong-lanceolate, acuminate at both ends, mostly somewhat petioled, sparingly serrate, upper face scabrous: heads nearly inch high and broad: bracts of the involucre linear-lanceolate and long-attenuate into almost filiform tips, externally hirsute, especially the margins, squarrose-spreading, often much surpassing the disk, outermost sometimes large and foliaceous: rays pale yellow, an inch or more long. — Fl. ii. 141; Ell. Sk. ii. 424; Torr. & Gray, l. c. *H. pubescens*, Bot. Reg. t. 524, but not that of Hort. Kew., &c. *H. squarrosus*, Nutt. Trans. Am. Phil. Soc. vii. 367. *H. spathulatus*, Ell. Sk. ii. 421, a form with mostly opposite leaves and less prolonged involucral bracts. — Moist woods, Illinois? and Virginia to Georgia and Alabama, most common along the mountains, in the lower country with leaves less pubescent beneath.

+ + + Leaves mostly scabrous both sides (in one sometimes soft tomentose-canescent beneath), the upper disposed to be alternate and not triplinerved, mostly petiolate and not broad: heads middle-sized.

++ Atlantic species: involucre loose or squarrose; its bracts linear-subulate or gradually attenuate from a narrowish base to a slender point, all nearly of the same length, equalling or surpassing the dull yellow disk: all producing slender creeping rootstocks and also forming one or more fleshy thickened roots (like tap-roots) at base of stem.

← **H. grosse-serrátus**, Martens. Stem very smooth and glabrous, commonly glaucous, 6 to 10 feet high, bearing numerous rather cymosely disposed and short-peduncled heads: leaves (not rarely some even of the uppermost opposite) slender-petioled, thinnish, oblong-lanceolate or narrower, or some of the cauline almost deltoid-lanceolate, gradually acuminate, sharply serrate (sometimes with long salient teeth), or upper merely denticulate, slightly scabrous above, whitish and minutely tomentulose or soft-puberulent beneath; larger cauline commonly 8 to 10 inches and the petiole an inch or two long: heads fully half-inch high, and deep yellow oblong rays over an inch long: bracts of the involucre mostly slender. — Sel. Sem. Hort. Lovan., & Linn. xiv. Suppl. 133; Torr. & Gray, Fl. ii. 326. — Dry plains and prairies, Ohio to Dakota, Missouri, and Texas. Eastward the smaller-leaved forms seem to pass into *H. giganteus*.

Var. **hypoleúcus**. Leaves almost silvery-canescent with fine and dense soft tomentum, the larger with either cuneate or truncate base. — Texas, *Drummond, Lindheimer, Wright*. (Var. γ, Torr. & Gray, in part.)

H. gigantéus, L. Stem hispidulous or scabrous, or below smooth, 3 to 10 feet high, commonly one or more of the roots becoming thick and tuber-like; the larger plants branching above, bearing scattered heads: leaves lanceolate or oblong-lanceolate, green and more or less scabrous both sides, tapering to base and summit, short-petioled or subsessile, minutely serrate or denticulate, occasionally nearly entire, commonly only 3 to 5 inches long: heads of the preceding or smaller: rays pale yellow, barely inch long. — Spec. ii. 905; Ait. Kew. iii. 249; Willd. Spec. iii. 2242; DC. l. c.; Torr. & Gray, Fl. ii. 325, excl. β. *H. altissimus*, L. Spec. ed. 2, ii. 1278; Jacq. Hort. Vind. t. 162. *H. gigas*, Michx. Fl. ii. 141. A low and mainly northern form is *H. tuberosus*, Parry in Owen Rep. Minnesota Surv. 614, and *H. subtuberosus*, Bourgeau in herb. Hook., "the Indian Potato of the Assiniboine tribe," the so-called "edible tubers" (which were also long ago noted by Douglas) being tuber-like thickened roots. — Moist or wet ground, Canada to Saskatchewan, and south to Alabama and Louisiana. Very variable: the var. *ambiguus*, Torr. & Gray, l. c., is intermediate between this species and *H. divaricatus*, probably a hybrid.

H. Maximiliáni, SCHRADER. Hispidulous-scabrous: stem stout, 2 or 3 (and even 10 to 12) feet high, below mostly rough-hispid: leaves almost all alternate, thickish, becoming rigid, very scabrous above, lanceolate, acute or acuminate at both ends, mostly subsessile, all entire or sparingly denticulate: heads comparatively large, short-peduncled, terminating somewhat simple stem or branches, and later in the axils of many of the cauline leaves: involucre of more rigid bracts: rays numerous, often inch and a half long, golden yellow: flowering late. — Ind. Sem. Hort. Gœtt. 1835; DC. Prodr. vii. 290; Torr. & Gray, Fl. ii. 325; Gray, Pl. Lindh. i. 41 (with var. *asperrimus*, which is merely a rougher form); Meehan, Nat. Fl. ii. t. 37. — Rich prairies and plains, west of the Mississippi, and from Saskatchewan and Minnesota to Texas.

++ ++ Pacific species: leaves mostly lanceolate, broader toward the base and tapering to an acute or acuminate apex, short-petioled or subsessile: involucre of narrow or small bracts: rays about inch long.

= Bracts of the involucre linear- or lanceolate-subulate, attenuate, fully equalling the disk. herbaceous, loose or soon squarrose-spreading: stem usually smooth and glabrous, except at the summit.

H. Nuttállii, TORR. & GRAY. Stem slender, 2 to 4 feet high, commonly simple: leaves lanceolate or the upper linear (3 to 6 inches long, 3 to 9 lines wide, in small plants not rarely all opposite), serrulate or entire: heads half-inch high: bracts of the involucre naked or somewhat hirsute at base: disk-corollas slightly pubescent toward the base: paleæ of the pappus long and narrow. — Fl. ii. 324. *H. Californicus*, Nutt. in herb., not DC. — In wet soil, Rocky Mountains, from western part of Wyoming and Utah to Oregon, Washington Terr., and interior of Brit. Columbia.

H. Paríshii, GRAY. Resembles the preceding, 6 to 15 feet high: leaves elongated-lanceolate, softly cinereous-puberulent or even canescent beneath, scabrous above: heads half-inch high and rays 10 to 18 lines long: bracts of the involucre linear-subulate, longer than the disk, villous toward the base: disk-corollas with a silky-villous ring or two tufts above the short proper tube: paleæ of the pappus slender-subulate. — Proc. Am. Acad. xix. 7. — S. E. California, in wet places and along streams at San Bernardino, *Parish*; fl. autumn.

H. Califórnicus, DC. Tall, 3 to 8 feet high, usually branching: leaves lanceolate, entire or serrate (the larger 4 to 10 inches long, sometimes an inch or two wide): heads mostly two-thirds inch high: rays over an inch long when well developed: bracts of the involucre slightly hirsute or naked: disk-corollas canescently puberulent toward the base: akenes very glabrous: paleæ of the pappus broadly lanceolate. — Prodr. v. 589; Torr. & Gray, Fl. ii. 325; Gray, Bot. Calif. i. 353. *H. giganteus*, var. *insulus*, Kellogg, Proc. Calif. Acad. v. 17. — California, along streams, from San Francisco Bay southward.

Var. **Utahénsis** (*H. giganteus*, var *Utahensis*, Eaton, Bot. King Exp. 169) seems rather to be a form of *H. Californicus*, with thin and smoother leaves, and involucre more hirsute. — Wahsatch Mts., Parley's Park, Utah, *Watson*.

Var. **Mariposiánus.** Leaves ample; upper cauline ovate or oblong-lanceolate, entire (7 or 8 inches long by 2 or more wide): pappus not rarely of 4 linear-lanceolate paleæ of nearly equal length, or two often reduced and short. — Banks of the Merced at Clark's Ranch, Mariposa Co., California, *Bolander*.

= = Bracts of the involucre broader and short, erect.

H. graciléntus, GRAY. Stem 2 to 5 feet high, rough-hispidulous, the slender branches glabrous or scabrous: leaves thickish, scabrous and commonly hispidulous both sides, sparingly denticulate or entire; lower cauline from broadly to ovate-lanceolate, triplinerved near the base, which is abruptly contracted into a short margined petiole; upper lanceolate: heads slender-peduncled, half-inch high: bracts of the involucre imbricated in about 3 ranks, thickish, ovate or oblong-lanceolate, acute or apiculate-acuminate, shorter than the disk, scabrous-puberulent, usually ciliolate: chaff of receptacle with puberulent obtuse or abruptly acutish tips, below often purplish: rays 12 to 16, about inch long. — Proc. Am. Acad. xi. 77, Bot. Calif. i. 616. — Low plains and along water-courses, San Diego Co. to San Bernardino Co., California, *Palmer, Parry* & *Lemmon, Parish*, &c.

++   ++   ++   Imperfectly known Pacific species, probably perennial, with foliaceous involucre.

H. Douglásii, Torr. & Gray. Stems branching, ascending, hispidulous: leaves alternate; upper rhomboid-oblong to spatulate-lanceolate, tapering into winged petioles, obtuse, entire, inch or two long: head half-inch high: bracts of the involucre almost all foliaceous, hispidulous; outer narrowly oblong, mostly obtuse, reflexed or spreading, longer than the disk, innermost shorter, erect, acute or somewhat acuminate: rays barely half-inch long: chaff of receptacle entire. — Fl. ii. 332. — California, *Douglas* (mentioned in Bot. Beech. 253); near Santa Clara, *Sinclair*, in Bot. Sulph. as "*H. Californicus.*"

+— +— +— +— Leaves all or most of them opposite, at least the cauline, or in *H. tuberosus*, &c., the upper alternate, all triplinerved or 3-nerved: Atlantic species.

++   Heads remarkably small, only 4 or 5 lines high and rather narrow, loosely paniculate: rays only 5 to 8, seldom inch long: stem and spreading branches slender: leaves scabrous above, puberulent or canescent-tomentulose beneath.

H. parviflórus, Bernh. Stem smooth and glabrous, 3 to 6 feet high: leaves thin, nearly membranaceous, ovate-lanceolate or narrower, cuneately or almost truncately contracted at base into a half-inch or inch long partly margined petiole, gradually attenuate-acuminate, serrulate, sometimes more serrate (4 to 7 inches long, the larger inch or two wide near the base), pale and when young tomentulose or puberulent beneath; bracts of the campanulate involucre subulate-lanceolate, shorter than the comparatively few-flowered disk, the tips loose or squarrose: rays 5 or 6, commonly half-inch but sometimes nearly inch long. — Spreng. Syst. iii. 617 (1826, & probably somewhat earlier), not of HBK. Nov. Gen. & Spec., 1820 (*H. micranthus*, Spreng.), which perhaps is not of the genus. *H. divaricatus*, Michx. Fl. ii. 141; Ell. Sk. ii. 428, not L. *H. strumosus,* var. *pallidus*, Ell. l. c., ex Torr. & Gray. *H. tracheliifolius,* Hook. Comp. Bot. Mag. i. 98. *H. microcephalus,* Torr. & Gray, Fl. ii. 229. — Moist woods and along streams, Pennsylvania to Illinois, Upper Georgia, Arkansas, and Louisiana.

Var. attenuátus. Leaves narrowly lanceolate, 5 inches long, at most half-inch wide, very scabrous above, therefore connecting with the following. — Dry woods, near Tallulah Falls, Georgia, *J. Donnell Smith.*

H. Schweinitzii, Torr. & Gray. Stem hispidulous or minutely strigose-pubescent, 2 to 5 feet high: leaves of thicker texture, shagreen-scabrous above, canescently tomentulose beneath, lanceolate (the larger 4 to 7 inches long, inch or less wide) and with more tapering less petioled base, serrulate or nearly entire: involucre hirsute: rays 6 to 8, half-inch long. — Fl. ii. 330; Chapm. Fl. 231. — Dry ground, W. North Carolina to Middle Georgia.

++   ++   Heads small, half-inch or less high, few or scattered, slender-peduncled: rays 6 to 10: whole plant glabrous and smooth! except perhaps the edges of the leaves and involucral bracts: involucre campanulate, of thickish smooth bracts; the outer lanceolate with gradually attenuate-subulate spreading tips; inner ovate-lanceolate or broader, somewhat acuminate, erect: akenes a little hairy at the summit: usually but not always one or two conspicuous acute squamellæ or short paleæ on each side between the lanceolate or ovate principal paleæ of the pappus, sometimes united with their base (like stipules), caducous with them.

H. longifólius, Pursh. Stem 3 to 7 feet high, simple: leaves elongated linear-lanceolate (3 to 8 inches long, quarter to half inch wide), thickish, mostly entire, sessile, lowest cauline and radical tapering into slender margined petioles: rays about 10, narrow, half-inch long: chaff of the receptacle glabrous, commonly 3-toothed, narrow: proper paleæ of the pappus 2 or 3, the squamellæ thin and small. — Fl. ii. 571; Ell. Sk. ii. 417; Torr. & Gray, Fl. ii. 431. *Leighia longifolia,* Nutt. Trans. Am. Phil. Soc. vii. 365. — W. Georgia, in wet soil, *Lyon,* &c. Little known; no sufficient specimens seen.

H. lævigátus, Torr. & Gray. No creeping rootstocks and no fleshy-thickened roots: stem 2 to 5 feet high, glaucous: leaves lanceolate, very acute, subsessile, thickish, pale beneath, sparsely serrulate or the upper entire: rays 6 to 8, broad, usually inch long, bright yellow: chaff of the receptacle entire, more or less pubescent on the back: squamellæ or intermediate paleæ of the pappus rather large and firm, half or a quarter the length of the lanceolate or ovate proper paleæ, sometimes wanting. — Fl. ii. 330; Gray, Man. 256. — Alleghany Mountains in Virginia and N. Carolina. Occurs in two forms; one slender, simple, 2 or 3 feet high, with narrow leaves 3 to 5 inches long, half-inch or less broad (this possibly may be *H. longifolius*): the other larger, 4 to 6 feet high, branching, with ampler leaves, the larger cauline ovate- or oblong-lanceolate and 2 or 3 inches wide, and rays over an inch long, and it would appear to pass into *H. strumosus* except for the remarkable smoothness. Bracts of the involucre minutely ciliolate.

++ ++ ++ Heads middle-sized, at least half-inch high: rays usually but not always more than 10, an inch or more long: plant multiplying by creeping rootstocks. (Species difficult of extrication, either confluent or mixed by intercrossing.)

= Cauline leaves all sessile and even somewhat connate by a more or less narrowed base, those of the flowering branches not rarely alternate, none more than serrulate, no lateral basal ribs.

H. doronicoídes, LAM. Minutely pubescent and somewhat scabrous: stem 3 to 7 feet high: leaves ovate-oblong, narrowed from below the middle to both ends, moderately so below, lightly or indistinctly triplinerved much above the base, 4 to 8 inches long: involucre of loose subulate-linear and slender pointed bracts, soft-pubescent or hirsute: rays 13 to 18, a third to half inch broad, sometimes inch and a half long: ovary and akene glabrous.— Dict. iii. 84; Torr. & Gray, Fl. ii. 327, in part, excl. syn. Vahl, &c., not of Gray, Man. *H. pubescens*, Hook. Bot. Mag. t. 2778, not Vahl. *H. cinereus?* var. *Sullivantii*, Torr. & Gray, l. c. 324, appears to be a form of this. — Dry Ground, Ohio to Missouri, &c.

= = Cauline leaves sessile or nearly so by a rounded or subcordate and 3-nerved base, thence gradually narrowed to the slender apex, of rather firm texture: heads and rays comparatively small.

H. divaricátus, L. Stem simple to the summit or nearly, a foot to a yard high, mostly slender, rigid, usually smooth and glabrous below and hispidulous-scabrous at summit, bearing few short-peduncled heads: leaves green and scabrous both sides, appressed-serrulate, all the cauline opposite and horizontally divaricate (whence the name), commonly 4 or 5 inches long, and at base an inch or two wide: head only half-inch high. bracts of the involucre lanceolate-subulate, usually hirsute-ciliate: rays 8 to 12, at most an inch long. — Spec. ii. 906 (excl. syn. Moris. Hist. sect. 6, t. 7. f. 66); Ait. Kew. iii. 250; Willd. Spec. iii. 2244; Torr. & Gray, Fl. ii. 329. *H. truncatus*, Schweinitz in Ell. Sk. ii. 416. *Chrysanthemum Virginianum*, &c., Moris. Hist. sect. 6, t. 3, f. 62! — Dry and sandy or gravelly soil, Canada and Saskatchewan to Florida and Louisiana.

= = = Cauline leaves short-petioled or upper subsessile, serrulate or serrate with small erect teeth, or the uppermost entire, all triplinerved from near the base.

H. hirsútus, RAF. Stem simple or branched at summit, 2 to 4 feet high, rigid, commonly smooth below, rough and hispidulous above: leaves oblong-lanceolate and ovate-lanceolate, subsessile or short-petioled with a roundish or broad abrupt and rarely subcordate or sometimes rather cuneate base, thence gradually tapering to the point in the manner of *H. divaricatus*, scabrous above, somewhat so and little paler beneath: bracts of the involucre usually broadly lanceolate and acuminate, ciliate, unequal, commonly erect and not surpassing the disk: rays 12 to 15, rather broad, fully inch long. — Ann. Nat. (1820). 14; DC. Prodr. v. 591; Torr. & Gray, Fl. ii. 329. *H. diversifolius* & *H. hispidulus?* Ell. Sk. l. c. — Dry or moist soil, Ohio to Wisconsin and south to Georgia and Texas.

Var. **trachyphýllus**, TORR. & GRAY, l. c., a form from Arkansas, with thick very rough leaves, and larger heads with squarrose involucre.

Var. **stenophýllus**, TORR. & GRAY, l. c., a small form, with narrow lanceolate leaves almost sessile by a somewhat contracted base. — *H. strumosus*, var.? *leptophyllus*, Torr. & Gray, l. c., may be the same with smoother stem. — Louisiana and Texas.

H. strumósus, L. Rootstocks long and slender, often branching, thickened often into a narrow fusiform tuber at the apex: stem usually branching, 3 to 6 feet high, glabrous and very smooth and often glaucous, but summit and branches not rarely hispidulous: leaves oblong- or ovate-lanceolate, or the lower sometimes ovate, acute or acuminate, slightly serrate or some of them entire, bright-green and somewhat papillose-scabrous above, whitish beneath (either with or without minute tomentum), abruptly contracted or more tapering into a margined petiole (the larger 5 to 8 inches long and 2 wide): heads rather small (half-inch high), but the rays ample, 9 to 15, commonly oblong, an inch to inch and a half long: bracts of the involucre rather broadly or ovate-lanceolate, acuminate, sometimes with attenuate spreading tips. rarely surpassing the disk, ciliate, either glabrous or pubescent on the back: pappus not rarely with intermediate squamellæ, either free or adnate to the base of the paleæ. — Spec. ii. 905; Ait. Kew. iii. 249; Torr. & Gray, l. c. *H. lævis*, Walt. Car. 215? *H. neglectus*, Otto, in Berlin Garden, is either a glabrous form of this, or is *H. lævigatus*. — Open woods and banks, Canada to Wisconsin, Georgia, and Arkansas.

Var. móllis, TORR. & GRAY, l. c. Leaves canescently tomentulose beneath, not rarely subcordate, commonly larger (upper cauline not rarely 6 to 8 inches long): involucre looser, the bracts mostly with prolonged attenuate tips: there are similar forms without the pubescence, except when young. — *H. mollis*, Willd. Spec. iii. 2240, excl. syn. Michx.; Hook. Bot. Mag. t. 3689, not LAM. *H. macrophyllus*, Willd. Hort. Berol. t. 70, & Enum. 920. — Mass. to Iowa; commoner westward.

H. tracheliifólius, WILLD. Resembles the two preceding: leaves thinner, nearly of the same rather dull green hue both sides, all distinctly short-petioled, lower more sharply serrate: involucre of the following, i. e. the bracts all loose and spreading, linear-attenuate, hirsute, surpassing the disk, sometimes much prolonged and attenuate-foliaceous. — Spec. iii. 2241, & Enum. 920. *H. prostratus*, Willd. l. c. 2242, a weak form, decumbent in cultivation. — Moist or dry ground, Penn.? and Ohio to Wisconsin and Illinois.

= = = = Cauline leaves more conspicuously petioled, prominently serrate, thinnish or soft, veiny, commonly broad, the upper disposed to be alternate: stems mostly branching: involucral bracts loose, hirsute-ciliate.

H. decapétalus, L. Rootstocks rather slender, branching, more or less tuberous-thickened at apex: stem smooth and glabrous below, 2 to 5 feet high; the branches slightly pubescent or scabrous: leaves usually membranaceous, ovate or oblong-ovate, acuminate, saliently serrate, green both sides, either smooth and glabrous or above papillose-scabrous and slightly scabrous below, 4 to 8 inches long, the truncate or somewhat cuneate base abruptly contracted into a winged or naked petiole: bracts of the involucre narrowly lanceolate-linear or linear, thin, often foliaceous and surpassing the disk: rays 8 to 10 or more, light yellow, only an inch long. — Spec. ii. 905; Ait. l. c.; Willd. l. c.; Hook. Bot. Mag. t. 3510; Torr. & Gray, l. c. *H. frondosus*, L. Amœn. Acad. iv. 290, & Spec. ed. 2, ii. 1277, merely a form with foliaceous involucre. *H. strumosus*, Willd. l. c. 2422; Ell. Sk. ii. 420. *H. tenuifolius*, Ell. l. c., thin-leaved form of shady places. — Banks of streams and moist woods, Canada to Michigan, Illinois, Kentucky, and Georgia, in the upper country.

Var. **multiflórus**? *H. multiflorus*, L. l. c.; Bot. Mag. t. 227, known only in cultivation, from early times; must have been derived from *H. decapetalus*. It has short and thick rootstocks, somewhat firmer leaves, on naked petioles, larger heads, more numerous bracts to the involucre, and 20 or more rays. The more common form of it in gardens is dwarf, and the disk filled with transformed ligulate flowers.

H. tuberósus, L. (JERUSALEM ARTICHOKE.) Stem usually pubescent or hirsute, 5 to 10 feet high, branching at summit: leaves mostly alternate on the branches, and sometimes on the upper part of the stem, ovate or subcordate, sometimes oblong, acuminate, thickish-membranaceous, dull green, minutely pubescent and occasionally cinereous beneath, soon scabrous above: bracts of the involucre lanceolate, attenuate-acuminate, hirsute, at least the margins toward the base: rays often inch and a half long, 12 to 20: bracts of the receptacle hirsute-pubescent on the back: akenes more or less pubescent at summit and margins, mostly long and slender: horizontal rootstocks enlarging at apex into either oval or fusiform fleshy tubers (in cult. large and oblong or roundish, sweet and edible). — L. Spec. ii. 905 (excl. habitat); Jacq. Hort. Vind. t. 161; Trumbull & Gray in Am. Jour. Sci. ser. 3. xiii. (May, 1877), 347; Decaisne in Fl. Serres, xxiii 1881. *H. doronicoides*, Torr. & Gray, Fl. ii. 327, in part; Gray, Man. 257, not LAM. — Moist alluvial ground, Upper Canada to Saskatchewan, and south to Arkansas and middle parts of Georgia. Was cultivated by the aborigines, and the tubers developed; now widely dispersed under cultivation. Among the various indigenous forms the following may be distinguished.

Var. **subcanéscens**. Mostly dwarf (about 2 feet high), comparatively small-leaved, rough-hispidulous or scabrous, but the lower face of the leaves whitish with soft and fine pubescence. — Plains of Minnesota, Dakota, &c., *Kennicott, Coues, Ward*, sometimes with well-developed tubers. Also, a larger form with narrower leaves, near St. Louis, Missouri, *Engelmann*, &c.

H. DEALBATUS. A foot or two high from a frutescent base, canescent with fine appressed pubescence: leaves ovate to oblong, obtuse, entire or repand, 3-nerved at the rounded or abruptly contracted base (about inch long), rather long-petioled; lower opposite, upper alternate: head solitary, terminating simple stems or few branches, slender-peduncled, barely half-inch high: involucre short-campanulate, canescent, of oblong-linear obtuse bracts, shorter than the fuscous disk: rays 4 or 5 lines long: akenes turgid, sericeous-pubescent. — Lower California,

Belding, 1875. At All-Saints Bay, 70 miles below the U. S. boundary, *Parry*, 1883; perhaps therefore within the U. S. A singular species, with aspect of a *Viguiera*, but a caducous pappus of two lanceolate paleæ and no squamellæ.

105. **FLOURÉNSIA**, DC. (*M. J. P. Flourens*, a distinguished physiologist.) — Founded on two homogamous northern Mexican species, of very distinct habit and character, shrubby, almost glabrous, somewhat resiniferous-viscid, much branched, with alternate entire leaves, either corymbed or paniculate short-peduncled heads from upper axils, and whitish or yellowish flowers. To these the founder added two Chilian radiate species, viz. *F. corymbosa*, which is a *Viguiera* (*V. Pæppigii*); and *F. thurifera* (*Helianthus thurifer*, Molina), which may probably remain as a subgenus, *Diomedia*, Bertero and Colla, not Cass. — DC. Prodr. v. 592, excl. no. 2; Gray, Proc. Am. Acad. xix. 7.

F. cérnua, DC., l. c. Very branching and leafy shrub, with the aromatic bitterness and odor of hops, 3 to 6 feet high : branches puberulent : leaves obovate and oblong, half to inch and a half long, acute at both ends, dull, obscurely veiny : heads seldom half-inch long, subsessile in the axils or terminating paniculate branchlets, soon nodding : involucre campanulate, shorter than the disk, of lanceolate erect imbricated bracts, with some outer and spreading foliaceous ones passing into leaves : tips of the short style-branches much dilated, wider than high : awns of the pappus rigid, half the length of the appressed-villous akene, the slender squamellæ not surpassing the villous hairs. — Gray, Pl. Wright. i. 114, & ii. 89. *Helianthus cernuus*, Benth. & Hook. Gen., ex Hemsl., but it is not really so referred, nor has it any likeness to that genus. — Arid hills and plains, W. Texas to Arizona, *Wright, Lemmon, &c.* (Adj. Mex., *Berlandier, Gregg*, &c.)

F. LAURIFÓLIA, DC. l. c., of N. E. Mexico, *Berlandier, Palmer*, is larger, with oblong and more veiny lucid leaves (2 to 4 inches long, on distinct petioles), corymbosely clustered heads of twice or thrice the size, &c.; may occur on the Lower Rio Grande.

106. **ENCÉLIA**, Adans. (*Christopher Encel*, wrote upon oak-galls.) — Herbs or some under-shrubby (all American, chiefly subtropical); with alternate or opposite leaves, commonly with rather showy radiate heads of flowers on naked peduncles; the rays mostly yellow, occasionally wanting; the disk yellow or brownish. Chaffy bracts of the receptacle usually soft and mainly scarious. — Benth. & Hook. Gen. ii. 378. *Encelia, Simsia* (Pers.), & *Armania* (Bertero); DC. Prodr., with *Geræa*, Torr. & Gray, & *Barrattia*, Gray & Engelm. Neglecting the pappus, which is inconstant, the four sections may be reduced to two.

§ 1. EUENCÉLIA. Akenes densely long-ciliate : upper and commonly most of the leaves alternate : petioles naked. — *Encelia*, Adans. Fam. ii. 128. *Pallasia*, L'Her. ex Ait., not L. f. *Geræa*, Torr. & Gray, &c.

* Shrubby or lignescent at base, with herbaceous flowering branches: leaves from ovate to oblong-lanceolate, mostly entire.

E. MICROPHÝLLA, Gray, Proc. Am. Acad. xv. 37, & xix. 7, of Northern Mexico, makes the nearest approach to *Flourensia*, and commonly has a biaristellate pappus.

E. ALBÉSCENS, Gray, l. c. viii. 658, of Sonora in Mexico, *Palmer*, appears to be more herbaceous than the following species ; the akenes less strongly villous on the edges, except next the summit, and the faces pubescent : pappus biaristellate. It may be expected in S. Arizona.

E. HALIMIFÓLIA, CAV. Ic. iii. 6, t. 210 (*Pallasia grandiflora*, Willd. Spec. iii. 2261), from "Nova Hispania," i. e. Mexico, probably from the Pacific side. This resembles *E. Californica*, and, being described as having green and glabrous leaves and ciliate involucral bracts, is very probably identified in a plant collected on the Xaqui River, Sonora, by *Palmer*, perhaps not far below the Mexican border of Arizona. It is probably also *E. conspersa*, Benth. Bot. Sulph., of Lower California.

— **E. califórnica,** NUTT. Woody only at base, 2 to 4 feet high, strong-scented, minutely pubescent and sometimes cinereous when young, at least the foliage glabrate and green: leaves from ovate to oblong-lanceolate, rarely denticulate or toothed, about 2 inches long: heads commonly solitary and large, the disk nearly inch broad, brownish or purplish : involucre white-villous. rays 16 to 20, an inch or more long, golden yellow · akenes obovate, with very shallow notch and no pappus ; the margins very long-villous. — Trans. Am. Phil. Soc. vii. 357 ; Torr. & Gray, Fl. ii. 317 ; Gray, Bot. Calif. i. 351. — Dry ground, California near the coast, from Santa Barbara to San Diego, thence east to the borders of Arizona, where is a smaller-flowered form, *E. conspersa*, Gray, Bot. Mex. Bound. 88, not Benth. ?

— **E. farinósa,** GRAY. Shrubby, except the nearly leafless flowering branches or corymbosely branched peduncles, 2 to 5 feet high. leaves (and the leafy branches) silvery-white with a close furfuraceous tomentum, ovate or ovate-oblong, obtuse, contracted at base into a rather long petiole · heads somewhat paniculate, smaller ; the disk only half-inch broad, yellowish · involucre short, barely pubescent rays 6 to 10, only half-inch long : akenes obovate, with a deep notch and no pappus. — Emory Rep. 143, & Bot. Calif. l. c.   *E. nivea*, Gray. Bot. Mex. Bound. 88, not Benth. — Dry hills, S. E. California and Arizona, first coll. by *Coulter.*

— **E. frutéscens,** GRAY. Shrubby below, 2 or 3 feet high, with widely spreading monocephalous branches, hispidulous-scabrous and at least the branches cinereous : leaves ovate or oblong, obtuse, half-inch or an inch long, abruptly petioled mostly from a rounded base: heads rather long-peduncled, variable in size : rays either none, few, or numerous, but short (quarter to half inch long) and 3-4-lobed : akenes very long-villous on the margins, with a small narrow notch at summit · pappus either none or of two delicate long-villous awns. — Proc. Am. Acad. viii. 657, & Bot. Calif. l. c.   *Simsia (Geræa) frutescens*, Gray, Bot. Mex. Bound. 89. — Gravelly hills and ravines, S. Utah, Arizona, and S. E. California; first coll. by *Fremont.*

* * Herbaceous perennial. leaves linear, entire.

E. scapósa, GRAY. Minutely scabrous-puberulent, a foot or more high : leaves all crowded at and near the base of the slender scapiform and simple monocephalous stem, rather rigid, entire, 2 or 3 inches long, a line or two wide : involucre loose : rays several, obovate or cuneiform, half-inch or less long, 3-toothed : akenes (immature) very villous all over, as also the pappus of two chaffy awns. — Proc. Am. Acad. xix. 7.   *Simsia? (Geræa) scaposa*, Gray, Pl. Wright. ii. 88. — New Mexico, and stony hills between the Mimbres and the Rio Grande, *Wright.*

* * * Herbaceous from an annual or biennial root (at least the first species): leaves apparently all alternate, somewhat dentate: awns of the pappus large and conspicuous, thick at base, continuous from the rather strong and very villous margins of the cuneate akene. — *Geræa*, Torr. & Gray, Proc. Am. Acad. v. 48.

— **E. eriocéphala,** GRAY. A foot or two high, hirsute with white hairs: stem simple or branched from the annual root, leafy below, nearly leafless toward the somewhat paniculate heads : leaves cuneate-obovate or ovate-oblong ; lower tapering into margined petioles, uppermost reduced to sparse subulate bracts : heads about half-inch high : bracts of the involucre linear-lanceolate, green, but the lower half and the margins very white with long villous pubescence : rays 12 or more, cuneate-obovate or spatulate, half-inch or more long, golden yellow : akenes cuneate, slightly emarginate between the thick-based awns. — Proc. Am. Acad. viii. 657 ; Bot. Calif. l. c.   *Geræa canescens*, Torr. & Gray, Proc. Am. Acad. v. 48.   *Simsia (Geræa) canescens*, Gray, Pl. Fendl. 85. — Low grounds and sand-hills, through the arid region of W. Arizona and adjacent parts of Nevada and S. E. California; first coll. by *Coulter*, then by *Fremont.*

Var. **paniculáta.** A greener and less hairy form, paniculately branched ; the numerous heads of only half the ordinary size. — S. Arizona, *Pringle.*

— **E. víscida,** GRAY. A foot or two high, branching, leafy up to the usually short simple peduncles, viscid-glandular and hirsutely villous : leaves thinnish ; cauline all ovate or oblong, obtuse, closely sessile and clasping by an auriculate or cordate base ; lower ones and base of stem not seen : heads nearly an inch high and broad : bracts of the viscid involucre oblong, obtuse, at length much shorter than the yellow disk : rays none : akenes narrowly cuneate, truncate between the awns. — Proc. Am. Acad. xi. 78, & Bot. Calif. ii. 616. — Mountains of San Diego Co., California, *Palmer, Parish.*

§ 2. SÍMSIA. Akenes naked, at least not strongly ciliate: leaves usually opposite or the upper alternate, broad, usually serrate, sometimes 3–5-lobed, not rarely auriculate-dilated at the insertion: herbs. — *Simsia*, Pers. *Barrattia*, Gray & Eugelm., merely wants the pappus.

　　* Root annual: petioles all naked at base: some uppermost leaves alternate.

E. exaristáta, GRAY. Stem 2 feet or more high, rather slender, minutely glandular-puberulent and sparsely villous-hirsute, naked at summit and bearing loosely paniculate heads: leaves ovate and oblong-ovate, barely serrate, rarely somewhat incised, on narrowly margined petioles: heads half-inch high, rather narrow: bracts of the involucre lanceolate; outer series villous-hirsute, more than half the length of the narrow and granulose-glandular inner ones: rays 4 to 9, not surpassing the disk: akenes very smooth and glabrous throughout, obovate, slightly emarginate at summit, destitute of pappus, or not rarely with two minute vestiges of awns. — Hemsl. Bot. Biol. Centr.-Am. ii. 183, & Proc. Am. Acad. xix. 8. *Simsia lagascœformis*, Gray, Pl. Wright. i. 107, not DC. *S. exaristata*, Gray, Pl. Wright. ii. 87. — Valleys along water-courses, Western Texas to Southern Arizona, *Wright, Thurber, Lemmon*. (Mex. ?)

　　** Root perennial, thick and fleshy-tuberous: leaves all opposite, even on the branches, on margined or narrowly winged petioles, these united at base on each side by a foliaceous appendage, the two often connate into an amplexicaul disk.

E. cálva, GRAY. Scabrous-pubescent and often hispidulous: stem 2 or 3 feet high, with opposite branches, terminating in long and naked monocephalous peduncles: leaves deltoid-ovate and subcordate, often hastately 3-lobed, irregularly dentate: involucre hemispherical, half-inch high, hirsute and hispid, outer bracts foliaceous and somewhat squarrose: rays 15 to 20, half-inch long: akenes wholly smooth and glabrous, obcordate-oval, without vestige of pappus. — Proc. Am. Acad. xix. 8. *Barrattia calva*, Gray & Eugelm. Proc. Am. Acad. i. 40. *Simsia (Barrattia) calva*, Gray, Pl. Lindh. ii. 228. — Rocky hills and edges of oak woods, Texas, *Lindheimer, Wright*, &c. (Adj. Mex., *Berlandier*.)

E. subaristáta, GRAY, l. c. Too closely like the preceding, sometimes more canescently hispid: akenes minutely pilose-pubescent, ciliolate toward the summit, bearing two rigid scabro-hispidulous awns, which are half the length of the akene or often reduced to mere rudiments. — *Simsia subaristata*, Gray, Pl. Fendl. 84. — S. W. Texas, *Wright, Palmer*. (Monterey, Mex., *Gregg*, &c.)

107. **HELIANTHÉLLA**, Torr. & Gray. (*Helianthus* with altered termination, the principal species resembling that genus.) — Perennial (N. American) herbs, of diverse habit, commonly simple-stemmed and entire-leaved: rays yellow: disk either yellow or purplish-brown. — Fl. ii. 233; Gray, Proc. Am. Acad. xix. 9.

§ 1. ENCELIÓPSIS. Silvery-canescent, scapose, with large heads (disk an inch broad and flat), thick-leaved: chaffy bracts of the receptacle soft and scarious: akenes flat, oblong-cuneate, very villous, with narrow callous margins and summit, the latter bordered between the short subulate awns by a very short fringe of membranaceously confluent squamellæ. Anomalous species.

H. nudicaúlis, GRAY. Cespitose, with a stout multicipital caudex, densely tomentulose-canescent: leaves all radical and rosulate-tufted, obovate or orbicular, obtuse, an inch or more long, abruptly contracted into a longer margined petiole: scapes naked, nearly a foot high, monocephalous: bracts of the involucre all canescent and lanceolate, numerous in 2 or 3 series, equal: rays 20 or more, linear, about inch long: disk-corollas also yellow; the short ovate teeth hispidulous-pubescent outside: immature akenes 4 lines long, including the short awns, which do not surpass the villosity. — Proc. Am. Acad. xix. 9. *Encelia (Geræa) nudicaulis*, Gray, Proc. Am. Acad. viii. 656. — S. W. Utah, *Capt. Bishop*. Caudelaria, Esmeralda Co, Nevada, *Shockley*.

H. argophýlla, GRAY, l. c. Said to be "2 or 3 feet high, leafy, with cauline leaves similar to the radical ones"; these very white with the dense silvery tomentum, rhomboid-obovate or cuneate and acute: mature akene 5 lines long. — *Tithonia argophylla*, Eaton, Bot. King Exp. 423. *Encelia (Geræa) argophylla*, Gray, l. c. — S. W. Utah, near St. George, *Palmer*. Incompletely known. Perhaps the two are not specifically different.

§ 2. HELIANTHÉLLA proper. Habit somewhat of *Wyethia*, leafy-stemmed: leaves from lanceolate to ovate, mostly triplinerved above the tapering base, and commonly venulose-reticulated, varying from opposite to alternate: rays broad: tube of the disk-corollas usually nearly half the length of the throat, and the short ovate lobes more or less puberulent: akenes flat, from cuneate-obovate and emarginate to slightly obcordate: style-appendages obtuse, mostly short and spatulate or oblong.

* Chaffy bracts of the receptacle soft and scarious: akenes with some long villous hairs on the margins and sometimes on the faces.

+ Heads showy, large or middle-sized, solitary, or some later ones in the axils of bract-like leaves below: bracts of the involucre loose and lanceolate-attenuate or linear, more or less foliaceous, conspicuously hirsute-ciliate: disk yellowish, with dark anthers.

H. quinquenérvis, GRAY. Somewhat hirsutely pubescent or almost glabrous: stems solitary or scattered, 2 to 4 feet high: leaves mostly opposite, oblong- or ovate-lanceolate, acuminate, 4 to 9 inches long, triplinerved below the middle and commonly with a lower pair at some distance, uppermost sessile, lower ones tapering into margined petioles, and the lowest (a foot or more long) into longer petioles: head mostly long-peduncled, ample, the disk a full inch in diameter: bracts of the involucre lanceolate, more or less foliaceous: rays 15 to 20, pale yellow, commonly inch and a half long: akenes cuneate-obovate and obscurely obcordate, 4 lines long, with margins and commonly a part of the faces long-villous: pappus of 2 slender awns, of half the length of the akene, and nearly thrice the length of the squamellæ, which form a conspicuous finely dissected fringe. — Proc. Am. Acad. xix. 10. *Helianthus quinquenervis*, Hook. Lond. Jour. Bot. vi. 247. *Helianthella uniflora*, Gray in Proc. Acad. Philad. 1863, 65; Porter & Coulter, Fl. Colorado, 71; Eaton in Bot. King Exp. 170; not Torr. & Gray, Fl., except as to one of Nuttall's specimens of *Leighia uniflora*, but not the original from Wyeth. — Rocky Mountains from Dakota and Montana to S. Colorado.

Var. **Arizónica**. Akenes obovate, even 5 lines long, with delicate awns rarely twice the length of the broader squamellæ. — Northern Arizona, *Woodhouse*, and S. W. Arizona, *Lemmon*.

H. Párryi, GRAY. Hispidulous-hirsute: stems numerous from a thickened root, a foot high, rather slender: leaves mostly alternate, more rigid, lanceolate and an inch or two long, or the lowest and radical oblong-spatulate and of double the size: heads and rays barely half the size of the preceding: pappus of fimbriately dissected squamellæ only, or with a pair of slender awns not surpassing these. — Proc. Acad. Philad. 1863, 68; Porter & Coult. Fl. Colorado, 71. — Rocky Mountains of Colorado and northern part of New Mexico, at 8,000 to 10,000 feet, *Parry, Hall*, &c.

H. MEXICÁNA, Gray, Proc. Am. Acad. xi. 37, of San Luis Potosí, Mexico, ranks between the preceding and the following, but is nearer the former; has solitary heads, dark brownish disk of 4 lines in diameter, broad rays half-inch long, and almost linear leaves.

+ + Heads small: involucre more imbricated: rays few and hardly surpassing the dark-purple disk.

H. microcéphala, GRAY. Hispidulous-scabrous: stems numerous from a greatly thickened root, a foot or less high, slender, somewhat paniculately or corymbosely branched at summit and bearing several heads: leaves rigid, all but the lower alternate: radical lanceolate-spatulate; upper cauline nearly linear and sessile, inch long: involucre somewhat campanulate, 3 or 4 lines high; its bracts linear-oblong, mostly obtuse: rays not over 3 lines long: immature akenes villous, at least at the summit: pappus of several slender squamellæ intermixed with the long hairs, longer than the breadth of the ovary, two marginal ones often extended and awn-like. — Proc. Am. Acad. xix. 10. *Encelia (Geræa) microcephala*, Gray, Proc. Am. Acad. viii. 657. — Borders of Colorado and adjacent New Mexico and Utah, north to Rabbit Valley, *Newberry, Brandegee, Ward*.

— *H. Nevadensis*, Greene

* * Chaffy bracts of the receptacle rather firm-chartaceous: stems a foot or two high.

— **H. Douglásii**, Torr. & Gray, extended. Hirsute-pubescent with spreading hairs, at least the upper part of the stem: leaves mostly opposite and oblong-lanceolate; upper sessile or nearly so: disk of the head an inch broad: involucre hirsute: rays an inch long: akenes obovate, more or less ciliate-fringed: pappus a pair of elongated awns with more or less chaffy-dilated base, or sometimes (as in the original specimen) reduced to this base, and with mostly conspicuous squamellæ. — *H. Douglasii*, Torr. & Gray, Fl. ii. 334. *H. lanceolata*, Torr. in Wilkes Ex. Exped. xvii. 354, hardly of Torr. & Gray. Fl. — Dry ground, W. Idaho and E. Oregon and Washington Terr., *Douglas* (awns of pappus reduced, perhaps not constantly), *Spalding, Cusick, Brandegee*. Ciliation of ovary and akene variable, sometimes wanting except near the summit.

— **H. uniflóra**, Torr. & Gray, l. c. Minutely pubescent or somewhat scabridous, or glabrate: leaves more commonly opposite, sometimes all alternate, oblong-lanceolate (2 to 5 inches long); lower short-petioled: involucre pubescent or slightly hirsute: rays a full inch long: akenes more or less ciliate: pappus a pair of long awns and rather conspicuous squamellæ. — Gray, Proc. Am. Acad. xix. 10. *H. lanceolata*, Torr. & Gray, l. c. (*Leighia lanceolata*, Nutt. Trans. Am. Phil. Soc. vii. 365, which is said by Nuttall to have three or more heads, but of which we have only two or three flowers, is probably of this species). *H. multicaulis*, Eaton in Bot. King Exp. 170, small form. *Helianthus uniflorus*, Nutt. Jour. Acad. Philad. vii. 37, & *Leighia uniflora*, Nutt. Trans. Am. Phil. Soc. l. c., by char. and genuine specimens. — Rocky Mountains, Montana and E. Idaho to S. Utah, *Wyeth, Burke, Watson, Ward*, &c.

— **H. Califórnica**, Gray. Minutely scabrous-puberulent or almost glabrous. stems slender, rarely bearing 2 or 3 small heads: leaves more commonly alternate, lanceolate, nearly all tapering into slender or distinct petioles: rays half-inch or more long, usually little surpassing the involucre: akenes obovate, wholly glabrous, the roundish summit slightly notched at maturity, minutely 2-aristellate and with very short squamellæ, but whole pappus often obsolete in age, margins very obscurely ciliolate near the summit. — Pacif. R. Rep. iv. 103; Bot. Calif. i. 352. — California, from Napa Valley to the Sierra Nevada, from the heads of the Sacramento to Mariposa Co.; first coll. by *Bigelow*.

§ 3. Pseudo-heliánthus. Habit of the narrow-leaved *Helianthi*: slender and leafy-stemmed: leaves all linear and one-nerved, with revolute margins, alternate, hispidulous-scabrous: bracts of the involucre linear-attenuate, hispid, squarrose-spreading: rays long and narrow: style-appendages of the disk-flowers long and slender, hirsute: chaffy bracts of the receptacle rather rigid, obscurely 3-toothed at the apex: akenes less flat, the lateral angles being usually developed, or even quadrangular.

H. grandiflóra, Torr. & Gray, l. c. Stem 3 or 4 feet high (the base unknown): leaves somewhat broadly linear (2 lines wide by 2 inches or more long), strongly papillose-scabrous above: head nearly three-fourths inch high and broad: rays 16 to 20, inch and a half long: immature akenes broadly oblong, glabrous below, the acute almost winged margins produced on each side at apex into a chaffy tooth, and one or both of these commonly extended into a chaffy persistent awn, the salient border connecting them villous and minutely multi-squamellate. — E. Florida, *Leavenworth, Burrows*. Mature akenes not seen.

H. tenuifólia, Torr. & Gray, l. c. Stem 2 feet or more high, more slender and simple from a narrow somewhat moniliform horizontal tuber: leaves nearly filiform: head one half smaller: rays 10 to 15, an inch or more long: akenes slightly pubescent, quadrangular and moderately or the outer very little compressed, the anterior and posterior angles narrowly and acutely margined, these two and sometimes the other angles surmounted by a subulate or triangular short persistent chaffy and pointed tooth, and with some minute intermediate squamellæ. — Sandhills and dry pine barrens on and near the Apalachicola River, Florida, *Chapman, Mohr*.

108. **ZEXMÉNIA**, Llave & Lex. (Anagram of Ximenez, the genus being likened to *Ximenesia*.) — Mexican genus of numerous species, two of them reaching U. S., perennial herbs or some rather shrubby; with mostly opposite

leaves, and heads of yellow flowers, of moderate or small size, in ours solitary on slender peduncles terminating the branches. Ray-akenes commonly triquetrous and 3-awned; those of the disk either much compressed or thicker, with winged or bordered margins, but the wings variable. Style-branches of the hermaphrodite flowers with acute hispid tips. — Nov. Veg. Descr. i. 13; Gray, Pl. Wright. i. 12; Benth. & Hook. Gen. ii. 373. — The genuine species (*Lasianthæa*, Zucc., *Lipochætæ Americanæ*, DC.) form a marked group: some others too nearly approach *Wedelia* on the one hand, and *Verbesina* on the other.

— **Z. brevifólia**, GRAY. Much branched and below shrubby, 2 or 3 feet high, strigose-scabrous or hispidulous, and the branches cinereous: leaves (alternate on the branches!) small, less than an inch long, ovate and oval, mostly entire, short-petioled: heads solitary on slender peduncles terminating the branches, half-inch high: involucre between hemispherical and campanulate, of broad mostly ovate bracts imbricated in 3 or 4 series, the outer looser and partly foliaceous: rays 5 to 9, small: corolla-lobes glabrous: akenes obovate, flat, some nearly marginless, some at maturity conspicuously callous-winged, slightly narrowed at summit between the wings or margin and the subulate-attenuate awns; between the bases of these the free or partly united squamellæ are conspicuous, yet sometimes obsolete in age. — Pl. Wright. i. 112, & Bot. Mex. Bound. 92. — Rocky banks, S. W. Texas, *Wright, Parry, Palmer*. (Adj. Mex.)

Z. híspida, GRAY. Herbaceous and branched or many-stemmed from a barely lignescent base or root, strigose-hispid, about 2 feet high: branches terminated by solitary long-peduncled heads: leaves sessile or nearly so, lanceolate or the lower rhomboid-lanceolate, acute or acuminate and with acute or cuneate base, irregularly more or less serrate, sometimes with a pair of coarser salient teeth or lobes above the base: involucre biserial; the outer bracts more loose and foliaceous, lanceolate from a broader base, as long as the oblong inner ones: rays 7 to 9, orange-yellow, barely half-inch long: corolla-lobes puberulent-ciliolate: akenes obovate, either narrowly or (when well developed) broadly winged, or sometimes winged only near the summit, appearing obcordate, the pappus in the centre of the notch, consisting of a somewhat elevated cupule of united firm squamellæ and one or two (or in the ray 3) variable awns, these occasionally abortive or little exceeding the squamellæ; usually an appressed fleshy scale or protuberance on each side of the base of the akene. — Proc. Am. Acad. xix. 10. *Wedelia hispida*, HBK. Nov. Gen. & Spec. iv. 214, t. 371 (poor details, from flowers only); Bot. Reg. t. 543 (details copied from HBK.); DC. Prodr. v. 539, excl. syn. Cav.; Benth. & Hook. Gen. ii. 370. *Stemmodontia scaberrima*, Cass. Dict. xlvi. 407. *Lipochæta (Catamenia) Texana*, Torr. & Gray, Fl. ii. 357; Gray, Pl. Lindh. ii. 229. *Zexmenia Texana*, Gray, Pl. Wright. i. 112. *Wirtgenia Texana*, Schultz Bip. in Seem. Bot. Herald, 304. — Dry ground, common in Texas. (Mex.)

Z. podocéphala. Herbaceous from a lignescent root, 2 or 3 feet high, rough-hirsute or hispidulous: stems with few and slender branches, terminated by solitary long-peduncled heads: leaves ovate, nearly sessile by a rounded base, obtuse or acute, serrate, thinnish, very veiny (the larger 3 or 4 inches long): head and involucre nearly of the preceding, but corolla-lobes hispidulous: akenes obovate, with narrow at length callous wings, more or less confluent with the rather long awns; the intermediate squamellæ small and distinct, absolutely wanting in the original specimens of *Wright*, on which was founded *Verbesina podocephala*, Gray, Pl. Wright. ii. 92, and of *Schott*, but obvious in specimens of *Rothrock* and *Lemmon*. — S. Arizona.

109. VERBESÍNA, L. partly, Less. (Unmeaning name.) — American herbaceous or more or less shrubby plants; with heads of yellow or rarely white flowers. — Benth. & Hook. Gen. Pl. ii. 379, with part of *Actinomeris*, *Verbesina*, *Ximenesia*, DC.

§ 1. VERBESINÁRIA, DC. Heads narrow, mostly small, cymosely clustered or paniculate: involucre imbricated in two or more series, the bracts not elongated-foliaceous: rays (rarely wanting) few or several, styliferous and usually fertile:

disk from flattish to low conical: awns of the pappus not hooked: ours all perennial herbs. — Gray, Proc. Am. Acad. xix. 11.

* Akenes wholly wingless: receptacle nearly flat: flowers yellow, the rays 1 to 5, lanceolate: leaves opposite.

V. occidentális, Walt. Green and minutely pubescent or glabrous, 4 to 7 feet high, with erect narrowly 4-winged branches, leafy up to the short peduncles of the corymbosely paniculate open cymes: leaves ovate and the uppermost oblong-lanceolate, acuminate, acutely serrate, the larger about 8 inches long, contracted into a margined petiole: involucre oblong, 4 or 5 lines high: akenes obovate-oblong, pubescent. — Car. 213. *V. Siegesbeckia*, Michx. Fl. ii. 134; DC. Prodr. v. 616; Torr. & Gray, Fl. ii. 358. *Siegesbeckia occidentalis*, L. Spec. ii. 900, & Pl. Gronov. *V. Phœthusa*, Cass. Dict. li. 476, & lix. 149; DC. Prodr. l. c., but there are no squamellæ. *Phœthusa Americana*, Gærtn. Fruct. ii. 425, t. 169, f. 3, hairs at summit of akene exaggerated, and awns missing. *P. borealis*, Spreng. Syst. iii. 591. *Coreopsis alata*, Pursh, Fl. ii. 567, therefore *Actinomeris alata*, Nutt. Gen. 181. — Borders of woods and banks, S. Penn. to Illinois and Florida.

* * Akenes or most of them broadly winged at maturity, but variable: receptacle convex to conical: flowers both of ray and disk white or whitish: the anthers blackish: rays 3 to 5, obovate, short: leaves alternate.

V. Virgínica, L. Minutely tomentose-pubescent or puberulent, 3 to 6 feet high: stem or branches winged or wingless: leaves green and glabrate or minutely hispidulous-scabrous above, cinereous to canescent beneath, ovate or the upper narrower, from denticulate to coarsely serrate, contracted below into a winged petiole: heads small, 3 or 4 lines high, crowded on the irregular branches of the compound paniculate naked cyme: bracts of the involucre lanceolate, rather obtuse, erect, pubescent: awns of the pappus slender, sometimes obsolete. — Spec. ii. 901; Walt. l. c.; Michx. l. c.; DC. l. c.; Torr. & Gray, l. c. 359. *V. paniculata*, Poir. Dict. viii. 456. *V. microptera*, DC. l. c.; akenes sometimes but not always imperfectly winged. *V. polycephala*, DC. l. c., rather robust form. *V. villosa*, Nutt. Trans. Am. Phil. Soc. vii. 370, a tomentose form. *V. Texana*, Buckley in Proc. Acad. Philad. 1861, 458. — Rich dry soil, Penn. ? and Illinois to Florida and Texas. (Mex.)

Var. **laciniáta**. Leaves variously and irregular sinuate- or laciniate-lobed, rarely almost to the midrib; the principal lobes 3 to 5. — *Siegesbeckia laciniata*, Poir. Dict. vii. 158. *Verbesina laciniata*, Nutt. Gen. ii. 170. *V. sinuata*, Ell. Sk. ii. 411; DC. l. c.; Torr. & Gray, l. c. — Along the coast, S. Carolina to Florida.

§ 2. PTERÓPHYTON. Heads (solitary or scattered) comparatively broad: involucre more or less imbricated, all or at least the inner bracts erect or appressed: disk convex to oval and the akenes all erect in fruit; the receptacle from convex to conical: rays several to numerous, either neutral or styliferous (even in the same species), but almost always infertile: akenes flat: awns of the pappus not hooked, often obsolete or wanting: perennial herbs. — Gray, Proc. Am. Acad. xix. 12. Part of *Pterophyton*, Cass. & of *Actinomeris*, Nutt.

* Stems wholly wingless and marginless: leaves long and linear, not decurrent: bracts of the involucre narrow, the outer loose and disposed to become foliaceous.

V. longifólia, Gray. Stems slender, smooth and glabrous, 2 or 3 feet high, very leafy, branching at summit and bearing several heads: leaves alternate or some 2–3-nate, sessile, scabrous, reticulate-veiny and with prominent midrib, 4 to 9 inches long, quarter to half inch wide: head hemispherical, half-inch high, with flattish disk, often subtended by one or two linear leaf-like bracts: involucral bracts linear: rays about 15, neutral, inch long: akenes obovate, smooth, with narrow wing, a shallow notch, and no awns or rarely a rudimentary one. — Proc. Am. Acad. xix. 12. *Actinomeris longifolia*, Gray, Pl. Wright. ii. 89. — Mountains of S. Arizona, *Wright, Rothrock*, &c.

* * Stems wholly wingless: leaves ovate to oblong, sessile, not decurrent, mostly opposite: bracts of the involucre broader and closer: rays not rarely styliferous.

V. Wríghtii, Gray, l. c. Scabrous and mostly hispidulous: stems stout, 1 to 3 feet high, somewhat branching, bearing few or solitary long-pedunculate showy heads: leaves from broadly ovate to oblong, thickish, serrate, triplinerved: heads hemispherical, three-fourths inch high: bracts of the involucre oval or oblong, obtuse, in 2 or 3 series: rays about 12, oval or oblong, sometimes inch long, rarely wanting: akenes obovate, smooth, with either broad or narrow wings, and only minute callous teeth for pappus, or some of the inner with short awns: receptacle low. — *Actinomeris Wrightii*, Gray, Pl. Fendl. 85, & Pl. Lindh. ii. 229; Rothr. in Wheeler Rep. vi. 162, t. 8. — Rocky ground, W. Texas to Arizona, *Wright, Thurber*, &c. (Adj. Mex.)

V. Wárei, Gray, l. c. Scabrous, somewhat hispidulous: stem slender, a foot or two high, simple, leafless at the peduncle-like summit, which bears one or two small heads: leaves 4 or 5 pairs, narrowly oblong, obtuse at both ends, obscurely serrulate, reticulate-veiny, lucid; the upper very small: bracts of the involucre oblong-lanceolate, shorter than the ovoid-conical fruiting disk: "rays 3 or 4," small: akenes oblong, with narrow or rather broad wings, connected by an obscure epigynous border: pappus of 2 minute teeth or none. — *Actinomeris pauciflora*, Nutt. in Am. Jour. Sci. v. 301, & (§ *Achæta*) Trans. Am. Phil. Soc. vii. 364; Torr. & Gray, l. c. But there is a *V. pauciflora* of Hemsley in Mexico. — Florida, in low pine barrens near the coast, *Ware, Chapman*.

V. nudicaúlis, Gray, l. c. Scabrous-hirsute and hispidulous, 2 or 3 feet high; the naked summit of the stem or branches bearing a few mostly pedunculate small heads: leaves in numerous pairs, dull green, elliptical-oblong, obtuse at both ends or the apex acutish, acutely and irregularly serrate, loosely pinnately veined: bracts of the involucre oblong-linear, short: disk in fruit merely convex: rays 7 to 12, linear, an inch or more long, the head only quarter-inch wide: wings of the akene often one or both wanting, sometimes rather conspicuous: pappus 2-aristellate or obsolete. — *Helianthus? aristatus*, Ell. Sk. ii. 428. *Actinomeris nudicaulis*, Nutt. Trans. Am. Phil. Soc. l. c.; Torr. & Gray, l. c. — Dry sandy woods, Georgia, Alabama, and Florida.

* * * Stems winged by decurrence of the more or less broad sessile leaves. Leaves in our species only pinnately-veined: stems 2 or 3 feet high, simple or with sparing flowering branches.

— **V. heterophýlla**, Gray, l. c. Hispidulous-scabrous, below somewhat hirsute: lower leaves nearly all opposite, approximate, oblong or oval, obtuse, sometimes acute, minutely serrate (1 to 3 inches long), indistinctly veiny, decurrent into wings; those toward the naked summit and on the slender flowering branches small, lanceolate, soon reduced to linear bracts: heads somewhat paniculate, barely half-inch high in fruit, then with strongly convex disk: bracts of the involucre barely in 2 series, small, lanceolate; those of the receptacle very similar, rigid: rays 5 to 10, linear: akenes obovate, narrowly winged, 2-aristellate. — *Actinomeris heterophylla*, Chapm. in Bot. Gazette, iii. 6. — Dry pine barrens, E. Florida, *Chapman, Garber, Curtiss*. Related to the preceding.

— **V. helianthoídes**, Michx. Pubescent, stouter: stem usually winged up to the short peduncles: leaves alternate, or rarely some of the lower opposite, ovate-lanceolate or ovate, acuminate, serrate, transversely veiny-scabrous above, canescently soft-pubescent beneath, at least when young: heads few, fully half-inch high; the disk and receptacle at maturity either strongly convex or conical: involucre of 2 or 3 series of erect lanceolate bracts: rays 8 to 15, inch or more long: akenes somewhat pubescent or scabrous, rather broadly winged, 2-aristellate. — Fl. ii. 134; Pursh, Fl. ii. 565. *Actinomeris helianthoides*, Nutt. Gen. ii. 181; Ell. Sk. ii. 413; DC. Prodr. v. 575, & vii. 290 (vars. *Nuttallii* & *Elliottii*); Torr. & Gray, l. c. *A. oppositifolia*, DC. Prodr. vii l. c., not of Fresenius? — Prairies and open woods, Ohio to Iowa, Georgia, and Texas; first coll. by *Michaux*.

§ 3. XIMENÉSIA. Heads (solitary or scattered) broad: involucre of spreading linear and foliaceous equal bracts: disk and receptacle merely convex: rays numerous and conspicuous, usually fertile: akenes flat; the awns not hooked: root annual. — *Ximenesia*, Cav.

— **V. encelioídes**, Benth. & Hook. A foot or two high, freely branching, pale and cinereous or sometimes canescent with fine and soft appressed pubescence: leaves mostly alternate, and the upper face green, from ovate or cordate to deltoid-lanceolate, variously serrate or laciniate-dentate, some with nearly naked, most with winged petioles, and these commonly with auriculate-dilated appendage at base: heads large, the disk three-fourths inch in diameter: rays 12 to 15, inch long, deeply 3-cleft at summit: akenes obovate, mostly broadly winged and with short setiform awns; the outermost often awnless and pubescent, sometimes rugose and thick-winged. — *Ximenesia encelioides*, Cav. Ic. ii. 60, t. 178; DC. Prodr. v. 627 (under several varieties); Torr. & Gray, l. c. *Pallasia serratifolia*, Smith in Rees Cycl — Low grounds, Texas and S. Colorado to Arizona: also Florida, where it was probably introduced. Now widely dispersed in warm regions and cult. (Mex.)

110. **ACTINÓMERIS**, Nutt., partly. (From ἀκτίς, a ray, and μερίς, a part.) — Tall perennials, of the Atlantic U. S.; the somewhat simple stems (4 to 8 feet high) leafy to the top, below mostly winged in the manner of *Verbesina* by decurrent prolongations from the base of the leaves; these alternate or some lower ones occasionally opposite, lanceolate or broader, acuminate at both ends, pinnately veined, serrate, thinnish: heads loosely corymbose-paniculate: flowers yellow or white, produced in late summer. — Gray, Proc. Am. Acad. xix. 11. *Actinomeris*, spec. 1, Nutt. Gen. ii. 181, & § 1, 2, Torr. & Gray. Fl. ii. 335.

A. squarrósa, NUTT. l. c. Heads with 2 to 8 irregular yellow rays; disk-flowers yellow: involucral bracts linear to narrowly spatulate: akenes mostly with broad and firm wings: pappus of 2 or in marginal akenes 3 awns. — Ell. Sk. ii. 413 (excl. var. *alba*); Torr. & Gray, l. c. (forms *alternifolia* and *oppositifolia*, the latter of rare occurrence); Meehan, Nat. Fl. i. t. 39. *A. alternifolia*, DC. Prodr. v. 575. *A. oppositifolia*, Fresenius, Ind. Sem. Hort. Francf. 1836, an occasional form. *Coreopsis alternifolia*, L. Spec. ii. 909; Jacq. Hort. Vind. t. 110. *C. procera*, Ait. Kew. iii. 258. *C. acuta*, Pursh, Fl. ii. 569? *Verbesina Coreopsis*, Michx. Fl. ii. 134. — Rich or alluvial soil, either moist or dry, W. New York to Iowa, south to Florida and Louisiana. Wholly wingless-stemmed specimens occasionally occur.

A. álba, TORR. & GRAY, l. c. Rather smoother: heads smaller: rays none; disk-flowers dull white: akenes oftener wingless or narrowly winged: awns of the pappus not rarely 3 or 4, and commonly some interposed small ones or aristellate squamellae! — *A. squarrosa*, var. *alba*, Nutt. l. c.; Ell. l. c. *A. alternifolia*, var. *alba*, DC. l. c. *Athanasia paniculata*, Walt. Car. 201. *Verbesina Coreopsis*, var. *alba*, Michx. l. c. — Alluvial soil, S. Carolina to Louisiana, near the coast: rare in herbaria. Specimens from Dr. Mellichamp, S. Carolina, all exhibit the squamellate-aristellate pappus, not before known in this genus.

111. **SYNEDRÉLLA**, Gaertn. (Συνεδρία, a sitting together, the heads in the original species being collected at the nodes.) — Tropical annuals; with branching stems, opposite and more or less serrate petioled leaves, and small heads of yellow flowers, the rays short. — Fruct. ii. 456, t. 171; Hook. Exot. Fl. t. 60. *Synedrella, Oligogyne* in part, & *Calyptrocarpus* (Less.), DC. Prodr. v. 629.

S. viális, GRAY. Diffuse or procumbent, slender, strigulose-hirsute or more hairy: leaves ovate, about inch long: heads only 3 lines long, solitary or scattered, some subsessile, others slender-peduncled: principal bracts of the involucre 4 or 5, ovate or oblong: rays 5 to 8, with oblong exserted ligule: akenes or many of them tuberculate-scabrous at maturity, some of the outer occasionally trigonal, mostly flattened, and with or more commonly without a coriaceous and thickish undulated wing-like border, the central ones narrower and marginless: pappus of 2 or sometimes 3 rigid diverging awns, with occasionally one or two additional teeth or squamellae, arising from an obscure border. — Proc. Am. Acad. xvii. 217. *Calyptrocarpus vialis*, Less. Syn. 221, & Linn. ix. 269. *Oligogyne Tampicana*, DC. Prodr. v. 629; Deless. Ic. Sel. ix. t 38; Gray, Pl. Wright. i. 111. *Zexmenia hispidula*, Buckley in Proc. Acad. Philad. 1861, 458. — Waysides and waste grounds, southern borders of Texas. (Mex., S. Am.?)

112. **COREÓPSIS**, L. TICKSEED. (Κόρις, a tick, and ὄψις, resemblance, from the form of the akene.) — Herbs, mostly Eastern North American and opposite-leaved, of various habit; with pedunculate heads terminating the branches; the rays commonly showy, yellow, particolored, or sometimes rose-

colored: fl. summer and autumn. — Gen. no. 981; Torr. & Gray, Fl. ii. 338; Benth. & Hook. Gen. ii. 384, partly.

C. ÁSPERA, Pursh, Fl. ii. 570, and C. FLEXICAÚLIS, Raf. in Med. Rep., are not identified, and probably not of the genus.

§ 1. CALLIÓPSIS. Style-tips truncate or obtusely short-conical: akenes not villous-ciliate: outer involucre small, short and calyculiform, except in the last species: rays obovate or cuneate, inclined to be palmately 3-4-toothed or lobed. — § *Calliopsis, Coreoloma,* & *Cosmella,* Torr. & Gray, Fl.

* Perennials, with rose-red rays and yellow disk-flowers: akenes oblong, nearly straight, smooth. — § *Cosmella,* Torr. & Gray.

C. rósea, NUTT. Nearly glabrous, a foot or less high from slender creeping rootstocks, branched, leafy, bearing numerous small and short-peduncled heads: leaves opposite, linear or nearly so and entire, or the lower 2-3-toothed or 3-parted: involucre 2 or 3 lines high: rays rose-color, coarsely 3-toothed or lobed: akenes with merely callous margins and an obscure entire border at summit. — Gen. ii. 179; Bart. Fl. Am. Sept. t. 12; Torr. & Gray, Fl. ii. 348; Torr. Fl. N. Y. i. t. 57. *Calliopsis rosea,* Spreng. Syst. iii. 611. — Grassy swamps, coast of Mass. to Delaware and Georgia.

C. nudáta, NUTT. l. c. Very smooth and glabrous: stem 2 to 4 feet high from a thick or tuberous rootstock, rush-like, below bearing some alternate terete and filiform-subulate leaves (the larger a foot long), above some scattered smaller ones, gradually reduced to bracts, the naked summit forking and bearing a few slender pedunculate heads: involucre 4 or 5 lines high: rays showy, obscurely lobed, bright purplish rose-color, inch long: akenes with fimbriately or pectinately dissected wings, and two short upwardly hirsute subulate awns. — Torr. & Gray, l. c.; Hook. f. Bot. Mag. t. 6419. — Pine-barren swamps, Florida.

* * Perennials, with yellow rays, dark purple disk-flowers, and mostly entire leaves: akenes oblong or elliptical, straight, with fimbriate border or dissected wings and a pair of awns. — § *Coreoloma,* Torr. & Gray, excl. first species. § *Rhabdocaulis* & *Eublepharis,* Nutt. Trans. Am. Phil. Soc. vii. 359, excl. sp.

+ Stems not rarely alternate-leaved throughout, strict; the summit or flowering branches (bearing solitary or scattered heads) naked and rush-like, their leaves being reduced to small subulate bracts: cauline thickish and rather fleshy, especially when near brackish water, all tapering or contracted at base.

C. gladiáta, WALT. Glabrous, or young leaves not rarely pilose-pubescent: stem terete, 2 to 4 feet high, all the upper part naked: principal leaves alternate, from broadly obovate-oval to lanceolate-linear, obtuse, scarious-edged; lower 3 to 6 inches long, with long margined petiole dilated and partly clasping at insertion: involucre 4 or 5 lines high: rays commonly inch long: mature akenes bordered by a strong pectinate fringe, and surmounted by 2 short rigid awns or teeth which may not surpass the fringe. — Car. 215; Nutt. Gen. l. c.; Ell. Sk. ii. 244; Torr. & Gray, Fl. ii. 347. *C. dichotoma,* Michx. Fl. ii. 137, mainly and by the char. — Moist pine barrens, S. Carolina to Florida, in the low country, commoner near the coast.

C. angustifólia, AIT. Wholly glabrous: stem slender, mostly quadrangular, 1 to 3 feet high: leaves narrower and smaller than in the foregoing, sometimes all opposite; lower spatulate-lanceolate and the upper spatulate-linear: heads and rays smaller, the latter about half-inch long: akenes with narrow lacerate fimbriate wings and slender setiform awns. — Ait. Kew. iii. 253; Torr. & Gray, l. c. *C. dichotoma* in part, Michx. l. c. *C. linifolia,* Nutt. Jour. Acad. Philad. vii. 75. *C.* (*Rhabdocaulis*) *linifolia* & *angustifolia,* Nutt. Trans. Am. Phil. Soc. l. c. — Moist pine barrens or swamps, N. Carolina to Florida and Texas.

+ + Stems leafy to near the summit, and the leaves opposite.

C. integrifólia, POIR. Nearly glabrous, 2 or 3 feet high: leaves ovate or oblong, entire, only inch and a half long, rounded at base; upper almost sessile; lower abruptly petioled: rays three-fourths inch long: akenes not seen: ovary minutely serrulate-hispidulous on the margin, minutely awned. — Suppl. ii. 353; DC. Prodr. v. 570; Torr. & Gray, Fl. ii. 347, mainly. — Carolina, *Bosc* (originals in herb. Poir., now of Cosson, & herb. DC.), S. Carolina, *Ravenel,* and Georgia, Decatur Co., *Chapman.* Too little known. Ligules said by Poiret to be linear-oblong and entire, which does not accord with our specimens, nor with the group.

* * * Annuals or biennials, one has been thought perennial, with opposite leaves, yellow or particolored rays, and dark-purple or brown disk-flowers: akenes short, with entire scarious wings or none, incurved at maturity, one or both faces sometimes becoming papillose or tuberculate-roughened, or in some remaining smooth: heads scattered-paniculate: herbage glabrous or nearly so, except in *C. Drummondii*. — § *Calliopsis*, Torr. & Gray, slightly extended.

╉ Rays pure yellow: pappus a pair of conspicuous slender awns (or these rarely abortive): leaves from entire to 3-parted or simply pinnately divided. — *Coreopsidium*, Torr. & Gray.

C. Leavenwórthii, Torr. & Gray. Annual, sometimes seemingly perennial, slender, 1 or 2 feet high: lower leaves or their 3 to 7 divisions from broadly linear to spatulate-lanceolate: rays barely half-inch long: wings of the akene on each side as wide as the body, equalled or surpassed by the distinct erect awns. — Moist ground in pine barrens, Florida; first coll. by *Leavenworth*.

Var. Gárberi. Very remarkable form, more robust, all the cauline leaves pinnately 5-7-parted or divided into shorter and broader divisions; the terminal one from obovate to lanceolate-oblong. — Tampa, Florida, *Garber*: perhaps in over-luxuriant condition.

╉ ╉ Rays with base or lower part brown-purple: pappus none or minute: leaves all 1-2-pinnately divided.

╉╉ Akenes winged.

C. Atkinsoniána, Dougl. Root "perennial" or annual, flowering in autumn, stem 2 to 4 feet high: lobes of the leaves linear or nearly so: akenes with a narrow wing, sometimes a mere scarious margin, and usually a pair of distinct short subulate teeth for pappus. — Lindl. Bot. Reg. t. 1376; Torr. & Gray, l. c. 346. *Calliopsis Atkinsoniana*, Hook. Fl. i. 311. — River banks, Oregon, Washington Terr., and east to Dakota; first coll. by *Douglas*.

C. cardaminefólia, Torr. & Gray. Root annual: stem a span to 2 feet high: lobes of the lower or radical leaves oval to lanceolate, sometimes linear; of the upper mostly linear: rays rarely half-inch long: akenes with a moderately broad wing, with which is sometimes connected two obscure teeth. — Fl. l. c., with var. *lineariloba*, the narrowest-leaved form. *Calliopsis cardaminefolia*, DC. Prodr. v. 568. — Low grounds, W. Louisiana and Texas to Kansas and New Mexico. (Adj. Mex.)

╉╉ ╉╉ Akenes wingless: pappus none or an obscure border: annuals.

C. tinctória, Nutt. Glabrous, 2 or 3 feet high: radical and some lower cauline leaves 2-pinnately divided into lanceolate or linear divisions: upper with 3 to 7 linear divisions: outer involucre short and close: rays from half to three-fourths inch long, sometimes base only, sometimes nearly all crimson-brown: akenes oblong, thinnish, moderately incurved. — Jour. Acad. Philad. ii. 114; Bart. Fl. Am. Sept. t. 45; Bot. Mag. t. 2512; Bot. Reg. t. 846; Brit. Fl. Gard. t. 72; Torr. & Gray, l. c. "*Calliopsis bicolor*, Reichenb. Mag. t. 70." *C. tinctoria*, DC. l. c. 568; Hook. Bot. Mag. t. 3511, var. *atropurpurea*. — Low ground, Saskatchewan and Minnesota to Louisiana, Texas, and Arizona. Common everywhere in gardens.

C. Drummóndii, Torr. & Gray, l. c. Low, pubescent with many-jointed lax hairs, sometimes glabrous: divisions of the radical and lower cauline leaves from roundish-ovate to oblong-lanceolate; of the uppermost sometimes linear: peduncles inclined to be solitary terminating stem and branches: outer involucre of loose and spreading more foliaceous bracts, little shorter than the inner: rays broad, sometimes inch long, brown-purple only at base: akenes oval or obovate, thick, much incurved at maturity, a cartilaginous margin bordering the inner face. — *C. diversifolia*, Hook. Bot. Mag. t. 3474, not DC. *Calliopsis Drummondii*, Don, Brit. Fl. Gard. ser. 2, t. 315. — Sandy soil, Texas; first coll. by *Drummond*. Common in cultivation.

Var. Wrightii. Lobes of the leaves narrower, linear and the broadest linear-oblong: heads smaller: akenes circinately incurved. — Pl. Wright. ii. 90. — Rocky hills on the San Pedro, W. Texas, *Wright*.

§ 2. LEÁCHIA. Style-tips hispid or hirsute and abruptly produced into a cusp or acute cone: akenes nearly orbicular, incurved at maturity, some or all of them becoming papillose or muriculate at maturity, often developing a callus at base and apex of the inner face (this varying greatly even in the same head): pappus of two small chaffy teeth, or none: outer involucre little shorter and more herbaceous than the inner: rays cuneate, palmately 3-5-lobed or toothed, mostly yellow, as also the disk-flowers: narrow chaffy bracts of receptacle attenuate-filiform at apex: heads usually showy, on long and simple peduncles: leaves all opposite, entire or pinnately 3-7-parted, mostly petioled. — *Leachia*, Cass. Dict. x., xxv., lix. *Coreopsides*, Mœnch. *Chrysomelea*, Tausch. § *Eucoreopsis*, *Leachia*, Torr. & Gray.

* Root annual: style-tips almost truncate and with a short conical point: rays with some brown-purple lines or spots toward the base: leaves long-petioled. Transition to preceding section.

C. coronáta, Hook. Sparsely hirsute-pubescent or mainly glabrous, a foot or two high, lax: leaves entire or the lower 3-5-parted, obovate and spatulate-oblong, the lateral divisions when present small: bracts of the outer involucre lanceolate or oblong-lanceolate: rays an inch or less long, bright yellow, with deeper or orange hue at base, above which are delicate brownish-purple markings, thus forming a sort of corona: akenes with a rather broad wing and a pappus of 2 minute squamellate teeth. — Bot. Mag. t. 3460 (not L.); Torr. & Gray, Fl. ii. 345. — E. Texas, *Berlandier, Drummond, Lindheimer,* &c. Rather common in ornamental cultivation.

* * Root apparently perennial: style-tips with conspicuous cusps: rays sometimes brown-purple at base: heads small: cauline leaves hardly petioled, very slender.

C. Harveyána. A foot or more high, smooth and glabrous: stems slender, branching above: leaves pinnately parted into 3 to 7 and upper often palmately parted into 3 to 5 filiform divisions (no broader than the rhachis); lowest cauline and radical petioled and the divisions narrowly linear: involucre about 3 lines high: bracts of the outer involucre narrowly lanceolate-linear, little shorter than the inner: rays 3 or 4 lines long; disk-flowers brownish in age: akenes orbicular (only a line long), outer narrowly winged (and the wing occasionally laciniate-dentate), mostly muricate-roughened; inner smooth and wingless or nearly so; callus small or none: pappus a pair of obtuse short squamellæ. — Arkansas, on cliffs near Fort Smith, *Prof. F. L. Harvey.*

* * * Root perennial, or in the first species sometimes annual: rays yellow throughout (the larger inch long): style-tips with conspicuous cusp: calli of the akene often very large: pappus a pair of small denticulate or fimbriolate squamellæ, which become subulate teeth, sometimes deciduous or obsolete; at least lower leaves slender-petioled: species apparently confluent.

← Wings of the akene thin-scarious, outspread, broad when well developed.

C. grandiflóra, Nutt. Glabrous except the hirsute-ciliate petioles, rarely sparsely pilose, a foot or two high: radical and some lower cauline leaves lanceolate or spatulate and entire; upper or sometimes all the cauline 3-5-parted or divided, the divisions lanceolate or linear, or even almost filiform, sometimes again 2-3-parted: heads, &c., nearly of the next, usually larger: akenes with more conspicuous squamellate or paleaceous pappus. — Hort. Barclay & Trans. Am. Phil. Soc. vii. 358; Sweet, Brit. Fl. Gard. t. 175; DC. Prodr. v. 572; Torr. & Gray, Fl. ii. 344, with the vars. *longipes* & *subintegrifolia*. *C. longipes*, Hook. Bot. Mag. t. 3586; DC. l. c. *C. Boykiniana* & *C. heterophylla*, Nutt. Trans. Am. Phil. Soc. l. c. — Low grounds, Georgia to S. Missouri and Texas. Variable species: involucre 5 to 7 lines high: rays half-inch to inch long: foliage diverse.

C. lanceoláta, L. Low, only a foot or two high, including the long and simple naked peduncles: leaves ordinarily a few pairs, oblong-spatulate to lanceolate or nearly linear, obtuse, thickish, all entire, or rarely 1 or 2 small lateral lobes: rays commonly inch long and half-inch broad, sometimes smaller: pappus very small or obsolete. — Spec. ii. 908 (Martyn, Hist. Pl. t. 26; Dill. Elth. t. 48); Michx. Fl. ii. 136; Torr. & Gray, Fl. ii. 344. *Leachia lanceolata*, &c., Cass. *Chrysomelea lanceolata*, Tausch. — In rich or sandy damp soil, W. Canada and Illinois, Virginia, &c., to Florida and Louisiana. The ante-Linnæan figures well represent the species; the type glabrous or nearly so, except hirsute ciliation: passes into

Var. **angustifólia**, Torr. & Gray, l. c. (var. *glabella*, Michx. l. c., partly); a low form, with narrow leaves (2 to 4 lines wide) all crowded on the abbreviated stems, and scapiform peduncles about a foot long. — Shore of L. Superior to Florida.

Var. **villósa**, Michx. l. c. Leaves spatulate-obovate to oblong-lanceolate or oblong, villous-hirsute with many-jointed hairs, as also lower part of the stem. — *C. crassifolia*, Ait. Kew. iii. 253; Ell. Sk. ii. 434. *C. oblongifolia*, Nutt. Jour. Acad. Philad. vii. 76. Illinois to Florida.

— **C. pubéscens,** ELL. Taller, 1 to 4 feet high, more leafy, from pubescent to nearly glabrous: leaves thickish, oblong, or the lower oval-obovate and the upper oblong-lanceolate, often all entire, some not rarely with 2 or even 4 small lanceolate lateral lobes or divisions: heads usually smaller than in the preceding: akenes similar. — Sk. ii. 441; Chapm. Fl. Suppl. 630. *C. auriculata,* Schk. Handb. t. 260; DC. Prodr. l. c.; Torr. & Gray, Fl. ii. 343, in part (γ & δ), and of old gardens. *Leachia trifoliata,* Cass.? — Virginia to S. Illinois, Missouri, and south to Florida. In the middle or low country southward only a slender form, usually with lateral lobes to upper leaves; in the mountains a larger plant in all its parts, with larger leaves 3 to 5 inches long, 1 or 2 inches wide, all entire, or a few 3-parted, the var. γ, Torr. & Gray, Fl. l. c.

+ + Wings of the akene narrow, strongly involute and callous-thickened at maturity.

— **C. auriculáta,** L. Low and weak, stoloniferous, below commonly villous-hirsute: stems a foot or so high, including the long and slender peduncle, often simple: leaves of few pairs, ovate to round-oval, only an inch or so long, entire and some with a pair of smaller basal lobes, all but the upper slender-petioled: head comparatively small: rays little more than half-inch long: akenes by involution of margins oblong and umbilicate. — Spec. ii. 908 (Pluk. Alm. t. 242, f. 4, and perhaps t. 83, f. 5; Moris. Hist. iii. sect. 6. t. 3, f. 45); Michx. Fl. ii. 138; Ell. l. c. (var. *diversifolia*), Torr. & Gray, l. c., as to typical form, but the akenes were then unknown. *C. diversifolia,* DC. Prodr. v. 571, excl. syn. — Wooded ground, Virginia and Kentucky to the borders of Florida.

§ 3. EUCOREÓPSIS. Style-tips produced into a cusp or acute cone: akenes straight or little incurved, oblong, with narrow wing or none; no calli on the inner face: rays mostly entire or slightly toothed (yet sometimes 2–3-cleft) at the apex, pure yellow: disk-corollas yellow (sometimes dull, rarely turning brown): leaves opposite, in some seemingly verticillate. — Torr. & Gray, Fl., excl. *Leachia.*

* Perennials, mostly low (a foot or two high), leafy to the summit: leaves sessile, palmately divided or cleft, but never serrate, not veiny: involucre becoming rigid, its bracts all united at the base; outer oblong-linear, erect, about the length of the inner: rays from oblong to lanceolate: chaff of the receptacle linear-filiform and persistent: akenes oblong, narrowly wing-margined: pappus 2-toothed or 2-aristellate, or obsolete: stems and branches striate-angled when dry. — § *Gyrophyllum,* Nutt.

+ + Leaves 3-cleft to or below the middle, but not to the base, which has a 3-nerved midrib.

— **C. palmáta,** NUTT. Glabrous, rigid: stem nearly simple: leaves cuneiform in outline; the undivided basal portion little wider than the rather broadly linear lobes, which are either simple or again 1–3-lobed, the margins scabrous: rays obovate-oblong: akenes oblong. — Gen. ii. 573; Torr. & Gray, Fl. ii. 342. *C. pauciflora,* Lehm. Ind. Sem. Hort. Hamb. 1833, & Linn. x. Suppl. 76. *C. præcox,* Fresenius, Ind. Sem. Hort. Francf. 1838. *Calliopsis palmata,* Spreng. Syst. iii. 611. — Plains and prairies, Winnipeg and Wisconsin to Illinois, Louisiana, and W. Texas; first coll. by *Nuttall.*

+ + Leaves divided to the base, the pair thus imitating a whorl of six, or the uppermost simple, rarely some of the lower also simple.

— **C. verticilláta,** L. Glabrous, slender: leaves 2–3-ternately dissected into very narrowly linear or nearly filiform lobes; heads small: rays narrowly oblong: disk-corollas dull yellow: akenes obovate-cuneiform. — Spec. ii. 907; Lam. Dict. ii. 108; Michx. l. c. (var. *tenuifolia*); Torr. & Gray, l. c. *C. tenuifolia,* Ehrh. Beitr. vii. 168; Willd. Spec. iii 2252; Schk. Handb. t. 260; Bart. Fl. Am. Sept. t. 73; DC. l. c. — Moist ground, Upper Canada and Maryland to upper parts of Carolina and Arkansas.

C. delphinifólia, LAM. Stouter than the preceding: divisions of the leaves fewer and wider; the middle one once or its midlobe again 3-parted, lateral ones 2-parted or simple; lobes all linear, 2 lines wide: disk-flowers brown! — Dict. ii. 108; DC. l. c., Torr. & Gray. l. c. *C. verticillata,* Ehrh. l. c.; Willd. l. c.; Bot. Mag. t. 156; Schk. Handb. t. 260. *C. verticillata,* var. *linearis,* Michx. l. c. — Pine woods, &c., Virginia to Alabama and the borders of Florida.

— **C. senifólia,** MICHX. Stem stouter and often taller (2 or 3 feet high): leaves divided into 3 commonly oblong-lanceolate or ovate-lanceolate and entire sessile divisions (of 1½ to 3½ inches in length and half-inch to inch wide), thus closely imitating a whorl of six : disk-flowers dull yellow : akenes obovate-elliptical, 2-toothed at summit by extension of the broadish wing, the teeth sometimes aristellate-pointed. — Fl. ii. 128; Pursh, Fl. ii. 568; Nutt. in Jour. Acad. Philad. vii. 77; Torr. & Gray, l. c. *C. major*, Walt. Car. 214 ?— Dry and usually sandy woodlands, N. Carolina to Florida. The typical form is softly and minutely pubescent. Passes into the following

Var. stelláta, Torr. & Gray, l. c. Smooth and glabrous throughout : divisions of the leaves from oblong to broadly lanceolate, sometimes rather attenuate at base (rarely, or in a monstrosity, the middle one 3-cleft !), sometimes a part or even all the leaves entire ! — *C. stellata* (Soland. in herb. Banks), Nutt. Jour. Acad. l. c. *C. senifolia*, Hook. Bot. Mag. t. 3484. *C. Œmleri*, Ell. Sk. ii. 435, the abnormal entire-leaved form. — Upper country and mountains, Virginia and Kentucky to Georgia. Passes, especially in the lower country, to

Var. rígida, Nutt. Divisions of the leaves lanceolate, mostly attenuate at base, from 2 to 3 or 4 lines wide, mostly quite glabrous. — Gen. ii. 180. *C. Wrayi*, Nutt. Jour. Acad. l. c. *C. delphinifolia*, var. *rigida*, Torr. & Gray, l. c. — S. Carolina to Florida, in the low and middle country. Narrowest-leaved forms closely approach *C. delphinifolia*.

* * Perennial, tall : leaves petioled and pinnately 3-5-divided, except the uppermost : otherwise nearly as the preceding. — § *Chrysostemma*, Torr. & Gray. *Chrysostemma*, Less., DC.

C. trípteris, L. Smooth and glabrous, or leaves minutely pubescent : stem strict, 4 to 8 feet high, simple, with corymbose or fastigiate flowering branches : leaflets lanceolate, rather obtuse, 2 to 5 inches long ; the pinnate veins connected by an obscure vein just within the scabrous margin : heads half-inch or less high, and oblong rays almost inch long : disk-flowers dull yellow turning brownish : akenes with narrow wings obscurely lacerate or denticulate at summit : no proper pappus : heads when bruised anise-scented. — Spec. ii. 908 (Moris. Hist. sect. 6, t. 3, f. 44) ; Michx. Fl. ii. 138 ; Torr. & Gray, l. c. *Anacis tripteris*, Schrank, ex DC. *Chrysostemma tripteris*, Less. Syn. 227 ; DC. Prodr. v. 568 ; Hook. Bot. Mag. t. 3553. — Near streams, in rich soil, Penn. to Wisconsin and Louisiana.

* * * Perennial, tall : leaves short-petioled, undivided, copiously pinnately veiny and serrate : style-tips conical-pointed ; akenes oblong-lanceolate, wingless, the narrow truncate apex destitute of teeth or any kind of pappus: flowering late. — § *Silphidium*, Torr. & Gray.

C. latifólia, Michx. Glabrous and smooth, or pubescent, 3 to 5 feet high, leafy and simple to near the top: leaves membranaceous, 6 to 9 inches long, ovate or oval, acuminate at both ends, short-petioled ; the long-mucronate teeth callous-tipped : heads several or numerous : involucre half-inch high, narrow ; bracts of the outer loose and herbaceous, linear, more or less shorter than the thin and narrowly oblong inner ones, hardly united at base : rays narrowly oblong, entire, over half-inch long : disk-corollas yellow, barely brownish in age : akenes nearly 4 lines long. — Fl. ii. 137 ; Torr. & Gray, Fl. ii. 341. — Higher mountains, N. Carolina to Georgia.

* * * * Annuals, late-flowering, some perhaps biennial, leafy-stemmed and branching : leaves thinnish, petioled, pinnately 3-7-parted or divided, rarely undivided, these or their divisions pinnately veiny and incised or serrate, the principal veins often running to the sinuses : heads numerous : bracts of the involucre mostly distinct to the base: the outer loose and spreading or reflexed, usually foliaceous, irregular, sometimes numerous : rays obovate-oblong, almost always entire, conspicuously many-nerved : disk-flowers dull yellow: anthers black : akenes wingless or obscurely margined, obovate or cuneate-oblong, unicostate on each face, straight, more or less 2-dentate or 2-aristate, sometimes the lateral ribs produced at summit into a tooth or awn. — § *Diodonta*, Torr. & Gray. *Diodonta*, Nutt. Trans. Am. Phil. Soc. vii. 360. (Transition to *Bidens*: differing only in the absence of retrorse barbs to the awns of the pappus, and some species hybridizing with those of *Bidens*.)

+— Heads radiate: rays golden yellow: bracts of the outer involucre about 8, not longer than the inner: akenes cuneate-oblong or obovate-cuneate, somewhat angulate-thickened on the faces, obscurely ciliate or naked-margined.

C. aúrea, Ait. Glabrous or nearly so, 1 to 3 feet high : leaves various, more commonly 3-7-divided, with lanceolate divisions or leaflets incisely serrate or lobed, or upper leaves undivided : rays half or two-thirds inch long. akenes broadly cuneate, only one or two lines long, slightly hairy, bearing two very short and rather divergent and blunt chaffy teeth, and rarely obscure ones from the lateral angles. — Ait. Kew. iii. 252; Torr. & Gray, Fl. ii. 339. *C. coronata*, L. Spec. ed. 2, ii. 1281, as to herb., but excl. syn. Plum. (from which the "foliis lineatis," &c., is taken) & Vaill.; Walt. Car. 215; name best not restored. *Diodonta mitis, aurea,* & *leptophylla,* Nutt. Trans. Am. Phil. Soc. l. c. 360. — Wet ground, Virginia to Florida. The original *C. aurea* is a form with some of the upper leaves lanceolate and entire, lowermost of 3 leaflets. Extreme forms are: var. *subintegra,* Torr & Gray, with all or most of the leaves undivided and lanceolate (*C. arguta,* Pursh, Fl. 567, & *C. ambigua,* Nutt. Jour. Acad. Philad. vii. 75); var. *leptophylla,* Torr. & Gray, l. c., with leaves or their few divisions elongated linear, only a line or two wide (*Diodonta leptophylla,* Nutt. Trans. Am. Phil. Soc. l. c.), a brackish coast form; and var. *incisa,* Torr. & Gray, l. c., with nearly all the leaves 3-7-divided and the divisions incised or coarsely serrate (*C. mitis,* Michx. Fl. ii. 138, & the *C. coronata* of herb. Linn.), the form which approaches or passes into the following.

C. trichospérma, MICHX. Glabrous or nearly so, a foot or two (rarely 3 to 5) high: leaves almost all 3-5-divided into lanceolate coarsely serrate or pinnately incised divisions: rays oval-obovate, two-thirds to three-fourths inch long: akenes narrowly cuneate-oblong, sparsely hairy or glabrate, about 4 lines long and barely a line wide, or the outer somewhat broader and shorter, bearing a pair of strong subulate pointed erect teeth, commonly equal in length to the breadth of the summit of the akene. — Fl. ii. 130; Torr. & Gray, Fl. ii. 340. *C. aurea,* Lindl. Bot. Reg. t. 1228. *Diodonta coronata,* Nutt. l. c. — Wet ground, coast of Mass. to Virginia and N. E. Georgia. Also shore of L. Erie to Illinois; where is a

Var. **tenuíloba.** Tall, much branched: divisions of the leaves from narrowly lanceolate to linear: akenes smaller (outer barely 3 lines long), and with shorter somewhat spreading teeth: approaching *C. aurea.* — Peat bogs, Indiana and Illinois, *Vasey, Stewart,* &c.

+ + Heads radiate: rays golden yellow, sometimes inch long: akenes obovate, very flat, with very thin margins hispid-ciliate: leaves all 3-7-divided or parted; the divisions serrate, incised, or some again cleft: herbage somewhat pubescent or glabrous. (Hybrids of these with *Bidens frondosa* or others are not uncommon.)

C. aristósa, MICHX. Stem 1 to 3 feet high: divisions of the leaves lanceolate, acuminate: bracts of the outer involucre 8 to 10, barely ciliate, not surpassing the inner: akene with a pair of slender upwardly scabrous awns of its own length, or these rarely wanting. — Fl. ii. 140; Torr. & Gray, l. c.; Hook. f. Bot. Mag. t. 6462. *C. aristata,* Willd. Spec. iii. 2253. *Diodonta aristosa,* Nutt. l. c. — Swamps, Michigan to Iowa, Missouri and W. Louisiana: Southwestward with the var. *mutica* (the awns wanting), there disposed to pass into the next.

C. involucráta, NUTT. Heads rather larger: bracts of the outer involucre 12 to 20, mostly surpassing the inner, slender, hispid on the back and margins: akenes with 2 short acute teeth. — Jour. Acad. Philad. vii. 74; Torr. & Gray, l c. *Diodonta involucrata,* Nutt. Trans. Am. Phil. Soc. l. c. — W. Illinois to Kansas and Texas.

+ + + Heads without rays, or rarely a rudimentary one, short-pedunculate: disk dull yellow: outer involucre of 3 to 5 irregular foliaceous bracts, some or most of them surpassing the head: herbage glabrous or nearly so: leaves slender-petioled.

C. bidentoídes, NUTT. Rather stout, 1 to 4 feet high, with ascending branches: leaves undivided, lanceolate, acuminate, serrate, tapering at base into the long petiole: heads oblong, half to three-fourths inch long, and outer involucral bracts sometimes inch and a half long, resembling uppermost leaves; bracts of inner involucre with somewhat petaloid margins and tips: akenes cuneate-linear, 4 or 5 lines long, more or less exceeding the two setiform upwardly hispidulous awns, rarely vestiges of awns from lateral nerves. — Torr & Gray, Fl. ii. 339. *Diodonta (Heterodonta) bidentoides,* Nutt Trans Am. Phil. Soc. vii 361. — Muddy shores of Delaware River and Bay, from above Philadelphia, first coll by *Nuttall.*

C. discoídea, TORR. & GRAY, l. c. Slender, with widely spreading branches, a foot or two high: leaves membranaceous, irregularly serrate; uppermost undivided and rhomboid-lanceolate; lower divided into 3 sessile or petiolulate leaflets: heads campanulate, a quarter-inch high: akenes narrowly cuneiform, 2 or 3 lines long, the two subulate teeth rather than awns mostly shorter than the width of its summit. — River borders and swamps, Connecticut and N. New York to Ohio, Virginia, and Texas.

113. **BÍDENS,** Tourn. BUR-MARIGOLD. (Lat. *bidens,* with two teeth or prongs: name from the adjective, i. e. *planta bidens,* therefore feminine.) — Herbs of wide distribution, chiefly American; with opposite either simple or compound

leaves, and solitary or paniculate heads of mostly yellow (sometimes white, rarely purple) flowers; in summer or autumn. — Linn. Gen. no. 932; DC. Prodr. v. 593; Benth. & Hook. Gen. ii. 387.

§ 1. PLATYCARPÆA, DC. Akenes flat, from obovate to cuneiform, not at all contracted at the summit, 2-4-awned: outer involucre foliaceous and spreading: veins of the leaves commonly terminating in the sinuses: ours annuals.

* Heads erect, rayless, or rarely with one to five small rays, these usually shorter than the disk and therefore inconspicuous: disk greenish yellow: leaves mostly petioled.

B. frondósa, L. (STICK-TIGHT.) Glabrous or somewhat hairy, branching, 2 to 6 feet high: leaves except the uppermost pinnately 3-5-divided into lanceolate or broader sharply serrate and pinnately veiny commonly petiolulate leaflets: outer involucre often very leafy: akenes obovate or oblong, more or less hairy (the hairs of the margin ascending except near the summit), 2-awned. — Spec. ii. 832; Torr. & Gray, Fl. ii. 351. — Shady or moist ground, preferring manured soil, Florida and Texas to Saskatchewan and Brit. Columbia, everywhere common, and with the habit of a naturalized weed. Near Philadelphia, along with this and *Coreopsis bidentoides*, occurs a form with *upwardly hispidulous* awns, doubtless a hybrid.

B. connáta, MUHL. Glabrous, a foot or two high, loosely branched: leaves either all undivided, oblong or broadly lanceolate, acuminate at both ends, sharply serrate, tapering into margined petioles or the upper sessile; or some with a pair of lateral divisions which are sessile and decurrent on the petiole: akenes oblong-cuneate or the outermost obovate, nearly glabrous but retrorsely hispid-ciliate, commonly 3-awned. — Willd. Spec. iii. 1718; DC. l. c.; Torr. & Gray, l. c. *B. tripartita*, Bigel. Fl. Bost. ed. 2, 294, not L. *B. petiolata*, Nutt. Jour. Acad. Philad. vii. 99, a thin-leaved small-headed form; while var. *comosa*, Gray, Man. 261, is a stout and larger-headed form with very leafy involucre. Slender forms imitate *Coreopsis discoidea.* — Wet ground, Canada to Illinois, Missouri, and Georgia.

* * Heads disposed to nod after anthesis, commonly with conspicuous rays: leaves all sessile and undivided; upper pairs somewhat connate round the stem: margins of the cuneate akenes and the rigid awns retrorsely aculeolate-hispid.

B. cérnua, L. Stem glabrous or setulose-hispid, from a span to a yard high: leaves oblong-lanceolate, coarsely and irregularly sharply serrate: heads conspicuously nodding after anthesis, commonly surpassed by the foliaceous outer involucre: rays ovate or oval, little surpassing the disk or wanting: akenes usually 4-awned. — Spec. ii. 832 (discoid); Willd. Spec. iii. 1716; Schk. Handb. t. 235; Fl. Dan. t. 841; Torr. & Gray, Fl. ii. 352, with var. *elata*, a large form of the Pacific coast. *B. quadriaristata*, var. *dentata*, Nutt. Trans. Am. Phil. Soc. l. c. 368. *Coreopsis Bidens*, L. l. c. 908, radiate form. — Wet ground, from Hudson's Bay and Saskatchewan to the Pacific coast, and in the Atlantic States south to Virginia and Missouri; at some stations seemingly introduced. (Eu., N. Asia.)

B. chrysanthemoídes, MICHX. Glabrous, often decumbent at base, a foot or two high: leaves lanceolate, rather minutely and evenly serrate: heads rather large, little or not at all nodding: outer involucre seldom surpassing the inner, conspicuously surpassed by the oval or broadly oblong (usually inch-long) rays: akenes 2-4- (more commonly 2-) awned. — Fl. ii. 136; Willd. Spec. iii. 1717; Torr. & Gray, l. c.; Sprague, Wild Flowers of Amer. 131, t. 30. *B. quadriaristata*, DC. l. c. *B. helianthoides*, HBK. Nov. Gen. & Spec. iv. 230. *Helianthus lævis*, L. Spec. ii. 906, viz. Pl. Gronov. Fl. Virg., ed. 1, 104 (not of ed. 2). *Coreopsis Bidens* (& *C. perfoliata?*), Walt. Car. 215. — Wet grounds, Canada to Florida, Arizona, and California. (Mex., S. Am.)

§ 2. PSILOCARPÆA, DC. (*Ceratocephalus*, Vaill.) Akenes narrow, linear-tetragonal; the outer almost always shorter and more truncate at apex than the inner, which generally taper upward, but are not distinctly rostrate: outer involucre seldom foliaceous or enlarged.

* Leaves mainly divided into 3 to 5 ovate merely serrate divisions or leaflets: rays when present white: annuals, at least with us, varying from pilose-pubescent to nearly glabrous: akenes 4 to 5

or even 6 lines long, in the same plant either smooth with sparing bristles, or the outer becoming tuberculose and rough.

◂ **B. leucántha,** WILLD. Leaves of rather firm texture, some undivided and ovate; these and the 3 or occasionally 5 ovate or oblong-ovate divisions evenly serrate, more or less lineately veiny: heads corymbosely paniculate on rather short peduncles: rays obovate, bright white, 5 to 8 lines long, rather showy. — Spec. iii. 1719; Torr. & Gray, l. c. *B. striata,* Sweet, Brit. Fl. Gard. t. 237; Hook. Bot. Mag. t. 3155. *Coreopsis leucanthema,* L. Amœn. Acad. iv. 291. *C. leucantha,* L. Spec. ed. 2, ii. 1282; Desc. Fl. Ant. t. 583. *C. coronata,* L. l. c. 1281, as to syn. Plum. t. 53, f. 2. — Common in S. Florida. (W. Ind., Mex.)

◂ **B. pilósa,** L. Stem sometimes tall, usually weak: leaves thin; leaflets 3 to 5, irregularly serrate, sometimes incised, or the lower divisions occasionally 3-lobed: heads fewer and scattered: rays commonly none, at most inconspicuous and yellowish-white. — Spec. ii. 832 (but the cited figure, Dill. Elth. t. 43, probably belongs to *B. frondosa*); Willd. l. c. *Coreopsis alba,* L. Spec. ii. 908 (Herm. Parad. t. 124? excl. syn. Pluk.). *Bidens Californica,* DC. Prodr. v. 599; Torr. & Gray, Fl. ii. 354. Variable, and the slender forms in warm countries seeming to pass into *B. bipinnata.* — S. California and Arizona, introduced? (Mex. to Chili, W. Ind., and all tropics.)

  * * Leaves all once to thrice 3-5-nately parted or divided into oblong or linear ultimate lobes: root in ours annual.

  +— Heads narrow: rays inconspicuous and yellowish or none: akenes long and slender, at least the central ones much surpassing the involucre.

  ++ Lobes of the thin leaves from oblong to lanceolate: heads slender-peduncled.

◂ **B. bipinnáta,** L. (SPANISH NEEDLES.) Primary and secondary divisions of the leaves rather ovate or deltoid-lanceolate in circumscription, and the lobes mostly acute: akenes all slender, the inner ones 5 to 9 lines long, outermost moderately shorter and thicker: awns 3 or 4, sometimes only 2. — Spec. ii. 832; Michx. ii. 135; Torr. & Gray, l. c. — Damp thickets and waste ground; a common and disagreeable weed, New England to Florida and Arizona. (Trop. Am., &c.)

B. Bigelóvii, GRAY. Lobes of the leaves linear-oblong, mostly obtuse: innermost akenes 5 or 6 lines long, 2-3-awned; outermost of half that length or less, stouter, 2-corniculate, or with a pair of short awns, or even with none. — Bot. Mex. Bound. 91. *B. tenuisecta,* in part, Gray, Pl. Wright. i. 109. — S. W. Texas to S. Arizona, first coll. by *Wright* and *Bigelow.*

  ++ ++ Lobes of the leaves linear.

◂ **B. tenuisécta,** GRAY. A foot or two high, branched from the base, sparsely hirsute or glabrous: leaves 2-3-ternately or pinnately dissected into narrow linear lobes (of a line or more in width): heads on naked rather long and stout peduncles, many-flowered, 4 or 5 lines high in flower: involucre hirsute, especially at base: akenes glabrous, 2-awned; inner 5 lines long, with tapering summit; outermost 3 lines long, stouter and with broad summit and usually short awns: rays yellow, mostly surpassing the disk. — Pl. Fendl. 86. — Along water-courses, Colorado, New Mexico, and Arizona; first coll. by *Fendler.*

B. Lemmóni. A foot or less high, slender, with short branches, smooth and glabrous: leaves twice ternately parted into entire obtuse rather broadly linear lobes (a line or two wide); the uppermost simply 3-5-parted, subtending the mostly sessile 5-9-flowered cylindraceous and glabrous heads: akenes nearly of the preceding, but the outer nearly like the inner: rays apparently none. — S. Arizona, in Apache Pass, *Lemmon.*

B. heterospérma, GRAY. Slender, glabrous, paniculately much branched: leaves once or twice ternately parted into filiform-linear (half-line wide) lobes: heads on slender peduncles, few-flowered, in flower barely 3 lines long: rays apparently none: akenes 2-3-awned, smooth; the inner 4 or 5 lines long, outermost only 2 lines long and their short awns caducous. — Pl. Wright. ii. 90. — S. Arizona: raised from seed coll. by *Wright* (thought to come from New Mexico): rediscovered in Apache Pass by *Lemmon.*

  +— +— Heads broader, many-flowered, and with comparatively large deep yellow rays: akenes all short, hardly surpassing the involucre.

B. prócera, DON. Erect and tall from an annual or biennial root, glabrous: leaves twice or thrice parted into narrow linear lobes (mostly of an inch or more in length and less than a line wide): heads corymbosely paniculate: outer involucre small and inconspicuous, merely spreading: rays oval (in ours half-inch long, in Mexican sometimes an inch): disk in fruit only 4 lines high, comparatively broad: outer akenes narrowly cuneate-oblong and only 2 lines long, innermost 3 lines long and cuneate-linear, apex not attenuate: awns 2, strongly barbed, of half or a third the length of the akene. — Bot. Reg. t. 684 (1822); DC. Prodr. v. 603; Gray, Proc. Am. Acad. xix. 16. *B. fœniculifolia*, DC. l. c. (probably); Gray, Pl. Wright. ii. 90; Rothr. Wheeler Rep. vi. 165. *B. feruliæfolia*, Hemsl. Bot. Biol. Centr.-Am. ii. 202, mainly, not Jacq. — Arizona, *Wright, Thurber, Lemmon*. (Mex.)

* * * Leaves some undivided, some 3-5-parted into lanceolate or linear divisions: root perennial.

B. heterophýlla, ORT. Glabrous or nearly so, often tall: leaves of firm texture and with ascending veins, mostly serrate with erect teeth, from oblong to lanceolate and tapering into a petiole, sometimes all undivided, commonly some 3-parted or the upper 5-parted into lanceolate or linear lobes: heads in flower 3 and in fruit about 5 lines high: rays broadly obovate, half to three-fourths inch long, deep yellow: akenes cuneate-linear, the inner (3 lines long) little longer than the outermost, 2-3-awned. — Dec. 99, t. 12; DC. Prodr. v. 597. *B. arguta*, HBK. Nov. Gen. & Spec. iv. 231; DC. l. c. *B. longifolia*, DC. l. c. — S. Arizona, along streams, *Pringle, Lemmon*. (Mex.)

Var. **Wrightii**. Base of stem and lower leaves unknown: upper leaves and (when divided) their divisions lanceolate-linear, entire or nearly so, the longer 4 or 5 inches long and 3 or 4 lines broad. — Pl. Wright. ii. 90 (as unnamed doubtful var.); Rothrock in Wheeler Rep. vi. 165. — S. Arizona, *Wright, Rothrock, Lemmon*. Seemingly an extreme form of a variable species.

§ 3. HYDROCARPÆA. Akenes almost terete, cartilaginous, truncate at both ends, bearing 3 to 6 very long and rigid acerose awns, which are smooth below, the upper part densely and retrorsely hispidulous: aquatic: submersed leaves filiformly dissected: rays conspicuous, yellow.

B. Béckii, TORR. Submersed stems much elongated in deep water, thickly beset with the almost capillary ternately multifid leaves; emersed summit bearing a few pairs of oblong-lanceolate serrate leaves, or the lower pinnatifid: head short-peduncled: bracts of the involucre oblong: rays obovate, over half-inch long: mature akenes half-inch and the rigid diverging awns an inch or less long, very persistent. — Torr. in Spreng. Neu. Entd. ii. 135, Syst. iii. 455, & Fl. N. Y. i. 388, t. 58. — In slow-flowing streams and ponds, Canada to New Jersey and Missouri; first coll. by *L. C. Beck*.

114. **CÓSMOS**, Cav. (Κόσμος, an ornament.) — Tropical American herbs, chiefly Mexican, too near *Bidens;* for there is one yellow-rayed species, and certain species with purple rays have hardly a beak to the akenes. — Ic. i. 9, t. 14, 79; DC. Prodr. v. 606; Benth. & Hook. Gen. ii. 387. *Cosmea*, Willd. Spec. iii. 2250.

C. CAUDÁTUS, HBK. Apparently annual: leaves twice pinnately parted into lanceolate acute lobes: rays rose-colored, seldom much surpassing the involucre: akenes fusiform, with beak longer than the body (in all nearly inch long), 2-awned. — HBK. Nov. Gen. & Spec. iv. 240; Torr. & Gray, Fl. ii. 349. — Key West, *Blodgett*. (Evidently introduced from W. Ind.)

C. DIPINNÁTUS, CAV. l. c. t. 14. Annual: leaves pinnately divided into narrowly linear or almost filiform lobes: heads very showy, the deep rose-colored rays commonly an inch or more long: akenes smooth and glabrous throughout, with abrupt beak very much shorter than the body, or in some flowers reduced to a mere neck: awns 1 to 3, short (Mexico); in var. EXARISTÁTUS, DC. l. c., the awns wholly wanting. — The var., S. Texas, near Marfa, on an abandoned ranch (base of stem becoming lignescent), *Havard*. (Mex.)

C. parviflórus, HBK. Annual, slender: heads smaller, with either white or rose-colored rays half to a quarter of an inch long: beak of the akenes slender, usually half the length of the body, scabrous, 2-3-awned: otherwise as preceding, into which it may pass. — Nov. Gen. & Spec. iv. 241. *C. bipinnatus*, var. *parviflorus*, Gray, Pl. Wright. ii. 90. *Coreopsis parviflora*, Willd. Spec. iii. 2551. *Bidens Humboldtii*, Schultz Bip. in Bot. Herald, 307. — S. W. Texas to Arizona. (Mex.)

115. HETEROSPÉRMUM, Cav. (*Heterosperma.* Ἕτεροϛ, other, in sense of unlike, σπέρμα, seed.) — Small or slender annuals (from the Mexican border southward), mainly glabrous, branching; with opposite pinnately or ternately dissected or sometimes undivided leaves, and small heads of yellow flowers; the 3 to 5 rays little exserted. — Cav. Ic. iii. 24, t. 267; HBK. Nov. Gen. & Spec. iv. 245, t. 383, 384; DC. Prodr. v. 632.

H. pinnátum, Cav. l. c. About a foot high; leaves pinnately 3-7-parted into linear divisions, which are either all entire or some of them again 2-3-parted: heads slender-peduncled, about 3 lines long: outer involucre of 3 to 5 linear foliaceous bracts, hispidly ciliate at base, and overtopping the thin and oval striate inner bracts: outer akenes oval, at maturity cymbiform or becoming oblong by inflexion of the callous wing, destitute of pappus; innermost commonly infertile, subulate, attenuate into a scabrous beak, bearing a pair of short deciduous awns. — Willd. Spec. iii. 2129; DC. Prodr. v. 632. *H. tagetinum*, Gray, Pl. Fendl. 87. & Pl. Wright. ii. 91, a form with simply pinnate leaves often marked with glandular spots, the awns sometimes wholly wanting or caducous. — W. Texas to Arizona. (Mex.)

116. LEPTÓSYNE, DC., extended. (Λεπτοσύνη. slenderness; a name applicable to the original, but not to most of the species here associated, except as to the leaves and their divisions.) — Herbaceous or suffruticose plants (of California and Arizona), smooth and glabrous; with alternate or opposite and usually rather fleshy ternately or pinnately divided or dissected leaves, and showy pedunculate heads, both disk and ray flowers bright yellow. Habit of *Coreopsis* (which it represents on the western side of the continent), but mostly with pistillate rays, and always with a ring on the tube of the disk-corollas or at its junction with the throat. — Gray, Proc. Am. Acad. xvii. 218. *Leptosyne*, DC. Prodr. v. 531, with *Agarista*, DC. l. c. 569. *Coreocarpus* & *Acoma*. Benth. Bot. Sulph. 28. 29, t. 16, 17. *Leptosyne* & *Pugiopappus*. Gray (Pacif. R. Rep. iv. 104), Bot. Calif. i. 354.

§ 1. EULEPTÓSYNE. Akenes callous-winged and commonly meniscoidal at maturity, a small or obscure saucer-shaped cup in place of pappus: rays pistillate and commonly fertile, obovate, more or less 3-lobed: style-tips of the disk-flowers capitellate either with or without a minute setiform cusp: low annuals, with all but the lowest leaves alternate, and long or scape-like monocephalous peduncles: bracts of the outer involucre linear or lanceolate, loose. — *Leptosyne*, DC. l. c.

L. Dougásii, DC. A span to a foot high, leafy only at or near the base: leaves once to thrice parted into nearly filiform divisions: rays half-inch or more long: ring of the disk-corollas usually distinctly bearded: akenes thickened at maturity (at least the more fertile outer ones) and corky-winged, also corky-ridged down the inner face, roughened nearly throughout with capitellate or clavate short and rigid bristles: pappus-cup somewhat conspicuous. — Prodr. v. 531: Torr. & Gray, Fl. ii. 355; Gray, Bot. Calif. i. 356. *L. Californica*, Nutt. Trans. Am. Phil. Soc. vii. 363, & *L. Newberryi*, Gray, Proc. Am. Acad. vii. 358. Bot. Calif. l. c.; state with young akenes or infertile inner ones thin-winged, and ring of corolla-tube less bearded. — California (from Monterey to San Diego and San Bernardino) and adjacent Arizona; first coll. by *Douglas:* flowering early.

L. Stillmáni, Gray. Stouter, more leafy below: lobes of the leaves linear, a line or more broad: ring of the disk-corollas beardless: akenes somewhat obovate, quite smooth and naked on the back, becoming papillose or tuberculate on the inner face, at least along the slightly ridged centre, the corky wing more or less rugose. — Bot. Mex. Bound. 92. & Bot. Calif. i. 356. — California, from San Francisco Bay northward and eastward; first coll. by *Stillman*.

§ 2. TUCKERMÁNNIA. Akenes plane, oblong, smooth and glabrous, with obscure wing-like margin: pappus none or sometimes the margins continued into an acute tooth or short naked awn: rays fertile, oblong, obscurely toothed at the apex: ring of disk-corollas beardless: perennial, with more fleshy leaves and thickened succulent stem or caudex: the heads large and showy. — *Tuckermannia*, Nutt. Trans. Am. Phil. Soc. l. c. 363.

L. marítima, GRAY. Stems low, fleshy-herbaceous from a thick base or caudex: branches terminating in monocephalous peduncles of a span to a foot in length: leaves bipinnately divided into narrowly linear lobes of a line or two in width: rays 16 to 20, an inch or more long, and disk commonly an inch in diameter. — Proc. Am. Acad. vii. 358, & Bot. Calif. l. c.; Regel, Rev. Hort. 1872, tab. *Tuckermannia maritima*, Nutt. l. c.; Torr. & Gray, Fl. ii. 355; Torr. Bot. Mex. Bound. 92, t. 31. *Coreopsis maritima*, Hook. f. Bot. Mag. t. 6241. — S. Coast of California, at San Diego and on the adjacent islands.

L. gigantéa, KELLOGG. Fleshy-woody stem 2 to 8 feet high, 1 to 5 inches thick, leafy at top: leaves twice or thrice pinnately divided into filiform lobes: heads smaller (disk half-inch in diameter) on short corymbosely clustered peduncles: inner bracts of the involucre with prominent midrib. — Proc. Calif. Acad. iv. 198; Gray, Bot. Calif. i. 356. — California, on the mountains near Sta. Barbara and San Miguel, and islands off the coast; first coll. by *Coulter*. May be a form of the preceding, but seemingly is quite distinct. (Guadalupe Island, *Palmer*.)

§ 3. PUGIOPÁPPUS. Akenes dimorphous; those of the ray- or outermost disk-flowers very like those of the preceding section (oval, flat, glabrous), either fertile or sterile; those of the disk also flat, but narrowly oblong, marginless, clothed at least on the margins with long and soft-villous hairs (which are bidentate at apex under a lens), bearing a conspicuous pappus of a pair of linear triquetrous paleæ: annuals with the habit and otherwise the character of *Euleptosyne;* the ample golden yellow rays multinervose, commonly styliferous, not rarely fertile, yet sometimes neutral or with mere included rudiment of style. — *Agarista*, DC. Prodr. v. 569; Torr. & Gray, Fl. ii. 337, not Don. *Pugiopappus*, Gray, Pacif. R. Rep. l. c., & Proc. Am. Acad. vi. 545, viii. 659.

L. Bigelóvii, GRAY. A foot or less high, with the habit of *L. Douglasii*, leafy only at base, and with long often scapiform peduncles: leaves once or twice ternately or quinately parted into narrow linear lobes: involucre half-inch or less high; its outer bracts linear or nearly so, inner oblong-ovate: rays obovate or quadrate-oblong, half to two-thirds inch long, 10-12-nerved: ring of disk-corollas beardless: ray-akenes oblong, with narrow callous-winged margin; disk-akenes elongated-oblong, very villous at the margins, sparsely so or naked on one or both faces, twice the length of the paleæ of the pappus. — *Pugiopappus Bigelovii* & *P. Breweri*, Gray, Proc. Am. Acad. viii. 660, & Bot. Calif. l. c., the former described from immature and incomplete specimens, in which the villosity of the disk-akenes was little developed. — Southern part of California, from San Buenaventura and Tejou to the Mohave Desert.

L. calliopsidea, GRAY. A foot or two high, rather stout and leafy, with peduncles a span long: lobes of the leaves narrowly linear, sometimes incised: heads rather large and broad: bracts of the outer involucre broadly ovate, thick, a little shorter than the narrowly ovate inner ones: rays broadly cuneate-obovate, commonly an inch long and three-fourths inch wide, 15-20-nerved: ring of the disk-corollas pubescent: ray-akenes broadly oval, distinctly thin-winged; disk-akenes cuneate-oblong, little longer than the paleæ of the pappus, very long villous on the margins and inner face. — *Agarista calliopsidea*, DC. Prodr. v. 569; Torr. & Gray, l. c. *Coreopsis calliopsidea*, Bolander, Cat. Pl. San Francisco. *Pugiopappus calliopsideus*, Gray Proc. Am. Acad. & Bot. Calif. l. c. *Leptosyne maritima*, Rev. Hortic. 1873, 330, tab. — Moist hillsides and plains, California, from the Sacramento southward.

Var. nána. A span or so high, with more scapiform peduncles, leaves crowded at base, heads and rays smaller, outer involucre comparatively shorter, and ray-akenes narrower or less margined. — San Bernardino Co. at Mohave Station, &c., *Lemmon*, *Pringle*.

§ 4. COREOCÁRPUS. Akenes nearly of *Euleptosyne*, but mostly with tuberculate rather than winged margins, and some of them bearing a pair of sometimes retrorsely hispid awns; those of the ray-flowers mostly fertile: style-branches of the disk-flowers produced into a subulate appendage: outer involucre of a few small inconspicuous bracts: annuals or suffruticose perennials; with branching stems, opposite leaves, and small cymose or paniculate heads on short slender peduncles. — *Coreocarpus* & *Acoma*, Benth. Bot. Sulph. l. c. *Coreocarpus*, Benth. & Hook. Gen. ii. 384.

L. Arizónica, GRAY. Stems 2 or 3 feet high, and paniculately branched from a woody base, rigid, slender: leaves 3–5-parted into mostly entire linear acute lobes: heads loosely cymose, 3 or 4 lines long: outer involucre of 1 to 3 small loose bracts; inner of 6 to 8 ovate ones in two ranks: rays 5 or 6, about 3 lines long: disk-corollas with a bearded ring: akenes narrowly oblong, with faces either smooth or papillose-muriculate, and margins beset with a wing which is wholly dissected into a pectinate tubercular fringe (in the manner of *Coreopsis*, § *Coreoloma*), the inner and less fertile or infertile marginless, some without pappus, others bearing either one or two short and setiform awns, which are either naked or sparingly denticulate, the denticulations spreading or a few of them recurved. — Proc. Am. Acad. xvii. 218. — Along streams in the Santa Catalina Mountains, Southern Arizona, *Lemmon*, *Pringle*.

L. PARTHENIOÍDES (*Coreocarpus parthenioides*, Benth. Bot. Sulph. t. 16), L. HETEROCÁRPA (*C. heterocarpus*, Gray, Proc. Am. Acad. v. 162), and L. DISSÉCTA (*Acoma dissectum*, Benth. l. c. t. 17) are species of Lower California, insufficiently known.

117. **THELESPÉRMA**, Less. (Θηλή, a nipple, σπέρμα, seed, from the papillosity of some of the akenes.) — Perennial, sometimes annual or suffrutescent herbs (of the Great Plains, and one on the Pampas of S. Amer.), smooth and glabrous; with habit of *Coreopsis*, opposite usually finely dissected leaves, and pedunculate heads; the rays normally golden yellow, disk-flowers yellow, sometimes purplish or brownish. — Less. in Linn. vi. 511; Gray in Kew Jour. Bot. i. 252, & Pl. Wright. i. 109. *Cosmidium*, Torr. & Gray, in Nutt. Trans. Am. Phil. Soc. l. c., & Fl. ii. 350.

* Lobes of the disk-corollas linear or lanceolate, longer than the short campanulate throat: style-appendages with cuspidate or subulate tips: pappus evident: chaff of receptacle falling with and partly embracing the akenes.

T. SCABIOSOÍDES, Less., of the Pampas in S. America, closely represents *T. gracile*, but has more filiform foliage and longer-awned pappus.

COSMIDIUM BURRIDGEÁNUM of the gardens is a hybrid of *T. filifolium* and *Coreopsis tinctoria*, acquiring its brown-purple rays from the latter.

T. filifólium, GRAY. A foot or two high from an annual or biennial root, loosely branching, leafy: leaves not rigid, bipinnately divided into filiform lobes no wider than the rhachis: bracts of the outer involucre 8, subulate-linear, almost equalling or more than half the length of the inner, which are connate only to the middle: rays broad, over half-inch long: disk usually purple turning brownish: outer akenes becoming coarsely papillose on the back; the stout triangular-subulate pappus-scales not longer than the width of the akene. — Kew Jour. Bot. i. 252, & Pl. Wright. i. 109. *Coreopsis trifida*, Lam. Ill. t. 704; Poir. Suppl. ii. 353, ex tab. *C. filifolia*, Hook. Bot. Mag. t. 3505. *Cosmidium filifolium*, Torr. & Gray, Fl. ii. 350. — Dry uplands and plains, Arkansas to Texas.

T. ambíguum, GRAY. A foot high, perennial and spreading by creeping rootstocks, rather rigid, usually more naked above or with longer peduncles: cauline leaves less compound; the lobes from filiform to narrowly linear; bracts of inner involucre connate to or above the middle: rays rarely wanting; otherwise as the preceding. — Proc. Am. Acad. xix. 16. *T. filifolium*, Gray, Pl. Wright. i. 109, & ii. 90, chiefly. — Plains and hills, W. Texas to New Mexico, Colorado, and Montana.

T. grácile, GRAY, l. c. More rigid, a foot or two high from a deep perennial root, less branched, naked above: leaves once or twice 3–5-nately divided or parted into filiform-linear or broader lobes, or some upper ones filiform and entire: bracts of the outer involucre 4 to 6, very short, ovate or oblong; of the inner one connate to above the middle, the edges of their lobes slightly scarious: disk mostly yellow, scarcely brownish after anthesis: akenes less papillose or roughened, the breadth of the summit exceeded by the subulate awns: rays usually none, rarely present and 2 or 3 lines long. — *Bidens gracilis*, Torr. Ann. Lyc. N. Y. ii. 215. *Cosmidium gracile*, Torr. & Gray, l. c. — Plains, Nebraska and Wyoming to W. Texas and Arizona. (Adj. Mex.)

* * Lobes of disk-corollas from ovate to oblong, decidedly shorter than the cylindraceous throat; the proper tube also shorter than in the foregoing: pappus shorter and more coroniform, destitute of retrorse bristles or hairs, or wanting.

+— Leafy-stemmed, branching, herbaceous to the ground: style-appendages subulate-tipped.

T. subsimplicifólium, GRAY. Stems slender, rigid, 1 to 3 feet high: leaves sometimes all entire and filiform (1¼ to 3 inches long), sometimes 3–5-parted into filiform entire lobes: outer bracts of the involucre oblong to linear, short: rays half-inch long: akenes short-fusiform: pappus 2 minute slightly hairy teeth, or obsolete. — Bot. Mex. Bound. 90. *T. simplicifolium*, Gray, Kew Jour. Bot. l. c. *Cosmidium simplicifolium*, Gray, Pl. Fendl. 86. — Rocky prairies, Texas to Arizona. (Adj. Mex.)

+— +— Low, branching from a lignescent base, very leafy below, sending up long and naked or scapiform peduncles: outer involucre short and small: akenes fusiform, more incurved at maturity.

T. subnúdum, GRAY. Rather stout: leaves thickish and rigid, once or twice ternately parted into linear or lanceolate lobes: peduncles 4 to 10 inches long: head rather large (half-inch high): rays sometimes none, sometimes ample (the larger two-thirds inch long and over half-inch wide): style-appendages subulate-tipped: pappus a minute 4–5-toothed naked crown, or obsolete. — Proc. Am. Acad. x. 72. Includes also *T. subsimplicifolium*, var. *scaposum*, Gray, coll. Parry, &c. — New Mexico, S. Utah, and N. Arizona, *Palmer, Parry, Ward*. Also apparently Green River, Wyoming, *Parry*, a plant referred to *T. gracile*.

T. lóngipes, GRAY. Fastigiately much branched at the woody base, very leafy: leaves 3–5-parted into filiform divisions which are usually no wider than the rhachis: peduncles filiform, wholly simple, 5 to 10 inches long: head small (quarter-inch high), rayless: style-appendages tipped with a very short cone: akenes barely 2 lines long, arcuate at maturity, falling free from the chaff: pappus quite obsolete. — Pl. Wright. i. 109; Rothrock in Wheeler Rep. vi. 164. — Dry hills and banks, W. Texas and Arizona, *Wright, Rothrock* (not showing the woody stems), *Lemmon*. (Mex., *Schaffner*.)

118. **BALDWÍNIA**, Nutt., in the form of *Balduina*. (*Dr. Wm. Baldwin*, collaborator with Elliott, died early.) — Apparently biennials or annuals (of S. Atlantic States), mostly glabrous or minutely puberulent: with alternate entire leaves, puncticulate in the manner of *Helenium* and veinless, and solitary or corymbosely paniculate heads of yellow flowers, or those of the disk sometimes purplish-tinged: fl. late summer and autumn. — Nutt. Gen. ii. 175; Ell. Sk. ii. 447; Benth. & Hook. Gen. ii. 391. *Baldwinia* and *Actinospermum* (Ell.), Torr. & Gray, Fl. ii. 388. (True affinity rather with the *Helenioideæ*.)

B. uniflóra, NUTT. l. c. Stem rather stout, simple or simply branched, 1 to 3 feet high from a perhaps "perennial" root, with terminal usually elongated peduncle bearing a solitary large head: leaves obtuse, spatulate, or the upper linear: bracts of the involucre numerous, in about 4 series, thickish, at first appressed: rays 20 to 30, cuneate-linear, 3-toothed at truncate apex, inch or more long: concreted chaff of receptacle truncate: akenes cylindraceous-obconical, with pappus of 7 to 9 narrowly oblong paleæ of nearly its length. — Ell. Sk. ii. 447. — Low pine barrens, S. Carolina to Florida and Louisiana; first coll. by *Bartram*.

B. multiflóra, NUTT. l. c. Slender, from an annual or biennial root, branching above, very leafy up to the several or numerous slender peduncles, glabrous or sometimes sparsely hirsute: leaves all narrowly linear: heads small (3 or in fruit 5 lines high): bracts of the involucre fewer and narrow: rays 8 or 10, cuneate, half-inch long, 3-4-lobed at summit: alveoli cuspidate-toothed at the angles: akenes stipitate, turbinate. the flat summit crowned with the pappus of about 12 radiate and orbicular-obovate paleæ. — *Actinospermum*, Ell. l. c. *A. angustifolium*, Torr. & Gray, l. c. *Buphthalmum angustifolium*, Pursh, Fl. ii. 564. — Sandhills, Georgia and Florida; first coll. by *Bartram*.

119. **MARSHALLIA**, Schreb. (*Humphry Marshall*. author of the earliest indigenous work on the sylva of N. America.) — Low and smooth nearly glabrous perennials (of S. Atlantic States); with fibrous roots, commonly simple stems, and solitary pedunculate (*Armeria*-like) heads of rose-purple or white glandular-puberulent flowers, with blue anthers, produced in spring or summer: peduncle puberulent: leaves alternate, entire, mostly 3-nerved, but not manifestly veiny. — Gen. ii. 810; Torr. & Gray, Fl. ii. 390. *Persoonia*. Michx. Fl. ii. 104, not Smith. *Trattenickia*, Pers. Syn. 403, not Willd. *Therolepta*, Raf.

* Leaves thickish, mostly obtuse, all but the upper tapering below into a slender sessile base or margined petiole; radical spatulate.

M. angustifólia, PURSH. Sometimes 2 feet high and branching above. cauline leaves linear, or the uppermost linear-subulate; radical spatulate: bracts of involucre narrow, mostly acute, rigid, head only half-inch high: corollas pale purple.: akenes minutely pubescent or at maturity glabrous, longer than the pappus. — Fl. ii. 520; Ell. Sk. ii. 316 (& var. *cyanauthera*); Torr. & Gray, l. c. *Athanasia graminifolia*, Walt. Car. 201. *Persoonia angustifolia*, Michx. l. c. *Trattenickia angustifolia*, Pers. l. c. — Low pine barrens, N. Carolina (and Tennessee ?) to Florida and Louisiana.

M. cæspitósa, NUTT. More tufted, a foot high or less, either leafy only at base and with scapiform peduncle, or sparsely leafy-stemmed and sparingly branching: leaves spatulate-linear, or somewhat lanceolate and the upper linear: bracts of involucre narrow-linear, acute or acutish: head two-thirds inch or more high: corollas pale rose-color or white: akenes obpyramidal, villous on the angles, shorter than the pappus. — Nutt. in DC. Prodr. v. 680; Hook. Bot. Mag. t. 3704; Torr. & Gray, l. c. — Calcareous soil, Arkansas to Texas: first coll. by *Berlandier* and *Nuttall*.

M. lanceoláta, PURSH. A foot or less high, commonly leafy only at base and with scapiform simple peduncle: leaves lanceolate, oblanceolate, or spatulate, 3 to 6 lines wide: bracts of involucre oblong-linear or lanceolate, obtuse: akenes elongated-turbinate, pubescent, much longer than the pappus. — Fl. ii. 519; Ell. l. c.; Torr. & Gray, l. c. *Persoonia lanceolata*, Michx. l. c. *Trattenickia lanceolata*, Pers. l. c. — Open dry woods, N. Carolina to Florida, preferring the upper districts.

Var. **platyphýlla**, M. A. CURTIS. Leafy-stemmed, sometimes 2 feet high, with spatulate-oblong leaves 2 to 6 inches long, all obtuse. — Chapm. Fl. 241. — Moist or wet ground, N. Carolina, &c., from the middle country westward.

* * Leaves thinner, conspicuously 3-nerved; cauline acuminate.

M. latifólia, PURSH, l. c. A foot or so high, leafy to the middle or more: cauline leaves oblong- or ovate-lanceolate, sessile by a merely narrowed base, gradually acuminate, 2 or 3 inches long: bracts of the involucre linear, acute or acutish, rigid. — Torr. & Gray, l. c. *Athanasia trinervia*, Walt. l. c. *Persoonia latifolia*, Michx. l. c., t. 43. *Trattenickia latifolia*, Pers. l. c. *Marshallia Schreberi*, Tratt. Arch. Gen. i. 108. — Moist soil, Virginia to Mississippi, along the middle country.

120. **GALINSÓGA**, Ruiz & Pav. (*M. Galinsoga*, a Spanish physician and botanist.) — Annuals of Tropical America, the common species now widely disseminated.

G. parviflóra, CAV. A foot or two high, loosely branching, slender, somewhat pubescent: leaves thin, ovate, acute, serrate, 3-nerved from near the base, petioled: heads 2 lines long, slender-peduncled from the summit of the branches, somewhat paniculate: rays whitish, barely exserted: disk-flowers yellow: pappus usually of 8 to 16 short paleæ. — Ic. iii. 41, t. 281; DC. Prodr. v. 677; Gray, Man. 264; Reichenb Ic. Fl. Germ. t. 983. — Open or waste grounds, perhaps indigenous to New Mexico and Arizona, an introduced weed about gardens in the Northern States. In indigenous plants of the Southern border (var. *Caracasana*, & var. *semicalva*, Gray, Pl. Wright. ii. 98) pappus of the ray much reduced or wanting. (Mex., S. Amer.)

121. **BLEPHARIPÁPPUS**, Hook. (Βλεφαρίς, the eyelash, πάππος, seed-down, from the fringed paleæ of the pappus.) — A single but variable species. (Transition to the *Madieæ*.)

——**B. scáber**, Hook. Annual, a span to a foot high, loosely branched, puberulent and scabrous, and with some hispid hairs, above more or less glandular: leaves alternate, narrowly linear, with revolute or involute margins when dry, entire; heads short-peduncled, terminating the paniculate branchlets, 3 to 5 lines high: both rays and disk-flowers white: anthers brownish-purple. — Fl. i. 316; Torr. & Gray, Fl. ii. 391; Gray, Bot Calif. i. 358. *Ptilonella scabra*, Nutt Trans. Am. Phil. Soc. l. c. 386. — Dry plains and mountains, interior of Oregon, Idaho, &c., to Nevada and the Sierra Nevada, California.

Var. **subcálvus**, Gray, Bot. Calif. l. c. Pappus both of ray and disk obsolete or reduced to hyaline vestiges. — Eastern borders of California, *Lemmon*, *Matthews*, &c.

——Var. **lǽvis**, Gray, l. c. Slender, with filiform branches, almost smooth: heads few-flowered. — California, *Bridges*. Taken for *Hemizonia* in Gen. Pl. ii. 395.

122. **MÁDIA**, Molina. TARWEED. (*Madi*, the Chilian name of the common species.) — Glandular and viscid herbs, mostly heavy-scented; with leaves entire or merely toothed, some or all of them alternate; heads axillary and terminal; the yellow flowers vespertine or matutinal, closing in sunshine: in summer. — Molina, Chil.; Cav. Ic. iii. 50, t. 298; Don in Bot. Reg.; Benth. & Hook. Gen. ii. 393. *Madaria* (DC.), *Madariopsis*, *Madorella*, *Amida*, *Anisocarpus*, & *Harpæcarpus*, Nutt. Trans. Am. Phil. Soc. l. c.

§ 1. MADÁRIA. Ligules exserted and conspicuous: disk-flowers sterile or partly fertile: disk-corollas pubescent, except in the first species: herbage hirsute, the upper part minutely glandular. — Gray, Proc. Am. Acad. ix. 188, & Bot. Calif. i. 358.

* Annual, low and slender, with mostly alternate leaves and small heads: pappus both to ray and disk-flowers!

M. **Yosemitána**, Parry. A span or more high: leaves linear, entire: heads slender-pedunculate, 2 lines high: ray-flowers 5 to 10, with ligules a line or two long: disk-flowers 3 to 10, sterile: corollas nearly glabrous: bracts of the involucre with short and narrow tips; of the receptacle 4 to 8, more or less connate by their margins: ray-akenes semi-obovate or slightly lunate, bearing an evident pappus in the form of a ciliolate crown: pappus of the disk-flowers of about 5 sparsely barbellate awns, nearly equalling the corolla. — Gray, Proc. Am. Acad. xvii. 219 — California; near Fresno, *Eisen*: at the foot of the upper Yosemite Fall, *Parry* (few-flowered form); near Auburn, *Marcus E. Jones*, a larger form, with 8 to 10 rays and about as many disk-flowers.

* * Perennial, taller, with larger heads and some or most of the leaves opposite, occasionally dentate: a manifest pappus to the disk-flowers, of plumose-lacerate or fimbriate paleæ. — *Anisocarpus*, Nutt. Trans. Am. Phil. Soc. vii. 388.

——M. **Nuttálli**, Gray. Stem slender, a foot or two high: leaves linear-lanceolate: heads sparsely paniculate, 4 lines high, usually slender-peduncled: involucral bracts 8 to 12, with short inconspicuous tips: exserted ligules 3 to 5 lines long: only ray-akenes fertile; these obovate-falcate, much compressed, with sides many-striate and nearly nerveless: pappus of sterile disk-flowers of small oblong paleæ. — Proc. Am. Acad. l. c. viii. 391, ix. 188, & Bot. Calif. i. 358. *Anisocarpus madioides*, Nutt. l. c.; Torr. & Gray, Fl. ii. 403. — Woods, from Monterey, California, to Brit. Columbia; first coll. by *Nuttall*.

——M. **Bolánderi**, Gray, l. c. Stem 2 to 4 feet high: leaves linear (the longer 7 to 10 inches long, 4 lines wide): heads half to three-fourths inch high: involucral bracts and rays 12 to 16; bracts of the receptacle linear and unconnected: ray-akenes linear-falcate, 1-2-nerved on the narrow faces, commonly with a rudiment of pappus: disk-akenes numerous, straighter, all the outer ones fertile, all with a pappus of slender paleæ, which are either little or much shorter than the corolla. — *Anisocarpus Bolanderi*, Gray, Proc. Am. Acad. vii. 360. — Woods of the Sierra Nevada, California, from Mariposa to Plumas Co.; first coll. by *Bolander*, and northward to Scott Mountains, *Pringle*.

* * * Annual, with showy heads, chiefly alternate leaves, and no pappus: pubescence viscid as well as hirsute or hispid.

M. radiáta, KELLOGG. Stem stout, 2 or 3 feet high: larger leaves broadly lanceolate, denticulate: bracts of the involucre 10 to 20, with short tips: rays as many, half-inch long, obtusely 3-toothed: disk-flowers very numerous on a nearly flat glabrous receptacle, all but the central ones fertile, somewhat clavate and 4-angular, straightish: ray-akenes narrowly obovate-falcate, flat, tipped with a minute reflexed beak! — Proc. Calif. Acad. iv. 190; Gray, Bot. Calif. i. 359. — California, near the mouth of the San Joachin River, *Bolander*.

M. élegans, DON. Stem less stout, a foot or two high, or in depauperate forms only a span or two, above sometimes copiously beset with stipitate viscid glands, sometimes these almost wanting: leaves linear or lanceolate, mostly entire: bracts of the involucre 5 to 15, with linear tips: rays acutely 3-lobed, yellow throughout or with a brown-red spot at base: disk-flowers more numerous than the rays, on a convex hirsute-fimbrillate receptacle, all sterile: fertile akenes obliquely obovate-cuneate, nearly nerveless, depressed-truncate and wholly beakless at summit. — Don in Bot. Reg. t. 1458; Gray, l. c. *M. viscosa*, var., Hook. Fl. ii. 24, not Cav. *Madaria elegans* & *M. corymbosa* (with var. *hispida?*), DC. Prodr. v. 692. *M. elegans*, Hook. Bot. Mag. t. 3548. *M. corymbosa*, Endl. Iconogr. t. 36. *M. racemosa*, Nutt. Trans. l. c. — Hills and plains, throughout California, Oregon, and the borders of Nevada; first coll. by *Douglas*.

§ 2. EUMÁDIA. Ligules inconspicuous or short, from twelve to one, or rarely none: disk-flowers few or numerous and fertile: the corollas pubescent: pappus none: receptacle flat, smooth: glandular and viscid heavy-scented annuals. — Gray, l. c. *Madia, Madariopsis, Madorella*, & *Amida*, Nutt. l. c.

M. satíva, MOLINA. Commonly robust, 1 to 3 feet high, pubescent with slender somewhat viscid hairs and beset with pedicellate very viscid glands: leaves from broadly lanceolate to linear: heads commonly short-peduncled or sessile and rather scattered, 5 or 6 lines high: rays 5 to 12, with honey-yellow ligules about 2 lines long: disk-akenes cuneate-oblong and quadrangular, being prominently one-nerved on the faces (2 lines long), those of the ray somewhat falcate-obovate, either with or without an obvious nerve on the sides. — Don in Bot. Reg. l. c.; DC. Notul. Jard. Genev. & Prodr. l. c.; Torr. & Gray, Fl. ii. 404. *M. sativa* (with false char.) & *M. mellosa* (which would have been the better name to use), Molina, Chil. ed. 1, 354. *M. viscosa*, Cav. Ic. iii. 50, t. 298. *M. mellosa*, Jacq. Hort. Schœub. iii. 29, t. 302. *M. stellata*, Fisch. & Meyer, Ind. Sem. Petrop., few-flowered form, like that figured by Jacquin. — Oregon and California. (Chili.)

Var. **congésta**, TORR. & GRAY, l. c. The common TARWEED near the coast, stout, branching, very sticky: heads mostly crowded or glomerate at the end of the branches, many-flowered; the rays from 8 to 12. — *M. capitata*, Nutt. l. c. Nearly *M. viscosa*, Cav. l. c. — Fields and waysides throughout the western portion of California and Oregon; probably an introduction from Chili, or the contrary.

Var. **racemósa**, GRAY, l. c. Slender, simple-stemmed, with fewer-flowered heads somewhat racemosely disposed: disk-akenes flatter and nerve less distinct. — *M. racemosa*, Torr. & Gray, l. c. *Madorella racemosa*, Nutt. l. c. — Oregon to Idaho, interior of California, and Nevada. Approaching the fewer-flowered Chilian *M. mellosa*, Jacq., &c., perhaps passing into the next.

M. dissitiflóra, TORR. & GRAY, l. c. Slender, a foot or two high, often loosely branching, moderately viscid: heads 3 or barely 4 lines high, scattered or loosely paniculate: rays 5 to 8: disk-flowers few: akenes shorter and broader (a line or two long), also thicker, not angled nor with the sides evidently nerved. — *M. sativa*, var. *dissitiflora*, Gray, l. c. *Madorella dissitiflora*, Nutt. l. c. *Sclerocarpus gracilis*, Smith in Rees Cycl.? — Not uncommon throughout Oregon and California.

━━ **M. glomeráta**, Hook. A foot or so high, rigid, very leafy, hirsute, glandular only toward the inflorescence: leaves narrowly linear: heads glomerate: rays 2 to 5 or sometimes none, not surpassing the about equal number of disk-flowers: akenes (2 lines or more long) narrow, those of the disk 4-5-angled; of the ray somewhat curved and 1-nerved on each face. — Fl. ii. 24; Gray, l. c. *Amida hirsuta* & *A. gracilis*, Nutt. l. c.; Torr. & Gray, l. c. — Rocky Mountains of Colorado to Saskatchewan, Washington Terr., Oregon, and the Sierra Nevada in California.

§ 3. Harpæcárpus. Ligules very short and inconspicuous, not surpassing the solitary fertile disk-flower, all destitute of pappus: corolla glabrous. — Gray, l. c. *Harpæcarpus*, Nutt. l. c. 389.

━━ **M. fílipes**, Gray, l. c. Slender annual, a span to a foot or more high, hirsute, glandular above, paniculately branched; the small heads (a line or two long) on long filiform peduncles: leaves narrowly linear: bracts of the involucre 4 to 8, lunate and strongly carinate in fruit, almost destitute of free tips, hispid-glandular: bracts of receptacle united into a 3-5-toothed cup: ray-akenes obovate-lunate, the tip somewhat pointed by a small epigynous disk: disk-akene straight and obliquely obovate. — *Sclerocarpus exiguus*, Smith in Rees Cycl.? *Harpæcarpus madarioides*, Nutt. l. c. *H. exiguus*, Gray, Bot. Mex. Bound. 101. — Open grounds, from S. California to British Columbia near the coast, and eastward to Idaho.

123. **HEMIZONÉLLA**, Gray. (Diminutive of *Hemizonia*.) — Little annuals of Pacific N. America; with somewhat the aspect and characters of the *Harpæcarpus* section of *Madia*, hirsute-pubescent and above glandular, diffusely branching: leaves linear, entire, opposite or some of the upper alternate: heads in the forks and cymosely clustered, terminating the branchlets, short-peduncled, small (a line or two in length); the very small corollas yellow. Involucre glandular-hispid on the back. — Proc. Am. Acad. ix. 189, & Bot. Calif. i. 360. *Hemizonia* § *Hemizonella*, Gray, Proc. l. c. vi. 548.

━━ **H. Durándi**, Gray. A span high: earliest heads slender-peduncled: akenes narrowly oblong-obovate or somewhat fusiform, manifestly obcompressed with the inner face slightly angulate, tipped with a short but conspicuous incurved beak. — *H. Durandi* & *H. parvula*, Gray, Proc. Am. Acad. ix. 189. *Hemizonia Durandi* & *H. parvula*, Gray, Proc. Am. Acad. vi. 549. — Dry ground, California, from the Yosemite Valley to Washington Territory; first coll. by *Pratten*.

H. mínima, Gray, l. c., with syn. An inch or two high: peduncles all shorter than the heads: ray-akenes obovate, less incurved, much obcompressed, the beak obsolete or a minute inflexed apiculation. — Dry sterile soil, California, through the eastern ranges of the Sierra Nevada, from Mariposa Co. northward, *Brewer*, *Matthews*, &c.

124. **HEMIZÓNIA**, DC. Tarweed. (Composed of ἥμι, half, ζώνη, girdle, from the half-enclosed ray-akenes.) — Californian herbs, nearly all annuals or biennials, usually glandular, viscid, and heavy-scented; with alternate or sometimes opposite leaves, and middle-sized or small heads of yellow or white flowers, the anthers commonly brownish. Fl. summer or later. — Torr. & Gray, Fl. ii. 396; Benth. & Hook. Gen. Pl. ii. 394; Gray, Proc. Am. Acad. ix. 190, xix. 17, & Bot. Calif. i. 361. *Hemizonia*, *Hartmannia*, in part, & *Calycadenia*, DC. Prodr. v. 692–695.

§ 1. Euhemizónia, Gray, l. c. Ray-akenes only fertile, obovate-triangular, with depressed terminal areola hardly eccentric, glabrous, smooth and even: disk-akenes abortive and without pappus: annuals, a foot or so high; with entire or merely denticulate and mostly linear leaves, and white or yellow flowers: rays 3-lobed. — *Hemizonia*, DC. (the typical species of both sections).

* Akenes rounded on the back and with a ventral angle, destitute or nearly so of basal stipe: rays exserted but rather short: chaffy bracts none or hardly any among the inner flowers: leaves narrow, quite entire, or rarely a few salient denticulations. (Ambiguous species, with the habit, but not the akenes, of *Madia*.)

H. Wheéleri, GRAY. Loosely branching, slender, green, slightly pubescent, minutely glandular above: heads scattered: rays 5 or 6, bright yellow; marginal bracts of the receptacle distinct. — Bot. Calif. i. 617; Rothrock in Wheeler Rep. vi. 361, t. 10. — Olancha Mountains, of the southern Sierra Nevada, California, *Rothrock*.

H. citriodóra. Simple-stemmed, with short-pedunculate corymbosely panicled heads, or loosely branched above and heads more scattered, "lemon-scented," cinereously villous-hirsute and above with small pedicellate glands interspersed: rays 8 or 9, greenish-yellow: marginal bracts of the receptacle lightly united into a cup. — *Madia citriodora*, Greene, Bull. Torr. Club, ix. 63. — Northern California, from Siskiyou Co., *Greene*, to Placer and Sacramento Co., *Bolander* (1865), *Mrs. Curran*. With specimens from the latter a less villous and more glandular form, *Madia anomala*, Greene, ined.

* * Akenes obovate-triangular, with a dorsal and two lateral angles, the ventral face broad and nearly plane, surface smooth and shining, base usually extended into a small inflexed stipe, with a whitish callous at its insertion (but sometimes the stipe short or obsolete and the callus at the very base of the akene): receptacle chaffy throughout: rays either white or light yellow in the same species, opening only in bright sunshine.

+ Heads terminating paniculate or usually corymbosely cymose branches.

H. congésta, DC. Soft-hirsute or villous, but not lanate, slightly glandular toward the clustered or scattered heads: bracts of the involucre with lanceolate foliaceous tips, little surpassed by the rays: marginal bracts of the receptacle either lightly connate or nearly distinct: inflexed stipe of the akene conspicuous. — Prodr. v. 692; Gray, l. c. (not of Pacif. R. Rep. iv. 109, which proves to be young *Lagophylla*). — California, near San Francisco, *Douglas, G. R. Vasey*. Specimens formerly referred to this still little-known species belong to the following.

H. luzulæfólia, DC. Villous, and below even sericeous-lanate, at least when young, above becoming very viscid-glandular and corymbosely or paniculately branched: lower leaves elongated, 3–5-nerved: bracts of the involucre with short and broadish herbaceous tips: marginal bracts of the receptacle united into a cup: rays 5 to 10, rather large, white, sometimes tinged with pink, or not rarely pale yellow (var. *lutescens*, Greene, Bull. Torr. Club, ix. 16): stipe of akene as in the preceding, or shorter, or obsolete. — Prodr. l. c.; Gray, l. c. *H. sericea*, Hook. & Arn. Bot. Beech. 356. *H. rudis*, Benth. Bot. Sulph. 31, a very branching and late-flowering form. — Dry open grounds, throughout western part of California, very common from San Francisco Bay to Monterey. Varying greatly.

+ + Heads disposed to be sessile along simple branches.

H. Cleveládi, GREENE. More slender, below villous with long spreading hairs, not lanate: leaves all narrowly linear, mostly one-nerved: heads smaller, nearly all after the terminal one subsessile in the axils or on short leafy branchlets, thus as it were spicately or racemosely disposed: rays white: akenes and flowers as in the smaller-headed form of the preceding. — Bull. Torr. Club, ix. 109. — California, from Mendocino Co. (*Kellogg*) to Lake Co., *Bolander, Cleveland*.

§ 2. HARTMÁNNIA. Gray, l. c. Ray-akenes opaque and often rugose or tuberculate (rarely smooth and shining), very gibbous, turgid, the terminal areola from the summit of the inner angle or face, and by gibbosity commonly intraapical, raised on a little beak (rostellum) or apiculation: flowers in all yellow: ours annuals. — *Hartmannia* (excl. spec.) & part of *Hemizonia*, DC. Prodr.

H. FRUTÉSCENS, Gray, Proc. Am. Acad. xi. 79, & Bot. Calif. i. 361, an outlying species from Guadalupe Island, off Lower California, is remarkable for having a woody-based stem, and is probably the only species that is really perennial. This and

H. STREÉTSII, Gray, l. c. xii. 162, from San Benito Island, Lower California, are the only known representatives of the genus beyond the limits of this Flora.

* Receptacle conical or convex, many-flowered, all the disk-flowers subtended by narrow and mostly quite distinct chaffy bracts, some of them not rarely fertile: ray-flowers usually numerous and in more than one series, with short and yellow ligules; their akenes obovate-triangular, with very oblique apiculation, usually smoothish: rigid and branching annuals; with some or all of the lower leaves incisely pinnatifid, and the uppermost clustered around the sessile heads.— *Hartmannia* § *Olocarpha*, DC. Prodr.

+— Leaves and bracts not pungent, but the upper gland-tipped.

— **H. macradénia**, DC. Stout, hirsute, viscid-glandular, very leafy: upper leaves linear, entire or laciniately dentate; those of the branchlets and axillary fascicles linear-subulate, truncately gland-tipped: some of these and most of those crowded around the sessile glomerate heads, also the bracts of the involucre and even those of the conical receptacle, beset with stipitate tack-shaped glands: heads fully half-inch in diameter: pappus none. — Prodr. v. 693; Hook. & Arn. Bot. Beech. 356; Torr. & Gray, Fl. ii. 400; Gray, Bot. Calif. i. 363. — Dry open ground, from the Bay of San Francisco southward. An unpleasantly scented *Tarweed*.

+— +— Upper leaves or their lobes and the bracts of the involucre rigid, pungently pointed, none gland-tipped.

═**H. Fitchii**, GRAY. Villous-hirsute, somewhat viscid, above beset with small scattered tack-shaped glands: leaves some (even of the lower) entire and elongated linear-acerose, very pungent, some of the lower once or twice pinnately parted: bracts of the involucre subulate; those of the receptacle pointless, soft, bearded with long villous hairs: disk-akenes sterile, with pappus of 8 to 12 linear paleæ, fringed or bearded at tip, somewhat united at base, nearly equalling their corolla. — Pacif. R. Rep. iv. 109, & Bot. Calif. l. c. — Common in California north and east of Sacramento; first coll. by *Rev. Mr. Fitch*.

--**H. Párryi**, GREENE. Sparsely or slightly hirsute, sometimes minutely viscid-glandular: leaves short; lower sparingly pinnatifid; upper subulate-acerose, as also the tips of the involucral bracts; those of the receptacle thin, villous on the margin, acute or obtuse, but neither pointed nor rigid: sterile disk-akenes with a pappus of 3 to 5 narrowly linear slender, pointed naked paleæ which equal the corolla. — Bull. Torr. Club, ix. 16. (Has been inexcusably confounded with the preceding and following.) — Not uncommon in California from Lake Co. to San Bernardino Co., *Torrey, Parry, Parish*, &c.

═**H. púngens**, TORR. & GRAY. Hirsute or hispid, sometimes only slightly so, hardly at all viscid or glandular: cauline leaves pinnatifid or the lower bipinnatifid, and the lobes short; those of the branchlets and fascicles entire, lanceolate or linear-subulate, with very pungent tips, those around the head little surpassing it: bracts of the receptacle also pungently pointed: pappus to disk-flowers none. — Fl. ii. 399; Bot. Calif. l. c. *Hartmannia pungens*, Hook. & Arn. Bot. Beech. 357; Hook. Ic. Pl. t. 334. — Dry hills and fields, from San Francisco Bay southward; first coll. by *Douglas*.

* * Receptacle flat or nearly so, naked among the disk-flowers, which are surrounded by a circle of connate or sometimes distinct bracts: rays golden yellow and with glandular usually slender tubes: some of the pubescence glandular or viscid: no large tack-shaped or terminal truncate glands.

+— Rays 12 to 24, oblong-cuneate; their akenes occupying more than one series, obscurely rugose: disk-flowers as numerous, with wholly sterile or abortive ovary, and small plurisquamellate pappus or none.

H. corymbósa, TORR. & GRAY, l. c. Erect, corymbosely branched above, hirsute, with or without short-pedicellate glands intermixed: lower or sometimes most of the cauline leaves pinnately parted into linear lobes; those of the branches narrowly linear: heads rather large (a third to half inch high): rays 15 to 25, oblong-cuneate: bracts of receptacle well united into a cup: akenes 4-5-nerved or angled (the nerve of the inner face indistinct or wanting), and with beak short and stout: disk-pappus setosely plurisquamellate. — *H. angustifólia*, Benth. Pl. Hartw., not DC. *H. macrocephala*, Nutt. Pl. Gamb. 174. *H. balsamifera*, Kellogg, Proc. Calif. Acad. ii. 64, t. 13. *Hartmannia corymbosa*, DC. Prodr. v. 694. — W. California, in low grounds, common from San Francisco Bay to San Luis Obispo; first coll. by *Douglas*.

H. angustifólia, DC. Diffuse, a span to a foot high, hirsutely pubescent and glandular, becoming viscid: cauline leaves all linear, small, entire: heads corymbosely paniculate or scattered; rays 12 to 15: bracts of the receptacle less united, or almost separate. akenes 3-nerved, with prominent upturned beak: disk-pappus minute and squamellate or nearly wanting. — Prodr. v. 692; Torr. & Gray, Fl. ii. 398; Gray, Bot. Calif. i. 362. *H. multicaulis*, Hook. & Arn. Bot. Beech. 355 ? *H. decumbens*, Nutt. Pl. Gamb. 175. — W. California, in open grounds, from San Francisco Bay southward; first coll. by *Douglas*.

Var. **Bárclayi**, GRAY, Proc. Am. Acad. ix. 190, & Bot. Calif. l. c. (excl. pl. *Brewer*), from Monterey, *Barclay*, has more conspicuous laciniate pappus to disk-flowers.

+ + Rays 8 to 20, broadly cuneate or quadrate: disk-flowers more numerous, with well-formed and often fertile ovary and a conspicuous pappus of coriaceous oblong obtuse paleae, which are hirsute at summit and margins, and even on the back: stems erect, paniculately branched, 2 feet or more high, very leafy.

H. floribúnda, GRAY. Minutely glandular-pubescent and viscid, not hirsute: cauline leaves all linear, small, entire: heads disposed to be racemose-paniculate on the branches: rays about 20; their akenes in more than one series, somewhat tuberculate-rugose, obscurely 4-angled, with very short straight beak: disk-akenes numerous, with pappus of 5 to 8 broadish paleae shorter than the proper tube of the corolla. — Proc. Am. Acad. xi. 79, & Bot. Calif. i. 616. — California, southern part of San Diego Co., *Palmer*, *Cleveland*.

H. paniculáta, GRAY. More diffusely branched, below commonly hirsute, the branchlets and heads viscid-glandular · cauline leaves laciniate-pinnatifid; those of the branches entire or 2-3-dentate, linear, small; of the flowering branchlets mostly very small and bract-like, erect: heads sparsely paniculate, barely 3 lines high: involucral bracts minutely densely glandular: rays about 8; their akenes coarsely rugose or pitted on the back: receptacular bracts connate or distinct: disk-flowers about 11; their well-formed akenes with a pappus of 8 or 10 oblong paleae which exceed the proper tube of the corolla. — Proc. Am. Acad. xix. 17. — Santa Barbara Co. to San Diego Co., *Brewer*, *Parish*, *Jared*. Includes plant of coll. Brewer, referred in Bot. Calif. to *H. angustifolia*, var. *Barclayi*.

+ + + Rays 5 (rarely 3, 4, or 6), broadly cuneate or quadrate: disk-flowers not over 6, surrounded by mostly 5 receptacular bracts, which are usually more or less connate; their akenes generally sterile, the paleae of their pappus not hirsute: stems paniculately branched, a foot or two high, some taller: lower cauline leaves pinnatifid; upper and rameal entire, small.

H. Kellóggii, GREENE. Hirsute, sparsely so above, bearing short-pedicelled loosely paniculate heads: cauline leaves mostly pinnately-parted or toothed: involucre quarter-inch high; the bracts hirsutely glandular on the back, broadly lanceolate: rays fully 3 lines long: bracts of the receptacle rather broad, well united into a cup: ray-akenes tuberculate-rugose (a line or more long), bearing a rather strongly lateral and slender curved (almost sigmoid) beak: sterile disk-akenes with pappus about equalling the tube of their corolla, composed of lacerately truncate paleae, which are mostly connate to near their summits. — Bull. Torr. Club, x. 41. — Central California near Antioch (*Kellogg*), and along the San Joaquin Valley, *Greene*.

H. Wrightii, GRAY. Hirsute below, 1 to 3 feet high, with widely-spreading branches, when much branched decumbent; the slender or filiform branchlets terminated by pedicellate heads: lower cauline leaves laciniate-pinnatifid; those of the branchlets mostly minute and very viscid-glandular, as is the involucre; its bracts ovate-lanceolate; those of the receptacle partly united: ray-akenes obscurely tuberculate-rugose, with short beak: sterile disk-akenes with pappus of 8 or 9 oblong firm paleae, their summit erose-laciniate. — Proc. Am. Acad. xix. 17. — S. California, about San Bernardino, *W. G. Wright*, *Parish*, *Parry*. Found also as a waif near San Francisco, *Greene*. Heads always scattered, and most of them on pedicels of fully their own length.

H. fasciculáta, TORR. & GRAY. More or less hirsute or hispid below, a span to 2 feet high, commonly with rather rigid ascending glabrate or viscid-glandular branches, bearing usually fasciculate-clustered sessile small heads: cauline leaves mostly pinnately parted or laciniate; uppermost on the branches subulate-linear and rather crowded about the heads or clusters: bracts of the involucre narrowly lanceolate, either glabrous or glandular-hispidulous: of the receptacle lightly united or nearly free: ray-akenes either smoothish or at length transversely rugose, apiculate with a small very short beak; disk-akenes chiefly sterile, with conspicuous pappus of 8 or 10 narrowly oblong or linear lacerate-tipped paleae. — Fl. ii. 397; Gray, l. c. *H. glomerata*, Nutt. Trans. Am. Phil. Soc. l. c. *Hartmannia fasciculata*, DC.

Prodr. l. c. — Dry ground, W. California, common from Monterey to San Diego; first coll. by *Coulter, Douglas*, &c. Passes into

Var. ramosíssima. Diffuse, sometimes decumbent: upper leaves mostly entire: heads less fascicled or all scattered: akenes at maturity rugose. — *H. ramosissima*, Benth. Bot. Sulph. 30; Gray, Bot. Mex. Bound. 100, & Bot. Calif. i. 362. — Same range, and to San Bernardino Co.

Var. Lóbbii (*H. Lobbii*, Greene, Bull. Torr. Club, ix. 109, founded on a single specimen, coll. by Lobb, thought to come from near Monterey) appears to be nothing more than a tall and slender form of this species, with stem 2 feet high, long and slender branches, very small and numerous leaves on the branchlets, rays reduced to 3 or 4, and disk-flowers to about the same number, each subtended and partly enclosed by a free bract. But specimens hardly sufficient.

* * * Receptacle flat: all the flowers subtended and akenes partly enclosed by bracts; the corolla-tubes glandular: ligules yellow and broad, 5 to 8: ray-akenes somewhat 5-nerved or angled, i. e. ventral face somewhat carinate-angled, with short upturned beak: disk-flowers 8 to 15, with akenes mostly sterile and destitute of pappus: slender virgately branched or paniculate annuals, with lowest cauline leaves commonly laciniate-dentate, the upper all small and linear, none of them at all pungently pointed, but those of the branchlets tipped with a sessile truncate gland.

H. Heermánni, GREENE. Viscid and somewhat pubescent or hirsute, heavy-scented, paniculately branched, 1 to 3 feet high, the minute leaves of the diffuse flowering branchlets rather scattered: involucre nearly hemispherical; its bracts (and rays) 5 to 9, viscid-pubescent and copiously beset with pedicellate glands; the terminal gland inconspicuous: beak and stipe of ray-akenes somewhat conspicuous: disk-flowers 10 to 15. — Bull. Torr. Club, ix. 15. *H. macradenia*, Durand, Pacif. R. Rep. v. 10, not DC. *H. ramosissima*, in part, Rothrock in Wheeler Rep. vi. 365. — Southern part of California, from Santa Barbara to Kern Co., &c., first coll. by *Heermann*.

H. virgáta, GRAY. Less pubescent and viscid or nearly glabrous, the stem or long branches virgate and bearing numerous racemosely or somewhat paniculately disposed heads on short densely foliolose branchlets; their leaves Heath-like, line long, all glandular-truncate: involucre campanulate or in age oblong; its mostly 5 bracts becoming coriaceous, with stout involute tip bearing a large truncate gland, the back nearly glabrous and sparsely beset with some stout pedicellate glands or gland-tipped processes: stipe of ray-akene hardly any, and its beak short: disk-flowers 7 to 10. — Bot. Mex. Bound. 100, & Bot. Calif. i. 363. — California, from Lake Co. to Los Angeles, &c.; first coll. by *Fremont*.

§ 3. CALYCADÉNIA, Gray, l. c. Ray-flowers few (1 to 7), with very broad palmately 3-lobed or parted ligule; their akenes mostly dull, obovoid-triangular and little oblique; the terminal areola scarcely if at all eccentric: disk-flowers surrounded by a circle of herbaceous bracts (forming a kind of inner involucre), which are connate into a cup or rarely separable; their akenes well formed and the outer not rarely fertile (then hairy), turbinate-quadrangular or slightly obcompressed, straight, bearing a conspicuous paleaceous pappus: annuals, with entire narrowly linear leaves, often becoming filiform by revolution of the margins; those of the axillary fascicles and clusters near the heads usually tipped with tack-shaped or when dry saucer-shaped conspicuous glands, which are either sessile or short-stipitate, sometimes similar glands along their backs or edges: heads as it were involucrate by some bract-like leaves. — *Calycadenia*, DC. Prodr. v. 695.

* Wholly destitute of tack-shaped glands, paniculately and diffusely much branched and heads scattered: rays 3-parted down to the slender tube, and disk-corollas cleft into oblong-linear lobes; both white: ray-akenes almost beaked. — *Osmadenia*, Nutt.

H. tenélla, GRAY. Slender, 6 to 18 inches high, sparsely hirsute-pubescent or hispid, and filiform branchlets minutely viscid-glandular: leaves almost filiform: involucre cylindraceous-campanulate: ray-flowers 3 to 5; their akenes rugose, short-stipitate and abruptly rostellate-apiculate: disk-flowers 5; their pappus of 4 or 5 lanceolate palea: tapering into stout rough awns, and as many intermediate short and lacerate-truncate ones. — Proc. Am. Acad. ix. 191, Bot. Calif. l. c. *Osmadenia tenella*, Nutt. Trans. Am. Phil. Soc. vii. 392. *Calycadenia tenella*, Torr. & Gray, Fl. ii. 402. — Found only near San Diego, California; first by *Coulter* and *Nuttall*.

　＊　＊　Tack-shaped or saucer-shaped glands borne at least by the leaves next the heads and those fascicled in the axils: stem strict or with ascending branches: disk-corollas long and narrow, 5-toothed: ray-akenes truncate at summit, and with a depressed or sometimes slightly protuberant terminal areola; no basal stipe: anthesis commonly (or perhaps always) vespertine or matutinal.

　+　Heads very few-flowered and narrow, spicately and sparsely scattered along flexuous simple branches: flowers white or rose-tinged.

H. pauciflóra, GRAY, l. c. A foot or less high, with spreading filiform branches, sparsely hirsute, glabrate: heads solitary and sessile in the axils of small remote leaves; these and the floral ones sparsely hispid near the base: ray-flowers solitary or 2, the ligule 3-parted: disk-flowers 3 in a 3-lobed cup; their pappus of 5 subulate-awned and 5 small truncate paleæ: ray-akenes glabrous: tack-shaped glands small and sparse, short-stalked. — *Calycadenia pauciflora*, Bot. Mex. Bound. 100. — California, from unrecorded station, *Fremont*. Also Lakeport, Lake Co., *Pringle*.

　+　+　Heads many-flowered, loosely paniculate or racemosely scattered along the slender spreading branches: flowers yellow: plant remarkably glabrous.

H. truncáta, GRAY, l. c. A foot or two high: leaves rather lucid and thickish, some of them hispidulous-scabrous, or the lower with a few bristles, and those next the heads occasionally setose-ciliate, otherwise very smooth: glands mostly only terminal, large and subsessile: heads oval-campanulate, 4 or 5 lines long: ray-flowers 5 to 8, with ovate-oblong boat-shaped involucral bracts and glabrous triangular-obpyramidal akenes: bracts of the receptacle 7 to 9, lightly connate to the top into a truncate cup, at length separable: disk-flowers 10 to 20; their pappus of 7 to 10 oblong and somewhat erose fimbriate pointless paleæ, much shorter than the akene, sometimes obsolete. — *Calycadenia truncata*, DC. Prodr. l. c.; Torr. & Gray, l. c. — California, from near San Francisco Bay northward into Oregon; first coll. by *Douglas*.

　+　+　+　Heads 8-15-flowered, in axillary and terminal short-pedunculate clusters on the strict stem or branches: pubescence all soft and short, grayish.

H. móllis, GRAY, l. c. A foot or two high, the stem only puberulent: leaves cinereous-pubescent; those of the fascicles and around the heads and the bracts tipped with a short-stalked dark gland, also some on the back: ray-flowers 3 to 5, with sometimes white sometimes yellow 3-parted ligules on a short slender tube: chaff of receptacle forming a 6-8-toothed cup: ray-akenes obpyramidal, glabrous: disk-flowers 5 to 10, with pappus of 5 or 6 subulate-awned paleæ nearly twice the length of the akenes, and one or two small pointless ones. — *H. angustifolia*, Durand, in Pacif. R. Rep. l. c., not DC. *Calycadenia mollis*, Gray, Proc. Am. Acad. vii. 360. — Sierra Nevada, California, in the foothills up to 4,000 feet in Merced Co. and Tuolumne Co.; first found with bright white rays, later with yellow also, by *Lemmon*, &c.

　+　+　+　+　Heads several-many-flowered, mostly glomerate or spicately paniculate on the strict stem or branches, in depauperate slender plants solitary in the axils: leaves rather rigid: pubescence setose-hirsute or hispid, at least on the margins of the upper leaves: lobes of the disk-corollas sometimes strongly and sometimes sparsely and obscurely hispidulous-glandular or barbellate on the outside.

H. Douglásii, GRAY, l. c., partly. Whitish-hirsute and hispid: tack-shaped glands not rare on the margins as well as the tips of many of the leaves, mostly none on the bracts of the involucre and receptacle: flowers yellow or white and purplish-tinged: akenes silky-villous, at least when young, but often glabrate: pappus a little shorter than the disk-corolla, of 10 or sometimes 12 narrow linear-lanceolate paleæ which are gradually attenuate into an awn-like point, as long as or longer than the akenes, or 2 or 3 of them not rarely shorter or pointless. — *Calycadenia villosa*, DC. l. c.; Torr. & Gray, l. c., founded on slender and too young specimens of coll. Douglas. *H. hispida*, Greene, Bull. Torr. Club, ix. 63; a robust form, 1 to 3 feet high, with yellow flowers; coll. near Atwater Station, Merced Co., by *Greene* and *Parry*. *H. spicata*, Greene, Bull. Torr. Club, ix. 16, a dwarf form, with white flowers; coll.

by the same at Milton. — Desiccated plains, from Lake Co. to Merced Co., California; first coll. by *Douglas*, in immature and depauperate specimens. As polymorphous as the next species.

H. multiglandulósa, Gray, l c. Hirsute or hispid, also puberulent: tack-shaped glands usually abundant on the back of the bracts of the involucre and of the receptacle: flowers white, sometimes purplish-tinged: ray-akenes glabrous or glabrate, short and broadly obpyramidal-obovate, glabrous or soon glabrate: pappus much shorter than the disk-corolla and shorter than the akenes, of 10 or rarely 12 unequal paleæ, 5 of them oblong- to lanceolate-subulate and attenuate at summit into an awn-like point, the others obtuse or erose-truncate. — *Calycadenia multiglandulosa* & *C. cephalotes*. DC. Prodr. v. 695. — Common in California, especially in the Great Valley and north of the Bay of San Francisco. Runs into many and various forms. The type of the species has the heads or clusters sessile and not much crowded in the axils of the leaves along the virgate stem or its basal branches: odor said to be disagreeable.

Var. **cephalótes**. Stouter, with heads densely glomerate at the summit of the stem and in approximate axils, sometimes appearing later in remoter axils: herbage heavy-scented. — *H. cephalotes*, Greene in Bull. Torr. Club, ix. 110. *Calycadenia cephalotes*, DC. l. c. A common form: odor said to be balsamic.

Var. **spársa**. Slender, lax, a span to a foot high: lower and sometimes all the leaves opposite: heads usually solitary in a few axils; the terminal glands on the bracts few. — *H. Fremonti*, Gray, Proc. Am. Acad. ix. 191. (*Calycadenia Fremonti*, Gray, Bot. Mex. Bound. 100.) *H. oppositifolia*, Greene, Bull. Torr. Club, l. c. — Valley of the Sacramento, *Fremont, Mrs. Bidwell, Parry*, &c.

§ 4. BLEPHARIZÓNIA, Gray. Ray-flowers 7 to 10, with 3-lobed ligules: disk-flowers 10 to 20; outer ones subtended by one or two series of linear receptacular bracts: akenes of disk disposed to be fertile and nearly like those of the ray, except in their pappus of about 20 short and stout densely plumose awns: ray-akenes elongated-turbinate, hardly oblique, sericeous-hirsute, about 10-nerved, with broad and depressed terminal areola, this obscurely coroniform-bordered. — Proc. Am. Acad. ix. 192, & Bot. Calif. i. 366.

H. plumósa, Gray, l. c. Strongly ill-scented annual, 2 to 5 feet high, paniculately branched, hirsute-pubescent, above most copiously beset with very viscid tack-shaped glands: cauline leaves linear, entire; those of the branchlets very small, oblong or oval, bract-like: heads racemosely paniculate, broad (4 or 5 lines long): involucral bracts short, very glandular: pappus in the original specimens nearly half the length of the disk-akenes. — *Calycadenia plumosa*, Kellogg, Proc. Calif. Acad. v. 49. — Banks and dried beds of streams, near Stockton, California; the original discoverer unknown; recently collected by *Lemmon*, &c.

Var. **subplumósa**. Pappus only one quarter the length of the disk-akenes, or even hardly longer than the diameter of their summit: heads more sparse, terminating loosely paniculate branches. — Near Stockton, apparently same habitat as that of the original species, *Parry, Mrs. Curran*.

125. **ACHYRACHÆNA**, Schauer. (Ἄχυρον, chaff, and *achænium*, the botanical name of the fruit of Compositæ, &c.: relates to the very chaffy pappus.) — Del. Sem. Hort. Vratisl. 1837; DC. Prodr. vii. 292; Torr. & Gray, Fl. ii. 392; Benth. & Hook. Gen. ii. 396. *Lepidostephanus*, Bartl. Ind. Sem. Hort. Gœtt. 1837. — Single species, a Californian annual.

A. móllis, Schauer, l. c. A span to a foot high, erect, villous-pubescent, slightly glandular-viscid: leaves alternate, or the lower opposite, long and narrowly linear, entire or the lower laciniate: heads solitary and long-peduncled, terminating the stem and fastigiate branches, an inch or less long: corollas whitish or yellowish and turning brownish: pappus and disk-akenes each quarter-inch long: in fruit and when mature and dry the akenes with their spreading pappus diverging, forming a globular silvery-chaffy head, resembling that of Thrift.

— *Lepidostephanus madioides,* Bartl. l. c. — Open grounds; fl. in spring, throughout the western part of California; first coll. by *Douglas.*

126. **LAGOPHÝLLA,** Nutt. (λαγός, a hare, φύλλον, foliage.) — Slender (Pacific N. American) herbs, paniculately much branched, usually more or less cinereous with sericeous pubescence (this so long and copious on the crowded upper leaves of the original species as to have suggested the generic name, from some likeness to a hare's foot): leaves narrow, entire or nearly so, the lower opposite, upper alternate, sometimes bearing small tack-shaped glands: heads small, with "pale yellow" or white and rose-tinged rays, apparently vespertine. Bracts and chaff promptly deciduous with the mature akenes, leaving the naked receptacle terminating and little thicker than the peduncle. — Trans. Am. Phil. Soc. l. c. 390; Torr. & Gray, Fl. ii. 402.

§ 1. HOLOZÓNIA. Perennial and spreading by creeping scaly rootstocks: pubescence all short: heads naked, scattered, mostly slender-peduncled: corollas white or purplish-tinged: chaff of receptacle connate into a 9–12-toothed cup: ray-akenes bearing a shallow entire or denticulate cupule in place of pappus (as sometimes in *Layia*): ovary of sterile disk-flowers occasionally bearing 2 to 5 nearly capillary naked bristles, which are very caducous, sometimes almost equalling the corolla. — *Holozonia*, Greene, Bull. Torr. Club, ix. 122, 146.

L. filipes, GRAY. Rootstocks elongated, rigid, partly sheathed by the approximate pairs of connate scales: stems diffusely branched: filiform branchlets and peduncles glabrous or sparsely glandular: cauline leaves linear, minutely soft-villous; those of the branchlets minute, oblong, commonly beset with short-stipitate dark glands: involucre loosely villous; its bracts little longer than the clavate-obovate obscurely 5-nerved akene, which bears a conspicuous white saucer-shaped cupule. — Pacif. R. Rep. iv. 109, Bot. Mex. Bound. 101, & Bot. Calif. i. 367. *Hemizonia filipes,* Hook. & Arn. Bot. Beech. 356; Torr. & Gray, Fl. ii. 359. *Holozonia filipes,* Greene, Bull. Torr. Club, l. c., where the peculiar characters were pointed out, and not unnaturally taken to be generic. — Rocky hills near streams, Napa Co. to Mendocino Co.; first coll. by *Douglas.*

§ 2. LAGOPHYLLA proper. Annuals: heads subtended by bracteal leaves which may sometimes imitate an outer involucre, disposed to be sessile and glomerate, or at length short-peduncled: no cupule or pappus to the akenes: chaff or bracts of the receptacle mostly quite distinct: stems below smooth and glabrous, or early glabrate.

＊ Green or barely cinereous, not canescent: heads loose or scattered: ligules much exserted, pale yellow?

L. dichótoma, BENTH. Stem a foot or two high, dichotomously paniculate; the branchlets puberulent: leaves sparse; cauline spatulate, occasionally dentate, strigulose-pubescent; of the branchlets short, hirsute-ciliate, as also the broadish bracts of the involucre, and with small and sparse or no glands: akenes obovate, much obcompressed, no nerve or keel to the ventral face. — Pl. Hartw. 317; Gray, Bot. Calif. i. 366. — Plains of Feather River, on the Sacramento, and Lake Co., California, *Hartweg, Fitch, Bigelow, Mrs. Curran.*

L. glandulósa, GRAY. Stem virgately paniculate, slender, a foot or two high: leaves cinereous-puberulent, linear or the radical spatulate-lanceolate, entire, sometimes even the lower as well as the small and scattered upper ones (also the branchlets) beset with small tack-shaped glands, sometimes these all but or quite absent: bracts of the involucre and the outer subtending bracts resembling the ordinary leaves, and inconspicuously if at all ciliate: akenes nearly of the following. — Proc. Am. Acad. xvii. 219. — Not rare from Butte Co. to Mariposa, *Mrs. Bidwell, G. R. Vasey, Lemmon, Mrs. Curran, Congdon.* Badly named, the glands inconstant in this, and occasionally seen in all the species.

* * Typical species: leaves canescent with soft silky pubescence: the short ones subtending the crowded heads conspicuously and densely ciliate with very soft villous hairs, and buck occasionally beset with sessile or short-stipitate glands: involucral bracts comose-ciliate at the sides (along the line of infolding): ligules short, pale yellow according to Nuttall, but certainly sometimes if not always purplish or rose-color: akenes clavately obovate-oblong, carinate down the ventral face: stems at length becoming naked below by the early fall of the older leaves. — *Lagophylla*, Nutt.

L. ramosíssima, NUTT. Slender, paniculately much branched, 6 to 30 inches high: leaves entire; radical and lowest cauline obovate-spatulate; upper lanceolate or linear, obtuse; uppermost linear-oblong: heads 3 lines long, glomerate in small and at length rather scattered irregular clusters: akenes only a line and a half long. — Trans. Am. Phil. Soc. vii. 390; Torr. & Gray, Fl. ii. 402; Gray, Bot. Calif. i. 367, mainly. *L. minima*, Kellogg, Proc. Calif. Acad. v. 53. — Dry ground, common through California, and to Washington Terr., Nevada, and W. Idaho; first coll. by *Nuttall*.

L. congésta, GREENE. Robust, a foot to a yard high, with short branches and larger heads in thick glomerules: akenes 2 lines long. — Bull. Torr. Club, x. 87. *Hemizonia congesta*, Gray in Pacif. R. Rep. iv. 109 (immature), not DC. — From Marin Co. to the Sierra Nevada and to Mendocino Co., California, *Bigelow, Torrey, Lemmon, Greene, Mrs. Curran*. Chaff of receptacle not found to be "united into a cup": perhaps only a gigantesque form of the preceding species.

127. **LÁYIA**, Hook. & Arn. (*Thomas Lay*, naturalist in Beechey's Voyage.) — Annuals, of California and adjacent parts; with chiefly alternate leaves, and branches terminated by usually showy heads of flowers, in spring and early summer: disk-corollas sparsely hispidulous or hirsute on the lobes, yellow: rays yellow or white. — Bot. Beech. 148 & 357 (not 182); Torr. & Gray, Fl. ii. 393; Gray, Pl. Fendl. 103, & Bot. Calif. i. 368. *Madaroglossa* & *Oxyura*, DC. Prodr. v. 693, 694. *Eriopappus*, Arn. in Lindl. Introd. Nat. Syst. ed. 2, 443. *Callichroa*, Fisch. & Meyer, Ind. Sem. Hort. Petrop. ii. 31. *Calliglossa*, Hook. & Arn. Bot. Beech. 356. *Calliachyris*, Torr. & Gray, in Bost. Jour. Nat. Hist. v. 110. — Certain species are so much alike in their whole aspect and structure that the technical characters which alone distinguish them may be expected to give way.

§ 1. MADAROGLÓSSA, Gray, Pl. Fendl. l. c. Pappus of about 10 to 20 stout bristles, which are long-plumose or villous below the middle: akenes all narrow and somewhat clavate, mostly with an obvious almost cupulate epigynous disk, at least in the ray: receptacle naked and pubescent among the disk-flowers: herbage hispid or hirsute, somewhat viscid, above beset with scattered stipitate blackish glands. — *Madaroglossa*, DC. l. c. *Layia*, Hook. & Arn.

* Rays bright white (sometimes tinged with rose), large and conspicuous, commonly half to three-fourths inch long, 3-lobed: lower leaves lanceolate or linear, laciniate-pinnatifid or incised, upper narrower and entire: pubescence more or less hispid or hirsute and with scattered short-stipitate dark glands, especially toward the heads: lobes of the disk-corollas with some sparse hispid hairs: pappus bright white.

L. glandulósa, HOOK. & ARN. A span to a foot or more high, diffusely branched: dark glands sometimes abundant, sometimes scarce: rays 8 to 13: villous hairs of the pappus-bristles copious, the outer straight and erect, the inner soon crisped and interlaced into a woolly mass. — Bot. Beech. 358; Torr. & Gray, l. c. *L. Neo-Mexicana*, Gray, Pl. Wright. ii. 98, a form with vestiges of pappus to ray-akenes. *Blepharipappus glandulosus*, Hook. Fl. i. 316. *Eriopappus glandulosus*, Arn. l. c. *Madaroglossa angustifolia*, DC. Prodr. v. 694, ex Hook. & Arn. — Barren ground, British Columbia to S. California and the Mexican border, and east to Idaho and New Mexico. Variable, sometimes with stems almost glabrous, sometimes with hairs of the pappus less copious.

Var. **rósea**, GRAY, Bot. Calif. i. 368, a rare state with rose-purple rays. — Ojai, California, *Peckham, Palmer*.

L. heterótricha, Hook. & Arn. l. c. Generally larger and more erect: dark glands copious: rays 10 to 18: long-villous hairs of the pappus-bristles less abundant, all erect, the inner woolly ones wanting.— Gray, l. c. *Madaroglossa heterotricha*, DC. l. c.; Hook. Ic. Pl. t. 326.— California, from the Lower Sacramento Valley southward.

* * Rays apparently white, but small and inconspicuous, little if at all surpassing the disk: pappus dull white.

L. carnósa, Torr. & Gray. Dwarf, barely a span high, diffusely branched from the base, somewhat pubescent: dark glands few or wanting: leaves succulent, spatulate to linear-oblong, an inch or less long, some sinuate-pinnatifid: pappus-bristles sparsely plumose with straight villous hairs: akenes of the ray also pubescent!— Fl. ii. 394; Gray, Bot. Calif. l. c. *Madaroglossa carnosa*, Nutt. Trans. Am. Phil. Soc. vii. 393.— Sands of the California sea-beach, from San Diego to Marin Co., *Nuttall, Parry, Bigelow*.

* * * Rays as well as disk-flowers yellow, or the former rarely white-edged.

+ Pubescence hirsute rather than hispid: inner hairs on the pappus woolly and interlaced in the manner of *L. glandulosa*, but mostly less densely so.

L. élegans, Torr. & Gray, l. c. Diffuse: stipitate glands small and sparse: leaves linear; the lower pinnately toothed or parted into linear lobes: rays 10 to 12, half-inch long: pappus white or whitish, its copious villous hairs much shorter than the aristiform bristles.— Gray, Pl. Fendl. 103, & Bot. Calif. i. 369. *Madaroglossa elegans*, Nutt. Trans. Am. Phil. Soc. vii. 393.— Common from Santa Barbara Co. southward to San Bernardino, California, *Nuttall, Cleveland, Parish*, &c.; also in the northern part of the State at Ukiah, *Kellogg*.

+ + Pubescence hispid; the stem often dark-spotted at the base of the papillæ of the stronger bristles: hairs on the pappus less copious, all straight and erect: stems and branches mostly upright.

L. hieracioídes, Hook. & Arn. l. c. Leaves from linear to oblong, mostly laciniate-dentate: rays 10 to 15, small and short, little surpassing the disk: pappus dull white or rusty.— Gray, l. c. *Madaroglossa hieracioides*, DC. l. c.— California, from San Mendocino Co. to Santa Barbara, &c.; first coll. by *Douglas*.

L. gaillardioídes, Hook. & Arn. Leaves more commonly laciniate-pinnatifid: heads usually larger: rays 10 to 20, orange-yellow, half to three-fourths inch long: pappus dull white or rusty.— Bot. Beech. 357 (*Tridax? gaillardioides* or *Layia*, Bot. Beech. 148); Torr. & Gray, Fl. ii. 393; Gray, Bot. Calif. i. 369.— Common in W. California, from Mendocino Co. to Tejon; first coll. by *Lay*.

§ 2. Callichróa, Gray, l. c. Pappus of 5 to 25 naked aristiform bristles, or rarely wanting: otherwise as in the preceding.— *Callichroa*, Fisch. & Meyer, Ind. Sem. Hort. Petrop. l. c.

L. pentachǽta, Gray. Somewhat hirsute and viscid-pubescent, hardly hispid, erect, a foot or two high, paniculately branched; stipitate glands minute and sparse: cauline leaves mostly pinnatifid and the lower laciniately bipinnatifid; the lobes narrowly linear: rays ample, half-inch or more long, golden- or orange-yellow: disk-akenes minutely pubescent or glabrate: pappus of 5 or rarely fewer rigid and smooth bristles, sometimes even wholly wanting in certain specimens apparently of very same parentage.— Pacif. R. Rep. iv. 108, t. 16; Bot. Calif. i. 369.— California, along the foothills of the Sierra Nevada, from Placer to Fresno Co., *Bigelow, Bolander, Parry*.

L. platyglóssa, Gray. Usually more hirsute and lower: stipitate glands small and sparse: cauline leaves linear, simply pinnatifid into short linear lobes, most of the upper entire: rays half-inch long, light yellow, commonly with white tips to the lobes: disk-akenes silky-hirsute: pappus of 15 to 20 upwardly scabrous stout awn-like bristles, only a little shorter than the corolla.— Pl. Fendl. l. c.; Bot. Calif. l. c. *Callichroa platyglossa*, Fisch. & Meyer, Ind. Sem. Hort. Petrop. l. c., & Sert. Petrop. t. 5; Don, Brit. Fl. Gard. ser. 2, t. 373; Hook. & Arn. l. c.; Hook. Bot. Mag. t. 3719; Torr. & Gray, Fl. ii. 395. *Madaroglossa (Callichroa) hirsuta* & *angustifolia*, Nutt. Trans. Am. Phil. Soc. l. c.— Common in low grounds throughout W. California, where it is called *Tidy-tips*.

Var. **breviséta**, Gray, Bot. Calif. i. 370. Pappus only half the length of corolla and of the young akene: cauline leaves mostly pinnatifid.— S. California, in the vicinity of Los Angeles, *Bigelow*.

§ 3. CALLIGLÓSSA, Gray, l. c. Pappus wanting or of few or several flattened awns or paleæ (instead of bristles), either naked or with long hairs only at base. — *Calliglossa*, Hook. & Arn., with *Oxyura*, DC. & Lindl.

* Rays pure white: only marginal receptacular bracts present: pappus aristiform: habit of *L. glandulosa*: a few small stipitate glands on the upper leaves and involucre.

L. Douglásii, HOOK. & ARN. Low, sparingly hirsute or hispid: radical leaves pinnatifid-dentate; upper linear and entire: rays rather short, broad, 3-cleft: lobes of disk-corolla hirsute outside: akenes narrow, those of the disk villous-pubescent: pappus of about 10 minutely scabrous linear-subulate flat paleæ, nearly equalling disk-corolla; their margins toward the base scantily beset with long and straight villous hairs. — Bot. Beech. 358; Gray, Proc. Am. Acad. ix. 194. — Gravelly banks, between the Dalles and Great Falls of the Columbia River, *Douglas* (the pappus in the specimen fulvous by discoloration); Austin, Nevada, 1882, *M. E. Jones* (pappus bright white). Probably only a form of *L. glandulosa* with more paleaceous and almost naked pappus.

* * Rays yellow at base, white or pale at summit: bracts of the involucre lanate-ciliate at the basal margins where infolded around the akene: both ray- and disk-akenes mostly oblong-obovate.

+ Pappus of 7 to 12 broadish naked paleæ: disk-akenes more or less villous-hirsute. — *Calliachyris*, Torr. & Gray.

L. Jonésii, GRAY. Somewhat hispidulous and viscid, a few small and sessile dark glands on and near the involucre: leaves hispidulous-ciliate, narrowly linear, simply pinnatifid, and upper ones 3-lobed or entire: heads rather small: rays only quarter-inch long: receptacular bracts only marginal: paleæ of the pappus ovate or oblong-ovate, acuminate, often erose-denticulate, not longer than the tube of the corolla. — Proc. Am. Acad. xix. 18. — San Luis Obispo, California, *M. E. Jones*.

L. Fremónti, GRAY. A foot high, minutely pubescent, not glandular: leaves not ciliate, nearly all pinnately parted into oblong-linear or spatulate short lobes: rays ample, half to three-fourths inch long: receptacular bracts to many of the flowers: paleæ of the pappus from ovate to oblong-lanceolate, tapering into a subulate awn, nearly equalling the corolla, the margins entire, accompanied by a few free long-villous hairs, which much exceed those of the surface of the akene. — Pl. Fendl. 103, Bot. Calif. i. 370. *Calliachyris Freemonti*, Torr. & Gray, Jour. Bot. Nat. Hist. Soc. v. 140. — California, upper valley of the Sacramento to Tuolumne Co.; first coll. by *Fremont*.

+ + Pappus subulate-aristiform and unequal and naked, or none: chaffy bracts to most of the disk-flowers: herbs loosely erect or diffuse (about a foot high), not glanduliferous, with herbage glabrous or minutely pubescent, but the margin of the leaves and yet more the base of the bracts strongly hispidulous-ciliate: lower leaves pinnately parted or lobed; upper entire: heads showy, with ample usually particolored rays.

L. Calliglóssa, GRAY, l. c. Akenes villous-pubescent or partly glabrate: pappus of usually several (10 to 18) very unequal and rigid subulate awns, which are somewhat scabrous or slightly hirsute near the dilated base, the marginal ones rather shorter than the corolla, the smaller hardly half as long. — *Oxyura chrysanthemoides*, Lindl. Bot. Reg. t 1850; Fisch. & Meyer, Hort. Petrop. t. 6. *Calliglossa Douglasii*, Hook. & Arn. Bot. Beech. 356. *Callichroa* (*Calliglossa*) *Douglasii*, Torr. & Gray, Fl. ii. 396. — California, common around San Francisco Bay; probably first coll. by *Douglas*.

Var. **oligochǽta**, GRAY, Bot. Calif. l. c. Pappus reduced to the two marginal awns (and these sometimes slender) and to some intermediate rudiments or small awns: leaves less lobed. — Petaluma, Santa Rosa, and elsewhere, north of Bay of San Francisco, *Newberry, Bolander*, &c.

L. chrysanthemoides, GRAY, l. c. Akenes wholly glabrous, broader, with no epigynous disk (the base of corolla covering the top of the ovary): no pappus: receptacle becoming convex: otherwise quite like the preceding species. — *Oxyura chrysanthemoides*, DC. (in Lindl. Syst. Nat. &) Prodr. v. 693, not of Bot. Reg. *Tollatia chrysanthemoides*, Endl. Gen. Suppl., & Walp. Repert. ii. 631. *Hartmannia ciliata*, DC. Prodr. v. 694. — California, not rare near San Francisco; first coll. by *Douglas*.

Tribe VI. HELENIOIDEÆ, p. 70.

128. CLÁPPIA, Gray. (*Dr. A. Clapp*, author of a Synopsis of the Medicinal Plants of the U. S.) — Bot. Mex. Bound. 93; Benth. & Hook. Gen. ii. 413, & Ic. Pl. xi., partly. (The excluded *C. aurantiaca*, Benth. Ic. Pl. t. 1104, is a *Dysodia*, apparently wanting the oil-glands.) — Single species.

C. suædæfólia, Gray, l. c. Suffruticose, a foot high, widely branching, not punctate nor glandular: leaves alternate, fleshy, terete, linear, entire, or the lower pinnately 3–5-parted, sessile; head (half-inch in diameter) pedunculate, terminating herbaceous branchlets: flowers doubtless yellow. — Benth. Ic. Pl. t. 1105. — S. Texas; on the Rio Grande at Laredo, *Berlandier*. Alkaline flats of the Pecos, *Havard*.

129. JAÚMEA, Pers. (*I. H. Jaume St. Hilaire*, a French botanist.) — Herbs or suffruticose plants (mainly S. American); with opposite entire leaves, and terminal pedunculate heads of yellow flowers. — Syn. Pl. ii. 397; Benth. & Hook. Gen. ii. 397 (including *Coinogyne*, Less., *Espejoa*, DC., *Chæthymenia*, Hook. & Arn., &c.); Gray. Bot. Calif. ii. 371. *Kleinia*, Juss., not L.

— **J. carnósa**, Gray. Procumbent or ascending perennial herb, fleshy, glabrous, leafy to the short-pedunculate head; leaves spatulate-linear, almost terete, about inch long: head half-inch long, fleshy; rays 6 to 10, linear, not surpassing the disk: receptacle conical: akenes glabrous, destitute of pappus. — Wilkes Exped. xvii. 360, & Bot. Calif. i. 372. *Coinogyne carnosa*, Less. in Linn. vi. 520; Torr. & Gray, Fl. ii. 410. — Salt marshes and sea-beaches, Brit. Columbia to California; probably first coll. by *Chamisso*.

130. VENEGÁSIA, DC. (*Michael Venegas*, a Jesuit missionary, early writer upon California.) — Prodr. v. 43; Benth. & Hook. Gen. ii. 397. — Single species, yellow-flowered.

— **V. carpesioides**, DC. l. c. Large perennial herb, with glabrous leafy branches: leaves alternate, slender-petioled, membranaceous, ovate and subcordate, mostly denticulate, veiny, somewhat puberulent or atomiferous: heads terminal and from upper axils, short-peduncled, inch broad, and the about 15 rays an inch long. — Gray, Bot. Calif. i. 372. *Partheniopsis maritima*, Kellogg, Proc. Calif. Acad. v. 100. — Rocky banks of streams, coast of California, from Santa Barbara southward; first coll. by *Douglas* and *Coulter*: fl. summer.

131. RIDDÉLLIA, Nutt. (*Prof. John L. Riddell*, author of a Synopsis of the Flora of Western States.) — Low and corymbosely branched woolly herbs (Texano-Arizonian); with alternate and spatulate or linear leaves, the cauline entire, and small heads of yellow flowers; the ligules large in proportion, becoming pale or whitish in age and thin-papery; fl. summer. In habit not unlike *Zinnia* § *Diplothrix* of the same regions. Bracts of the involucre distinct, but connivent-erect, and connected by the intricate wool so as to seem connate. — Trans. Am. Phil. Soc. n. ser. vii. 371; Gray, Pl. Fendl. 94, & Bot. Calif. i. 372. *Psilostrophe*, DC. Prodr. vii. 261.

   ＊ Rays at maturity half-inch long: akenes and pappus glabrous, or the former with few and short scattered hairs: perennial.

— **R. tagetína**, Nutt. l. c. Loosely or somewhat villosely lanate, sometimes glabrate in age, rather widely branched: radical and even lower cauline leaves often laciniate-pinnatifid: heads numerous, mostly cymosely clustered and short-peduncled: paleæ of the pappus oblong-lanceolate, entire, usually obtuse, half or three-fourths the length of the disk-corolla. — Torr. in Emory Rep. t. 5; Gray, Pl. Fendl. 94. — W. Texas to E. Colorado and Arizona; first coll. by *James*.

Var. sparsiflóra. Heads more scattered and slender-peduncled: paleæ of the pappus linear-lanceolate, mostly acute. — S. Utah, *Bishop, Mrs. Thompson.*

R. Coóperi, GRAY. Canescent with close and matted tomentum, no villous hairs, or the wholly entire narrow leaves glabrate: stems much branched from a ligneous base: heads scattered, slender-peduncled: paleæ of the pappus from broadly oblong to lanceolate, erose-laciniate at summit or nearly entire, less than half the length of the disk-corolla. — Proc. Am. Acad. vii. 358, & Bot. Calif. ii. 373. — Gravelly plains and banks, S. E. California to S. Utah and Arizona; first coll. by *Dr. Cooper.*

* * Rays at maturity only a quarter of an inch in length: akenes and pappus long-villous: biennial or annual?

R. arachnoídea, GRAY. Loosely lanate: stem and branches rather strict: foliage of *R. tagetina*: heads clustered, short-peduncled: arachnoid hairs even longer than the somewhat turbinate akenes: paleæ of the pappus subulate-lanceolate, their margins and apex more or less deliquescent into long and arachnoid hairs. — Pl. Fendl. 94. *Psilostrophe gnaphalioides*, DC. l. c. — Western Texas along the Rio Grande, *Wright*, &c. (Adj. Mex., *Berlandier, Gregg,* &c.)

132. **BAILÉYA,** Harvey & Gray. (*Jacob Whitman Bailey*, the pioneer in microscopical research in U. S.) — Soft and densely floccose-woolly annuals or biennials, of the Texano-Arizonian district; with alternate leaves, the lower once or twice pinnatifid, and terminal long-pedunculate solitary heads of yellow flowers, the large and persistent rays deflexed in age: fl. summer. — Pl. Fendl. 105; Proc. Am. Acad. ix. 195.

B. pauciradiáta, HARV. & GRAY, l. c. Villosely and floccosely lanate, a foot or so high, loosely paniculately branched, leafy: leaves sparingly laciniate-pinnatifid or the upper entire, linear: heads small, short-peduncled: involucre quarter-inch high and broad: ligules 5 or 6, roundish-oval, 3 or 4 lines long: disk-flowers 10 to 25: akenes subclavate, with slightly narrowed summit, strongly many-nerved, muriculate-scabrous, obscurely resinous-atomiferous. — Gray, Bot. Calif. i. 373. — Sandy deserts, S. E. California and adjacent Arizona, *Coulter, Schott, Cooper, Parish.* Still rare.

B. multiradiáta, HARV. & GRAY, l. c. Densely floccosely white-tomentose, at length much branched from the base and leafy: radical and lower leaves spatulate or broader, mostly laciniate-pinnatifid or sparingly bipinnatifid; uppermost small, spatulate-linear, entire: heads on slender and often long peduncles: involucre mostly half-inch broad: ligules 25 to 50, cuneate-oblong or at length broader and nearly quadrate, 5 or 6 lines long: disk-flowers very numerous: akenes oblong-prismatic and obscurely striate, broadest at the truncate apex, minutely scabrous and resinous-atomiferous. — Torr. in Emory Rep. t. 6; Rothrock in Wheeler, Rep. vi. 175. *B. pleniradiata* & *B. multiradiata*, Pl. Fendl. l. c., Bot. Calif. l. c.; the former the commoner form, branching and leafy, with more numerous and smaller heads. — Plains, from W. Texas to S. Utah, Arizona, and the borders of S. E. California; first coll. by *Coulter.* (Adj. Mex.)

Var. nudicaúlis. More simple-stemmed, or branched only from a stout (biennial?) base: leaves more divided: peduncles elongated, sometimes scapiform; head larger. — *B. multiradiata,* Harv. & Gray, l. c., mainly. — Same range, or more southern. (Adj. Mex.)

133. **WHITNÉYA,** Gray. (*Josiah D. Whitney*, Director of California Geological Survey.) — Proc. Am. Acad. vi. 549, ix. 195, & Bot. Calif. i. 374. — Single species, yellow-flowered; perhaps most related to *Arnica.*

W. dealbáta, GRAY, l. c. Low perennial herb, from filiform rootstocks, with aspect of *Arnica*, canescent with minute and close tomentum: stems simple or sparingly branched, bearing 2 to 4 pairs of opposite entire leaves, and solitary few slender monocephalous peduncles: radical leaves obovate or oblong-spatulate, obtuse, 3-nerved, 2 or 3 inches long: upper small, lanceolate: head half-inch high: rays inch or more long. — Sierra Nevada, California, in Mariposa Co., at 5,000 feet or higher; first coll. by *Brewer* and *Bolander.*

134. LAPHÁMIA, Gray. (*Dr. Increase Allen Lapham,* of Wisconsin, died in 1875.) — Low suffruticulose perennials, Texano-Arizonian, growing in crevices of rocks, mostly with petioled and dentate or laciniate small leaves, the upper alternate. rarely all opposite : small heads of yellow (rarely white?) flowers. either cymosely disposed or singly terminating the branches : fl. spring and summer. — Pl. Wright. i. 99. t. 9 ; Benth. & Hook. Gen. ii. 308. excl. spec.

L. PENINSULARIS, Greene, Bull. Calif. Acad. i. 8, is an extra-limital species (with rather large and radiate heads and no pappus) from Lower California.

§ 1. PAPPÓTHRIX, Gray, l. c. Pappus of about 20 unequal rigid hispidulous bristles, hardly as long as the somewhat quadrangular-compressed akene, shorter than the corolla : rays none : disk-flowers 12 to 15 ; the corolla with short proper tube and cylindraceous throat : bracts of the involucre 5 to 8, linear-oblong, nearly plane : stems slender, a span or more high and much branched from the stout woody base : leaves mostly opposite, as broad as long, abruptly slender-petioled : short-peduncled heads rather scattered.

L. rupéstris, GRAY, l. c. Pubescent, slightly viscid, leafy to summit : leaves (half-inch long) sometimes crenately sometimes strongly and acutely dentate or almost laciniate: pappus much exceeding short proper tube of the corolla. — S. W. Texas, *Wright, Bigelow*.

L. cinérea, GRAY. Tomentose-canescent : leaves more orbicular, almost entire: pappus hardly surpassing the proper tube of the corolla, which is more than half the length of the short-cylindraceous throat : akenes sometimes 4-nerved.— Bot. Mex. Bound. 82. — Rocks along Escondido Creek, S. W. Texas. *Bigelow*.

§ 2. LAPHÁMIA proper, Gray, l. c. Pappus of a solitary very slender bristle (very rarely a pair from the same angle), or none: akenes flatter : disk-flowers 15 to 20: their corollas with longer and glandular tube. — *Monothrix*, Torr. in Stansb. Exped. 389. t. 7.

* Involucre 15-20-flowered, of nearly as many plane and linear pubescent bracts: leaves nearly orbicular in outline, palmately lobed or dissected, not punctate, the lower opposite.

L. Lemmóni, GRAY. Depressed and diffuse, much branched, hardly a span high, villously pubescent, leafy throughout : leaves a quarter or third of an inch in diameter and with petiole of equal length, obtusely 3-lobed and the lobes coarsely crenate-dentate : heads (3 or 4 lines long) short-peduncled . rays none : akenes canescently puberulent : pappus a very delicate bristle, or occasionally a pair from the same angle, little surpassing the proper tube of the corolla, or often none.— Proc. Am. Acad. xvi. 191. — Southern Arizona, near Camp Lowell, *Lemmon*.

Var. **pedáta,** GRAY, l. c. Leaves pedately parted and cut into narrow lobes. — With the other form, also on the Chiricahua Mountains, *Lemmon*.

* * Involucre 15-25-flowered, rather narrow, glabrous, of thinnish nearly plane bracts, 2 or 3 lines long: herbage merely puberulent: leaves mostly angulate-toothed or incised, the lower opposite: heads commonly corymbosely cymose and pedunculate.

L. halimifólia, GRAY. Stems a span or more high and crowded on a thick woody caudex : leaves coriaceous, resinous-punctate or atomiferous, somewhat viscid, broadly ovate or rhombic, seldom inch long, laciniately dentate, abruptly long-petioled : rays 4 to 6, with broad and short ligules little longer than the tube: pappus none. — Pl. Wright. l. c. 99, t. 9.— S. W. Texas, *Wright, Bigelow*.

L. angustifólia, GRAY. Leaves lanceolate or rhombic-lanceolate, tapering into margined petioles, laciniately 1-5-toothed or lobed : heads less numerous, scattered : rays none : otherwise much like the preceding species.—Pl. Wright. l. c. & ii. 81. — S. W. Texas, on high and rocky hills of the Pecos, *Wright, Havard*, the latter's specimens connecting with var. *laciniata,* Gray. Bot. Mex. Bound. 82, which proves to be only a form with long and weak stems, hanging from rocks on the Rio Grande, *Bigelow, Schott*.

L. Lindheímeri, Gray. Stems a foot or less high from a thick woody base: leaves thinner, oblong or ovate, glabrous, few-toothed or some entire, contracted at base into a short petiole: heads loosely cymose: rays 3 to 6, very short, sometimes none: pappus a single slender bristle equalling the proper tube of the corolla. — Pl. Wright. i. 101. — Rocky banks of the Guadalupe, near New Braunfels, Texas, *Lindheimer*.

* * * Involucre 35-50-flowered, of numerous carinate-concave bracts, somewhat puberulent or glandular on the back: herbage minutely puberulent: leaves thickish.

+— Flowers said to be white: leaves mostly opposite, numerous up to the heads, dentate.

L. Pálmeri, Gray. Scabrous-puberulent: leaves broadly ovate or deltoid-rotund, rigid, coarsely 5-7-dentate or laciniate-lobed, half-inch long, veiny, abruptly short-petioled: heads somewhat crowded on the fastigiate flowering branches, little surpassing the upper leaves: involucre campanulate, about 35-flowered; its bracts linear, somewhat pubescent: rays none: pappus a bristle of the length of the akene and a little shorter than the corolla. — Proc. Am. Acad. xiii. 372. — Cañons at Beaver-dam, N. W. Arizona, pendulous from rock-crevices, *Palmer*, who notes that the flowers are "creamy white."

+— +— Flowers yellow: leaves small, 2 to 4 lines long, mostly orbicular, more entire, the upper alternate, scattered: heads solitary and naked, terminating the loose branchlets, nearly hemispherical.

L. megacéphala, Watson. Base of stem and lower leaves unknown; those of flowering branches all very small, alternate, short-petioled: involucre about 50-flowered; its bracts lanceolate-linear, minutely glandular: rays none: pappus none. — Am. Nat. vii. 301. — S. Nevada, *Wheeler*.

L. Stansbúrii, Gray. Stems slender and lax from a woody base: lower leaves opposite and on petioles of their own length; upper alternate, also slender-petioled: involucre 35-40-flowered, its bracts fewer and broader, lanceolate-oblong, nearly glabrous: rays 6 to 10, conspicuous, oblong: pappus a bristle somewhat shorter than the disk-corolla. — Pl. Wright. i. 101; Eaton in Bot. King Exp. 164. *Monothrix Stansburii*, Torr. in Stansb. Rep. 389, t. 7. — Rocks on Stansbury Island, &c., Salt Lake, Utah; first coll. by *Stansbury*.

§ 3. Díthrix. Pappus a pair of stouter naked bristles, one from each angle of the akene: head only 6–8-flowered.

L. bisetósa, Torr. Hispidulous-puberulent, minutely resinous-atomiferous and punctate: stems 1 to 3 inches high from the woody base: leaves mostly alternate, coriaceous, spatulate-ovate, obscurely few-toothed (quarter-inch long including the petiole): heads solitary and sessile: rays none: involucre (3 lines long) with bracts broadly linear, slightly pubescent, carinate-concave at base: flowers proportionally large: corolla (whitish or pale yellow?) with glandular tube one-third the length of the campanulate cylindraceous throat: akenes hispidulous-puberulent, the narrow marginal nerves naked: rigid awns rather shorter than the akene, more than half the length of the corolla. — Gray, Pl. Wright. ii. 106. — On the Rio Grande, Texas, in a cañon below Presidio del Norte, *Parry*.

135. **PERÍTYLE**, Benth. (Περί, around; τύλη, a callus; the akenes callous-margined.) — Californian and Mexican herbs, the genuine species mostly annuals; with petiolate dentate or palmately-lobed leaves, lower opposite, upper alternate, and small or middle-sized pedunculate heads terminating the branches: disk-flowers yellow (or sometimes white?): rays when present yellow or white. — Bot. Sulph. 23 & 119, t. 15; Gray, Pl. Fendl. 77, Proc. Am. Acad. ix. 194, & Bot. Calif. i. 396.

P. incána, Gray, Proc. Am. Acad. xi. 78, from Guadalupe Island, off Lower California, is an outlying anomalous species: all the others are as follows.

§ 1. Crown of the pappus an entire or undulate firm and shallow border: akene hardly ciliate: suffruticulose: transition to *Laphamia*.

P. disséeta, Gray. Dwarf, 3 or 4 inches high from the woody base, cinereous-pubescent, very leafy: leaves with blade (quarter-inch long) equalled by the petiole, round-cordate in outline, pedately cleft or parted and dissected into short linear lobes: heads subsessile, 3 or 4 lines high: involucre campanulate, of numerous narrow linear bracts: rays none: disk-flowers about 20 (perhaps white): akenes linear-oblong, minutely cinereous-hirsute, and the cartilaginous margins somewhat more hirsute; a short scabrous awn from one angle, of nearly half its length, or this wanting: style-branches slender-subulate, not short and obtuse, as said in Proc. Am. Acad. ix. 195. — *Laphamia dissecta*, Torr. in Pl. Wright. ii. 81. — Rocks at Presidio del Norte on the Rio Grande, between Texas and Mexico.

§ 2. Genuine species: pappus a crown of hyaline lacerate squamellæ, either somewhat united at base or distinct, rarely obsolete.

  * Suffruticulose perennial, with commonly dissected leaves: rays and perhaps disk-flowers also white.

P. coronopifólia, GRAY. Cinereous-puberulent, many-stemmed from the woody base, a foot or less high, slender, leafy: leaves small, somewhat pedately or pinnately once or twice divided or parted into linear or narrow spatulate lobes, or some coarser and merely trifid: heads disposed to be paniculate, 3 lines high: rays as long, broadly oblong, coarsely 3-toothed at apex: style-tips slender-subulate: akenes narrowly oblong, glabrate on the faces, densely hirsute-ciliate: awns 2, little shorter than the corolla. — Pl. Wright. ii. 82, & Bot. Mex. Bound. 82. — Rocks on mountain-sides, New Mexico and Arizona; first coll. by *Wright*. Varies with roundish merely incisely-cleft leaves.

  * * Herbaceous, chiefly and perhaps all with annual root, loosely branching, and bearing scattered pedunculate heads: leaves often palmately cleft.

  + Akenes thin-margined, hispidulous or hirsutely ciliate: crown of pappus minute or obsolete and awns wanting: style-appendages short, acute. (Perhaps extra-limital.)

P. Fitchii, TORR. Viscid-pubescent: leaves and involucre nearly of the following species: akenes unknown: ovaries apparently destitute of pappus. — Pacif. R. Rep. iv. 100. — "California, *Rev. A. Fitch*," in herb. Torr. Probably from the islands: imperfect, seemingly winter specimens. (To this apparently is to be joined var. PALMERI, *P. Emoryi* of coll. Palmer, no. 44, which has the whole aspect and foliage of *P. Californica*, var. *nuda*, but akenes narrowly oblong, somewhat falcately oblique, with a short pappus of numerous squamellæ united into an erose-denticulate crown. — Guadalupe Island off Lower California.)

  + + Akenes callous-margined and densely ciliate with long beard: pappus-crown more conspicuous: awns rarely wanting.

  ++ Style-branches with short and obtuse or acute minutely hirsute appendages: rays 6 to 12, short, the oblong or broader ligule little longer than the tube, perhaps always white.

P. Califórnica, BENTH. Somewhat hirsutely pubescent, also viscid and glandular: leaves broadly ovate or roundish-cordate, incisely lobed or more deeply 3–5-cleft and the lobes coarsely dentate: heads fully 3 or 4 lines high and broad: bracts of the involucre narrowly oblong: akenes oblong, densely hispid-villous on the margins, crowned with conspicuous squamellæ, and with a single more or less barbellate awn of about the length of the akene. — Bot. Sulph. 23, t. 15. *P. Emoryi*, Torr. in Emory Rep. (1848), 142; Gray, Bot Calif. i. 397, form with usually more rounded lobed and incised leaves. — Desert-region of the Mohave and Gila, S. E. California and W. Arizona. (Lower California, Guadalupe Island, &c. Now found by many collectors.)

Var. **núda**, GRAY, Bot. Calif. l. c., under *P. Emoryi*. Awn of the pappus none: otherwise as in the *P. Emoryi* form. — *P. nuda*, Torr. Pacif. R. Rep. iv. 100. — With the aristate form and commoner. (Lower Calif.)

P. plumígera, GRAY. Flowering branches only seen, small-leaved, viscid-glandular: heads much smaller than in the preceding (narrowish, barely 3 lines high): akenes oval-oblong, the margins very densely long-villous: awn solitary, longer than the akene, sparsely barbellate-hispid. — Pl. Fendl. l. c. — "California," probably Arizona, *Coulter*. Possibly a late-flowering form of the preceding.

P. microglóssa, BENTH. Merely puberulent, obscurely glandular above: leaves broadly ovate with subcordate or truncate base, or upper somewhat hastate, incisely dentate, often 3–5-lobed: heads 3 lines high: akenes obovate or obovate-oblong, with broad summit, villous-ciliate margins, and a pair of delicate awns, which barely equal the breadth of the akene and are twice or thrice the length of the crown of squamellæ. — Bot. Sulph. 119; Hemsl. Biol.

Centr.-Am. Bot. ii. 210. *P. Californica*, Gray, Proc. Am. Acad. v. 159, not Benth. *P. Acmella*, Gray, Pl. Fendl. 77, & Bot. Calif. l. c., with *P. Californica*, mainly. *Spilanthes Pseudo-Acmella*, Hook. & Arn. Bot. Beech. 150. *Baltimia* § *Dichetophora* sp., Benth. & Hook. Gen. ii. 269. — California, from Monterey? southward, *Lay & Collie, Coulter, Parish.* (Mex.)

Var. effúsa. Very much branched from the annual root, paniculately floriferous: leaves and heads smaller (the former half-inch or so, the latter only 2 lines high): akenes correspondingly small, narrowly obovate-oblong. — Santa Catalina Mountains, S. Arizona, *Pringle*.

++ ++ Style-branches tipped with setaceous-filiform acute hispidulous appendages: rays with narrow ligules, or wanting in one species: disk-corollas slender, with long and narrow throat: akenes oblong: pappus of a rather conspicuous crown of squamellæ and one long and delicate awn: heads 5 lines high: bracts of the involucre linear: perhaps perennials or with lignescent base, not improbably all of one species.

P. leptoglóssa, GRAY. Minutely puberulent or glabrate, not at all glandular: leaves roundish-subcordate, coarsely and doubly crenate-dentate (half to three-fourths inch long): rays oblong-linear, 4 lines long: akenes (a line long) linear-oblong, with comparatively short hispid ciliation, the setiform awn shorter than the disk-corolla. — Pl. Fendl. 77; Bot. Calif. l. c. — " California," *Coulter*, more probably from Arizona.

P. Párryi, GRAY. Minutely pubescent and obscurely viscid: leaves reniform-cordate, crenately dentate and often lobed (the larger inch broad): rays oblong, barely 2 lines long: akenes (a line and a half long) oblong, strongly hirsute-ciliate: awn of the pappus nearly equalling the disk-corolla. — Pl. Wright. ii. 106. — S. border of Texas, or on the Mexican side, in a cañon of the Rio Grande below Presidio, *Parry*. Also mountains on the Texan side, *Havard*.

P. aglóssa, GRAY, l. c. Somewhat puberulent, obscurely viscid: leaves roundish, with subcordate or truncate base, mostly 3–5-cleft and coarsely dentate (the larger 2 inches broad): bracts of the involucre very narrowly linear: rays none: akenes narrowly oblong, with rather short and dense hirsute ciliation: awn of the pappus equalling the disk-corolla. — Cañon of the Rio Grande, with or near the preceding, *Parry*.

136. **PERÍCOME,** Gray. (Περί, around, and κόμη, a tuft of hairs; a coma of long hairs all round the margin of the akenes.) — Pl. Wright. ii. 82; Benth. & Hook. Gen. ii. 406. — The latter authors indicate a Mexican radiate species, of anomalous character, which they associate with the typical

P. caudáta, GRAY, l. c. Rather tall widely branching perennial herb, strong-scented, very minutely puberulent: leaves opposite, long-petioled, green and membranaceous, minutely somewhat resinous-atomiferous, triangular-hastate (2 to 5 inches long), with sparingly crenate-dentate or entire margins, caudately long-acuminate, as also in less degree are the basal angles: heads numerous in terminal corymbiform cymes, half-inch or less high; flowers golden yellow, conspicuously longer than the glabrous involucre: akenes linear-oblong; the flat faces glabrous, the nerviform margins densely villons-bearded: pappus a crown of hyaline squamellæ which are more or less connate and fimbriate-lacerate at summit, the fringe dissected into bristles or hairs somewhat simulating those of the margin of the akene; also sometimes a slender awn from one or both margins of the akene. — Rocky cañons, &c., S. Colorado, New Mexico, and Arizona; first coll. by *Wright, Bigelow*, &c. Fl. late summer and autumn.

137. **EATONÉLLA,** Gray. (*Prof. Daniel Cady Eaton*, author of Ferns of N. America, the *Compositæ* of King's Expedition, &c., grandson of *Amos Eaton* for whom was named the genus *Eatonia*.) — Very floccose-lanate annuals, of California and adjacent Nevada; with mostly alternate leaves and small sessile heads of yellow or white flowers: fl. spring or early summer. — Bot. Calif. i. 379, as subgenus under *Actinolepis*: Proc. Am. Acad. xix. 19.

E. nívea, GRAY, l. c. Depressed in a small tuft from a slender root, an inch or so high, subcaulescent, densely leafy, white with long and loose wool: leaves obovate-spatulate, entire, equalling or surpassing the sessile heads; involucre of about 8 narrowly oblong bracts, subtending as many ray-flowers: ligules hardly exceeding the disk: disk-corollas 5-toothed: akenes all compressed and with only marginal callous nerves, linear-oblong, the dark faces polished and shining, the comose long and soft villous hairs of the margin bright white: pappus a pair of comparatively large opaque paleæ, of broadly ovate or quadrate form (the insertion of the two occupying the whole circumference of the akene), sparingly laciniate-dentate or erose at summit, and the middle produced into a subulate naked awn which nearly equals the 4-toothed corolla. — *Burrielia nivea*, D. C. Eaton, Bot. King Exp. 174, t. 18. *Actinolepis* (*Eatonella*) *nivea*, Gray, Bot. Calif. i. 379. — Sterile hills of the eastern side of the Sierra Nevada; in the Pah-Ute Mts., Nevada, *Watson*, and Surprise Valley, E. California, *Lemmon*.

E. Congdóni, GRAY. A span or two high, loosely branching, sparsely leaved, floccosely lanate: leaves oblong linear, sparsely sinuate-dentate or repand: heads short-peduncled or nearly sessile at the summit of the stem: involucre of 5 or 6 oval-oblong herbaceous bracts: ray-flowers none: disk-corollas 4-toothed: akenes oval (the faces at first pubescent, at length glabrate), the outermost triangular-obcompressed, the others compressed and flat: pappus of 2 to 4 very thin and hyaline erose-laciniate awnless paleæ, not exceeding the long villosity, forming a crown. — Proc. Am. Acad. xix. 30. — California, at Deer Creek, Tulare Co., *Congdon*, and on the San Joaquin, *Parry*.

138. **MONOLÓPIA**, DC. (Μονόλοπος, single husk, alluding to the uniserial involucre.) — Annual herbs, Californian, clothed with floccose wool; with alternate (or only lower sometimes opposite) sessile leaves, and comparatively large pedunculate heads of golden yellow flowers terminating the stem and few branches. — Prodr. vi. 74; Hook. Ic. Pl. t. 343. 344. *Spiridanthes*, Fenzl in Endl. Gen. Suppl. ii. 105.

§ 1. MONOLÓPIA proper. Ray-corollas with ample coarsely 3-4-toothed or lobed ligule, and bearing at base on the opposite side of the style a roundish denticulate appendage: leaves undivided, strictly sessile or partly clasping by a broadish base. — Bot. Calif. i. 383.

M. májor, DC. l. c. A foot or two high, rather stout and simple; the floccose white wool tardily deciduous: leaves from linear to lanceolate-oblong, repand-serrate to entire: bracts of the broad (half-inch high) involucre united to above the middle, the lobes triangular-ovate: ligules 6 to 10 lines long: akenes glabrous or nearly so at maturity. — Hook. Ic. Pl. t. 344. & Bot. Mag. t. 3839; Gray, Bot. Calif. l. c. *Hologymne Douglasii*, Fisch. & Meyer, Ind. Sem. Hort. Petrop. viii. 64. — Common in low ground, through W. California.

Var. **lanceoláta**, GRAY, Bot. Calif. l. c. A mere form, with bracts of involucre distinct to near the base. — *M. lanceolata*, Nutt. Pl. Gamb. 175. Near Los Angeles, &c.

M. grácilens, GRAY. A foot or more high, slender, loosely paniculately branched, bearing scattered small heads; involucre only quarter-inch high; its oval or ovate bracts distinct to the base: akenes only a line long. — Proc. Am. Acad. xix. 20. — California in the coast ranges, near New Almaden and Santa Cruz, *Bolander, Torrey, Isaman, Pringle*.

§ 2. PSEUDO-BÁHIA. Ray-corollas destitute of internal appendage, barely 3-toothed at apex: leaves all alternate, commonly laciniately cleft, narrowed at base into more or less of a petiole. — Bot. Calif. l. c.

M. minor, DC. l. c. Loosely lanate, a span or more high: cauline leaves 3-5-cleft into linear lobes: heads 3 lines high: bracts of the involucre about 10, somewhat in 2 series, oblong, separate to below the middle: ovary glabrous. — Hook. Ic. Pl. t. 343. — California, *Douglas*. Not since detected.

M. Heermánni, DURAND. Whitened with a close and fine flocculent tomentum, which is deciduous, the foliage glabrate and green in age, a span or two high, branching: leaves pinnatifid or pinnately parted into linear lobes or divisions, or some of the cauline bipinnatifid: heads 3 or 4 lines high: involucral bracts distinct nearly to base: akenes sericeous-puberulent or glabrate. — Pl. Pratten. in Jour. Acad. Philad. ser. 2, iii. 93. *M. bahiæfolia*, var. *pinnatifida*, Gray, Bot. Calif. i. 383. — Foothills of the Sierra Nevada, California, from Calaveras to Tulare Co., *Heermann, Pratten, Congdon*. Also near Auburn, *Bolander*.

M. bahiæfólia, BENTH. Smaller than the foregoing, and with similar flocculent tomentum; the simple monocephalous stems only 2 inches high: leaves small (at most half-inch long), spatulate to linear, entire, or lower ones 3-lobed: head hardly 3 lines high: involucral bracts distinct to the middle: immature akenes sparsely pubescent. — Pl. Hartw. 317; Gray, Bot. Calif. i. 383, excl. var. — Valley of the Sacramento, *Hartweg*. Probably depauperate specimens.

139. **LASTHÉNIA**, Cass. (Λασθενία, a courtesan, who was a pupil of Plato: name given, by some freak of the founder, to a genus of three Western American plants.) The Chilian *L. obtusifolia* has comparatively few-flowered nearly or quite homogamous heads, and a less developed receptacle. — Low and slender annuals, mostly quite glabrous and slightly succulent; with opposite and linear or narrowly lanceolate mostly entire leaves, their sessile bases connate round the stem; the yellow-flowered heads pedunculate, terminating the stem and branches. — "Opusc. Phyt. iii. 88"; DC. in Lindl. Bot. Reg. t. 1780, 1823, & Prodr. v. 664; Torr. & Gray, Fl. ii. 377.

§ 1. LASTHENIA proper. Pappus paleaceous: heads discoid; the ligules not surpassing the involucre or the short glabrous disk-corollas, therefore wholly inconspicuous. — *Rancagua*, Pœpp. & Endl. Nov. Gen. & Spec. i. 15, t. 24, 25.

L. glabérrima, DC. l. c. Somewhat fleshy: stems ascending, a span to a foot long: heads on long peduncles which are enlarged at summit, nodding after anthesis: leaves elongated-linear: involucre about 15-toothed: corollas all shorter than the minutely puberulent oblong-linear akenes: pappus of 5 to 10 rigid paleæ, two or three of them with subulate or short-awned points, the others erose or laciniate. — Torr. & Gray, l. c.; Gray, Bot. Calif. i. 384. — Wet meadows near brackish water, along the coast of California and Oregon.

§ 2. HOLOGÝMNE. Pappus wanting: rays large, conspicuously exserted: disk-corollas fully as long as the akene; their lobes sparsely papillose-barbellate outside, as in *Monolopia*. — Gray, Bot. Calif. l. c. *Hologymne*, Bartl. Ind. Sem. Gœtt. 1837, 1839. *Xantho*, Remy, in Ann. Sci. Nat. ser. 3, xii. 191. *Lasthenia*, Lindl. Bot. Reg. l. c.

L. glabráta, LINDL. Somewhat fleshy, sometimes slightly pubescent: stems erect: leaves shorter: peduncles somewhat enlarged under the erect head: involucre more hemispherical: ligules 3 to 6 lines long: akenes narrowly obovate-oblong with acutish edges, smooth and glabrous — DC. Prodr. v. 665; Hook. Bot. Mag. t. 3730; Torr. & Gray, l. c. *L. glabrata* & *L. californica* (a smaller form, mistaken for preceding species which DC. had so named), Lindl. Bot. Reg. t. 1780, 1823. *Hologymne glabrata*, Bartl. l. c. *Monolopia glabrata*, Fisch. & Meyer, Sert. Petrop. 1835. — Moist grounds throughout W. California.

Var. **Coúlteri**. A smaller form: akenes smaller and narrower, with obtuse edges, sprinkled with minute rough points or glands. — Saline marshes, S. California, *Coulter* (no. 338), *Brewer, Cleveland, Pringle*.

140. **BURRIÉLIA**, DC., partly. (*Andrés Marcos Burriel*, a Spanish Jesuit and historian, who, in 1758, wrote a History of California, and edited the account by Venegas of the establishment of its missions.) — Prodr. v. 664; Benth. & Hook. Gen. ii. 398: now reduced to one of the three original species. Perhaps too near the following somewhat earlier-published genus.

B. microglóssa, DC. l. c. Slender annual, a span high, hirsute: leaves an inch long and barely a line wide, entire: involucre 3 lines high, equalling the yellow flowers. — Low ground, from San Francisco Bay to San Bernardino, California; fl. spring; first coll. by *Coulter* and *Douglas*.

141. BAÉRIA, Fisch. & Meyer. (In honor of the eminent Russian zoölogist, *Karl Ernst von Baer.*) — Californian annuals, or one perennial species; with opposite and entire or pinnately dissected and sessile leaves, sometimes connate at the base; and slender-pedunculate heads of yellow flowers terminating the branches: fl. spring and early summer. — Ind. Sem. Hort. Petrop. ii. 29. Jard. Petrop. t. 6, & Sert. Petrop. t. 7; Don in Sweet, Brit. Fl. Gard. ser. 2, t. 395; Benth. & Hook. Gen. ii. 399; Gray, Proc. Am. Acad. ix. 196, Bot. Calif. i. 375, & Proc. Am. Acad. xix. 21. *Burrielia* in part, DC. Prodr. v. 664; Torr. & Gray, Fl. ii. 378. excluding the typical species. *Dichæta* & *Ptilomeris*, Nutt. Trans. Am. Phil. Soc. n. ser. vii. 382, 383. *Dichæta* & *Hymenoxys*, Torr. & Gray, Fl. ii. 379, 380.

§ EUBAÉRIA. Pappus of uniform (or mainly uniform) and entire awned or pointed paleæ or chaffy-based awns, or wanting (even present or absent in the very same species): receptacle muricate-roughened: ligules mostly conspicuous: leaves linear and entire, except in one species. — *Burrielia*, Torr. & Gray, l. c., excl. *B. microglossa.*

* Akenes slenderly subclavate-linear: style-tips abruptly terminated by a conspicuous narrow-subulate appendage which usually surpasses the broad basal portion: receptacle slender-subulate and elongated in the manner of *Burrielia:* heads small: involucral bracts and oval rays 5, or sometimes only 4.

B. leptálea. Wholly glabrous: stems filiform, a span high: leaves nearly filiform, quarter to half inch long: involucre 2 lines high: ligules mostly as long: anther-tips filiform: pappus of 2 or 3 scabrous flattened awns with gradually dilated base. — *Burrielia leptalea*, Gray, Proc. Am. Acad. vi. 546, & Bot. Calif. i. 375. — Monterey Co., valley of the Nacismento, *Brewer*, and the Salinas, *Greene*. Known only as an exiguous vernal plant: probably also occurs in a larger form.

B. débilis, GREENE in herb. Minutely pubescent: stems weak, 6 to 10 inches long: leaves flaccid, linear, the largest inch and a half long: involucre 2 or 3 lines high: ligules hardly over a line long: anther-tips ovate-lanceolate: pappus of 3 or 4 firm ovate-lanceolate and awned paleæ, or in some heads none, then the akene with narrower apex. — Plains of Fresno and mountains of Kern Co., *Greene*.

* * Akenes more clavate. with scanty aristiform pappus or commonly none, then less truncate or slightly contracted at summit, either glabrous or minutely papillose-glandular in same species: style-tips capitate and mostly with a small apiculation: receptacle conical: large-flowered and with some hirsute pubescence, a foot or so high unless depauperate: rays and plane involucral bracts 7 to 12. — *Baeria*, Fisch. & Meyer, l. c.

B. macrántha, GRAY. Apparently perennial, rather stout, with peduncles 4 to 8 inches long: leaves more or less 3-nerved and obtuse, 2 or 3 lines wide (the lower 4 to 8 inches long), hispidly ciliate, at least toward the base: head about half-inch high and broad: involucre of about 12 hirsute-pubescent thickish herbaceous bracts: ligules half to three-fourths inch long. — Proc. Am. Acad. xix. 21. *B. chrysostoma*, var. *macrantha*, Gray, Proc. Am. Acad. ix. 196, & Bot. Calif. l. c. *Burrielia chrysostoma*, var. *macrantha*, Gray, Pacif. R. Rep. iv. 106. — Coast of California, north of San Francisco Bay to Humboldt Co., *Andrews*, *Bigelow*, *Bolander*.

Var. pauciaristáta, GRAY, l. c. Clearly perennial, often only 6 inches high: leaves shorter, hispid-ciliate: ligules only 4 or 5 lines long: pappus when present of 1 to 3 subulate chaffy awns rather than paleæ, little shorter than the akene. — Coast of Mendocino Co., *Bolander*, *Pringle*.

B. chrysóstoma, FISCH. & MEYER, l. c. Annual, slender: leaves narrowly linear (a line or less wide): heads 3 or 4 lines high: bracts of the broad involucre 7 to 12, in depauperate plants sometimes fewer: ligules 3 or 4 lines long: pappus (perhaps always) none. — Don, Brit. Fl. Gard. ser. 2, t. 395; DC. Prodr. v. 254; Gray, Bot. Calif. i. 375, excl. var. *Burrielia chrysostoma*, Torr. & Gray, Fl. ii. 379. — Moist ground, common almost throughout California.

* * * Akenes more cuneate and broad at the summit, usually but not always pappose, more or less 4-angular, not glandular, mostly canescent-hispidulous: receptacle conical or cylindraceous: heads middle-sized or smaller.

+ Some hirsute or strigulose pubescence, but no woolliness: style-tips capitate, without any obvious apiculation: plants slender, a span to a foot high according to situation and season: pappus-paleæ 1 to 5.

B. grácilis, GRAY. Bracts and rays 10 to 12, when depauperate 5 or 6: ligules 2 or 3 lines long: akenes almost equalled by the pappus; this in type specimens of 3 or 4 awns from small lanceolate paleæ. — Proc. Am. Acad. ix. 146, & xix. 21. *Burrielia gracilis*, DC. Prodr. v. 664. *B. hirsuta*, Nutt. Trans. Am. Phil. Soc. n. ser. vii. 381, a state destitute of pappus. — Common in W. California, especially southward, and W. Arizona: variable. Extreme variations are

Var. **aristósa**, GRAY, l. c., with awns very gradually and slightly widened downward, or in some flowers wanting. — *Burrielia gracilis*, Hook. Bot. Mag. t. 3758.

Var. **tenérrima**, with pappus-awns of the preceding, but usually fewer: depauperate form: bracts and rays only 5 or 6. — Probably *Burrielia tenerrima*, DC. Prodr. v. 664.

Var. **paleácea**, GRAY, l. c., with awns more or less abruptly dilated at base into a conspicuous oval or ovate palea, occasionally wanting, rarely one or two of the paleæ awnless. — *Burrielia longifolia* & *B. parviflora*, Nutt. l. c.

B. cúrta, GRAY. Bracts and rays 8 or 10: pappus of 4 or 5 ovate or oblong pointless paleæ (or rarely of a single one), in length not exceeding the breadth of the akene, or in some plants obsolete or wanting: leaves all filiform-linear: heads 2 or 3 lines high and wide. — Proc. Am. Acad. xix. 21. — Southeastern California, near San Bernardino, *W. G. Wright, Lemmon*.

+ + Glabrous except some fine deciduous woolliness: leaves and involucral bracts more or less fleshy-thickened, heads about 4 lines high, many-flowered: style-tips ovate or capitate, and with a conical or subulate apiculation or appendage: pappus of firm ovate or deltoid paleæ abruptly attenuate-awned, about equalling the akene.

B. Clevelándi, GRAY. Leaves linear, a line wide, obtuse, entire: involucral bracts 8 to 12, plane: pappus-paleæ only 2, slender-awned. — Proc. Am. Acad. xix. 22. — San Diego, *Cleveland*. Too little known.

B. carnósa, GREENE. Leaves filiform, entire: involucral bracts about 7, with a strong carinate midrib: pappus of 4 or 5 subulate-awned paleæ. — Bull. Torr. Club, x. 86. — Bay of San Francisco, in salt marsh at Vallejo, *Greene*.

B. platycárpha, GRAY. Leaves narrowly linear to filiform, some laciniate-pinnatifid: involucral bracts 6 or 7, manifestly 3-nerved at base, middle nerve at length carinate-thickened: pappus-paleæ 5 to 7, slender-awned. — Proc. Am. Acad. ix. 196, xix. 22. *Burrielia platycarpha*, Gray, Bot. Mex. Bound. 97. — Lower Sacramento and Byron Springs, *Stillman, Greene*.

§ 2. DICHÆTA. Pappus of two forms both in ray and disk, i. e. of truncate or muticous paleæ alternating with awned ones or naked awns, or wanting in some species: receptacle, &c., of § 1: involucral bracts more obviously carinate-concave in middle, the concavity partly embracing subtended akene, disposed to be deciduous with it at maturity of the fruit: heads not large (3 or 4 lines high): leaves from entire to laciniate-pinnatifid in the same species. — Gray, Proc. Am. Acad. ix. 196, & xix. 22. *Dichæta*, Nutt. Trans. Am. Phil. Soc. vii. 383; Torr. & Gray, l. c.

B. marítima, GRAY. Low and diffuse, glabrate: leaves oblong-linear, inch long, entire, or lowest sparingly laciniate-toothed: head rather narrow: involucral bracts and short orbicular rays 6 or 8: pappus of 3 to 5 slender-subulate awns and at least as many small and narrow laciniate squamellæ or paleæ. — Proc. Am. Acad. ix. 196. *Burrielia maritima*, Gray, Proc. Am. Acad. vii. 358. — Farallones Islands, off San Francisco, *Gruber*.

B. Fremónti, Gray, l. c. Erect, slender, a span or two high, somewhat hirsute-pubescent: leaves some narrowly linear and entire, the others palmately or pedately 3-5-parted above into linear lobes: bracts of the broad involucre 10 to 12: rays as many or fewer, with oval ligules seldom surpassing the disk: pappus of about 4 slender awns and as many or more numerous narrow small paleæ, or rarely none. — *Dichæta Fremontii,* Torr. in Pl. Fendl. 102. *Burrielia (Dichæta) Fremontii,* Benth. Pl. Hartw. 317. — Lower valley of the Sacramento to San Francisco Bay; first coll. by *Fremont.*

B. uliginósa, Gray, l. c. A span to a foot or more high, at length loosely branched and diffuse, villous-tomentose when young, commonly glabrate: leaves linear or ligulate (the larger 4 to 10 inches long), laciniate-pinnatifid and the linear segments sometimes again cleft, or the upper occasionally entire; involucral bracts and oblong exserted rays 10 to 13: pappus sometimes none, commouly of 2 or 3 stout chaffy awns, and as many or twice as many shorter broad and truncate laciniate-fimbriate paleæ. — *Dichæta uliginosa,* Nutt. Trans. Am. Phil. Soc. l. c. 383; Torr. & Gray, Fl. ii. 380. — Wet ground or in shallow water, San Francisco Bay to Santa Barbara; first coll. by *Coulter.*

Var. **ténera,** Gray. Depauperate, on drier soil, 2 to 6 inches high: leaves linear, entire, or some of the lowermost laciniate: rays oval or oblong. little or not at all exceeding the disk: paleæ and awns each usually 2, the former very broad and quadrate, or splitting into 2 or 3. — Proc. Am. Acad. xix. 22. *Dichæta tenella,* Nutt. l. c.; Torr. & Gray, l. c. — With the ordinary form: also at Tulare Station, *Parry.*

§ 3. **PTILÓMERIS.** Pappus wholly of awned or of muticous and commonly erose paleæ, or sometimes wanting: receptacle not muricate-roughened, rather scrobiculate: involucral bracts in fruit carinate at centre outside, plicate-concave within, and at length deciduous with the subtended akene, as in § *Dichæta:* heads of the same: leaves all pinnately or lower ones bipinnately parted into linear and attenuate divisions: not woolly, mostly somewhat glandular, diffuse. — Proc. Am. Acad. xix. 23. *Hymenoxys* (Hook.), Torr. & Gray, Fl. ii. 380, not Cass. *Ptilomeris,* Nutt. Trans. Am. Phil. Soc. vii. 382. *Actinolepis* § *Ptilomeris,* Benth. & Hook. Gen. ii. 399; Gray, Proc. Am. Acad. ix. 197, & Bot. Calif. i. 378. *Baeria* § *Ptilomeris,* Gray, Proc. Am. Acad. xix. 21, 23.

   * Rays 6 to 8, oblong, short-exserted: involucral bracts ovate-oval: receptacle either acutely or obtusely conical, glabrous: heads small (barely 3 lines high): plants a span high, minutely pubescent, obscurely if at all glandular, with filiform-linear divisions to the leaves: the two following perhaps forms of one species.

B. affínis, Gray. Pappus of 8 or 10 oblong or lanceolate paleæ with laciniate-setulose margins, fully equalling the corolla tube, some or most of them produced into an awn almost equalling disk-corolla, or in the ray blunt and awnless. — Proc. Am. Acad. xix. 23. *Ptilomeris affinis,* Nutt. Pl. Gamb. 174. — S. California, from Los Angeles to San Bernardino, *Gambel, Nevin, Parish.*

B. tenélla, Gray, l. c. Pappus of 6 to 10 short and firm quadrate or broadly cuneate paleæ, with the truncate muticous summit denticulate or nearly entire, not surpassing the tube of the corolla. — *Ptilomeris tenella,* Nutt. Pl. Gamb. 173. *Actinolepis tenella,* Gray, Bot. Calif. i. 378. mainly. — Los Angeles, California, *Gambel, Parry.*

   * * Rays 10 to 15, elongated-oblong, exserted: involucral bracts oblong-lanceolate: receptacle acutely conical, minutely and sparsely pubescent: plants minutely glandular-pubescent, diffusely branched, a span to near a foot high, perhaps all varieties of one, the difference being mainly in the pappus.

B. coronária, Gray, l. c. Pappus of 8 to 12 lanceolate or oblong denticulate paleæ, all tapering into awns, little shorter than disk-corollas, or some in the ray awnless: rays nearly half-inch long. — *Ptilomeris coronaria* & *P. aristata,* Nutt. in Trans. Am. Phil. Soc. vii. 382. *Shortia Californica,* Nutt. in garden catalogues. *Hymenoxys Californica,* Hook. Bot. Mag. t. 3828; Torr. & Gray, l. c., with var. *coronaria. Actinolepis (Ptilomeris) coronaria,* Gray, Proc. Am. Acad. ix. 198, & Bot. Calif. l. c. — California, *Nuttall.* Not since collected, but common in cultivation, especially in France.

B. anthemoídes, Gray, l. c. More glandular, and with somewhat more filiform divisions to the leaves: pappus wanting. — *Ptilomeris (Ptilopsis) anthemoides*, Nutt. l. c. *Hymenoxys calva*, Torr. & Gray, Fl. l. c. *Actinolepis (Ptilomeris) anthemoides*, Gray, Bot. Calif. l. c. — San Diego, California, *Nuttall*, and near Julian City, *Bolander*.

B. mútica, Gray, l. c. Like the preceding, probably the pappose state of it: pappus of 6 to 8 quadrate-oblong paleæ, the obtuse or truncate summit erose. — *Ptilomeris mutica*, Nutt. l. c. *Hymenoxys mutica*, Torr. & Gray, l. c. *Actinolepis (Ptilomeris) mutica*, Gray, Bot. Calif. l. c. — San Diego, California, *Nuttall*, *Cleveland*.

142. **SYNTRICHOPÁPPUS**, Gray. (Σύν, θρίξ, πάππος, bristles of pappus united.) — Low and small Californian and Arizonian winter annuals, floccose-woolly, mostly alternate-leaved, branched from the base; with short-peduncled heads terminating the branches; flowers all yellow or rays sometimes rose-red. — Pacif. R. Rep. iv. 106, t. 15, Bot. Calif. i. 394, & Proc. Am. Acad. xix. 20.

S. Fremónti, Gray, l. c. About a span high, loosely floccose: leaves spatulate or linear-cuneate, often 3-lobed at summit: involucre 3 lines high, of about 5 broadly oblong bracts: rays 5, rather large: flowers all golden yellow: pappus bright white. — Desert plains, S. E. California, adjacent Nevada, S. Utah, and Arizona; first coll. by *Fremont*.

S. Lemmóni, Gray. Smaller, slender, lightly woolly, glabrate in age: leaves spatulate or linear, entire: involucre of 6 to 8 narrowly oblong bracts: rays small, rose-purple and white or white-edged; disk-corollas pale yellow: pappus none. — Proc. Am. Acad. xix. 20. *Actinolepis Lemmoni*, Gray, Proc. Am. Acad. xvi. 102. — S. E. California, on the Mohave Desert, *Lemmon*. Summit of Cajon Pass, *Parish*.

143. **ERIOPHÝLLUM**, Lag. ("Εριον, wool, φύλλον, foliage, the plants woolly.) — Mostly floccose herbs, rarely suffruticose (of W. N. America and probably in northern parts of Mexico); with alternate or partly opposite leaves, and peduncled or sometimes sessile heads; the flowers wholly yellow, or one or two with rose-purple rays, one rayless. — Nov. Gen. & Spec. 28; Dougl. in Bot. Reg. t. 1167; Gray, Proc. Am. Acad. xix. 24. *Eriophyllum* & *Phialis*, Spreng. Gen. 631. *Trichophyllum*, Nutt. Gen. ii. 166; Hook. Fl. i. 315. *Bahia*, DC. Prodr. v. 656, in part; Torr. & Gray, Fl. ii. 374, partly, not Lag. *Actinolepis*, DC. Prodr. v. 655.

§ 1. ACTINÓLEPIS. Low and diffuse winter-annuals, with short-peduncled or sessile heads only 2 or 3 lines high: involucral bracts few, distinct to the base, herbaceous or chartaceous in age: anther-tips from ovate-lanceolate to linear-subulate. — Gray, Proc. Am. Acad. xix. 24. *Actinolepis*, DC., Benth. & Hook. Gen. ii. 399; Gray, Proc. Am. Acad. ix. 198, & Bot. Calif. i. 377, excl. § *Ptilomeris*.

    * Heads sessile or nearly so in the forks, or at summit of branches, then subtended by a leaf or glomerate, 2 lines high, wholly yellow-flowered: receptacle flat or barely convex: anther-tips ovate-lanceolate, obtuse: leaves small, spatulate, commonly 3-lobed or 3-toothed at summit. — *Actinolepis*, DC. l. c., founded on specimens with infertile disk-flowers.

E. multicaúle, Gray, l. c. Whitened with rather close cottony wool, sometimes denudate in age: stems slender, at length much branched, a span high, most of the internodes exceeding the leaves: rays 3 to 5, obovate, a line long: akenes glabrate: pappus of 10 to 15 rather firm narrowly subulate or almost aristiform paleæ, or sometimes wanting in all or some of the disk-flowers, especially when these are infertile; then their style is only minutely forked at the apex. — *Actinolepis multicaulis*, DC. Prodr. v. 656; Hook. Ic. t. 325; Torr. Bot. Mex. Bound. t. 33. — Southern California to Arizona, from Santa Barbara to Tucson, in low ground; first coll. by *Coulter* and *Douglas*.

E. Prínglei, Gray, l. c. More loosely and copiously woolly, depressed, inch or two high, flowering almost from the base: rays none: flowers all fertile: akenes villous: pappus of about 10 much larger wholly silvery-scarious oblong-lanceolate and pointless erose paleæ. — Gravelly plains from the Mohave Desert in S. E. California to Tucson, Arizona, *Palmer, Lemmon, Pringle, Parish.*

* * Heads pedunculate, terminating the branches, 3 or 4 lines high: receptacle convex or conical: plants 3 to 5 inches high, erect and at length diffuse, with mostly entire leaves.

+- Rays about 5, inconspicuous: disk-flowers not numerous: anther-tips ovate-oblong, obtuse.

E. nubígenum, Greene. Densely white-woolly: leaves lanceolate-spatulate (about half-inch long): heads short-peduncled, narrow: involucre of 5 oblong bracts: rays with oval ligule, hardly exceeding the disk-flowers, yellow: receptacle with conical centre: pappus of about 10 oblong or narrow nerveless and obtuse erose thinnish paleæ, half the length of the corolla, one third that of the akene. — Gray, Proc. Am. Acad. xix. 25. — On Cloud's Rest, above the Yosemite, at 9,000 feet, *Mrs. Curran.*

+- +- Rays 5 to 9, exserted and ample, oval or oblong: disk-flowers more numerous: anther-tips narrow and slender: receptacle high-convex or obtusely low-conical.

E. Wallácei, Gray. Thickly clothed with cottony wool: leaves obovate or spatulate, occasionally 2-3 toothed at apex: pappus of 6 to 10 short-oval or obovate obtuse and pointless nerveless paleæ, of firm texture and opaque: style-tips somewhat subulate-conical: corollas all yellow. — Proc. Am. Acad. xix. 25. *Bahia Wallacei*, Gray, Pacif. R. Rep. iv. 105. *Actinolepis Wallacei*, Gray, Proc. Am. Acad. ix. 198. — Plains, from San Diego Co., California (first coll. by *W. A. Wallace*), to adjacent Arizona and S. Utah. A var. with pale purple and white rays (*Bahia rubella*, Gray, Bot. Mex. Bound. 95), S. E. California, *Parry*.

E. lanósum, Gray, l. c. More thinly and floccosely woolly: leaves spatulate-linear, entire: pappus of about 5 oblong and rather firm nerveless and obtuse paleæ and as many alternating paleaceous awns of double the length: style-tips obtuse and sometimes with a minute cuspidate apiculation: rays white or rose-color. — *Burrielia* (*Dichæta*) *lanosa*, Gray, Pacif. R. Rep. iv. 107. *Actinolepis lanosa*, Gray, Proc. Am. Acad. ix. 198. — Dry plains, S. E. California to Arizona and S. Utah; first coll. by *Bigelow*.

§ 2. Trichophýllum. Larger, erect: heads when clustered small, when solitary commonly rather large: involucral bracts of firm texture: rays and disk-flowers golden yellow; tube of corolla commonly glandular or hairy: anther-tips ovate, mostly obtuse: akenes linear or cuneate-linear, glabrous or nearly so: pappus of short opaque and firm nerveless and pointless paleæ, sometimes very small, rarely obsolete or wanting. — Gray, Proc. Am. Acad. xix. 25. *Trichophyllum*, Nutt. l. c. *Phialis*, Spreng. l. c., but involucre seldom gamophyllous.

* Suffruticose or suffrutescent, leafy to the top, branching: heads small, compactly corymbosely cymose, short-peduncled: ligules roundish-oval, only a line or two long: pappus of oblong-linear paleæ much shorter than the akene: leaves mostly lobed or divided, and the margins revolute.

E. stæchadifólium, Lag. l. c. Canescent with close-pressed pannose tomentum, at length partly denudate, 1 to 4 feet high from a woody base: leaves once or twice pinnately parted into linear divisions and rhachis, or the upper linear with a pair of lateral lobes, or some of them entire, upper face soon glabrate and green; heads 3 or 4 lines high, numerous in rather loose paniculate clusters: involucre cylindraceous-campanulate: its bracts 8 to 10, linear-spatulate to narrowly oblong, thinnish: receptacle convex, alveolate-toothed: rays 6 to 8: paleæ of the pappus 8 to 12, the four over the angles of the akene rather longer. — *Helenium stæchadifolium*, Spreng. Syst. iii. 574. *Bahia artemisiæfolia*, Less. in Linn. vi. 253; DC. Prodr. v. 657; Torr. & Gray, l. c. *B. stæchadifolia*, DC. l. c. 656 (with wrong habitat, the plant of Hænke coming from Monterey, California); Nutt. Trans. Am. Phil. Soc. l. c. (with var. *Californica*). Lagasca's original appears to have been a branch of the form with uppermost leaves entire. — Coast of California, from San Francisco to Santa Barbara.

E. confertiflórum, Gray, l. c. Similarly white-woolly, a foot or two high, with slender and more strict stems naked at summit: leaves small, of mostly cuneate outline, pinnately or somewhat ternately once or twice 3-7-parted into narrow linear divisions: heads 2 lines high, several or numerous in a compact cymose cluster, mostly short-peduncled or subsessile: involucre oval or obovoid-oblong, of about 5 broadly oval thin-coriaceous bracts: receptacle convex or low-conical in the centre, not alveolate: rays 4 or 5 : paleæ of the pappus 8 to 14. — *Bahia confertiflora*, DC. l. c. 657 ; Torr. & Gray, l. c. — Hills, California, common from near the coast to the Sierra Nevada.

Var. trifidum, GRAY, l. c. A form with small short leaves, simply 3–5-cleft into oblong or short-linear lobes. — *B. trifida*, Nutt. Trans. Am. Phil. Soc. l. c.; Torr. & Gray, l. c. *B. confertiflora*, var. *trifida*, Gray, Bot. Calif. l. c. — With the ordinary form. Autumnal specimens, coll. by *Parish*, on the San Bernardino Mountains, are tomentose with longer and looser wool.

Var. laxiflórum, GRAY, l. c. Heads loosely fastigiate-cymose and mostly slender-peduncled. — *Bahia tenuifolia*, DC. l. c. — California, *Douglas* (herb. DC.), *Coulter*. An ambiguous form with larger heads and rays, coll. at San Bernardino, *Parish*.

* * Herbaceous, commonly and perhaps always perennials: heads larger, mostly solitary or scattered and conspicuously pedunculate: receptacle from convex or low conical to flat (even in the same species): ligules 6 to 13, from quarter to half inch long, oblong or oval: leaves variable.

+— Akenes glabrous, glabrate, or sparsely appressed-pilose, not glandular.

— — **E. cæspitósum,** DOUGL. Floccosely white-woolly, many-stemmed from the root : leaves in age with upper face often glabrate; lower ones from spatulate or cuneate to roundish in outline, from incisely 3–5-lobed to pinnately parted, or the upper varying to linear and entire: involucral bracts 8 to 12, oblong or oval: tube of disk-corollas mostly hirsute-glandular and longer than the pappus, which is variable, sometimes very short, sometimes obsolete. — Lindl. Bot. Reg. t. 1167 (but the gamophyllous involucre of the figure is seldom found); Gray, l. c. *Actinella lanata*, Pursh, Fl. ii. 560. *Helenium lanatum*, Spreng. Syst. iii. 574. *Trichophyllum lanatum*, Nutt. Gen. ii. 167; Hook. Fl. i. 315. *Bahia lanata*, DC. Prodr. v. 657; Torr. & Gray, Fl. ii. 375, incl. var. *tenuifolia* (which is not *B. tenuifolia*, DC., but merely the most slender form of the present species). — Moist or dry ground, common from Montana to Brit. Columbia, and thence to S. California, under very various forms, which are indefinable as species. Taking as the type the original of Pursh and Nuttall, with rather slender stems a foot or more high, principal leaves somewhat palmately pinnatifid into narrow divisions, or incisely cleft, and heads rarely half-inch high, the main divergent forms are: —

Var. latifólium, GRAY, l. c. The opposite extreme in foliage: stems commonly 2 feet long, branched and lax when growing in shade : leaves thin, dilated, from rhombic or cuneate to oblong-lanceolate, 3–5-lobed and incised or dentate, the lobes from oblong to broadly lanceolate: peduncles comparatively short: rays 9 to 13 : corolla-tube either sparsely or densely hirsute with gland-tipped hairs, much longer than the pappus, the rounded paleæ of which do not exceed the breadth of the narrowly oblong-cuneate or narrower glabrous akenes, commonly very short and forming a kind of crown, sometimes quite obsolete (as occurs in other forms also). — *Bahia arachnoidea*, Fisch. & Lallement, Ind. Sem. Hort. Petrop. 1842; Gray, Pl. Fendl. 100, & Bot. Calif. i. 382. *B. latifolia*, Benth. Bot. Sulph. 30. *Eriophyllum cæspitosum*, Bot. Reg. t. 1167, is nearly this. — California, near the coast, in or near Redwood forests, from Humboldt Co. to Santa Cruz. *Bahia lanata*, var. *brachypoda*, Gray, Bot. Calif. l. c., is a sea-shore form of this, with leaves thickish under exposure, heads clustered and remarkably short-peduncled, and pappus larger. Forms connecting with var. *integrifolium* occur in the Sierra, in groves of *Sequoia gigantea*.

Var. achillæoídes, GRAY, l. c. Leaves pinnately parted or cleft, with the 3 to 5 divisions mostly narrow and laciniately incised or pinnatifid : heads somewhat corymbosely collected and rather short-peduncled : involucre hemispherical, 3 or 4 lines high ; rays and involucral bracts 9 to 13: akenes sparsely pubescent or glabrate. — *Bahia achillæoides*, DC. l. c. *B. lanata*, var. *achillæoides*, Gray, Bot. Calif. l. c. — California, near the coast.

Var. grandiflórum, GRAY, l. c. Rather strict and stout, densely woolly : leaves all linear or the lower narrowly lanceolate or spatulate, laciniate-serrate or entire, or some parted into a few narrowly linear divisions : heads solitary and long-peduncled : involucre half-inch high, hemispherical, densely woolly, of 10 to 13 bracts : rays as many, large : akenes usually somewhat pubescent : corolla-tube sparsely hirsute-glandular. — *Bahia lanata*, Benth. Pl. Hartw. 317. *B. lanata*, var. *grandiflora*, Gray, Bot. Calif. l. c. — California, valley of the Sacramento. (Guadalupe Island, off Lower California.)

—— **Var. leucophýllum,** Gray, l. c. Smaller, a span to a foot high, rather strict: leaves narrow, entire or sparingly cleft or parted: heads solitary, long-peduncled: involucre campanulate, 4 or 5 lines high, of about 8 oblong bracts: pappus in the typical plant of narrow lanceolate paleæ, four of them twice the length of the others, but this is inconstant. — *Bahia leucophylla,* DC. l. c. — Brit. Columbia to N. California, and east to Idaho.

—— **Var. integrifólium,** Gray, l. c. Low, often dwarf, cespitose-tufted, 3 to 10 inches high: leaves from narrowly spatulate or oblanceolate and entire to more dilated and 3-lobed at summit, or at base and on sterile shoots cuneate and incisely lobed: heads rather long-peduncled: involucre, &c., of the preceding, sometimes smaller and of only 6 bracts: paleæ of the pappus mostly of same length, about equalling the very glandular but not hirsute corolla-tube: akenes glabrous, rarely somewhat glandular-atomiferous near the summit. — *Trichophyllum integrifolium,* Hook. Fl. i. 316. *T. multiflorum,* Nutt. Jour. Acad. Philad. vii. 37. *Bahia integrifolia,* DC. l. c.; Gray, Bot. Calif. l. c. *B. multiflora,* Nutt. Trans. Am. Phil. Soc. l. c. *B. leucophylla,* Torr. & Gray, l. c., in part. *B. cuneata,* Kellogg in Proc. Calif. Acad. v. 49, a form passing into the preceding — Rocky Mountains in Montana and Wyoming to Brit. Columbia and along the higher portions of the Sierra Nevada, California, south to San Bernardino Co.

+ + Akenes like the corolla-tube glandular: stems less than a foot high, slender.

E. grácile, Gray, l. c. Loosely floccose-woolly: leaves so far as known all very narrowly linear and entire (an inch or two long, half-line wide): head on a long slender peduncle: involucre nearly 4 lines high, campanulate, of about 10 oblong bracts: rays about 8: receptacle nearly flat, alveolate-dentate: akenes slender, 2 lines long: paleæ of the pappus oblong or quadrate, exceeding the breadth of the akene. — *Bahia gracilis,* Hook. & Arn. Bot. Beech. 353; Torr. & Gray, l. c.; Gray, Bot. Calif., in part. — S. Idaho, on Snake River, *Tolmie.* Not since seen.

E. Watsóni, Gray, l. c. Canescent with fine and close tomentum, fastigiately branched: leaves cuneate or spatulate in outline, with tapering slender base or petiole, 3-lobed at summit: involucre 3 lines high, short-campanulate, of 6 or 7 oval bracts: rays 5 to 7: receptacle conical, naked: akenes shorter and thicker: pappus a crown of truncate laciniate-dentate paleæ, decidedly shorter than the breadth of the akene. — *Bahia leucophylla,* Eaton, Bot. King Exp. 173, in part. *B. gracilis,* Gray, Bot. Calif. l. c., in part. — N. Nevada, at Robert's Station, at 6,000 feet, *Watson.*

* * * Annuals, with leaves apparently all alternate, and small pedunculate heads terminating the lax slender branches: receptacle conical: pappus a crown of small paleæ, not longer than the breadth of the summit of the akene, sometimes very short or obsolete: style-tips conical.

E. ambíguum, Gray, l. c. Somewhat loosely floccose-woolly, or denudate: stems branching from the decidedly annual root, 3 to 10 inches high: leaves from spatulate to linear-lanceolate (an inch or less long), entire, or 3-toothed or lobed, especially the broader sometimes dilated-cuneate lowermost: involucre campanulate, 3 lines high, of 6 to 9 oblong-lanceolate bracts, which are either distinct to the base or lightly coherent for two thirds their length: rays 5 to 9, oblong or oval: tube of the corolla glandular-hirsute: akenes pubescent or the inner ones glabrous. — *Lasthenia* (*Monolopia*) *ambigua,* Gray, Proc. Am. Acad. vi. 547. *Bahia Wallacei,* Gray in Jonr. Bot. Nat. Hist. vii. 145, not of Pacif. R. Rep. *B. parviflora* & *B.* (*Pseudo-Monolopia*) *ambigua,* Gray, Bot. Calif. i. 382. — S. E. California, near Tejon, *Xantus, Van Horn, Parry,* and near Hot Springs, San Bernardino Co., *Parish.*

144. **BÁHIA,** Lag. (*Juan Francisco Bahi,* Professor of Botany at Barcelona.) — Suffruticose or mostly herbaceous plants (of Rocky Mountain district, Mexico, and Chili), not lanate but in some canescent; with opposite or sometimes alternate leaves, and small or middle-sized pedunculate heads of yellow flowers terminating the branches. — Lag. Nov. Gen. & Spec. 30; Gray, Proc. Am. Acad. xix. 26. *Stylesia,* Nutt. Trans. Am. Phil. Soc. n. ser. vii. 377, founded on the original *Bahia.* Species of *Bahia,* Less. Syn. 238; DC. Prodr. v. 656; Benth. & Hook. Gen. ii. 402. *Achyropappus,* HBK. Nov. Gen. & Spec. iv. 257. t. 390, not Bieb. Species of *Schkuhria* & of *Villanova,* Benth. & Hook. Gen. 403, 404.

Amauria, Benth. Bot. Sulph., of Lower California, insufficiently known, is perhaps an epappose *Bahia*.

§ 1. Suffruticose (*B. ambrosioides*, Lag., of Chili) or herbaceous from a perennial sometimes lignescent root: paleæ of the pappus 4 to 8, obovate or spatulate, with rounded or truncate scarious summit, and thickened base or imperfect costa: leaves dissected or cleft, the lower opposite.

B. oppositifólia, NUTT. A span or two high, fastigiately branched and many-stemmed, herbaceous to the base, very leafy up to the short-peduncled heads, cinereous with fine close pubescence: leaves mostly opposite, petioled, palmately or pedately 3–5-parted into linear divisions little broader than the margined petiole: head 4 or 5 lines high: bracts of the involucre oblong or oval, comparatively close (the outer obscurely carinate-one-nerved): rays 5 or 6, oval, hardly surpassing the disk-flowers: akenes slender, glandular: pappus half the length of the corolla-tube, the paleæ narrowly obovate, with strongly opaque centre evanescent near the summit. — Torr. & Gray, Fl. ii. 376; Gray, Pl. Fendl. 99. *Trichophyllum oppositifolium*, Nutt. Gen. ii. 167. — Sterile hills and plains, Nebraska to Colorado and the borders of New Mexico.

B. absinthifólia, BENTH. About a foot high from an herbaceous root or barely suffrutescent base, diffusely branched, tomentulose-canescent, and with sparsely corymbose-paniculate heads on slender peduncles: leaves opposite and the upper alternate, pedately or sometimes pinnately 3–5-parted into narrowly linear or lanceolate divisions and lobes: involucre more lax; its bracts oblong-spatulate or lanceolate with narrowed base: rays 9 to 12, lanceolate-oblong, much exceeding the disk: akenes slender, pubescent: pappus nearly equalling the proper corolla-tube, its paleæ more dilated and broadly thin-scarious above. — Pl. Hartw. 18. — Arizona, near Tucson, *Palmer*, *Lemmon*. (Forms almost as slender and narrow-leaved as the plant of Northern Mexico.)

Var. **dealbáta**, GRAY. Perhaps more lignescent at base, more whitened with fine pannose tomentum: leaves less divided, commonly only 3-cleft into lanceolate or linear-oblong lobes, or some lower ones oblong-lanceolate and entire. — Pl. Wright. i. 121; Proc. Am. Acad. xix. 27. *B. dealbata*, Gray, Pl. Fendl. 99. — Dry plains, W. Texas to Arizona. (Adj. Mex.)

§ 2. Herbaceous from a perennial caudex: leaves all alternate and entire, coriaceous: paleæ of the pappus about 10, linear-lanceolate, and with a distinct excurrent or percurrent costa. — § *Platyschkuhria*, Gray.

B. nudicaúlis, GRAY. Cinereous-puberulent and glabrate, upper part of the scapiform stem and involucre minutely glandular, a span or two high: leaves nearly all radical, oval or spatulate-oblong (an inch or more long), tapering into a slender petiole: heads solitary or few and somewhat corymbosely paniculate, nearly half-inch high: involucre hemispherical, of about 10 oblong bracts: rays 6 to 9, oblong: pappus fully half the length of the cuneate-linear sparsely hairy akene; the thin margins of the paleæ of the pappus erose, and the short-excurrent awn barbellate-hispidulous. — Proc. Am. Acad. xix. 27. — *Schkuhria* (*Platyschkuhria*) *integrifolia*, Gray in Am. Nat. viii. 213, & Proc. Am. Acad. ix. 198, excl. var. — Wind River Mountains, N. W. Wyoming, *Parry*.

B. oblongifólia, GRAY, l. c. Smaller: stems sparsely leafy almost to the 3-cephalous naked inflorescence: leaves narrowly oblong: head only 4 lines high, narrow: paleæ of the pappus firmer, smoother, and with entire edges, little shorter than the glabrate akene. — *Schkuhria integrifolia*, var. *oblongifolia*, Gray in Am. Nat. l. c. — On the San Juan and Rio Colorado, near their junction, S. E. Utah or adjacent Colorado, *Newberry*, *Brandegee*.

§ 3. Annuals, with once or twice palmately or pedately divided leaves: akenes mostly hirsute along the slender attenuate base, at least on the angles. — *Achyropappus*, HBK.

* Leaves mainly opposite, at least all the lower ones, their divisions narrowly linear: pappus of broad and very obtuse paleæ, scarious above, callous-thickened and opaque at base, as in § 1: ray-flowers occasionally wanting. — Gray, Proc. Am. Acad. xix. 27. *Achyropappus*, HBK., DC.

B. Bigelóvii, Gray. Slender, a foot high, diffuse, strigose-puberulent: leaves 3-parted and the divisions sometimes 2-3-parted into linear-filiform segments and lobes: peduncles elongated, filiform: involucre hemispherical, 2 lines high; its bracts 8 or 9, oval and tapering to both ends, viscidly hirsute: rays as many, oblong: tube of disk-corollas hirsute with viscid hairs; throat broadly cyathiform: paleæ of the pappus broadly cuneate-obovate, half the length of the corolla-tube, callous-thickened only at base. — Bot. Mex. Bound. 96. *Schkuhria (Achyropappus) Bigelovii*, Gray, Proc. Am. Acad. ix. 199; Proc. Am. Acad. xix. 1. c. — S. W. Texas, in the valley of the Limpio, *Bigelow*.

B. Neo-Mexicána, Gray. A span or more high, minutely puberulent: leaves 3-7-parted into narrow linear divisions; uppermost little shorter than the slender peduncles: involucre of about 10 sparingly pubescent spatulate bracts: rays none : disk-corollas small, with glandular tube, almost equalled by the obovate paleæ of the pappus, which are much thickened at and near the base. — Proc. Am. Acad. xix. 27. *Schkuhria Neo-Mexicana*, Gray, Pl. Fendl. 96, & Proc. Am. Acad. ix. 199. — Northern New Mexico and S. Colorado, *Fendler, Bigelow, Parry*, &c.

* * Leaves mainly opposite, with linear divisions: flowers perhaps white: paleæ of the pappus lanceolate and with a complete costa.

B. Woodhoúsii, Gray. Low, cinereous-puberulent: peduncles hardly longer than the heads: leaves thickish, 3-parted, and the middle divisions sometimes with a pair of lateral lobes: involucre 3 lines high; its bracts 8 or 9, oblong-obovate, obtuse: rays 8 or 9, with oblong ligules a line or two long, hardly surpassing the disk-flowers: paleæ of the pappus 8 to 10, with hyaline margins and a strong opaque costa, which reaches the acute apex. — Proc. Am. Acad. xix. 28. — *Achyropappus Woodhousii*, Gray, Proc. Am. Acad. vi. 546. *Schkuhria Woodhousii*, Gray, Proc. Am. Acad. ix. 199. — Northern part of New Mexico, *Dr. Woodhouse*.

* * * Leaves all or mostly alternate, naked-petioled, 2-3-ternately divided or parted; the divisions from linear-spatulate to obovate, comparatively short: heads loosely cymose-paniculate at the naked summit of the erect stems, hemispherical, yellow-flowered, with oblong or obovate exserted rays: paleæ of the pappus oblong to narrowly lanceolate, with a distinct procurrent or excurrent costa, or the pappus wanting: akenes tetragonal-clavate, or those of the ray slender obpyramidal with 4 sides.

+ Tube of the disk corollas glandular but not hirsute; lobes ovate or oblong, shorter than the dilated throat: pappus present.

B. pedáta, Gray. A foot or two high, cinereous-puberulent: leaves pedately divided, commonly into 3 petiolulate obovate or cuneate segments, of which the lateral are 2-parted and the middle 3-7-lobed; the lobes obovate or broadly oblong, short : heads 5 lines high : bracts of the involucre oblong, obtuse, shorter than the disk: rays about 12, oblong : paleæ of the pappus 10 to 12, spatulate-oblong, with costa vanishing near the obtuse or retuse summit. — Pl. Wright. i. 132, & Proc. Am. Acad. xix. 28. *Schkuhria (Achyropappus) pedata*, Gray, Proc. Am. Acad. ix. 199. — S. W. Texas, on the Limpio and Rio Grande, *Wright, Bigelow*.

B. biternáta, Gray. More pubescent and slender: leaves ternately dissected into linear and obtuse or somewhat spatulate segments and lobes, the primary ones slightly petiolulate : heads 4 lines high : bracts of the involucre obovate : rays 8 or 10, broadly obovate : paleæ of the pappus 12 to 14, longer and narrower, about equalling the corolla-tube, those of the outer flowers obovate and obtuse, with costa evanescent below the apex ; of the inner flowers longer, elongated-lanceolate, and with costa excurrent into an awn-like cusp; in intermediate flowers of intermediate character. — Pl. Wright. ii. 95, & Proc. Am. Acad. l. c. *Schkuhria (Achyropappus) biternata*, Gray, Proc. Am. Acad. ix. 199. — Borders of W. Texas and adjacent New Mexico, *Wright, Bigelow, Thurber*. May pass into the preceding.

+ + Tube of disk-corollas viscid-hirsute: the limb cleft into narrow lobes which are much longer than the throat and little shorter than the tube: pappus none.

B. chrysanthemoídes, Gray. Taller and stouter, 1 to 4 feet high, puberulent or below glabrous, above with the flowering branches and short peduncles glandular-pubescent and viscid: leaves 1-3-ternately divided or parted ; the lobes from oblong and obtuse to nearly linear: heads 5 or 6 lines high and broad: bracts of the involucre 16 to 20, crowded, from oblong-lanceolate to obovate-oblong, most of them conspicuously acuminate: rays as many, obovate-oblong: akenes obscurely striate on the four narrow faces, the whole apex covered by the base of the corolla. — Proc. Am. Acad. xix. 28. *Amauria? dissecta*, Gray, Pl. Fendl. 104. *Villanova chrysanthemoides*, Gray, Pl. Wright. ii. 96. — Along mountain water-courses, Colorado to S. Arizona; first coll. by *Fremont*.

145. **AMBLYOPÁPPUS**, Hook. & Arn. ('Αμβλύς, blunt, πάππος, pappus.) — Jour. Bot. iii. 321; Benth. & Hook. Gen. ii. 406. *Aromia*, Nutt. Trans. Am. Phil. Soc. vii. 395. *Infantea*, Remy, in Gay, Fl. Chil. iv. 257, t. 48. — Low annual (of Chili, and *Schkuhria pusilla*, Wedd., is perhaps a second species in Bolivia), probably introduced into California.

A. **pusillus**, Hook. & Arn. l. c. A span or two high, nearly glabrous, balsamic-viscid, paniculately or corymbosely branched, with small short-peduncled heads terminating the branches: leaves linear and alternate, entire or lower pinnately 3–5-parted and opposite: involucre 2 lines high, equalling the yellowish flowers. — *Aromia tenuifolia*, Nutt. l. c. *Infuntea Chilensis*, Remy, in Gay, l. c. — Around San Diego, California, and southward. (Chili.)

146. **SCHKÚHRIA**, Roth. (*Christian Schkuhr*, of Wittenberg.) — Slender and paniculately much branched annuals (Mexican and Andean), somewhat pubescent, never tomentose; the small pedunculate heads of yellow (rarely purplish) flowers terminating the branchlets: leaves alternate, or the lower opposite, pinnately 3–7-parted or uppermost entire, the divisions and rhachis filiform. Herbage sometimes minutely resinous-atomiferous and the leaves impressed-punctate. — Roth, Catalecta Bot. i. 116; Cass.; Less., &c. *Tetracarpum*, Mœnch, Meth. Suppl. 241. *Schkuhria & Hopkirkia*, DC. Prodr. v. 654, 660. *Schkuhria*, Benth. & Hook. Gen. ii. 403, in part, excl. *Achyropappus*, &c. — Our species, and *S. Wislizeni* of Northern Mexico, form a section (the genus *Hopkirkia*, DC.), with leaves more commonly only 3-parted and on the branches entire, heads only 3–5-flowered, with a single ray-flower or none: obpyramidal akenes in length only about double the width of the summit, their angles very densely long-villous, some hairs also on the faces: scarious tips of involucral bracts purple-tinged: stems diffusely corymbose-paniculate.

S. **Hopkirkia**, GRAY. Pappus equalling the corolla; its paleæ all alike, ovate-oblong, with percurrent costa projecting as a cusp: faces of the akene conspicuously 3-nerved. — Pl. Wright. ii. 94. *Hopkirkia anthemidea*, DC. Prodr. v. 660. — S. Arizona, *Wright, Lemmon*. (Northern Mex., *Hænke*.)

S. **Wrightii**, GRAY, l. c. Pappus shorter than the corolla; its paleæ all obovate and obtuse or erose-truncate, destitute of costa, merely thickened at very base: akenes rather less thick and faces less striate. — S. Arizona, *Wright, Thurber, Lemmon*.

147. **HYMENÓTHRIX**, Gray. (From ὑμήν, membrane, θρίξ, bristle, the pappus a combination of awn and thin palea.) — Herbs of Arizona and vicinity, glabrous or somewhat pubescent; with probably annual or perhaps perennial root, branching stems of 1 to 3 feet high, alternate leaves once to thrice parted into linear divisions or lobes, and numerous corymbosely cymose heads (about one-third inch high); the corollas yellow or white and purple, strikingly different in the two species.

H. **Wislizéni**, GRAY. Glabrous: lobes of the leaves often spatulate-linear and broadish: heads radiate: involucre of comparatively narrow acutish and yellow-tinged bracts, hardly any accessory ones: corollas yellow; those of the disk with oblong lobes only half the length of the narrowly obconical throat: style-tips pointless: akenes rather slender, barely pubescent: pappus-awns narrowly margined below, naked and hispidulous above. — Pl. Fendl.

102, & Pl. Wright. ii. 97; Rothrock in Wheeler Rep. vi. 168. — River-bottoms, &c., S. Arizona and New Mexico; first coll. by *Wislizenus*. (Adj. Mex.)

H. Wríghtii, GRAY. Leaves with very narrow linear or almost filiform divisions, the lower cauline hirsute: heads broader: involucre of obovate-oblong and very obtuse purple-tinged bracts, and a few smaller narrow accessory ones: rays none: disk-corollas white or purplish, 5-parted almost down to the narrow tube into oblong-linear widely spreading lobes: style-tips with a slender-subulate cusp: akenes broader, villous: pappus of broader paleæ and smoother awned tips — Pl. Wright. ii. 97; Torr. in Sitgreaves Rep. t. 6; Rothrock, l. c. — Along streams, S. Arizona, *Wright, Thurber,* &c. (Lower California, *Orcutt.*)

148. HYMENOPÁPPUS, L'Her. (From ὑμήν, membrane, πάππος, pappus, the latter of hyaline paleæ.) — North American and North Mexican herbs (chiefly of the prairies and plains), perennial, biennial, or some perhaps winter annuals, mostly floccose-tomentose and with sulcate-angled erect stems, alternate 1-2-pinnatifid or parted leaves, the lower sometimes entire, and corymbosely cymose or solitary pedunculate middle-sized heads of white or yellow flowers. Leaves in some species evidently impressed-punctate. When the corolla is deeply cleft the nerves of its lobes are deeply intramarginal. Fl. spring. — "L'Her. Diss. cum icon."; Michx. Fl. ii. 103; Cass. Dict. lv. 266, 279; DC. Prodr. v. 658; Gray, Proc. Am. Acad. xix. 29.

* Lobes of the white corolla as long as the short-campanulate or crateriform throat; the tube long and slender, much exceeding the short pappus: stamens with even the filaments mostly exserted: akenes merely pubescent, clavate-obpyramidal, with much thickened summit and stipitiform base: involucre of comparatively lax and partly white-petaloid bracts: heads corymbiform-cymose and rather numerous, on short peduncles: comparatively Eastern species, biennials, 1 to 3 feet high.

+ Pappus of very small obovate or roundish nerveless paleæ forming a crown, much shorter than the breadth of the summit of the merely pubescent akene, often minute, even obsolete: floccose or pannose tomentum thin, sometimes deciduous.

H. scabiosǽus, L'HER. Leafy to the top, thinly tomentose: radical leaves pinnately parted or occasionally entire, cauline irregularly 1-2-pinnately parted into broadly or narrowly linear lobes: heads about 5 lines high: the broad involucre somewhat radiate-expanded, its mainly white bracts roundish-obovate, at first surpassing the disk: akenes short-pubescent. — Michx. Fl. ii. 104; Torr. & Gray, Fl. ii. 372. *Rothia Carolinensis,* Lam. Jour. Hist. Nat. i. 16, t. 1, & Ill. t. 667. — Sandy pine-barrens, Middle Florida to S. Carolina, and west to Illinois and Texas.

H. corymbósus, TORR. & GRAY. More slender, smaller, and glabrate, naked above: lower leaves 2-pinnately and the small upper ones mostly simply parted into narrowly linear acute divisions and lobes: heads 3 or 4 lines high: bracts of the involucre much smaller, shorter than the flowers, obovate-oblong, the petaloid summit only greenish-white: akenes puberulent. — Fl. ii. 372. — Prairies, Nebraska to Arkansas and Texas. The var. *Nuttallii,* Torr. & Gray, as to plant in herb. Torr., belongs here, but the *H. tenuifolius* of Nuttall in other herbaria is Pursh's species.

+ + Pappus of larger spatulate-obovate paleæ, in length nearly equalling the breadth of the summit of the villous-pubescent akene, partly traversed by a callous-thickened axis or obscure costa.

H. artemisiæfólius, DC. Pannosely or somewhat floccosely white-tomentose, or somewhat denudate in age: leaves from simply pinnatifid or lyrately few-lobed, and sometimes quite entire (lanceolate or oblong), to bipinnately parted into broadly linear or narrowly oblong obtuse divisions and lobes: heads 4 lines high: bracts of the involucre obovate-oblong, about equalling the disk-flowers, dull white, lower half green. — Prodr. v. 658; Torr. & Gray, Fl. ii. 372. — Texas: first coll. by *Berlandier.*

* * Lobes of the corolla more or less shorter than the throat: pappus conspicuous, of spatulate or narrow paleæ, which have a manifest costa or thicker opaque axis, this evanescent near or below the obtuse or retuse apex: akenes villous: involucre greener, less petaloid.

+— Stems leafy, from a biennial root a foot or two high: heads rather numerous and corymbosely cymose, on rather short slender peduncles: corolla-tube slender, throat short, and lobes rather long.

H. flavéscens, GRAY. Densely white-tomentose, sometimes glabrate in age: leaves once or twice or even thrice pinnately parted; the divisions or lobes from narrowly to rather broadly linear: heads 4 or 5 lines high: bracts of the involucre roundish-obovate or ovate, with greenish-white or barely yellowish margins: corolla from yellowish to yellow, and short-campanulate throat almost equalled by the lobes: akenes rather short-villous: paleæ of the pappus spatulate, usually only half the length of the slender corolla-tube. — Pl. Fendl. 97, & Pl. Wright. i. 121, ii. 94 (excl. the last var.); Rothrock in Wheeler Exped. vi. 167, where one form is printed "*H. canescens.*" *H. robustus*, Greene, Bull. Torr. Club, ix. 63, stout specimens of the form with finely much divided leaves and somewhat reduced pappus. — Sandy plains and valleys, W. Texas and New Mexico to Arizona. (Adj. Mex.)

H. tenuifólius, PURSH. Lightly tomentose, or soon glabrate and green: leaves rather rigid, once or twice (or radical thrice) pinnately parted into very narrowly linear or filiform divisions, their margins soon revolute: heads only 3 or 4 lines high: involucre more erect and close; its bracts oblong-obovate, greenish with whitish apex and margins: corolla dull white; its lobes moderately shorter than the throat: paleæ of the pappus shorter than the corolla-tube, oblong-spatulate: akenes long-villous. — Fl. ii. 742; Nutt. Gen. ii. 139; DC. Prodr. v. 658; Torr. & Gray, l. c. — Plains, from Nebraska to Arkansas, Texas, and apparently also in Utah.

+— +— Stems clustered on a perennial caudex, leafy below, naked above, bearing few or solitary comparatively large heads.

H. filifólius, HOOK. Tomentose-canescent, or somewhat denudate and glabrate: stems a span to a foot high, sometimes scapiform: leaves nearly of *H. tenuifolius*, or of more filiform rigid divisions: heads a third to half inch high: bracts of the involucre oblong or obovate-oblong, largely green or else white-woolly, the tips whitish or purplish-tinged: corolla yellowish-white or sometimes clear yellow, its reflexed lobes or teeth very much shorter than the throat: akenes very long-villous: paleæ of the pappus equalling or much shorter than the tube of the corolla, but commonly equalled by the villosity of the akene. — Fl. i. 317, but the pappus is not "extremely minute." *H. filifolius* & *H. luteus*, Nutt. Trans. Am. Phil. Soc. l. c.; Torr. & Gray, Fl. l. c. *H. tenuifolius*, Eaton in Bot. King Exp. 173. — Rocky Mountain plains, from Nebraska and Montana to New Mexico, mountains of Arizona, and southern borders of California. The forms referable to *H. luteus* are more white-tomentose, have shorter and more crowded lobes to the leaves, and southward have almost scapiform stems. Northeastern forms are greener, more leafy, and with smaller heads, approaching *H. tenuifolius*.

* * * Lobes of the honey-colored or yellow corolla much shorter than the throat: akenes broad, the faces almost destitute of nerves: pappus obsolete or wanting: root perennial: fl. July-Oct.

H. Mexicánus, GRAY. Densely floccose-tomentose, sometimes denudate in age, a foot or two high from a thick root or caudex: radical leaves from lanceolate to spatulate, and from entire to pinnately parted, the lobes entire; upper cauline leaves linear or lanceolate, often entire: heads few or several and loosely corymbose-paniculate, 4 lines high: bracts of the involucre oval or ovate, green with yellowish tips: akenes slightly pubescent and glabrate. — Proc. Am. Acad. xix. 29. *H. flavescens*, var.? Gray, Pl. Wright. ii. 94. — Mountain ravines, New Mexico, *Wright, Greene, Rusby*. (Mountains near San Luis Potosi, Mexico, these certainly perennial, *Schaffner*.)

149. FLORESTÍNA, Cass. (Probably dedicated to a female friend.) — Slender annuals (of Mexico and its northern borders), leafy-stemmed, loosely paniculately branched, pubescent and above beset with stipitate glands: all but the lowest leaves alternate, petiolate, simply palmately or pedately divided into entire segments, rarely entire: heads loosely paniculate, quarter-inch high: flowers white or flesh-color, in summer. — Bull. Philom. 1815, & Dict. xvii. 155, t. 86; DC. Prodr. v. 655, excl. spec. — Consists of the Mexican *F. pedata*, Cass., and the following.

F. trípteris, DC. l. c. Lowest leaves commonly ovate or oblong and entire; others of 3 oval or oblong or the upper linear leaflets: tips of involucral bracts and flowers usually dull white: anther-tips acutish. — Gray, Pl. Wright. i. 121. — S. Texas; first coll. by *Berlandier*. (Adj. Mex.)

150. POLÝPTERIS, Nutt. (Πολύς, many, and πτέρις, meant for πτερόν, wing; many-winged or feathered, i. e. the pappus.) — Southeastern N. American herbs (entering Mexico). more or less scabrous-pubescent; with undivided and mostly entire petiolate leaves, all or the upper alternate, and loosely corymbose-cymose or paniculate and pedunculate heads of rose-purple or flesh-colored flowers, in summer and autumn. — Gen. ii. 139; Ell. Sk. ii. 314 (not of DC., which was a *Gaillardia*); Nutt. Trans. Am. Phil. Soc. ser. 2, 377; Gray, Proc. Am. Acad. xix. 30. Part of *Palafoxia*, Less., DC., &c.

§ 1. Heads homogamous, middle-sized or small: bracts of the involucre herbaceous up to the small sphacelate colored tip: corollas 5-parted nearly down to the slender tube: akenes narrowly obpyramidal: root annual. (Nearest to *Florestina*.)

P. callósa, Gray, l. c. Slender, paniculately branched, a foot or two high: leaves linear, slightly petioled: peduncles glandular: involucre turbinate, 10–12-flowered, quarter-inch high, of 8 or 10 linear-oblong bracts: akenes minutely pubescent or glabrous: paleæ of the pappus all short, obovate or roundish, with costate-thickened centre seldom reaching the obtuse or erose and retuse apex, occasionally minute or wholly wanting. — *Stevia callosa*, Nutt. Jour. Acad. Philad. ii. 121; Bart. Fl Am. Sept. t. 46. *Florestina callosa*, DC. Prodr. v. 655. *Palafoxia callosa*, Torr. & Gray, Fl. ii. 369. — Low or dry ground, Arkansas to Texas: first coll. by *Nuttall*.

P. Texána, Gray, l. c. Stouter: leaves from lanceolate-linear to lanceolate-oblong (at least below), distinctly petioled: peduncles less glandular: involucre campanulate or broader, 20–30-flowered, 3 to 5 lines high, of 8 to 12 spatulate-oblong bracts: paleæ of the pappus from oblong-ovate to oblong-lanceolate, with slender nearly complete or slightly excurrent costa, sometimes almost as long as the akene, in the outer flowers often much shorter. — *Palafoxia Texana*, DC. Prodr. v. 124; Torr. & Gray, Fl. l. c. — River-banks, Texas; first coll. by *Berlandier*. (Adj. Mex.)

§ 2. Heads heterogamous, larger, with palmately 3-lobed rays: disk-corollas parted not quite to the filiform tube: bracts of the involucre herbaceous up to the small and narrow sphacelate colored tip: akenes slender: root annual.

P. Hookeriána, Gray, l. c. Stouter, 1 to 4 feet high, above glandular-pubescent and somewhat viscid: leaves from narrowly to broadly lanceolate, mostly 3-nerved below: involucre many-flowered, broad, half-inch or more high, of 12 to 16 lanceolate bracts in two series, the outer looser and often wholly herbaceous, inner with purplish tips: ray-flowers 8 to 10, the deeply 3-cleft rose-red rays half-inch long, but sometimes reduced or abortive: pappus in the ray a crown of 6 to 8 short and obtuse rather rigid spatulate paleæ; in the disk of narrowly lanceolate thin paleæ, traversed by an excurrent costa, attenuate at apex into a slender point or short awn, nearly of the length of the akene. — *Stevia sphacelata* (Nutt.). Torr. in Ann. Lyc. N. Y. ii. 214. *Palafoxia Texana*, Hook. Ic. Pl. t. 148, not DC. *P. Hookeriana*, Torr. & Gray, Fl. ii. 368; Hook. f. Bot. Mag. t. 5549, with var. *subradiata*, a reduced state. — Sandy plains of Nebraska to Texas. (Adj. Mex.)

§ 3. Heads homogamous, rather large: corollas with the base of 5-parted limb forming a short-campanulate throat: involucre more imbricated and whitish-scarious, glabrous: akenes slender: root perennial. — *Polypteris*, Nutt.

P. integrifólia, Nutt. Not glandular: stems 2 to 5 feet high, fastigiately corymbose at summit, almost glabrous: leaves scabrous, lanceolate and obtuse, upper ones linear, lowest spatulate-oblong to obovate: heads fully half-inch high, many-flowered: principal bracts of the involucre obovate-spatulate, very obtuse, thin, mainly whitish, some outer or accessory bracts narrower and shorter, partly herbaceous: corollas white or flesh-color: paleæ of the pappus little shorter than the akenes, linear-lanceolate, gradually attenuate, more or less pointed by the excurrent tip of the strong costa. — Gen. ii. 139; Ell. Sk. ii. 314, not DC. *Paleolaria fastigiata*, Less. Syn. 156. *Palafoxia fastigiata*, DC. Prodr. v. 125. *P. integrifolia*, Torr. & Gray, Fl. ii. 269. — Pine barrens, Georgia and Florida; first coll. by *Dr. Baldwin*.

151. **PALAFÓXIA**, Lag. (*José Pulafox*, noted Spanish general.) — Herbaceous or suffruticose plants (of Mexico and the U. S. borders); with branching stems, rather large scattered or loosely cymosely disposed pedunculate heads of flesh-colored or whitish flowers; the leaves linear to oblong, alternate, entire, the lower short-petioled. — Nov. Gen. & Spec. 26 (Elench. Hort. Madr. 1815); Gray, Proc. Am. Acad. xix. 31. *Palafoxia* in part, Less., DC., Benth. & Hook.

P. LATIFÓLIA, DC. Prodr. v. 125, of Southern Mexico, is unknown to us, and by its opposite cordate leaves and obovate involucral bracts is probably of some other genus.

* Anomalous species, connecting with *Polypteris*.

P. Feáyi, GRAY. A foot or two high, suffruticose at base, very leafy to near the summit, minutely scabrous· leaves short (little over inch long), oblong or ovate-oblong and rounded at both ends, or uppermost lanceolate and acutish, thickish, 3-nerved at base: heads corymbosely cymose, over half-inch high: involucre campanulate, about half the length of the flowers; its bracts spatulate-linear, at apex truncate-obtuse and somewhat purplish-sphacelate: corollas with oblong lobes fully half the length of the cylindraceous throat: pappus shorter than the corolla-tube and several times shorter than the glabrate akene, of 8 oblong rigid pointless lacerately scarious-edged paleæ (comparable with those of some outermost flowers of the following).— Proc. Am. Acad. xii. 59, xix. 31.— Coast of S. Florida, in sandy soil, *Feay, Chapman, Curtiss*, no. 1507.

* * Genuine species, with narrow and paniculately scattered heads, narrowly linear involucral bracts, these in age usually concave and applied to the subtended akenes.

P. lineáris, LAG. l. c. Flowering as an annual, but becoming perennial and frutescent, strigose-cinereous and partly hirsute or hispid, slender flowering branches sometimes glanduliferous: leaves linear, or lower ones lanceolate, more or less canescent: heads about inch long, 15-30-flowered (or by depauperation 10-12-flowered)· corolla-lobes oblong-linear, half the length of the throat: pappus of 4 (sometimes 5) linear hyaline paleæ with strong and rigid excurrent costa, and little shorter than the slender akenes, or sometimes 2 to 4 additional and shorter blunt ones, or in the outer flowers all reduced, short, and of firmer texture, with imperfect costa, or abortive.— DC. Prodr. v. 124; Hook. Bot. Mag. t. 2132. *Ageratum lineare*, Cav. Ic. iii. 3, t. 205. *Paleolaria carnea*, Cass. Bull. Philom. 1816, & Dict. *P. leucophylla*, Gray, Proc. Am. Acad. viii. 291, & Bot. Calif. i. 388, a shrubby form with reduced pappus, from seeds of which were raised plants having nearly the ordinary pappus of the species, which, although flowering as an herb with seemingly annual root along the Mexican border, was originally described as shrubby.— On the Colorado near Fort Yuma, &c., S. California, and Arizona. (Mex.)

152. **RIGIOPÁPPUS**, Gray. (From ῥίγιος, stiffened, and πάππος, pappus.)— Proc. Am. Acad. vi. 548, Bot. Calif. i. 387; Benth. & Hook. Gen. ii. 406. — Single but variable species.

R. leptócladus, GRAY, l. c. Slender annual, a span to a foot high, minutely hirsute-pubescent to almost glabrous, paniculately or subcorymbosely branched: branches commonly filiform, elongated, and leafless below, smooth, simple or proliferous, bearing solitary heads: leaves all alternate, very narrowly linear, sessile, erect, entire, those of the branches near the heads small and subulate: involucre 3 lines high; flowers yellow but often changing to purple or whitish: paleæ rather than awns of the pappus from half to two-thirds the length of the akene, 3 to 5, occasionally only 2 or 1, or rarely wanting. — Dry ground, interior region of Washington Terr. to the middle of California and Nevada; first coll. by *Lyall*.

Var. longiaristátus. A small form: involucre only 2 lines high: pappus of (mostly 3) more slender awns, subulate-dilated at base, much longer than the corolla, rather longer than the akene. — Rattlesnake Bar, California, *Mrs. Curran.*

153. CHÆNÁCTIS, DC. (χαίνω, to gape, and ἀκτίς, ray, the enlarging orifice and limb of the marginal corollas in most species simulating a kind of ray.) — Herbaceous or rarely suffrutescent (Western N. American); with alternate mostly pinnately dissected leaves, and pedunculate solitary or sometimes cymosely disposed heads of yellow, white, or flesh-colored flowers. Pappus more commonly shorter or of fewer paleæ in the outer flowers. Akenes pubescent, rarely glabrate. — Prodr. v. 659; Benth. & Hook. Gen. ii. 401; Gray, Proc. Am. Acad. vi. 545, x. 73.

§ 1. CHÆNACTIS proper. Pappus of entire or merely erose persistent paleæ, rarely obsolete: akenes more or less tetragonal or terete, slender.

  * Corollas yellow, the marginal ones with enlarged throat and limb, somewhat unequally or as if palmately 5-lobed: annuals, mostly winter annuals, flowering in spring.
  + Pappus of 4 (rarely if ever "5 or 6") nearly equal narrowly oblong or oblong-lanceolate acutish paleæ, at least the inner attaining to the throat of the corolla.

C. lanósa, DC. Floccosely white-woolly when young, flowering from near the base with (3 to 8 inches) long naked peduncles, the earliest scapiform: leaves thickish, simply pinnately parted into a few narrowly linear (rarely again parted) lobes no wider than the rhachis, or uppermost entire: heads half-inch high: involucral bracts nearly linear: marginal flowers moderately ampliate, not surpassing the disk. — Prodr. l. c.; Torr. & Gray, Fl. ii. 370; Gray, Bot. Calif. ii. 389. — California, common from Monterey southward to San Bernardino, &c.

C. glabriúscula, DC. Taller, stouter, more caulescent, a foot or more high, thinly floccose, at length denudate, branching above, and with stout sometimes elongated peduncles bearing solitary heads of two-thirds to three-fourths inch high: leaves with more numerous and irregular lobes: bracts of the involucre broader, thickish, glabrate, obtuse: marginal corollas with much ampliate and more palmate limb, surpassing the disk. — Prodr. l. c.; Gray, l. c. *C. denudata*, Nutt. Pl. Gamb. 177. The var. *megacephala*, Gray, Pacif. R. Rep. iv. 104, is merely a larger form. — California, from valley of the Sacramento southward.

C. tenuifólia, NUTT. Somewhat white-tomentulose when young, glabrate, loosely branched, often diffuse, bearing scattered or paniculately disposed heads (a third of an inch high) on short slender peduncles: leaves once or twice pinnately parted into irregular and small linear or oblong or sometimes nearly filiform lobes: involucral bracts narrow, rather rigid: limb of marginal corollas short, not surpassing the disk. — Trans. Am. Phil. Soc. l. c. 375; Torr. & Gray, l. c.; Gray, Bot. Calif. l. c. *C. filifolia*, Gray, Pl. Fendl. 98, the most slender-leaved form. On the sea-shore occurs an opposite extreme, with primary divisions of the leaves pinnatifid into very short and thickish lobes. — Coast of California, from Santa Barbara to San Diego; also San Bernardino.

  + + Pappus of very obtuse mostly unequal paleæ, or obsolete.

C. heterocárpha, GRAY. Lightly floccose, soon denudate, a span or two high, simple or sparingly branched: leaves pinnately or sometimes bipinnately parted into irregular and unequal rather crowded and short divisions and lobes: heads half-inch high, mostly on rather long peduncles terminating stem and branches: bracts of the involucre broadly linear or sometimes wider: limb of the marginal flowers conspicuously ampliate, surpassing the disk: pappus of inner flowers of 4 elliptical-oblong paleæ fully half the length of the corolla, and with 4 or fewer alternate outer and roundish very short ones, but these occasionally wanting; in the outermost flowers all shorter or very short. — Pl. Fendl. 98. & Bot. Calif. l. c. Var. *tanacetifolia*, Gray, l. c. (*C. tanacetifolia*, Gray, Proc. Am. Acad. vi. 543), proves to be only a stunted and condensed form. — California, from the Upper Sacramento and Lake Co. to San Bernardino Co.; first coll. by *Hartweg.*

C. Névii, GRAY, Proc. Am. Acad. xix. 30. Dwarf, rather stout, puberulent, or leaves nearly glabrous: peduncles short: marginal corollas little ampliate: pappus of a few minute dentiform vestiges: otherwise resembles the preceding, so far as an insufficient specimen shows. — Coll. in Idaho, 1876, *Nevius*.

* * Corollas white or pale flesh-colored.

+— Marginal ones with throat and limb manifestly enlarged, and unequally 5-lobed or even palmately ligulate: bracts of the involucre linear, obtuse or acutish: pappus of 4 paleæ: winter annuals.

C. Fremónti, GRAY, l. c. Glabrate, the slight woolliness caducous, or glabrous, except the puberulent or hispidulous peduncles, a foot or less high, rather stout: leaves thickish, narrowly linear, many entire, some with 2 to 5 similar linear lobes: heads half or two-thirds inch high, terminating rather simple erect branches: bracts of the involucre thickish, rather acute, with prominent midrib: marginal corollas comparatively large and conspicuous, ligulately palmate, not rarely developing a cuneate almost equally 4-5-cleft ligule (of 3 lines in length): paleæ of the pappus linear-lanceolate, nearly equalling disk-corolla, with manifestly thickened axis at base forming a vanishing costa. — Desert of the Mohave and Lower Colorado, California, and adjacent Nevada and Arizona, *Fremont* (imperfect specimen), *Newberry*, *Parish*, *Lemmon*, &c. Partly confounded in Bot. Calif. with the next.

C. stevioídes, HOOK. & ARN. Floccose-tomentose, glabrate in age, seldom a foot high, freely and loosely branched, bearing numerous somewhat cymosely disposed heads (of half-inch in height) on short slender peduncles: leaves 1-2-pinnately parted into short linear lobes, uppermost rarely entire: bracts of involucre narrowly linear, obtuse, with obscure midrib: marginal corollas with moderately ampliate unequally 5-lobed limb, not surpassing the disk: paleæ of the pappus scarcely thickened at base, those of the inner flowers oblong-lanceolate and shorter than the corolla, of the outer ones ovate or oblong, often unequal, sometimes much shorter. — Bot. Beech. 353; Torr. & Gray, Fl. ii. 371; Eaton in Bot. King Exp. 172. — Dry interior region, Utah and S. Idaho, to eastern side of Sierra Nevada and through Arizona; first coll. by *Tolmie*.

C. brachypáppa, GRAY. Resembles the preceding: leaves perhaps thicker: heads broader: involucral bracts with prominent midrib: paleæ of pappus alike in inner and outermost flowers, quadrate or slightly cuneate, very truncate, not longer than the short proper corolla-tube, barely one fourth the length of the akene. — Proc. Am. Acad. viii. 390, & Bot. Calif. i. 389. — S. E. Nevada, in the Pahranagat Mountains, *Miss Searls*.

+— +— Marginal corollas little enlarged, nearly regular: receptacle commonly with a few fimbrillæ or bracts in the form of setiform awns: bracts of the receptacle very narrowly linear, cuspidately or setaceously acuminate: pappus of 4 paleæ: winter annuals, minutely puberulent, with no woolliness.

C. carphoclínia, GRAY. A foot or less high, diffusely much branched, slender, bearing numerous scattered heads (barely half-inch high) on short filiform peduncles: leaves 1-2-pinnately parted into almost filiform lobes: involucre 30-40-flowered: awns on the receptacle 5 to 10 among and nearly equalling the flowers, rigid, persistent: paleæ of the pappus ovate-lanceolate, acute or acuminate, and little or moderately shorter than the inner corollas, or in the outer much shorter, occasionally very short. — Bot. Mex. Bound. 94, & Bot. Calif. l. c. — Arid districts, W. Arizona and S. Utah to S. E. California; first coll. by *Gen. Thomas*.

C. attenuáta, GRAY. More slender, with narrow 15-20-flowered heads: ray-corollas hardly at all enlarged: hardly any fimbrillæ on the receptacle: paleæ of the pappus very short, broadly obovate-cuneate and truncate: otherwise nearly like the preceding. — Proc. Am. Acad. x. 73, & Bot. Calif. l. c. — Ehrenberg, Arizona, *Janvier*, through *Canby*.

+— +— +— Marginal corollas not larger than the others (or only slightly so in *C. Xantiana*), receptacle quite naked: involucral bracts pointless, narrowly linear, rather loose, the midrib obvious: pappus of 4 conspicuous paleæ and usually 2 to 4 small alternating outer ones: leaves simply pinnately parted, with divisions entire or merely 1-2-toothed: winter annuals.

C. Xantiána, GRAY. Stout, often a foot or more high, tomentulose when young, some glabrate: ascending simple branches terminated by large (three-fourths to inch long) solitary many-flowered heads on thick often fistulous peduncles: leaves with a few narrowly linear distant lobes, or some entire: corollas with short oval or oblong lobes a little bearded externally, or in the margin rather broader and more spreading, but equal: anthers partly exserted (in the manner of the genus): pappus of 4 lanceolate paleæ little shorter than the corolla, and of as many very short obovate or obcordate ones. — Proc. Am. Acad. vi. 545,

x. 74, & Bot. Calif. i. 390, with var. *integrifolia*, which is more slender, fewer-flowered, and usually entire-leaved. *C. glabriuscula*, var. *megacephala*, Gray, Jour. Bot. Nat. Hist. Soc., vii. 146, not Pacif. R. Rep. — Eastern California and adjacent Nevada, from Tejon to Carson, &c., *Dr. Horn, Anderson, Lemmon*.

C. macrántha, EATON. A span high, rather simply branched from the base, canescently tomentulose, partly glabrate: leaves short, with linear or oblong-linear lobes usually approximate: heads 12–20-flowered, mostly short-peduncled, or the earlier on longer naked peduncles from near the base of the stem: bracts of the involucre thinnish, more or less tomentose: corollas half to three-fourths inch long, narrow, externally puberulent, all alike; the 5 short teeth linear-oblong, ascending or barely spreading: anthers wholly included in the throat, the tips lanceolate: pappus of 4 linear-oblong paleæ barely half the length of the corolla, and 2 to 4 very short cuneate-oblong ones, but these occasionally obsolete or wanting. — Bot. King Exp. 171, t. 18: Gray, Bot. Calif. l. c. — Hills in the desert region, W. Nevada to S. Utah and the Mohave in California; first coll. by *Watson*.

+ + + + Marginal corollas not distinctly larger than nor different from the others (the lobes if slightly larger still regular): bracts of many-flowered involucre linear or somewhat spatulate, obtuse, sometimes one or two loose and shorter outer ones: pappus of 8 to 14 mostly equal and large obtuse paleæ: biennial, perennial, or suffrutescent plants: fl. summer. — *Macrocarphus*, Nutt.

C. Douglásii, HOOK. & ARN. Canescent with a fine somewhat floccose or pannose tomentum, or sometimes early glabrate, a span to a foot or more high from a biennial or more enduring root: leaves mostly of broad outline and bipinnately parted into crowded short and very obtuse divisions and lobes: heads from half to three-fourths inch long, in larger plants several or numerous and corymbosely cymose: paleæ of the pappus from linear-ligulate to narrowly oblong and from half to three-fourths the length of the corolla, or in marginal flowers shorter and broader. — Gray, Proc. Am. Acad. x. 74, & Bot. Calif. l. c. *C. Douglasii* & *C. achilleæfolia*, Hook. & Arn. Bot. Beech. 354; Torr. & Gray, Fl. l. c.; Torr. in Stansb. Rep. t. 6. *Hymenopappus Douglasii*, Hook. Fl. i. 316; DC. Prodr. v. 658: *Macrocarphus Douglasii* & *M. achilleæfolius*, Nutt. Trans. Am. Phil. Soc. n. ser. vii. 376. — Dry plains and mountains, Montana to New Mexico, west to Washington Terr. and California. From S. E. California, *Palmer*, an incomplete specimen of a peculiar large and glabrate form, with sparser divisions to the leaves, and shorter spatulate-oblong paleæ of pappus. Very variable species.

Var. **alpina**. Dwarf, 3 to 5 inches high, consisting of a rosette or thick tuft of leaves with very approximate divisions, and naked or scapiform stems, bearing mostly solitary heads, surmounting the subterranean branches of a multicipital perennial caudex or rootstock. — Alpine region of the Rocky and Cascade Mountains in Colorado and Wyoming, of the Sierra Nevada, California, and north to Washington Terr. Seems distinct from the following.

C. Nevadénsis, GRAY. Very dwarf, in small tufts surmounting filiform branches of subterranean rootstocks, mostly growing in volcanic scoriæ or ashes: leaves small (half to barely inch long), densely white-woolly, crowded, obovate or flabelliform-cuneate in outline, once or twice pinnatifid or parted into obovate or spatulate-linear lobes: peduncles inch or less long, bearing a solitary rather narrow head. — Bot. Calif., 390. *Hymenopappus Nevadensis*, Kellogg, Proc. Calif. Acad. v 46. — Alpine region of the Sierra Nevada, California, from Shasta and Lassen to the sources of the San Joaquin, *Kellogg, Muir, Lemmon*, &c.

C. santolinoídes, GREENE, in herb. Subcanlescent perennial: leaves all crowded on short tufted shoots from a slightly ligneous crown, white-tomentose, linear in outline, with broadish rhachis thickly beset with small (line or so long) oblong obtusely few-lobed and crispate divisions: peduncles scapiform, 4 to 6 inches high, simple or once or twice forked, glandular and viscid: head half-inch high, rather narrow: pappus of 8 or 10 linear-ligulate paleæ, a little shorter than the corolla. — San Bernardino Mountains, above Bear Valley, S. E. California, *Parish*.

C. suffrutéscens, GRAY. Canescently tomentose, a foot or more high from decumbent woody stems: leaves pinnately parted into 5 to 7 narrowly linear entire or rarely 1–2-toothed divisions: heads solitary or scattered, on slender peduncles, three-fourths inch high: pappus of 10 to 13 linear or narrowly ligulate-oblong paleæ a little shorter than the corolla, or in the outermost flowers considerably shorter. — Proc. Am. Acad. xvi. 100. — California, on the rocky banks of the Sacramento, below Mount Shasta, *Lemmon* (perhaps a mistake as to habitat); S. E. California, south of San Jacinto Mountains, *Parish.*

§ 2. ACARPHÆA. Pappus of deciduous and fimbriate paleæ, or wanting: akenes obovate- or linear-clavate, hardly angled, blackish: involucre viscid: corollas whitish or ochroleucous, all alike or nearly so, the marginal not obviously ampliate: annuals. — *Acarphæa*, Gray, Pl. Fendl. 98; characterized anew in Proc. Am. Acad. xix. 30.

C. artemisiæfólia, GRAY. A foot or two high, paniculately branched, furfuraceous-pubescent, somewhat viscid, above glandular-hirsute, especially the naked summit and peduncles and involucre of the loosely cymose-paniculate heads: leaves 2-3-pinnately divided or parted into short linear or oblong lobes: involucre broadly campanulate, half-inch high, many-flowered; its bracts lanceolate-linear, acute: akenes linear-clavate, flattened, hardly at all angled, the sides minutely impressed-striate; epigynous disk small and obscurely annulate. — Proc. Am. Acad. x. 74, & Bot. Calif. l. c. *Acarphæa artemisiæfolia*, Gray, Pl. Fendl. 98, & Bot. Mex. Bound. 95, t. 32. — San Diego Co., California; first coll. by *Coulter.*

C. thysanocárpha, GRAY. Slender and low annual, paniculately branched, viscid-puberulent, with some early deciduous villosity, sparsely leafy up to the subsessile small heads: leaves narrowly linear, entire: involucre barely 3 lines high, of few linear-oblong and viscidulous bracts, 7-10-flowered: akenes clavate-obovate, obscurely angled: pappus about half the length of the corolla, of 8 or 9 nearly equal thin spatulate paleæ which are crosely fimbriate quite down to their unguiculate base, deciduous. — Proc. Am. Acad. xix. 30. — Sierra Nevada in Kern Co.? California, at 9,800 feet, *Rothrock*, no. 345. Apparently depauperate or unseasonable specimens of a peculiar plant; coll. Sept.

154. HÚLSEA, Torr. & Gray. (The late *Dr. G. W. Hulse*, U. S. Army.) — Herbs, of the Sierra Nevada and its continuations, viscid-pubescent and balsamic-scented, most of the species when young floccose-woolly; with alternate mostly sessile entire or dentate or pinnatifid leaves, and solitary or scattered large heads of yellow flowers, or rays sometimes purple; in summer. — Bot. Mex. Bound. 98; Pacif. R. Rep. vi. 77, t. 13; Bot. Calif. i. 385.

* More or less floccose-woolly when young, and denudate in age: upper leaves reduced in size and bract-like on the naked flowering branches or peduncles: root perennial, or in the first species perhaps biennial.

H. Califórnica, TORR. & GRAY, l. c. Robust, 2 feet or more high, leafy, bearing several paniculately disposed heads, when young whitened by long and soft loose wool: leaves entire or nearly so; lower spatulate or lingulate, uppermost ovato-lanceolate to linear: involucre two-thirds inch high and broad; its bracts very numerous, linear, gradually acute, villose-lanate: rays very many, with linear ligule half-inch long: paleæ of the pappus quadrate-oblong and somewhat equal, or the two over the principal angles longer, crose-denticulate at summit. — Gray, Bot. Calif. i. 386. — S. California, in mountains of San Diego Co., *Parry*, and (near Campo, June, 1880), *Parish, G. R. Vasey.*

H. vestíta, GRAY. Commonly a foot or less high from a rosette of panuosely white-tomentose spatulate leaves (either entire or lyrately dentate, tardily somewhat denudate); the flowering stems sometimes scapiform and monocephalous, commonly sparsely leaved below and bearing two or three slender pedunculate heads: involucre half-inch high, of mostly broadly lanceolate viscid-pubescent bracts: rays little surpassing the disk-flowers, sometimes shorter, or even wanting, yellow or changing to reddish: pappus of conspicuous and silvery quadrate crose-toothed paleæ, either nearly equal or two rather longer. — Proc. Am. Acad. vi. 547, & Bot. Calif. i. 387. (Forms have been distributed under the names of *H. Parryi*, Gray, and *H. callicarpha*, S. Watson.) — S. E. California; volcanic hill south of Mono Lake, *Bigelow*, low, scapiform, with large head. San Jacinto Mountain, San Diego Co., 1882, *Parish*. Mohave country, San Bernardino Co., *Parry*, 1876, form with dentate or almost pinnatifid leaves. Also a more leafy and branched form, 2 feet high, with more deciduous wool and rather longer rays, *Parish.*

Var. pygmǽa. Depressed, rising only 2 inches high, the head subsessile in the tuft of leaves: rays saffron or rose-colored. — San Bernardino Co., on the summit of Greyback Mountain, *Lemmon, W. G. Wright*, and Bear Valley, *Parish*.

H. álgida, GRAY. A span or two high from a deep perennial rootstock, the villous or cottony wool caducous, viscid pubescence remaining: stem simple, stout, terminated by a solitary short-peduncled large head: leaves linear-lingulate, irregularly dentate, sometimes with large salient teeth; lower crowded (2 to 5 inches long, quarter to half inch wide), upper gradually smaller and sparser: involucre almost inch high and broad; its bracts linear, attenuate-acute, lax, villose-lanate and viscid: rays very numerous, linear, nearly half-inch long, yellow: pappus short, not exceeding the breadth of the akene, equalled by its hairs; the paleæ deeply fimbriate-lacerate. — Proc. Am. Acad. vi. 547, Bot. Calif. i. 386. — California, on the higher summits of the Sierra Nevada, from Mount Dana southward, *Brewer, Bolander, Muir*, on Mount Whitney up to 13,700 feet, *Rothrock*.

H. nána, GRAY. A span high from long branching rootstocks rising through volcanic ashes and scoriæ, villous-lanate when young, viscid-pubescent: leaves crowded around base of the thickish (inch or two long, or sometimes very short) monocephalous peduncle, oblong-spatulate, pinnatifid or incised, mostly tapering into a margined petiole: involucre half-inch or more high, of lanceolate bracts: rays about 30, yellow, broadly linear, nearly half-inch long: paleæ of the pappus (either broad or apparently splitting into narrower ones) usually longer than the breadth of the akene, equalled by its villous hairs, incisely or fimbriately lacerate. — Pacif. R. Rep. vi. 76. t. 13, Bot. Calif. l. c. — Volcanic peaks of the Cascade Mountains, Oregon, *Newberry, Cusick*, to Washington Terr., *Suksdorf*.

Var. Lárseni, GRAY, Bot. Calif. l. c. More woolly even in age, and leaves somewhat scattered on the flowering stems, even up to the head: rays smaller. — California, in volcanic ashes on peaks of northern part of the Sierra Nevada, such as Shasta and Lassen; first coll. by *Lemmon* and *Larsen*.

* * Apparently quite destitute of floccose wool from the first, but with some long and soft many-jointed and viscidulous hairs: stems mostly simple, equably leafy to the top, bearing solitary or somewhat racemosely disposed short-pedunculate heads: paleæ of the pappus conspicuous, oblong or narrower, the two over the angles longer.

H. heterochróma, GRAY. Rather stout, sometimes over 2 feet high from an annual root: leaves oblong, saliently dentate: involucre two-thirds or three-fourths inch high, of linear-lanceolate attenuate-acute bracts: rays very numerous, 3 or 4 lines long, rose-purple, sometimes inconspicuous or obsolete: tube of disk-corollas hirsute: shorter paleæ of the pappus truncate-lacerate. — Proc. Am. Acad. vii. 369, & Bot. Calif. l. c. — California, from the Yosemite, *Bolander*, to the mountains of San Bernardino Co., *Lemmon, Parish*.

H. brevifólia, GRAY, l. c. Slender, a foot high from an annual or possibly perennial root, the stem or simple branches bearing a solitary comparatively small and narrow head: leaves small (the largest inch and a half long), spatulate-oblong, denticulate: involucre half-inch high, of linear rather loose bracts: rays only 10 or 12, 3 or 4 lines long, light yellow: paleæ of the pappus rather entire. — California, along the Merced in and near the Yosemite Valley, *Bolander*, &c.

155. TRICHOPTÍLIUM, Gray. (Θρίξ, τριχός, hair, and πτίλον, feather or plumage, the pappus-paleæ feathery-dissected.) — Single species, yellow-flowered winter annual; fl. spring.

T. incísum, GRAY. Diffusely branched, low and spreading, loosely floccose-woolly, also somewhat pubescent and glandular: leaves oblong-rhomboidal or cuneate-lanceolate, incisely and acutely dentate, alternate or the lower opposite: heads scarcely half-inch high, on slender peduncles terminating stem and branches. — Bot. Mex. Bound. 97, Pacif. R. Rep. v. t. 5, & Bot. Calif. i. 395. *Psathyrotes incisa*, Gray, Pl. Thurb. 322. — Arid district of the Mohave, Lower Colorado, and Gila, W. Arizona and S. E. California; first coll. by *Fremont*.

156. BLENNOSPÉRMA, Less. (βλέννα, mucus, σπέρμα, seed; the akenes developing copious mucus when wetted; that is, the club-shaped papillæ then swell up through imbibition, open at the apex, or else split into two valves, and emit a pair of uncoiling filaments of extreme tenuity, in the manner of

Crocidium, to which this anomalous genus is perhaps most related.) — Low and small annuals, of two species, one Chilian, the other Californian. — Less. Syn. 267; Torr. & Gray, Fl. ii. 272; Remy in Gay, Fl. Chil. iv. t. 48; Benth. & Hook. Gen. ii. 404. *Apalus*, DC. Prodr. v. 507. *Coniothele*, DC. l. c. 531.

— B. Califórnicum, Torr. & Gray, l. c. A span or two high, at length diffusely branched, glabrous or nearly so, with pedunculate heads terminating the branches: leaves alternate, pinnately parted into narrowly linear usually entire lobes; heads a third to half inch in diameter when expanded: flowers pale yellow, with ligules 2 or 3 lines long, or the alternate ones sometimes destitute of corolla: disk-flowers shorter than the involucre: style-branches of fertile flowers broad. — *Coniothele Californica*, DC. Prodr. v. 531. — Moist ground, Upper Sacramento to San Diego, California: fl. summer; first coll. by *Douglas*.

157. ACTINÉLLA, Pers., Nutt. (Changed from *Actinea*, from ἀκτίς, ray.) — Low herbaceous or rarely suffruticose plants (all American); the herbage usually impressed-punctate and often resinous-atomiferous, bitter-aromatic, generally Chamomile-scented; leaves all alternate and narrow or with narrow lobes; the heads of yellow flowers commonly slender-pedunculate. — Pers. Syn. ii. 469 (*Actinea*, Juss. Ann. Mus. Par. ii. 425, t. 61, a S. American form, somewhat approaching *Helenium*, but not to be combined with *Cephalophora*, which is a reduced rayless *Helenium*); Nutt. Gen. ii. 173, & Trans. Am. Phil. Soc. n. ser. vii. 378; Torr. & Gray, Fl. ii. 381; Gray, Proc. Am. Acad. xix. 31. *Hymenoxys*, Cass. Dict. lv. 278; DC. Prodr. v. 661. *Actinella, Hymenoxys*, and a part of *Cephalophora*, Benth. & Hook. Gen. ii. 413–415.

A. (Plateiléma) Pálmeri, Gray, Proc. Am. Acad. xviii. 109, xix. 31, is an outlying species, of Northern Mexico, remarkable for its few and broad and nearly herbaceous involucral bracts, convex receptacle, and truncate laciniate paleæ of the pappus.

§ 1. Euactinélla. Involucre of numerous herbaceous or nearly membranaceous (not rigid) nearly equal and similar bracts, distinct to the base: receptacle obtusely conical or hemispherical: heads mostly solitary on long or scapiform peduncles, rarely sessile in the cluster of leaves: rays inclined to persist and turn pale: akenes silky-villous: pappus of 5 to 7 hyaline paleæ. — Gray, l. c.

* Winter annual or at most biennial, caulescent, entire-leaved: receptacle conical.

— A. linearifólia, Torr. & Gray. Slender, a span to a foot high, sometimes strict and nearly simple, generally diffusely branched, villous-pubescent and glabrate: leaves linear or the lowest somewhat spatulate: peduncles filiform, a span long; head 3 lines high; rays 4 lines long: paleæ of the pappus ovate, abruptly acuminate-awned. — Fl. ii. 383. *Hymenoxys linearifolia*, Hook. Ic. t. 146; DC. Prodr. vii. 243. — Texas and borders of Louisiana, in sandy soil; first coll. by *Drummond*. (Adj. Mex.)

* * Perennials, mostly with multicipital caudex, commonly lanate in the axils of the radical leaves.

+ Leaves except in one form of the first species quite entire, all on the crowns of the caudex, which bear a simple scapiform peduncle (or none): receptacle obtusely or low conical: involucre villous-lanate: paleæ of the pappus hyaline from broadly ovate to oblong, mostly traversed by an indistinct costa, and usually produced at apex into an awn: well-formed heads 4 to 6 lines high, and rays as long.

— A. scapósa, Nutt. In the typical form somewhat like the preceding in aspect, especially when leafy along the base of the scape, loosely villous and glabrate, rather sparsely cæspitose, the branches of the caudex being slender and often ascending: leaves linear to lanceolate or some of the earlier ones spatulate, not rarely laciniate-lobed: scape a span to a foot high. — Trans. Am. Phil. Soc. l. c.; Torr. & Gray, Fl. ii. 382. *Cephalophora (Actinella) scaposa*, DC. Prodr. v. 663. *Guillardia Raymeriana*, Scheele in Linn. xxii. 161. *A. lanuginosa*, Buckley in Proc. Acad. Philad. 1861, 459. — Rocky prairies, &c., Texas to New Mexico; first coll. by *Berlandier*. (Adj. Mex.)

—— **Var. lineáris**, NUTT. l. c. Leaves all narrowly linear and entire, more rigid. — Texas to New Mexico, and the borders of Colorado: broader-leaved and dwarfer forms very like glabrate *A. acaulis*. (Adj. Mex.)

— — **A. acaúlis**, NUTT. Densely cespitose, the branches of the caudex short, thick, and crowded, canescently villous or sericeous, sometimes more naked: leaves thickish, all entire, from spatulate to nearly linear, commonly short (half-inch to 2 inches long), densely crowded on the caudex: scape half-inch to 6 inches high: rays 3 to 5 inches long (rarely wanting). — Gen. ii. 173; Torr. & Gray, l. c.; Eaton in Bot. King Exp. 174. *A. lanata*, Nutt. Trans. Am. Phil. Soc. l. c.; Torr. & Gray, l. c., a loosely villous form. *Galardia acaulis*, Pursh, Fl. ii. 743. *Cephalophora (Actinella) acaulis*, DC. l. c. — Rocky Mountains and the bordering plains and hills, Dakota to Montana, and south to New Mexico, W. Nevada, and Arizona. Passes into

Var. glábra, GRAY, Man. ed. 5, 363. Leaves green, spatulate-linear, from sparingly villous or glabrate to nearly glabrous, even to the base and axils. — *A. glabra* & *A. Torreyana*, Nutt. l. c.; Torr. & Gray, Fl. ii. 382. — Rocky hills and bluffs, Wyoming Terr. to New Mexico and Utah. Also on an ancient mound at Joliet, Illinois, *Scammon*, *W. Boott*, probably adventive.

A. depréssa, TORR. & GRAY. Pulvinate-cespitose: leaves densely crowded on the very thick dense branches of the caudex, spatulate-linear, half-inch long. either sericeous-canescent or glabrate: head strictly sessile, immersed among the long-villous bases of the leaves. — Pl. Fendl. 100, with var. *pygmœa*, a diminutive silky-canescent form. — Mountains of W. Colorado or E. Utah, *Fremont, Ward*, and the small variety, Raton Mountains, *Gordon*. Perhaps a state of *A. acaulis*.

+ + Leaves all quite entire, crowded on the caudex, also scattered along the simple or sparingly branched stems: peduncles slender: heads, &c., of the preceding subdivision.

A. argéntea, GRAY. Commonly rather stout, a span to a foot high, silvery-canescent with appressed silky pubescence: lower leaves spatulate and oblanceolate, uppermost linear: heads 4 or 5 lines high and rays 5 or 6 lines long, but sometimes of less than half this size: paleæ of the pappus 5, from broadly ovate or obovate to oblong, with manifest costa produced into an awn which usually about equals the disk-corolla. — Pl. Fendl. 100; Rothrock in Wheeler Rep. vi. 173. — Hills of New Mexico; first coll. by *Fendler*.

A. leptóclada, GRAY. A span or two high, more slender, sparsely and more loosely silky-villous, glabrate, the narrower (sometimes all narrow-linear) leaves and lower part of the stems not rarely glabrous: heads usually smaller than of the foregoing. — Pacif. R. Rep. iv. 107. — New Mexico and S. W. Colorado to Arizona? *Bigelow, Newberry, Brandegee*, &c.

+ + + Leaves mostly parted or dissected into narrow linear lobes, crowded on the thick comparatively simple caudex and scattered on the short flowering stems: heads large; bracts of the involucre herbaceous but very woolly, loose: receptacle hemispherical: paleæ of the pappus 5 or 6, elongated-lanceolate, attenuate into a subulate but hardly awned point, somewhat shorter than the disk-corolla.

A. Brandegéi, T. C. PORTER. Leaves glabrate, with 2 or 3 lobes toward the upper part, or some entire, narrowly linear, only 2 or 3 on the somewhat scapiform simple flowering stem (of a span or more in height): head therefore conspicuously pedunculate, half-inch high and wide: involucral bracts lanceolate: rays 12 to 16, 3 or 4 lines long. — Gray, Proc. Am. Acad. xiii. 373. *A. grandiflora*, var. *glabrata*, Porter & Coulter, Fl. Colorad. 76. — Alpine region of the Sangre de Christo and adjacent mountains of S. Colorado, *Parry* (1867), *Brandegee*, *Gray & Hooker*.

A. grandiflóra, TORR. & GRAY. A span or two high, very stout, floccose-woolly, tardily somewhat glabrate in age: stem simple or branching below, leafy: leaves with petiole scarious-dilated at base, lower ones 2–3-ternately or quinately parted, upper with 3 to 5 simple lobes: involucre about an inch broad, very woolly; its bracts linear: rays 30 or more, over half-inch long: plants generally growing singly and the caudex on a perpendicular root, as if biennial. — Bost. Jour. Nat. Hist. Soc. v. 110; Gray, Am. Jour. Sci. xxxiii. 240. — Alpine region of the Rocky Mountains from Montana to Colorado; first coll. by *Fremont*.

A. CHRYSANTHEMOÍDES and A. INSIGNIS, Gray, Proc. Am. Acad. xix. 32. of Mexico (large and tall species, with much divided leaves, and at most biennial roots), rank next to *A. grandiflora*.

§ 2. Hymenoxys. Involucre double or of two distinct series of coriaceous or rigid erect bracts, the outer more commonly connate at base: leafy-stemmed herbs; ours all with ray-flowers. — Gray, Pl. Fendl. 101, & Proc. Am. Acad. xiii. 373, & xix. 32. *Hymenoxys*, Cass. Dict. lv. 278 (founded on a rayless South Amer. species; DC. Prodr v. 661; Benth. & Hook. Gen. ii. 415.

    * Simple-stemmed and monocephalous or nearly so, perennial: pappus-paleæ elongated and pointed.

A. Bigelóvii, Gray. Habit of *A. leptoclada*, loosely woolly, tardily glabrate: stems strict and slender, 6 to 15 inches high from an apparently multicipital caudex, terminated by a single long-pedunculate head, rarely with one or two branches: leaves all very narrowly linear, rigid, not perceptibly punctate, some of them with a pair of subulate lobes, the others quite entire: head half-inch high: involucre hemispherical; its bracts all lanceolate, acute, coriaceous, about 12 in each series, distinct; those of the inner a little longer, scarious-margined, and attenuate-cuspidate: rays about 12, half-inch long: receptacle obtusely low-conical: paleæ of the pappus about 10, subulate-lanceolate, nearly equalling disk-corolla, with more or less evident costa, gradually attenuate into an aristiform cusp. — Pl. Wright. ii. 97, & Bot. Mex. Bound. 99. — Pine forests in the mountains of New Mexico, *Bigelow*, *Newberry*, *Palmer*, *Greene*.

    * * Stems branching above and bearing numerous or several heads.

    + Most of the leaves entire, some 3-cleft: pappus of about 5 broad and truncate paleæ.

A. Rúsbyi, Gray. Greene and glabrous or nearly so: stems a foot or more high from a lignescent perhaps biennial root, strict, fastigiately branched at summit into a cyme of many small (3 lines high) heads: leaves rigid, linear; upper cauline all entire; lower and the long and narrow radical ones some entire, some 3-cleft: outer involucre shorter than inner, of 7 or 8 thickish subulate-lanceolate coriaceous bracts, connate only at very base: ligules 2 or 3 lines long, quadrate: paleæ of the pappus rather firm, quadrate or broadly cuneate and truncate, without costa, not surpassing the proper tube of the disk-corolla. — Proc. Am. Acad. xix. 33. — Grassy slopes of the Mogollon Mountains, New Mexico, *Rusby*.

    + + Leaves all or mostly 1-3-ternately parted.

    ++ Paleæ of the pappus oblong or roundish, pointless, mostly thin and hyaline: heads rather large: outer involucre more or less cupulate: rays laciniate at apex: root apparently perennial.

A. Vaséyi, Gray. Nearly glabrous, a foot high: leaves once or twice 3-parted into linear lobes: heads rather numerous and fastigiately cymose: involucre narrowly campanulate, 4 lines high; outer nearly equalling the inner, united high up into a 7-9-toothed cup: paleæ of the pappus oblong or broadly lanceolate, more or less obtuse, about half the length of the disk-corolla. — Proc. Am. Acad. xvii. 219, xix. 33. — Organ Mountains, New Mexico, *G. R. Vasey*.

A. Coóperi, Gray. Puberulent, 2 feet or more high, paniculately branched above and with more scattered heads: lower leaves twice ternately or quinately and the upper simply 3-5-nately parted into (mostly inch long) nearly filiform lobes of hardly more width than the rhachis: involucre almost hemispherical; outer of 6 to 10 bracts which are united only toward the base: paleæ of the pappus ovate or quadrate-oblong, with very obtuse erose summit, not half the length of the disk-corolla. — Proc. Am. Acad. vii. 394, xix. 33.

    ++ ++ Paleæ of the pappus ovate to lanceolate, attenuate into a slender point or awn, fully half the length of disk-corolla.

A. biénnis, Gray. Mostly stout and a foot or more high from a tap root, probably never more than biennial, cinereous-puberulent, sometimes more hoary, sometimes green and glabrate: leaves simply 3-5-parted into narrow linear lobes: heads loosely cymose, hemispherical: involucre about 4 lines high; bracts of its outer series 12 to 14, plane or barely carinate-thickened at base, nearly distinct: rays about as many, half to full inch long when well developed, narrowly cuneate: paleæ of the pappus ovate-lanceolate, from acuminate to cuspidate. — Proc. Am. Acad. xiii. 373; also part of *A. Richardsonii*, Bot. Calif. i. 394. &c. *A. Richardsonii*, var. *canescens*, Eaton, Bot. King Exp. 175, a hoary and dwarf form. — Arid mountain districts of Utah, Nevada, and on the borders of California and Arizona, *Watson*, *Ward*, *Lemmon*, *Palmer*, &c.

A. Richardsónii, NUTT. A span to a foot high, in tufts from a multicipital perennial caudex, obscurely puberulent or nearly glabrous, woolly in the axils of radical leaves, fastigiately cymose, polycephalous: upper leaves mostly once and lower twice ternately parted into long and simple filiform-linear lobes, rather rigid: involucre campanulate, 2 or 3 lines high, 6–9-angled; the 6 to 9 bracts of the outer strongly carinate, united for the lower quarter or third: rays broadly or sometimes narrowly cuneate, 2 to 4 lines long: paleæ of the pappus attenuate-acuminate. — Trans. Am. Phil. Soc. vii. 379; Torr. & Gray, l. c.; Gray, Pl. Fendl. 101, with var. *floribunda*, a tall and full-flowered form. *Picradenia Richardsonii,* Hook. Fl. i. 317, t. 103. — Plains, Saskatchewan and E. Oregon to Utah and New Mexico.

A. odoráta, GRAY. Diffuse and at length much branched from an annual root, a span to 2 feet high, with scattered small heads terminating leafy branches: leaves once to thrice ternately parted into filiform lobes, not rigid: involucre campanulate, rigid; outer of 7 or 8 oblong bracts, united at base: paleæ of the pappus aristately attenuate. — Pl. Fendl. 101, Pl. Wright. i. 122, & Proc. Am. Acad. xix. 33. *Hymenoxys odorata*, DC. Prodr. v. 661; Deless Ic. iv. t. 42. *Philozera multiflora,* Buckley, Proc. Acad. Philad. 1861, 459. — Open ground, Texas to S. California; also sparingly in Kansas, where it is probably naturalized. (Mex.)

158. HELÉNIUM, L. SNEEZE-WEED.

(Ancient Greek name of Elecampane, or some other plant, which was said to be named after the wise *Helenus,* son of Priam.) — N. American and Mexican herbs, erect, mediocre or tall; with alternate simple leaves, which are sometimes decurrent, commonly resinous-atomiferous (therefore bitter-aromatic) and impressed-punctate, and with pedunculate heads of usually yellow or occasionally brownish-tinged flowers, produced in summer or autumn. — DC. Prodr. v. 667; Benth. & Hook. Gen. ii. 413, with the synonymy (except *Amblyolepis,* and adding *Hecubæa*); Gray, Proc. Am. Acad. ix. 202. *Helenium. Leptopoda* (Nutt.), and *Hecubæa,* DC. Prodr., to which *Cephalophora* (Cass.), § 1, DC., should be added.

§ 1. OXYLEPIS. Rays fertile, numerous, long and narrow: disk-corollas with moderately long proper tube: pappus of elongated paleæ: bracts of the involucre numerous in two series, tardily reflexed in fruit: leaves not decurrent on the stem. — Gray, Proc. Am. Acad. ix. 205. *Dugaldea,* Cass. *Oxylepis.* Benth. Pl. Hartw. 87.

H. Hoopésii, GRAY. Slightly tomentose or pubescent when young, soon glabrate: stem stout, 1 to 3 feet high, from a strong perennial root, leafy, bearing several or sometimes solitary large heads: leaves thickish, entire, oblong-lanceolate, or the lower spatulate with long tapering base, somewhat nervose: rays becoming inch long, tardily reflexed: disk half to three-fourths inch high, hemispherical: receptacle in fruit ovoid-hemispherical: paleæ of the pappus ovate-lanceolate, long attenuate-acuminate, a little shorter than the corolla. — Proc. Acad. Philad. 1863, 65, Proc. Am. Acad. ix. 205, & Bot. Calif. i. 392. — Rocky Mountains, Montana to New Mexico, Arizona, and Sierra Nevada, California; first coll. by *Thomas Hoopes.*

§ 2. EUHELÉNIUM. Rays fertile (rarely sterile, occasionally wanting), with cuneate or oblong soon drooping rays: disk-corollas with proper tube very short or reduced to a mere ring: paleæ of the pappus not dissected: involucre comparatively simple and small, of slender linear or subulate often unequal bracts, soon reflexed: plants from glabrous to puberulent, leafy-stemmed, mostly branching. — Gray, l. c.

* Root annual: leaves all filiform-linear, not decurrent on the stem or branches.

H. tenuifólium, NUTT. Glabrous, slender, fastigiately much branched, very leafy up to the slender peduncles: leaves mostly entire: rays often half-inch long, much surpassing the globular disk: receptacle depressed-hemispherical (a line and a half in diameter): paleæ of the pappus ovate, abruptly tipped with a longer awn which equals the villous akene and is little shorter than the disk-corolla. — Jour. Acad. Philad. vii. 66 ; Hook. Comp. Bot. Mag. i. 98; Torr. & Gray, Fl. ii. 385; Meehan, Native Fl. ii. t. 10. — River bottoms, &c., Arkansas to Mississippi, Florida, and Texas: becoming a naturalized weed throughout Southern Atlantic States.

—— **Var. bádium**, Gray. Disk dull purplish brown (instead of yellow): lower leaves sometimes pinnately parted, the radical into short linear or even somewhat oblong entire or few-toothed lobes. — Proc. Am. Acad. xviii. 108. — Texas, *E. Hall, Reverchon, Palmer.*

* * Root annual, or at most biennial: leaves broader, at least some of them decurrent and forming wings on the stem and branches: rays in several species occasionally particolored with brownish-red.

+ Paleæ of the pappus obtuse or at least pointless, destitute of costa.

++ Rays present: disk and receptacle in fruit elongated.

—— **H. quadridentátum**, Labill. Loosely paniculate: lower leaves incisely pinnatifid; upper lanceolate, entire: heads with oval disk becoming oblong, half-inch long, surpassing the rays; receptacle cylindraceous-oblong; disk-corollas more commonly 4-toothed: pappus of very short roundish-oval paleæ. — Act. Soc. Nat. Hist. Par. i. 22, t. 4; Lam. Ill. t. 688; Lindl. Bot. Reg. t. 598 ; DC. Prodr. v. 666. *H. quadripartitum*, Link, Enum. ii. 338 ? *Rudbeckia alata*, Jacq. Ic. Rar. t. 593. *Tetrodus quadridentatus*, Cass. Dict. lv. 272. — Low ground, Carolina to Texas, near the coast; adventive in ballast-heaps to Philadelphia. (Mex.)

++ ++ Rays present: disk globular.

H. élegans, DC. Strict, slender: leaves narrowly lanceolate and entire, or lowermost broader and sometimes slightly toothed; heads of the smallest (2 or 3 lines high), with brownish or purplish disk, equalled or surpassed by the pure yellow or particolored or sometimes largely brownish-purple rays: receptacle barely hemispherical: pappus minute, the roundish-ovate paleæ decidedly shorter than the breadth of the akene. — Prodr. v. 667. *H. microcephalum*, var. *bicolor*, Torr. & Gray, Fl. ii. 385. *H. microcephalum*, Curtiss, distrib. 1513. — Moist ground, W. Louisiana and Texas; first coll. by *Berlandier.* (Adj. Mex.)

H. microcéphalum, DC. Freely branching: leaves lanceolate or oblong, the lower denticulate or repand-toothed : heads with yellow or fuscous disk (3 or 4 lines high) much surpassing or sometimes equalled by the rays: receptacle conical-ovate : paleæ of the pappus ovate, short, but nearly half the length of the akene. — Prodr. v. 667 ; Gray, Proc. Am. Acad. ix. 202, in part. *H. heterophyllum*, DC. l. c. as to Berland. 2113 from Reynosa, not of char. *H. Texanum*, Buckley in Proc. Acad. Philad. 1861, 460, to which aristate pappus is wrongly assigned. — Moist ground, Texas; first coll. by *Berlandier.* (Adj. Mex.)

H. amphíbolum, Gray. Stouter, freely branching: upper leaves lanceolate to linear and entire; lower varying to oblong and toothed or laciniate-pinnatifid: heads with fuscous-purplish globose disk (3 or 4 lines in diameter), equalled or surpassed by the yellow rays: receptacle more than hemispherical: paleæ of the pappus roundish and very small, as in *H. elegans.* — Proc. Am. Acad. ix. 202. *H. Mexicanum*, DC. Prodr. v. 666, by the char., not HBK. — Southern borders of Texas, on the Rio Grande, at Presidio and Eagle Pass, *Havard.* (Adj. Mex.)

H. ooclínium, Gray. Freely branching, rather stout: leaves lanceolate, usually more or less dentate or denticulate: heads with yellowish and fuscous ovate-globose disk (3 or 6 or rarely 4 lines high), longer than the yellow rays: receptacle ovoid-conical (acutish or obtuse): paleæ of the pappus comparatively large, ovate, obtuse, often almost the length of the akene, sometimes with traces of a costa or of a mucro. — Proc. Am. Acad. ix. 202. — Southern borders of Texas, along the Rio Grande, at Presidio and (with smaller heads) at Eagle Pass, *Havard.* (Northern Mex.)

++ ++ ++ Rayless.

H. Thúrberi, Gray. Slender, puberulent, freely branched, 2 or 3 feet high: leaves mostly linear-lanceolate, entire, the lowest broader and denticulate or rarely laciniate: heads globose-ovoid, 3 or 4 lines high, fuscous: receptacle relatively large, broadly ovate: pappus of ovate obtuse paleæ, about one third the length of the corolla and of the akene. — Proc. Am. Acad. xix. 32. — S. Arizona, *Coulter* (359, distributed as of California), *Thurber* (wrongly referred to *H. puberulum* in Bot. Mex. Bound.), *Pringle, Lemmon.*

╋ ╋ Paleæ of the pappus acuminate, mucronately cuspidate, or awned, the costa commonly manifest: heads with globose disk and semi- or sub-globose receptacle: herbage puberulent.

— **H. Mexicánum**, HBK. (*H. varium*, Schrader), by some said to be perennial, has paleæ of the pappus from apiculate to aristellate-acuminate. To it may belong Coulter's no. 357 (specimen too incomplete), ticketed "California," but probably belonging to his Mexican collection.

H. pubérulum, DC. Mostly tall, freely branching, and with long monocephalous peduncles: leaves lanceolate or the lower broader, all entire: heads about half-inch in diameter: rays one, two, or sometimes three lines long, equalling or exceeding the small involucre, rarely obsolete: paleæ of the pappus ovate, short-awned, not half the length of the corolla. — Prodr. v. 667; Torr. & Gray, Fl. ii. 385; Gray, Bot. Calif. i. 393. *H. pubescens*, Hook. & Arn. Bot. Beech. 355, not Ait. *H. Californicum*, Link. Ind. Sem. Berol. 1840? *H. decurrens*, Vatké, Ind. Sem. Berol. 1875. *H. Mexicanum*, Gray, Pacif. R. Rep. iv. 107, probably. *Cephalophora decurrens*, Less. in Linn. vi. 517; DC. Prodr. v. 663. — Moist or wet ground, California, common.

H. laciniátum, Gray. A foot or two high, more cinereous: leaves lanceolate or linear, pinnatifid-dentate or laciniate, or the upper entire: heads 4 or 5 lines in diameter: rays as in the preceding: involucre commonly more conspicuous: paleæ of the pappus more than half the length of the corolla.— Proc. Am. Acad. ix. 203, & Bot. Calif. l. c.— S. E. California and adjacent Arizona, *Coulter*, &c. (Adj. Mex.)

* * * Root perennial: rays sterile, either neutral or with abortive style and akene: ligules equalling or exceeding the globular disk: receptacle ovate: leaves mostly narrowly decurrent on the stem and branches: paleæ of the pappus aristate-acuminate, hardly half the length of the disk-corolla: heads on short slender peduncles.

H. nudiflórum, Nutt. Somewhat puberulent, 1 to 3 feet high, with leafy branches and corymbosely disposed heads: leaves from narrowly lanceolate to oblong, entire, or the radical obovate or spatulate and dentate: rays half to three-fourths inch long, either pure yellow or partly (sometimes wholly) brown-purple, once or twice the length of the brownish or purplish disk: receptacle ovate, in age acutish, but sometimes rounder and very obtuse. — Gray, Proc. Am. Acad. ix. 203. excl. syn. *H. parviflorum*. *H. nudiflorum* & *H. micranthum*, Nutt. Trans. Am. Phil. Soc. l. c. 384. *H. quadridentatum*, Hook. Comp. Bot. Mag. i. 98. *H. atropurpureum*, Kunth, Ind. Sem. Berol. 1845, 21, purple-rayed state. *H. Seminariense*, Featherman in Louisiana Univ. Rep. 1871. *Leptopoda brachypoda*, Torr. & Gray, Fl. ii. 388; Curtiss. distrib., a very slender and small-rayed form. — Low ground, N. Carolina and Illinois to Arkansas and Texas; and naturalized eastward. Hybridizes with *H. autumnale*.

H. parviflórum, Nutt. l. c. Glabrate or glabrous, much branched and with scattered small heads: leaves broadly lanceolate, with contracted base, sparingly denticulate, very narrowly decurrent on the branches: disk and rays yellow, the former 3 or 4 lines in diameter; the latter 3 to 5 lines long. styliferous: receptacle short-ovate. — Georgia, *Nuttall* (a specimen named by him is ticketed Alabama); in a swamp near Macon, *J. Donnell Smith*. Seemingly quite distinct. Simple-stemmed and low specimens with larger heads, Delaware Co., Penn., verge rather to *H. autumnale*.

* * * * Root perennial: rays fertile and conspicuous: stem or branches more or less winged by the decurrent leaves: receptacle from half to two-thirds spherical: pappus with the paleæ acuminate-aristate, not rarely somewhat lacerate or with one or two setiform teeth.

╋ Heads corymbose at summit of very leafy stem and branches; the disk globose: leaves mostly serrate or denticulate: flowering late.

H. autumnále, L. Nearly glabrous or minutely pubescent: stem narrowly winged, 2 to 6 feet high: leaves lanceolate to ovate-oblong: heads about half-inch in diameter, usually equalled by the rays: pappus commonly half or two-thirds the length of disk-corolla. — Spec. ii. 866; Lam. Ill. t. 688; Schknhr, Handb. t. 230; Bart. Fl. Am. Sept. t. 26; Hook. Bot. Mag. t. 2994; Torr. & Gray, Fl. ii. 384. *H. longifolium*, Smith in Rees Cycl. *H. pumilum*, Willd. Enum. Suppl. 60, may be a common dwarf form. *H. pubescens*, Ait. Kew. iii. 287. *H. canaliculatum*, Lam. Journ. Hist. Nat. ii. 213, t. 35, & *H. tubuliflorum*, DC. Prodr. v. 666, a state with tubulose ligules. *H. altissimum* & *H. comiantatum*, Link. Ind. Sem. Berol. 1840. *H. grandiflorum*, Nutt. Trans. Am. Phil. Soc. vii. 384, larger-flowered form. *H. montanum*, Nutt. l. c. — Wet ground, Canada to Georgia, Texas, and westward to Brit. Columbia and Arizona; the var. *grandiflorum*, with rays three-fourths inch long, only in the northwest.

╋ ╋ Heads solitary or few on long (sometimes foot long) peduncles, terminating the stem or lax branches; disk depressed-globose or almost hemispherical: leaves en ire.

H. Bigelóvii, GRAY. Almost glabrous: stem simple or loosely branched, 2 or 3 feet high: leaves from narrowly to oblong-lanceolate, the radical oblong-spatulate, elongated: peduncles mostly slender: disk of the head three-fourths globose at maturity, two-thirds to three-fourths inch in diameter, equalled by the rays: paleæ of the pappus ovate-lanceolate or subulate and awn-pointed, considerably shorter than the corolla. — Pacif. R. Rep. iv. 107, Proc. Am. Acad. ix. 205, & Bot. Calif. i. 393. — Wet ground, California, not rare from Lake Co. to San Bernardino Co.; first coll. by *Bigelow*.

H. Bolánderi, GRAY. Somewhat furfuraccous-pubescent: stems stout, often simple, 1 or 2 feet high: leaves oblong to ovate-lanceolate, or lowest obovate: peduncles thick, commonly upwardly enlarged and fistulous: disk of the head decidedly broader than high, inch or more wide: rays often inch long: paleæ of pappus lanceolate or subulate, with slender awn, almost equalling disk-corolla. — Proc. Am. Acad. vii. 358, ix. 204, & Bot. Calif. l. c. — Low grounds near the coast, northeastern part of California, *Bolander*, &c.

§ 3. LEPTÓPODA. Rays neutral, very numerous, mostly surpassing the linear bracts of the involucre, cuneate, 3–5-cleft, yellow, as are also mostly the flowers of the broad disk: pappus of thin-scarious wholly nerveless sometimes lacerate or fimbriate paleæ: simple-stemmed perennials (sometimes biennials?), from slender or filiform rootstocks; virgate stem continued into an unusually long solitary peduncle, the apex of which is mostly turbinate-thickened under the large and broad head: leaves narrowly or not at all decurrent. — Gray, Proc. Am. Acad. ix. 204. *Leptopoda*, Nutt. Gen. ii. 174; Ell. Sk. ii. 445; Torr. & Gray, Fl. ii. 386, excl. § 2.

* Receptacle ovate-conical and the disk semi- to sub-globose: involucre soon reflexed and the rays (over half-inch long) drooping in the manner of *Euhelenium*: nearly glabrous, with somewhat elongated-lanceolate mostly entire cauline leaves, but no conspicuous radical tuft: akenes more or less hairy on the ribs.

H. Curtísii, GRAY. Stem slender, 2 or 3 feet high: disk of the head half-inch in diameter, surpassed by the rays: paleæ of the pappus almost entire, obovate, muticous, about one third the length of the disk-corolla. — Proc. Am. Acad. l. c. *Leptopoda integrifolia,* M. A. Curtis in Torr. & Gray, Fl. ii. 387, under *L. brevifolia,* var. — Near Raleigh, N. Carolina, *M. A. Curtis*.

H. fimbriátum, GRAY, l. c. Less slender: disk two-thirds or three-fourths inch broad, equalled by the rays: paleæ of the pappus broad, dissected from summit to beyond the middle into many capillary bristles. — *Gaillardia fimbriata,* Michx. Fl. ii. 142. *L ptopoda fimbriata,* Torr. & Gray, Fl. l. c. — Low pine barrens, Florida and Texas.

* * Receptacle and disk depressed-hemispherical or flatter: involucre and rays merely horizontal or tardily recurved: flowering stem usually from a rosulate cluster of radical leaves: cauline leaves gradually diminished upward, the uppermost usually bracteiform and subulate, all somewhat fleshy.

H. Nuttállii, GRAY, l. c. A foot or more high, with nearly the foliage of the preceding and head of the following: ovary and akene glabrous and glandular-atomiferous: paleæ of the pappus oval or oblong, more or less erose or lacerate, muticous, or some of them aristellate. — *Leptopoda Helenium,* Nutt. Gen. ii. 174, excl. syn.? *L. denticulata,* Nutt. Trans. Am. Phil. Soc. vii. 372. *L. decurrens,* Macbride in Ell. Sk. ii. 446, form with denticulate leaves. — Damp ground, S. Carolina to Florida and Louisiana.

Var. **incisum,** GRAY, l. c. Leaves incised or sinuate-pinnatifid in the manner of the following. — *Leptopoda incisa,* Torr. & Gray, Fl. ii. 387. — Georgia, *Le Conte*.

H. vernále, WALT. Somewhat puberulent or tomentulose and viscidulous, a foot or two high: principal leaves in a radical tuft, spatulate-lanceolate or narrower, 4 to 6 inches long, from repand-denticulate to incisely pinnatifid: upper cauline small, linear-subulate and bract-like: disk of the head two-thirds or three-fourths inch broad, yellow: akenes pubescent: paleæ of the pappus obovate or spatulate, with lacerate or fimbriate-toothed summit. — Walt. Car. 210; Gray, Proc. Am. Acad. ix. 205. *Leptopoda puberula,* Macbride in Ell.

l. c.; Torr. & Gray, l. c., with var. *pinnatifida. L. pinnatifida*, Schweinitz; Nutt. Trans. Am. Phil. Soc. l. c. — Pine barren swamps. N. Carolina to Florida.

H. brevifólium, GRAY, l. c. More glabrous: leaves shorter and entire or nearly so, lower and radical spatulate: head smaller, with brownish or purplish disk: akenes pubescent: paleæ of the pappus nearly entire. — *Leptopoda brevifolia*, Nutt. Trans. Am. Phil. Soc. l. c.; Torr. & Gray, l. c., excl. var. — Pine barren swamps, N. Carolina to Alabama.

159. **AMBLYÓLEPIS**, DC. (Composed of ἀμβλύς. blunt, and λεπίς. scale; from the pappus.) — Prodr. v. 667. — Single species, exhaling the odor of Melilot in drying: fl. all summer.

A. setígera, DC. l. c. Annual, a foot or so high, sometimes glabrous and very smooth, sometimes villous with very long hairs rising from minute papillæ, especially along the margins of the leaves: stem loosely branching below, terminated by long monocephalous peduncles: leaves membranaceous, bright green, entire; radical oblong-spatulate with long tapering base: cauline oblong or ovate, with rounded or subcordate half-clasping base and mucronate-acuminate tip: head large: flowers all golden yellow: rays almost inch long. 3–4-lobed: paleæ of the pappus 5, about half the length of the akene, broadly ovate, silvery-scarious, entire and nerveless, very obtuse, or in some outer flowers short-acuminate. — Gray, Pl. Wright. i. 121. — Prairies of Texas; first coll. by *Berlandier*. (Adj. Mex., *Palmer*.)

160. **GAILLÁRDIA**, Fougéroux. (*M. Gaillard de Merentonneau*.) — N. American herbs (and one extra-trop. S. Amer.), chiefly of the Atlantic side; with alternate sometimes resinous-atomiferous and impressed-punctate leaves, and ample and showy Scabious-like heads on terminal or sometimes scapiform peduncles: the flowers often fragrant, yellow or reddish-purple; in summer. — Mem. Acad. Sci. Par. 1786, 5. t. 1, 2; DC. Prodr. v. 651; J. Gay in Ann. Sci. Nat. ser. 2. xii. 56. *Galardia*, Lam. Dict. ii. (1786), 590, & Ill. t. 708; Michx. Fl. ii. 142; Nutt. Gen. ii. 175. *Calonea*, Buchoz. Ic. (1786). t. 126, ex DC. *Virgilia*, L'Her. & Smith, not Lam. *Guntheria*. Spreng. Syst. iii. 356.

§ 1. Style-branches tipped with short (in ours naked) appendage of only once to thrice the length of the penicillate tuft: lobes of disk-corolla short and obtuse: rays sometimes fertile, often none: akenes villous all over: winter annuals or at most biennials. — *Guntheria*, Spreng. Syst. iii. 356, 449. and *Cercostylis*, Less. Syn. 239; an extra-tropical S. American species. *Agassizia*, Gray & Engelm. Proc. Am. Acad. i. 50, & Jour. Bot. Nat. Hist. vi. 229.

G. COMÓSA, Gray, Proc. Am. Acad. xviii. 109, xix. 34, of Coahuila, Mexico, is a third species of this section: it has truly fertile rays, exceedingly long hairs to the akene which nearly cover the short-awned pappus and at length almost equal the disk-corolla, and very short soft fimbrillæ to the receptacle; the head on a naked scape.

G. símplex, SCHEELE. Leaves all in a radical cluster, or a few near the base of the simple (foot or two long) monocephalous scape, commonly spatulate, from pinnatifid to coarsely dentate or some entire: head globose in fruit: involucre of about 2 series of short and narrow bracts: flowers heliotrope-scented: rays none or imperfect and irregular and styliferous, or but few fully developed and neutral: villous hairs of the akene little surpassing the base of the large paleæ of the pappus, these 6 to 11, their slender awns at length surpassing disk-corolla. — Scheele in Linn. xxii. 160. *G. tuberculata*, Scheele, l. c. 349, is apparently the subcaulescent and more radiate form. *Agassizia suavis*, Gray & Engelm. l. c. — Rocky prairies of Texas; first coll. by *Lindheimer* and *Wright*.

§ 2. Style-branches tipped with a long hispid or hispidulous filiform appendage: rays neutral, in first species sometimes wanting. — *Gaillardia*, Foug.. DC., &c.

* More or less pubescent or hirsute with many-jointed hairs, leafy-stemmed: leaves not coriaceous: bracts of the involucre (at least the outer and larger) mainly foliaceous and spreading, lanceolate or narrower: disk-flowers apt to turn brown or dark-purple: villous hairs covering the akene mainly at its base or below the broad summit: paleæ of the pappus slender-awned.

+ Fimbrillæ of the receptacle obsolete or reduced to very short soft teeth: corolla-lobes caudately acuminate from a short broadish base.

G. lanceoláta, Michx. Minutely or somewhat cinereously pubescent, not hirsute, about 2 feet high from an annual or perhaps perennial root, virgately branched: leaves rather small, from spatulate-lanceolate to linear, entire or slightly or sparsely serrate: outer bracts of the involucre lax and herbaceous to the base: rays rather few and sparse, half or two-thirds inch long, 3-cleft into narrow lobes and with slender tapering base, sometimes obsolete or wanting: flowers sweet-scented: the disk commonly dark and the rays yellow or copper-colored with dark veins. — Fl. ii. 142; Gray, l. c.; Torr. & Gray, Fl. ii. 365, not DC. *Galardia bicolor*, Pursh, Fl. ii. 572 (excl. syn. and it cannot be *Lysera Caroliniana*, Walt.); Nutt. l. c.; Ell. Sk. ii. 449. *Polypteris integrifolia*, DC. Prodr. v. 659, excl. syn. — Dry pine barrens, S. Carolina to Florida, Kansas, and Texas.

+ + Fimbrillæ of the receptacle setiform or subulate-aristiform, mostly surpassing the akenes.

++ Lobes or teeth of disk-corolla subulate-acute and usually tipped with a seta or cusp, externally beset with long and beaded hairs: rays usually numerous and when well developed contiguous or overlapping, short-cuneate at base: pappus aristate even in the ray-flowers: bracts of involucre callous at base, more or less hirsute, as also the herbage.

G. aristáta, Pursh. Perennial; often 2 feet or more high: leaves of firm texture, lanceolate or broader, or lower spatulate, from entire to laciniate-dentate or sinuate-pinnatifid: rays all yellow, in the largest heads inch and a half long: setiform fimbrillæ sometimes little shorter than disk-corollas. — Fl. ii. 573; Lindl. Bot. Reg. t. 1186; Hook. Bot. Mag. t. 2940, & Fl. i. 315; DC. Prodr. v. 652; Gray, l. c.; Torr. & Gray, l. c. *G. bicolor*, Hook. Fl. l. c., excl. syn. *G. bicolor*, var. *aristata*, Nutt. Gen. ii. 175; Hook. Bot. Mag. sub t. 3368. *G. rustica*, Cass. Dict. xviii. 20; Desf. Cat. *G. lanceolata*, DC. l. c., excl. syn. — Plains, Saskatchewan to Brit. Columbia and Oregon, south to S. Colorado, New Mexico, and even the borders of California?

G. pulchélla, Foug. l. c. Annual, a foot or less high, diffusely branched at base: leaves softer, from entire to pinnatifid: rays two-colored, lower part red-purple or darker, the upper or teeth yellow, at most inch long: lobes of disk-corolla more attenuate: fimbrillæ rather stouter, hardly surpassing the mature akenes. — Cass. Dict. xviii. 19; DC. l. c.; Gay, l. c.; Torr. & Gray, l. c. *G. bicolor*, Lam. Dict. ii. 506, & Ill. t. 708; Sims, Bot. Mag. t. 1602 (as to figure); Hook. Bot. Mag. t. 3551 (var. *Drummondii, integerrima*). *G. Drummondii*, DC. Prodr. v. 652. *Virgilia helioides*, L'Her.; Smith. Exot. Bot. i. t. 37. — Plains, Louisiana, Arkansas, and Texas to Arizona. (Adj. Mex.)

Var. picta. Form with somewhat succulent leaves, when growing near the sea-shore: fimbrillæ of the receptacle shorter and stouter, more or less subulate. — *G. bicolor*, var. *Drummondii*, Hook. Bot. Mag. t. 3368. *G. picta*, Don, Brit. Fl. Gard. ser. 2, t. 267; Gay, l. c. — Texas, in low grounds. Common in gardens.

++ ++ Teeth of the disk-corolla short and broad, obtuse, pointless or obscurely apiculate: involucre more or less callous at base.

= Akenes destitute of villous hairs (glabrous or glabrate) at the upper part, and not overtopped by the basal villi: fimbrillæ of the receptacle setiform, equalling or surpassing the akenes: leaves undivided.

G. amblýodon, Gay. Annual, a foot or two high, leafy to the top, mostly hirsute: leaves oblong or the lower spatulate, all sessile by an auriculate base, denticulate or the upper entire; bracts of the involucre hirsute-ciliate, outer with conspicuous erect callous base; rays numerous and contiguous, oblong-cuneate, throughout brownish red or maroon-color, an inch or less long: ray-pappus awnless. — Ann. Sci. Nat. l. c.; Torr. & Gray, Fl. ii. 367; Gray, Chloris Am. Bor. (Mem. Am. Acad. iii.) 32, t. 4; Meehan, Nat. Fl. ii. t. 46. — Sandy prairies of Texas; first coll. by *Drummond*.

G. Mexicána, Gray. A foot or less high from a perennial root, with the habit of *G. lanceolata*, minutely pubescent, naked above, with long rather rigid peduncles: leaves lanceolate, rather small, entire, or the lowest cauline and radical sparingly dentate or laciniate: head rather small (disk barely half-inch in diameter): rays rather sparse and narrow, half-inch or less long, yellow and brownish: teeth of disk-corolla oblong: akenes with rather short and scanty villosity, surpassed by the numerous setiform fimbrillæ. — Proc. Am. Acad. xix. 34. *G. pulchella*, var., Gray, Pl. Wright. i. 120, along with plants mentioned as *G. lanceolata*. — Hills of the Rio Frio, S. W. Texas, *Wright*. (Adj. Mex. to San Luis.)

= = Akenes densely long-villous all over: fimbrillæ subulate-setaceous: rays yellow: peduncles scapiform or from short leafy stems, 5 to 10 inches long: some or even all the leaves pinnatifid, but very variable.

G. pinnatífida, Torr. Perennial, cinereous-pubescent: leaves sometimes linear or with linear lobes, sometimes spatulate and sinuate or even entire: pappus-paleæ lanceolate. — Ann. Lyc. N. Y. ii. 214; Torr. & Gray, l. c. — Plains, W. Texas to Colorado and Arizona; first coll. by *James*. (Adj. Mex.)

G. Arizónica. Annual, greener: leaves less frequently pinnatifid and with only oblong lobes: pappus-paleæ obovate-oblong, very obtuse or retuse. — High plains of S. Utah and Arizona, *Palmer, Parry, Greene, Pringle*. Has been confounded with the preceding.

* * Glabrous or nearly so, thick-leaved, impressed-punctate, low, perennial from a stout multicipital caudex: rays and disk-flowers both yellow: bracts of involucre more coriaceous, mostly ovate or oblong and with short herbaceous tips: teeth of disk-corolla short, ovate, obtuse: akenes moderately villous all over.

G. spathuláta, Gray. Hardly a foot high, leafy-stemmed, branched from the base: leaves spatulate, entire, inch long, uppermost gradually smaller: head barely half-inch in diameter: rays few and small: pappus with awns surpassing disk-corolla: fimbrillæ setaceous-attenuate, shorter than the akenes. — Proc. Am. Acad. xii. 59. — Rabbit Valley, Utah, *Ward*.

G. acaúlis, Gray. Leaves all clustered on the thick caudex, ovate and obovate, somewhat spatulate, contracted into slender petioles, entire or sparingly dentate: scapes a span to a foot high: head larger: rays more numerous, over half-inch long, rather narrow and with narrow lobes: pappus with short awns not equalling the disk-corolla: fimbrillæ subulate, shorter than the akenes. — Proc. Am. Acad. xi. 73. — S. W. Utah and adjacent Arizona, at Mokiak Pass, &c., *Parry, Palmer*.

161. **SARTWÉLLIA,** Gray. (In honor, now in memory, of *Dr. Henry P. Sartwell*.) — Annuals (of the Texano-Mexican border), glabrous, a foot or two high, leafy, fastigiately branched, and bearing very numerous small heads (only 2 lines high) of yellow flowers in corymbiform cymes; the leaves all narrowly linear or filiform, entire, rather fleshy, opposite, slightly connate at base. — Pl. Wright. i. 122, t. 6, & Proc. Am. Acad. xix. 34. — Two species.

S. Flavériæ, Gray, l. c. Leaves nearly filiform: pappus a truncate cupule. — S. W. Texas, on the Pecos, &c., *Wright, Thurber, Havard*.

S. Mexicána, Gray, Proc. Am. Acad. xviii. 107, xix. 34 (coll. *Palmer* in Northern Mexico), has less filiform leaves, and a pappus of nearly distinct paleæ, with which as many longer delicate awns alternate.

162. **FLAVÉRIA,** Juss. (From *flavus*, yellow; plants used to dye yellow.) — Glabrous herbs (mainly tropical-American), mostly annuals; with small and fascicled or glomerate heads of yellowish or yellow flowers, and opposite sessile leaves, the broader ones 3-nerved. Akenes mostly smooth and glabrous. — Gen. 186: Benth. & Hook. Gen. ii. 409. *Flaveria & Broteroa (Brotera,* Spreng.), DC. Prodr. v. 635. *Vermifuga,* Ruiz & Pav. Prodr. 114, t. 24.

§ 1. Involucre 4–15-flowered, composed of 3 to 5 principal bracts.

* Heads more or less clustered in broad and open naked-pedunculate compound terminal corymbiform cymes: leaves somewhat fleshy: involucre of 5 bracts: corollas except in the last species nearly or quite glabrous.

F. chloræfólia, Gray. Glaucous, 1 to 3 feet high: leaves entire, from ovate-oblong to lanceolate, broadest (half to fully an inch broad) and connate or connate-perfoliate at base:

heads about 12-flowered, 3 lines long: no ray. (A few flowers once seen with a pappus of 4 thin paleae!)— Pl. Fendl. 88, & Pl. Wright. 114. — Low grounds, on and near the Rio Grande, S. W. Texas, *Wright, Parry, Bigelow.* (Adj. Mex., *Wislizenus, Gregg, Palmer.*)

F. longifólia, GRAY, l. c. Rather stout, 1 to 3 feet high, pale: leaves from linear to lanceolate, broadest or not narrowed at the closely sessile base, 2 to 5 inches long, entire or with rare spinulose denticulations: heads in very ample cymes, 10-15-flowered, often 3 lines long: no ray: bracts of the involucre broad. — *Gymnosperma? oppositifolium,* DC. Prodr. v. 312. — Not yet found on the Texan side of the Rio Grande. (Adj. Mex.)

F. lineáris, LAG. Rather slender, a foot or two high: leaves from narrowly linear to lanceolate, or sometimes lower oblong-lanceolate (and inch broad), all contracted above the somewhat connate bases, sometimes denticulate: heads smaller and more glomerate, 5-8-flowered, commonly uniligulate. — Nov. Gen. & Spec. 33; Torr. & Gray, Fl. 360. *F. maritima,* HBK. Nov. Gen. & Spec. iv. 285. *F. tenuifolia,* Nutt. Jour. Acad. Philad. vii. 81. *Selloa nudata,* Nutt. in Am. Jour. Sci. v. 300; therefore *Gymnosperma nudatum,* DC. Prodr. v. 312. — Coast and Keys of S. Florida; first coll. by *Ware.* (Cuba & Bahamas.)

* * Heads in closer subsessile or short-pedunculate or foliose-involucrate chiefly terminal glomerules: involucre of mostly 3 bracts, narrow, 3-5-flowered or some only 2-flowered, commonly uniligulate: disk-corollas sparsely hirsute at base.

F. angustifólia, PERS. Erect, a foot or two high: leaves from linear to lanceolate, serrulate or entire, lightly 3-nerved, sessile by broadish or little contracted base: ligule somewhat exceeding the disk. — Syn. ii. 489; DC. Prodr. v. 635; Gray, Pl. Fendl. 88. *Milleria angustifolia,* Cav. Ic. iii. 12, t. 223. — Alkaline ground, S. W. Texas to E. Colorado and New Mexico. (Mex.)

F. CONTRAYERBA, Pers., is S. American, spreading to W. Indies, and possibly to within our borders, has mostly oblong-lanceolate leaves contracted at base and conspicuously 3-nerved, more glomerate heads, and ligule not exceeding the disk or wanting.

§ 2. Involucre 1-2-flowered, of 1 to 3 unequal bracts: heads densely glomerate. — *Broteroa,* DC., corrected from *Brotera,* Spreng. in Schrad. Jour. Bot. (1800), ii. 186, t. 5.

F. repánda, LAG. l. c. Divergently branched annual: leaves obovate to oblong-lanceolate with narrowed petiole-like base, strongly 3-nerved, acutely serrate: glomerules of many confluent heads, sessile in the forks and involucrate at end of the branches, outermost heads commonly of a single short-ligulate flower. — *F. Contrayerba,* Gray, Pl. Wright. i. 114, not Pers. *Brotera Contrayerba,* Spreng. l. c. *B. trinervata,* Pers. Syn. ii. 498. *B. Sprengelii,* Cass. Dict. xxxiv. 304. *Nauenbergia trinervata,* Willd. Spec. iii. 2393. *Broteroa trinervata,* DC. Prodr. v. 636. — S. W. borders of Texas, *Wright.* (Mex., &c.)

163. **POROPHÝLLUM**, Vaill. (Πόρος, a passage or pore, φύλλον, leaf, the foliage or involucre appearing as if punctate on account of the translucent oil-glands.) — Herbaceous or suffrutescent plants (of the warmer parts of America), usually glaucous; with alternate or opposite undivided leaves, and pedunculate heads of yellow or purplish flowers. Oil-glands present in the involucre when wanting in the leaves, in the form of dots or stripes. — L. Hort. Cliff. 494; Adans. Fam. ii. 122; DC. Prodr. v. 647, excl. § 2, 3. *Kleinia,* Jacq. Stirp. Am. 215, t. 127, not L.

* Annual, with broad crenate-repand leaves on slender petioles: bracts of cylindrical involucre 5: corollas purplish, with filiform tube several times longer than the throat and limb: akenes filiform or slender-fusiform.

P. macrocéphalum, DC. A foot or two high: leaves roundish-oval to oblong (or some of the lowest narrower), about the length of the petiole: peduncles enlarged above, clavate and fistulous: head inch long: bracts of involucre obtuse: akenes much longer than the pappus. — Prodr. v. 468; Gray, Pl. Wright. ii. 93. — Rocky hills and ravines, Arizona, *Wright, Thurber,* &c. (Mex.)

* * Perennial (as to N. American species), with narrow entire sessile leaves, glaucescent, much branched, 1 to 3 feet high.

P. grácile, BENTH. Liguescent at base, with slender striate branches: odor pungent, "Fennel-like": leaves narrowly linear with tapering base, or uppermost filiform or subulate, or all filiform: involucre cylindraceous, half-inch long; its bracts 5, oblong or linear-oblong, obtuse, scarious-margined, often slightly purple-tinged: corollas dull white and purple, with tube as long as the narrowish throat and short triangular-lanceolate lobes: akenes attenuate at apex, rather longer than the pappus. — Bot. Sulph. 29; Gray, Bot. Calif. i. 399. *P. Greggii*, Gray, Pl. Wright. i. 120, in part, & ii. 94. — Arid plains, S. W. Texas to San Diego Co., California. (Lower Calif., Adj. Mex.)

P. scopárium, GRAY. Shrubby at base, with slender rush-like branches: leaves thick and firm, linear-subulate and filiform, narrow at base, mucronate-apiculate: involucre campanulate, 4 or 5 lines high, of 7 to 9 broadly lanceolate greenish bracts, one third to half shorter than the mature pappus: corollas yellow, with very short obtuse teeth, and narrow throat much longer than the proper tube (i. e. below the insertion of the stamens): akenes not attenuate at apex, fully equalled by the pappus. — Pl. Wright. i. 120, ii. 94, & Proc. Am. Acad. xviii. 108. *P. Greggii*, Gray, Pl. Wright. l. c., as to pl. Gregg. a stouter form. — Rocky banks and plains, S. W. Texas and New Mexico; first coll. by *Wright*. W. borders of Texas, *Havard*. (Adj. Mex.)

P. AMPLEXICAULE. Engelm. in Pl. Wright. l. c., of adjacent Mexico, is stouter, less branched, with solitary and larger heads, and fleshy-coriaceous leaves lanceolate, tapering from a partly clasping base, all but the uppermost opposite: bracts of the involucre 8 to 10, half-inch long.

164. CHRYSACTÍNIA, Gray. (Χρυσός, gold, ἀκτίς, ray. from the golden yellow rays, which distinguish the genus from the preceding.) — Pl. Fendl. 93, & Pl. Wright. i. 119. — Single species, with resinous-aromatic odor.

C. Mexicána, GRAY, l. c. Fruticulose, about a foot high from a stout base, much branched, very leafy: leaves alternate, Heath-like, thick or almost terete, short-linear or filiform, with narrowed base, cuspidate-mucronate, entire, with abundant round oil-glands: heads on slender peduncles terminating the branches, a third of an inch high: bracts of the involucre lanceolate, hardly longer than the akenes, usually bearing a single large and prominent infra-apical oil-gland: disk-corollas with short proper tube and long cylindrical throat (in the way of *Porophyllum scoparium* and *P. amplexicaule*): akenes shorter than the pappus. — Rocky ground, W. Texas and adjacent New Mexico; first coll. by *Gregg*. (Mex.)

165. NICOLLÉTIA, Gray. (Memory of *J. N. Nicollet*, astronomer and explorer of the region between Upper Mississippi and Missouri Rivers.) — Two low annuals; with alternate leaves irregularly pinnately parted, and leafy branches terminated by large heads of purple or flesh-colored flowers, or disk-corollas at first yellow. — Rep. Fremont 2d Exped. 315, Pl. Wright. i. 119. & Bot. Calif. i. 398.

N. occidentális, GRAY, l. c. Stout, somewhat fleshy, a foot or two high: branches leafy up to the head: leaves with numerous or several short lanceolate-subulate or linear setosely tipped lobes: involucre three-fourths inch long, of 8 to 12 bracts: ligules oblong, little surpassing the disk. — Sandy banks and plains of the Mohave Desert region, S. E. California; first coll. by *Fremont*, who made his earliest exploration under Nicollet.

N. Edwárdsii, GRAY. More slender, a span or two high: leaves attenuate-linear, few-lobed: heads somewhat naked-pedunculate: involucre turbinate at base, half-inch long, of 8 or 9 bracts: ligules much exserted, elongated-oblong, dentate or denticulate at the truncate summit, commonly half-inch long. — Pl. Wright. i. 119, t. 8. & Bot. Mex. Bound. 93. — Sandy banks and plains, S. W. Texas and adjacent New Mexico, *Bigelow*, *Parry*. (Adj. Mex., *Dr. Edwards*, *Palmer*. Lower Calif., *Streets*.)

166. DYSÓDIA, Cav., as *Dyssodia*. (Δυσωδία, an ill smell.) — Herbs or suffrutescent plants, of N. America and Mexico, mostly strong-scented; with alternate or opposite leaves, and solitary or rarely somewhat paniculate heads of yellow or orange flowers, sometimes turning purplish or reddish. — Anal. Cienc. Nat. vi. 334; Lag. Nov. Gen. & Spec. 29 (mainly); Cass. Dict. xxv. 396; DC. Prodr. v. 639 (excl. § 5 and incl. *Clomenocoma* & *Lebetina*, Cass.); Benth. & Hook. Gen. ii. 409 (but not excl. § *Gymnolæna*, DC.); Gray, Proc. Am. Acad. xix. 37. *Bœbera*, Willd. Spec. iii. 2125.

D. ANTHEMIDIFÓLIA, Benth. Bot. Sulph., of Lower California, is of a peculiar section (*Bœberastrum*, Gray, l. c.), with simple and more open involucre, broad conspicuous rays, style-branches nearly of *Bœbera*, and pappus with the paleaceous part more conspicuous, the lower bristles on the sides much reduced in size.

§ 1. EUDYSÓDIA, Gray. l. c. Involucre calyculate with some external loose bracts: style-branches (at least in our species) tapering into slender-subulate appendages: teeth of the corolla usually narrow: heads comparatively large, pedunculate, and terminating naked branches; perennials; ours obviously frutescent at base, very glabrous, and with glabrous akenes which are shorter than the pappus; this of rather scanty bristles; the receptacle minutely if at all fimbrillate.

D. Coóperi, GRAY. Stout, a foot or two high: leaves all alternate, sessile, thickish, short (the larger less than inch long), from broadly ovate to lanceolate, acute, spinulose-dentate, many with a pair of stipule-like small lobes at base, mostly glandless: head broad, inch high: principal bracts of the involucre 20 to 30, distinct, subulate-acuminate; accessory ones small and subulate: rays little surpassing the disk, orange or turning purplish. — Proc. Am. Acad. ix. 201, & Bot. Calif. i. 398. — Dry ravines of the Mohave Desert, S. E. California, *Cooper, Palmer, Lemmon, Parish*.

D. porophylloides, GRAY. Stems 1 to 3 feet high from a woody base, with numerous spreading slender branches: leaves partly alternate, 3-5-parted; the lower petioled and with cuneate to lanceolate entire or incised divisions; upper sessile and the divisions linear-subulate, not setigerous: head narrower, half to three-fourths inch high: principal bracts of the involucre 14 to 20, linear, abruptly acute or mucronate, commonly slightly united below: rays few and inconspicuous, yellow. — Pl. Thurb. in Mem. Amer. Acad. v. 322, & Bot. Calif. l. c. — Dry hills and mesas, S. E. California and Arizona; first coll. by *Thurber*.

D. SPECIÓSA, Gray, Proc. Am. Acad. v. 163, is a species of Lower California, allied to the preceding, with opposite trifoliate leaves, and mostly petiolulate leaflets.

D. CANCELLÁTA, Gray, Proc. Am. Acad. xix. 38 (*Lebetina*, Cass), is a species very closely related to *D. porophylla*, DC. (and with similar abruptly short-appendiculate style-appendages), but the pappus is anomalous in having an outer series of short and blunt and wholly naked paleæ. Common in the northern part of Mexico, reaching so near the Texan border that it may be expected within it.

§ 2. BŒBERA, DC., excl. spec. Involucre regularly calyculate with accessory bracts: style-branches with very short conical tips: corolla-teeth short, ovate: paleæ of the pappus multicapillary: akenes pubescent: receptacle merely pubescent or puberulent: rather low herbs (the two Mexican species perennials, with naked-peduncled conspicuously radiate heads); all with opposite pinnately divided leaves, and some pubescence.

D. chrysanthemoides, LAG. Much branched and ill-scented annual, leafy up to the subsessile or short-pedunculate small heads. leaves 1-2-pinnately parted into linear lobes: involucre purplish-tinged or greenish, campanulate, of 8 or 10 scarious-tipped oblong bracts, and some linear loose accessory ones: rays few and inconspicuous, not surpassing the disk. — Nov. Gen. & Spec. 29; DC. l. c.; Torr. & Gray, Fl. ii. 362. *D. glandulosa*, Cav. *D. fastigiata*, DC. l. c., excl. syn. *Tagetes papposa*, Vent. Hort. Cels. t. 36; Michx. Fl. ii. 132. *Bœbera chrysanthemoides*, Willd. Spec. iii. 2125. *B. glandulosa*, Pers. Syn. ii. 459. — Alluvial soil, Minnesota to Louisiana and southwest to Arizona; now spreading eastward in the Atlantic States as a weed. (Mex.)

167. HYMENATHÉRUM, Cass. ('Υμήν, membrane, ἀθήρ, awn, the paleæ of the pappus awned.) — Low herbs or suffruticulose plants (chiefly of the Mexican borders), of various habit, mostly pleasant-scented; with alternate or opposite leaves, and small or barely middle-sized usually radiate heads of yellow flowers. — Cass. Bull. Philom. 1817, 1818. & Dict. xxii. 313; Gray, Pl. Fendl. 88, & Pl. Wright. i. 115; Benth. & Hook. Gen. ii. 410. *Hymenatherum* (excl. § 2), *Dysodia* § *Aciphyllæa*, & *Gnaphaliopsis*, DC. Prodr. Now adding *Thymophylla*, Lag. (slightly earlier published name, but obscure), & *Lowellia*, Gray, with muticous pappus. Vide Proc. Am. Acad. xix. 40.

§ 1. ACIPHYLLÆA, Gray. Paleæ of the simple pappus numerous (18 to 20), above resolved into about 5 or the alternate ones into 3 capillary bristles, like those of *Dysodia* (to which it makes transition): heads sessile or nearly so at the end of the woody branchlets: leaves entire, opposite. — Pl. Wright. i. 115.

H. acerósum, GRAY, l. c. Shrubby, a span to a foot high from a thick base, rigid, exceedingly branched: branches barely puberulent: leaves filiform-acerose, usually with shorter ones fascicled in most of the axils: heads 3 or 4 lines high: involucre with copious large oil-glands, subtended by uppermost pair of leaves or by a few shorter subulate foliaceous bracts: rays oblong. — *Dysodia?* (*Aciphyllæa*) *acerosa*, DC. Prodr. v. 641. *Aciphyllæa acerosa,* Gray, Pl. Fendl. 91. — W. borders of Texas to Arizona toward the Mexican boundary, *Wright,* &c. (Mex.)

§ 2. DYSODIÓPSIS, Gray, l. c., excl. spec. Paleæ of the simple pappus only 10, rigid, not longer than the thickish akene, much shorter than disk-corolla, some entire with a single awn, others with 3 aristate-subulate tips: heads loosely foliose-calyculate: leaves alternate.

H. tagetoídes, GRAY, l. c. A rigid annual, and becoming perennial, glabrous, a foot or so high, fastigiately branched at summit: leaves narrowly linear, 2 or 3 inches long, rigid, laciniately and spinulosely dentate or almost pinnatifid: heads indistinctly peduncled, less than half-inch high: involucre rigid; its bracts obviously imbricated, but connate almost to the tip: rays oblong, conspicuous. — *Dysodia tagetoides,* Torr. & Gray, Fl. ii. 361. — Low prairies, Texas; first coll. by *Drummond.*

§ 3. EUHYMENATHÉRUM, Gray, l. c. Paleæ of the pappus 10 to 20, all or the inner ones 1–3-aristate, and the awns about equalling or surpassing the disk-corolla: heads naked at base, or with some small and scanty subulate accessory bracts. (See also § 4.)

* Rays inconspicuous and few, with ligule not surpassing the disk or the double and dimorphous pappus; thus of 10 rigid paleæ in each series, inner with stout awns.

H. Neo-Mexicánum, GRAY. A slender erect annual, a foot or less high, glabrous, fastigiately branched above: leaves mostly pinnately parted into a few linear-filiform entire divisions: lower opposite, upper alternate: heads short-peduncled: involucre turbinate, 3 or 4 lines high, of 5 to 7 oblong connate bracts, subtended by 2 to 4 filiform-subulate bractlets: akenes appressed-villous at the attenuate base, shorter than the inner pappus, the oblong-lanceolate paleæ of which are cleft into 3 scabrous rigid awns, the middle one longer; those of the short outer pappus oblong-spatulate, retuse. — Proc. Am. Acad. xix. 40. *Adenophyllum Wrightii,* Gray, Pl. Wright. ii. 92. — Hillsides, New Mexico, *Wright.*

* * Rays exserted and conspicuous, oblong: awns of the pappus capillary or slender-setiform.

+ Most of the paleæ of the nearly homomorphous pappus 3-awned, lateral awns shorter: glabrous leafy-stemmed herbs, either annuals or slender-rooted subperennials: upper leaves all alternate.

H. polychǽtum, GRAY. Low, diffusely much branched from an annual root, leafy to near the numerous short-peduncled heads: leaves not rigid, pinnately parted into several short-filiform obtuse and pointless divisions: involucre barely 3 lines high, 10–16-toothed:

pappus of 18 or 20 very narrow paleæ, of 2 or 3 lengths, the smaller attenuate into a short single awn, the larger into a much longer capillary awn, with a delicate short one at each side of its base, or rarely an additional pair of setæ. — Pl. Wright. i. 116. — Prairies, S. W. Texas and New Mexico, *Wright*. (Chihuahua, adj. Mex., *Thurber*.)

H. Wrightii, GRAY. Erect or diffuse, a foot or less high from a firm annual or perhaps perennial root: branches rather simple, bearing few or solitary heads on peduncles 1 to 3 inches long: leaves not rigid, narrowly linear or almost filiform (an inch or more long), setulose-mucronate, many entire, some with 1 to 3 small subulate lobes: involucre fully 3 lines high, 16-20-toothed : paleæ of the pappus 10, all slenderly 3-awned from a short lanceolate base; lateral awns with subulate base, half the length of middle one. — Pl. Fendl. 89, Pl. Lindh. ii. 229, & Pl. Wright. l. c. — Prairies of Texas, *Wright*, *Lindheimer*, &c.

H. tenuilobum, DC. Diffusely branched and spreading from a seemingly annual but sometimes more enduring root: branches a span to a foot long : heads on filiform (1 to 4 inches long) peduncles: leaves rather rigid, all pinnately parted into 7 to 11 subulate-filiform setulose-mucronate divisions (of only 2 to 4 lines in length) : involucre 3 lines high, about 12-toothed : paleæ of the pappus 10, more rigid, all nearly similar and bearing two lateral and a middle longer stouter awn, the latter hardly longer than the lanceolate paleaceous portion (which, however, sometimes splits away from the awn on each side), rarely one or two smaller lateral setæ or cusps. — Prodr. v. 642. *H. tenuifolium*, Gray, Pl. Wright. l. c., not Cass. — S. Texas along and near the Rio Grande, *Berlandier*, *Wright*, &c. (Adj. Mex.)

+ + All 10 paleæ of the pappus nearly similar and tapering into a single short awn, and the larger mostly 2-setulose; leaves acerose.

H. Thúrberi, GRAY. Habit and character of a more leafy-stemmed form of *H. pentachætum*: paleæ of the pappus not distinctly in two series, all narrowly lanceolate, alternate shorter ones subulate-awn-pointed, the others with awn rather shorter than the paleæ, and a pair of obscure or more manifest setulose teeth at its base. — Proc. Am. Acad. xix. 41. *H. tenuifolium*, var. ? Gray, Pl. Wright. ii. 93. — Texas or adjacent New Mexico, on the Mexican border near El Paso, *Wright*. Corralitas, *Thurber*. (Apparently also Northern Mexico, *Parry*.)

+ + + Pappus manifestly double and dimorphous, of 10 scarious paleæ; the 5 outer shorter, spatulate or oblong, obtuse and pointless; inner lanceolate or oblong, bearing a single awn, of equal or greater length, between a pair of cusps or subulate or sometimes aristellate teeth.

++ Low and diffuse suffruticulose perennials, minutely cinereous-pubescent or glabrate, not woolly, much branched from the base: leaves rigid or rigidulous, pinnately parted into few or several mostly filiform or acerose entire divisions, subulate or setulose-mucronate at tip: heads on elongated filiform peduncles.

H. Hartwégi, GRAY. A span or two high, nearly herbaceous and glabrous: leaves chiefly opposite, of few rather long filiform-acerose divisions: heads numerous: involucre rather narrow, 2 lines high, almost naked at base: outer paleæ of pappus subcoriaceous, with truncate summit obscurely denticulate. — Pl. Wright. i. 117. *H. Berlandieri*, Benth. Pl. Hartw. 18, not DC. — W. Texas to S. Arizona, *Wright*, *Lemmon*. (Mex.)

H. pentachǽtum, DC. Decidedly suffruticulose, low, diffuse, cinereous-puberulent, sometimes glabrate and rather shining, sometimes the foliage canescent with short and fine spreading pubescence : leaves rigid, upper alternate, divisions slender subulate-acerose : involucre from broadly campanulate to hemispherical, 2 or 3 lines high : outer paleæ of the pappus thinnish, usually erose at summit. — Gray, Pl. Wright. i. 117. *H. pentachætum* (the outer pappus overlooked) & *H. Berlandieri*, DC. Prodr. v. 642. — Dry hills, Texas (first coll. by *Berlandier*) to Arizona and S. Utah : very variable. (Mex.)

H. Treculii, GRAY. Diffuse, nearly herbaceous, almost glabrous, with loose elongated leafy branches and very scattered heads : leaves perhaps rather succulent, pectinately parted into linear-subulate equal short (2 or 3 lines long) divisions, which are rather narrower than the rhachis : involucre (3 lines high) and pappus of the preceding. — Proc. Am. Acad. xix. 42. — S. E. Texas, *Trécul*, in herb. Mus. Paris.

++ ++ Low and densely floccose-lanate and soft-leaved annual. — *Gnaphaliopsis*, DC.

H. Gnaphaliópsis, GRAY. Depressed or diffusely spreading, clothed even to the involucre with dense white wool in the manner of a Cudweed, leafy up to the sessile or short-peduncled solitary heads: leaves mostly alternate, spatulate, entire, barely half-inch long: involucre quarter-inch high ; rays oval, short : receptacle flat and wholly naked : paleæ of inner pappus narrowly lanceolate. — Pl. Feudl. 90 (as *H. gnaphaloides*) & 115; Pl. Wright. l. c. *Gnaphaliopsis micropoides*, DC. Prodr. vii. 258. — Hills and plains, S. Texas, *Wright, Havard*, &c. (Adj. Mex.; first coll. by *Berlandier*.)

§ 4. THYMOPHÝLLA. Gray. Paleæ of the pappus 5 to 12, truncate and muticous (yet in one species occasionally some are short-awned!), somewhat coriaceous, distinct or cupulately connate. — Proc. Am. Acad. xix. 42. *Thymophylla*, Lag. Nov. Gen. & Spec. 25 ; Gray, Pl. Fendl. 91, & Pl. Wright. i. 119, t. 7 ; Benth. & Hook. Gen. ii. 410 (as *Thymophyllum*).

* Fruticulose plants, with habit and character of the acerose-leaved genuine species of *Hymenatherum*, but white tomentose, and rays in one species wanting.

H. SETIFÓLIUM, Gray, l. c. (*Thymophylla setifolia*, Lag. l. c., on which the long imperfectly known genus was founded), may possibly reach our limits. It has a canescent involucre, no rays, and normally a pappus of 5 or 6 distinct quadrate paleæ. But in some specimens of Parry and Palmer's no. 516 occurs an inner alternating series of longer and narrower aristate paleæ, — completely invalidating Lagasca's genus.

H. Gréggii, GRAY. A span or two high in dense tufts: branches thickly leafy up to the filiform glabrate peduncles : leaves white-tomentose, short, Heath-like ; lower 3-7-parted, upper entire, setaceous : involucre campanulate, glabrous, naked at base : rays 10 to 12, short, but distinctly exserted, sometimes wanting ; paleæ of the pappus united into an entire truncate cup. — Proc. Am. Acad. xix. 42. *Thymophylla Greggii*, Gray, Pl. Fendl. 92 (specimen apparently rayless), & Pl. Wright. i. 109, t. 7. radiate. — S. W. Texas, on the Pecos, &c., growing in large bunches, *Wright*. (Adj. Mex., *Gregg*.)

* * Annual, wholly glabrous, wholly resembling *H. polychætum* and its near allies, except the pappus. — *Lowellia*, Gray.

H. aúreum, GRAY. A span or two high, erect or diffuse, much branched, bearing numerous short-peduncled heads : leaves mostly alternate, pinnately parted into 7 to 9 linear-filiform pointless divisions : involucre broadly campanulate, 3 lines high : rays about 12, oblong, 3 lines long: pappus of 6 or 8 quadrate or oblong and erose-truncate paleæ, in length little exceeding the breadth of the akene. — Proc. Am. Acad. xix. 42. *Lowellia aurea*, Gray, Pl. Fendl. 89, & Pl. Wright. i. 118. — Plains of Colorado to W. Texas ; first coll. by *Fendler*, next by *Wright*.

168. TAGÉTES, Tourn. (A name of the early botanists for the "*French*" or "*African Marigold*" of the gardens, *T. patula*, and its larger form, *T. erecta*, L. Fuschius says it was applied by *Apuleius* to the Tansy ; some derive the word from *Tages*, an Etruscan deity.) — Mexican and S. American herbs, mostly annuals, strong-scented, branching ; with opposite and sometimes alternate leaves, in one species nearly entire, in most pinnately divided, having copious oil-glands, bearing large and showy or small and comparatively inconspicuous heads of mostly yellow or orange flowers, in cultivation some flame-colored or reddish. — Inst. 488, t. 278 ; L. Gen. ; DC. Prodr. v. 642 ; Gray, Proc. Am. Acad. xix. 42.

—— T. LÚCIDA, Cav., the species with simple and narrow sessile leaves, and cymose heads with 2 or 3 rather showy rays, may yet be found within our Mexican border. One of our two indigenous species has handsome exserted rays, the other has inconspicuous rays and the most slender heads in the genus.

T. Lemmóni, GRAY. Nearly glabrous, 2 or 3 feet high from a *perennial* root, lignescent at base, rather slender, fastigiately branched, bearing numerous cymosely disposed heads on slender short peduncles: leaves all opposite, 3-7-foliolate ; leaflets lanceolate-linear or sometimes lanceolate-oblong, with attenuate base, serrulate, not setiferous (an inch or two long), sometimes a minute lower pair : involucre turbinate-campanulate, 4 lines high : rays 6 to 8, nearly half-inch long, obovate-oblong : lobes of the disk-corolla nearly beardless : pappus much shorter than the akene, of 1 to 3 subulate and one or two shorter truncate paleæ. — Proc. Am. Acad. xix. 40, 42. — S. Arizona, in the Huachuca Mountains, *Lemmon*.

—— **T. micrántha**, Cav. Slender, diffusely much branched, Anise-scented, a span to a foot high from an annual root, with loosely paniculate slender-peduncled heads: leaves linear-filiform, 3-5-parted, or some of the lowest undivided, not serrate: involucre fusiform, about half-inch long, few-flowered; rays 1 to 3, the oval pale yellow or white ligules only a line long: akenes slender, glabrate, longer than the pappus of 2 oval or truncate thin paleæ and 2 longer awns. — Ic. iv. 31, t. 352; DC. Prodr. v. 646; Gray, Pl. Wright. ii. 93. — Dry ground, New Mexico and Arizona, *Wright, Rothrock*, &c. (Mex.)

169. **PÉCTIS**, L. (Πεκτέω, to comb, the leaves of most species of the genus pectinately setiferous. It is an ancient Latin name of some plant, appropriated to this genus by Linnæus.) — Herbs, all American, mostly low and spreading, usually glabrous, heavy-scented; with narrow opposite leaves conspicuously dotted with round oil-glands; and with mediocre or small heads of yellow flowers, occasionally turning purplish, slender rigid bristles fringing at least the base of the leaves, or rarely quite wanting.

§ 1. EUPÉCTIS. Pappus of a few paleæ or slender awns with or without a dilated or chaffy base, or in some (and occasionally in all) of the akenes reduced to a paleaceous crown, or to a few squamellæ, or obsolete: base of the leaves copiously setiferous. — Gray, Proc. Am. Acad. xix. 44. *Pectis* § *Pectidopsis* & *Eupectis*, Gray, Pl. Wright. i. 83. *Pectidopsis, Pectis* (excl. spec.), and part of *Lorentea*, DC. Prodr. v. 98–101.

* Pappus paleaceous, conspicuous: the paleæ (in ours 4 or 5 or 6 in the disk, 2 or 3 in the ray) mostly prolonged into awns or subulate points: bracts of involucre 4 to 6, broad or broadish: ours annuals.

—— **P. prostráta**, Cav. Procumbent or prostrate: leaves oblanceolate or spatulate-linear (seldom inch long): heads sessile or nearly so: disk-flowers 5 or 6: paleæ of the pappus ovate-lanceolate or narrower, thin, often unequal, short-awned. — Ic. t. 324; DC. Prodr. v. 100; Gray, Pl. Wright. i. 83. *Chthonia prostrata*, Cass. Dict. ix. 173. — S. W. Texas to Arizona. (Mex., W. Ind.?)

P. ciliáris, L. Erect or diffuse, sometimes a foot high: leaves linear-oblanceolate or narrower, commonly inch long: heads nearly sessile: disk-flowers 4 to 8: paleæ of the pappus lanceolate-subulate tapering into a slender awn, more rigid and equal than in the preceding. — Spec. ed. 2, 1250. — Coast and keys of S. Florida, *Blodgett, Garber*. (W. Ind.)

—— **P. linifólia**, Less. Erect, diffusely branched, slender, a span to a foot high: leaves narrowly linear, inch long: heads on minutely bracteate filiform (commonly inch long) peduncles: involucre (2 lines long) of narrower bracts: paleæ of the pappus ovate or ovate-lanceolate, abruptly long-awned, or some nearly awnless. Varies with peduncles not longer than the head. — Less. in Linnæa, vi. 709 (excl. syn.); DC. l. c. 99 (excl. syn.); Griseb. F. Brit. W. Ind. 378, not L. — S. Florida, *Blodgett, Garber, Curtiss*, &c. (W. Ind.)

* * Pappus pauciaristate; viz. of 1 to 5 or 6 upwardly scabrous (usually slender and setiform but rigid) awns, at most dilated only at very base, with or without a short chaffy crown of connate or separate squamellæ, sometimes reduced to this, the awns being absent: bracts of the short-cylindraceous rather many-flowered involucre linear, at length with involute margins partly surrounding outer akenes: low and much branched annuals, with slender narrow-linear leaves, bearing a few bristles next the base.

+— Heads subsessile or short-peduncled, more or less fastigiate or cymose at the end of the branches: bracts of the involucre about 8.

—— **P. tenélla**, DC. A span or more high: pappus of 3 to 6 slender awns, not much shorter than the akene: no squamellæ or crown. — Prodr. v. 99; Gray, Bot. Mex. Bound. 73. — S. Texas, *Berlandier, Thurber, Havard*. (Adj. Mex.)

—— **P. angustifólia**, Torr. A span or two high, lemon-scented: pappus a crown of 4 or 5 mostly connate squamellæ, and not rarely one or sometimes two slender usually short awns.

— Ann. Lyc. N. Y. ii. 214; Gray, Pl. Fendl. 61, Pl. Wright. i. 82, & Bot. Mex. Bound. l. c. *P. fastigiata*, Gray, Pl. Feudl. l. c. *Pectidopsis angustifolia*, DC. l. c.; Torr. & Gray, Fl. ii. 62. — Dry hills and plains, Texas to Colorado and Arizona; first coll. by *James*. (Adj. Mex.)

+ + Heads scattered or solitary, on filiform (one or two inches long) peduncles terminating stem and diffuse branches: involucral bracts obtuse, at length more or less infolding ray-akenes: plants a span or two high: leaves shorter than or equalling the internodes, an inch or two long.

P. filipes, GRAY. Lemon-scented, at length much branched: leaves narrowly linear: involucre narrow, of 5 narrowly oblong bracts; ligules 2 lines long: pappus both in disk and ray-flowers of 2 or 3 (rarely one) rigid subulate awns, shorter than or equalling disk-corolla, with thickened bases and usually very short and blunt interposed squamellæ, sometimes all coroniform-concreted, or some disk-flowers destitute of pappus. — Pl. Fendl. 62, & Pl. Wright. ii. 69. *P. Jaliscana*, Hemsl. Bot. Biol. Centr.-Am. in part, not of Hook. & Arn. (as *Taliscana*) as was supposed. — Common from S. W. Arizona (probably not California) to S. W. Texas; first coll. by *Coulter*.

P. Rúsbyi, GREENE, in herb. "Mint-scented" · leaves linear, thickish: involucre campanulate, of 7 to 9 oblong very obtuse bracts: ligules 3 lines long: pappus in ray-flowers of two or more slender awns: in disk-flowers a conspicuous crown of numerous setaceous-pointed squamellæ more or less concreted at base: flowers more numerous than in the preceding. — Beaver Creek, N. Arizona, *Rusby*.

§ 2. PECTÓTHRIX. Pappus of numerous capillary bristles (at least in most of the disk-flowers) and no paleæ. — Gray, l. c.

* Pappus barbellate-setose, uniserial, but occasionally reduced to a crown of squamellæ: annual.

P. pappósa, GRAY. A span to a foot high, diffusely or divaricately much branched from the base: leaves very narrowly linear, elongated (the larger 2 inches long, barely a line wide), very few bristles at their base: peduncles once to thrice the length of the heads: involucre of 7 to 9 linear bracts: akenes sparsely hispidulous with short capitate bristles: pappus shorter than the disk-corolla, normally of 12 to 18 unequal capillary bristles, which are strongly but rather sparsely barbellate, sometimes (especially in the ray) reduced to a setulose or squamellate crown, or quite obsolete. — Pl. Fendl. 62; Pl. Wright ii. 69. *P. tenella*, Rothrock in Wheeler Rep. vi. 171, not DC. — S. California, Arizona, Utah, and New Mexico: first coll. by *Coulter*.

* * Pappus of merely scabrous capillary bristles, with shorter outer ones more or less in a separate series: ours perennial.

P. lóngipes, GRAY. Low and much branched from a perennial root, forming spreading or depressed tufts: leaves crowded, linear, conspicuously setiferous at base: peduncles elongated, often scape-like, 3 or 4 inches long: involucre campanulate, 3 lines high, many-flowered, of 12 or 13 linear bracts: rays as many, 3 or 4 lines long: pappus of the ray-flowers setosely 2-aristate: of the disk multisetose, i. e. of 20 or 30 upwardly denticulate capillary bristles, the longer equalling disk-corolla, and of some small more attenuate outer ones. — Pl. Wright. ii. 70; Rothrock, l. c., but root not annual. — S. Arizona; first coll. by *Wright*, *Thurber*, &c.

§ 3. PECTÍDIUM. Pappus when developed of few (1 to 4 or 5) subulate and rigid mostly corneous and persistent awns, or reduced to a crown or vestige in some flowers. — § *Pectidium* & *Heteropectis*, Gray. Pl. Wright. i. 83.

* Diffuse low annuals, puberulent, with setiferous leaves: corneous divergent awns of the pappus retrorsely barbed, in the way of *Bidens*. — *Heteropectis*, Gray, l. c.

P. Coúlteri, GRAY. A span or less high, very diffuse, slender: leaves narrowly linear, setiferous merely at base, only half-inch long: peduncles twice or thrice the length of the head: involucre cylindraceous, 2 lines high, 10-20-flowered, of 5 linear-oblong bracts: serrulate-hispid awns of the pappus 3 or 4 in the ray, 2 or 3 in the disk. — Pl. Fendl. 62. — Arizona (probably not California), *Coulter*, *Palmer*.

P. MULTISÉTA. Benth. Bot. Sulph. 20 (Lower California, *Hinds*, *Xantus*). is larger, with broader and serrulate leaves, all the teeth and the apex bristle-tipped; outer akenes with 2 or 3 retrorsely hispid awns; inner with a single awn and a short crown.

* * Erect and comparatively tall annual, with leaves sparingly if at all setiferous at base: akenes all with 2 or 3 corneous and subulate diverging smooth awns (rarely an ascending denticulation): rays small, turning purplish. — *Pectidium*, DC.

P. punctáta, Jacq. A foot or more high, paniculately branched, very smooth : leaves linear, with copious small oil-glands : heads slender-peduncled, quarter-inch long : involucre cylindrical, few-flowered, of 4 or 5 narrow bracts, involute in age. — Stirp. Amer. t. 128; L. Spec. ed. 2, 1250; Griseb. l. c. *P. linifolia*, L. Amœn. Acad. v. 407, & Spec. l. c., founded on pl. of Browne and Sloane. *Pectidium punctatum*, Less. in Linn. vi. 706; DC. l. c. 98. — S. W. Arizona, *Palmer*, *Smart*, *Lemmon*. Not yet seen from Florida, where it would rather be expected. (W. Ind., S. Calif., Galapagos.)

* * * Erect and rather tall perennial, with leaves wholly destitute of bristles: pappus in some flowers of one or two conspicuous erect and smooth paleaceous awns or rigid aristiform paleæ, and 2 or 3 rigid squamellæ, or sometimes all reduced to a crown of corneous squamellæ, or nearly obsolete: rays conspicuous, turning purplish.

P. imbérbis, Gray. Wholly smooth and glabrous: stems a foot or two high, paniculately branched, rather rigid and junciform, sometimes few-leaved: leaves narrowly linear, quite entire, sparingly punctate with oil-glands: heads half-inch long, slender-pedunculate: involucre cylindrical, of 5 or 6 linear obtuse bracts, with margins strongly involute in age: rays 5: disk-flowers 5 to 7, with lobes of corolla bearing a large dark gland. — Pl. Wright. ii. 70; Rothrock in Wheeler Rep. vi. 172. — S. Arizona, Sanoita Valley, [*Wright*, *Rothrock*, &c. (Adj. Mex.)

Tribe VII. ANTHEMIDEÆ, p. 77.

170. LEUCÁMPYX, Gray. (Λευκάμπυξ, with white head-band; the circle of bracts of the head white-bordered.) — Benth. & Hook. Gen. ii. 422; Porter & Coulter, Fl. Colorad. 77; Rothrock in Wheeler Rep. vi. 175, t. 12. — Single species.

L. Newbérryi, Gray in Porter & Coulter, l. c. Perennial herb, a foot or two high, with the aspect and some of the characters of *Hymenopappus* (except the rays), flocculent-woolly, glabrate in age: leaves 2-3-pinnately parted into filiform-linear segments: heads few or several at the naked summit of the stem : involucre nearly half-inch broad : rays three-fourths inch long, obscurely 3-lobed at summit, at first yellow, soon changing to cream-color or white : akenes 2 lines long, turning black. — Cañons, &c., S. W. Colorado, *Newberry*, *Porter*, *Brandegee*. Also W. New Mexico, *Loew*.

171. ÁNTHEMIS, L. Chamomile. (Ancient Greek and Latin name of Chamomile.) — Herbs, usually with pinnately dissected leaves, and rather large heads on peduncles terminating the branches; disk-flowers yellow; rays white, rarely yellow, fertile, except in the first species. A large Old-World genus, one or two species naturalized.

— **A. Cótula**, L. (Mayweed.) Annual weed, of the section Maruta, has receptacle of the head conical, destitute of bracts near the margin, bristly ones at the centre: rays mostly neutral, white, sometimes abortive: akenes 10-ribbed, rugose or tuberculate: stem low: leaves finely 3-pinnately dissected: herbage unpleasantly strong-scented, acrid. — Spec. ii. 894; Barton, Veg. Mat. Med. t. 14. *Maruta fœtida*, Cass. Dict. xxix. 174. *M. Cotula*, DC. Prodr. vi. 13. — Common in waste grounds and along roadsides; fl. late summer and autumn. (Nat. from Eu.)

— **A. arvénsis**, L. (Field Chamomile.) Annual weed, not unpleasantly scented: leaves 1-2-pinnately parted into linear-lanceolate lobes: heads rather long-peduncled: bracts of involucre obtuse, whitish-scarious: receptacle conical; its bracts lanceolate, acuminate: rays white: akenes with a very short slightly toothed margin in place of pappus. — Engl. Bot. t. 602; Fl. Dan. t. 1179; DC. Prodr. vi. 6. — Old fields, sparingly established in the Atlantic States, Oregon, &c. (Nat. from Eu.)

A. nóbilis, L., the officinal CHAMOMILE, a low perennial, with pleasant aromatic filiformly dissected foliage, not uncommon in gardens, is said to be occasionally spontaneous, but rarely.

A. tinctória, L., — an erect herb, rather stout, with large heads, yellow rays, or occasionally pale or partly white, and quadrangular akenes, — has sometimes escaped from gardens.

172. ACHILLÉA, Vaill. YARROW. (After *Achilles*.) — Perennial herbs; with small and corymbosely cymose heads of white, yellow, or sometimes rose-colored flowers, at least in the ray; disk commonly yellow. — Linn. Gen. no. 661. *Ptarmica* & *Millefolium*, Tourn. *Ptarmica* & *Achillea*, DC. — Many Old-World species, very few American, all perennial.

§ 1. Heads rather narrow: receptacle at length elevated. — *Achillea*, DC.

A. Millefólium, L. (MILFOIL or YARROW.) From villous-lanate to glabrate: stems simple, a foot or two (on high mountains a span) high: leaves elongated and narrow in outline, sessile, bipinnately dissected into numerous small and linear to setaceous-subulate divisions; heads numerous, crowded in a fastigiate cyme: involucre oblong; its bracts pale or sometimes fuscous-margined, or even wholly brownish; rays 4 or 5, about the length of the involucre, white occasionally rose-color. — Very variable; in grassy fields of Atlantic States green and more or less glabrate, and with open foliage (perhaps introduced from Europe); northward and on mountains mostly lanate (var. *lanata*, Koch), with divisions of the narrow leaves much crowded; including *A. gracilis* & *A. occidentale*, Raf. in DC. Prodr. vi. 24; *A. tomentosa*, Pursh, Fl. ii. 319. *A. lanulosa*, Nutt. Jour. Acad. Philad. vii. 20; *A. setacea*, Schwein. in Long Exped. ii. 119. Form with dark involucre, *A. Millefolium*, var. *nigrescens*, E. Meyer, Pl. Labrad.; *A. borealis*, Doug. Veg. Sitch. 149. *Ptarmica borealis*, DC. — Common from Labrador to Alaska, south to Texas and California. (All N. hemisphere.)

§ 2. Heads broader: involucre campanulate: receptacle low. — *Ptarmica*, Tourn., DC.

A. multiflóra, Hook. Villous-pubescent, soon glabrate: stem strict, 2 feet high: leaves linear, closely pectinate-pinnatifid into lanceolate-subulate minutely denticulate lobes, the sinuses extending fully half-way to the midrib: heads in rather a close cyme, rays 10 or 12, very short and small, white. — Fl. i. 318; Torr. & Gray, Fl. ii. 409. *A. Ptarmica*, Richards. in Frankl. Journ. 33. — Saskatchewan to Fort Franklin and Behring Strait; first coll. by *Richardson* and *Drummond*.

A. Ptármica, L. (SNEEZEWORT.) A foot or two high, loosely branching above, bearing more loosely disposed and pedunculate heads: leaves glabrous, linear, finely and closely serrate: rays 8 to 12, comparatively large, roundish, white. — Fl. Dan. 643; Engl. Bot. 757; Pursh, Fl. ii. 552. *Ptarmica vulgaris*, Blackw. Herb. t. 236; DC. Prodr. vi. 23. — "Open dry swamps, Canada and New York," *Pursh*. The latter habitat unsupported. New Brunswick, apparently indigenous in Restigouche and Kent Counties, *Fowler*. Locally naturalized in Mass. and Michigan. (Eu., N. Asia.)

173. MATRICÁRIA, Tourn., L. (Name given by the herbalists, from *mater* or *matrix*, to herbs of reputed medicinal virtues.) — Herbs, chiefly of Europe and Asia; with finely once to thrice dissected leaves, and pedunculate heads, the disk-flowers yellow, those of the ray white, or occasionally (and in one of our species constantly) wanting. — Benth. & Hook. Gen. ii. 427.

§ 1. Akenes obpyramidal, with 3 strong and thick (lateral and facial) ribs. — *Tripleurospermum*, Schultz Bip. *Chamaemelum*, Visiani; Boiss., in part.

M. inodóra, L. Nearly scentless, annual, an arctic form apparently biennial or perennial: leaves 2-3-pinnately divided into filiform or narrow linear lobes: heads large: rays half to three-fourths inch long: receptacle at length ovate; pappus a minute entire or 4-toothed border. — Fl. Suec. ed. 2. 297; DC. Prodr. vi. 52; Torr. & Gray, Fl. ii. 412. *Chrysanthemum inodorum*, L. Spec. ed. 2. 1253; Fl. Dan. t. 696; Schk. Handb. t. 253. *Pyrethrum inodorum*, Smith, Engl. Bot. t. 676; Hook. Fl. i. 320. *Tripleurospermum inodorum*, Schultz Bip. Tanacet. 31. — Arctic sea-coast to Alaska and the Hudson Bay country, commonly in a dwarf and monocephalous form with blackish involucre (var. *nana*, Torr. & Gray, l. c., *Chrysanthemum grandiflorum*, Hook., *Pyrethrum inodorum*, var. *nanum*, Hook. Fl. l. c.), occasionally wanting the ray, var. *eligulata*, Seem. Bot. Herald, 33. The common taller and branching European form is naturalized in some parts of Canada and Maine. (Eu., Asia.)

§ 2. Akenes more terete, with 3 to 5 slender often unequal or indistinct ribs, the surface commonly developing mucilage when wetted.

M. Chamomílla, L. Annual, a foot or two high, quite resembling *Anthemis Cotula*, aromatic: heads 3 lines high, and rays of the same length: bracts of the involucre oblong, fuscous: receptacle ovate-conical or oblong in age: akenes small, with an obscure border and usually no distinct pappus; the inner face unequally 5-ribbed. — Curt. Fl. Lond. v. t. 63; Schk. Handb. t. 253. — Waste grounds, S. New York and New Jersey. (Nat. from Eu.)

Var. coronáta, Gay, ex Boiss. Akenes of the ray and commonly most of the disk furnished with a conspicuous thin-scarious cleft and toothed and sometimes unilateral pappus, not rarely surpassing the tube of the corolla. — *M. coronata*, Gay in Koch, Fl. Germ. ed. 2, 416. *M. Courrantia*, DC. l. c. 72; Webb, Phyt. Canar. ii. t. 89. *M. pyrethroides*, DC. l. c., from Mex. *Courrantia chamomilloides*, Schultz Bip. in Webb, Phyt. Canar. ii. 278. — Cult. fields, S. Texas, *Bigelow*. (Adj. Mex.)

M. discoídea, DC. Annual, somewhat aromatic, glabrous, a span to a foot high, very leafy: leaves 2-3-pinnately dissected into short and narrow linear lobes: heads all short-peduncled: bracts of the involucre broadly oval, white-scarious with greenish centre, hardly half the length of the well-developed greenish yellow ovoid disk: receptacle high-conical: akenes oblong, somewhat angled, with an obscure coroniform margin at summit, this occasionally produced into one or two conspicuous oblique auricles of coriaceous texture. — Prodr. vi. 50; Torr. & Gray, Fl. ii. 413. *M. tanacetoides*, Fisch. & Meyer, Ind. Sem. Petrop. vii. 52. *Santolina suaveolens*, Pursh, Fl. ii. 520. *Artemisia matricarioides*, Less. in Linn. vi. 210. *Tanacetum matricarioides*, Less. Syn. 265. *T. suaveolens*, Hook. Fl. i. 327, t. 110. *T. pauciflorum*, DC. l. c. 131, not Richards. *Cotula matricarioides*, Boug. Veg. Sitch. 150. *Lepidotheca* (*Lepidanthus*) *suaveolens*, Nutt. Trans. Am. Phil. Soc. n. ser. vii. 397. — Open ground, W. California to Unalaska and Behring Island, east to Montana, and becoming naturalized in the Atlantic States near railroad stations. (N. Asia; nat. in N. Eu.)

174. **CHRYSÁNTHEMUM**, Tourn., L. (Old Greek name, Χρυσάνθεμον, i. e. golden flower.) — *Chrysanthemum* & *Leucanthemum*, Tourn. *Pyrethrum*, Gærtn., &c. — Mostly an Old-World genus, only a small portion of the species with yellow rays: fl. summer.

C. segétum, L. (Corn-Chrysanthemum or Corn-Marygold of Europe), is a ballast-weed at New York and Philadelphia, and is in fields at Oakland, California. This and *C. coronárium*, L., are genuine Chrysanthemums, annuals, with golden yellow rays as well as disk-flowers, and 3-sided or 3-winged ray-akenes.

C. Sinénse and C. Índicum, L., of China and Japan, are the parents of the autumn-flowering perennial Chrysanthemums of gardens and houses, and form a peculiar section of the genus.

C. ? nánum, Hook. Fl. i. 320, is *Blennosperma Californicum*.

§ 1. **Pyréthrum**, Benth. & Hook. Herbaceous or suffruticulose perennials; with comparatively large and broad heads, either solitary or loosely corymbose: rays usually conspicuous: akenes all equably 5-10-costate. — *Pyrethrum*, Gærtn. *Pyrethrum*, *Leucanthemum*, *Plagius*, &c., DC. *Tanacetum* in part, Schultz Bip.

* Rays described as yellow, but perhaps white, short: leaves bipinnately dissected into many small linear lobes.

C. bipinnátum, L. Slender, a span to a foot high from a creeping rootstock, villous or glabrate, bearing usually a solitary head of half-inch diameter: rays obovate, little surpassing the merely convex disk: pappus a short crown. — Spec. ii. 890; founded on Gmel. Fl. Sibir. ii. 205, t. 85, f. 1. *Pyrethrum bipinnatum*, Willd. Spec. iii. 2160; DC. Prodr. vi. 60. *Tanacetum Kotzebuense*, Bess., ex DC. l. c. 131. *T. bipinnatum*, Schultz Bip. Tanacet. 48. —

Cape Espenberg, Arctic Amer., *Eschscholtz.* Yukon Valley, Alaska, *L. M. Turner*, a glabrate form. (E. Asia to Russ. Lapland.)

* * Rays white, elongated: heads solitary, mostly long-peduncled: leaves undivided or merely pinnatifid.

C. LEUCÁNTHEMUM, L. (OX-EYE DAISY, WHITEWEED.) Glabrous, a foot or two high from a creeping base or rootstock, simple or sparingly branched: cauline leaves spatulate, and the upper gradually narrower, becoming small and linear, pinnately dentate or incised, partly clasping at base; radical broader, petioled: head broad and flat: rays inch long: pappus none. — Fl. Dan. 994; Engl. Bot. 601. *Leucanthemum vulgare*, Lam. Fl. Fr. ii. 137; DC. Prodr. vi. 46. — Common weed in pastures and meadows through Atlantic States, &c.; here and there met with in similar situations quite to the Pacific. Occurs occasionally with abortive, deformed, or tubular and laciniate rays. (Nat. from Eu.)

C. árcticum, L. Nearly or quite glabrous, rather fleshy, a span to a foot high: leaves cuneate, with long tapering base or petiole, crenately toothed or incised at summit, sometimes 3-5-lobed; uppermost small and linear, nearly entire: bracts of the involucre broad, brown-margined: rays nearly inch long: pappus none. — Spec. ii. 889; Pursh, Fl. ii. 526; Willd. Hort. Berol. t. 33. *Leucanthemum arcticum*, DC. l. c.; Torr. & Gray, Fl. ii. 412. — Coasts, Hudson's Bay and arctic shores to Arctic Alaska and islands. (Kamtschatka and Japan to Lapland.)

C. integrifólium, RICHARDS. Villous when young: stem simple and scapiform from a leafy tufted base, 2 to 4 inches high: leaves linear or slightly spatulate, entire: bracts of involucre oblong, blackish: rays less than half-inch long: pappus none. — App. Frankl. Journ. ed. 2, 33; Hook. in Parry Voy. & Fl. i. 319, t. 109. *Leucanthemum integrifolium*, DC. l. c.; Torr. & Gray, l. c. — Arctic sea-coast; first coll. by *Parry* and *Richardson*. (Arctic Asia.)

* * * Rays white, broad: heads globular-depressed, comparatively small, loosely corymbose-cymose: leaves bipinnately parted or cleft.

C. PARTHÉNIUM, PERS. (FEVERFEW.) Much branched, puberulent, leafy: leaves thin, pinnately parted, and the oval or oblong divisions pinnatifid or incised and toothed: rays oval or obovate, 2 or 3 lines long (in cultivation sometimes all the disk-flowers changed into rays): pappus a minute crown. — Benth. Brit. Flora, ed. 4, 250. *Matricaria Parthenium*, L.; Fl. Dan. t. 624. *M. odorata*, Lam. *Pyrethrum Parthenium*, Smith. *Leucanthemum Parthenium*, Godron; Gray, Man. *Tanacetum Parthenium*, Schultz Bip. — Roadsides and waste grounds, sparingly in Atlantic States; escaped from cult. (Nat. from Eu.)

§ 2. GYMNÓCLINE, Benth. & Hook. (*Gymnocline*, Cass.). Consists of perennial species, with small and corymbosely disposed rather narrow heads, resembling *Achillea* except in the naked receptacle, and when discoid or nearly so making transition to *Tanacetum*. An outlying member of this group is

C. BALSÁMITA, L., with its rayless or discoid form, var. TANACETOIDES, Boiss. (COSTMARY, MINT-GERANIUM, of the gardens), is beginning to escape to roadsides in a few places. It is known by its sweet-scented herbage, barely serrate oblong leaves, and yellowish flowers; when the rays appear they are white. (Adv. from Asia.)

175. **SOLÍVA,** Ruiz & Pav. (*Dr. Salvador Soliva*, of Spain.) — Small and depressed herbs, mostly if not all annuals and S. American; with mainly alternate and petioled pinnately dissected leaves; the heads of greenish flowers sessile in the axils or forks. — Prodr. 113, t. 24. *Solivæa*, Cass. Dict. xxix. 177. *Gymnostyles*, Juss. Ann. Mus. Par. iv. 258. t. 61.

S. séssilis, RUIZ & PAV. Villous, or the leaves glabrate; these twice divided; primary divisions 3 to 5, petiolate, parted into 3 to 5 narrow lanceolate lobes: heads depressed: akenes broadly obovate, thin-winged, the wings entire or sometimes panduriform-excised near the base, spinulose-pointed at summit, in some the wings reduced to an acute margin: persistent style long and stout. — Syst. 215; DC. Prodr. vi. 143. *S. daucifolia*, Nutt. Trans. Am. Phil. Soc. l. c. 403; Torr. & Gray, Fl. ii. 425; Gray, Bot. Calif. i. 406. *Gymnostyles Chilensis*, Spreng. Syst. iii. 500. — Moist ground, coast of California, from Santa Barbara to Mendocino Co. (Chili, whence probably introduced.)

S. NASTURTIIFÓLIA, DC. l. c. Much depressed, spreading, small: leaves glabrate, pinnately parted into 5 to 9 oblong divisions of about a line in length; these entire or the lower few-toothed: heads globular: akenes small, very numerous, villous at apex, cuneate, the margins much thickened and tuberculate-rugose: style short and slender. — Torr. & Gray, l. c. *Gymnostyles nasturtiifolia*, Juss. Ann. Mus. iv. 262, t. 61, f. 2. *G. stolonifera*, Nutt. Gen. ii. 185; Ell. Sk. ii. 473. — A humble weed, near dwellings, coast of N. Carolina to Georgia. (Nat. from Buenos Ayres.)

176. CÓTULA, L. (Κοτύλη, a small cup or disk.) — Low herbs of the southern hemisphere, one or two naturalized in the northern, strong-scented; leaves alternate, lobed or dissected; flowers yellow: ours more or less perennial by creeping base, or annual. — Benth. & Hook. Gen. ii. 428.

C. CORONOPIFÓLIA, L. Somewhat succulent, nearly glabrous: ascending stems often a foot high: leaves lingulate-linear, laciniate-pinnatifid, or uppermost entire, the base clasping or sheathing: head much depressed, a third to half inch broad: female flowers a single row, on flattened pedicels which lengthen in fruit, their akenes bordered with a thick spongy wing and notched at both ends: disk-akenes with wing reduced to a thickened border. — Lam. Ill. t. 700; Dill. Elth. t. 23; DC. Prodr. vi. 28. — Wet ground, thoroughly established on the coast of California, and on some water-courses in the interior: a rare ballast-weed on the Atlantic coast. (Nat. from S. Afr.)

C. AUSTRÁLIS, Hook. f. Slender, diffusely branched, somewhat pubescent: leaves 2-pinnately dissected into linear lobes: heads small: female flowers in 2 or 3 rows, their akenes distinctly pedicelled; those of the disk less so. — Fl. N. Zeal. i. 128; Gray, Bot. Calif. i. 405. *Strongylosperma australe*, Less.; DC. l. c. 82. — Waste ground, coast of California. *Kellogg, Cleveland.* Oregon, *E. Hall.* (Sparingly nat. from Australia.)

177. TANACÉTUM, Tourn. TANSY. (Name of the old herbalists, of quite uncertain derivation.) — Chiefly perennials, of the northern hemisphere, strong-scented, alternate-leaved, yellow-flowered. Disk-flowers 5-toothed. — Torr. & Gray, Fl. ii. 414.

§ 1. Robust erect perennials, leafy to the summit: leaves 2–3-pinnately dissected into very numerous divisions and lobes; also with interposed small ones on the main rhachis: pappus coroniform-dentate: receptacle flat, quite naked. — § *Eutanacetum* & *Omalotes*. DC. Prodr. vi. 128, 83.

T. VULGÁRE, L. (COMMON TANSY.) Acrid-aromatic, glabrous or somewhat pubescent, 2 or 3 feet high: divisions and lobes of the leaves decurrent-confluent, the teeth cuspidate-acuminate: heads numerous and crowded in the corymbiform cymes, 3 to 5 lines broad, depressed-hemispherical: ray-corollas terete, inconspicuous, with oblique 3-toothed limb. — Escaped from gardens to roadsides, &c., in Atlantic States and Canada. (Nat. from Eu.)

T. Huronénse, NUTT. Comparatively sweet-aromatic, villous when young, sometimes glabrate, commonly a foot high: leaves with fewer interposed segments on the rhachis; lobes and teeth narrowly oblong to linear, mucronate or acuminate: heads much fewer (1 to 5) and larger; the disk convex, half-inch broad: corollas of female flowers with a flattish tube and a 3–5-lobed limb, which not rarely expands into a cuneate rather obvious ligule (thus making a transition to *Chrysanthemum* and showing relationship to *C. bipinnatum*). — Gen. ii. 141; Torr. & Gray, Fl. ii. 414. *T. Douglasii*, DC. Prodr. vi. 128. *T. pauciflorum*, Richards. App. Frankl. Journ. ed. 2, 30; Hook. Fl. i. 327, not DC. *T. boreale*, Nutt. Trans. Am. Phil. Soc. n. ser. vii. 401, not Fischer in DC., which is rather a form of *T. vulgare*. — Banks of streams, &c., N. Maine (*Goodale*), New Brunswick, and Lake Superior to Hudson's Bay, west to Washington Terr. and Oregon on the coast.

T. camphorátum, LESS. Pleasantly camphoric-aromatic, villous-tomentose, at least when young, glandular, robust, 1 or 2 feet high: pinnæ and segments of the leaves much crowded; the latter oval or short-oblong, entire or crenately few-lobed, rounded-obtuse, at most callose-apiculate, usually with revolute margins: heads several in a corymbiform cluster, short-pedunclcd, hemispherical, the flat or at length low-convex disk half-inch broad : disk-corollas with flattened tube and 3 small lobes, not surpassing the disk-flowers, regularly 3-5-toothed, not at all liguliform.— Linnæa. vi. 521. *T. Huronense*, Nutt. Trans. Am. Phil. Soc. l. c., in part ; Torr. & Gray, Fl., in part : Gray, Bot. Calif. i. 402. *T. elegans*, Decaisne in Fl. Serres, ser. 2, xii. t 1191. *Omalanthus camphoratus*, Less. Syn. 260 : Hook. Fl. i. 321, as to pl. Calif. only. *Omalotes camphorata*, DC. Prodr. vi. 83. — Sea-beaches, San Francisco, California ; first coll. by *Menzies*, next by *Chamisso*.

§ 2. Low perennials: slender stems more naked above, bearing rather small globular heads ; leaves less dissected or entire : corolla of disk-flowers not oblique nor dilated at summit, regularly 2-4-toothed ; akenes usually utricular ; pappus obsolete or none : receptacle convex or conical.— § *Sphæromeria*, Torr. & Gray, l. c. 415. *Sphæromeria*, Nutt. Trans. Am. Phil. Soc. n. ser. vii. 402.

* Herbaceous to the thickened root: leaves dissected ; receptacle densely fimbrillate-hirsute.

T. potentilloides, GRAY. Silvery-sericeous ; stems decumbent or ascending, a span to a foot long, the naked summit bearing a few slender-peduncled somewhat corymbiform-paniculate heads (of 3 or 4 lines in diameter) : radical leaves 2-3-pinnately and cauline 1-2-pinnately parted into rather few mostly linear lobes : bracts of the involucre roundish-ovate or obovate.— Proc. Am. Acad. ix. 204 : Bot. Calif. l. c. *Artemisia potentilloides*, Gray, Proc. Am. Acad. vi. 551.— Eastern ranges of the Sierra Nevada, California and adjacent Nevada, *Lemmon, Anderson*.

* * Suffrutescent at base, erect : leaves simply or pedately cleft or entire ; receptacle not hirsute, sometimes conical : heads only 2 lines broad. — *Sphæromeria*, Nutt.

T. diversifólium, EATON. Glabrous, very smooth, 8 to 15 inches high, leafy ; leaves some narrowly linear and entire, 1-nerved, some pinnately or pedately parted : heads several or rather numerous in a corymbiform cyme, slender-peduncled : female flowers 8 or 10, with 3-4-toothed corolla. — Bot. King Exp. 180. t. 19. — Utah, in the American Fork Cañon ; first coll. by *Watson*.

T. cánum, EATON. Silvery with minute close tomentum, a span or two high : lower leaves cuneate and 3-lobed or 3-cleft ; upper linear-lanceolate. mostly entire : heads few or several, very short-peduncled or in clusters of 2 or 3 terminating the short branches of the cyme : female flowers 4 to 8, with a truncate obscurely toothed corolla. — Bot. King Exp. 179, t. 19, f. 8-14 ; Rothrock in Wheeler Rep. vi. 366. — Nevada, in a cañon of E. Humboldt Mountains, *Watson*. Olanche Mountain, S. E. California, *Rothrock*.

T. Nuttállii, TORR. & GRAY. Silvery-canescent, loosely cespitose, a span high : leaves short, mostly broadly cuneate with tapering base, obtusely 3-5-lobed at the broad summit ; those of the flowering stems usually oblong or linear and entire : heads few, somewhat paniculate or loosely clustered, some of them slender-pedunculate : involucre very scarious. — Fl. ii. 415. *Sphæromeria argentea*, Nutt. Trans. Am. Phil. Soc. vii. 402. — Rocky Mountains of N. Wyoming, *Nuttall, Parry*.

T. capitátum, TORR. & GRAY, l. c. Silvery-canescent, densely cespitose, a span high ; leaves simply or pedately 3-5-parted into linear lobes, or some of them only 3-cleft at summit : flowering stems scapiform or 2-4-leaved : heads 10 or more, sessile in a globose glomerule. — *Sphæromeria capitata*, Nutt. Trans. Am. Phil. Soc. l. c. — Rocky Mountains of N. Wyoming. *Nuttall, Parry*.

178. **ARTEMÍSIA**, Tourn., L. WORMWOOD, SAGE-BRUSH, MUGWORT. (Ancient name of Mugwort. in memory of *Artemisia*. wife of Mausolus.) — Herbs and low shrubs (chiefly of the northern hemisphere, and most abundant in arid regions), bitter-aromatic ; with alternate leaves, and small paniculately disposed heads. commonly nodding. at least when young ; the flowers yellow or whitish, or turning brownish. usually sprinkled with resinous globules. Anthers commonly tipped with subulate-acuminate appendages. in the manner of *Ambrosia*, but not inflexed. — Besser in DC. Prodr. vi. 93. *Artemisia, Abrotanum, & Absinthium*, Tourn.

§ 1. DRACÚNCULUS, Bess. — Heads heterogamous; the disk-flowers hermaphrodite but sterile, their ovary abortive, and style mostly entire, peltate-penicillate at tip: receptacle not hairy. — *Oligosporus*, Cass.

* Akenes and flowers beset with long cobwebby and crisped hairs: spinescent undershrub. — *Picrothamnus*, Nutt.

A. spinéscens, EATON. Stout and densely branched, rigid, 4 to 18 inches high, villous-tomentose: leaves small, pedately 5-parted and the divisions 3-lobed; lobes spatulate: heads globose, racemosely glomerate on short and leafy branchlets, which persist as slender spines: bracts of the involucre 5 or 6, broadly obovate: female flowers 1 to 4; hermaphrodite-sterile flowers 4 to 8, their corolla ventricose-campanulate from a narrow base. — Bot. King Exp. 180, t. 19, f. 15-21; Gray, Bot. Calif. i. 404. *Picrothamnus desertorum*, Nutt. Trans. Am. Phil. Soc. n. ser. vii. 417; Torr. & Gray, Fl. ii. 289. — Whole desert region of Wyoming, Utah, Nevada, and Idaho, reaching the borders of California; first coll. by *Douglas* (incomplete specimens), then *Nuttall*.

* * Akenes nearly glabrous: receptacle except in last species hemispherical or ovate: no spines.

+ Biennial herb; leaves all filiform.

A. caudáta, MICHX. Glabrous, with one or more strict stems, 2 to 6 feet high: leaves 1-3-pinnately divided into slender filiform lobes: heads small (a line in diameter), very numerous in an ample elongated thyrsus. — Fl. ii. 129; Nutt. Gen. ii. 144; Torr. & Gray, Fl. ii. 417. — Sandy ground, Canada to Texas near the coast, Illinois to Saskatchewan and Kansas.

+ + Perennial herbs, the last two or three species sometimes frutescent at base: heads many-flowered.

++ Leaves dissected.

A. Canadénsis, MICHX. A foot or two high from a perennial (or sometimes biennial?) root: glabrous or mostly with at least the radical and sometimes all the leaves either sparsely or canescently silky-pubescent: leaves mostly 2-pinnately divided into narrow linear or almost filiform but plane lobes, of thickish texture: heads 1 or 2 lines long, very numerous in a compound oblong or pyramidal virgate panicle (in reduced specimens northward fewer in a simple panicle): involucre greenish, glabrous or rarely pubescent. — Fl. ii. 129 (northern form with heads 2 lines broad); Nutt. l. c.; Torr. & Gray, l. c.; Besser, Dracunc. 90, & DC. l. c., partly (mixed with *A. caudata*). *A. peucedanifolia*, Juss. Herb.; Besser, l. c. 29 (*A. Canadensis ferulaceo-folio*, Vaill.), spec. Herb. Tourn., DC. l. c. 99, excl. pl. Mitch. *A. campestris*, Pursh, Fl. ii. 521 ? excl. syn.; Richards. App. Frankl. Journ. *A. desertorum*, in part, Bess. in Hook. Fl. *A. commutata*, Bess. Dracunc. 68, & in DC. Prodr. vi. 98. *A. Pacifica*, Nutt. Trans. Am. Phil. Soc. l. c. 399 ? — Rocky banks and plains, N. New England to Hudson's Bay, west to the Pacific in Washington Terr., and south in the Rocky Mountain region to New Mexico, Utah, and Arizona. (N. W. Asia.)

A. boreális, PALL. A span or two high from a stout caudex: stems simple: leaves silky-pubescent or silky-villous; radical and lower 1-2-ternately or pinnately divided into linear lobes; uppermost linear and entire or 3-parted: heads (2 lines broad) comparatively few, crowded in a narrow (rarely compound) spiciform thyrsus with leaves interspersed: involucre pilose or glabrate, pale-fuscous to brownish. — It. iii. 129, t. hh, f. 1; Bess. Dracunc. 78, & in DC. l. c. 98; Torr. & Gray, l. c. *A. spithumœa*, Pursh, Fl. ii. 522. *A. violacea*, Ledeb. lc. Fl. Alt. t. 475. — Arctic America, Labrador to Alaska, and Rocky Mountains to Colorado in the alpine region. (Greenland, N. Asia.)

Var. **Wormskióldii**, BESS. Taller, 10 to 16 inches high, with more numerous heads in looser or compound narrower thyrsus. — Dracunc. 83, & Hook. l. c. 327. *A. Grœnlandica*, Wormsk. Fl. Dan. t. 1585, small specimen. — Hudson's Bay and mountains of Lower Canada (where it seemingly passes into *A. Canadensis*, in coll. *Allen*) to Washington Terr. and N. Alaska. (Greenland, N. E. Asia.)

A. pedatífida, NUTT. Cespitose, with a stout lignescent caudex, very dwarf, canescent throughout with a fine and close pubescence: leaves chiefly crowded in radical tufts and on the base of the (inch or two high) rather naked flowering stems, once or twice 3-parted into narrowly spatulate or nearly linear obtuse entire divisions: heads (hardly 2 lines broad) few, loosely spicately or racemosely disposed, canescently pubescent: heads 12-15-flowered; the hermaphrodito-sterile flowers with style barely 2-lobed at summit and no ovary. — Trans.

Am. Phil. Soc. vii. 399; Torr. & Gray, Fl. ii. 419. — Arid grounds in the Rocky Mountains of Wyoming, Montana, and Idaho, *Nuttall*, *Fremont* (without flowers), *Parry*. Has been wrongly referred to the following section of the genus.

— **A. pycnocéphala,** DC. A foot or two high, either herbaceous or with a woody base, densely silky-villous, even to the involucre, robust: leaves 1-3-pinnately parted into rather few and short linear or spatulate lobes: heads numerous (2 lines broad), glomerate in an elongated and interrupted spiciform leafy thyrsus. — Gray, Bot. Calif. i. 404. *A. pycnocephala* & *A. pachystachya*, DC. l. c. 99 & 114; Torr. & Gray, l. c. *A. pycnostachya*, Nutt. l. c., error in name. *Oligosporus pycnocephalus*, Less. in Linn. vi. 524. — Sea-shores. California, from Monterey to Humboldt Co.; first coll. by *Chamisso*.

++ ++ Leaves mostly entire, occasionally some 3-cleft, or the lowest even more divided: base of stems rather lignescent.

A. glaúca, Pall. Minutely silky-pubescent or canescent, sometimes glabrate and glaucous: stems strict, a foot or two high: leaves rather short, from linear- to oblong-lanceolate: heads nearly of the next, into which it probably passes. — Willd. Spec. iii. 1331; Bess. Dracunc. 55, & DC. l. c. *A. glauca*, var. *fastigiata*, Bess. l. c. *A. dracunculoides*, var. *incana*, Torr. & Gray, Fl. ii. 416. — Saskatchewan and Minnesota, *Drummond*, *Nicollet*, *Kennicott*.

— **A. dracunculoídes,** Pursh. Glabrous, wanting the scent and taste of *A. Dracunculus*, which it much resembles: stems 2 to 4 feet high, either virgately or paniculately branched: leaves narrowly or sometimes more broadly linear: heads very numerous in a compound and crowded or open and diffuse panicle. — Pursh, Fl. ii. 742; Torr. & Gray, Fl. ii. 416. *A. Dracunculus*, Pursh, Fl. ii. 521. *A. cernua*, Nutt. Gen. ii. 143. *A. inodora*, Hook. & Arn. Bot. Beech. 150. *A. Nuttalliana*, Bess. in Hook. Fl., &c., shorter-leaved form, with lower leaves more freely 3-cleft. — Plains, Missouri to Saskatchewan and Brit. Columbia, and from Texas to Arizona and California. Polymorphous.

A. Lewísii, Torr. & Gray, Fl. ii. 417, appears to be a fictitious species. The plant referred to *A. Santonica* by Pursh is wholly obscure. The specimen in herb. Michaux, with no indication of source, which Besser made a var. *Americana* of *A. variabilis*, Tenore, is without much doubt European. The plant of Engelmann, referred to by Besser in Linnæa, xv. 111, is an imperfect specimen, probably of *A. Canadensis*.

++ ++ ++ Suffruticose: heads very small and numerous, few-flowered.

— **A. filifólia,** Torr. Minutely canescent, even to the 3-6-flowered involucre, 1 to 3 feet high, with virgate rigid branches, very leafy: leaves all slender filiform, commonly 3-parted; the upper and those in axillary fascicles entire: heads crowded in an elongated leafy panicle: receptacle small, not pilose. — Ann. Lyc. N. Y. ii. 211; Torr. & Gray, Fl. ii. 417: Torr. in Marcy Rep. t. 12. *A. Plattensis*, Nutt. Trans. Am. Phil. Soc. vii. 397. — Plains, Nebraska to New Mexico and western borders of Texas; first coll. by *James*.

§ 2. EUARTEMÍSIA. Heads heterogamous; the disk-flowers hermaphrodite and fertile, with 2-cleft style. — § *Abrotanum* & *Absinthium*, Bess.

* Akenes obovoid or oblong, wholly destitute of pappus: receptacle beset with long woolly hairs. — § *Absinthium*, Bess.

A. scopulórum, Gray. Herbaceous, a span or two high from a stout multicipital caudex, silky-canescent: stems simple, bearing 3 to 12 spicately or racemosely disposed hemispherical (rarely solitary) heads: radical and few lower cauline leaves pinnately 5-7-divided, and divisions 3-parted into spatulate-linear lobes; uppermost simply 3-5-parted or entire: involucre 2 lines broad, villous, 18-30-flowered; its bracts brown-margined: corollas hirsute at summit. — Proc. Acad. Philad. 1863, 66; Eaton, Bot. King Exp. 184. — Alpine region of the Rocky Mountains in Colorado, Utah, and Wyoming: first coll. by *Parry*, *Hall* & *Harbour*. Var. *monocephala*, Gray, l. c., is merely a form with single head.

— **A. frígida,** Willd. Herbaceous from a suffrutescent base, silky-canescent and silvery, about a foot high: stems simple or branching, bearing numerous racemosely disposed heads in an open panicle: leaves mainly twice ternately or quinately divided or parted into linear crowded lobes, and usually a pair of simple or 3-parted stipuliform divisions at base of the petiole: heads globular, barely 2 lines in diameter: involucre pale, canescent, its outer bracts narrow and herbaceous: corollas glabrous. — Spec. iii. 1838 (Gmel. Fl. Sibir. t. 63); Pursh. Fl. ii. 521; Ledeb. Ic. Fl. Alt. t. 462; Bess. in Hook. Fl. i. 321. *A. sericea*, Nutt. Gen. ii.

24

143. **A. virgata**, Richards. in Frankl. Journ. — Plains and mountains, Saskatchewan to Minnesota and W. Texas, west to Idaho, Nevada, New Mexico, &c. (N. Asia.)

A. Absínthium, L. (WORMWOOD.) Frutescent, paniculately branched, 2 or 3 feet high, bearing numerous small heads in leafy panicles: leaves 2-3-pinnately parted into lanceolate or oblong obtuse and entire or sparingly incised lobes: involucre canescent, of one or two loose and narrow herbaceous bracts and several roundish and scarious: corollas glabrous. — Spec. ii. 848; Engl. Bot. t. 1230. *Absinthium vulgare*, Lam. Fl. Fr.; Gærtn. Fr. t. 164. — Roadsides, escaped from gardens, Newfoundland to New England. Also Moose Factory, Hudson's Bay. (Nat. from Eu.)

* * Akenes broad or broadish and truncate at summit, commonly bearing a minute or even a conspicuous squamellate or coroniform-dentate pappus, therefore having the character of Tanacetum, but the heads paniculate: receptacle glabrous or barely pubescent. (Here belongs *A. Australis*, Less., of Hawaiian Islands, as well as the anomalous *A. Chinensis*, L.) — *Crossostephium*, Less. *Artemisia* § *Tanaceum*, Nutt.

A. Califórnica, LESS. Shrubby, with habit of *A. Abrotanum*, 4 or 5 feet high, paniculately branched, minutely canescent or cinereous: leaves 1-2-pinnately parted into few filiform lobes not wider than the rhachis, or uppermost entire: heads very numerous in leafy panicles: involucre hemispherical, many-flowered, about 2 lines broad: akenes 3-5-ribbed, with a minute squamellate crown at the broad summit. — Linn. vi. 523; Hook. & Arn. Bot. Beech. 150; Torr. & Gray, Fl. ii. 424; Gray, Bot. Calif. i. 403. *A. Fischeriana*, Besser, Abrot. 21; DC. Prodr. vi. 105. *A. abrotanoides, Fischeriana, & foliosa*, Nutt. Trans. Am. Phil. Soc. vii. 397, 399. California, along the coast, from San Francisco Bay southward and to San Bernardino Co.; first coll. by *Menzies*.

* * * Akenes obovoid or oblong, with small epigynous disk, wholly destitute of pappus: receptacle not villous. — § *Abrotanum*, Bess.

A. Abrótanum, L. (SOUTHERNWOOD), cultivated in old gardens, has become spontaneous in a few places from New York southward.

A. prócera, Willd., a less shrubby and finer-leaved species, has escaped from gardens at Buffalo, New York.

+ Annuals and biennials.

A. Annua, L. A tall and much branched glabrous species, native to Asia, with a very ample and loose panicle of small heads, and leaves 2-pinnately divided into oblong deeply pinnatifid segments. — Naturalized in waste places around Nashville, Tennessee.

A. biénnis, WILLD. Wholly glabrous, inodorous and nearly insipid: stem strict, 1 to 3 feet high, leafy to the top, bearing close glomerules of small heads in the axils from toward the base of the stem to the somewhat naked and spiciform summit: leaves 1-2-pinnately parted into lanceolate or broadly linear laciniate or incisely toothed lobes; or the uppermost small, sparingly pinnatifid and less toothed. — Phytogr. 1794, 11, & Spec. iii. 1842 (excl. hab. New Zeal.); Pursh, Fl. ii. 522; Nutt. Gen. ii. 144; Bess. in Hook. Fl.; DC. Prodr. vi. 120; Torr. & Gray, l. c. *A. Hispanica*, Jacq. Ic. Rar. i. t. 172, not Lam. — Open grounds, Hudson's Bay to Oregon and Colorado; also in Utah and S. California; common also from Ohio and Tennessee to Missouri, probably by immigration, now spreading to the seaboard. (Kamtschatka, N. India.)

+ + Perennials, some fruticulose.

++ Heads many-flowered, collected in a single capitate glomerule or dense cluster: dwarf, arctic, with leaves mainly in radical tufts. (Nearly related species.)

A. Senjavinénsis, BESS. Cespitose-proliferous, very densely villous with long hairs, which on the radical tufts conceal the foliage: leaves much crowded in the tufts, and scattered on the flowering stems, cuneate or oblong, simply 3-5-cleft into oblong or lanceolate lobes: heads in a dense villous glomerule, fuscous: involucral bracts sphacelate: corolla glabrous. — Abrot. 65 (as *Semavinensis*), Suppl. in Bull. Mosc. ix. 64, & DC. Prodr. vi. 116. *A. androsacea* (characteristic name), Seem. Bot. Herald. 34, t. 6 (founded on *A. glomerata*, Hook. & Arn. Bot. Beech. 125, not Ledeb.); Hook. f. Arct. Pl. 331. — Kotzebue Sound, *Beechey*. (Adj. Asia, Arakamtchetchene Island, *Wright*.)

A. glomeráta, LEDEB. Silky-canescent with mostly close short pubescence: leaves usually twice or thrice ternately parted and cleft into lanceolate or spatulate lobes: heads cymose-glomerate, fuscous or pale: flowers sparsely pilose, at least the summit of the corolla. —

Mem. Acad. Petrop. v. 564; Bess. Abrot. 63; DC. Prodr. vi. 116; Torr. & Gray, Fl. ii. 423. *A. globularia*, Ledeb. Fl. Ross. ii. 588, in part. *A. leontopodioides* & *A. corymbosa* (form with heads pedunculate), Fisch. in Bess. Abrot. & DC. l. c. — Arctic Alaska, *Seemann*, *Muir*. (Adj. Asia.)

A. globulária, Cham. Canescently pubescent: leaves once or twice ternately parted into linear or broader lobes: heads globular- or somewhat racemiform-capitate, both involucre and flowers dark purplish-brown, the latter glabrous. — Cham. in Bess. l. c.; DC. l. c. *A. Senjavinensis*, Ledeb. Fl. Ross. ii. 588, at least in part, not Bess. Perhaps an extreme arctic form of *A. Norvegica*, as was suspected by Maxim. Diagn. Pl. Jap., & Dec. xi. 534. — Arctic Alaska and islands. St. Paul's Island, *Mrs. M. Intyre*. (Adj. Asia.)

++ ++ Heads many-flowered, broad (2 to 5 lines in diameter), several or rather numerous and loosely racemose or paniculate on mostly simple stems of a foot or less in height: subarctic and subalpine, with dissected leaves and no cottony tomentum.

A. Richardsoniána, Bess. A span to near a foot high, with rather slender ascending stems from a cespitose caudex: leaves silvery-canescent with fine very close-pressed pubescence; radical twice ternately or quinately divided or parted into oblong-linear or narrower lobes (of only 2 or 3 lines in length); cauline sparse, mostly trifid: heads comparatively small (2 lines high), several or rather numerous in a strict and simple racemiform inflorescence, fuscous: corollas pilose or sometimes glabrous. — Suppl. 64. & DC. l. c. 117; Torr. & Gray, Fl. ii. 423. *A. arctica* & *A. cæspitosa*, Bess. in Hook. Fl. i. 323, 324. — Arctic coast to Bear Lake (*Richardson*, &c.), northern Rocky Mountains, and Mount Ranier, Washington Terr., *Tolmie*. (From the char. probably *A. heterophylla*, Bess. Abrot., which is said to be *A. trifurcata*, Steph. in Spreng. Syst. iii. 488, and to occur in Arct. Amer. as well as Arct. Asia to Kamtschatka.)

A. Norvégica, Fries. Rather stout, 5 to 25 inches (commonly a foot) high, from villous or sericeous-pubescent to glabrate: leaves twice 3–7-nately parted into linear or lanceolate or more dilated segments: heads large (commonly 4 or 5 lines broad), loosely racemose or racemose-paniculate, most of them long-pedunculed: bracts of the involucre broadly brown-margined: corollas yellow or turning brown, loosely pilose, rarely almost glabrous. — Fries in Liljeb. Fl. 1815, Novit. ed. 1, 56, ed. 2, 265; Reichenb. Ic. Crit. i. 74, t. 89: Bess. Abrot. 76; DC. l. c. *A. rupestris*, Fl. Dan. t. 801, not L. *A. Chamissoniana*, var. *saxatilis*, Bess. l. c. & Hook. Fl. i. 324. *A. Richardsoniana*, Gray, in Am. Jour. Sci. ser. 2, xxxiii. 239, not Bess. *A. arctica*, Gray, in Proc. Acad. Philad. 1863, 66. — Alpine and subalpine region of the Rocky Mountains, from lat. 62° to S. Colorado, Utah, and the Sierra Nevada, California. (N. E. Eu.)

Var. **Pacífica.** Robust, glabrous or glabrate up to the heads, sometimes two feet high: leaves broader; their divisions from lanceolate to cuneate, commonly laciniate. — *A. longepedunculata*, Rudolphi, ex Bess. Abrot. 77. *A. arctica*, Less. in Linn. vi. 213; Hook. & Arn. Bot. Beech. 125; DC. Prodr. vi. 119; Torr. & Gray, Fl. ii. 423. *A. Chamissoniana*, Bess. in Hook. Fl. l. c. (mainly). & Abrot. 77, t. 4, of which the largest and coarsest-leaved form is his var. *Ochotensis!* — Arctic coast to the Aleutian Islands, &c., in various forms. (Adj. E. Asia.)

A. Párryi, Gray. Rather stout, a foot or less high, wholly glabrous, leafy up to the loosely paniculate inflorescence of numerous short-pedunculed heads: leaves 2–3-pinnately parted into mostly linear thickish lobes: involucre 2 or 3 lines broad, its bracts greenish with brownish margins and with the corollas glabrous. — Proc. Am. Acad. vii. 361. — Mountains of Colorado, at Sangre de Cristo Pass, 11,000 feet, *Parry*, *Brandegee*.

++ ++ ++ Heads many-flowered, large and broad (4 lines long), in a racemose-glomerate and thyrsoid inflorescence, white-tomentose as well as the herbage.

A. Stelleriána, Bess. A foot or two high from a creeping lignescent base, robust, densely white-tomentose, the tomentum of the stem cottony: leaves obovate or spatulate in outline, sinuately or incisely pinnatifid; lobes obtuse: corolla glabrous: akenes a line and a half long, oblong, not contracted at summit; the coat utricular. — Abrot. 79, t. 5; DC. l. c. *A. Chinensis*, Pursh, Fl. ii. 521, not L. &c. — This may be what Pursh saw in herb. Lambert, from N. W. America, probably from Pallas. It is indigenous from Kamtschatka to Japan, and not improbably on the American coast. Singularly, it grows wild in large tufts on Lynn Beach, Massachusetts! Also of Sweden, Fl. Dan. t. 3045.

++ ++ ++ ++ Heads comparatively small (1 to 3 lines high and broad), variously paniculate, 12-many-flowered: flowers glabrous: herbs, or occasionally suffrutescent at base, mostly whitened (at least when young and on the lower face of the leaves) with cottony tomentum.

= Tall, with numerous amply paniculate heads, strict stems, and undivided elongated-lanceolate or linear leaves (the lowest sometimes cleft), 3 to 7 inches long: involucre oblong.

A. serráta, NUTT. Stems 6 to 9 feet high, very leafy: leaves green and glabrous above, white-tomentose beneath, lanceolate or uppermost linear, all serrate with sharp narrow teeth, pinnately veined, the earliest sometimes pinnately incised: heads rather few-flowered, less than 2 lines long, greenish, hardly pubescent. — Gen. ii. 143. *A. Ludoviciana*, var. *serrata*, Torr. & Gray, Fl. ii. 420. — Prairies and low grounds, Illinois to Dakota; first coll. by *Nuttall*.

A. longifólia, NUTT. l. c. Stem 2 to 5 feet high: leaves entire, at first tomentulose, but usually glabrate above, white tomentose beneath, linear or linear-lanceolate (1 to 5 lines wide), entire ; veins obsolete: heads usually canescent, 2 or 3 lines long. — Torr. & Gray, Fl. ii. 419, not Bess. Rocky banks, Minnesota and Nebraska to Saskatchewan and Montana; first coll. by *Nuttall*, or by Lewis & Clarke, if perhaps *A. integrifolia* of Pursh.

= = Moderately tall or sometimes low: leaves various, more or less cleft or divided, or when entire comparatively short, not filiform or very narrowly linear. Species of very difficult discrimination.

a. Involucre canescently lanate-tomentose.

A. Ludoviciána, NUTT. A foot to a yard high, simple or with virgate branches, sometimes paniculate, completely and somewhat flocculently white-tomentose, or upper face of leaves sometimes early glabrate and green: leaves from linear-lanceolate to oblong, sometimes nearly all undivided and entire; commonly the lower with a few coarse teeth or incisions, or 2-3-cleft, or irregularly 3-5-parted into lanceolate or linear entire lobes: heads glomerately paniculate, not over 2 lines long; involucre campanulate or in fruit ovoid, 12-20-flowered. — Torr. & Gray, Fl. ii. 420 (excl. var. *serrata*); Bess. Revis. Artem. in Linn. xv. 104; Gray, Bot. Calif. i. 404. *A. Ludoviciana* (with incised or subpinnatifid leaves) & *A. gnaphalodes* (with undivided leaves), Nutt. Gen. ii. 143. *A. integrifolia*, Pursh, l. c., at least in part, not L. *A. Purshiana*, Bess. Abrot. 59, & Hook. Fl. i. 323. *A. Douglasiana*, Bess. l. c., an entire-leaved less white-tomentose Western form. *A. Hookeriana*, Bess. l. c.; the plant taken to be this, of "Rocky Mts., Saskatchewan, &c., *Drummond*," in herb. Hook., but not ticketed, is a tall and large-leaved form. — Plains and banks, Saskatchewan to Texas, east to Illinois and Upper Michigan, and west to Brit. Columbia, California, and Arizona. The *Wild Sage* of Lewis & Clarke, at least in part. (Adj. Mex.)

b. Involucre not lanate (at least when fully developed), from pilose-pubescent or minutely canescent to glabrate or glabrous: divisions of the leaves broad or narrow, but not filiform.

A. Mexicána, WILLD. Intermediate between preceding and following, paniculately branched, 2 to 4 feet high, less tomentose: leaves narrow-lanceolate to linear, commonly attenuate, some 3-5-cleft or parted; radical cuneate, incisely pinnatifid or trifid · heads very numerous in an ample loose panicle, many pedicellate, 1 to 2 lines long: involucre campanulate, arachnoid-canescent or glabrate, largely scarious, 10-20-flowered. — Spreng. Syst. iii. 490; Less. in Linn. v. 163; DC. Prodr. vi. 114; Bess. Revis. l. c. 106. *A. Indica*, var. *Mexicana*, Bess. Abrot. 56. *A. vulgaris*, var. *Americana*, Bess. in Linn. xv. 105. *A. vulgaris*, var. *Mexicana*, Torr. & Gray, Fl. ii. 421. *A. Ludoviciana*, in part, Gray, Pl. Wright. ii. 98. *A. cuneifolia?* & *A. Lindheimeriana*, Scheele in Linn. xxii. 162, 163. *A. Ludoviciana*, var. *Mexicana*, forma *tenuifolia*, Gray, Pl. Wright. ii. 98, from New Mexico, &c., is a very narrow-leaved variety, with strict panicle. — Dry plains, Arkansas and Texas to Arizona and S. W. Nevada. (Mex.)

A. vulgáris, L. (MUGWORT.) Paniculately branched: leaves white with cottony tomentum beneath, green and soon glabrate or glabrous above, usually bipinnately cleft or parted and laciniate, and the lobes lanceolate or coarser; upper sometimes linear: heads numerous and glomerate-paniculate, 2 lines long: involucre mostly oblong-campanulate, scarious, sparingly arachnoid but usually glabrate. — Michx. Fl. ii. 128; Pursh, Fl. ii. 522; Torr. & Gray, Fl. l. c., excl. var. *Mexicana*. — The common European form is apparently indigenous at Hudson's Bay, &c., and is naturalized in Canada (*A. Indica, Canadensis*, Bess. in Hook. Fl.) and Atlantic States. (Eu., Asia.)

Var. Tilésii, LEDEB. Robust, leafy to the very summit: heads glomerate, fuscous: involucre broadly campanulate, arachnoid-cottony when young, but glabrate, many-flowered: leaves coarsely cleft and laciniate, the lobes lanceolate, attenuate-acute. — Fl. Ross. ii. 586. *A. Tilesii.* Ledeb. Mem. Acad. Petrop. v. 568; Bess. Abrot. 70; Less. in Linn. vi. 214; DC. l. c.; Torr. & Gray, l. c. — Arctic coast to Unalaska. (Adj. E. Asia.)

Var. Califórnica, BESS. Less branched or simple-stemmed, with more naked panicle: heads of var. *Tilesii* or smaller, or at maturity sometimes oblong, glabrate. — Bess. in Linn. xv. 91 (founded on *A. integrifolia*, Less.); Torr. & Gray, l. c.; Gray, Bot. Calif. i. 404. *A. heterophylla*, Nutt. Trans. Am. Phil. Soc. vii. 400. *A. Tilesii*, var. *elatior*, Torr. & Gray, Fl. ii. 422. — Northern Rocky Mountains to Alaska, south to the coast of California and in the Sierra Nevada.

A. franserioídes, GREENE. Habit of *A. vulgaris*, glabrous throughout, or minutely and obscurely cinereous-puberulent: stem rather stout, 2 or 3 feet high: leaves comparatively ample, green above, pale and barely cinereous beneath; lower bipinnately and upper simply pinnately parted into lanceolate-oblong obtuse entire or 2-3-cleft divisions and lobes: heads numerous, loosely racemose on the branches of the leafy elongated panicle, 2 or 3 lines broad; involucre greenish, glabrous, low-hemispherical, 30-40-flowered. — Bull. Torr. Club, x. 42. *A. discolor*, Torr. & Gray in Pacif. R. Rep. ii. 126; Rothrock in Wheeler Rep. vi. 176, not Dougl. — Ronbidean's Pass, Mountains of S. Colorado, *Gunnison*. Pinos Altos Mountains, New Mexico, *Greene*. Mount Graham, Arizona, *Rothrock*.

A. discolor, DOUGL. A foot high, mostly slender, from a lignescent slender caudex, glabrous or glabrate except the lower face of the leaves: these white with close cottony tomentum (which is rarely deciduous), 1-2-pinnately parted into narrow linear or lanceolate entire or sparingly laciniate divisions and lobes: heads glomerate in an interrupted spiciform or virgate panicle, 1 or 2 lines high: involucre hemispherical-campanulate, greenish and scarious, glabrous or soon becoming so, 20-30-flowered. — Dougl. in herb. Hook.; Bess. Suppl. & DC. Prodr. vi. 109; Torr. & Gray, l. c.; Gray, Bot. Calif. i. 404. *A. Ludoviciana* & *A. Michauxiana*, Bess. Abrot. 38, 71, & in Hook. Fl. l. c., not Nutt. — Mountains of Brit. Columbia and Montana to Utah, Nevada, and the Sierra Nevada in California.

Var. incómpta. A stouter form, with coarser and less dissected leaves, having mostly broader (sometimes short-oblong) lobes, or the upper entire. — *A. incompta*, Nutt. Trans. Am. Phil. Soc. vii. 400. — Rocky Mountains from Montana and Wyoming to Washington Terr., Nevada, and the Sierra Nevada in California.

= = = = Not tall, sometimes low, herbaceous or suffrutescent at base: leaves or their divisions narrowly linear, simple, small: heads 15-20-flowered, in a narrow thyrsoid or spiciform panicle.

A. Lindleyána, BESS. A foot or two, rarely only a span high, slender, with thin flocculent tomentum soon deciduous, or persisting on the lower face of the mostly entire leaves (these inch or less long, a line or much less wide, the lower occasionally with 2 or 3 small lobes): heads barely 2 lines high, loosely spicate on the simple stem or paniculate branches of the inflorescence: involucre sparingly pubescent or glabrate, pale fuscous — Abrot. 35, & in Hook. l. c., described from herb. Lindl. *A. pumila*, Nutt. Trans. Am. Phil. Soc. l. c. 399, a dwarf state. — Sandy banks of the Columbia River and its tributaries, Idaho, Oregon, and Washington Terr., *Douglas, Nuttall, Hall* (distrib. as *A. discolor?*), *Brandegee*. Also on the sands of the sea-shore near the Straits of Juan de Fuca, *Douglas*.

A. Wríghtii, GRAY. Cinereous or canescent with minute pubescence, or radical shoots sometimes white-tomentose, 10 to 20 inches high, very leafy up to the strict virgate panicle: leaves pinnately 5-7-parted into very narrow linear and by revolution filiform entire divisions: heads numerous and crowded: involucre minutely cinereous-canescent, glabrate in age. — Proc. Am. Acad. xix. 48. — Plains of S. Colorado and adjacent New Mexico, *Wright* (no. 1279. Pl. Wright. ii. 98, mention only), *Palmer, Greene, Rothrock* (no. 539), *Brandegee*.

= = = = Pinnately parted leaves mostly attenuate-filiform: heads simply and loosely racemose-spicate.

A. Prescottiána, BESS. Much branched from the base, a foot or two high, slender, glabrous or early glabrate, lower leaves cuneate-linear and incised or cleft at apex, slightly tomentose beneath; most of the cauline pinnately parted into 5 to 7 delicate filiform divisions (of an inch or less long): involucre glabrous, hemispherical, about 15-flowered. — Abrot. 72, & in Hook. l. c. — "Quicksand River, near the Grand Rapids of the Columbia," *Douglas*.

Described by Besser from herb. Lindl., here from herb. Hook. A peculiar and little known species, to which Douglas had applied the appropriate-name of *A. leptophylla*.

++ ++ ++ ++ Heads small and narrow, very few-flowered: flowers glabrous: stems woody at base: habit of the following section.

A. Bigelóvii, Gray. Silvery-canescent throughout, a foot high: leaves from oblong- to linear-cuneate, mostly 3-toothed at the truncate apex, about half-inch long: heads very numerous and crowded in the oblong or virgate thyrsiform panicle, tomentoso-canescent, containing only one or two hermaphrodite and as many female flowers, all fertile. — Pacif. R. Rep. iv. 110. — Rocky banks and cañons, Colorado, on the Upper Canadian and Arkansas, common where the latter leaves the mountains; first coll. by *Bigelow*.

§ 3. SERIPHÍDIUM, Bess. Heads homogamous, the flowers all hermaphrodite and fertile: receptacle not hairy. — Gray, Proc. Am. Acad. xix. 49.

* Anomalous species of Southwestern border, tall, mainly herbaceous, 3 to 5 feet high, with ample and naked compound panicles; the heads nodding in anthesis, as is common in the genus.

A. Parishii, Gray. Frutescent, cinereous-puberulent: leaves linear and entire, below passing into elongated slender-spatulate and with 3-toothed apex: panicle a foot or two long, loose: heads mostly pedicellate (2 lines long): involucre oblong-campanulate, canescent, 6-7-flowered: akenes sparsely arachnoid-villous! — Proc. Am. Acad. xvii. 220. — Interior of Los Angeles Co., California, *Parish*.

A. Pálmeri, Gray. Wholly or nearly herbaceous, obscurely puberulent; but leaves white beneath with close cottony tomentum, pinnately 3-5-parted into long narrowly linear entire lobes, their margins revolute: heads glomerate on the branches of the open panicle, hemispherical, less than 2 lines in diameter: involucre greenish, about 20-flowered; many of the flowers subtended by scarious-hyaline bracts of the receptacle! — Proc. Am. Acad. xi. 79, & Bot. Calif. i. 618. — Jamul Valley, 20 miles south of San Diego, on the borders of California and Lower California, *Palmer*, *Miss Bird*.

* * SAGE-BRUSH or SAGE-BUSHES, low shrubs, or fruticulose, canescent or silvery with very fine and close tomentum: heads glomerate or strict in the paniculate or spiciform inflorescence, not nodding even when young: corollas sometimes turning reddish.

+— *Foliose-spicate:* heads solitary in the axils, surpassed by the rigid leaves.

A. rígida, Gray. A span to a foot high from a thick woody base or short stem, producing a profusion of rigid and slender rather simple fastigiate branches, leafy to the very top: leaves also rigid, silvery-canescent, filiform-linear, 3-5-parted or cleft, or some of the upper and fascicled ones entire (even the lower rarely inch long), most of them subtending a sessile head: involucre oblong to campanulate, 5-12-flowered, less than 2 lines long; bracts oval, hyaline-margined. — Proc. Am. Acad. xix. 49. *A. trifida*, var. *rigida*, Nutt. Trans. Am. Phil. Soc. vii. 398. — On high rocky ridges, N. E. Oregon and adjacent Idaho, *Nuttall* (without flowers), *Cusick*.

+— +— More naked-paniculate or thyrsoid, at least the upper heads or clusters exceeding the subtending leaves; these not rigid.

++ Heads comparatively small and few-flowered, mostly oblong, one or two lines long: involucral bracts rather firm in texture, well imbricated, the outer successively shorter: leaves seldom over an inch long, mostly shorter.

A. arbúscula, Nutt. Dwarf, a span or rarely a foot high, with a stout base and slender flowering branches: leaves short, cuneate or flabelliform, 3-lobed or parted, with the lobes obovate to spatulate-linear, sometimes again 2-lobed; those subtending the heads usually entire and narrow: panicle strict and comparatively simple and naked, often spiciform and reduced to few rather scattered sessile heads: involucre 5-9-flowered. — Trans. Am. Phil. Soc. l. c.; Torr. & Gray, Fl. ii. 418; Eaton, Bot. King Exp. 182; Gray, Bot. Calif. i. 405. — High mountains and elevated arid plains, Wyoming and Utah to Idaho and the Sierra Nevada, California. Two forms, passing into each other (both coll. by *Nuttall*, &c.); one with involucre more campanulate, 7-9-flowered; in the other oblong and only 4-5-flowered; sometimes the inflorescence simply spiciform, sometimes freely naked-paniculate.

A. tridentáta, Nutt. l. c. Larger, 1 to 6 (or even 12) feet high, much branched: leaves cuneate, obtusely 3-toothed or 3-lobed, or even 4-7-toothed, at the truncate summit, uppermost cuneate-linear: heads densely paniculate: involucre 5-8-flowered, its outer or accessory tomentose-canescent bracts short and ovate. — Torr. & Gray, l. c.; Eaton, l. c. — Plains and also on the drier mountains, Montana to Colorado, Washington Territory, and eastern slope of the Sierra Nevada, California, immensely abundant, the characteristic *Sage-brush* or *Sage-wood* of the region.

Var. angustifólia, GRAY. Leaves all narrow; lower spatulate-linear, barely 3-toothed at the roundish summit; upper entire and more linear, a line or less wide: heads small: shrub 3 or 4 feet high, with foliage too like that of the following species, but involucre of *A. tridentata.* — Proc. Am. Acad. xix. 49. — Arid plains, S. Idaho and W. New Mexico to the Mohave Desert and the southern borders of San Diego Co., California.

A. trífida, NUTT. l. c. A foot or two high, sometimes lower, much branched: leaves 3-cleft and 3-parted: the lobes and the entire upper leaves narrowly linear or slightly spatulate-dilated: heads numerous in the contracted leafy panicle, or spicately disposed on its branches: involucre 3-5-flowered, rarely 6-9-flowered, its outer or accessory bracts oblong to short-linear or lanceolate. — Torr. & Gray, Fl. ii. 419 (excl. var.); Eaton, l. c. — Plains and valleys, Wyoming and Utah to Washington Terr. and the Sierra Nevada, California.

++ ++ Heads somewhat larger and broader, glomerate-paniculate, 7-14-flowered: involucre short-campanulate; inner bracts more scarious: stems low, suffruticose.

= Pubescence looser, furfuraceous-tomentose: inner bracts of the involucre narrow.

A. Bolánderi, GRAY. A foot or two high: leaves all narrowly linear, half a line wide, acutish, entire, or some with one or two slender lobes: heads numerous, densely glomerate-paniculate, 14-flowered, mostly equalled or surpassed by one or two linear-subulate herbaceous accessory bracts. — Proc. Am. Acad. xix. 50. — *A. trifida*, in part, Gray, Bot. Calif. i. 405. — Mono Pass, in the eastern part of the Sierra Nevada, California, *Bolander.*

= = Canescent pubescence minute and very close: bracts of the involucre broad.

A. cána, PURSH. A foot or two high, freely branched, silvery-canescent: leaves lanceolate-linear or narrower, somewhat tapering to both ends, an inch or two long, entire, rarely with 2 or 3 acute teeth or lobes: margins not revolute: heads glomerate in a leafy contracted panicle, 6-9-flowered, rarely 5-flowered, usually with one or two linear subulate accessory bracts. — Fl. ii. 521; Bess. in Hook. Fl. & DC. Prodr. vi. 105: Torr. & Gray, l. c. *A. Columbiensis*, Nutt. Trans. Am. Phil. Soc. l. c. — Plains, Saskatchewan to Montana, Dakota, and Colorado: common only northward.

A. Rothróckii, GRAY. A foot or less high, less canescent or cinereous: leaves (inch or less long) from cuneate and obtusely 3-lobed at dilated summit to spatulate-lanceolate or the upper linear, sometimes all entire: heads 2 or 3 lines long, glomerate-paniculate, 9-12-flowered: proper bracts of the involucre all ovate or oval, glabrate. — Bot. Calif. i. 618; Rothrock in Wheeler Rep. 366, t. 13; Gray, Proc. Am. Acad. xix. 50. *A. trifida,* Gray, l. c. 405, in part. — California, in the eastern and southern part of the Sierra Nevada, *Rothrock, Bolander, &c.*, and S. Utah, *Ward, Parry.*

TRIBE VIII. SENECIONIDEÆ. p. 79.

179. TUSSILÁGO, Tourn. COLTSFOOT. (*Tussis* and *ago*, allays cough.) — Single species, indigenous to Europe and Asia, naturalized in N. America.

T. FÁRFARA, L. Low perennial herb, cottony-tomentose; with extensively creeping root-stocks, sending up in earliest spring a scape beset with alternate lanceolate bracts, and terminated by a head of yellow flowers: later developing rounded- or angulate-cordate irregularly dentate leaves on long and stout radical petioles, glabrate in age. — Wet grounds, a common weed in N. Atlantic States and Canada. (Nat. from Eu.)

180. PETASÍTES, Tourn. BUTTER-BUR. SWEET COLTSFOOT. (Πέτασος, a broad-brimmed hat, alluding to the large and broad leaves.) — Perennial herbs, of the northern temperate zone; with thickish and mostly creeping rootstocks, sending up scapiform and foliose-bracteate simple flowering stems, and ample radical leaves on strong petioles, cottony-tomentose or glabrate; the flowers whitish or purplish, in spring. — Gærtn. Fruct. ii. 406, t. 166; Grenier & Godr. Fl. Fr. ii. 89; Reichenb. Ic. Fl. Germ. t. 896-901; Benth. & Hook. Gen. ii. 438. *Nardosmia* (Cass.) & *Petasites*, DC. Prodr. v. 205, 206.

§ 1. No ligule to female flowers: an introduced plant. — *Petasites*, DC.

P. VULGÁRIS, DESF. Rootstock very stout: leaves at maturity very large, round-cordate, angulate-dentate and denticulate: heads racemosely disposed: flowers purplish. — *Tussilago Petasites*, L. — In cult. and waste grounds, spreading in the vicinity of Philadelphia, *C. E. Smith*. (Nat. from Eu.)

§ 2. Female flowers with distinct ligules: rootstocks in ours slender and creeping: leaves developing with or soon following the whitish blossoms, in spring. — *Nardosmia*, Cass.; so named from the fragrant flowers of the original species.

P. sagittáta, GRAY. Leaves from deltoid-oblong- to reniform-hastate, from acute to rounded-obtuse, repand-dentate, very white-tomentose beneath, when full grown 7 to 10 inches long: heads short-racemose becoming corymbose: ligules equalling or shorter than the disk. — Bot. Calif. i. 407. *Tussilago sagittata*, Pursh, Fl. ii. 332. *Nardosmia sagittata*, Hook. Fl. i 307, and apparently a part of *N. frigida*, Hook. — Wet ground, Hudson's Bay to Fort Franklin, west to the Rocky Mountains in Brit. Columbia, and south to those of Colorado.

P. frígida, FRIES. Leaves small (1 to 3 or 4 inches long), rounded- or oblong-cordate to reniform-hastate, sometimes even truncate at base, angulately or more deeply and sinuately lobed, the lobes entire: heads few, corymbose. — "Syll. 20," & Sum. Veg. Scand. 182. *Tussilago frigida*, L.; Fl. Dan. t. 61, not of Pursh, whose plant from Canada and New England is either fictitions or the succeeding species. *T. corymbosa*, R. Br. in Parry Voy. & Richards. App. Frankl. Journ. *Nardosmia angulosa*, Cass. Dict. xxxiv 188. *N. frigida* & *N. corymbosa*, Hook. l. c., at least mainly. — Arctic coast and west to Kotzebue Sound, the Aleutian Islands, &c. (N. Eu. & Asia.)

P. palmáta, GRAY. Leaves (7 to 10 or even 18 inches broad) round-reniform in outline, palmately 7-11-cleft to beyond the middle or deeper; the lobes oblong-lanceolate to oblong-cuneate, laciniate-dentate: scape multibracteate, bearing rather numerous heads. — Bot. Calif. i. 407. *Tussilago palmata*, Ait. Kew. ii. 188, t. 2; Pursh, l. c. *Nardosmia palmata*, Hook. l. c.; Torr. & Gray, l. c. — Wet woodlands, Newfoundland and Labrador, Canada, New England, and Wisconsin to Brit. Columbia and California. (E. Asia.)

181. CACALIÓPSIS, Gray. (Κακαλία, ancient Greek name of Coltsfoot? and ὄψις, likeness; from resemblance, if not to the ancient *Cacalia*, at least to that of Tournefort.) — Proc. Am. Acad. xix. 50. — Single known species.

C. Nardósmia, GRAY, l. c. Robust perennial, a foot or two high, floccose-woolly, at length glabrate: leaves considerably resembling those of *Petasites palmata*, alternate, long-petioled, all but 2 or 3 radical, orbicular-cordate or flabellate, 5-9-cleft or rarely parted; the lobes or divisions rather broad, incisely lobed or dentate: heads (an inch high) few or several, pedunculate, corymbosely or racemosely disposed at the naked summit of the stem: corolla honey-yellow: flowers honey-scented. — *Cacalia Nardosmia*, Gray, Proc. Am. Acad. vii. 361. *Adenostyles Nardosmia*, Gray, l. c. viii. 631, & Bot. Calif. i. 301, following Benth. & Hook. — Open pine woods, California from Mendocino Co. northward (*Bolander, Kellogg, Greene*) to Oregon and Washington Terr., *Suksdorf, Howell*.

182. LÚINA, Benth. (Anagram of *Inula*, which this genus approaches.) — Hook. Ic. Pl. t. 1139; Benth. & Hook. Gen. ii. 438. — Single species.

L. hypoleúca, BENTH. l. c. Herbaceous and simple-stemmed from a stout woody rootstock, white with appressed tomentum: stems hardly a foot high, equably leafy up to the corymbiform cyme of several small heads: leaves ovate or oval, alternate, sessile, entire, inch or less long, nervose-veiny and reticulated, the upper face soon glabrate and green, involucre 4 lines high, nearly equalling the light yellow corollas. — Gray, Proc. Am. Acad. ix. 206, & Bot. Calif. i. 409. — Cascade Mountains, on the border of Brit. Columbia, *Lyall*. Yakima Co., Washington Terr., *Brandegee*.

Var. **Califórnica**, GRAY, l. c. More densely woolly, and upper face of the leaves tardily glabrate: corolla-lobes shorter. — W. California, on Chimney Rock, Mendocino Co. (and, according to the ticket, behind Santa Cruz), *Kellogg*.

183. **PEUCEPHÝLLUM**, Gray. (Πεύκη, the Fir, φύλλον, leaf, from some likeness in foliage.) — Bot. Mex. Bound. 74; Benth. & Hook. Gen. ii. 438. *Psathyrotes* § *Peucephyllum*, Gray, Proc. Am. Acad. ix. 206. — Single species.

P. **Schóttii**, GRAY, l. c. Shrub 2 to 10 feet high, glabrous but resinous-viscid and balsamic, very much branched, rigid (the stem at base often 3 inches in diameter, including the rough bark); branches and branchlets very leafy up to the terminal heads: leaves alternate and some fascicled in the axils, nearly terete, half-inch to inch long, as it were acerose but bluntish and not very rigid, minutely impressed-punctate; the lower sometimes 3-parted: heads barely half-inch high: corollas dull yellowish, with the teeth becoming fuscous; anthers included or half-exserted. — *Psathyrotes Schottii*, Gray, Proc. Am. Acad. ix. 206, & Bot. Calif. i. 409. — Desert region of S. E. California and adjacent Arizona; first coll. by *Schott* and *Newberry;* later by *Parry, Lemmon, Pringle*, &c.

184. **PSATHYRÓTES**, Gray. (Ψαθυρότης, brittleness, from the brittle stems and branches.) — Low and pubescent or scurfy winter annuals (of Nevada and Arizona); with round-cordate or ovate petioled leaves, and rather small heads of yellowish flowers, sometimes turning purplish. — Pl. Wright. ii. 100; Proc. Am. Acad. vii. 363; also Proc. Am. Acad. ix. 206, & Bot. Calif. i. 409, excl. § 2. *Tetradymia* § *Polydymia*. Torr. in Emory Rep. 1848, 145. *Bulbostylis* (*Psathyrotus*), Nutt. Pl. Gamb. in Jour. Acad. Philad. n. ser. i. 179.

§ 1. Divaricately much branched, spreading or depressed, leafy; with solitary heads in the forks, either erect or nodding on short or slender peduncles: corollas more or less woolly at summit: style-branches glabrous, or with some very minute pubescence at or toward the tip.

P. **ramosíssima**, GRAY. Lanate, at least the stems and branches, and the young leaves covered with dense and somewhat scurfy white tomentum: leaves long-petioled, roundish, subcordate or almost cuneate at base, coarsely crenate (half-inch wide): outer bracts of the involucre 5, spatulate-obovate, much larger than the inner, the upper part spreading and foliaceous: corollas plainly yellow: akenes short-turbinate, densely long-villous. — Proc. Am. Acad. vii. 363, & Bot. Calif. l. c. *P. annua*, Gray, Pl. Thurb. 323, & Bot. Mex. Bound. 102, in part. *Tetradymia* (*Polydymia*) *ramosissima*, Torr. in Emory Rep. l. c. — Gravelly hills and rocks, along the Mohave and Gila, S. E. California and throughout adjacent Arizona; first coll. by *Emory*.

P. **ánnua**, GRAY. Furfuraceous-canescent or cinereous: leaves more dentate, seldom cordate, commonly wider than long: outer bracts of the involucre ovate-oblong or narrower, less foliaceous, rather shorter than the inner, erect: corollas more slender, pale yellow, changing sometimes to purplish: akenes oblong-turbinate, densely villous: pappus rather less copious. — Pl. Wright. ii. 100, Proc. Am. Acad. vii. 364, & Bot. Calif. l. c. *Bulbostylis* (*Psathyrotus*) *annua*, Nutt. l. c. — Saline plains, Nevada, eastern borders of California, S. Utah, and adjacent Arizona; first coll. by *Gambel*.

P. **pilífera**, GRAY. Minutely furfuraceous-tomentose: leaves dilated rhombic-obovate or roundish with cuneate base, entire; their margin and sometimes upper face and long petiole beset with very long and soft (probably viscid) many-jointed hairs: heads narrower: outer bracts of cylindraceous involucre oblong-linear, herbaceous only at summit: young akenes oblong, short-hirsute: style-branches dorsally somewhat pubescent for some distance below the truncate tip. — Proc. Am. Acad. xix. 50. — Southern Utah, near Kanab, *Mrs. Thompson, Parry*.

§ 2. Scapose, erect: corollas nearly glabrous throughout: style-branches flatter, very obtuse, externally minutely hirsute over most of the back.

P. scapósa, GRAY. Leaves all at or near the base, ovate or roundish, almost entire, short-petioled, at first loosely white-tomentose, at length glabrate: scapes or naked peduncles several, 3 or 4 inches high, bearing 3 to 7 corymbosely disposed heads, glandular-pubescent, as also the campanulate involucre: bracts of the latter all somewhat herbaceous; outer ones broadly linear or barely oblong, equalling and not unlike the inner: akenes oblong-turbinate, hirsute: pappus about half the length of the corolla. — Pl. Wright. ii. 100, t. 13. — Borders of Texas, New Mexico, and Chihuahua, near El Paso, on the Rio Grande, *Wright*. (Adj. Mex.)

185. **BARTLÉTTIA**, Gray. (*John R. Bartlett*, Commissioner of the Mexican Boundary Survey, in which this plant was discovered.) — Pl. Thurb. in Mem. Amer. Acad. v. 324; Bot. Mex. Bound. 102. — Single species.

B. scapósa, GRAY, l. c. Slender winter-annual, almost glabrous, flowering almost from the base by monocephalous scapes of 6 to 9 inches high, and later by similar peduncles terminating sparsely leafy branching stems: leaves slender-petioled, roundish or subcordate, membranaceous, repand-dentate, some 3-5-lobed: head half-inch or less high: involucre pubescent: flowers yellow: pappus rather fragile, little longer than the akene. — New Mexico, near El Paso, perhaps only below the Mexican boundary, *Thurber*, *Schott*, *G. R. Vasey*. (Adj. Mex.)

186. **CROCÍDIUM**, Hook. (Diminutive formed from κρόκη, loose thread or wool, alluding to the wool which usually persists in the axils of the leaves.) — Fl. Bor.-Am. i. 335, t. 118; Torr. & Gray, Fl. ii. 448; Benth. & Hook. Gen. ii. 440. — Single species; fl. early spring.

C. multicaúle, HOOK. l. c. Small winter annual, a span or two high, flocculent-woolly when young, soon mostly glabrate, producing many simple stems from the tuft of obovate or spatulate few-toothed sessile or short-petioled radical leaves: cauline leaves small, lanceolate to linear; head slender-pedunculate, rather small, but showy; the ray and disk deep golden yellow. — Plains and hills, British Columbia and Idaho to the northern part of California; first coll. by *Douglas*.

187. **HAPLOÉSTHES**, Gray. ('Απλόος, simple, ἐσθής, garment, the involucre of unusually few pieces.) — Pl. Fendl. 109, Pl. Wright. i. 125, & Bot. Mex. Bound. 102. — Single species.

H. Gréggii, GRAY, l. c. Somewhat fleshy, herbaceous or suffrutescent, a foot or two high, fastigiately branched, glabrous, leafy up to the loose cymes of a few slender-pedunculate naked heads: leaves all opposite, very narrowly linear or filiform, entire; the lower connate at base: heads 2 or 3 lines high: flowers yellow: ligules 1 or 2 lines long. — Saline soil, S. E. Colorado and W. Texas to the Mexican border, *Wright*, *Bigelow*, *Parry*, &c. (Adj. Mex.; first coll. by *Gregg*.)

188. **LEPIDOSPÁRTUM**, Gray. (Λεπίς, a scale, and σπάρτον, the Broom plant.) — Proc. Am. Acad. xix. 50. — Single species.

L. squamátum, GRAY, l. c. A rigid Broom-like shrub, 4 or 5 feet high; seedling plants floccose-tomentose and with spatulate entire alternate leaves of half-inch or more in length; but the primary branches and whole subsequent growth glabrous or nearly so, and beset with small and thickish appressed green scales in place of leaves: heads terminal or more commonly spicate-paniculate on the slender branchlets, 3 to 5 lines long: involucre very glabrous, 10-18-flowered: corollas pale yellow. — *Linosyris squamata*, Gray, Proc. Am. Acad. viii. 290. *Tetradymia* (*Lepidosparton*) *squamata*, Gray, Proc. Am. Acad. ix. 207, & Bot. Calif. i. 408; var. *Breweri* & var. *Palmeri* are mere varying forms. *Carphephorus junceus*, Durand in Pacif. R. Rep. v. 8, not Benth. Has been mistaken also for a *Baccharis*. — Dry hills and arid plains, from Los Angeles Co., California, to Arizona; first coll. by *Heermann* and by *Brewer*.

189. **TETRADÝMIA**, DC. (Τετράδυμος, four together, the heads of the principal species only 4-flowered.) — Low and rigid shrubs (of the arid interior of N. America), sometimes spinescent, canescently tomentose; with alternate and sometimes fascicled narrow and entire leaves, rather large cymose or clustered heads of yellow flowers, and a copious white pappus. — Prodr. vi. 540; Deless. Ic. Sel. iv. t. 60; Torr. & Gray, Fl. ii. 447.

§ 1. EUTETRADÝMIA. Involucre 4-flowered, of 4 or 5 bracts; pappus extremely copious; akenes either very villous, glabrate, or glabrous, varying even in the same species; undershrubs, a foot or two high.

T. canéscens, DC. Permanently canescent with a dense close tomentum, unarmed, fastigiately branched; leaves from narrowly linear to spatulate-lanceolate, an inch or less long; heads half to three-fourths inch long, most of them short-pedunculate. — Prodr. l. c.; Deless. Ic. iv. t. 60. — Hills and plains, along with *Artemisia tridentata*, N. Wyoming and Brit. Columbia to New Mexico, Arizona, and eastern borders of California. Passes freely into

Var. **inérmis**, GRAY, Bot. Calif. i. 408. A form with shorter and crowded branches, shorter leaves more inclined to spatulate and lanceolate, and smaller heads — *T. inermis*, Nutt. Trans. Am. Phil. Soc. vii. 415; Torr. & Gray, l. c. — The commonest and almost the only form eastward.

T. glabráta, GRAY. Whitened with looser at length deciduous tomentum, unarmed; branches more slender, spreading; leaves at length naked and green, primary ones slender-subulate, cuspidate, on young shoots appressed, half-inch long; those of fascicles in their axils spatulate-linear, fleshy, pointless; heads mostly short-pedunculate; involucre often glabrate; akenes as far as known very villous. — Pacif. R. Rep. ii. 122, t. 5; Eaton, Bot. King Exp. 193; Gray, Bot. Calif. i. 408. — Common in Utah and to the eastern borders of California and S. E. Oregon; first coll. by *Beckwith*.

T. Nuttállii, TORR. & GRAY. Pubescence and foliage of *T. canescens*, var. *inermis*, bearing rigid divergent spines in place of primary leaves; leaves of the axillary fascicles mostly spatulate; heads more glomerate. — Fl. l. c.; Eaton, l. c. *T. spinosa*, Nutt. l. c., not Hook. & Arn. — Utah and Wyoming or S. Idaho, *Nuttall, Watson*.

§ 2. LAGOTHÁMNUS, Torr. & Gray, l. c. Involucre 5–9-flowered, of 5 or 6 broader bracts; proper pappus less copious, reduced nearly or quite to a single series of bristles, which are covered by a false pappus of the extremely long very soft and white woolly hairs which densely clothe the akene; shrubs 2 to 4 feet high, at least the branches densely white-tomentose. — *Lagothamnus*, Nutt. Trans. Am. Phil. Soc. vii. 416.

T. spinósa, HOOK. & ARN. Branches divaricate, rigid, bearing rigid and straight or recurved spines in place of primary leaves; secondary leaves fascicled in their axils, small, fleshy, linear-clavate, glabrous or glabrate; heads scattered, pedunculate, fully half-inch long; pappus of comparatively rigid capillary bristles, somewhat surpassing the wool of the akene. — Bot. Beech. 360; Torr. & Gray. l. c.; Gray, Bot. Calif. l. c. — *Lagothamnus microphyllus* & *L. ambiguus*, Nutt. Trans. Am. Phil. Soc. vii. 416. — S. Wyoming and Utah to Idaho, E. Oregon, and along the southeastern borders of California to border of Arizona.

T. comósa, GRAY. Branches erect, elongated; primary leaves linear, soft, floccose-tomentose; the earlier 2 or 3 inches long and 2 lines wide, plane; those of the branches often filiform and deciduous, some of the upper changed to long and soft spines; fascicled secondary leaves wanting, or fewer and like those of *T. spinosa*; heads corymbose or glomerate at the summit of the branches; pappus finer and more scanty, concealed by the long wool of the akene. — Proc. Am. Acad. xii. 60; Bot. Calif. ii. 458. — N. W. borders of Nevada (*Lemmon*), San Bernardino and San Diego Counties, California, *Parry, Lemmon, Parish, Cleveland*.

190. RAILLARDÉLLA, Gray. (Diminutive of *Raillardia*, an allied Hawaiian genus of shrubs.) — Perennial and mostly scapose herbs of the Sierra Nevada, California, intermediate between the *Senecionideæ* and the *Helenioideæ*. Leaves entire, narrow; cauline alternate or none: head solitary, with yellow flowers; in summer. — Proc. Am. Acad. vi. 550 (§ of *Raillardia*), & in Benth. & Hook. Gen. ii. 442; Bot. Calif. i. 416.

§ 1. Genuine species, with creeping rootstocks, producing rosulate clusters of spatulate-lanceolate or narrower thickish leaves, and occasionally one or two small ones near the base of the otherwise naked elongated simple scape, which is terminated by the solitary (commonly inch long) head: pappus-bristles 15 to 20 or more, conspicuously short-plumose, white: no hirsute pubescence, but involucre and upper part of scape glandular.

R. argéntea, GRAY, l. c. Rootstocks extensively creeping, somewhat lignescent: leaves silvery with a silky tomentum, inch or two long: scape 2 to 4 inches high: head narrow, in depauperate specimens 7-8-flowered, but usually about 15-flowered: no rays. — High Sierra Nevada (9,000 to 11,000 feet) from the San Bernardino Mountains to Lassen, *Brewer, Greene, Lemmon*, &c.

R. scapósa, GRAY, l. c. Somewhat pubescent, but no tomentum, glabrate: leaves 1 to 4 inches long: scape 4 to 10 inches high: involucre cylindraceous, in depauperate plants 10-12-flowered, in others 20-30-flowered: corollas light yellow. — Sierra Nevada above and east of the Yosemite, at 8,000 to 9,000 feet; first coll. by Brewer and *Bolander;* the latter found some specimens with incipient rays, connecting with

Var. Eiseni. A small form: heads with 3 or 4 deformed rays. — *R. Eiseni*, Kellogg in herb. Calif. Acad. — Mountains of King's River, Fresno Co., *G. Eisen.*

R. Pringlei, GREENE. Rootstock stout and branching: leaves glabrous and smooth, thickish, some obscurely denticulate, 3 or 4 inches long, 3 or 4 lines wide: scape 10 to 18 inches high: involucre campanulate, about 40-flowered, of correspondingly numerous and more distinct bracts: flowers orange-yellow, 6 to 10 of them conspicuously radiate: pappus-bristles rather fewer (15 to 18) and rather less plumose than in the foregoing. — Bull. Torr. Club, ix. 17. — High mountains of N. California, west of Mount Shasta, *Pringle*.

§ 2. Anomalous species, hirsute, leafy-stemmed, perhaps some of the central flowers infertile.

R. Muírii, GRAY. About a foot high, roughish-hirsute, leafy below, sparsely so and bearing stipitate glands toward the summit: leaves inch long, lanceolate-linear, acute, closely sessile; radical ones unknown: heads terminal and one or two lateral, half-inch high, wholly discoid: involucre campanulate, hirsute, its narrow bracts distinct to the base: akenes oblong with tapering base: pappus of 11 or 12 somewhat more aristiform and rather less plumose bristles than in preceding species. — Bot. Calif. ii. 618. — In the Sierra Nevada, probably southward, but station unknown, *Muir*. Too little known.

191. ÁRNICA, L. (Thought to be a corruption of *Ptarmica*.) — Perennial herbs, of the northern temperate and arctic zones; with erect stems, either quite simple or branching above, opposite leaves (or upper occasionally alternate), and comparatively large long-pedunculate heads of yellow flowers; the rays usually elongated, rarely wanting. Anthers yellow except in the last species. Fl. summer. — Gærtn. Fr. t. 173; Schkuhr, Handb. t. 248; Torr. & Gray, Fl. ii. 449.

   * Radical leaves roundish and sessile in an ample rosulate cluster. Atlantic U. S.

A. nudicaúlis, NUTT. Hirsute: stem robust, 1 to 3 feet high, simple and bearing few heads, or loosely paniculate with many: leaves denticulate or nearly entire; radical 2 to 5 inches long; cauline only one or two remote pairs up to the inflorescence, small, oval, closely sessile: rays half-inch long. — Gen. ii. 164; Ell. Sk. ii. 333; DC. Prodr. vi. 318; Torr. &

Gray, l. c. *A. Claytoni*, Pursh, Fl. ii. 527. *Doronicum acaule*, Walt. Car. 205. *D. nudicaule*, Michx. Fl. ii. 121. — Pine barrens, &c., Penn. to Florida.

* * Radical leaves mostly cordate at base, on slender or sometimes winged petioles: rootstocks slender and creeping. Pacific and Rocky Mountain species.

+ Rays wanting or rarely some rudiments: cauline leaves sometimes by disjunction alternate, some of them petioled, irregularly dentate: heads rather numerous, paniculate.

A. parviflóra, GRAY. A foot high, slender, pubescent, even the peduncles but slightly glandular; leaves narrowly deltoid or oblong, truncate or abrupt at base, an inch or two long: involucre 4 or 5 lines high, about 20-flowered; its linear bracts sparsely pubescent: akenes not pubescent, minutely glandular. — Proc. Am. Acad. vii. 363, & Bot. Calif. i. 415. — California, in Humboldt Co., *Bolander*. Also at some station north of San Francisco Bay, *G. R. Vasey*.

A. discoídea, BENTH. A foot or two high, stouter, more or less villous and viscid: radical and lowest cauline leaves from ovate with truncate or abruptly cuneate base to cordate, not rarely wing-petioled: involucre half-inch high, 30-50-flowered, usually very villous and glandular; its bracts lanceolate or linear: akenes pubescent. — Pl. Hartw. 319; Gray, Bot. Calif. l. c., with a part of *A. cordifolia*. — Wooded hills in the coast ranges of California, from San Luis Obispo Co. northward to Washington Terr.; first coll. by *Hartweg*. Northwardly seems to pass into *A. cordifolia*.

+ + Rays conspicuous and elongated, rarely wanting: cauline leaves all opposite, in one or two or at most three pairs, broad, usually membranaceous, dentate or denticulate.

A. cordifólia, HOOK. A foot or two, or when alpine a span or two high, pubescent, or the stems hirsute and peduncles villous; lower cauline as well as radical leaves long-petioled, deeply cordate, yet sometimes only ovate; upper cauline small, sessile: heads few, in smaller plants solitary: involucre two-thirds inch long, pubescent or villous: rays commonly inch long: akenes more or less hirsute. — Fl. i. 331; Torr. & Gray, Fl. ii. 430. *A. macrophylla*, Nutt. Trans. Am. Phil. Soc. l. c. 407. *Senecio Cumingii*, Klatt in Abh. Nat. Hist. Gesellsch. xv. 9, is either this or the next. — Woods and high mountains, Brit. Columbia, and mountains near Saskatchewan, to those of Colorado, Utah, Nevada, and eastern borders of the Sierra Nevada, California.

Var. eradiáta is an ambiguous form; with smaller and rayless heads, and oblong-ovate at most subcordate leaves. — E. Oregon, Montana, &c. Transition to *A. Parryi*.

A. latifólia, BONG. Minutely pubescent or commonly glabrous, with smaller heads than the preceding: only radical leaves cordate or subcordate and petioled; cauline 2 or 3 pairs, equal, ovate or oval, usually sharply dentate, closely sessile by a broad base, or lowest with contracted base: akenes commonly glabrate or glabrous. — Veg. Sitch. 147; Torr. & Gray, l. c. *A. Menziesii*, Hook. Fl. i. 331, t. 111. — Pine woods, Alaska and Brit. Columbia to Oregon, and Rocky Mountains to Colorado and Utah; first coll. by *Menzies*.

Var. viscídula. Viscidly pubescent: cauline leaves less broad at base: heads rather larger: akenes pubescent. — High Sierra Nevada, California, *Greene, Pringle*. And a very similar plant from Sitka.

* * * No cordate leaves; radical leaves petioled, tapering or sometimes abrupt at base: rootstocks usually creeping and slender. Western and Northern species.

+ Leafy to the top: cauline leaves very seldom less than 4 pairs, and the upper not conspicuously diminished: heads several or few, or in smaller plants solitary.

++ Heads all with rays half-inch or more long: plants a foot or two high: the species confluent.

A. amplexicaúlis, NUTT. Slightly pubescent or almost glabrous: leaves from ovate to lanceolate-oblong, acute or acuminate, all the cauline sessile by a half-clasping base, saliently and very acutely dentate: akenes hirsute-pubescent. — Trans. Am. Phil. Soc. vii. 408; Torr. & Gray, l. c. — Woods and shaded rocks, Oregon to Brit. Columbia, *Nuttall, Lyall, Wallace*, &c. Broad-leaved forms much resembling the preceding, except in more leafy stems and want of cordate radical leaves: narrower-leaved forms nearly pass into the succeeding.

A. Chamissónis, LESS. From tomentulose- or villous-pubescent to nearly glabrous: leaves oblong or oblong-lanceolate, denticulate or dentate, acute or obtuse; lowest tapering into a margined petiole, upper broad at base (sometimes ovate-lanceolate) and somewhat clasping: akenes hirsute-pubescent. — Less. in Linn. vi. 238; DC. Prodr. vi. 317; Torr. &

Gray, l. c. partly; Gray, Bot. Calif. l. c. *A. mollis*, Hook. Fl. i. 231 ; Torr. & Gray, Fl. l. c.; Torr. Fl. N. Y. t. 60, a form with comparatively few and mostly broad leaves. *A. lanceolata*, Nutt. Trans. Am. Phil. Soc. l. c. *A. latifolia*, Gray, Bot. Calif. ii. 458, & i. 415, in part, almost glabrous broad-leaved form. — Unalaska and Sitka to the Sierra Nevada, California, and mountains of Utah and Colorado; east to L. Superior, Mount Washington, and the mountains of Lower Canada. A form with comparatively narrow leaves, N. Maine and Lower Canada, *Goodale, Allen*, &c.

A. longifólia, EATON. Many-stemmed in a tuft, minutely puberulent: cauline leaves elongated-lanceolate, tapering to both ends, entire or denticulate, somewhat nervose (3 to 6 inches long), lower with narrowed bases connate-vaginate ; heads corymbosely disposed, short-peduncled: akenes minutely glandular, not hairy. — Bot. King Exp. 186. — On rocks, in the mountains, at 9,000 feet, from above Summit (*Jones, Pringle*) to Kern Co. (*Rothrock*) in California, Clover Mountains, Nevada (*Watson*), and Wahsatch Mts. (*Jones*) in Utah.

A. foliósa, NUTT. Tomentose-pubescent, strict: leaves lanceolate, denticulate, nervose; upper partly clasping by narrowish base, lower with tapering bases connate: heads short-peduncled, rarely solitary: akenes hirsute-pubescent or glabrate. — Trans. Am. Phil. Soc. vii. 407 (excl. var. *nana*); Gray, Bot. Calif. i. 416. *A. Chamissonis*, Torr. & Gray, l. c., in part. *A. montana*, Hook. Fl. l. c., in part. — Wet meadows and mountain-sides, Saskatchewan to Oregon, N. California, and southward along the Sierra Nevada, and in the Rocky Mountains south to Colorado.

Var. **incána**, GRAY, Bot. Calif. l. c. White with very soft floccose tomentum. — Wet meadows, mostly in water, in the Sierra Nevada, California and adjacent Nevada; first coll. by *Brewer* and *Torrey*.

++ ++ Heads rayless: stems leafy even on the flowering branches.

A. viscósa, GRAY. A foot or less high, fastigiately branching, very viscid-pubescent: leaves small (inch or less long), ovate-oblong, entire, closely sessile, but not connate at base: involucre 4 lines high, considerably shorter than the (25 or 30) flowers: corollas pale yellow: akenes glandular-hirsute. — Proc. Am. Acad. xiii. 374, & Bot. Calif. ii. 458. — N. California, on Mt. Shasta at 8,000 feet, *Gray & Hooker*.

+– +– Less leafy: cauline leaves one or two (rarely three) pairs, and the upper mostly small.

++ ++ Heads rayless, mostly 3 to 5 and rather short-peduncled at the naked summit of the stem.

A. Párryi, GRAY. A foot or less high, slender, simple, somewhat hirsutely pubescent and above glandular: leaves membranaceous, commonly denticulate; radical oval to ovate-oblong (1 to 3 inches long), abruptly or cuneately contracted at base into a short margined petiole; cauline remote: involucre hirsute and glandular, half-inch or less high: occasionally some outermost corollas ampliate: akenes glabrous or with a few sparse hairs. — Am. Nat. viii. 213. *A. angustifolia*, var. *discoidea, latifolia*, Gray in Am. Jour. Sci. ser. 2, xxxiii. 238. *A. angustifolia*, var. *eradiata*, Gray, Proc. Acad. Philad. 1863, 68. — Rocky Mountains, from Colorado (on the borders of alpine region) to Wyoming, in the Wahsatch, Utah, and west to Oregon and Washington Terr.; first coll. by *Parry*.

++ ++ Heads conspicuously radiate, solitary or very few, mostly long-peduncled.

= Anthers yellow, as in all the preceding species: tube of disk-corollas hirsute.

A. Nevadénsis, GRAY. Half a foot high, puberulent, sometimes cinereous: leaves all oval or oblong, mostly obtuse, entire or a few small denticulations (inch or two long), obscurely triplinerved or 3-nerved at base; radical roundish to obovate, either abruptly contracted or tapering into slender petiole: involucre half-inch high: akenes minutely pubescent and glabrate. — Proc. Am. Acad. xix. 55. — Sierra Nevada, California. Lassen's Peak, *Mrs. Austin*, cinereous form, with rays almost inch long, bears resemblance to *Whitneya*. Peaks south of Summit, at 9,000 feet, *Pringle*, greener, roundish-leaved, with rays half-inch long.

A. alpína, OLIN. A span to 18 inches high, pubescent, hirsute, or at summit villous, strict, simple and monocephalous, occasionally 3-cephalous: leaves thickish, from narrowly oblong to lanceolate, or the radical oblong-spatulate and small uppermost linear, entire or denticulate, 3-nerved; bases of the cauline hardly at all connate: akenes hirsute-pubescent, rarely glabrate. — "Murr. Syst. Veg., 1774" (according to Fries, but not found there), "Olin, Monogr. Arnic. Upsaliæ, 1799," ex Fries, Summ. Veg. Scand. 187; Wahl. Fl. Suec. ii. 530; Gray, Bot. Calif. i. 416. *A. angustifolia*, Vahl, Fl. Dan. t. 1524; DC. Prodr. vi. 317; Torr.

& Gray, Fl. ii. 449. *A. plantaginea* & *A. fulgens*, Pursh, Fl. ii. 527. — Labrador and north to the arctic coast, west to the Aleutian Islands, south to the Sierra Nevada, California, and to Colorado in the Rocky Mountains; the southern forms comparatively large and broad-leaved. (N. Eu., Greenland.)

Var. **Lessingii**, Torr. & Gray, l. c., perhaps from Kotzebue Sound, is a thinner-leaved form, of lax habit; the akenes only sometimes glabrous, and the anthers not "black-ish." — *A. alpina*, Less. in Linn. vi. 235; Herder, Pl. Radd. ii. 110. (N. E. Asia.)

== Anthers black: leaves broad: head large, solitary. High Northwestern species.

A. **Unalaschénsis**. Less. Robust, a span or two high, hirsute or villous: leaves oblong, mostly acutish and obviously serrate or denticulate with subulate callous teeth: disk-corollas glabrous or nearly so: akenes slightly hairy or glabrate. — Linn. vi. 235; Herder, l. c. — Unalaska, and other Aleutian Islands, Behring Island, &c.; first coll. by *Chamisso*.

A. **obtusifólia**, Less. l. c. Taller or longer-pedunculate, pubescent or glabrate: leaves oblong or spatulate, very obtuse, almost or quite entire, nervose: disk-corollas "glabrous" or upper part of the tube hispidulous: akenes glabrous: resembles *A. montana*. — Unalaska, *Chamisso*. Shumagin Islands, *Harrington*.

192. **SENÉCIO**, Tourn. Groundsel. (Old Latin name of Groundsel, from *senex*, old man, in allusion to the hoary pappus.) — One of the largest known genera, very widely dispersed over the world, most of the species (all of ours) herbs; with alternate leaves, and yellow-flowered heads of middle or rarely larger size: fl. spring and summer. Minute short hairs or papillæ on the akenes of most species swell and emit a pair of spiral threads when wetted. Before wetting the akenes may be really or apparently glabrous, and after wetting become canescent. — Less. Syn. 391; DC. Prodr. vi. 340; Benth. & Hook. Gen. ii. 446, partly. — Arrangement wholly artificial.

S. Canadénsis, L., Spec. ii. 869, and Cineraria Canadensis, L. Spec. ed. 2, ii. 1244 (to which Nutt., Gen. ii. 165, gave the name of *S. Kalmii*), were said to be of "Canada, *Kalm*." They are not so indicated in the Linnæan herbarium: both are probably South European specimens. The first belongs to *S. artemisiæfolius*, Pers.; the second is a thinner-leaved form of *Cineraria maritima*, L., the *S. Cineraria*, DC. — Cineraria Carolinensis, Walt. Car. 207, is undeterminable.

S. ciliátus, Walt. Car. 208, is probably only *Erigeron Canadense*, L.

S. floccíferus, DC. Prodr. vi. 426, is *Malacothrix obtusa*, Benth.

S. Cinerária, DC., the Dusty-Miller of house-cultivation, has been found wild on the beach of San Francisco Bay, California, at Alameda.

S. Jacobǽa, L., of Europe, and some other species, occasionally occur as ballast waifs.

S. Pálmeri, Gray, a peculiar frutescent species of Guadalupe Island, off Lower California, is quite beyond our limits.

§ 1. Perennials (one or two suffruticose); with pubescence, if any, of a tomentose character, mostly floccose and when deciduous leaving the surface smooth and naked, never viscid nor obviously hirsute.

* Heads an inch or distinctly over half-inch high, very many-flowered.
+ Disk-corollas deeply 5-toothed: heads of the largest.

S. **Rugélia**, Gray. Lightly floccose-tomentose when young, soon glabrate: stems simple, a foot high from a creeping rootstock, bearing 3 to 5 naked slender-pedunculate somewhat racemosely disposed heads: leaves membranaceous; radical and lowest cauline ovate, denticulate, 2 to 5 inches long, long-petioled; others small and few, bract-like, sessile: involucre not calyculate, of about 12 linear-lanceolate thickish glabrous bracts: rays none: pappus rather sordid. — Proc. Am. Acad. xix. 54. *Rugelia nudicaulis*, Shuttlew. in coll. Rugel; Chapm. Fl. 246. — Woods, Smoky Mountains, N. Carolina and Tennessee, *Rugel, Buckley*. Style-branches capitellate-truncate and pubescent at summit, and a few obscure minute hairs on the back.

S. Pseudo-Árnica, LESS. Floccosely white-tomentose, more or less glabrate in age: stem stout, 6 to 30 inches high, equably very leafy to top, bearing solitary or several corymbosely disposed heads on stout bracteolate peduncles: leaves oblong-lingulate or the lower spatulate, denticulate or dentate, 5 to 8 inches long, sessile by a partly clasping auriculate base: involucre calyculate by few or several slender-subulate loose accessory bracts: rays numerous, half-inch or more long: pappus dull white. — Less. in Linn. vi. 240; Hook. Fl. i. 334, t. 113; DC. Prodr. vi. 358; Torr. & Gray, Fl. ii. 446. *Arnica maritima,* L. Spec. ii. 884; Pursh, Fl. ii. 528. *A. Doronicum,* Pursh, l. c. — Sea-beaches, &c., Newfoundland, New Brunswick, and border of Maine to Labrador, and west to the Aleutian Islands. (N. Asia.)

+ + Disk-corollas merely 5-toothed. Rocky-Mountain and more Western species.

++ Heads radiate.

= Alpine species of the Rocky Mountains.

S. Soldanélla, GRAY. Apparently glabrous from the first, a span high, somewhat succulent: leaves mostly radical and long-petioled, from round-reniform to spatulate-obovate, denticulate or entire; cauline one or two or none: head solitary, erect, two thirds to nearly a full inch high: involucral bracts lanceolate and a very few calyculate ones: rays 6 to 10, oblong, quarter-inch long. — Proc. Acad. Philad. 1863, 67; Porter & Coulter, Fl. Colorad. 83. — High alpine region, mountains of Colorado, *Parry, Hall & Harbour, Coulter,* &c.

S. ampléctens, GRAY. Lightly floccose-woolly at first, soon glabrate, a foot or so high, few-several-leaved, terminated by one or two long-pedunculate nodding heads: leaves thinner than in the foregoing, from denticulate to conspicuously and sharply dentate; radical obovate to spatulate, tapering into a winged petiole; cauline as large or larger (4 to 6 inches long), oblong or narrower, half-clasping or more, the upper by a broad base: involucre over half-inch high, of linear bracts and a few loose calyculate ones: rays linear, inch long or more, acute or acutely 2-3-toothed at tip. — Am. Jour. Sci. ser. 2, xxxiii. 240, & Proc. Acad. Philad. l. c. — Alpine and subalpine region, Rocky Mountains, Colorado; first coll. by *Parry.*

Var. taraxacoídes, GRAY. Only a span or two high, with fewer and smaller cauline leaves; these and the radical commonly spatulate and with tapering base, not rarely laciniately subpinnatifid: head smaller, even down to half-inch, and with rays of only the same length. — Proc. Acad. Philad. 1863, 67; Eaton, Bot. King Exp. 192. — High alpine, in the mountains of Colorado and Nevada; first coll. by *Parry.* The most dwarf forms are very unlike the type.

= = Not alpine: scapiform stem low, strict and strictly monocephalous.

S. Actinélla, GREENE. Floccosely white-tomentose, glabrate in age: simple stem 6 to 10 inches high, bearing several small and appressed linear bract-like leaves and an erect head of two thirds of an inch in height: radical leaves in a rosulate tuft, obovate-spatulate, denticulate, subcoriaceous, an inch or more long including the cuneate narrowed base or short winged petiole: involucral bracts subulate-linear: rays 9 to 12, rather conspicuous, broadly linear. — Bull. Torr. Club, x. 87. — N. Arizona, near Flagstaff, *Rusby.*

= = = Not alpine, with leafy stems a foot to a yard high, and several or few or sometimes solitary erect heads. (Here *S. Clarkianus,* if the heads were a little larger.)

S. Whippleánus, GRAY. Probably floccose when young, sprinkled with less deciduous araneose hairs; stem robust, apparently 3 or 4 feet high, naked above, with an ample loose cyme: leaves ample (6 or 8 inches long), sinuately or laciniately pinnatifid, the lobes few and irregular; cauline sessile: peduncles mostly elongated, naked: involucral bracts fleshy-thickened, oblong-linear, abruptly acuminate; a very few loose and small slender calyculate bracts: rays half-inch long. — Proc. Am. Acad. xix. 54, without char. *S. eurycephalus,* var. *major,* Gray, Pacif. R. Rep. (Bot. Whipp.) iv. 111. — Lower Sierra Nevada, at Murphy's, Calaveras Co., California, *Bigelow.* Further specimens needed. The broad heads nearly three-fourths inch high.

S. Mendocinénsis, GRAY. Lightly arachnoid-floccose, soon glabrate: stem robust, 2 or 3 feet high, leafy below, naked above, bearing a corymbiform cyme of several heads on sparsely setaceous-bracteolato peduncles: leaves somewhat succulent, irregularly repand-denticulate to dentate; radical and lower 3 to 6 inches long, oval to oblong-lanceolate, tapering into margined petioles; upper lanceolate from a broad sessile base, above reduced to subulate bracts; involucral bracts linear-subulate, and with several loose and slender calyculate ones: rays oblong, seldom half-inch in length. — Proc. Am. Acad. vii. 362, & Bot. Calif. l. c. 413. — Plains, Mendocino to Humboldt Co., California, *Bolander, Kellogg, Harford*.

S. Greénei, GRAY. Lightly floccose-tomentose, seldom a foot high, simple, bearing 1 to 3 short-peduncled heads: leaves (about inch long) coarsely dentate; radical roundish, with abrupt or somewhat cuneate base, coarsely crenate-dentate, slender-petioled; cauline few, sessile, upper lanceolate and entire, sometimes all small and bract-like: heads two-thirds inch long: bracts of involucre linear, no outer calyculate ones: rays deep orange, half-inch or more long: style-tips of disk-flowers conspicuously penicillate-margined and with a central cusp. — Proc. Am. Acad. x. 75, & Bot. Calif. i. 412. — Wooded mountain-side, near the Geysers in Lake Co., California, *Greene*.

S. megacéphalus, NUTT. About a foot high, loosely floccose-woolly, tardily glabrate, leafy: leaves entire, lanceolate, or the radical spatulate-lanceolate and tapering into a petiole, and uppermost cauline attenuate, thickish (obscurely glandular under the wool?): heads 1 to 3, short-peduncled (8 lines to an inch high); involucre calyculate by some very loose and setaceous-subulate elongated accessory bracts; sometimes the true bracts and peduncles bear a few hirsute hairs besides the loose wool: rays over half-inch long. — Trans. Am. Phil. Soc. l. c. 410; Torr. & Gray, Fl. ii. 438. — Mountains of Idaho, *Nuttall, Watson,* and Rocky Mountains, at 5,000 to 8,000 feet, near British Boundary, *Lyall, Canby.*

++ ++ Heads rayless, nodding: some sparse crisped hairs in place of tomentum: caudex hardly any; the root a cluster of fibres.

S. Bigelóvii, GRAY. Robust, 2 or 3 feet high, leafy up to near the racemiform or simply paniculate inflorescence, pubescent with some sparse crisped hairs when young, and with mere traces of arachnoid caducous wool, at length glabrate: leaves from elongated-oblong to lanceolate, denticulate or more dentate, acute or acuminate; radical and lower cauline 3 to 6 inches long, abrupt at base and naked-petioled, or tapering into a winged petiole or partly clasping base; upper lanceolate with partly clasping base: heads in small plants few or solitary, in larger ones several, nodding on their peduncles: involucre very broadly campanulate: its bracts lanceolate, thickish; a few small and loose subulate accessory bractlets at base. — Pacif. R. Rep. iv. 111; Porter & Coulter, Fl. Colorad. 83; Rothrock in Wheeler Rep. 178. With var. *Hallii,* Gray, Proc. Acad. Philad. l. c. (more sessile-leaved), and var. *monocephalus,* Rothrock. l. c. (smallest form). — Mountains of Colorado, New Mexico, and Arizona, at 8,000 to 10,000 feet; first coll. by *Bigelow*.

* * Heads middle-sized or small (half-inch or less).

+ Nodding on the paniculate pedicels in anthesis, rayless, a few loose setaceous or subulate bractlets at their base: very early glabrate or quite glabrous leafy-stemmed plants: leaves at most dentate, all either petioled or attenuate at base.

S. Rúsbyi, GREENE. Stem 2 to 4 feet high: leaves very obscurely pruinose-puberulent under a lens, ovate-lanceolate, callous-denticulate; the lower (3 to 6 inches long) with abrupt or truncate base and winged petiole with dilated and somewhat auriculate half-clasping insertion; upper cuneately contracted into the winged petiole; the small uppermost closely sessile, attenuate-acuminate: heads (4 or 5 lines long) less nodding than in the next, almost hemispherical. — Bull. Torr. Club, ix. 64, at least as to pl. Rusby. — New Mexico, in the Mogollon Mountains. *Rusby.* Apparently in Santa Catalina Mountains, Arizona, *Lemmon,* but specimens insufficient. Nearly related to the following: root nearly of the preceding.

S. cérnuus, GRAY. Quite glabrous, usually more slender, 2 or 3 feet high: leaves lanceolate or the larger oblong-lanceolate, entire, denticulate, rarely with a few scattered coarser teeth, all tapering at base into a barely margined petiole, or upper into a narrowed not clasping base: heads (4 to almost 6 lines long) several or numerous in the panicle, most of them decidedly nodding: involucre narrow-campanulate: flowers pale yellow. — Am. Jour. Sci. ser. 2, xxxiii. 10; Porter & Coulter, Fl. Colorad. 82. — Mountains of Colorado, wholly below the alpine region; first coll. by *Parry.*

+ + Heads erect, mostly radiate, occasionally rayless in same species.

++ Stem frutescent below.

S. Lemmóni, GRAY. Loosely much branched, early glabrate and smooth: main stems decidedly woody: branches slender, spreading, very leafy below, nearly naked at summit, bearing several or numerous loosely cymose slender-pedunculate heads: leaves somewhat succulent, lanceolate, irregularly and sparsely dentate with salient teeth, attenuate below and with a dilated cordate-clasping base, or the lower tapering into a naked petiole; uppermost small, linear, entire: heads 4 or 5 lines high: rays about 12; disk-flowers 20 or more. — Proc. Am. Acad. xvii. 220. — Santa Cataliua Mountains, S. Arizona, *Lemmon*.

++ ++ Stems herbaceous, numerously leafy to the top: leaves all rounded-subcordate and angulately somewhat lobed, palmately veined and reticulate-venulose, petioled: heads small and numerous in a compound cyme.

S. Hartwégi, BENTH. Flocculent-tomentulose when young, or nearly glabrous: stems 2 or 3 feet high from a somewhat tuberous rootstock: leaves chartaceo-membranaceous (2 to 4 inches broad, and petiole inch or two long), the margin with 7 to 9 short angulate lobes or coarse teeth, and sinuses denticulate: veinlets minutely reticulated: heads 3 or 4 lines long, crowded: involucre narrow-campanulate, 12-20-flowered; its bracts lanceolate, short: rays few. — Pl. Hartw. 18, form with leaves tomentulose beneath. *S. Seemanni*, Schultz Bip. in Seem. Bot. Herald, 311, glabrous form. — Cañons, S. Arizona, near Fort Huachuca, *Lemmon*. (Mex.; of a Mexican type unlike any other N. American.)

++ ++ ++ Stems numerously and nearly equably leafy to the top: leaves pinnately veined, not conspicuously reticulated, from entire to laciniate-dentate, never divided or dissected, nor narrowly linear: glabrous, or very early glabrate and smooth, seldom a vestige of wool at anthesis.

= Low, alpine: heads subsolitary, radiate.

—— **S. Fremónti**, TORR. & GRAY. Many-stemmed from a thickish caudex, a span to a foot high: leaves thickish, from rounded-obovate or spatulate to oblong (inch or sometimes 2 inches long), obtuse, obtusely or acutely dentate, sometimes even pinnatifid-dentate, lower abruptly contracted into a winged petiole; uppermost sessile by broadish base: heads half-inch high, short-peduncled, subtended by a few short loose bractlets: rays 3 to 5 inches long. — Fl. ii. 445. — Alpine region of the Rocky Mountains (first coll. by *Fremont*), from near Brit. boundary to S. Colorado, Utah, and Lassen's Peak, California: passing to

Var. occidentális, GRAY. More slender, with rounder leaves and heads longer-peduncled; in high alpine stations becoming very dwarf, and flowering almost from the ground. — Bot. Calif. i. 618. — Sierra Nevada, California, at 10,000 to 12,000 feet, *Rothrock*, &c. Also Rocky Mountains of N. Wyoming and Montana, at 7,000 to 8,000 feet, *Lyall*, *Parry*, very dwarf.

= = Rather low, with numerous cymosely paniculate and small heads, always rayless.

S. rapifólius, NUTT. About a foot high: leaves ovate or oblong, throughout very sharply and unequally dentate, rather fleshy; radical tapering into a petiole, cauline mostly clasping by a broad subcordate base: heads 3 lines high, about 15-flowered: involucral bracts 8 to 10, narrowly oblong. — Trans. Am. Phil. Soc. vii. 409: Torr. & Gray, Fl. ii. 441. — Rocky Mountains, Wyoming, about the sources of the Platte, *Nuttall*, *Fremont*, &c.

= = = Tall, with corymbosely cymose and radiate heads: involucre setaceously few-bracteolate, campanulate or narrower: leaves nearly membranaceous.

—— **S. trianguláris**, HOOK. Rather stout: stem simple, 2 to 5 feet high, bearing several or somewhat numerous heads in a corymbiform open cyme: leaves all more or less petioled and thickly dentate (sometimes minutely so, sometimes with long lanceolate-subulate and very salient teeth), deltoid-lanceolate, or the lower triangular-hastate or deltoid-cordate, and uppermost lanceolate with cuneate base: heads about half-inch high: involucre campanulate, mostly 25-30-flowered; the oblong-linear rays 6 to 12. — Fl. i. 332, t. 115; Torr. & Gray, Fl. ii. 441; Eaton, Bot. King Exp. 189; Gray, Bot. Calif. i. 414. *S. longidentatus*, DC. Prodr. vi. 428. — Wooded districts in wet ground, Saskatchewan to Washington Terr., south to the higher mountains of Colorado and through the Sierra Nevada, California.

S. Huachucánus, GRAY. Two or three feet high, somewhat branching: leaves ovate- to oblong-lanceolate, acuminate, minutely denticulate; lower cauline (4 to 6 inches long) tapering into a winged petiole, upper partly clasping by a broad subcordate base: heads fastigiately cymose, small, about 4 lines high: involucre cylindraceous-campanulate, 15-18-flowered: the small rays 3 or 4. — Proc. Am. Acad. xix. 54. — High bluffs near Fort Huachuca, S. Arizona, *Lemmon*.

—— **S. sérra**, HOOK. Strict, 2 to 4 feet high, very leafy, sometimes simple and bearing rather few somewhat large (half-inch long) heads, commonly branching at summit, then bearing numerous corymbosely paniculate smaller heads: leaves (4 to 6 inches long) all lanceolate and tapering to both ends, sessile by a narrow base (or the lowest oblong-spatulate and tapering into a short petiole), usually with whole margin thickly serrate or serrulate with very acute salient teeth : involucre oblong-campanulate, 20-30-flowered : rays 5 to 8, oblong-linear, sometimes fully half-inch long. — Fl. i. 332 (Torr. & Gray, Fl. ii. 441, as to name only, the char. taken from *S. longidentatus*, DC., wrongly referred, and syn. belonging to *S. triangularis*); Gray, Proc. Acad. Philad. 1863, 68, under *S. Andinus?* — Mountains, from Wyoming to Idaho and S. Colorado; first coll. by *Douglas*. The form with the very serrate leaves of the original of Douglas, but with much fewer and larger heads, mountains of Colorado, *Fremont, Hall & Harbour, Parry, Rothrock* (under *S. Andinus*). Passes into

Var. integriúsculus. Heads smaller (usually only 3 or 4 lines high) and narrower, fewer-flowered: leaves minutely serrate or denticulate, or the upper entire, sometimes all entire or nearly so, generally shorter and smaller, or broader and not acuminate. — *S. Andinus*, Nutt. Trans. Am. Phil. Soc. l. c.; Torr. & Gray, l. c.; Gray, Bot. Calif. l. c. *S. lanceolatus,* Torr. & Gray, Fl. ii. 440, an entire-leaved form. — Common from Wyoming to E. Oregon, and in the mountains of Nevada and the borders of California; perhaps first coll. by *Nuttall*.

++ ++ ++ ++ Stem not numerously but somewhat equably leafy up to the inflorescence : leaves all entire or denticulate : involucre fleshy-thickened !

S. crássulus, GRAY. A foot or less high, glabrous apparently from the first: stem rather stout, 5-7-leaved, bearing 3 to 8 pedunculate rather large (fully half-inch high) and thick heads: leaves oblong-lanceolate, of rather firm texture, apiculate-acute, 2 to 5 inches long; radical and lowest cauline spatulate or obovate-oblong, narrowed into a short winged petiole; upper sessile by partly clasping or decurrent base : involucre broadly campanulate, 40-50-flowered, of 12 or more lanceolate to oblong fleshy-thickened but thin-edged bracts, the base also much thickened, the whole becoming conical and multangular in fruit: rays about 8, oblong. — Proc. Am. Acad. xix. 54. *S. integerrimus,* Gray, Am. Jour. Sci. l. c., & Proc. Acad. Philad. 1863, 67, not Nutt. *S. lugens,* var. *Hookeri,* Eaton, Bot. King Exp. 188, in part. — Subalpine, Rocky Mountains of Colorado (first coll. by *Parry*) to the Wahsatch in Utah (*Watson*), and in N. Wyoming, *Parry*.

++ ++ ++ ++ Stems either few-leaved or with the upper leaves (and sometimes most of the cauline) reduced in size; the inflorescence therefore naked: none with narrow linear leaves (except one scapose species).

= Plant tall and simple-stemmed, with a coarsely fibrous cluster of roots, perhaps not perennial : leaves fleshy-coriaceous, all entire or barely denticulate.

S. hydróphilus, NUTT. Very glabrous and smooth, sometimes glaucous : stem robust, 2 to 4 feet high, strict : leaves lanceolate, with strong midrib and obsolete veins; radical oblanceolate and stout-petioled, sometimes a foot long and nearly two inches wide; upper cauline sessile or partly clasping : heads numerous in a branching corymbiform cyme, 5 lines high, short-pedicelled : involucre narrowly campanulate, slightly bracteolate; its bracts 8 to 12 : disk-flowers 15 to 30; rays 3 to 6 and small, sometimes none. — In water or very wet ground, especially in brackish water. Montana to Brit. Columbia, south to Colorado, and west to San Francisco Bay, California.

= = Plants mostly in clumps or tufts, or from tufted or creeping rootstocks.

a. Stems commonly robust, from a foot or rarely less to 3 or even 5 feet high, bearing mostly numerous heads in a cyme : involucre sparingly calyculate: leaves from entire to dentate, only in the last species at all laciniate, none really cordate nor with permanent tomentum. Western species, none truly alpine.

1. Glaucous or glaucescent, apparently quite glabrous throughout from the very first: heads many-flowered.

S. Clevelándi, GREENE. Stems rather rigid and slender, a foot or two high from firm creeping rootstocks: leaves subcoriaceous, entire, obtuse, with veins almost obsolete, spatulate or rarely obovate; radical and lower cauline an inch or two long, tapering into much longer slender petioles; upper cauline few and smaller, with shorter petioles : heads 4 or 5 lines high: involucral bracts subulate-linear: rays 6 to 8 and short, sometimes fewer, occasionally none. — Bull. Torr. Club, x. 87. — Springy ground, Lake Co., California, *Cleveland, Pringle.*

S. Toluccánus, DC. Prodr. vi. 428; Gray, Proc. Am. Acad. xviii. 110; apparently a common Mexican species.

Var. microdóntus. About 2 feet high from a short rootstock or caudex: leaves thickish and firm; radical obovate to oblong, obscurely veiny, mostly acute, numerously denticulate, 2 to 6 inches long, tapering into shorter wing-margined petioles; cauline sessile, few and oblong-lanceolate toward the base of the stem, or commonly only one or two small and bract-like ones on a scapiform stem, these subtending the rather few-headed branches of the cyme: heads nearly half-inch high: involucral bracts linear: rays 6 to 10, conspicuous. — Pinos Altos Mountains, New Mexico, *Greene.* Mountains of S. Arizona, *Pringle, Lemmon.* Agrees with a specimen of coll. Seemann, N. W. Mexico, but not well with *S. Toluccanus,* var. *modestus,* Schultz Bip. in Bot. Herald, 211. Approaches very smooth forms of *S. lugens.*

2. Not glaucous, usually more or less woolly-pubescent when young, and the wool sometimes tardily deciduous, often quite glabrate and green at flowering time: heads many-flowered: rays 8 to 12, conspicuous.

S. integérrimus, NUTT. Leaves oblong-lanceolate, or the radical elongated-oblong, quite entire or denticulate; upper ones reduced and bract-like, attenuate-subulate from a dilated base: heads several, umbellately cymose, commonly half-inch high: involucral bracts narrow, acute or acuminate. — Gen. ii. 165, & Trans. Am. Phil. Soc. l. c.; Torr. & Gray, Fl. ii. 439. — Dakota to Wyoming and Saskatchewan; first coll. by *Nuttall.*

S. lúgens, RICHARDS. Lightly floccose-woolly when young, in the typical form early glabrate and bright green. stem 6 inches to 2 feet high, few- and small-leaved and naked above, terminated by a cyme of several or rather numerous heads (these about 5 lines high): radical and lower cauline leaves spatulate, varying to oval or oblong, either gradually or abruptly contracted at base into a winged or margined short petiole, usually repand- or callous-denticulate; upper cauline lanceolate or reduced and bract-like: bracts of the campanulate involucre lanceolate, with obtuse or acutish commonly blackish-sphacelate tips: rays 10 or 12, conspicuous. — App. Frankl. Journ. ed. 2, 31; Hook. Fl. i. 332, t. 114; Torr. & Gray, Fl. ii. 439; Eaton, Bot. King Exp. 188, var. *Hookeri,* chiefly, & var. *Parryi. S. campestris,* Hook. f. Arct. Pl. 294, 332, partly. *Cineraria pratensis,* Herder, Pl. Radd. ii. 127, in part. — Low or moist grounds, Subarctic America to Kotzebue Sound? through the whole Rocky Mountains to New Mexico, and west to California. In various forms.

Var. foliósus, GRAY, Bot. Calif. i. 413. Floccose wool usually persistent up to flowering, and vestiges remaining to near maturity: stem seldom over a foot high, stouter, more leafy to near the inflorescence: leaves comparatively large, oblong to broadly lanceolate: heads often very numerous and crowded in the corymbiform cyme, then narrower: tips of involucral bracts conspicuously blackish. — *S. exaltatus,* Nutt., var. *minor,* Gray, Am. Jour. Sci. l. c. 406. *S. lugens,* var. *exaltatus,* Eaton, l. c. — Mountains of Colorado and Utah, from base up to 10,000 or 12,000 feet; first coll. by *Parry.*

Var. exaltátus, GRAY, l. c. Lightly floccose when young, and not rarely with looser and more persistent scattered hairs: stem stout, 1 to 3 or even 4 or 5 feet high: leaves thickish; radical longer-petioled, from spatulate-lanceolate to obovate or ovate, the broader ones abrupt and sometimes even subcordate at base; cauline occasionally laciniate-dentate: heads mostly numerous in the cyme. — *S. exaltatus,* Nutt. Trans. Am. Phil. Soc. l. c. 410; Torr. & Gray, l. c. — Wet ground, Brit. Columbia and Idaho to California, where it connects with the next species.

Var. ochroleúcus. Rays yellowish-white: otherwise like broader-leaved forms of the preceding, some radical leaves subcordate. — *S. cordatus,* Nutt. l. c. 411, probably, but color of flowers not noted. — Open woods, on Columbia River, Klikitat Co., Washington Terr., *Suksdorf.*

3. Like the preceding, but with fewer-flowered heads and fewer or no rays: upper leaves occasionally incised.

S. aronicoídes, DC. Robust, lightly floccose when young, and usually with some deciduous villosity, 1 to 3 feet high: leaves variable, from broadly ovate to oblong, repand-denticulate to coarsely dentate, or cauline sometimes pinnatifid-laciniate: heads mostly smaller than in preceding, often only 10–12-flowered: rays when present only one or two and short. — Prodr. vi. 426; Torr. & Gray, Fl. ii. 441; Gray, Bot. Calif. i. 414. — *S. exaltatus,* var.

uniflosculosus, Gray, Pacif. R. Rep. iv. 111. — Low grounds, common in California; first coll. by *Douglas*. Connects with *S. lugens*, var. *exaltatus*.

   b. Stems low and simple, bearing a solitary or few comparatively large heads: involucre not at all calyculate: leaves entire or merely dentate; radical and lower ones spatulate to obovate. Arctic-alpine species, loosely cottony-woolly, tardily glabrate.

S. Hoókeri, TORR. & GRAY. Perhaps a less arctic variety of the next, bearing 3 to 5 closely corymbose heads, or a var. of *S. campestris* of the Old World, but ovaries and akenes glabrous. — Fl. ii. 438. *S. integrifolius*, Hook. Fl. i. 334, excl. syn. *S. campestris*, Hook. f. Arct. Pl. 395, partly. *Cineraria integrifolia*, Richards. l. c. — Arctic and Subarctic America and high-northern Rocky Mountains, *Richardson*, &c.

S. frígidus, LESS. A span or two high, 3–5-leaved, bearing a solitary head, sometimes 2 or 3: leaves spatulate, or the radical rounded-obovate and cauline lanceolate from a broad or narrow sessile base, these sometimes dentate: involucre half-inch high, usually villous with some purplish hairs, especially at the thickened base or summit of the peduncle: rays rather numerous, becoming half-inch long: ovaries and akenes glabrous or sparsely hairy. — Less. in Linn. vi. 239; Hook. Fl. i. 334, t. 112; Torr. & Gray, Fl. ii. 445. *Cineraria frigida*, Richards. l. c.; Hook. & Arn. Bot. Beech. 126; Herder, l. c. 124. *C. atropurpurea*, Ledeb. ex DC., &c. — Newfoundland? and Labrador, Arctic coast to Kotzebue Sound, &c. (N. E. Asia.)

   c. Stems low, only 2 to 6 inches high, scapiform: leaves clustered on the rootstock or caudex, entire or crenate; those of the scape few and very small, reduced to mere bracts: involucre slightly calyculate. Rocky Mountain species, chiefly alpine or subalpine.

   1. Leaves linear, not thick: akenes papillose-hirtellous.

S. Thúrberi, GRAY. Leaves densely tufted on the branches of the multicipital caudex, about inch long, barely a line wide toward the apex, tapering into a slender base, entire or nearly so, tomentose-canescent, tardily glabrate: scapes glabrate, 4 to 6 inches high, bearing 2 to 5 heads; these 4 or 5 lines high: rays 7 to 10, 3 lines long. — Proc. Acad. Philad. 1863, 68. *S. canus*, var. *pygmæus*, Gray, Bot. Mex. Bound. 103. — Mountain-sides, Santa Rita del Cobre, New Mexico, *Thurber, Bigelow*.

   2. Leaves thick and coriaceous, tapering into a petiole, crowded on the multicipital caudex, nearly veinless, even the midrib obscure: akenes glabrous.

S. werneriæfólius, GRAY. Woolly and canescent, tardily glabrate: leaves quite entire, erect or ascending, from spatulate-linear (2 or 3 inches long, including the petiole-like base, by 2 or 3 lines wide) to elongated-oblong (inch long and half-inch wide) and short-petioled, the margins sometimes revolute: scape a span high, rather stout, bearing 2 to 8 heads; these 4 or 5 lines high; rays 10 or 12, oblong, 2 lines long, rarely few or wanting. — Proc. Am. Acad. xix. 54. *S. aureus*, var. *werneriæfolius*, Gray, Proc. Acad. Philad. 1863, 68; Porter & Coulter, Fl. Colorad. 81. — Mountains of Colorado, alpine, in coniferous woods near the upper limit of trees, and in the alpine region, mostly on the upper waters of Clear Creek, *Hall & Harbour, Greene, Coulter*, &c.

S. petræus, KLATT. Glabrous or early glabrate: leaves from orbicular-obovate or oval (a quarter to half an inch long) to cuneate-oblong (largest inch long), entire or 3–7-crenate-toothed at the broad summit, abruptly petioled: scapes 1 to 3 inches high, bearing solitary or several clustered heads; these 4 or 5 lines high: rays 6 to 10, golden yellow, 3 lines long. — Abhand. Nat. Gesellsch. Halle, xv. (1881). *S. aureus*, var. *alpinus*, Gray, Am. Jour. Sci. n. ser. xxxiii. 11; Porter & Coulter, l. c. *S. aureus*, var. *borealis*, mainly, Gray, Bot. Calif. i. 412. — Alpine region of the Rocky Mountains in Colorado (first coll. by *Parry*), of Utah (*Ward*), and highest peaks of the Sierra Nevada, California, *Brewer*, &c. Approaches the preceding on one hand, and *S. aureus*, var. *borealis*, on the other.

   3. Leaves round-cordate, crenate, purple-tinged beneath, slender-petioled, more or less clustered at the base of the scape: akenes glabrous: plants very glabrous.

S. renifólius, PORTER. Two inches high from filiform creeping rootstocks: leaves thickish, resembling those of *Ranunculus Cymbalaria*, rounded-subcordate or reniform, only about half-inch wide, coarsely 5–7-crenate: scape or peduncle little surpassing the leaves, bearing a solitary comparatively large (half-inch long) head: rays about 8, oblong, 4 lines long. — Porter & Coulter, Fl. Colorad. 83. — High alpine region on Whitehouse Mountain, in Central Colorado, at 13,000 feet, *J. M. Coulter*.

S. Cardamíne, Greene. Scapes a span or two high, slender, bearing solitary or 2 or 3 small (about 4 lines high) heads, and below one or two very small oblong-cordate clasping pinnatifid-dentate bract-like leaves: radical leaves orbicular-cordate, repand-crenate, thinnish, inch or two in diameter, on long slender petioles: rays about 8, pale yellow. — Bull. Torr. Club, viii. 98. — New Mexico, on the higher slopes of the Mogollon Mountains, *Greene*.

d. Stems low (2 to 6 inches high) and slender, 1-2-cephalous, few-leaved: leaves mostly lyrate-pinnatifid. High northern species.

S. resedifólius, Less. Glabrous or soon glabrate: stems simple: earlier radical leaves roundish or subcordate, crenate or crenately lobed, later ones lyrate-pinnatifid, slender-petioled, all or the terminal lobes crenate-incised: heads 4 or 5 lines high: involucre very obscurely bracteolate: rays 5 lines long: style-branches commonly with slender cusp: akenes either papillose-hirsute or glabrous. — Less. in Linn. vi. 243; Hook. Fl. i. 333, t. 117; Torr. & Gray, Fl. ii. 445. *Cineraria lyrata,* Ledeb. Fl. Alt. iv. 102; Reichenb. Ic. Bot. Crit. ii. t. 101. — From Great Bear Lake, &c., near the Arctic Circle, to Kotzebue Sound and the Aleutian Islands. (N. Asia.)

Var. Columbiénsis. Heads rayless: stems often sparingly branched and 2-4-leaved. — Muckluug River, British Columbia, *Mr. Mackay*.

e. Stems a foot or two high (or in reduced forms lower), bearing some leaves and corymbosely cymose (only when depauperate solitary) heads: involucre sparingly or inconspicuously calyculate, or nearly naked at base: foliage various. Not arctic nor alpine, except perhaps one variety of *S. aureus:* usually some floccose tomentum, at least when young.

1. Leaves all entire, rarely a tooth or a few obscure denticulations, and narrowed at base.

S. fastigiátus, Nutt. Cinereous with a fine and close pannose tomentum, or glabrate: stems strict, simple, 1 or 2 feet high, terminated by a fastigiate cyme of several heads, or sometimes with branches terminated by single and rather larger heads: leaves lanceolate or spatulate-lanceolate, obtuse, about 2 inches long; upper often linear; lower cauline and the sometimes oblong radical tapering into slender petioles: heads 4 or 5 lines high: rays conspicuous: akenes glabrous. — Trans. Am. Phil. Soc. vii. 410; Torr. & Gray, Fl. ii. 439. — Plains of Oregon, Washington Terr., and adjacent Idaho; first coll. by *Nuttall*.

Var. Laýneæ. Stems disposed to branch, and the branches to bear 2 or 3 or sometimes solitary heads, of half-inch in height: leaves mostly apiculate-acute. — *S. Layneæ,* Greene in Bull. Torr. Club, x. 87. — Sweetwater Creek, El Dorado Co., California, *Mrs. K. Layne-Curran*.

2. Leaves from entire or serrate to pinnatifid in the same species, none pinnately divided: rays occasionally wanting. Species of perhaps impossible limitation.

S. cánus, Hook. Permanently canescent with pannose tomentum, or at length flocculent, but rarely at all glabrate: stems from a span to a foot or rarely 2 feet high: leaves sometimes all undivided or even entire, the radical and lower from spatulate to oblong or roundish-oval (half-inch to thrice that length) and slender-petioled, sometimes laciniate-toothed or pinnatifid (either the upper or lower ones, or both): heads 4 or 5 lines high: akenes very glabrous (in figure of Hooker hispidulous on the angles): style-tips usually with central cusp. — Fl. i. 333, t. 116; Torr. & Gray, l. c.; Gray, Bot. Calif. i. 412. *S. integrifolius,* Nutt. Gen. ii. 165. *Cineraria integrifolia minor,* Pursh, Fl. ii. 528. *S. Purshianus,* Nutt. Trans. Am. Phil. Soc. vii. 412. *S. Howellii,* Greene, Bull. Torr. Club, viii. 98. — Rocky banks, Saskatchewan and Dakota to the mountains of Colorado, west to Brit. Columbia, Oregon, Nevada, and the Sierra Nevada as far as Kern Co., California. — A notable and dubious form, low and stout, with comparatively large heads and always undivided leaves, abounds in the mountains of Colorado, at the upper limit of trees.

S. tomentósus, Michx. Canescent or cinereous with a close or at length floccose and more or less deciduous wool: stems rather stout, commonly 2 feet high: leaves thickish, oblong, crenate or sometimes entire; the larger radical ones ample, 5 or 6 inches long, on elongated stout petioles and with stout midrib; cauline similar and smaller or lyrate-pinnatifid, often few and small: heads, &c., of the next species: akenes always hispidulous, at least on the angles. — Fl. ii. 119; Ell. Sk. ii. 329; Torr. & Gray, Fl. ii. 443. *S. integrifolius,* var. *heterophyllus,* Nutt. Gen. ii. 165. *Cineraria heterophylla,* Pursh, Fl. ii. 528. — Open or sparsely wooded moist ground, Delaware to Florida and Arkansas; first coll. by *Michaux*.

S. aúreus, L. Very early glabrate, usually quite free from wool at flowering (in spring or early summer) and a foot or two high from small rootstocks: radical leaves mostly rounded and undivided, and cauline lanceolate and pinnatifid or laciniate: most polymorphous species, of which the typical form is bright green, 1 to 3 feet high, surculose by slender rootstocks: leaves thin; principal radical ones roundish, cordate or truncate at base, crenate-dentate (1 to 3 inches in diameter), on long slender petioles; lower cauline similar, with 2 or 3 lobelets on the petiole, or lyrately divided or lobed; others more laciniate-pinnatifid and lobes often incised; uppermost sparse and small, with closely sessile or auriculate-dilated incised base: heads rather numerous, 4 or 5 lines high: rays 8 to 12, conspicuous, rarely wanting: akenes quite glabrous. — Spec. ii. 870; Michx. Fl. ii. 820; Ell. Sk. ii. 331; DC. Prodr. vi. 432; Torr. & Gray, Fl. ii. 442; Sprague, Wild Flowers, 77, t. 15, the normal form. *S. gracilis,* Pursh, Fl. ii. 529; DC. l. c., a slender or depauperate form. *S. fastigiatus,* Schwein. in Ell. l. c. — Swamps and wet banks, usually in shaded ground, Newfoundland to Florida, Texas, and to Brit. Columbia and the Sierra Nevada, California.

Var. **obovátus,** Torr. & Gray, l. c. Radical leaves of thicker texture, rotund with abrupt or truncate base, or obovate and cuneate-contracted into a short margined petiole, or the earliest in the rosulate tufts almost sessile and humifuse: otherwise as in the typical form. — *S. obovatus,* Muhl. in Willd. Spec. iii. 1999; Pursh, l. c.; Ell. l. c. *S. Elliottii,* Torr. & Gray, Fl. ii. 443, a form with the early radical leaves more plantagineous and very short-petioled. — More open and moist grounds, Canada to Indiana and Georgia, in the upper country, characteristically developed southward.

Var. **Balsámitæ,** Torr. & Gray, l. c. Less glabrate, not rarely holding more or less wool until fruiting: depauperate stems a span or two, larger fully 2 feet high: principal or earliest radical leaves oblong, sometimes oval, commonly verging to lanceolate, inch or two long, serrate, contracted into slender petioles; the succeeding lyrately pinnatifid: heads usually rather small and numerous: akenes almost always hispidulous-pubescent on the angles. — *S. Balsamitæ,* Muhl. l. c.; Pursh, l. c. *S. Pluttensis,* Nutt. Trans. Am. Phil. Soc. l. c. 413, a robust and larger-leaved western form, verging toward *S. tomentosus. S. aureus,* var. *lanceolatus,* Oakes in Hovey's Mag., & Torr. & Gray, l. c., an attenuated form of this, or of the type, growing in shady swamps. *S. pauperculus,* Michx. Fl. ii. 120, depauperate form. — Rocky or nearly dry ground, Canada to Texas, and northwestward to Brit. Columbia.

Var. **compáctus.** A span or two high, in close tufts, rather rigid, when young whitened with fine tomentum, glabrate in age: radical leaves oblanceolate or attenuate-spatulate, entire or 3-toothed at apex, or pinnatifid-dentate, an inch or more long, thick and firm at maturity; cauline lanceolate or linear, entire or pinnatifid: heads rather numerous and crowded in the cyme, rather small: ovaries papillose-hispidulous on the angles. — *S. aureus,* var. *borealis,* Gray, Pl. Wright. i. 125, & Proc. Acad. Philad. 1863, 68, in part. — N. W. Texas (*Wright*) to the base of the mountains in Colorado, *Hall & Harbour, Greene, &c.;* mostly in saline soil.

Var. **boreális,** Torr. & Gray, l. c. A foot down to a span high, at summit bearing either numerous or few heads; these not rarely rayless: leaves thickish; radical from roundish with abrupt or even truncate base to cuneate-obovate and cuneate-spatulate, half-inch to inch long, slender-petioled; cauline seldom much pinnatifid: akenes glabrous. — *S. elongatus, pauciflorus,* & *Cymbalaria?* Pursh, Fl. ii. 529, 530. *S. aureus,* var. *foliosus,* &c., Hook. l. c. *S. aureus,* var. *borealis* & var. *discoideus,* Torr. & Gray, l. c. *S. cymbalarioides* & *S. debilis,* Nutt. Trans. Am. Phil. Soc. l. c. 408, 412. — Labrador to Brit. Columbia, Oregon, the high Sierra Nevada in California, and mountains of Nevada, Utah, and Colorado, where are forms undistinguishable from the following.

Var. **cróceus,** Gray. A span to a foot or two high, glabrous or early glabrate: leaves somewhat succulent; radical oblong to roundish, sometimes lyrate; cauline very various: heads usually numerous in the cyme: flowers saffron-colored or orange, at least the rays, or these sometimes wanting. — Proc. Acad. Philad. 1863, 68; Porter & Coulter, Fl. Colorad. 82; Eaton, Bot. King Exp. 190, & *S. Fendleri* of the same. *S. aureus,* var. *multilobatus,* Gray, Bot. Calif. i. 411, in part. — Wet ground, high mountains of Colorado, Utah, Nevada, north to Montana, and sparingly in the Sierra Nevada; first coll. by *Parry,* &c.

Var. **subnúdus.** Wholly glabrous or glabrate, slender, a span or two high, bearing 2 or 3 small cauline leaves and a solitary head, or not rarely a pair: radical leaves few, spatulate or obovate, sometimes roundish, half-inch or less long, occasionally lyrate: cauline incised or sparingly pinnatifid: rays conspicuous. — *S. subnudus,* DC. Prodr. vi. 428; Nutt.

l. c.; Torr. & Gray, l. c. Here perhaps *S. Cymbalaria*, Pursh, Fl. ii. 530. — Wet ground on mountains, Wyoming to Brit. Columbia, Oregon, and sparingly in California. The most depauperate form.

S. Féndleri, GRAY. Very canescent with pannose or floccose wool, in age tardily glabrate; stems rather stout, 5 to 15 inches high, leafy, the larger plants branching: leaves oblong-lanceolate or narrower; radical sometimes almost entire, more commonly like the cauline sinuately pectinate-pinnatifid or even pinnately parted, the short oblong divisions incisely 2-4-lobed; heads mostly numerous and crowded, small (3 or 4 lines high): rays rather numerous: akenes and ovaries glabrous. — Pl. Fendl. 108, Pacif. R. Rep. iv. 111, & Proc. Acad. Philad. l. c. — Dry ground, mountains of New Mexico and Colorado, at 6,000 to 8,000 feet, *Fendler, Bigelow, Parry*, &c.

S. Neo-Mexicánus, GRAY. More or less canescent with looser tomentum, in age glabrate: stems robust, a foot or two high (often from a simple thickish caudex), few-leaved, simple or often branching above, and bearing loose cymes of comparatively large (often half-inch) heads: leaves thickish (inch or two long); radical oblong-obovate to spatulate, with cuneate or tapering base, sometimes coarsely few-toothed only at summit, many lyrate-pinnatifid, with few or several pairs of small lateral lobes; cauline similar or more pinnatifid, and the lobes incisely few-toothed: rays 12 to 16, in larger heads half-inch long: akenes sometimes hispidulous-papillose, sometimes quite glabrous. — Proc. Am. Acad. xix. 55. Has been variously referred to *S. Fendleri*, to doubtful forms of *S. aureus*, &c. — Mountains and wooded hills of New Mexico, *Fendler, Wright, Thurber, Henry, Greene*, &c. Arizona, *Lemmon, Pringle*. San Bernardino Mountains, California, *Parish*.

S. Arizónicus, GREENE. Lightly and loosely floccose-woolly when young, early glabrate and green: stems a foot or two high, sometimes from a thick perpendicular caudex: leaves mainly in the radical tuft, thickish, ovate to oblong-obovate (commonly 2 or 3 inches long), dentate with mucronate teeth, often with rounded or subcordate, but some with cuneate base, with or without one or two pairs of small lobes on the petiole; lower cauline leaves one or two and usually lyrate-pinnatifid, upper very small and bract-like: heads loosely cymose, 5 or 6 lines high: rays 9 to 12, conspicuous. — Bull. Torr. Club, x. 87. — Arizona, *Palmer, Pringle* (referred to a form of *S. aureus*), *Rusby*.

3. Leaves all or mainly bipinnately dissected into narrow lobes. Atlantic species.

S. Millefólium, TORR. & GRAY. Early glabrate: stems slender, a foot or two high, bearing a corymbose cyme of rather numerous heads: these 3 lines high: radical and cauline leaves similar (or the earliest less dissected), the very numerous lobes linear-oblong or narrow (1 to 3 lines long), thickish: small upper leaves narrow and more simply dissected: rays few, a line or two long. — Fl. ii. 444. — Sides of precipitous mountains, North and South Carolina, especially at Table Mountain, S. Carolina, and vicinity; first coll. by *Fraser*.

4. Leaves mostly once pinnately divided or parted, and again lobed or incised. Pacific species.

S. Bolánderi, GRAY. Glabrous or early glabrate: stems weak and slender, 6 to 30 inches high from slender creeping rootstocks: leaves thin and membranaceous, mostly petioled; early radical orbicular, subcordate, palmately 5-9-lobed or crenate-incised; others pinnately divided into 5 to 9 distinct leaflets, or upper lobes confluent with rounded terminal one, all obtusely incised: heads several, loosely cymose, 4 or 5 lines high: rays 5 to 8, rather long. — Proc. Am. Acad. vii. 362, & Bot. Calif. i. 411. — Sandstone bluffs and in Redwoods, Mendocino Co., California, *Bolander, Rattan*, to Cascade Mountains, Oregon, *Kellogg, Howell*.

S. eurycéphalus, TORR. & GRAY. Floccose-woolly when young, sometimes early glabrate: stems robust, 1 to 3 feet high, corymbosely branching above, bearing several or numerous loosely cymose heads: leaves irregularly pinnately parted or the lower divided, radical mostly lyrate; divisions of the cauline from cuneate to linear-lanceolate, variously lobed or incised, mucronately tipped: heads hardly at all calyculate, fully half-inch high, commonly as broad, but sometimes half smaller: rays 10 to 12, the larger half-inch long. — Gray, in Pl. Fendl. 109, & Bot. Calif. i. 411, excl. var. *major*, Pacif. R. Rep. iv. 111. — Low grounds, California north of the Bay of San Francisco, and on Monte Diablo; first coll. by *Fremont* and *Hartweg*.

S. eremóphilus, RICHARDS. Stems freely branching, leafy up to the inflorescence: leaves mostly oblong in outline, laciniately pinnatifid or pinnately parted, the lobes usually incised or acutely dentate: heads numerous in corymbiform cymes, 4 or 5 lines high, short-peduncled; involucre campanulate or narrower, minutely bracteolate; proper bracts commonly purple-tipped: rays 7 to 9, 2 or 3 lines long: akenes either minutely papillose-cinereous or glabrous. — App. Frankl. Journ. ed. 2, 31; Hook. Fl. i. 334; Torr. & Gray, Fl. ii. 444; Eaton, Bot. King Exp. 192. — Shady moist ground, from Mackenzie River and Saskatchewan, along the Rocky Mountains to those of New Mexico, Utah, and Arizona; first coll. by *Richardson*. In cañons of S. Arizona, a form with narrowest and even linear lobes to the leaves, coll. *Lemmon*.

++ ++ ++ ++ ++ ++ Stems leafy, numerously or somewhat equally so up to the top, all pinnately lobed or parted, or when entire narrowly linear.

= Leaves comparatively broad, pinnatifid and laciniate: early glabrate if not glabrous.

S. Clarkiánus, GRAY. Stems strict and simple, 3 or 4 feet high, striate-angled: leaves lanceolate; cauline 4 to 7 inches long, sessile, simply pinnatifid or laciniate-dentate; the salient lobes or teeth lanceolate or triangular, very acute: heads several, cymose or somewhat paniculate, fully half-inch high, short-peduncled: involucre of subulate-linear bracts, and several more slender loose calyculate ones: rays 4 or 5 lines long, narrow. — Proc. Am. Acad. vii. 362, & Bot. Calif. i. 413. — Moist ground, in the Sierra Nevada, California, at 8,000 to 9,000 feet, Yosemite to Kern Co., *Bolander, Rothrock*, &c.; first coll. at Clark's Ranch.

= = Leaves or their divisions from linear to filiform, or broader toward the base of the stems.

S. Douglásii, DC. Lignescent and sometimes decidedly shrubby at base, many-stemmed, a foot or two or southward even 5 or 6 feet high, either white-tomentose or glabrate and green: leaves thickish, sometimes all entire and elongated-linear (mostly 2 to 4 lines long and 1 or 2 lines wide), more commonly pinnately parted into 3 to 7 linear or nearly filiform entire divisions: heads several or numerous and cymose, from a third to half an inch high, obscurely bracteolate, the proper bracts linear: rays 8 to 16, a third to half an inch long: akenes canescent with a fine strigulose pubescence. — Prodr. vi. 429; Torr. & Gray, Fl. ii. 443; Gray, Bot. Calif. i. 411. *S. Regiomontanus*, DC. l. c. (Monterey, California), & probably *S. stœchadiformis*, DC. *S. longilobus*, Benth. Pl. Hartw. 18; Gray, Fl. Fendl. 108. *S. filifolius*, Nutt. Trans. Am. Phil. Soc. vii. 414. *S. Riddellii*, *S. filifolius*, & *S. spartioides*, Torr. & Gray, l. c. *S. fastigiatus?* Gray, Pl. Wright. ii. 99, a peculiar and abnormal broader-leaved form. — Open plains and hills, Nebraska to Texas, S. Utah, Arizona, S. California, and northward near the Pacific coast to Lake Co.

§ 2. Perennial? viscidly pubescent: heads conspicuously radiate.

S. Párryi, GRAY. Rather stout, a foot or two high, branching, sparsely leafy to the inflorescence, pubescent with short and spreading and some longer viscid hairiness: root not seen: leaves irregularly dentate, oblong or the lowest spatulate, auriculate-clasping at base: heads cymose or somewhat paniculate, about half-inch long: involucre sparsely calyculate: akenes strigulose-canescent. — Bot. Mex. Bound. 103. — S. E. California on the San Bernardino and San Francisco Mountains, *Lemmon, Greene*. First coll. within the Mexican lines, on the Rio Grande in Chihuahua, below San Carlos, *Parry*.

§ 3. Annuals or biennials.

* Indigenous species, of Southern range: heads conspicuously radiate: akenes seldom glabrous.

S. ampulláceus, HOOK. Lightly floccose or araneose-woolly when young, glabrate and smooth: stem mostly stout, a foot or two high, leafy to near the summit: leaves all undivided, repand-dentate or entire (1 to 6 inches long), ovate or oblong; lowest obovate with tapering wing-petioled base; upper mostly clasping with broad base: heads rather numerous in naked loose cymes: involucre (4 lines high) calyculate-bracteolate, cylindraceous, becoming thickened and conoidal after anthesis: rays 7 to 9, oblong: akenes canescent. — Bot. Mag. t. 3487; Torr. & Gray, Fl. ii. 440; Gray, Pl. Lindh. i. 42. — Sandy prairies, Texas; first coll. by *Drummond*.

S. Califórnicus, DC. Early glabrate if not glabrous, slender, a foot or so high: leaves lanceolate, linear, or the lower oblong, varying from denticulate to pinnatifid, the lobes short and obtuse, all but the lowest auriculate-sessile or clasping at base (one or two inches long): heads several and loosely paniculate or cymose at the naked summit of the stem: involucre broadly campanulate, 3 or 4 lines high, nearly naked at base: rays oblong, 3 or 4 lines long: akenes canescent. — Prodr. vi. 426; Torr. & Gray, l. c.; Gray, Bot. Calif. i. 410. *S. Coro-*

nopus, Nutt. Trans. Am. Phil. Soc. l. c.; Torr & Gray, l. c.; a form with leaves deeply and even doubly pinnatifid. — Low ground, California, from Santa Barbara southward. (Lower Calif.)

S. multilobátus, Torr. & Gray. Early glabrate and smooth, a foot or two high from a winter-annual or biennial root, naked and often branching above, bearing numerous corymbosely cymose heads: radical and lower cauline leaves lyrate, and the divisions dentate; upper pinnately parted, their mostly numerous divisions narrowly cuneate, incised or 2-3-lobed at the apex: involucre 3 lines high, nearly or quite naked at base: rays 3 or 4 lines long: akenes slightly hispidulous or glabrate. — Pl. Fendl. 109, excl. var. pl. Coulter, which is probably *S. Douglasii* S. *Tampicanus*, Gray, Pl. Wright. ii. 89, perhaps also i. 109. *S. aureus*, var. *multilobatus*, Gray, Bot. Calif. l. c., partly. — S. Utah, Arizona, and western borders of Texas, *Fremont, Wright, Palmer*, &c.

S. lobátus, Pers. (Butter-weed.) Lightly floccose-tomentose when very young, early glabrous, very smooth, soft-succulent or tender: stem fistulous, 1 to 3 feet high, sometimes depauperate and slender, commonly branching, and bearing compound or paniculate cymes: leaves lyrately parted or divided, irregular and variable; divisions from roundish to cuneate or oblong, obtusely sinuate-lobed or toothed: involucre barely 3 lines high, nearly naked at base: rays 6 to 12: akenes minutely hispidulous on some of the angles. — Syn. ii. 436; Ell. Sk. ii. 332; Torr. & Gray, l. c. *S. lyratus*, Michx. Fl. ii. 120, not L., &c. *S. glabellus*, Poir. Dict. vii. 102. *S. Carolinianus*, Spreng. Syst. iii. 559. *S. Mississippianus*, DC. Prodr. vi. 427. *S. densiflorus*, Martens, Bull. Acad. Brux. viii. 67. *S. Schweinitzianus*, Nutt. in Trans. Am. Phil. Soc. vii. 411. *S. imparipinnatus*, Klatt in Naturf. Gesellsch. Halle. xv. — Wet grounds, in the low country, N. Carolina to Texas, common. (Adj. Mex.)

* * Indigenous, of northern range: heads obviously radiate: akenes glabrous: pappus elongated.

S. palústris, Hook. Loosely woolly or villous with long and many-jointed hairs, in age sometimes glabrate: stem 6 to 20 inches high from an annual or biennial root, leafy, usually stout: leaves broadly lanceolate, from sinuate-dentate to pinnatifid-laciniate, cauline sessile by a cordate or auriculate partly clasping base: heads crowded in a glomerate or corymbiform cyme, in flower only 4 lines long, and with short light-yellow rays, in fruit with pappus half-inch or more long: involucre naked at base. — Fl. i. 334; DC. Prodr. vi. 363; Torr. & Gray, Fl. ii. 438. *S. Kalmii*, Less. in Linn. vi. 244, not Nutt., which is only a changed name for *Cineraria Canadensis*, L. *Cineraria palustris*, L. Spec. ed. 2, 1243; Fl. Dan. t. 573; Schkuhr, Handb. t 246. *C. congesta*, R. Br. in Parry, Voy., Richards., &c., only an arctic and woolly condensed form, var. *congestus*, Hook. l. c. — Wet ground, N. Wisconsin, Iowa, and Minnesota to the Arctic sea-coast, N. Alaska, &c. (N. Asia, Eu.)

* * * Naturalized annual weeds from Europe: rays none or minute.

S. sylváticus, L. Slender, glabrate or somewhat pubescent, a span to a foot or more high: leaves usually pinnatifid: heads 3 or 4 lines high, narrow, nearly naked at base, bearing a few rays with inconspicuous ligule not surpassing the disk: akenes canescent. — Engl. Bot. t. 748; Fl. Dan. t. 869. — Waste grounds, of sparing occurrence in Nova Scotia and coast of California. (Nat. from. Eu.)

S. vulgáris, L. (Groundsel.) Stouter, more branchy and leafy to the top, glabrate: leaves incisely pinnatifid, the oblong or roundish lobes and the sinuses sharply toothed: heads thicker, 4 or 5 lines high: tips of the involucral bracts and the short calyculate ones at base blackish: rays none: akenes canescently puberulent. — Engl. Bot. t. 747; Fl. Dan. t. 513; Pursh, Fl. ii. 528. — Waste grounds and cult. fields, not rare on both the Northern Atlantic and Pacific coasts. (Nat. from Eu.)

S. viscósus, L. Coarser, viscid-pubescent, strong-scented: leaves once or twice pinnatifid: heads rather larger, more pedunculate: involucre sparingly and slenderly bracteolate at base, its bracts not black-tipped: rays with inconspicuous ligule: akenes glabrous. — Engl. Bot. t. 32; Fl. Dan. t. 1230. — Waste grounds on coast of New England, near Providence and Boston. (Nat. from Eu.)

193. CACÁLIA, L. Indian Plantain.

(Ancient Greek name of some Senecioneous plant, perhaps Coltsfoot.) — Perennial herbs, not fleshy (some shrubby in the tropics), natives of America and Asia in the northern hemisphere, with aspect mostly unlike *Senecio*. Leaves petioled. Our species all smooth, glabrous, and akenes glabrous: fl. summer. — L. Gen. ed. 4, 362 (partly); DC. Prodr. vi. 327 (with *Psacalium*, & excl. § 3, 4); Gray, Proc. Am. Acad. xix. 51.

§ 1. Involucre in our of rather many bracts, calyculate with some small loose ones, and many-flowered: corolla-lobes shorter than the throat: receptacle plane.

C. suavéolens, L. Nearly glabrous: stem striate-angled, 3 to 5 feet high, leafy up to the corymbiform cyme of numerous heads; leaves hastate and on margined or winged petioles, or uppermost merely truncate or cuneate at base, acutely and often doubly dentate: proper bracts of the involucre about 12: flowers 25 to 30: corolla-lobes fully half the length of the throat: style-branches capitellate-truncate. — Spec. ii. 835; Walt. Car. 195; Michx. Fl. ii. 96; Schkuhr, Handb. t. 236; Torr. & Gray, Fl. ii. 434. *Senecio suaveolens*, Ell. Sk. ii. 328. — Moist and shaded ground, W. New England to Michigan and Illinois, and along the mountain region to W. Florida.

C. HASTÁTA, L., which reaches Kamtschatka, is said to have been collected in Sitka by four collectors (see Herder in Pl. Radd. iii. 108); but Stewart's plant, named by Herder, is *Prenanthes alata*, and probably the others likewise.

§ 2. Involucre of about 5 narrowly oblong or linear bracts and as many flowers: receptacle commonly with a fleshy projection or 2 or 3 thickish fimbrillæ in the centre: corolla-lobes longer than the throat: heads numerous in corymbose cymes. — § *Conophora*, DC.

* Leaves merely lobed, pedately ribbed, veiny: plants glabrous and smooth.

C. reniformis, Muhl. Green, not at all glaucous: stem angled, 4 to 9 feet high: leaves slightly angulate-lobed, repand-dentate, ample; radical dilated-reniform, often 2 feet wide; upper cauline subcordate or flabelliform: corolla parted down almost to the proper tube. — Muhl. in Willd. Spec. iii. 1735 (where the heads are wrongly said to be many-flowered); Pursh, Fl. ii. 518; Torr. & Gray, Fl. ii. 435. — Rich and damp woods, Penn. to Carolina and Tennessee along the mountains.

C. atriplicifólia, L. Glaucous: stem terete, 3 to 6 feet high, naked at summit: leaves of firmer texture, lobed or incised, but not dentate; radical from round-reniform to subcordate-ovate (larger 6 inches broad); cauline angulate-cordate or triangular, or with cuneate base and 3 to 7 laciniate lobes, to rhombic-lanceolate and entire in the uppermost: cymes open: corolla-lobes fully twice the length of the throat. — Spec. ii. 835; Walt. l. c.; Michx. l. c.; Pursh, l. c.; Schkuhr, Handb. t. 236; Torr. Fl. N. Y. i. 401, t. 59. *C. atriplicifolia*, etc., Moris. Hist. iii. sect. 7, t. 15, f. 7. *C. gigantea*, Nees & Schauer, Ind. Sem. Vratisl. 1841, & Linnæa, xvi. 216. *Senecio atriplicifolius*, Hook. Fl. i. 332, with var. *reniformis*. — Moist or dry ground, W. Canada and New York to Florida, west to Michigan and Illinois.

C. diversifólia, Torr. & Gray. Not glaucous: stem striate, 2 or 3 feet high: corolla-lobes a little longer than the oblong-campanulate throat: otherwise nearly as in the preceding, into which it may pass. — Fl. ii. 435. — River swamps in Middle Florida, Chapman. S. Carolina, Ravenel.

* * Leaves from sinuately dentate to entire, 3-7-nerved or triplinerved: plants glabrous and smooth: style-tips with or without a short setiform central cusp.

+— Corolla-lobes moderately longer than the oblong-campanulate throat.

C. Floridána, Gray. Not glaucous: stem 3 or 4 feet high, rigid, striate-angled: leaves thickish, ovate or oblong, obtuse, cuneate-contracted at base into a margined petiole, 3-5-nerved from or near the base, obtusely dentate (cauline 2 or 3, and radical 5 or 6 inches long): cymes open, irregular. — Proc. Am. Acad. xix. 52. — Coast of Florida, *Palmer, Chapman*.

+— +— Corolla parted down almost to the proper tube: stems comparatively naked above, bearing loose fastigiate-corymbose cymes.

C. ováta, Ell. Somewhat glaucous: stem terete: 3 to 4 feet high: leaves thinnish, from oval, or radical broadly ovate, to oblong or upper cauline oblong-lanceolate, obtuse or acute, entire or with a few irregular teeth: uppermost sessile; lower and radical nervose at base and triplinerved above it, the nerves commonly diverging. — Ell. Sk. ii. 310; Torr. & Gray,

l. c. — Damp woods, Georgia and W. Florida to Louisiana. It is impossible to determine whether this or the next is Walter's *C. ovata*.

C. tuberósa, NUTT. Green, not glaucous: stem 2 to 5 feet high from "a napiform root" or stock, striate-angled: leaves thickish, from oval to oblong-lanceolate, entire or denticulate, or rarely repand-dentate, conspicuously 5-7-nerved from base, and the nerves parallel and continued to the apex; radical plantaginous, 3 to 8 inches long, contracted or tapering at base into (sometimes foot long) petioles; lower cauline similar, upper comparatively few and small. — Gen. ii. 138; Torr. & Gray, Fl. ii. 436. *C. paniculata* & *C. pteranthes*, Raf. Ann. Nat. 1820, 14. *C. ovata*, Walt. Car. 196? from char., not Ell. — Wet prairies, &c., W. Canada and Wisconsin to Alabama.

C. lanceoláta, NUTT. Somewhat glaucous: stem terete, 2 or 3 feet high, slender: leaves all lanceolate and lightly 3-5-nerved, or even linear and 1-3-nerved, thickish, entire, sometimes 2 or 3 laciniate teeth or small lobes: heads and cymes of the preceding or fewer. — Gen. l. c.; Ell. l. c.; Torr. & Gray, l. c. *C. hastata?* Walt. l. c. 195? — Wet pine barrens, &c., S. Carolina to Florida and Louisiana.

* * * Leaves decompound: stem and branches slightly pubescent: corolla divided down to the proper tube into linear lobes somewhat exceeding it in length.

C. decompósita, GRAY. Stem slender, 3 feet high, floccose-woolly at base, naked and paniculately branched above, bearing numerous small (4 or 5 lines high) heads in open corymbiform cymes: leaves large (radical 2 feet high including the petiole), 3 or 4 times pinnately divided into linear chiefly entire lobes, the primary and secondary divisions more commonly alternate: involucre about half the length of the (5 or 6) flowers. — Pl. Wright. ii. 99. *Senecio Grayanus*, Hemsl. Biol. Centr.-Am. Bot. ii. 241. — Mountains of S. Arizona, *Wright, Lemmon*.

194. **ERECHTÍTES**, Raf. FIREWEED. (Name of a Groundsel by Dioscorides.) — Coarse and homely annuals (Eastern American, and some in New Zealand and Australia); with rank smell, alternate leaves, and cymosely or paniculately disposed heads of whitish or dull yellow flowers. — DC. Prodr. vi. 294; Benth. & Hook. Fl. ii. 443. *Neoceis*, Cass.

E. hieracifólia, RAF. Glabrous or with some hirsute pubescence: stem commonly stout, 1 to 6 feet high, sulcate, leafy to top: leaves of tender texture, lanceolate or broader, sessile, acute, acutely dentate, or some incised or pinnatifid, upper commonly with auriculate partly clasping base: heads half-inch high, cylindraceous, rather fleshy, setaceously bracteolate: pappus white. — DC. Prodr. l. c.; Torr. & Gray, Fl. ii. 434. *E. (hieracifolia,) prealta, elongata*, &c., Raf. Fl. Ludov. & in DC. *Senecio hieracifolius*, L. Spec. ii. 866. *Cineraria Canadensis*, Walt. Car. 207? — Moist woods and copses, a common weed in enriched soil, and especially where woods have been recently burned away (fl. late summer), Newfoundland and Canada to Louisiana. (Extends to S. Amer.)

TRIBE IX. **CYNAROIDEÆ**, p. 81.

195. **SAUSSÚREA**, DC. (*Theodore*, and his father *Horace Benedict Saussure*, eminent Genevese naturalists.) — Perennials of the northern temperate and arctic zones; with middle-sized heads of purple or violet-blue flowers. — Ann. Mus. Par. xvi. t. 10-13, & Prodr. vi. 532; Benth. & Hook. Gen. ii. 471. — Ours all have the distinct and deciduous outer pappus of true *Saussurea*: fl. late summer.

S. alpína, DC. l. c. Low, 2 to 12 inches high, with few cymose-glomerate heads, loosely arachnoid-tomentose and glabrate: leaves from narrowly to oblong-lanceolate or even broader, all narrowed at base, denticulate, sometimes entire: bracts of the involucre chartaceo-membranaceous, acutish or acute, outer shorter: usually some setose chaff of the receptacle among the flowers. — Torr. & Gray, Fl. ii. 452; Reichenb. Ic. Fl. Germ. t. 816,

&c.; Herder, Pl. Radd. iii. 36. *S. angustifolia*, DC. l. c. *S. monticola*, Richards. App. Frank. Journ. ed. 2, 29. *S. multiflora*, Richards. l. c., ed. 1. — Mackenzie River to Arctic coast and Kotzebue Sound. (Eu., N. Asia.)

Var. **Ledeboúri.** More glabrate ; leaves from sinuately or laciniate-dentate to entire : involucre looser; its bracts mostly attenuate-acuminate, less unequal, or the outermost prolonged to the height of the inner : chaff of the receptacle either sparse or wanting. — *S. alpina*, Hook. Fl. i. 303, in part. *S. Ledebouri*, Herder, l. c. 41. *S. subsinuata, nuda*, & *Tilesii*, Ledeb. lc. Fl. Alt. t. 60, 61, 62. *S. subsinuata*, Seem. Bot. Herald, 35. t. 7. *S. acuminata*, Turcz. in DC. l. c. 636, exactly *S. nuda*, Ledeb. l. c. — Northern Rocky Mountains in the alpine region to Kotzebue Sound and Alaskan islands ; in this country the commoner form and manifestly passing into *S. alpina*. (Adj. Asia.)

— **S. Americána,** Eaton. Tall, 2 or 3 feet high, leafy, lightly arachnoid when young, soon glabrate, bearing numerous corymbosely cymose heads ; leaves membranaceous, denticulate or dentate, ovate and oblong-ovate, acute or acuminate; radical and lower cauline subcordate and on slender margined petioles (4 inches long) ; upper sessile with acute base; uppermost lanceolate : heads half to three-fourths inch long : involucre cylindraceous or somewhat turbinate, pubescent. 10-17-flowered ; its bracts thin-coriaceous, 5–6-ranked, all pointless and obtuse ; outer successively shorter, ovate : corollas "dark blue " or " purple " : receptacle bearing more or less copious setiform chaff ["naked" according to Eaton]. — Lot. Gazette, vi. 283. — Mountains of Eastern Oregon, *Cusick.* and Simcoe Mountains, Washington Terr., *T. J. Howell.* Related to the W. Asiatic *S. latifolia*, Ledeb., and *S. grandifolia*, Maxim., especially to the latter, which *has* an equally copious outer pappus.

196. **ÁRCTIUM, L.** Burdock. ("Αρκτος, a bear, from the rough involucre ?) — Coarse and rank biennials, of the Old World, unarmed, except the hooked tips of the involucral bracts forming the bur; with large and roundish mostly cordate leaves, the lower on stout petioles, and middle-sized heads of pink or purplish flowers, in summer. — Benth. & Hook. Gen. ii. 466. *Lappa*, Tourn., Juss., Gærtn., DC., &c.

— **A. Láppa, L.** Plant 3 to 5 feet high, with somewhat cymosely disposed heads : leaves mostly green and glabrous above, whitish with cottony down beneath : in the larger form, var. majus (*Lappa major*, Gærtn., *Arctium majus*, Schkuhr), the bur an inch or more in diameter, its bracts all spreading and glabrous or nearly so. — Common in waste or manured ground, near dwellings. (Nat. from Eu.)

Var. tomentósum (*A. Bardana*, Willd., *Lappa tomentosa*, Lam.), a more woolly form ; with bracts of involucre cottony-webbed. — Rare in N. America.

— Var. mínus (*A. minus*, Schkuhr, *Lappa minor*, DC.), with smaller and only slightly webby heads; these more paniculate, and innermost bracts or awns of the bur erect. Varies with laciniate leaves. — Not uncommon. All the forms are vile weeds.

197. **CÁRDUUS, Tourn., L.**, partly. Plumeless Thistle. (Ancient Latin name of Thistle.) — Old World genus, one species locally naturalized.

— **C. nútans, L.** (Musk Thistle.) Biennial, 1 to 3 feet high, green : stem sinuately and interruptedly winged : head solitary, nodding : corollas crimson-purple. — Fl. Dan. 675; Reichenb. Ic. Germ. 877. — On the Susquehanna near Harrisburg, Penn. (Nat. from Eu.)

C. críspus, C. acanthoídes, and C. pycnocéphalus, L., occasionally appear as ballast-weeds or waifs at seaports.

C. pectinátus, L. Mant. 279, grown in the Upsal Garden, from unknown source, said by Willdenow to come from Pennsylvanian seeds, but doubtless not American, is referred by Sprengel to *C. defloratus.*

198. **CNÍCUS, Tourn., L.**, partly. Plumed Thistle. (Latin name of Safflower, changed from κνῆκος, of Dioscorides, applied by the herbalists and early botanists to Thistles.) — Stout herbs (of the northern hemisphere) ; with sessile leaves, commonly with prickly teeth and tips, and large or middle-sized

heads ; the flowers red, purple, or rose-color, rarely white or yellowish, in summer. Many hybridize ! — L. Gen. ed. 6, 409 (where the char. is *pappus plumosus*, and in Spec. ed. 2, two years earlier, *C. benedictus* is referred to *Centaurea*) ; Willd. Spec. iii. 1662 ; Benth. & Hook. Gen. ii. 468. *Cirsium*, DC. Fl. Fr. ed. 3, iv. 110, & Prodr. vi. 634, not Tourn.

§ 1. Naturalized from Europe : one species with diœcious heads.

C. arvénsis, Hoffm. (Canada Thistle). Perennial and spreading by creeping rootstocks, a foot or two high, corymbosely branching, usually glabrate and green ; stem and branches wingless : leaves lanceolate, pinnatifid and toothed, furnished with abundant weak prickles ; heads loosely cymose, less than inch high, diœcious ; in male plant ovate-globular, and flowers (rose-purple) well exserted ; in female oblong-campanulate and flowers less projecting : bracts of involucre all appressed, short, and with very small weak prickly points : only abortive anthers to the female flowers. — Fl. Germ. iv. 180 ; Pursh, Fl. ii. 506. *Serratula arvensis*, L. Spec. ii. 820 ; Fl. Dan. t. 644. *Carduus arvensis*, Curt. Fl. Lond. t. 57 ; Engl. Bot. t. 975. *Cirsium arvense*, Scop. Fl. Carn. ; DC. Prodr. vi. 643 ; Torr Fl. N. Y. i. 408, t. 61 ; Reichenb. Ic. Fl. Germ. t. 842. *Breea arvensis*, Less. Syn. 9. — Meadows, pastures, and waste grounds, from Newfoundland through the Northern and Middle Atlantic States : too common weed. (Nat. from Eu.)

C. lanceolátus, Hoffm. l. c. (Common Thistle of fields.) Biennial, 3 or 4 feet high, with large heads (almost 2 inches high) terminating somewhat leafy branchlets, cottony-tomentose when young, becoming green, more or less villous or hirsute : leaves lanceolate, deeply pinnatifid and with lanceolate lobes, rigidly prickly ; upper face strigose-setulose ; base decurrent on the stem into interrupted prickly wings : bracts of involucre arachnoid-woolly, lanceolate and mostly attenuate into slender and rigid prickly-pointed spreading tips : flowers rose-purple, hermaphrodite. — Willd. Spec. iii. 1666 ; Pursh, l. c. *Carduus lanceolatus*, L. ; Engl. Bot. t. 107 ; Fl. Dan. t 1173. *Cirsium lanceolatum*, Scop. l. c. ; DC. l. c. ; Reichenb. Ic. Fl. Germ t. 826. — Pastures and waste grounds, Newfoundland and Canada to Georgia (very common northward) ; also in Oregon. (Nat. from Eu.)

§ 2. Indigenous species, all but one Alaskan species endemic, all or mostly biennials.

 * Bracts of the ovoid or hemispherical involucre appressed-imbricated and the outer successively shorter, all with loose and dilated fimbriate or lacerate white-scarious tips. — *Echenais*, Cass., DC.

C. Americánus, Gray. A foot or two high, branching above : branches bearing solitary or scattered naked heads : leaves white-tomentose beneath, lanceolate or broader, sinuately pinnatifid, or some merely dentate, others pinnately parted, weakly prickly : heads erect, inch high · principal bracts of the involucre naked-edged or merely fimbriate-ciliate (not setose-spinuliferous) below, and the dilated scarious apex as broad as long, fimbriate-lacerate, tipped with barely exserted cusp or mucro ; innermost with lanceolate nearly entire scarious tips : flowers ochroleucous : stronger pappus-bristles dilated-clavellate at tip. — Proc. Am. Acad. xix. 56, without char. *C. carlinoides*, var. *Americanus*, Gray, Proc. Am. Acad. x. 48, & Bot. Calif. i. 420, excl. syn. Nutt., &c. *Echenais carlinoides*, var. *nutans*, Gray, Proc. Acad. Philad. 1863, 69. — Lower mountains of Colorado and New Mexico to the coast of California. (A hybrid with *C. undulatus* ? with red-purple flowers and purplish tips to involucral bracts, is from Pinos Altos Mountains, New Mexico, Greene.)

 * * Bracts of the involucre mostly loose, not appressed-imbricated nor rigid, tapering gradually from a narrow base to a slender-prickly or muticous apex ; outer not very much shorter than the inner, wholly destitute of dorsal glandular ridge or spot,

  +— Some with scarious or fringed tip or margins, at least the innermost, slightly or not at all prickly-pointed (except accessory leafy ones) : leaves not decurrent on the stem, moderately prickly : Rocky Mountain and Western species.

C. Párryi, Gray. Green, lightly arachnoid and villous when young, 2 feet or so high : leaves lanceolate, sinuate-dentate : heads (inch high) several and spicately glomerate or more racemosely paniculate, more or less bracteose-leafy at base : accessory and outer proper bracts or some of them pectinately fimbriate-ciliate down the sides, innermost with more or less dilated or margined mostly lacerate-fimbriate tips: corollas pale yellow; the lobes longer than the throat: pappus of fine soft bristles, none of them obviously clavellate. — Proc. Am. Acad. x. 47; Rothrock in Wheeler Rep. vi. 180. — Rocky Mountains in Colorado and Utah, at about 8,000 feet; first coll. by *Parry*. Appears to hybridize with *C. eriocephalus*, &c.

C. remotifólius, GRAY, l. c. Loosely arachnoid-woolly when young, 3 to 8 feet high: leaves from sinuately to deeply pinnatifid, more or less whitened by the loose tomentum beneath even in age: heads (inch and a half high) pedunculate, scattered, naked or nearly so at base: involucre lightly arachnoid and glabrate; the bracts attenuate, the outer into a weak small prickle; the inner or some of them with a scarious (from broadly subulate to ovate-lanceolate) entire or sparingly lacerate tip: corolla ochroleucous, its lobes much shorter than the throat; pappus of coarser bristles, the strongest with conspicuously clavellate tips. — *Carduus remotifolius*, Hook. Fl. i. 302. *Cirsium remotifolium*, DC.; Torr. & Gray, Fl. ii. 460. *C. stenolepidum*, Nutt. Trans. Am. Phil. Soc. vii. 419. — Along streams, Columbia River, from the Yakima district, Washington Terr., to the coast, and to Mendocino Co., California. Here no. 559 *Kellogg & Harford* (not "Hall & Harbour"), doubtfully referred to *C. Americanus* in Bot. Calif. i. 421, a form most approaching the latter species.

+ + None of the involucral bracts with fimbriate or scarious-dilated tips, or obscurely so in the first species.

++ Proper bracts nearly all tipped with a slender acicular prickle, also somewhat viscidly long-woolly; leaves narrow, well armed with prickles: stem a foot or two high, leafy: pappus-bristles not clavellate-tipped. Rocky-Mountain species.

C. Hookeriánus, GRAY, l. c. Arachnoid white-woolly, hardly glabrate, stout: leaves pinnatifid; the short lobes rather distant, sparsely prickly; base little or not at all decurrent: heads few and sessile in a terminal cluster, or scattered, inch and a half high, somewhat bracteose-leafy at base: proper bracts tapering from a broadish base into a rather rigid subulate prickly point: corollas white or whitish. — *Carduus discolor*, var. *fl. albis*, Hook. Fl. i. 302. *Cirsium Hookerianum*, Nutt. Trans. Am. Phil. Soc. vii. 418. — Upper wooded and subalpine region of the Rocky Mountains, north of lat. 48°, *Douglas, Bourgeau*, &c.

C. eriocéphalus, GRAY, l. c. Loosely arachnoid-woolly and partly glabrate, very leafy: leaves pinnatifid into very numerous and crowded and numerously prickly short lobes, the base decurrent on the stem into prickly wings: heads (inch long) several, sessile, and crowded in a leaf-subtended at first nodding glomerule; the subtending leaves and the involucral bracts densely long-woolly (or the inmost bracts glabrous), all very slender-prickly: corollas light yellow or yellowish. — *Cirsium eriocephalum*, Gray, Proc. Acad. Philad. 1863, 69; Eaton, Bot. King Exp. 196, excl. var. — Alpine region of the Rocky Mountains, at the head of Clear Creek and its tributaries; first coll. by *Parry*.

++ ++ Proper bracts of the involucre tapering into an almost innocuous weak and short prickle or soft point: leaves green both sides, glabrate, mostly membranaceous, not decurrent on the stem, except the lower of the last species. Pacific species, with middle-sized or small heads.

C. édulis, GRAY, l. c. Stem robust and somewhat succulent, 3 to 6 feet high, pubescent, leafy to the top: leaves oblong or narrower, from slightly to deeply sinuate-pinnatifid, weakly prickly-ciliate: heads (the larger inch and a half high) scattered or few in a cluster, usually bracteose-leafy at base: involucre conspicuously arachnoid-woolly when young, partly glabrate in age: corollas dull purple or whitish; the lobes much shorter than throat, filiform in the dried state and capitellate-callous at apex! — Bot. Calif. i. 420. *Cirsium edule*, Nutt. l. c.; Torr. & Gray, l. c. — Low grounds, British Columbia to W. California.

C. Hállii, GRAY. Glabrate and green: stem slender, 2 or 3 feet high, moderately leafy: leaves pinnatifid, the lobes and teeth rather strongly prickly: heads solitary and pedunculate, or 2 or 3 in a small terminal cluster (inch or more high), more or less bracteose-leafy at base: involucre sparingly arachnoid when young, soon glabrate, the attenuate tips of all but the outermost innocuous: corollas rose-purple, varying to white; the lobes linear, plane, obtuse. — Proc. Am. Acad. xix. 56. — Oregon, *Hall* (310, was referred to *C. edulis*), to S. California (San Bernardino Co., *Lemmon*, &c.) and S. Utah, *Mrs. Thompson*.

C. Kamtscháticus, MAXIM. Glabrate and green, leafy up to the naked and short-pedunculate (inch high) heads: leaves oblong-ovate or oval, from barely dentate to incisely pinnatifid, 6 to 10 inches long, weakly prickly; lower decurrent on the stem into narrow prickly wings: involucral bracts all attenuate-subulate from a narrow base, arachnoid-pubescent when young or glabrate: corolla-lobes narrowly linear, apiculate: larger pappus bristles clavellate. — Mel. Biol. ix. 310. *Cirsium Kamtschaticum*, Ledeb. in DC. Prodr. vi. 644, & Fl. Ross. ii. 736. — Atkha, one of the Aleutian Islands, *Lieut. Turner*. Said to be "7 feet high": corollas whitish: anther-tips slender, as in pl. Kamts., and longer than in var.? *Grayanus*, Maxim., of Japan. (Kamtschatka to Japan.)

++ ++ ++ Proper bracts of the involucre not at all prickly, but the large (2 inches high) heads conspicuously and numerously bracteose-leafy at base. Atlantic species.

C. horridulus, Pursh. Arachnoid when young, glabrate with age, 1 to 3 feet high, the larger plants branching and bearing several heads: leaves elongated-lanceolate, not decurrent, pinnatifid, strongly prickly: head about 2 inches high, surrounded by a whorl of 8 to 30 linear or lanceolate numerously and strongly prickly leaves, which usually equal in length the involucre of gradually attenuate weak-pointed minutely scabrous bracts: flowers pale yellow, rarely purple (var. *Elliottii*, Torr. & Gray). — Fl. ii. 507; Ell. Sk. ii. 272; Gray, Proc. Am. Acad. x. 40. *C. spinosissimus*, Darlingt. Fl. Cest. ed. 2, 438. *Cirsium horridulum*, Michx. Fl. ii. 90; DC. Prodr. vi. 651. *C. megacanthum*, Nutt. in Trans. Am. Phil. Soc. vii. 421, large form. *Carduus spinosissimus*, Walt. Car. *C. horridulus*, Pers. Syn. ii. 390. — Sandy or gravelly soil, New England, near the coast, to Florida and Texas.

* * * Bracts of the involucre moderately unequal or the lower not rarely about equalling the upper, more rigid and imbricated at base, but most of them with more or less herbaceous spinescent-tipped spreading upper portion, and no glandular dorsal ridge. Rocky Mountain and Pacific species.

+- Heads (only inch high) few or several and sessile in a terminal cluster: stem leafy to the top.

C. Eatóni, Gray. A foot or so high, mostly simple, loosely arachnoid-woolly or glabrate: leaves pinnatifid or pinnately parted into short lobes, mostly very prickly, either green and glabrate or remaining whitish-woolly beneath: involucre rather narrow, from arachnoid-ciliate to glabrate or apparently glabrous; its principal bracts erect, with broadish appressed base, abruptly attenuate into the subulate-acerose slightly herbaceous spinescent portion, outermost little shorter than the inner: corolla whitish, its lobes considerably shorter than the throat. — Proc. Am. Acad. xix. 56. *Cirsium eriocephalum*, var. *leiocephalum*, *C. foliosum*, & *C. Drummondi* in part, Eaton in Bot. King Exp. 195, 196. — Mountains of Utah (Uintah and Wahsatch) and of Colorado, from 8,000 to 11,000 feet, also in Humboldt Mountains, Nevada, *Watson, Jones, Hall & Harbour*, &c.

+- +- Heads solitary terminating the stem or branches (involucre usually long-woolly when young, but sometimes glabrate), hemispherical,

++ Middle-sized: flowers white or pale purple: anther-tips deltoid.

C. Andréwsii, Gray. Probably tall, branching; the loose wool deciduous except from the heads: stem strongly striate: upper leaves laciniate-pinnatifid and with narrowly lanceolate prickly lobes: bracts of the involucre with coriaceous oblong-ovate base, greenish at short upper part, where it is abruptly contracted into an aristiform spinescent appendage: corollas apparently whitish; the lobes fully twice the length of the throat. — Proc. Am. Acad. x. 45, & Bot. Calif. i. 420. — W. California, *Andrews*, station unknown.

C. Califórnicus, Gray, l. c. Tall and branching, with white wool more or less deciduous: leaves from sinuately to deeply pinnatifid, moderately prickly: principal bracts of the involucre with somewhat foliaceous and subulate spinescent summit, sometimes very conspicuous, sometimes smaller and attenuate more directly into the prickle: corollas cream-color, white, or rarely purple; lobes shorter than the throat. — *Cirsium Californicum*, Gray, Pacif. R. Rep. iv. 112. — California, from the Stanislaus (where first coll. by *Bigelow*) to San Diego and San Bernardino and adjacent Arizona. A variety of forms here assembled, some with larger heads and more leafy-bracted involucre passing to the next.

++ ++ Large heads, the larger fully 2 inches high and broad: slender corolla-lobes considerably longer than the throat: herbage and commonly squarrose involucre copiously white-woolly, sometimes glabrate in age: anther-tips narrow and acuminate.

C. Neo-Mexicánus, Gray, l. c. Stout, 2 to 4 feet high: spinescent rigid tips to the principal involucral bracts half to nearly full inch long: corollas from white to pale purple: node on the style generally manifest and obscurely bearded: otherwise as the next, into which it seems to pass. — *Cirsium Neo-Mexicanum*, Gray, Pl. Wright. ii. 101. *C. canescens*,

Gray, Pl. Fendl. 110, not Nutt. — Plains of S. Colorado, New Mexico, and Arizona; first coll. by *Fendler, Wright*, &c.

C. occidentális, GRAY, l. c. Mostly stout, 2 to 5 feet high, very white with thick coating of cottony wool; leaves from sinuate-dentate to pinnatifid, not very prickly: involucral bracts sometimes narrow and herbaceous-acerose from a little dilated base, sometimes with broader more coriaceous base, or the outer with lanceolate-subulate tips: corollas red or crimson (the longer inch and a half long): style destitute of node. — *Carduus occidentalis*, Nutt. Trans. Am Phil. Soc. vii. 418. *Cirsium Coulteri*, Gray, Pl. Wright. ii. 110; Eaton in Bot. King Exp. 195. — S. Oregon and W. California to San Diego and to the Mohave; first coll. by *Coulter*. Varies much in the size of the heads; these in some plants only inch and a half long, narrower, and involucre glabrate, its outer bracts successively shorter, with lanceolate-subulate squarrose green tips; approaching *C. Californicus* and also the following section.

  * * * * Bracts of the involucre regularly and chiefly appressed-imbricated in numerous ranks; the outer successively shorter, not herbaceous-tipped or appendaged, except that the innermost (which are all muticous or innocuous) are in one or two species obviously scarious-tipped.

  +— Heads oblong or cylindraceous, showy (1½ to 2 inches long): flowers bright red or crimson-pink: involucral bracts comparatively large, not at all glandular on the back; inner ones all erect and purplish-tinged. Arizonian and Californian.

  ++ White with cottony wool, which is tardily if at all deciduous, 1 to 3 feet high.

C. Andersóni, GRAY, l. c. Slender, rather lightly and loosely woolly; leaves lightly prickly, sinuate-pinnatifid, rather sparse: heads naked-pedunculate: involucral bracts comparatively loose and erect, all gradually attenuate from a narrow base; outermost tipped with a small weak prickle: corolla bright pink-red; its slender lobes about equalling the throat: style considerably prolonged above the very obscure node. — Dry hills, E. California, adjacent Nevada, and S. W. Idaho; common along the Sierra south to the Yosemite and Kern Co.; first coll. by *Anderson*.

C. Arizónicus, GRAY, l. c. More densely white-woolly, branching and leafy: leaves sinuate or pinnatifid; lobes prickly-pointed: heads more numerous, less peduncled: involucral bracts well imbricated, soon glabrate; outer coriaceous, ovate-oblong to ovate-lanceolate, abruptly contracted into a rigid prickle of rarely over their own length, inner attenuate: corolla crimson-purple or carmine; its lobes twice the length of the throat: style produced at tip to only 4 or 6 times its diameter above the manifest node. — *Cirsium undulatum*, var., Gray, Pl. Wright. ii. 101. — Sandy or gravelly places, Arizona and S. W. Utah; first coll. by *Wright* and by *Thurber*.

  ++ ++ Green and glabrous or very early glabrate, 3 or 4 feet high.

C. Rothróckii, GRAY. Stout, branching, leafy to the top: leaves from incisely pinnatifid to pinnately parted, conspicuously prickly: heads rather thicker than in the foregoing: involucre similar, but longer prickly (prickles sometimes even three-fourths inch long): corolla and style similar, or node of the latter less evident. — Proc. Am. Acad. xvii. 220 (form noted by Rothrock in Wheeler Rep. under *C. Arizonicus*). — Cañons of S. Arizona, *Rothrock, Lemmon*.

  +— +— Heads broad, mostly large: flowers from rose-purple to white: involucre glabrous or early glabrate, the light arachnoid wool caducous; its bracts rather large, chartaceous or coriaceous, not at all glandular on the back, outer tipped with a short weak prickle or innocuous cusp, innermost wholly unarmed and not rarely scarious-tipped.

  ++ Eastern species: leaves equally green both sides: anther-tips broadish.

C. púmilus, TORR. Somewhat villous-pubescent: stem stout, mostly simple, a foot or two high (rarely taller) and bearing 1 to 3 large heads: leaves oblong or lanceolate, commonly pinnatifid, copiously prickly and setose-ciliate: heads full 2 inches high, often leafy-bracteose at base, arachnoid when young: involucral bracts mostly lanceolate: corollas rose-purple, occasionally white, with lobes shorter than throat: flowers distinctly fragrant. — Compend. 282; Bigel. Fl. Bost. ed. 2, 292; Gray, Proc. Am. Acad. x. 40; Sprague, Wild Flowers, 138, t. 32. *Carduus odoratus*, Muhl. Cat. 70. Darlingt. Fl. Cest. ed. 1, 85. *C. pumilus*, & var. *hystrix*, Nutt. Gen. ii. 130. *Cirsium pumilum*, Spreng. Syst. iii. 375; DC. Prodr. vi. 651; Torr. & Gray, l. c. — Open ground, Mass., near the coast, to Penn. and New Jersey.

++ ++ Western species: leaves either green both sides or deciduously white-woolly beneath: involucral bracts plane: anther-tips narrow, very acute.

C. quercetórum, GRAY. Lightly villous-arachnoid when young, soon glabrate: stem stout, a foot or less high, bearing few or several thick heads: leaves mostly petiolate (the larger a foot long), pinnately parted and the oblong divisions often 3–5-cleft, strongly or weakly prickly; involucral bracts thickish-coriaceous, closely imbricated in numerous ranks; outer only mucronately cuspidate or with short prickle (outermost only about 3 lines long); innermost obscurely scarious at tip: corollas purplish or whitish, the lobes equalling or longer than the throat. — Proc. Am. Acad. x. 40, & Bot. Calif. i. 418. — Dry hills, at Oakland and vicinity, California, *Kellogg, Bolander*, &c.

C. Drummóndii, GRAY, l. c. Green and somewhat villous-pubescent, or when young lightly arachnoid-woolly (at least the lower face of the leaves), either stemless and bearing sessile heads in a cluster on the crown, or caulescent and even 2 or 3 feet high, with solitary or several loosely disposed heads: leaves from sinuate or almost entire to pinnately parted, moderately prickly: larger heads fully 2 inches high: involucral bracts thin-coriaceous or chartaceous, mostly acuminate, weak-prickly pointed or innocuous, innermost with more scarious and sometimes obviously dilated and erose-fimbriate tips: corollas either white or sometimes rose-purple, with lobes usually shorter than the throat. — *Carduus pumilus*, Hook. Fl. i. 302, excl. syn. *Cirsium Drummondii*, Torr. & Gray, Fl. ii. 459. — From Fort Franklin, near the Arctic Circle, to the Saskatchewan, along the Rocky Mountains to Colorado and Utah, west to Oregon, and south along the Sierras to S. California. Polymorphous and of very wide range.

Var. **acauléscens**, GRAY, l. c. Smaller, with heads (solitary or several on the crown, encircled by the radical leaves) only inch and a half long, or less, and proportionally narrow: outer involucral bracts with a longer but rather weak prickle. — *Cirsium acaule*, var. *Americanum*, Gray, Proc. Acad. Philad. 1863, 68. — Mountains of Colorado to the Sierra Nevada in S. California.

C. foliósus, GRAY, l. c. More woolly, usually also villous when young: stem stout, leafy to the cluster of a few sessile heads, a span or two high: leaves commonly elongated, linear-lanceolate, laciniately deutate, arachnoid-tomentose beneath: heads broad, inch and a half high, leafy-bracteose: involucre nearly of the preceding: corollas pale or white, with lobes equalling or longer than the throat. — *Carduus foliosus*, Hook. Fl. i. 303. *Cirsium foliosum*, DC. Prodr. vi. 654. — Prairies of the northern Rocky Mountains, *Drummond*. Idaho, *Burke, Spalding*.

C. scariósus. White with cottony tomentum, at least the lower face of the leaves: stem about a foot high: leaves of lanceolate outline, mostly pinnately parted into lanceolate long-prickly lobes; upper face sometimes villous, sometimes only cottony and early glabrate: heads nearly of preceding, 2 or 3 in a sessile cluster, or solitary on short leafy branches: innermost bracts of involucre commonly with more conspicuous erose or entire scarious tips: corollas pale or white. — *Cirsium scariosum*, Nutt. Trans. Am. Phil. Soc. vii. 420. — Rocky Mountain plains, Wyoming and Utah, *Nuttall, Ward, Palmer*, &c. Has been referred to *C. Americanus* and (in Proc. Am. Acad. xix. 56) to *C. foliosus*.

++ ++ ++ Species of Mexican border, with dense white tomentum, smaller and obscurely carinate outer involucral bracts, and blunt very scarious tips to the inner: anther-tips very acute.

C. Wheéleri, GRAY. Stem slender, 2 or 3 feet high, white with close cottony wool, as is the lower face of the leaves: these narrowly lanceolate or linear, sparingly laciniate-pinnatifid, glabrate and green above, slightly prickly: head solitary, nearly 2 inches high, naked at base: outer involucral bracts firm-coriaceous, much appressed, carinate-thickened down the middle of the back, abruptly tipped with a small weak prickle; inner with conspicuous scarious or scarious-edged and erose tip or appendage: corolla crimson-purple; its lobes much longer than throat. — Proc. Am. Acad. xix. 56. — Rocky Cañon, south of Camp Apache, *Arizona, Rothrock*, in Wheeler Exped., where it was referred to *C. undulatus*.

+ + + Heads large or comparatively small: flowers usually rose or flesh-colored: involucral bracts closely appressed, coriaceous or thickish, commonly with a glandular or viscid ridge, short line, or a broader spot on the back near the summit.

++ Canescent, at least the lower face of the leaves white-tomentose, very rarely glabrate in age: heads naked, solitary or scattered.

= = **Leaves** pinnately parted into narrow and linear mostly entire divisions; anther-tips attenuate-subulate.

C. Pítcheri, TORR. A foot or two high, with herbage persistently white-tomentose throughout: lower leaves a foot or so long, with divisions (2 to 4 inches long, 2 or 3 lines wide) either entire or some again pinnately parted into shorter lobes, weakly prickly-tipped; the winged rhachis not wider than the divisions: heads few or solitary, 2 inches high: involucre glabrate; the bracts rather small, viscid down the back, tipped with small short prickle: corollas ochroleucous. — Torr. in A. Eaton, Man. ed. 5, 180; Gray, Proc. Am. Acad. x. 42. *Cirsium Pitcheri*, Torr. & Gray, Fl. ii. 456. — Sand-banks on the shores of the Great Lakes from the head of Lake Michigan northwestward, and in Dakota, *Suckley*; first coll. by Dr. *Pitcher*.

= = **Leaves** from undivided to pinnately parted, the lobes lanceolate or broader, disposed to be white-tomentose above as well as below: prickle on cusp of the principal involucral bracts more or less rigid and pungent.

    a. Bracts of the involucre minutely scabrous-ciliolate.

C. Gráhami, GRAY. Stem 3 to 8 feet high: leaves elongated-lanceolate (larger ones a foot or more long), from repand-dentate to sinuate and pinnatifid (sometimes delicately, sometimes strongly prickly), upper face at length glabrate and green: heads 1½ to 2 inches high: involucre glabrate and greenish; the bracts lanceolate-subulate, tipped with a short rigid cusp rather than prickle, the margins at least of the principal ones minutely scabrous-ciliolate: corollas crimson-red: anther-tips attenuate-subulate. — Proc. Am. Acad. xix. 57. *C. undulatus*, var. *Grahami*, Gray, Proc. Am. Acad. x. 43. *Cirsium Grahami*, Gray, Pl. Wright. ii. 102; Hook. Bot. Mag. t. 2885. — Wet ground, Arizona, *Wright, Thurber, Lemmon*.

    b. Bracts of the involucre smooth and naked, or else tomentose on the margins.

C. ochrocéntrus, GRAY. Resembles the next following species, usually taller, even to 6 or 8 feet high, the white tomentum mostly persistent: leaves commonly but not always deeply pinnatifid and armed with long yellowish prickles: heads 1 or 2 inches high: principal bracts of the involucre broader and flatter, the viscid line on the back narrow or not rarely obsolete, tipped with a prominent spreading yellowish prickle: corollas purple, rarely white. — Proc. Am. Acad. xix. 57. *C. undulatus*, var. *ochrocentrus*, Gray, Proc. Am. Acad. x. 43. *Cirsium ochrocentrum*, Gray, Pl. Fendl. 110. — Plains, &c., W. Texas to Colorado, the eastern Sierra Nevada, and Arizona. (Adj. Mex.)

C. undulátus, GRAY. A foot or two high, persistently white-tomentose: leaves rarely pinnately parted, moderately prickly: heads commonly inch and a half high: principal bracts of the involucre mostly thickened on the back by the broader glandular-viscid ridge, comparatively small and narrow, tipped with an evident spreading short prickle: corollas rose-color, pale purple, or rarely white; its lobes equalling or surpassing the throat in length: anther-tips attenuate-subulate. — Proc. Am. Acad. x. 42, excl. var. *ochrocentrus*, & var. *Grahami*. *Carduus undulatus*, Nutt. Gen. ii. 130. *C. discolor*, Hook. Fl., in part. *C. Douglasii*, DC. Prodr. vi. 643, excl. habitat. *Cirsium Hookerianum*, Hook. Lond. Jour. Bot. vi. 253, not Nutt. — Plains, &c., from Lake Huron and Minnesota to Saskatchewan, west to Oregon, south to Kansas and New Mexico.

    Var. **canéscens,** GRAY, l. c., is merely a form with smaller heads, sometimes not over an inch high, the leaves varying from ciliately spinulose-dentate to deeply pinnatifid. — *Cirsium canescens* & *C. brevifolium*, Nutt. Trans. Am. Phil. Soc. vii. 421. — Minnesota to New Mexico and S. Utah.

    Var. **megacéphalus,** GRAY, l. c. Stouter form, usually broader-leaved, with broad heads 2 inches or more high — Minnesota and Texas (where coll. by *Berlandier*) to Idaho.

C. Bréweri, GRAY, l. c. Usually both very white-tomentose and tall (5 to 10 feet high): leaves mostly elongated-lanceolate, conspicuously prickly: heads paniculate, sometimes very numerous, subsessile, merely inch high, or when solitary inch and a half high: bracts of the globular involucre much appressed, firm-coriaceous, the tip externally bearing an oval or oblong greenish viscid-glandular spot; outer ones ovate to oblong, abruptly tipped with a rather slender spreading prickle: corollas pale purple or whitish, the lobes shorter than the throat: anther-tips deltoid, merely acute. — Springy soil, Sierra Nevada from Lake Tahoe and Mendocino Co., California (first coll. by *Anderson* and *Brewer*), to E. Oregon, *Cusick*, &c. Also, less white-woolly, San Juan, Monterey Co., *Brewer*, leading to the var.

Var. Vaséyi. Perhaps a distinct species, only arachnoid-tomentose and greenish, even glabrate in age. — California, in Plumas and Sierra Co., *Lemmon, Mrs. Ames.* A remarkably glabrate form, with involucral bracts obscurely glandular, and tipped with very short prickle, growing in dry soil exposed to the sun, Tamalpais, *G. R. Vasey.* Also a robust form, equally glabrate and green, with the glandular spot on the involucral bracts conspicuous and narrow: in salt marshes, Suisin Bay, *Greene.*

= = = Leaves in the same species from undivided to pinnately parted, and the lobes from ovate to lanceolate, upper face soon glabrate and green: involucral bracts tipped with weak setiform prickles or sometimes hardly any: anther-tips subulate, very acute: corolla flesh-colored, rarely white.

C. altíssimus, WILLD. Stem branching, 3 to 10 feet high: leaves in the typical form ovate-oblong or narrower, sometimes with merely spinuloso-ciliate slightly toothed margins, sometimes laciniate-cleft or sinuate, or lower ones deeply sinuate-pinnatifid, weakly prickly: heads one and a half to two inches high: involucral bracts firm-coriaceous, abruptly tipped with a spreading setiform prickle, the short outermost ovate or oblong: roots fascicled and not rarely tuberous-thickened below the middle, in the manner of *Dahlia.* — Willd. Spec. iii. 1671; Ell. Sk. ii. 268; Gray, Proc. Am. Acad. x. 42. *Carduus altissimus,* L. Spec. ii. 824. *Cirsium altissimum,* etc., Dill. Elth. i. 81, t. 69. *C. altissimum & C. diversifolium,* DC. Prodr. vi. 640. — Borders of woods, and in open ground, common from New York to Wisconsin, Florida, and Texas.

Var. filipéndulus, GRAY, Proc. Am. Acad. xix. 56. Smaller, 2 or 3 feet high: roots tuberiferous: leaves commonly deeply pinnatifid: heads few, only inch and a half high. — *Cirsium filipendulum,* Engelm. in Gray, Man. ed. 5, 273. *C. Virginianus,* var. δ? Torr. & Gray, l. c. — Prairies and Live-oak thickets, Texas and Colorado. (Adj. Mex.)

Var. díscolor, GRAY, l. c. Stem 2 to 6 feet high, freely branching: leaves nearly all deeply pinnatifid into lanceolate lobes, or those of upper leaves linear: heads fully inch and a half high. — *C. discolor,* Muhl. in Willd. Spec. iii. 1670; Ell. l. c.; Bigel. Fl. Bost. ed. 2, 292. *Carduus discolor,* Nutt., Darlingt., &c. *Cirsium discolor,* Spreng. Syst. iii. 373; DC. l. c.; Torr. & Gray, l. c. — Borders of fields and thickets, Canada and New England to Illinois and Georgia.

C. Virginiánus, PURSH. Stem slender, 2 or 3 feet high, simple or branching: leaves narrow, varying as in the preceding: heads more naked-pedunculate, only an inch long: involucral bracts small and narrow, thinner, tapering into a very weak short spreading bristle-like prickle, sometimes hardly any: flowers rose-purple. — Fl. ii. 506; Ell. l. c. *Carduus Virginianus,* L. l. c.; Jacq. Obs. iv. t. 99; Nutt. l. c. *Cirsium Virginianum,* Michx. Fl. ii. 90; DC. l. c.; Torr. & Gray, Fl. ii. 457, excl. last var. *C. Texanum,* Buckley in Proc. Acad. Philad. 1862, imperfect specimen, apparently of this species. — Pine woods and dry banks, Virginia to Texas.

++ ++ Green or with only light and thin arachnoid tomentum, this at length mostly deciduous: involucre innocuous or nearly so. Atlantic species.

= Heads only inch high, loosely somewhat paniculate: principal bracts of the involucre conspicuously viscid-glandular on the back, more or less cuspidate-tipped: stems branching, 2 to 8 feet high.

C. Nuttállii, GRAY, l. c. Early glabrate: stem slender, below winged by decurrence of the leaves: those when young lightly arachnoid beneath and often villous with jointed hairs above, deeply pinnatifid and with narrow lobes, slender-prickly: heads rather narrow: involucre nearly glabrous, of very small and narrow thinnish bracts, the lower ones acicular-mucronate: corollas white or pale purple. — *Carduus glaber,* Nutt. Gen. ii. 129? but if so, hardly from New Jersey. *Cnicus glaber,* Ell. Sk. ii. 270. *Cirsium Nuttallii,* DC. Prodr. vi. 651. — Dry ground, S. Carolina to Florida, toward the coast. Nearly related to *C. Virginianus.*

C. Wrìghtii, GRAY, l. c. Robust and tall, with thin arachnoid wool tardily deciduous from the ample (foot or more long) sinuate or pinnatifid weakly prickly leaves: heads in a naked panicle, hemispherical: bracts of the involucre small: outer ones subulate, cuspidate-tipped: corollas white, or possibly flesh-color: larger pappus-bristles strongly clavellate at tip. — *Cirsium Wrightii,* Gray, Pl. Wright. ii. 101. — Near springs, S. W. Texas and E. Arizona, *Wright.*

== == Heads large, oblong or cylindraceous, commonly solitary and pedunculate: involucral bracts comparatively large, gradually acuminate into a mucronate cusp or weak and short prickle, glabrate, the viscid dorsal ridge narrow: corollas purple: leaves when young canescently floccose-woolly beneath, oblong-linear or narrowly lanceolate.

C. repándus, ELL. A foot or two high, leafy: leaves mostly undulate-lobulate, rather densely prickly at margins: heads inch and a half long: involucre narrow-campanulate. — Sk. ii. 269; Gray, l. c. *Cirsium repandum*, Michx. Fl. ii. 89; DC. Prodr. vi. 651. *Carduus repandus*, Pers. Syn. ii. 386. *C. Virginianus*, Walt. Car. 195 ? — Dry pine barrens, N. Carolina to Florida.

C. Lecóntei, GRAY. Stem slender but rigid, commonly simple and bearing a single conspicuously pedunculate head (of full 2 inches in height): leaves sparsely dentate or pinnatifid-lobulate, with scattered prickles: involucre cylindraceous. — Proc. Am. Acad. x. 39. *Cnicus Virginianus*, Hook. Comp. Bot. Mag. i. 48. *Cirsium Lecontei*, Torr. & Gray, Fl. ii. 459. — Wet pine barrens, Georgia to Florida and Louisiana; first coll. by L. Conte.

== == == Heads inch and a half high, rather broad: involucre arachnoid-woolly; its principal bracts broad and pointless. Atlantic species.

C. múticus, PURSH. Obscurely arachnoid when young and with some villosity: stem 3 to 8 feet high, branching above: leaves deeply pinnatifid, sparsely weak-prickly, glabrate: involucre sometimes glabrate in age: bracts with broad and short viscid ridge or spot just beneath the obtuse or acutish sometimes mucronulate apex, lowest ovate or oblong and very short, innermost linear: flowers rose-purple. — Gray, l. c. *C. glutinosus*, Bigel. Fl. Bost. ed. 2, 291, not Lam. *Carduus muticus* and perhaps *C. glaber*, Nutt. Gen. ii. 129. *Cirsium muticum*, Michx. Fl. ii. 89; DC. l. c.; Torr. & Gray, Fl. ii. 458, excl. syn. of the var.?, which is a more rigid form, growing in open ground. *C. Bigelovii*, DC. l. c. — Low ground and shady swamps, Newfoundland to Saskatchewan, Florida, and Louisiana.

199. **ONOPÓRDON**, Vaill. COTTON THISTLE. (Old Greek name, meaning *Asses' Thistle*.) — Large and stout biennials of the Old World, one sparingly naturalized; fl. late summer. — DC. Prodr. vi. 617. *Onopordum*, L.

O. ACÁNTHIUM, L. White with cottony wool: stem 3 to 9 feet high, branching, winged throughout by decurrence of the large oblong sinuate-lobed and prickly leaves: wings sinuate, very prickly: heads pretty large: involucre globular, arachnoid or partly glabrate; bracts rigid, subulate and prickly tipped, squarrose: corollas light purple or paler: pappus fuscous, scabrous, not twice the length of the slightly rugose akene. — Fl. Dan. t. 909; Engl. Bot. t. 907. — Waste grounds near dwellings and roadsides in Atlantic States, not abundant. (Nat. from Eu.)

200. **SÍLYBUM**, Vaill. MILK THISTLE. (Σίλυβος, ancient Greek name of an edible-stemmed Thistle, perhaps the present plant.) — Single species.

S. MARIÁNUM, GAERTN. Prickly-leaved biennial or annual, glabrate or nearly glabrous; with ample sinuate or pinnatifid green leaves, blotched with white along the veins: corollas rose-purple, deeply cleft. — Escaped from gardens in a few places, also a ballast-weed, disposed to be naturalized southward, especially in California: fl. summer. (Adv. from Eu.)

201. **CENTAURÉA**, L. STAR THISTLE. &c. (Κενταύρειον, plant of the Centaurs, name applied by the herbalists to two or three widely different genera.) — An immense genus in the Old World, one species only indigenous to N. America, two or three in Chili. — *Centaurea* & *Carbenia* (Adans.). Benth. & Hook. Gen. Pl. ii. 477, 482.

§ 1. CARBÉNIA. Akenes terete, strongly many-striate, with lateral scar, the corneous margin at summit 10-dentate: pappus double, each of 10 aristiform bristles, outer longer and naked, inner short and fimbriolate: anthers with elongated cartilaginous terminal appendages, which are connate to their blunt tips:

head surrounded by large and leafy accessory bracts. — *Carbeni*, Adans. Fam. ii. 116. *Cnicus*, Gærtn., DC., not L.

C. BENEDÍCTA, L. (BLESSED THISTLE.) Low and branching annual, hirsute or pubescent: leaves prominently reticulated, sinuate-pinnatifid or laciniate-dentate, the teeth or margins weakly prickly; lower attenuate at base; upper narrowly oblong, partly clasping by broad base: heads sessile, inch and a half high, equalled by the oblong involucral leaves: proper involucre of thin-coriaceous bracts in few ranks, all or most of them abruptly tipped with an aristiform or spinescent and pectinately prickly spreading appendage: receptacle very densely setose with long and soft capillary bristles: corollas light yellow: longer bristles of the pappus alternating with inner and with the teeth of the akene. — Spec. ed. 2, ii. 1296; Sibth. Flora Græca, t. 906. *Cnicus benedictus*, L. Spec. ed. i. 826; Gærtn. Fruct. ii. t. 162; DC. Prodr. vi. 606; Torr. & Gray, Fl. ii. 455. — Waste grounds, at seaports and elsewhere near dwellings, in the Southern Atlantic States and in California; not common. (Nat. from Eu.)

§ 2. CENTAUREA proper. Akenes more or less compressed or quadrangular: pappus of indefinite (either scanty or numerous) bristles or narrow paleæ: involucre globular or ovoid.

* Old World species, sparingly naturalized, with comparatively small heads: scar or insertion of akene lateral.

+ Bracts of the involucre (or some of them) armed with a rigid spine or prickle, and also more or less spinulose along its sides or base: cartilaginous appendages terminating the anthers commonly elongated and connate: ours annuals, none with the marginal corollas enlarged. — *Calcitrapa*, Juss.

C. CALCÍTRAPA, L. (STAR THISTLE.) Low, much branched, diffusely spreading, green, glabrate or hairy: leaves narrow, laciniate-pinnatifid; uppermost somewhat involucrate-crowded at base of the sessile heads: principal bracts of the involucre becoming corneous, armed with a widely spreading very long and rigid spine, which bears 2 or 3 spinules on each side at base: corollas purple or purplish: pappus wanting. — Engl. Bot. t. 125; Torr. & Gray, Fl. ii. 454. — Sparingly established at seaports from New York southward, chiefly as a mere ballast-weed. (Nat. from Eu.)

C. SOLSTITIÁLIS, L. Erect, a foot or two high, canescent with cottony wool: radical leaves lyrate-pinnatifid; cauline lanceolate and linear, mostly entire, decurrent on the branches in narrow wings: heads naked, somewhat pedunculate: intermediate bracts of the globular involucre tipped with a long spreading spine, having one or two spinules at base; outermost bearing a few small palmate prickles; innermost only scarious-tipped: corollas yellow: pappus double; outer of short and squamellate, inner of longer bristles. — Engl. Bot. t. 243; Reichenb. Ic. Fl. Germ. t. 795; Gray, Bot. Calif. i. 421. — Near San Francisco and San Diego, California, sparingly introduced. (Nat. from Eu.)

C. MELITÉNSIS, L. Erect, 2 to 4 feet high, paniculately branched, cinereous-pubescent, somewhat woolly at first: radical leaves lyrate-pinnatifid; cauline lanceolate or linear, mostly entire, narrowly decurrent on the branches: heads smaller, sessile or 1-2-leaved at base: principal bracts of involucre bearing a spreading slender spine of about their own length, which is pectinately spinulose towards its base; innermost with simply spinescent tip; outermost usually with the central spine reduced and the spinules palmate: corollas yellow: pappus of very unequal rigid bristles or squamellæ: akene lightly costate. — Sibth. Flora Græca, t. 909; Reichenb. Ic. Fl. Germ. xv. t. 796; Gray, Bot. Calif. l. c. — Fields, California and Arizona, rather common. (Nat. from Eu.)

+ + Bracts of the involucre unarmed, most of them terminated by a scarious discolored fimbriate-ciliate or lacerate appendage. — *Jacea, Platylophus, Cyanus*, &c., Cass.

++ Perennials, with rose-purple flowers: pappus obsolete.

C. NÍGRA, L. (KNAPWEED, HARDHEADS.) A foot or two high, branching, roughish-pubescent: leaves lanceolate and entire, or lower sparingly toothed: most of the involucral bracts with strongly pectinately ciliate-fringed blackish appendages, these only conspicuous: flowers all hermaphrodite, marginal ones not enlarged or rarely so. — Fl. Dan. t. 606; Engl. Bot. t. 278. — Fields, Newfoundland to E. New England. (Nat. from Eu.)

— C. Jácea, L. Heads usually larger: brownish appendages of the involucral bracts merely lacerate: marginal flowers neutral and with enlarged palmate corollas, forming conspicuous false rays: otherwise like the preceding. — Fl. Dan. t. 519; Reichenb. Ic. Fl. Germ. xv. t. 754, 755. — Charlotte, Vermont, *Pringle*. Near New York, &c., as a ballast-weed. (Nat. from Eu.)

++ ++ Annual, with blue flowers, varying to white or purple: pappus of unequal bristles about the length of the akene.

— C. Cýanus, L. (Bluebottle.) Slender, branching, a foot or two high, whitened when young with floccose wool: leaves linear, entire, or lower toothed, sometimes pinnatifid: heads naked on slender peduncles: involucral bracts rather narrow, fringed with short scarious teeth: marginal flowers neutral, with much enlarged radiatiform corollas. — Engl. Bot. t. 277; Reichenb. l. c. t. 768. — Escaped from gardens sparingly in the Atlantic States. (Nat. from. Eu.)

* * American species: heads large: scar or insertion of akene obliquely basal: bracts of involucre unarmed, the appendage conspicuously pectinate-fimbriate: anther-appendages distinct. — *Plectocephalus*, Don.

— C. Americána, Nutt. Annual, nearly glabrous: stem stout, commonly simple, 2 to 6 feet high, striate-sulcate, thickened under the naked head: leaves entire or mostly so, oblong-lanceolate, mucronate: involucre inch or inch and a half in diameter; its very numerous bracts all with conspicuously fringed scarious appendages: flowers rose-color or flesh-color; the hermaphrodite ones forming a disk of 1 to 3 inches in diameter; the neutral marginal ones (with their very narrow lobes an inch long) forming an ample ray: style filiform, entire to the minutely 2-dentate stigmatic tip: pappus of copious similar but unequal bristles longer than the akene. — Jour. Acad. Philad. ii. 117; Barton, Fl. Am. Sept. t. 50; Reichenb. Ic. Exot. t. 132; Fl. Serres. iv. t. 327; Meehan, Nat. Flowers, ser. 2, ii. t. 17. *C. Nuttallii*, Spreng. Syst. iv. 298. *C. Mexicana* & *C. Americana*, DC. Prodr. vi. 575. *Plectocephalus Americanus*, Don, Brit. Fl. Gard. ser. 2. t 51. — Plains of Arkansas and Louisiana to Arizona; first coll. by *Nuttall*. (Adj. Mex.)

Tribe X. MUTISIACEÆ, p. 82.

202. HECASTOCLÉIS, Gray. ("Εκαστος, each, κλείω, to shut up, each flower in an involucre of its own). — Proc. Am. Acad. xvii. 220. — Single species.

H. Shockléyi, Gray, l. c. Low and glabrous shrub, with rigid branches, and rigid leaves of two sorts; cauline small, linear-lanceolate or subulate, cuspidate-tipped, and on the sides usually a few spiniform teeth, also fascicled on axillary spurs; floral ones 3 or 4 in a whorl or cluster, larger (half-inch or more long) and oval or ovate, papyraceous, reticulated, margined with sparse slender prickles, forming a loose external involucre around a fascicle of few or several sessile heads (those about 5 lines long and fusiform): flower apparently dull white. — Esmeralda Co., W. Nevada, in an arid desert region, *W. S. Shockley*. By the style and habit evidently Mutisiaceous rather than Cynaroideous.

203. GOCHNÁTIA, HBK. (*F. C. Gochnat*, of Strasburg.) — American shrubby plants; with coriaceous leaves usually entire and tomentose beneath, and white or whitish flowers. — Nov. Gen. & Spec. iv. 19, t. 309. *Gochnatia* & *Moquinia* (at least in part), DC.; Benth. & Hook. Gen. ii. 490.

G. hypoleúca, Gray. Rigid shrub, 6 to 8 feet high: leaves oblong or oval, very short-petioled, commonly inch or more long, glabrous and bright green above, finely white-tomentose beneath (like an Olive-leaf) as also the branchlets: heads in sessile somewhat thyrsoid-paniculate fascicles, half-inch or less long: involucre cylindraceous, 5–7-flowered: bracts ovate and oblong, outermost very short: flowers white, all hermaphrodite! — Proc. Am. Acad. xix. 57. *Moquinia hypoleuca*, DC. Prodr. vii. 23. — Southern Texas, between the Rio Frio and the Nueces, *Palmer*. (Adj. Mex.; first coll. by *Berlandier*.)

204. CHAPTÁLIA, Vent. (*J. A. C. Chaptal*, an eminent chemist.) — Perennial herbs (all American), chiefly stemless, low, and floccose-tomentose; with leaves in a radical tuft, persistently canescent beneath, glabrate above; scapes naked; heads at first nodding; flowers white or purplish, or the rays rose-purple: fl. spring and summer.

§ 1. Akenes of female flowers merely attenuate into a neck; those of hermaphrodite flowers all abortive: scapes elongated. — *Chaptalia*, DC.

C. tomentósa, Vent. Leaves spatulate or oblanceolate, thickish, entire or retrorsely denticulate, white beneath with dense matted tomentum: scapes a span to a foot high: rays broadly linear, commonly purple: akenes glabrous. — Hort. Cels. t. 61; Pursh, Fl. ii. 577; Sims, Bot. Mag. t. 2257; DC. Prodr. vii. 41; Torr. & Gray, Fl. ii. 464. *Perdicium semiflosculare*, Walt. Car. 204. *Tussilago integrifolia*, Michx. Fl. ii. 121. *Gerbera Walteri*, Schultz Bip. in Seem. Bot. Herald, 313. — Moist pine barrens, N. Carolina to Florida and E. Texas.

§ 2. Akenes of all the flowers fertile, and with slender usually filiform beak: corollas of hermaphrodite flowers sometimes hardly bilabiate, of innermost female flowers somewhat so: scapes elongated. — *Leria*, DC.

C. nútans, Hemsl. Leaves obovate or oblong, sometimes lyrate-sinuate, thin, beneath white with more cottony or even arachnoid and partly deciduous tomentum: scapes a foot or two high: rays small and narrow, little exserted: akenes pubescent or glabrate, the beak as long as the body. — Bot. Biol. Centr.-Amer. ii. 255. *Tussilago nutans*, L. Amœn. Acad. v. 406 (Plum. ed. Burm. t. 41, f. 1). *Leria lyrata*, Cass. Dict. xxvi. 102. *L. nutans*, DC. Ann. Mus. Par. xix. 68, & Prodr. l. c. 42. *Gerbera nutans*, Schultz Bip. l. c. — Wooded grounds, Texas to New Mexico and Arizona. (Mex., W. Ind., S. Am.)

205. PERÉZIA, Lag. (*Lorenzo Perez*, of Toledo, pharmacist and writer on materia medica in the sixteenth century.) — Perennial herbs, all American (Texan, Californian, and southward, chiefly along the Andes), not lanate, except at the base of the stem, mostly with reticulated leaves, often setulose-ciliate or spinulose; heads solitary or cymose or paniculate; the corollas rose-purple to white, rarely blue, never yellow. — Amœn. Nat. i. 31; Gray, Pl. Fendl. 110, & Pl. Wright. i. 126; Benth. & Hook. Gen. ii. 500. *Perezia*, *Clarionea* (Lag. incd.), *Homoianthus*, *Dumerilia* (Less., not Lag., nor DC. Ann. Mus.), *Proustia* § *Thelecarpæa*, & *Acourtia* (Don), DC. Prodr., &c. *Drosia*, Cass. — § Euperezia (*Perezia*, Lag. l. c., *Clarionea* & *Homoianthus*, DC.), of S. American species, is distinguished by radiate heads, the corollas of marginal flowers having elongated and conspicuously liguliform outer lip, the two lobes of the inner much shorter and smaller.

§ Acoúrtia, Gray, Proc. Am. Acad. xix. 58, has flowers nearly or quite homomorphous, the marginal corollas with 3-toothed outer lip hardly ever longer than the two lobes of the inner: flowers commonly fragrant: involucre usually naked at base: leaves coriaceous or papyraceous, reticulated: usually a tuft of wool at base of the stem. — *Acourtia*, Don in Trans. Linn. Soc. xvi. 203; DC. Prodr. vii. 65. *Perezia*, Llav. & Lex.; Less.; DC. l. c. 62. *Dumerilia*, Less. & DC. l. c. 66, not Lag., nor Cass. Of few Chilian, numerous Mexican, and the following Texano-Californian species.

* A span or two high: heads (half-inch to inch long) single or few, 20–30-flowered: flowers purple.

P. runcináta, Lag. Acaulescent, scabrous-puberulent or glabrate: rootstocks apparently short, sending down tuberous-thickened fascicled roots: radical leaves runcinate-pinnatifid,

4 to 8 inches long, thin-papyraceous; lobes rounded, copiously fringed with spinulose teeth, margined-petioled: scapes naked, equalling the leaves, bearing solitary or a few pedunculate heads; bracts of the involucre rather few in three series, lanceolate, setaceous-acuminate: pappus rather sordid. — Lag. in herb. ex Don; Gray, Pl. Fendl. 110, Pl. Wright. l. c. *Clarionea runcinata*, Don in Trans. Linn. Soc. xvi. 207; DC. l. c. — Dry ground, E. & S. Texas, *Wright, Hall*, &c. (Adj. Mex.)

— **P. nána**, Gray. Leafy-stemmed, glabrous: rootstocks slender, creeping: first leaves small and scale-like; principal cauline leaves firm-chartaceous, orbiculate, dilated-obovate, or ovate (inch or two long), coarsely spinulose-dentate, sessile or partly clasping the slender stem: heads mostly sessile, solitary and terminal. bracts of involucre 3 or 4 series, thinnish, acutish; the short outer ones ovate, innermost lanceolate, mucronulate: pappus white. — Pl. Fendl. 111. — Dry plains and rocky bluffs, S. W. Texas to Arizona, *Wright, Palmer*, &c. (Mex., first coll. by *Gregg*.)

✱ ✱ Taller, 1 to 3 feet high, branching, especially above, leafy up to the corymbiform polycephalous inflorescence: leaves closely sessile by sagittate-cordate or sometimes truncate base, densely and spinulosely denticulate: heads 5-15-flowered, narrow, half-inch or less long, subsessile and fasciculate-crowded or short-pedicelled, quite naked at base: involucral bracts thinnish, not very many, in only three series: flowers rose-purple and sometimes white in the same species: pappus white, soft.

+— Involucre 8-15-flowered; its bracts not attenuate-acuminate.

— **P. Wríghtii**, Gray. Glabrous throughout, or obscurely puberulent, but smooth: leaves thin, oblong to nearly ovate (larger 4, smaller 1 or 2 inches long), often unequally or doubly dentate: heads 8-12-flowered: involucral bracts all pointless and obtuse, or the narrow innermost barely acutish: corollas pale rose to whitish. — Pl. Wright. i. 127, ii. 102; & Proc. Am. Acad. xix. 60. *P. Arizonica*, Gray, Bot. Calif. i. 422, a form of drier districts, rather more rigid, the involucral bracts all rounded-obtuse. *P. Coulteri*, Gray, Proc. Am. Acad. xv. 40, as to pl. Parry & Palmer, no. 234. — Rocky hills and ravines, S. W. Texas to S. Arizona; first coll. by *Coulter*, then by *Wright*. (Mex., *Schaffner, Parry & Palmer*.)

— **P. microcéphala**, Gray. Scabro-puberulent and minutely resinous-glandular: leaves more chartaceous, oblong, commonly obtuse, finely and closely denticulate· heads 10-15-flowered, larger than in preceding (over half-inch long when well developed): involucral bracts scaberulous on the back, abruptly acute or mucronate-acuminate: corollas rose-color. — Pl. Wright. i. 127, & Bot. Calif. i. 422. *Acourtia microcephala*, DC. Prodr. vii. 66. — California, on hills back of Monterey? (*Douglas*), Santa Barbara, and San Diego.

+— +— Involucre 5-6 flowered; bracts attenuate-acuminate: fully developed heads half-inch long.

P. Thúrberi, Gray. Scabro-puberulent, viscidulous-glandular: leaves firm-chartaceous, oblong-ovate, denticulate and partly doubly dentate (larger 5 to 8 inches long): involucral bracts lanceolate, gradually tapering to a very acute point, scaberulous externally: corollas sometimes deep rose-color, sometimes white. — Pl. Thurb. in Mem. Am. Acad. v. 324, & Proc. Am. Acad. xix. 59. — S. Arizona, on rocky hills, *Thurber, Lemmon*.

206. **TRÍXIS**, P. Browne. (Τριξύς, threefold, the corolla being trifid.) — American, chiefly subtropical, fruticose or perennial herbaceous plants; with entire or merely denticulate leaves, and paniculately or corymbosely cymose heads, of moderate size; the corollas yellow or sometimes whitish. — Hist. Jam. 312; Lag. Amœn. Nat. i. 35. *Perdicium*, L., in part.

- **T. angustifólia**, DC. Suffruticose, fastigiately or corymbosely much branched, a foot or two high, sericeous-puberulent, from subcanescent to glabrate, somewhat resinous-atomiferous, leafy up to the heads: leaves sessile, rather rigid, from broadly to very narrowly lanceolate, entire or denticulate with sparse mucrouiform teeth (2 or 3 inches long): heads simply fascicled or singly terminating leafy branchlets, half-inch and more long, 9-12-flowered, subtended by a few lanceolate or linear bracteiform leaves which do not exceed the 8 or 10 linear-lanceolate and equal proper bracts of the involucre; these in age gibbous and indurated at base: receptacle copiously villous: corollas golden yellow; outer lip of the marginal ones quarter-inch long: pappus barely fulvous. — Prodr. vii. 69; Gray, Pl. Wright. i. 128, ii. 102. *T. frutescens*, Gray, Bot. Mex. Bound. 103, vars. *T. Californica*, Kellogg in Proc. Calif. Acad. ii. 182, fig. 53, with some seeming monstrosities. *T. corymbosa*, Gray in Coll. Pringle, &c.; but that species should have petiolate leaves and loosely corymbose heads. — Hills and cañons, S. W. Texas to Arizona, *Wright*, &c. Founded on Mexican specimens with narrow leaves revolute when dry. (Mex.)

Var. **latiúscula**. Leaves lanceolate, plane, commonly glabrate and greener, from 4 to nearly 12 lines wide, thence varying into the narrow-leaved form. — Gray, Pl. Wright. ii. 102. *T. suffruticosa*, Wats. Bot. Calif. ii. 459. — Cañons, S. New Mexico to San Diego Co., California, *Wright*, *Palmer*, *Greene*, *Lemmon*, &c.

T. FRUTÉSCENS, P. Browne, which the broad-leaved forms of the preceding species nearly approach, was collected by *Berlandier* near Matamoras, but has not yet come from Texas.

TRIBE XI. CICHORIACEÆ, p. 83.

207. PHALACRÓSERIS, Gray. (Φαλακρός, bald-headed, and σέρις, the Greek name of some kind of Cichoriaceous plant). — Proc. Am. Acad. vii. 364; Bot. Calif. i. 423. — Single species.

P. **Bolánderi**, GRAY, l. c. Glabrous and acaulescent perennial, with thickish root: leaves lanceolate, entire, clustered on the caudex, slightly succulent: scape perfectly naked, a span to a foot high: solitary head half-inch high: flowers deep yellow, in summer. — California, in wet mountain meadows of the higher Sierra Nevada, Mariposa Co.; first coll. by *Torrey* and by *Bolander*.

208. ATRICHÓSERIS, Gray. ("Αθριξ, without hair, and σέρις, a Cichoriaceous plant.) — *Malacothrix* § *Anathrix*, Gray, Proc. Am. Acad. ix. 213, & Bot. Calif. i. 435. — Single species.

A. **platyphýlla**. Winter annual, wholly glabrous, somewhat glaucous: leaves all or chiefly in a rosulate radical tuft, broadly cuneate or obovate, mostly rounded at summit, sessile, spinuloso-denticulate, somewhat veiny (inch or two long); those of stem reduced to very small scattered bracts: stem slender, a foot or two high, at summit deliquescent into a diffuse cymose panicle of few or numerous slender-pedunculate heads: involucre quarter-inch high, about half the length of the corollas (these white or with purple base): akenes 2 lines long, at maturity nearly equalling the narrow and open bracts of the involucre, white, sometimes with 4 or 5 very thick corky ribs and much smaller alternate ones, sometimes more terete and obscurely costate, the truncate summit wholly destitute of the border of *Malacothrix*, its areola small: receptacle rather fleshy, scrobiculate. — *Malacothrix? platyphylla*, Gray, l. c. — Gravelly deserts of the Mohave, S. W. California, to the southern borders of Utah, *Cooper*, *Palmer*, *Parry*, *Parish*.

209. LÁMPSANA, Tourn. (Ancient Greek name, of obscure derivation; but the λαμψάνα of Dioscorides and the *Lapsana* of Pliny, whose orthography was followed by Linnæus, were Cruciferous plants.) — Yellow-flowered and leafy-stemmed branching annuals of the Old World, one sparingly naturalized: fl. summer.

L. COMMÚNIS, L. (NIPPLEWORT.) A foot or two high, hirsutely pubescent or glabrate: leaves ovate, repand-dentate, or lower lyrate and uppermost oblong: heads loosely paniculate: involucre 2 or 3 lines high. — Roadsides, in a few places, Penn. to New England, more abundant in Canada, also on the Columbia River. (Nat. from Eu.)

210. APÓGON, Ell. ('Απώγων, beardless, i. e. no pappus.) — Low annuals of the Southern Atlantic States, glaucescent, mostly glabrous, a span to a foot high, branching from the base, bearing scattered rather small heads on slender peduncles: flowers yellow, in spring and early summer. Leaves variable, lanceolate or lower oblong, from entire or repand to dentate, or radical lyrate-pinnatifid, uppermost closely sessile, often seemingly opposite. — Sk. ii. 267; DC. Prodr. vii. 78; Torr. & Gray. Fl. ii 466.

— **A. húmilis**, ELL. l. c. Peduncles naked, or rarely with some obscure glandular-bristly hairs under the head: this in fruit only 2 lines high: corollas pure yellow, little longer than involucre: akenes oblong-obovate. — DC. l. c.; Torr. & Gray, l. c., in part. *A. lyratum*, Nutt. Jour. Acad. Philad. vii. 71, & Trans. Am. Phil. Soc. n. ser. vii. 424. *Serinia cæspitosa*, Raf. Fl. Ludov. 149, cited in DC. l. c. 261, should be either this or the next. — Open ground, S. Carolina to Texas and Arkansas.

A. grácilis, DC. l. c. Sometimes slender and strict, not rarely more robust than the preceding, often some bristly hairs on the stem and lower leaves: peduncles usually glandular-hispid some way below the head; this commonly 3 lines high in fruit: corollas orange, conspicuously exserted, twice the length of the involucre: akenes rather thicker and obtuser at apex, sometimes an obscure vestige of pappus! — *A. humilis*, Nutt. Trans. Am. Phil. Soc. l. c., not Ell. — Rocky prairies, &c., Texas; first coll. in a very slender form by *Berlandier*. Grows with the preceding, keeping distinct.

A. Wrightii. Resembling slender and narrow-leaved form of the preceding (such as Berlandier's original specimens): rather diffuse: heads equally small: akenes larger and thicker (over half-line long), little contracted at either end, and with comparatively large areola (yet less than the full breadth of the akene), this bordered by obscure vestige of pappus. Possibly a hybrid between *A. gracilis* and *Krigia occidentalis*. — E. Texas, *Wright*, in fruit.

211. **KRÍGIA**, Schreb. (*David Krig*, or *Krieg*, an early collector in Maryland and Delaware.) — Low herbs of Atlantic U. S., glabrous or somewhat hispidulous; with small or middle-sized heads of yellow flowers, terminating slender naked peduncles or scapes; these not rarely glandular-hispidulous at summit: fl. in spring or summer. — Gen. Pl. 532. Benth. & Hook. Gen. ii. 507. *Krigia* & *Cynthia*, Don; Torr. & Gray, Fl. ii. 467, 468.

§ 1. CÝMBIA, Torr. & Gray. l. c. Acaulescent annuals: bracts of the involucre 5 to 8, oblong-lanceolate, in fruit becoming broader and firmer, erect and navicular-carinate, with a conspicuous midnerve, or sometimes 2–3-nerved: akenes turbinate, mostly 5-paleaceous and 5-aristate.

K. occidentális, NUTT. Scapes a span or more high, commonly glandular-hispidulous, at least toward the summit: leaves obovate to lanceolate, entire, lyrately lobed or pinnatifid: heads 2 or 3 lines high: akenes transversely rugulose: paleæ of the pappus conspicuous, rounded-obovate; bristles or rather awns alternating with these and over the stronger angles of the akene sometimes equalling it in length, sometimes not surpassing the paleæ, sometimes (var. *mutica*, Torr. & Gray) obsolete or wanting. — Jour. Acad. Philad. vii. 104, & Trans. Am. Phil. Soc. vii. 427; Torr. & Gray, Fl. ii. 468. *K. nervosa*, Hook. Ic. Pl. iii. t. 227. & *K. bellioides*, Scheele in Linn. xxv. 257, normal form, with pappus-awns double the length of the paleæ. — Prairies of Arkansas and Texas; first coll. by *Nuttall*.

§ 2. EUKRÍGIA, Torr. & Gray, l. c. Acaulescent and subcaulescent winter annual; bracts of the involucre 9 to 18, thin, remaining narrow and nearly nerveless, reflexed after the fall of the narrowly turbinate somewhat 5-angular akenes: pappus of 5 to 7 (commonly 5) roundish short paleæ, and of as many alternating nearly capillary long bristles. — *Krigia*, Schreb., &c.

— **K. Virgínica**, WILLD. Varying much in size; often sparsely hispidulous: scapes 2 or 3 inches or at length a foot or more high, slender, not rarely caulescent below: leaves from spatulate-obovate to lanceolate or linear, from few-toothed or entire to pinnately parted: heads 3 or 4 lines high: pappus-bristles fully twice the length of the akene. — Spec. iii. 1618. *K. Virginica, dichotoma*, & *Caroliniana*, Nutt. Gen. ii. 127. *K. leptophylla*, DC. Prodr. vii. 88, slender form. *Hyoseris Virginica*, L. Spec. ii. 809; Lam. Jour. Hist. Nat. i. 22, t. 12; Walt. Car. 193; Michx. Fl. ii. 88. *Hyoseris Caroliniana*, Walt. l. c.! Sandy ground, Canada to Florida and Texas: fl. from spring to autumn.

§ 3. CYNTHIA. Caulescent or acaulescent perennials, glaucescent, comparatively large-flowered: involucre of the preceding section: akenes less turbinate, of 10 to 15 smaller and more squamellate oblong paleæ and 15 or 20 slender capillary bristles. — *Cynthia*, Don in Edinb. Phil. Jour. xii. 305; DC. Prodr. vii. 89; Torr. & Gray, l. c. *Adopogon*, Neck. Elem. i. 55.

K. Dandélion, NUTT. Scapigerous, or at length leafy-stemmed only next the ground: crown bearing oval or globose tubers on filiform stolons: leaves lanceolate or almost linear, from denticulate to laciniate-lobed or pinnatifid: scapes 6 to 18 inches high, naked: head about half-inch high. — Gen. ii. 127; Ell. Sk. ii. 267. *Tragopogon Dandelion*, L. Spec. ed. 2, ii. 1111. *Hyoseris major*, Walt. Car. 194. *H. angustifolia*, Michx. Fl. ii. 87. *Troximon Dandelion*, Pers. Syn. ii. 360. *Cynthia Dandelion* & *C. Boscii*, DC. Prodr. vii. 89. *C. lyrata*, Nutt. Jour. Acad. Philad. vii. 69. *Krigia Caroliniana*, Hook. Comp. Bot. Mag. i. 100, a slender form. — Moist ground, Maryland to Florida, Arkansas, and Texas.

K. montána, NUTT. l. c. Caulescent or subcaulescent from short cespitose rootstocks, not tuberiferous: peduncles simple and naked, a span to a foot long: leaves from oblong to linear, from entire to pinnatifid, thickish: head smaller than of the preceding. — *Hyoseris montana*, Michx. Fl. ii. 87. *Cynthia Dandelion*, var. γ, Torr. & Gray, Fl. ii. 469. *C. Dandelion*, Meehan, Nat. Flowers, ser. 2, ii. t. 35. — Crevices of rocks, Alleghany Mountains (Blue Ridge), N. and S. Carolina and Georgia; first coll. by *Michaux*.

K. amplexicaúlis, NUTT. l. c. Caulescent, not tuberiferous, glaucous: stem a foot or two high, 1–3-leaved, bearing one or two or few somewhat umbellate heads on moderately long peduncles: leaves oblong or oval, obtuse, entire, repand and denticulate, or radical somewhat lyrately lobed; these contracted into winged petioles; cauline partly clasping by a broad base: heads a third of an inch high. — *Tragopogon Virginicum*, L. Spec. ii. 789. *Hyoseris amplexicaulis*, Michx. Fl. ii. 87. *H. biflora*, Walt. Car. 194? *H. prenanthoides*, Willd. Spec. iii. 1618. *Cynthia Virginica*, Don, l. c.; Torr. & Gray, l c. *C. amplexicaulis*, Beck, Bot. 168; Darl. Fl. Cestr. 441. *C. Griffithii*, Nutt. Jour. Acad. Philad. vii. 69, with lower leaves runcinate-lyrate. *Luthera Virginica*, Schultz Bip. in Linn. x. 257. — Moist banks, New York to Minnesota and Colorado, south to Georgia.

212. **CICHÓRIUM**, Tourn. SUCCORY, CHICCORY, ENDIVE. (Arabic name Latinized.) — Old World herbs, fl. summer.

C. INTYBUS, L. (CHICCORY) Deep-rooted perennial, more or less hirsute, at least below, with rigid stout branches: radical leaves runcinate; cauline oblong or lanceolate, commonly dentate; those of flowering branches mostly reduced and scale-like, subtending solitary or clustered sessile heads, or some heads raised on a fistulous peduncle: flowers showy, matutinal, closing by midday, sky-blue, varying occasionally to purple or white. — Roadsides, common in E. New England, and in a few places westward. (Nat. from Eu.)

213. **STEPHANOMÉRIA**, Nutt. (Στεφάνη, a coronal or wreath, μέρος, a division; no particular application.) — W. N. American perennials or annuals, mostly smooth and glabrous; with branching or rarely virgate and often rigid or rush-like stems, small or merely scale-like leaves on the flowering branches, and usually paniculate small or middle-sized heads of rose-colored or flesh-colored flowers, open only in early morning. — Trans. Am. Phil. Soc. vii. 427; Torr. & Gray, Fl. ii. 722; Benth. & Hook. Gen. ii. 533 (excl. *Rafinesquia*): Gray, Bot. Calif. i. 427. ' *Jamesia*, Nees in Pl. Neuwied Trav. 516, not Torr. & Gray.

§ 1. ALLÓSERIS, Gray. Heads large for the genus, about 12-flowered: involucre somewhat imbricated, the outer bracts being of 2 or 3 lengths: receptacle alveolate, and the short alveoli fimbriolate-hirsute: pappus-bristles 12–20, short-plumose for their whole length, sordid or almost fuscous. — Proc. Am. Acad. vi. 552, Bot. Calif. l. c., & Proc. Am. Acad. xix. 60.

— **S. cichoriácea,** Gray l c. Perennial, 1 to 4 feet high, comparatively stout, when young sometimes tomentulose leaves resembling those of Chiccory, lanceolate, sparsely denticulate to runcinate-laciniate involucre half-inch high: heads sessile along naked branches: mature akenes short-linear, smooth, lightly and acutely 5-angled. — Rocky hills and cañons through the southern portions of California, *Dr. Horn, Parish, Pringle.*

§ 2. STEPHANOMERIA proper. Heads 3-20-flowered: receptacle quite naked: involucre slightly imbricated by having one or two intermediate bracts, especially in the earlier species, or only calyculate at base: pappus setose and plumose throughout or only above the middle, the lower part of the bristle either slender to base or sometimes paleaceous-dilated. — Gray, l. c. 61.

* Heads fully half-inch high, 10-20-flowered, somewhat corymbosely disposed,
+ Terminating leafy stems and branches: pappus sordid or gravish, of 10 or 12 rather long-plumose bristles: akenes smooth and even, with slender ribs or angles: plants a span to a foot high from perennial roots: involucre obscurely imbricated, 10-12-flowered.

S. Párryi, Gray. Rather stout, widely branched from the base: leaves thickish, deeply runcinately pinnatifid; those of the flowering branchlets rather numerous up to the head, small, somewhat spinulose-lobed: pappus-bristles rather stout, naked (and often united in twos or threes) at base. — Proc. Am. Acad. xix. 61. — Arid districts, near St George, S. Utah, *Parry.* Borders of the Mohave Desert, S. E. California, *Palmer, Pringle.*

— **S. lactucina,** Gray. Rather slender, with erect branches, leafy up to the nearly naked peduncles. leaves linear or narrowly lanceolate, entire or with a few salient teeth: pappus-bristles slender and plumose to the base. — Proc. Am. Acad. vi 552; Bot. Calif. l. c. — Woods of the Sierra Nevada, California, from Mariposa Co. to Shasta, *Newberry, Brewer, Bolander,* &c.

+ + Heads naked-paniculate: pappus bright white: involucre merely calyculate.

— **S. Thúrberi,** Gray. Simple-stemmed from a probably biennial root, a foot or two high: leaves mainly at and near the base, runcinate-pinnatifid, inch or two long; those of the naked stem and few corymbosely-paniculate branches reduced to linear-subulate or inconspicuous bracts heads rather few: involucre narrow, 16-20-flowered. bristles of the pappus 20 to 30, soft and slender, very plumose to base. — Pl. Thurb. in Mem. Am. Acad. v. 325, & Bot. Mex. Bound. 105. — New Mexico and adjacent Arizona, *Thurber, Bigelow, Henry, Greene,* &c.

S ELATA, Nutt. Pl. Gamb. 173, — said to be probably perennial and blue-flowered, simple-stemmed, 3 or 4 feet high, with very narrow linear leaves, about 10-flowered heads, involucre (6-8-phyllous) and branches sprinkled with resinous dots, and plumose white pappus, coll. at Santa Barbara, California, — remains quite obscure.

* * Heads quarter to third inch high, or sometimes higher, narrow, mostly 5-flowered (flowers from 3 to 6, occasionally 8 or 9), and with about the same number of involucral bracts: mature akenes either smooth and even between the ribs, or rugose, or tubercular-thickened, sometimes in the same species. — *Jamesia,* Nees, l. c.

+ Perennials, paniculately or fastigiately branched from thick and tortuous roots or a lignescent base, with striate and rush-like branches, small-leaved or nearly leafless above: pappus-bristles not at all squamellate-appendaged or dilated at base.

— **S. runcináta,** Nutt. Comparatively stout and rigid, a foot or two high, with spreading branches: heads mostly 4 or 5 lines high and scattered along the branches: lower leaves runcinate-pinnatifid, commonly lanceolate; upper linear or reduced to scales: pappus dull white, plumose only to near the base. — Torr. & Gray, Fl. ii. 472; Gray, Pl. Fendl. 112. *S. runcinata* & *S. heterophylla,* Nutt. Trans. Am. Phil. Soc. l. c., at least in part and by char., but poor specimens, seemingly confused with next. *Prenanthes runcinata,* James in Long Exped. *P.? pauciflora,* Torr. Ann. Lyc. N. Y. ii. 210. — Plains, Nebraska to Wyoming, N. W. Texas, Arizona, and S. California; first coll. by *James.*

— **S. minor,** Nutt. l. c. More slender and with ascending branches bearing usually terminal and smaller heads: cauline leaves all slender, often filiform: pappus white, very plumose down to base. — Torr. & Gray, l. c. *Prenanthes? tenuifolia,* Torr. l. c. *Lygodesmia minor,* Hook. Fl. i. 205, t. 103 A. *Jamesia pauciflora,* Nees in Neuwied Trav. 516 (16). — Plains and mountains, from borders of Brit. America to those of Texas, Arizona, the Sierra Nevada in California, and Washington Terr. Generally of more northern range than the foregoing, not throughout distinguishable, perhaps has been rightly combined with it.

S. myrióclada, EATON. Very slender stems and tortuous filiform branches very numerous and fastigiately crowded in an erect tuft, a foot or two high, terminated by scattered small heads: leaves linear and very small: involucre 2 and 3 lines long (of 4 or 5 as well as "3" narrow bracts) and 3-5-flowered: akenes pluristriate at maturity: pappus white, its bristles naked or merely hirsute below the middle or at the base. — Bot. King Exp. 198, t. 20. — Dry rocky ridges, Thousand Spring and Goose Creek Valleys, Nevada, *Watson.* Hawthorne, Nevada, *M. E. Jones.*

+ + Biennial, or probably perennial with long and slender subterranean shoots: pappus bright white; the bristles long-plumose to base, which is not at all paleaceous-dilated.

S. Wrightii, GRAY. A foot or two high, slender, with single corymbosely paniculate stems: cauline leaves mostly filiform and entire; those of the radical tuft linear to spatulate and laciniate-pinnatifid: heads nearly half-inch long, 5-flowered, sparse, pedunculate, terminating slender branches: akenes smooth on the salient ribs and narrow intervals, contracted at summit: pappus long-plumose. — Proc. Am. Acad. xix. 60. *S. runcinata,* var., Gray, Pl. Wright. ii. 103, no. 1301. — W. Texas, in pebbly bed of Howard's Creek, *Wright* (without the elongated root or shoot), and adjacent New Mexico, *Bigelow.* Apparently same from N. Arizona, *Rusby,* seemingly perennial from long and filiform subterranean shoots.

+ + + Annual, strictly erect: pappus white; the bristles plumose to base, not paleaceous-dilated.

S. virgáta, BENTH. Stem rigid, 1 to 4 feet high: heads 3 or 4 lines long, mostly subsessile or short-peduncled, spicately or thyrsoidly disposed along the naked upper part of virgate stem or similar branches, but sometimes more loosely paniculate on open branchlets: upper leaves linear, small and entire; lower oblong or spatulate, often sinuate or pinnatifid; involucre 4-8-flowered, originally described as "8-10-flowered"· akenes subclavate or oblong, rugose-tuberculate between the narrow ribs: pappus moderately plumose. — Bot. Sulph. 32; Gray, Proc. Am. Acad. l. c. *S. paniculata,* chiefly, Eaton, Bot. King Exp. 198, t. 20, f. 5; Gray, Bot. Calif. i. 428. Possibly (from habitat not improbably) *S. elata,* Nutt. Pl. Gamb. 173; but flowers not blue, and no resinous dots on involucre and branchlets. — California, common from San Bernardino and San Diego Co., to Oregon, east to Nevada and Utah.

+ + + + Annual, strictly erect: pappus grayish or fuscous; its bristles short-plumose nearly or quite to the more or less paleaceous or squamelliferous base.

S. paniculáta, NUTT. Stem erect from an annual root, a foot or two high, bearing numerous narrow 3-5-flowered heads in an elongated narrow or more open panicle, or else more strictly disposed on virgate branches: leaves linear or the lower lanceolate: akenes nearly of the preceding: pappus decidedly different. — Trans. Am. Phil. Soc. vii. 428; Torr. & Gray, Fl. ii. 473. — Plains of Idaho, and probably Northern Nevada, to E. Oregon, *Nuttall, Hall, Cusick,* &c.

+ + + + + Annuals or biennials: bristles of the white or whitish pappus plumose above but naked below the middle, at base more or less dilated or abruptly paleaceous, or else with one or two adnate squamellæ or bristly teeth at or near insertion: akenes thick-ribbed and tuberculate-rugose at maturity: stems paniculately and often divergently branched, bearing scattered squamulose-peduncled heads. — § *Hemiptilium,* Gray, Bot. Calif., in part only.

S. exígua, NUTT. A foot or two high, with slender branches and branchlets, but stem not rarely robust (therefore ill named from depauperate specimens): radical and lower cauline leaves pinnatifid or bipinnatifid, those of the branches mainly reduced to short scales: involucre 3 to 5 lines long, with commonly 5 flowers, "3 or 4" when depauperate, rarely 6 or 8 in strong plants: bristles of the pappus 9 to 18, their more or less dilated and paleaceous or thickened bases commonly a little connate in 4 or 5 phalanges and often 1-2-setulose on each side. — Trans. Am. Phil. Soc. l. c. 428; Torr. & Gray, Fl. ii. 473 (attenuated form); Eaton, Bot. King Exp. 198, t. 20, f. 6, 7; Gray, Bot. Calif. i. 428. *Hemiptilium Bigelovii,* Gray, Bot. Mex. Bound. 105, a stout form. — Interior of Wyoming to the Upper Rio Grande on the border of Texas, west to Nevada and E. California.

S. pentachǽta, EATON. A span or two or even 2 or 3 feet high, like the preceding, or divaricately branched from the base: pappus of 5 or sometimes 7 bristles, all distinct to the base, which is little dilated, plumose only above the middle. — Bot. King Exp. 199, t. 20,

f. 8-10; Gray, Proc. Am. Acad. xix. 63. — Desert region, W. Nevada, *Watson*, *Shockley*. Edge of desert at San Felipe, San Diego Co., California, *Parish*.

§ 3. HEMIPTÍLIUM, Gray, l. c., xix. 63. Heads 5-flowered, small: receptacle naked: involucre merely calyculate: pappus of 4 to 6 narrow and rigid paleæ (rather than awns), not longer than the akene, sparsely short-plumose toward the summit, fuscous. — *Hemiptilium*, Gray. Bot. Mex. Bound. 105, excl. spec.

S. Schóttii, GRAY. Probably annual, with habit of *S. paniculata* or *S. exigua*, slender: loosely paniculate, 3 lines long: involucre of 4 or 5 thinnish bracts and 2 or 3 small calyculate ones: ligules barely 3 lines long: akenes less than 2 lines long, rather narrow, 4-5-angled, tapering very slightly from truncate summit to base, minutely scabrous between the smooth angles. — Bot. Calif. i. 427. — *Hemiptilium Schottii*, Gray, Bot. Mex. Bound. l. c. — Arizona, on the Gila River, *Schott*. Not since collected.

214. **CHÆTADÉLPHA,** Gray. (Χαίτη, bristles, and ἀδελφή, sister, the bristles or awns of pappus as it were 5-adelphous.) — Proc. Am. Acad. ix. 218; Rothrock in Wheeler Rep. 182, t. 15. — Single species.

C. Wheéleri, GRAY, l. c. Much branched from a perennial root, flexuous and fastigiate, with aspect of *Stephanomeria*, or more of *Lygodesmia*, a foot or two high: leaves narrowly linear, entire, uppermost reduced to subulate scales: heads solitary terminating the branchlets: involucre half-inch and more high, somewhat exceeded by the pappus. — W. Nevada, on the borders of Arizona, *Wheeler*. Near Pyramid Lake, *Lemmon*.

215. **RAFINÉSQUIA,** Nutt. (*Constantine S. Rafinesque Schmalz*, a noted botanist.) — Glabrous and branching slightly succulent and Sonchus-like winter annuals (Californian and New Mexican), leafy; with pinnatifid leaves, reduced on the flowering branches to herbaceous bracts: the heads rather large, with showy white or rose-tinged flowers, mostly matutinal. — Nutt. Trans. Am. Phil. Soc. vii. 429; Gray, Pl. Wright. ii. 103, & Bot. Calif. i. 429.

R. Califórnica, NUTT. l. c. Mostly robust, 2 or 3 feet high, paniculately branching, bearing numerous heads: leaves oblong (larger 4 to 6 inches long); cauline partly clasping: involucre thickened at base (half to three-fourths inch high), of 12 to 15 principal bracts and some spreading calyculate ones: ligules comparatively short, white: beak of akenes very slender, as long as the body: pappus dull white. — Torr. Bot. Mex. Bound. t. 34, figure not good. — Moist or shaded ground, common in California toward the coast: a smaller-flowered form in N. W. Arizona, *Palmer*.

R. Neo-Mexicána, GRAY. A foot or less high, more slender, bearing few but larger and more showy heads and much smaller leaves, the lower of these often runcinate: involucre narrow, more cylindraceous, sometimes inch long, little thickened at base, of fewer bracts: ligules large and conspicuous (half-inch and more long), white or tinged with flesh-color: beak of akene more gradually tapering, therefore stouter, rather shorter than the body: pappus bright white, of firmer bristles, the plume somewhat arachnoid. — Pl. Wright. l. c. — Sand-hills, &c., in the desert region, S. E. California to S. Utah and New Mexico on the Rio Grande; first coll. by *Wright*.

216. **TRAGOPÓGON,** GOAT'S-BEARD, SALSIFY. (Τράγος, goat, πώγων, beard.) — Old World biennials or rarely perennials, glabrous; with long taproot; entire and grass-like nervose leaves clasping at base; long and stout peduncles commonly thickened and fistulous under the large head; the flowers yellow or purple, closing at noon or earlier. — Two species sparingly naturalized, one of them cultivated.

T. PORRIFÓLIUS, L. (SALSIFY, OYSTER-PLANT.) Commonly 2 or 3 feet high: peduncle strongly clavate-thickened and fistulous for 2 or 3 inches beneath the head, which becomes

3 inches high: flowers violet-purple, mostly surpassed by the involucre: outermost akenes squamellate-muricate. — Sparingly in fields and near dwellings, as an escape from cultivation in the Atlantic States, a naturalized weed in California and Oregon. (Nat. from Eu.)

T. PRATÉNSIS, L. (GOAT'S-BEARD.) A foot or two, or the larger form a yard high: leaves with broader base: peduncles little enlarged except close under the head: flowers yellow, equalling the involucre, sometimes longer. — Sparingly found in fields, &c., New England to New Jersey and Wisconsin. (Nat. from Eu.)

217. ANISÓCOMA, Torr. & Gray. ("Άνισος, unequal, κόμη, tuft of hair; from the pappus.) — Jour. Bost. Soc. Nat. Hist. v. 111, t. 13; Eaton, Bot. King Exp. 197; Gray, Bot. Calif. i. 430. — Single species.

A. acaúle, TORR. & GRAY, l. c. Low winter annual, glabrous, except a dense white tomentum on the edges of the pinnately lobed and often runcinate leaves: these all in a rosulate radical cluster (inch or two long): scapes numerous, naked, a span high: head about inch high: ligules conspicuous, light yellow. — *Pterostephanus runcinatus*, Kellogg, Proc. Calif. Acad. iii. 20, f. 4, badly characterized. — Dry plains and hills, of the eastern part of the Sierra Nevada, from Sierra Co. to the Mohave, California, and adjacent Nevada; first coll. by *Fremont*.

218. HYPOCHŒRIS, L. (A name of Theophrastus for some plant of this tribe.) — Old World and S. American herbs; with yellow flowers; one species sparingly introduced.

H. GLABRA, L. Nearly glabrous; a rosulate tuft of oblong-spatulate sinuate-dentate leaves from an annual root, sending up branching scapes a span to a foot high, bearing a few middle-sized heads: involucral bracts lanceolate: outermost akenes truncate, inner slender-beaked: bristles of the somewhat sordid pappus arachnoid-plumose, but naked at tip, also some fine and shorter naked ones in an outer series. — Fields, E. California. (Nat. from Eu.)

H. RADICATA, L., which is hirsute and has all the akenes rostrate, is an occasional ballast-weed, at Philadelphia and New York.

219. MICRÓSERIS, Don. (Μικρός, little, σέρις, Endive or Lettuce; not an apposite name for our larger species.) — W. and S. American (but almost all Californian) annuals, biennials, or some perennials, glabrous or merely furfuraceous-puberulent, acaulescent or subcaulescent; with heads of yellow flowers terminating naked scapes or elongated simple peduncles, commonly nodding before expansion. Foliage very variable. — Don in Phil. Mag. xi. 388 (1832); Gray, Proc. Am. Acad. ix. 207, & Bot. Calif. i. 423. *Bellardia*, Colla in Mem. Acad. Taurin. xxxviii. 40, t. 34. *Lepidonema*, Fisch. & Meyer, Ind. Sem. Petrop. 1835. *Fichtea*, Schultz Bip. in Linn. x. 255. *Calais*, DC. Prodr. vii. 85; Gray, Pacif. R. Rep. iv. 121. *Phyllopappus*, F. Muell. in Linn. xiv. 507. *Uropappus* & *Scorzonella*, Nutt. Trans. Am. Phil. Soc. vii. 426. *Microseris* & *Scorzonella*, Benth. & Hook. Gen. ii. 506, 533.

§ 1. PTILÓPHORA, Gray, l. c. Pappus of 15 to 20 white and soft plumose bristles with paleaceous base: akenes linear-columnar, of same diameter from base to summit: stems more or less branching and leaf-bearing: perennials, with fusiform biennial roots.

M. nútans, GRAY. Slender, a foot or so high: fusiform roots either fascicled or solitary: leaves from entire and spatulate-obovate to pinnately parted into narrow linear lobes: heads 8-20-flowered, slender-peduncled: involucre cylindraceous, of 8 to 10 linear-lanceolate gradually acuminate principal bracts and a few short loose calyculate ones: bristles of pappus several times longer than the oblong scale at the base. — Proc. Am. Acad. ix. 208. *Scorzo-*

nella nutans (Geyer), Hook. Lond. Jour. Bot. vi. 253. *Ptilophora nutans*, Gray, Pl. Fendl. 113. *Calais* (*Ptilophora*) *nutans*, Gray, Pacif. R. Rep. iv. 112. *Stephanomeria intermedia*, Kellogg, Proc. Calif. Acad. v. 39. — Wet grassy grounds, borders of Brit. Columbia and Montana to S. W. Colorado, Utah, and the higher Sierra Nevada, California: fl. spring.

M. májor, Gray, l. c. Stouter, often more than 2 feet high, apparently thicker-rooted: heads larger, sometimes inch high: involucre of more lanceolate bracts imbricated in three lengths, pappus rather less plumose: leaves oblong-lanceolate, entire or sparingly laciniate. — *Ptilophora major*, Gray, Pl. Fendl. 113. *Calais major*, Gray, Pacif. R. Rep. l. c. 114. — Idaho, Utah, &c.; first coll. by *Spalding*.

Var. **laciniáta**, Gray, l. c. Lower: leaves pinnately parted and laciniate. — *Calais graciloba*, Kellogg, Proc. Calif. Acad. l. c. — W. Idaho, *Spalding*. Mendocino Co., California, *Kellogg*.

§ 2. Scorzonélla, Gray, l. c. Pappus (somewhat sordid) of 5 to 10 attenuate bristles with paleaceous base, or of short paleæ bearing long bristles, these either subplumose or naked: akenes truncate at summit, slightly attenuate downward only: involucre loosely imbricated in 2 or 3 series, many-flowered: perennials or biennials, often branching from base and somewhat leafy-stemmed: root fusiform. — *Scorzonella*, Nutt., Benth. & Hook. l. c.

* Pappus-bristles barbellate or short-plumose, not more than 4 times the length of the entire lanceolate palea: akenes slender.

M. sylvática, Gray, l. c. A foot or so high, not rarely simple and monocephalous: leaves from broadly lanceolate to linear, laciniate-pinnatifid: head almost inch high: involucral bracts mostly abruptly acuminate from an oval or oblong base: awn-like bristles of the pappus (commonly 10) almost or quite plumose. — *Scorzonella sylvatica*, Benth. Pl. Hartw. 320. *Calais* (*Anacalais*) *sylvatica*, Gray, Pacif. R. Rep. iv. 113. — California, in woods of the Sacramento and its tributaries, *Hartweg, Bigelow, Kellogg & Harford*.

Var. **Stillmáni**, Gray, l. c. Sometimes several-stemmed from the base and with leaves pinnately parted into long linear lobes: involucral bracts lanceolate and more gradually slender-acuminate: pappus-bristles (5 to 10) merely barbellate under a lens: akenes smooth, obscurely costate. — California, on the Sacramento, &c., *Stillman, Bigelow*. Near San Francisco, *Samuels, G. R. Vasey*.

* * Pappus-bristles (10, sometimes 8) naked, barely denticulate or scabrous, entire or nearly so, several or many times longer than the small palea: akenes columnar or rather slender: heads large or middle-sized, with elongated ligules: involucre not reflexed at the fall of the akenes, its bracts with loose and conspicuously acuminate or attenuate tips: stem sometimes scapiform, often few-leaved and branching. — *Scorzonella*, Nutt. (Foliage variable, as in the genus generally: species too nearly connected.)

+ Heads and leaves of the largest.

M. prócera, Gray. Glaucous: stem robust, 2 or 3 feet high, branching: leaves chiefly oblong and apiculate-acuminate, denticulate or entire, rarely laciniate-lobed; larger cauline 6 to 8 inches long and one or two wide; radical oblong-lanceolate, commonly a foot long: involucre a full inch or more high and broad; outer bracts broadly ovate with abrupt acumination; innermost lanceolate-acuminate: akenes nearly 3 lines long: paleæ of the pappus lanceolate or oblong-lanceolate, acute, about one fourth the length of the awn. — Proc. Am. Acad. xix. 64. *M. laciniata*, var. *procera*, Gray, Proc. l. c. ix. 209, & Bot. Calif. i. 424. *Calais glauca*, var. *procera*, Gray, Proc. Am. Acad. vii. 364. — Hillsides, W. California, from Sonoma Co. northward to Klamath, Oregon, *Torrey, Bolander, Kellogg, Kronkhite*.

+ + Heads and leaves less ample and stem less robust, a foot or less to two feet high.

M. laciniáta, Gray. Rather stout, glancescent: leaves laciniate-pinnatifid into attenuate lobes, or sometimes entire; radical 4 to 10 inches long: involucre an inch or less high; its bracts as in the preceding, or nearly all with broad dilated base: akenes columnar, 2 lines long: paleæ of the pappus deltoid or triangular-ovate, not longer than breadth of the akene, abruptly tipped by an awn or bristle 8 or 9 times longer. — Proc. Am. Acad. ix. 209, excl. var. *procera*. *Hymenonema? laciniatum*, & perhaps also *H.? glaucum*, Hook. Fl. i. 301. *Scorzonella laciniata*, Nutt. Trans. Am. Phil. Soc. vii. 426. *Calais* (*Scorzonella*) *laciniata*,

Gray, Pacif. R. Rep. iv. 112 — Alluvial ground, Washington Terr. and Oregon to N. W. California; first coll. by *Douglas*. Some forms are to be distinguished from the smaller of the preceding by the pappus only.

—— **M. leptosépala**, Gray. Slender, a foot or less high: leaves from linear to narrowly lanceolate, and from entire to attenuate-pinnatifid: heads comparatively small and fewer-flowered: involucre only half-inch high, of fewer bracts reduced almost to two series; outer ovate or oblong with abrupt acumination, inner lanceolate-attenuate: akenes more slender: paleæ of the pappus (often only 8) ovate-lanceolate or narrower, a quarter or fifth of the length of the akene, tapering from base gradually into the awn. — Proc. Am. Acad. ix. 209, Bot. Calif. l. c. in part only, & Proc. Am. Acad. xix. 64. *Scorzonella leptosepala* (& *S. glauca*), Nutt. l. c. *Hymenonema? glaucum*, Hook. l. c.? *Calais laciniata*, Gray, Proc. Am. Acad. viii. 208, pl. 11all. — Low grounds and meadows, Washington Terr. & Oregon, along the Columbia River, *Garry? Nuttall, Hall, Suksdorf.*

M. Bolánderi, Gray. Slender, a foot or two high: leaves from narrowly linear-lanceolate to somewhat spatulate, entire or with a few small salient linear lobes; radical a span to a foot long including the margined petiole: involucre half to two-thirds inch high; its bracts all gradually lanceolate-attenuate from a broadish base, or some small outermost ovate and abruptly acuminate: paleæ of the pappus (8 to 10) little exceeding the breadth of the akene, broadly ovate, mostly obtuse, abruptly tipped by the long slender awn. — Proc. Am. Acad. xix. 64. *Calais Bolanderi*, Gray, Proc. Am. Acad. vii. 365. *Microseris leptosepala*, Gray, Proc. Am. Acad. ix. 209, in part, & Bot. Calif. i. 425, as to Californian plant. — Swamps, Mendocino and Humboldt Co., N. W. California, *Bolander, Kellogg, Rattan, Pringle*. Apparently same from Seattle, Washington Terr., *Mrs. Summers*.

* * * Pappus-bristles or awns 5, not over thrice the length of the palea, rising from an apical cleft, rather strongly denticulate.

M. Párryi, Gray. Furfuraceous-puberulent: leaves all radical, lanceolate, a span long, from entire to laciniate-pinnatifid: scape 7 inches high: involucre campanulate, half-inch high, of ovate and oblong and hardly acuminate bracts: ligules little exserted. — Proc. Am. Acad. ix. 209, & Bot. Calif. l. c. *Calais Parryi*, Gray, Pacif. R. Rep. iv. 122, & Bot. Mex. Bound. 104. — S. California, near San Diego, *Parry*. Known only from scanty and imperfectly developed specimens; perhaps not of this section.

§ 3. CÁLAIS. Pappus of 5 scarious awn-tipped paleæ (in one species sometimes fewer or deciduous, and with only minute palea at base of the capillary awn): involucre either sparingly imbricated or merely calyculate, of thinnish bracts, the principal ones lanceolate: acaulescent or subcaulescent annuals, with leaves very variable in all the species, some narrow and entire, some laciniate-toothed, some very commonly pinnatifid or parted into slender lobes. — *Calais*, DC. Prodr. vii. 85.

* Paleæ of pappus acutely bifid or bidentate at apex by early splitting away from the base of the awn, which thus rises from the notch: akenes more or less slender and narrower upward, but not rostrate, for the slender seed reaches nearly to the apex: acaulescent, with slender scapes and middle-sized or smaller heads. — *Calais* § *Calocalais*, DC. l. c.

+— Awn very long in proportion to the paleæ of the pappus.

M. macrochǽta, Gray, l. c. Subcaulescent: scapes or peduncles sometimes even 2 feet high: involucre narrow, 8 or 10 lines high; its bracts attenuate-acuminate, outermost fully half the length of the inner: forming akenes decidedly contracted toward summit: pappus-paleæ thin, small, cleft quite to the middle from the first, bearing a long and weak capillary awn. — *Calais macrochæta*, Gray, Pl. Fendl. 112, & Pacif. R. Rep. iv. 113. — S. W. Idaho (*Spalding*), and California along the coast from San Francisco (*Bigelow*) to San Diego (*Nuttall, Cleveland*); but mature fruit still unknown.

+— +— Awn short in proportion to the lanceolate paleæ, which about equal the akene: larger heads inch or more high: involucral bracts less acuminate, the few outer of variable length: scapes or scapiform peduncles rising from a span to even 2 feet in height.

——**M. linearifólia**, Gray, l. c. Peduncle more or less thickened upward and in strong plants fistulous under the oblong head: leaves sometimes almost villous-pubescent when young:

akene attenuate above almost into a beak: pappus silvery white; very delicate awn barely half the length of the more deeply notched palea. — *Calais linearifolia*, DC. l. c., excl. syn. *Uropappus linearifolius* & *U. grandiflorus*, Nutt. Trans. Am. Phil. Soc. vii. 425. — Open low grounds, throughout California, to Nevada, Arizona, and New Mexico.

M. Lindléyi, GRAY, l. c. Peduncle little or not at all thickened upward: akene gradually and slightly attenuate toward the summit: pappus sordid: paleæ bearing a stronger and more exserted awn from a shallow notch, sometimes those of marginal flowers villous outside, and their akenes pubescent. — *Calais Lindleyi*, DC. l. c., excl. syn. *Uropappus Lindleyi* & *U. heterocarpus*, Nutt. l. c. — California, from San Francisco Bay to San Diego; apparently less common than the preceding.

* * Paleæ of the sordid or merely whitish pappus entire, surmounted by the awn, conspicuous except in the second species: scapes slender, a span to 18 inches high: akenes mostly upwardly scabrous on the 10 equal ribs, occasionally the outermost villous. — *Calais § Eucalais*, DC., &c.

+ Akenes attenuate-fusiform, the upper and slightly narrowed half not occupied by the seed!

M. attenuáta, GREENE. Leaves mostly pinnately parted into narrow linear lobes: scapes a span or two high: involucre half-inch high, barely calyculate at base: pappus of the length of the akenes (nearly 4 lines); the paleæ oblong-lanceolate, about half the length of the awn, externally either lightly or conspicuously villous. — Proc. Am. Acad. xix. 65 — California, at Berkeley and hills north of Monte Diablo, *Greene*.

+ + Akenes from slender-subclavate to turbinate, the cell filled by the seed.

++ Paleæ of the pappus very small (not over half-line long) or obsolete, glabrous, and the slender bristles fragile or deciduous.

M. aphantocárpha, GRAY, l. c. Involucre merely calyculate: akenes (1½ to 2 lines long) oblong-clavate, with usually some constriction or rounding of the summit, shorter than the rigid capillary awns of the pappus: these merely deltoid-dilated at base, or hardly so. — *Calais aphantocarpha*, Gray, Proc. Am. Acad. vi. 552. — California, in Alameda and Contra Costa Co.; first coll. by *Brewer*.

— **Var. tenélla**, GRAY, l. c. Depauperate in the original specimens: bristles of pappus 2 to 5, sometimes with manifest broadly-ovate palea at base, sometimes deciduous. — *Calais (Aphanocalais) tenella*, Gray, Pacif. R. Rep. iv. 114, t. 17. — Same range, and along the Napa and Lower Sacramento; first coll. by *Bigelow & Fitch*.

++ ++ Paleæ of the pappus conspicuous, persistent.

= Akene (with truncate summit) little over a line long and the long-awned paleæ of the pappus of not more than half its length, both glabrous: head small.

M. élegans, GREENE, in herb. A span to a foot high, slender: head in fruit less than half-inch high: akenes tapering gradually from very summit to base: paleæ of the pappus ovate-deltoid or sometimes rather narrower, either obscurely emarginate or more attenuate into a slender awn of about 4 times the length: sometimes a minute villosity covering the truncate summit of the akene. — California, common in Contra Costa Co., *Hall, Lemmon, Parry, Greene*, &c. Between *M. aphantocarpha* and *M. Bigelovii*.

= = Akenes 2 or 3 lines long, some of the outermost not rarely villous: paleæ of the pappus seldom shorter, in the same species and even in the same head disposed to be either glabrous or scabro-puberulent, or externally villous.

M. acumináta, GREENE. A span or two high: head in fruit about inch high: akenes apparently all glabrous, slenderly somewhat fusiform-turbinate (obscurely contracted between middle and summit), almost 3 lines long, not half the length of the pappus: paleæ narrowly lanceolate, gradually attenuate into a rather shorter awn. — Bull. Torr. Club, x. 58. *Calais Douglasii*, Gray, Pacif. R. Rep. iv. 113, not DC. — California, on Mark West's Creek, *Bigelow* (has passed for the following species; pappus nearly glabrous). Foot-hills of the Sierra Nevada, *Mrs. Curran*, with paleæ minutely silky-villous externally.

— **M. Bigelóvii**, GRAY. A span to a foot or more high: head half-inch or more high: involucre inclined to be somewhat imbricated: akenes oblong-turbinate, not contracted under the truncate summit, only 2 lines long; outermost sometimes villous: paleæ of the pappus oblong- to ovate-lanceolate, mostly if not always glabrous, or scaberulous, varying considerably in size, only half or a third the length of the awn. — Proc. Am. Acad. ix. 209. *Calais Bige-*

M. Howellii, Gray.

lovii, Gray, Pacif. R. Rep. iv. 164, t. 17. — Common in the district around San Francisco Bay, California, and south to Tulare Co.; first coll. by *Bigelow*.

M. Douglásii, GRAY. Rather coarser: scapes 8 to 20 inches high: head broad: akenes oblong-turbinate, thickish, obviously contracted under the summit, nearly 3 lines long; outermost usually white-villous: paleæ of the pappus ovate to orbicular (2 lines high and often as wide), firm-scarious, commonly imbricated or convolutely overlapping, abruptly acute or retuse at the apex, a half or a third the length of the awn, sometimes glabrous, sometimes densely villous outside. — *M. Douglasii* & *M. cyclocarpha*, with var. *eriocarpha*, Gray, Proc. Am. Acad. ix. 210. *Calais Douglasii*, DC. Prodr. vii. 85; Hook. & Arn. Bot. Beech. 361. *C. cyclocarpha*, Gray, Pacif. R. Rep. iv. 115, t. 18. *C. eriocarpha*, Gray, Proc. Am. Acad. vi. 552. — W. California, from Humboldt Co. to San Francisco Bay, and southward; perhaps first coll. by *Douglas*.

M. platycárpha, GRAY, l. c. A span to a foot high, slender: head half-inch or less high: proper bracts of involucre rather few and broad (oblong): akenes turbinate, tapering gradually from the broad summit to base; outermost densely short-villous: paleæ of the pappus ovate, 2 lines long, somewhat longer than the akene, abruptly acuminate into a short awn or cusp. — *Calais platycarpha*, Gray, Pacif. R. Rep. l. c. — Hills around San Diego and San Luis Rey, *Parry*, *Cleveland*, *Pringle*, &c. (Lower Calif., *Parry*, &c.)

§ 4. NOTHOCÁLAIS. Pappus of 20 to 24 narrowly linear-lanceolate silvery-white paleæ, occupying two or more series, with obscure mid-nerve, very gradually attenuate into a slender awn: akenes attenuate-fusiform: seed not reaching to the tapering summit: bracts of the oblong-campanulate involucre narrowly lanceolate, nearly equal, in about two series: perennial from a thick caudex. Intermediate between *Microseris* and *Troximon!* — Gray, Proc. Am. Acad. xix. 65.

M. troximoídes, GRAY. Acaulescent or nearly so: leaves tufted on the caudex, rather fleshy, narrowly linear-lanceolate, entire or undulate, 4 to 6 inches long: scapes a span to a foot high: involucre three-fourths inch high: ligules somewhat elongated: mature akenes half-inch long: pappus somewhat longer, its almost setiform paleæ a quarter of a line wide below. — Proc. Am. Acad. ix. 211. — Wooded hills and open plains, Montana and Idaho (first coll. by *Spalding*), Washington Terr. and Oregon to N. W. California.

220. **LEÓNTODON**, L. partly, Juss. HAWKBIT. (λέων, lion, ὀδούς, tooth, from the toothed leaves.) — Low perennials of the Old World, one naturalized in N. E. America, belonging to section OPORINIA, Koch, having simple pappus of a single series of plumose bristles, and the unopened heads not drooping.

L. AUTUMNÁLIS, L. (FALL DANDELION.) Short rootstock or caudex præmorse: leaves lanceolate, more or less pinnatifid, somewhat pubescent with simple hairs: scapes 5 to 15 inches high, sometimes simple, commonly once to thrice forked: peduncles clavate-thickened under the pubescent much calyculate involucre: akenes all alike. — *Apargia autumnalis*, Hoffm. Fl. Germ. iv. 113; Schk. Handb. t. 220; Pursh, Fl. ii. 497. *Oporinia autumnalis*, Don in Ediub. Phil. Jour. vi. 309; DC. Prodr. vii. 108. — In grassy grounds, Newfoundland to E. New England and sparingly to Penn.; fl. June to Nov. (Nat. from Eu.)

L. HÍSPIDUS, L., with double pappus, the outer of short naked bristles, and L. nÍRTUS, L., with both kinds of bristles plumose, and a paleaceous crown to outermost akenes, are sparing ballast-weeds at the ports of New York and Philadelphia.

220ᵃ. **PICRIS**, L. (Greek name for some bitter herb of this suborder, from πικρός, bitter.) — Leafy-stemmed and coarse herbs, chiefly biennials or annuals, and of the Old World, with aspect of the larger kinds of *Hieracium*, rough-bristly, yellow-flowered. — Benth. & Hook. Gen. ii. 511. *Picris* & *Helmintha*, Juss. Gen. 170.

P. HIERACIOÍDES, L. Rather tall, hispid, and some of the bristles minutely glochidiate, corymbosely branched: leaves lanceolate or broader, with partly clasping base, irregularly dentate: heads half-inch or more high: onter bracts of the involucre broader and spreading: akenes oblong, with 5 broad ribs and little or no beak: pappus of unequal sparsely plumose bristles, deciduous in a ring. — Lam. Ill. t. 648; Reichenb. Ic. Fl. Germ. t. 1375. — Introduced in a few places (as in Illinois, *Hall*), and as a ballast-weed. (Nat. from Eu.)

Var. **Japónica**, Regel. Very hispid with dark bristles, even to the involucre. — *P. Japonica*, Thunb. Fl. Jap. 299. *P. Kamtschatica*, Ledeb. Mem. Acad. 1814, & Fl. Alt. iv. 159. *P. Davurica*, Fischer & Horuem Hort. Hafn. Suppl. 155. — Sitka, *Mertens*, according to Herder. (Occurs on Behring Island, off Kamtschatka, as well as on the mainland, Japan, &c.)

——P. (Helmintha) echioides, L., of the Old World., is a ballast-weed of occasional appearance near New York and Philadelphia: it is known by the ovate and subcordate foliaceous outer bracts of the involucre, 3 to 5 in number, and by the narrow inner ones becoming thickened at base in age; also by the slender beak to the akene and a densely plumose pappus.

221. **PINAROPÁPPUS**, Less. (Πιναρός, dirty, πάππος, pappus, this being sordid or fuscous.) — Syn. 143; DC. Prodr. vii. 99. — Single species.

P. róseus, Less. l. c. Glabrous and glaucescent deep-rooted perennial: stems scapiform with a few minute bracts, and monocephalous, or leafy below with a few naked branches, slender, rather rigid: leaves lanceolate and entire, and some pinnatifid: involucre over half-inch high: ligules conspicuous, rose-tinged or almost white. — *Troximon Ræmerianum*, Scheele in Linn. xxii. 165. — High and rocky prairies, Texas, *Lindheimer*, *Wright*, &c. (Mex.)

222. **CALYCÓSERIS**, Gray. (Κάλυξ, a cup, alluding to the shallow cup at summit of akene, σέρις, a Cichoriaceous plant.) — New Mexican and Californian winter annuals, low, branching from the base, glabrous below and glaucescent; with leaves pinnately parted into narrow linear lobes, and showy rather large heads terminating the branches; the ligules elongated; peduncles sparsely or copiously hispid with tack-shaped glands. Fl. spring. — Pl. Wright. ii. 104, t. 14, Bot. Mex. Bound. 106, & Bot. Calif. i. 431.

C. Wrightii, Gray. Flowers rose-color: akenes with thick and broad somewhat rugulose ribs and thickish beak. — Pl. Wright. l. c. t. 14. — New Mexico from the Rio Grande to Arizona and S. Utah; first coll. by *Wright*.

C. Párryi, Gray. Flowers yellow: akenes more slender, 5-angled by the acute ribs, with narrower beak and smaller apical cup. — Bot. Mex. Bound. l. c.; Bot. Calif. l. c. — San Diego Co., California, to S. Nevada and adjacent Utah; first coll. by *Parry*.

223. **MALACÓTHRIX**, DC., extended. (Μαλακός, soft, θρίξ, hair.) — W. N. American herbs, leafy-stemmed or sometimes scapose; with pedunculate heads usually nodding before anthesis: flowers yellow or white, sometimes becoming purplish-tinged; in spring and early summer. — DC. Prodr. vii. 192; Torr. & Gray. Fl. ii. 485; Gray, Pl. Fendl. 113; Benth. & Hook. Gen. ii. 518; Gray, Proc. Am. Acad. ix. 213, & Bot. Calif. i. 432, excl. § 3.

§ 1. Malacólepis, Gray, l. c. Involucre very broad, of silvery-scarious bracts, with only a linear central portion green, regularly imbricated in several series; the short outer ones orbicular; inner from oval to oblong-lanceolate: receptacle bearing slender persistent bristles: corollas white, purplish-tinged in fading: broad-leaved annual.

M. Coúlteri, Gray, l. c. A foot or two high, rather stout, glabrous: leaves oblong or spatulate, upper cauline ovate or cordate and clasping, sparsely laciniate-dentate: heads terminating loose branches, short-peduncled, hemispherical, over half-inch high: akenes acutely about 15-ribbed and 4-5-angled, the summit obscurely denticulate by projection of the ribs: one or two stouter pappus-bristles more persistent. — S. California, from the Mohave desert to San Luis Obispo, &c.; first coll. by *Coulter*.

§ 2. MALACÓTHRIX proper. Involucre of narrow and acute or acuminate bracts, only narrowly scarious-margined, much less imbricated: bristles on the receptacle sparing, or fragile and deciduous, rarely none. — *Leptoseris, Leucoseris,* & *Malacomeris,* Nutt.

* Annuals: flowers light yellow, sometimes purplish in fading.

+– Simply scapose, with solitary large head, about 3-serially imbricated involucre, and herbage long-woolly when young. — *Malacothrix,* DC.

M. Califórnica, DC. Leaves once or partly twice laciniately pinnatifid into narrow linear or almost filiform lobes, when young woolly with long and loose very soft hairs (whence the generic name), as also is the base of the broadly campanulate (two-thirds inch high) involucre; the outer bracts slender-subulate: delicate bristles of the receptacle generally present: akenes narrow, lightly striate-costate, the acutish base with a small concave callus: outer pappus of 2 persistent bristles and between them some minute pointed teeth: scape a foot or less high, bractless or nearly so. — Prodr. vii. 192; Gray, l. c., excl. var. *glabrata,* Eaton. — Open grounds, California, from the Sacramento valley to San Diego; first coll. by *Douglas.*

+– +– Subcaulescent or more leafy-stemmed, more or less branching, early glabrate or glabrous: involucral bracts nearly or wholly of two lengths; the outer (or calyculus) short, proportionally broader and loose. — *Leptoseris,* Nutt.

++ Heads comparatively large, and on elongated or the earlier on scapiform peduncles: leaves and their divisions long and slender, nearly as in the preceding species.

M. glabráta. Erect, or with ascending branches from the base, these leafy, often again branching and bearing a few lateral as well as terminal heads: involucre fully half-inch high, glabrous, or outer bracts sometimes tomentulose canescent when young: flowers, &c., as in *M. Californica.* — *M. Californica,* var. *glabrata,* Eaton, Bot. King Exp. 201; Gray, Bot. Calif. l. c. *M. Torreyi,* Gray, Proc. Am. Acad. ix. 213, as to "slender narrow-leaved form." — Dry eastern portion of the Sierra Nevada in California and Nevada, to S. E. Utah and Arizona; first coll. by *Anderson.*

++ ++ Heads smaller, with broadish campanulate involucre seldom less than half-inch high, short-peduncled on the leafy usually spreading branches: lower leaves oblong, rather short, pinnatifid, with short and dentate lobes; teeth and lobes commonly callous-mucronate: plants a span to a foot high.

M. Torréyi, Gray. Akenes linear-oblong, 5-angled by as many salient often almost wing-like ribs, a much less prominent pair in each interval: outer pappus of 2 to 5 or sometimes 8 stouter persistent bristles, between the thickish bases of which are minute teeth: bracts of the involucre acuminate: peduncles commonly with some sparse gland-tipped hairs. — Proc. Am. Acad. ix. 213, & Bot. Calif. i. 433. *M. sonchoides,* Torr. in Stansb. Rep. 392; Gray, Pl. Wright. ii. 105, in part; Eaton, Bot. King Exp. 201, not Torr. & Gray. — Low grounds, Utah to W. Nevada and S. E. Oregon, probably to California; first coll. by *Stansbury.*

M. sonchoídes, Torr. & Gray. Akenes linear-oblong, 15-striate-costate, somewhat angled by 5 moderately stronger ribs, the summit with a 15-denticulate white border: no persistent bristles: involucral bracts rather broader, merely acute: branches more diffuse: rhachis of the principal leaves as well as lobes dentate. — Fl. ii. 486; Gray, l. c. *M. obtusa,* Eaton, l. c., in part. *Leptoseris sonchoides,* Nutt. Trans. Am. Phil. Soc. vii. 428. — Plains of W. Nebraska to New Mexico, Nevada, and adjacent California and Arizona; first coll. by *Nuttall.*

M. Féndleri, Gray. Akenes cylindrical, equably 15-costate, dark-colored; the summit bordered by a shallow cupulate crown, its margin entire, white within: no persistent pappus-bristles or only one. — Pl. Wright. ii. 104, Bot. Mex. Bound. 106, & Proc. Am. Acad. ix. 213. — E. New Mexico to S. E. California, *Fendler, Bigelow, Wright, Lemmon,* &c.

++ ++ ++ Heads small, numerous and loosely paniculate on slender erect and rather naked stem and branches: involucre seldom over 3 lines high, narrower, fewer-flowered: the tips of the bracts commonly sphacelate or purplish.

M. XÁNTI, Gray, l. c., the only outlying species of the genus (Cape San Lucas, Lower California, *Xantus*), is 2 feet high, with leaves mainly radical and lyrate-pinnatifid, panicle very naked, narrow involucre 4 lines high, akenes obtusely 15-ribbed, five ribs moderately stronger, cupulate apex obtusely 5-toothed, outer pappus of 3 to 5 very slender persistent bristles. Heads larger than in either of the following.

M. Clevelándi, Gray. Akenes oblong-linear, minutely striate-costate, 4 or 5 of the ribs slightly more prominent: outer pappus of one persistent bristle and a conspicuous circle of narrow white setulose teeth: leaves narrow, only some of the radical pinnatifid. — Bot. Calif. i. 433. — From Antioch (*Mrs. Curran*) to Santa Barbara and San Diego in California (first coll. by *Cleveland*); also mountains of Arizona.

M. obtúsa, Benth. Akenes obovate-oblong, obtusely angled by 5 rather prominent ribs, the others delicate or obscure, the apex somewhat contracted and its border entire: no persistent pappus-bristles: remains of tomentum in axils of leaves, &c.: radical leaves thickish, spatulate-oblong, sinuate-dentate or pinnatifid; the teeth or lobes short-oblong, sometimes very obtuse: corollas (white?) in dried specimens purplish-tinged. — Gray, l. c. *M. obtusa*, & *M. parviflora*, Benth. Pl. Hartw. 321. *Senecio flocciferus*, DC. Prodr. vi. 426. — California, from Monterey to Humboldt Co. and in the Yosemite; first coll. by *Douglas* and *Hartweg*.

* * Suffrutescent-perennial: "flowers yellow." — *Malacomeris*, Nutt.

M. incána, Torr. & Gray. Low, white-tomentose: leaves in tufts on short basal shoots, pinnatifid, with short lobes: flowering branches scape-like, a few inches high, bearing one or two rather large heads: involucre broadly campanulate: no persistent pappus-bristles. — Fl. ii. 486. *Malacomeris incana*, Nutt. Trans. Am. Phil. Soc. vii. 435. — Island in the bay at San Diego, California, *Nuttall*, who only has collected it, and in imperfect specimens.

* * * Somewhat suffrutescent and leafy paniculately branching perennials: flowers white (changing to rose-color?): involucre broadly campanulate (nearly half-inch high), many-flowered; the loose calyculate bracts numerous, subulate, passing into similar bractlets on the peduncle: receptacle obscurely dentate-alveolate, no bristles detected: no persistent exterior pappus-bristles. — *Leucoseris*, Nutt.

M. saxátilis, Torr. & Gray. Minutely tomentose when young, soon glabrate, somewhat succulent, a foot or two high: leaves lanceolate or the lower spatulate, either entire or laciniate-pinnatifid: heads terminating the paniculate branches: akenes narrowly oblong, 10-15-costate, at maturity somewhat 4-5-angled by the stronger ribs: apex slightly contracted, bearing a very short multidenticulate white border. — Gray, Proc. Am. Acad. & Bot. Calif. l. c. *M. saxatilis* & *M. commutata*, Torr. & Gray, l. c., excl. syn. *Senecio flocciferus. Leucoseris saxatilis* & *L. Californica*, Nutt. Trans. Am. Phil. Soc. vii. 440, 441. *Hieracium? Californicum*, DC. Prodr. vii. 235. *Sonchus? Californicus*, Hook. & Arn. Bot. Beech. 361. — Coast of California at Santa Barbara and southward; first coll. by *Coulter*. Passes on the mountains and in the interior district into

Var. tenuifólia. Early glabrate or glabrous: stems slender, not succulent, 2 to 4 feet high, with long and slender loosely-paniculate branches, bearing slender-pedunculate heads (of equal or smaller size): leaves narrowly lanceolate to linear, or on branchlets almost filiform. — *M. tenuifolia*, Torr. & Gray, l. c.; Gray, Bot. Calif. l. c. *Leucoseris tenuifolia*, Nutt. l. c. — Mountain-sides and cañons, Santa Barbara to San Diego, also Tejon, San Bernardino, and Arizona; first coll. by *Coulter*.

224. GLYPTOPLEÚRA, Eaton. (Γλυπτός, carved, πλευρά, side, from the sculpturing of the akenes.) — Winter annuals of the Utah-Nevada desert, many-stemmed and depressed, forming flat and leafy tufts, only an inch or two high; with thickish and oblong runcinate leaves on margined petioles: heads rather large for the size of the plant: fl. spring. — Bot. King Exp. 207, t. 20; Gray, Proc. Am. Acad. ix. 209, & Bot. Calif. i. 431.

G. margináta, Eaton, l. c. Corollas white, turning pink in fading. little exserted: lobes and mostly whole margin of the leaves densely scarious-fringed, this white border mainly cut into short obtuse teeth, only pectinate-setiform on the leaves subtending the heads. — Western borders of Nevada, from the Truckee to Candelaria (*Watson, Lemmon, Shockley*), and to the Mohave desert in California, *Parish*.

G. setulósa, Gray, l. c. Corollas yellow changing to pink, much exserted (half to three-fourths inch): white margin of the leaves less conspicuous, mainly composed of distinct subulate or acicular white teeth. — St. George, S. Utah, to the Mohave desert, *Parry, Palmer, Parish*, &c.

225. APARGÍDIUM, Torr. & Gray. (Likeness to *Apargia*, a sort of Dandelion.) — Fl. ii. 474; Gray, Bot. Calif. i. 439. — Single species.

A. boreále, Torr. & Gray, l. c. Glabrous and slender perennial: leaves wholly radical, linear-lanceolate, entire or nearly so, thinnish: scapes at length a foot high: involucre half to three-fourths inch high: corollas deep yellow, conspicuous. — *Apargia borealis*, Bongard, Veg. Sitch. 146. *Leontodon boreale*, DC. Prodr. vii. 102. *Microseris borealis*, Schultz Bip., ex Herder in Pl. Radd. iii. (4), 28. — Wet meadows and bogs, Alaskan Islands (*Mertens*, &c.) to Mendocino Co., California. Mature akenes not yet seen.

226. HIERÁCIUM, Tourn. Hawkweed. (The Greek and Latin name, from ἱέραξ, a hawk.) — A huge European genus, and with a moderate number of peculiar American species; perennial herbs, often with toothed but never deeply lobed leaves; heads in ours from small to barely middle-sized, paniculate, rarely solitary; the flowers yellow, in one species white, produced in summer and autumn, usually open through the day. — Frœlich in DC. Prodr. vii. 198; Fries, Symb. Hist. Hier. (1848), & Epicrisis Hier. (1862); Benth. & Hook. Gen. Pl. ii. 516; Gray, Proc. Am. Acad. xix. 65. Sections after Fries.

H. KÁLMII, L. The original in the Linnæan herbarium is some wholly undetermined plant, probably not at all from Pennsylvania, nor from America, certainly not of this genus.

§ 1. PILOSÉLLA, Fries. Involucre not distinctly calyculate nor regularly much imbricate: pappus a single series of delicate bristles: akenes oblong, truncate: natives of the Old World.

H. AURANTÍACUM, L. Somewhat stoloniferous from the tufted rootstocks, long-hirsute and above setose-hispid as well as setulose-glandular, the involucre especially with dark hairs: leaves radical and near the base of the simple scape or peduncle: heads (four lines high) in a naked cymose cluster: flowers deep orange-color to flame-color: pappus whitish. — Jacq. Fl. Austr. t. 410; Fl. Dan. t. 1112. — Escaped from gardens to roadsides and fields in several places, New England and New York. (Nat. from Eu.)

H. PRÆÁLTUM, VILL. Glaucous, 2 feet or more high: stems scapiform, leafy only near the base, and there (as also the lanceolate leaves) sparsely beset with bristly hairs: heads rather numerous in an open cyme: involucre about three lines high. — A form of this appears to be established, along fences and field borders, near Evans Mills and Carthage, N. New York, *L. F. Ward*. (Nat. from Eu.)

§ 2. ARCHIERÁCIUM, Fries. Involucre of the comparatively large heads irregularly more or less imbricated: pappus of more copious and unequal bristles: akenes columnar, truncate: chiefly natives of the Old World.

* Stem scapiform, or only with a leaf or two above the base.

H. MURÓRUM, L. The form called *H. præcox*, Schultz Bip., or nearly: leaves thin, oval or oblong, obtuse, incisely dentate toward the subordate base; scapiform stem a foot or less high, bearing few or several cymose heads: involucre 4 or 5 lines high, dark-glandular. — Open woodlands near Brooklyn, New York, *Merriam*. Also apparently in Lower Canada. (Nat. from Eu.)

H. vulgátum, FRIES. Habit of the preceding, or more leafy: leaves from oblong to broadly lanceolate, mostly acute at both ends, decurrent on the petiole: heads few, rather smaller than in the foregoing. — Novit. ii. 258, Symb. Hier. 115, & Epicr. 98; Reichenb. Ic. Fl. Germ. xix. t. 1526, 1527. *H. sylvaticum*, Smith (that of L. is rather *H. murorum*); Fl. Dan. t. 1113; Schlecht. in Linn. x. 87. *H. molle*, Pursh, Fl. ii. 503, not Jacq. — Labrador, *Kohlmeister*, &c. Canada, on shores of the Lower St. Lawrence (*Macoun*), there perhaps introduced. (Greenland, Eu., N. Asia.)

H. ALPÍNUM, L., which has only a single large and dark-haired head, is in Greenland only, beyond our range.

* * Stem leafy to the top (a foot to a yard high), bearing short-pedunculate broad heads : involucre half-inch high, or sometimes smaller: no stolons or running rootstocks: no cluster of leaves at base of the developed stems; cauline leaves all closely sessile; receptacle conspicuously fimbrillate-dentate: ligules not ciliate.

H. umbellátum, L. A foot or two high, strict, bearing a few somewhat umbellately disposed heads: leaves narrowly or sometimes broadly lanceolate, nearly entire, sparsely denticulate, occasionally laciniate-dentate, all narrow at base: involucre usually livid, glabrous or nearly so; outermost bracts loose or spreading. — Fl. Dan. t. 680; Fl. Lond. vi. t. 58; Richards. App. Frankl. Journ. ed. 2, 29 ? in part; Fries, l. c. *H. Canadense*, var. *angustifolium*, Torr. & Gray, Fl. ii. 476, in part. *H. macranthum*, Nutt. Trans. Am. Phil. Soc. vii. 446. *H. rigidum ?* Fries in Epicr. 134. — N. shore of Lake Superior to the Rocky Mountains, and northward. (Kamtschatka, N. Asia, Eu.)

H. Canadénse, Michx. Taller, robust, with corymbosely or paniculately cymose heads: leaves from lanceolate to ovate-oblong, acute, sparsely and acutely dentate or even laciniate, at least the upper partly clasping and broad or broadish at base: involucre usually pubescent when young, glabrate, occasionally glandular; the narrow outermost bracts loose: pappus sordid. — Fl. ii. 86; Torr. & Gray, l. c. *H. virgatum, fasciculatum,* & *macrophyllum,* Pursh, Fl. ii. 504. *H. Kalmii*, Spreng. Syst. iii. 646; Bigel. Fl. Bost.; Torr. Compend., &c., not L. *H. scabriusculum*, Schwein. App. Long Exp. *H. prenanthoides*, Hook. Fl. i. 300, not Vill. *H. helianthifolium*, Frœlich in DC. l. c. 225. *H. corymbosum*, Fries, Symb. Hier. 185, as to pl. Newfoundl. ? also *H. auratum*, Fries, l c. 181, & Epicr. 124; these being thin-leaved forms of shady places. — Open woods in dry soil, Newfoundland ? and New England to Penn., north to the Mackenzie River, west to Oregon and Brit. Columbia, northwardly passing into *H. umbellatum*. (Greenland, N. Eu., if also *H. crocatum*, Fries.)

§ 3. STENOTHÉCA, Torr. & Gray, l. c. Involucre a series of equal bracts and a few short calyculate ones, usually narrow and few-many-flowered: pappus of more or less scanty equal bristles: akenes in a few species slender or tapering to the summit. (Name therefore more applicable to the involucre than to the akenes.) — Fries, l. c. *Stenotheca*, Monnier, Ess. Hierac. 71, there restricted to species with attenuate akenes. Species of *Pilosella*, Schultz Bip. in Flora, 1862, 433-440.

* Atlantic species, all yellow-flowered and with sordid pappus.

+– Akenes columnar, at maturity not at all attenuate upward: panicle not virgate.

++ Heads 15-20-flowered, narrow, effusely paniculate, on divergent or divaricate slender pedicels: stem leafy, sometimes almost leafless in depauperate plants.

H. paniculátum, L. Slender, 1 to 3 feet high, usually leafy up to the sparse compound panicle, nearly smooth and glabrous (except some villosity at base of stem), not glandular: leaves thin, lanceolate or broader, tapering to both ends, sparingly denticulate or salient-dentate: peduncles and pedicels filiform, an inch or more long: involucre 3 or 4 lines long, of 8 to 14 narrow principal bracts. — Spec. ii. 802; Michx. Fl. ii. 86; Torr. & Gray, Fl. ii. 478. — Open dry woods, Canada and New England to upper parts of Georgia and Alabama. *H. venosum*, var. *caulescens*, Arvet-Touvet, and *H. Sullivantii*, Arvet-Touvet, Spicil. Hier. (1881), 11, are seemingly depauperate forms of this.

++ ++ Heads 15-40-flowered, narrow-campanulate or oblong, on erect or ascending slender pedicels, in a naked and very loose corymbiform-paniculate cyme.

H. venósum, L. (RATTLESNAKE-WEED.) Slender: stem leafless from a depressed radical rosette, or 1-2-leaved above it, a foot or two high, glabrous or nearly so, branching above into a lax corymbiform cyme of few or several heads: leaves obovate to spatulate-oblong, mostly denticulate, subsessile, commonly purple-veined and sparsely setose-villous: involucre 4 lines long, 15-35-flowered (or even only 12-flowered), of 10 to 14 principal bracts and very few bractlets, either glabrous or with the peduncles beset with some small glandular hairs: akenes short, strictly columnar, even when young. — Spec. ii. 800 (founded on the syn., but the "scapo crassissimo" of Gronovius unaccountable); Willd. Spec. iii. 1570; Torr. & Gray, l. c.; Fries, l. c. *H. Gronovii*, L. l. c. 802, as to herb. & descr. (but not the Gronovian plant); Willd. l. c.; Michx. l. c., in part, the var. *subcaulescens*, Torr. & Gray, l. c. *H. subnudum*,

Frœl. in DC. vii. 218, chiefly. *Stenotheca venosa*, Monnier, Ess. Hier. 72. — Open pine woods and sandy barrens, Canada and Saskatchewan to Georgia and Kentucky.

H. Mariánum, WILLD. Larger, 2 or 3 feet high, few–several-leaved, pilose-hirsute below, branching at summit into a very open cymose panicle of several or numerous 20–43-flowered heads: leaves obovate-oblong with tapering base; radical erect or ascending, attenuate below into petioles, rarely at all purplish-veiny: peduncles and pedicels commonly minutely whitish-tomentulose, also usually the base of the involucre, at least when young, and beset with few and sparse or more copious glandular bristles: akenes slender-columnar, with tapering summit when forming, but not so at maturity.— Spec. iii. 1572, partly (& as to syn. *H. Marianum*, &c., Pluk. Mant. 102, t. 420, f. 2, whence the name); Frœl. in DC. Prodr. vii. 217. *H. Gronovii*, var. *subnudum*, in part, & some of *H. scabrum*, Torr. & Gray, Fl. ii. 447. *H. Carolinianum*, Fries, Symb. Hier. 145, & Epicr. 151. *H. Rugelii*, Arvet-Touvet, Spicil. Hier. (1881), 11, by the char.— Dry and open woods and clearings, New England to Penn. and Georgia. Various forms almost fill the interval between the preceding and the following species.

++ ++ ++ Heads 40–50-flowered, thickish (and the tumid-campanulate involucre 4 or 5 lines high), on shorter and rather rigid spreading pedicels, and somewhat crowded in a convex or barely flat-topped cyme: no rosulate tuft of radical leaves at flowering time.

H. scábrum, MICHX. Robust, 2 or 3 feet high, mostly leafy up to the inflorescence, hirsutely hispid below, glandular-hispid above: whole inflorescence and mostly base of involucre densely beset with dark glandular bristles and with some fine grayish tomentum: leaves obovate to spatulate-oblong, obtuse, denticulate, pubescent or hirsute, sessile by a narrow base: akenes exactly columnar.— Fl. ii. 86; Pursh, Fl. ii. 504; Torr. & Gray, Fl. ii. 476; Fries, l. c. *H. Marianum*, Willd. l. c., in part (as to one specimen); Bigel. Fl. Bost. ed. 2, 288; Ell. Sk. ii. 263. — Dry open woods, Canada to Lake Superior, Missouri, and to Georgia.

+— +— Akenes fusiform or with tapering summit: heads 15–30-flowered, on short and ascending pedicels disposed in a narrow thyrsiform or almost virgate panicle: glandular-bristly hairs on peduncles and cylindraceous involucre either scanty or numerous: radical leaves generally present at flowering time, and destitute of colored veins, oblong-obovate, all more or less long-pilose or setiferous, especially along the midrib beneath.

H. Gronóvii, L. Stem strict, 1 to 3 feet high, leafy (3–12-leaved) below, continued nearly through the virgate or thyrsiform panicle: pubescence mainly soft-setose, the stronger bristles from papillæ: cauline leaves oval or oblong, closely sessile mostly by a broad base; lowest and radical obovate or spatulate with attenuate base or short petiole: involucre 3 or 4 lines long, 15–20-flowered: akenes fusiform, with gradually tapering beak-like summit: pappus dirty whitish.— Spec. ii. 802, as to pl. Gronov. (excl. remarks and pl. herb., which are of *H. venosum*); Michx. Fl. (var *foliosum*); Monnier, Ess. Hier. 30; Torr. & Gray, Fl. ii. 677, not of Willd., Frœl. in DC., &c. *H. Marianum*, Fries, Symb. Hier. 147, & Epicr. 152, not Willd., except perhaps in small part. *Stenotheca Mariana*, Monnier, l. c. 72? *S. subnuda*, Monnier, l. c. t. 2, f. 5; depauperate form (var. *subnudum*, Torr. & Gray), with narrow panicle reduced to a few heads. *H. Gronovii*, var. *hirsutissimum*, Torr. & Gray, l. c., is the most setose-hirsute form, with narrow panicle a foot or more long: and from that character, either this or the next must be *H. Pennsylvanicum*, Fries, Symb. Hier. 150, & Epicr. 156; yet the akenes described are like those of *H. Marianum*, Willd.— Sandy ground, and open dry woods, Canada! to Florida, Missouri, and Louisiana.

H. longípilum, TORR. Stouter, leafy to near the middle of the stem, and with linear-lanceolate or subulate bracts up to the narrow panicle: pubescence mainly setose and most abundant; the bristles from a small papilla, upright, commonly half-inch to even an inch long, fulvous or rufous, denticulate: leaves spatulate-oblong or upper lanceolate, thickish, the radical commonly present in a tuft at flowering time: involucre 5 or 6 lines long, 20–30-flowered, oblong-campanulate, and with short peduncles more or less tomentulose as well as glandular: akenes fusiform, but much less tapering upward than in the preceding: pappus at maturity (fuscous.— Hook. Fl. i. 298 (note); Torr. & Gray, Fl. ii. 477; Fries, l. c. *H. barbatum*, Nutt. Jour. Acad. Philad. vii. 70, & Trans. Am. Phil. Soc. vii. 446, not Tausch. — Open woods and prairies, Michigan to Nebraska and Texas.

Var. **spathulátum** (*Pilosella spathulata*, Schultz Bip. in Flora, 1862, conjectured by the author to be a variety of *Hieracium scabrum*), collected on Tuscarora Mountain, in the

Alleghanies of Penn., *Porter* and *Traill Green*, seems to be a depauperate form of the present species, with stem naked and leafless except near the base, and bristly hairs not so long: but heads in the specimens barely in blossom, and akenes unknown.

* * Rocky Mountain and Pacific species. (Involucre in most cases less obviously double than in the Eastern species; the calyculate bracts sometimes unequal or emulating the interior, or else obsolete.)

+ Crinite-hirsute with long and whitish or yellowish shaggy denticulate hairs, especially on both sides of the entire leaves, on the branching leafy stems and panicle, and commonly but not always on the involucre also: flowers yellow: akenes columnar and short, not at all narrowed upward, at most a line and a half long, shorter than the sordid pappus.

H. Scoúleri, Hook. Robust, a foot or two high: long and soft setose hairs commonly from small papillæ: leaves lanceolate or spatulate-lanceolate (3 to 6 inches long): panicle irregular or branching: heads half-inch high: involucre somewhat furfuraceous and glandular, also sparsely or copiously beset with long bristly hairs: pappus whitish. — Fl. i. 298, & Torr. & Gray, Fl. ii. 478, partly (some specimens of coll. Scouler distributed being *H. cynoglossoides*, and the plant from " Pennsylvania, *Schweinitz*," of Hooker, being *H. Gronovii*); Eaton, Bot. King Exp. 199. — Montana to Oregon and Brit. Columbia, southeast to the Wahsatch Mountains, Utah.

H. hórridum, Fries. Low (a span to a foot high), in tufts, branched from the caudex: softer villous hairs not from papillæ: leaves lingulate-lanceolate or spatulate-oblong, lowest petioled: panicle corymbiform-cymose, of numerous small and rather narrow heads: involucre 3 lines high, sometimes nearly naked, oftener beset with scattered and long bristly hairs: pappus fuscous. — Epicr. Hier. 154; Arvet-Touvet, l. c. 19. *H. Breweri*, Gray, Proc. Am. Acad. vi. 553, & Bot. Calif. i. 440. — On rocks, in the higher Sierra Nevada, California, from Shasta to San Bernardino Co.; first coll. by *Bridges*, next by *Brewer*.

H. RELÍCINUM, Fries, Epicr. 153, would seem to be only a taller and simpler-stemmed form of the preceding, with widely open panicle and long-hirsute involucre. Described from a specimen in herb. DC., from mountains of California, *Bridges*.

+ + Crinitely long-villous with soft-woolly and blackish smooth hairs, which involve the heads, &c., but are wanting to lower leaves; no stellular pubescence and no glands: flowers yellow: pappus fuscous.

H. triste, Cham. A span or two high: stem simple, few-leaved, bearing solitary or mostly 2 to 4 somewhat racemosely disposed heads: radical leaves obovate to spatulate, entire, green and glabrate, or with sparse pale hairs; cauline oblong, upper ones and stem more or less villous-lanate: heads half-inch high: livid involucre and peduncles densely clothed with the very long dark-brown or partly grayish soft wool: akenes short-columnar. — Cham. in herb. Willd.; Spreng. Syst. iii 640; Frœl. in DC. Prodr. vii. 209; Torr. & Gray, Fl. ii. 458, partly; Fries, l. c. — Aleutian Islands to Behring Strait; first coll. by *Chamisso* and *Eschscholtz*.

+ + + Dark-hirsute (verging to naked) and somewhat glandular (also whitish with short stellular-tomentum) on the involucre: leaves and lower part of scapiform stems not even pilose (but glabrous or at most puberulent): flowers yellow: pappus sordid.

H. grácile, Hook. Pale green, in tufts: leaves nearly all in radical clusters, obovate- to oblong-spatulate (1 to 3 inches long) and attenuate into petioles, entire or repand-denticulate: stems or scapes slender, 8 to 18 inches high, cinereous-tomentulose above, bearing few or several racemosely disposed livid heads, the lower linear-bracteate: involucre about 4 lines high, usually blackish-hairy at base in the manner of the preceding, but the hairs much shorter than the head, also (as on the peduncles) some more setulose and glandular ones; akenes short-columnar. — Fl. i. 298; Fries, l. c., not of Frœl., which is later. *H. arcticum*, Frœl. in DC. Prodr. vii. 209. *H. Hookeri*, Steud. Nomen. ed. 2, 763. *H. triste*, in part, Torr. & Gray, Fl. ii. 478. *H. triste*, var. *gracile*, Gray, Bot. Calif. i. 441. — Alaska (Norfolk Sound, ex Frælich), Brit. Columbia, Northern Cascade and Rocky Mountains, and south to those of Utah and Colorado. Passes into

Var. detónsum. A span to nearly a foot high, with rather smaller heads: dark hirsute hairs wholly wanting, or only some smaller ones on the involucre. — *H. triste*, var. *detonsum*, Gray, Bot. Calif. l. c. — Mountains of Brit. Columbia to those of Colorado, and alpine region in the Sierra Nevada, California, at some stations accompanying the typical form.

+ + + + Not crinite (yet sometimes scattered bristles on the involucre and panicle), but at least the radical leaves and base of stem sparsely or even thickly setose-hirsute with long spreading hairs.

++ Flowers white: stems leafy and in larger plants loosely branching, depauperate or subalpine plants even scapose: involucre 18–30-flowered: akenes linear-columnar (only a line and a half long), not at all narrowed upward: pappus sordid: leaves entire or denticulate.

H. albiflórum, Hook. A foot to a yard high, smaller plants with simple and larger with compound open corymbiform-paniculate cyme. leaves oblong, thin (2 to 4 or larger 5 to 6 inches long), upper with usually narrowed sessile base, lower tapering into petiole: involucre narrow-campanulate, 4 or 5 lines high, of linear-lanceolate bracts, pale or livid, mostly glabrous or nearly so, not rarely a few bristly hairs. — Fl. i. 298; Torr. & Gray, Fl. ii. 479; Fries, Symb. Hier. 143; Gray, Bot. Calif. i. 440. — Open dry woods, Rocky Mountains, from lat. 56° to Colorado and Utah, and Brit. Columbia to mountains of S. California; first coll. by *Drummond*, *H. Vancouverianum*, Arvet-Touvet, Spicil. Hier. 10 (at least specimens coll. Lyall distributed from Kew as "H. Sconleri"), is of this species, and doubtless white-flowered.

++ ++ Flowers yellow: stems more or less leafy, except in var. of *H. cynoglossoides*: involucre 15–30-flowered, oblong-campanulate, of rather numerous narrow and acute or acutish bracts: akenes columnar, not at all tapering upward, not over a line and a half long: pappus from sordid to dull white.

= Leaves or many of them salient-dentate: pappus whitish.

H. argútum, Nutt. A foot or two high, slender, hirsute with long shaggy hairs at base of stem, glabrous or merely puberulent above and throughout the very lax diffuse naked panicle: leaves numerous at and near the base of the stem, broadly lanceolate (or radical oblong-spatulate), acute or acuminate, tapering into margined petioles, larger ones 4 inches long, half or two-thirds inch wide, each margin with 4 or 5 salient triangular teeth; upper ones linear and entire, much reduced in size (1 to 3 liues wide): peduncles elongated and with the involucre more or less dark-glandular, sometimes a few scattered dark hairs. — Trans. Am. Phil. Soc. vii. 447. — Hills behind Santa Barbara, California, *Nuttall* (specimen not seen), *Rothrock*, who found it in Bartlett's Cañon, young, color of flowers uncertain; Santa Lucia Mountains, *Parry*, an almost naked-stemmed form with radical leaves merely denticulate, the involucre and peduncles less glandular and more scurfy-puberulent; corollas certainly yellow. Also coll. by *Hænke?* if *Pilosella arguta*, Schultz Bip. in Flora, 1862, 438.

H. Parishii, Gray. Leafy up into the narrowly oblong panicle, puberulent above, with no glandular hairs or stipitate glands: lower leaves shaggy-hirsute (along with base of stem), elongated-lanceolate (5 to 8 inches long, half-inch or more wide), tapering to the base or margined petiole, with 5 to 8 salient teeth to each margin; upper leaves linear-lanceolate, entire, those subtending lower branches of panicle (2 inches long) little shorter than they: peduncles seldom much longer and often shorter than the heads: involucre pale, granulose-puberulent. — Proc. Am. Acad. xix. 67. — Rock-crevices, San Bernardino Mountains, S. E. California, *Parish*.

= = Leaves all entire, or merely repand, or slightly denticulate.

H. Rúsbyi, Greene. Leafy-stemmed, 2 feet or more high, bearing numerous compound-paniculate heads: stem hirsute below, above smooth and glabrous up to the rather short-peduncled heads: leaves all elongated-oblong; cauline little diminished in size upward (3 or 4 inches long), quite entire, mostly half-clasping at base: involucre 3 lines high, pale, barely puberulent: akenes short-columnar, blackish: pappus sordid. — Bull. Torr. Club, ix. 64; Gray, Proc. Am. Acad. xix. 69. — Mogollon Mountains, New Mexico, *Rusby*.

Var. Wrightii, Gray, l. c. More robust and branching: bristles of the stem truly hispid from papilliform base: branches and even peduncles setulose-hispidulous, and the latter obscurely glandular: sometimes a few small bristles near the tips of the involucral bracts: pappus dull white. — *Crepis ambigua*, Gray, Pl. Wright. i. 129, not Pl. Fendl. — W. borders of Texas, between the Limpio and the Rio Grande, *Wright*.

H. cynoglossoídes, Arvet. Stem a foot or less high (either from naked base or more commonly a radical tuft of leaves), simple, 2–several-leaved, bearing few or several cymosely disposed heads, setose-hirsute or hispid at base, either hispidulous or glabrous above: leaves lanceolate to spatulate-oblong, at least the lower conspicuously setose-hirsute; upper sometimes glabrous; involucre 4 or 5 lines high, glandular, sometimes as also peduncles glandular-hispidulous; akenes rather short-columnar; pappus whitish. — Gray, Proc. Am. Acad. xix. 68. *H. Scouleri*, Hooker, in herb. & distrib., partly; Torr. & Gray, Fl. ii. 478 & Gray, Bot. Calif. l. c., mainly, not Hook. Fl. *H. cynoglossoides*, Arvet-Touvet, Spicil. Hier. 20, founded on undeveloped specimen of Parry's N. Wyoming coll. no. 188. E. Hall's Oregon coll. 523 consists of this and *H. Scouleri* mixed. — N. W. Wyoming and Montana (*Porter, Parry*) to Washington Terr. and Oregon (first coll. by Tolmie, &c.); also Siskiyou Co., California, *Greene*, passing to

Var. nudicaúle, GRAY, l. c. Leaves all in the radical tuft, or only one or two very small and bracteiform on the (8 to 12 inch high) glabrous scape. — Northern Sierra Nevada, California, *Lemmon, Mrs. Austin*, the latter on Lassen's Peak.

++ ++ ++ Flowers apparently yellow, only 5 to 15 in the narrow and diffusely paniculate heads: involucre cylindraceous, not at all glandular (4 lines high), of 7 to 9 broadish-linear and obtuse principal bracts and 2 or 3 short ones: akenes comparatively large, fully 2 lines long, chestnut-brown, slightly or at maturity not perceptibly tapering to the summit: pappus dark-fuscous: leaves obovate-spatulate, all in a radical tuft at base of the loosely branching (span to foot high) scapes.

H. Bolánderi, GRAY. Radical leaves sparsely or densely long-hirsute, no other pubescence, scapes and involucre smooth and glabrous. — Proc. Am. Acad. vii. 365, Bot. Calif. i. 440, & Proc. Am. Acad. xix. 68. — Mountains of California, Humboldt Co., *Bolander*. Near headwaters of the Sacramento, *Pringle*. Sierra Co., *Lemmon*. Only Bolander's specimen has akenes narrowed upward.

H. Gréenei, GRAY. Radical leaves villous-hirsute, also canescent-tomentose on both sides with stellular pubescence: scape with peduncles and involucre cinereous-tomentose. — Proc. Am. Acad. xix. 69. — Pine woods of Scott's Mountains, Siskiyou Co., N. California, *Greene*.

++ ++ ++ ++ Flowers yellow, 20 to 30 in the oblong heads: akenes fusiform, tapering gradually to a narrow summit, fully as long as the white or whitish and softer pappus: stems scapiform, bearing one or two small leaves toward the base and subulate bracts subtending peduncles or simple branches of the panicle: leaves of radical tuft obovate to spatulate, obtuse, entire or minutely denticulate, contracted into short wing-margined petioles. — § *Chionoracium*, Schultz Bip. in Bonplandia, 1861. *Crepidispermum*, Fries, Symb. *Heteropleura*, Schultz Bip. in Flora, 1861. 434. (Transition to *Crepis*.)

H. Prínglei, GRAY. Strictly scapose, completely destitute of setose hairs and of glands: leaves wholly rosulate, very villous-lanate both sides, obovate (2 or 3 inches long): scape very slender, a foot or more high, minutely soft-pubescent, as also the involucre, loosely paniculate above, bearing few (4 or 5 lines long) and scattered heads: forming akenes somewhat narrowed upward: young pappus soft, bright-white. — Proc. Am. Acad. xix. 69. — S. Arizona, on Santa Rita Mountains, *Pringle, Lemmon*. Specimens too young.

H. Féndleri, SCHULTZ BIP. Subscapose, not rarely one or two leaves toward base of the simple or paniculately branching stem, sparsely setose-hirsute, not at all lanate, not glandular or only obscurely so on the peduncles; radical leaves spatulate or broader; cauline verging to lanceolate, reduced above to linear bracts: heads few and racemiform-paniculate, or more numerous and corymbosely disposed, rather long-pedunculate: involucre half-inch high, of 16 to 24 linear bracts and some short ones. puberulent or glabrate, with or without scattered setose hairs: akenes 2½ to fully 3 lines long, tapering from near the base to summit (at maturity the alternate nerves usually thicker than the others), sometimes reddish, at length commonly blackish: pappus copious, soft, sordid-whitish. — Bonplandia, ix. 173; Gray, l. c. *Crepis ambigua*, Gray, Pl. Fendl. 114. — New Mexico, *Fendler, Wright, G. R. Vasey*. Colorado, *Parry, Hall & Harbour*.

← **Var. díscolor**, GRAY, l. c. Radical leaves (sometimes large, roundish, and over 2 inches broad) purple beneath: pappus nearly pure white. — Santa Rita and Huachuca Mountains, S. Arizona, *Lemmon, Pringle*, the latter distributed as *Hieracium erythrospermum*, Greene, ined., which is the following.

← **Var. Mogollénse**, GRAY, l. c. Leaves narrower, hardly if at all purple-tinged: bristly hairs disposed to be shorter: peduncles minutely and sparsely glandular-setulose: involucre smaller (only 5 lines high): immature akenes reddish: pappus pure white. — *H. brevipilum*, Greene in Bull. Torr. Club, ix. 64, first distributed as "*Hieracium erythrospermum*." — Mogollon Mountains, New Mexico, *Rusby*.

++ ++ ++ ++ ++ Flowers white or flesh-colored: akenes slender-columnar, hardly narrowed upward, about the length of the *bright white soft pappus*: stem leafy. (Transition to *Crepis*.)

H. cárneum, GREENE. Wholly glabrous and smooth except below: stem slender, 2 feet or more high, loosely paniculate-branched, glaucescent, its base and the oblong or lanceolate subsessile radical leaves beset with long villous-setiform hairs: cauline leaves narrowly-lanceolate to linear, entire, very smooth, some of the lower sparsely piliferous: heads scattered in the corymbiform or irregular panicle; involucre campanulate, 4 or 5 lines high, pale, of narrow linear-lanceolate bracts, 15-20-flowered: corollas light rose-color: akenes 2 lines long. — Bot. Gazette, vi. 184; Gray, l. c. 69. — Mountains of New Mexico, *Greene*. Also coll. by *Bigelow* or *Wright*. Huachuca Mountains, S. Arizona, *Lemmon*.

H. Lemmóni, GRAY. Villously or hirsutely setose throughout up to the racemiform close thyrsus: stem simple, 2 feet or more high, very leafy: leaves thinnish, lanceolate-oblong, denticulate with callous or glandular teeth; cauline partly clasping, acute; lowest oblong-spatulate, 4 to 7 inches long, tapering into winged petioles; those of radical cluster wanting: heads numerous and crowded in the oblong thyrsus, 4 lines high, 12-20-flowered: involucre glabrous or nearly so, not glandular, not longer than the canescently puberulent peduncles; its principal bracts narrowly linear, greenish-livid, obtuse: corollas short, seemingly white: akenes hardly 2 lines long, slender, obscurely if at all narrowed upward when mature but obviously so when younger: pappus less copious than in the preceding, bright white. — Proc. Am. Acad. xix. 70. — S. Arizona, at Bear Spring, Cave Cañon, near Fort Huachuca, *Lemmon*. A species of Mexican type, of the group *Thyrsoidea* of Fries.

H. ABSCISSUM, LESS., a Mexican species (with habit of *H. Lemmoni*, but less leafy), probably also including *H. thyrsoideum*, Fries, is said, in Fries, Epicrisis, 150, to come from "Texas ad Malpays de la Joyas" (an unrecognized locality), and from "Alabama."

227. **CRÉPIS**, L. (Name used by Pliny for some now unknown plant, from κρηπίς, a boot or sandal.) — Chiefly a European genus, of annuals or perennials, with soft white pappus and narrow-necked or beaked akenes, some with truncate or merely upwardly attenuate akenes; the involucre apt to be thickened at base, and leaves to be pinnatifid. Flowers in all ours yellow. — Torr. & Gray, Fl. ii. 487; Benth. & Hook. Gen. ii. 513.

* Annuals or hardly biennials, sparingly introduced from Europe: akenes beakless or nearly so: bracts of involucre thickening and becoming more or less rigid at base after anthesis.

C. vírens, L. A foot or two high, erect or ascending: leaves from dentate to laciniate-pinnatifid, spatulate to lanceolate; cauline with sagittate somewhat clasping base; heads slender-peduncled, small: involucre 3 or 4 lines high: akenes oblong, 10-striate, smooth, slightly and about equally contracted at both ends. — Vill. Fl. Delph. iii. 142. *C. polymorpha*, Wallr.; DC. Prodr. vii. 162, mainly. *Maluedthrix crepoides*, Gray, Pacif. R. Rep. xii. 49, & *Crepis Cooperi*, Gray, Proc. Am. Acad. ix. 214, a small and diffuse somewhat naked-stemmed form, with scattered heads. — At landings and near towns on the Columbia River, Oregon and Washington Terr., probably at first a ballast-weed. (Nat. from Eu.)

C. tectórum, L. Usually more slender: leaves narrow, less or not at all sagittate at base: akenes fusiform, with gradually attenuate summit, upwardly scabrous on the ribs. — A ballast-weed at New York Harbor. In fields at Lansing, Michigan. (Nat. from Eu.)

C. biénnis, L. Generally larger, more pubescent or hirsute, leafy-stemmed: leaves runcinate-pinnatifid, or some of the lower spatulate and barely dentate; cauline with sagittate-dentate base: involucre 4 to 6 lines high, broadly campanulate, somewhat canescently pubescent and hispidulous: akenes oblong with narrower summit, 13-striate, smooth. — Engl. Bot. t. 149; DC. Prodr. vii. 163 (excl. var. *Americana*); Reichenb. Ic. Fl. Germ. t. 1439. — Waste grounds, Vermont, *Pringle*. (Nat. from Eu.)

* * Perennials, indigenous westward or northward: akenes beakless or short-beaked.

+ Low or depressed, branched from base, glaucescent and wholly glabrous, bearing numerous clustered and narrow short-peduncled heads: involucre cylindrical, 8-14-flowered, of 8 to 10 smooth and narrowly linear obtuse equal bracts, in a single series (unchanged in fruit except by thickened midrib close to the base in *C. nana*), and 3 or 4 short calyculate ones at base: akenes

Crepis. COMPOSITÆ. 431

narrow, 10-striate, the summit with a more or less dilated disk bearing the soft deciduous pappus. — *Youngia*, Ledeb., &c., not Cass.

C. nána, RICHARDS. Forming depressed tufts on slender creeping rootstocks: leaves chiefly radical (inch or two long, including petiole or attenuate base), obovate to spatulate, entire, repand-dentate, or lyrate, commonly equalling the clustered scapes or stems: heads in fruit half-inch high or nearly: akenes linear, unequally costate, obscurely contracted under the moderately dilated pappiferous disk. — App. Frankl. Journ. ed. 2, 92; Hook. App. Parry Voy. 397, t. 1, & Fl. i. 297; Torr. & Gray, Fl. ii. 488. *Hieracium*, etc., Gmel. Fl. Sibir. ii. 20, t. 7. *Prenanthes pygmæa*, Ledeb. in Mem. Acad. Petrop. v. 553. *P. polymorpha*, Ledeb. Fl. Alt. iv. 144. *Burkhausia nana*, DC. Prodr. vii. 156. *Youngia pygmæa*, Ledeb. Fl. Ross. ii. 838. — Arctic coast and islands, and alpine mountain summits south to Colorado and the Sierra Nevada in California. (N. Asia.)

C. élegans, HOOK. Many-stemmed from a perennial tap-root, a span to a foot high, diffusely branched: leaves entire or nearly so; radical spatulate, cauline from lanceolate to linear: heads smaller or narrower than in the preceding: akenes linear-fusiform, minutely scabrous on the equal narrow ribs, attenuate into a short slender beak, which is discoid dilated at summit. — Fl. i. 297; DC. Prodr. vii. 172; Torr. & Gray, l. c. *Burkhausia elegans*, Nutt. Trans. Am. Phil. Soc. vii. 435. — Saskatchewan district to Dakota and Montana; first coll. by *Drummond*.

  ← ← More robust and taller, with scapiform or few-leaved stems and larger heads: akenes thicker, not discoid-dilated at the insertion of the pappus.

  ++ No furfuraceous or canescent pubescence: foliage mostly glabrous: involucre campanulate, many-flowered; its bracts lanceolate or linear, acute, little thickened below after flowering: thick root possibly biennial, probably perennial: heads few or several and loosely corymbosely cymose: pappus not remarkably copious. — *Crepidium*, Nutt.

C. glaúca, TORR. & GRAY. Usually scapose, a foot or two high, glaucescent or glaucous: radical leaves from obovate-spatulate to lanceolate, from entire to laciniate-pinnatifid: involucre 4 lines high, glabrons or nearly so, as also the peduncles: akenes oblong, with slightly narrowed summit, strongly and evenly 10-costate. — Fl. ii. 438; Eaton, Bot. King Exp. 203; Gray, Bot. Calif. i. 436. *Crepidium glaucum*, Nutt. Trans. Am. Phil. Soc. vii. 436. — Moist and saline ground, Saskatchewan and Nebraska, Utah and Nevada. Probably Arizona (*Rothrock*), but specimen too young and leafy, and peduncles sparingly hispidulous-glandular. *Crepidium caulescens*, Nutt. l. c., is probably a somewhat leafy-stemmed form.

C. runcináta, TORR. & GRAY, l. c. Not glaucous or slightly so, a foot or two high: radical leaves obovate-oblong to oblong-lanceolate, from repand to runcinate-pinnatifid with short lobes or teeth; cauline none, or small and narrow at the forks: involucre half-inch high or smaller, pubescent, often hirsute, sometimes (with peduncles and upper part of scape) glandular-hispidulous: akenes narrowly oblong, moderately narrowed upward, somewhat evenly 10-costate. — *C. biennis*, var., Hook. Fl. i. 297, not L. *C. biennis*, var. *Americana*, DC. Prodr. vii. 163. *Hieracium runcinatum*, James in Long Exped. i. 453; Torr. in Ann. Lyc. N. Y. ii. 209. *Crepidium runcinatum*, Nutt. l. c. — Saskatchewan to Montana and south to Colorado and Utah, in subalpine swamps; first coll. by *James*.

C. Andersóni, GRAY. Not glaucous, a foot or more high; leaves laciniately pinnatifid or dentate, but not runcinate: involucre half to three-fourths inch high, cinereous-pubescent, of broader and firmer bracts, more imbricated, ontermost oblong- to ovate-lanceolate: akenes fusiform, unequally 8-10-costate, tapering into a short but manifest beak. — Proc. Am. Acad. vi. 553, & Bot. Calif. i. 436. — Eastern Sierra Nevada, California and adjacent Nevada, in low grounds: a form with a cauline leaf or two in uplands; first coll. by *Anderson*.

  ++ ++ Furfuraceous- or cinereous-pubescent, at least the foliage, sometimes also hirsute, deep-rooted perennials, more or less leafy-stemmed: akenes oblong to fusiform, beakless, 10-12-costate: pappus of very copious bristles, persistent: bracts of involucre at length with more or less thickened or carinate midrib, at least the base: leaves usually laciniate-pinnatifid. — *Crepis* § *Leptotheca* & *Psilochenia*, Nutt. Trans. Am. Phil. Soc. vii. 437, but false character of akenes of the latter, and outer flowers not sterile. Species difficult.

  = Principal bracts of the narrow involucre and flowers 5 to 8: no hirsute pubescence: pappus moderately copious and soft.

C. acumináta, NUTT. l. c. Minutely cinereously-puberulent below, but green: stem slender, 1 to 3 feet high, 1-3-leaved, bearing a fastigiate or corymbiform cyme of numerous small heads: leaves elongated, slender-petioled, oblong-lanceolate in outline, laciniate-pinnatifid, tapering to both ends, the apex usually into a lanceolate or linear tail-like prolongation (of 2 or 3 inches in length); the lobes also mostly linear-lanceolate, rarely short: involucre narrow-cylindraceous, a third to half inch long, rarely over 6-flowered, their inflorescence smooth and glabrous: the few calyculate bractlets minute and often tomentulose: akenes at maturity rostrate, considerably longer than the pappus, lightly striate-costate, moderately attenuate at summit. — Torr. & Gray, Fl. ii. 489; Torr. in Stansbury Rep. 392, t. viii. (akene too rostrate); Eaton, Bot. King Exp. 204, hardly of Gray, Bot. Calif. — Dry ground, Moutana and Wyoming to E. Oregon, southward to Utah and mountains of S. E. California; first coll. by *Nuttall.*

C. intermédia. Habit and foliage of the preceding, or less tall, more cinereous-puberulent, usually with fewer heads: involucre half-inch or more long, canescently puberulent; its bracts in age more carinate by thickened midrib, the calyculate ones less minute: akenes acutely 10-costate at maturity, oblong-fusiform, slightly attenuate upward, longer than or equalling the pappus. — *C. acuminata,* Gray, Bot. Calif. l. c., partly. Rocky Mountains in Colorado to the Sierra Nevada, California, and north to the interior of Washington Terr. and borders of British Columbia. Appears to pass both into preceding and following.

Var. grácilis. A very slender form, with rhachis and apical prolongation as well as lobes of the leaves attenuate-linear. — *C. occidentalis,* var. *gracilis,* Eaton, Bot. King Exp. 203, mainly.

Var. pleurocárpa. Leaves runcinately dentate, or subpinnatifid, or some entire, not prolonged at apex: akenes merely oblong, hardly narrowed upward, shorter than the pappus, very saliently 10-costate. — Mountains about headwaters of the Sacramento, N. California, *Pringle,* coll. 1881, taken as a well-marked species: but the coll. 1882, distributed as "*C. pleurocarpa,* Gray," accords both to leaves and akenes with *C. intermedia.*

= = Principal bracts of involucre 9 to 24 and flowers 10 to 30: pappus exceedingly copious and pluriserial, rather harsher.

C. occidentális, NUTT. Often hirsute as well as canescent, rather robust, a span to a foot or so high, commonly leafy-stemmed and branching: leaves oblong-lanceolate or broader in outline, variously laciniate-pinnatifid or incised, apex seldom much prolonged: heads few or several, mostly on stout peduncles: involucre half to two-thirds inch high, oblong-cylindraceous to campanulate, canescent: akenes (4 or 5 lines long, longer than the pappus) usually with tapering summit and acute ribs. — Jour. Acad. Philad. vii. 29; Torr. & Gray, Fl. 488; Gray, Bot. Calif. i. 435. *Psilochenia occidentalis,* Nutt. Trans. Am. Phil. Soc. l. c. — Plains of Nebraska and Wyoming to Washington Terr., and south to the mountains of Colorado and California. Varies widely, as into

Var. costáta, GRAY, Bot. Calif. l. c. Low and stouter form, with broader heads, and no hirsute pubescence: involucre oblong, of 10 to 14 bracts: akenes thicker, oblong, sometimes hardly at all narrowed at summit and more saliently costate. — Utah, on Stansbury Island, *Watson.*

Var. Nevadénsis, KELLOGG. Stout, a span or two high, hirsute as well as canescent, or canescent only: leaves broad, disposed to be laciniately bipinnatifid: heads solitary or few, on stout peduncles: involucre campanulate; its principal bracts 16 to 20: akenes gradually narrowed to summit. — *C. occidentalis,* var. *Nevadensis* & var. *subcaulis,* Kellogg, Proc. Calif. Acad. v. 50. Var. *costata* in part, & var. *Nevadensis,* Gray, Bot. Calif. l. c. — High Sierra Nevada, California, *Kellogg, Lemmon,* &c.

Var. crinita, GRAY, Bot. Calif. l. c. Stout, a span to a foot high, barbately and above somewhat viscidly hirsute even to the involucre; this from broadly campanulate to oblong, 20-30-flowered: akenes (as far as seen) oblong, strongly costate, obscurely narrowed at summit. — Sierra Nevada from Sierra Co., *Lemmon, Mrs. Ames,* to Siskiyou Co., *Greene.* Also collected by the Wilkes Expedition, in Washington Territory, or perhaps rather N. California.

228. **PRENANTHES,** Vaill. (Πρηνής, drooping, ἄνθη, blossom.) — Perennial herbs, the original a European species, with loosely paniculate heads, few-nerved akenes, and soft bright white pappus. But the American species all belong to the following well-marked subgenus,

§ 1. NÁBALUS. Endl., with more contracted inflorescence, dull-colored flowers, more nerved akenes (only in the last species tapering at summit), and stiffer sordid pappus. (North American & North Asiatic.) — *Nabalus*, Cass. Dict. xliii. 281; Hook. Fl. i. 293; Torr. & Gray, Fl. ii. 480. *Harpalyce*, Don, in Edinb. Phil. Jour. vi. 305. not DC. Fl. late summer and autumn.

* Heads 20-35-flowered, comparatively broad, corymbosely paniculate: leaves mostly wing-petioled.

P. crepidínea, MICHX. Minutely pubescent or partly glabrous: stem stout, 5 to 9 feet high, branching above, leafy up to the short branches of inflorescence: leaves ample, ovate-deltoid, or radical hastate and uppermost oblong, acutely or laciniately dentate: involucre half to two-thirds inch long, oblong-campanulate, sparsely hirsute: flowers ochroleucous: akenes finely 12-15-costate, four or five of the ribs stronger: pappus sordid. — Fl. ii. 84. *Harpalyce crepidina*, Don ex Steud. *Nabalus crepidineus*, DC. Prodr. vii. 241; Torr. & Gray, l. c. 483. — Rich soil, Penn. and western borders of New York to Illinois and Kentucky; first coll. by Michaux.

* * Heads 8-15-flowered, narrow, crowded or sometimes scattered in an elongated racemiform or thyrsoid-virgate inflorescence which terminates the simple (1 to 5 feet high) stem: cauline leaves sessile; radical and lower tapering into winged petioles, not cordate or deltoid; all simply pinnately veined: root usually fusiform-thickened or tuberous, simple or palmately branched.

+ Thyrsus hirsute or pubescent: heads little or not at all drooping, on pedicels much shorter than the involucre, 12-14-flowered: akenes at maturity about 15-nerved, somewhat angled by four or five of the nerves being stronger: stems leafy up to the strict thyrsus: leaves ordinarily only denticulate, lower spatulate-oblong to obovate.

P. áspera, MICHX. l. c. Minutely scabrous-pubescent or below puberulent: upper leaves lanceolate, not clasping: thyrsus a foot or two long: involucre roughish-hirsute: flowers yellowish cream-color. — *P. Illinoensis*, Pers. Syn. ii. 366; Pursh, Fl. ii. 500. *Chondrilla Illinoensis*, Poir. Suppl. ii. 331. *Nabalus Illinoensis*, DC. Prodr. vii. 242. *N. asper*, Torr. & Gray, l. c. — Prairies and barrens, Ohio and Kentucky to Iowa and Louisiana; first coll. by Michaux.

P. racemósa, MICHX. l. c. Leaves and stem glabrous and glaucous: upper cauline leaves lanceolate to ovate, partly clasping, the broader ones by cordate or auriculate base: thyrsus a span to 2 feet long: involucre rather loosely hirsute: flowers purplish. — *Harpalyce racemosa*, Don ex Steud.; Beck, Bot. 168. *Nabalus racemosus*, DC. l. c.; Torr. & Gray, l. c. — Moist or low ground, N. Maine and Canada, also New Jersey, to Saskatchewan and the Rocky Mountains, south to Colorado; first coll. by Michaux.

Var. pinnatifida. Large: leaves all lyrately or laciniately pinnatifid. — *N. racemosus*, var., Torr. & Gray, l. c. — Hackensack Marshes, New Jersey, *Carey*.

+ + Thyrsus and whole plant smooth and glabrous: heads pendulous and more pedicellate, in a looser racemiform thyrsus, 8-12-flowered: akenes about 5-nerved or angled, the intermediate nerves obscure.

P. Mainénsis. About two feet high, leafy up to and into the panicle: leaves nearly those of *P. racemosa*, but thinner and less glaucous; the radical ovate, commonly with abrupt or rounded base; upper subtending clusters of the interrupted narrow thyrsus: heads all drooping both before and after anthesis, resembling those of the following species. — Shore of the St. John's River, at St. Francis, N. Maine, *Pringle*. Growing with or near *P. racemosa*. And a looser form of the latter, "very common on the St. John's River," *Goodale*, is somewhat between the two; so that this may be a hybrid of *P. racemosa* with *P. serpentaria*.

P. virgáta, MICHX. l. c. Glaucescent, very smooth, 2 to 4 feet high, very strict: radical and lower leaves oblong-lanceolate, deeply sinuate-pinnatifid or pinnately parted, and divisions sometimes lobed or few-toothed; upper not clasping, decreasing to linear-lanceolate and entire, and to small subulate bracts of the naked and slender (1 or 2 feet long) racemiform inflorescence: flowers whitish or pale flesh-color: pappus sordid-stramineous. — Willd. Spec. iii. 1533; Pursh, l. c.; Ell. Sk. ii. 258. *P. autumnalis*, &c., Gronov. Fl. Virg.; Walt. Car. 193. *P. simpler*, Pursh, l. c. *Harpalyce virgata*, Don ex Steud.; Beck, l. c. *Nabalus virgatus*, DC. l. c.; Torr. & Gray, l. c. — Moist ground in pine barrens, New Jersey to Florida, in the low country.

* * * Heads 5-18-flowered, sometimes racemose, usually paniculate, commonly pendulous: leaves diverse, but the cauline nearly all petioled; lower and radical or some of them cordate, or hastate, or truncate at base: root mostly fusiform-thickened or tuberous, as in the preceding: akenes obscurely or minutely striate and sometimes 3-4-costate or angled.

+ Involucre cylindraceous, distinctly calyculate with very short and ovate to triangular-subulate appressed scale-like bracts: principal bracts with their covered margins white-scarious in dried specimens. Species variously called WHITE LETTUCE, LION'S-FOOT, RATTLESNAKE-ROOT, GALL-OF-THE-EARTH.

++ Pappus cinnamon-brown: stem tall, generally purplish.

P. álba, L. Glabrous, often glaucescent, 2 to 5 feet high, rather stout: leaves sometimes all deltoid-hastate and nearly dentate, on slender winged petioles, or uppermost oblong with tapering base, or most of them 3-5-lobed or parted: inflorescence thyrsoid-paniculate: involucre 8-12- (rarely 5-) flowered, commonly purplish-tinged: flowers dull white: pappus reddish-brown. — L. as to Hort. Cliff. & Syn. Pluk., not of herb.; Michx. Fl. ii. 83, in part (not of herb. proper); Sims, Bot. Mag. t. 1079; Pursh, Fl. ii. 499; Bigel. Fl. Bost. ed. 2, 286. *P. rubicunda,* Willd. Spec. iii. 2537, excl. syn.; Pursh, l. c. *P. suavis,* Salisb. Parad. Lond. t. 85. *P. Miamensis, ovata,* & *proteophylla,* Riddell, Syn. W. Pl., to be divided between this and following species. *Harpalyce alba,* Don ex Steud.; Beck, l. c. *Nabalus trifidiolatus,* Cass. Dict. xxxiv. 95. *N. suavis,* DC. l. c. 241. *N. albus,* Hook. Fl. ii. 294, chiefly; Torr. & Gray, Fl. ii. 480, excl. var. — Open oak-woods and sandy or gravelly ground, Canada and New England to Saskatchewan, Illinois, and the upper country of Georgia, &c.

++ ++ Pappus sordid straw-color or whitish: leaves diversely variable, assuming all the forms of the preceding species.

P. serpentária, PURSH. Commonly 2 feet high, glabrous or a little hirsute-pubescent: stem sometimes purple-spotted, rather stout: inflorescence corymbosely thyrsoid-paniculate; the heads mostly glomerate at summit of ascending or spreading flowering-branches or peduncles: involucre green, rarely purplish-tinged, 8-12-flowered: flowers purplish, greenish white, or ochroleucous. — Fl. ii. 499, t. 24; Ell. Sk. ii. 261. *P. alba,* L. Spec., as to Pl. Gronov.; Walt. Car. 193; Ell. Sk. ii. 259. *Harpalyce serpentaria,* Don, l. c.; Beck, l. c. *Nabalus trilobatus,* Cass. Dict. l. c.? *N. serpentarius* (Hook. l. c.), *N. trilobatus,* & *N. Fraseri,* DC. l. c. (*N. glaucus, Prenanthes glauca,* Raf. Fl. Lud. 57, & *Esopon glaucum,* Raf. l. c. 149, has no foundation.) — Open grounds, commonly in sandy or sterile soil, New Brunswick and Canada to Florida.

Var. nána. Stem more simple and strict, 6 to 16 inches high, smooth and glabrous: inflorescence contracted; often sessile or subsessile clusters of heads in the axils of most of the cauline leaves: involucre livid-greenish. — *P. alba,* var. *nana,* Bigel. Fl. Bost. l. c. *Nabalus nanus* (also *N. serpentarius,* var. *lævis*), DC. l. c.; Torr. & Gray, l. c. A form with leaves pedately parted into narrow divisions is probably *N. trifoliolatus,* Cass. — Mountains of N. New England to Canada, S. Labrador, and Newfoundland.

Var. barbáta. Sometimes hirsutulous-pubescent: leaves from oblong to deltoid-hastate, from denticulate to sinuate-lobed, upper not rarely sessile: involucre sometimes sparingly and sometimes copiously beset with bristles. — *P. crepidinea,* Ell. Sk. ii. 259, not Michx. *Nabalus integrifolius* (Cass. l. c. 96, with "subsagittate" leaves), & *N. Fraseri* in part, DC. l. c. *N. Fraseri,* var. *integrifolius* & var. *barbatus,* Torr. & Gray, Fl. ii. 481. *N.* (& *Prenanthes*) *Roanensis,* Chickering in Bot. Gazette, v. 155, vi. 191, a mountain form, a span to a foot high, with most of the leaves hastate-deltoid, and bristles on the involucre copious and conspicuous. — Mountains of N. and S. Carolina to Alabama; apparently first coll. by *Fraser,* then by *Macbride:* the high mountain form by *Chickering* on Roan Mountain. Occasionally a few of these setose hairs are found on the involucre of ordinary *P. serpentaria,* and in this variety some heads are almost destitute of them.

P. altíssima, L. Commonly 3 to 7 feet high, slender, not glaucous, glabrous or nearly so (in open ground sometimes purple-stemmed and hispidulous!): leaves thin, disposed to be hastate, deltoid, or cordate, sometimes ovate, and denticulate or dentate: lower not rarely from 3-lobed to pedately 5-parted: panicle elongated and loose, very commonly subsessile clusters in the axils of many cauline leaves: involucre narrow-cylindrical, greenish, always glabrous, 5-6-flowered: flowers greenish-ochroleucous. — Spec. ii. 797, from char., syn. Pluk. (Alm. t. 317) & Vaill., and perhaps herb. (*P. alba,* L. herb., specimen from Kalm, is either this or

P. serpentaria.) *Harpalyce altissima,* Don, l. c.; Beck. l. c. *Nabalus altissimus,* Hook. Fl. i. 294; DC. l. c.; Torr. & Gray, l. c., with named varieties, *ovatus, cordatus* (*N. cordatus,* Hook., *Prenanthes corduta,* Willd. Hort. Berol. t. 25), *deltoideus* (*N. deltoideus* & *N. cordatus,* DC. l. c., *Prenanthes deltoideus,* Ell. l. c.), & *dissectus,* all too transitional for preservation. — Woods and shaded banks, Newfoundland to Saskatchewan, Pennsylvania, and to the mountains of Georgia.

+ + Involucre campanulate-oblong, of livid or greenish bracts nearly or quite destitute of scarious margins, imperfectly calyculate by 2 or 3 irregular and loose linear accessory bracts, less pendulous than in the preceding: pappus sordid-whitish: plants glabrous or a little pubescent.

P. Boóttii. A span or two high, simple, several-leaved, bearing 7 to 15 racemosely disposed heads: leaves deltoid-oblong and obtuse, or somewhat hastate and the upper acute, on margined petioles, uppermost lanceolate, all entire or denticulate, the lamina only an inch or so long: involucre half-inch long, of 10 to 15 obtuse and rather unequal proper bracts, 10-18-flowered: flowers dull white. — *Prenanthes alba,* var. *nana,* Bigel. Fl. Bost. l. c., in part. *Nabalus Boottii,* DC. Prodr. vii. 241; Torr. & Gray, l. c. 482. — Alpine region, mountains of Maine, New Hampshire, Vermont, and N. New York; first coll. by *Boott* and *Bigelow*.

P. aláta. A foot or two high, the larger plants branching: leaves hastate-deltoid, acute or acuminate, sharply and irregularly dentate, abruptly contracted or some of the upper cuneately decurrent into winged petioles, or small uppermost narrower and sessile by a tapering base: heads loosely and somewhat corymbosely paniculate: involucre of 8 to 10 bracts, 7-15-flowered: flowers purplish: akenes slender, 3 or almost 4 lines long, at least sometimes with tapering summit! — *Sonchus hastatus,* Less. in Linn. vi. 99; Bong. Veg. Sitch. 146. *Nabalus alatus,* Hook. Fl. i. 294, t. 102; Torr. & Gray, Fl. ii. 483. *Mulgedium hastatum,* DC. Prodr. vii. 250. — Unalaska and other Aleutian Islands to Oregon; first coll. by *Chamisso,* &c. (Adj. Asia.)

Var. **sagittáta.** Leaves sagittate or hastate, with the basal lobes mostly slender and prolonged: heads in a virgate panicle: involucre narrower, pale green (not livid), very glabrous, subtended by more numerous slender calyculate bracts: immature akenes little over 2 lines long, not tapering at summit. — Rocky Mountains, N. Montana, in Jocko Cañon, *Watson.* Upper Flathead, *Canby & Sargent.*

229. LYGODÉSMIA, Don. (Λύγος, a pliant twig, and δέσμη, bundle, from the vimineous fasciculate stems of the typical species.) — N. American herbs, mostly smooth and glabrous; with usually rush-like rigid or tough stems, linear or scale-like leaves, and terminal or scattered heads which are always erect: the flowers pink or rose-color, produced in spring or summer. — Don, in Edinb. Phil. Jour. vi. 305; DC. Prodr. vii. 198; Torr. & Gray, Fl. ii. 484; Benth. & Hook. Gen. 530; Gray, Proc. Am. Acad. ix. 217. Genus somewhat polymorphous.

§ 1. Erect perennials, with striate-angled junciform stems and branches, not spinescent, and terminal solitary heads: akenes slender, terete, almost filiform, slightly tapering to summit, 4–8-nerved or at maturity nerveless: pappus soft and copious, whitish or sordid.

L. júncea, Don, l. c. Fastigiately much branched from the deep-rooted base, about a foot high: leaves persistent, small, somewhat nervose: lower lanceolate-linear from a broadish base, inch or two long; upper reduced to small subulate scales: involucre at most half-inch long, 5-flowered: ligules a quarter or third of an inch long. — Hook. Fl. i. 295. t. 103; Torr. & Gray, l. c. *Prenanthes juncea,* Pursh, Fl. ii. 498; Nutt. Gen. ii. 123. — Plains of the Saskatchewan and Minnesota to the Rocky Mountains, New Mexico, and even in Nevada, *Watson.*

L. grandifióra, Torr. & Gray. Stems separate or few from the root, simple below, a span to a foot high; the larger plants leafy, corymbosely branched above, and bearing few or numerous short-pedunculate heads: leaves all entire, of firm and thickish texture, linear-attenuate, 2 to 4 inches long, only the very uppermost reduced to scales: involucre fully three-fourths inch long, 5-10-flowered: ligules of equal length, showy, rose-red. — Fl. ii. 485.

L. juncea, var. *dianthopsis*, Eaton, Bot. King Exp. 200, the well-developed and taller form, leafy to the top. *Erythremia grandiflora*, Nutt. Trans. Am. Phil. Soc. vii. 445, dwarf form. — Gravelly hills or slopes, W. Wyoming (first coll. by *Nuttall*), Utah about Salt Lake (first coll. by *Stansbury*), and S. Utah.

L. aphýlla, DC. l. c. Stems mostly solitary from the root, slender and quite rush-like, 2 feet or so high, naked or nearly so, once or twice forked above, and bearing solitary long-peduncled heads: leaves rather fleshy, chiefly radical or near the ground, filiform, elongated, entire, or with one or two rare teeth ; upper reduced to minute scales at the forks : involucre (mostly 10-flowered) and rose-colored ligules each two-thirds to three-fourths inch long. — Torr. & Gray, l. c.; Chapm. Fl. 251. *Prenanthes aphylla*, Nutt. Gen. ii. 123 ; Ell. Sk. ii. 261. *Erythremia aphylla*, Nutt. Trans. Am. Phil. Soc. l. c. 446. — Dry pine barrens, S. Georgia and Florida; first coll. by *Baldwin*.

Var. Texána, TORR. & GRAY, l. c. Stouter : leaves more numerous, from filiform and usually with 2 or 3 lateral lobes to linear (2 lines broad) and sparingly pinnately lobed, also some smaller leaves on the stem: some Texan specimens nearly like those of Florida. — Rocky hills and plains, Texas ; first coll. by *Berlandier, Drummond*, &c.

§ 2. Diffuse and spinescent perennial, with flexuous branches not striate-angled : pappus rigidulous, whitish, of unequal bristles. — § *Pleiacanthus*, Nutt.

L. spinósa, NUTT. l. c. Stems slender and rigid, low, much branched from an indurated and matted-woolly base, otherwise glabrous : branchlets divergent, spinescent, bearing minute scales in place of leaves and lateral very short-peduncled heads : lower cauline leaves linear, entire, thickish, above soon reduced to scales : involucre 3-5-flowered ; its proper bracts not more numerous, rather loose, lanceolate ; the unequal and more imbricated calyculate ones comparatively broad and large : akenes much shorter than the pappus, not at all narrowed upward, 4-5-costate. — Torr. & Gray, l. c.; Eaton, Bot. King Exp. 200; Gray, Bot. Calif. 441. — Gravelly hills and plains in the arid district, S. Idaho to S. Nevada and the eastern borders of California; first coll. by *Nuttall*.

§ 3. Paniculately branched annuals, not spinescent : pappus white and soft.

L. rostráta, GRAY. Stem erect, 1 to 3 feet high, striate, leafy, corymbose-paniculate : leaves narrowly linear, attenuate to both ends, entire, obscurely 3-nerved ; cauline 3 to 7 inches long, barely 2 lines wide; uppermost slender-subulate : heads numerous, on scaly-bracteolate erect peduncles : involucre over half-inch high, 8-9-flowered, of as many very narrowly linear bracts and a few short calyculate ones : rays small and narrow, probably purplish : akenes slender-fusiform, 4 or 5 lines long, distinctly attenuate at summit but not truly rostrate, 5-8-striate, longer than the soft rather dull-white pappus. — Proc. Am. Acad. ix. 217. *L. juncea*, var. *rostrata*, Gray, Proc. Acad. Philad. 1863, 69. — Plains along the eastern base of the Rocky Mountains, from the Saskatchewan (Cypress Hills, *Macoun*) to Wyoming and Colorado, where first coll. by *Hall & Harbour*.

L. exígua, GRAY, l. c. A span or two high, effusely paniculate from the base, bearing numerous small heads terminating short-filiform divergent branchlets or peduncles : branches not striate : radical leaves spatulate or oblong (about inch long), from nearly entire to runcinate-pinnatifid ; cauline small and entire, soon reduced to minute bracts : involucre oblong, 2 lines high, 4-5-flowered, of as many narrowly oblong bracts and one or two very small calyculate ones : akenes not 2 lines long, gradually tapering from the truncate summit to base, broadly 4-5-costate, or rather narrowly 4-5-sulcate, somewhat longer than the bright white pappus. — *Prenanthes exigua*, Gray, Pl. Wright. ii. 105. — Stony hills, S. E. New Mexico, *Wright*. S. W. Utah, *Parry, Palmer*. Mohave Desert, S. E. California, *Parish*.

230. **TRÓXIMON**, Nutt. (Probably from τρώγω, τρώξομαι, to chew, of no obvious application to this, or to the factitious genus, partly *Krigia*, partly *Scorzonera*, for which Gaertner coined the name.) — N. American with one or two S. American herbs, acaulescent or nearly so, with a cluster of sessile or subsessile radical leaves, and simple scapes bearing a head of yellow or rarely purple flowers, in summer. Occasionally in one species some chaffy bracts among the

flowers. — Nutt. in Fras. Cat. & Gen. ii. 127 : Benth. & Hook. Gen. ii. 522 ; Gray, Proc. Am. Acad. ix. 215, Bot. Calif. i. 437. & Proc. Am. Acad. xix. 71. *Troximon* & *Macrorhynchus* (Less.), DC., Torr. & Gray, Fl. ii. 489. 491.

§ 1. EUTRÓXIMON. Akenes more or less linear, beakless, or tapering gradually into a short and thickish beak, on which the nerves or ribs of the body are prolonged to the apex: pappus rigidulous: perennial from a strong caudex.

* No beak to the akene, its moderately short contracted summit of the same texture as the body and equally 10-costate: involucral bracts somewhat equal, all tapering to a slender acumination, the outer from an oblong or ovate-lanceolate base, glabrous: corolla yellow: root perennial.

T. alpéstre, GRAY. Dwarf from an elongated rootstock or caudex, glabrous: leaves diverse (2 or 3 inches long), narrowly spatulate or lanceolate and pinnately lobed or incised, or parted into narrow linear divisions: scape 2 or 3 inches high, weak: involucre campanulate, 7 or 8 lines high; the bracts in about 2 series: akenes 2½ lines long, equalled by the slender uniform pappus-bristles. — Proc. Am. Acad. xix. 70, 71. — On Mount Paddo, Washington Terr., *Suksdorf*, 1880. Summit of Cascade Mountains, Oregon, *L. F. Henderson*.

T. cuspidátum, PURSH. Glaucescent, tomentulose when young, a span or two or the scape at length a foot high from a thickened caudex: leaves entire, elongated linear-lanceolate and upwardly linear-attenuate, thickish, often nervose, mostly tomentulose-ciliate (2 to 5 lines wide): involucre about inch high; its bracts in 2 or 3 series: akenes 3 or 4 lines long when mature, rather shorter than the unequal rigidulous pappus. — Fl. ii. 472; Torr. & Gray, l. c.; Gray, Man. 277, & Proc. Am. Acad. ix. 215. *T. marginatum*, Nutt. l. c. — Prairies of W. Illinois and Wisconsin to Dakota; first coll. by *Nuttall* and *Bradbury*. Stronger pappus-bristles gradually and slightly widened toward the base.

* * Akenes with apex tapering gradually into a rather stout and nerved beak which is shorter than the body. — § *Nothotroximon*, Gray, Bot. Calif. l. c., partly.

T. barbellulátum, GREENE in herb. Slender, not glaucous: leaves linear-lanceolate, laciniate-pinnatifid into a few short and narrow lobes, or some entire: involucre narrow, over half-inch high, rather few-flowered: its 10 to 12 bracts nearly equal, lanceolate, acuminate, glabrous: flowers yellow: akenes with the beak (of fully half the length of the fusiform body) 3 lines long, about the length of the soft distinctly barbellulate pappus. — Castle Lake, near Mount Shasta, California, *C. H. Dwinelle*, from *Greene*.

T. glaúcum, NUTT. Usually a foot or two high, rather stout, pale or glaucous, either glabrous or with loose pubescence: leaves linear to lanceolate, from entire to sparingly dentate or sometimes laciniate, 4 to 12 inches long: involucre commonly an inch high and many-flowered; its bracts lanceolate or broader: outer series shorter, often pubescent, or even villous: akenes with the beak 5 or 6 lines long, longer than the pappus, the copious and rather rigid bristles of which are (as in most species) only denticulate-scabrous. — Nutt. in Fras. Cat. & Gen. ii. 128; Pursh, l. c.; Sims, Bot. Mag. t. 1667; Torr. & Gray, l. c. *Macrorhynchus glaucus*, Eaton, Bot. King Exp. 204. — Grassy plains, Saskatchewan and Dakota to Brit. Columbia, and mountains of Utah and Colorado; first coll. by *Nuttall* and *Bradbury*.

Var. **parviflórum**. A small and slender form: leaves only 2 to 6 inches long: scape a span to a foot high: head smaller and narrower. — *T. parviflorum*, Nutt. Trans. Am. Phil. Soc. vii. 434. *Macrorhynchus cynthioides*, Hook. Lond. Jour. Bot. vi. 256 ! — Plains of Nebraska and Wyoming to the mountains of New Mexico.

Var. **laciniátum**, GRAY, Bot. Calif. l. c. Dwarf (a span or two high), with the small heads of the preceding variety, varying to larger, glabrous or glabrate, when young often cinereous-pubescent throughout: rays sometimes purplish externally or in fading: leaves mostly of lanceolate outline and laciniate-pinnatifid. — Mountains of Colorado and New Mexico to the higher Sierra Nevada, California. Larger forms pass into the next.

Var. **dasycéphalum**, TORR. & GRAY. Commonly robust, with large and broad heads; the involucre inch broad as well as high, and from villous to cinereous-pubescent, sometimes early glabrate: receptacle not rarely bearing a few chaffy bracts among the flowers: leaves from elongated-lanceolate to oblong-spatulate (the broadest even inch and a half wide), from entire to laciniate or rarely pinnatifid: scape from a span to 2 feet high. — *Ammogeton scorzoneræfolium*, Schrad. Ind. Sem. Hort. Goett. 1833 ; DC. Prodr. vii. 98. *Troximon glaucum*, Richards. App. Frankl. Jour.; Hook. Bot. Mag. 3462. *T. pumilum*, Nutt.

Trans. Am. Phil. Soc. l. c., a dwarf form. *T. taraxacifolium*, Nutt. l. c., a larger form. — Dakota to Saskatchewan and to near Arctic coast, south to the mountains of Colorado, west to the Sierra Nevada and Washington Terr. on the mountains. Passes through smoother and narrowish-leaved forms to the type of this polymorphous species.

§ 2. MACRORHÝNCHUS. Akenes with a slender and mostly filiform nerveless beak and soft pappus. — *Macrorhynchus*, Torr. & Gray, Fl. l. c. *Trochoseris*, Endl. Gen., & Pœpp. & Endl. Nov. Gen. & Spec. iii. 56, t. 263. *Troximon* in part, *Stylopappus*, *Cryptopleura*, & *Kymapleura*, Nutt. Trans. Am. Phil. Soc. vii. 430, 434.

* Perennials, with akene acute or tapering at summit

+— Into a beak not longer or little longer than the cylindraceous or narrowly fusiform body.

T. aurantiacum, HOOK. Loosely soft-pubescent and glabrate: leaves from linear-lanceolate to spatulate, thinnish, entire, or sparingly laciniate-dentate, occasionally pinnatifid: scape from a span to a foot or more high: involucre oblong to campanulate, 7 to 9 lines high; its bracts from broadly to narrowly lanceolate and acute, or outer and looser ones oblong and obtuse: flowers orange, commonly changing to brownish red or purple: akenes thickish, 3 or 4 lines long, and the firm beak only 2 or 3 lines long: pappus somewhat rigidulous. — Fl. i. 300, t. 104. *T. roseum*, Nutt. Trans. Am. Phil. Soc. l. c., a small form. *Macrorhynchus aurantiacus*, Fisch. & Meyer, Ind. Sem. Hort. Petrop. 1837 ? *M. troximoides*, Torr. & Gray, Fl. ii. 491. — Mountain prairies and banks of streams, northern Rocky Mountains to Brit. Columbia and Oregon, perhaps California, and mountains of Colorado.

Var. **purpúreum**, GRAY. Leaves apparently thickish, laciniate, and with the purple-tinged involucre very glabrous or glabrate: "flowers purple." — Proc. Am. Acad. xix. 72. *Macrorhynchus purpureus*, Gray, Pl. Fendl. 114. — Along Santa Fé Creek, New Mexico, *Fendler*. A similar form in mountains of Colorado.

T. grácilens, GRAY. Resembles slender forms of preceding: leaves mostly entire, flaccid, from lanceolate to nearly linear, or some narrowly spatulate: scape 10 to 18 inches high: head and involucral bracts narrow: flowers deep orange: akenes fusiform-linear, 3 or 4 lines long; the very slender beak 4 or 5 lines long: pappus soft, but not flaccid. — Proc. Am. Acad. xix. 71. — Cascade Mountains of Oregon and Washington Terr., *Lyall*, *Nevius*, *Suksdorf*, *Brandegee*. Rocky Mountains in N. Wyoming, *Forwood*.

Var. **Greénei**, GRAY, l. c. A dubious form, smaller: leaves narrowly linear, with a few linear lobes. — N. California, in Scott Mountains, Siskiyou Co., in dry open ground at about 7,000 feet, *Greene*.

T. Nuttállii, GRAY. Resembles broad-leaved forms of *T. glaucum*, robust: leaves thickish, from spatulate to lanceolate, from sparingly dentate to pinnatifid, a span to near a foot long (the thick midrib nervose when dry): scape 6 to 20 inches high: head broad, an inch or more high: involucre more or less pubescent: flowers yellow: thickish akene and beak each 3 or 4 lines long. — Proc. Am. Acad. ix. 216, & Bot. Calif. i. 438 (excl. pl. Nevius). *Stylopappus elatus*, Nutt. l. c. 433. *Macrorhynchus elatus*, Torr. & Gray, l. c. *M. grandiflorus*, Eaton, Bot. King Exp. 206. *Troximon aurantiacum*, Gray, Bot. Calif. l. c., as to Calif. plant. — Low or moist ground, Oregon, and the Sierra Nevada in California to S. Utah; perhaps first coll. by *Nuttall*.

T. apargioides, LESS. Low and tufted from a multicipital lignescent caudex, glabrate: leaves linear or narrowly lanceolate, entire or with a few salient teeth or lobes, or pinnatifid with sparse linear divisions: scapes a span or two high: head half-inch high: involucre campanulate; outer bracts at least pubescent: akenes and beak each 1¼ to 2 lines long: pappus soft, dull white. — Linnæa, vi. 594; Gray, Bot. Calif. l. c., partly. — Sandy soil on and near the coast, San Francisco Bay, &c., California; first coll. by *Chamisso*.

+— +— Beak slender-filiform or almost capillary, 2 to 4 times the length of the short-fusiform or oblong akene (this rarely over 2 lines long): pappus soft and fine, rather flaccid: flowers all yellow. — *Stylopappus*, Nutt. l. c.

++ Pappus about the length of the beak, whitish.

T. húmile, GRAY. Leaves hirsutely pubescent, from spatulate and repand-dentate or lyrate-pinnatifid to lanceolate or broader in outline and pinnately parted into linear lobes: scapes from a span to a foot high, slender: involucre permanently villous with apparently somewhat viscid hairs: ligules exserted: closed head in fruit from half-inch to hardly inch high: filiform beak only about twice the length of the whitish akene. — Proc. Am. Acad. xix. 72. *Leontodon hirsutum*, Hook. Fl. i. 296, therefore *Taraxacum hirsutum*, Torr. & Gray, Fl. ii. 494, ex char. *Borkhausia Lessingii*, Hook. & Arn. Bot. Beech. 145, excl. syn. *Macrorhynchus Lessingii*, Hook. & Arn. l. c. 361, excl. syu., for it is not Lessing's plant described in Linnæa. *M. humile*, Benth. Pl. Hartw. 320, a small form. *M. Harpodii*, Kellogg, Proc. Calif. Acad., a larger form. *Troximon apargioides* in part, Gray, Bot. Calif. l. c. — California near the coast, from Monterey to Washington Terr. Variable in size, the flowering head sometimes nearly as large as in *T. grandiflorum*.

++ ++ Pappus much shorter than the almost capillary beak, usually bright white.

— **T. laciniátum**, Gray, l. c. Smooth and glabrous, or with sparse soft pubescence: leaves elongated-lanceolate, laciniate-dentate or commonly deeply pinnatifid into linear lobes: scapes a foot or two high: involucre glabrous or glabrate, or base of the outer of the lanceolate bracts tomentose: closed head in fruit not over inch high: akene 2 and beak 5 to 7 lines long. — *Stylopappus laciniatus* (original specimen, and one like it from Vancouver's Island, *Lyall*, small and with small immature heads, but apparently of the species) and especially var. *longifolius*, Nutt. l. c. *Macrorhynchus laciniatus*, & var., Torr. & Gray, l. c. *Troximon grandiflorum*, var. *tenuifolium* & var. *laciniatum*, Gray, Bot. Calif. l. c. — Low ground, Brit. Columbia to Oregon, aud California to San Francisco Bay or nearly.

T. grandiflórum, Gray. Leaves hirsutely or cinereons-pubescent, or glabrate, spatulate to lanceolate, sinuate-dentate to laciniate-pinnatifid, or even pinnately parted: scapes stout, a foot or two high : involucre broad, usually well imbricated; the bracts lanate or tomentose when young, often glabrate in age: ligules short: head in fruit an inch to inch and a half high: akene 2 and capillary beak 6 to 8 lines long. — Proc. Am. Acad. ix. 216, & Bot. Calif. l. c., excl. vars. *Stylopappus grandiflorus*, Nutt. l. c. *Macrorhynchus grandiflorus*, Torr. & Gray, l. c. — Plains and moist hillsides, Washington Terr. to S. California, toward the coast. Some forms seem to pass into the preceding.

* * Perennial, with habit of the last preceding species : akene abruptly long-beaked from a broad truncate summit.

— **T. retrórsum**, Gray, l. c. Villous-tomentose when young: leaves pinnately parted into linear-lanceolate usually retrorse lobes, the terminal lobe long and narrow, all callous-tipped : scapes about a foot high: involucre narrowly oblong, 1½ to 2 inches high when mature; its narrow linear bracts hardly surpassed by the soft white pappus: ligules short: akene 3 lines and filiform beak about an inch long. — *Macrorhynchus retrorsus*, Benth. Pl. Hartw. 30; Gray, in Wilkes Exp. xvi. 373. *M. angustifolius*, Kellogg, Proc. Calif. Acad. v. 47. — Open pine woods, California, from Mendocino and the Upper Sacramento to mountains of San Bernardino; first coll. by *Pickering & Brackenridge*, then by *Hertweg*. Also in S. W. Idaho, *Nevius*.

* * * Annuals, slender, mostly low, occasionally subcaulescent: flowers yellow. — *Macrorhynchus*, Less. Syn. 137, but "achenium plano-obcompressum" is erroneous. *Kymapleura* in corrig. (*Macrorhynchus* in text) & *Cryptopleura*, Nutt. Trans. Am. Phil. Soc. vii. 430.

T. heterophýllum, Greene. Somewhat villosely or hirsutely pubescent, or glabrate: leaves from spatulate to linear-lanceolate, denticulate to pinnatifid: scapes a span or two (rarely a foot) high: involucre oblong-campanulate, half to three-fourths inch high; its bracts erect, lanceolate or narrower; outer decidedly shorter than the glabrous inner ones, more or less pubescent with simple or gland-tipped hairs (not villous): akenes various but at most 2 lines long, usually fusiform; filiform beak fully 3 lines long, mostly longer than the white or whitish pappus. — Bull. Torr. Club, x. 88; Gray, Proc. Am. Acad. xix. 72. *T. Chilense*, Gray, Proc. Am. Acad. ix. 216, & Bot. Calif. i. 439. *Macrorhynchus heterophyllus* & *Cryptopleura Californica*, Nutt. l. c. *M. Californicus* & *M. heterophyllus*, Torr. & Gray, Fl. ii. 493. *M. Chilensis*, Hook. Lond. Jonr. Bot. vi. 256. — Open and low ground throughout California, at least near the coast, to Brit. Columbia, and east to Utah. — Varies mainly in the akenes; these generally glabrous, occasionally outer ones pubescent or hirsute; sometimes all alike and from 10-striate to acutely 10-costate; sometimes the outer ones more acutely or even alately costate, and passing into the following forms described by Nuttall, even taken as of different and peculiar genera, but they are rather conditions than varieties.

⎯⎯ **Var. Kymapleúra,** GREENE, l. c. Outermost and sometimes all of the akenes thicker and blunter or truncately obtuse by the development of the ribs into wings, which become sinuously undulate, covering the whole surface. — *M.* (*Kymapleura*) *heterophyllus,* Nutt. l. c., changed in corrig. to *Kymapleura heterophylla.* — Common in California, with other forms.

⎯⎯ **Var. Cryptopleúra,** GREENE, l. c. Some marginal akenes becoming utricular and lightly nerved, enlarging to almost a line in diameter. — *Cryptopleura Californica,* Nutt. l. c. — With other forms, less common.

231. TARÁXACUM, Haller. DANDELION, *i. e.* DENT DE LION. (Ταράσσω, to stir up, alluding to medicinal virtues.) — Perennials, of the northern hemisphere, sending up in spring, from a rosulate cluster of runcinate-pinnatifid or lyrate radical leaves, naked fistulous scapes, which elongate with and after anthesis of the showy head of yellow flowers: involucre reflexed at maturity of the fruit, which, with the expanded pappus, raised on the elongated beak, is displayed in a globose body. The common and only North American, but very polymorphous species, is the following.

⎯⎯ **T. officinále,** WEBER. Root vertical: leaves from spatulate-oblong to lanceolate, from irregularly dentate to runcinate-pinnatifid: akenes oblong-obovate or narrower, squamulose or spinellose-muricate toward the summit, abruptly contracted into a conical or pyramidal apex, which is prolonged into a filiform beak of twice or thrice the length of the akene. In the ordinary form of the fields the involucral bracts are obscurely or not at all corniculate, and the calyculate bracts are linear, elongated, and recurved ; leaves usually lobed. — Weber (not Wiggers) Prim. Pl. Holst. 56; Vill. Dauph.; Koch, Fl. Germ., &c. *T. Dens-leonis,* Desf. Fl. Atl. ii. 228 ; DC. Prodr.; Torr. & Gray, Fl. ii. 494. *Leontodon Taraxacum,* L. *L. officinalis,* Withering. *L. vulgare,* Lam. — Common everywhere in fields and yards, an introduction from Europe: perhaps nowhere here indigenous, but it comes even from Modoc Co., California. (Eu., Asia, &c.)

Var. **alpínum,** KOCH. Outer involucral bracts ovate to broadly lanceolate, spreading, none conspicuously corniculate. — *Leontodon alpinus,* Hoppe. *Taraxacum latilobum,* DC. Prodr. vii. 494 ? — Labrador to Brit. Columbia, and southward along higher mountains to Colorado, Utah, and California.

Var. **glaucéscens,** KOCH. Outer involucral bracts lanceolate to linear, loosely erect or spreading, inner ones and sometimes outer with a corniculate appendage below the tip : leaves generally glaucescent. — *T. corniculatum* and *T. ceratophorum,* DC. l. c. *Leontodon ceratophorum,* Ledeb. Ic. Fl. Alt. t. 34. — Unalaska, &c. (Adj. Asia, Greenland.)

Var. **lívidum,** KOCH. Outer involucral bracts ovate to ovate-lanceolate, all apt to be dark-colored in drying, obscurely or not at all corniculate : leaves from denticulate to runcinate-dentate, sometimes pinnatifid. — *T. palustre,* DC., &c. *T. lanceolatum,* Poir. *T. montanum,* Nutt. Trans. Am. Phil. Soc. vii. 430, not Meyer & DC. *Leontodon lividus,* Waldst. & Kit. — Rocky Mountains, south to New Mexico, north to Arctic coast and islands, and the Aleutian Islands, in various forms. (N. Asia, Eu., Greenland.)

Var. **scopulórum.** Minute: leaves and scape an inch or less long: head 3 or in fruit even 5 lines high, narrow, few-flowered : outer involucral bracts lanceolate, rather loose; inner somewhat corniculate. — *T. lævigatum,* Gray, Proc. Acad. Philad. 1863, 70. — Highest alpine region of the Rocky Mountains in Colorado, *Hall & Harbour, Brandegee.*

T. PHYMATOCÁRPUM, J. Vahl in Fl. Dan. t. 2297, of Greenland, is near var. *lividum,* but the akene is broad and its beak shorter.

232. PYRRHOPÁPPUS, DC. (Πυῤῥός, flame-colored. πάππος, pappus.) — Atlantic N. American and adjacent Mexican herbs; with leafy or sometimes scapiform stems, undivided or pinnatifid leaves, and rather large slender-pedunculate heads of golden yellow flowers, produced in late spring and summer. Principal bracts of the involucre always more or less corniculate behind the tip, in the manner of certain forms of Dandelion. — Prodr. vii. 144 (excl. S. African sp.); Torr. & Gray, Fl. ii. 495; Benth. & Hook. Gen. ii. 523.

* Scapose, monocephalous, perennial by roundish tubers.

P. scapósus, DC. l. c. Hirsutulous-pubescent, low and simple: globular tuber (three-fourths inch in diameter) sending up a slender caudex, bearing at the surface of the ground a cluster of pinnatifid leaves and scapes of a span or two high: the latter simple and naked, sometimes a bract or small leaf near the base: head seldom an inch high in fruit: calyculate bracts of involucre short and small, subulate; principal ones obscurely corniculate at tip; flowers citron-yellow: pappus fulvous. — *P. grandiflorus,* Nutt. Trans. Am. Phil. Soc. l. c. 430; Torr. & Gray, l. c.; Engelm. & Gray, Pl. Lindh. i. 42. *Barkhausia grandiflora,* Nutt. Jour. Acad. Philad. vii. 69. — Prairies of Arkansas and Kansas; first coll. by *Pitcher,* Texas; first coll. by *Berlandier.*

* * More or less leafy-stemmed and branching: heads moderately long-pedunculate.

+— Leaves diversely pinnatifid, laciniate, sinuate-dentate, or some upper ones entire.

P. Caroliniánus, DC. Annual or biennial, freely branching, 2 to 5 feet high, nearly glabrous, but peduncles and involucre mostly cinereous-puberuleut: upper leaves when undivided usually elongated lanceolate and gradually attenuate to the tip: flowers very bright yellow: fruiting heads fully inch high: calyculate bracts setaceous-subulate, loose, half or a third the length of the principal ones; these conspicuously corniculate at the apex: pappus rufous. — Torr. & Gray, l. c.; Nutt. Traus. Am. Phil. Soc. l. c., with var. *maximus.* *P. multicaulis,* Curtiss, Distrib. N. Am. Pl. 1623, not DC. *Leontodon Carolinianum,* Walt. Car. 192. *Scorzonera pinnatifida,* Michx. Fl. ii. 89. *Chondrilla laevigata,* Pursh, Fl. ii. 497. *Barkhausia Caroliniana,* Nutt. Gen. ii. 126; Ell. Sk. ii. 251. — Dry ground, Maryland to Florida, Arkansas, and Texas.

P. multicaúlis, DC. l. c. A foot or two high from a thickened apparently perennial root (but flowering first season), less leafy, at length many-stemmed from base and diffuse or ascending: leaves seldom large: head in fruit two-thirds to three-fourths inch high: calyculate bracts of involucre short and subulate: pappus rufous or fulvous. — Texas (first coll. by *Berlandier*), New Mexico (*Newberry, Greene, Rusby*), and Arizona (*Lemmon*), the latter a dwarf and very narrow-leaved form. (Mex., where *P. pauciflorus* and even *P. Sessilums* are probably forms of it.)

+— +— Leaves all undivided, narrow: stems junciform.

P. Rothróckii, Gray. Glabrous, or involucre obscurely puberulent: stems 1 to 3 feet high, slender, erect from a thickened perennial root: leaves narrowly lanceolate or linear, entire or merely denticulate (3 to 9 inches long, 1½ to 4 lines wide); radical ones spatulate-lanceolate: calyculate bracts of involucre short and subulate: head in fruit only two-thirds inch high: pappus sordid-whitish. — Proc. Am. Acad. xi. 80; Rothrock in Wheeler Rep. vi. 181, t. 14. — Mountains of S. Arizona, *Rothrock, Lemmon.*

233. **CHONDRÍLLA,** L. (Name by Dioscorides, of unexplained meaning, for some gummiferous plant.) — Old World herbs. perennials or biennials; with virgate or rush-like stems and branches, leafy below, and small heads of yellow flowers; one species introduced.

C. júncea, L. Hirsute towards the base, 1 to 3 feet high, glabrous above: lower leaves runcinate; upper linear and entire, those on the long slender branches reduced to linear-subulate bracts: heads scattered or in small clusters and nearly sessile along the branches: akenes somewhat clavate, bearing a circle of scales at base of the filiform beak. — Old fields and banks, S. Maryland and N. Virginia, common about Washington. (Nat. from Eu.)

234. **LACTÚCA,** Tourn. Lettuce. (Ancient Latin name. from *lac,* milk, referring to the milky juice.) — Mostly tall herbs (of the northern hemisphere); with leafy stems, and paniculate middle-sized or small heads of yellow, blue, or sometimes whitish flowers, in summer. Involucre in ours glabrous and smooth. — Benth. & Hook. Gen. ii. 524, excl. § 5, 6. *Lactuca* & *Mulgedium,* Cass., DC., &c.

§ 1. SCARÍOLA, DC. Akenes very flat, orbicular to oblong, abruptly produced into a filiform beak of softer texture, which bears the soft white pappus on its dilated apex: involucre cylindraceous or in fruit conoidal, glabrous: ours biennials or sometimes annuals.

* Introduced: heads 6–12-flowered: akenes several-nerved, margined.

— **L. Scariola, L.** Strict, 2 to 6 feet high, glaucous-green, glabrous except lower part of stem, which has stiff hristles: leaves becoming vertical by a twist, lanceolate to oblong, with spinulose-denticulate margins, sometimes sinuate-toothed, sometimes pinnatifid; midrib beneath beset with weak prickles rather than bristles; base sagittate-clasping: panicle open: heads small: flowers pale yellow: beak about the length of the obovate-oblong striate-nerved akene. — Waste ground, becoming common in Atlantic States near towns and habitations. (Nat. from Eu.)

* * Indigenous: heads 12–20-flowered: akenes blackish, obscurely scabrous-rugulose, lightly one-nerved on the middle of each face, sometimes with obscure nerves toward the distinct thin margins; the beak a little shorter or longer than the body: most of the cauline leaves partly clasping by a sagittate or auriculate base.

+— Involucre irregularly calyculate, but little imbricated, hardly over half-inch long. Species seemingly confluent.

— **L. Canadénsis, L.** (FIRE-WEED, WILD LETTUCE, TRUMPET-WEED.) Glabrous, glaucescent: stem strict, 4 to 9 feet high, very leafy up to the elongated narrow panicle: leaves mostly sinuate-pinnatifid, 6 to 12 inches long, with margins entire or sparingly dentate, and midrib naked or rarely some sparse bristles: involucre half-inch or less high: flowers pale yellow: akenes broadly oval, rather longer than the beak. — Spec. ii. 796; Gray, Man. 280. *L. Caroliniana*, Walt. Car. 193? *L. longifolia*, Michx. Fl. ii. 85. *L. elongata*, Muhl. in Willd. Spec. iii. 1523; Pursh, Fl. ii. 252; Hook. Fl. i. 296; Torr. & Gray, l. c., var. *longifolia*. *Galathenium elongatum*, Nutt. Trans. Am. Phil. Soc. vii. 443. *Sonchus pallidus*, Willd. Spec. iii. 1521; Pursh, l. c., founded wholly on char. of *Lactuca Canadensis*, L. — Rich moist grounds, Nova Scotia and Canada to Saskatchewan, south to the upper part of Georgia. Specimens from a grain-field in Sierra Valley, California, probably introduced with grain.

— **L. integrifólia, Bigel.** Glabrous, less leafy, 3 or 4 feet high, loosely branched above, or heads loosely paniculate: leaves oblong-lanceolate, acuminate (larger 7 to 10 inches long, 1½ to 3 inches broad), whitish beneath, denticulate, sometimes quite entire, all undivided, midrib naked: involucre barely half-inch long: flowers yellow or purplish-tinged: akenes oval, longer than the beak. — Fl. Bost. ed. 2, 287; DC. l. c. 137, not Nutt. *L. sagittifolia*, Ell. Sk. ii. 253. *L. elongata*, var. *integrifolia*, Torr. & Gray, l. c. *L. Canadensis*, var. *integrifolia*, Gray, Man. l. c. *Galathenium integrifolium* and partly *G. salicifolium*, Nutt. l. c. — Open grounds, New England to Illinois and Georgia.

— **L. hirsúta, Muhl.** Stems 2 or 3 feet high, rather few-leaved, often reddish, the naked summit paniculate-branched or bearing a loose panicle of heads, the base commonly hirsute: leaves hirsute on both faces, or glabrous except the hirsute or hispid midrib, mostly runcinate-pinnatifid, with narrow rhachis; cauline 3 or 4 inches long: involucre rather over half-inch long: flowers yellow-purple or dull red, or sometimes whitish: akenes oblong-oval, about the length of the beak. — Cat. & in Nutt. Gen. ii. 124. *L. sanguinea*, Bigel. Fl. Bost. ed. 2, 287. *L. elongata*, var. *sanguinea* & var. *albiflora*, Torr. & Gray, l. c. *L. Canadensis*, Gray, Man. l. c. *Galathenium sanguineum*, & *G. Floridanum*, Nutt. l. c. — Dry and open ground, E. Massachusetts to Louisiana and Texas.

— **L. graminifólia, Michx.** Perhaps perennial, glaucescent and glabrous, or merely hispid on the midrib beneath, or hirsute in the manner of the foregoing species: stem slender, 2 or 3 feet high, terminating in a naked loose panicle of comparatively large heads: leaves elongated-linear or linear-lanceolate (4 to 12 inches long, 2 to 5 lines wide), rather rigid, entire, or with spreading or deflexed lobes, or the radical spatulate-lanceolate and pinnatifid: involucre 6 or 7 lines long, with outer bracts broader and more imbricated: flowers purple or pale blue, varying to white or yellow: akenes elliptical-oblong, longer than the beak. — Fl. ii. 85; Ell. Sk. ii. 253; DC. Prodr. vii. 134; Torr. & Gray, l. c. *L. elongata*, var. *graminifolia*, Chapm. Fl. 252. *L. graminea*, Spreng. Syst. iii. 659. *Galathenium graminifolium* & *G. salicifolium* in part, Nutt. l. c. — Dry and fertile soil, S. Carolina to Florida and Texas; also New Mexico and Arizona.

+ + Involucre more imbricated, commonly three-fourths inch high; outermost and intermediate bracts ovate and ovate-lanceolate.

L. Ludoviciána, DC. Glabrous, leafy to the open panicle, 2 to 5 feet high: leaves all oblong and auriculate-clasping, 3 or 4 inches long, sinuate-pinnatifid, somewhat spinulosely dentate, more or less bristly-ciliate, more or less hispidulous-setose on the midrib beneath: peduncles squamose-bracteolate: flowers yellow; akenes oblong-oval, about equalled by the filiform beak. — Prodr. vii. 141; Torr. & Gray, l. c. *Sonchus Ludovicianus*, Nutt. Gen. ii. 125. *Galathenium Ludovicianum*, Nutt. Trans. Am. Phil. Soc. l. c. — Moist or dry banks of lakes and streams, Dakota, *Nuttall, Geyer*. Iowa, *Arthur*. Black Hills of the Platte, *Hayden*. Rio Limpio, S. W. Texas, *Bigelow*.

§ 2. LACTUCÁSTRUM. Akenes lanceolate-oblong, flat, marginless, tapering into a beak nearly like that of the preceding section, but not longer than the breadth of the body: root perennial: involucre well imbricated, glabrous.

L. pulchélla, DC. A foot or two high, very glabrous, glaucescent, leafy up to the open corymbiform panicle: leaves from linear-lanceolate to narrowly oblong, entire or runcinate-dentate, or some lower ones pinnatifid; cauline sessile, with base not auriculate-clasping, disposed to be vertical: branches of the loose panicle and peduncles squamose-bracteolate: involucre two-thirds inch high, 15-20-flowered; its outer bracts ovate-lanceolate: flowers bright blue or violet-purple; akenes barely 2 lines long, striate-nervose; the tip of short beak soft and usually whitish. — Prodr. vii. 134; Gray, Bot. Calif. i. 442. *L. integrifolia*, Nutt. Gen. l. c., not Bigel. *Sonchus pulchellus*, Pursh, Fl. ii. 502. *S. Sibiricus*, Richards.; Hook. Fl. i. 293, not L. *Mulgedium pulchellum*, Torr. & Gray, Fl. ii. 497. *M. pulchellum* & *M. heterophyllum*, Nutt. Trans. Am. Phil. Soc. vii. 441. — Alluvial ground, Upper Michigan to the Hudson's Bay region in lat. 60°, south to New Mexico, west to Brit. Columbia and mountains of Nevada and adjacent California.

§ 3. MULGÉDIUM. Akenes thickish, oblong, with some strong ribs and nerves, contracted at summit into a stout short beak mainly of the texture of the body, or into a mere (even obscure) neck under the dilated pappiferous apex: involucre (glabrous, 15-25-flowered) and habit of § *Scariola*, or more branching: glabrous biennials or annuals, with or without some hairs or weak bristles on the midrib and veins beneath, commonly with bluish flowers. — Here characterized for the American species only of *Mulgedium*, Cass. (*Agathyrsus*, Don), leaving the older name, *Cicerbita*, Wallr., for the Old World species of less affinity to true *Lactuca*.

* Flowers light blue: pappus bright white: cauline leaves on margined or winged petioles, not clasping nor auriculate at insertion: heads loosely paniculate.

L. Floridána, Gærtn. Stem 3 to 7 feet high: leaves deeply lyrate-pinnatifid; lobes simply or doubly dentate, lateral ones ovate, terminal dilated-deltoid and acuminate: involucre half-inch long: akenes acuminate into a manifest beak. — Gærtn. Fruct. ii. 262, name, but the akenes figured, t. 158, probably from herb. Banks, are of *L. leucophæa*. *Sonchus Floridanus*, L. Spec. ii. 795; Willd. Spec. iii. 1520; Michx. Fl. ii. 85, in part; Ell. Sk. ii. 225. *Mulgedium lyratum*, Cass. Dict. xxxiii. 297. *M. Floridanum*, DC. Prodr. vii. 249; Torr. & Gray, Fl. ii. 498, excl. vars. *Agathyrsus Floridanus*, Beck, Bot. 171. *Galathenium Floridanum*, Nutt. Trans. Am. Phil. Soc. l. c. 443. — Alluvial ground and along streams, Penn. to Illinois, Florida, and Texas.

L. acumináta, Gray. Leaves from ovate-oblong to oblong-lanceolate, acuminate at both ends, or cauline not rarely sagittate or hastate, sharply and sometimes doubly serrate, occasionally some of the lower cleft at base, forming a pair of lateral lobes: involucre 5 lines high: akenes beakless and with hardly a neck: otherwise nearly like the preceding. — Proc. Am. Acad. xix. 73. *L. villosa*, Jacq. Hort. Schœnbr. iii. t. 367; Beck, Bot. 170, but the plant mostly glabrous or nearly so. *Sonchus acuminatus*, Willd. Spec. iii. 1521; Ell. l. c. *S. Floridanus*, Michx. l. c., in part. *Mulgedium acuminatum*, DC. l. c.; Torr. & Gray, l. c. — Borders of woodlands, New York to Illinois and Florida.

* * Flowers bluish to yellowish or whitish: pappus sordid or fuscous: upper cauline leaves sessile by a mostly narrowed but auriculate or partly clasping base: heads in a pyramidal more crowded panicle. — *Mulgedium* § *Agalma*, DC. l. c., in part.

L. leucophǽa, Gray, l. c. Stem 3 to 12 feet high, stout, leafy up to the panicle: leaves ample, sinuately or runcinately pinnatifid, coarsely and irregularly or doubly dentate: involucre oblong, 5 lines high. akenes narrowed at summit into a short but manifest neck. — *L. Canadensis flore leucophœo*, Tourn. *Sonchus alpinus*, L., as to char. (& of Smith, lc. Ined. t. 21), & *S. Canadensis*, L., as to habitat, owing to transposition by Linnæus. *S. spicatus*, Lam. Dict. iii. 401, excl. syn. Walt. *S. racemosus*, Lam. l. c. 400. *S. biennis*, Mœnch, Meth. 545. *S. leucophœus*, Willd. Spec. iii. 1520, excl. syn. Walt. *S. Floridanus*, Ait. Kew. iii. 116, from fruit of which is probably that of *Lactuca Floridana*, Gærtn. t. 158. *S. acuminatus*, Bigel. Fl. Bost. ed. 2, 290. *S. pallidus*, Torr. Compend. 279. *Agathyrsus leucophœus*, Beck, Bot. 170. *Mulgedium leucophœum*, DC. l. c., Nutt. l. c. (§ *Leucomela*); Torr. & Gray, Fl. ii. 499. *M. multiflorum*, DC. l. c. *Sonchus multiflorus*, Desf. Cat. (and so *Galathenium multiflorum*, Nutt. l. c.) is, from sessile cauline leaves, probably this species. — Moist grounds and border of woods, Newfoundland to Canada, Iowa, mountains of Carolina and Tennessee, and northwestward to coast of Oregon and Brit Columbia.

Var. integrifólia. Leaves undivided (simulating those of *L. acuminata*, but sessile), or the lower sinuate-pinnatifid. — *Mulgedium leucophœum*, var. *integrifolia*, Torr. & Gray, l. c. — Ohio, *Lea*. Canton, Illinois, *Wolf*.

L. MACROPHÝLLA, *Sonchus macrophyllus*, Willd., is not known in this country, and is doubtless an Old World species.

L. ALPÍNA, *Sonchus alpinus*, L., is not American. For an account of the early confusion between this and *L. leucophœa*, see the latter species, *supra*, and Torr. & Gray, Fl. ii. 500.

235. SÓNCHUS, Tourn. SOW-THISTLE. (The ancient Greek name.) — Herbs of the Old World, some species now widely diffused, the following naturalized in N. America. Stems leafy: leaves somewhat spinulosely or ciliately dentate: flowers yellow, in summer: pappus white.

* Coarse annual weeds, of cultivated soil and around dwellings; with mostly runcinately or lyrately pinnatifid leaves, of tender texture, beset with soft spinulose serratures; upper cauline auriculate-clasping, and lobes ovate or oblong: heads about half-inch high, somewhat corymbose-paniculate, on short peduncles; these sometimes setose-glandular: akenes flat, thin-edged, oblong-obovate.

S. OLERÁCEUS, L. Leaves with soft or hardly spinulose teeth; auricles of the cauline ones acute: akenes striate-nerved and transversely rugulose-scabrous. — Common in yards and gardens. (Nat. from Eu.)

S. ASPER, VILL. Teeth of the leaves longer and more prickly; auricles of the clasping base rounded: akenes smooth, 3-nerved on each side, margined. — Torr. & Gray, Fl. ii. 501, with syn. *S. Carolinianus*, Walt. Car. 192; Ell. Sk. ii. 255. *S. spinulosus*, Bigel. Fl. Bost. ed. 2, 292. — More common westward and southward, widely dispersed, even to remote districts.

* * Slender annual; with leaves pinnately parted into narrow lobes.

S. TENÉRRIMUS, L. A foot or two high, with rather few and scattered pedunculate heads, glabrous: lobes of the leaves mostly linear or narrowly lanceolate, somewhat spinulosely denticulate: akenes narrow, thickish, rugose-scabrous. — *S. tenuifolius*, Nutt. Trans. Am. Phil. Soc. vii. 438. — San Diego, California, *Nuttall, Orcutt*. (Nat. from Eu.)

* * * Strong-rooted perennial, with deep yellow flowers: akenes thickish.

S. ARVÉNSIS, L. Rootstocks creeping: stems 2 feet high, naked at summit, bearing few or several and corymbosely paniculate showy heads, leaves runcinate-pinnatifid or some undivided, denticulate-spinulose, cauline partly clasping at base: peduncles and involucre more or less glandular-bristly: head almost inch high: akenes oblong, about 10-costate, rugulose on the ribs. — On shores and banks of streams, in several places in N. Atlantic States, and Salt Lake City, Utah. (Nat. from Eu.)

ADDITIONS.

24. PENTACHÆTA, p. 120, after P. aurea, add:

P. Lýoni. Hirsute, at least the margins of the plane linear or spatulate-linear leaves, 4 to 7 inches high, with the sparing ascending branches leafy up to the head or short peduncle: involucre hirsute; its bracts linear-lanceolate and of nearly equal length, green, with narrow scarious margins: pappus-bristles 9 to 11 or commonly 12!— San Pedro, Los Angeles Co., California, in clayey soil near Palos Verdes Mountain, *W. S. Lyon*. An anomalous species, evidently allied to *P. aurea*, notwithstanding the involucre and the more numerous pappus-bristles so repugnant to the generic name.

27. CHRYSOPSIS, § AMMODIA, p. 124, add:

C. Wrightii. Pubescent with fine soft hairs: bracts of the involucre all partly herbaceous, and the inner nearly equalling the flowers: corollas with limb slightly hairy outside: stigmatic portion of the style-branches not much longer than broad, several times shorter than the subulate-linear appendage: outer pappus scanty and obscure; inner extremely copious: otherwise like *C. Breweri*. — S. California, on the San Bernardino Mountains, at 11,500 feet, *W. C. Wright*.

82. FRANSERIA, p. 251, after F. deltoidea, add:

F. cordifólia. Cinereous-puberulent, woody at base, branching above into a narrow and loose panicle: leaves all long-petioled, cordate, obscurely 3–5-lobed, crenately serrate, an inch or two long, thin: fertile involucres granulose-puberulent, the few subulate spines rather shorter than the small body. — Arizona, in the mountains near Tucson, *Pringle, Parish*. (Adj. Mex., *Pringle*.)

137ᵃ. CROCKÉRIA, Greene, Nov. Gen. ined., next to Eatonella, p. 72, 323. (Dedicated by the discoverer to *Charles Crocker, Esq.*, of San Francisco, one of the most liberal and enlightened promoters of botanical investigation in California and adjacent regions.) — Habit, involucre, flowers, and receptacle essentially of *Lasthenia* § *Hologymne*. Akenes oval-obovate, very flat, the plane sides nerveless, glabrous; margins with a distinct filiform nerve, and very densely ciliate with short and pyriform or clavate rather rigid more or less glandular hairs; apex truncate. Pappus none.

C. chrysántha, GREENE, in Bull. Calif. Acad. ined. A span or two high from a slender annual root, nearly glabrous, not at all woolly: leaves all opposite, linear, entire: heads a quarter-inch high: involucre nearly hemispherical, shorter than the disk; the 12 to 14 ovate bracts cupulate-connate to the middle: ray- and numerous disk-flowers golden yellow, and quite like those of *Lasthenia glabrata*. To refer the plant to that genus seems impracticable. — Valley of the San Joaquin, California, in alkaline soil near Lake Tulare, April 15, 1884, *E. L. Greene*.

167. HYMENATHERUM, p. 357, add:

§ 2ᵃ. HETEROCHRÓMEA. Paleæ of the simple pappus 10, little shorter than the slender akene and the disk-corolla, lanceolate, resolved above into 5 or 7 awns, the central one longer, and the lateral successively shorter: rays white!

H. concínnum. Depressed and spreading from the annual root, mostly glabrous, glaucescent: leaves chiefly alternate, thickish, pinnately parted into narrowly linear obtuse and pointless divisions: heads sessile and clustered at summit of the short leafy branchlets: involucre 12-14-toothed, nearly naked at base: rays 10 or 12, the showy oblong ligules (2 lines long) bright white; the disk-flowers yellow. — Arizona, on the borders of Sonora, 1884, *Pringle.* — A handsome species, anomalous for its heterochromous flowers; and in other respects serving to connect the first two sections with true *Hymenatherum*.

192. SENECIO, at end of genus, p. 394, add:

* * * * Indigenous winter annual: heads rayless or with a few minute rays.

S. Mohavénsis. Glabrous, branching from the base, rather slender, leafy to the loose polycephalous panicle: leaves ovate or oblong, sinuate-dentate or sparingly incised, cauline all more or less cordate-clasping or auriculate: heads slender-peduncled, 4 lines high: involucre narrow-campanulate, 18-20-flowered; calyculate bracts few and inconspicuous: ray-flowers when present with corolla commonly biligulate, not surpassing the disk-flowers: akenes canescent. — S. E. California on or near the Mohave and Colorado Rivers, *Lemmon*. (Also within the borders of Sonora, Mex., *Pringle.*)

226: HIERACIUM, to H. Marianum, p. 426, add:

Var. spathulátum. A mountain form leaves all or mainly radical, unusually setose-hirsute or long-villous: scapiform stem simple, 10 to 16 inches high, bearing few rather short-pedicelled heads. — *H. longipilum*, var. *spathulatum*, at foot of p. 426, which is to be cancelled. *Pilosella spathulata*, Schultz Bip. in Flora, 1862, 439 — On Tuscarora or Two-top Mountain, Franklin Co., Pennsylvania, *Porter* and *Traill Green* (1845 and 1884), flowering and fruiting in June.

Page 355. *Nicolletia occidentalis* proves to be a deep-rooted perennial, according to Lemmon and Parish. Possibly *N. Edwardsii* is also perennial.

Page 415. *Rafinesquia Californica* not rarely has pale rose-colored ray-flowers.

ENUMERATION OF GENERA AND SPECIES.

*** The figures give the whole number of species under each genus; those appended in parentheses represent the introduced (naturalized or adventive) species.

Order CAPRIFOLIACEÆ, p. 7.

1. Adoxa 1
2. Sambucus 5
3. Viburnum . . . 14
4. Triostenm 2
5. Linnæa 1
6. Symphoricarpos . . 7
7. Lonicera 15
8. Diervilla 2

Genera, 8. Indigenous species, 47.

Order RUBIACEÆ, p. 19.

1. Exostema 1
2. Pinckneya 1
3. Bouvardia 2
4. Houstonia 13
5. Oldenlandia . . . 3
6. Pentodon 1
7. Hamelia 1
8. Catesbæa 1
9. Randia 1
10. Genipa 1
11. Cephalanthus . . 1
12. Morinda 1
13. Guettarda . . . 2
14. Erithalis 1
15. Chiococca . . . 1
16. Psychotria . . . 2
17. Strumpfia . . . 1
18. Ernodea 1
19. Mitchella . . . 1
20. Kelloggia . . . 1
21. Mitracarpus . . . 1
22. Richardia . . . 1
23. Crusea 3
24. Spermacoce . . . 5
25. Diodia 2
26. Galium . . . (4) 37

Genera, 26. Indigenous species, 82; Naturalized, 4 = 86.

Order VALERIANACEÆ, p. 42.

1. Valeriana 8   2. Valerianella . . (1) 13

Genera, 2. Indigenous species, 21; Naturalized, 1 = 22.

Order DIPSACACEÆ, p. 47.

1. Dipsacus . . . (2) 2

Indigenous species, none; Naturalized, 2.

Order COMPOSITÆ, p. 48.

Tribe I. Vernoniaceæ.

1. Stokesia 1
2. Elephantopus . . . 3
3. Vernonia 10

Tribe II. Eupatoriaceæ.

4. Stevia 6
5. Sclerolepis 1
6. Trichocoronis . . . 2
7. Ageratum . . (1) 3
8. Hofmeisteria . . . 1
9. Mikania 2
10. Eupatorium . . . 39
11. Carminatia . . . 1
12. Kuhnia 2
13. Brickellia . . . 30
14. Carphochæte . . 1
15. Liatris 15
16. Garberia 1
17. Carphephorus . . 5
18. Trilisia 2

Tribe III. Asteroideæ.

19. Gymnosperma . . 1
20. Xanthocephalum . 2
21. Gutierrezia . . . 5
22. Amphiachyris . . 2
23. Grindelia 12
24. Pentachæta . . . 4
25. Bradburia . . . 1
26. Heterotheca . . . 2
27. Chrysopsis . . . 13
28. Acamptopappus . 2
29. Xanthisma . . . 1
30. Aplopappus . . . 43
31. Bigelovia 31
32. Solidago 78
33. Brachychæta . . 1
34. Lessingia 6
35. Bellis 7
36. Aphanostephus . . 4
37. Greenella . . . 2
38. Keerlia 2
39. Chætopappa . . 3
40. Monoptilon . . . 1
41. Dichætophora . . 1
42. Boltonia 3
43. Townsendia . . . 16
44. Corethrogyne . . 3
45. Psilactis 2
46. Eremiastrum . . 1
47. Sericocarpus . . 4
48. Aster 124
49. Erigeron . . . (1) 71
50. Conyza 1
51. Baccharis . . . 19

Tribe IV. Inuloideæ.

52. Pluchea 3
53. Pterocaulon . . . 2
54. Micropus 2
55. Stylocline 3
56. Psilocarphus . . 2
57. Evax 4
58. Filago . . . (2) 5
59. Antennaria . . . 12
60. Anaphalis . . . 1
61. Gnaphalium . . 15
62. Inula . . . (1) 1
63. Adenocaulon . . 1

Tribe V. Helianthoideæ.

64. Plummera . . . 1
65. Dicranocarpus . . 1
66. Guardiola . . . 1
67. Polymnia . . . 2
68. Melampodium . . 4
69. Acanthospermum (2) 2
70. Silphium 11
71. Berlandiera . . . 4
72. Chrysogonum . . 1
73. Lindheimera . . 1
74. Engelmannia . . 1
75. Parthenium . . . 6
76. Parthenice . . . 1
77. Iva 11
78. Oxytenia 1
79. Dicoria 2
80. Hymenoclea . . . 2
81. Ambrosia . . . 8
82. Franseria . . . 11
83. Xanthium . . (2) 3
84. Zinnia 5
85. Sanvitalia . . . 2
86. Heliopsis . . . 4
87. Tetragonotheca . 3
88. Sclerocarpus . . 1
89. Eclipta 1
90. Melanthera . . . 3
91. Varilla 2
92. Isocarpha . . . 1
93. Spilanthes . . . 1
94. Echinacea . . . 2
95. Rudbeckia . . . 21
96. Lepachys . . . 4
97. Wedelia 1
98. Borrichia . . . 2
99. Balsamorrhiza . 8
100. Wyethia . . . 12
101. Gymnolomia . . 5
102. Viguiera . . . 7
103. Tithonia . . . 1
104. Helianthus . . . 40
105. Flourensia . . . 1
106. Encelia 9
107. Helianthella . . 10
108. Zexmenia . . . 3
109. Verbesina . . . 9
110. Actinomeris . . 2
111. Synedrella . . . 1
112. Coreopsis . . . 28
113. Bidens 14
114. Cosmos . . (2) 3
115. Heterospermum . 1
116. Leptosyne . . . 7
117. Thelesperma . . 6
118. Baldwinia . . . 2
119. Marshallia . . . 4
120. Galinsoga . . . 1
121. Blepharipappus . 1
122. Madia 9
123. Hemizonella . . 2
124. Hemizonia . . . 25
125. Achyrachæna . . 1
126. Lagophylla . . 5
127. Layia 13

ENUMERATION OF GENERA AND SPECIES. 449

TRIBE VI. HELENIOIDEÆ.

128. Clappia 1	142. Syntrichopappus . 2	156. Blennosperma . 1
129. Jaumea 1	143. Eriophyllum . . 11	157. Actinella . . . 15
130. Venegasia . . . 1	144. Bahia 10	158. Helenium . . . 20
131. Riddellia . . . 3	145. Amblyopappus . 1	159. Amblyolepis . . 1
132. Baileya 2	146. Schkuhria . . . 2	160. Gaillardia . . . 9
133. Whitneya . . . 1	147. Hymenothrix . . 2	161. Sartwellia . . . 1
134. Laphamia . . . 10	148. Hymenopappus . 7	162. Flaveria . . . 5
135. Perityle . . . 9	149. Florestina . . . 1	163. Porophyllum . . 3
136. Pericome . . . 1	150. Polypteris . . . 4	164. Chrysactinia . . 1
137. Eatonella . . 2	151. Palafoxia . . . 2	165. Nicolletia . . . 2
137ª. Crockeria . . . 1	152. Rigiopappus . . 1	166. Dysodia . . . 3
138. Monolopia . . . 5	153. Chænactis . . . 18	167. Hymenatherum . 14
139. Lasthenia . . . 2	154. Hulsea 6	168. Tagetes 2
140. Burriellia . . . 1	155. Trichoptilium . 1	169. Pectis 12
141. Baeria 17		

TRIBE VII. ANTHEMIDEÆ.

170. Leucampyx . . 1	173. Matricaria . . (1) 3	176. Cotula . . . (2) 2
171. Anthemis . . (2) 2	174. Chrysanthemum (3) 6	177. Tanacetum . (1) 8
172. Achillea . . . 3	175. Soliva . . . (1) 2	178. Artemisia . . (2) 42

TRIBE VIII. SENECIONIDEÆ.

179. Tussilago . . (1) 1	185. Bartlettia . . . 1	190. Raillardella . . 4
180. Petasites . . (1) 4	186. Crocidium . . . 1	191. Arnica 15
181. Cacaliopsis . . . 1	187. Haploesthes . . 1	192. Senecio . . . (3) 57
182. Luina 1	188. Lepidospartum . 1	193. Cacalia 9
183. Peucephyllum . 1	189. Tetradymia . 5	194. Erechtites . . . 1
184. Psathyrotes . . 4		

TRIBE IX. CYNAROIDEÆ.

195. Saussurea . . . 2	198. Cnicus . . . (2) 37	200. Silybum . . (1) 1
196. Arctium . . (1) 1	199. Onopordon . . (1) 1	201. Centaurea . . (7) 8
197. Carduus . . (1) 1		

TRIBE X. MUTISIACEÆ.

202. Hecastocleis . . 1	204. Chaptalia . . . 2	206. Trixis 1
203. Gochnatia . . . 1	205. Perezia 5	

TRIBE XI. CICHORIACEÆ.

207. Phalacroseris . . 1	217. Anisocoma . . 1	226. Hieracinm . . (3) 28
208. Atrichoseris . . 1	218. Hypochœris . (1) 1	227. Crepis . . . (3) 11
209. Lampsana . . (1) 1	219. Microseris . . . 19	228. Prenanthes . . 10
210. Apogon 3	220. Leontodon . . (1) 1	229. Lygodesmia . . 6
211. Krigia 5	220ª. Picris 1	230. Troximon . . . 13
212. Cichorium . . (1) 1	221. Pinaropappus . . 1	231. Taraxacum . . 1
213. Stephanomeria . 13	222. Calycoseris . . 2	232. Pyrrhopappus . 4
214. Chætadelpha . . 1	223. Malacothrix . . 11	233. Chondrilla . . (1) 1
215. Rafinesquia . . 2	224. Glyptopleura . . 2	234. Lactuca . . . (1) 10
216. Tragopogon . (2) 2	225. Apargidium . . 1	235. Sonchus . . (4) 4

Genera, 237. Indigenous species, 1551; Naturalized, 59 = 1610.

INDEX.

NAMES of orders are in CAPITALS; of suborders, tribes, &c., in SMALL CAPITALS; of admitted genera and species, in ordinary Roman type; of synonyms, in *Italic* type.

Abrotanum, 369.
Absinthium, 369.
— *vulgare*, 370.
Acamptopappus, 54, 124.
  Shockleyi, 124.
  sphærocephalus, 124.
Acantholæna, 250.
Acanthospermum, 60, 239.
  australe, 239.
  Brasilum, 239.
  hispidum, 240.
  humile, 240.
  xanthioides, 239.
Acanthoxanthium, 253.
Acarphæa, 342.
  artemisiæfolia, 342.
Achæta, 288.
Achillea, 78, 363.
  borealis, 363.
  gracilis, 363.
  lanulosa, 363.
  Millefolium, 363.
  multiflora, 363.
  occidentalis, 363.
  Ptarmica, 363. *setacea* 363
  tomentosa, 363.
Achyrachæna, 69, 312.
  mollis, 312.
Achyrœa, 124.
Achyropappus, 332.
  Woodhousei, 333.
Aciphyllæa, 357.
  acerosa, 357.
Acmella occidentalis, 258.
  repens, 258.
Acoma dissectum, 301.
Acosmia, 263.
Acourtia, 408.
  microcephala, 409.
Actinea, 344.
Actinella, 75, 344.
  acaulis, 345.
  argentea, 345.
  biennis, 346.
  Bigelovii, 346.
  Brandegei, 345.
  chrysanthemoides, 345.
  depressa, 345.
  glabra, 345.
  grandiflora, 345.
  insignis, 345.
  lanata, 330, 345.
  lanuginosa, 344.
  leptoclada, 345.
  linearifolia, 344.
  odorata, 347.
  Richardsonii, 347.
  Richardsonii, 346.
  Rusbyi, 346.
  scaposa, 344.
  Torreyana, 345.
  Vaseyi, 346.
Actinophoria, 129.
Actinolepis, 328.
  anthemoides, 328.
  coronaria, 327.
  lanosa, 329.
  Lemmoni, 328.
  multicaulis, 328.
  mutica, 328.
  nivea, 328.
  tenella, 327.
  Wallacei, 329.
Actinomeris, 67, 289.
  alata, 287.
  alba, 289.
  alternifolia, 289.
  helianthoides, 288.
  heterophylla, 288.
  longifolia, 287.
  nudicaulis, 288.
  oppositifolia, 288, 289.
  pauciflora, 288.
  squarrosa, 289.
  Wrightii, 289.
Actinospermum
  angustifolium, 303.
ADENOCAULEÆ, 59.
Adenocaulon, 59, 237.
  bicolor, 237.
Adenostyles Nardosmia 376.
Adenophyllum Wrightii, 357.
Adopogon, 412.
Adoxa, 7, 8.
  Moschatellina, 8.
Æthulia uniflora, 92.
Agassizia, 351.
  suavis, 351.
Agarista, 299, 300.
  calliopsidea, 300.
Agathyrsus, 443.
  Floridanus, 443.
  leucophæus, 443.
Ageratum, 51, 93.
  altissimum, 101.
  cœlestinum, 93.
  conyzoides, 93.
  corymbosum, 93.
  lineare, 338.
  littorale, 93.
  maritimum, 93.
  Mexicanum, 93.
  punctatum, 92.
  Wrightii, 92.
Alarconia angustifolia, 268.
  helenioides, 267.
  helianthoides, 267.
Aldama uniserialis, 256.
Alloseris, 412.
Alymnia, 238.
Amauria, 332.
  dissecta, 334.
Amblylepis, 76, 351.
  setigera, 351.
Amblyopappus, 73, 334.
  pusillus, 334.
Ambrosia, 63, 248.
  absinthifolia, 249.
  acanthicarpa, 251.
  aptera, 249.
  artemisiæfolia, 249.
  bidentata, 249.
  cheiranthifolia, 249.
  confertiflora, 250.
  coronopifolia, 250.
  crithmifolia, 250.
  elatior, 249.
  fruticosa, 250.
  glandulosa, 250.
  heterophylla, 249.
  hispida, 250.
  integrifolia, 249.
  Lindheimeriana, 250.
  longistylis, 249.
  longistylis, 250.
  paniculata, 249.
  Peruviana, 250.
  Pitcheri, 246.
  psilostachya, 250.
  pumila, 250.
  tenuifolia, 250.
  tomentosa, 251. *trifida*
AMBROSIEÆ, 62.
Amellastrum, 172.
Amellus Carolinianus, 256.
  spinulosus, 130.
  villosus, 123.
Amida gracilis, 306.
  hirsuta, 306.
Ammodia, 124.
  Oregana, 124.
Ammogeton
  scorzoneræfolium, 438.
Amphiachyris, 53, 116.
  dracunculoides, 116.
  Fremonti, 116.
Amphipappus Fremonti, 116.
Anacalais, 417.

452 INDEX.

Anacis tripteris, 294.
Anaphalis, 59, 233.
— margaritacea, 233.
Ancistrocarphus, 227.
— *filagineus*, 228.
Angelandra, 244.
Anisocarpus Bolanderi, 305.
— *madioides*, 304.
Anisocoma, 85, 416.
— acaule, 416.
Anotis lanceolata, 26.
Antennaria, 58, 231.
— alpina, 232.
— *alpina*, 231.
— argentea, 232.
— Carpathica, 231.
— dimorpha, 231.
— flagellaris, 231.
— Geyeri, 231.
— *hyperborea*, 233.
— *Labradorica*, 232.
— luzuloides, 232.
— *margaritacea*, 233.
— microcephala, 232.
— *monocephala*, 232.
— *parvifolia*, 233.
— plantaginifolia, 233.
— racemosa, 233.
— stenophylla, 231.
ANTHEMIDEÆ, 77, 362.
Anthemis, 78, 362.
— arvensis, 362.
— Cotula, 362.
— nobilis, 363.
— tinctoria, 363.
Apalus, 344.
Apargia autumnalis, 420.
— *borealis*, 424.
Apargidium, 86, 424.
— borealis, 424.
Aparine, 35.
Aphanostephus, 55, 168.
— Arizonicus, 168.
— Arkansanus, 164.
— humilis, 164.
— *pilosus*, 164.
— ramosissimus, 168.
— ramosus, 164.
— *Riddellii*, 163.
Aphantochæta, 119.
— exilis, 120.
Aplodiscus, 142.
Aplodiscus, 125.
Aplopappus, 54, 125.
— acaulis, 132.
— *alpigenus*, 201.
— apargioides, 127.
— armerioides, 132.
— aureus, 129.
— *baccharoides*, 160.
— Berberidis, 126.
— Bloomeri, 134.
— Brandegei, 132.
— canescens, 123.
— carthamoides, 126.
— cervinus, 134.
— ciliatus, 125.
— croceus, 128.
— cuneatus, 133.
— *discoideus*, 143.
— divaricatus, 130.
— ericoides, 133.
— *florifer*, 168.
— Fremonti, 128.
— gracilis, 130.

Greenei, 135.
gymnocephalus, 205.
Hænkei, 170.
Hallii, 129.
hirtus, 127.
Hookerianus, 131.
integrifolius, 128.
inuloides, 128.
lanceolatus, 129.
lanceolatus, 127.
lanuginosus, 131.
laricifolius, 133.
linearifolius, 132.
linearifolius, 222.
Lyallii, 131.
Macronema, 135.
Menziesii, 143.
mollis, 135.
monactis, 133.
multicaulis, 129.
nanus, 134.
Nuttallii, 125.
paniculatus, 127.
Parryi, 131.
pinifolius, 134.
phyllocephalus, 130.
pygmæus, 131.
racemosus, 126.
ramulosus, 223.
resinosus, 124.
rubiginosus, 130.
sphærocephalus, 124.
spinulosus, 130.
squarrosus, 125.
stenophyllus, 132.
suffruticosus, 135.
tenuicaulis, 129.
tortifolius, 173.
uniflorus, 128.
Watsoni, 134.
Whitneyi, 127.
Apogon, 84, 410.
— gracilis, 411.
— humilis, 411.
— *lyratus*, 411.
— Wrightii, 411.
Archieracium, 424.
Arctium, 82, 397.
— *Bardana*, 397.
— Lappa, 397.
— *majus*, 397.
— *minus*, 397.
Arctogeron, 200.
Argyrochæta bipinnatifida, 244.
Armania, 281.
Arnica, 81, 380.
— alpina, 382.
— amplexicaulis, 381.
— *angustifolia*, 382.
— Chamissonis, 381.
— *Chamissonis*, 382.
— *Claytoni*, 380.
— cordifolia, 381.
— discoidea, 381.
— Doronicum, 384.
— foliosa, 382.
— *fulgens*, 383.
— *lanceolata*, 382.
— latifolia, 381.
— *latifolia*, 382.
— longifolia, 382.
— *macrophylla*, 381.
— maritima, 384.
— *Menziesii*, 381.
— mollis, 382.

— *montana*, 382.
— Nevadensis, 382.
— nudicaulis, 380.
— obtusifolia, 383.
— Parryi, 382.
— parviflora, 381.
— *plantaginea*, 383.
— Unalaschensis, 383.
— viscosa, 382.
Arnicella, 128.
Aromia tenuifolia, 334.
Arrhenachne, 221.
Arrow-wood, 10, 225.
Artemisia, 78, 367.
— *abrotanoides*, 370.
— *Abrotanum*, 370.
— *Absinthium*, 370.
— androsacea, 370.
— annua, 370.
— arbuscula, 374.
— arctica, 371.
— biennis, 370.
— Bigelovii, 374.
— Bolanderi, 375.
— borealis, 368.
— cæspitosa, 371.
— Californica, 370.
— *campestris*, 368.
— cana, 375.
— Canadensis, 368.
— *capillifolia*, 97.
— caudata, 368.
— *cernua*, 369.
— *Chamissoniana*, 371.
— Chinensis, 371.
— *Columbiensis*, 375.
— *commutata*, 368.
— corymbosa, 371.
— *cuneifolia*, 372.
— *desertorum*, 368.
— discolor, 373.
— *Douglasiana*, 372.
— dracunculoides, 369.
— *Dracunculus*, 369.
— filifolia, 360.
— *Fischeriana*, 370.
— *foliosa*, 370.
— frauseriodes, 378.
— frigida, 369.
— glauca, 369.
— globularia, 371.
— glomerata, 370.
— *gnaphalodes*, 372.
— *Grænlandica*, 368.
— *heterophylla*, 371, 373.
— *Hispanica*, 370.
— *Hookeriana*, 372.
— incompta, 373.
— *Indica*, var., 372.
— *inodora*, 369.
— *integrifolia*, 372, 373.
— *leptophylla*, 374.
— *leontopodioides*, 271.
— Lewisii, 369.
— *Lindheimeriana*, 372.
— Lindleyana, 373.
— *longepedunculata*, 371.
— longifolia, 372.
— Ludoviciana, 373.
— *matricarioides*, 364.
— Mexicana, 372.
— Norvegica, 371.
— *Nuttalliana*, 369.
— *pachystachya*, 369.
— *Pacifica*, 368.

INDEX. 453

Palmeri, 374.
Parishii, 374.
Parryi, 371.
  pedatifida, 368.
  peucedanifolia, 368.
  Plattensis, 369.
  potentilloides, 367.
  Prescottiana, 373.
  procera, 370. *procerior 97*
  pumila, 373.
  Purshiana, 372.
  pycnocephala, 369.
  pycnostachya, 369.
  Richardsoniana, 371
  rigida, 371. *371*
  Rothrockii, 375.
  rupestris, 371.
  Santonica, 369.
  scopulorum, 369.
  Semavinensis, 370.
  Senjavinensis, 370.
  Senjavinensis, 371.
  sericea, 369.
  serrata, 372.
  spinescens, 368.
  spithamæa, 368.
  Stelleriana, 371.
  Tilesii, 373.
  tridentata, 374.
  trifida, 375.
  trifida, 374, 375.
  trifurcata, 371.
  variabilis, 369.
  violacea, 368.
  virgata, 370.
  vulgaris, 372.
  Wrightii, 373.
-Artichoke, 82, 280.
-*Artorhiza*, 265.
Asperula odorata, 35.
-Aster, 56, 172.
  abbreviatus, 194.
  acuminatus, 199.
  acuminatus, 194.
  adnatus, 180.
  adscendens, 191.
  adscendens, 192, 193.
  adulterinus, 189.
  æstivus, 188, 189, 192.
  alatus, 180.
  albus, 198.
  alpigenus, 201.
  alpinus, 178. *Americanus 171*
  amethystinus, 185.
  amœnus, 195.
  amplexicaulis, 178, 180, 183.
  amplus, 194.
  amygdalinus, 196.
  Andersonii, 201.
  Andinus, 191.
  angustus, 204.
  anomalus, 181.
  annuus, 172, 219.
  arenarioides, 201.
  argenteus, 179.
  argutus, 189.
  artemisiæflorus, 187.
  asperrimus, 178.
  asperugineus, 212.
  asperulus, 181.
  attenuatus, 183.
  auritus, 180.
  azureus, 181.
  Baldwinii, 182.
  bellidiflorus, 188.

bicolor, 146.
biennis, 179, 185, 206.
biflorus, 176.
bifrons, 187.
Bigelovii, 205.
blandus, 195.
blephariphyllus, 202, 205.
Bloomeri, 178.
borealis, 188.
Bostoniensis, 185.
bracteolatus, 191.
brumalis, 189.
cærulescens, 188.
cæspitosus, 190.
Californicus, 208.
campestris, 178.
Canbyi, 193.
canescens, 206.
coricifolius, 202.
carneus, 188.
carnosus, 202.
Caroliuianus, 179.
Chamissonis, 190.
Chapmani, 201.
Chilensis, 190.
chrysanthemoides, 206.
ciliatus, 172.
ciliatus, 180, 185.
ciliolatus, 182.
Collinsii, 171.
Coloradoensis, 205.
commutatus, 185.
concinnus, 183.
concinnus, 178.
concolor, 180.
confertus, 183, 195.
consanguineus, 196.
conspicuus, 177.
conyzoides, 171.
cordifolius, 182.
cordifolius, 174, 182.
coridifolius, 186.
cornifolius, 197.
Cornuti, 194.
corymbosus, 174.
Curtisii, 177.
Cusickii, 195.
cyaneus, 183.
denudatus, 191.
dichotomus, 197.
diffusus, 186.
diffusus, 186, 199.
discoideus, 144.
divaricatus, 172, 174, 197, 199, 203.
divergens, 186.
diversifolius, 181.
Douglasii, 192.
Douglasii, 193, 194.
dracunculoides, 188.
Drummondii, 182.
dumosus, 185.
dumosus, 184, 185.
Durandi, 190.
elegans, 200.
elegans, 176, 184, 199.
Elliottii, 194.
elodes, 130.
eminens, 188, 189, 190.
Engelmanni, 199.
ericæfolius, 198.
ericoides, 184.
ericoides, 185, 186.
eryngiifolius, 173.
Espenbergensis, 176.

exilis, 203.
exscapus, 168.
falcatus, 185.
falcatus, 191.
Fendleri, 178.
filaginifolius, 170.
firmus, 195.
flexuosus, 202.
floribundus, 189.
foliaceus, 193.
foliolosus, 186.
fragilis, 186, 187.
Fremonti, 191.
frondosus, 204. *glabell*
glabriusculus, 200.
glacialis, 201, 208, 209, 211.
glaucescens, 183.
gracilentus, 183.
gracilis, 176.
graminifolius, 216.
grandiflorus, 178.
grandiflorus, 174.
Greenei, 188.
gymnocephalus, 205.
Hallii, 191.
hebecladus, 185.
Herveyi, 175.
hesperius, 192.
heterophyllus, 182.
hiemalis, 189.
hirsuticaulis, 187.
hirtellus, 182.
hispidus, 195.
horizontalis, 187.
humilis, 197.
hyssopifolius, 172.
imbricatus, 201.
incanus, 206.
infirmus, 197.
integrifolius, 177.
junceus, 188.
junceus, 188.
Kingii, 178.
Kumleini, 179.
lævigatus, 183, 189.
lævis, 183.
Lamarckianus, 188.
lanceolatus, 188.
latifolius, 175.
laxifolius, 188, 189, 192.
laxus, 188, 189.
leucanthemus, 187.
ledifolius, 199.
ledophyllus, 200.
Lemmoni, 199.
linariifolius, 197.
Lindleyanus, 182.
linifolius, 172, 204.
longifolius, 188.
longifolius, 189, 190.
lucidus, 195.
lutescens, 199.
macrophyllus, 175.
Marylandicus, 171.
Menziesii, 190.
microphyllus, 180.
mirabilis, 175.
miser, 172, 183, 187.
modestus, 179.
montanus, 176, 179.
multiceps, 179.
multiflorus, 185.
multiflorus, 186, 191.
mutabilis, 172.
mutabilis, 183, 189.

454 INDEX.

mutatus, 179.
Neesii, 190.
nemoralis, 199.
Novæ-Angliæ, 178.
Novi-Belgii, 189.
nudiflorus, 176.
Nuttallii, 178, 191.
obliquus, 188.
oblongifolius, 178.
obovatus, 197.
occidentalis, 192.
onustus, 188, 189.
Oolentangiensis, 181.
Oreganus, 192.
pallens, 194.
Palmeri, 293.
paludosus, 174.
paniculatus, 187.
paniculatus, 181, 182, 189.
Parryi, 200.
parviflorus, 207.
parviflorus, 187.
patens, 180.
patentissimus, 180.
Pattersoni, 205.
Pattersoni, 206.
patulus, 194.
pauciflorus, 202.
pauciflorus, 184.
pendulus, 186.
Pennsylvanicus, 183.
peregrinus, 196.
phlogifolius, 180.
phyllolepis, 179.
pilosus, 184, 185.
politus, 183.
polyphyllus, 184.
Porteri, 184.
præaltus, 188, 189.
præcox, 182, 194.
prenanthoides, 194.
P. escottii, 176.
pulchellus, 201.
pulcherrimus, 197.
puniceus, 195.
puniceus, 194.
purpuratus, 183.
pygmæus, 196.
racemosus, 186.
radula, 176.
radula, 177, 190.
radulinus, 177.
ramulosus, 185, 191.
recurvatus, 188.
Reevesii, 184.
reticulatus, 197.
Richardsonii, 176.
rigidus, 178, 197.
rigidulus, 188.
riparius, 202.
roseus, 178.
rubricaulis, 183.
sagittifolius, 182.
sagittifolius, 181.
salicifolius, 188.
salicifolius, 188, 189.
saluginosus, 196, 208, 209.
Sayanus, 179.
scaber, 181.
scabrosus, 172.
scandens, 179.
scoparius, 185.
scopulorum, 198.
secundiflorus, 186.
sericeus, 179.

serotinus, 189.
Shastensis, 174.
Shortii, 181.
Sibiricus, 176.
simplex, 188, 192.
solidagineus, 171.
solidaginoides, 171.
Sonoræ, 202.
sparsiflorus, 186, 202.
spathulatus, 191.
speciosus, 176.
spectabilis, 178.
spectabilis, 190.
spinosus, 203.
spinulosus, 174.
spurius, 178.
squarrosus, 180.
squarrulosus, 189.
stenomeres, 198.
stenophyllus, 188.
strictus, 176, 188.
subasper, 188.
subspicatus, 193.
subulatus, 204.
subulatus, 203.
surculosus, 176.
tanacetifolius, 206.
tardiflorus, 194.
tardiflorus, 189.
tenuifolius, 202.
tenuifolius, 184, 186, 187, 188.
thyrsiflorus, 190.
Tilesii, 196.
tomentellus, 170.
tortifolius, 172.
Townshendii, 205.
Tradescanti, 187.
Tradescanti, 186, 188, 194.
Tripolium, 202.
turbinellus, 183.
umbellatus, 196.
Unalaschensis, 179, 190, 208.
undulatus, 181.
undulatus, 180.
uniflorus, 199.
urophyllus, 182.
ver nus, 172, 216.
versicolor, 183.
villosus, 184.
vimineus, 186.
vimineus, 183, 194, 195.
virgatus, 183.
virgatus, 184.
virgineus, 189.
Watsoni, 201.
Wrightii, 173.
Xylorrhiza, 202.
ASTEROIDEÆ, 162, 114.
Astranthium integrifolium, 163
Astropolium, 203.
Athanasia graminifolia, 303.
hastata, 257.
paniculata, 289.
trinervia, 303.
Atrichoseris, 84, 410.
platyphylla, 410.
Aurelia, 117.
amplexicaulis, 118.
decurrens, 119.

BACCHARIDEÆ, 57.
Baccharis, 57, 221.
Alamani, 225.
angustifolia, 222.
Bigelovii, 224.

brachyphylla, 223.
cærulescens, 225.
consanguinea, 222.
Douglasii, 224.
Emoryi, 222.
fœtida, 226.
glomeruliflora, 222.
glomeruliflora, 222.
glutinosa, 224.
Hankei, 224.
halimifolia, 222.
Havardi, 224.
juncea, 221.
pilularis, 222.
pilularis, 223.
Plummeræ, 224.
ptarmicæfolia, 224.
pteronioides, 223.
ramulosa, 223.
salicifolia, 222.
salicina, 222.
salicina, 223.
sarothroides, 223.
Seemanni, 221.
sergiloides, 223.
Texana, 222.
thesioides, 224.
veneto, 143.
viminea, 225.
viscosa, 226.
Wrightii, 222.
Baeria, 72, 324.
affinis, 327.
anthemoides, 328.
carnosa, 326.
chrysostoma, 325.
Clevelandi, 326.
coronaria, 327.
curta, 326.
debilis, 325.
Fremonti, 327.
gracilis, 325.
leptalea, 325.
macrantha, 325.
maritima, 326.
microglossa, 324.
mutica, 328.
Palmeri, 326.
platycarpha, 326.
tenella, 327.
uliginosa, 327.
Bahia, 73, 331.
absinthifolia, 332.
achillæoides, 330.
ambigua, 331.
ambrosioides, 332.
arachnoidea, 330.
artemisiæfolia, 329.
Bigelovii, 333.
biternata, 333.
chrysanthemoides, 333.
confertiflora, 330.
cuneata, 331.
dealbata, 332.
gracilis, 331.
integrifolia, 331.
lanata, 330.
latifolia, 330.
leucophylla, 331.
multiflora, 331.
Neo-Mexicana, 333.
nudicaulis, 332.
oblongifolia, 332.
oppositifolia, 332.
parviflora, 331.

INDEX. 455

——rubella, 329.
stæchadifolia, 329.
tenuifolia, 330.
trifida, 330.
—— Wallacei, 329, 331.
Woodhousii, 333.
Bahiopsis lanata, 271.
Baileya, 71, 318.
—— multiradiata, 318.
—— panciradiata, 318.
—— pleniradiata, 318.
- Baldwinia, 68, 302.
—— multiflora, 302.
—— uniflora, 302.
- Balsamorrhiza, 66, 265.
Bolanderi, 266.
Careyana, 265.
—— deltoidea, 266.
—— glabrescens, 266.
—— helianthoides, 266.
hirsuta, 266.
—— Hookeri, 266.
incana, 266.
macrophylla, 266.
—— sagittata, 265.
— Barkhausia Caroliniana, 441.
elegans, 431.
grandiflora, 441.
— Lessingii, 439.
nana, 431.
Barrattia, 283.
calva, 283.
Bartlettia, 80, 378.
scaposa, 378.
—Bedstraw, 35.
—Bellardia, 416.
—Bellis, 55, 163.
—— integrifolia, 163.
—— Mexicana, 163.
—— perennis, 163.
xanthocomoides, 163.
—Berlandiera, 61, 242.
—— incisa, 243.
longifolia, 242.
—— lyrata, 243.
+++ Texana, 242.
—— tomentosa, 243.
Berthelotia, 235.
—Betckea, 44, 46, 47.
—— major, 47.
—— samolifolia, 47.
—Bezanilla, 228.
—Bidens, 68, 235.
arguta, 298.
—— Beckii, 298.
Bigelovii, 297.
—— bipinnata, 297.
— Californica, 297.
—— cernua, 296.
—— connata, 296.
ferulæfolia, 238.
fœniculifolia, 238.
—— frondosa, 296.
—— gracilis, 302.
—— helianthoides, 236.
heterophylla, 238.
heterosperma, 297.
Humboldtii, 298.
Lemmoni, 297.
—— leucantha, 297.
longifolia, 298.
—— nivea, 297.
—— petiolata, 296.
—— pilosa, 297.
procera, 297.

—— quadriaristata, 296.
—— striata, 297.
—— tenuisecta, 297.
—— tripartita, 296.
Bigeluvia, 54, 135.
acradenia, 142.
albida, 137.
arborescens, 141.
Bigelovii, 137.
Bolanderi, 136.
brachylepis, 141.
ceruminosa, 138.
Cooperi, 141.
coronopifolia, 142.
—— coronopifolia, 143.
—— depressa, 137.
diffusa, 141.
—— Douglasii, 137.
—— dracunculoides, 139.
Drummondii, 142.
Engelmanni, 137.
—— graveolens, 139.
Greenei, 138.
—— Hartwegi, 143.
—— Howardi, 136.
—— intricata, 203.
lanceolata, 140.
leiosperma, 139.
juncea, 138.
Menziesii, 143.
—— Missouriensis, 139.
Mohavensis, 138.
—— nudata, 141.
paniculata, 138.
—— Parishii, 141.
—— Parryi, 136.
pluriflora, 142.
pulchella, 137.
—— rupestris, 133.
—— spathulata, 133.
—— teretifolia, 138.
—— tridentata, 143.
—— uniligulata, 154.
Vaseyi, 140.
—— veneta, 142.
—— viscidiflora, 140.
—— Wrightii, 142.
— Biotia, 174.
—— commixta, 175, 176.
—— corymbosa, 175.
—— latifolia, 175.
—— Schreberi, 175.
—Blazing Star, 103.
—Blennosperma, 60, 343.
Californicum, 343.
—Blepharipappus, 69, 304.
—— glandulosus, 314.
—— scaber, 304.
Blepharizonia, 312.
—Blepharodon, 129.
—Bluebottle, 407.
—Bluets, 24.
—Bœbera, 356.
—— chrysanthemoides, 356.
—— glandulosa, 356.
Bœberastrum, 356.
Bolophia, 245.
alpina, 245.
Bolophytum, 245.
—Boltonia, 56, 166.
—— asteroides, 166.
—— diffusa, 166.
—— glastifolia, 166.
—— latisquama, 166.
—Bombycilæna, 227.

—Boneset, 99.
—Borreria, 33.
—— Domingensis, 34.
—— micrantha, 34.
—— parciflora, 34. podocep
—— nubulata, 33.
—Borrichia, 66, 265.
—— arborescens, 265.
—— frutescens, 265.
— Bouvardia, 19, 23.
angustifolia, 24.
—— coccinea, 24.
hirtella, 24.
—— Jacquini, 24.
ovata, 23.
—— quaternifolia, 24.
splendens, 24.
—— ternifolia, 24.
—— triphylla, 23.
—Brachyachyris, 115.
—— Euthamiæ, 115.
—Brachyactis, 204.
—— ciliata, 204.
frondosa, 204.
—Brachychæta, 53, 54, 161.
—— cordata, 161.
Brachycome xanthocomoides, 166.
—Brachyris, 115.
Californica, 115.
—— divaricata, 115.
—— dracunculoides, 116.
—— Euthamiæ, 115.
—— microcephala, 115, 116.
—— ovatifolia, 161.
—— paniculata, 115.
—— ramosissima, 116.
Bradburia, 53, 120.
hirtella, 120.
—Brauneria, 258.
—Breea arvensis, 398.
—Brickellia, 51, 103.
atractyloides, 104.
—— baccharidea, 106.
—— betonicæfolia, 107.
—— brachyphylla, 108.
—— Californica, 106.
—— cordifolia, 105.
—— Conferi, 105.
—— Cusickii, 124.
—— cylindracea, 107.
—— dentata, 106.
Fendleri, 96.
floribunda, 105.
frutescens, 108.
—— grandiflora, 105.
—— Greenei, 104.
—— hastata, 104.
—— incana, 104.
—— laciniata, 106.
Lemmoni, 107.
—— linearifolia, 140.
linifolia, 104.
microphylla, 106.
Mohavensis, 104.
multiflora, 108.
—— oblongifolia, 104.
—— oliganthes, 107.
parvula, 107.
Pringlei, 107.
reniformis, 106.
Riddellii, 108.
Rusbyi, 106.
simplex, 105.
spinulosa, 108.
squamulosa, 108.

tenera, 106.
Wislizeni, 107.
Wrightii, 106.
Wrightii, 105.
Brotera Contrayerba, 354.
— *Sprengelii,* 354.
— *trinervata,* 354.
Broteroa trinervata, 354.
Bulbostylis, 104.
  annua, 377.
  Cavanillesii, 106.
  deltoides, 100.
  microphylla, 106.
  oliganthes, 107.
Buphthalmum angustifolium, 303.
  arborescens, 265.
  frutescens, 265.
  helianthoides, 255.
  repens, 265.
  sagittatum, 266.
Burdock, 397.
Bur-Marigold, 295.
Burrielia, 72, 324.
  chrysostoma, 325.
  Fremonti, 327.
  gracilis, 326.
  hirsuta, 325.
  lanosa, 329.
  leptalea, 325.
  longifolia, 326.
  maritima, 327.
  microglossa, 324.
  nivea, 323.
  parviflora, 326.
  platycarpha, 326.
  tenerrima, 326.
Bush Honeysuckle, 18.
Butter-bur, 375.
Butter-weed, 394.
Button Snakeroot, 109.

Cacalia, 81, 394.
  atriplicifolia, 395.
  cordifolia, 94.
  decomposita, 398.
  diversifolia, 397.
  Floridana, 397.
  gigantea, 397.
  hastata, 397.
  hastata, 398.
  lanceolata, 398.
  Nardosmia, 376.
  ovata, 397.
  ovata, 398.
  paniculata, 398.
  pteranthes, 396.
  suaveolens, 397.
  tuberosa, 396.
Cacaliopsis, 79, 376.
  Nardosmia, 376.
Cachinilla, 225.
Cœnotus, 220.
Calais, 418.
  aphantocarpha, 419.
  Bigelovii, 419.
  Bolanderi, 418.
  cyclocarpha, 420.
  Douglasii, 419, 420.
  eriocarpha, 420.
  glauca, 417.
  graciloba, 417.
  laciniata, 417, 418.
  Lindleyi, 419.
  linearifolia, 418.
  macrochæta, 418.

major, 417.
nutans, 417.
Parryi, 418.
platycarpha, 420.
sylvatica, 417.
tenella, 419.
Calea aspera, 257.
  oppositifolia, 257.
Calliachyris Fremonti, 316.
Callistrum, 175.
Callichroa, 315.
  Douglasii, 316.
  platyglossa, 315.
Calliglossa, 316.
Calliopsis, 290.
  Atkinsoniana, 291.
  bicolor, 291.
  cardaminefolia, 291.
  Drummondii, 291.
  palmata, 293.
  rosea, 290.
  tinctoria, 291.
Calocalais, 418.
Calonea, 351.
Calostelma, 109.
Calycadenia, 310.
  cephalotes, 312.
  Fremonti, 312.
  mollis, 311.
  multiglandulosa, 312.
  pauciflora, 311.
  plumosa, 312.
  tenella, 311.
  truncata, 311.
  villosa, 311.
Calycoseris, 86, 421.
  Parryi, 421.
  Wrightii, 421.
Calymandra, 229.
  candida, 230.
Calyptrocarpus cialis, 289.
Canada Thistle, 390.
CAPRIFOLIACEÆ, 7.
Caprifolium, 16.
  bracteosum, 18.
  ciliosum, 16.
  Douglasii, 17.
  flavum, 17.
  Fraseri, 17.
  hispidulum, 18.
  glaucum, 18.
  gratum, 18.
  occidentale, 16.
  parviflorum, 18.
  pubescens, 17.
  sempervirens, 16.
Carbenia, 405.
Cardoon, 82.
CARDUINEÆ, 82.
Carduus, 82, 397.
  acanthoides, 397.
  altissimus, 404.
  arvensis, 398.
  crispus, 398.
  discolor, 397, 403, 404.
  Douglasii, 403.
  foliosus, 402.
  glaber, 404, 405.
  horridulus, 400.
  lanceolatus, 398.
  muticus, 405.
  nutans, 397.
  occidentalis, 401.
  odoratus, 401.
  pectinatus, 397.

pumilus, 401, 402.
pycnocephalus, 397.
remotifolius, 399.
repandus, 405.
spinosissimus, 400.
undulatus, 403.
Virginianus, 404, 405.
Carminatia, 51, 103.
  tenuiflora, 103.
Carphephorus, 52, 112.
  atriplicifolius, 113.
  bellidifolius, 113.
  corymbosus, 113.
  junceus, 113.
  junceus, 378.
  Pseudo-Liatris, 113.
  tomentosus, 113.
Carphochæte, 52, 109.
  Bigelovii, 109.
Catesbæa, 20, 28.
  parviflora, 28.
Catamenia, 286.
Centaurea, 82, 405.
  Americana, 407.
  benedicta, 406.
  Calcitrapa, 406.
  Cyanus, 407.
  Jacea, 407.
  Melitensis, 406.
  Mexicana, 407.
  nigra, 406.
  Nuttallii, 407.
Centauridium Drummondii, 125.
CENTAURINEÆ, 82.
Centrocarpha aristata, 260.
  grandiflora, 261.
  triloba, 259, 260.
Centrospermum humile, 240.
  xanthioides, 239.
Cephalanthus, 20, 29.
  occidentalis, 29.
  salicifolius, 29.
Cephalophora acaulis, 345.
  decurrens, 349.
  scaposa, 344.
Ceratocephalus, 296.
Cercomeris, 249.
Cercostylis, 351.
Chænactis, 74, 339.
  achillæfolia, 341.
  attenuata, 340.
  artemisiæfolia, 342.
  brachypappa, 340.
  carphoclinia, 340.
  Douglasii, 341.
  filifolia, 330.
  Fremonti, 340.
  glabriuscula, 339.
  glabriuscula, 341.
  heterocarpha 339.
  lanosa, 339.
  macrantha, 341.
  Nevadensis, 341.
  Nevii, 339.
  santolinoides, 341.
  stevioides, 340.
  suffrutescens, 341.
  tanacetifolia, 339.
  tenuifolia, 339.
  thysanocarpha, 342.
  Xantiana, 340
Chænolobus pycnostachyus, 226.
  virgatus, 226.
Chætadelpha, 84, 415.
  Wheeleri, 415.

INDEX. 457

— *Chætanthera asteroides*, 165.
— Chætopappa, 55, 165.
    asteroides, 165.
    modesta, 165.
    Parryi, 165.
— *Chæthymenia*, 317.
— *Chamædaphne*, 31.
— *Chomæmelum*, 363.
— Chamomile, 362, 363.
— Chaptalia, 83, 408.
    nutans, 408.
    tomentosa, 408.
— Chiccory, 412.
— Chiococca, 21, 3.
    parvifolia, 30.
    racemosa, 30.
— *Chionoracium*, 429.
— Chondrilla, 87, 441.
    Ill.noensis, 433.
    juncea, 441.
    lævigata, 441.
Chorisiva, 247.
— Chrysactinia, 77, 355.
    Mexicana, 355.
— Chrysanthemum, 78, 364.
    arcticum, 365.
    Balsamita, 365.
    bipinnatum, 364.
    Carolinianum, 166.
    grandiflorum, 364.
    Indicum, 364.
    inodorum, 363.
    integrifolium, 365.
    Leucanthemum, 365.
    nanum, 364.
    Parthenium, 365.
    segetum, 364.
    Sineuse, 361.
— *Chrysastrum*, 144.
Chrysocoma acaulis, 91.
    coronopifolia, 97.
    dracunculoides, 139.
    gigantea, 90.
    graminifolia, 90, 161.
    graveolens, 139.
    nauseosa, 139.
    tomentosa, 89.
    virgata, 141.
— Chrysogonum, 61, 243.
    Diotostephus, 243.
    Virginianum, 243.
Chrysoma, 161.
    pumila, 160.
    solidaginoides, 161.
    uniligulata, 154.
— *Chrysomelea*, 292.
    lanceolata, 292.
— Chrysopsis, 53, 121.
    acaulis, 132.
    albo, 198.
    alpina, 198.
    amygdalina, 196.
    argentea, 121.
    aspera, 121.
    Bolanderi, 123.
    Breweri, 124.
    cæspitosa, 132.
    canescens, 213.
    coronopifolia, 206.
    decumbens, 122.
    dentata, 122.
    divaricata, 130.
    echioides, 123.
    falcata, 123.
    hirtella, 210.

hispida, 123.
humilis, 197.
hyssopifolia, 122.
gossypina, 122.
graminifolia, 121.
Lamarckii, 130.
linariifolia, 197.
Mariana, 122.
mollis, 123.
oligantha, 121.
Oregana, 124.
pilosa, 124.
pinifolia, 121.
scabra, 121.
scabrella, 122.
sessiliflora, 123.
trichophylla, 122.
villosa, 122.
Wrightii, 145.
— *Chrysostemma triypteris*, 294.
— Chrysothamnopsis, 136.
Chrysothamaus, 137.
— Chrysothamaus, 136.
    depressus, 137.
    dracunculoides, 139.
    lanceolatus, 140.
    pulchellus, 137.
    pumilus, 140.
    speciosus, 137, 139.
    viscidiflorus, 140.
— *Chthonia prostrata*, 360.
— CICHORIACEÆ, 83, 410.
— Cichorium, 84, 412.
    Intybus, 412.
— Cinchona Caribæa, 23.
    Caroliniana, 23.
    Jomaicensis, 23.
— CINCHONACEÆ, 19.
— Cineraria atropurpurea, 389.
    campestris, 388.
    Canadensis, 383, 394, 396.
    Corolinensis, 383.
    congesta, 394.
    frigida, 389.
    heterophylla, 390.
    integrifolia, 389, 390.
    Lewisii, 211.
    lyrata, 390.
    maritima, 383.
    palustris, 394.
— Cirsium, 398.
    acaule, 402.
    altissimum, 404.
    arvense, 398.
    Bigelovii, 405.
    brevifolium, 403.
    Californicum, 400.
    canescens, 400, 403.
    Coulteri, 401.
    discolor, 404.
    diversifolium, 404.
    Drummondii, 400, 402.
    edule, 399.
    eriocephalum, 399, 400.
    filipendulum, 404.
    foliosum, 402.
    Grahami, 403.
    Hookerianum, 403.
    horridulum, 400.
    Kamtschaticum, 400.
    lanceolatum, 398.
    Lecontei, 405.
    megacanthum, 400.
    muticum, 405.
    Neo-Mexicanum, 400.

— Nuttallii, 404.
— *ochrocentrum*, 403.
    Pitcheri, 403.
    pumilum, 401.
    remotifolium, 399.
    repandum, 405.
    scariosum, 402.
    stenolepidum, 399.
    Texanum, 404.
    undulatum, 401, 403.
    Virginianum, 404.
    Wrightii, 404.
Clappia, 70, 317.
    aurontiaca, 317.
    suædæfolia, 317.
— Clarionea, 408.
    runcinata, 409.
— Clavigera, 104.
    dentata, 108.
    Riddellii, 108.
    trachyphylla, 108.
— Cleavers, 35.
— Clomenocoma, 356.
— Clot-bur, 252.
— Cnicus, 82, 397.
    altissimus, 404.
    Americanus, 398.
    Andersoni, 401.
    Andrewsii, 400.
    Arizonicus, 401.
    arvensis, 398.
    benedictus, 406.
    Breweri, 403.
    Californicus, 400.
    carlinoides, 398.
    Drummondii, 402.
    Eatoni, 400.
    edulis, 399.
    eriocephalus, 399.
    foliosus, 402.
    glaber, 404.
    glutinosus, 405.
    Grahami, 403.
    Hallii, 399.
    Hookerianus, 399.
    horridulus, 400.
    Kamtschaticus, 399.
    lanceolatus, 398.
    Lecontei, 405.
    muticus, 405.
    Neo-Mexicanus, 400.
    Nuttallii, 404.
    occidentalis, 401.
    ochrocentrus, 403.
    Parryi, 398.
    Pitcheri, 403.
    pumilus, 401.
    quercetorum, 402.
    remotifolius, 399.
    repandus, 405.
    Rothrockii, 401.
    scariosus, 402.
    spinosissimus, 400.
    undulatus, 403.
    Virginianus, 404.
    Virginianus, 405.
    Wheeleri, 402.
    Wrightii, 404.
— Cockle-bur, 252.
Cœlestina, 93.
    ageratoides, 93.
    cærulea, 93, 102.
    corymbosa, 93.
    maritima, 93.
— COFFEACEÆ, 20.

458　INDEX.

- *Coinogyne carnosa*, 317.
- *Coleosanthus*, 104.
- Coltsfoot, 375.
- Compass-plant, 242.
- COMPOSITÆ, 48.
- Cone-flower, 259.
- *Coniothele Californica*, 344.
- Conoclinium, 102.
  - *betonicum*, 102.
  - *cœlestinum*, 102.
  - *dichotomum*, 102.
  - *dissectum*, 102.
  - *rigidum*, 95.
- Conophora, 395.
- Conyza, 57, 221.
  - *Altaica*, 204.
  - *ambigua*, 221.
  - *amplexicaulis*, 226.
  - *angustifolia*, 226.
  - *asteroides*, 171.
  - *bifoliata*, 172.
  - *bifrons*, 226.
  - *camphorata*, 226.
  - *Carolinensis*, 226.
  - *Coulteri*, 221.
  - *Coulteri*, 220.
  - *linifolia*, 171.
  - *Marylandica*, 226.
  - *pycnostachya*, 226.
  - *sinuata*, 221.
  - *subdecurrens*, 220, 221.
  - *uliginosa*, 226.
  - *virgata*, 226.
- CONYZEÆ, 57.
- Conyzopsis, 204.
- Coral-berry, 13.
- Coreocarpus *dissectus*, 301.
  - *heterocarpus*, 301.
  - *parthenioides*, 301.
- *Coreoloma*, 290.
- COREOPSIDEÆ, 67.
- *Coreopsides*, 292.
- *Coreopsidium*, 291.
- Coreopsis, 68, 289.
  - *acuta*, 289.
  - *aluta*, 287.
  - *alba*, 297.
  - *alternifolia*, 289.
  - *angustifolia*, 290.
  - *angustifolia*, 273.
  - *aristata*, 295.
  - aristosa, 295.
  - aspera, 290.
  - Atkinsoniana, 291.
  - *aurea*, 294.
  - *aurea*, 295.
  - auriculata, 293.
  - *auriculata*, 293.
  - *Bidens*, 296.
  - bidentoides, 295.
  - *Boykiniana*, 292.
  - calliopsidea, 300.
  - cardaminefolia, 291.
  - coronata, 292.
  - *coronata*, 294, 295.
  - *crassifolia*, 292.
  - delphinifolia, 293.
  - *dichotoma*, 290.
  - discoidea, 295.
  - *diversifolia*, 291, 293.
  - Drummondii, 291.
  - flexicaulis, 290.
  - gladiata, 290.
  - grandiflora, 292.
  - Harveyana, 292.

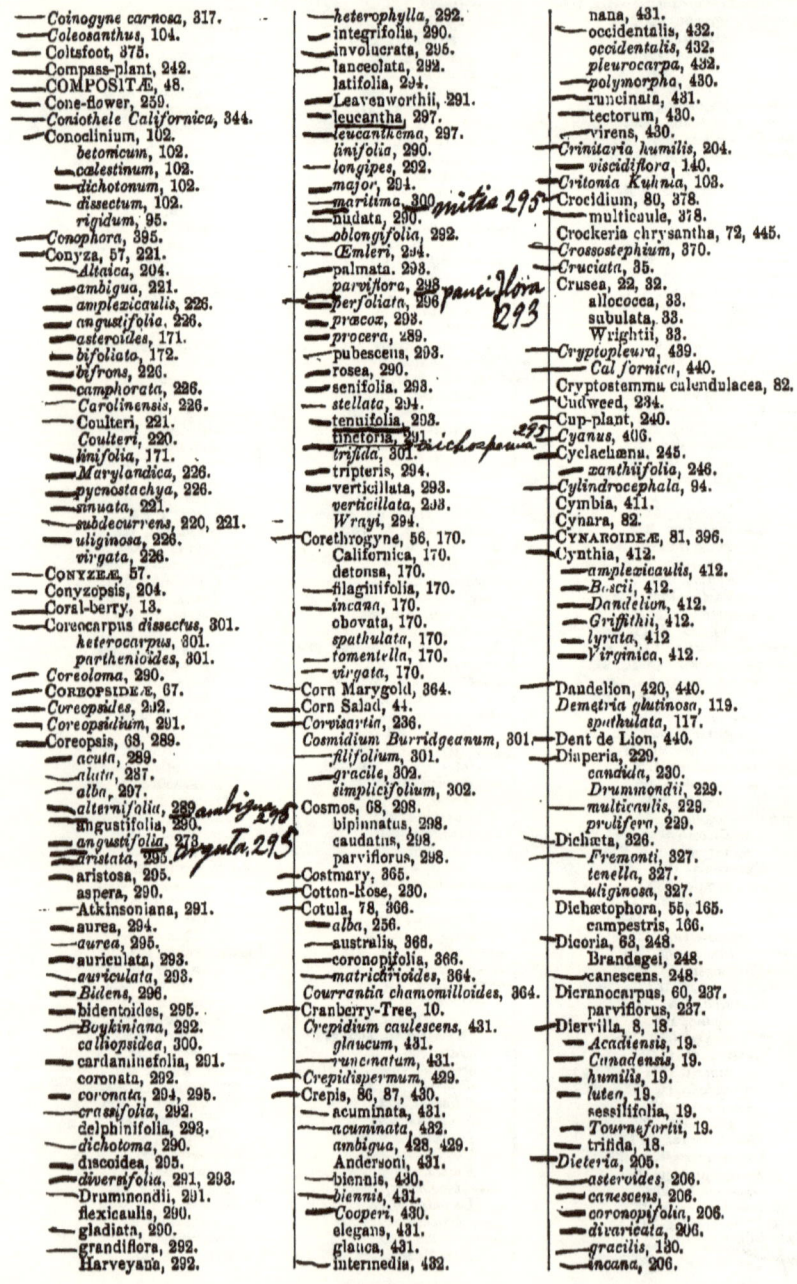

- *heterophylla*, 292.
- integrifolia, 290.
- involucrata, 295.
- lanceolata, 292.
- latifolia, 294.
- Leavenworthii, 291.
- leucantha, 297.
- *leucanthema*, 297.
- *linifolia*, 290.
- *longipes*, 292.
- *major*, 294.
- *maritima*, 300. *mitis 295*
- nudata, 290.
- *oblongifolia*, 292.
- *Œmleri*, 294.
- palmata, 293.
- *parviflora*, 298. *pauciflora*
- *perfoliata*, 296.
- *præcox*, 293. *293*
- procera, 289.
- pubescens, 293.
- rosea, 290.
- senifolia, 293.
- *stellata*, 294.
- tenuifolia, 293.
- tinctoria, 291. *trichosperma 295*
- *trifida*, 301.
- tripteris, 294.
- verticillata, 293.
- *verticillata*, 293.
- *Wrayi*, 294.
- Corethrogyne, 56, 170.
  - Californica, 170.
  - detonsa, 170.
  - filaginifolia, 170.
  - *incana*, 170.
  - obovata, 170.
  - spathulata, 170.
  - *tomentella*, 170.
  - *virgata*, 170.
- Corn Marygold, 364.
- Corn Salad, 44.
- *Corvisartii*, 236.
- Cosmidium Burridgeanum, 301.
  - *filifolium*, 301.
  - *gracile*, 302.
  - simplicifolium, 302.
- Cosmos, 68, 298.
  - bipinnatus, 298.
  - caudatus, 298.
  - parviflorus, 298.
- Costmary, 365.
- Cotton-Rose, 230.
- Cotula, 78, 366.
  - *alba*, 366.
  - australis, 366.
  - coronopifolia, 366.
  - *matricarioides*, 364.
- *Courrantia chamomilloides*, 364.
- Cranberry-Tree, 10.
- *Crepidium caulescens*, 431.
  - *glaucum*, 431.
  - *runcinatum*, 431.
- *Crepidispermum*, 429.
- Crepis, 86, 87, 430.
  - acuminata, 431.
  - *acuminata*, 432.
  - ambigua, 428, 429.
  - Andersoni, 431.
  - biennis, 430.
  - *biennis*, 431.
  - *Cooperi*, 430.
  - elegans, 431.
  - glauca, 431.
  - intermedia, 432.

- nana, 431.
- *occidentalis*, 432.
  - *occidentalis*, 432.
  - *pleurocarpa*, 432.
  - *polymorpha*, 430.
  - *runcinata*, 431.
  - *tectorum*, 430.
  - virens, 430.
- *Crinitaria humilis*, 204.
  - *viscidiflora*, 140.
- *Critonia Kuhnia*, 103.
- Crocidium, 80, 378.
  - multicaule, 378.
- Crockeria chrysantha, 72, 445.
- *Crossostephium*, 370.
- *Cruciata*, 35.
- Crusea, 22, 32.
  - allococca, 33.
  - subulata, 33.
  - Wrightii, 33.
- *Cryptopleura*, 439.
  - *Californica*, 440.
- Cryptostemma calendulacea, 82.
- Cudweed, 234.
- Cup-plant, 240.
- *Cyanus*, 406.
- Cyclachænu, 245.
  - *xanthiifolia*, 246.
- *Cylindrocephala*, 94.
- Cymbia, 411.
- Cynara, 82.
- CYNAROIDEÆ, 81, 396.
- Cynthia, 412.
  - *amplexicaulis*, 412.
  - *Bascii*, 412.
  - *Dandelion*, 412.
  - *Griffithii*, 412.
  - *lyrata*, 412.
  - *Virginica*, 412.
- Dandelion, 420, 440.
- Demetris glutinosa, 119.
  - spathulata, 117.
- Dent de Lion, 440.
- Diaperia, 229.
  - candida, 230.
  - *Drummondii*, 229.
  - *multicaulis*, 229.
  - *prolifera*, 229.
- Dichæta, 326.
  - *Fremonti*, 327.
  - *tenella*, 327.
  - *uliginosa*, 327.
- Dichætophora, 55, 165.
  - campestris, 166.
- Dicoria, 63, 248.
  - Brandegei, 248.
  - canescens, 248.
- Dicranocarpus, 60, 237.
  - parviflorus, 237.
- Diervilla, 8, 18.
  - *Acadiensis*, 19.
  - *Canadensis*, 19.
  - *humilis*, 19.
  - *lutea*, 19.
  - sessilifolia, 19.
  - *Tournefortii*, 19.
  - trifida, 18.
- *Dieteria*, 205.
  - *asteroides*, 206.
  - *canescens*, 206.
  - *coronopifolia*, 206.
  - *divaricata*, 206.
  - *gracilis*, 130.
  - *incana*, 206.

INDEX. 459

pulverulenta, 206.
sessiliflora, 206.
setigera, 208.
spinulosa, 140.
viscosa, 206.
Diodia, 22, 34.
glabra, 34.
hirsuta, 34.
teres, 35.
tetracocca, 33.
tetragona, 35.
tricocca, 33.
Virginiana, 35.
Virginica, 35.
Diodonta, 294.
aristosa, 295.
bidentoides, 295.
coronata, 295.
involucrata, 295.
leptophylla, 295.
mitis, 295.
Diomedea, 281.
Diotostephus repens, 243.
Diplopappus albus, 198.
amygdalinus, 196, 197.
canescens, 214.
cornifolius, 197.
dubius, 219.
ericoides, 133, 198.
filifolius, 213.
graminifolius, 121.
grandiflorus, 214.
hispidus, 123.
incanus, 206.
leucophyllus, 170.
linariifolius, 197.
linearis 213.
Marianus, 122.
robovatus, 197.
occidentalis, 215.
rigidus, 197.
scaber, 121.
sericeus, 121.
spinulosus, 130.
trichophyllus, 122.
umbellatus, 196.
villosus, 123.
Diplostelma bellioides, 165.
Diplostephium acuminatum, 199.
alpinum, 198.
amygdalinum, 196.
boreale, 197.
cornifolium, 197.
dichotomum, 197.
linariifolium, 197.
obovatum, 197.
umbellatum, 196.
Diplothrix, 253.
acerosa, 254.
DIPSACACEÆ, 47.
Dipsacus, 47.
fullonum, 48.
sylvestris, 48.
Distasis concinna, 210.
heterophylla, 165.
modesta, 165.
Distreptus, 88.
Dithrix, 320.
Dockmackie, 10.
Dœllingeria, 196.
amygdalina, 197.
cornifolia, 197.
obovata, 197.
ptarmicoides, 198.
Donia, 117.

ciliata, 125.
glutinosa, 118, 119.
inuloides, 117.
lanceolata, 129.
squarrosa, 118.
uniflora, 128.
Doronicum acaule, 381.
glutinosum, 119.
lævifolium, 216.
nudicaule, 381.
ramosum, 219.
Dracopis, 263.
amplexicaulis, 263.
Dracunculus, 368.
Dufresnia, 44.
Dugaldea, 347.
Duhamelia, 28.
Dumerilia, 408.
Dunmtia Achyranthes, 257.
Dysodia, 77, 355.
acerosa, 357.
anthemidifolia, 356.
cancellata, 356.
chrysanthemoides, 356.
Cooperi, 356.
fastigiata, 356.
glandulosa, 356.
speciosa, 356.
tagetioides, 357.
Dysodiopsis, 357.

Eatonella, 72, 322.
Congdonii, 323.
nivea, 322.
Echenais carlinoides, 398.
Echinacea, 65, 258.
angustifolia, 258.
atrorubens, 259.
pallida, 258.
purpurea, 258.
sanguinea, 258.
serotina, 258.
Echinomeria apetala, 274.
Eclipta, 64, 256.
alba, 256.
brachypoda, 256.
erecta, 256.
integrifolia, 163.
procumbens, 256.
prostrata, 256.
Egletes Arkansana, 164.
humilis, 164.
ramosissima, 164.
Elecampane, 236.
Elephantopus, 50, 88.
Carolinianus, 88.
elatus, 88.
nudatus, 88.
nudicaulis, 88, 89.
scaber, 88, 89.
tomentosus, 88.
Elephantosis, 88.
Encelia, 67, 281.
albescens, 281.
argophylla, 284.
Californica, 282.
calva, 282.
conspersa, 281, 282.
exaristata, 283.
eriocephala, 282.
farinosa, 282.
frutescens, 282.
halimifolia, 281.
microcephala, 284.
microphylla, 281.

nivea, 282.
nudicaulis, 283.
scaposa, 282.
subaristata, 283.
Enceliopsis, 283. viscida.
Endive, 412.
Engelmannia, 61, 244.
pinnatifida, 244.
Texana, 244.
Enula Campana, 236.
Erechtites, 81, 396.
elongata, 398.
hieracifolia, 398.
prealta, 398.
Ereicotis, 27.
Eremiastrum, 56, 171.
bellioides, 171.
Ericameria, 132.
diffusa, 141.
erecta, 134.
microphylla, 133.
nana, 134.
resinosa, 134.
Erigerastrum, 196.
Erigeridium, 216.
Erigeron, 56, 207.
acris, 219.
alpinus, 219.
alpinus, 207, 220.
ambiguus, 219, 221.
Andersonii, 201.
annuus, 218.
argentatus, 212.
Arizonicus, 218.
armeriæfolius, 220.
asper, 209.
asperugineus, 212.
Bellidiastrum, 217.
bellidifolius, 216.
Beyrichii, 219.
Bigelovii, 210.
Bloomeri, 214.
Bonariensis, 208, 220.
Brandegei, 210.
Brazoensis, 218.
Breweri, 215.
cæspitosus, 214.
cæspitosus, 212, 214.
camphoratus, 226.
Canadensis, 226.
canescens, 213, 214.
canus, 212.
Carolinianus, 161.
Chrysopsidis, 210.
ciliatus, 204.
cinereus, 218.
compositus, 211.
concinnus, 210.
corymbosus, 214.
Coulteri, 209.
decumbens, 215.
decumbens, 215.
delphinifolius, 219.
delphinifolius, 219.
discoideus, 221.
divaricatus, 221.
divaricatus, 218.
divergens, 218.
divergens, 217.
Douglasii, 215.
Drœbachensis, 220.
Eatoni, 214.
elongatus, 220.
eriocephalus, 207.
eriophyllus, 220.

filifolius, 213.
flagellaris, 217.
florifer, 168.
foliosus, 215.
glabellus, 209.
glabratus, 220.
glandulosus, 212.
glandulosus, 121.
glaucus, 208.
grandiflorus, 208.
grandiflorus, 209.
heterophyllus, 219.
hirsutus, 210.
hispidus, 208.
Howellii, 209.
hyssopifolius, 216.
incomptus, 218.
inornatus, 215.
integrifolius, 216, 219.
Kamtschaticus, 220.
lanatus, 207.
leiomerus, 211.
Lemmoni, 216.
linifolius, 220.
lonchophyllus, 220.
longipes, 217.
macranthus, 200.
maritimus, 208.
miser, 216.
modestus, 218.
Muirii, 207.
multiflorus, 203.
nanus, 212.
Neo-Mexicanus, 219.
nervosus, 121.
Nevadensis, 214.
nudicaulis, 216.
nudiflorus, 218.
ochroleucus, 213.
ochroleucus, 214.
Oreganus, 216.
paniculatus, 221.
Parishii, 212.
pedatus, 211.
peucephyllus, 213.
Philadelphicus, 217.
Philadelphicus, 217, 219.
pilosus, 122.
poliospermus, 216.
Pringlei, 211.
pulchellus, 207, 209, 216, 217.
pumilus, 210.
pumilus, 213.
purpureus, 217.
pusillus, 221.
quercifolius, 217.
quercifolius, 218.
racemosus, 220.
radicatus, 211.
repens, 217.
Rusbyi, 217.
salsuginosus, 208.
scoposus, 217.
speciosus, 209.
stenophyllus, 213.
stenophyllus, 213, 215.
strictus, 221.
strigosus, 219.
strigosus, 210.
subdecurrens, 220.
subdecurrens, 221.
supplex, 215.
tener, 212.
tenuis, 218.
trifidus, 211.

uniflorus, 207.
ursinus, 211.
Utahensis, 213.
vernus, 216.
Eriocarpum grindelioides, 126.
Eriopappus glandulosus, 314.
Eriophyllum, 72, 73, 328.
ambiguum, 331.
cæspitosum, 330.
confertiflorum, 329.
gracile, 331.
multicaule, 328.
nubigenum, 329.
Pringlei, 329.
stæchadifolium, 329.
Wallacei, 329.
Watsoni, 331.
Erithalis, 21, 30.
fruticosa, 30.
odorifera, 30.
Ernodea, 21, 31.
littoralis, 31.
Erythremia aphylla, 436.
grandiflora, 436.
Esopon glaucum, 434.
Espejoa, 317.
Espeletia amplexicaulis, 267.
helianthoides, 266.
sagittata, 266.
Eublepharis, 290.
Eucephalus, 199.
albus, 198.
elegans, 200.
ericoides, 198.
glaucus, 200.
EUINULEÆ, 59.
EUPATORIACEÆ, 50, 91.
Eupatoriophalacron, 256.
Eupatorium, 51, 94.
ageratifolium, 100.
ageratoides, 101.
album, 98
altissimum, 99.
altissimum, 101.
ambiguum, 97, 98.
amœnum, 98.
aromaticum, 101.
Berlandieri, 100, 101.
betonicum, 102.
Bigelovii, 96.
Brickellia, 105.
Bruneri, 96.
calocephalum, 95.
cassinifolium, 98.
ceanothifolium, 101.
cœlestinum, 102.
compositifolium, 97.
concinnum, 95.
conyzoides, 95.
cordatum, 101.
cordiforme, 101.
coronopifolium, 97.
crassifolium, 97, 102.
Cubense, 100.
cuneatum, 100.
cuneifolium, 100.
dissectum, 102.
divergens, 95.
dubium, 96.
falcatum, 96.
Fendleri, 96.
foeniculaceum, 97.
foeniculoides, 97.
floribundum, 95.
Fraseri, 101.

fusco-rubrum, 96.
glandulosum, 98.
glastifolium, 114.
glaucescens, 98.
grandiflorum, 105.
Greggii, 102.
Hartwegi, 102.
heteroclinium, 95.
hyssopifolium, 98.
incarnatum, 101.
ivæfolium, 85.
lævigatum, 96.
lanceolatum, 99.
leptophyllum, 97.
leucolepis, 98.
Lindheimerianum, 100.
linearifolium, 98.
luteum, 102.
maculatum, 95, 96.
Marrubium, 99.
Maximiliani, 95.
melissoides, 101.
mikanioides, 97.
multinerve, 100.
obovatum, 99.
obscurum, 95.
occidentale, 100.
odoratum, 95, 101.
oliganthes, 107.
Oreganum, 101.
ovatum, 99.
Parryi, 96.
parviflorum, 97, 98.
pauperculum, 102.
perfoliatum, 99.
pilosum, 97.
pinnatifidum, 97.
pubescens, 99,
punctatum, 96.
purpureum, 95.
pycnocephalum, 100.
racemosum, 97.
resinosum, 100.
Rothrockii, 102.
rotundifolium, 99.
Sabeanum, 95.
sagittatum, 94.
salviæfolium, 100.
scabridum, 99.
scandens, 94.
Schiedeanum, 100.
semiserratum, 98.
semiserratum, 97.
serotinum, 97.
sessilifolium, 99.
solidaginifolium, 97.
Sonoræ, 100.
speciosum, 109.
stigmatosum, 98.
suaveolens, 101.
ternifolium, 96.
teucrifolium, 99.
tortifolium, 98.
trifoliotum, 95.
truncatum, 99, 100.
urticæfolium, 101.
verbenæfolium, 99.
verticillatum, 95.
villosum, 100.
violaceum, 102.
Wrightii, 100.
Euphrosyne ambrosiæfolia, 246.
xanthiifolia, 246.
Eurybia commixta, 175.
corymbosa, 174.

INDEX. 461

Jussiæi, 175.
—*macrophylla*, 175.
Euthamia, 160.
— *graminifolia*, 160.
— *occidentalis*, 160.
— *tenuifolia*, 161.
Euthamiopsis, 141.
Evax, 58, 228.
— candida, 230.
— caulescens, 228.
— multicaulis, 229.
— prolifera, 229.
Everlasting, 233, 234.
Exostema, 19, 23.
— Caribæum, 23.
Exostemma, 23.

Fedia, 44.
— *amarella*, 45.
— *chenopodifolia*, 45.
— *Fagopyrum*, 45.
— *longifolia*, 46.
— *Nuttallii*, 46.
— *olitoria*, 44.
— *patellaria*, 46.
— *radiata*, 45.
— *stenocarpa*, 45.
— *umbilicata*, 45.
— *Woodsiana*, 45.
Feverwort, 12.
Fichtea, 416.
FILAGINEÆ, 57.
Filaginopsis Drummondii, 229.
— *multicaulis*, 229.
Filago, 58, 230.
— Arizonica, 230.
— Californica, 230.
— depressa, 230.
— Gallica, 230. *Germanica 230*
— *parvula*, 230.
— repens, 231.
— Texana, 231.
Fireweed, 396.
Flaveria, 76, 353.
— *angustifolia*, 354.
— chloræfolia, 353.
— Contrayerba, 354.
— linearis, 354.
— *longifolia*, 354.
— *maritima*, 354.
— repanda, 354.
— *tenuifolia*, 354.
FLAVERIEÆ, 76.
Fleabane, 207.
Florestina, 74, 336.
— *callosa*, 337.
— tripteris, 337.
Flourensia, 66, 281.
— cernua, 281.
— *corymbosa*, 281.
— laurifolia, 281.
— thurifera, 281.
Fly Honeysuckle, 15.
Frauseria, 63, 250.
— *albicaulis*, 251.
— ambrosioides, 252.
— *artemisioides*, 252.
— bipinnatifida, 251.
— Chamissonis, 251.
— *chenopodiifolia*, 251, 252.
— cordifolia, 445.
— *cuneifolia*, 251.
— deltoidea, 251.
— discolor, 251.
— dumosa, 251.

— eriocentra, 251.
— Hookeriana, 250.
— ilicifolia, 252.
— *Lessingii*, 251.
— *montana*, 251.
— *pumila*, 250. *tenuifolia 250*
— tomentosa, 251.
Fuller's Teasel, 48.

Gaillardia, 76, 351.
— acaulis, 353.
— *acaulis*, 345.
— amblyodon, 352.
— aristata, 352.
— Arizonica, 353.
— bicolor, 352.
— comosa, 351.
— *Drummondii*, 352.
— lanceolata, 352. *fimbriata*
— Mexicana, 352. *350*
— picta, 352.
— pinnatifida, 353.
— pulchella, 352.
— *pulchella*, 353.
— *Rœmeriana*, 344.
— rustica, 352.
— simplex, 351.
— spathulata, 353.
— tuberculata, 351.
Galardia, 351. *a bicolor 350*
Galatella graminifolia, 216.
— nemoralis, 199.
— *obtusifolia*, 171.
Galathenium elongatum, 442.
— *Floridanum*, 442, 443.
— *graminifolium*, 443.
— *integrifolium*, 442.
— *Ludovicianum*, 443.
— *multiflorum*, 444.
— *salicifolium*, 442, 443.
— *sanguineum*, 442.
Galinsoga, 69, 303.
— parviflora, 303.
GALINSOGEÆ, 68.
Galium, 23, 35.
— *acutissimum*, 40.
— Andrewsii, 41.
— Anglicum, 36.
— angustifolium, 39.
— Aparine, 36.
— Arkansanum, 38.
— asperrimum, 39.
— asprellum, 39.
— *Bermudianum*, 37.
— *Bermudiense*, 37.
— bifolium, 36.
— Bloomeri, 40.
— Bolanderi, 41.
— boreale, 38.
— *boreale*, 37.
— *brachiatum*, 37, 39.
— Brandegei, 38.
— Californicum, *ericoides 37*
— circæzans, 37.
— *Claytoni*, 38.
— concinnum, 38.
— cuspidatum, 39.
— Fendleri, 41.
— hispidulum, 42.
— *hispidum*, 42.
— *hypotrichium*, 40.
— Kamtschaticum, 37.
— lanceolatum, 37.
— latifolium, 38.
— *Littellii*, 37.

— margaricoccum, 41.
— Matthewsii, 40.
— *micranthum*, 39.
— Mollugo, 35.
— multiflorum, 40.
— Nuttallii, 41.
— obovatum, 37.
— obtusum, 38.
— *Parisiense*, 36.
— *parviflorum*, 39.
— *Pennsylvanicum*, 39.
— pilosum, 37.
— proliferum, 37.
— pubens, 40.
— punctatum, 37.
— puncticulosum, 40.
— *purpureum*, 37.
— Rothrockii, 39.
— rubioides, 38.
— *septentrionale*, 38.
— *spinulosum*, 39.
— stellatum, 40.
— strictum, 38.
— suaveolens, 39.
— suffruticosum, 41.
— *Texanum*, 36.
— Texense, 36.
— tinctorium, 38.
— Torreyi, 37.
— tricorne, 36.
— trifidum, 38.
— triflorum, 39.
— uncinulatum, 36.
— uniflorum, 41.
— *Vaillantii*, 36.
— verum, 35.
— virgatum, 36.
— Wrightii, 39.
Gall-of-the-earth, 434.
Gamochæta, 236.
Garberia, 52, 112.
— fruticosa, 112.
Gardenia florida, 29.
Gurdenia clusiæfolia, 29.
— *Randia*, 29.
Genipa, 20, 29.
— clusiæfolia, 29.
Georgia Bark, 23.
Geræa, 281, 282.
— *canescens*, 282.
Gerbera nutans, 408.
— *Walteri*, 408.
Glyptopleura, 86, 423.
— marginata, 423.
— setulosa, 423.
GERBEREÆ, 83.
GNAPHALIEÆ, 58.
Gnaphaliopsis micropoides, 359.
Gnaphalium, 59, 234.
— alpinum, 232.
— *Americanum*, 236.
— Arizonicum, 234.
— *Californicum*, 235.
— *Carpathicum*, 232.
— *Chamissonis*, 236.
— *Chilense*, 235.
— *conoideum*, 234.
— decurrens, 235.
— *dimorphum*, 231.
— dioicum, 233.
— *falcatum*, 236.
— *filaginoides*, 230.
— *gossypinum*, 235.
— *hyemale*, 236.
— leucocephalum, 235.

luteo-album, 235.
margaritaceum, 233.
microcephalum, 234.
Norvegicum, 236.
obtusifolium, 234.
palustre, 235.
Pennsylvanicum, 236.
plantagineum, 233.
plantaginifolium, 233.
polycephalum, 234.
purpureum, 236.
pusillum, 236.
ramosissimum, 235.
rectum, 235.
spathulatum, 236.
spicatum, 236.
Sprengelii, 234.
Sprengelii, 235.
stachydifolium, 236.
strictum, 235.
supinum, 236.
sylvaticum, 236.
sylvaticum, 236.
uliginosum, 238.
ustulatum, 235.
virgatum, 226.
Wrightii, 234.
Goat's-beard, 415, 416.
Gochnatia, 83, 407.
hypoleuca, 407.
GOCHNATIEÆ, 83.
Golden-rod, 148.
Greenella, 55, 164.
Arizonica, 164.
discoidea, 164.
Grindelia, 53, 116.
angustifolia, 117.
Arizonica, 118.
arguta, 118.
coronopifolia, 117.
costata, 117.
cuneifolia, 118.
discoidea, 119.
glutinosa, 119.
grandiflora, 118.
hirsutula, 117.
humilis, 119.
humilis, 119.
integrifolia, 117.
integrifolia, 118.
inuloides, 117.
lanceolata, 118.
latifolia, 119.
microcephala, 117, 118.
nana, 119.
nuda, 118.
Oregana, 118.
Pacifica, 119.
robusta, 119.
rubricaulis, 117.
squarrosa, 118.
squarrosa, 119.
stricta, 117.
subdecurrens, 118.
Texana, 118.
virgata, 118.
Groundsel, 383, 394.
Guardiola, 60, 237.
platyphylla, 237.
Guettarda, 21, 29.
ambigua, 30.
Blodgettii, 30.
elliptica, 30.
scabra, 30.
Gum-plant, 116, 119.

Guntheria, 351.
Gutierrezia, 53, 115.
Berlandieri, 116.
Californica, 115.
divaricata, 115.
eriocarpa, 116.
Euthamiæ, 115.
Lindheimeriana, 116.
linearifolia, 115.
linearifolia, 115.
microcephala, 115.
microphylla, 115.
sphærocephala, 115.
Texana, 116.
Gymnocline, 365.
Gymnolæna, 356.
Gymnolomia, 66, 269.
encelioides, 269.
multiflora, 269.
Porteri, 269.
tenuifolia, 269.
triloba, 269.
Gymnopsis, 269.
uniserialis, 256.
Gymnosperma, 52, 114.
corymbosum, 114.
multiflorum, 114.
nudatum, 354.
oppositifolium, 354.
scoparium, 114.
Gymnostyles Chilensis, 365.
nasturtiifolia, 366.
stolonifera, 366.
Gymena, 225.
dentata, 226.
viscida, 226.
Gyrophyllum, 293.

Halea Ludoviciana, 256.
repanda, 256.
Texana, 256.
Hamelia, 20, 28.
coccinea, 28.
patens, 28.
Haploesthes, 80, 378.
Greggii, 378.
Haplopappus, 125.
Hardheads, 406.
Harpæcarpus exiguus, 306.
madarioides, 306.
Harpalium, 270, 271.
rigidum, 275.
Harpalyce, 433.
alba, 434.
altissima, 435.
crepidina, 433.
racemosa, 433.
serpentaria, 434.
virgata, 433.
Hartmannia, 307.
ciliata, 316.
corymbosa, 308.
fasciculata, 309.
pungens, 308.
Hawkbit, 420.
Hawkweed, 424.
Hecastocleis, 83, 407.
Shocklevi, 407.
Hectorea villosissima, 121.
Hecubæa, 347.
Hedyotis acerosa, 27.
angustifolia, 26.
auricularia, 28.
Boscii, 27.
cærulea, 24.

calycosa, 26.
ciliolata, 26.
gentianoides, 24.
glomerata, 28.
Halei, 28.
humifusa, 26, 27.
lanceolata, 28.
longifolia, 26.
minima, 25.
rotundifolia, 25.
rubra, 25.
serpyllifolia, 24.
stenophylla, 26, 27.
umbellata, 26.
Virginica, 28.
Heleastrum, 173.
album, 198.
paludosum, 174.
HELENIEÆ, 72.
HELENIOIDEÆ, 70, 317.
Helenium, 76, 347.
altissimum, 349.
amphibolum, 348.
atropurpureum, 349.
autumnale, 349.
Bigelovii, 350.
Bolanderi, 350.
brevifolium, 351.
Californicum, 349.
canaliculatum, 349.
commutatum, 349.
Curtisii, 350.
decurrens, 349.
elegans, 348.
fimbriatum, 350.
grandiflorum, 349.
Hoopesii, 347.
lanatum, 330.
longifolium, 349.
Mexicanum, 349.
Mexicanum, 348, 349.
micranthum, 349.
microcephalum, 348.
montanum, 349.
nudiflorum, 349.
Nuttallii, 350.
ooclinium, 348.
parviflorum, 349.
puberulum, 340.
pubescens, 349.
pumilum, 349.
quadridentatum, 348.
quadridentatum, 349.
quadripartitum, 348.
Seminariense, 349.
tenuifolium, 347.
Thurberi, 348.
tubuliflorum, 349.
varium, 349.
vernale, 350.
Helepta grandiflora, 255.
Helianthella, 67, 283.
argophylla, 283.
Californica, 285.
Douglasii, 285.
grandiflora, 285.
latifolia, 270.
Mexicana, 284.
microcephala, 284.
nudicaulis, 283.
Parryi, 284.
quinquenervis, 284.
tenuifolia, 285.
uniflora, 285.
uniflora, 284.

HELIANTHOIDEÆ, 59, 237.
Helianthus, 66, 271.
— *altissimus*, 276.
— angustifolius, 273.
— annuus, 272.
argophyllus, 272.
aristatus, 288.
— atrorubens, 274.
— *atrorubens*, 274, 275.
Bolanderi, 272.
Californicus, 277.
— canescens, 276.
— *cernuus*, 281.
— ciliaris, 274.
cinereus, 275.
cinereus, 279.
— *crassifolius*, 275.
cucumerifolius, 273.
dealbatus, 280.
— debilis, 273.
— decapetalus, 280.
— *dentatus*, 270.
— *diffusus*, 274.
— divaricatus, 279.
— *divaricatus*, 278.
— *diversifolius*, 279.
doronicoides, 279.
— *doronicoides*, 280.
Douglasii, 278.
Dowellianus, 275.
— *erythrocarpus*, 272.
exilis, 273.
— Floridanus, 273.
— *frondosus*, 280.
— giganteus, 276.
— *giganteus*, 273, 277.
— *gigas*, 276.
gracilentus, 277.
— grosse-serratus, 276.
— heterophyllus, 274.
— *heterophyllus*, 275.
— hirsutus, 279.
— *hispidulus*, 279.
Hookerianus, 268.
— *integrifolius*, 272.
— lætiflorus, 275.
lævigatus, 278.
— *lævis*, 255, 279, 296.
— lenticularis, 272.
Lindheimerianus, 273.
longifolius, 278.
longifolius, 268.
— *macrocarpus*, 272.
— *macrophyllus*, 280.
— Maximiliani, 277.
— *microcephalus*, 278.
— *Missouriensis*, 275.
— *Missouricus*, 274.
— mollis, 276.
— *mollis*, 280.
— *multiflorus*, 272, 280.
— *neglectus*, 279.
— Nuttallii, 277.
— occidentalis, 275.
— orgyalis, 273.
— *ovatus*, 272.
Parishii, 277.
— parviflorus, 276.
— *patens*, 272.
pauciflorus, 272.
— petiolaris, 272.
— *præcox*, 273.
— *prostratus*, 280.
— *pubescens*, 276, 279.
— pumilus, 275.

quinquenervis, 284.
— radula, 274.
— rigidus, 274.
scaberrimus, 272.
— *scaberrimus*, 275.
Schweinitzii, 278.
— *silphioides*, 274.
— *sparsiflorus*, 274.
spathulatus, 276.
squarrosus, 276.
— strumosus, 279.
— *strumosus*, 278, 279, 280.
— subtuberosus, 276.
— *tenuifolius*, 280.
tephrodes, 271.
thurifer, 281.
tomentosus, 276.
— trachelifolius, 280.
— *trachelifolius*, 278.
tricuspis, 275.
— *truncatus*, 279.
— *tubæformis*, 272.
— tuberosus, 280.
— *tuberosus*, 276.
Heliochroa amœna, 285.
— *elatior*, 258.
— *furcata*, 258.
— *Linnæana*, 258.
— Heliomeris multiflora, 269.
— *tenuifolia*, 269.
Heliopsis, 64, 254.
— annua, 255.
— *Balsamorrhiza*, 266.
— buphthalmoides, 255.
— *buphthalmoides*, 255.
— *canescens*, 255.
gracilis, 255.
— lævis, 254.
parviflora, 255.
— scabra, 255.
terebinthacea, 266.
— Helmintha, 420.
Helogyne, 93.
— Hemiachyris Texana, 116.
— Hemiambrosia, 250.
— Hemiptilium, 415.
— *Bigelovii*, 414.
Schottii, 415.
— *Hemizanthidium*, 250, 251.
— Hemizonella, 69, 306.
— *Durandi*, 306.
minima, 306.
— *parvula*, 306.
Hemizonia, 69, 306.
angustifolia, 308.
— *angustifolia*, 311.
— *balsamifera*, 308.
cephalotes, 312.
— *citriodora*, 307.
Clevelandi, 307.
congesta, 307.
congesta, 314.
— corymbosa, 308.
decumbens, 309.
Douglasii, 311.
— fasciculata, 309.
filipes, 313.
— Fitchii, 308.
floribunda, 309.
Fremonti, 312.
frutascens, 307.
— *glomerata*, 309.
Heermanni, 310.
hispida, 311.
Kelloggii, 309.

Lobbii, 310.
luzulæfolia, 307.
— macradania, 308.
— *macradenia*, 310.
— *macrocephala*, 308.
mollis, 311.
— *multicaulis*, 309.
— multiglandulosa, 312.
— *oppositifolia*, 312.
paniculata, 309.
— Parryi, 308.
pauciflora, 311.
plumosa, 312.
— pungens, 308.
— *ramosissima*, 310.
rudis, 307.
sericea, 307.
spicata, 311.
Streetsii, 307.
— tenella, 310.
— truncata, 311.
virgata, 310.
Wheeleri, 307.
Wrightii, 309.
— Herba Impia, 230.
Hesperastrum, 174.
— Hesperevax, 228.
— Heterochæta, 207.
Heterochromea, 446.
HETEROCHROMEÆ, 54.
— *Heterodonta*, 295.
Heterogyne, 253.
Heteropectis, 361.
— *Heterophanta*, 231.
— *Heteropleura*, 429.
Heterosperma, 299.
Heterospermum, 68, 299.
pinnatum, 299.
tagetinum, 299.
— Heterotheca, 53, 120.
— *Chrysopsidis*, 121.
— *floribunda*, 121.
— grandiflora, 121.
— *Lamarckii*, 120.
— *Lamarckii*, 130.
— *latifolia*, 121.
— *leptoglossa*, 121.
— *scabra*, 121.
Hieracium, 86, 424.
abscissum, 430.
— albiflorum, 428.
alpinum, 424.
— *arcticum*, 427.
argutum, 428.
— aurantiacum, 424.
— *aureum*, 425.
— *barbatum*, 426.
Bolanderi, 429.
brevipilum, 429.
Breweri, 427.
Californicum, 423.
— *Canadense*, 425.
— *Canadense*, 425.
carneum, 430.
— *Caroliniamum*, 426.
— *corymbosum*, 425.
— *cynoglossoides*, 428.
— *erythrospermum*, 429.
— *fasciculatum*, 425.
Fendleri, 429.
— gracile, 427.
Greenei, 429.
— Gronovii, 426.
— *Gronovii*, 425, 426, 427.
— *helianthifolium*, 425.

464 INDEX.

—*Hookeri*, 427.
horridum, 427.
—Kalmii, 424.
Kalmii, 425.
Lemmoni, 430.
— longipilum, 426, 446.
—*macranthum*, 425.
—*macrophyllum*, 425.
Marianum, 426, 446.
—*Marianum*, 426.
—*molle*, 424.
— murorum, 424.
— paniculatum, 425.
Parishii, 428.
—*Pennsylvanicum*, 426.
præaltum, 424.
præcox, 424.
—*prenanthoides*, 425.
Pringlei, 429.
—*pusillus*, 207.
relicinum, 427.
—*rigidum*, 425.
Rugelii, 426.
—*runcinatum*, 431.
Rusbyi, 428.
—*scabriusculum*, 425.
— scabrum, 426.
scabrum, 426.
—Scouleri, 427.
—*Scouleri*, 429.
—*subnudum*, 425.
—*Sullivantii*, 425.
—*sylvaticum*, 424.
thyrsoideum, 430.
triste, 427.
—*triste*, 427.
—*umbellatum*, 425.
—*Vancouverianum*, 428.
venosum, 425.
—*venosum*, 425.
—*virgatum*, 425.
—*vulgatum*, 424.
—High Cranberry, 10.
—High-water Shrub, 247.
—Hobble-bush, 9.
Hofmeisteria, 51, 93.
pluriseta, 93.
—Hologymne, 324.
Douglasii, 323.
glabrata, 324.
Holozonia, 313.
filipes, 313.
—HOMOCHROMEÆ, 52.
—*Homoianthus*, 408.
—*Homopappus*, 126.
argutus, 127.
glomeratus, 127.
inuloides, 128.
—*multiflorus*, 129.
paniculatus, 127.
—*racemosus*, 127.
spathulatus, 149.
—*squarrosus*, 125.
—Honeysuckle, 141.
Hopkirkia anthemoides, 334.
—Horse-Gentian, 12.
—Houndstongue, 118.
—Houstonia, 20, 24.
acerosa, 27.
— angustifolia, 26.
— cærulea, 24.
— ciliolata, 26.
—*coccinea*, 24.
fasciculata, 27.
—*fruticosa*, 26.

humifusa, 25.
—*Linnæi*, 24, 25.
—*longifolia*, 26.
— minima, 25.
— patens, 24.
— pubescens, 26.
— purpurea. 26.
—*pusilla*, 24.
— rotundifolia, 25.
rubra, 25.
—*rupestris*, 26.
— serpyllifolia, 24.
— subviscosa, 25.
—*tenuifolia*, 26.
—*varians*, 26.
—Wrightii, 26.
—Hulsea, 75, 342.
algida, 343.
brevifolia, 343.
Californica, 342.
callicarpha, 342.
heterochroma, 343.
nana, 343.
Parryi, 342.
vestita, 342.
—Hydrocarpæa, 298.
—Hymenatherum, 77, 357.
— acerosum, 357.
aureum, 359.
—*Berlandieri*, 358.
concinnum, 446.
Gnaphaliopsis, 358.
—*gnaphalodes*, 359.
Greggii, 359.
Hartwegi, 358.
Neo-Mexicanum, 357.
—pentachætum, 358.
—polychætum, 357.
setifolium, 359.
—tagetoides, 357.
—*tenuifolium*, 358.
—tenuilobum, 358.
Thurberi, 358.
Treculii, 358.
Wrightii, 358.
—Hymenoclea, 63, 248.
— monogyra, 248.
— Salsola, 248.
—*Hymenonema glaucum*, 417, 418.
laciniatum, 417.
—Hymenopappus, 74, 335.
— artemisiæfolius, 335.
corymbosus, 335.
—*Douglasii*, 341.
—flifolius, 336.
—flavescens, 336.
—*luteus*, 336.
Mexicanus, 336.
—*Nevadensis*, 341.
—*robustus*, 336.
scabiosæus, 335.
— tenuifolius, 336.
—*tenuifolius*, 336.
Hymenothrix, 73, 334.
Wislizeni, 334.
—Wrightii, 335.
—*Hymenoxys*, 346.
Californica, 327.
linearifolia, 344.
mutica, 328.
odorata, 347.
—*Hyoseris amplexicaulis*, 412.
— *angustifolia*, 412.
— *biflora*, 412.
— *Caroliniana*, 411.

—*major*, 412.
— *montana*, 412.
— *prenanthoides*, 413.
—Hypochœris 85, 416.
— glabra, 416.
— radicata, 416.

—Ianthe, 197.
—Indian Currant, 13.
—Indian Plantain, 394.
Infantea Chilensis, 334.
—Innocence, 24.
—Inula, 59, 236.
—*argentea*, 121.
—*divaricata*, 130.
—*ericoides*, 198.
—*falcata*, 122.
—*glandulosa*, 122.
—*glutinosa*, 119.
—*gossypina*, 122.
—*graminifolia*, 121.
—Helenium, 236.
—*Mariana*, 122.
—*scabra*, 121.
—*serrata*, 117.
—*subaxillaris*, 121.
—INULOIDEÆ, 57, 225.
—Iron-weed, 89.
Isocarpha, 65, 257.
oppositifolia, 257.
—*Isocoma vernonioides*, 143.
—Isopappus, 130.
—*divaricatus*, 130.
—*Hookerianus*, 131.
—Iva, 62, 245.
— ambrosiæfolia, 246.
— angustifolia, 247.
—*annua*, 246.
— axillaris, 247.
— ciliata, 246.
dealbata, 246.
—*foliosa*, 247.
— frutescens, 247.
Hayesiana, 247.
— imbricata, 246.
— microcephala, 247.
— *monophylla*, 249.
Nevadensis, 247.
—*paniculata*, 246.
—*xanthiifolia*, 246.
—*Ixora Americana*, 24.
—*ternifolia*, 24.

—Jacea, 406.
—*Jamesia*, 412, 413.
—*pauciflora*, 413.
—Jaumea, 70, 317.
— carnosa, 317.
JAUMIEÆ, 70.
—Jerusalem Artichoke, 280.
—Joe-Pye Weed, 95.

Kalliactis, 265.
—Keerlia, 55, 164.
— bellidifolia, 164.
— effusa, 165.
— ramosa, 164.
—*skirrobasis*, 164.
Kelloggia, 22, 31.
— galioides, 32.
—*Kleinia*, 317, 354.
—Knapweed, 406.
—*Knoxia*, 31.
—Krigia, 84, 411.
— amplexicaulis, 412.

INDEX.

bellioides, 411.
— Caroliniana, 411.
— *Caroliniana*, 412.
— Dandelion, 412.
— dichotoma, 411.
— leptophylla, 411.
— montana, 412.
nervosa, 411.
occidentalis, 411.
— Virginica, 411.
- Kuhnia, 51, 103.
— *Critonia*, 103.
— eupatorioides, 103.
— *frutescens*, 103.
— *glutinosa*, 99, 103.
— *leptophylla*, 103.
macrantha, 103.
Maximiliani, 103.
— *paniculata*, 103.
— rosmarinifolia, 103.
Schaffneri, 103.
suaveolens, 103.
— spec. Raf., 103.
Kuhnioides, 113.
Kymapleura, 439.
— *heterophylla*, 440.

LABIATIFLORÆ, 50; 82.
Lactuca, 87, 441.
— acuminata, 443.
alpina, 444.
— Canadensis, 442.
— *Canadensis*, 444.
— Caroliniana, 442.
— elongata, 442, 443.
— Floridana, 443.
— *Floridana*, 444.
— *graminea*, 443.
— graminifolia, 442.
— hirsuta, 442.
— integrifolia, 442.
— *integrifolia*, 443.
— leucophæa, 444.
— *longifolia*, 441.
Ludoviciana, 443.
macrophylla, 444.
— pulchella, 443.
— *sagittifolia*, 442.
— *sanguinea*, 442.
— Scariola, 442.
— *villosa*, 444.
— Lactucastrum, 443.
- Lagophylla, 70, 313.
congesta, 314.
dichotoma, 313.
filipes, 313.
glandulosa, 313.
— ramosissima, 314.
Lagatea, 200.
— Lagothamnus, 379.
ambiguus, 379.
microphyllus, 379.
— Lampsana, 84, 410.
— communis, 410.
Laphamia, 71, 319.
angustifolia, 319.
bisetosa, 320.
cinerea, 319.
dissecta, 321.
halimifolia, 319.
Lemmoni, 319.
Lindheimeri, 320.
megacephala, 320.
Palmeri, 320.
peninsularis, 319.

rupestris, 319.
Stansburii, 320.
— *Lappa major*, 397.
— *minor*, 397.
tomentosa, 397.
— *Lapsana*, 410.
— *Lasianthæa*, 286.
— Lasthenia, 72, 324.
ambigua, 331.
Californica, 324.
— *glaberrima*, 324.
glabrata, 324.
obtusifolia, 324.
Laurestinus, 9.
— Layia, 70, 314.
Calliglossa, 316.
carnosa, 315.
chrysanthemoides, 316.
Douglasii, 316.
— elegans, 315.
— Fremonti, 316.
gaillardioides, 315.
glandulosa, 314.
heterotricha, 315.
Jonesii, 316.
— *Neo-Mexicana*, 314.
pentachæta, 315.
— platyglossa, 315.
— Leachia, 291.
— *lanceolata*, 292.
— *trifoliata*, 293.
Lebetina, 356.
— Leighia, 270.
bicolor, 273.
— *lanceolata*, 285.
longifolia, 278.
— *uniflora*, 284, 285.
Leontodon, 85, 420.
alpinus, 440.
autumnalis, 420.
boreale, 424.
— Carolinianum, 441.
ceratophorum, 440.
— *hirsutum*, 439.
— hirtus, 420.
— hispidus, 420.
lividus, 440.
officinalis, 440.
— *Taraxacum*, 440.
— *vulgare*, 440.
— Lepachys, 66, 263.
— *angustifolia*, 264.
— columnaris, 264.
peduncularis, 264.
— pinnata, 263.
— *pinnatifida*, 264.
serrata, 264.
— Lepidanthus, 364.
— Lepidaploa, 89.
— Lepidonema, 416.
— Lepidospartum, 80, 378.
squamatum, 378.
— *Lepidostephanus madioides*, 313.
— *Lepidotheca suaveolens*, 364.
— *Leptoclinium fruticosum*, 112.
— Leptogyne, 225.
— Leptopoda, 350.
— *brachypoda*, 349.
brevifolia, 350, 351.
decurrens, 350.
denticulata, 350.
— *fimbriata*, 350.
Helenium, 350.
incisa, 350.
— *integrifolia*, 350.

pinnatifida, 351.
— *puberula*, 350.
— Leptoseris, 422.
Californica, 423.
saxatilis, 423.
sonchoides, 422.
— Leptosyne, 68, 299.
— Arizonica, 301.
Bigelovii, 300.
— Californica, 299.
calliopsidea, 300.
Douglasii, 299.
gigantea, 300.
heterocarpa, 301.
— maritima, 300.
— Newberryi, 299.
parthenioides, 301.
Stillmani, 299.
— Leptatheca, 431.
Leria lyrata, 408.
— *nutans*, 408.
— Lessingia, 54, 161.
Germanorum, 162.
— *glandulifera*, 162.
— leptoclada, 162.
nana, 163.
— ramulosa, 162.
virgata, 162.
Lettuce, 441.
Leucampyx, 78, 362.
Newberryi, 362.
Leucanthemum arcticum, 365.
integrifolium, 365.
— Parthenium, 365.
— *vulgare*, 365.
— Leucacoma, 172.
— *Leucopsidium Arkansanum*, 164.
— *humile*, 164.
— Leucopsis, 123.
— Leucoseris, 422.
tenuifolia, 423.
— Leysera Caroliniana, 352.
Liatris, 52, 109.
— acidota, 110.
— aspera, 110.
bellidifolia, 113. *borealis*
Boykini, 110.
— brachystachys, 111.
Chapmanii, 112.
— corymbosa, 113.
— cylindracea, 109.
— *cylindrica*, 110.
— *dubia*, 111.
— elegans, 109.
— *flexuosa*, 110.
fruticosa, 112.
Garberi, 112.
— gracilis, 110.
— graminifolia, 111.
— *graminifalia*, 110.
heterophylla, 110.
intermedia, 109.
— *lævigata*, 112.
— *lanceolata*, 111.
— macrostachya, 111.
mucronata, 110.
— *odoratissima*, 113.
— *oppositifolia*, 95.
— *paniculata*, 113.
pauciflora, 112.
— *pauciflosculosa*, 112.
— *pilosa*, 110.
— *propinqua*, 111.
pumila, 111.
— punctata, 110.

30

INDEX.

pycnostachya, 110.
  radians, 109.
  resinosa, 110, 111.
  acariosa, 110.
  secunda, 112.
  sessiliflora, 111.
  sphæroidea, 110.
  spicata, 111.
  squamosa, 113.
  squarrosa, 109.
  squarrulosa, 110.
  stricta, 110.
  tenuifolia, 112.
  tomentosa, 113.
  umbellata, 90.
  virgata, 111.
  Walteri, 113.
LiguliflorÆ, 50, 83.
Lindheimera, 61, 244.
  Texana, 244.
Linnæa, 7, 13.
  borealis, 13.
Linosyris, 136.
  albicaulis, 139.
  arborescens, 141.
  Bigelovii, 138.
  Bolanderi, 136.
  ceruminosa, 138.
  coronopifolia, 142.
  dentata, 143.
  depressa, 137.
  Drummondii, 142.
  graveolens, 139.
  heterophylla, 142.
  hirtella, 142.
  Howardi, 136.
  humilis, 204.
  lanceolata, 140.
  Mexicana, 143.
  Nevadensis, 136.
  Parryi, 136.
  pluriflora, 142.
  ramulosa, 223.
  serrulata, 140.
  Sonoriensis, 141.
  squamata, 378.
  teretifolia, 138.
  Texana, 222.
  viscidiflora, 138, 139, 140.
  Wrightii, 142.
Linosecomia glauca, 274.
Lion's-foot, 434.
Lipochæta Texana, 286.
Logfia subulata, 230.
Lonicera, 8, 14.
  alba, 30.
  albiflora, 18.
  Breweri, 15.
  cærulea, 15.
  Californica, 18.
  Canadensis, 15.
  Caroliniana, 16.
  ciliata, 15.
  ciliosa, 16.
  ciliosa, 18.
  Diervilla, 19.
  dioica, 17.
  Douglasii, 17.
  dumosa, 18.
  flava, 17.
  flava, 17.
  hirsuta, 17.
  hispidula, 18.
  glauca, 17.
  grata, 17.

  Goldii, 17.
  intermedia, 16.
  interrupta, 18.
  involucrata, 16.
  Ledebourii, 16, media 17
  microphylla, 17.
  Mociniana, 16.
  oblongifolia, 15.
  occidentalis, 16.
  parviflora, 17.
  pilosa, 18.
  pubescens, 17.
  sempervirens, 16.
  subspicata, 18.
  Sullivantii, 17.
  Symphoricarpos, 13.
  Tartarica, 16.
  Utahensis, 15.
  velutina, 15.
  villosa, 15, 17.
  Virginiana, 16.
Loniceræ, 7.
Lophochæna, 264.
Lorentea, 254, 360.
Lowellia aurea, 359.
Luina, 79, 376.
  hypoleuca, 376.
Luthera Virginica, 412.
Lygodesmia, 87, 435.
  aphylla, 436.
  exigua, 436.
  grandiflora, 435.
  juncea, 435.
  juncea, 436.
  minor, 413.
  rostrata, 436.
  spinosa, 436.

Machæranthera, 204.
  canescens, 205, 206.
  grandiflora, 125.
  parviflora, 207.
  Shastensis, 174.
  tanacetifolia, 206.
Macrocarphus
  achilleæfolius, 341.
  Douglasii, 341.
Macrohoustonia, 24.
Macronema, 135.
  discoidea, 135.
  suffruticosa, 135.
Macrorhynchus, 438.
  angustifolius, 439.
  aurantiacus, 438.
  Californicus, 439.
  Chilensis, 440.
  cynthioides, 437.
  elatus, 438.
  glaucus, 437.
  grandiflorus, 438, 439.
  Harfordii, 439.
  heterophyllus, 439.
  humile, 439.
  laciniatus, 439.
  Lessingii, 439.
  purpureus, 438.
  retrorsus, 439.
  troximoides, 438.
Madaria, 304.
  corymbosa, 305.
  elegans, 305.
  racemosa, 305.
Madariopsis, 305.
Madaroglossa, 314.
  angustifolia, 314, 315.

  carnosa, 314.
  elegans, 315.
  heterotricha, 315.
  hieracioides, 315.
  hirsuta, 315.
Madder, 35.
Madia, 69, 304.
  anomala, 307.
  Bolanderi, 304.
  capitata, 305.
  citriodora, 307.
  dissitiflora, 305.
  elegans, 305.
  filipes, 306.
  glomerata, 306.
  mellosa, 305.
  Nuttallii, 304.
  radiata, 305.
  sativa, 305.
  stellata, 305.
  viscosa, 305.
  Yosemitana, 304.
MadieÆ, 69.
Madorella dissitiflora, 305.
  racemosa, 305.
Malacolepis, 421.
Malacomeris, 423.
  incana, 423.
Malacothrix, 86, 421.
  Californica, 422.
  Californica, 422.
  Clevelandi, 423.
  commutata, 423.
  Coulteri, 421.
  crepoides, 423.
  Fendleri, 422.
  glabrata, 422.
  incana, 423.
  obtusa, 423.
  obtusa, 422.
  parviflora, 423.
  platyphylla, 410.
  saxatilis, 423.
  sonchoides, 422.
  sonchoides, 422.
  tenuifolia, 423.
  Torreyi, 422.
  Torreyi, 422.
  Xanti, 422
Mallostoma acerosa, 27.
Margacola parvula, 93.
Marsh Elder, 247.
Marshallia, 68. 303.
  angustifolia, 303.
  cæspitosa, 303.
  lanceolata, 303.
  latifolia, 303.
  Schreberi, 303.
Maruta Cotula, 362.
  fœtida, 362.
Matricaria, 78, 362.
  asterioides, 166.
  Chamomila, 364.
  coronata, 364.
  Courrantia, 364.
  discoidea, 364.
  glastifolia, 166.
  inodora, 363.
  odorata, 365.
  Parthenium, 365.
  pyrethroides, 364.
  tanacetoides, 364.
Matthiola, 29.
  scabra, 30.
Mayweed, 362.

INDEX. 467

Megalastrum, 173.
MELAMPODIEÆ, 60.
Melampodium, 60, 238.
— australe, 239.
— cinereum, 239.
cupulatum, 239.
hispidum, 239.
leucanthum, 239.
longicornu, 239.
ramosissimum, 239.
Melananthera, 257.
Melanthera, 65, 257.
angustifolia. 257.
deltoidea, 257.
hastata, 257.
lanceolata, 257.
Linnæi, 257.
microphylla, 257.
panduræformis, 257.
triloba, 257.
Micropus. 5 226.
amphibolus, 227.
angustifolius, 227.
Californicus, 227.
Grayana, 227.
minimus, 229.
Microseris. 85, 416.
acuminata, 419.
aphantocarpha, 419.
attenuata, 419.
Bigelovii, 419.
Bolanderi, 418.
borealis, 424.
cyclocarpha, 420.
Douglasii, 420.
elegans, 419.
laciniata, 417.
leptosepala, 418.
leptosepala, 418.
Lindleyi, 419.
linearifolia, 418.
macrochæta, 418.
nutans, 416.
Parryi, 418.
platycarpha, 420.
sylvatica, 417.
troximoides, 420.
Mikania, 51, 94.
artemisioides, 97.
convolvulacea, 94.
cordifolia, 94.
gonoclada, 94.
menispermoidea, 94.
pubescens, 94.
rubiginosa, 94.
scandens, 94.
suaveolens, 94.
Milfoil, 363.
Milk Thistle, 405.
Milleria angustifolia, 354.
MILLERIEÆ, 59.
Mint Geranium, 365.
Mist-flower, 102.
Mitchella, 22, 31.
repens, 31.
undulata, 31.
Mitracarpium, 32.
Mitracarpum, 32.
Mitracarpus, 22, 32.
breviflorus, 32.
linearis, 32.
Molina viscosa, 225.
Monachæna, 247.

Monolopia, 72, 323.
babiæfolia, 324.
glabrata, 324.
gracilens, 323.
Heermanni, 323.
lanceolata, 323.
major, 323.
minor, 323.
Monoptilon, 55, 165.
bellidiforme, 165.
Monothrix Stansburii, 320.
Moquinia hypoleuca, 407.
Morinda, 21, 29.
Roloc, 29.
Mugwort, 367, 372.
Mulgedium, 443.
acuminatum, 444.
Floridanum, 443.
hastatum, 435.
heterophyllum, 443.
leucophæum, 444.
lyratum, 443.
multiflorum, 444.
pulchellum, 443.
MUTISIACEÆ, 82, 407.
Myrstiphyllum, 30.

Nabalus, 433.
alatus, 436.
albus, 434.
altissimus, 435.
asper, 433.
Boottii, 435.
cordatus, 435.
crepidineus, 433.
deltoideus, 435.
Fraseri, 434.
glaucus, 434.
Illinoensis, 433.
integrifolius, 434.
nanus, 434.
racemosus, 433.
Roanensis, 434.
serpentarius, 434.
trifoliolatus, 434.
trilobatus, 434.
virgatus, 434.
Nardosmia angulosa, 376.
corymbosa, 376.
frigida, 376.
palmata, 376.
sagittata, 376.
NASSAUVIEÆ, 83.
Nauenbergia trinervata, 354.
Neocis, 398.
Nicolletia, 77, 355, 446.
Edwardsii, 355.
occidentalis, 355.
Nippleworth, 410.
Nothocalais, 420.

Obeliscaria columnaris, 264.
pinnata, 264.
pulcherrima, 264.
Tagetes, 264.
Odontocarpha, 115.
Ogilva, 230.
Oldenlandia, 20, 27.
acerosa, 27.
angustifolia, 27.
Boscii, 27.
cærulea, 24.
glomerata, 27.
Greenei, 27.
Halei, 28.

humifusa, 26.
pentandra, 28.
purpurea, 26.
rotundifolia, 25.
rubra, 25.
serpyllifolia, 24.
subviscosa, 25.
uniflora, 28.
Oligogyne Tampicana, 289.
Oligosporus, 368.
pycnocephalus, 369.
Oligotrichium, 218.
Olocarpha, 308.
Omalanthus camphoratus, 367.
Omalotes camphoratus, 367.
Omalotheca, 231, 236.
supina, 236.
Onopordon, 82, 405.
acanthium, 405.
Onopordum, 405.
Oporinia, 420.
autumnalis, 420.
Oritrophium, 175.
Orthomeris, 198.
Osmadenia tenella, 311.
Osmia, 94.
Osteospermum Uvedalia, 238.
Ox-eye Daisy, 365.
Oxylepis, 347.
Oxytenia, 62, 248.
acerosa, 248.
Oxytripolium, 203.
Oxyura chrysanthemoides, 316.
Oyster-plant, 415.
Palafoxia, 74, 338.
callosa, 337.
fastigiata, 338.
Feayi, 338.
Hookeriana, 337.
integrifolia, 338.
latifolia, 338.
leucophylla, 338.
linearis, 338.
Texana, 337.
Paleolaria carnea. 338.
fastigiata, 338.
Pallasia, 281.
grandifolia, 281.
serratifolia, 289.
Pappochroma, 206.
Pappotbrix, 319.
Partheniastrum, 244.
Parthenice, 62, 245.
mollis, 245.
Parthenium, 62, 244.
alpinum, 245.
argentatum, 245.
Hysterophorus, 244.
incanum, 244.
integrifolium, 245.
lobatum, 244.
lyratum, 244.
ramosissimum, 245.
Partridge-berry, 31.
Patrinia ceratophylla, 43.
longifolia, 43.
Pectidium, 361.
punctatum, 362.
Pectidopsis angustifolia, 361.
Pectis, 77, 360.
angustifolia, 360.
ciliaris, 360.
Coulteri, 361.
fastigiata, 361.

INDEX.

filipes, 361.
imberbis, 362.
Jaliscana, 361.
linifolia, 360, 362.
longipes, 361.
multiseta, 361.
papposa, 361.
prostrata, 360.
punctata, 362.
Rusbyi, 361.
tenella, 360.
tenella, 361.
Pectothrix, 361.
Pentachæta, 53, 119.
alsinoides, 120.
aphantochæta, 120.
aurea, 120.
exilis, 120.
gracilis, 120.
Lyoni, 445.
Pentodon, 6, 28.
Halei, 28.
Pentotis, 28.
Peramibus hirtus, 259.
Perdicium, 409.
semiflosculare, 408.
Perezia, 83, 408.
Arizonica, 409.
Coulteri, 409.
microcephala, 409.
nana, 409.
runcinata, 408.
Pericome, 71, 322.
caudata, 322.
Perityle, 71, 320.
Acmella, 322.
aglossa, 322.
Californica, 321.
Californica, 321, 322.
coronopifolia, 321.
dissecta, 321.
Emoryi, 321.
Fitchii, 321.
incana, 320.
leptoglossa, 322.
microglossa, 322.
nuda, 321.
Parryi, 322.
plumigera, 321.
PERITYLEÆ, 71.
Persoonia angustifolia, 303.
lanceolata, 303.
latifolia, 303.
Petasites, 79, 375.
frigida, 376.
palmata, 376.
sagittata, 376.
vulgaris, 376.
Peucephyllum, 79, 377.
Schottii, 377.
Phænactis, 207, 208, 209.
Phæthusa Americana, 287.
borealis, 287.
Phalcrocline, 225.
Phalacroloma, 207, 218.
acutifolium, 219.
Beyrichii, 219.
obtusifolium, 219.
Phalacroseris, 84, 410.
Bolanderi, 410.
Phialis, 328, 329.
Philozera multiflora, 347.
Phyllacrocephala, 94.
Phyllactis obovata, 42.
Phyllopappus, 416.

Phyllopappus, 124.
Phyllotheca, 123.
Picradenia Richardsonii, 347.
Picris, 85, 420.
Davurica, 421.
echioides, 421.
hieracioides, 420.
Japonica, 421.
Kamtschatica, 421.
Picrothamnus desertorum, 368.
Picrotus, 246.
Pilosella, 424.
arguta, 428.
spathulata, 426.
Pinaropappus, 85, 421.
roseus, 421.
Pinckneya, 19, 23.
pubens, 23.
pubescens, 23.
Pinknea pubescens, 23.
Pityopsis, 121.
argentea, 121.
falcata, 122.
graminifolia, 121.
pinifolia, 122.
Plagius, 364.
Plateilema, 344.
Platycarpæa, 296.
Platylophus, 406.
Platyschkuhria, 332.
Pleiocephalus Americanus, 407.
Plectritis, 46.
brachystemon, 47.
capitata, 47.
congesta, 47.
macrocera, 46.
major, 47.
samolifolia, 47.
Pluchea, 57, 225.
bifrons, 226.
borealis, 225.
camphorata, 226.
fœtida, 226.
glabrata, 226.
Marylandica, 226.
petiolata, 226.
purpurascens, 226.
PLUCHEINEÆ, 57.
Plummera, 59, 237.
floribunda, 237.
Polyactidium, 207, 219.
Polyactis, 207, 219.
Polydymia, 377.
Polymnia, 60, 238.
Canadensis, 238.
Caroliniana, 243.
Tetragonotheca, 255.
Uvedalia, 238.
Polymniastrum, 238.
Polypappus sericeus, 225.
Polypteris, 74, 337.
callosa, 337.
Hookeriana, 337.
integrifolia, 337.
integrifolia, 352.
Texana, 337.
Porophyllum, 76, 354.
amplexicaule, 355.
gracile, 355.
Greggii, 355.
macrocephalum, 354.
Prenanthes, 87, 432.
alata, 435.
alba, 434.
alba, 434, 435.

altissima, 434.
ophylla, 436.
aspera, 433.
autumnalis, 433.
Boottii, 435.
cordata, 435.
crepidinea, 433.
crepidinea, 434.
deltoidea, 435. *die*
exigua, 436.
glauca, 434.
Illinoensis, 433.
juncea, 435.
Mainensis, 433.
Miamensis, 434.
ovata, 434.
pauciflora, 413.
polymorpha, 431.
proteophylla, 434.
pygmæa, 431.
racemosa, 433.
Roanensis, 434.
rubicunda, 434.
runcinata, 413.
serpentaria, 434.
simplex, 433.
suavis, 434.
tenuifolia, 413.
Prionopsis, 125.
Chapmanii, 174.
ciliata, 125.
Proustia, 408.
Psacalium, 397.
Psathyrotes, 80, 377.
annua, 377.
incisa, 343.
pilifera, 377.
ramosissima, 377.
scaposa, 378.
Pseudo-Bahia, 323.
Pseudo-Monolepis, 331.
Psilactis, 56, 170.
asteroides, 171.
brevilingulata, 171.
Coulteri, 171.
Psilocarpæa, 296.
Psilocarphus, 58, 228.
brevissimus, 228.
globiferus, 228.
Oreganus, 228.
tenellus, 228.
Psilochæna, 431.
occidentalis, 432.
Psilostrophe gnaphaliodes, 318.
Psychotria, 21, 30.
chimarroides, 31.
lanceolata, 31.
nervosa, 31.
oligotricha, 31.
rufescens, 31.
tenuifolia, 31.
undata, 30.
Psychotrophum, 30.
Ptarmica borealis, 363.
vulgaris, 363.
Pterocaulon, 57, 226.
pycnostachyum, 226.
virgatum, 226.
Pteronia Caroliniana, 109.
Pterophyton, 287.
Ptilomeris, 327.
affinis, 327.
anthemoides, 328.
aristata, 327.
coronaria, 327.

INDEX. 469

mutica, 328.
tenella, 327.
—Ptilonella scabra, 304.
—Ptilophora, 416.
major, 417.
—nutans, 417.
Pugiopappus, 300.
Bigelovii, 300.
Breweri, 300.
calliopsideus, 300.
—Pulicaria annua, 219.
—Pyrethrum, 364.
bipinnatum, 364.
—inodorum, 363.
—Parthenium, 365.
Pyrochæta, 170.
—Pyrrhopappus, 87, 440.
—Carolinianus, 441.
grandiflorus, 441.
multicaulis, 441.
—multicaulis, 441.
—pauciflorus, 441.
Rothrockii, 441.
scaposus, 441.
—Sesseanus, 441.
Pyrrocoma arguta, 127.
—carthamoides, 126.
—foliosa, 128.
glomerata, 127.
—grindelioides, 125.
—Menziesii, 143.
paniculata, 127.
—racemosa, 127.
radiata, 126.

—Rafinesquia, 85, 415.
—Californica, 415, 446.
—Neo-Mexicana, 415.
—Ragweed, 248.
—Raillardella, 81, 380.
argentea, 380.
Eiseni, 380.
Muirii, 380.
Pringlei, 380.
scaposa, 380.
—Raillardia, 380.
—Rancagua, 324.
—Randia, 20, 28.
— aculeata, 28.
— clusiæfolia, 29.
— latifolia, 29.
— mitis, 29.
Xalapensis, 29.
—Ratibida columnaris, 264.
— sulcata, 264.
—Rattlesnake-root, 433, 434.
—Relbunium, 35, 40, 41.
— microphyllum, 41.
— polyplocum, 41.
—Rhabdacaulis, 290.
—Rhinactina, 260.
—Rhyncholepis, 227.
—Richardia, 22, 32.
—scabra, 32.
—Richardsonia scabra, 32.
— Riddellia, 71, 317.
—arachnoidea, 318.
Cooperi, 318.
— tagetina, 317.
—RIDDELLIEÆ, 71.
— Rigiopappus, 74, 338.
— leptocladus, 338.
Roioc, 29.
—Rosalesia, 104.
— Rosin-weed, 240, 241.

Rothia Carolinensis, 335.
Rubia, 35.
—Brownei, 42.
—peregrina, 42.
tinctoria, 35.
—Walteri, 42.
RUBIACEÆ, 19.
—Rudbeckia, 65, 250.
— alata, 348, apetala, 274
alismæfolia, 261.
—amplexicaulis, 263.
—amplexifolia, 263.
—angustifolia, 278.
—apetala, 274.
—aristata, 259.
aspera, 261.
— atrorubens, 259.
bicolor, 260.
bupleuroides, 259.
Californica, 262.
— chrysomela, 261.
— columnaris, 264.
digitata, 262, 264.
— discolor, 260, 261.
fulgida, 260.
— fulgida, 261.
glabra, 262.
— globosa, 264.
— gracilis, 260.
grandiflora, 261.
Heliopsidis, 261.
Heliopsidis, 261.
heterophylla, 263.
— hirta, 260.
— hispida, 258.
— laciniata, 262.
lævigata, 262.
— maxima, 262.
— Mohrii, 259.
— mollis, 261.
— montana, 263.
nitida, 262.
— occidentalis, 263.
— odorata, 260.
— oppositifolia, 255.
— pallida, 258.
— perfoliata, 258.
pinnata, 264.
Porteri, 259.
— purpurea, 258.
— quinata, 262.
— radula, 274.
rupestris, 260.
— serotina, 258, 260.
spathulata, 261.
— spathulata, 263.
— speciosa, 261.
— speciosa, 258.
— strigosa, 260.
— subtomentosa, 260.
— Tagetes, 260.
— tomentosa, 260, 264.
— triloba, 259.
— Rugelia nudicaulis, 383.

— Sage-brush, 367, 374.
— Sage-bush, 374.
— Salsify, 415.
— SAMBUCEÆ, 7.
— Sambucus, 7, 8.
— Canadensis, 8.
— glauca, 9.
— humilis, 9.
melanocarpa, 8.
— Mexicana, 9.

— nigra, 9.
— pubens, 9.
— pubescens, 9.
— racemosa, 8.
— velutina, 9.
— Santolina suaveolens, 364.
— Sanvitalia, 64, 254.
— Aberti, 254.
acinifolia, 254.
angustifolia, 254.
— ocymoides, 254.
procumbens, 254.
— trayiæfolia, 254.
villosa, 254.
Sartwellia, 76, 353.
Flaveriæ, 353.
Mexicana, 353.
— Saussurea, 82, 396.
acuminata, 397.
— alpina, 396.
— Americana, 897.
— angustifolia, 397.
Ledebouri, 397.
— monticola, 397.
nuda, 397.
subsinuata, 397.
Scabiosa atropurpurea, 47.
Scabious, 47.
— Scariola, 442.
Schkuhria, 78, 384.
Bigelovii, 333.
bifurcata, 333.
Hopkirkia, 334.
integrifolia, 332.
— Neo-Mexicana, 333.
pedata, 333.
pusilla, 334.
Wislizeni, 334.
Woodhousii, 333.
— Wrightii, 334.
Sclerocarpus, 64, 256.
— exiguus, 306.
— gracilis, 305.
uniserialis, 256.
— Sclerolepis, 51.
— verticillata, 51.
— Scorzonella, 416, 417.
— glauca, 418.
laciniata, 417.
— leptosepala, 418.
— nutans, 417.
sylvatica, 417.
— Scorzonera pinnatifida, 441.
— Sellou glutinosa, 114.
— nudata, 354.
— Senecio, 81, 383.
Actinella, 384.
— amplectens, 384.
ampullaceus, 393.
— Andinus, 387.
Arizonicus, 392.
aromioides, 388.
— atriplicifolius, 397.
— aureus, 391.
— aureus, 389, 394.
— Balsamitæ, 391.
Bigelovii, 385.
— Bolanderi, 392.
Californicus, 393.
— campestris, 388, 389.
— canus, 390.
canus, 389.
Cardamine, 390.
— Carolinianus, 394.
— cernuus, 385.

ciliatus, 221, 383.
Cineraria, 383.
Clarkianus, 393.
Clevelandi, 387.
Coronopus, 393.
crassulus, 387.
Cumingii, 381.
Cymbalaria, 391, 392.
cymbalarioides, 391.
debilis, 391.
densiflorus, 394.
Douglasii, 393.
Elliottii, 391.
elongatus, 391.
eremophilus, 392.
eurycephalus, 392.
eurycephalus, 384.
exaltatus, 388.
fastigiatus, 390.
fastigiatus, 391, 393.
Fendleri, 392. *Fendleri* 391
filifolius, 383.
flocciferus, 383, 422.
Fremonti, 386.
flabellus, 394. *gracilis* 391
Grayanus, 397.
Greenei, 385.
Hartwegi, 386.
hieraciifolius, 397.
Hookeri, 389.
Howellii, 390.
Huachucanus, 386.
hydrophilus, 387.
imparipinnatus, 394.
integerrimus, 388.
integerrimus, 387.
integrifolius, 389, 390.
Jacobæa, 383.
Kalmii, 383, 394.
lanceolatus, 387.
Layneæ, 390.
Lemmoni, 385.
lobatus, 396.
longidentatus, 386.
longilobus, 393.
lugens, 388.
lugens, 387.
lyratus, 394.
megacephalus, 385.
Mendocinensis, 384.
Millefolium, 392.
Mississippianus, 394.
Mohavensis, 446.
multilobatus, 394.
Neo-Mexicanus, 392.
obovatus, 391.
Palmeri, 383.
palustris, 394.
Parryi, 393.
pauciflorus, 391.
pauperculus, 391.
petræus, 389.
Plattensis, 391.
Pseudo-Arnica. 384.
Purshianus, 390.
rapifolius, 386.
Regiomontanus, 393.
renifolius, 389.
resedifolius, 390.
Riddellii, 393.
Rugelia, 383.
Rusbyi, 385.
Schweinitzianus, 394.
Seemanni, 386.
serra, 386.

— Soldanella, 384.
— spartioides, 393.
— *stæchadiformis*, 393.
— *suaveolens*, 395.
subnudus, 391.
sylvaticus, 394.
Tampicanus, 394.
Thurberi, 389.
Toluccanus, 388.
tomentosus, 390.
triangularis, 386.
viscosus, 394.
vulgaris, 394.
werneriæfolius, 389.
Whippleanus, 384.
SENECIONIDEÆ, 79, 375.
Sericocarpus, 56, 171.
— conyzoides, 171.
— *Oregonensis*, 172.
— rigidus, 172.
— solidagineus, 171.
— tortifolius, 172.
Sericophyllum, 121.
Serinia cæspitosa, 411.
Seriphidium, 374.
Serratula arvensis, 398.
— Carolinensis, 91.
— compta, 111.
— glauca, 89.
— Noveboracensis, 89.
— pilosa, 111.
— præalta, 89.
— scariosa, 110.
— speciosa, 109.
— spicata, 111.
— squarrosa, 109.
Seven-Years' Apple, 29.
Shortia Californica, 327.
Silphium, 61, 240.
— asperrimum, 240.
— Asteriscus, 241.
— *Asteriscus*, 243.
— *atropurpureum*, 241.
— albiflorum, 242.
— *betonicæfolium*, 242.
— compositum, 241.
— *confunctum*, 240.
— *ornatum*, 240.
— dentatum, 241.
— elatum, 242.
— erythrocaulon, 240.
— gracile, 241.
— *gummiferum*, 242.
— *Hornemanni*, 240.
— integrifolium, 240.
— *lævigatum*, 240, 241.
— lanceolatum, 241.
— laciniatum, 242.
— *laciniatum*, 241.
— nudicaule, 241.
— *Nuttallianum*, 243.
— *punctum*, 243.
— perfoliatum, 240.
— pinnatifidum, 242.
— *radula*, 240.
— reticulatum, 241, 243.
— *reniforme*, 242.
— *scabrum*, 240, 241.
— *speciosum*, 242.
— scaberrimum, 240.
— spicatum, 242.
— *subacaule*, 243.
— tetragonum, 240.
— trifoliatum, 241.
— ternatum, 241.

— *ternifolium*, 241.
— terebinthinaceum, 242.
— *tomentosum*, 242.
Simsia, 283.
calva, 283.
— canescens, 282.
exaristata, 283.
— frutescens, 282.
lagascæformis, 283.
lanceolata, 285.
scaposa, 282.
Snowberry, 13, 30.
Sneezewort, 363.
Solidago, 54, 143.
— alba, 146.
— altissima, 151, 153, 157.
— ambigua, 143, 146.
amplexicaulis, 153.
— *amplexicaulis*, 160.
— angulata, 152.
— angusta, 145.
angustifolia, 150.
argentea, 145.
— arguta, 154.
— arguta, 145, 155.
— aspera, 153.
— asperata, 152, 153.
— asperula, 153.
— axillaris, 145.
— azurica, 149.
bicolor, 146.
Bigelovii, 146.
Boottii, 154.
brachyphylla, 154.
Buckleyi, 147.
cæsia, 145.
— Californica, 158.
— Californica, 147.
— Canadensis, 157.
carinata, 149.
— carnosa, 149.
Chapmani, 151.
— ciliaris, 155.
— cinerascens, 158.
Cleliæ, 143.
— compacta, 147.
— conferta, 158.
— confertiflora, 148.
— confertiflora, 144.
confinis, 149.
— cordata, 161.
corymbosa, 159.
— corymbosa, 144, 148.
Curtisii, 156.
— decemflora, 158, 159.
diffusa, 152.
— Drummondii, 159.
dubia, 143.
— elata, 144, 157.
Elliottii, 153.
elliptica, 143.
elliptica, 153.
elongata, 157.
— elongata, 153, 156.
— eminens, 157.
— erecta, 144, 146, 152.
— fistulosa, 151.
— flabellata, 145.
— flabelliformis, 145.
flavovirens, 150.
flexicaulis, 145. 146
— fragrans, 143, 156.
— Frankii, 152.
fuscata, 144.
— Gattingeri, 156.

INDEX. 471

— *genistoides*, 150.
— *gigantea*, 156.
— *glaberrima*, 155.
— *glabra*, 156.
— glomerata, 147.
— *glutinosa*, 148.
 gracilis, 145.
 gracillima, 150.
— *graminifolia*, 161.
— *grandiflora*, 144, 159.
 Guiradonis, 151.
— *hirsuta*, 146.
— *hirta*, 153.
— *hispida*, 146, 158.
 Houghtoni, 160.
— humilis, 148.
— *humilis*, 153.
 incana, 158.
 integerrima, 149.
 integrifolia, 149.
— juncea, 154.
— *lævigata*, 149.
 lanata, 146.
— lanceolata, 160.
— lancifolia, 145.
— *lateriflora*, 143, 153, 187.
— latifolia, 145.
 latifolia, 143.
 latissimifolia, 143.
 Leavenworthii, 156.
— *leiocarpa*, 147.
 lepida, 157.
— leptocephala, 161.
— *limonifolia*, 149.
 Lindheimeriana, 147.
— *linoides*, 150, 154.
 lithospermifolia, 144, 149.
 livida, 144, 145.
— *longifolia*, 157.
— macrophylla, 147.
— *macrophylla*, 145.
 Marshalli, 158.
— *Mexicana*, 149.
— Missouriensis, 155.
— *Missouriensis*, 156.
 mollis, 157, 158.
— monticola, 146.
— *Muhlenbergii*, 155.
— *multiflora*, 153.
— multiradiata, 147.
 nana, 158.
— neglecta, 154.
— nemoralis, 158.
— nitida, 160.
— *Noveboracensis*, 143, 149.
— *nutans*, 157.
 obovata, 150.
— occidentalis, 160.
— odora, 150.
— Ohioensis, 159.
— patula, 152.
 pauciflora, 144.
 paucifloculosa, 161.
 petiolaris, 144.
— *petiolaris*, 146, 147, 149, 152, 158.
— pilosa, 151.
— *pilosa*, 153.
— *Pitcheri*, 156.
— *procera*, 157.
 pubens, 146.
— puberula, 150.
— *puberula*, 158.
— *pulverulenta*, 150.
— *puncticulata*, 151.

— *pyramidata*, 151.
— radula, 158.
 recurvata, 144, 145.
— *recurvata*, 153.
— *reflexa*, 157.
— *retrorsa*, 151.
— Riddellii, 160.
— *rigida*, 159.
— *rigidula*, 153.
— *rotundifolia*, 159.
— rugosa, 153.
 rupestris, 156.
 salicina, 152.
— *Sarothræ*, 115.
— *scaberrima*, 159.
— *scabra*, 152, 153, 157.
 scabrida, 157.
 Schraderi, 145.
— sempervirens, 149.
— *sempervirens*, 149, 152.
— serotina, 156.
— *serotina*, 155.
 Shortii, 156.
 sparsiflora, 159.
 spathulata, 148.
 speciosa, 152.
 speciosa, 147.
 spectabilis, 151.
— *sphacelata*, 161.
 spiciformis, 149.
 spithamæa, 147.
— squarrosa, 144.
— *squarrosa*, 144.
 stricta, 149.
— *stricta*, 148, 152, 157.
 tenuifolia, 161.
 Terra-Novæ, 154.
— *thyrsoidea*, 147.
 Tolmieana, 151.
 tortifolia, 151.
 uliginosa, 151.
— *uliginosa*, 154.
— ulmifolia, 153.
— *ulmifolia*, 159.
— velutina, 158.
— verna, 152.
— *verrucosa*, 143, 155.
— *villosa*, 151, 153.
— *viminea*, 146, 149.
— Virgaurea, 148.
— *Virgaurea*, 147, 148.
— *virgata*, 150.
— Virginiana, 153.
 Soliva, 78, 365.
 daucifolia, 365.
 nasturtifolia, 366.
 sessilis, 366.
 Solivæa, 365.
— Sonchus, 87, 444.
— *acuminatus*, 444.
— *alpinus*, 444.
— *arvensis*, 444.
— asper, 444.
— *biennis*, 444.
 Californicus, 423.
— Canadensis, 444.
— *Carolinianus*, 444.
— *Floridanus*, 443, 444.
 hastatus, 435.
— *leucophæus*, 444.
 Ludovicianus, 443.
 macrophyllus, 444.
— *multiflorus*, 444.
— oleracens, 444.
— *pallidus*, 442, 444.

— *pulchellus*, 443.
— *racemosus*, 444.
— *Sibiricus*, 443.
— *spicatus*, 444.
— *spinulosus*, 444.
 tenerrimus, 444.
 tenuifolius, 444.
— Sow-Thistle, 444.
 Sparganophorus verticillatus, 92.
 Spermacoce, 22, 33.
— *Chapmanii*, 34.
— *diodina*, 35.
— glabra, 34.
— *hirta*, 34.
 involucrata, 33.
— parviflora, 34.
— podocephala, 34.
 Portoricensis, 34.
— *pygmæa*, 34.
— *subulata*, 33.
— tenuior, 34.
 tetracocca, 33.
— *Virginiana*, 35.
 Sphæromeria argentea, 367.
— *capitata*, 367.
 Spilanthes, 65, 258.
— Nuttallii, 258.
— *occidentalis*, 258.
 Pseudo-Acmella, 322.
— *repens*, 258.
— Spilanthus, 258.
— *Spiridanthes*, 328.
— *Stæhelina elegans*, 109.
— *Starkea pinnata*, 130.
 Starwort, 172.
— Star Thistle, 405.
 Staurospermum, 32.
 Stemmodontia scaberrima, 286.
— STELLATÆ, 23.
— Stenactis, 207, 219.
— *ambigua*, 219.
— *annua*, 219.
 Beyrichii, 219.
— *dubia*, 219.
 glauca, 208.
— *speciosa*, 209.
— *strigosa*, 219.
— *verna*, 216.
— Stenotheca, 425.
— *Mariana*, 426.
— *nubnuda*, 426.
— *venosa*, 426.
— Stenotus, 131.
— *acaulis*, 132.
— *armerioides*, 132.
— *cæspitosus*, 132.
— *florifer*, 168.
 linearifolius, 132.
— *multicaulis*, 129.
— *pygmæus*, 131.
— *Stephananthus*, 221.
— Stephanomeria, 84, 412.
— *cichoriacea*, 413.
 elata, 413.
— *exigua*, 414.
— *heterophylla*, 413.
— *intermedia*, 417.
— lactucina, 413.
— minor, 413.
 myrioclada, 414.
 paniculata, 414.
 Parryi, 413.
 pentachæta, 414.
— runcinata, 413.
— Thurberi, 413.

472 INDEX.

Schottii, 415.
virgata, 414.
Wrightii, 414.
— Stevia, 51, 91.
amabilis, 91.
— angustifolia, 92.
— callosa, 337.
— canescens, 92.
— ivæfolio, 92.
Lemmoni, 92.
macella, 91.
micrantha, 91.
Plummeræ, 92.
punctata, 92.
salicifolia, 92. *serrata 92*
Sphacelata, 337.
virgata, 92.
Stick-tight, 286.
— Stokesia, 50, 88.
cyanea, 88.
Strigia, 103.
Strangylasperma australe, 366.
Strumpfia, 21, 31.
maritima, 31.
Stylesia, 331.
Stylimnus, 225.
— Stylocline, 58, 227.
acaulis, 229.
— filaginea, 228.
— gnaphalioides, 227.
micropoides, 227.
— Stylopappus, 439.
elatus, 438.
grandiflorus, 439.
— laciniatus, 439.
Succory, 412.
Sunflower, 271.
Suprago, 109.
Sweet Scabious, 47.
Symphiotrichium unctuosum, 189.
— Symphoria conglomerata, 13.
— elongata, 14.
— glomerata, 13.
— heterophylla, 14.
— occidentalis, 13.
— racemosa, 13.
Symphoricarpos, 7, 13.
ciliatus, 14.
longiflorus, 14.
— mollis, 14.
— montanus, 14.
— occidentalis, 13.
— oreophilus, 14.
— parviflorus, 13.
— racemosus, 13.
— rotundifolius, 14.
— spicatus, 13.
— vulgaris, 13.
— Synedrella, 67, 289.
— vialis, 289.
Syntrichopappus, 73, 328.
Fremonti, 328.
Lemmoni, 328. *Syntrip vacciflora, 31*
— Tagetes, 77, 359.
Lemmoni, 359.
— lucida, 359.
— micrantha, 360.
— papposa, 359.
TAGETINEÆ, 76.
— Tanacetum, 78, 366.
bipinnatum, 364.
— boreale, 366.
camphoratum, 366.
canum, 367.

capitatum, 367.
diversifolium, 367.
— Douglasii, 366.
elegans, 367.
— Huronense, 366.
Huronense, 367.
Kotzebuense, 364.
— matricarioides, 364.
Nuttallii, 367.
— Parthenium, 365.
pauciflorum, 364, 366.
— potentilloides, 367.
— suaveolens, 364.
— vulgare, 366.
Taraxacum, 87, 440.
ceratophorum, 440.
corniculatum, 440.
— Dens-Leonis, 440.
— hirsutum, 439.
lævigatum, 440.
lanceolatum, 440.
latilobum, 440.
montanum, 440.
officinale, 440.
palustre, 440.
phymatocarpum, 440.
— Tarweed, 304, 306.
— Teasel, 47.
Tessaria borealis, 225.
— Tetracarpum, 334.
— Tetradymia, 80, 379.
— canescens, 379.
— comosa, 379.
glabrata, 379.
inermis, 379.
Nuttallii, 379.
ramosissima, 377.
spinosa, 379.
squamata, 378.
Tetragonosperma
— lyratifolium, 256.
— Tetragonotheca, 64, 255.
helianthoides, 255.
Ludoviciana, 256.
— Texana, 256.
Fetrodus quadridentatus, 348.
— Thelecarpæa, 408.
Thelesperma, 68, 301.
ambiguum, 301.
— filifolium, 301.
— gracile, 302.
— longipes, 302.
simplicifolium, 302.
subnudum, 302.
subsimplicifolium, 302.
— Therolepta, 303.
— Thistle, 397.
Thoroughwort, 94, 99.
— Thymelea, 31.
— Thymophylla, 359.
— Greggii, 359.
setifolia, 359.
— Thymophyllum, 359.
Tickseed, 289.
Tinker's-weed, 12.
— Tithonia, 66, 271.
argophylla, 284.
Thurberi, 271.
tubæformis, 271.
Tulatia chrysanthemoides, 316.
Townsendia, 56, 166.
Arizonica, 169.
condensata, 167.
eximia, 167.
Fendleri, 167.

— florifer, 167.
Fremontii, 169.
glabella, 169.
— grandiflora, 167.
incana, 169.
Mexicana, 169.
— Parryi, 167.
Rothrockii, 168.
scapigera, 168.
— sericea, 168.
spathulata, 169.
strigosa, 169.
strigosa, 168.
Watsoni, 168.
Wilcoxiana, 168.
Wrightii, 173.
— Tragacanthes, 97.
Tragopogon, 85, 415.
— Dandelium, 412.
— porrifolius, 415.
— pratensis, 416.
Virginicum, 412.
Trattenickia angustifolia, 303.
— lanceolata, 303.
latifolia, 303.
Trichocoronis, 51, 92.
rivularis, 93.
Wrightii, 92.
Trichophyllum, 329.
— integrifolium, 331.
— lanatum, 330.
— multiflarum, 331.
— oppositifolium, 332.
Trichoptilium, 75, 343.
incisum, 343.
Tridax gaillardioides, 315.
Trillisa, 52, 113.
— odoratissima, 113.
— paniculata, 114.
Trimorphæa, 219.
— vulgaris, 220.
Triosteospermum, 12.
Triosteum, 7, 12.
— angustifolium, 12.
— majus, 12.
— minus, 12.
— perfoliatum, 12.
Tripleurospermum
— inodorum, 363.
Triplopappus, 196.
— Tripolium angustum, 204.
carici folium, 202.
conspicuum, 201, 203.
— divaricatum, 203.
frondosum, 204.
imbricatum, 201.
occidentale, 192.
Origanum, 192.
paludosum, 174.
— subulatum, 202, 203, 204.
Trixis, 83, 409.
— augustifolia, 409.
— Californica, 409.
— corymbosa, 410.
frutescens, 410.
— frutescens, 410.
suffruticosa, 410.
— Trochaseris, 438.
— Troximon, 86, 87, 436.
alpestre, 437.
— apargioides, 438.
— apargioides, 439.
— aurantiacum, 438.
— aurantiacum, 438.
— barbellulatum, 437.

INDEX. 473

Chilense, 439.
cuspidatum, 437. *Dandelion*
glaucum, 437. 412
gracilens, 438.
grandiflorum, 439.
— *grandiflorum*, 439.
heterophyllum, 439.
— humile, 438.
— *laciniatum*, 439.
marginatum, 437.
— Nuttallii, 438.
parviflorum, 437.
pumilum, 438.
retrorsum, 439.
Rœmerianum, 421. *Rosenw* 438
taraxicifolium, 438.
Trumpet Honeysuckle, 16.
Trumpet Weed, 95.
TUBULIFLORÆ, 49.
Tuckermannia, 300.
gigantea, 300.
— *maritima*, 300.
Tulocarpus, 237.
Tussilago, 79, 375.
corymbosa, 376.
Farfara, 375.
frigida, 376.
integrifolia, 408.
— *nutans*, 408.
— *palmata*, 376.
Petasites, 376.
— *sagittata*, 376.
Twin-flower, 13.

Uropappus, 416.
— *grandiflorus*, 419.
— *heterocarpus*, 419.
— *Lindleyi*, 419.
— *linearifolius*, 419.
Uvedalia, 238.

— *Vaccinium album*, 16.
— Valeriana, 42.
— Arizonica, 43.
capitata, 43.
— *ciliata*, 43.
— dioica, 43.
— edulis, 43.
— *locusta*, 44, 45.
— pauciflora, 43. *radiata* 45
— scandens, 44.
— Sitchensis, 43.
— sorbifolia, 44.
— sylvatica, 43.
VALERIANACEÆ, 42.
Valerianella, 43, 44.
amarella, 45.
— anomala, 47.
aphanoptera, 47.
— *cærulea*, 44.
— chenopodifolia, 45.
— congesta, 47.
Fagopyrum, 45.
— longiflora, 46.
— macrocera, 46.
Nuttallii, 46.
— olitoria, 44.
patellaria, 46.
— radiata, 45.
— *rhombicarpa*, 44.
— samolifolia, 47.
— stenocarpa, 45.
— *triquetra*, 45.
— *umbilicata*, 45.
Woodsiana, 45.

Vanilla-plant, 113.
Varilla, 65, 257.
Mexicana, 257.
Texana, 257.
Venegasia, 70, 317.
— carpesioides, 317.
Verbesina, 67, 286.
— *alba*, 286.
Coreopsis, 289.
— encelioides, 288.
— helianthoides, 288.
— heterophylla, 288.
laciniata, 287.
longifolia, 287.
microptera, 287.
nudicaulis, 288.
— occidentalis, 287.
— *paniculata*, 287.
— *Phæthusa*, 287.
podocephala, 286.
— *polycephala*, 287.
Siegesbeckia, 287.
— *Texana*, 287.
Warei, 288.
— *villosa*, 287.
— Virginica, 287.
Verbesinaria, 286.
VERNESINEÆ, 64.
Vermifuga, 253.
Vernonia, 50, 80.
— altissima, 90.
— angustifolia, 90.
— Arkansana, 89.
— Baldwinii, 90.
— corymbosa, 90.
— fasciculata, 90. *v. glauca* 89
— Jamesii, 90.
— Lettermani, 90.
— Lindheimeri, 91.
— Noveboracensis, 89.
oligophylla, 91.
— *ovalifolia*, 90.
— *præalta*, 89, 90.
— scaberrima, 91.
— *sphæroidea*, 90.
— *tomentosa*, 89.
VERNONIACEÆ, 50, 88.
Viburnum, 7, 9.
— acerifolium, 10.
— *acerifolium*, 10.
— *alnifolium*, 10.
— cassinoides, 11.
— *cassinoides*, 12.
— densiflorum, 10.
— dentatum, 10.
— *dentatum*, 10.
— *edule*, 10.
— ellipticum, 10. *grandifolium* 10
— *grandiflorum*, 10.
— *lævigatum*, 12.
— *Lantana*, 10.
— lantanoides, 9.
— Lentago, 12.
— molle, 11. *nitidum* 12
— *nudum*, 11.
— obovatum, 12.
— *opuloides*, 10.
— Opulus, 10.
— *Oxycoccus*, 10.
— pauciflorum, 10.
— prunifolium, 12.
— pubescens, 11. (*not italics*)
— *pyrifolium*, 11, 12.
— *Rafinesquianum*, 11.
— *scabrellum*, 11.

æquamatum, 11.
Tinus, 10.
— *trilobum*, 10.
— *villosum*, 11.
Viguiera, 66, 270.
— *brevipes*, 270.
canescens, 270.
— cordifolia, 270.
deltoidea, 271.
— *dentata*, 270.
— helianthoides, 270.
laciniata, 270.
lanata, 271.
— *microcline*, 270.
nivea, 271.
Parishii, 271.
reticulata, 271.
tephrodes, 271.
— *Texana*, 270.
tomentosa, 271.
— *triquetra*, 270.
Villanova *bipinnatifida*, 244.
— *chrysanthemoides*, 334.
Virga-aurea, 144.
Virgaurea, 144.
Virgilia, 351.
— *helioides*, 352.

Wayfaring Tree, 9.
Wedelia, 66, 264.
— carnosa, 265.
— *hispida*, 266.
Weigela, 18.
White Lettuce, 434.
Whiteweed, 365.
Whitneya, 71, 318.
dealbata, 318.
Wild Coffee, 12.
Wirtgenia *Texana*, 286.
Wolf-berry, 13.
Woodbine, 14.
Woodruff, 35.
Woodvillea, 207, 208.
calendulacea, 208.
Wormwood, 367, 370.
Wyethia, 66, 267.
— amplexicaulis, 267.
angustifolia, 268.
Arizonica, 267.
coriacea, 268.
glabra, 267.
helenioides, 267.
helianthoides, 267.
longicaulis, 267.
mollis, 268.
ovata, 268.
reticulata, 268.
robusta, 268.
scabra, 268.
Wythe-rod, 11.

Xanthidium ambrosioides, 252.
— *discolor*, 251.
rhombophyllum, 251.
— *tenuifolium*, 250.
Xanthiopsis, 252.
Xanthisma, 54, 124.
— Texanum, 124.
Xanthium, 63, 252.
— *Americanum*, 252.
— Canadense, 252.
— *Carolinense*, 252.
— *echinatum*, 252.
— *macrocarpum*, 252.
— *maculatum*, 252.

spinosum, 253.
　strumarium, 252.
Xantho, 324.
Xanthocephalum, 53, 114.
　Benthamianum, 114.
　gymnospermoides, 114.
　sericocarpum, 114.
　Wrightii, 114.
Xanthocoma, 114.
Ximenesia, 288.
　encelioides, 289.
Xylorrhiza glabriuscula, 200.
　villosa, 200.
Xylosteon, 14.
Xylosteum ciliatum, 16.
　involucratum, 16.
　oblongifolium, 15.
　Solonis, 15.
　Tartaricum, 15.
　villosum, 15.

Yarrow, 363.
Youngia pygmæa, 431.

Zexmenia, 67, 285.
　brevifolia, 286.
　hispida, 286.
　hispidula, 289.
　podocephala, 286.
　Texana, 286.

Zinnia, 63, 253.
　acerosa, 254.
　anomala, 253.
　bicuspis, 253.
　grandiflora, 253.
　intermedia, 253.
　juniperifolia, 253.
　leptopoda, 253.
　multiflora, 253.
　pauciflora, 253.
　pumila, 253.
　revoluta, 253.
　tenuiflora, 253.
ZINNIEÆ, 63.

www.ingramcontent.com/pod-product-compliance
Lightning Source LLC
Chambersburg PA
CBHW051855300426
44117CB00006B/401